Perspective of
QUANTUM MECHANICS

For BSc Physics [Honours / Major], MSc [Physics] and BTech students of Indian universities

Perspective of
QUANTUM MECHANICS

Prof. (Dr) S P Kuila MSc, PhD [IIT, Kharagpur]
Professor and Head, Engineering Physics Department
Bengal Institute of Technology and Management, Santiniketan, WB and
Former Professor of Physics, Abhedananda Mahavidyalaya, Sainthia, Birbhum, WB

New Central Book Agency (P) Ltd

Co-Published by

Bijay Publications
E-mail : bijaypublications@gmail.com

Exclusively Distributed by

Bijay Pustak Bhandar (P) Ltd
90/6A M. G. Road, Bhawani Dutta Lane, Kolkata-700007
✆ : (033) 4605-7856, 8777898576, 8274916104
E-mail : bijaypustakbhandar@gmail.com

NCBA

REGD OFFICE
8/1 Chintamoni Das Lane, Kolkata 700 009, India
email: ncbakolkata@gmail.com

OVERSEAS
NCBA (UK) Ltd, 149 Park Avenue North, Northampton, NN3 2HY, UK
email: ncbauk@yahoo.com

CORPORATE OFFICE
212 Shahpur Jat, New Delhi 110 049
email: ncbaexports@ncbapvtltd.com

BIJAY PUBLICATIONS
90/6A M. G. Road, Bhawani Dutta Lane, Kolkata-7
E-mail : bijaypublications@gmail.com

BIJAY PUSTAK BHANDAR (P) LTD
90/6A M. G. Road, Bhawani Dutta Lane, Kolkata-7
E-mail : bijaypustakbhandar@gmail.com

PERSPECTIVE OF QUANTUM MECHANICS ● Prof. (Dr) S P Kuila

First Published: 2008
Reprinted: 2010, 2011, 2012, 2013, 2015, 2016, 2018
Thoroughly Revised and Updated Second Edition: 2019
Reprinted: 2021, 2023

PUBLISHER • TYPESETTER • PRINTER
New Central Book Agency (P) Ltd
8/1 Chintamoni Das Lane, Kolkata 700 009

CO-PUBLISHED BY
Bijay Publications
90/6A M G Road, Bhawani Dutta Lane, Kolkata 700007

TECHNICAL EDITOR
Dipan Roy

PROJECT SUPERVISOR
Abhijit Sen

ISBN: 978 81 7381 567 6

Price ₹ 875.00

Preface to the Second Edition

The second edition of the book, PERSPECTIVE OF QUANTUM MECHANICS, has been thoroughly revised and upgraded with the addition of questions from different standard universities throughout the country. Four new chapters, viz., Theory of Scattering (Chapter 12), Approximate Methods in Quantum Mechanics (Chapter 13), Time Dependent Perturbation (Chapter 14), and WKB Approximation and Variational Method (Chapter 15) have been added at the request of university teachers in order to cover the syllabi prescribed for the Postgraduate courses. The present edition is of ample help to the students pursuing masters degree in Physics as well as those who intend to clear GATE, JAM, JEST, CSIR-NET, SET, IAS and IPS examinations.

The book covers the papers of Physics C-III of semester II and C-IX of semester IV as well as papers on generic elective on Modern Physics and Quantum Mechanics of the UGC's new syllabus prescribed for the students pursuing Physics Honours course of the Indian universities.

The overwhelming response already accorded to the book and the constant flow of suggestions from the students and teachers of the different parts of the country are expected to continue for further improvement of the text.

1st January, 2019 S P Kuila

Preface to the First Edition

QUANTUM MECHANICS forms the foundation of Physics and as such is indispensable formalism in the study of physical properties of matter, especially when sub-atomic phenomena are concerned. Though many a book has been written on the subject, a major part of these are either extremely insufficient or highly mathematical, thereby appearing unsuitable for the undergraduate students. Keeping this point in mind, a long experience spanning over three decades has been exploited to present a suitable book on the subject to the Physics General and Honours students of different Indian and South Asian universities. The book has been written primarily on the basis of my class lecture notes in the most coherent and exhaustive manner, so that the students can grasp the subject with minimum labour and time. The critical points have been discussed with utmost lucidity and clarity in a self-developed technique.

The book has been enriched with large number of solved problems apart from appropriate illustrations and examples in each chapter. Moreover, exercise included at end of the each chapter comprises of four different types of questions, viz., a. Objective Type, b. Short Answer Type, c. Broad Answer Type and d. Numerical Problems to test the depth of understanding of the subject.

It is expected that the book will be able to quench the thirst of knowledge of the undergraduate as well as postgraduate students in Quantum Mechanics and meet all the requirements for the university examinations as well as various competitive examinations.

I take this opportunity to thank numerous students and colleagues working in different colleges, institutes and universities throughout India for their encouragement and constructive suggestions during the preparation of the manuscript. I would also like to thank Mr Amitabha Sen, Mr Soumen Pal, Mr Abhijit Sen and Mr Prabhat Jas of New Central Book Agency for their active cooperation in publishing the book in due time.

I would also like to thank Mr N R Datta, Chairman-cum-Managing Trustee, MEMD Trust, Santiniketan (WB) for his keen interest in the spread of higher education in West Bengal.

Lastly, it would be gross injustice if I do not express my sincere thanks to my wife Mrs Jyotsna Kuila and son Sri Sandip Kuila for the constant encouragement during the preparation of the manuscript.

25 December 2007

S P Kuila

Engineering Physics Department

Bengal Institute of Technology and Management

Santiniketan, WB

Contents

Chapter 1

Foundation of Quantum Mechanics

1.1 Introduction

The motion of objects either directly observable or made observable by simple instruments like microscope can be successfully explained by classical mechanics based on the principle of *Newton's laws of motion.* When the classical concepts were applied to the particles of atomic dimensions, e.g., electrons, they failed to explain the actual behaviour. Thus, classical mechanics is unable to describe the behaviour of atomic phenomenon, like the motion of electrons in an atom.

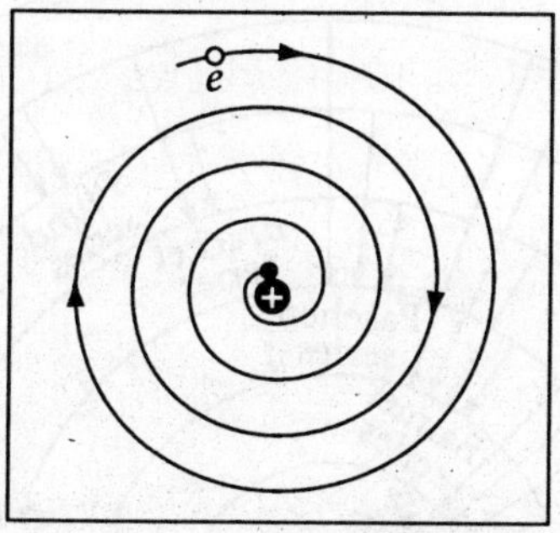

Fig 1.1 Spiraling path of the electron

According Rutherford's model of atom, an atom consists of a heavy positive nucleus with light negative electrons revolving round the nucleus in circular orbits. According to classical idea the electrostatic attractive force would bring the electron closer to the nucleus.

Again, as the electron revolves round the nucleus, it experiences a centripetal acceleration. Now, according classical electromagnetic theory, *a charged particle, whenever it undergoes an acceleration or deceleration, it radiates electromagnetic energy*. Therefore, the energy of the electron should decrease continuously and it should come closer and closer until it collapses with the nucleus (Fig 1.1). This leads to the instability of the atom which is contrary to experience. This shows the *failure* of classical theory to explain the stability of the atom. The failure of classical theory is again reflected from the fact that it could not account the discrete set of spectral lines of hydrogen spectrum represented by

$$\bar{\nu} = R_H \left(\frac{1}{n_f^2} - \frac{1}{n_i^2} \right) \tag{1.1}$$

where $\bar{\nu}$ is the wave number, R_H is the *Rydberg constant*, n_i and n_f are integers such that $n_i > n_f$.

Actually the observed spectral series are as follows:

Lyman series: $\bar{\nu} = R_H \left(\frac{1}{1^2} - \frac{1}{n_i^2} \right), \; n_i = 2, 3, 4, \ldots$

Balmer series: $\bar{\nu} = R_H \left(\frac{1}{2^2} - \frac{1}{n_i^2} \right), \; n_i = 3, 4, 5, \ldots$

Paschen series: $\bar{\nu} = R_H \left(\frac{1}{3^2} - \frac{1}{n_i^2} \right), \; n_i = 4, 5, 6, \ldots$

Bracket series: $\bar{\nu} = R_H \left(\frac{1}{4^2} - \frac{1}{n_i^2} \right), \; n_i = 5, 6, 7, \ldots$

Pfund series: $\bar{\nu} = R_H \left(\frac{1}{5^2} - \frac{1}{n_i^2} \right), \; n_i = 6, 7, 8, \ldots$

Thus, the excited atoms of hydrogen emit electromagnetic radiation of certain definite wavelength in a continuous fashion (Fig 1.2). Classical theory failed to explain the origin of this spectrum of hydrogen atom. So we conclude that classical theory does not hold good in the region of atomic dimension.

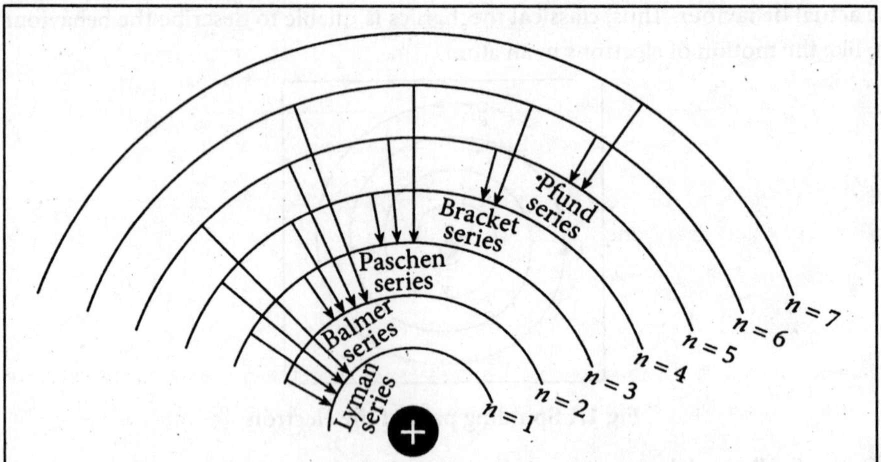

Fig 1.2 Origin of spectral series of hydrogen

The failure of classical theory led to the development of new mechanics called *Quantum Mechanics*. The mathematical theory of this theory was developed by *Erwin Schrödinger in* 1926. In this mechanics earlier ideas of classical wave theory are combined with *Louis de Broglie's wave-particle relationship*. The mathematical theories of *Wave Mechanics* and *Matrix Mechanics* appear to be different at the first sight, but they are equivalent. A more complete theory of particles, *the Quantum Field Theory* has been formulated in 1947.

1.2 Inadequacy of classical theory to explain the spectrum of blackbody radiation

Anybody which has the capacity of absorbing radiation of all wavelengths incident on it, is called *a perfect blackbody*. A perfect blackbody is the best possible emitter at any given temperature. The radiation emitted by such a body is known as *blackbody radiation* (or full radiation or temperature radiation).

Using infrared spectrometer and bolometer the intensities of the blackbody radiation of different wavelengths were measured by *Lummer and Pringsheim* (1899). The results of their experiment are as shown in Fig 1.3.

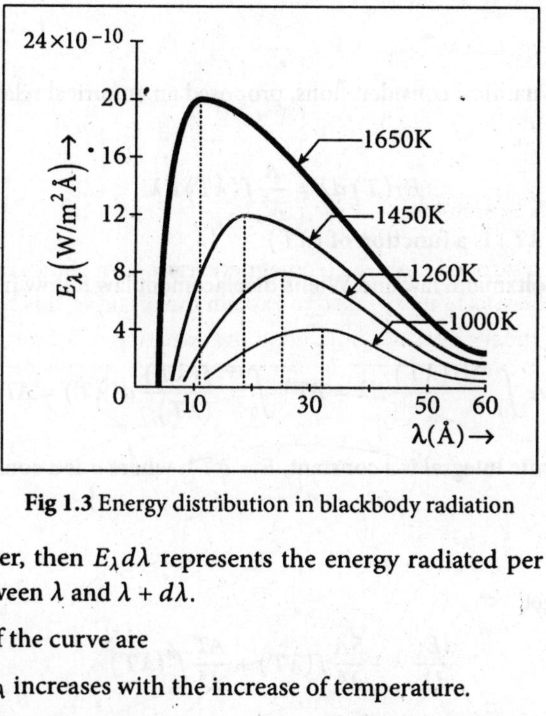

Fig 1.3 Energy distribution in blackbody radiation

If E_λ is the emissive power, then $E_\lambda d\lambda$ represents the energy radiated per unit area per second for wavelength in the range between λ and $\lambda + d\lambda$.

The chief characteristics of the curve are

(a) For every wavelength, E_λ increases with the increase of temperature.

(b) At a constant temperature, E_λ increases as λ increases till it becomes the maximum at a certain wavelength and, then E_λ decreases as λ increases further. At a higher wavelength λ_m at which E_λ is maximum, shifts towards shorter wavelength such that

$$\lambda_m T = \text{constant} \tag{1.2}$$

where T is the absolute temperature of the emitter. Relation (1.2) is known as the *Wien's displacement law*.

(c) E_m and T are connected by

$$\frac{E_m}{T} = \text{constant} \tag{1.3}$$

(d) The area under the curve and λ-axis at a particular temperature T represents the total radiation emitted per square metre per second over all wavelengths, i.e.,

$$E = \int_0^\infty E_\lambda d\lambda = \sigma T^4 \quad (1.4)$$

where σ is the *Stefan's constant* and $\sigma = 5.6697 \times 10^{-8}\ \mathrm{W/m^2\,K^4}$. Eq. (1.4) is the *Stefan's law* regarding total radiation emitted by a blackbody at absolute temperature T.

1.3 Theories of blackbody radiation

Various workers attempted to offer theoretical interpretation of the spectral distribution of energy in the spectrum of blackbody radiation on the basis of classical mechanics and electromagnetic theory with partial success. Ultimately in 1901, *Max Planck* proposed *Quantum Theory* of radiation and could offer a satisfactory explanation of the spectrum of blackbody radiation.

1.3.1 Wien's theory

Wien in 1893, from thermodynamical considerations, proposed an empirical relationship between E_λ and λ for a given temperature T as

$$E_\lambda(T)\,d\lambda = \frac{A}{\lambda^5} f(\lambda T)\,d\lambda \quad (1.5)$$

where A is a constant and $f(\lambda T)$ is a function of (λT).

We can show that Stefan-Boltzmann law and Wien's displacement law follow from Wien's distribution law.

Putting $x = \lambda T$ in eq. (1.5)

$$E = \int_0^\infty E_\lambda d\lambda = A\int_0^\infty \frac{f(\lambda T)}{\lambda^5} d\lambda = AT^4 \int_0^\infty \frac{f(\lambda T)}{(\lambda T)^5} d(\lambda T) = AT^4 \int_0^\infty \frac{f(x)}{x^5} dx$$

$\because \int_0^\infty \frac{f(x)}{x^5} dx$ being a definite integral is a constant, $E = \sigma T^4$, where σ is a constant. This relation is called *Stefan-Boltzmann law.*

Wien's displacement law

Differentiating eq. (1.5), we get

$$\frac{dE_\lambda}{d\lambda} = -\frac{5A}{\lambda^6} f(\lambda T) + \frac{AT}{\lambda^5} f'(\lambda T)$$

If $E_\lambda = (E_\lambda)_{\max}$ at $\lambda = \lambda_m$, then

$$\left(\frac{dE_\lambda}{d\lambda}\right)_{\max} = 0 \text{ or, } Tf'(\lambda_m T) - T\frac{5f(\lambda_m T)}{\lambda_m} = 0 \text{ or, } \lambda_m f'(x_m) - 5f(x_m) = 0$$

where $x_m = \lambda_m T$. The above equation is in a single variable x_m and can have only one solution.

$\therefore x_m = \lambda_m T = \text{constant}.$

This is known as *Wien's displacement law* which explains the shift of emission peaks towards lower value of λ with the increase of T.

Wien's distribution formula explains the experimental results fairly well for low values of λT, but for higher values, it gives E_λ lower than the experiment values.

1.3.2 Rayleigh-Jeans law

Applying the law of equipartition of energy to the electromagnetic vibrations, *Lord Rayleigh* with a contribution from *J.H. Jeans* (in 1900) obtained a formula for energy density inside an enclosure with perfectly reflecting walls. According to this formula the energy density in the frequency range ν to $\nu + d\nu$ at temperature T is

$$U_\nu d\nu = \frac{8\pi\nu^2 kT}{c^3} d\nu \tag{1.6}$$

where U_ν is the energy per unit volume per unit frequency range at ν, k is the Boltzmann constant and c is the velocity of light in vacuum.

Since,

$$c = \nu\lambda \text{ or, } \nu = \frac{c}{\lambda} \text{ and } d\nu = -\frac{c}{\lambda^2} d\lambda$$

$$U_\lambda d\lambda = -U_\nu d\nu = -\frac{8\pi}{c^3}\left(\frac{c}{\lambda}\right)^2 kT\left(-\frac{c}{\lambda^2}\right) d\lambda \text{ or, } U_\lambda d\lambda = \frac{8\pi kT}{\lambda^4} d\lambda \tag{1.7}$$

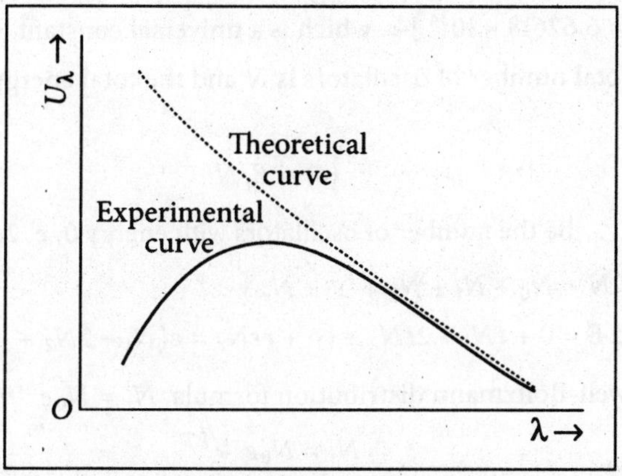

Fig 1.4 Deviation of Rayleigh-Jeans law from experimental result

Equation (1.6) gives *Rayleigh-Jeans law* in terms of wavelength. This law can explain the experimental results in the long wavelength side, but fails in the short wavelength side [Fig 1.4], since $U_\lambda \to \infty$ as $\lambda \to 0$ for a constant temperature. This is contrary to experimental findings.

Ultraviolet catastrophe

The total energy of radiation per unit volume of the enclosure for all wavelengths from 0 to ∞ is

$$U = \int_0^\infty U_\lambda d\lambda = \int_0^\infty \frac{8\pi kT}{\lambda^4} d\lambda = 8\pi kT\left[-\frac{1}{3\lambda^3}\right] = \infty \text{ [from eq. (1.7)]}$$

Hence, for a given quantity of radiant energy all the energy will be confined to vibrations of small wavelength. But, experimental results show that $U_\lambda d\lambda \to 0$ as $\lambda \to 0$ [Fig 1.4]. This serious disagreement between theory and experiment is known as *ultraviolet catastrophe*. This indicates the limitations of the classical mechanics on the basis of which the principle of equipartition is deduced.

1.3.3 Planck's law of radiation

Max Planck in Germany (1901) put forward a new postulate regarding the nature of vibration of the linear harmonic oscillator (LHO) which is in thermal equilibrium with the electromagnetic radiation within the radiation enclosure. According to Planck:

1. An oscillator has a discrete set of energies which are integral multiples of a finite quantum of energy $\epsilon = h\nu$, here h is *Planck's constant* and ν is the frequency of the oscillator. So the energy of the oscillator can have only values like

$$E = n\varepsilon = nh\nu \tag{1.8}$$

where $n = 0, 1, 2, 3, \ldots$. This shows that the lowest state (ground state) energy is zero.

2. So long the oscillator has energy $E = nh\nu$, it cannot emit or absorb energy. The emission or absorption of energy occurs only when the oscillator jumps from one energy state to another. If the oscillator jumps from the state of quantum number n_2 to a lower state of quantum number n_1, then the energy emitted is

$$\Delta E = E_2 - E_1 = (n_2 - n_1)\, h\nu \text{ or, } \Delta E = h\nu \text{ if } n_2 = n_1 + 1 \tag{1.9}$$

When an oscillator absorbs a quantum of energy $h\nu$, it jumps to the next higher state of energy. Planck estimated the value of $h = 6.62618 \times 10^{34}$ J-s, which is a universal constant.

If in an enclosure the total number of oscillators is N and the total energy is E, then the average energy per oscillator is

$$\langle \varepsilon \rangle = E/N \tag{1.10}$$

If $N_0, N_1, N_2, \ldots, N_r, \ldots$ be the number of oscillators with energy $0, \varepsilon, 2\varepsilon, \ldots, r\varepsilon, \ldots$ respectively, then

$$N = N_0 + N_1 + N_2 + \cdots + N_r + \ldots \tag{1.11}$$

$$\text{and } E = 0 + \varepsilon N_1 + 2\varepsilon N_2 + \cdots + r\varepsilon N_r = \varepsilon(N_1 + 2N_2 + \cdots + rN_r) \tag{1.12}$$

According to the Maxwell-Boltzmann distribution formula, $N_r = N_0 e^{-r\varepsilon/kT}$.

$$\therefore N_1 = N_0 e^{-\varepsilon/kT}$$

$$N_2 = N_0 e^{-2\varepsilon/kT}$$

$$\ldots\ldots\ldots\ldots\ldots$$

Using these values in eqs (1.11) and (1.12), we get

$$\begin{aligned} N &= N_0 + N_0 e^{-\varepsilon/kT} + N_0 e^{-2\varepsilon/kT} + \cdots + N_0 e^{-r\varepsilon/kT} + \ldots \\ &= N_0 \left[1 + e^{-\varepsilon/kT} + e^{-2\varepsilon/kT} + \cdots + N_r e^{-r\varepsilon/kT} + \ldots\right] \end{aligned} \tag{1.13}$$

$$\begin{aligned} \text{and } E &= \varepsilon\left[N_0 e^{-\varepsilon/kT} + 2N_0 e^{-2\varepsilon/kT} + \cdots + rN_0 e^{-r\varepsilon/kT} + \ldots\right] \\ &= N_0\varepsilon\left[e^{-\varepsilon/kT} + 2e^{-2\varepsilon/kT} + \cdots + r e^{-r\varepsilon/kT} + \ldots\right] \end{aligned} \tag{1.14}$$

Putting $y = e^{-\varepsilon/kT}$, we have from eqs (1.13) and (1.14)

$$N = N_0\left(1 + y + y^2 + \cdots + y^r + \ldots\right) = \frac{N_0}{1-y} \tag{1.15}$$

$$\text{and } E = N_0\varepsilon\left(y + 2y^2 + \cdots + ry^r + \ldots\right) = \frac{N_0\varepsilon y}{(1-y)^2} \tag{1.16}$$

Now, from eq. (1.10)

$$\langle \varepsilon \rangle = \frac{E}{N} = \frac{N_0 \varepsilon \frac{y}{(1-y)^2}}{\frac{N_0}{(1-y)}} = \frac{\varepsilon y}{1-y} = \frac{\varepsilon e^{-\varepsilon/kT}}{1-e^{-\varepsilon/kT}}$$

$$\text{or, } \langle \varepsilon \rangle = \frac{\varepsilon}{e^{\varepsilon/kT}-1} = \frac{h\nu}{e^{h\nu/kT}-1} \tag{1.17}$$

But, $h\nu = \frac{h}{2\pi} 2\pi\nu = \hbar\omega$, where $\hbar = \frac{h}{2\pi}$ and $\omega = 2\pi\nu$.

$$\therefore \langle \varepsilon \rangle = \frac{\hbar\omega}{e^{\hbar\omega/kT}-1} \quad [\because \varepsilon = h\nu = \hbar\omega] \tag{1.18}$$

If $h\nu \ll kT$, then

$$e^{\frac{h\nu}{kT}} \simeq 1 + \frac{h\nu}{kT} \left(\because \frac{h\nu}{kT} \ll 1 \right)$$

so that

$$\langle \varepsilon \rangle = \frac{h\nu}{1 + \frac{h\nu}{kT} - 1} = kT \tag{1.19}$$

Hence, if $h\nu \ll kT$, then $\langle \varepsilon \rangle$ reduces to the classical limit. It can be shown that the number of oscillators per unit volume in the frequency range ν to $\nu + d\nu$ is

$$N(\nu)\, d\nu = \frac{8\pi\nu^2}{c^3} d\nu$$

where c is the velocity of light in vacuum.

Now, the energy per unit volume of the radiation enclosure in the frequency range ν to $\nu + d\nu$ is

$$U_\nu d\nu = N(\nu)\, d\nu \times \langle \varepsilon \rangle = \frac{8\pi\nu^2}{c^3} \times \frac{h\nu}{e^{h\nu/kT}-1} d\nu$$

$$\text{or, } U_\nu d\nu = \frac{8\pi h\nu^3}{c^3} \frac{1}{e^{h\nu/kT}-1} d\nu \tag{1.20}$$

The relation (1.20) is called *Planck's radiation law* which explains the distribution of energy in the spectrum of blackbody radiation over whole range of wavelength.

Again, since $\nu = \frac{c}{\lambda}$, $\therefore |d\nu| = \left| -\frac{c}{\lambda^2} d\lambda \right|$.

$$\therefore U_\lambda d\lambda = \frac{8\pi h}{c^3} \left(\frac{c^3}{\lambda^3} \right) \frac{1}{e^{hc/\lambda kT}-1} \left(\frac{c}{\lambda^2} d\lambda \right) = \frac{8\pi hc}{\lambda^5} \frac{1}{e^{hc/\lambda kT}-1} d\lambda \tag{1.21}$$

Equation (1.21) represents *Planck's radiation law in terms of wavelength.*

It can be shown that *Wien's law, Rayleigh-Jeans law* and *Stefan's law* are the special cases of *Planck's law* of radiation.

(a) ***Wien's formula:*** In the low wavelength region $hc/\lambda kT \to \infty$, hence from eq. (1.21) we can write:

$$U_\lambda d\lambda = \frac{8\pi hc}{\lambda^5} \frac{1}{e^{hc/kT}} d\lambda = \frac{8\pi hc}{\lambda^5} e^{-hc/kT} d\lambda \tag{1.22}$$

If E_λ represents the intensity of radiation of wavelength λ, then it can be easily shown that

$$E_\lambda = \frac{cU_\lambda}{4} \text{ or, } U_\lambda = \frac{4E_\lambda}{c}$$

From the above eq. (1.22)

$$E_\lambda d\lambda = \frac{2\pi hc^2}{\lambda^5} e^{-hc/kT}\, d\lambda = \frac{A}{\lambda^5} f(\lambda T)\, d\lambda \tag{1.23}$$

where $A = 2\pi hc^2$ and $f(\lambda T) = e^{-hc/kT}$.

Equation (1.23) is the *Wien's radiation law* in terms of wavelength λ.

(b) ***Wien's displacement law***: For a particular temperature T, if λ_m be the wavelength corresponding to the maximum energy density, then $\left[\frac{dU_\lambda}{d\lambda}\right]_{\lambda_m} = 0.$

Now, since

$$U_\lambda = \frac{8\pi hc}{\lambda^5} \frac{1}{e^{hc/kT} - 1}$$

$\therefore \ln U_\lambda = \ln(8\pi hc) - 5\ln\lambda - \ln\left(e^{hc/\lambda kT} - 1\right).$

Differentiating w.r.t. λ, we get

$$\frac{1}{U_\lambda}\frac{dU_\lambda}{d\lambda} = 0 - \frac{5}{\lambda} - \left[\frac{1}{e^{hc/\lambda kT} - 1}\right] e^{hc/\lambda kT}\left(-\frac{hc}{\lambda^2 kT}\right) = \frac{1}{\lambda}\left[-5 + \frac{hc}{\lambda kT}\left(\frac{e^{hc/\lambda kT}}{e^{hc/\lambda kT} - 1}\right)\right]$$

At $\lambda = \lambda_m$, $\frac{dU_\lambda}{d\lambda} = 0.$

$$\therefore -5 + \frac{hc}{\lambda_m kT}\left(\frac{e^{hc/\lambda kT}}{e^{hc/\lambda kT} - 1}\right) = 0$$

Putting $\frac{hc}{\lambda_m kT} = x$, we get

$$-5 + \frac{xe^x}{e^x - 1} = 0 \text{ or, } \frac{xe^x}{e^x - 1} = 5$$

Solving, we get $x = 4.9651$, $\therefore \frac{hc}{\lambda_m kT} = 4.9651.$

$$\therefore \lambda_m T = \frac{hc}{4.9651 \times k} = \frac{6.62 \times 10^{-34} \times 3 \times 10^8}{4.9651 \times 1.38 \times 10^{-23}}$$

$$\text{or, } \lambda_m T = 2.898 \times 10^{-3} \text{ m K} \tag{1.24}$$

This is *Wien's displacement law*.

(c) ***Rayleigh-Jeans law***: In the long wavelength region

$$\frac{hc}{\lambda kT} \to 0 \text{ and } e^{\frac{\lambda c}{\lambda kT}} \simeq 1 + \frac{hc}{\lambda kT}$$

Then from

$$U_\lambda d\lambda = \frac{8\pi hc}{\lambda^5} \frac{1}{e^{hc/\lambda kT} - 1} d\lambda$$

we get

$$U_\lambda d\lambda = \frac{8\pi hc}{\lambda^5}\frac{1}{hc/\lambda kT}d\lambda = \frac{8\pi kT}{\lambda^4}d\lambda \tag{1.25}$$

This is *Rayleigh-Jeans law*.

(d) ***Stefan-Boltzmann law***: *J. Stefan* in 1879, deduced empirically that the total radiant energy of all wavelengths emitted per unit area per second by a heated body is proportional to the fourth power of its absolute temperature. *Boltzmann* in 1884, deduced the same law by considering blackbody radiation as the working substance of a Carnot cycle. That is why the law is known as *Stefan-Boltzmann law*.

The energy density of the total radiation for all possible wavelengths in a blackbody enclosure at temperature T is

$$U = \int_0^\infty U_\lambda d\lambda = \int_0^\infty \frac{8\pi hc}{\lambda^5}\frac{1}{e^{hc/\lambda kT}-1}d\lambda$$

Let

$$x = \frac{hc}{\lambda kT}, \therefore \lambda = \frac{hc}{kTx} \text{ and } d\lambda = -\frac{hc}{kTx^2}dx.$$

When $\lambda \to 0, x \to \infty$ and when $\lambda \to \infty, x \to 0$.

$$\therefore U = \int_\infty^0 \frac{8\pi hc}{1}\left(\frac{kTx}{hc}\right)^5 \frac{1}{e^x-1}\left(-\frac{hc}{kTx^2}dx\right) = -\int_\infty^0\left(\frac{8\pi k^4T^4}{h^3c^3}\right)\frac{x^3}{e^x-1}dx = \frac{8\pi k^4T^4}{h^3c^3}\int_0^\infty \frac{x^3}{e^x-1}dx$$

But, $\int_0^\infty \frac{x^3}{e^x-1} = \frac{\pi^4}{15}$ [$\because$ the integral is a standard integral]

$$\therefore U = \frac{8\pi k^4T^4}{h^3c^3}\frac{\pi^4}{15} = \frac{4}{c}\left(\frac{2}{15}\frac{\pi^5k^4}{h^3c^2}\right)T^4$$

The total emissive power E is related to U at the same temperature T by $E = \frac{cU}{4}$ or, $U = \frac{4E}{c}$.

$$\therefore E = \left(\frac{2\pi^5k^4}{15h^3c^2}\right)T^4 \text{ or, } E = \sigma T^4 \tag{1.26}$$

where $\sigma = \frac{2\pi^5k^4}{15h^3c^2} = 5.67\times10^{-8}\ \text{W/m}^2\,\text{K}^4$.

Equation (1.26) is called the *Stefan-Boltzmann law*.

Planck's conception of discrete energy was successful in explaining the distribution of energy in the spectrum of blackbody radiation which led to the formulation of *Quantum Mechanics*.

1.4 Specific heat of solids

Einstein (1907) first applied *Planck's quantum hypothesis* to resolve the discrepancies of the classical theory of specific heat of solids. According to classical *Dulong-Petit law* the gram-molecular specific heat of all solids has the same value, viz., ~ 6 calorie/°C/mole at above room temperature. Classical theory assumes that each atom behaves as a three-dimensional oscillator vibrating about their mean position of rest. Corresponding to each Cartesian component of vibration, there are two degrees of freedom, viz., translational and rotational.

Again, according to the theorem of equipartition of energy, the mean kinetic energy associated with each degree of freedom is $\frac{1}{2}kT$, where k is the Boltzmann constant and T is the absolute temperature.

Hence, the energy associated with each component is $\frac{1}{2}kT + \frac{1}{2}kT = kT$. So the mean energy associated with the above oscillator is

$$\langle \varepsilon \rangle = 3kT \tag{1.27}$$

The total energy per gm-mole of the solid is

$$E = N\langle \varepsilon \rangle = 3NkT = 3RT \tag{1.28}$$

where N is the Avogadro number and $R = Nk$, the universal gas constant. The specific heat of the solid is, therefore,

$$C_v = \frac{dE}{dT} = 3R \sim 6\text{cal/°C/mole} \tag{1.29}$$

This is *Dulong-Petit law* of specific heat of solid which holds good at higher temperature only. Experiment on Ag, Ge and Si shows that C_v decreases with the decrease of T and tends to zero at $T \to 0$ [Fig 1.5]. The discrepancy was resolved by Einstein and, then Debye by using the quantum theory proposed by Max Planck.

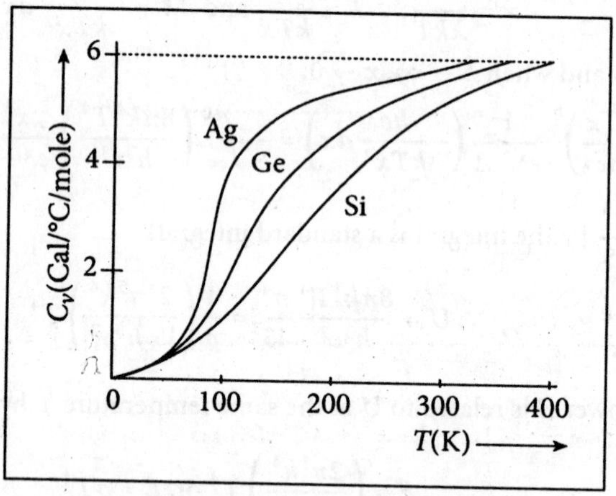

Fig 1.5 Variation of C_v with temperature

Einstein replaced the idea of classical harmonic oscillator by quantum harmonic oscillator which can have only discrete energy values and retained all assumptions of classical theory. The main assumptions of Einstein's theory were as follows:

(a) The atoms of a crystal are regarded as identical and independent harmonic oscillators.

(b) A solid having N atoms is equivalent to having $3N$ one-dimensional harmonic oscillators.

(c) Due to the identical environment each oscillator vibrates with the same natural frequency ω.

(d) As the oscillators are quantum mechanical, they can have discrete energy spectrum.

(e) Any quantum state may contain any number of oscillators.

(f) The oscillators are distinguishable and hence obey Maxwell-Boltzmann distribution of energy.

In view of Planck's quantum theory the discrete energy values of an oscillator having frequency ν are

$$\varepsilon_n = nh\nu = n\hbar\omega \tag{1.30}$$

where $n = 0, 1, 2, 3, \ldots$ is the quantum number. Using wave mechanical consideration the energy of a harmonic oscillator is

$$\varepsilon_n = \left(n + \tfrac{1}{2}\right) h\nu_0 = \left(n + \tfrac{1}{2}\right) \hbar\omega_0 \tag{1.31}$$

Here, $\frac{1}{2}\hbar\omega_0$ is the zero-point energy contribution to the internal energy of the oscillator.

Now, the average energy of an oscillator is given by

$$\langle\varepsilon\rangle = \frac{\sum_{n=0}^{\infty} \varepsilon_n e^{-\varepsilon_n/kT}}{\sum_{n=0}^{\infty} e^{-\varepsilon_n/kT}} \tag{1.32}$$

$$= \frac{\sum_{n=0}^{\infty} \left(n + \frac{1}{2}\right) \hbar\omega_0 e^{\left[-\left(n+\frac{1}{2}\right)\frac{\hbar\omega_0}{kT}\right]}}{\sum_{n=0}^{\infty} e^{\left[-\left(n+\frac{1}{2}\right)\frac{\hbar\omega_0}{kT}\right]}}$$

$$= \frac{\hbar\omega_0 \sum_{n=0}^{\infty} \left(n + \frac{1}{2}\right) e^{\left[\left(n+\frac{1}{2}\right)x\right]}}{\sum_{n=0}^{\infty} e^{\left[\left(n+\frac{1}{2}\right)x\right]}} \tag{1.33}$$

where $x = -\dfrac{\hbar\omega_0}{kT}$.

$$\therefore\ \langle\varepsilon\rangle = \frac{\hbar\omega_0\left(\frac{1}{2}e^{\frac{x}{2}} + \frac{3}{2}e^{\frac{3x}{2}} + \frac{5}{2}e^{\frac{5x}{2}} + \ldots\right)}{\left(e^{\frac{x}{2}} + e^{\frac{3x}{2}} + e^{\frac{5x}{2}} + \ldots\right)} = \hbar\omega_0 \frac{d}{dx} \ln\left[e^{\frac{x}{2}} + e^{\frac{3x}{2}} + e^{\frac{5x}{2}} + \ldots\right]$$

$$= \hbar\omega_0 \frac{d}{dx} \ln\left[e^{\frac{x}{2}}\left(1 + e^x + e^{2x} + \ldots\right)\right]$$

But, $\ln\left(1 + e^x + e^{2x} + \ldots\right) = -\ln\left(1 - e^x\right)$.

$$\therefore\ \langle\varepsilon\rangle = \hbar\omega_0 \frac{d}{dx}\left[\frac{x}{2} - \ln\left(1 - e^x\right)\right] = \hbar\omega_0\left[\tfrac{1}{2} + \frac{e^x}{1 - e^x}\right] = \hbar\omega_0\left[\frac{1}{2} + \frac{1}{e^{-x} - 1}\right]$$

$$= \hbar\omega_0\left[\frac{1}{2} + \frac{1}{e^{\hbar\omega_0/kT} - 1}\right] = \frac{1}{2}\hbar\omega_0 + \frac{\hbar\omega_0}{e^{\frac{\hbar\omega_0}{kT}} - 1} \tag{1.34}$$

Again, if we use $\varepsilon_n = n\hbar\omega_0$ in eq. (1.32), we get

$$\langle\varepsilon\rangle = \frac{\hbar\omega_0}{e^{\frac{\hbar\omega_0}{kT}} - 1} \tag{1.35}$$

We note from eq. (1.34) that for $T = 0$, $\langle\varepsilon\rangle = \frac{1}{2}\hbar\omega_0$. This means that the atoms are having vibrational energy $\frac{1}{2}\hbar\omega_0$ even at $T = 0$.

Since, $\frac{1}{2}\hbar\omega_0$ is temperature-independent, it has no contribution to specific heat of solids.

Thus, the internal energy is

$$E = 3N\langle\varepsilon\rangle = \frac{3}{2}N\hbar\omega_0 + \frac{3N\hbar\omega_0}{e^{\hbar\omega_0/kT} - 1} \tag{1.36}$$

Molar specific heat

$$C_v = \left(\frac{dE}{dT}\right)_V = 3Nk\left(\frac{\hbar\omega_0}{kT}\right)^2 \frac{e^{\frac{\hbar\omega_0}{kT}}}{\left(e^{\hbar\omega_0/kT}-1\right)^2} \tag{1.37}$$

Now, the *characteristic temperature* (θ_c), also known as the *Einstein temperature* (θ_E) is given by

$$\theta_E = \frac{\hbar\omega_0}{k} \tag{1.38}$$

∴ eq. (1.37) reduces to the form:

$$C_v = 3Nk\left(\frac{\theta_E}{T}\right)^2 \frac{e^{\theta_E/T}}{\left(e^{\theta_E/T}-1\right)^2} \tag{1.39}$$

This shows that C_v of a solid is a function of temperature.

Special cases

1. ***When the temperature (T) is large***: For T larger such that

$$kT \gg \hbar\omega_0 \text{ or, } T \gg \frac{\hbar\omega_0}{k} \text{ or, } T \gg \theta_E$$

we get

$$e^{\hbar\omega_0/kT} - 1 = 1 + \frac{\hbar\omega_0}{kT} + \cdots - 1 = \frac{\hbar\omega_0}{kT} \tag{1.40}$$

Equation (1.34), then reduces to $\langle\varepsilon\rangle = \frac{1}{2}\hbar\omega_0 + kT \simeq kT$.

This shows that at high temperature the average energy $\langle\varepsilon\rangle$ is the same as that obtained from classical theory.

From eq. (1.37)

$$C_v = \left(\frac{dE}{dT}\right)_V = 3Nk\left(\frac{\hbar\omega_0}{kT}\right)^2 \frac{\left(1+\frac{\hbar\omega_0}{kT}\right)}{\left(\frac{\hbar\omega_0}{kT}\right)^2} \quad \left[\because e^{\hbar\omega_0/kT} - 1 = \frac{\hbar\omega_0}{kT}\right]$$

$$= 3Nk\left(1 + \frac{\hbar\omega_0}{kT}\right)$$

As $T \to \infty$, we get $C_v = 3Nk = 3R$.

This is the *Dulong-Petit law* as obtained from classical theory.

2. ***When the temperature is low***: When $T \to 0$

$$kT \ll \hbar\omega_0 \text{ or, } T \ll \frac{\hbar\omega_0}{k} \text{ or, } T \ll \theta_E. \ \therefore\ e^{\hbar\omega_0/kT} - 1 \simeq e^{\hbar\omega_0/kT}$$

From eq. (1.33)

$$\langle\varepsilon\rangle = \frac{1}{2}\hbar\omega_0 + \hbar\omega_0 e^{-\frac{\hbar\omega_0}{kT}} \tag{1.41}$$

So the average energy decreases exponentially with temperature at low temperature region.

$$\therefore C_v = \left(\frac{dE}{dT}\right)_V = 3Nk\left(\frac{\hbar\omega_0}{kT}\right)^2 e^{-\frac{\hbar\omega_0}{kT}} \tag{1.42}$$

$$\text{or, } C_v = 3Nk\left(\frac{\theta_E}{T}\right)^2 e^{-\frac{\theta_E}{T}} \tag{1.43}$$

Hence, for $T \ll \theta_E$, $C_v \propto e^{-\frac{\theta_E}{T}}$.

This shows a rapid decrease in C_v with the decrease of T. For most of the solids C_v is found to vary as T^3.

Fig 1.6 shows a theoretical plot of C_v against T/θ_E using eq. (1.43). For comparison some experimental point for diamond are also shown. Though the Einstein's theory agrees well with experiment over a wide range of temperature, it shows a more rapid decrease of C_v in the low temperature region (for liquid HeT), where $C_v \propto T^3$.

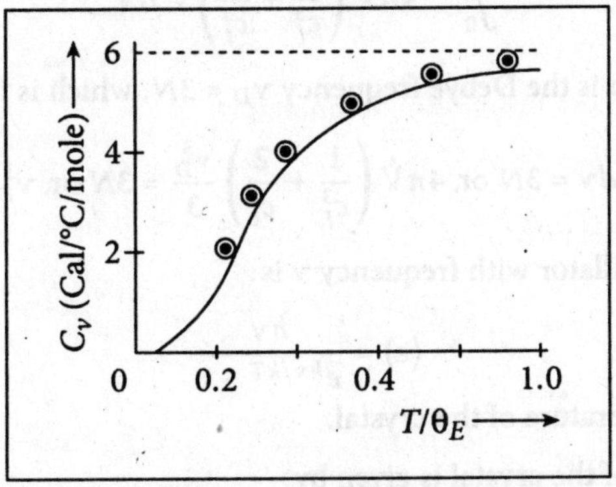

Fig 1.6 Plot of C_v against (T/θ_E)

It is true that Einstein's theory using quantum idea offers a better explanation to the variation of specific heat with temperature, still it fails to account the variation of C_v at very low temperature. This is due to the wrong assumption that all the atomic oscillators oscillate with the same natural frequency ω_0.

Later on *P. Debye* improved Einstein's theory by considering a whole range of frequencies up to a maximum cut-off frequency ν_m for the solids. According to Debye, C_v at low temperature is given by

$$C_v = 3R\frac{12\pi^4}{15}\left(\frac{T^3}{\theta_D}\right) \tag{1.44}$$

where $\theta_D = \dfrac{h\nu_m}{kT}$ is a constant for a given solid and is called the *Debye temperature.* This is *Debye T^3-law* which agrees well with experiment at low temperature.

1.4.1 Debye's theory of specific heat of solids

Debye, in 1912, modified Einstein's theory of specific heats of solids on the basis of the following assumptions:

1. Any solid is capable of vibrating elastically in many different modes, the frequency varying from one mode to another. The number of modes of vibration of the solid are, however limited to $3N$, where N is the number of atoms in the solid.
2. If a continuum solid is subjected to elastic vibration, two types of vibrations are produced, viz., *longitudinal vibrations* and *transverse vibrations.* The number of modes of longitudinal vibrations and transverse vibrations per unit volume in the frequency range ν to $\nu + d\nu$ are respectively

$$\frac{4\pi\nu^2 d\nu}{c_l^3} \text{ and } \frac{2\times 4\pi\nu^2 d\nu}{c_t^3}$$

where c_l and c_t are the velocities of longitudinal and transverse waves. The factor 2 before $\frac{4\pi\nu^2 d\nu}{c_t^3}$ is due to the fact that the transverse vibration has two states of polarisation at right angles to each other. So the number of independent modes of vibration in the frequency range ν and $\nu + d\nu$ in volume V is

$$4\pi V\left(\frac{1}{c_l^3}+\frac{2}{c_t^3}\right)\nu^2 d\nu$$

Hence, the total number of independent modes of vibration is given by

$$\int_0^{\nu_D} 4\pi V\left(\frac{1}{c_l^3}+\frac{2}{c_t^3}\right)\nu^2 d\nu$$

The upper limit of integration is the Debye frequency $\nu_D = 3N$, which is finite by assumptions.

$$\therefore \int_0^{\nu_D} 4\pi V\left(\frac{1}{c_l^3}+\frac{2}{c_t^3}\right)\nu^2 d\nu = 3N \text{ or, } 4\pi V\left(\frac{1}{c_l^3}+\frac{2}{c_t^3}\right)\frac{\nu_D^3}{3} = 3N \text{ or, } \nu_D^3 = \frac{9N}{4\pi V}\left(\frac{1}{c_l^3}+\frac{2}{c_t^3}\right)^{-1} \quad (1.45)$$

3. The average energy of an oscillator with frequency ν is

$$\langle\varepsilon\rangle = \frac{h\nu}{e^{h\nu/kT}-1} \quad (1.46)$$

where T is the absolute temperature of the crystal.

Now, the vibrational energy of the crystal is given by

$$E = \int_0^{\nu_D} 4\pi V\left(\frac{1}{c_l^3}+\frac{2}{c_t^3}\right)\frac{h\nu^3}{e^{h\nu/kT}-1}d\nu = 4\pi Vh\left(\frac{1}{c_l^3}+\frac{2}{c_t^3}\right)\int_0^{\nu_D}\frac{\nu^3 d\nu}{e^{h\nu/kT}-1}$$

$$\text{or, } E = \frac{9Nh}{\nu_D^3}\int_0^{\nu_D}\frac{\nu^3 d\nu}{e^{h\nu/kT}-1}$$

$$\because 4\pi V\left(\frac{1}{c_l^3}+\frac{2}{c_t^3}\right) = \frac{9N}{\nu_D^3} \text{ [by eq. (1.45)]} \quad (1.47)$$

Now, let $\frac{h\nu}{kT} = x$ and $\frac{h\nu_D}{kT} = x_m$, then

$$\nu = \frac{xkT}{h}, \therefore d\nu = \frac{kT}{h}dx$$

From eq. (1.47)

$$E = \frac{9Nh}{\nu_D^3}\left(\frac{kT}{h}\right)^4\int_0^{x_m}\frac{x^3\,dx}{e^x-1} = 9N\left(\frac{kT}{h\nu_D}\right)^3 kT\int_0^{x_m}\frac{x^3\,dx}{e^x-1} \quad (1.48)$$

$$= 9NkT\left(\frac{T}{\theta_D}\right)^3\int_0^{\theta_D/T}\frac{x^3\,dx}{e^x-1} \quad (1.49)$$

where $\theta_D = \frac{h\nu_D}{k}$ is called the *Debye temperature.*

Now, the specific heat is

$$C_v = \left(\frac{dE}{dT}\right)_V = 9Nk\left(\frac{T}{\theta_D}\right)^3\int_0^{\theta_D/T}\frac{e^x x^4}{(e^x-1)^2}dx = 3R\left(\frac{\theta_D}{T}\right)F_D \text{ [for } N = N_a] \quad (1.50)$$

where

$$F_D = 3\left(\frac{T}{\theta_D}\right)^4 \int_0^{\theta_D/T} \frac{e^x x^4}{(e^x-1)^2}\,dx \tag{1.51}$$

is the *Debye function.*

Special cases

1. ***When the temperature T is high:*** For $T \gg \theta_D$, $x = \dfrac{\theta_D}{T}$ is small as compared to unity.

So $e^x - 1 \simeq 1 + x + \dfrac{x^2}{2} + \cdots - 1 \simeq x$.

$\therefore$ from eq. (1.49)

$$E = 9NkT\left(\frac{T}{\theta_D}\right)^3 \int_0^{\theta_D/T} x^2\,dx = 3NkT \tag{1.52}$$

$$\text{and } C_v = \left(\frac{dE}{dT}\right)_V = 3Nk = 3R \text{ [for } N = N_a \text{ the Avogadro number]} \tag{1.53}$$

Equation (1.53) shows that Debye's theory reduces to classical Dulong-Petit law at high temperature.

2. ***When temperature T is low:*** For $T \ll \theta_D$, $x_m = \dfrac{\theta_D}{T} \to \infty$.

$\therefore$ eq. (1.49) reduces to

$$E = 9NkT\left(\frac{T}{\theta_D}\right)^3 \int_0^{\infty} \frac{x^3\,dx}{e^x - 1}$$

But, $\displaystyle\int_0^{\infty} \frac{x^3\,dx}{e^x-1} = \frac{\pi^4}{15}$ [Standard integral].

$$\therefore E = 9NkT\left(\frac{T}{\theta_D}\right)^3 \frac{\pi^4}{15} = \frac{3}{5}\pi^4 Nk\frac{T^4}{\theta_D^3} \tag{1.54}$$

Equation (1.54) shows that the vibrational energy of the crystal is proportional to T^4, which is analogous to *Stefan's law.*

Now,

$$C_v = \left(\frac{dE}{dT}\right)_V = \frac{12}{5}\pi^4 Nk\left(\frac{T}{\theta_D}\right)^3$$

$$\text{or, } C_v = \frac{12}{5}\pi^4 R\left(\frac{T}{\theta_D}\right)^3 \text{ [for } N = N_a\text{, the Avogadro number]} \tag{1.55}$$

Hence, at low temperatures $C_v \propto T^3$. This is known as *Debye* T^3*-law.*

Debye law is able to explain the experimental results with excellent agreement. Fig 1.7 shows the results for silver at $\theta_D = 225\,\text{K}$.

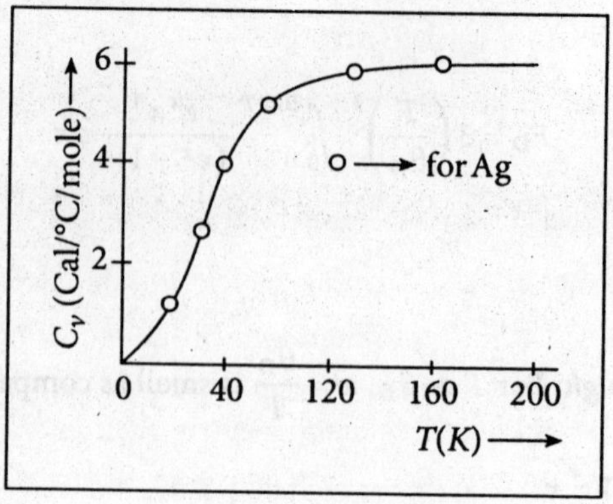

Fig 1.7 C_v vs. T plot

1.4.2 Comparison between Einstein's equation and Debye equation

According to Einstein all oscillators in the solid oscillate with a single frequency, $\nu_0 = \frac{\omega_0}{2\pi}$, which means all the atomic oscillators are excited in unison.

Debye assumed that there is a spectrum of frequency from $\nu = 0$ to $\nu = \nu_D$ (the maximum value of frequency) as the atoms are not excited to the same extent.

Hence, in Einstein's theory there is no upper limit of frequency, but in case of Debye's theory there is an upper limit called *cut-off frequency*.

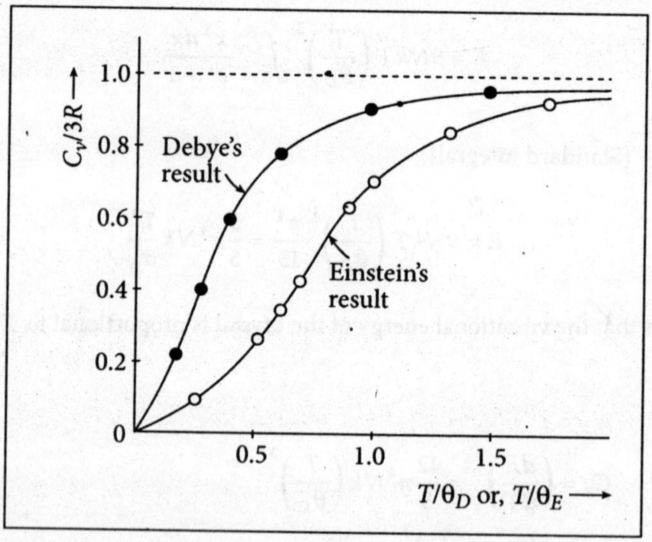

Fig 1.8 Shows the difference of Einstein's result and Debye's result on C_v

Fig 1.8 shows the temperature variation of C_v for Diamond (θ_E = 1326 K) and Al (θ_D = 396 K). From this figure it is clear that Debye's theory gives result which are much more closer to experimentally observed values than Einsteins values.

1.4.3 Discussion of Debye's equation

Though for most of the solids Debye's equation gives satisfactory results, it is based on assumptions:

(a) isotropy of the solid

(b) absence of dispersion of sound waves in the crystal medium

(c) degeneracy of the allowed modes.

In this model the frequency distribution function is proportional to ν^2 with a sharp cut-off at $\nu = \nu_D$. More rigorous calculations show that this is true only near $\nu = 0$, where the solid behaves as an elastic continuum. At all other frequencies, the frequency spectrum differs widely from Debye's assumptions. For good conductors electronic specific heat is also to be considered along with that due to lattice vibration. *Born* and *Karman* have developed the lattice theory of atomic heat capacity which agrees better with experimental results than Debye's model. *C. V. Raman* on the basis of X-ray reflection has shown that ν_D has significance only at lowest temperature.

1.4.4 Drawbacks of the Debye theory

Though Debye theory successfully explains the results in case of many solids, it suffers from the following limitations:

(a) This theory cannot be applied to crystals consisting more than one type of atoms.

(b) The cut-off frequency ν_D is assumed to be the same both for longitudinal and transverse vibrations which is not true.

Also the total number of vibrational modes is assumed to be $3N$, but a continuous solid should have infinite frequencies.

(c) The theory assumes that θ_D is independent of temperature, but practically it is found to vary up to 10 % with temperature.

(d) Debye's model is valid for long wavelengths only which indicates the presence of low frequencies only.

(e) The theory does not take into account of the interaction among the atoms and the contribution of electronic specific heat.

Note: If $D(\omega)\,d\omega$ be the number of modes of vibrations between frequency intervals ω and $\omega + d\omega$, then the vibrational energy of the crystal for these modes is

$dE = \langle\varepsilon(\omega)\rangle\, D(\omega)\, d\omega$ and the total energy of the crystal is

$$E = \int_0^{\omega_m} \langle\varepsilon(\omega)\rangle\, D(\omega)\, d\omega \tag{i}$$

Now, the possible energy level of a harmonic oscillator is $\varepsilon_n = (n + \frac{1}{2})\,\hbar\omega$.

The average value of ε_n at thermal equilibrium can be shown is

$$\langle\varepsilon\rangle = h\nu\left[\frac{1}{2} + \frac{1}{e^{h\nu/kT} - 1}\right] \tag{ii}$$

Also

$$D(\omega) = \frac{3V\omega^2}{2\pi^2 v_0^2} \tag{iii}$$

where v_0 = the phase velocity and is constant. v_0 may be taken to be equal to the velocity of long elastic waves propagating through the crystal. Using (ii) and (iii) in (i), we get

$$E = \frac{3\hbar V\omega_m^4}{16\pi^2 v_0^3} + \frac{3\hbar V}{2\pi^2 v_0^3}\int_0^{\omega_m} \frac{\omega^2}{e^{\hbar\omega/kT}-1}\,d\omega \tag{iv}$$

Putting $x = \dfrac{\hbar\omega}{kT}$

$$x_m = \frac{\hbar\omega_m}{kT} = \frac{\hbar v_0}{kT}\left(\frac{6N\pi^2}{V}\right)^{1/3} \tag{v}$$

$$\text{and } \theta_D = \frac{\hbar\omega_m}{k} = \frac{\hbar v_0}{k}\left(\frac{6N\pi^2}{V}\right)^{1/3} \tag{vi}$$

Using (v) and (vi) in (iv) along with $\omega_m = v_0\left(\dfrac{6\pi^2 N}{V}\right)^{1/3}$, we get

$$E = \frac{9}{8}Nk\theta_D + 9NkT\left(\frac{T}{\theta_D}\right)^3\int_0^{x_m}\frac{x^3}{e^x-1}\,dx \tag{vii}$$

Hence,

$$C_v = \left(\frac{\partial E}{\partial T}\right)_V = 9Nk\left(\frac{T}{\theta_D}\right)^3\int_0^{\theta_D/T}\frac{x^4e^x}{(e^x-1)^2}\,dx \tag{viii}$$

As $T \to 0$, $\chi_m \to \infty$

$$\therefore \int_0^{\infty}\frac{x^3}{e^x-1}\,dx = \frac{\pi^4}{15}$$

Then from (vii)

$$E = \frac{9}{8}Nk\theta_D + \frac{3}{5}\pi^4 NkT\left(\frac{T}{\theta_D}\right)^3$$

At $T = 0$

$$E = \frac{9}{8}Nk\theta_D = \frac{9}{8}R\theta_D \text{ and } C_v = \frac{12}{5}\pi^4 Nk\left(\frac{T}{\theta_D}\right)^3 \tag{ix}$$

1.5 Photoelectric effect

The phenomenon in which electrons are emitted from the metal surface (e.g., Na, K, Cs, etc.) under the action of light (photon) is called the *photoelectric effect*. The metals that exhibit the photoelectric effect are called *photosensitive materials* and the emitted electrons are called *photoelectrons.*

The photoelectric effect was first observed by *Heinrich Hertz* in 1887 and was studied in detail by *Philipp Lenard,* a student of Hertz, Hallwachs, Elster and Geitel.

Lenard's experiment

Lenard's experimental arrangement is as shown in Fig 1.9. Two plane metal plates *C* and *D* are mounted in an evacuated tube. Plate *C* is coated with photosensitive metal and maintained at a negative potential whereas *A* another metal plate and kept at zero potential. The beam of negative ions emitted from *C* can pass through a hole in *A*. UV radiation is incident on *C* through the quartz window *W* fitted to the side tube *G* and negative ions are emitted. These are attracted to *A* and a part of it passing through the hole in *A* is collected at *D*. The resulting electric current can be detected by an electrometer *E*.

The tube is now placed in a magnetic field which is perpendicular to the plane of the paper. The path of the negative ions is bent due to the action of the magnetic field into circular arc and the ions are collected by *E*. The corresponding current is recorded by another electrometer not shown in the diagram.

If *V* be the potential difference between *C* and *A*, then the KE of the emitted negative ions is

$$\tfrac{1}{2}mv^2 = eV \tag{1.56}$$

where *m* is the mass of the ion and *v* is its velocity.

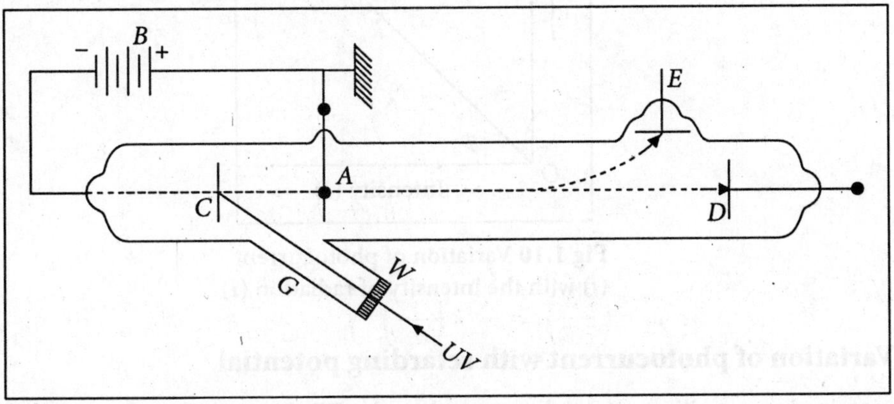

Fig 1.9 Lenard's experiment on photoelectricity

If *R* is the radius of curvature of the ion track and *B* the magnetic field, then

$$Bev = \frac{mv^2}{R} \tag{1.57}$$

e being the charge on the ion. From eqs (1.56) and (1.57)

$$\frac{e}{m} = \frac{2V}{B^2R^2} \tag{1.58}$$

Measuring e/m of the negative ions Lenard proved experimentally that these are nothing, but electrons.

Further, if a positive potential be applied to *C* the electric current is found to decrease with the increase of positive potential. Ultimately a stage comes when at a particular p.d. (V_S) the current becomes zero. Then this V_S is called *stopping potential*. Electrons of all velocities up to the maximum v_m are stopped from reaching *A*. Then

$$\tfrac{1}{2}mv_m^2 = eV_S \tag{1.59}$$

It was found that

(a) If the intensity of UV is increased, it will cause no electrons to reach *A*. But, if *C* is made negative w.r.t. *A*, then on increasing the intensity of UV, the current increases.

(b) For constant intensity if the frequency of the incident radiation be increased, then stopping potential increases.

(c) The photocurrent is independent on frequency of light, but depends on the intensity.

The experimental observations are summarised below.

1.5.1 Variation of photocurrent with the intensity of the incident light

The PD between C and A is increased so that the current is the maximum for a monochromatic light. If now the intensity of the light is increased, then the photocurrent (i) is also found to increase linearly as shown in Fig 1.10. The straight line graph shows that the number of photoelectrons emitted per second by the metal surface is directly proportional to the intensity of the incident radiation.

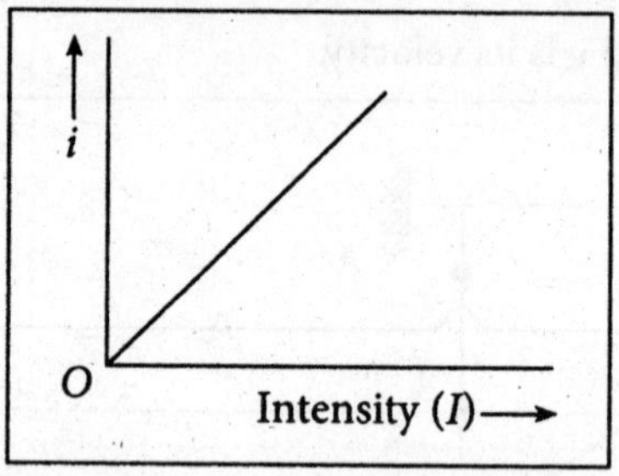

Fig 1.10 Variation of photocurrent (i) with the intensity of radiation (I)

1.5.2 Variation of photocurrent with retarding potential

If C is maintained at a positive potential w.r.t. A, then the PD between C and A is called the *retarding potential.* Keeping the intensity (I_1) of the incident monochromatic light fixed, if the retarding potential be increased, then at a particular value of the *retarding potential* (V) the photocurrent (i) reduces to zero [Fig 1.11]. Then V_S is called the *stopping potential.* The experiment is repeated with light of same frequency but higher intensity (I_2). The curves show that for a given metal the stopping potential V_S is independent of the intensity of the incident radiation.

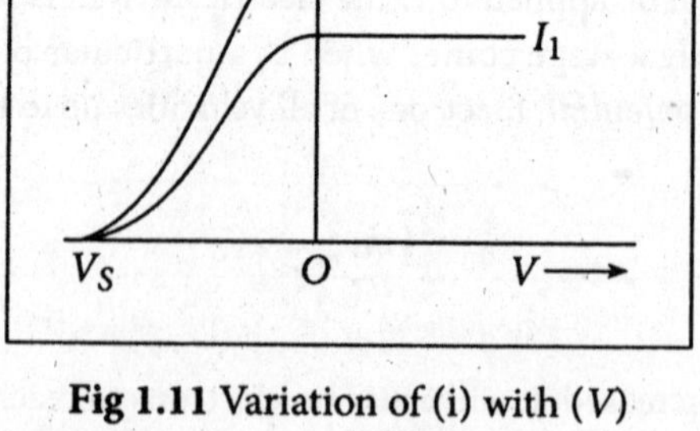

Fig 1.11 Variation of (i) with (V)

This is due to the fact that the photoelectrons are emitted with KE ranging from zero to a certain maximum value. V_S just stops the photoelectrons of maximum KE from reaching the collector in accordance with (1.59).

1.5.3 Variation of stopping potential with the frequency of the incident radiation

The cathode surface is illuminated by light of known frequency ν_1 and an appropriate intensity (I_m). Now, the magnitude of the retarding PD is increased till the photoelectric current is reduced to zero at the stopping potential V_{S_1}. The experiment is, then repeated with the same intensity of the incident light, but of frequencies ν_2 and ν_3. The results are shown in Fig 1.12. It is observed that *the stopping potential increases with the increase of frequency.*

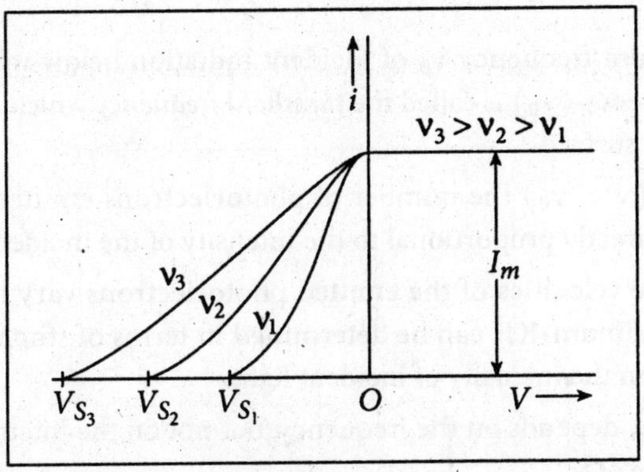

Fig 1.12 Plot of i against V showing the variation of stopping potential (V_S) with frequency (ν) of radiation

1.5.4 The variation of the maximum kinetic energy with frequency

According to the experimental observation, the maximum KE $\frac{1}{2}mv_m^2 = eV_S$ increases linearly with frequency ν after a certain minimum frequency ν_0 of the incident radiation [Fig 1.13]. Fig 1.13 shows the graph of eV_S against frequency ν for three different metals. The slope of the three straight lines is the same and equal to the Planck's constant h.

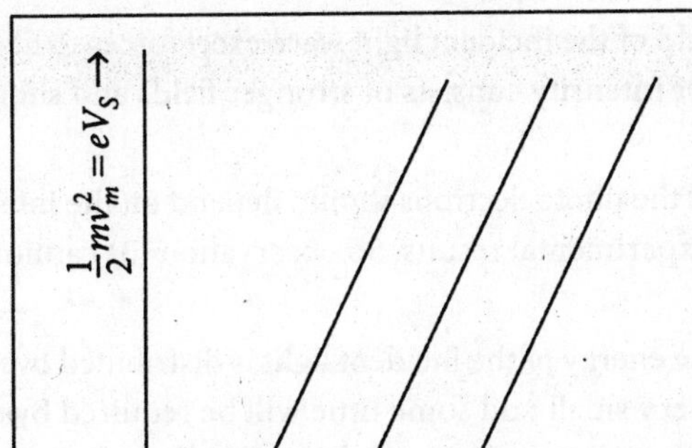

Fig 1.13 Plot of eV_s against ν

However, the intercepts on the ν-axis are different. *The minimum frequency ν_0 that causes the emission with zero KE from the surface is called the threshold frequency.* No emission occurs with frequency below ν_0.

1.5.5 Instantaneous nature of photoelectric effect

Photocurrent is observed in the circuit immediately on illumination of the cathode without any time lag, i.e., *the photoelectric effect is instantaneous.*

1.6 Characteristics of photoelectric effect

The following characteristics (or, laws) of photoelectric emission were derived from the experimental observations of Lenard:

1. There is a certain minimum frequency ν_0 of incident radiation below which there is no emission of photoelectrons. This frequency (ν_0) is called the *threshold frequency* which depends on the material and the nature of the emitting surface.
2. For a given frequency $\nu(\nu > \nu_0)$ the number of photoelectrons emitted per second for a constant accelerating potential is directly proportional to the intensity of the incident radiation.
3. For frequency $\nu > \nu_0$ the velocities of the emitted photoelectrons vary between zero and a definite maximum (v_m). The maximum KE can be determined in terms of stopping potential. The stopping potential is independent on the intensity of incident light.
4. The maximum velocity v_m depends on the frequency, but not on the intensity of the incident light. v_m increases with the increase of ν.
5. The emission of photoelectrons is *instantaneous* no matter how feeble the incident light may be. The time lag between the incidence of photon and emission of electrons ~ 10^{-9} s.

1.7 Failure of the electromagnetic wave theory

1. According to wave theory, photoelectric effect should occur for any frequency of light, provided that its intensity can cause vibrations of electrons to the amplitude required so as to free them from the material. But, experimentally for each metal, there exists a *threshold frequency* below which no emission should occur. Thus, e.m. theory fails to explain observation (1).
2. The electric and magnetic field of the incident light wave exert forces to liberate electrons from the metal surface. So the light of higher intensity consists of stronger fields and should give higher velocity to the photoelectrons.

   In other words, the energy of the photoelectrons should depend on the intensity of the incident radiation. But, this is just contrary to experimental results. So observation (3) cannot be accounted for by classical e.m. theory of radiation.
3. According to wave theory, the energy of the incident light is distributed over the wavefront. So the energy incident on any electron is very small and some time will be required by an electron to absorb enough energy to escape from the surface. But, experiment shows that there is no time lag between the irradiation of the surface and the start of the emission. So classical e.m. theory fails to account observation (5).

   The wave theory thus fails to account the characteristics of photoelectric effect.

1.8 Einstein's photoelectric equation

In 1901, *Max Planck* proposed his quantum theory which explained the long standing problem regarding the distribution of energy in the spectrum of blackbody radiation. According to this theory, the emission and absorption of radiation occur discretely in the basic unit, called *quantum of energy,* i.e., energy of a quantum $\varepsilon = h\nu$, where h is the Planck's constant and ν is the frequency of radiation. So the emitted or absorbed energy is $E = n\varepsilon = nh\nu$, where $n = 1, 2, 3, \ldots.$

Einstein's hypothesis

Einstein applied Planck's quantum theory to offer a satisfactory explanation of the photoelectric phenomenon. He made the following assumptions:

1. Planck proposed that the emission and absorption of radiation takes place in discrete quanta (photon) of energy $\varepsilon = h\nu$. Einstein extending Planck's idea postulated that radiation travels in discrete quanta with the speed of light.
2. The incident photon on the metal surface can transfer the whole of its energy content to one electron without any time lag. The probability of absorption of two or more photons at the same time is negligible.
3. A certain minimum energy called the *work function* W_0 is required to release an electron from the metal surface.

The photoelectric equation

When a radiation of frequency ν is incident on the metal surface in vacuum, it can either be reflected from the metal surface or it can be absorbed completely by an electron. In the latter case, the electron can be knocked out of the metal surface. The emission of the electron from the metal surface is possible only if $h\nu > W_0$. The balance of energy $(h\nu - W_0)$ is utilised to impart kinetic energy of the emitted electron. Some electrons, emitted from the interior of the metal, expand a fraction of their acquired energy in collision with the atoms. The loss of energy is a variable quantity leading the emitted energy (on velocity) of the photoelectron to vary from zero to a certain maximum [Fig 1.14]. So the maximum kinetic energy of the emitted electron is

$$\tfrac{1}{2}mv_m^2 = h\nu - W_0 \tag{1.60}$$

where m is the mass and v_m, the maximum velocity of the emitted electron.

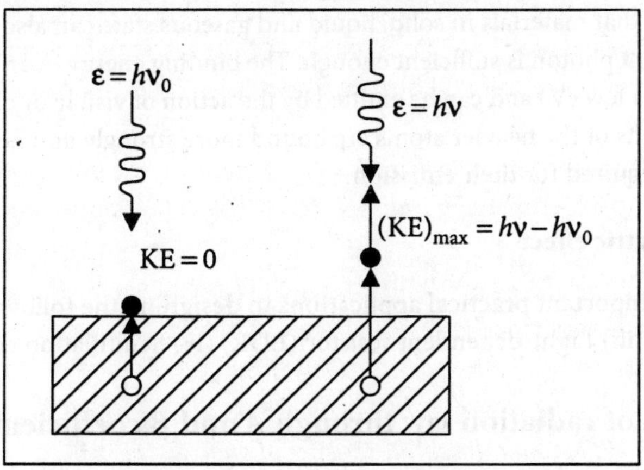

Fig 1.14 Mechanism of photoelectric emission

If the whole energy of the photon is utilised to release the electron from the metal surface, then

$$h\nu - W_0 = 0 \text{ or, } W_0 = h\nu = h\nu_0 \tag{1.61}$$

Here, $\nu = \nu_0$ is called the *threshold (or critical) frequency.*

From eq. (1.60)

$$\tfrac{1}{2}mv_m^2 = h\nu - h\nu_0 = h(\nu - \nu_0) \tag{1.62}$$

Equations (1.60) and (1.62) are the well-known *photoelectric equation of Einstein.*

Explanation of photoelectric phenomenon

1. If $\nu < \nu_0$, then from eq. (1.62), $\tfrac{1}{2}mv_m^2$ is negative. This means that no electrons could be emitted if the frequency of the incident photon ν be less than ν_0. This establishes the existence of threshold frequency ν_0. This explains observation (1).
2. The maximum kinetic energy, according to eq. (1.62) depends only on the frequency ν of the incident photon and not on the intensity of the incident radiation. The maximum kinetic energy varies linearly with ν, when $\nu > \nu_0$. This explains observation (4).
3. An increase in intensity of the incident radiation for a fixed frequency $\nu(\nu \geqslant \nu_0)$ is equivalent to increase in the number of photon and not energy.

   Hence, the number of emitted electrons with the increase in intensity will increase, but their maximum kinetic energy will remain unchanged. Thus, observation (2) is explained.
4. If $\nu \geqslant \nu_0$, then the incident photon a concentrated packet of energy, is absorbed at an instant ($\tau \sim 10^{-9}$ s) by the electron and is emitted instantaneously. This explains the instantaneous emission of the photoelectrons [observation (5)].

Hence, Einstein's quantum theory can explain successfully all the photoelectric characteristics.

It is to be kept in mind that materials in solid, liquid and gaseous state can also show photoelectric effect if the energy of the incident photon is sufficient enough. The binding energy of an electron in the outermost orbit of an atom is low (~ a few eV) and can be emitted by the action of visible or ultraviolet photon. But, the electrons in the inner orbits of the heavier atoms are bound more strongly and hence high-energy photons of X-rays or, γ-rays are required for their emission.

Application of photoelectric effect

Photoelectric effect has important practical applications in designing the following devices: (i) Photocell (ii) Photomultiplier tube (iii) Light-dependent resistor (LDR) (iv) Scintillation counter (v) Solar cells.

Einstein's derivation of radiation law through *A* and *B* coefficients

The radiation law was derived by Einstein by considering statistical approach and quantum ideas.

Let an atom be placed in a uniform field of energy density U_ν for frequency ν. The absorption or emission of radiation between the two states E_1 and E_2 in the atom can take place due to transitions between the states as follows:

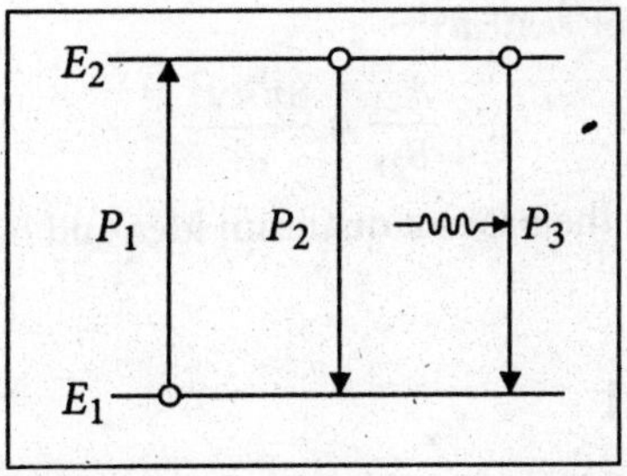

Fig 1.15 Absorption and emission of radiation between two states E_1 and E_2 in the atom

1. The transition of an atom from E_1 to E_2 can take place through the absorption of a quantum of radiation from the field and the probability of this transition is $P_1 = B_{12} U_\nu$.
2. The transition from E_2 to E_1 can take place in the following two ways:
   (a) due to spontaneous emission of a quantum independent of the incident radiation and the probability of this transition is $P_2 = A_{21}$.
   (b) due to induced emission of a quantum of radiation assisted by the incident radiation with a probability $P_3 = B_{21} U_\nu$.

The transition from E_1 to E_2 in accordance with *Boltzmann law* in thermal equilibrium is proportional to $e^{-E_1/kT}$ and that from E_2 to E_1 is proportional to $e^{-E_2/kT}$, where k is the *Boltzmann constant* and T is the temperature of the medium.

In equilibrium

$$P_2 + P_3 = P_1 \text{ or, } (A_{21} + B_{21} U_\nu)\, e^{-E_2/kT} = B_{12} U_\nu e^{-E_1/kT} \tag{1.63}$$

where A and B are constants and depend on the atomic states E_1 and E_2, but independent of U_ν. A and B are called *Einstein's A and B coefficients*.

For large T, i.e., for high value of U_ν, $B_{21} U_\nu \geqslant A_{21}$.

$$\therefore B_{21} U_\nu = B_{12} U_\nu \text{ or, } B_{21} = B_{12} \tag{1.64}$$

This is *the principle of detailed balance* for induced transitions.

Putting $B_{21} = B_{12}$ in eq. (1.63), we get

$$(A_{21} + B_{21} U_\nu)\, e^{-E_2/kT} = B_{21} U_\nu e^{-E_1/kT}$$

$$\text{or, } A_{21} e^{-E_2/kT} = B_{21} U_\nu \left(e^{-E_1/kT} - e^{-E_2/kT}\right)$$

$$\text{or, } U_\nu = \frac{A_{21} e^{-E_2/kT}}{B_{21}\left(e^{-E_1/kT} - e^{-E_2/kT}\right)} = \frac{A_{21}/B_{21}}{e^{(E_2-E_1)/kT} - 1} \tag{1.65}$$

$$\text{or, } U_\nu = \frac{A_{21}/B_{21}}{e^{h\nu/kT} - 1} \tag{1.66}$$

$\because E_2 - E_1 = h\nu$, according to quantum theory Planck's law of blackbody radiation is

$$U_\nu = \frac{8\pi h\nu^2}{c^3} \frac{1}{e^{h\nu/kT} - 1} \tag{1.67}$$

Thus, comparing eqs (1.66) and (1.67), we get

$$\frac{A_{21}}{B_{21}} = \frac{8\pi h\nu^2}{c^3} \tag{1.68}$$

The above results are obtained on the basis of quantum idea and hence establishes the validity of quantum theory.

1.9 The Compton effect

When a beam of monochromatic X-rays of wavelength λ is scattered by a light element (like carbon), it is observed that the scattered X-rays have maximum intensities at two wavelengths λ and $\lambda'(\lambda' > \lambda)$. The X-radiation of wavelength λ is called *the unmodified radiation* whereas the radiation of wavelength λ' is called *the modified radiation.*

The phenomenon was discovered by *A.H. Compton* of USA in 1922 and is known as the *Compton effect.* The wavelength difference $\Delta\lambda = \lambda' - \lambda$ is called the *Compton shift* and is independent of the wavelength of the incident beam and the nature of the scatter, but depends on the angle of scattering.

1.9.1 The experiment

The experimental arrangement is shown in Fig 1.16 below:

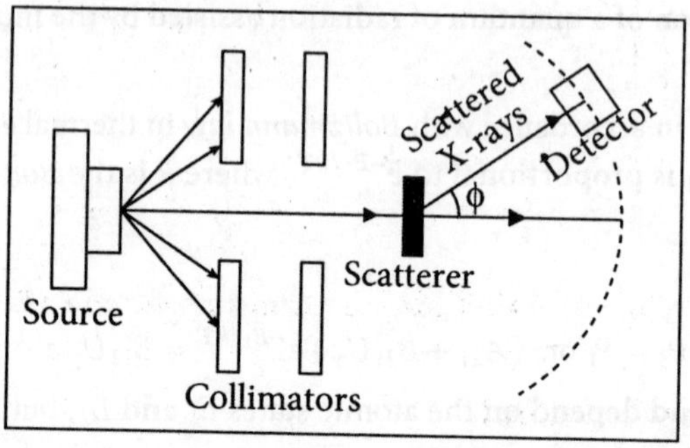

Fig 1.16 Experimental arragement for the study of Compton scattering

A beam of monochromatic X-rays were allowed to fall on a block of graphite and the intensity of the scattered X-rays were measured as a function of the wavelength of X-rays at different scattering angles using a detector (e.g., an ionisation chamber). For each angle of scattering ϕ (other than $\phi = 0$) two peaks appear corresponding to the scattered X-ray photons with two different wavelengths. The intensity (I) variation of the scattered X-rays with wavelength (λ) for different angle of scattering (ϕ) is shown in Fig 1.17.

1.9.2 Failure of classical theory

In view of classical wave theory, X-rays are electromagnetic waves. If ν be the frequency of the incident X-rays on the material, then they cause electrons in the material to oscillate at the same frequency ν. The oscillating electron, then reradiate X-rays of the same frequency ν.

Hence, the scattered X-radiation should consist of single frequency ν (or, wavelength λ).

Again, the electrons should radiate X-rays uniformly in all possible directions and hence, the wavelength of the scattered radiation should not vary with the scattering angle ϕ, which is a contrary to experiment conducted by Compton.

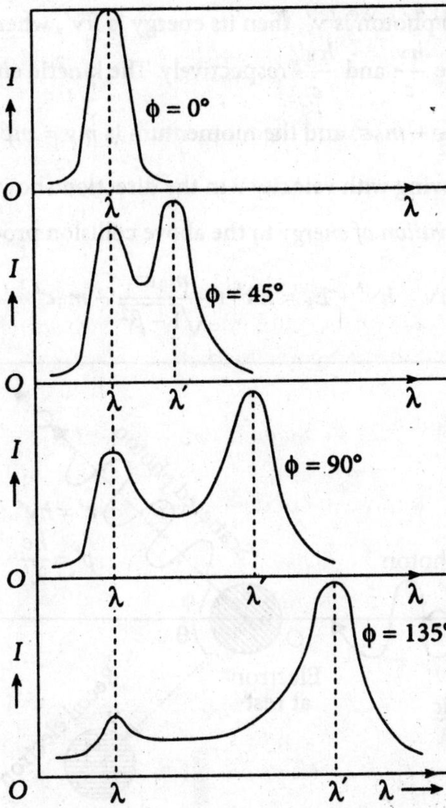

Fig 1.17 Compton shift ($\Delta\lambda$) at various angles of scattering.

1.9.3 Theory of the Compton effect

Compton applied *Einstein's quantum theory of light* with the assumption that the incident photons possess momentum in addition to energy. In order to derive an expression for the Compton shift ($\Delta\lambda$), Compton assumed the following:

1. The beam of incident monochromatic X-radiation of frequency ν consists of a stream of photons each of which having energy $\varepsilon = h\nu$ travels along the direction of the beam with velocity c, where c is the velocity of light in free space. Each photon of energy ε has a momentum $p = \frac{\varepsilon}{c} = \frac{h\nu}{c}$, where h is the Planck's constant.

2. The scattering is due to the elastic collision between the photon and the free electron at rest.

   Hence, the conservation laws of energy and momentum hold good in the scattering process.

We consider a beam of monochromatic X-rays of frequency ν and wavelength λ be incident on a target electron at O of an element of low atomic number (e.g., carbon).

Let the incident photon after suffering elastic collision with the free electron is scattered through an angle ϕ and the electron recoils in a direction θ with the direction of incidence [Fig 1.18].

If the frequency of the scattered photon is ν', then its energy is $h\nu'$, where $\nu' < \nu$. The momenta of the incident and scattered photons are $\frac{h\nu}{c}$ and $\frac{h\nu'}{c}$ respectively. The kinetic energy gained by the electron is $E_k = mc^2 - m_0c^2$ or, $E_k = \frac{m_0c^2}{\sqrt{1-\beta^2}} - m_0c^2$ and the momentum is $mv = m\beta c$, where $\beta = v/c$ and m is the mass of the electron when it is moving with velocity v in the direction shown in Fig 1.18.

Now, applying the *law of conservation of energy* to the above collision process, we get

$$h\nu = h\nu' + E_k = h\nu' + \frac{m_0c^2}{\sqrt{1-\beta^2}} - m_0c^2 \tag{1.69}$$

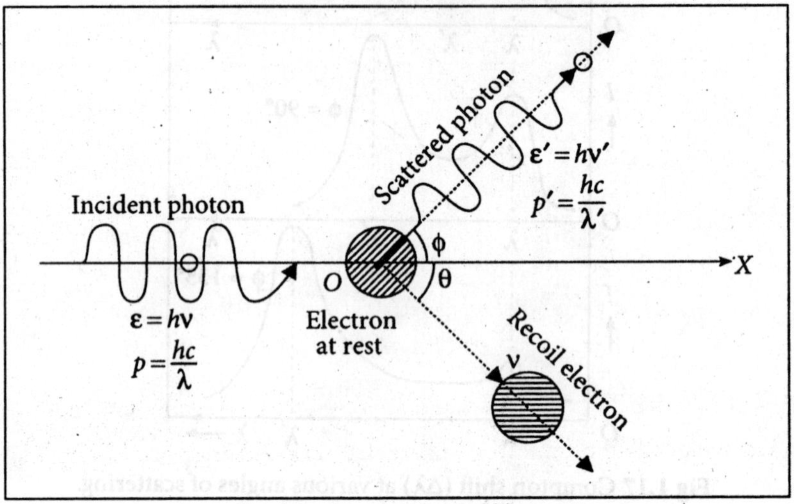

Fig 1.18 Compton effect

Applying the *law of conservation of linear momentum along and perpendicular to the direction of incidence*, we get

$$\frac{h\nu}{c} = \frac{h\nu'}{c}\cos\phi + \frac{m_0\beta c}{\sqrt{1-\beta^2}}\cos\theta \tag{1.70}$$

$$\text{and } 0 = \frac{h\nu'}{c}\sin\phi - \frac{m_0\beta c}{\sqrt{1-\beta^2}}\sin\theta \tag{1.71}$$

Equations (1.70) and (1.71) can be expressed as

$$\frac{m_0\beta c}{\sqrt{1-\beta^2}}\cos\theta = \frac{h\nu}{c} - \frac{h\nu'}{c}\cos\phi \tag{1.72}$$

$$\text{and } \frac{m_0\beta c}{\sqrt{1-\beta^2}}\sin\theta = \frac{h\nu'}{c}\sin\phi \tag{1.73}$$

Squaring and adding eqs (1.72) and (1.73), we get

$$\frac{m_0^2\beta^2c^2}{1-\beta^2} = \frac{h^2\nu^2}{c^2} + \frac{h^2\nu'^2}{c^2} - \frac{2h^2\nu\nu'}{c^2}\cos\phi \tag{1.74}$$

From eq. (1.69)

$$\left(\frac{m_0c^2}{\sqrt{1-\beta^2}}\right)^2 = (h\nu - h\nu' + m_0c^2)^2 = h^2\nu^2 + h^2\nu'^2 + m_0^2c^4 - 2h^2\nu\nu' + 2hm_0c^2(\nu - \nu') \quad (1.75)$$

Dividing both sides of the above equation by c^2, we get

$$\frac{m_0^2c^2}{1-\beta^2} - m_0^2c^2 = \frac{m_0^2\beta^2c^2}{1-\beta^2} = \frac{h^2\nu^2}{c^2} + \frac{h^2\nu'^2}{c^2} - \frac{2h^2\nu\nu'}{c^2} + 2m_0h(\nu - \nu') \quad (1.76)$$

Now, subtracting eq. (1.74) from eq. (1.76), we get

$$0 = 2m_0h(\nu - \nu') - \frac{2h^2\nu\nu'}{c^2}(1 - \cos\phi) \text{ or, } \frac{\nu - \nu'}{\nu\nu'} = \frac{h}{m_0c^2}(1 - \cos\phi)$$

$$\text{or, } \frac{c}{\nu'} - \frac{c}{\nu} = \frac{h}{m_0c}(1 - \cos\phi) \quad (1.77)$$

Putting $\frac{c}{\nu'} = \lambda'$ and $\frac{c}{\nu} = \lambda$, we get

$$\lambda' - \lambda = \frac{h}{m_0c}(1 - \cos\phi) \quad (1.78)$$

$$\text{or, } \Delta\lambda = \lambda_c(1 - \cos\phi) \quad (1.79)$$

where $\Delta\lambda = \lambda' - \lambda$ is called *the Compton shift* and $\lambda_c = \frac{h}{m_0c}$, *the Compton wavelength*. It is evident from eq. (1.79) that $\Delta\lambda$ is independent on the scattering material as well as on the wavelength λ of the incident radiation.

However, $\Delta\lambda$ depends on the angle of scattering ϕ.

Putting $h = 6.626 \times 10^{-34}$ J-s, $m_0 = 9.11 \times 10^{-31}$ kg and $c = 2.998 \times 10^8$ m/s, we get

$$\lambda_c = \frac{h}{m_0c} = \frac{6.626 \times 10^{-34}}{9.11 \times 10^{-31} \times 2.998 \times 10^8} = 0.02426 \text{ Å} \quad (1.80)$$

Thus, the Compton wavelength λ_c is a universal constant.

Now, from eqs (1.79) and (1.80), Compton shift

$$\Delta\lambda = 0.02426(1 - \cos\phi) \quad (1.81)$$

Equation (1.81) has excellent agreement with the experimental results establishing the fact that *Compton effect is a quantum phenomenon.*

In conclusion, the Compton shift $\Delta\lambda$

(a) depends only on the scattering angle ϕ

(b) is independent on the nature of the scattering material

(c) is independent on the wavelength of the incident radiation

(d) when larger, the smaller the rest mass of the scatterer.

The variation of $\Delta\lambda$ with the scattering angle ϕ is shown in Fig 1.19.

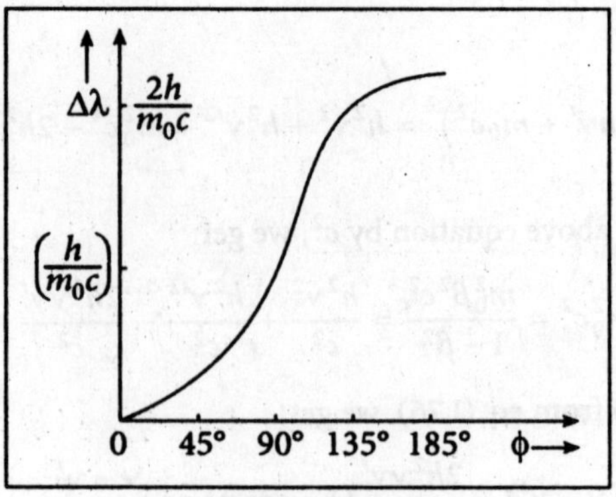

Fig.1.19 Variation of Δλ with ϕ

1.9.4 Relation between ϕ and θ

We have from Compton theory

$$\frac{c}{\nu'} - \frac{c}{\nu} = \frac{h}{m_0 c}(1 - \cos\phi)$$

$$\text{or, } \frac{1}{\nu'} - \frac{1}{\nu} = \frac{h}{m_0 c^2}(1 - \cos\phi) \tag{1.82}$$

$$\text{or, } \frac{\nu}{\nu'} = 1 + \frac{h\nu}{m_0 c^2}(1 - \cos\phi) \tag{1.83}$$

$$\text{or, } \frac{\nu}{\nu'} = 1 + \alpha(1 - \cos\phi) \tag{1.84}$$

where $\alpha = d\frac{h\nu}{m_0 c^2}$.

Also

$$\frac{m_0 \beta c}{\sqrt{1-\beta^2}} \cos\theta = \frac{h\nu}{c} - \frac{h\nu'}{c}\cos\phi \tag{1.85}$$

$$\text{and } \frac{m_0 \beta c}{\sqrt{1-\beta^2}} \sin\theta = \frac{h\nu'}{c}\sin\phi \tag{1.86}$$

Dividing eq. (1.86) by eq. (1.85), we get

$$\tan\theta = \frac{\nu' \sin\phi}{\nu - \nu'\cos\phi} = \frac{\sin\phi}{\nu/\nu' - \cos\phi} \tag{1.87}$$

From eq. (1.84)

$$\nu' = \frac{\nu}{1 + \alpha(1 - \cos\phi)} = \frac{\nu}{1 + 2\alpha \sin^2\frac{\phi}{2}} \tag{1.88}$$

From eqs (1.87) and (1.88)

$$\tan\theta = \frac{2\sin\frac{\phi}{2}\cos\frac{\phi}{2}}{1+2\alpha\sin^2\frac{\phi}{2}-\cos\phi} = \frac{2\sin\frac{\phi}{2}\cos\frac{\phi}{2}}{2\sin^2\frac{\phi}{2}+2\alpha\sin^2\frac{\phi}{2}}$$

$$\text{or, } \tan\theta = \frac{2\sin\frac{\phi}{2}\cos\frac{\phi}{2}}{2\sin^2\frac{\phi}{2}(1+\alpha)} = \frac{\cos\frac{\phi}{2}}{(1+\alpha)\sin\frac{\phi}{2}} \text{ or, } \tan\theta = \frac{\cot\frac{\phi}{2}}{(1+\alpha)}$$

$$\text{or, } \cot\theta = (1+\alpha)\tan\frac{\phi}{2} \tag{1.89}$$

Equation (1.89) gives a relation between the angle of scattering (ϕ) and the angle of recoil (θ).

1.9.5 Kinetic energy of the recoil electron

The kinetic energy of the recoil electron is given by

$$E_K = h\nu - h\nu' = h\nu\left[1 - \frac{1}{1+2\alpha\sin^2\frac{\phi}{2}}\right]\left[\because \nu' = \frac{\nu}{1+2\alpha\sin^2\frac{\phi}{2}}\right]$$

$$= h\nu\left[\frac{2\alpha\sin^2\frac{\phi}{2}}{1+2\alpha\sin^2\frac{\phi}{2}}\right] \text{ or, } E_K = h\nu\frac{\alpha(1-\cos\phi)}{1+\alpha(1-\cos\phi)} \tag{1.90}$$

In terms of the angle of recoil θ

$$E_K = h\nu\frac{2\alpha\cos^2\theta}{(1+\alpha)^2 - \alpha^2\cos^2\theta} \tag{1.91}$$

The value of E_K was experimentally determined by *Compton* and *Simon* (in 1925) and *Bless* (in 1927) which agrees well with the theoretical values obtained by Compton.

Compton effect thus proves that *radiant energy is quantised and is a decisive phenomenon which supports quantisation of energy.*

1.9.6 Limitations of Compton theory

1. In the experiment of Compton effect, X-radiation of incident wavelength λ along with the X-rays of wavelength λ' were observed. But, Compton theory does not offer any interpretation regarding the existence of unmodified X-rays of wavelength λ.

   The existence of scattered X-rays of wavelength λ can be understood as follows:

   The incident X-ray photons collide with (i) loosely bound outer electrons (ii) tightly bound inner electrons of the atom. The tightly bound electron is not detached from the atom and as such the entire atom has to undergo recoil. In such cases, m_0 in the expression of $\Delta\lambda$ should be replaced by M, the total mass of the atom which reduces the value of $\Delta\lambda \to 0$ and goes undetected.

   Example: For a graphite target $M = 12 \times 1840 \times m_0$.

$$\therefore \Delta\lambda = \frac{2h}{Mc}\sin^2\frac{\phi}{2} = \frac{2}{12\times 1840}\left(\frac{h}{m_0 c}\right)\sin^2\left(\frac{180°}{2}\right) = 9.058\times 10^{-5}\times 0.02426$$

$$\text{or, } \Delta\lambda = 2.197\times 10^{-6}\ \text{Å} \tag{1.92}$$

   which is quite negligible.

2. Experimentally, it has been observed that the intensity of modified X-rays (wavelength λ') is greater than that of the unmodified X-rays (wavelength λ) for lighter elements. But, reverse is the case for heavier elements. Compton theory does not offer any interpretation of the above facts.

1.10 Compton scattering and photoelectric effect

For the occurrence of Compton effect, the energy of the incident photon $(h\nu)$ must be larger than the binding energy E_B of the electron with the nucleus (i.e., $h\nu > E_B$), so that the electron recoils and a low-energy scattered photon appears. It can be shown that the incident photon cannot be absorbed completely by the target atom so that scattered photon does not appear. If this would be possible, then

$$h^2\nu^2 = m^2v^2c^2 \tag{1.93}$$

$$\text{and } h^2\nu^2 + 2m_0c^2h\nu = \left(m^2 - m_0^2\right)c^4 \text{ [putting } \nu' = 0 \text{ in eq. (1.75)]} \tag{1.94}$$

$$\text{or, } h^2\nu^2 + 2m_0c^2h\nu = m^2v^2c^2 \left[\because m^2 = \frac{m_0^2}{1-\beta^2} \text{ and } \beta = \frac{v}{c}\right]$$

Equations (1.93) and (1.94) cannot be satisfied simultaneously.

Hence, we conclude that in Compton scattering, the appearance of the scattered photon is a must and the complete absorption of the incident photon is impossible for the sake of conservation of energy and momentum.

In case of photoelectric effect, the energy of the incident photon $h\nu \sim E_B$. With the incidence of the photon of energy $h\nu$ the binding between the electron and the ion breaks down. The resulting ion recoils with a definite energy and momentum.

Now, the kinetic energies of the electron and the ion neglecting relativistic effect can be expressed as

$$E_e = \frac{1}{2}m_0v_e^2 = \frac{p_e^2}{2m_0} \text{ and } E_0 = \frac{1}{2}\mu_A v_i^2 = \frac{p_i^2}{2\mu_A}$$

where μ_A is the mass of the ion and m_0 is the rest mass of the electron. $\mu_A \gg m_0$, so, $E_e \gg E_i$, even if $p_i \simeq p_e$. This implies that the ion can recoil with momentum p_i so that conservation of momentum holds good. The corresponding KE is negligible.

However, the photoelectric equation of Einstein accounts the energy balance. Photoelectric effect and Compton scattering can take place simultaneously when electromagnetic radiation passes through a matter and both the processes have their relative magnitude as energy loss is concerned. If $h\nu$ is much higher, then *pair production* can take place in which a photon is converted into an electron-positron pair.

1.11 The wave-particle duality

It is evident from different experimental results that light (radiation) plays a dual role, in some experiment it behaves as wave, while in other experiment as particle.

In optical phenomena like interference, diffraction, reflection, refraction and polarisation of light which are the result of interaction of light with light, it behaves as a continuous wave. These phenomena can be explained by the wave theory or the electromagnetic theory of radiation.

There is another class of phenomena, like blackbody radiation, the photoelectric effect and Compton effect where light behaves as a flux of photons. These arise due to the interaction of radiation with matter and cannot be explained on the basis of wave theory or electromagnetic theory of radiation. To explain these phenomena the radiant energy is considered as a stream of energy packets called *quanta* or *photons each of energy* $\varepsilon = h\nu$. The frequency and wavelength are relevant concept to wave whereas quantum is relevant to particle. So *radiation possesses dual character,* but both characters, *wave* and *particle* are never exhibited simultaneously in any experiment.

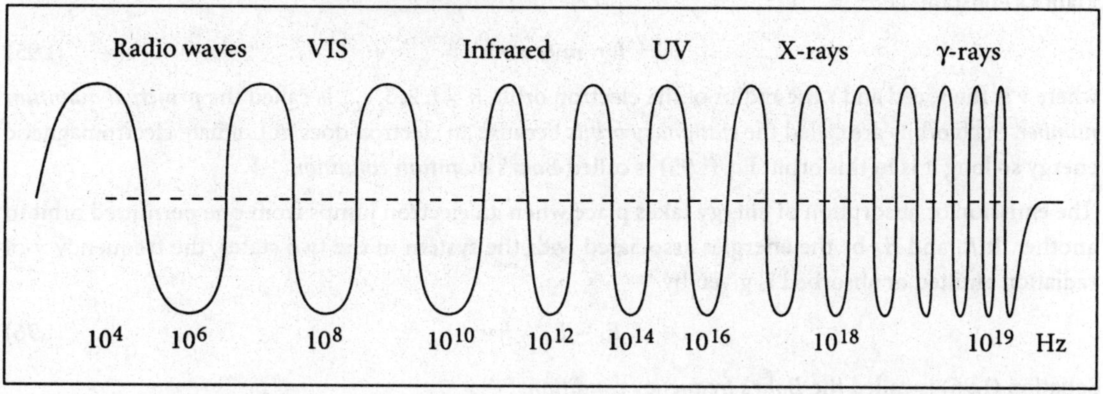

Fig 1.20 Wave nature is prominent in the low-frequency region whereas particle nature is prominent in the high-frequency region

The manifestation of wave-particle dualism becomes transparent if we consider the entire electromagnetic spectrum. At lower frequency there is radio waves ($\lambda \sim$ a few hundred metres) and spreads over a large volume of space [Fig 1.20]. So the energy available at any point is insignificantly small and as such the particle nature cannot be observed.

Again, at the higher frequency end of the electromagnetic spectrum there are ultraviolet X-rays and γ-rays ($\lambda \sim$ a few angstrom) the waves are located in a very small region of space due to their very short wavelength and the energy at a particular point assumes a large value. Consequently, the wave property is less noticeable as compared to particle property.

Hence, in low-frequency region the wave nature is prominent whereas in high-frequency region the particle nature dominates. The visible region is the transition region where both wave and particle aspects are observed.

According to quantum theory, energy and momentum of a photon are given by $\varepsilon = h\nu$ and $p = h/\lambda$. Here, ε and p are characteristics of particle and ν and λ are those of waves. Hence, *wave-particle dualism* is apparent from the above relations.

Again, the amplitude (A) of a light wave determines the probability that a photon can be found at a particular point in space. The probability (P) to observe photons is $P \propto |A|^2$. The wave-particle dualism is well expressed by this relation.

1.12 Bohr's theory of hydrogen atom

In 1913, *Niels Bohr* of Denmark applied Planck-Einstein's quantum theory to *Rutherford atom model* and successfully explained the origin of spectral lines in hydrogen atom. To overcome the difficulties associated with Rutherford atom model, Bohr proposed the following postulates:

1. In hydrogen atom the electron moves in a circular orbit around the nucleus at its centre. The nucleus being very heavy as compared to the electron is practically at rest.

2. Out of infinite number of circular orbits, only those orbits are permissible for the electron in which the total angular momentum of the electron is equal to an integral multiple of $\hbar$ $\left(\hbar = \frac{h}{2\pi}\right)$, where h is the Planck's constant, i.e.,

$$L = mvr = n\hbar \tag{1.95}$$

where v is the speed and r the radius of the electron orbit, $n = 1, 2, 3, \ldots$ is called the *principal quantum number*. Such orbits are called the *stationary orbits* because an electron does not radiate electromagnetic energy so long it is in this orbit. Eq. (1.95) is called *Bohr's quantum condition*.

3. The emission or absorption of energy takes place when an electron jumps from one permitted orbit to another. If E_i and E_f be the energies associated with the system in the two states, the frequency ν of radiation emitted or absorbed is given by

$$E_i \sim E_f = h\nu \tag{1.96}$$

Equation (1.96) is called the *Bohr's frequency condition*.

Postulate (2) is a violation of the laws of classical mechanics according to which an electron could revolve in any orbit of any radius. Postulate (2) is also a violation of the laws of classical electrodynamics according to which the revolving electron (having acceleration $f = v^2/r$) would loose energy through the emission of electromagnetic waves due to its centripetal acceleration.

Now, in view of wave mechanics an electron moving in a circular orbit of radius r with a constant speed v is associated with a *de Broglie wavelength*

$$\lambda = \frac{h}{p} = \frac{h}{mv} \tag{1.97}$$

According to postulate (2)

$$mvr = n\hbar = \frac{nh}{2\pi} \text{ or, } 2\pi r = n\frac{h}{mv} = n\frac{h}{p} = n\lambda \tag{1.98}$$

Thus, the circumference of the orbit is an integral multiple of wavelength associated with the electron. This means that the wave at any point will be in phase with the wave that would result after the electron completes one revolution. Such waves interfere constructively causing the formation of a stationary wave, i.e., a suitable orbit would result (Fig 1.21).

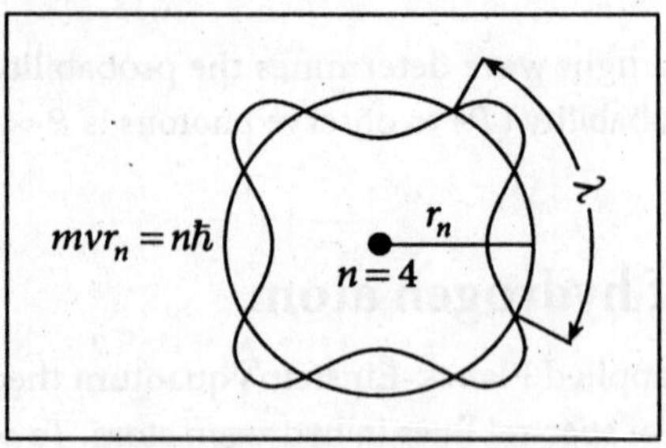

Fig 1.21 Constructive interference of electron wave

If $2\pi r \neq n\hbar$, then the wave at any point would be out of phase with the wave resulting after one orbit is completed. When the wave resulting after the electron has undergone many revolutions, is added to it, the wave train will cancel and the electron wave will vanish (Fig 1.22).

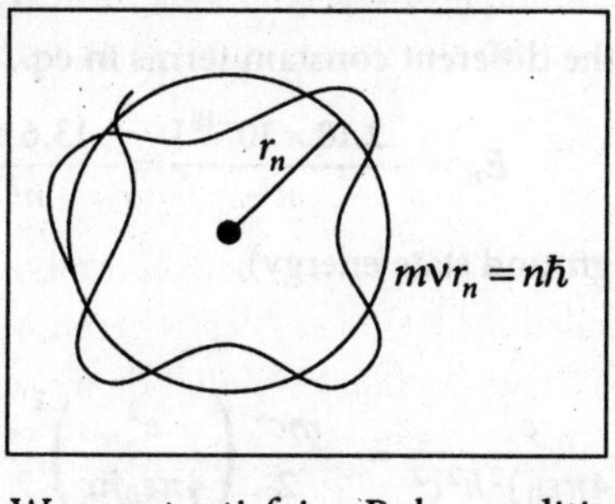

Fig 1.22 Wave not satisfying Bohr condition

1.12.1 Energy levels

Since, the electron in hydrogen atom revolves in a circular orbit of radius r around the nucleus of charge $+Ze = +e$ ($Z = 1$ for hydrogen) with speed v, the centripetal force mv^2/r must be equal to the electrostatic attraction $\dfrac{e^2}{4\pi\varepsilon_0 r^2}$ between them.

$$\therefore \frac{mv^2}{r} = \frac{e^2}{4\pi\varepsilon_0 r^2} \tag{1.99}$$

From eq. (1.95)

$$v = \frac{n\hbar}{mr} \tag{1.100}$$

$\therefore$ from eq. (1.99)

$$r = \frac{e^2}{4\pi\varepsilon_0 m v^2} = \frac{e^2}{4\pi\varepsilon_0 m \times \frac{n^2\hbar^2}{m^2r^2}} \text{ or, } r = \frac{e^2 m r^2}{4\pi\varepsilon_0 n^2\hbar^2}$$

$$\text{or, } r = \frac{4\pi\varepsilon_0 n^2\hbar^2}{me^2} \tag{1.101}$$

So r can take only certain allowed values.

From eqs (1.100) and (1.101)

$$v = \frac{e^2}{4\pi\varepsilon_0 n\hbar} \tag{1.102}$$

Now, the total energy of the electron is

$$E = \text{KE} + \text{PE} = \frac{1}{2}mv^2 + \left(-\frac{e^2}{4\pi\varepsilon_0 r}\right) = \frac{e^2}{2\times 4\pi\varepsilon_0 r} - \frac{e^2}{4\pi\varepsilon_0 r} \left[\because mv^2 = \frac{e^2}{4\pi\varepsilon_0 r}\right]$$

$$\text{or, } E = -\frac{1}{2}\frac{e^2}{4\pi\varepsilon_0 r} \tag{1.103}$$

The energy E is *negative* which means that the *electron is bound to the nucleus* and to separate it from the nucleus work has to be done.

Now, from eqs (1.101) and (1.103)

$$E = -\frac{1}{2}\frac{e^2}{4\pi\varepsilon_0} \times \frac{me^2}{4\pi\varepsilon_0 n^2\hbar^2} = -\frac{me^4}{(4\pi\varepsilon_0)^2 2\hbar^2}\left(\frac{1}{n^2}\right) \tag{1.104}$$

Using the standard values of the different constant terms in eq. (1.104), we get for the nth quantum state

$$E_n = -\frac{2.18 \times 10^{-18}\,\text{J}}{n^2} = -\frac{13.6\,\text{eV}}{n^2} \tag{1.105}$$

For $n = 1$, $E_1 = -13.6$ eV (the ground state energy).

Again,

$$E_1 = -\frac{mc^2}{2}\frac{e^4}{(4\pi\varepsilon_0)^2\hbar^2 c^2} = -\frac{mc^2}{2}\left(\frac{e^2}{4\pi\varepsilon_0\hbar c}\right)^2 = -\frac{mc^2}{2}\alpha^2 = -13.6\,\text{eV} \tag{1.106}$$

Here, $\alpha = \dfrac{e^2}{4\pi\varepsilon_0\hbar c} \sim \dfrac{1}{137}$ is called the *fine structure constant.*

The first Bohr orbit radius

$$r_1 = a_0 = \frac{\hbar^2 4\pi\varepsilon_0}{me^2} = \left(\frac{\hbar}{mc}\right)\frac{4\pi\varepsilon_0\hbar c}{e^2} = \frac{\hbar}{mc}\left(\frac{1}{\alpha}\right)$$

$$\text{or, } a_0 \simeq 9.53 \times 10^{-10}\,\text{m} \tag{1.107}$$

For the nth orbit

$$E_n = -\frac{mc^2\alpha^2}{2n^2} = +\frac{E_1}{n^2} \tag{1.108}$$

$$\text{and } r_n = a_0 n^2 = \frac{\hbar}{mc}\frac{n^2}{\alpha} \tag{1.109}$$

We note from eqs (1.108) and (1.109) that E decreases as $1/n^2$ and r_n increases as n^2. Hence, *an electron far away from the nucleus is less bound to the nucleus than those nearer to it.*

Also the orbits of large n values are more closely spaced.

Bohr's theory explained the experimentally observed discrete nature of hydrogen spectrum. According to Bohr's frequency condition, the frequency (ν) of the emitted photon when an electron jumps from an orbit of energy E_i to an orbit of energy E_f is given by

$$\nu = \frac{E_i - E_f}{h} = \frac{mc^2\alpha^2}{2h}\left(\frac{1}{n_f^2} - \frac{1}{n_i^2}\right) \quad [n_i > n_f]$$

$$\text{or, } \frac{c}{\lambda} = \frac{mc^2\alpha^2}{2h}\left(\frac{1}{n_f^2} - \frac{1}{n_i^2}\right)$$

$$\text{or, } \frac{1}{\lambda} = \frac{mc\alpha^2}{2h}\left(\frac{1}{n_f^2} - \frac{1}{n_i^2}\right) = \frac{mc\alpha^2}{4\pi\hbar}\left(\frac{1}{n_f^2} - \frac{1}{n_i^2}\right) \left[\because \hbar = \frac{h}{2\pi}\right]$$

or, the wave number

$$\bar{\nu} = \frac{1}{\lambda} = R_H\left(\frac{1}{n_f^2} - \frac{1}{n_i^2}\right) \tag{1.110}$$

where

$$R_H = \frac{mc\alpha^2}{4\pi\hbar} = \frac{mc}{4\pi\hbar}\left(\frac{e^2}{4\pi\varepsilon_0\hbar c}\right)^2 \text{ or, } R_H = \frac{me^4}{(4\pi\hbar)^3\varepsilon_0^2 c} \tag{1.111}$$

R_H is called the *Rydberg constant* for hydrogen having value $R_H = 1.09737 \times 10^7\ \text{m}^{-1}$, which agrees well with the experimental value. For an atom having Z electrons, Rydberg constant is given by

$$R_Z = \frac{me^4Z^2}{(8\pi\hbar)^3\varepsilon_0^2 c} = R_H Z^2. \tag{1.112}$$

1.12.2 Origin of spectral series

The origin of the discrete series of spectral series can be understood from Bohr's theory as follows. We have the wave number of a spectral line as

$$\tilde{\nu} = R_H\left(\frac{1}{n_f^2} - \frac{1}{n_i^2}\right)$$

1. ***Lyman series***: If $n_f = 1$ and $n_i = 2, 3, 4, \ldots$, we get a series of spectral lines having wave number $\tilde{\nu}_{LI}$, $\tilde{\nu}_{LII}$, $\tilde{\nu}_{LIII}$, ... as follows:

$$\tilde{\nu}_{LI} = R_H\left(\frac{1}{1^2} - \frac{1}{2^2}\right) = \frac{3}{4}R_H \quad \text{or, } \lambda_{LI} = \frac{4}{3R_H}$$

$$\tilde{\nu}_{LII} = R_H\left(\frac{1}{1^2} - \frac{1}{3^2}\right) = \frac{8}{9}R_H \quad \text{or, } \lambda_{LII} = \frac{9}{8R_H}$$

$$\tilde{\nu}_{LIII} = R_H\left(\frac{1}{1^2} - \frac{1}{4^2}\right) = \frac{15}{16}R_H \quad \text{or, } \lambda_{LIII} = \frac{16}{15R_H}, \text{ etc.}$$

These lines are all in *ultraviolet region*.

2. ***Balmer series***: If $n_f = 2$, $n_i = 3, 4, 5, \ldots$, we get a series of spectral lines as follows:

$$\tilde{\nu}_{BI} = R_H\left(\frac{1}{2^2} - \frac{1}{3^2}\right) = \frac{5}{36}R_H, \quad \therefore \lambda_{BI} = \frac{36}{5R_H}$$

$$\tilde{\nu}_{BII} = R_H\left(\frac{1}{2^2} - \frac{1}{4^2}\right) = \frac{12}{16}R_H = \frac{3}{4}R_H, \therefore \lambda_{BII} = \frac{4}{3R_H}$$

$$\tilde{\nu}_{BIII} = R_H\left(\frac{1}{2^2} - \frac{1}{5^2}\right) = \frac{21}{50}R_H, \quad \therefore \lambda_{BIII} = \frac{50}{21R_H}, \text{ etc.}$$

These are all in the *visible region*.

3. ***Paschen's series***: If $n_f = 3$ and $n_i = 4, 5, 6, \ldots$, then we get Paschen series:

$$\tilde{\nu}_{PI} = R_H\left(\frac{1}{3^2} - \frac{1}{4^2}\right) = \frac{7}{192}R_H, \quad \therefore \lambda_{PI} = \frac{192}{7R_H}$$

$$\tilde{\nu}_{PII} = R_H\left(\frac{1}{3^2} - \frac{1}{5^2}\right) = \frac{16}{225}R_H, \quad \therefore \lambda_{PII} = \frac{225}{16R_H}$$

$$\tilde{\nu}_{PIII} = R_H\left(\frac{1}{3^2} - \frac{1}{6^2}\right) = \frac{27}{36}R_H = \frac{3}{4}R_H, \therefore \lambda_{PIII} = \frac{4}{3R_H}, \text{ etc.}$$

The series appear in *infrared region*.

4. ***Brackett series***: If $n_i = 4$ and $n_f = 5, 6, 7, \ldots$, we get Brackett series:

$$\nu_{\text{Br I}} = R_H\left(\frac{1}{4^2} - \frac{1}{5^2}\right) = \frac{9}{400}R_H \text{ or, } \lambda_{\text{Br I}} = \frac{400}{9R_H}$$

$$\nu_{\text{Br II}} = R_H\left(\frac{1}{4^2} - \frac{1}{6^2}\right) = \frac{5}{36}R_H \text{ or, } \lambda_{\text{Br II}} = \frac{36}{5R_H}$$

$$\nu_{\text{Br III}} = R_H\left(\frac{1}{4^2} - \frac{1}{7^2}\right) = \frac{33}{784}R_H \text{ or, } \lambda_{\text{Br III}} = \frac{784}{33R_H}, \text{ etc.}$$

These lines appear in *infrared region*.

5. ***Pfund series***: If $n_f = 5$ and $n_i = 6, 7, 8, \ldots$, we get Pfund series

$$\bar{\nu}_{\text{FI}} = R_H\left(\frac{1}{5^2} - \frac{1}{6^2}\right) = \frac{11}{900}R_H, \quad \therefore \lambda_{\text{FI}} = \frac{900}{11R_H}$$

$$\bar{\nu}_{\text{FII}} = R_H\left(\frac{1}{5^2} - \frac{1}{7^2}\right) = \frac{24}{1225}R_H, \quad \therefore \lambda_{\text{FII}} = \frac{1225}{24R_H}$$

$$\bar{\nu}_{\text{FIII}} = R_H\left(\frac{1}{5^2} - \frac{1}{8^2}\right) = \frac{39}{16000}R_H, \quad \therefore \lambda_{\text{FIII}} = \frac{16000}{39R_H}, \text{ etc.}$$

All the lines appear in *infrared region.*

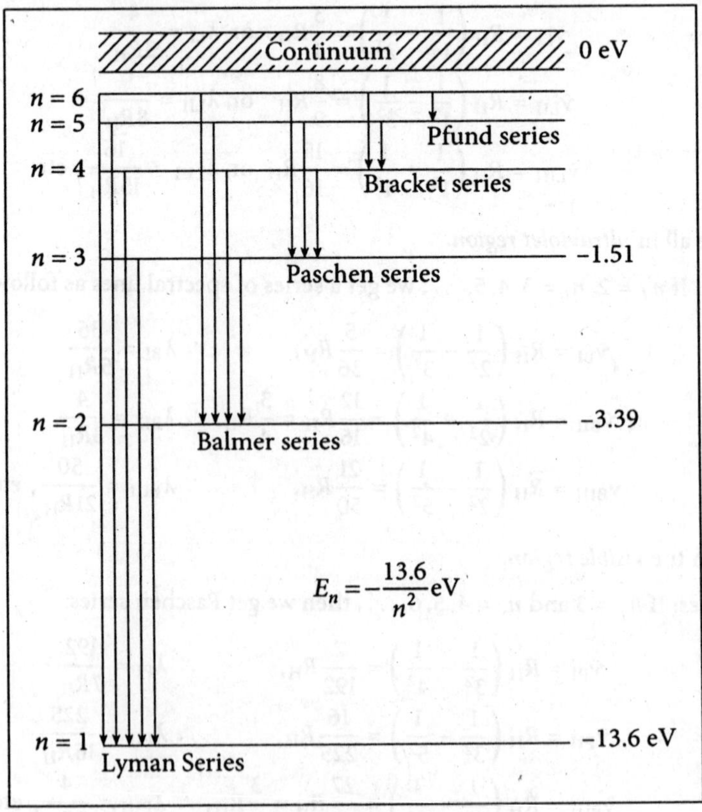

Fig 1.23 Energy level diagram

1.12.3 Energy level diagram

The quantised allowed energy values are called *energy level* or *energy states of a system*. The plot of these energy values as horizontal lines is the energy level diagram.

The level with $n = 1$ is called the *ground state*, the atom is in the normal state. As n increases the energy E becomes less and less negative and the states are called the *excited states*. For $n \to \infty$, $E_n \to 0$ and the electron is completely removed from the atom. This corresponds to series limit and ionisation of the atom. The levels for $E \gg 0$ form a continuum [Fig 1.23].

1.13 Bohr's theory and motion of the nucleus

In Bohr's theory it has been assumed that the nucleus of the hydrogen atom is so heavy that it remains at rest at the centre of the circular orbit and the electron revolves around the nucleus. This is not true as the mass of the nucleus is only 1836 times the mass of an electron. With a finite mass of the nucleus, both the nucleus as well as the electron revolve about their common centre of gravity with the same angular velocity ω [Fig 1.24].

If r_1 and r_2 be the distances of the electron and the nucleus from the CM of the system, then $r = r_1 + r_2$ and $Mr_2 = mr_1$,

M is the mass of the nucleus.

$$\therefore r_1 = \frac{Mr}{M+m} \text{ and } r_2 = \frac{mr}{M+m}$$

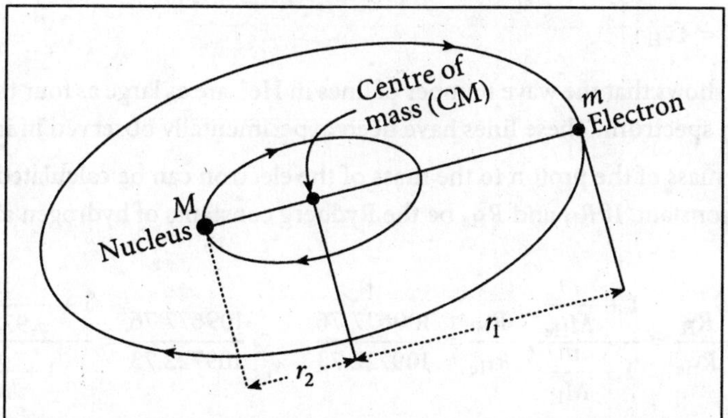

Fig 1.24 Motion of the electron relative to the moving nucleus

The kinetic energy of the system is

$$E_k = \frac{1}{2}MV^2 + \frac{1}{2}mv^2 = \frac{1}{2}M\omega^2 r_2^2 + \frac{1}{2}mr_1^2\omega^2 = \frac{\omega^2}{2}[Mr_2^2 + mr_1^2]$$

$$= \frac{\omega^2}{2}\left[M\left(\frac{mr}{M+m}\right)^2 + m\left(\frac{Mr}{M+m}\right)^2\right] = \frac{r^2\omega^2}{2}\frac{mM}{M+m} = \frac{1}{2}\mu r^2\omega^2 \tag{1.113}$$

where $\mu = \dfrac{mM}{m+M}$ is *the reduced mass* of the system and $\mu \to m$ as $M \to \infty$.

Now, all the expressions of Bohr's theory will be valid, if we replace m by μ. The wave number, then becomes

$$\bar{\nu} = \frac{\mu c\alpha^2}{2h}\left(\frac{1}{n_f^2} - \frac{1}{n_i^2}\right) = \frac{c\alpha^2}{2h}\left(\frac{mM}{m+M}\right)\left(\frac{1}{n_f^2} - \frac{1}{n_i^2}\right)$$

$$= \frac{mc\alpha^2}{2h}\left(\frac{1}{1+\frac{m}{M}}\right)\left(\frac{1}{n_f^2} - \frac{1}{n_i^2}\right) = \frac{mc\alpha^2}{4\pi\hbar^2}\left(\frac{1}{1+\frac{m}{M}}\right)\left(\frac{1}{n_f^2} - \frac{1}{n_i^2}\right)$$

$$\text{or, } \bar{\nu} = R_N\left(\frac{1}{n_f^2} - \frac{1}{n_i^2}\right) \tag{1.114}$$

where

$$R_N = \frac{mc\alpha^2}{4\pi\hbar^2\left(1+\frac{m}{M}\right)} = \frac{R_H}{1+\frac{m}{M}} \tag{1.115}$$

Since, $\mu = \frac{mM}{M+m}$, it is evident that $\mu < m$. When $M \to \infty$, $\mu \to m$ and $R_N \to R_H$. Equations (1.114) and (1.115) have important and interesting conclusions.

(a) Ions having hydrogen-like structure, i.e., a nucleus of charge $+2e$ and a single electron must give rise to spectral lines similar to hydrogen atom. For example, consider the case of He^+ having nuclear charge $+2e$. The wave number of the spectral lines are

$$\bar{\nu}_{He} = R_N\left(\frac{1}{n_f^2} - \frac{1}{n_i^2}\right) = Z^2R_H\left(\frac{1}{n_f^2} - \frac{1}{n_i^2}\right) = 4R_H\left(\frac{1}{n_f^2} - \frac{1}{n_i^2}\right) \quad [\because Z = 2 \text{ for } He^+]$$

$$\text{or, } \bar{\nu}_{He} = 4\bar{\nu}_H \tag{1.116}$$

Equation (1.116) shows that the wave number of lines in He^+ are as large as four times the corresponding line of hydrogen spectrum. These lines have been experimentally observed in ionised helium.

(b) The ratio of the mass of the proton to the mass of the electron can be calculated by knowing the value of the Rydberg constant. If R_H and R_{He} be the Rydberg constants of hydrogen and helium respectively, then

$$\frac{R_H}{R_{He}} = \frac{1+\frac{m}{M_{He}}}{1+\frac{m}{M_H}}, \quad \frac{R_H}{R_{He}} = \frac{109677.76}{109722.73}, \quad \therefore \frac{109677.76}{109722.73} = \frac{1+\frac{m}{3.9717M_H}}{1+\frac{m}{M_H}}$$

Solving, we get

$$\frac{M_H}{m} = 1840 \tag{1.117}$$

Thus, $M_H = 1840m$, i.e., the mass of a proton is 1840 times the mass of an electron which is in excellent agreement with values determined by other methods.

(c) *Wavelength difference of hydrogen and deuterium lines*: We have the wavelength of hydrogen (λ_H) and deuterium (λ_D) are given by

$$\frac{1}{\lambda_H} = \bar{\nu}_H = R_H\left(\frac{1}{n_f^2} - \frac{1}{n_i^2}\right), \quad \frac{1}{\lambda_D} = \bar{\nu}_D = R_D\left(\frac{1}{n_f^2} - \frac{1}{n_i^2}\right)$$

$$\therefore \frac{\lambda_D}{\lambda_H} = \frac{R_H}{R_D} = \frac{\dfrac{R_H}{1+\dfrac{m}{M_H}}}{\dfrac{R_H}{1+\dfrac{m}{M_D}}} = \frac{1+\dfrac{m}{M_D}}{1+\dfrac{m}{M_H}}, \therefore \frac{\lambda_H - \lambda_D}{\lambda_H} = \frac{m}{M_H} - \frac{m}{2M_H} = \frac{m}{2M_H}$$

[Since, $M_D = 2M_H$ and $m \ll M_H$, or $\dfrac{m}{M_H} \ll 1$]

$$\therefore \Delta\lambda = \lambda_H - \lambda_D = \frac{m}{2M_H}\lambda_H = \frac{\lambda_H}{3672}\left[\because \frac{m}{M_H} = 1836\right] \tag{1.118}$$

Franck-Hertz experiment

J. Franck and *G. Hertz* of Germany discovered (in 1914) experimentally that the discrete energy levels in atom do really exist and are the same as those suggested by the observation of line spectra. This experimental verification of Bohr's theory constitutes another triumph of the quantum hypothesis proposed by Max Planck.

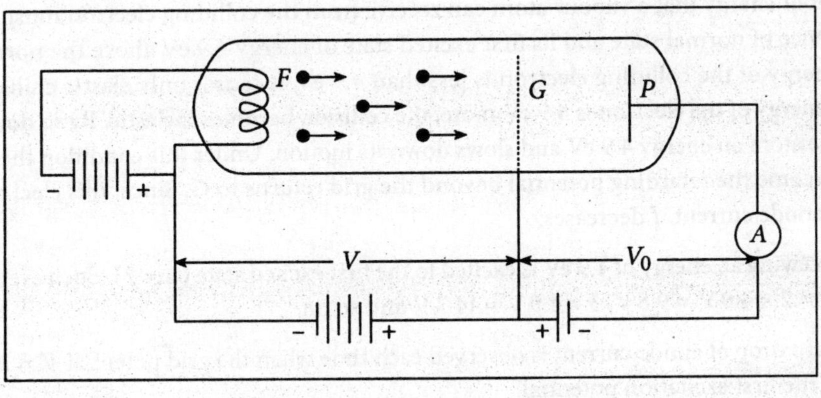

Fig 1.25 Franck-Hertz experiment

The schematic representation of the experimental arrangement is shown in Fig 1.25 in which vapours of various elements are bombarded by electrons in the tube. A small PD V_0 between the grid (G) and the collector (P) prevents electrons having energies less than a certain minimum from contributing to the current I recorded by the ammeter (A). With the increase of V more and more electrons reach the plate P and the current I increases.

The variation of I with V is studied and the results are plotted taking V along x-axis and I along y-axis [Fig 1.26]. The current first increases and after reaching a maximum at about 4.9 V, it decreases sharply with the further increase of V. It reaches a minimum and, then again increases. The current assumes the maximum values for V = 9.8 V, 17.7 V, . . . and so on.

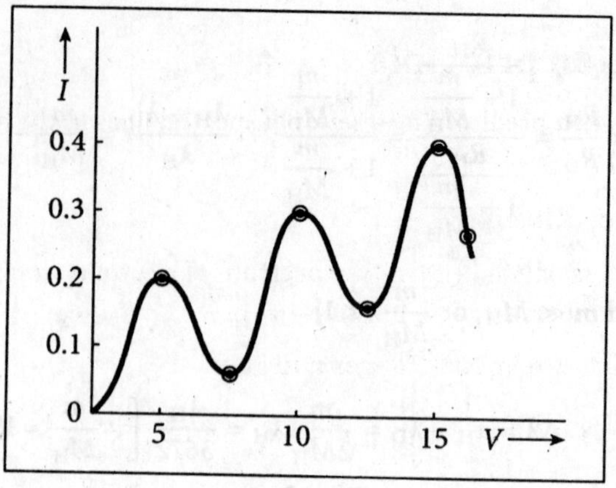

Fig 1.26 A series of critical potentials for an atomic species

Explanation of the results

When V is low, the collision between a vapour atom and an electron is elastic. As a result the electron does not lose energy, but changes its direction of motion, without exciting the atom. When the electron reaches G, it acquires sufficient KE to reach P. With the increase of V more and more electrons reach P, and I increases.

The amount of energy that a vapour atom can receive from the colliding electron must be equal to the energy difference of normal state and its first excited state of energy 4.9 eV above the normal state. So as long as the energy of the colliding electron is less than 4.9 eV, it causes only elastic collision. When the accumulated energy of the electron is 4.9 eV more, the collision becomes inelastic. By so doing the electron transfers to the atom an energy 4.9 eV and slows down its motion. Under this condition the electron being unable to overcome the retarding potential beyond the grid returns to G. Since, this electron is unable to contribute to anode current, I decreases.

The atom receiving an energy of 4.9 eV is excited to the first excited state ($n = 2$). Such events are repeated again when $V = 2 \times 4.9\text{ V} = 9.8\text{ V}$, $3 \times 4.9\text{ V} = 14.7\text{ V}$ and so on.

Hence, a sharp drop of anode current is observed each time when the grid potential $V = nV_1$ ($n = 1, 2, 3, \ldots$), V_1 being the first excitation potential.

The mercury vapour in the tube emits UV light at $\lambda = 2537$ Å. This is due to the transition of the atom from $n_i = 2$ to $n_f = 1$ state.

Now,

$$\lambda = \frac{c}{\nu} = \frac{hc}{E_f - E_i} = \frac{hc}{4E} \text{ or, } \lambda = \frac{hc}{\Delta E} = \frac{(6.62 \times 10^{-34}\text{J-s}) \times (3 \times 10^{8}\text{m/s})}{(4.9\text{ V}) \times (1.60 \times 10^{-19}\text{C})} = 2533\text{ Å}. \quad (1.119)$$

The excellent agreement between theory and experiment lends support to the Bohr's model.

Hence, *Franck-Hertz experiment offers a direct proof to the existence of discrete energy levels as proposed by Bohr.* Franck and Hertz received Nobel Prize in Physics in 1925 for this experiment.

1.14 Ritz combination principle

With the help of this principle, Ritz predicted new series of spectral lines in the hydrogen spectrum before their discovery.

Statement of Ritz combination principle

If ν_1 and ν_2 be the wave numbers of two lines in the spectrum of a certain atom, it often happens that the lines also occurs at the wave numbers $(\nu_1 + \nu_2)$ or $(\bar{\nu}_1 - \bar{\nu}_2)$.

The wave numbers for the first two lines of Balmer series are

$$\bar{\nu}_{H_\alpha} = R_H\left(\frac{1}{2^2} - \frac{1}{3^2}\right) \text{ and } \bar{\nu}_{H_\beta} = R_H\left(\frac{1}{2^2} - \frac{1}{4^2}\right)$$

Combining, we get

$$\bar{\nu}_{H_\beta} - \bar{\nu}_{H_\alpha} = R\left(\frac{1}{3^2} - \frac{1}{4^2}\right) \tag{1.120}$$

This represents the first line of a new series in the *infrared region* of hydrogen spectrum which was discovered by *Paschen*. The difference in wave number of H_γ and H_α similarly gives the second line of *Paschen series*.

The validity of the principle can be tested as follows:

The first line of *Paschen series* $(n_f = 3, n_i = 4)$ and *Balmer series* $(n_f = 2, n_i = 3)$ are

$$\bar{\nu}_{43} = 5.331 \times 10^5 \text{ m}^{-1}$$

$$\bar{\nu}_{32} = 15.233 \times 10^5 \text{ m}^{-1}$$

$$\therefore \bar{\nu}_{43} + \bar{\nu}_{32} = 5.331 \times 10^5 + 15.233 \times 10^5 \text{ m}^{-1} = 20.564 \times 10^5 \text{ m}^{-1}$$

Again, for the second line of *Balmer series* $(n_f = 2, n_i = 4)$

$$\bar{\nu}_{43} = 20.565 \times 10^5 \text{ m}^{-1}$$

Note: Due to the operation of selection rule and quantum condition, each combination is not allowed.

Thus, the combination of the first time of *Paschen series* and *Balmer series* gives the second line of *Balmer series*. This proves the validity of the principle.

Ritz combination principle and alkali spectra

The alkali spectra arise due to the atoms having one electron in the outermost orbit. The alkali spectra metal, viz., Na, K, Li, Cs all have one electron in their outermost orbit. Hydrogen atom and the singly ionised atoms of Be, Mg, Ca, etc. have also one electron in their outermost orbit. The spectra emitted by such atoms can be written using Ritz combination principle.

The general expression for the wave number $(\bar{\nu})$ of alkali spectra is given by

$$\bar{\nu} = R\left[\frac{1}{(p-\alpha)^2} - \frac{1}{(q-\beta)^2}\right] \tag{1.121}$$

where p is a first term and q is a running term. α and β are characteristic constants. The wave number of the different series in alkali spectra are

1. ***Principal series***: In this case, $p = 1$ and $q \geqslant 2$.

$$\therefore \tilde{\nu}_P = R\left[\frac{1}{(1-\alpha)^2} - \frac{1}{(q-\beta_p)^2}\right] \tag{1.122}$$

2. ***Sharp series***: For this case, $p = 2$ and $q \geqslant 2$.

$$\therefore \tilde{\nu}_S = R\left[\frac{1}{(2-\alpha)^2} - \frac{1}{(q-\beta_p)^2}\right] \tag{1.123}$$

3. ***Diffuse series***: For this case, $p = 2$ and $q \geqslant 3$.

$$\therefore \tilde{\nu}_D = R\left[\frac{1}{(2-\alpha)^2} - \frac{1}{(q-\beta_d)^2}\right] \tag{1.124}$$

4. ***Fundamental series***: For this case, $p = 3$ and $q \geqslant 4$.

$$\therefore \tilde{\nu}_F = R\left[\frac{1}{(3-\alpha)^2} - \frac{1}{(q-\beta_f)^2}\right] \tag{1.125}$$

Here, β_p, β_s, β_d and β_f are the characteristic constants of the respective series.

1.15 Bohr's correspondence principle

The classical physics accounts successfully the dynamical behaviour of macroscopic bodies including the atoms. Kinetic theory of matter was also formulated using the laws of classical mechanics.

Again, for large quantum numbers (n) in Bohr's theory, the successive allowed energy values differ very little from each other and assume continuous nature.

Also the spectral lines tend to the series limit and become much closer to form a continuous appearance.

Neils Bohr, therefore, (in 1923) felt the need of enunciating his famous *correspondence principle* as follows.

The behaviour of an atomic system as predicted by quantum theory tends asymptotically to that expected in classical theory in transitions involving states of large quantum numbers.

Now, the frequency of revolution of an electron in a circular orbit of radius r with a speed v is

$$\nu_{Cl} = \frac{1}{T} = \frac{1}{\left(\frac{2\pi r}{v}\right)} = \frac{v}{2\pi r} \text{ or, } \nu_{Cl} = \frac{me^4}{2\pi n^3\hbar^3} \tag{1.126}$$

since

$$v = \frac{e^2}{n\hbar} \text{ and } r = \frac{n^2\hbar^2}{me^2}$$

Again, the frequency of the radiation emitted when an atom jumps from n_f to n_i quantum states having energies E_f and E_i respectively is

$$\nu_Q = \frac{E_f - E_i}{2\pi\hbar} = \frac{me^4}{4\pi\hbar^3}\left[\frac{1}{n_i^2} - \frac{1}{n_f^2}\right]$$

If $n_i = n$ and $n_f = n + 1$, then

$$\nu_Q = \frac{me^4}{4\pi\hbar^3}\left[\frac{1}{n^2} - \frac{1}{(n+1)^2}\right] = \frac{me^4}{4\pi\hbar^3}\frac{2n+1}{n^2(n+1)^2}$$

For $n \to \infty$, $\dfrac{2n+1}{n^2(n+1)^2} \to \dfrac{2n}{n^4} = \dfrac{2}{n^3}$.

$$\therefore \nu_Q = \frac{me^4}{4\pi\hbar^3} \times \frac{2}{n^3} = \frac{me^4}{2\pi n^3 \hbar^3} \tag{1.127}$$

Hence, for $n \to \infty$, $\nu_{Cl} = \nu_Q$. This principle offers important guidelines to form laws of atomic mechanics.

1.15.1 Success of Bohr's theory

Bohr's theory of hydrogen spectra has the following successes:

1. It provides a satisfactory explanation to the origin of spectral lines.
2. Empirically determined value of Rydberg constant and that evaluated from Bohr's theory agree well.
3. The value of the ratio of mass of the electron to the mass of the proton agrees excellently with the value obtained by other methods.
4. The theory predicts new undiscovered spectral series lines which have been discovered by the spectroscopist later on.
5. The calculated value of the radius of the first Bohr orbit, $a_0 = \dfrac{h^2}{4\pi^2 me^2} = 0.528$ Å agrees closely with the value obtained by other methods.
6. It helped H.C. Urey to discover heavy hydrogen (deuterium).

1.15.2 Failure of Bohr's theory

Bohr's theory suffers from the following failures:

1. The theory does provide any idea regarding the distribution of electrons in an atom.
2. In this theory the equilibrium of the electron is governed by classical laws, but emission of radiation is explained by quantum rules.
3. The theory does not give any idea about the intensities and the state of polarisation of spectral lines. It is unable to explain the *multiplicity* and *fine structure* of spectral lines.
4. The dynamical problems of atoms with more than one electron cannot be treated with its help.
5. The selection rules cannot be calculated with the help of this theory.
6. The electron orbits according to Bohr's theory are circular. In fact, they may be elliptical in case of heavier atoms.

1.16 The Sommerfeld relativistic atom model

The hydrogen spectral lines when observed through spectroscopes of low resolving power appears to be single lines. But, when observed through spectroscopes of high resolving power, the single line appears to consist of groups of separate and distinct lines located very close to each other. This is known as *fine structure of spectral lines.* Now, the origin of more than one spectral line can be explained only when several electron orbits of slightly different energies exist for a particular quantum number n. But, in view of Bohr's theory only one electron orbit is possible for each value of n. So Bohr's theory is unable to account the fine structure of spectral lines.

A. Sommerfeld in 1915 introduced *elliptic orbits for the electrons* and the *concept of relativistic variation of mass of the electron* to explain the origin of fine structure of spectral lines.

1.16.1 Elliptical orbits

Periodic motion under the action of an inverse square law force leads to elliptic orbits. It was therefore, postulated by Sommerfeld that in an atom an electron moves in an elliptic orbit with the central nucleus located at a focus. In an elliptic orbit, an electron can be located by two coordinates r (the radius vector) and ϕ (the azimuthal angle) as shown in Fig 1.27. If p_r and p_ϕ be the resolved components of the momentum p of the electron along radial direction and perpendicular to it, then Sommerfeld assumed that

$$p_r\, dr = n_r h \tag{1.128}$$

$$\text{and } p_\phi\, d\phi = n_\phi h \tag{1.129}$$

where n_r and n_ϕ are called the *radial quantum number* and *azimuthal quantum number* respectively.

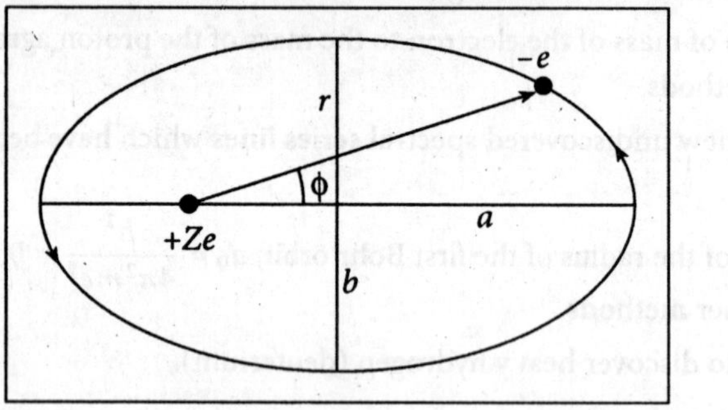

Fig 1.27 Sommerfeld elliptic orbit of the electron with the nucleus at one focus

1.16.2 Allowed elliptical orbits

According to classical mechanics the angular momentum p_ϕ remains constant during the motion.

$$\therefore \int_0^{2\pi} p_\phi\, d\phi = p_\phi \cdot 2\pi = n_\phi h \text{ or, } p_\phi = \frac{n_\phi h}{2\pi} \tag{1.130}$$

For elliptical orbits, the radius vector is variable. The momentum

$$p_r = m\left(\frac{dr}{dt}\right) = m\left(\frac{dr}{d\phi}\right)\left(\frac{d\phi}{dt}\right) = m\left(\frac{dr}{d\phi}\right)\left(\frac{p_\phi}{mr^2}\right) = \frac{p_\phi}{r^2}\left(\frac{dr}{d\phi}\right)$$

$$\therefore \oint p_r\, dr = \oint \left(\frac{p_\phi}{r^2}\right)\left(\frac{dr}{d\phi}\right) dr = n_r h \text{ or, } p_\phi \oint \left(\frac{1}{r}\frac{dr}{d\phi}\right)^2 d\phi = n_r h \tag{1.131}$$

Again, the equation of an ellipse is

$$\frac{1}{r} = \frac{1 + \varepsilon\cos\phi}{a(1-\varepsilon)^2} \tag{1.132}$$

where ε is the eccentricity and a the semi-major axis of the ellipse.

$$\therefore \frac{1}{r^2}\frac{dr}{d\phi} = \frac{\varepsilon\sin\phi}{a(1-\varepsilon)^2} \tag{1.133}$$

From eqs (1.132) and (1.133), we get

$$\frac{1}{r}\frac{dr}{d\phi} = \frac{\varepsilon\sin\phi}{1+\varepsilon\cos\phi} \tag{1.134}$$

Using eq. (1.134) in eq. (1.131), we get

$$p_\phi \oint \left(\frac{\varepsilon \sin\phi}{1+\varepsilon\cos\phi}\right)^2 d\phi = n_r h \text{ or, } p_\phi \int_0^{2\pi} \frac{\varepsilon^2 \sin^2\phi}{(1+\varepsilon\cos\phi)^2} d\phi = n_r h \tag{1.135}$$

Let $u = \varepsilon \sin\phi$ and $v = \dfrac{1}{(1+\varepsilon\cos\phi)2\pi}$, then

$$p_\phi \int_0^{2\pi} \frac{\varepsilon^2 \sin^2\phi}{(1+\varepsilon\cos\phi)^2} d\phi = \int_0^{2\pi} u\, dv = [uv]_0^{2\pi} - \int_0^{2\pi} v\, du \text{ [integrating by parts]}$$

$$= \left[\frac{\varepsilon\sin\phi}{1+\varepsilon\cos\phi}\right]_0^{2\pi} - \int_0^{2\pi} \frac{\varepsilon\cos\phi}{1+\varepsilon\cos\theta} d\theta = -\int_0^{2\pi} \frac{\varepsilon\cos\phi}{1+\varepsilon\cos\phi} d\phi$$

$$= \int_0^{2\pi} \left(\frac{1}{1+\varepsilon\cos\phi} - 1\right) d\phi = 2\pi\left(\frac{1}{\sqrt{1-\varepsilon^2}} - 1\right)$$

$$\therefore n_r h = 2\pi p_\phi \left(\frac{1}{\sqrt{1-\varepsilon^2}} - 1\right) = n_\phi h \left(\frac{1}{\sqrt{1-\varepsilon^2}} - 1\right) \left[\because p_\phi = \frac{n_\phi h}{2\pi}\right]$$

$$\therefore \frac{1}{\sqrt{1-\varepsilon^2}} = 1 + \frac{n_r}{n_\phi} = \frac{n_\phi + n_r}{n_\phi} = \frac{n}{n_\phi} \text{ (say)}$$

where $n = n_r + n_\phi$ is an integer as n_r and n_ϕ are both integers. Here n is known as the *principal* or *total quantum number.*

Since, $b = a\sqrt{1-\varepsilon^2}$, where b is the semi-minor axis of the ellipse, we have

$$\frac{b}{a} = \sqrt{1-\varepsilon^2} = \frac{n_\phi}{n} \tag{1.136}$$

This is *the condition of quantisation of the orbits.* Thus, the allowed elliptical electron orbits are those for which the ratio of minor to major axes is the ratio n_ϕ/n.

As $b \leqslant a$, $n_\phi \leqslant n$. When $b = a$, $\varepsilon = 0$, in this case, the orbit is a circle for which $n_\phi = n$. The maximum value of azimuthal quantum number is n and the radial quantum number $n_r = 0$.

If $n_\phi = 0$, then $p_\phi = 0$, which means the electron executes simple harmonic motion through the nucleus which is absurd.

Hence, $n_\phi = 0$ is not acceptable. The minimum value of $n_\phi = 1$. So $n_\phi = 1, 2, 3, \ldots, n$.

As $n_r = n - n_\phi$ the corresponding values of n_r are $n_r = (n-1), (n-2), (n-3), \ldots, 2, 1, 0$.

1.16.3 Energy

Now, the total energy of the electron is given by

$$E_n = \text{KE} + \text{PE} = \frac{1}{2}m\left(\dot{r}^2 + r^2\dot{\phi}^2\right) - \frac{Ze^2}{4\pi\varepsilon_0 r}$$

But, $p_r = m\dot{r}$ and $p_\phi = mr^2\dot{\phi}$.

$$\therefore E_n = \frac{1}{2m}\left(p_r^2 + p_\phi^2/r^2\right) - \frac{Ze^2}{4\pi\varepsilon_0 r} = \frac{p_\phi^2}{2mr^2}\left[\left(\frac{1}{r}\frac{dr}{d\phi}\right)^2 + 1\right] - \frac{Ze^2}{4\pi\varepsilon_0 r}$$

$$\text{or, } \left(\frac{1}{r}\frac{dr}{d\phi}\right)^2 = \frac{2mE_n}{p_\phi^2}r^2 + \frac{2mZe^2}{4\pi\varepsilon_0 p_\phi^2} - 1$$

Squaring both sides of eq. (1.134), we get

$$\left(\frac{1}{r}\frac{dr}{d\phi}\right)^2 = \frac{\varepsilon^2 \sin^2\phi}{(1+\varepsilon\cos\phi)^2} \tag{1.137}$$

$$\because r = \frac{a(1-\varepsilon^2)}{1+\varepsilon\cos\phi} \text{ [from eq. (1.132)]}$$

$$\therefore 1 + \varepsilon\cos\phi = \frac{a(1-\varepsilon^2)}{r}$$

$$\text{or, } \varepsilon^2\sin^2\phi = \varepsilon^2 - \varepsilon^2\cos^2\phi = \varepsilon^2 - \left[\frac{a(1-\varepsilon^2)}{r} - 1\right]^2 = (\varepsilon^2 - 1) - \frac{a^2(1-\varepsilon^2)^2}{r^2} + \frac{2a(1-\varepsilon^2)}{r} \tag{1.138}$$

From eqs (1.137) and (1.138), we get

$$\left(\frac{1}{r}\frac{dr}{d\phi}\right)^2 = \frac{r^2}{a^2(1-\varepsilon^2)^2}\left[(\varepsilon^2-1) - \frac{a^2(1-\varepsilon^2)^2}{r^2} + \frac{2a(1-\varepsilon^2)}{r}\right] = -\frac{r^2}{a^2(1-\varepsilon^2)} + \frac{2r}{a(1-\varepsilon^2)} - 1 \tag{1.139}$$

Comparing the coefficients of r^2 and r of eqs (1.137) and (1.138), we get

$$\frac{2mE_n}{p_\phi^2} = -\frac{1}{a^2(1-\varepsilon^2)} \tag{1.140}$$

$$\text{and } \frac{2mZe^2}{4\pi\varepsilon_0 p_\phi^2} = \frac{2}{a(1-\varepsilon^2)} \tag{1.141}$$

Dividing eq. (1.140) by eq. (1.141), we get

$$\frac{4\pi\varepsilon_0 E_n}{Ze^2} = -\frac{1}{2a} \text{ or, } E_n = -\frac{Ze^2}{8\pi\varepsilon_0 a}. \tag{1.142}$$

From eq. (1.142)

$$a = \frac{4\pi\varepsilon_0 p_\phi^2}{mZe^2(1-\varepsilon^2)} \text{ or, } a = \frac{n_\phi^2 h^2}{4\pi^2}\cdot\frac{4\pi\varepsilon_0}{mZe^2}\cdot\frac{n^2}{n_\phi^2} = \frac{\varepsilon_0 n^2 h^2}{m\pi Ze^2}\left[\because p_\phi = \frac{n_\phi h}{2\pi}, 1-\varepsilon^2 = \frac{n_\phi^2}{n}\right] \tag{1.143}$$

From eqs (1.142) and (1.143)

$$E_n = -\left(\frac{1}{n^2}\right)\frac{mZ^2e^4}{8\varepsilon_0^2 h^2} \tag{1.144}$$

For hydrogen $Z = 1$

$$E_n = -\left(\frac{1}{n^2}\right)\frac{me^4}{8\varepsilon_0^2 h^2}. \tag{1.145}$$

This energy is exactly the same as the energy of an electron in the nth quantum state. E_n depends only on n, but not on n_ϕ. The introduction of elliptic orbit does not give rise to new energy levels and thus it was unable to explain the origin of fine structure of spectral lines.

The equality of energy values for a given n, but different n_ϕ is called *degeneracy.*

For $n = 4$ the possible values of n_ϕ and n_r are as follows:

(a) $n_\phi = 4,\ n_r = 0\ :\ \varepsilon = 0$ (circle) (b) $n_\phi = 3,\ n_r = 1\ :\ \varepsilon = \frac{\sqrt{7}}{4}$ (ellipse)

(c) $n_\phi = 2,\ n_r = 2\ :\ \varepsilon = \frac{2\sqrt{3}}{4}$ (ellipse) (d) $n_\phi = 1,\ n_r = 3\ :\ \varepsilon = \frac{\sqrt{15}}{3}$ (ellipse)

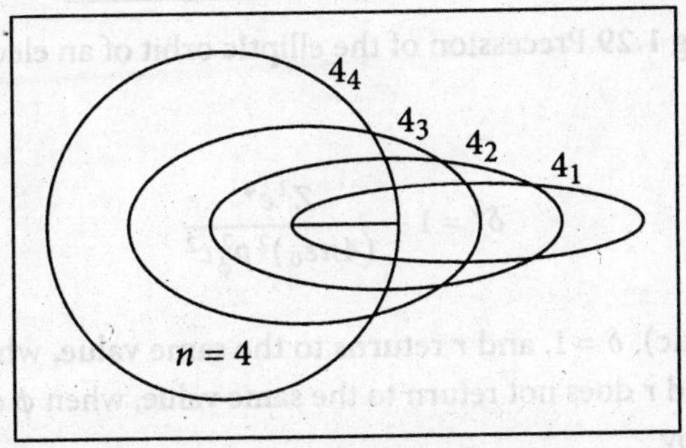

Fig 1.28 Sommerfeld's elliptic orbits for $n = 4$

Thus, there are four orbits with different values of ε, but with the same values of energy. It means that the introduction of elliptical orbits gives no new energy levels and hence no new transitions.

1.16.4 Relativistic correction to Sommerfeld elliptical orbits

Since, the velocity of an electron in the innermost Bohr orbit is 0.007 of the velocity of light, the electron in the elliptic orbit must undergo a variation of mass according to the relation

$$m = \frac{m_0}{\sqrt{1 - v^2/c^2}}$$

where m_0 is the rest mass, m the mass when the electron moves with velocity v and c the velocity of light in free space.

Secondly, the velocity of the electron is the maximum when the electron is at the perihelion and minimum when the electron is at the aphelion of the elliptical path. According to the theory of relativity, the variation of velocity introduces the variation of mass. The above two facts justify the introduction of the relativistic correction to the electron mass.

The application of relativistic correction leads to the fact that there is a slow *precession* of the major axis of the plane about an axis through its one of the foci. The equation to the path of the electron is given by

$$r = \frac{a(1 - \varepsilon^2)}{1 + \varepsilon \cos \delta\phi} \tag{1.146}$$

which represents a *Rosette curve* [Fig 1.29].

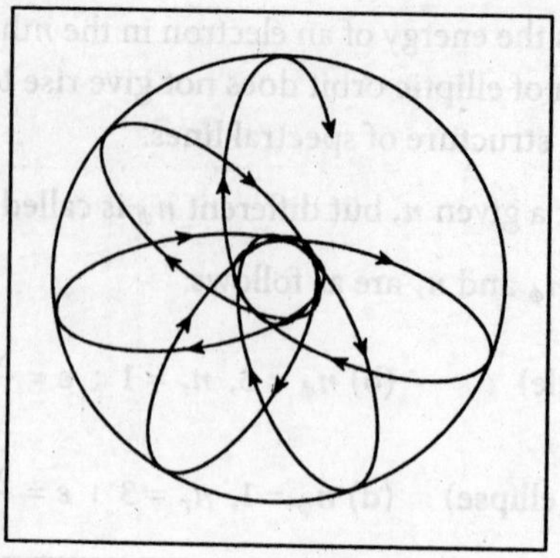

Fig 1.29 Precession of the elliptic orbit of an electron

The value of

$$\delta^2 = 1 - \frac{Z^2 e^4}{(4\pi\varepsilon_0)^2 p_\phi^2 c^2} \tag{1.147}$$

When $c \to 0$ (non-relativistic), $\delta = 1$, and r returns to the same value, when ϕ varies from 0 to 2π. But, when $c \to$ finite value, $\delta < 1$ and r does not return to the same value, when ϕ changes by 2π. r returns to the same values when ϕ changes by

$$\frac{2\pi}{\delta} > 2\pi \; (\because \delta < 1)$$

So the axis of the elliptical orbit precesses through an angle, $\Delta\phi = \dfrac{2\pi}{\delta} - 2\pi$.

Applying relativistic correction, the energy of the electron in a particular quantum state represented by n and n_ϕ is found to be

$$E_n = -\frac{\mu Z^2 e^4}{8\varepsilon_0^2 n^2 h^2}\left[1 + \frac{Z^2\alpha^2}{n^2}\left(\frac{n}{n_\phi} - \frac{3}{4}\right)\right] \tag{1.148}$$

where α is the *finite structure constant* given by

$$\alpha = \frac{e^2}{2\varepsilon_0 ch} = \frac{1}{137} \tag{1.149}$$

Equation (1.148) shows that E_n does not only depend on n, but also on n_ϕ. So transitions corresponding to the various possible n_ϕ values can occur.

Since, the energy differences of these values are very small, the frequencies of the resulting spectral lines will be very close. Thus, a group of closely associated lines appear instead of a single line as in Bohr's theory [Fig 1.30]. The selection rule for the transition is

$$\Delta n_\phi = \pm 1 \tag{1.150}$$

which suggests a finite number of close lines.

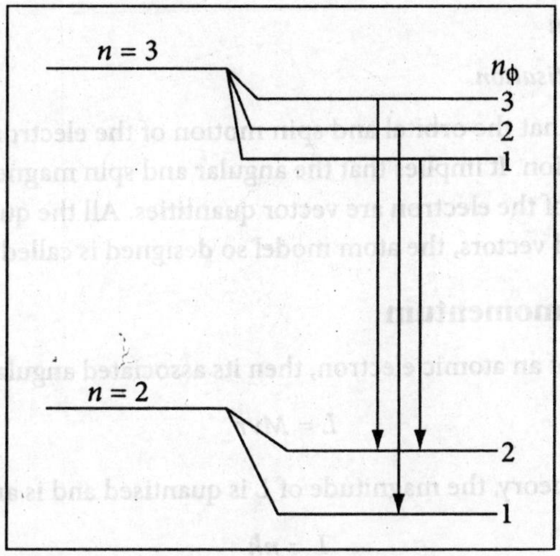

Fig 1.30 Fine structure of levels of one-electron atom

1.16.5 Limitations of old quantum theory

The quantum theory based on Bohr's quantum condition and Wilson-Sommerfeld quantum rule is known as *old quantum theory*. It is unable to explain some problems, e.g., energy states of hydrogen atom, particle in infinite square well, harmonic oscillator, etc. The main limitations of the old quantum theory are as follows:

(a) It is not applicable to non-periodic system.

(b) It cannot explain the relative intensities of the spectral lines.

(c) The spectral lines of hydrogen molecule and normal helium atom cannot be explained by this theory.

(d) The theory does not account the spin motion of the electron and the Pauli's exclusion principle.

(e) Even after applying the selection rule $\Delta n_\phi = \pm 1$, Sommerfeld theory cannot give the actual number of spectral lines.

(f) Sommerfeld method of elliptical orbits gives the correct total (n) of possible azimuthal quantum numbers, but the actual values are incorrect. Spectroscopic studies reveals that the azimuthal quantum number can take the value zero so that the total values are $0, 1, 2, \ldots, (n-1)$. So a new azimuthal quantum number l is introduced where $l = n_\phi - 1$.

A satisfactory explanation of the above difficulties has been provided by new *quantum mechanics*.

1.17 Vector atom model

The old quantum theory formulated by *Bohr-Sommerfeld could not explain the anomalous Zeeman effect, Paschen-Back effect, Stark effect, etc.* which were discovered in subsequent years. To provide satisfactory explanation of the above effects, a new model called *Vector Atom Model* was proposed. The main contributors to the Vector Atom Model are *Bohr, Sommerfeld, Uhlenbeck, Goudsmit, Pauli, Stern and Gerlach.*

1.17.1 The concept of vector atom model

The vector atom model is based on the following two concepts:

(a) *The concept of electron spin*

(b) *The concept of space quantisation.*

Quantum theory suggests that the orbital and spin motion of the electron are to be quantised both in magnitude as well as in direction. It implies that the angular and spin magnetic momenta associated with the orbital and spin motions of the electron are vector quantities. All the quantities which determine the state of the atom are quantised vectors, the atom model so designed is called the *vector atom model.*

1.17.2 Orbital angular momentum

If r be the radius of the orbit of an atomic electron, then its associated angular momentum L is

$$L = Mvr \tag{1.151}$$

In accordance with Bohr's theory, the magnitude of L is quantised and is an integral multiple of $\hbar$. Thus,

$$L = n\hbar \tag{1.152}$$

In Sommerfeld's theory, the angular momentum is represented by p_ϕ, where $p_\phi = n_\phi \hbar$.

In vector atom model n_ϕ is represented by l and is called the *orbital quantum number.*

$\therefore p_\phi = l\hbar$, where $l = 0, 1, 2, \ldots, (n-1)$.

The quantum states with different l values and their designation are shown in Table 1.1.

Table 1.1

Orbital quantum number	0	1	2	3	4	5
Designation of the states	*s*	*p*	*d*	*f*	*g*	*h*

The angular momentum is a vector quantity which is directed along the axis of rotation and is perpendicular to the plane of the orbit of the electron as shown in Fig 1.31.

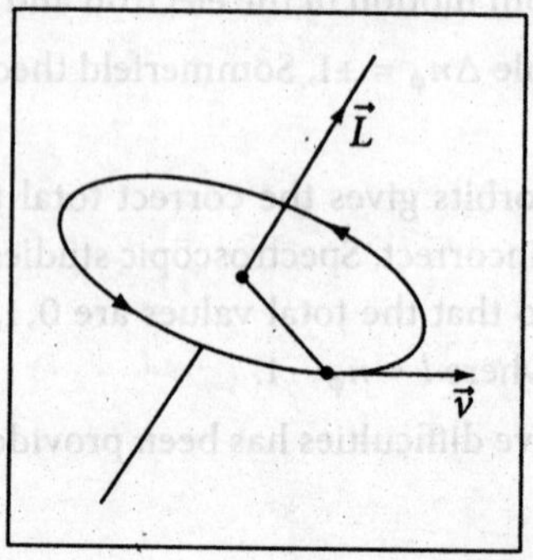

Fig 1.31 Orbiting electron produces angular momentum

1.17.3 Orbital magnetic moment

The orbital motion of the electron is equivalent to a current loop which produces a magnetic field perpendicular to the plane of the orbit. The direction of the magnetic field is given by the left-hand rule as shown in Fig 1.32. The magnetic field is characterised by a magnetic dipole moment. The magnetic moment associated with the current loop is $\mu_l = IA$.

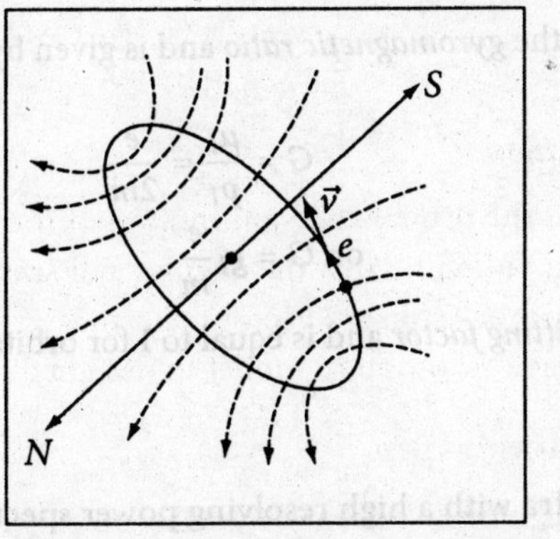

Fig 1.32 Orbiting electron produces magnetic field.

But, $I = e/T$ and $A = \pi r^2$, where r is the radius of the electron orbit and T is the time period of revolution of the electron.

$$\therefore \mu_l = \frac{e}{T}\pi r^2$$

If v be the orbital velocity of the electron, then $\dfrac{2\pi r}{v} = T$.

$$\therefore \mu_l = e\pi r^2\left(\frac{v}{2\pi r}\right) = \frac{evr}{2} \text{ or, } \mu_l = \frac{e}{2m}(mvr) = \frac{e}{2m}p_l \; [\because p_l = mvr]$$

$$\text{or, } \mu_l = \left(\frac{e}{2m}\right) l\hbar \tag{1.153}$$

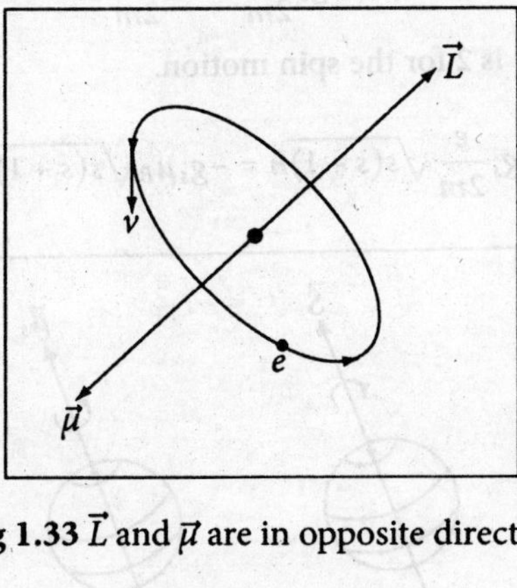

Fig 1.33 $\vec{L}$ and $\vec{\mu}$ are in opposite direction

But,

$$\frac{e\hbar}{2m} = \mu_B \text{ the Bohr magneton} \tag{1.154}$$

$$= 9.273 \times 10^{-24}\,\text{A m}^2$$

$$\therefore \mu_l = \mu_B l$$

The ratio of $\frac{\mu_l}{p_l} = G$ is called the *gyromagnetic ratio* and is given by

$$G = \frac{\mu_l}{p_l} = \frac{e}{2m} \tag{1.155}$$

$$\text{or, } G = g_l \frac{e}{m} \tag{1.156}$$

where g_l is called the *Landé splitting factor* and is equal to 1 for orbital motion of the electron.

1.17.4 Electron spin

The investigation of alkali spectra with a high resolving power spectroscope revealed that many spectral lines consists of a group of lines closed to each other. To explain this multiple structure of the spectral lines the Dutch Physicists *G.E. Uhlenbeck* and *S.A. Goudsmit* put forward in 1925 the hypothesis of *spinning electron*. According to this hypothesis, the electron revolves about its own mechanical axis in a way similar to top motion. The spin angular momentum is given by

$$\vec{S} = s\hbar \tag{1.157}$$

where s is called the *spin quantum number* having only one value, i.e., $s = \frac{1}{2}$.

1.17.5 Magnetic moment due to electron spin

A magnetic field [Fig 1.34] is generated due to the spin motion of the electron. The direction of the spin angular momentum $(\vec{S})$ and the associated spin magnetic moment $(\vec{\mu}_s)$ are also shown in Fig 1.34. The magnetic moment $\vec{\mu}_s$ is given by

$$\mu_s = g_s \frac{e}{2m} \vec{S} = 2 \frac{e}{2m} \vec{S} \tag{1.158}$$

where the Landé splitting factor g_s is 2 for the spin motion.

$$\text{or, } \mu_s = -g_s \frac{e}{2m} \sqrt{s(s+1)}\hbar = -g_s \mu_B \sqrt{s(s+1)} = -\frac{g_s \mu_B \vec{S}}{\hbar} \tag{1.159}$$

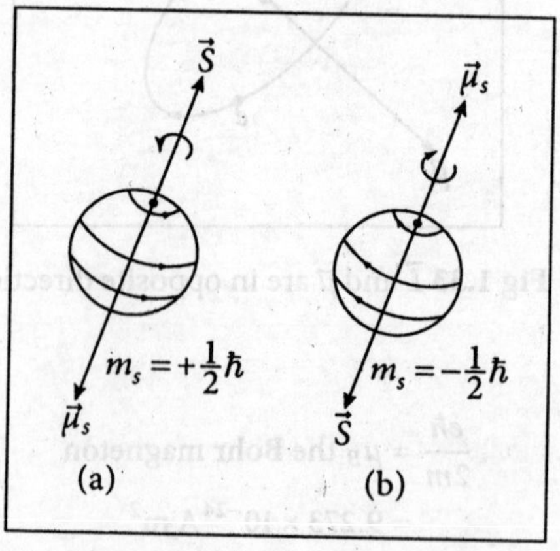

Fig 1.34 Spin up (a) and spin down (b)

1.17.6 Space quantisation

Classically, the orbits of atomic electrons can orient in all possible directions in space, but quantum theory allows only certain *discrete orientations* in space out of infinite possible orientations. This is known as *space quantisation.*

To specify the quantisation of orbits in space, a reference direction is required. Generally, a magnetic field is imagined to be along the z-axis and the allowed orientations of the orbit are determined relative to that magnetic field direction. The space quantisation of an electron orbit is specified by the projection of its total angular momentum on to reference direction, such projections being themselves quantised. The spin motion of the electrons also follows space quantisation.

(a) ***Orientation of the orbits:*** When the electron orbit of an atom is placed in a strong magnetic field $\vec{B}$, its angular momentum vector L interacts with the magnetic field through the magnetic moment vector μ. L can take only such orientations in space for which the projection of L_Z along $\vec{B}$ will be integral multiple of $\hbar$ [Fig 1.35] such that $L_Z = m_l\hbar$, where m_l is called the *orbital magnetic quantum number.* The possible values of m_l are $l, l-1, l-2, \ldots, 0, \ldots, -l$, i.e., there are $(2l+1)$ values of m_l. Fig 1.35 shows $(2l+1)$ values of m_l for $l = 3$. The angle between L_Z and L is given by

$$\cos\theta = \frac{L_Z}{L} = \frac{m_l\hbar}{l\hbar} = \frac{m_l}{l} \tag{1.160}$$

Fig 1.35 Space quantisation

Since, according to wave mechanics l is replaced by $\sqrt{l(l+1)}$.

$$\therefore \cos\theta = \frac{m_l}{\sqrt{l(l+1)}} \tag{1.161}$$

The torque due to the magnetic field causes the vector L to precess about the direction of B maintaining the same orientation θ as shown in Fig 1.36.

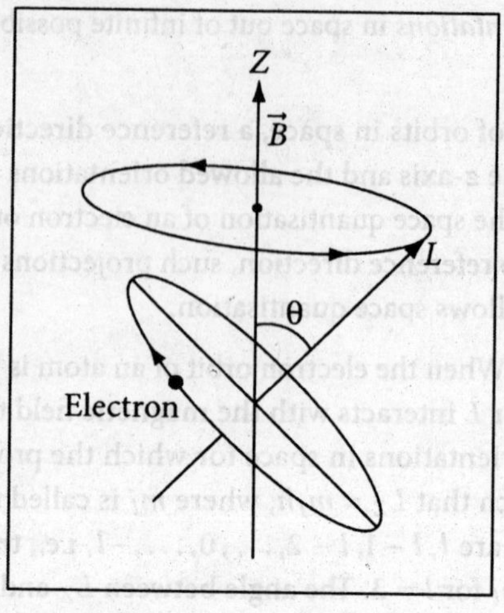

Fig 1.36 $\vec{L}$ **precesses about** Z**-direction**

The magnetic energy of the rotating electron will be

$$\Delta E = \mu B \cos\theta \tag{1.162}$$

But,

$$\cos\theta = \frac{m_l}{l} \text{ and } \mu = \frac{e\hbar}{2mc} l$$

$$\therefore \Delta E = \frac{e\hbar}{2mc} l B \frac{m_l}{l}$$

$$\Delta E = \frac{e\hbar B m_l}{2mc} \tag{1.163}$$

But,

$$\Delta E = m_l h \nu_L \text{ or, } \nu_L = \frac{eB}{4\pi mc} \tag{1.164}$$

ν_L is called the *Larmor precessional frequency.*

(b) ***Spin of electron***: The spin angular momentum vector $\vec{S}$ of the electron can assume only two possible orientations relative to the magnetic field $\vec{B}$ [Fig 1.37]. The projection of $\vec{S}$ along $\vec{B}$ is denoted by m_s and is called the *magnetic spin quantum number*. m_s can have only two values $+\frac{1}{2}$ and $-\frac{1}{2}$ corresponding to *spin-up* and *spin-down* orientations.

Thus, the projection on the Z-component of S, i.e., S_Z is given by

$$S_Z = m_s \hbar \tag{1.165}$$

m_s can take $(2s + 1) = (2 \times \frac{1}{2} + 1) = 2$ discrete values in space w.r.t. the magnetic field $\vec{B}$ [Fig 1.37].

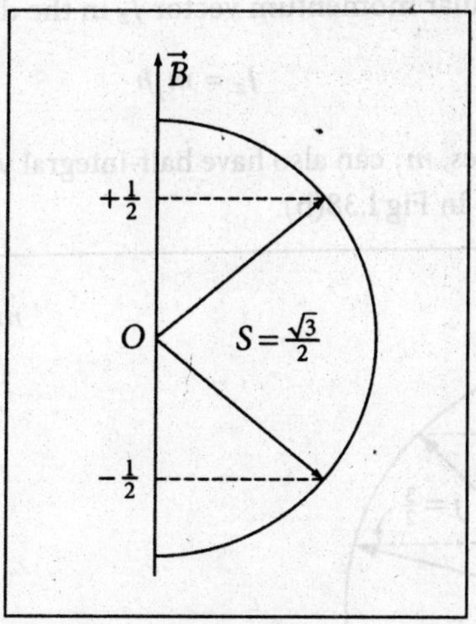

Fig 1.37 Space quantisation of spin angular momentum for a single electron

1.17.7 Total angular momentum of electron

An electron possesses an orbital angular momentum $l\hbar$ and a spin angular momentum $s\hbar$. The total angular momentum is given by $p_j = j\hbar$, which is the vector sum of $l\hbar$ and $s\hbar$ [Fig 1.38(a)]; j is called the *total angular momentum quantum number*.

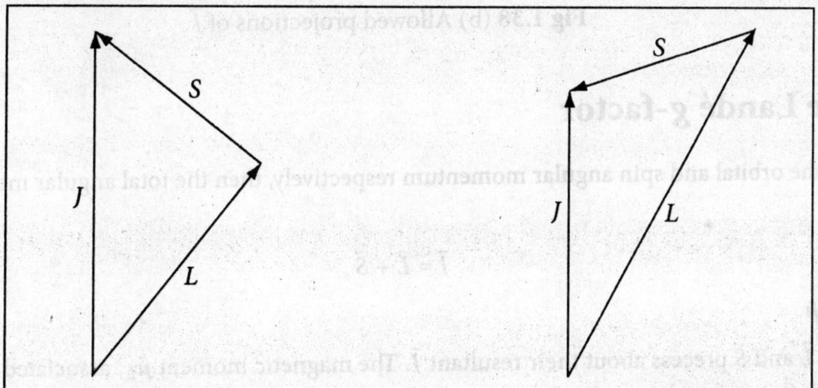

$j = l + s = 1 + \frac{1}{2} = \frac{3}{2}$, when $l = 1$ and $s = +\frac{1}{2}$

$j = l - s = 1 - \frac{1}{2} = \frac{1}{2}$, when $l = 1$ and $s = -\frac{1}{2}$

Fig 1.38 (a) Two ways in which $\vec{L}$ and $\vec{S}$ can be added to form $\vec{J}$

For each value of the orbital quantum number l there are two possibilities as below

$$\left.\begin{aligned} j\hbar &= l\hbar + s\hbar \\ \text{and } j\hbar &= l\hbar - s\hbar \end{aligned}\right\} \tag{1.166}$$

$$\therefore j = l + s \text{ and } j = l - s \tag{1.167}$$

The projection of the total angular momentum vector J_z in the direction of $\vec{B}$ is

$$J_z = m_j \hbar \tag{1.168}$$

As j can have half-integral values, m_j can also have half-integral values. m_j can have $(2j+1)$ values from $+j$ to $-j$ excluding zero as shown in Fig 1.38(b).

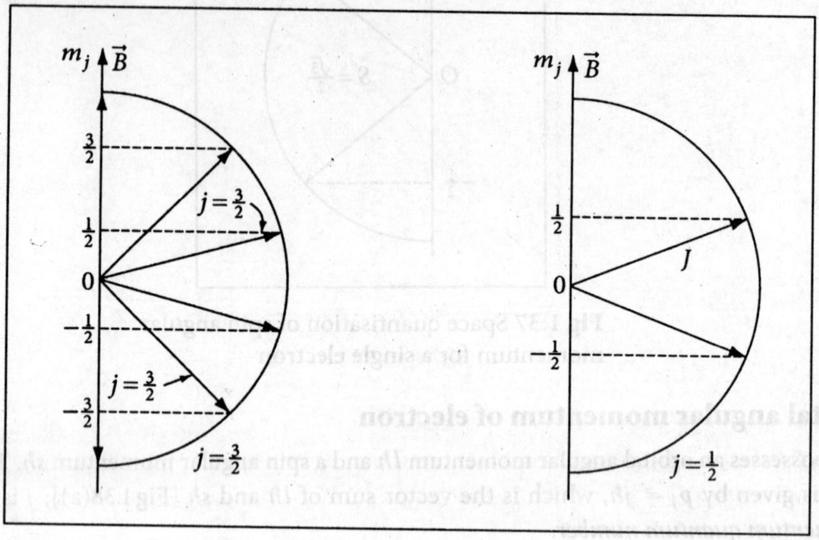

Fig 1.38 (b) Allowed projections of $\vec{J}$

1.18 The Landé g-factor

If $\vec{L}$ and $\vec{S}$ be the orbital and spin angular momentum respectively, then the total angular momentum of the electron

$$\vec{J} = \vec{L} + \vec{S} \tag{1.169}$$

all in units of $\hbar$.

The vectors $\vec{L}$ and $\vec{S}$ precess about their resultant $\vec{J}$. The magnetic moment $\vec{\mu}_L$, associated with $\vec{L}$ and $\vec{\mu}_S$, associated with $\vec{S}$ are given by

$$\vec{\mu}_L = \frac{e\hbar}{2mc}\vec{L} \tag{1.170}$$

$$\text{and } \vec{\mu}_S = \frac{e\hbar}{2mc}\vec{S} \tag{1.171}$$

$\vec{\mu}_L$ and $\vec{\mu}_S$ precess about $\vec{J}$. The total magnetic moment $\vec{\mu} = \vec{\mu}_L + \vec{\mu}_S$ will also precess about $\vec{J}$, but $\vec{\mu}$ will have different direction from $\vec{J}$ [Fig 1.39]. $\vec{\mu}_L$ and $\vec{\mu}_S$ are in the opposite direction of $\vec{L}$ and $\vec{S}$.

Again, since the gyromagnetic ratio for spin motion (μ_s/S) is twice that for the orbital motion (μ_l/L), so $\vec{L}$ and $\vec{\mu}_L$ are of the same length but length of μ_s is twice that of $\vec{S}$. That is why $\vec{\mu}$ and $\vec{J}$ are in different directions.

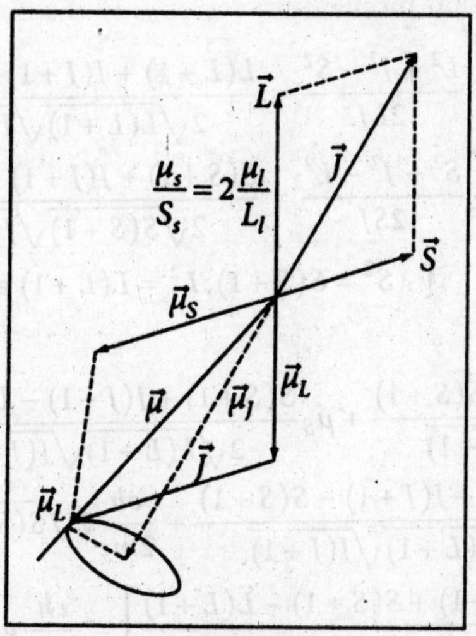

Fig 1.39 Origin of *g*

Since, the vectors $\vec{L}$ and $\vec{S}$ precess about $\vec{J}$, $\vec{\mu}_L$ and $\vec{\mu}_S$ must also precess about $\vec{J}$. If each of the vectors is resolved along and perpendicular to $\vec{J}$, then the value of the perpendicular components averaged over a period of motion will be zero as it is constantly changing direction, i.e., $\mu_\perp = 0$. Then $\vec{\mu}_J$ will be the sum of the components of $\vec{\mu}_L$ and $\vec{\mu}_S$ along $\vec{J}$.

$$\therefore \mu_J = \mu_L \cos\left(\vec{L}, \vec{J}\right) + \mu_S \cos\left(\vec{S}, \vec{J}\right) \tag{1.172}$$

where $\cos\left(\vec{L}, \vec{J}\right)$ is the cosine of the angle between $\vec{L}$ and $\vec{J}$ and $\cos\left(\vec{S}, \vec{J}\right)$ is that between $\vec{S}$ and $\vec{J}$. The orientations of $\vec{L}$ and $\vec{S}$ relative to $\vec{J}$ are shown in Figs 1.40(a) and (b).

$$\therefore S^2 = L^2 + J^2 - 2\vec{L}\cdot\vec{J} \text{ and } L^2 = S^2 + J^2 - 2\vec{S}\cdot\vec{J}$$

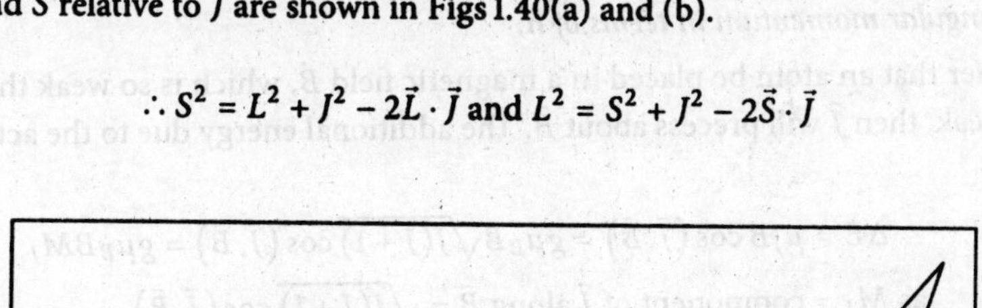

Fig 1.40 Orientations of $\vec{L}$ and $\vec{S}$ w.r.t. $\vec{J}$

Now,

$$\cos\left(\vec{L},\vec{J}\right) = \frac{L^2+J^2-S^2}{2LJ} = \frac{L(L+1)+J(J+1)-S(S+1)}{2\sqrt{L(L+1)}\sqrt{J(J+1)}}$$

$$\cos\left(\vec{S},\vec{J}\right) = \frac{S^2+J^2-L^2}{2SJ} = \frac{S(S+1)+J(J+1)-L(L+1)}{2\sqrt{S(S+1)}\sqrt{J(J+1)}}$$

$$[\because S^2 = S(S+1), L^2 = L(L+1) \text{ and } J^2 = J(J+1)]$$

Now, from eq. (1.172)

$$\mu_J = \mu_L \frac{L(L+1)+J(J+1)-S(S+1)}{2\sqrt{L(L+1)}\sqrt{J(J+1)}} + \mu_S \frac{S(S+1)+J(J+1)-L(L+1)}{2\sqrt{L(L+1)}\sqrt{J(J+1)}}$$

$$= \frac{e\hbar}{2mc}\sqrt{L(L+1)}\frac{L(L+1)+J(J+1)-S(S+1)}{2\sqrt{L(L+1)}\sqrt{J(J+1)}} + \frac{e\hbar}{2mc}2\sqrt{S(S+1)}\frac{S(S+1)+J(J+1)-L(L+1)}{2\sqrt{S(S+1)}\sqrt{J(J+1)}}$$

$$= \frac{e\hbar}{2mc}\sqrt{J(J+1)}\left[1+\frac{J(J+1)+S(S+1)-L(L+1)}{2J(J+1)}\right] = \frac{e\hbar}{2mc}gJ(J+1) \tag{1.173}$$

where

$$g = 1 + \frac{J(J+1)+S(S+1)-L(L+1)}{2J(J+1)} \tag{1.174}$$

g is called *the Landé g-factor* or *the splitting factor* which determines the splitting of the energy levels in the presence of a weak external magnetic field. The splitting is determined by the values of L, S and J.

Since, Bohr magneton, $\mu_B = \dfrac{e\hbar}{2mc}$, we have from eq. (1.169), $\mu_J = \mu_B gJ$ so that

$$g = \frac{\mu_J}{\mu_B J} \tag{1.175}$$

Hence, g, *the Landé splitting factor, is the ratio of the magnetic moment of the system expressed in Bohr magnetons to its angular momentum in terms of* $\hbar$.

Now, we consider that an atom be placed in a magnetic field $\vec{B}$, which is so weak that the coupling of $\vec{L}$ and $\vec{S}$ does not break, then $\vec{J}$ will precess about $\vec{B}$. The additional energy due to the action of the magnetic field is

$$\Delta E = \mu_J B\cos\left(\vec{J},\vec{B}\right) = g\mu_B B\sqrt{J(J+1)}\cos\left(\vec{J},\vec{B}\right) = g\mu_B BM_J \tag{1.176}$$

$$\text{as } M_J = \text{component of } \vec{J} \text{ along } \vec{B} = \sqrt{J(J+1)}\cos\left(\vec{J},\vec{B}\right) \tag{1.177}$$

$\dfrac{e\hbar}{2mc}B$ is a unit of energy used for expressing the splitting of the energy levels placed in a magnetic field and is called a *Lorentz unit*.

1.19 Zeeman effect

Quantum theory provides a successful account of the experimentally observed Zeeman effect.

Zeeman effect is the phenomenon in which each spectral line in presence of a strong magnetic field (~ 10,000 gauss) is observed to spit into *two* components when viewed in longitudinal direction (i.e., along the magnetic field) and three components when viewed transverse to the magnetic field. This is known as *normal Zeeman effect*.

When the magnetic field is comparatively weak, each spectral line splits into *more than three* components. This is known as *anomalous Zeeman effect.*

Each spectral line emitted by the source S placed in the magnetic field $\vec{B}$ when viewed along and perpendicular to $\vec{B}$ is found to split up as shown in Fig 1.41.

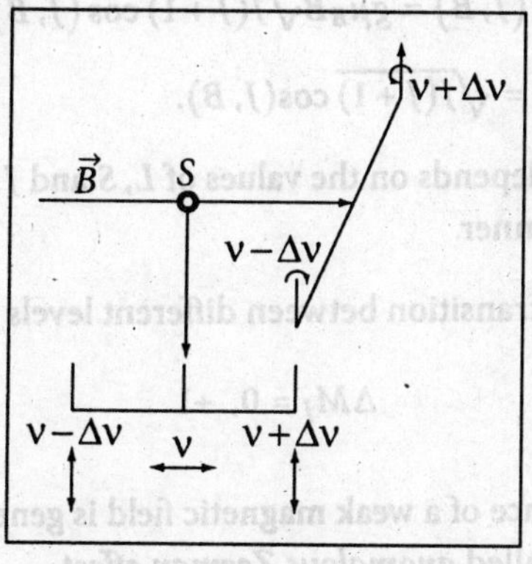

Fig 1.41 Normal Zeeman effect

1.19.1 Origin of normal Zeeman effect

In a strong magnetic field each energy level of an atom with given n and l splits up into $(2l + 1)$ sub-levels. These sub-levels have different values of the orbital magnetic momentum quantum number m_l. Using the selection rule

$$\Delta m_l = 0, \ \pm 1 \tag{1.178}$$

we get three spectral lines originating from the transition between two levels with given values of n and l [Fig 1.42].

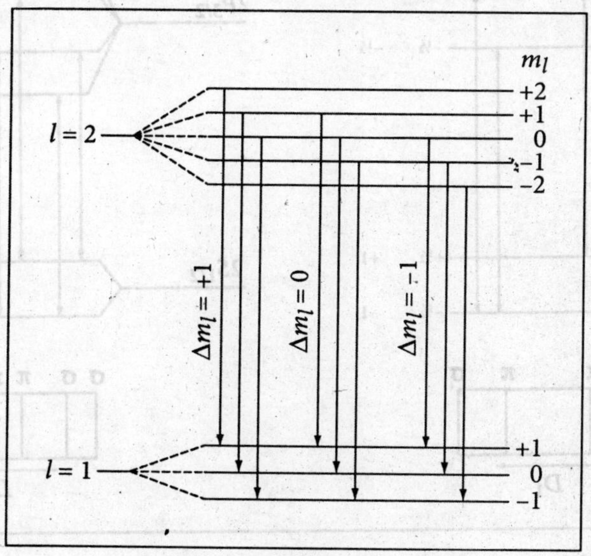

Fig 1.42 Normal Zeeman effect

1.19.2 Origin of anomalous Zeeman effect

From the results of Art. 1.19 it can be shown that the splitting of atomic energy levels in a magnetic field of flux density $\vec{B}$ is given by

$$\Delta E = \mu_J B \cos(\vec{J}, \vec{B}) = g\mu_B B\sqrt{J(J+1)}\cos(\vec{J}, \vec{B}) = g\mu_B B M_J \tag{1.179}$$

where M_J = component of J along $\vec{B} = \sqrt{J(J+1)}\cos(J, B)$.

The value of g given by eq. (1.174) depends on the values of L, S and J showing that different energy levels in presence of $\vec{B}$ split in different manner.

The selection rules governing the transition between different levels are

$$\Delta M_J = 0,\ \pm 1 \tag{1.180}$$

The number of split levels in presence of a weak magnetic field is generally more than two or three unlike normal Zeeman effect. Hence, it is called *anomalous Zeeman effect.*

The D_1-line (λ = 5890 Å) and D_2-line (λ = 5896 Å) of sodium D-line splits into four and six components as shown in Fig 1.43.

The origin of D_1- and D_2-line are as $3^2P_{1/2} \rightarrow 3^2S_{1/2}$ gives D_1-line and $3^2P_{3/2} \rightarrow 3^2S_{1/2}$ gives D_2-line.

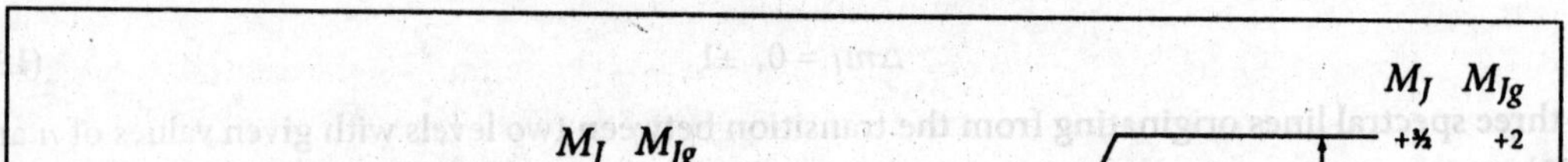

Fig 1.43 Anomalous Zeeman effect of D_1- and D_2-line

Now, for $2S_{1/2}$-state, $g = 2\ (\because L = 0,\ S = \frac{1}{2},\ J = \frac{1}{2})$
and $M_J = +\frac{1}{2}$
for $2P_{1/2}$-state, $g = \frac{2}{3}\ (\because L = 1,\ S = \frac{1}{2},\ J = \frac{1}{2})$
and $M_J = +\frac{1}{2}$
for $2P_{3/2}$-state, $g = \frac{4}{3}\ (\because L = 1,\ S = \frac{1}{2},\ J = \frac{3}{2})$
and $M_J = +\frac{3}{2},\ \pm\frac{1}{2}$.

Thus, for the above levels $gM_J = (\pm 1, \pm\frac{1}{3})$ and $(\pm 2, \pm\frac{2}{3})$ respectively. Fig 1.43 shows levels and the allowed transitions.

For elements like Mg, Ca, Sr, Cd, etc. for which the ground state is $^1S_0$ the spectral lines in weak magnetic field show normal Zeeman effect.

1.20 Stern and Gerlach's experiment

In 1921, *Otto Stern* and *W. Gerlach* of Germany performed an experiment in which atomic beams were deflected in a nonuniform magnetic field. The experiment provides a direct proof of (i) the existence of two spin states $(\pm\frac{1}{2})$ of the electron (ii) the postulate of space quantisation.

Atomic beam of silver was used in Stern and Gerlach's experiment. Other workers used atomic beams of hydrogen, lithium, sodium, potassium, copper and gold.

An electron in hydrogen atom is associated with intrinsic spin magnetic moment (μ_s) due to its spin motion. If the atom be placed in a *uniform magnetic field* $(\vec{B})$, there will be no net force acting on its electron, but a torque $\vec{\tau} = \vec{\mu}_s \times \vec{B}$ will act on it. Instead, if the atom be placed in a *nonuniform magnetic field,* the electron experiences a force. The force acting on the electron in a nonuniform magnetic field suggests that the electron is associated with a spin and hence spin magnetic moment $(\vec{\mu}_s)$.

We suppose that the electron is a small magnet whose N-pole and S-pole are separated by a distance d. The magnetic field $\vec{B}(x, y, z)$ changes with position as it is nonuniform and can be expressed as

$$\vec{B}(x, y, z) = \hat{i}B_x(x, y, z) + \hat{j}B_y(x, y, z) + \hat{k}B_z(x, y, z)$$

Let the magnet be situated along the z-axis with its S-pole at the origin [Fig 1.44]. The magnetic moment $\mu = md$, where m is the magnetic pole strength and μ is directed from S to N.

The force on the S-pole, $F_1 = mB_z(0, 0, 0)$ is along the negative direction of z-axis.

The force on the N-pole is

$$F_2 = mB_z(0, 0, d) = m\left[B_z(0, 0, 0) + d\frac{\partial B}{\partial Z}\bigg|_{x=y=z=0} + \ldots\right]$$

along the positive direction of z-axis.

Then net force on NS is along the positive z-axis

$$F_z = F_2 - F_1 = m\left[B_z(0, 0, 0) + d\frac{\partial B}{\partial Z}\bigg|_{x=y=z=0} + \ldots\right] - mB_z(0, 0, 0) = md\frac{\partial B}{\partial Z}\bigg|_{x=y=z=0} = \mu\frac{\partial B}{\partial Z}\bigg|_{x=y=z=0} \tag{1.181}$$

where $\mu = md$, the Z-component of the magnetic moment.

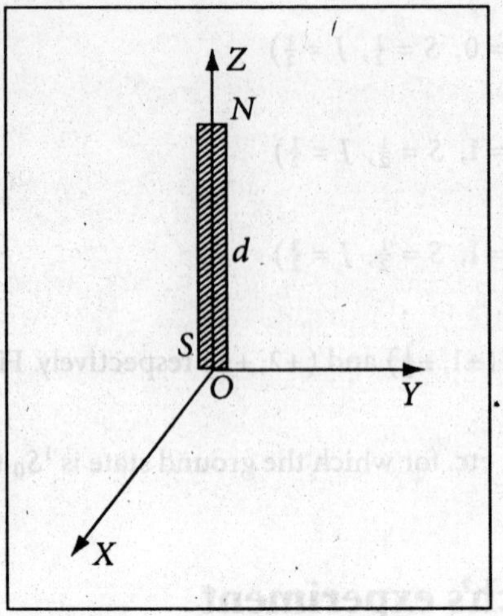

Fig 1.44 Force on a magnetic dipole in a magnetic field

According to the classical idea, $\mu_Z = \mu \cos\theta$, where θ is the angle between μ and the z-axis.

Hence, F_z can have all values $+\mu\dfrac{\partial B}{\partial Z}$ and $-\mu\dfrac{\partial B}{\partial Z}$. When a magnetic dipole passes through a *non-homogeneous magnetic field*, it must be deflected from its rectilinear path. The magnitude of deflection being a function of the extent of homogeneity of $\vec{B}$, therefore, if a beam of magnetic dipole passes through a magnetic field, a continuous range of deflection would be observed leading to the broadening of the beam.

But, quantum mechanically, all settings of the magnetic moment relative to $\vec{B}$ are not possible, instead, a certain discrete number of settings are allowed leading to the splitting of the beam into a number of discrete components. In the *Stern-Gerlach's experiment,* this splitting of the beam was really demonstrated.

1.20.1 Experimental arrangement

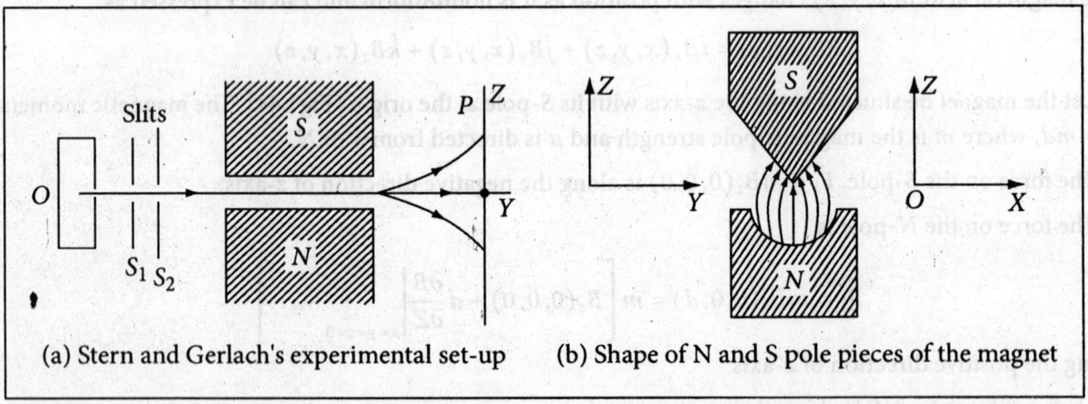

(a) Stern and Gerlach's experimental set-up (b) Shape of N and S pole pieces of the magnet

Fig 1.45 Stern and Gerlach's experimental arrangement

The experimental set-up is shown in Fig 1.45(a). The silver atoms are obtained by heating silver metal in a furnace O situated in vacuum. An atomic beam of silver is obtained by collimating it through a set of slits S_1, S_2 into an inhomogeneous magnetic field. The inhomogeneous magnetic field is produced by the pole pieces N and S of a powerful magnet. The S-pole is of the shape of a knife-edge and the N-pole forms a groove as shown in Fig 1.45(b).

The magnetic field $\vec{B}$ is along the z-axis and the beam passes along the y-axis. The beam is received on the photographic plate P placed perpendicular to the knife-edge. The whole arrangement is kept in an highly evacuated chamber so that the silver atoms can traverse the field without any collision.

Results. The traces obtained on the photo plate using silver are as shown in Fig 1.46. A straight line trace is obtained without the application of magnetic field. A double trace is obtained with some irregularities in presence of $\vec{B}$. The double trace is found to converge because field gradient decreases transversely.

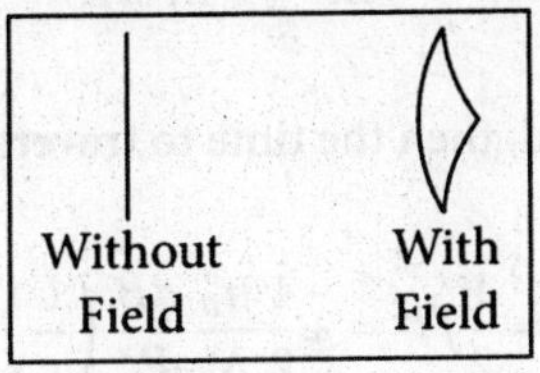

Fig 1.46 Traces obtained on the photo-plate

1.20.2 Explanation

In absence of the magnetic field the atomic beam is undeflected and travels towards the photo plate P. But, in presence of the inhomogeneous magnetic field each atom of the beam experiences a translators force in a direction perpendicular to the path of atom beam. Under this situation the atoms strike the photo plate P at a point off the axis.

The normal state of a silver atom is $^2S_{1/2}$. For the ground state $l = 0$, $s = \frac{1}{2}$. As $l = 0$, $\mu_l = 0$ and the magnetic moment is due to spin motion only.

The Landé splitting factor g is

$$g = 1 + \frac{j(j+1) + s(s+1) - l(l+1)}{2j(j+1)} = 1 + \frac{\frac{1}{2}(\frac{1}{2}+1) + \frac{1}{2}(\frac{1}{2}+1) - 0(0+1)}{2 \times \frac{1}{2}(\frac{1}{2}+1)}$$
$$= 2 \; (\because j = 0 + \tfrac{1}{2} = \tfrac{1}{2}, s = \tfrac{1}{2} \text{ and } l = 0) \tag{1.182}$$

Hence, if the silver atom be placed in a nonuniform field ($\vec{B}$), it would take two specific positions given by

$$\cos\theta = \frac{J_z}{J} = \frac{m_j\hbar}{\sqrt{j(j+1)}\hbar} = \frac{m_j}{\sqrt{j(j+1)}} \tag{1.183}$$

The component of the magnetic moment μ_j along the direction of $\vec{B}$ is given by

$$\mu_j \cos\theta = 2\sqrt{j(j+1)}\mu_B \times \frac{m_j}{\sqrt{j(j+1)}} = 2m_j\mu_B$$

For $j = l + s = 0 + \frac{1}{2} = \frac{1}{2}$, $m_j = +\frac{1}{2}$ or $-\frac{1}{2}$.

If we use $J = j\hbar$, then $\cos\theta = \frac{J_z}{J} = \frac{m_j\hbar}{j\hbar} = \frac{m_j}{J}$.

Using $m_j = \pm\frac{1}{2}$, $\cos\theta = \frac{\pm 1/2}{1/2} = 1$, or, -1, i.e., $\theta = 0°$ or $180°$.

Also for

$$m_j = \tfrac{1}{2}, \mu_j = 2 \times \tfrac{1}{2} \times \mu_B \text{ or, } \mu_j = \mu_B \tag{1.184}$$

The value of $\theta = 0°$ or $180°$ indicates that μ_B will be directed either along $\vec{B}$ or opposite to it.

If the silver atom enters the nonuniform magnetic field $\vec{B}$ along the y-axis with velocity v and dB/dZ the field gradient along the z-axis, then the force on the atom along the z-axis is

$$F_Z = \pm\mu_B \frac{dB}{dZ}$$

The acceleration is

$$f = \frac{F_Z}{M} = \pm\frac{\mu_B}{M}\frac{dB}{dZ}$$

where M is the mass of the silver atom.

If L = the length of the path in the field, then the time to traverse is $t = L/v$. The displacement along the Z-direction is

$$d = \frac{1}{2}ft^2 = \pm\frac{1}{2}\frac{\mu_B}{M}\frac{dB}{dZ}\left(\frac{L}{v}\right)^2. \tag{1.185}$$

The most probable velocity of the silver atom, $v = \sqrt{\dfrac{2kT}{M}}$.

$$\therefore d = \pm\frac{1}{2}\frac{\mu_B}{M}\left(\frac{dB}{dZ}\right)\frac{L^2 M}{2kT} = \pm\frac{1}{4}\frac{\mu_B}{kT}L^2\left(\frac{dB}{dZ}\right) \tag{1.186}$$

where k = the Boltzmann constant; T = the absolute temperature of the silver atom. The experimentally measured value of d agrees well with the calculated value. Thus, the experiment provides a brilliant confirmation of the principle of quantisation of spin magnetic moment.

The ground states of H, Li, Na, K, Cu, Ag, Au are represented by $^2S_{1/2}$, which means $2J + 1 = 2$ (as $l = 0$, $s = \frac{1}{2}, \therefore J = l + s = \frac{1}{2}$). They all if subjected to a nonuniform magnetic field give two traces.

Also for them $g = 2$ and $m_s = \pm\frac{1}{2}. \therefore \mu = \pm 1\mu_B$.

But, the elements like Zn, Cd, Hg have two S electrons outside the closed shells, hence their normal state is $^1S_0$. This means that $J = 0$ and the atom has no resultant angular momentum and no magnetic moment. Thus, they produce no effect in Stern-Gerlach experiment.

1.21 The Harmonic oscillator

(a) ***Classical theory***: A one-dimensional harmonic oscillator consists of a particle of mass m bounded to the equilibrium position at $x = 0$ by a restoring force $F = -kx$, where k is the force constant and x is the displacement from 0. The particle is constraint to execute motion along the x-axis.

If the motion of the particle be described by the simple law, $F = -kx$, then the oscillator is termed as the *harmonic oscillator.*

The potential energy (V) of the particle is

$$V = -\int_0^\infty F dx = \int_0^\infty kx\,dx = \tfrac{1}{2}kx^2 \tag{1.187}$$

Again, the differential equation of motion of the particle is

$$m\frac{d^2x}{dt^2} = -kx \tag{1.188}$$

The solution of eq. (1.188) is

$$x = a\cos 2\pi\nu(t - t_0) \tag{1.189}$$

where A and t_0 are constants. ν is the natural frequency.

The fundamental frequency of the oscillator is

$$\nu = \frac{1}{2\pi}\sqrt{\frac{k}{m}}$$

Now, the velocity of the particle at any instant is

$$\frac{dx}{dt} = -2\pi\nu a \sin 2\pi\nu(t - t_0)$$

The kinetic energy

$$T = \frac{p^2}{2m} = \frac{1}{2}m\left(\frac{dx}{dt}\right)^2$$

∴ total energy of the oscillator is

$$E = T + V = \frac{1}{2}m\left(\frac{dx}{dt}\right)^2 + \frac{1}{2}kx^2 = 2m\pi^2\nu^2a^2\left[\sin^2 2\pi\nu(t - t_0) + \cos^2 2\pi\nu(t - t_0)\right] = \frac{1}{2}ka^2 \tag{1.190}$$

Equation (1.190) shows that the energy E of the oscillator is continuous and all positive values of E are possible.

(b) ***Old quantum theory***: According to Wilson-Sommerfeld quantisation rule

$$\oint p_x\, dx = nh \tag{1.191}$$

where $n = 0, 1, 2, \ldots$.

The phase integral $\oint p_x\, dx$ represents an area in the phase space.

Since, the particle executes simple harmonic motion in one dimension, the phase integral represents an ellipse in phase space having two coordinates x and p_x.

$$\therefore \oint p_x\, dx = \text{area of the ellipse} = \pi ab \tag{1.192}$$

where a and b are the semi-major and semi-minor axes respectively.

Now, total energy of the particle is given by

$$E = \frac{p_x^2}{2m} + \frac{1}{2}kx^2 \tag{1.193}$$

This represents an ellipse in x, p_x space.

Now, when $p_x = 0$, we get

$$E = \frac{1}{2}ka^2 \text{ or, } a = \sqrt{\frac{2E}{k}}$$

and when $x = 0$

$$E = \frac{b^2}{2m} \text{ or, } b = \sqrt{2mE}$$

Using the values of a and b in eq. (1.192), we get

$$\oint p_x\,dx = \pi ab = \pi\sqrt{\frac{2E}{m}}\cdot\sqrt{2mE} = \frac{2\pi E}{2\pi\nu} = \frac{E}{\nu}\quad\left[\because \nu = \frac{1}{2\pi}\sqrt{\frac{k}{m}}\right] \tag{1.194}$$

Comparing eqs (1.191) and (1.194), we get

$$\frac{E}{\nu} = nh \text{ or, } E = nh\nu \tag{1.195}$$

So quantum mechanically the energy levels of a harmonic oscillator are not continuous, but an integral multiple of $h\nu$ [Fig 1.47].

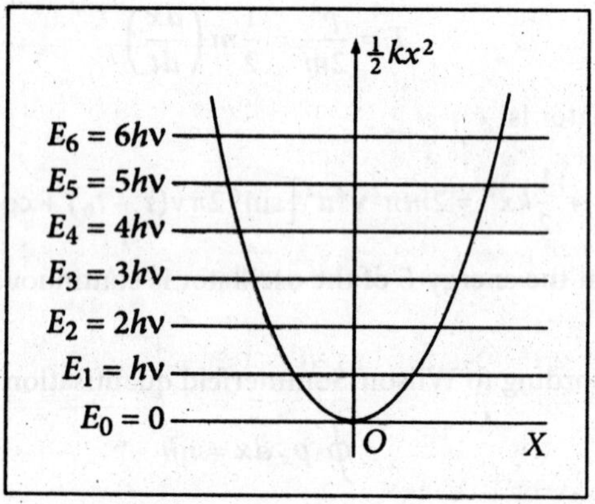

Fig 1.47 Energy levels of a harmonic oscillator

For emission and absorption of light of frequency ν the selection rule is $\Delta n = \pm 1$.

(c) ***The particle in a box:*** Wilson-Sommerfeld quantisation rule can be applied to the case of a particle in a box. It will be seen that the phase integral is quantised and the energy of the particle is also quantised.

We consider a particle of mass m constraint to move in one dimension (say, x). At $x = 0$ and $x = a$ the particle suffers perfectly elastic collisions with the rigid walls as shown in Fig 1.48(a). We assume that no other force acts on the particle.

The phase diagram is shown in Fig 1.48(b). The phase integral for the periodic motion is expressed as

$$J = \oint p_x\,dx = p\oint dx = 2ap \tag{1.196}$$

where p is the constant and represents the magnitude of momentum of the particle.

But, the quantisation demands that

$$J = 2ap = nh \tag{1.197}$$

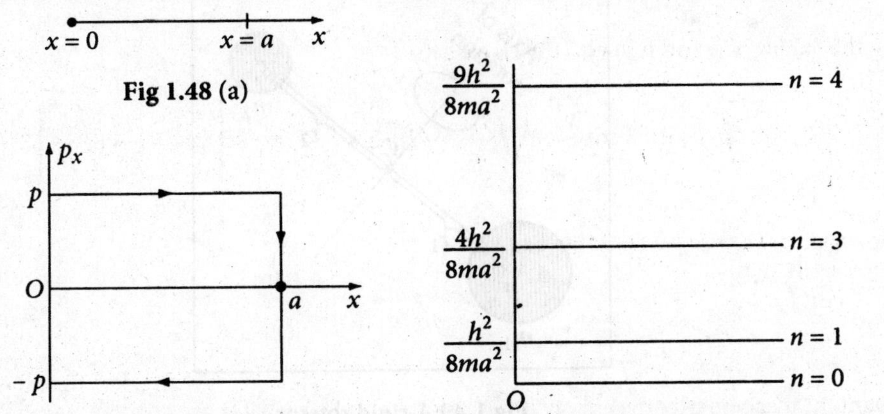

Fig 1.48 (a)

Fig 1.48 (b) Particle in a box

Fig 1.48 (c) Energy level diagram for a particle in a box

The allowed values of the momentum of the particle is given by the above equation. The total energy of the particle is the kinetic energy and is given by

$$E_n = \frac{p^2}{2m} = \frac{n^2h^2}{8ma^2} \quad (n = 0, 1, 2, \dots) \tag{1.198}$$

Equation (1.198) shows that the particle has a discrete set of energy values [Fig 1.48(c)]. For an electron it can be shown that $E_1 = 0.6 \times 10^{-10}$ erg and $E_2 = 2.4 \times 10^{-10}$ erg. So the difference of energy between the two levels is 1.8×10^{-10} erg or 113 eV which is large in the atomic scale of energies. This justifies the need of introduction of quantum theory.

1.22 The rigid rotator

The spectrum emitted by diatomic molecules (like HCl, KBr, etc.) can be explained on the basis of rigid rotator model. A rigid rotator consists of two atoms of masses m_1 and m_2 at a certain distance from one another. The system is capable of rotating about an axis passing through the centre of mass C and perpendicular to the line joining m_1 and m_2. The rotator can take any orientation in space because the rotation is assumed to be free.

If I be the moment of inertia of the system about the axis of rotation and ω, the angular velocity, then the kinetic energy of the rotator is

$$E = \tfrac{1}{2}I\omega^2$$

The phase integral of the rotator is

$$\oint p_\phi \, d\phi = 2\pi p_\phi \tag{1.199}$$

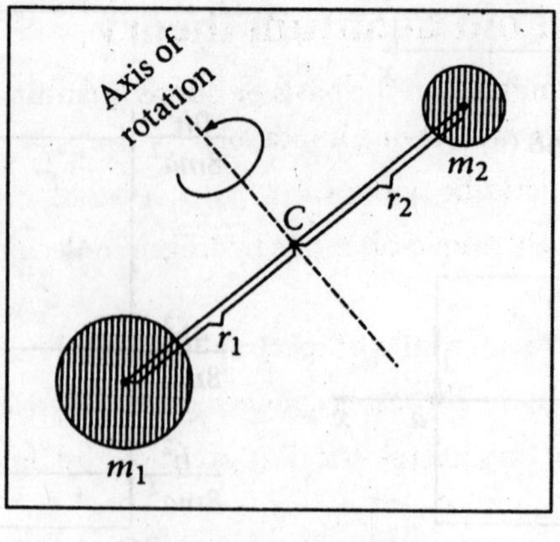

Fig 1.49 A rigid rotator

Again, according to the quantisation rule of Wilson-Sommerfeld

$$\oint p_\phi \, d\phi = nh \tag{1.200}$$

where $n = 0, 1, 2, 3, \ldots$.

Comparing eqs (1.199) and (1.200), we get

$$2\pi p_\phi = nh \text{ or, } p_\phi = \frac{nh}{2\pi} \tag{1.201}$$

This shows that the angular momentum of the rigid rotator is an integral multiple of $\frac{h}{2\pi}$, i.e., the angular momentum is quantised.

The energy of the rotator is given by

$$E = \frac{p_\phi^2}{2I} = \frac{1}{2I}\left(\frac{nh}{2\pi}\right)^2 = \frac{n^2h^2}{8\pi^2 I} \tag{1.202}$$

Equation (1.205) shows that $E \propto n^2$, i.e., the energy levels are discrete, but spaced unequally [Fig 1.50]. This leads to the support of quantum theory.

Fig 1.50 Energy levels of a rigid rotator

1.23 Inadequacy of old quantum theory

The old quantum theory formulated on the basis of Bohr's quantum condition and Wilson-Sommerfeld quantisation rule suffers from the following limitations:

1. It is not applicable to non-periodic systems.
2. The spectral lines of relatively simple cases like hydrogen molecule and normal helium are not possible to be explained with its help.
3. It cannot explain the relative intensities of spectral lines.
4. The theory does not account the processes connected with electron spin and Pauli's exclusion principle.
5. The discrete and non-radiating energy states as proposed by Bohr was purely empirical without any theoretical proof.

New quantum theory based on *Wave Mechanics* or *Matrix Mechanics*, however successfully resolved the above difficulties.

1.24 Answers of some important conceptual questions

1.1. *What led to the discovery of quantum theory?*

Ans. Classical theory based on *Newton's Laws of Motion* failed to explain the experimental results on *blackbody radiation, photoelectric effect, Compton effect, stability of atoms* and many other experimental results of *modern physics*. The failure of classical theory to account the above experimental results led to the discovery of a new theory, *the quantum theory*.

Max Planck in 1901 proposed the *quantum theory* which could successfully explain the distribution of energy in the spectrum of blackbody radiation according to which the energy of the blackbody radiation is discrete [i.e., energy $E = nh\nu$, where $n = 1, 2, 3, \ldots$, h is the *Planck's constant* and ν is the frequency of radiation] instead of continuous as an classical theory. This marvellous idea of Planck opened the door of *modern physics*.

1.2. *What do you mean by blackbody and blackbody radiation?* **[CU (H) 2002, 1995, 1992]**

Ans. Anybody which has the capacity of absorbing radiation of all wavelengths incident on it, is called a *blackbody* or a *perfect blackbody*. It neither reflects nor transmits any of the incident radiation and, therefore, appears black whatever be the colour of the radiation. *A perfect blackbody is an ideal concept.*

The radiation emitted by a blackbody consists of all wavelengths and is called a blackbody radiation (or, full radiation).

The heat radiation in an isothermal enclosure is termed as blackbody radiations. The radiation within such an enclosure can be observed through a *small hole* in it which acts as a perfect absorber for radiation that enters the hole and hence the hole acts as a *blackbody*.

1.3. *What are the characteristics of blackbody radiation?*

Ans. The following are the characteristics of blackbody radiation:

(a) Blackbody radiation consists of nearly all wavelengths (i.e., $\lambda = 0$ to α).

(b) Blackbody radiation resembles in many ways a perfect gas. Perfect gas consists of molecules of KE ranging from 0 to ∞. Blackbody radiation also consists of photons of energy $\varepsilon = 0$ to ∞.

(c) Blackbody radiation is *isotropic* and *homogeneous*.

(d) The energy density of blackbody radiation depends only on the temperature of the emitter.

(e) The character of the emitted and absorbed blackbody radiation is the same.

1.4. *Define emissive power and absorptive power of a blackbody. What is Kirchhoff's law of blackbody radiation?* **[CU (H) 1992]**

Ans. **Emissive power:** It is defined as the *energy emitted normally from unit area of the surface of the emitter per unit time through unit solid angle.*

If $u_\lambda dx$ be the energy radiated normally from an elementary area dS in elementary solid angle dw in time dt, then

$$e_\lambda d\lambda = \frac{u_\lambda d\lambda}{dS\, dw\, dt}$$

Here, e_λ represents the *emissive power* in the wavelength range λ to $\lambda + d\lambda$.

Absorptive power: It is defined as the ratio of absorbed radiation (Q'_λ) to that incident (Q_λ) on the body, i.e., absorptive power, $a_\lambda = Q'_\lambda / Q_\lambda$.

Kirchhoff's law: *It is defined as the ratio of the emissive power to the absorptive power of a body at any temperature and for the same wavelength and is equal to the emissive power of a perfect blackbody, i.e.,* $e_\lambda / a_\lambda = E_b$, *where* E_b = *emissive power of blackbody.*

1.5. *What are the basic differences between Wien's law and Rayleigh-Jeans law of blackbody radiation?*

Ans. Following are the basic differences between the above two laws:

(a) *Wien's law* was deduced by considering blackbody radiation to be longitudinal wave. But, in the derivation of *Rayleigh-Jeans law* radiation was supposed to be transverse electromagnetic wave.

(b) According to *Wien's theory* the energy density of wavelength λ is

$$E_\lambda \propto \frac{1}{\lambda^5} e^{-C_2/\lambda T}$$

where C_2 is a constant and T is the absolute temperature of the emitter.

But, according to *Rayleigh-Jeans law*, $E_\lambda \propto 1/\lambda^4$.

(c) Wien's law can explain the distribution of energy in the short wavelength region of the blackbody energy distribution curve and fails in the long wavelength region.

Rayleigh-Jeans law is valid only in the long wavelength region and leads to *ultraviolet catastrophe* in the short wavelength region of the blackbody spectrum.

1.6. *What is ultraviolet catastrophe?*

Ans. According to Rayleigh-Jeans law of blackbody radiation the energy in the narrow wavelength region λ to $\lambda + d\lambda$ is given by

$$E_\lambda\, d\lambda = \frac{8\pi kT}{\lambda^4} d\lambda$$

where k is the Boltzmann constant, E_λ is the energy density per unit wavelength and T is the absolute temperature of the emitter.

According to the above law, $E_\lambda \to \infty$ as $\lambda \to 0$, i.e., the whole energy of the blackbody radiation is concentrated in the shortwavelength at ultraviolet end of the spectrum. But, experimental observation shows that $E_\lambda \to 0$ as $\lambda \to 0$. This serious disagreement between theory and experiment is known as *ultraviolet catastrophe*. This is due to the failure of the *law of equipartition of energy* on the basis of which *Rayleigh-Jeans law* was derived. This indicates the inadequacy of *classical theory* to explain the distribution of energy in the spectrum of blackbody radiation.

1.7. *Bring out the basic differences in the formulation point of view of Wien's law, Rayleigh-Jeans law and Planck's law of radiation.*

Ans. Following are the basic differences in the formulation point of view of the above three laws:

(a) *Wien's law* and *Rayleigh-Jeans* law were derived on the classical approach that energy of blackbody radiation is *continuous* whereas *Planck's law* assumes that energy is *discrete*.

(b) Wien assumed thermal radiation to be *longitudinal wave* whereas Rayleigh-Jeans assumed it to be *transverse electromagnetic wave*.

But, Planck assumed that energy of thermal radiation is *discrete* and consists of photons. Photons are packets of concentrated energy $\varepsilon = h\nu$, where ν is the frequency of radiation and h is *Planck's constant*. The photons are unbreakable in any situation.

Planck's law of radiation corrected the radical mistake in *Wien's law* and *Rayleigh-Jeans law* and removed *ultraviolet catastrophe*.

1.8. *State Stefan-Boltzmann law.*

Ans. Stefan-Boltzmann law. The law states that *the total thermal energy (Q) emitted per unit area per second from a perfect blackbody is proportional to the fourth power of its temperature (T), i.e.,* $Q = \sigma T^4$, *where* $\sigma = 5.672 \times 10^{-5}$ ergs/cm^2/k^4 *is the Stefan's constant.*

If a body at temperature T be surrounded by another blackbody at absolute temperature T_1, then the amount of heat lost per second per unit area of the former is $Q = \sigma\left(T^4 - T_1^4\right)$.

1.9. *What is photoelectric effect? How does classical theory fail to explain this effect?*

[cf. CU (H) 2005; NBU (H) 2003]

Ans. See Arts. 1.5 and 1.7.

1.10. *Discuss how the photon concept explains the observed facts of photoelectricity.*

Ans. See explanation of photoelectric phenomenon (Art 1.8).

1.11. *Why is it important to have a clean surface, while doing photoelectric experiment?* **[HPU 1995]**

Ans. The presence of impurity layer on the photosensitive metal surface increases the work function of the metal surface. The incident photon, then will have to expend more energy to cross the impurity layer. Under such condition the encounter of the *photon* with the *electron* may not cause photoelectron emission.

1.12. *Why the photoelectron does not have as much energy as the quantum of light which causes its ejection?*

Ans. The KE of the ejected photoelectron according to *Einstein's theory* is

$$\tfrac{1}{2}mv^2 = h\nu - h\nu_0 = h\nu - W_0$$

where $h\nu$ is the energy of the incident photon and W_0 is the *work function*.

Since, the incident photon has to expend an energy (W_0 = the work function), the emitted photoelectron cannot have as much energy as the incident photon.

1.13. ***Does every incident photon on the photocathode cause emission of an electron?*** **[HPU 1997]**

Ans. **The incident photons which have energy ($h\nu$) less than the work function W_0 or frequency (ν) less than the *threshold frequency* (ν_0) are unable to eject photoelectrons. Moreover, there is a finite probability of absorption of photons by the electron even though they have $h\nu > W_0$ and $\nu > \nu_0$ and as such every incident photon cannot liberate a photoelectron.**

1.14. ***What is quantum yield?***

Ans. It is defined as the ratio of the number of emitted electrons (N_{ei}) to that of the incident photons **(N_{ph}). So quantum yield η is always less than unity, i.e., $\eta < 1$, as every incident photon cannot cause emission of an electron, i.e., $N_{el} < N_{ph}$.**

η has great importance in designing the *photoelectric devices.*

1.15. ***Explain why an electron in Compton effect cannot be scattered at an angle greater than*** **90°.**

[NBU (H) 2003; CU (H) 1996]

Ans. **The relation between the angle of recoil of the electron (θ) and the angle of scattering (ϕ) of the photon in Compton effect is**

$$\tan\theta = \frac{\cot\frac{\phi}{2}}{(1 + h\nu/m_0c^2)}$$

If (i) $\phi = 0$, then $\tan\theta = \infty$ and $\theta = \frac{\pi}{2}$ (ii) $\phi = \pi$, then $\tan\theta = 0$ and $\theta = 0$.

Since, $0 \leqslant \phi \leqslant \pi$ the value of θ must lie within 0 and $\pi/2$.

Hence, the electron cannot be scattered at an angle greater than 90°.

1.16. ***At what angle of scattering of the photon in Compton effect, the Compton shift will be the maximum?***

[WBUT 2005]

Ans. **The Compton shift is given by $\Delta\lambda = \lambda_0(1 - \cos\phi)$, where λ_0 is the Compton wavelength and ϕ is the angle of scattering of the photon.**

If $\phi = \pi$, then $\cos\phi = -1$ and $\Delta\lambda = 2\lambda_0$.

Since, the minimum value of $\cos\phi$ is -1 for $\phi = \pi$, the Compton shift will be the maximum for $\phi = \pi$.

1.17. ***Is it possible for a photon to transfer its whole energy to the electron in Compton effect?***

Ans. **The KE of the recoil electron in Compton effect is given by**

$$E_k = h\nu \frac{2\alpha\cos^2\theta}{(1+\alpha)^2 - \alpha^2\cos^2\theta}$$

where $\alpha = h\nu/mc^2$. θ is the angle of recoil and ν is the frequency of the incident photon.

For $E_k = (E_k)_{max}$, $\cos^2\theta$ must be maximum, i.e., 1.

$$\therefore (E_k)_{max} = \frac{2\alpha h\nu}{1+2\alpha} = \left(\frac{2\alpha}{1+2\alpha}\right)h\nu$$

$(E_k)_{max}$ is the maximum energy that the photon can transfer to the electron, since the electron was at rest before collision.

Since, $\left(\frac{2\alpha}{1+2\alpha}\right) < 1$, $(E_k)_{max} < h\nu$.

So the maximum energy that the photon can transfer to the electron is a *fraction* of its energy $h\nu$.

Hence, the photon can never transfer its whole energy to the electron.

1.18. *Show that a free electron at rest cannot absorb a photon in Compton effect.*

Ans. If p_{ph} and p_{el} be the momentum of a photon and a free electron, then if possible, let $p_{ph} = p_{el}$ or, $h\nu/c = p_{el}$.

If now E_{ph} and E_{el} be the energy of the photon and that of the electron, then

$$E_{ph} = E_{el} \text{ or, } h\nu = \sqrt{(p_{el}c)^2 + (m_0c^2)^2} \text{ or, } h\nu/c = \sqrt{p_{el}^2 + m_0^2c^2} \text{ or, } p_{ph} > p_{el}$$

Here, the conservation law of momentum breaks down.

Hence, *a free electron at rest cannot absorb a photon*. This justifies that *for Compton effect to occur the electron must be free.*

1.19. *Can a tightly bound inner electron of an atom cause Compton scattering?*

Ans. The tightly bound inner electron is not detached from the atom and as such the entire atom has to undergo recoil. In such case, the Compton shift is given by

$$\Delta\lambda = \frac{h}{\mu c}(1 - \cos\phi)$$

where μ is the mass of the entire atom at rest and ϕ is the angle of scattering.

Since, $\mu \gg m_0$, $\Delta\lambda \to 0$.

Hence, *a tightly bound inner electron of an atom cannot cause Compton effect.*

1.20. *Why Compton effect cannot be observed with visible light?* **[KU (H) 2013; WBUT 2000]**

Ans. Compton effect is the phenomenon of scattering of photon with free (or loosely bound) electron. Now, the energy of a photon of visible light ($\lambda = 4000 \times 10^{-10}$ m to 7500×10^{-10} m) is extremely smaller than the binding energy of an electron even for the lightest element hydrogen. The energy of a photon of $\lambda = 6000 \times 10^{-10}$ m is 2 eV whereas the binding energy of an electron in hydrogen atom is 13.6 eV. That is why the electron cannot be regarded as a free electron and as such Compton effect does not occur with visible light.

On the other hand, the energy of an X-ray photon is a few thousand times greater than that of a visible light photon. So the outer electrons of light elements (like carbon) can be regarded as free electrons and as such Compton effect is possible with X-rays photon.

Further, the maximum value of Compton shift for carbon $\Delta\lambda \simeq 1.1 \times 10^{-6}$ Å and to observe Compton shift with visible light using with optical instrument it must have resolving power

$$\frac{\lambda}{\Delta\lambda} = \frac{6000\text{ Å}}{1.1 \times 10^{-6}\text{ Å}} \simeq 5.45 \times 10^9$$

This value of resolving power is much more the limit of resolution of usual optical instrument and as such Compton shift remained unobserved.

1.21. *What is the limit of applicability of classical theory? Discuss.*

Ans. The dimension of Planck's constant ($h = 6.626 \times 10^{-34}$ J-s) is = [energy] × [time] = [length] × [momentum].

The dimension of a physical quantity is called *action* and so the action of Planck's constant is the fundamental quantum of action.

If the value of dimension of any physical system is of the order of the dimension of h, then the system is to be dealt with *quantum mechanics*. On the other hand, the system is to be dealt with classical mechanics if the value of the dimension of action is greater than that of h.

1.22. *Give a comparative study of photoelectric effect and Compton effect.*

Ans. The following gives a comparative study of photoelectric effect and Compton effect:

(a) Both the photoelectric effect and Compton effect result due to the action of a photon with an electron.

(b) The laws of conservation of energy and momentum are valid in both the effects.

(c) In photoelectric effect the energy of the incident photon is *completely* transferred to the free electron in metal, but in Compton effect only a *fraction* of the energy of the incident photon is transferred to the loosely bound electron.

(d) In photoelectric effect, if the energy of the incident photon (ε) be greater than the work function (W_0) of the metal, then the electron is ejected from it leaving a positive charge. But, in Compton effect the electron is recoiled after the scattering of the photon.

(e) The threshold energy in photoelectric effect is lower than that of the Compton effect.

1.23. *How could Bohr stabilize Rutherford atom?*

Ans. According to Rutherford atom model, the negatively charged electrons resolve around a positively charged heavy nucleus in circular orbits. The electrons being in accelerated motion, would undergo continuous radiation of energy and ultimately fall into the nucleus after transversing a spiraling path around the nucleus. Thus, the stability of the atom is lost.

In order to stabilise the atoms Bohr assumed that the electrons cannot resolve around the nucleus in any arbitrary orbit. Out of infinite number of orbits of continuously varying radius, only those orbits are possible for the electrons in which the total angular momentum (L) is an integral multiple of $\hbar(= h/2\pi)$, i.e., $L = n\hbar$, $n = 1, 2, 3, \ldots$, where h is the Planck's constant. These orbits are called *stationary orbits*, since the electron revolving in these orbits does not radiate energy. By so doing Bohr discarded the famous principle of classical electrodynamics that *a charged particle under acceleration or deceleration must radiate electromagnetic energy.*

1.24. *What is the significance of negative energy of the electron in hydrogen atom?*

Ans. Total energy of the electron of charge e and mass m revolving in the nth stationary orbit is given by

$$E_n = -\frac{me^4}{32\pi^2\varepsilon_0^2\hbar^2n^2} \quad \left[\hbar = \frac{h}{2\pi}\right]$$

where m, e are the mass and charge of the electron, ε_0 is the permittivity of the free space and n is the orbit number. Using standard values of m, e, ε_0 and $\hbar$, E_n comes out to be

$$E_n = -13.6 \text{ eV}$$

which is *negative* and *discrete.* The significance of the negative energy of the electron is that energy is to be supplied to take an electron away from the nucleus. The energy of the electron (E_n) is zero when it is removed to infinity from the nucleus. If $E_n > 0$, the electron will not revolve in a closed orbit around the nucleus and become free.

1.25. ***Is Rydberg's constant is a constant for all nuclei? Discuss.*** **[BU 1992]**

Ans. **If R_z and R_∞ be the Rydberg's constants of an atom of atomic number Z and of infinite mass respectively, then**

$$R_z = \frac{R_\infty}{1 + m/M_z}$$

where m is the mass of the electron and M_z the mass of the nucleus of atomic number Z.

Since, M_z is different for different nuclei, Rydberg's constant is not a universal constant, but different for different nuclei.

1.26. ***State Bohr's correspondence principle. What is its importance in atomic physics?***

Ans. **See Art. 1.15.**

1.27. ***State Ritz's combination principle and discuss its importance in spectroscopy.***

Ans. **See Art. 1.14.**

1.25 Solved problems

» ***Example*** **1.1. Luminosity of Rigel star in Orton Constellation is 17000 times that of the sun. If the surface temperature of the sun is 6000 K, calculate the temperature of the star.**

***Solution*: If the temperature of the surrounding is very very less than that of the radiation emitting body, then Stefan-Boltzmann law takes the form: $E = \sigma T^4$, where σ is the Stefan's constant and T is the absolute temperature of the body.**

If E_1 and E_2 be the energy radiated per unit area per second and T_1 and T_2 be the temperature of the star and the surface temperature of the sun respectively, then $E_1 = \sigma T_1^4$, $E_2 = \sigma T_2^4$, $\therefore E_1/E_2 = T_1^4/T_2^4$.

But, $E_1/E_2 = 17000$ and $T_2 = 6000\,K$.

$$\therefore 17000 = \frac{T_1^4}{(6000)^4} \quad \text{or,} \quad T_1^4 = (6000)^4 \times 17000.$$

$$\therefore T_1 = 6000 \times (17000)^{1/4} = \mathbf{68520\,K}.$$

» ***Example*** **1.2. If each square cm of the sun's surface radiates energy at the rate of 1.5×10^3 cal/sec/cm^2 and Stefan constant is 5.7×10^{-8} joule/sec/m^2 K^{-4}, calculate the temperature of the sun's surface in degree Centigrade.**

***Solution*: From Stefan-Boltzmann's law, $E = \sigma(T^4 - T_0^4)$.**

If $T \gg T_0$, then $E = \sigma T^4$.

The energy radiated by the sun is $E = 1.5 \times 10^3$ cal/cm^2/sec $= 1.5 \times 10^3 \times 4.2$ joules $(10^2\,\text{m})^2\,\text{sec}^{-1}$

$= 1.5 \times 4.2 \times 10^7$ joules/m^2/sec

and $\sigma = 5.7 \times 10^{-8}$ joule/sec/m^2(K)$^{-4}$.

Now, from $E = \sigma T^4$

$$1.5 \times 4.2 \times 10^7 = 5.7 \times 10^{-8} \times T^4$$

$$\therefore T = \left[\frac{1.5 \times 4.2}{5.7} \times 10^{15}\right]^{1/4} = 5764\,\text{K} = \mathbf{5492\,°C}$$

» ***Example* 1.3. Calculate the temperature of the sun from the following data: $\lambda_m T = 0.287$, $\lambda_m = 4753$ Å.**

***Solution*: The sun's temperature**

$$T = \frac{0.287}{\lambda_m} = \frac{0.287}{4753 \times 10^{-8}} = \mathbf{5766\ K}$$

» ***Example* 1.4. From Wien's displacement law, we have $\lambda_m T = \frac{hc}{5k} = b$.**

(a) Calculate the constant b. What is the unit of b?

(b) Radiation from the moon gives $\lambda_m = 4700$ Å and 14×10^{-6} m. What conclusion can you draw from this?

[$h = 6.6 \times 10^{-34}$ J-s, $k = 1.37 \times 10^{-23}$ J/K, $c = 3 \times 10^8$ m/s].

***Solution*:**

(a) We have $b = \lambda_m T$, the unit of b is mK.

Now, $b = \frac{hc}{5k} = \frac{6.6 \times 10^{-34} \times 3 \times 10^8}{5 \times 1.37 \times 10^{-23}} = \mathbf{0.002890\ mK}$. The unit of b is mk.

(b) When

$$\lambda_m = 4700\ \text{Å} = 4700 \times 10^{-10}\ \text{m},\ T_1 = \frac{0.002890}{4700 \times 10^{-10}} = 6166\ \text{K}$$

and when

$$\lambda_m = 14 \times 10^{-6}\ \text{m},\ T_2 = \frac{0.002890}{14 \times 10^{-6}} = \mathbf{207\ K}$$

As the sun's radiation is reflected from the moon, $T_1 = 6166$ K represents the temperature of the sun. $T = 207$ K represents the temperature of the moon.

» ***Example* 1.5. At what wavelength λ_m is the energy density maximum according to Wien's law? Estimate the temperature of the sun if λ_m for the sun is 4900 Å.** **[CU (H) 1995]**

***Solution*: According to Wien's displacement law, the energy density is the maximum at temperature T if $\lambda_m T = 0.2896$ cm K.**

$$\therefore T = \frac{0.2896}{\lambda_m} = 5910\ \text{K}$$

where $\lambda_m = 4900$ Å $= \mathbf{4900 \times 10^{-8}}$ **cm.**

» ***Example* 1.6.**

(a) Calculate the energy in electron-volt of a photon of wavelength 1 Å.

(b) What is the momentum of this photon? [*Given*: $h = 6.62 \times 10^{-34}$ J-s.]

***Solution*:**

(a) If h be the Planck's constant, c the velocity of light in free space and λ the wavelength of a photon, then its energy $\varepsilon = hc/\lambda$.

Here, $h = 6.62 \times 10^{-34}$ J-s, $c = 3 \times 10^8$ m/s, $\lambda = 1$ Å $= 10^{-10}$ m

$$\therefore \varepsilon = \frac{6.62 \times 10^{-34} \times 3 \times 10^8}{10^{-10}} = 19.86 \times 10^{-16}\ \text{J}$$

$$= \frac{19.86 \times 10^{-16}}{1.6 \times 10^{-19}}\ \text{eV} = \mathbf{12412.5\ eV}\ [\because 1\ \text{eV} = 1.6 \times 10^{-19}\ \text{J}]$$

(b) Momentum of the same photon

$$p = \frac{h}{\lambda} \text{ or, } p = \frac{6.62 \times 10^{-34}}{10^{-10}} = \mathbf{6.62 \times 10^{-24}\ kg\ m/s}$$

» ***Example* 1.7.** Calculate the number of photons emitted per second by a 100 watt sodium lamp. [Assume λ = 5893 Å of sodium light.]

Solution: Energy of a photon

$$\varepsilon = h\nu = \frac{hc}{\lambda} \text{ or, } \varepsilon = \frac{(6.63 \times 10^{-34}\ \text{J-s}) \times (3 \times 10^{8}\ \text{m/s})}{(5893 \times 10^{-10}\ \text{m})} = 3.375 \times 10^{-19}\ \text{J}$$

Now, the rate of energy emission from a 100 watt lamp is 100 J/s.

Since, the energy of a photon is 3.375 J, the number of photons emitted per second is

$$\frac{100}{3.375 \times 10^{-19}} = \mathbf{2.96 \times 10^{20}}$$

» ***Example* 1.8.** Calculate the frequency of the photon produced when an electron of 20 keV is brought to rest in one collision with a heavy nucleus.

Solution: If whole of the KE of the electron is used up to produce the photon, then $K + m_0c^2 = h\nu + m_0c^2$, where K is the KE of the electron and $h\nu$ the energy of the photon, m_0c^2 the rest mass of the electron.

$\therefore K = h\nu$ or, 20×10^3 eV = $(4.136 \times 10^{-15}$ eV s)ν.

$\therefore \nu = \mathbf{4.84 \times 10^{18}\ Hz}\ \left[\because \nu = \frac{k}{h}\right]$.

» ***Example* 1.9.** A radio station operates at a frequency of 103.7 MHz with a power output of 200 kW. Determine the rate of emission of quanta from the station.

Solution: The energy of each quantum is $\varepsilon = h\nu$ or, $\varepsilon = (6.63 \times 10^{-34}\ \text{J-s}) \times (103.7 \times 10^{6}/\text{s}) = 6.88 \times 10^{-26}$ J.

So the rate of emission of quanta $= \dfrac{\text{number of quanta}}{\text{time}} = 200 \times 10^3\ \text{J/s} \times \dfrac{1\ \text{quantum}}{6.88 \times 10^{-26}\ \text{J}}$

$= \mathbf{29.07 \times 10^{29}\ quanta/s.}$

» ***Example* 1.10.** The kinetic energies of photoelectrons range from zero to 4.0×10^{-19} J, when light of wavelength 3000 Å falls on a metal surface. What is the stopping potential? What is the threshold wavelength?

Solution: Maximum kinetic energy of photoelectrons is

$$K_{max} = 4.0 \times 10^{-19}\ \text{J} \times \frac{1\ \text{eV}}{1.6 \times 10^{-19}\ \text{J}} = 2.5\ \text{eV}$$

If V_s be the stopping potential, then

$$eV_s = K_{max} \text{ or, } V_s = \frac{2.5\ \text{eV}}{e} = 2.5\ \text{V}$$

Again,

$$eV_s = h\nu - h\nu_0 = \frac{hc}{\lambda} - \frac{hc}{\lambda_{th}} \text{ or, } 2.5\ \text{eV} = \frac{12.4 \times 10^3\text{eV Å}}{3000\ \text{Å}} - \frac{12.4 \times 10^3\text{eV Å}}{\lambda_{th}} \text{ or, } \lambda_{th} = \mathbf{7590\ Å}$$

» ***Example* 1.11.** In an experiment, tungsten cathode which has a threshold wavelength 2300 Å is irradiated by ultraviolet light of wavelength 1800 Å. Calculate

(a) the maximum energy of emitted photoelectrons and

(b) work function of tungsten. [*Given*: $h = 6.6 \times 10^{-34}$ J-s, 1 eV $= 1.6 \times 10^{-19}$ J and velocity of light, $c = 3 \times 10^8$ m/s.] **[BU (H) 2001]**

Solution: We have Einstein's photoelectric equation $E_{max} = \frac{1}{2}mV^2_{max} = h(\nu - \nu_0)$, where ν = frequency of the incident radiation, ν_0 = the threshold frequency.

λ and λ_0 are the corresponding wavelengths.

$$\therefore E_{max} = h\left[\frac{c}{\lambda} - \frac{c}{\lambda_0}\right] = hc\left[\frac{1}{\lambda} - \frac{1}{\lambda_0}\right]$$

Here, $h = 6.6 \times 10^{-34}$ J-s, $\lambda = 1800$ Å $= 1800 \times 10^{-10}$ m, $\lambda_0 = 2300$ Å $= 2300 \times 10^{-10}$ m.

$$\therefore E_{max} = 6.6 \times 10^{-34} \times 3 \times 10^8 \left[\frac{1}{1800 \times 10^{-10}} - \frac{1}{2300 \times 10^{-10}}\right]$$

$$= \frac{6.6 \times 10^{-34} \times 3 \times 10^8}{10^{-8}}\left[\frac{1}{18} - \frac{1}{23}\right] = 2.391 \times 10^{-19}\ \text{J}$$

$$= \frac{2.391 \times 10^{-19}}{1.6 \times 10^{-19}}\ \text{eV} = 1.485\ \text{eV}$$

The work function of tungsten is

$$W_0 = h\nu_0 = \frac{hc}{\lambda_0} = \frac{6.6 \times 10^{-34} \times 3 \times 10^8}{2300 \times 10^{-10}} = 8.608 \times 10^{-19}\ \text{J} = \frac{8.608 \times 10^{-19}}{1.6 \times 10^{-19}}\ \text{eV} = \mathbf{5.38\ eV}$$

» ***Example* 1.12.** What is the energy in ergs of a photon of wavelength 6000 Å? **[BU (H) 2003]**

Solution: The energy of a photon of wavelength λ is $E = \frac{hc}{\lambda}$, where $h = 6.6 \times 10^{-34}$ J-s, $c = 2.998 \times 10^8$ m/s and wavelength $\lambda = 6000$ Å $= 6000 \times 10^{-10}$ m.

$$\therefore E = \frac{6.6 \times 10^{-34} \times 2.998 \times 10^8}{6000 \times 10^{-10} \times 1.602 \times 10^{-19}} = \mathbf{0.687\ eV}$$

» ***Example* 1.13.** The work function of a tungsten surface is 5.4 eV. When the surface is illuminated by light of wavelength 175 nm the maximum photoelectron energy is 1.75 eV. Find Planck's constant from these data. **[WBUT (Question Bank)]**

Solution: If λ be the wavelength of the incident photon and W be the work function of metal, then its energy E is

$$E = h\nu = \frac{hc}{\lambda} = \left(\tfrac{1}{2}mv^2\right) + W$$

$$\therefore h = \frac{\lambda}{c}\left[\left(\tfrac{1}{2}mv^2\right)_{max} + W\right] = \frac{175 \times 10^{-9}}{3 \times 10^8}\left[(1.75 + 5.4) \times 10^{-19}\right]$$

$$= 58.3 \times 10^{-17} \times 11.36 \times 10^{-19} = 662.28 \times 10^{-36}$$

$\therefore$ Planck's constant $\boldsymbol{h \simeq 6.62 \times 10^{-34}}$ **J-s.**

» ***Example* 1.14.** The threshold wavelength for photoelectric emission in tungsten is 230 nm. What wavelength of incident light must be used in order to eject electrons with a maximum velocity of 5×10^5 m/s. [Mass of the electron $= 9.1 \times 10^{-31}$ kg] **[WBUT (Question Bank)]**

Solution: We have from the photoelectric equation

$$h\nu = \left(\frac{1}{2}mv^2\right)_{\max} + h\nu_0 \text{ or, } h\frac{c}{\lambda} = \left(\frac{1}{2}mv^2\right)_{\max} + h\frac{c}{\lambda_0}$$

$$\text{or, } (6.626\times10^{-34})\times(3\times10^8)\left(\frac{1}{\lambda}-\frac{1}{\lambda_0}\right) = 113.75\times10^{-21}$$

$$\text{or, } \frac{19.88\times10^{-26}}{\lambda} = \frac{19.88\times10^{-26}}{230\times10^{-9}} + 113.75\times10^{-21} \left[\because \tfrac{1}{2}mv^2 = \tfrac{1}{2}(9.1\times10^{-31})(5\times10^5)^2 = 113.75\times10^{-21}\right]$$

$$\text{or, } \lambda = \frac{19.88\times10^{-26}}{798.05\times10^{-21}} = 0.0203\times10^{-9}\text{ m}$$

$\therefore$ **the required wavelength = 2030 Å.**

» ***Example* 1.15.** The work function of aluminium is 4.2 eV. Calculate the kinetic energy of the fastest and slowest photoelectrons, the stopping potential and the cut-off wavelength when light of wavelength 2000 Å falls on a clean aluminium surface. [$h = 6.63\times10^{-34}$ J-s, $c = 3\times10^8$ m/s, 1 eV = 1.6×10^{-19} J and 1 Å = 10^{-10} m]

Solution: According to Einstein's photoelectric equation $K_{\max} = h\nu - W_0$, $K_{\max}$ is the KE of the fastest electron, W_0 = work function and ν = the incident frequency.

Now,

$$h\nu = \frac{hc}{\lambda} = \frac{(6.63\times10^{-34})\times(3\times10^8)}{(2000\times10^{-10})} = 9.94\times10^{-19}\text{ J} = \frac{9.94\times10^{-19}}{1.6\times10^{-19}} = 6.2\text{ eV}$$

$\therefore K_{\max} = 6.2 - 4.2 =$ **2.0 eV**.

The kinetic energy of the slowest photoelectron is **zero.**

Now, the stopping potential is equal to the maximum KE of the photoelectrons, i.e., **2.0 eV.**

$\therefore V_s = 2.0$ V.

$$\text{Now, cut-off frequency, } \nu_0 = \frac{W_0}{h} = \frac{4.2\times1.6\times10^{-19}}{6.63\times10^{-34}}\text{ J} = \mathbf{1.0\times10^{15}/s}$$

$$\text{cut-off wavelength, } \lambda_0 = \frac{c}{\nu_0} = \frac{3.0\times10^8}{1.0\times10^{15}} = 3.0\times10^{-7}\text{ m} = \mathbf{3000\text{ Å}}.$$

» ***Example* 1.16.** A ray of UV light of wavelength λ = 3000 Å falls on a surface of a material whose work function is 2.28 eV and ejects an electron. What is the velocity of the emitted photoelectron?

Solution: We have the work function, $W_0 = 2.28\text{ eV} = 2.28\times1.6\times10^{-19}\text{ J} = 3.648\text{ J}$.

Wavelength of the incident light, $\lambda = 3000\text{ Å} = 3000\times10^{-10}$ m.

$$\therefore \text{frequency, } \nu = \frac{c}{\lambda} = \frac{3\times10^8}{3000\times10^{-10}} = 10^{15}\text{ Hz.}$$

KE of the emitted photoelectron is $\frac{1}{2}mv^2 = h\nu - W_0 = 6.62\times10^{-34}\times10^{15} - 3.648\times10^{-19}$

$$= 2.972\times10^{-19}\text{ J.}$$

$\therefore$ velocity of the emitted electron is

$$v = \sqrt{\frac{2\times2.972\times10^{-19}}{9\times10^{-31}}} = \mathbf{8.13\times10^5\text{ m/s}}$$

» ***Example* 1.17.** What is the maximum kinetic energy of photoelectrons released from the surface of lithium by electromagnetic radiation whose electric field E is given by $E = A(1+\sin\omega t)\cos\omega_0 t$, where A is a constant, $\omega = 6\times10^{14}$/s, $\omega_0 = 3.6\times10^{15}$/s the work function of lithium is 2.4 eV.

***Solution*:** Here, $E = A\cos\omega_0 t + A\sin\omega t\cos\omega_0 t = A\cos\omega_0 t + \frac{A}{2}\{\sin(\omega+\omega_0)t - \sin(\omega_0-\omega)t\}$.

The threshold frequency

$$\omega_{th} \simeq 2.4\times1.6\times\frac{10^{-19}}{\hbar} \text{ or, } \omega_{th} \simeq 2.4\times1.6\times\frac{10^{-19}}{10^{-27}} \simeq \frac{2.4\times1.6\times10^{15}}{1.054} = 3.64\times10^{15}$$

The maximum photoelectric response will depend on the difference between the threshold frequency ω_{th} and frequencies ω_0, $(\omega_0+\omega)$ and $(\omega_0-\omega)$.

The maximum KE $= \hbar(\omega_0+\omega-\omega_{th})$

$$\text{or, } E_{max} = \hbar\left(3.6\times10^{15}+0.6\times10^{15}-3.64\times10^{15}\right) \simeq \hbar\times0.56\times10^{15} \text{ ergs}$$
$$= 1.054\times10^{-27}\times0.54\times10^{15} = \mathbf{0.569\times10^{12}\ erg}$$

» ***Example* 1.18.** A metallic surface when illuminated with light of wavelength λ_1, emits electrons with energies up to a maximum value E_1 and when illuminated with light of wavelength $\lambda_2(\lambda_2<\lambda_1)$, it emits electrons with energies up to a maximum value E_2. Prove that Planck's constant h and work function W_0 of the metal are

$$h = \frac{(E_2-E_1)\lambda_1\lambda_2}{c(\lambda_1-\lambda_2)} \text{ and } W_0 = \frac{E_2\lambda_2-E_1\lambda_1}{\lambda_1-\lambda_2}$$

***Solution*:** The maximum kinetic energy of the photoelectrons is given by

$$E_1 = h\nu_1 - W_0 \text{ or, } E_1 = \frac{hc}{\lambda_1} - W_0 \quad \text{(i)}$$

Similarly, the maximum KE

$$E_2 = \frac{hc}{\lambda_2} - W_0 \quad \text{(ii)}$$

Subtracting (i) from (ii), we get

$$E_2 - E_1 = hc\left(\frac{1}{\lambda_2}-\frac{1}{\lambda_1}\right) = hc\left(\frac{\lambda_1-\lambda_2}{\lambda_1\lambda_2}\right) \text{ or, } \mathbf{h} = \frac{(E_2-E_1)\lambda_1\lambda_2}{c(\lambda_1-\lambda_2)} \text{ (Proved)} \quad \text{(iii)}$$

Again,

$$E_1\lambda_1 = hc - W_0\lambda_1 \quad \text{(iv)}$$
$$\text{and } E_2\lambda_2 = hc - W_0\lambda_2 \quad \text{(v)}$$

Subtracting (4) from (5), we get

$$E_2\lambda_2 - E_1\lambda_1 = W_0(\lambda_1-\lambda_2) \text{ or, } \mathbf{W_0} = \frac{E_2\lambda_2-E_1\lambda_1}{\lambda_1-\lambda_2} \text{ (Proved)} \quad \text{(vi)}$$

» ***Example* 1.19.** Find the average number of photons per cubic metre in a monochromatic beam of radiation of frequency 10 kHz and of intensity 10^4 W/m^2.

***Solution*:** Let n = no. of photons per unit volume, each having energy $h\nu$.

We consider a length l of the beam of cross-sectional area A, then in the volume Al total energy of the photons is $(nAl)\times h\nu$.

The time taken by this energy to pass through the length l is $t = \frac{l}{c}$, where c = velocity of light in free space.

$\therefore$ intensity I of the beam is given by

$$I = \frac{nAlh\nu}{At} = \frac{nlh\nu}{l/c} = nh\nu c$$

$$\therefore n = \frac{I}{h\nu c} = \frac{10^4}{6.63 \times 10^{-34} \times 10^4 \times 3 \times 10^8} = \frac{10^{26}}{19.89} = \mathbf{5.03 \times 10^{24}\ photons/m^3}$$

» ***Example* 1.20.** Ultraviolet light of wavelength 350 nm and intensity 1.00 W/m^2 is directed at a potassium surface.

(a) Find the maximum KE of the photoelectrons.

(b) If 0.50 % of the incident photons produce photoelectrons, how many are emitted per second if the potassium surface has an area of 1.00 cm^2?

***Solution*:**

(a) Energy of the photon

$$E_p = \frac{1.24 \times 10^{-6}\ \text{eV m}}{(350\ \text{nm}) \times (10^{-9}\ \text{m/nm})} = 3.5\ \text{eV}\ [\because 1\ \text{nm} = 10^{-9}\ \text{m}]$$

Work function of potassium is W_0 = 2.2 eV.

So $KE_{max} = h\nu - W_0$ = 3.5 eV − 2.2 eV = **1.3 eV.**

(b) The photon energy in joule is 5.68×10^{-19}.

$\therefore$ the number of photons that reach the surface per second is

$$n_p = \frac{E/t}{E_p} = \frac{\left(\frac{P}{A}\right) A}{E_p} = \frac{(1.00\ \text{W/m}^2) \times (1.00 \times 10^{-4}\ \text{m}^2)}{5.68 \times 10^{-19}\ \text{J/photon}}$$

$\therefore$ the rate of emission of photoelectrons is $n_e = (0.0050)\, n_p$ = **8.8 × 10^{11} photoelectrons/s.**

» ***Example* 1.21.** Show that the maximum recoil energy of a free electron of rest mass m_0 when struck by a photon of wavelength λ is given by

$$E_{\max} = \frac{2m_0c^2\lambda_c^2}{\lambda^2 + 2\lambda_c\lambda}$$

where λ_c is the Compton wavelength of the electron.

***Solution*:** The recoil electron in the Compton scattering has kinetic energy

$$E = h(\nu - \nu') = hc\left(\frac{1}{\lambda} - \frac{1}{\lambda'}\right)$$

It appears that E will be $E_{\max}$ when λ' is the maximum.

$$\therefore E_{\max} = hc\left[\frac{1}{\lambda} - \frac{1}{\lambda'_{\max}}\right] \qquad \text{(i)}$$

But, $\lambda' = \lambda + \frac{h}{m_0 c}(1 - \cos\phi) = \lambda + \lambda_c(1 - \cos\phi)$.

If $\phi = 180°$, then $\cos\phi = -1$ and $\lambda'_{\max} = \lambda + 2\lambda_c$.

∴ from (i)

$$E_{max} = hc\left[\frac{1}{\lambda} - \frac{1}{\lambda + 2\lambda_c}\right] = \frac{2hc\lambda_c}{\lambda(\lambda + 2\lambda_c)}$$

$$\text{or, } E_{max} = \frac{2m_0c\lambda_c \times c\lambda_c}{\lambda^2 + 2\lambda_c\lambda} = \frac{2m_0c^2\lambda_c^2}{\lambda^2 + 2\lambda_c\lambda}\left[\because \lambda_c = \frac{h}{m_0c}, \therefore h = \lambda_c 2m_0c\right] \text{ (Proved)}$$

» ***Example* 1.22. X-rays of wavelength 1.0 Å are scattered from a carbon block. Find the wavelength of the scattered beam in a direction making 90° with the incident beam. What is the kinetic energy of the recoiling electron? [$h = 6.63 \times 10^{-34}$ J-s, $c = 3.0 \times 10^8$ m/s and 1 eV = 1.6×10^{-19} J]**

***Solution*: The Compton shift is given by**

$$\Delta\lambda = \lambda' - \lambda = \frac{2h}{m_0c}\sin^2\frac{\phi}{2} = \frac{h}{m_0c}(1 - \cos\phi)$$

When $\phi = 90°$, $\Delta\lambda = 0.0241(1 - 0) = 0.024$ Å.

Now, $\lambda' = \lambda + \Delta\lambda = 1.0 + 0.024 = 1.024$ Å.

Now, the energy imparted to the recoiling electron

$$= \frac{hc}{\lambda} - \frac{hc}{\lambda'} = \frac{hc(\lambda' - \lambda)}{\lambda\lambda'} = \frac{hc\Delta\lambda}{\lambda\lambda'} = \frac{(6.63 \times 10^{-34}) \times (3 \times 10^8) \times (0.024 \times 10^{-10}\text{ m})}{(1.0 \times 10^{-10}\text{ m}) \times (1.024 \times 10^{-10}\text{ m})}$$

$$= 4.73 \times 10^{-17}\text{ J} = \frac{4.73 \times 10^{-17}}{1.6 \times 10^{-19}} = \mathbf{296\ eV}$$

» ***Example* 1.23. A beam of γ-radiation having photon of energy 510 keV is incident on a foil of aluminium. Calculate the wavelength of radiation at 90° and also the energy and direction of the emission of the corresponding electron.**

***Solution*: Energy of a photon of frequency ν is $\varepsilon = h\nu = \frac{hc}{\lambda}$.**

Here, $\varepsilon = 510\text{ keV} = 510 \times 10^3 \times 1.6 \times 10^{-19}$ J.

$$\therefore \lambda = \frac{hc}{\varepsilon} = \frac{6.6 \times 10^{-34} \times 3 \times 10^8}{510 \times 10^3 \times 1.6 \times 10^{-19}} = \mathbf{2.426 \times 10^{-12}\ m}$$

The wavelength of the scattered radiation is

$$\lambda' = \lambda + \frac{h}{m_0c}(1 - \cos\phi) = (2.426 \times 10^{-12}) + \frac{6.6 \times 10^{-34}(1 - \cos 90°)}{9.1 \times 10^{-31} \times 3 \times 10^8}$$

$$= 2.426 \times 10^{-12} + 2.417 \times 10^{-12} = \mathbf{4.843 \times 10^{-12}\ m}$$

The energy of the recoil electron is

$$E = h\nu - h\nu' = h(\nu - \nu') = \frac{hc(\lambda' - \lambda)}{\lambda\lambda'}$$

$$= \frac{6.60 \times 10^{-34} \times 3 \times 10^8 \times (4.843 \times 10^{-12} - 2.426 \times 10^{-12})}{2.426 \times 10^{-12} \times 4.843 \times 10^{-12}}$$

$$= \mathbf{4.072 \times 10^{-14}\ J}$$

The direction of the recoil electrons is given by

$$\tan\theta = \frac{\tan\phi/2}{\left(1+\frac{h}{10^{-12}c}\right)} = \frac{\tan 45°}{\left[1+\frac{6.6\times10^{-34}}{(2.426\times10^{-12})\times9.1\times10^{-31}\times(3\times10^{8})}\right]}$$

$$= \frac{1}{1+0.9985} = \frac{1}{1.9985} = 0.5008$$

$\therefore \theta = \tan^{-1} 0.5008 = 26°36'$.

» ***Example* 1.24.** An X-ray photon of frequency 3×10^{19} Hz collides with a free electron at rest and is scattered at 60°. Find the direction in which the electron will move. [*Given*: $h/m_0c = 0.02426$ Å.]

Solution: Here, $\nu = 3\times10^{19}$ Hz, $\phi = 60°$, $h = 6.63\times10^{-34}$ J-s.

Now, the angle of scattering ϕ and the angle of recoil θ are given by $\cot\theta = (1+\alpha)\tan\frac{\phi}{2}$, where

$$\alpha = \frac{h\nu}{m_0c^2} = \left(\frac{h}{m_0c}\right)\frac{\nu}{c} = 0.02426\times10^{-10}\times\frac{3\times10^{19}}{3\times10^{8}} = 0.2426$$

$$\therefore \cot\theta = (1+0.2426)\tan 30° = 1.2426\times\frac{1}{\sqrt{3}} = \frac{1.2426}{1.7320}$$

$$\therefore \tan\theta = \frac{1.7320}{1.2426} = 1.3938$$

$\therefore \theta = 54.34°$.

» ***Example* 1.25.** Show that a free electron at rest cannot absorb a photon.

Solution: If p_{ph} and p_{el} be the momentum of a photon and that of a free electron, then if possible, let

$$p_{ph} = p_{el} \text{ or, } \frac{h\nu}{c} = p_{el}$$

If E_{ph} and E_{el} be the energies of a photon and a free electron, then

$$E_{ph} = E_{el} \text{ or, } h\nu = \sqrt{(p_{el}c)^2 + (m_0c^2)^2} \text{ or, } \frac{hc}{c} = \sqrt{p_{el}^2 + m_0^2c^2} > p_{el}$$

Thus, the conservation law of momentum breaks down.

Hence, the result.

Note: This shows that Compton scattering must occur with free electrons.

» ***Example* 1.26.** A photon of energy ε is scattered from a free electron of mass m_0 and observed at scattering angles greater than $\pi/3$. Show that the energy ε' of the scattered photon is less than $2m_0c^2$, however large the value of ε may be.

Solution: The energy of the scattered photon is

$$\varepsilon' = \frac{\varepsilon}{1+2\alpha\sin^2(\phi/2)} = \frac{\varepsilon}{1+\frac{2\varepsilon}{m_0c^2}\sin^2(\phi/2)}$$

For scattering angle $\phi > \pi/3$, $\sin^2(\phi/2) > 1/4$ as $\sin\phi$ is an increasing function of ϕ. For any value of $\phi > (\pi/3)$, $\sin^2\phi > 1/4 = k/4$, where $k \geqslant 1$ according as $\phi \geqslant \pi/3$.

∴ the energy of the scattered photon is

$$\varepsilon' = \frac{\varepsilon}{1+\dfrac{2\varepsilon}{m_0c^2}\cdot\dfrac{k}{4}} = \frac{\varepsilon}{1+\dfrac{k\varepsilon}{2m_0c^2}} = \frac{2m_0c^2}{\dfrac{2m_0c^2}{E}+k}$$

Since, $\frac{2m_0c^2}{E}$ is a positive quantity and $k \geqslant 1$, $\frac{2m_0c^2}{E} + k$ is always greater than 1 and ε' is always less than $2m_0c^2$, however large the value of ε may be.

» ***Example* 1.27.** A photon of energy E_0 is incident on a stationary electron target and the angle of Compton scattering of the photon is ϕ. Show using non-relativistic kinematics, that the kinetic energy of recoil of the electron is

$$\frac{E_0^2(1-\cos\phi)}{mc^2+E_0(1-\cos\phi)}$$

where m is the mass of the electron. **[BU (H) 1995]**

***Solution*:** Let a photon of the energy $E_0 = h\nu$ be incident on a stationary electron of mass m. After scattering, the scattered photon moves in a direction making an angle ϕ and the recoil electron in a direction making an angle θ with the direction of incidence [Fig 1.51].

From the conservation of energy principle, $E_0 = h\nu' + T$.

$$\therefore T = E_0 - h\nu' \quad \text{(i)}$$

where $T = \frac{1}{2}mv^2$ is the KE of the recoil electron and $h\nu'$ is the energy of the scattered photon.

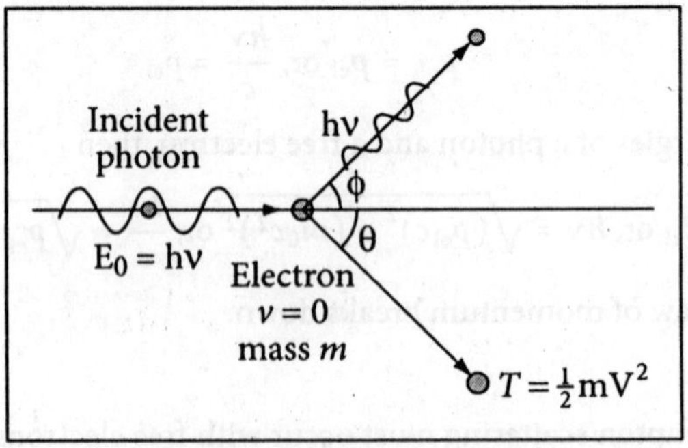

Fig 1.51 Compton scattering

Conservation of momentum along and perpendicular to the direction of incidence gives

$$\frac{E_0}{c} = \frac{h\nu'}{c}\cos\phi + mv\cos\theta$$

$$mv\cos\theta = \frac{E_0}{c} - \frac{h\nu'}{c}\cos\phi \quad \text{(ii)}$$

$$\text{and } 0 = \frac{h\nu'}{c}\sin\phi - mv\sin\phi \quad \text{(iii)}$$

Squaring and adding (2) and (3), we get

$$m^2v^2 = \left(\frac{E_0}{c} - \frac{h\nu'}{c}\cos\phi\right)^2 + \left(\frac{h\nu'}{c}\sin\phi\right)^2 = \frac{E_0^2}{c^2} - \frac{2E_0h\nu'}{c^2}\cos\phi + \left(\frac{h\nu'}{c}\right)^2$$

$$\text{or, } 2m\left(\frac{1}{2}mv^2\right) = \frac{1}{c^2}[E_0^2 - 2E_0h\nu'\cos\phi + (h\nu')^2]$$

$$\text{or, } 2mT = \frac{1}{c^2}[E_0^2 - 2E_0(E_0 - T)\cos\phi + (E_0 - T)^2]$$

$$= \frac{1}{c^2}[E_0^2 - 2E_0^2\cos\phi + 2E_0T\cos\phi + E_0^2 - 2E_0T + T^2]$$

$$= \frac{1}{c^2}[2E_0^2 - 2E_0^2\cos\phi + 2E_0T\cos\phi - 2E_0T + T^2]$$

$$\text{or, } 2mc^2T = 2E_0^2(1-\cos\phi) - 2E_0T(1-\cos\phi) + T^2$$

$$\text{or, } [2mc^2 + 2E_0(1-\cos\phi)]T = 2E_0^2(1-\cos\phi) + T^2$$

$$\text{or, } 2[mc^2 + E_0(1-\cos\phi)]T = 2E_0^2(1-\cos\phi) + T^2$$

$$\text{or, } T = \frac{2E_0^2(1-\cos\phi) + T^2}{2[mc^2 + E_0(1-\cos\phi)]}$$

$$T = \frac{E_0^2(1-\cos\phi)}{mc^2 + E_0(1-\cos\phi)} \quad [\text{where } T^2 \ll 2E_0^2(1-\cos\phi)]$$

» ***Example* 1.28.** **The Debye temperature of diamond is 2000 K. Calculate the mean velocity of sound in diamond, given the density and atomic mass of diamond as 3500 kg/m³ and 12 amu respectively. If the interatomic spacing is 1.54 Å, estimate the frequency of the dominant mode of lattice vibration.**

***Solution*:** The Debye temperature is defined as

$$\theta_D = \frac{h\nu_D}{k}, \text{ where } \nu_D = \frac{9N}{4\pi V}\left(\frac{1}{c_l^3} + \frac{2}{c_t^3}\right)^{-1}$$

Putting $c_l = c_t = c_s$,

$$\nu_D = c_s\left(\frac{3N}{4\pi V}\right)^{1/3} \quad \text{or, } \nu_D = \frac{hc_s}{k}\left(\frac{3N}{4\pi V}\right)^{1/3}$$

For one mole of atoms

$$\nu_D = \frac{hc_s}{k}\left(\frac{3N_a\rho}{4\pi M}\right)$$

where N_a = Avogadro number, M = molecular wt and ρ = density of the crystal.

$$\therefore c_s = \frac{k\theta_D}{h}\left(\frac{4\pi M}{3N_a\rho}\right)^{1/3}$$

But, here $\theta_D = 2000\,\text{K}$, $M = 12\,\text{kg}$, $\rho = 3500\,\text{kg/m}^3$.

$$\therefore c_s = \frac{1.38\times10^{-23}\times2000}{6.626\times10^{-34}}\left(\frac{4\pi\times12}{3\times6.023\times10^{26}\times3500}\right)^{1/3} = \mathbf{1.2\times10^4\ m/s}$$

Thus, the mean velocity of sound in diamond c_s = **1.2 × 10⁴ m/s.**

Now, the magnitude of the dominant photon wavelength λ_d at $T \ll \theta_D$ is

$$\lambda_d = \left(\frac{\theta_D}{T}\right) a \; [a = \text{interatomic distance}]$$

$$\therefore \lambda_d = \frac{2000}{298} \times 1.54 \times 10^{-10} = 1.03 \times 10^{-9} \text{ m}$$

The frequency of the corresponding lattice vibration is

$$\nu_d = \frac{c_s}{\lambda_d} = \frac{1.2 \times 10^4}{1.03 \times 10^{-9}} = \mathbf{1.16 \times 10^{13}\ Hz}$$

» ***Example* 1.29.** Calculate specific heat of silver at 20 K for which Debye temperature (θ_D) is 210 K.

Solution: We have

$$C_v = 3R\frac{12\pi^4}{15}\left(\frac{T}{\theta_D}\right)^3$$

$$\text{or, } C_v = 3 \times 1.986 \times \frac{12\pi^4}{15}\left(\frac{20}{210}\right)^3 = \mathbf{0.4010\ cal/K/mole}$$

This agrees well with the experimental value of $C_v = 0.3995$ cal/K/mole. [If C_v be calculated using Einstein's theory, then it would be 0.0181 cal/K/mole.

Hence, C_v differs widely from the observed value of $C_v = 0.3995$ cal/K/mole.]

» ***Example* 1.30.** Using the physical constants given below, calculate the following for hydrogen atom: (a) Velocity of an electron in the ground state (b) Radius of Bohr orbit in the ground state (c) Time taken by the electron to traverse the first Bohr orbit (d) Rydberg constant.

[Electronic charge, $e = 1.6\times10^{-19}$ C, Electronic mass $m = 9.1\times10^{-31}$ kg, Planck's constant, $h = 6.6\times10^{-34}$ J-s, Speed of light, $c = 3 \times 10^8$ m/s, Permittivity of free space, $\varepsilon_0 = 8.86 \times 10^{-12}$ C^2/N-m^2].

Solution:

(a) The velocity of the electron in the ground state ($n = 1$) is given by

$$v = \frac{e^2}{2h\varepsilon_0} = \frac{(1.6 \times 10^{-19})^2}{2 \times (6.6 \times 10^{-34}) \times (8.86 \times 10^{-12})} = \mathbf{2.19 \times 10^6\ m/s}$$

(b) The radius of the Bohr's orbit in the ground state, i.e., radius of the first Bohr orbit is given by

$$r_1 = \frac{h^2\varepsilon_0}{\pi m e^2} = \frac{(6.6 \times 10^{-34})^2 \times (8.86 \times 10^{-12})}{3.14 \times (9.1 \times 10^{-31}) \times (1.6 \times 10^{-19})^2} = 0.528 \times 10^{-10} \text{ m} = \mathbf{0.528\ Å}$$

(c) Time taken by the electron to traverse the first Bohr orbit is given by

$$T = \frac{2\pi r_1}{v} = \frac{2 \times 3.14 \times 0.528 \times 10^{-10}}{2.19 \times 10^6} = \mathbf{1.51 \times 10^{-16}\ sec}$$

(d) Rydberg constant for hydrogen atom is

$$R_H = \frac{me^4}{8\varepsilon_0^2 ch^3} = \frac{(9.1 \times 10^{-31}) \times (1.6 \times 10^{-19})^4}{8 \times (8.86 \times 10^{-12})^2 \times (3 \times 10^8) \times (6.6 \times 10^{-34})^3} = \mathbf{1.101 \times 10^7/m}$$

» ***Example* 1.31.** Calculate the frequency of the electron in the first Bohr orbit of hydrogen atom.

Solution: The velocity of the electron in the first Bohr orbit is $v = \dfrac{e^2}{2h\varepsilon_0}$.

The orbital frequency ν is given by

$$\nu = \frac{\text{electron velocity}}{\text{orbital circumference}} = \frac{v}{2\pi r} = \frac{e^2}{2h\varepsilon_0} \times \frac{1}{2\pi} \times \frac{\pi m e^2}{\varepsilon_0 h^2} = \frac{me^4}{4\varepsilon_0^2 h^3}$$

$$= \frac{(9.1 \times 10^{-31}) \times (1.6 \times 10^{-19})^4}{4 \times (8.854 \times 10^{-12})^2 \times (6.625 \times 10^{-34})^3} = \mathbf{6530 \times 10^{12}\ Hz}$$

» ***Example* 1.32.** **The wavelength of the first member of the Balmer series in hydrogen spectrum is 6563 Å; calculate the wavelength of the first member of Lyman series in the same spectrum.**

***Solution*:** The wave number for the first member of Balmer series is given by

$$\tilde{\nu}_1 = \frac{1}{\lambda_1} = R_H\left(\frac{1}{2^2} - \frac{1}{3^2}\right) = \frac{5}{36}R_H \qquad \text{(i)}$$

Again, the wave number for the first member of Lyman series is

$$\tilde{\nu}_2 = \frac{1}{\lambda_2} = R_H\left(\frac{1}{1^2} - \frac{1}{2^2}\right) = \frac{3R_H}{4} \qquad \text{(ii)}$$

From (i) and (ii)

$$\frac{\lambda_2}{\lambda_1} = \frac{5}{27} \text{ or, } \lambda_2 = \frac{5}{27}\lambda_1 \text{ or, } \lambda_2 = \frac{5 \times 6563}{27} = \mathbf{1215.37\ Å}$$

» ***Example* 1.33.** **Find the wavelength of the photon that is emitted when a hydrogen atom undergoes a transition from $n_i = 5$ to $n_f = 2$.**

***Solution*:** From Bohr's theory, the energy of the electron, while it is in the nth orbit is

$$E_n = -\frac{13.6\,\text{eV}}{n^2}$$

$$\therefore E_2 = -\frac{13.6\,\text{eV}}{2^2} = -3.40\,\text{eV}$$

$$E_5 = \frac{-13.6\,\text{eV}}{5^2} = -0.544\,\text{eV}$$

Again, from the Bohr's postulate, the energy of the emitted photon in the present case is

$$E_\gamma = -0.544\,\text{eV} - (-3.40\,\text{eV}) = 2.86\,\text{eV}$$

The wavelength of the same photon is

$$\lambda = \frac{hc}{E_\gamma} = \frac{12.4 \times 10^3\,\text{eV Å}}{2.86\,\text{eV}} = \mathbf{4340\ Å}$$

» ***Example* 1.34.** **Determine the wavelengths of hydrogen lines lying in the optical spectrum (3800 to 7700 Å).**

***Solution*:** The wavelength for hydrogen atom when an electron jumps from n_ith to n_fth orbit is given by

$$\frac{1}{\lambda} = \tilde{\nu} = R_H\left(\frac{1}{n_f^2} - \frac{1}{n_i^2}\right) = \left(1.097 \times 10^{-3}\,\text{Å}^{-1}\right) \times \left(\frac{1}{n_f^2} - \frac{1}{n_i^2}\right)$$

where $R_H = 1.097 \times 10^{-3}\ \text{Å}^{-1}$.

When $n_f = 1$, it can be shown that the wavelengths of Lyman series are

$$\frac{1}{\lambda} = \left(1.097 \times 10^{-3}\,\text{Å}^{-1}\right)\left(\frac{1}{1^2} - \frac{1}{n_i^2}\right), n_i = 2, 3, 4, \ldots$$

The longest wavelength corresponding to $n_i = 2$ is given by

$$\frac{1}{\lambda_{max}} = \left(1.097 \times 10^{-3}\ \text{Å}^{-1}\right)\left(1 - \frac{1}{2^2}\right)$$

$\therefore \lambda_{max} = 1215$ Å.

The shortest wavelength corresponds to $n_i = \infty$

$$\frac{1}{\lambda_{min}} = \left(1.097 \times 10^{-3}\ \text{Å}^{-1}\right)\left(\frac{1}{1^2} - \frac{1}{\infty^2}\right)$$

$\therefore \lambda_{min} = 912$ Å.

So none of these lines lie in the optical region.

For $n_f = 2$, the longest wavelength corresponds to $n_i = 3$.

So

$$\frac{1}{\lambda_{max}} = \left(1.097 \times 10^{-3}\ \text{Å}^{-1}\right)\left(\frac{1}{2^2} - \frac{1}{3^2}\right)$$

$\therefore \lambda_{max} = 6563$ Å.

The shortest wavelength for $n_f = 2$ is given by

$$\frac{1}{\lambda_{min}} = \left(1.097 \times 10^{-3}\ \text{Å}^{-1}\right)\left(\frac{1}{2^2} - \frac{1}{\infty^2}\right)$$

$\therefore \lambda_{min} = 3646$ Å.

Hence, some of the wavelengths in Balmer series ($n_f = 2$) lie in the optical region. To determine these wavelengths, we use $\lambda = 3800$ Å and find out n_i.

$$\therefore \frac{1}{3.8 \times 10^3\ \text{Å}} = \left(1.097 \times 10^{-3}\ \text{Å}^{-1}\right)\left(\frac{1}{4} - \frac{1}{n_i^2}\right) \quad \text{or, } n_i = 9.9$$

$\therefore$ the lines in the optical region are given by

$$\frac{1}{\lambda} = \left(1.097 \times 10^{-3}\ \text{Å}^{-1}\right)\left(\frac{1}{4} - \frac{1}{n_i^2}\right), n_i = 3, 4, 5, \ldots, 9$$

Now, the shortest wavelength of the Paschen series ($n_f = 3$) is given by

$$\frac{1}{\lambda} = \left(1.097 \times 10^{-3}\ \text{Å}^{-1}\right) \times \left(\frac{1}{3^2} - \frac{1}{\infty^2}\right)$$

$\therefore \lambda = 8200$ Å.

All other series will give rise to lines which are outside the optical region.

» ***Example* 1.35.** According to Bohr's theory, how many revolutions will an electron make in the first excited state of hydrogen if the lifetime of that state is 10^{-8} s?

Solution: The radius r_2 and the velocity v_2 of the electron for the state $n = 2$ are given by

$$r_2 = 4r_1 = 4(0.529\ \text{Å}) = 2.12\ \text{Å} = 2.12 \times 10^{-10}\ \text{m}[\because r_1 = 0.529\ \text{Å}]$$

$$\text{and } v_2 = \frac{v_1}{2} = \frac{c}{2(137)} = \frac{3 \times 10^8 \text{m/s}}{2(137)} = 1.10 \times 10^6 \text{m/s} \left[\because v_1 = \frac{c}{137}\right]$$

Then the angular velocity is

$$\omega = \frac{v_2}{r_2} = \frac{1.10 \times 10^6 \text{m/s}}{2.12 \times 10^{-10} \text{ m}} = 0.52 \times 10^{16} \text{ rad/s}$$

∴ the total no. of revolutions is

$$N = \frac{\omega t}{2\pi} = \frac{(0.52 \times 10^{16} \text{ rad/s})\,(10^{-8} \text{ s})}{6.28 \text{ rad/rev}} = \mathbf{8.3 \times 10^6 \text{ revolutions}}$$

» ***Example* 1.36. Show that in a Bohr atom, the *n*th orbit contains *n* complete de Broglie (electron) waves.** **[BU (H) 1994, 1997]**

***Solution*: According to Bohr's theory, the radius of the *n*th Bohr orbit is given by**

$$r_n = \frac{n^2 h^2 \varepsilon_0}{\pi m e^2}$$

The circumference of the *n*th Bohr orbit is

$$2\pi r_n = \frac{2n^2 h^2 \varepsilon_0}{m e^2}$$

But, the wavelength of the de Broglie wave for the electron is

$$\lambda = \frac{h}{mv} = \frac{h}{m}\left(\frac{2nh\varepsilon_0}{e^2}\right) \left[\because v_n = \frac{e^2}{2nh\varepsilon_0}\right]$$

The number of complete de Broglie waves in the *n*th Bohr orbit is

$$\frac{\dfrac{2n^2h^2\varepsilon_0}{me^2}}{\dfrac{h}{m}\left(\dfrac{2nh\varepsilon_0}{e^2}\right)} = n$$

» ***Example* 1.37. What is the velocity of the electron in the ground state of Bohr's hydrogen atom in terms of speed of light? What is this called?**

***Solution*: We have the velocity of the electron in the *n*th Bohr orbit is**

$$v_n = \frac{Ze^2}{2\varepsilon_0 nh} = \frac{e^2}{2\varepsilon_0 nh} \quad [\because Z = 1]$$

∴ velocity of the electron in the 1st Bohr orbit is $v_1 = \dfrac{e^2}{2\varepsilon_0 h}[\because n = 1]$.

If v_1 is expressed in terms of speed of light, then

$$v_1 = \frac{e^2}{2\varepsilon_0 ch} = \frac{(1.60 \times 10^{-19})^2}{(2 \times 8.85 \times 10^{-12}) \times (3 \times 10^8)\,(6.62 \times 10^{-34})} \simeq \frac{1}{137} = \alpha \text{ (say)}$$

This is called the ***fine structure constant.***

» ***Example* 1.38. Electrons of energies 10.20 eV and 12.09 eV can cause radiation to be emitted from hydrogen atoms. Calculate in each case, the principal quantum number of the orbit to which electron in the hydrogen atoms is raised and the wavelength of the radiation emitted if it drops back to the ground state.**

***Solution*: The energy of an electron in the *n*th Bohr orbit is**

$$E_n = -\frac{13.6}{n^2} \text{ eV} \quad [n \text{ is the principal quantum number}]$$

For $n = 1$, $E_1 = -13.6\,\text{eV}$

$n = 2$, $E_2 = -3.4\,\text{eV}$

$n = 3$, $E_3 = -1.51\,\text{eV}$.

Now, energy required by an electron to transit from $n = 1$ to $n = 2$ level is $(13.6 - 3.4)$ eV = 10.2 eV and the energy required for the transition from $n = 1$ to $n = 3$ level is $(13.6 - 1.51)$ eV = 12.09 eV.

This shows that the electron will be raised to $n = 2$ and $n = 3$ from $n = 1$ corresponding to energies 10.20 eV and 12.09 eV respectively.

Now, when the electron jumps from $n = 2$ to $n = 1$ level, we get the first member of Lyman series.

Hence,

$$\frac{1}{\lambda_1} = R_H\left[\frac{1}{1^2} - \frac{1}{2^2}\right] = \frac{3}{4}R_H \text{ or, } \lambda_1 = \frac{4}{3R_H} = \frac{4}{3 \times 10.97 \times 10^6} = 1216\,\text{Å}$$

When the electron jumps from $n = 3$ to $n = 1$ level

$$\frac{1}{\lambda_2} = R_H\left[\frac{1}{1^2} - \frac{1}{3^2}\right] = \frac{8R_H}{9}$$

$$\therefore \lambda_2 = \frac{9}{8R_H} = \frac{9}{8 \times 10.97 \times 10^6} = 1026\,\text{Å}$$

***Example* 1.39.** The amount of energy required to ionise a hydrogen atom in its ground state is 13.58 eV. How much energy is needed to excite it from the ground state to the $n = 4$ level? **[BU (H) 2003]**

Solution: Ionisation energy of the hydrogen atom is $(E_H)_1 = 13.58\,\text{eV}$.

Energy of the atom in the $n = 4$ level is

$$(E_H)_4 = \frac{13.58}{4^2} = \frac{13.58}{16} = 0.849\,\text{eV}$$

$\therefore$ energy needed to excite the atom from ground state to $n = 4$ state is $\Delta E = (E_H)_1 - (E_H)_4 = 13.58 - 0.849 =$ 12.731 eV.

» ***Example* 1.40.** Given the ground state energy of H-atom to be -13.58 eV, estimate the ground state energy of a doubly ionised Li atom (Li^{++}). (Neglect the effect of nuclear mass.)

Solution: Here, the ground state energy of the H-atom, $(E_H)_1 = -13.58\,\text{eV}$.

$\therefore$ the ground state energy of the doubly ionised Li (Li^{++}) is $(E_{4i}^{++}) = 2^2(E_H)_1$ [$\because$ for Li^{++}, $Z = 2$]

$= 4 \times 13.58 = $ **-54.32 eV.**

» ***Example* 1.41.** The ground state energy of an electron in the hydrogen atom is -13.6 eV. Compute the energy of (a) $n = 3$ state in He^+ (b) $n = 2$ state in Li^{++}.

Solution: The energy of an electron in the nth state of a hydrogen-like atom is

$$E_n = -\frac{RchZ^2}{n^2}$$

Here, $Rch = 13.6$.

$$\therefore E_n = -\frac{13.6Z^2}{n^2}$$

In case of hydrogen atom, $Z = 1$.

$$\therefore (E_n)_H = -\frac{13.6}{n^2}$$

For the ground state, $n = 1$.

$$\therefore (E_1)_H = -13.6\,\text{eV}$$

Similarly,

$$(E_2)_H = -\frac{13.6}{4} = -3.4\,\text{eV} \text{ and } (E_3)_H = -\frac{13.6}{9} = -1.5\,\text{eV}$$

(a) For He^+, $Z = 2$, so that for $n = 3$

$$(E_3)_{He} = 4 \times (E_3)_H = \mathbf{-6.0\,eV}$$

(b) For Li^{++}, $Z = 3$, so that for $n = 2$

$$(E_2)_{Li} = 9 \times (E_2)_H = \mathbf{-30.6\,eV}$$

***Example* 1.42.** Electrons of energy 12.2 eV are fired at hydrogen atoms in a gas-discharge tube. Determine the wavelengths of the lines that can be emitted by the hydrogen.

Solution: The maximum amount of energy that a hydrogen atom can absorb is equal to the electron energy 12.2 eV. This absorption of energy can raise the electron from $n = 1$ state to nth state of energy E_n (say).

$\therefore E_n = E_1 + 12.2\,\text{eV} = -13.6\,\text{eV} + 12.2\,\text{eV} = -1.4\,\text{eV}$.

But,

$$E_n = -\frac{E_1}{n^2} \text{ or, } -1.4\,\text{eV} = -\frac{13.6\,\text{eV}}{n^2} \text{ or, } n = 3.14$$

Since, n is an integer, the highest state that the electron can reach corresponds to $n = 3$. So the possible transitions that the electron can make are as follows:

$3 \to 2$, $2 \to 1$ and $3 \to 1$. The corresponding wavelengths are given as under

For $3 \to 2$: $\dfrac{1}{\lambda} = \left(1.097 \times \dfrac{10^{-3}}{\text{Å}}\right)\left(\dfrac{1}{2^2} - \dfrac{1}{3^2}\right)$ or, $\lambda = \mathbf{6563\,Å}$.

For $2 \to 1$: $\dfrac{1}{\lambda} = \left(1.097 \times \dfrac{10^{-3}}{\text{Å}}\right)\left(\dfrac{1}{1^2} - \dfrac{1}{2^2}\right)$ or, $\lambda = \mathbf{1215\,Å}$.

For $3 \to 1$: $\dfrac{1}{\lambda} = \left(1.097 \times \dfrac{10^{-3}}{\text{Å}}\right)\left(\dfrac{1}{1^2} - \dfrac{1}{3^2}\right)$ or, $\lambda = \mathbf{1026\,Å}$.

***Example* 1.43.** For hydrogen show that when $n \gg 1$ the frequency of the emitted photon in a transition from n to $(n-1)$ equals the orbital frequency of the electron. **[BU (H) 1994]**

Solution: The orbital frequency of the electron in the nth quantum state is

$$\nu_{el} = \frac{\omega}{2\pi} = \frac{v_n}{2\pi r_n} = \frac{2\pi\varepsilon_0 e^2/nh}{2\pi n^2h^2/4\pi^2\varepsilon_0 me^2} = \frac{4\pi^2\varepsilon_0^2 me^4}{n^3h^3} \quad \text{(i)}$$

The frequency of the emitted photon for the transition $n \to (n-1)$ is

$$\nu_Q = \frac{c}{\lambda} = cR_H \cdot \left[\frac{1}{(n-1)^2} - \frac{1}{n^2}\right] = cR_H \frac{2n-1}{n^2(n-1)^2}$$

For $n \gg 1$

$$\nu_Q \simeq cR_H \frac{2n}{n^4} = c\frac{2\pi^2\varepsilon_0^2 e^4 m}{h^3 c}\left(\frac{2}{n^3}\right) = \frac{4\pi^2\varepsilon_0^2 m e^4}{n^3 h^3} \quad \text{(ii)}$$

Equations (i) and (ii) show that $\nu_{el} = \nu_Q$ for $n \gg 1$. This forms the content of *Bohr's correspondence principle.*

» ***Example* 1.44.** A positronium atom is a system consisting of a positron and an electron. Calculate the reduced mass, the Rydberg constant and wavelength of the first Balmer line for positronium. [*Given*: $m = 9.1 \times 10^{-31}$ kg, $R_H = 1.09737 \times 10^{-3}$ Å and $H_\alpha = 6563$ Å.]

Solution: A positronium atom is a system consisting of a positron and an electron of the same mass, but opposite charge of equal magnitude.

The reduced mass of the positronium is

$$\mu = \frac{mm}{m+m} = \frac{m}{2} = \frac{9.1 \times 10^{-31}}{2} = 4.55 \times 10^{-31} \text{ kg}$$

Now, the Rydberg constant, $R_P = \mu e^4/8\varepsilon_0 c h^3$ is half of R_H.

$$\therefore R_P = \frac{R_H}{2} = 0.54868 \times 10^{-3}/\text{Å}$$

For hydrogen

$$\frac{1}{\lambda_{H_\alpha}} = R_H\left(\frac{1}{2^2} - \frac{1}{3^2}\right)$$

For positronium

$$\frac{1}{\lambda_{P_\alpha}} = R_P\left(\frac{1}{2^2} - \frac{1}{3^2}\right)$$

$$\therefore \frac{\lambda_{P_\alpha}}{\lambda_{H_\alpha}} = \frac{R_H}{R_P} = 2 \text{ or, } \lambda_{P_\alpha} = 2 \times \lambda_{H_\alpha} = 2 \times 6563 = \mathbf{13126\ Å}$$

» ***Example* 1.45.** A muon is an unstable elementary particle whose mass is $207m_e$ and whose charge is either $+e$ or $-e$. A negative muon (μ^-) can be captured by a nucleus to form a muonic atom.

(a) If a proton captures a μ^-. Find the radius of the first Bohr orbit. **[BU (H) 1993, 1997, 2001]**

(b) Find the ionisation energy of the atom.

Solution:

(a) Here, mass of the muon $m = 207m_e$ and mass of the nucleus $M = 1836m_e$, where m_e is the mass of the electron.

The reduced mass of the muonic atom is

$$\mu = \frac{mM}{m+M} = \frac{(207m_e) \times (1836m_e)}{207m_e + 1836m_e} = 186m_e$$

Now, according to Bohr's theory, the radius of the nth Bohr orbit is

$$r_n = \frac{n^2 h^2 \varepsilon_0}{\pi m_e e^2}$$

For $n = 1$

$$r_1 = \frac{h^2 \varepsilon_0}{\pi m_e e^2} = 5.29 \times 10^{-11} \text{ m or, } a_0 = r_1 = \mathbf{5.29 \times 10^{-11} \text{ m}}$$

Hence, r_1' corresponding to the reduced mass μ is

$$r_1' = \left(\frac{m_e}{m'}\right) r_1 = \left(\frac{m_e}{1836 m_e}\right) a_0 = \mathbf{2.85 \times 10^{-13} \text{ m}}$$

This shows that the muon is 186 times closer to the proton than an electron would be. So a muonic hydrogen atom is much smaller than an ordinary hydrogen.

(b) Now, the energy of the electron in the first Bohr orbit is $E_1 = -13.6\,\text{eV}$.

$$\therefore E_1' = \left(\frac{m}{m_e}\right) E_1 = 186 E_1 = -2.53 \times 10^3 \text{ eV} = \mathbf{-2.53\,keV}$$

Hence, the ionisation energy of the muonic atom is 2.53 keV, which is 186 times that of an ordinary hydrogen atom.

***Example* 1.46. Use vector atom model to determine the possible values of the total angular momentum of an f-electron.**

***Solution*: For an f-electron, the orbital quantum number, $l = 3$.**

Also its spin quantum number, $s = \pm\frac{1}{2}$.

$\therefore$ the total angular momentum quantum number of the f-electron is $j = l \pm s = 3 \pm \frac{1}{2} = \frac{7}{2}, \frac{5}{2}$.

The total angular momentum vector J is given by

$$J = \sqrt{j(j+1)}\,\hbar = \frac{3\sqrt{7}}{2}\hbar, \frac{\sqrt{35}}{2}\hbar$$

***Example* 1.47. Determine the energy difference between the electrons that are *aligned* and *anti-aligned* with a uniform magnetic field of 0.8 T when a beam of electrons moves perpendicular to the field.**

***Solution*: We know, the potential energy of an electron in a magnetic field $\vec{B}$ is**

$$E_B = -\vec{\mu}_s \cdot \vec{B} = -\mu_{sx} B_x - \mu_{sy} B_y - \mu_{sz} B_z$$

where the spin magnetic moment $\vec{\mu}_s = \hat{i}\mu_{sx} + \hat{j}\mu_{sy} + \hat{k}\mu_{sz}$ and $\vec{B} = \hat{i}B_x + \hat{j}B_y + \hat{k}B_z$.

If $\vec{B}$ is along z-axis, then $B_x = B_y = 0$.

$$\therefore E_B = -B\mu_{sz} = -B\left(-\frac{e\hbar}{m}\right) m_s$$

Hence, the energy difference is

$$\Delta E_B = B\frac{e\hbar}{m}\Delta m_s = (0.8\,\text{T})\left(2 \times 5.79 \times 10^{-5}\,\frac{\text{eV}}{\text{T}}\right) \times \left[\frac{1}{2} - \left(-\frac{1}{2}\right)\right] = \mathbf{9.26 \times 10^{-5}\ eV}$$

***Example* 1.48. Find out $\vec{L} \cdot \vec{S}$ in terms of J, L and S.**

***Solution*: We have $\vec{J} \cdot \vec{J} = |\vec{J}|^2$, where $\vec{J} = \vec{L} + \vec{S}$.**

$\therefore |\vec{J}|^2 = (\vec{L}+\vec{S})\cdot(\vec{L}+\vec{S}) = \vec{L}\cdot\vec{L} + 2\vec{L}\cdot\vec{S} + \vec{S}\cdot\vec{S} = |\vec{L}|^2 + 2\vec{L}\cdot\vec{S} + |\vec{S}|^2 = \frac{1}{2}[|\vec{J}|^2 - |\vec{L}|^2 - |\vec{S}|^2]$ or, $\vec{L}\cdot\vec{S} = \frac{1}{2}[|\vec{J}|^2 - |\vec{L}| - |\vec{S}|^2]$.

Again, $|\vec{J}|^2 = J(J+1)\hbar^2$, $|\vec{L}|^2 = L(L+1)\hbar^2$, $|\vec{S}|^2 = S(S+1)\hbar^2$.

$\therefore \boldsymbol{\vec{L} \cdot \vec{S} = \frac{1}{2}[J(J+1) - L(L+1) - S(S+1)]\hbar^2}$.

» ***Example* 1.49.** **Find the precessional frequency of an electron orbit when placed in a magnetic field of 6 Wb/m^2.**

***Solution*:** We have the frequency ν_L (Larmor precession)

$$\nu_L = \frac{eB}{4\pi m}$$

Here, $B = 6\ \text{Wb/m}^2$, $e = 1.6 \times 10^{-19}$ C, $m = 9.1 \times 10^{-31}$ kg.

$$\therefore \nu_L = \frac{1.6 \times 10^{-19} \times 6}{4 \times 3.14 \times 9.1 \times 10^{-31}} = \mathbf{8.399 \times 10^{10}\ Hz}$$

» ***Example* 1.50.** **A sample of an element is placed in a 0.300 T magnetic field and suitably excited. How far apart are the Zeeman components of 450 nm spectral lines of this element?**

***Solution*:** We know, the separation of the Zeeman components

$$\Delta\nu = \frac{eB}{4\pi m}$$

But, $\nu = c/\lambda$.

$$\therefore d\nu = -\frac{c\,d\lambda}{\lambda^2}$$

$$\therefore |\Delta\lambda| = \frac{\lambda^2 \Delta\nu}{c} = \frac{eB\lambda^2}{4\pi mc} = \frac{(1.60 \times 10^{-19}\ \text{C}) \times (0.300\ \text{T}) \times (4.5 \times 10^{-7}\ \text{m})^2}{(4\pi) \times (9.11 \times 10^{-31}\ \text{kg}) \times (3.00 \times 10^8\ \text{m/s})}$$

$$= 2.83 \times 10^{-12}\ \text{m} = \mathbf{0.00283\ nm}$$

» ***Example* 1.51.** **D_1 and D_2 lines of sodium are produced by the transitions $3\,^2P_{1/2} \to 3\,^2P_{1/2}$ and $3\,^2P_{3/2} \to 3\,^2S_{1/2}$. Calculate the Landé g-factors for the different levels and the consecutive sub-levels for each level due to a magnetic field of 10^4 gauss.**

***Solution*:** We have Landé splitting factor

$$g = 1 + \frac{J(J+1) + S(S+1) - L(L+1)}{2J(J+1)}$$

For $^2P_{3/2}$ state, $L = 1$, $S = \frac{1}{2}$, $J = \frac{3}{2}$
$^2P_{1/2}$ state, $L = 1$, $S = \frac{1}{2}$, $J = \frac{1}{2}$
$^2S_{1/2}$ state, $L = 0$, $S = \frac{1}{2}$, $J = \frac{1}{2}$

$$\therefore g\left(\text{for } ^2P_{3/2} \text{ state}\right) = 1 + \frac{\frac{3}{2}\cdot\frac{5}{2} + \frac{1}{2}\cdot\frac{3}{2} - 1\cdot 2}{2\cdot\frac{3}{2}\cdot\frac{5}{2}} = \frac{4}{3}$$

$$g\left(\text{for } ^2P_{1/2} \text{ state}\right) = 1 + \frac{\frac{1}{2}\cdot\frac{3}{2} + \frac{1}{2}\cdot\frac{3}{2} - 1\cdot 2}{2\cdot\frac{1}{2}\cdot\frac{3}{2}} = \frac{2}{3}$$

$$g\left(\text{for } ^2S_{1/2} \text{ state}\right) = 1 + \frac{\frac{1}{2}\cdot\frac{3}{2} + \frac{1}{2}\cdot\frac{3}{2} - 0 \times 1}{2\cdot\frac{1}{2}\cdot\frac{3}{2}} = 2$$

Now, splitting between consecutive sub-levels is $\Delta E_B = gh\nu_L \Delta M_J = gh\nu_L$ [$\because \Delta M_J = 1$]

$$\Delta E_B\left(\text{for } ^2P_{3/2} \text{ state}\right) = \tfrac{4}{3} \times 5.8 \times 10^{-5} = \mathbf{7.73 \times 10^{-5}\ eV}$$
$$\Delta E_B\left(\text{for } ^2P_{1/2} \text{ state}\right) = \tfrac{2}{3} \times 5.8 \times 10^{-5} = \mathbf{3.87 \times 10^{-5}\ eV}$$
$$\Delta E_B\left(\text{for } ^2S_{1/2} \text{ state}\right) = 2 \times 5.8 \times 10^{-5} = \mathbf{11.6 \times 10^{-5}\ eV}$$

» ***Example* 1.52.** Determine the maximum separation of a beam of hydrogen atoms that moves a distance of 20 cm with a speed of 2×10^5 m/s perpendicular to a magnetic field whose gradient is 2×10^2 T/m. [Neglect the magnetic moment of the proton.]

***Solution*:** Hydrogen atoms have zero orbital angular momentum in the ground state.

The force on a hydrogen atom is $F_z = \mu_{sz}\dfrac{dB}{dZ}$.

For the electron, $g_s = 2$, $\mu_{sz} = -\left(\dfrac{e}{m}\right)m_s\hbar$.

$$\therefore |\bar{F}_z| = \frac{e\hbar}{m}|m_s|\frac{dB}{dz} = \frac{e\hbar}{2m}\frac{dB}{dz} = \left(9.27 \times 10^{-24}\,\frac{\text{J}}{\text{T}}\right) \times \left(2 \times 19^2\,\frac{\text{T}}{\text{m}}\right) = 1.85 \times 10^{-21}\ \text{N}$$

Now, for constant acceleration, the separation of hydrogen beam is

$$\Delta z = \frac{1}{2}a_z t^2 = \frac{1}{2}\left(\frac{F_z}{m_{\text{H}}}\right)\left(\frac{\Delta y}{v}\right)^2$$

where a_z is the acceleration along z-direction, Δy, length of the magnets and v the velocity of hydrogen beam.

$$\therefore \Delta z = \frac{1}{2}\left(\frac{1.85 \times 10^{-21}\ \text{N}}{1.67 \times 10^{-27}\ \text{kg}}\right) \times \left(\frac{0.20\ \text{m}}{2 \times 10^5\ \text{m/s}}\right) = \mathbf{5.54 \times 10^{-7}\ m}$$

$\therefore$ total separation is $2\Delta z = 2 \times 5.54 \times 10^{-7}$ m $= \mathbf{1.11 \times 10^{-6}\ m.}$

» ***Example* 1.53.** Find the energies of the spectral lines emitted in the transition $2P_{1/2} \to 1S_{1/2}$ when a hydrogen atom is placed in a week magnetic field.

***Solution*:** For $1S_{1/2}$, $n = 1$, $l = 0$, $s = \frac{1}{2}$, $j = \frac{1}{2}$ and

$$g_j = \frac{1}{2j(j+1)}[3j(j+1) + s(s+1) - l(l+1)] = \frac{1}{2 \times \frac{1}{2} \times (\frac{1}{2}+1)}[3 \times \tfrac{1}{2}(\tfrac{1}{2}+1) + \tfrac{1}{2}(\tfrac{1}{2}+1) - 0(0+1)]$$
$$= \frac{1}{3/2}[3 \times \tfrac{1}{2} \times \tfrac{3}{2} + \tfrac{1}{2} \times \tfrac{3}{2}] = \tfrac{2}{3}[\tfrac{9}{4} + \tfrac{3}{4}] = \tfrac{2}{3} \times \tfrac{12}{4} = 2$$

$\therefore g_j = 2.$

Thus, the energy level $\left(1S_{1/2}\right)$ spits into two levels. The energy of each level changes due to the presence of the magnetic field. This change in energy is

$$\Delta_s = E_\mu = \frac{e\hbar}{2m}m_j B.$$

Thus, the energy changes in the above two levels are

$$\Delta_s = \pm\frac{e\hbar}{2m}B\begin{bmatrix} + \text{ sign for } m_j = \frac{1}{2} \\ - \text{ sign for } m_j = -\frac{1}{2} \end{bmatrix}$$
$$= \pm\mu_B B$$

For the $2P_{1/2}$ level, $l = 1, s = \frac{1}{2}, j = \frac{1}{2}$ and

$$g_j = \frac{3\cdot\frac{3}{4}+\frac{3}{4}-2}{\frac{3}{2}} = \frac{2}{3}$$

The corresponding energy shift is

$$\Delta_p = E_B = \frac{e\hbar}{2m}\frac{m_j}{2}B.$$

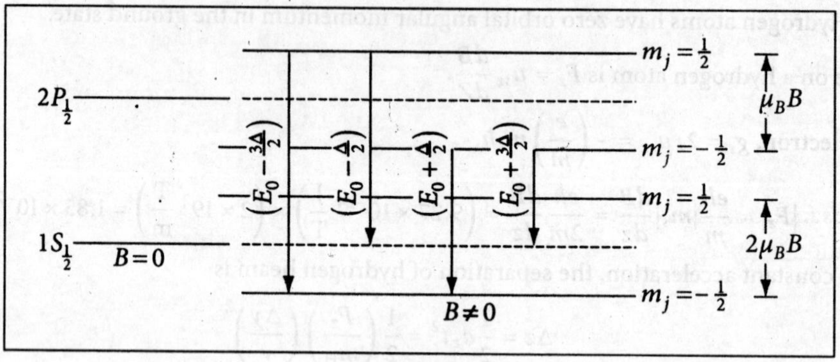

Fig 1.52

The energy shifts for the two levels are

$$\Delta_p = \pm\frac{e\hbar}{2m}\frac{B}{2}\begin{bmatrix} + \text{ sign for } m_j = \frac{1}{2} \\ - \text{ sign for } m_j = -\frac{1}{2} \end{bmatrix}$$

$$= \pm\frac{\mu_B B}{2}$$

In the absence of B the transition energy is $E_0 (= 10.2\,\text{eV})$, then the transition energies in the presence of B are $\left[E_0 - \frac{3\Delta}{2}\right], \left[E_0 - \left(-\frac{\Delta}{2}\right)\right], \left[E_0 + \frac{\Delta}{2}\right]$ and $\left[E_0 + \frac{3\Delta}{2}\right]$, where $\Delta = \mu_B B$ [Fig 1.52].

» ***Example* 1.54.** A positronium is a bound system composed of positron and electron. Determine the (a) Rydberg constant (b) ionisation potential of positronium.

***Solution*:**

(a) The mass of a positron is the same as the mass of an electron (m). So the Rydberg constant is

$$R_P = \frac{R_H}{1+\frac{m}{M_H}} = \frac{R_H}{1+\frac{m}{m}} = \frac{R_H}{2} = \mathbf{0.5485 \times 10^{-3}/\text{Å}}\ [\because R_H = 1.0960 \times 10^{-3}/\text{Å}]$$

(b) $\frac{1}{\lambda} = R_P\left(\frac{1}{n_i^2} - \frac{1}{n_f^2}\right)$ or, $\frac{hc}{\lambda} = h\nu = hcR_P\left(\frac{1}{n_i^2} - \frac{1}{n_f^2}\right)$.

The ionisation is required to excite the positronium from $n_i = 1$ to $n_f = \infty$.

$\therefore E_{ion} = hcR_P = (12.40 \times 10^3\ \text{eV Å})(0.5485 \times 10^{-3}/\text{Å}) = \mathbf{6.8\ eV}.$

EXERCISES

1(A): Objective (or MCQ) Type Questions

Select the correct answer:

1.1. Quantum nature of light emerged in an attempt to explain

(a) radioactivity
(b) interference of light
(c) blackbody radiation
(d) pair production.

1.2. Wien's displacement law shows that the intensity maximum shifts with increase of temperature towards

(a) higher wavelength
(b) lower wavelength
(c) does not shift
(d) only intensity increases.

1.3. Wien's radiation law could explain the distribution of thermal radiation in the spectrum of blackbody radiation in the region of

(a) low wavelength
(b) high wavelength
(c) middle wavelength
(d) none of them.

1.4. Rayleigh-Jeans law could explain the energy distribution in the spectrum of blackbody radiation in the region of

(a) low wavelength
(b) high wavelength
(c) entire wavelength
(d) X-rays.

1.5. The failure of classical wave theory to account the distribution of energy in the spectrum of blackbody radiation was due to the assumption that radiation energy is

(a) continuous
(b) discrete
(c) a mixture of continuous and discrete
(d) electromagnetic.

1.6. All the radiation laws can be shown to be special cases of

(a) Wien's law
(b) Rayleigh-Jeans law
(c) Planck's law
(d) Stefan-Boltzmann's law.

1.7. Photoelectrons are not emitted by photosensitive materials unless

(a) the frequency of incident light exceeds a certain maximum frequency
(b) intensity of incident light exceeds a certain minimum value
(c) the wavelength of incident light exceeds a certain minimum value
(d) the velocity of incident light exceeds a certain minimum value.

1.8. It is observed that

(a) the number of photoelectrons emitted is proportional to the intensity of light
(b) velocity of photoelectrons is directly proportional to the wavelength of light

(c) the velocity of photoelectrons is inversely proportional to the wavelength of light

(d) the KE of the photoelectrons is independent on the work function of the metal.

1.9. If the intensity of the incident light is doubled, then

(a) the frequency of the emitted photons is doubled

(b) the number of photons is doubled

(c) the number of photons remains unchanged

(d) the work function of the metal is reduced.

1.10. Ultraviolet light of wavelength 3500 Å falls on a potassium surface having a work function of 2 eV, the maximum energy of the photoelectrons will be of the order of

(a) 1 eV (c) 100 eV

(b) 10 eV (d) 1500 eV. **[IAS(Pre) 1993]**

1.11. Ultraviolet radiation of 6.2 eV falls on all aluminium surface (work function 4.2 eV). The kinetic energy of the fastest electron emitted is approximately,

(a) 3×10^{-21} J (c) 3×10^{-17} J

(b) 3×10^{-10} J (d) 3×10^{-15} J. **[IAS(Pre) 2000]**

1.12. The work function of a metal is W_0 and the frequency of the incident light is ν. There is no emission of photoelectrons when

(a) $\nu < W_0/h$ (c) $\nu > W_0/h$

(b) $\nu = W_0/h$ (d) $\nu \geqslant W_0/h$.

1.13. In Compton scattering, the incident photon loses maximum energy to the electron when the photon is scattered at

(a) 0° (c) 90°

(b) 45° (d) 180°. **[IAS(Pre) 1993, 1996, 1999, cf. 2006]**

1.14. The photoelectric emission depends on

(a) the work function of the emitter (c) frequency of the incident light

(b) intensity of the incident light (d) velocity of the incident light.

1.15. The Compton shift depends on

(a) angle of scattering (c) wavelength of the incident X-rays

(b) material of the target (d) none of them.

1.16. The magnitude of Compton shift is the maximum when the angle of scattering ϕ is

(a) 0° (c) π

(b) $\frac{\pi}{2}$ (d) 45°

1.17. In Compton effect, the whole of the photon energy cannot be handed over to the free electron at rest because, then

(a) conservation law of angular momentum breaks down

(b) conservation law of linear momentum breaks down

(c) conservation law of energy breaks down

(d) conservation law of mass-energy breaks down.

1.18. Compton shift for household object is not observed because their

(a) velocity is zero

(b) mass is very large

(c) charge is zero

(d) rotation is around the axis of the earth.

1.19. Compton effect is observed if

(a) the target electron is free

(b) the target electron is bound to the nucleus

(c) binding energy is very small compared to the energy of the incident photon

(d) none of them.

1.20. The KE of the recoil electron in Compton effect is derived from the

(a) lattice energy of the material

(b) energy of the incident photon

(c) thermal energy of the free electrons

(d) rest mass of the electron.

1.21. The maximum change in wavelength of radiation in Compton scattering is approximately,

(a) 0.005 Å

(b) 0.05 Å

(c) 0.5 Å

(d) 1 Å. **[IAS 1993]**

1.22. In Compton scattering on incoming photon of wavelength $\lambda = \frac{h}{m_0 c}$ [h = Planck's constant, m_0 = rest mass of the electron, c = velocity of light] is catered by an electron at rest. If the photon is scattered backwards at 180°, the ratio of momentum of scattered photon to recoil electron is

(a) 6.1

(b) 1.6

(c) 5 : 3

(d) 5.8. **[cf. IAS(Pre) 2004]**

1.23. Bohr postulated in his atom model quantisation of

(a) energy

(b) angular momentum

(c) spin

(d) linear momentum.

1.24. Bohr atom model

(a) assumes that only certain values of angular momentum are possible for orbital electrons in hydrogen atom

(b) explains the spectra of all elements

(c) assumes that the electrons have wave property

(d) assumes that the nucleus has infinite mass.

1.25. In hydrogen atom, the angular momentum of an electron can be integral multiple of

(a) $\frac{h}{4\pi}$ (c) $\frac{2\pi}{h}$

(b) $\frac{h}{2\pi}$ (d) $\frac{h}{\pi}$.

1.26. The first excitation potential of a given atom is 10.2 eV . Then its ionisation potential will be

(a) 13.6 eV (c) 12.5 eV

(b) 20.4 eV (d) 30.6 eV .

1.27. In Bohr's model of hydrogen atom

(a) the radius of the mth orbit is proportional to m^2

(b) the magnitude of the potential energy is greater than its kinetic energy

(c) total energy of the electron in the nth orbit is proportional to n

(d) Lyman series lies in the infrared region.

1.28. If the radius of the first Bohr orbit is 5.29 Å, then the radius of the second Bohr orbit will be

(a) 1.87 Å (c) 10.58 Å

(b) 2.64 Å (d) 21.16 Å.

1.29. Though Bohr's theory could explain the origin of spectral lines, it cannot explain the fine structure of because

(a) the theory assumes the principle of classical electromagnetic theory

(b) the theory discarded that the electron orbits are circular in shape

(c) it did not account the relativistic mass variation of electron

(d) it concerns to the simplest atom of hydrogen.

1.30. The orbits of electrons in hydrogen atom are

(a) equispaced (c) closely spaced for smaller value of n

(b) closely spaced for higher value of n (d) closely spaced for medium value of n.

1.31. Franck-Hertz experiment established that

(a) the electron orbits are discrete

(b) the electron orbits are continuous

(c) the electron revolving in a certain orbit possesses a definite amount of energy

(d) the electrons in the outermost shell have greater energy.

1.32. Introduction of elliptical electron orbits by Wilson-Sommerfeld leads to

(a) the degeneracy of electron orbits

(b) satisfactory explanation of fine structure of spectral lines

(c) different orbits with same value of principal quantum number n having the same energy

(d) assumption that an electron executes motion through the nucleus in a straight line.

1.33. Wilson-Sommerfeld theory explains the origin of fine structure of spectral lines by assuming

(a) the relativistic mass variation of the electron
(b) the electron orbits to be elliptical
(c) the angular momentum to be quantised
(d) that the nucleus is at rest.

1.34. The vector atom model is based on

(a) the idea of space quantisation and the hypothesis of spinning electron
(b) the principle of vector algebra
(c) discrete nature of energy and angular momentum
(d) the principle that magnetic moment is associated with both orbital angular momentum and spin angular momentum.

1.35. The spin motion of electron was first proposed by

(a) Schrödinger
(b) Pauli
(c) Bohr
(d) Uhlenbeck and Goudsmit.

1.36. For a particular value of angular momentum quantum number l the possible orientations of the electron orbit in space are

(a) l
(b) $(2l+1)$
(c) $(l+1)$
(d) $(2l-1)$.

1.37. Anomalous Zeeman effect is observed if the source is placed in

(a) strong magnetic field
(b) weak magnetic field
(c) no magnetic field
(d) nonuniform magnetic field.

1.38. The Zeeman splitting of spectral lines is proportional to

(a) magnetic field B
(b) Landé splitting factor g
(c) the principal quantum number n
(d) the specific change e/m of the electron.

1.39. In Stern-Gerlach experiment, the atomic beam is passed through a

(a) uniform magnetic field
(b) nonuniform magnetic field
(c) uniform electric field
(d) vacuum.

1.40. Stern-Gerlach experiment proved

(a) the idea of space quantisation and spinning electron
(b) the idea of spinning electron
(c) the idea of space quantisation
(d) the idea of spin magnetic moment.

1.41. Quantum mechanically, the energy states of a simple harmonic oscillator are

(a) continuous
(b) discrete
(c) partially continuous
(d) none of them.

1.42. Rigid rotor model is applied to explain the origin of molecular spectra of
(a) polyatomic molecules
(b) diatomic molecules
(c) monatomic molecule
(d) a mixture of diatomic and triatomic molecules.

1.43. The distribution of radiant energy against wavelength is fully explained for the whole spectral range by which one of the following laws?
(a) Wien's law
(b) Rayleigh-Jeans law
(c) Planck's law
(d) Stefan-Boltzmann law. **[IFS 2008]**

1.44. The distribution function for a photon is given by
(a) $e^{-\frac{\varepsilon}{kT}}$
(b) $\frac{1}{e^{\frac{\varepsilon}{kT}}+1}$
(c) $\frac{1}{e^{\frac{\varepsilon}{kT}}-1}$
(d) $\frac{1}{e^{-\frac{\varepsilon}{kT}}+1}$.

1.45. The average energy of Planck oscillator is given by
(a) kT
(b) $\frac{3kT}{2}$
(c) $\frac{h\nu}{e^{h\nu/kT}-1}$
(d) $\frac{kT}{2}$. **[IAS(Pre) 2004]**

1.46. Formula for total energy density at temperature T kelvin using Planck's law of radiation with usual symbol is
(a) $\frac{8\pi^5k^4T^4}{15C^3h^3}$
(b) $\frac{8\pi^5k^2T^2}{15C^3h^3}$
(c) $E = \frac{8\pi^5k^4T^4}{15Ch^3}$
(d) $E = \frac{8\pi^5k^4T^4}{15C^2h^2}$. **[IAS(Pre) 2002]**

1.47. Light coming from the discharge tube filled with hydrogen falls on the photoelectric cell. The work function of the surface of the cathode is 4 eV. Which one of the following values of the anode voltage (in V) w.r.t. the cathode will make the photocurrent zero?
(a) −4
(b) −6
(c) −8
(d) 10. **[IAS(Pre) 1995]**

1.48. Ultraviolet light of wavelength 3500 Å falls on potassium surface having a work function of 2 eV. The maximum energy of the photoelectrons will be of the order of
(a) 1 eV
(b) 10 eV
(c) 100 eV
(d) 1500 eV. **[IAS(Pre) 1993]**

1.49. In a photoelectric effect measurement the slopping potential for a given metal is found to be V_0 volts when radiation of wavelength λ_0 is used. If the radiation of wavelength $2\lambda_0$ is used with some metal then the stopping potential will be observed to be (symbols have their usual meaning):
(a) $\frac{V_0}{2}$
(b) $2V_0$
(c) $V_0 + \frac{hc}{2e\lambda_0}$
(d) $V_0 - \frac{hc}{2e\lambda_0}$.

1.50. The wavelength of the photoelectric threshold for silver is λ_0. The energy of the electron ejected from the surface of silver by an incident light of wavelength $\lambda(\lambda < \lambda_0)$ will be

(a) $hc(\lambda_0 - \lambda)$

(b) $\frac{hc}{\lambda_0 - \lambda}$

(c) $\frac{h}{c}\left(\frac{1}{\lambda} - \frac{1}{\lambda_0}\right)$

(d) $hc\frac{(\lambda_0 - \lambda)}{\lambda_0 \lambda}$. **[IAS(Pre) 1997]**

1.51. Consider the following statement statements: The maximum kinetic energy of photoelectron depends on: (i) Frequency of incident radiation (ii) Nature of photometer (iii) Intensity of incident radiation (iv) Which of these statements are correct?

(a) (i) and (ii)

(b) (ii) and (iii)

(c) (i) and (iii)

(d) (i), (ii) and (iii). **[IAS(Pre) 2003]**

1.52. In photoelectric effect with incoming radiation of frequency ν with $h\nu$ = 8 eV, electrons of energy 3 eV are emitted from a metal surface. The energy of the electrons emitted from this surface when radiation of frequency 1.2ν is incident, is

(a) 4.2 eV

(b) 5.2 eV

(c) 3.6 eV

(d) 4.6 eV. **[IAS(Pre) 2004, cf. 2007]**

1.53. What is the kinetic energy T of a photoelectron from the K shell of an atom, if E_k is the K-ionisation energy?

(a) $T = h\nu + E_k$

(b) $T = h\nu - E_k$

(c) $T = \sqrt{h^2\nu^2 - E_k^2}$

(d) $T = \sqrt{h^2\nu^2 + E_k^2}$.

(ν is the frequency of the photon). **[IAS(Pre) 2006]**

1.54. Upon which of the following parameters does the maximum energy of the photo electrons depend?

(i) The frequency of the radiation

(ii) The intensity of the radiation

(iii) The work function of the photo materials.

Select the answer using the codes given below:

(a) (i) and (ii) only

(b) (i) and (iii) only

(c) (ii) and (iii) only

(d) (i), (ii) and (iii). **[IAS(Pre) 2008]**

1.55. In Compton scattering the incident photon loses maximum energy to the electron when the photon is scattered at

(a) 0°

(b) 45°

(c) 90°

(d) 180°. **[IAS(Pre) 1993, 1996, 1999 cf. 2006]**

1.56. Which of the following statements connecting Compton effect is not correct?

(a) The wavelength of the scattered photon is greater than or equal to wavelength of the incident photon.

(b) The electron can acquire a kinetic energy equal to the energy of the incident photon.

(c) The energy of the incident photon is equal to the kinetic energy of the electron plus the energy of the scattered photon.

(d) The kinetic energy acquired by the electron is the largest when the incident and scattered photons move in opposite directions. **[GATE 2003]**

1.57. The quantity $\frac{h}{m_0 c}$, with m_0 as the electron has the dimension of

(a) length (c) mass

(b) time (d) temperature.

ANSWERS

1.1. (c).	**1.11.** (b).	**1.21.** (b).	**1.31.** (a).	**1.41.** (b).	**1.51.** (a).
1.2. (b).	**1.12.** (a).	**1.22.** (b).	**1.32.** (a).	**1.42.** (b).	**1.52.** (b).
1.3. (a).	**1.13.** (b).	**1.23.** (b).	**1.33.** (a).	**1.43.** (c).	**1.53.** (d).
1.4. (b).	**1.14.** (c).	**1.24.** (d).	**1.34.** (a).	**1.44.** (c).	**1.54.** (d).
1.5. (a).	**1.15.** (a).	**1.25.** (b).	**1.35.** (d).	**1.45.** (c).	**1.55.** (b).
1.6. (c).	**1.16.** (c).	**1.26.** (a).	**1.36.** (b).	**1.46.** (a).	**1.56.** (b).
1.7. (a).	**1.17.** (a).	**1.27.** (a).	**1.37.** (b).	**1.47.** (a).	**1.57.** (b).
1.8. (a).	**1.18.** (b).	**1.28.** (d).	**1.38.** (a).	**1.48.** (a).	
1.9. (b).	**1.19.** (a).	**1.29.** (c).	**1.39.** (b).	**1.49.** (d).	
1.10. (c).	**1.20.** (a).	**1.30.** (b).	**1.40.** (a).	**1.50.** (d).	

1(B): Short Answer Type Questions

1.1. Why classical electromagnetic theory fails to account the distribution of energy in the spectrum of blackbody radiation?

1.2. Show that Wien's law and Rayleigh-Jeans law of blackbody radiation are only the special cases of Planck's law.

1.3. What is ultraviolet catastrophe?

1.4. What are the basic postulates of Planck's law of radiation?

1.5. What are the laws of photoelectric effect?

1.6. What do you mean by (a) work function (b) threshold frequency and stopping potential?

1.7. Light of a frequency greater than the threshold frequency is incident on a photosensitive surface, but no photoelectrons are observed. Explain the possible reasons.

1.8. What are the three processes of interaction of radiation with matter? Indicate the relative energy of the incident photon in the above processes.

1.9. In the measurement of energy of photoelectrons what results are expected if

(a) keeping the frequency and intensity of light constant, the target material is changed;

(b) keeping target material and intensity constant, the frequency is changed;

(c) keeping frequency and target material constant, the intensity of incident light is changed.

1.10. What are the basic postulates of Einstein's photoelectric effect?

1.11. Obtain Einstein's photoelectric equation.

1.12. Can a photon hand over its cent percent energy to the photo electron? What is quantum yield?

1.13. What is Compton effect? How does it support the quantum nature of radiation?

1.14. Discuss the dependence of Compton wavelength on the frequency of the incident photon, nature of the scattering material and the angle of scattering.

1.15. Why an electron in the innermost shell of an atom is unable to show Compton effect?

1.16. Give a comparative study of photoelectric effect and Compton effect.

1.17. Indicate the postulates of Compton effect.

1.18. If the absorption of a photon by a free electron is ruled out, then how does photoelectric emission occur at all?

1.19. The existence of a cut-off frequency in the photoelectric effect is regarded as the most potent objection to a classical wave theory. Explain.

1.20. Choose appropriate clues to complete the following statements:

(a) The value of the maximum kinetic energy of photoelectrons depends on —— and ——, but not on ——.
(intensity of radiation, frequency of radiation, material of the photocell)

(b) When the intensity of radiation is doubled, the number of photoelectrons emitted per second by the cathode is ——, but the maximum kinetic energy of electrons is —— (unaltered/doubled).

1.21. In photoelectric effect, it is found that for a given monochromatic incident light, photoelectrons are emitted with a spread of velocities. Why?

1.22. Compton scattering formula suggests that objects viewed from different angles should reflect light of different wavelengths. Why do we not observe a change in the colour of objects as we change the viewing angle?

1.23. In Compton effect, show that the fractional change in energy of the photon is equal to $\frac{h\nu'}{m_0c^2}(1-\cos\phi)$, where ν' is the frequency of the scattered photon and ϕ is the angle of scattering.

1.24. How does an electron in hydrogen atom get from $4a_0$ to a_0 (2 → 1 transition) without being anywhere in between? [a_0 = the radius of the first Bohr orbit]

1.25. Why are the decreases in current in the Franck-Hertz experiment not sharp?

1.26. Show that for all Bohr orbits, the ratio of the magnetic dipole moment of the electronic orbit to its orbital angular momentum has the same value.

1.27. Justify the application of relativistic mechanics to the motion of an electron in an elliptic orbit.

1.28. Obtain classically the relation between magnetic moment $\vec{\mu}$ and the orbital angular momentum $\vec{L}$ of an electron.

1.29. Is it possible to do a Stern-Gerlach type experiment on free electrons to measure the spin magnetic moment of an electron?

1.30. For $l = 3$, calculate the possible angles that L makes with the z-axis.

1.31. A spectral line is placed in a weak magnetic field. What kind of Zeeman effect do you expect in the following cases: (a) $^1P \rightarrow {}^1S$ (b) $^2D_{5/2} \rightarrow {}^2P_{3/2}$ (c) $^3D_1 \rightarrow {}^3P_0$?

1.32. Into how many components will a beam of atoms split in an experiment similar to that of Stern and Gerlach if the atoms are in the states (a) $^2P_{3/2}$ (b) $^3D_1$ (c) $^3F_4$ (d) $^1S_0$?

1.33. The ground state of chlorine is $^2P_{3/2}$. Find the magnetic moment in the units of Bohr magneton.

1.34. Estimate the Zeeman splitting for hydrogen spectral lines in its ground state in a magnetic field of 1 tesla.

1.35. Did Bohr postulate the quantisation of energy in his atom model? If not, what really was quantised?

1.36. Give a qualitative discussion of Bohr's correspondence principle.

1.37. Apply Wilson-Sommerfeld quantisation rule to obtain permitted energy levels of a particle in a one-dimensional potential well with walls of infinite height at $x = 0$ and $x = a$. **[BU (H) 1994]**

1.38. What are the angles between $\vec{L}$ and z-axis for $l = 2$? **[BU (H) 2000]**

1.39. Show that Planck's constant has the dimension of angular momentum.

1.40. A muonic atom contains a nucleus of charge Ze and a negative muon μ^- moving about it. The μ^- has a charge $-e$ and mass $207m_e$, calculate the radius of the first Bohr orbit.

1.41. Which one has a lower ground state energy, normal H atom or He^+?

1.42. Into how many components a beam of normal silver atoms split when passed through an inhomogeneous magnetic field?

1.43. What is the number of levels into which the $^3P_{1/2}$ level of sodium atom will split when the atom is subjected to a weak magnetic field?

1.44. State Bohr's correspondence principle. **[BU (H) 1999]**

1.45. Write note on Planck's radiation law. **[WBUT 2002]**

1.46. Show that Planck's constant has the dimension of angular momentum. **[BU (H) 2000]**

[*Hints*: $\because E = h\nu$, $\therefore [h] = \frac{[E]}{[\nu]}$

But, $[E] = [F][L] = [MLT^{-2}][L] = [ML^2T^{-2}]$ and $[\nu] = [T^{-1}]$

$$\therefore [h] = \frac{[ML^2T^{-2}]}{[T^{-1}]} = [ML^2T^{-1}] = [MLT^{-1}][L] = [p][L] = [\Omega]$$

where $\Omega = pl$ is the angular momentum.]

1.47. The maximum KE of photoelectrons emitted by photons of wavelength 1800 Å from a certain metal is found to be 1.5 eV. What is the work function of the metal? **[BU (H) 2004]**

[*Hints*: Use the relation: $W_0 = h\nu - \frac{1}{2}mv_{\max}^2 = \frac{hc}{\lambda} - \frac{1}{2}mv_{\max}^2$.]

1.48. Show that the circumference of an allowed Bohr orbit contains an integral number of de Broglie wavelength. **[BU (H) 1994, 1997]**

1.49. Is it possible for a free electron to transfer all its energy to a free electron? Give reasons.

1.50. How is a wave packet formed? Show that the velocity of particle is represented by group velocity and not phase velocity. **[KU (H) 2012]**

1.51. Name one experiment in each of the following:

(a) Particle like behaviour of electromagnetic radiation.

(b) Wave like behaviour of electron. **[BU (H) 2011]**

1.52. Draw and briefly describe an experimental device which can demonstrate the photoelectric effect. **[BU (H) 2010]**

1.53. Explain the existence of stationary orbits in Bohr's theory on the basis of matter waves. **[KU (H) 2012]**

1.54. What are the essential features of photoelectric effect? Discuss the failure of classical electromagnetic theory to explain the features of photoelectric effect. How the photon concept explain all the features of photoelectric effect. **[KU (H) 2010]**

1.55. What is Compton effect? Deduce the relation between the increase in wavelength and the angle of scattering. **[WBCS 2002, cf. 1998, 1999]**

1.56. 60 keV γ-rays are Compton scattered by electrons in a metal. Assuming the electrons to be initially at rest deduce the maximum energy of scattered electrons. **[IAS(Main) 85]**

[*Ans.* $T = 11.43$ keV; $\varepsilon = 521.43$ keV]

1.57. With incident X-rays of wavelength 1.618 Å, Compton scattering is observed at 90° from the incident beam. Calculate (i) the wavelength of Compton radiation and (ii) the angle at which the recoil electron may be observed. **[IAS(Main) 1993]**

[*Ans.* 44.57°]

1.58. Derive an expression for the Compton sift $(\Delta\lambda)$ and show that it is independent of the wavelength of the incident radiation. Establish the parallelism between Compton and Random scattering. **[IAS(Main) 1998]**

1.59. What are Compton effect and Compton wavelength? Determine Compton shift. Show that the maximum Compton shift is twice the Compton wavelength. **[IAS(Main) 1999]**

1.60. X-ray photons of wavelength 2.0 pm incident on free electrons. They are scattered at an angle of 60° from the incident direction. Determine the following:

(a) Momentum of the incident photon in keV/c.

(b) Compton shift.

(c) Kinetic energy and recoil angle of the electron. **[IAS(Main) 2000]**

[*Ans.* $p = 621 = \frac{keV}{c}$; $\Delta\lambda = 1.2pm$; $T = 233$ keV; $\phi = 37.99°$]

1(C): Broad Answer Type Questions

1.1. Describe briefly an arrangement for studying the photoelectric phenomenon. Point out clearly the important features of the experimental results. How did Einstein explain them? Point out precisely the difference between the Planck's hypothesis and Einstein's theory. **[CU (H); BU (H) 2003]**

1.2. What are Wien's displacement law and Wien's radiation law? How does the latter differs from Rayleigh-Jeans law of blackbody radiation? Point out their inadequacy to explain the energy distribution in the spectrum of blackbody radiation.

1.3. What is photoelectric effect? How does one explain the fact that the maximum kinetic energy of the ejected electrons is independent of the intensity of the incident light? **[BU (H) 2001]**

1.4. An incident X-ray photon of frequency ν_0 is scattered by a free electron at rest through an angle ϕ. Using relativistic expression of electron energy, show that

(a) the change in the wavelength of the photon is given by

$$\Delta\lambda = \frac{h}{m_0 c}(1 - \cos\phi)$$

where m_0 = the rest mass of the electron.

(b) the KE of the recoil electron is

$$T = h\nu_0\left[\frac{(1-\cos\phi)\alpha}{1+\alpha(1-\cos\phi)}\right]$$

where $\alpha = \frac{h\nu_0}{m_0 c^2}$. [BU (H) 1998]

1.5. A photon of energy E_0 is incident on a stationary electron target and the angle of Compton scattering of the photon is ϕ. Show, using non-relativistic kinetics that the KE of recoil of the electron is

$$\frac{E_0^2(1-\cos\phi)}{mc^2 + E_0(1-\cos\phi)}$$

where m is the mass of the electron. [BU (H) 1995]

1.6. (a) What is Compton effect? Do you observe Compton effect with visible light? Give reasons for your answer.

(b) Show that the ratio of KE T of the recoil electron and the energy E of the incident photon in Compton effect is given by

$$\frac{T}{E} = \frac{\frac{2h\nu}{m_0 c^2}\sin^2\frac{\theta}{2}}{1 + \frac{2h\nu}{m_0 c^2}\sin^2\frac{\theta}{2}}$$

where ν is the frequency of the incident photon, θ is the angle through which the photon is scattered and m_0 is the rest mass of the electron. The other symbols have their usual significances.

[BU (H) 1993]

1.7. Draw neat diagrams showing the variation of:

(a) Photoelectric current vs retarding potential for three intensities $I_1 > I_2 > I_3$ of incident light having the same frequency.

(b) Photoelectric current vs retarding potential for three frequencies $\nu_1 > \nu_2 > \nu_3$ of incident light having the same intensity.

Elaborate how these observations can be explained by the quantum theory of light rather than by the classical electromagnetic theory. [BU (H) 1997]

1.8. Discuss how classical approaches failed to account for the spectral distribution of energy density in a blackbody radiation. How did Planck overcome this difficulty? Discuss Wien's displacement law from Planck's law.

1.9. Mention the salient points of the Bohr model of hydrogen atom and obtain an expression for the inverse of wavelength of the Balmer series. [BU (H) 1997]

1.10. Calculate the maximum shift in the wavelength of the scattered photon and the maximum kinetic energy of the recoil electron when an X-ray photon of energy E_0 is scattered by a free electron at rest. [BU (H) 2003]

1.11. (a) Give an account of Bohr's theory of the spectrum of a hydrogen atom. How does one incorporate the correction due to finite nuclear mass?

(b) Discuss Franck-Hertz experiment and explain how it confirmed the concept of discrete states of atoms.

(c) Show that the circumference of a circular Bohr orbit accommodates an integral number of de Broglie wavelength of the orbiting electrons. [BU (H) 2001]

1.12. If the electron in the hydrogen atom has mass m and the proton has a mass M, then from Bohr's theory obtain expressions for (a) the allowed energy levels of the atom (b) the allowed radii of the electron. [BU (H) 1999]

1.13. Using Bohr's postulates, deduce expressions for the energy levels and orbital radii of positronium which is a bound system comprising of an electron and a positron. What is the ionisation energy of positronium? [BU (H) 1998]

1.14. (a) Show that in the ground state of the hydrogen atom, the average magnetic moment due to the orbital motion of the electron is zero.

(b) Describe how the Stern-Gerlach experiment demonstrated the existence of the two-spin states of the electron.

1.15. (a) Explain the significance of Landé g-factor. Show that

$$g = 1 + \frac{J(J+1) + S(S+1) - L(L+1)}{2J(J+1)}$$

(b) The ground state of lithium atom is ${}^2S_{1/2}$. Find out the magnetic moment of this atom in the ground state in the unit of Bohr magneton. [BU (H) 1995, 1996, 1997]

(c) Draw a diagram showing the allowed transitions between the ${}^2P_{3/2}$ and ${}^2S_{1/2}$ terms in a weak magnetic field for the corresponding spectral lines. Calculate the shifts of the Zeeman components in $\mu_B \frac{B}{\hbar}$ units.

1.16. Discuss Bohr's theory of hydrogen atom and obtain formula for the Balmer series. Sketch an energy level diagram of the hydrogen atom. What is Bohr's correspondence principle. Describe Franck-Hertz experiment and explains its connection with Bohr's theory. [BU (H) 1996]

1.17. Give an account of Bohr's theory of the spectrum of hydrogen atom explaining the basic assumptions and calculate the quantised energy states of the same on this basis. What are the shortcomings? Briefly discuss the later improvements of the theory.

1.18. (a) Discuss Wilson-Sommerfeld quantisation rules and show that both Planck's and Bohr's quantum conditions are special cases of these rules.

(b) Show that in Sommerfeld's theory of elliptic orbits, the different values of n_ϕ correspond to ellipses of different eccentricities. What are the permissible values of n_ϕ? Why is $n_\phi = 0$ excluded? How is n_ϕ modified by the quantum mechanical theory?

1.19. (a) What are the rules of space quantisation? Discuss the origin of Zeeman effect. What is Larmor frequency?

(b) Explain the origin of anomalous Zeeman effect. Why is this effect more commonly observed than normal Zeeman effect? **[WBUT 2004]**

(c) Discuss the Zeeman splitting of sodium *D*-lines.

1.20. What is photoelectric effect? How does classical theory fail to explain this effect? **[WBUT 2004; NBU (H) 2003; BU (H) 2005]**

1.21. What is new in Einstein's concept? Write down Einstein's photoelectric equation and show how the photon concept explain all the observed facts.

1.22. What are the conservation laws satisfied in a Compton scattering experiment? Derive an expression for the Compton wavelength shift. **[NBU 2002; CU (H) 1996]**

1.23. A beam of monochromatic X-rays of wavelength λ is scattered by a light element. Obtain an expression for the wavelength of the scattered radiation. **[CU (H) 1994]**

1.24. What do you mean by a blackbody and blackbody radiation? **[CU (H) 2002]**

1.25. What was revolutionary in Planck's postulate regarding energy taken up by an oscillating dipole acting as a link between radiation and matter? Following Planck, calculate the average energy of such an oscillator. When does that value approach the classical limit? **[CU (H) 1992]**

1.26. (a) Derive an expression for Compton shift in wavelength for a photon scattered from a free electron at an angle θ.

(b) At which angle the shift will be maximum? **[WBUT 2005]**

1.27. Explain the basic features of the normal Zeeman effect.

Deduce the theory of splitting of spectral lines subjected to magnetic field. **[WBUT 2004]**

1.28. Obtain using vector atom model, the expression for Zeeman shift in energy in a weak magnetic field for one-electron atom. **[BU (H) 2006]**

1.29. (a) Derive an expression for the weak field Zeeman splitting of the energy level of a one-electron atom in the $S = 0$ state.

Show how the $^1P_1$ energy level is split due to a weak magnetic field.

(b) In the ground state of a sodium atom, the quantum numbers of the valence electron are $n = 3, l = 0, j = \frac{1}{2}$. What is the spectroscopic designation of the state. **[BU (H) 1999]**

1.30. (a) Elucidate the observations in the photoelectric effect which could not be explained by classical physics.

(b) Write and explain Einstein's photoelectric equation.

(c) Consider the emission of photoelectrons from a photo-cathode of work function 3×10^{-19} J by the action of light of frequency $7 \times 10^{14}/\text{s}$.

Write down the equation for stopping potential V_0 (negative and hence estimate Planck constant if $V_0 = 1$ volt. **[BU (H) 2013]**

1.31. (a) Consider the Compton scattering of an X-ray photon by an electron at rest. Derive an expression for the wavelength shift of the photon by this scattering.

(b) X-ray photon whose energy is much higher than the rest mass energy of an electron is back scattered by an electron at rest. Show that the energy of the scattered photon is half the rest mass energy of the electron. What is the electron's recoil KE? **[BU (H) 2014]**

1.32. (a) Obtain expressions for the quantised radii and energies of the electron in the Bohr model of the hydrogen atom taking into consideration a finite mass of the proton.

(b) A positronium is a bound state of an electron and a positron which is electron's antiparticle. Obtain the radius and energy of the ground state of positronium in terms of the Bohr radius (a_0) and the Rydberg constant (R) respectively. **[BU (H) 2014]**

1.33. (a) Calculate the energy values of different orbits of hydrogen atom according to Bohr theory.

(b) Find numerically the time taken by the electron to traverse the first Bohr orbit.

(c) Calculate the change in the expressions for energy levels in the hydrogen atom if the mass of the nucleus is taken into account.

(d) Show that the idea of Bohr's stationary orbits can be obtained from the de Broglie's hypothesis. **[BU (H) 2011]**

1.34. A photon collides with a free electron at rest. After collision the scattered photon makes an angle ϕ with the direction of the incident photon. Assuming the collision to be elastic in nature from relativistic considerations show that the energy of the recoiled electron is

$$I = h\nu_0 \frac{\alpha(1-\cos\phi)}{1+\alpha(1-\cos\phi)}$$

where ν_0 is the frequency of the incident photon and $\alpha = h\nu_0/m_0c^2$, m_0 is the rest mass of the electron.

1.35. (a) Draw and briefly describe an experimental device which can demonstrate the photoelectric effect.

(b) Draw graphs showing the experimental observations on *photoelectric current* against *retarding potential* for lights having

(i) the same frequency ν but three different intensities I_1, I_2, I_3 $(I_1 > I_2 > I_3)$

(ii) the same intensity I but three different frequencies ν_1, ν_2, ν_3 $(\nu_1 > \nu_2 > \nu_3)$.

Explain that the observations depicting in your graphs cannot be explained by classical wave theory of light but can be explained by the quantum theory of light.

(c) What is meant by the work function of a metal? **[BU (H) 2010, 2008]**

1.36. (a) Clearly state the postulates of Bohr's atomic model.

(b) Deduce the formula for the radii of the Bohr's orbit and the energies associated with those orbits.

(c) Show that the circumference of the Bohr orbits are integral multiples of the de Broglie wavelength of the electron. **[BU (H) 2009]**

1.37. Draw graphs of observed photoelectric current vs. retarding potential for the frequencies $\nu_1 > \nu_2 > \nu_3$ for the incident light and explain why observations cannot be understood by classical theory of light but can be interpreted in terms of Einstein's quantum theory of light. **[BU (H) 2008]**

1.38. (a) What are the difficulties in explaining the photoelectric effect with the wave theory of light?

(b) Use Einstein's explanation to write down the photoelectric equation. **[BU (H) 2007]**

1.39. (a) State the postulate of Bohr's theory of atomic spectra and show how well this explains the spectra.

(b) Indicate briefly how the theory was modified to accommodate nuclear mass. **[BU (H) 2007]**

1.40. An X-ray photon of wavelength λ is scattered by an electron initially at rest in laboratory frame of reference. Find the expression for the change in the wavelength of the electron of the photon as a result of scattering in terms of scattering angle. **[BU (H) 2009]**

1.41. (a) Arrive at the expressions for the angles of scattered photon and recoiled electron in the Compton scattering. Hence, show that while photon is scattered at angles between 0° to 180° the recoil electron is scattered at angles between 0° to 90°.

A 2 MeV gamma photon is scattered through an angle of 180° by a free electron at rest. What is the recoil KE of the electron?

(b) What will be observation if a photon is scattered by a molecule in stead of a free electron in Compton scattering experiment? **[BU (H) 2012]**

1.42. (a) Write down Einstein's equation in photoelectric effect explaining the terms involved.

(b) Explain the following in photoelectric effect:

(i) dependence of photoelectric current on the intensity of incident radiation

(ii) dependence of stopping potential on the wavelength of incident radiation. **[BU (H) 2012]**

1.43. (a) What is Compton effect? Obtain an expression for the change in wavelength of the incident high energy photon after collision with an atomic electron assuming it to be free and stationary.

(b) Why is the Compton line relatively more intense in case of light atoms?

(c) What is the essential difference between the coherent and incoherent scattering? **[BU (H) 2009]**

1.44. (a) What is stopping potential related to the maximum energy of the photon electron?

(b) Explain: (i) It is impossible for a photon to transfer its entire energy to the free electron. (ii) How do you account for the presence of unmodified component of X-rays in the study of the Compton effect? **[NBU (H) 2008]**

1.45. (a) State the postulates of Bohr's atomic model. Obtain radii, speed, kinetic, potential and total energy of various orbits. Hence, find the frequencies of transitions. **[cf. CU (H) 2013]**

(b) Explain the statement of Bohr's correspondence principle mentioning its significance. Give an example of the principle.

1.46. (a) Point out modifications of Bohr's theory due to finite mass of nucleus. Point out the inadequacies of Bohr's theory. How was it modified by Sommerfeld? **[cf. CU (H) 2014]**

(b) Draw qualitatively the possible orbits of electron in hydrogen atom for $n = 4$. Calculate the eccentricities of the orbit. **[cf. KU (H) 2011]**

1.47. (a) Explain the origin of D_1 and D_2 lines of Sodium.

(b) Discuss Zeeman splitting of sodium D-lines in a week magnetic field.

1.48. (a) State and explain the postulate of Bohr's theory of hydrogen atom. Obtain an expression for the radius of electron and the velocity in the nth orbit.

(b) An electron transition occurs from $n = 4$ to $n = 2$ energy level in hydrogen atom. Find the wavelength of emitted radiation in the energy of the electron in ground state (13.6 eV). To which series does this spectral line belong? **[Kakatiya U. 2016]**

1.49. (a) Explain Sommerfeld's elliptical orbit theory.

(b) Explain the variation of Rydberg constant with nuclear mass. **[Kakatiya U. 2016]**

1.50. (a) What is meant by Bohr magneton? Derive an expression for it.

(b) State Pauli's exclusion principle. Illustrate on the principle for distribution of electrons on *K* and *L* shells. **[Kakatiya U. 2016]**

1.51. (a) Explain Compton effect. Derive an expression for Compton-shift.

(b) Monochromatic X-rays of wavelength 0.15 Å undergoes Compton effect from a carbon block. Calculate the wavelength scattered through 135°.

1.52. (a) Discuss the theory of hydrogen atom taking into account the relativistic mass correction and spin orbit interaction.

(b) Explain the variation of Rydberg constant with the mass of a nucleus bringing the concept of reduced mass. **[Kakatiya U. 2016]**

1.53. (a) Explain Planck's quantum theory.

(b) List the experimental observations on photoelectric effect. Explain how Planck's quantum theory could give a complete explain of the same. **[Kakatiya U. 2016]**

1(D): Numerical Problems

1.1. The stopping potential for photoelectrons emitted from a sodium photo-cathode illuminated by light of wavelength $\lambda = 300$ nm is 1.85 V, while that for light of $\lambda = 400$ nm is 0.32 V. Determine the value of Planck's constant, work function of sodium and threshold wavelength for sodium. **[BU (H) 1993]**

1.2. The threshold wavelength for photoelectric emission in tungsten is 2300 Å. What wavelength of light must be used to eject electrons with a maximum energy of 1.5 eV? **[BU (H) 1997]**

[*Ans.* 2.3×10^{-7} m]

1.3. Find the energy of the photon which can impart a maximum energy of 50 keV to the electron. **[BU (H) 1998]**

1.4. What is the energy in ergs of a photon of wavelength 6000 Å? **[BU (H) 2003]**

[*Ans.* 3.309×10^{-12} erg]

1.5. X-rays of wavelength $\lambda = 0.124$ Å is scattered from a block of graphite. What is the wavelength of the X-rays scattered in the backward direction? **[BU (H) 2001]**

1.6. The photoelectric threshold wavelength for tungsten is 2300 Å. Calculate the kinetic energy in electron-volt of the ejected electrons from the surface by ultraviolet light of wavelength 1800 Å. **[BU (H) 2001]**

[*Ans.* 1.499 eV]

1.7. Find the stopping potentials for the photoelectrons emitted from the photo-cathode of rubidium having work function 2.1 eV when it is irradiated successively with light of wavelength 3650 Å, 4340 Å and 4860 Å. [*Ans.* 1.3 eV, 0.76 eV, 0.45 eV]

1.8. Calculate percentage change in wavelength in 0.15 Å photon which undergoes a 120° scattering from an electron. [*Ans.* 24.3 %]

1.9. Calculate the maximum energy in eV, that can be transferred to an electron in a Compton experiment when the incident quanta are X-rays of wavelength 0.50 Å. [*Ans.* 4.7 eV]

1.10. An electron that undergoes a *head-on* collision with an X-ray photon has a potential of 70 kV. If the electron was initially at rest what are the wavelengths of the initial and scattered X-ray photons?

[*Ans.* 0.0716 Å, 0.1201 Å]

1.11. The photoelectric work function of a metal is 6.27 eV. What is the threshold frequency of photo-emission? [Use the necessary constants from memory.]

1.12. Find the number of photons of yellow light of wavelength 6000 Å to produce 1 erg of energy. Assume the values of h and c. [*Ans.* 3×10^{11}]

1.13. NaCl has the same structure as KCl. The Debye temperatures of NaCl and KCl are 281 K and 230 K respectively. If the lattice heat capacity of NaCl at 5 K is 1.6×10^{-2} J/K/mole, estimate the heat capacities of KCl at 5 K and 3 K. [*Ans.* 2.9×10^{-2} J/K/mole, 6.26×10^{-3} J/K/mole]

1.14. Show that the zero-point energy of a solid according to Debye model is $\frac{9}{8} R\theta_D$.

1.15. In a Franck-Hertz experiment, hydrogen atoms are bombarded with electrons accelerated through a potential difference of 12.5 V. What are the wavelengths of the emission lines observed? [BU (H) 1998]

1.16. The Stern-Gerlach experiment is performed by using an atomic hydrogen beam. Assume that the length of the magnet is 0.5 m and the field gradient $\frac{\Delta B}{\Delta Z} = 0.8$ Wb/m^3.

Also let the velocity of the incident atoms be 500 m/s and let the width of the split beam just at the end of the magnet be 10^{-3} m. Calculate the magnetic moment of a hydrogen atom. [BU (H) 1998]

1.17. A muon is a particle of mass $206m_e$ and charge equal to that of an electron. If a negative muon is captured by a phosphorus nucleus ($Z = 5$), what is the energy of a photon emitted by a transition of the muon between $n = 3$ level and $n = 2$ level? [BU (H) 1995]

1.18. A muonic atom contains a nucleus of charge Ze and a negative muon (mass = 207 times the electron mass) moving about it. Derive expressions for the radius of the first Bohr orbit and binding energy of such an atom with $Z = 1$. [Take the proton mass to be 1836 times the electron mass.] [BU (H) 1993]

[*Ans.* 2.85×10^{-13} m, 2.53 keV]

1.19. In a Stern-Gerlach experiment, the magnetic field gradient is 5 V/m^2/mm with pole piece 0.07 m long. A narrow beam of silver atoms from an oven at 1250 K passes through the magnetic field. Calculate the separation of the beams as they emerge from the magnetic field. [*Ans.* 6.5×10^{-3} m]

1.20. A photon of frequency ν is back-scattered at 180° by an electron which was at rest.

(a) Write the equations for conservation of momentum and energy of this process.

(b) From the equation, calculate the momentum of the recoil electron in terms of ν and m.

[*Ans.* $P = \dfrac{4(h\nu)^2 + 4h\nu mc^2}{4h\nu c + 2mc^2}$]

1.21. What will be the separation between the adjacent normal Zeeman components for emitted radiation of 4500 Å in a magnetic field of 0.4 T? [*Ans.* 3.78×10^{-2} Å]

1.22. What magnetic flux density B is required to observe the normal Zeeman effect if a spectrometer can resolve spectral lines separated by 0.5 Å at 5000 Å? [*Ans.* 2.48 T]

1.23. In a normal Zeeman experiment, the calcium 4226 Å line splits into three lines separated by 0.25 Å in a magnetic field of 3 T. Determine e/m for the electron from these data.[*Ans.* $e/m = 1.76 \times 10^{11}$ C/kg]

1.24. Calculate the Compton wavelength (in Å) of an electron. **[WBUT 2005]**

1.25. In a Stern-Gerlach experiment, silver atoms traverse a distance of 0.1 m through an inhomogeneous magnetic field with gradient of 60 T/m. If the separation observed on the collector plate is 0.15 mm, determine the velocity of the silver atoms. [The mass of the silver atom is 1.79×10^{-25} kg.]

[*Ans.* 455 m/s]

1.26. Calculate the wavelength separation between two component lines which could be observed in the normal Zeeman effect with the following data:

Magnetic field strength, $B = 3000$ Oe, $e/m = 1.689 \times 10^{11}$ C/kg and $\lambda = 6000$ Å.

[*Ans.* 1.009×10^{-9} m]

1.27. A system consisting of an electron and positron revolving about their common centre of mass is known as a *positronium*. Calculate the distance between the two particles in the ground state. Also calculate the Rydberg constant and ionisation potential. [*Ans.* 1.6 Å; 0.54893×17/m; 6.8 eV]

1.28. Calculate the radius of the first Bohr orbit of the hydrogen atom. What will be the radius of the Bohr orbit if the electron is replaced by a negative muon of mass $212m_e$ having electronic charge?

[*Ans.* 0.529×10^{-8} cm, 2.5×10^{-11} cm]

1.29. Suppose, the wavelength of incident light in a photoelectric experiment changes from 300 nm to 400 nm. Find the corresponding change in the stopping potential. **[BU (H) 2005]**

Solution: The change in stopping potential is

$$(V_{S_1} - V_{S_2}) = \frac{h}{e}(\nu_1 - \nu_2) = \frac{hc}{e}\left(\frac{1}{\lambda_1} - \frac{1}{\lambda_2}\right)$$

$$\text{or, } \Delta V_S = \frac{6.602 \times 10^{-34} \times 3 \times 10^8}{1.602 \times 10^{-19}}\left(\frac{1}{300} - \frac{1}{400}\right) \times 10^9 = 100 \text{ V}$$

1.30. If the photoelectric threshold wavelength of sodium is 5420 Å, find the maximum speed of the photoelectrons ejected by incident of wavelength 4000 Å.

[*Hints*: Here, $\lambda_0 = 5420$ Å, $\therefore \nu_0 = \dfrac{c}{\lambda_0}$ and $W_0 = h\nu_0 = \dfrac{hc}{\lambda_0}$.]

Now, from $\dfrac{1}{2}mv_{max}^2 = h\nu + W_0 = \dfrac{hc}{\lambda} + \dfrac{hc}{\lambda_0}[\lambda = 4000 \text{ Å}], \therefore v_{max} = \left[\dfrac{2hc}{m}\left(\dfrac{1}{\lambda} + \dfrac{1}{\lambda_0}\right)\right]^{1/2}$.

1.31. Give the numerical values of the radius and the energy of the $n = 1$ orbit in the singly ionised helium atom. [*Given*: the corresponding values in hydrogen are 0.53×10^{-10} m and g −1.3 eV respectively.] **[BU (H) 2014]**

1.32. A metal is irradiated with lights of various frequencies and electrons are observed to be emitted. Show graphically the behaviour of the kinetic energy (k) of emitted electrons against the frequencies (ν) of the incident light. **[BU (H) 2014]**

1.33. Find the wavelength of the emitted photon when a hydrogen atom goes from the $n = 10$ state to the ground state. **[BU (H) 2013]**

1.34. Express energy of a 7000 Å photon in joule.

1.35. The maximum KE of photoelectrons is 1.3 eV when ultraviolet light of wavelength 350 nm is directed at a potassium surface. Find the work function of potassium. **[BU (H) 2007]**

1.36. A monochromatic X-ray beam whose wavelength is 1.5 pm is scattered through 45° of a target by Compton scattering. Find the wavelength of the scattered beam after deriving the necessary formula. **[BU (H) 2013]**

1.37. A photon of wavelength 10 Å collides with an electron at rest and is scattered at an angle of 90°. Calculate the change in wavelength of the incident photon and KE of the recoil electron.
[*Given*: $m = 9.1 \times 10^{-28}$ gm, $h = 6.66 \times 10^{-27}$ erg-sec, $c = 3 \times 10^{10}$ cm/s.]

1.38. A retarding potential of 6.2 eV can just stop the fastest photoelectrons emitted from caesium, whose work function is 1.8 eV. Find the wavelength of the most energetic incident photon. **[KU (H) 2015]**

1.39. The speed of an electron (9.1×10^{-31} kg) in the first Bohr orbit is 2×10^6 m/s. Find the de Broglie wavelength. **[KU (H) 2015]**

1.40. Find the surface area of a blackbody that radiates 1.0 W energy at 400 °C. **[KU (H) 2014]**

1.41. A beam of electrons bombards a sample of hydrogen. Through what potential difference must the electrons be accelerated so that the first line of Balmer series is emitted? **[BU (H) 2015]**

1.42. Using Sommerfeld's quantisation rule obtain permitted energy levels for a particle in a one-dimensional potential well with walls of infinite height at $x = 0$ and $x = a$. **[BU (H) 1999]**

1.43. Using Wilson-Sommerfeld's quantisation rule find the possible energies of a linear harmonic oscillator. **[KU (H) 1998; cf. BU (H) 1997, 1999]**

1.44. What is the value of the square of the angular momentum (L^2) for the electron in the state $4F$? In this case, what are the possible values of L_x? **[KU (H) 1998, 2000]**

[*Ans.* $L^2 \to 12\hbar$; $L_x \to -3h, -2h, -h, 0, h, 2h, 3h$]

1.45. Explain whether the level $^4D_{1/2}$ can exhibit Zeeman splitting. **[BU (H) 2007]**
[$g = 0$, no Zeeman splitting]

Chapter 2

Wave Nature of Matter

2.1 Introduction

We have seen that radiation exhibits the well-known phenomenon of *interference* and *diffraction* which are the result of interaction of radiation with radiation.

Again, in some phenomena like blackbody radiation, photoelectric effect and Compton effect radiation behaves as a set of discrete particles called *photons*. These phenomena are the result of interaction of radiation and matter. This is also evident from the principle of symmetry. According to this principle nature manifests in two forms, viz., matter and energy.

Since, radiant energy shows wave and particle nature, matter also possesses the same dual nature, viz., particle and wave under suitable conditions.

The wave-particle duality was first predicted by the French theoretical physicist *Pierre Louis de Broglie in* 1924. The theoretical concept was experimentally verified by *C.J. Davisson and L.H. Germer in* 1927 *and also by G.P. Thomson in the same year.* The wave properties of matter can be reconciled with particle properties by the superposition of waves of different wavelengths to form a group of waves or a *wave packet.* A wave packet can represent a particle in motion as a moving particle at any instant confined to a small region of space.

2.2 Origin of de Broglie's concept of matter wave

The origin of de Broglie's concept of dual nature of radiation, wave and particle, was the following:

1. ***Nature prefers symmetry***: *Matter* and *energy* are the two fundamental forms in which nature manifests herself. As radiant energy possesses wave and particle nature, matter also must possess the same dual nature, viz., particle and wave.

2. ***The close parallelism between mechanics and optics*:** According to the *Maupertian's principle of least action,* a moving particle always chooses that path for which the action is the minimum, i.e.,

$$\int_{P_1}^{P_2} p\,ds = \text{a minimum or, } \Delta \int_{P_1}^{P_2} p\,ds = 0. \tag{2.1}$$

According to *Fermat's principle of least time,* a light ray chooses always that path for which the time of transit is a minimum, i.e.,

$$\int_{P_1}^{P_2} \mu ds = \text{a minimum or, } \Delta \int_{P_1}^{P_2} \mu ds = 0. \tag{2.2}$$

Here, $p = mv$ is the momentum of the particle of mass m and velocity v.

Again, $\mu = A + \frac{B}{\lambda^2} + \frac{C}{\lambda^3} + \ldots$ is the refractive index of the medium for light of wavelength λ. The close parallelism of eq. (2.1) representing particle aspect and eq. (2.2) representing wave aspect shows that there is an intimate connection between *matter* and *wave.*

2.3 Phase and group velocities

Phase velocity

Let us consider a plane harmonic wave travelling along the positive direction of x-axis. If A be the amplitude and v the velocity, then

$$\psi(x,t) = A \sin \omega \left(t - \frac{x}{v}\right) \tag{2.3}$$

ω is the angular frequency given by $\omega = 2\pi\nu$ and ν being the cyclic frequency.

Now, the phase ϕ of the wave at any time t at a distance x from the origin is

$$\phi(x,t) = \omega\left(t - \frac{x}{v}\right)$$

Differentiating w.r.t. t, we get

$$\frac{\partial \phi}{\partial t} = \omega\left(1 - \frac{1}{v}\frac{dx}{dt}\right)$$

If the phase ϕ at a point is independent of time, then $\frac{\partial \phi}{\partial t} = 0.$

$$\therefore \omega\left[1 - \frac{1}{v}\left(\frac{dx}{dt}\right)_\phi\right] = 0 \text{ or, } 1 - \frac{1}{v}\left(\frac{dx}{dt}\right)_\phi = 0 \; [\because \omega \neq 0]$$

$$\text{or, } v = \left(\frac{dx}{dt}\right)_\phi$$

Here, $\left(\frac{dx}{dt}\right)_\phi$ represents the velocity with which the displacement of a given phase moves in the forward direction and is called the *phase velocity* denoted by v_p.

Hence, the phase velocity (v_p) of a wave is the velocity with which the displacement of a given phase moves forward. It is also the velocity of a monochromatic wave with which the *crest* or *trough* of the wave travels in a medium. In a *dispersive medium* v_p is different for different wavelengths.

Now, eq. (2.3) can be written as

$$\psi(x,t) = A\sin\left(\omega t - \frac{\omega x}{v_p}\right) = A\sin(\omega t - kx) \tag{2.4}$$

where $k = \frac{\omega}{v_p}$ is the wave propagation constant and can be expressed as

$$k = \frac{\omega}{v_p} = \frac{2\pi\nu}{v_p} = \frac{2\pi}{\lambda}\ [\because v_p = \nu\lambda] \tag{2.5}$$

The phase velocity (v_p) of the propagating monochromatic wave can be written as

$$v_p = \frac{\omega}{k} \tag{2.6}$$

The phase velocity of the wave may or may not vary, while travelling in a medium. If the medium be dispersive, then the phase velocity of the wave varies with wavelength or frequency. In such a medium, waves of different wavelengths propagate with different phase velocities. But, in a *non-dispersive medium*, the phase velocity does not vary with frequency or wavelength.

Group velocity

A wave group may be described by the superposition of individual waves of different wavelengths whose interference with one another results in the variation of amplitude defining the wave group shape. If the medium is *non-dispersive*, then the wave group travels with the common phase velocity. But, in a *dispersive* medium the different individual waves do not advance together. So the wave group has a velocity different from the phase velocities of the waves which make it up.

Hence, the velocity with which the wave group (or, wave packet) travels in a dispersive medium is called the *group velocity* (v_g). The group velocity and phase velocity of a wave group are thus different.

2.3.1* Expression for group velocity

Let us suppose, the wave group is formed due to the superposition of two monochromatic waves having the same amplitude, but differing by an amount $\Delta\omega$ in angular frequency and Δk in wave number [Fig 2.1]. The two-component waves may be represented by

$$\psi_1 = A\sin(\omega t - kx) \tag{2.7}$$

$$\psi_2 = A\sin[(\omega + \Delta\omega)t - (k + \Delta k)x] \tag{2.8}$$

The resultant displacement ψ at any time t at a distance x from the origin is

$$\begin{aligned}\psi &= \psi_1 + \psi_2 \\ &= A\sin(\omega t - kx) + A\sin[(\omega + \Delta\omega)t - (k + \Delta k)x] \\ &= 2A\sin\tfrac{1}{2}[(2\omega + \Delta\omega)t - (2k + \Delta k)x]\cos\tfrac{1}{2}(\Delta\omega t - \Delta kx)\end{aligned} \tag{2.9}$$

As $\Delta\omega$ and Δk are small compared to ω and k respectively, we write:

$$2\omega + \Delta\omega \simeq 2\omega$$

$$\text{and } 2k + \Delta k \simeq 2k$$

Equation (2.9) reduces to

$$\psi = 2A\cos\tfrac{1}{2}(\Delta\omega t - \Delta kx)\sin(\omega t - kx) \tag{2.10}$$

Equation (2.10) represents the analytical expression for the *wave group* (i.e., wave packet). From this equation it is clear that the sine factor represents a carrier wave whose phase velocity

$$v_p = \frac{\omega}{k}$$

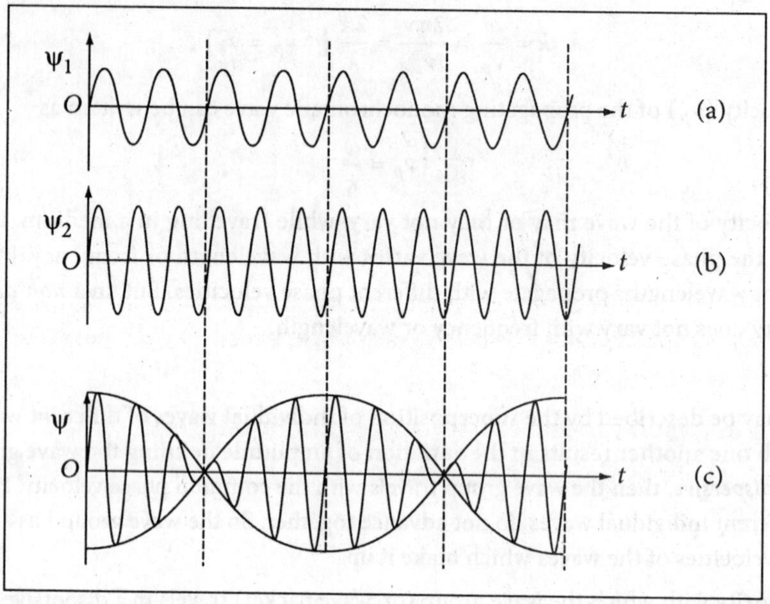

Fig 2.1 Superposition of two sinusoidal waves of slightly different frequencies [(a) and (b)]. Curve (c) represents the resultant motion.

Again, the amplitude of the resultant wave is

$$R = 2A\cos\tfrac{1}{2}(\Delta\omega t - \Delta k x) \tag{2.11}$$

Now, the group velocity (v_g) is the velocity with which the maximum amplitude moves. At $x = 0$ and $t = 0$

$$R_{\max} = 2A$$

If at $x = x'$ and $t = t'$ the amplitude is again $R_{\max}$, then from eq. (2.11)

$$\tfrac{1}{2}(\Delta\omega t' - \Delta k x') = 0 \text{ or, } \Delta k x' = \Delta\omega t' \text{ or, } \frac{x'}{t'} = \frac{\Delta\omega}{\Delta k}$$

When $\Delta k \to 0$

$$\lim_{\Delta k \to 0} \frac{x'}{t'} = \frac{d\omega}{dk} \text{ or, } v_g = \frac{d\omega}{dk} \tag{2.12}$$

The centre of the wave group moves with velocity $\dfrac{d\omega}{dk}$ and is called *the group velocity* v_g.

2.3.2 Relation between v_g and v_p

We have the group velocity

$$v_g = \frac{d\omega}{dk} = \frac{d}{dk}(v_p k) \left[\because v_p = \frac{\omega}{k}\right]$$

$$= v_p + k\frac{dv_p}{dk} = v_p + k\frac{dv_p}{d\lambda}\cdot\frac{d\lambda}{dk}$$

But,

$$\frac{d\lambda}{dk} = \frac{d}{dk}(\lambda) = \frac{d}{dk}\left(\frac{2\pi}{k}\right) = -\frac{2\pi}{k^2} \left[\because k = \frac{2\pi}{\lambda}\right]$$

$$\therefore v_g = v_p + k\frac{dv_p}{d\lambda}\left(-\frac{2\pi}{k^2}\right) = v_p - \frac{2\pi}{k}\frac{dv_p}{d\lambda} \text{ or, } v_g = v_p - \lambda\frac{dv_p}{d\lambda} \tag{2.13}$$

Equation (2.13) shows that for a non-dispersive medium, i.e., $dv_p/d\lambda = 0$, $v_g = v_p$, i.e., the group velocity and phase velocities are equal. But, for a dispersive medium in which $dv_p/d\lambda \neq 0$, $v_g < v_p$, i.e., *the group velocity is less than the phase velocity.*

2.3.3 Relation between group velocity and particle velocity

The group velocity (v_g) of a moving wave group is

$$v_g = \frac{d\nu}{d\left(\frac{\nu}{v_p}\right)} = \frac{d\nu}{d\left(\frac{1}{\lambda}\right)} \quad [\because v_p = \nu\lambda]$$

$$= \frac{d\nu}{dv}\bigg/\frac{d}{dv}\left(\frac{1}{\lambda}\right), \text{ } v \text{ being the velocity of the particle conserved}$$

$$\text{or, } v_g = \frac{d}{dv}\left\{\frac{m_0c^2}{h\sqrt{1-v^2/c^2}}\right\}\bigg/\frac{d}{dv}\left(\frac{m_0 v}{h\sqrt{1-v^2/c^2}}\right)$$

$$\therefore \nu = \frac{E}{h} = \frac{mc^2}{h} = \frac{m_0}{h\sqrt{1-v^2/c^2}} = \frac{m_0c^2}{h\sqrt{1-v^2/c^2}} \left[\because m = \frac{m_0}{\sqrt{1-v^2/c^2}}\right]$$

$$\text{and } \lambda = \frac{h}{p} = \frac{h}{mv} = \frac{h}{m_0 v/\sqrt{1-v^2/c^2}} = \frac{h\sqrt{1-v^2/c^2}}{m_0 v}$$

$$\therefore v_g = v \tag{2.14}$$

Hence, the group velocity (v_g) of matter waves associated with a moving body is equal to the velocity (v) of the particle.

The equality $v_g = v$ shows that instead of associating a single monochromatic de Broglie wave with the particle we should associate a group of packet of de Broglie waves of nearly equal wavelength around the de Broglie wavelength $\lambda = h/p$ of the particles concerned.

2.3.4 Formation of wave packet

We know, a wave packet comprises a group of waves, each with slightly different velocity and wavelength, with phases and amplitudes. The waves are so chosen that they interfere constructively over only a small region of space where a particle can be located, outside it they interfere destructively so that the amplitude reduces to zero.

The one-dimensional wave packet is represented by the curve of Fig 2.2 which travels along the positive x-direction with its own group velocity (v_g). The individual component waves possess an average velocity (u), called the *phase velocity*.

A plane wave travelling along the positive direction of x-axis can be mathematically represented by $\psi(x,t) = Ae^{-i(\omega t - kx)}$ or, $\psi(x,t) = Ae^{i(kx-\omega t)}$, where $k = 2\pi/\lambda$ is the wave propagation velocity, λ the wavelength and ω the frequency.

If a good number of such plane waves with propagation constant (k) slightly different from the average value (k_0) simultaneously propagate along the positive x-direction in a dispersive medium, then the wave packet of small extension Δx is formed [Fig 2.2].

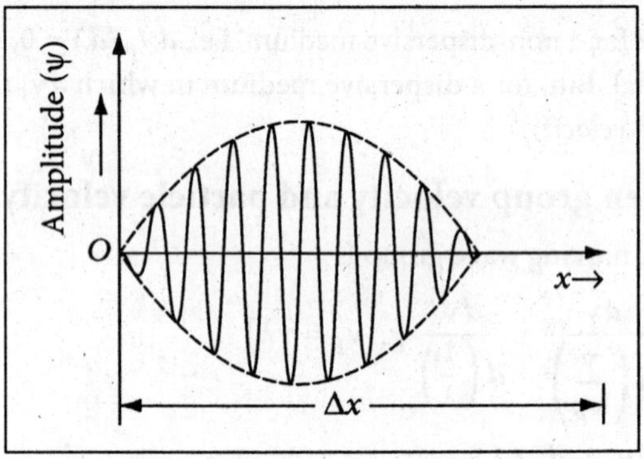

Fig 2.2 One-dimensional wave packet

According to Fourier's theorem, the wave packet can be expressed as

$$\psi(x,t) = \frac{1}{\sqrt{2\pi}} \int_{-\infty}^{+\infty} A(k)\, e^{i(kx-\omega t)}\, dk \tag{2.15}$$

The amplitude $A(k)$ is a function of k and is called the *Fourier transform* of the function $\psi(x,t)$. $A(k)$ is expressed as

$$A(k)\, e^{-i\omega t} = \frac{1}{\sqrt{2\pi}} \int_{-\infty}^{+\infty} \psi(x,t)\, e^{-ikx}\, dx \tag{2.16}$$

$$\text{or, } A(k) = \frac{1}{\sqrt{2\pi}} \int_{-\infty}^{+\infty} \psi(x,t)\, e^{-i(kx-\omega t)}\, dx \tag{2.17}$$

Since, $A(k)$ is independent of time t

$$\frac{dA(k)}{dt} = 0 \tag{2.18}$$

If Δx be the approximate spread of the wave packet, then it can be shown that $\Delta k \geqslant 1/\Delta x$. Eq. (2.15) can represent a wave packet if $A(k)$ is finite over a small range between $\left(k_0 - \frac{\Delta k}{2}\right)$ and $\left(k_0 + \frac{\Delta k}{2}\right)$ about k_0. Now, over a small range of k the variation $\omega(k)$ is small and can be expanded by Taylor's theorem as below:

$$\omega(k) = \omega(k_0 + k - k_0) = \omega(k_0) + (k - k_0)\left(\frac{d\omega}{dk}\right)_{k_0} + \frac{(k-k_0)^2}{21}\left(\frac{d^2\omega}{dk^2}\right)_{k_0} + \ldots$$

Neglecting the second and higher order terms

$$\omega(k) = \omega(k_0) + (k - k_0)\left(\frac{d\omega}{dk}\right)_{k_0} \tag{2.19}$$

$$\text{or, } \omega = \omega_0 + (k - k_0)\left(\frac{d\omega}{dk}\right)_{k_0} \tag{2.20}$$

where $\omega(k) = \omega$ and $\omega(k_0) = \omega_0$.

With this value of ω, $(kx - \omega t)$ becomes

$$(kx - \omega t) = kx - \left[\omega_0 + (k - k_0)\frac{d\omega}{dk}t\right] = k_0 x - \omega_0 t + kx - k_0 x - (k - k_0)\frac{d\omega}{dk}t$$

$$= (k_0 x - \omega_0 t) + (k - k_0)x - (k - k_0)\frac{d\omega}{dk}t = (k_0 x - \omega_0 t) + (k - k_0)\left[x - \left(\frac{d\omega}{dk}\right)t\right] \tag{2.21}$$

Equation (2.15), then takes the form:

$$\psi(x,t) = \frac{1}{\sqrt{2\pi}}\int_{-\infty}^{+\infty} A(k)\, e^{i[(k_0 x - \omega_0 t) + (k-k_0)\{x - (\frac{d\omega}{dk})t\}]}\, dk$$

$$= \frac{1}{\sqrt{2\pi}} e^{i(k_0 x - \omega_0 t)} \int_{-\infty}^{+\infty} A(k)\, e^{i(k-k_0)\, x\{x - (\frac{d\omega}{dk})t\}}\, dk \tag{2.22}$$

Equation (2.22) represents a modulated wave of wavelength $\lambda_0 = 2\pi/k_0$ and frequency $\omega_0/2\pi$. x and t determine the value of the integral. It is evident that the value of $\psi(x,t)$ depends on x and t through $\left(x - \frac{d\omega}{dk}t\right)$. Thus, the wave packet travels undistorted if

$$x - \frac{d\omega}{dk}t = 0 \text{ or, } \frac{x}{t} = v_g = \frac{d\omega}{dk} \tag{2.23}$$

v_g is called the *group velocity* of the waves.

The factor $e^{i(k_0 x - \omega_0 t)}$ in eq. (2.21) represents the phase velocity $v_p = \frac{\omega_0}{k_0}$.

A three-dimensional wave packet is represented by

$$\psi(\vec{r},t) = \frac{1}{(2\pi)^{3/2}}\int_{-\infty}^{+\infty} A(\vec{k}) e^{i(\vec{k}\cdot\vec{r} - \omega t)}\, d^3\vec{k} \tag{2.24}$$

where $d^3\vec{k} = dk_x\, dk_y\, dk_z$ the volume element in k-space. The corresponding Fourier transform is

$$A(\vec{k}) = \frac{1}{(2\pi)^{3/2}}\int_{-\infty}^{+\infty} \psi(\vec{r},t)\, e^{-i\vec{k}\cdot\vec{r}}\, d^3\vec{r} \tag{2.25}$$

where $d^3\vec{r} = dx\, dy\, dz$, the volume element in three-dimensional Cartesian coordinate space.

2.4 de Broglie's relation for a photon

According to *Planck-Einstein's equation,* the energy of a photon of frequency ν is given by

$$E = h\nu \tag{2.26}$$

where h = Planck's constant having value 6.6256×10^{-34} J-s.

Again, the total relativistic energy of a particle of rest mass m_0 and momentum p is given by

$$E^2 = p^2c^2 + m_0^2c^4 \tag{2.27}$$

where $c = 3 \times 10^{10}$ m/s, the velocity of light in vacuum.

For a photon, $m_0 = 0$.

$\therefore$ from eq. (2.27)

$$p = \frac{E}{c} \tag{2.28}$$

From eqs (2.26) and (2.28)

$$pc = h\nu = \frac{hc}{\lambda} \text{ or, } \lambda = \frac{h}{p} \tag{2.29}$$

Here, λ is the wavelength of the wave associated with a photon and p is its momentum regarding it as a particle.

de Broglie in 1924, proposed that the wave-particle duality is not limited to photon. He showed that the relation holds good for all sub-atomic particles like electrons, protons, neutrons, atoms, molecules, etc. while they are in motion.

If m be the mass and v the velocity, then according to de Broglie, the wavelength of the associated wave is

$$\lambda = \frac{h}{p} = \frac{h}{mv} \tag{2.30}$$

According to kinetic theory of gases

$$E_k = \frac{3}{2}kT \text{ or, } E_k = \frac{p^2}{2m}$$

where T is the absolute temperature.

$$p = \sqrt{2mE_k}. \therefore \lambda = \frac{h}{\sqrt{2mE_k}} = \frac{h}{\sqrt{3mkT}} \tag{2.31}$$

The frequency

$$\nu = \frac{E}{h} \tag{2.32}$$

where E is the relativistic energy and $m = \dfrac{m_0}{\sqrt{1 - v^2/c^2}}$ is the relativistic mass of the particle.

Also

$$\because p = \frac{h}{\lambda} = \frac{h}{2\pi}\frac{2\pi}{\lambda} = \hbar k \tag{2.33}$$

where $\dfrac{h}{2\pi} = \hbar$ and $\dfrac{2\pi}{\lambda} = k$.

$$\therefore E = h\nu = \frac{h}{2\pi}2\pi\nu = \hbar\omega \tag{2.34}$$

From eq. (2.30) we note that $\lambda \propto 1/m$, when v is constant. For macroscopic bodies, $m \to \infty, \therefore \lambda \to 0$.

For example, consider the case of a cricket ball of mass 500 gm flying with a velocity of 13.9 m/s, its de Broglie wavelength is

$$\lambda = \frac{6.62 \times 10^{-34}\ \text{J-s}}{0.5\ \text{kg} \times 13.9\ \text{m/s}} = 10^{-34}\ \text{m} = 10^{-24}\ \text{Å}$$

which is insignificant compared to the size of the ball. So for macroscopic objects de Broglie wavelength is insignificant. That is why we do not observe wave character for the planets moving around the sun.

On the other hand, for microscopic objects λ is large and significant. For example, an electron of mass 9.1×10^{-31} kg moving with velocity 4.47×10^{6} m/s, its de Broglie wavelength is

$$\lambda = \frac{h}{mv} = \frac{6.62 \times 10^{-34}\ \text{J-s}}{9.1 \times 10^{-31}\ \text{kg} \times 4.47 \times 10^{6}\ \text{m/s}} = 9.3 \times 10^{-10}\ \text{m} = 9.3\ \text{Å}$$

which is large compared to the size of the electron. It may, therefore, be treated more as a wave than a particle.

2.5 Nature of de Broglie waves

The conception of de Broglie wave (matter wave) associated with a particle of known momentum has been experimentally verified by the American physicists *C.J. Davisson* and *L.H. Germer* in 1927. The de Broglie hypothesis was further confirmed by the experiments conducted independently by G.P. Thomson in England and by *Kikuchi* in Japan in 1927. de Broglie, however gave no idea regarding the physical nature of the wave associated with a material particle in motion.

The relation $\lambda = h/p$ represents the wave associated with a moving particle of momentum p, which is monochromatic of wavelength λ. Now, two questions arise:

(a) How can a wave which spreads out over a large region of space represent a particle which is highly localised?

(b) If somehow a wave represents a particle what exactly is that which is waving the particle wave?

We now try to search answers for the above questions.

Let us now suppose that a particle be represented by the wave $\psi = A\sin(kx - \omega t)$.

It represents a pure sine wave extending from $-\infty$ to $+\infty$. It is characterised by a wavelength λ and momentum $p = h/\lambda$ and completely non-localised.

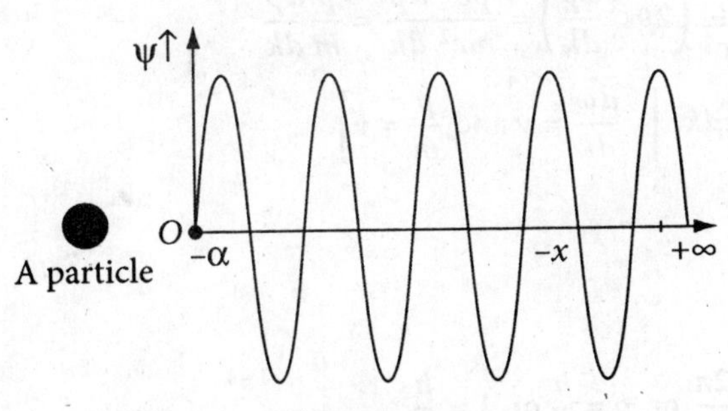

Fig 2.3 (a) A monochromatic wave

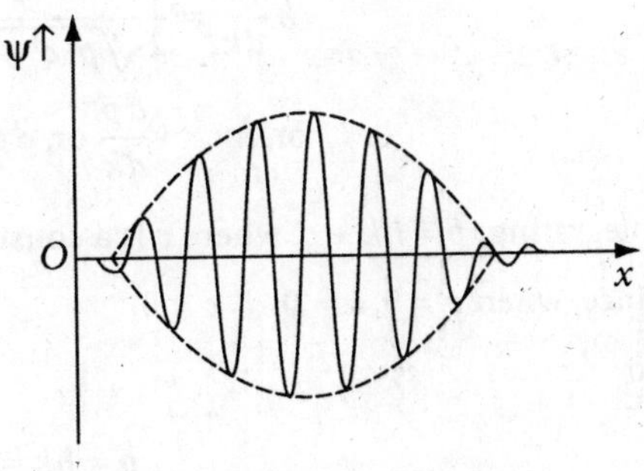

Fig 2.3 (b) A wave packet

Since, a particle (e.g., an electron) is completely localised, therefore, a monochromatic wave cannot represent a particle. It implies that de Broglie waves are not monochromatic waves, but could be a combination of several monochromatic waves. It is known that a superposition of several monochromatic waves with same amplitude and phase, but differing slightly in frequency travelling through the same part of a dispersive medium produce wave packet. Such wave packets are localised in a small part of space and capable of representing a particle. The separation between two successive amplitudes in a wave packet (λ_0) is characteristic of a wave and has a particle like localisation in space. But, a particle such as an electron or a proton retains its particle nature like mass and charge which are never divided in any diffraction experiment. For this reason, a particle in motion cannot be supposed to be just a wave packet. Therefore, *de Broglie wave packets are only theoretical conceptions which are employed to explain the experimental findings.*

2.6 de Broglie wavelength of material particles

To derive an expression of de Broglie wavelength for a material particle in motion following assumptions are made:

1. The energy of a moving particle is
$$E = h\nu \text{ [as in eq. (2.25)]} \tag{2.35}$$
where h is the Planck's constant and ν is the frequency of the wave associated with the moving particle.
2. The moving particle is associated with a wave packet of small extension.

Therefore, the particle velocity = the group velocity of the waves, i.e.,
$$v = v_g = \frac{d\omega}{dk} \tag{2.36}$$

Equation (2.35) can be written as $E = \frac{h}{2\pi} \cdot 2\pi\nu = \hbar\omega$; also $E = mc^2$, m being the mass of the particle and c the velocity of light in free space.

$$\therefore \hbar\omega = \sqrt{p^2c^2 + m_0^2c^4} \left[\because E = \sqrt{p^2c^2 + m_0^2c^4}\right] \tag{2.37}$$

where m_0 is the rest mass and p the momentum of the particle.

Differentiating eq. (2.37) w.r.t. k, we get

$$\hbar\frac{d\omega}{dk} = \frac{1}{2}\frac{1}{\sqrt{p^2c^2 + m_0^2c^4}}\left(2pc^2\frac{dp}{dk}\right) = \frac{pc^2}{mc^2}\frac{dp}{dk} = \frac{p}{m}\frac{dp}{dk}$$

$$\text{or, } \hbar v = v\frac{dp}{dk} \text{ or, } dp = \hbar dk, \left[\because \frac{d\omega}{dk} = v \text{ and } \frac{p}{m} = v\right]$$

Integrating, $p = \hbar k + c$, where c is a constant.

Since, when $p = 0, k = 0, \therefore c = 0$.

So

$$p = \hbar k = \frac{h}{2\pi}\frac{2\pi}{\lambda} \text{ or, } p = \frac{h}{\lambda} \text{ or, } \lambda = \frac{h}{p}$$

This is the expression for the *de Broglie wavelength* of a material particle of momentum p.

2.7 Relation between phase velocity and particle velocity

In ω be the angular frequency and k the wave propagation constant, then the phase velocity of the wave is

$$v_p = \frac{\omega}{k}$$

The momentum of the associated particle is $p = \hbar k$ and the total energy of the particle is $E = \hbar\omega$.

$$\therefore v_p = \frac{\omega}{k} = \frac{E}{\hbar} \cdot \frac{\hbar}{p} = \frac{E}{p}$$

Again, from the relativistic mass-energy relation, energy, $E = mc^2$ and momentum, $p = mv$.

$$\therefore v_p = \frac{mc^2}{mv} = \frac{c^2}{v} \tag{2.38}$$

Equation (2.38) represents the relation between phase velocity and particle velocity of a de Broglie wave.

Equation (2.38) shows that as $v < c$, $v_p > c$, i.e., *the phase velocity of a de Broglie wave is always greater than the velocity of light in free space.* But, according to the special theory of relativity, no energy can be transmitted with velocity greater than c. This suggests that the phase velocity v_p has no physical significance.

2.8 Relation between v_p and λ of de Broglie wave

For a de Broglie wave, the phase velocity of a particle of mass m and momentum p is given by $v_p = E/p$.

The total relativistic energy of the same particle is $E = \sqrt{p^2c^2 + m_0^2c^4}$, where m_0 is the rest mass of the particle.

$$\therefore v_p = \frac{\sqrt{p^2c^2 + m_0^2c^4}}{p} = \frac{pc}{p}\sqrt{1 + \frac{m_0^2c^2}{p^2}} = c\sqrt{1 + \frac{m_0^2c^2}{p^2}} \text{ or, } v_p = c\sqrt{1 + \left(\frac{m_0^2c^2}{h^2}\right)\lambda^2} \tag{2.39}$$

Equation (2.39) shows that $v_p > c$, and in free space, this is consistent with the relation $v_g = c^2/v_p$, since $v_g = v$ the particle velocity and v is always less than c. Further, v_p is a function of de Broglie wavelength λ even in vacuum. But, for light wave, v_p does not depend on λ. *Here, is the difference between de Broglie wave and light wave.*

2.9 de Broglie wavelength of a relativistic particle of mass m and kinetic energy K

If E_k be the kinetic energy of a particle of rest mass m_0, then its total relativistic energy is given by

$$E = E_k + m_0c^2 \text{ or, } \sqrt{p^2c^2 + m_0^2c^4} = E_k + m_0c^2$$

Squaring both sides

$$p^2c^2 + m_0^2c^4 = E_k^2 + 2E_km_0c^2 + m_0^2c^4 \text{ or, } p = \frac{\sqrt{E_k(E_k + 2m_0c^2)}}{c}$$

$\therefore$ de Broglie wavelength

$$\lambda = \frac{h}{p} = \frac{hc}{\sqrt{E_k(E_k + 2m_0c^2)}} \text{ or, } \lambda = \frac{h}{\sqrt{2m_0E_k\left(1 + \frac{E_k}{2m_0c^2}\right)}} \tag{2.40}$$

de Broglie wavelength λ can be calculated for any particle (e.g., electron, proton, etc.) if the kinetic energy E_k and rest mass m_0 are given.

2.10 The de Broglie model of the atom

Though Bohr's theory of atomic structure could explain the experimental observations relating to the atomic behaviour, the three ad hoc postulates of the theory remained without valid justification for a long time. de Broglie established that *his hypothesis of matter wave could provide the justification for the postulate regarding quantisation of angular momentum in Bohr's theory.*

Along with the electron, the associated de Broglie waves propagate along the circular path again and again. The wave after going round one full circular path must meet itself.

Otherwise, the wave would be out of phase with itself after one orbit and the wave would be annihilated by destructive interference. So the wave should produce standing wave profile in the orbit to avoid the radiation loss of electron energy. Just like the standing waves produced in a stretched string have discrete frequencies, an integral number of wavelengths fit into the circumference of the orbit. Thus, the condition of formation of standing waves in an orbit is

$$2\pi r_n = n\lambda \ (n = 1, 2, 3, \dots) \tag{2.41}$$

where r_n is the radius of the nth orbit. Such standing waves are shown in Fig 2.4 below:

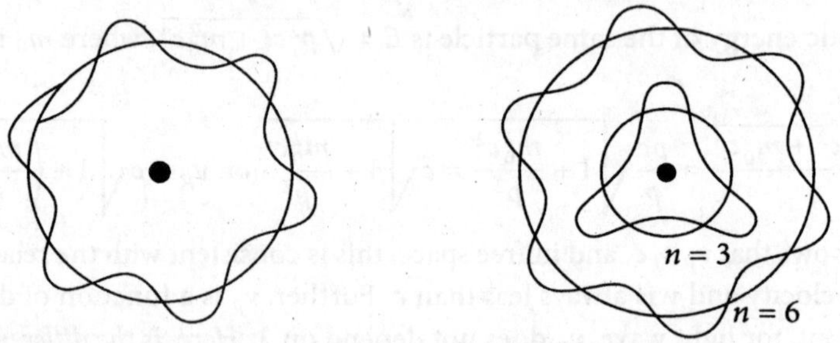

Fig 2.4 (a) The orbit does not contain integral multiple of λ, hence not allowed

Fig 2.4 (b) The allowed orbits accommodate integral multiple of electron waves

Again, since according to de Broglie's hypothesis $\lambda = h/mv$, we get from (2.41)

$$2\pi r_n = \frac{nh}{mv} \text{ or, } mvr_n = \frac{nh}{2\pi} = n\hbar \text{ or, } L = n\hbar \tag{2.42}$$

where $L = mvr_n$ is the angular momentum of the electron moving in the nth orbit with a speed v. de Broglie thus demonstrated that the quantisation of angular momentum is a direct consequence of wave nature of the electron.

Again, if λ be the wavelength of the electron wave in the first Bohr orbit, then $\lambda = h/mv$.

The speed of the electron is $v = \dfrac{e}{\sqrt{4\pi\varepsilon_0 mr}}$.

$$\therefore \lambda = \frac{h}{e}\sqrt{\frac{4\pi\varepsilon_0 r}{m}}$$

Putting $r = 5.3 \times 10^{-11}$ m, we get $\lambda = 3.3$ Å.

But, the length of the circumference of the first Bohr orbit is $2\pi r = 3.3$ Å.

This shows that one complete electron wave fits into the first Bohr orbit of hydrogen atom. In view of de Broglie's hypothesis, the quantum number n is equal to the number of de Broglie wavelengths that can fit into the circumference of the allowed orbits. Thus, the electron in an atom can be picturised either as *a particle moving in an orbit whose angular momentum is quantised or as a standing wave around the nucleus of an atom.*

Again, from Bohr's theory, the circumference of the nth orbit of the electron is $2\pi r_n$. The angular momentum of the electron

$$mv_n r_n = n\hbar \text{ or, } r_n = \frac{n\hbar}{p} \; [\because p = mv_n]$$

$$\therefore 2\pi r_n = 2\pi \times \frac{n\hbar}{p} = n\lambda \; [\because \hbar = h/2\pi \text{ and } \lambda = h/p]$$

This shows that the circumference of the nth Bohr orbit of the electron contains integral (n) *multiple of de Broglie wavelength.*

2.11 Characteristics of de Broglie waves

de Broglie waves have the following characteristics:

1. Lighter is the particle, greater is the de Broglie wavelength, i.e., $\lambda \propto \frac{1}{m}$.
2. Smaller is the velocity of the particle, greater is the wavelength of the associated wave.
3. $\lambda = \infty$, when $v = 0$, i.e., the waves are produced due to the motion of the particle. The waves are charge-independent and thus different from electromagnetic waves.
4. The velocity of the matter wave is not constant, but depends on the velocity of the particle. The velocity of electromagnetic wave is constant.
5. Both the wave and particle aspects of a moving body cannot appear simultaneously in the same experiment. We can say, that waves have particle-like properties and particles have wave-like properties and the concepts are inseparably linked.
6. The wave nature of matter is associated with the uncertainty in position of the particle.

2.12 Bragg's law for diffraction

We consider a set of parallel planes of a crystal having inter-planar distance d. Let a collimated beam of monochromatic X-rays of wavelength λ be incident on the atomic planes at a glancing angle θ.

Let the monochromatic beam of X-rays be incident on two consecutive planes AB and CD. The rays PO and QO' are reflected at O and O' respectively along OR and $O'S$ as shown in Fig 2.5. Then $\angle POA = \angle ROB = \theta$; also $\angle QO'C = \angle SO'D = \theta$.

OM and ON are dropped perpendiculars from O on QO' and $O'S$. The path difference of the beams POR and $QO'S$ is $\Delta = MO' + O'N = d\sin\theta + d\sin\theta = 2d\sin\theta$.

Now, for constructive interference of the beams $\Delta = n\lambda$, where λ is the wavelength of X-rays and n is an integer called the *order no. of diffraction.*

$$\therefore 2d \sin\theta = n\lambda \tag{2.43}$$

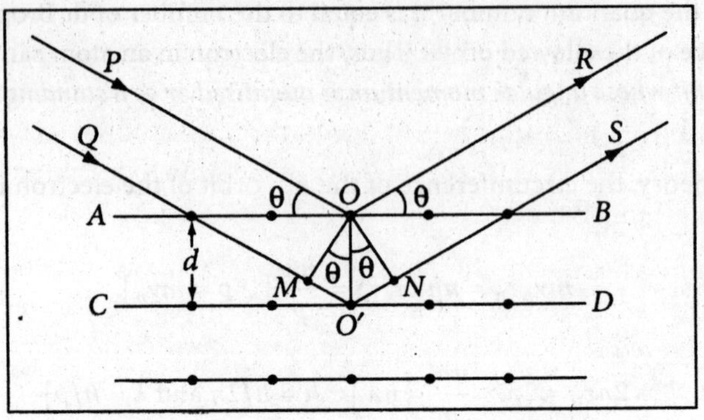

Fig 2.5 Diffraction of X-rays by crystal lattice

Equation (2.43) represents *the Bragg's law of X-ray diffraction.* It is evident that for a given value of d, λ and n the X-rays will be reflected along a direction making an angle θ with the incident plane of atoms.

2.13 Experimental proof of de Broglie's hypothesis

2.13.1 Davisson and Germer's experiment

In 1927, two American physicists *C.J. Davisson* and *L.H. Germer* first performed an experiment which conclusively demonstrated the wave character of electrons. They also succeeded in measuring the wavelength of slow electrons accelerated by a potential difference of 30 to 1000 volts by the method of diffraction.

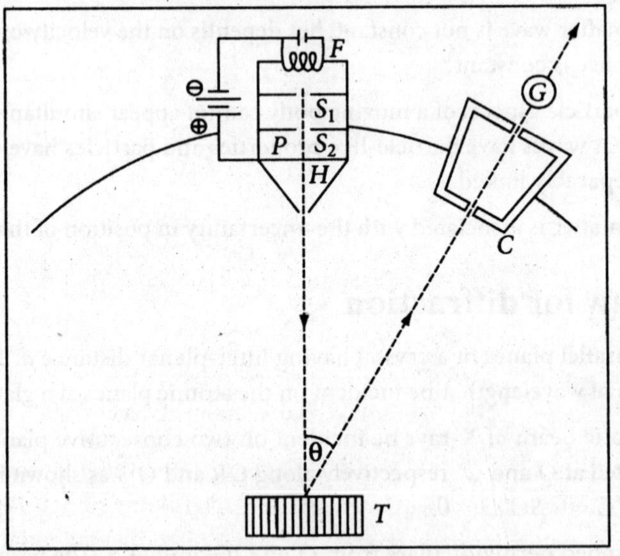

Fig 2.6 Experimental arrangement of Davisson and Germer

The arrangement of the experimental set-up is shown in Fig 2.6. The electron beam is produced by heating a tungsten filament F and is accelerated through a known potential difference V maintained between the filament F and the plate P. The electron beam is collimated by passing through narrow slits S_1 and S_2. The collimated beam, then finally emerges through a fine hole H. This beam of electrons, then is directed in a high vacuum to fall normally on a large single crystal of nickel T. The scattered electrons from T in different directions are collected by a Faraday cup C. The scattered intensity is measured by a sensitive galvanometer G. The collector C is such that all the electrons having the same energy as the incident beam can enter into it. The ejected electrons from the nickel crystal due to the impact of the incident electrons are being of low energy cannot enter the collector C. The intensity of the scattered electron beam can be measured by rotating C about an axis through the crystal.

Results. The results of measurements are shown in Fig 2.7 where the scattered intensity of the electron beam is plotted as a function of the angle of scattering in a polar graph. The radius vector of this graph is proportional to the scattered intensity. The angle between ordinate and the radius vector represents the angle of scattering (θ).

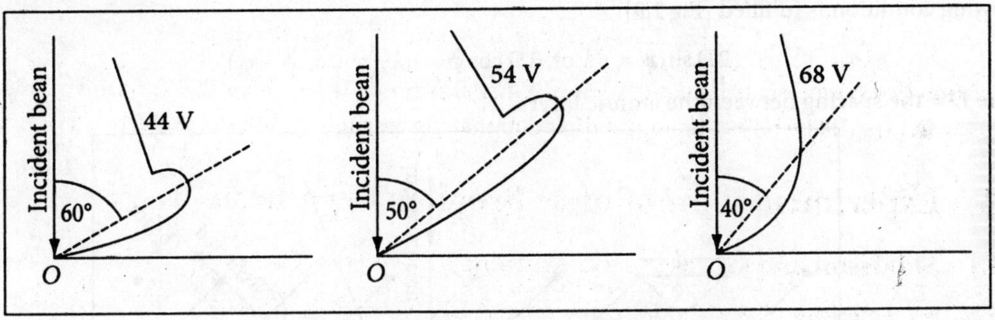

Fig 2.7 Results of Davisson and Germer's experiment

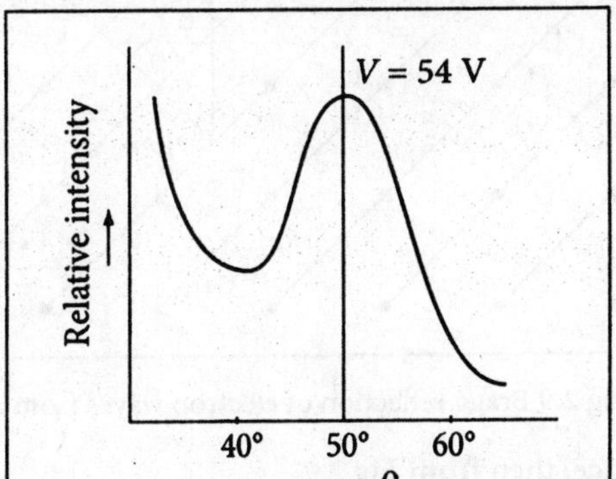

Fig 2.8 Plot of relative intensity against θ

The electron beam is accelerated to different energies by applying different accelerating potentials between F and H. At low value of the accelerating potential, the graph is fairly smooth, but at 44 V a spur (or hump) appears on the curve at scattering angle $\theta = 60°$. As the accelerating potential is gradually increased, the length of the spur increases. The spur becomes maximum at accelerating potential 54 V and scattering angle

$\theta = 50°$. On further increasing the accelerating potential, the spur decreases and disappears at 68 V at a scattering angle 40°. The graph becomes fairly smooth with further increase of accelerating voltage.

The plot of scattered intensity of electron beam against the angle of scattering is shown in Fig 2.8.

Observations and calculations

Since, there is a preferential reflection of 54 eV electron beam at a scattering angle of 50°, it is evident that electron beam undergoes diffraction from the single crystal of nickel.

Since, the atoms or group of atoms in the crystal lattice are arranged in a regular manner with spacing comparable to the wavelength of X-rays, the crystal can be used to diffract X-rays following the Bragg's relation

$2d \sin \theta = n\lambda$, where d is the spacing, λ is the wavelength of X-rays, θ is the glancing angle of incidence for order n.

We now consider an electron beam incident normally on the crystal surface at an angle ϕ with the normal to the atomic planes, then the reflected de Broglie waves associated with the electrons will reinforce if the following condition is fulfilled [Fig 2.9]

$$2D \sin \psi = n\lambda \text{ or, } 2D \cos \phi = n\lambda \quad \left(\because \psi = \tfrac{\pi}{2} - \phi\right) \tag{2.44}$$

where D = the spacing between the atomic layers.

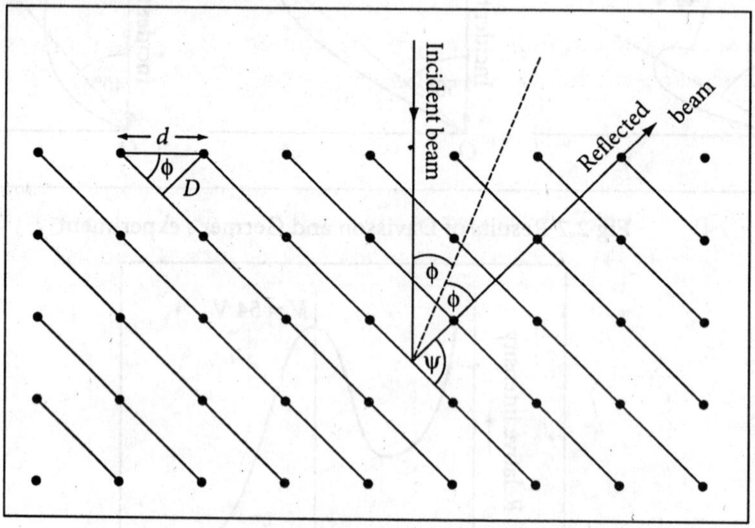

Fig 2.9 Bragg reflection of electron waves from a crystal

If d = the interatomic distance, then from Fig 2.9

$$D = d \sin \phi$$

$$\therefore n\lambda = 2(d \sin \phi) \cos \phi = d \sin 2\phi$$

$$\text{or, } n\lambda = d \sin \theta \tag{2.45}$$

since $\theta = 2\phi$ the angle between the incident and reflected beams and $n = 1, 2, 3, \ldots$. The de-Broglie wavelength of the electrons can be determined from eq. (2.45) using $d = 2.15$ Å and $\theta = 50°$ in the following way:

$$\lambda = 2.15 \sin 50° \text{ (for } n = 1)$$

$$\text{or, } \lambda = 1.65 \text{ Å} \tag{2.46}$$

Again, to calculate the de Broglie wavelength associated with an electron, we have the kinetic energy acquired by an electron in traversing a potential difference of V is $\frac{1}{2}m_0v^2 = eV$, where m_0 = rest mass of the electron of charge e, v = velocity of the electron in traversing through the potential difference of V.

$\therefore v = \sqrt{2Ve/m_0}$.

The de Broglie wavelength associated with the electron is

$$\lambda = \frac{h}{m_0 v} = \frac{h}{m_0 \times \sqrt{2Ve/m_0}} = \frac{h}{\sqrt{2m_0 eV}}$$

Putting $h = 6.62 \times 10^{-34}$ J-s, $m_0 = 9.1 \times 10^{-31}$ kg and $e = 1.6 \times 10^{-19}$ C, we get

$$\lambda = \frac{6.62 \times 10^{-34}}{\sqrt{V}(2 \times 1.6 \times 10^{-19} \times 9.1 \times 10^{-31})^{1/2}} \text{ or, } \lambda = \frac{12.27}{\sqrt{V}} \text{ Å} \tag{2.47}$$

In the present case, $V = 54$ V

$$\therefore \lambda = \frac{12.27}{\sqrt{54}} = 1.67 \text{ Å} \tag{2.48}$$

Thus, the value of λ calculated from diffraction experiment and from de Broglie's relation agrees well establishing the correctness of de Broglie hypothesis of matter wave.

2.13.2 G.P. Thomson's experiment

In 1928, the wave nature of electron was confirmed through G.P. *Thomson's electron diffraction experiment.* A beam of electrons of velocity v was passed through a thin gold foil and the diffraction pattern obtained was found to be exactly similar to that obtained from X-rays of wavelength, $\lambda = h/mv$.

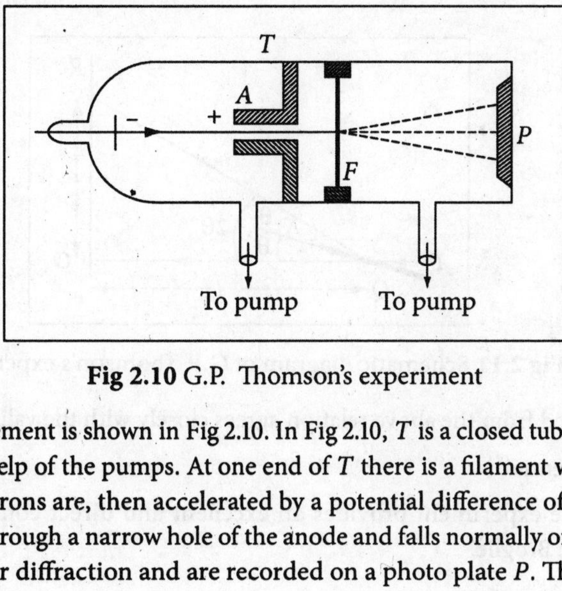

Fig 2.10 G.P. Thomson's experiment

The experimental arrangement is shown in Fig 2.10. In Fig 2.10, T is a closed tube in which the air pressure is made very low with the help of the pumps. At one end of T there is a filament which is heated to produce electron beams. These electrons are, then accelerated by a potential difference of 50,000 V to the anode A. The electron beam passes through a narrow hole of the anode and falls normally on a gold foil G of thickness 10^{-8} m. The electrons suffer diffraction and are recorded on a photo plate P. The diffraction pattern was exactly similar to those obtained by diffraction of X-rays by powdered crystals in *Debye-Scherrer method*. The diffraction pattern was found to consist of a series of concentric rings about a central spot.

To be sure that the pattern was due to the electron beam and not due to the X-rays produced due to the impact of cathode rays on the film, the experiment was repeated by introducing a magnetic field between the film and the photographic plate. The diffraction pattern in presence of the magnetic field was found to be exactly similar except that the whole pattern is moved to one side due to the field. This proves that the diffraction pattern was due to the charged particles (electrons), all of them having the same velocity. On removing the film, a central spot is recorded on the photo plate due to the direct electron beam. The experiment was repeated with beam of protons, neutrons and molecules and all are found to show wave character under suitable conditions.

Calculation of λ

The gold foil being polycrystalline, there are some atomic planes set at correct angle to give Bragg diffraction. If the number of crystals distributed at random be large, the Bragg reflection will form a series of concentric rings due to the intersection of the diffraction cone with the photographic plate.

In Fig 2.11, let PQ be the incident electron beam passing through the gold film at Q. The beam after suffering Bragg reflection at some crystal the reflected beam goes along QR meeting the photo plate at R. If O be the position of the central spot and $OR = r$, then $\frac{r}{l} = \tan 2\theta \simeq 2\theta$, where l is the distance QO and $\angle OQR = 2\theta$, the angle of scattering and θ is the glancing angle.

Again, if d be the distance between two parallel lattice planes and λ, the wavelength of the de Broglie wave associated with the electron, then according to Bragg's law, $2d \sin\theta = n\lambda$ [n is the order of diffraction.]

Again,

$$2d \sin\theta \simeq 2d\theta = d \tan 2\theta \; [\because \theta \text{ is small}]$$

$$\text{or, } d \tan 2\theta = 2d \sin\theta = n\lambda \text{ or, } d\frac{r}{l} = n\lambda \text{ or, } \lambda = \frac{rd}{nl}$$

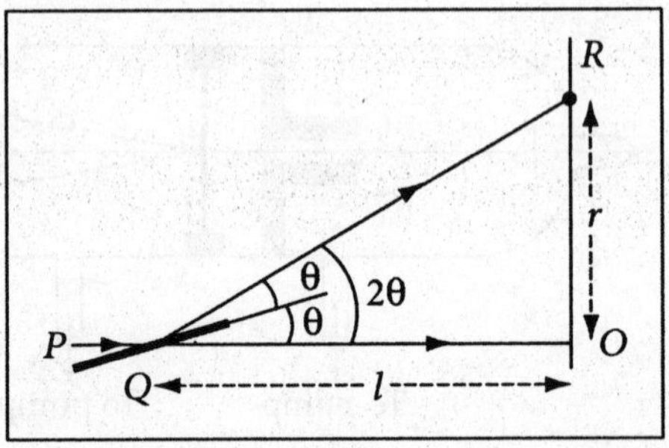

Fig 2.11 Schematic diagram of G.P. Thomson's experiment

The value of λ calculated from the above relation agrees closely with the value calculated from $\lambda = \frac{12.27}{\sqrt{V}}$, V is the accelerating potential.

Thus, we note that the experiment provides an excellent and direct confirmation of wave nature of electrons proposed by de Broglie.

2.14 de Broglie wavelength of high energy electrons

If the energy of the electron is so high that mass of the electron varies with velocity, then the simple expression of de Broglie wavelength of the electron ($\lambda = h/mv$) needs correction.

If p be the momentum and m_0 the rest mass of the electron, then the relativistic expression for energy is $E = \sqrt{p^2c^2 + m_0^2c^4}$, where c is the velocity of light in free space.

The kinetic energy of the electron is

$$E_K = E - m_0c^2 = \sqrt{p^2c^2 + m_0^2c^4} - m_0c^2$$

$$\text{or, } p^2c^2 + m_0^2c^4 = E_K^2 + m_0^2c^4 + 2m_0c^2E_K$$

$$\text{or, } p = \frac{\sqrt{E_K(E_K + 2m_0c^2)}}{c}$$

Now, the de Broglie wavelength of the electron is

$$\lambda = \frac{h}{p} = \frac{hc}{\sqrt{E_K(E_K + 2m_0c^2)}}$$

If V be the accelerating potential in volt, then $E_K = eV$.

$$\therefore \lambda = \frac{hc}{\sqrt{eV(eV + 2m_0c^2)}} \text{ or, } \lambda = \frac{hc}{e\sqrt{V\left(V + \frac{2m_0c^2}{e}\right)}} \tag{2.49}$$

If the rest mass energy of the electron $m_0c^2 = 0.51\,\text{MeV} = 0.51 \times 10^6$ eV, then

$$\frac{2m_0c^2}{e} = 1.02 \times 10^6 \text{ volts}$$

$$\therefore \lambda = \frac{300hc}{e\sqrt{V(V + 1.02 \times 10^6)}} \quad \left[\because V \text{ volt} = \frac{V}{300} \text{ esu}\right]$$

$$\text{or, } \lambda = \frac{300 \times 3 \times 10^{10} \times 6.62 \times 10^{-27} \times 10^8}{4.8 \times 10^{-10}\sqrt{V(V + 1.02 \times 10^6)}} \text{ Å} = \frac{12.41 \times 10^3}{\sqrt{V(V + 1.02 \times 10^6)}} \text{ Å} \tag{2.50}$$

If $E_K \ll m_0c^2$, then $\lambda = \frac{12.27}{\sqrt{V}}$ Å.

2.15 Diffraction of neutrons

Neutrons are nuclear particles of mass equal to that of a proton and charge zero. They play important role in determining the position of light atoms in crystals. In accordance with de Broglie hypothesis, the neutrons are associated with wave nature which suffers diffraction, while passing through a crystal.

The velocity of thermal neutrons is distributed over a wide range according to *Maxwell's law of velocity distribution*. If m be the mass of a neutron at equilibrium temperature T, then its most probable velocity is

$$v = \sqrt{\frac{2kT}{m}} \tag{2.51}$$

where k is the Boltzmann constant.

Taking $T = 300\,\text{K}$ (27 °C), $k = 1.38 \times 10^{-23}$ J/K and $m = 1.676 \times 10^{-27}$ kg.

$$v = \sqrt{\frac{2 \times 1.38 \times 10^{-23} \times 300}{1.676 \times 10^{-27}}} = 10^3 \sqrt{\frac{2 \times 1.38 \times 3}{1.676}} = 2.22 \times 10^3 \text{ m/s}$$

The most probable energy of these neutrons at 300 K is

$$E = \frac{1}{2}mv^2 = \frac{1}{2}m \times \frac{2kT}{m} = kT \tag{2.52}$$

$$= 1.38 \times 10^{-23} \times 300 = 4.14 \times 10^{-21}\text{ J}$$

$$= \frac{4.14 \times 10^{-21}}{1.6 \times 10^{-19}} \; [\because 1\text{ eV} = 1.6 \times 10^{-19}\text{ J}]$$

$$= 2.6 \times 10^{-2}\text{ eV}$$

Now, for thermal neutron at $T = 300\,\text{K}$, since the most probable velocity $v = \sqrt{\dfrac{2kT}{m}}$, its associated de Broglie wavelength is given by

$$\lambda = \frac{h}{mv} = \frac{h}{m \times \sqrt{\dfrac{2kT}{m}}} \text{ or, } \lambda = \frac{h}{\sqrt{2mkT}}$$

Taking $h = 6.63 \times 10^{-34}$ J-s, $m = 1.676 \times 10^{-27}$ kg, $T = 300\,\text{K}$ and $k = 1.38 \times 10^{-23}$ J/K, we get

$$\lambda = \frac{6.63 \times 10^{-34}}{\sqrt{2 \times 1.676 \times 10^{-27} \times 1.38 \times 10^{-23} \times T}} = \frac{6.63 \times 10^{-9}}{\sqrt{2 \times 1.676 \times 1.38 \times T}}\text{ m} = \frac{30.8 \times 10^{-10}}{\sqrt{T}}\text{ m}$$

$$\therefore \lambda = \frac{30.8}{\sqrt{T}}\text{ Å} \tag{2.53}$$

For $T = 300\,\text{K}$

$$\lambda = \frac{30.8}{\sqrt{300}} = \frac{30.8}{17.32} = 1.778\text{ Å}$$

The de Broglie wavelength of thermal neutron is thus of the same order as the interatomic distance in a crystal. *This suggests that the neutron beam can show diffraction and interference phenomena like X-rays of the same λ. Elsässer* (1936) first suggested the diffraction of neutrons from a crystal and the phenomenon was first experimentally demonstrated (in 1937) by *Halban* and *Preiswerk*.

Neutron diffraction has the following important practical applications:

1. Determination of the positions of light atoms in the crystal.
2. Determination of magnetic moment distribution in a magnetic material.
3. Invention of electron microscope.

2.16 Wave-particle duality

Max Planck in 1901 showed that the emission and absorption of electromagnetic radiation take place in *discrete quanta* each having energy $\varepsilon = h\nu$, where h is the Planck's constant and ν is the frequency of the radiation. Thus, the particle nature of radiation was established.

On the other hand, de Broglie in 1924 established the wave character of radiation consisting of *quanta*. According to him, a material particle of momentum p is associated with a wave of wavelength λ, where $\lambda = h/p$.

Thus, radiation and matter both have *dual character*. This phenomenon is called *wave-particle duality*. It is to be kept in mind that the particle nature and wave nature are *complimentary* but are *mutually exclusive*.

It is difficult to form a picture of *reality* behind this aspect of duality of matter and waves. What is the nature of this entity which can assume the properties of a particle (having momentum and kinetic energy) and also those of waves having wavelength and frequency?

To resolve the wave-particle contradiction, *Neils Bohr* suggested the *principle of complementarity*.

2.17 Bohr's principle of complimentarily

Max Planck in 1901 established the particle nature (quanta) of radiation through the interpretation of photo-electric effect. The wave character of radiation quanta was established by de Broglie in 1924. Though radiation possesses dual character, viz., *particle* and *wave*, they are *complementary* and *mutually exclusive*. This means the same experiment cannot reveal the *particle interference, diffraction* and polarisation experiment, only wave nature is revealed and no particle nature.

Similarly, in *photoelectric effect, Compton effect,* we observe the particle nature and no wave nature.

This mutual exclusiveness of particle and wave nature of radiation is called *Bohr's principle of complementarity.*

2.18 Double slit experiment with electrons: Nature of matter waves

The diffraction experiments with electrons show that a beam of electrons behaves very much like a beam of photons.

We consider the double slit experiment with electrons similar to Young's experiment in optics [Fig 2.12(a)]. A parallel of mono-energetic electrons be incident on two parallel slits each of width a and b, and separated by b, where a and b have the dimensions of de Broglie wavelength of the electrons, $\lambda = h/p$.

When one slit is shut off, a single slit diffraction pattern as shown in Fig 2.12(b) will be observed. If both the slits are open, the pattern having the appearance shown in Fig 2.12(c) will be observed. The interference effect between the two-slits shows up by the appearance of the interference fingers within the envelope of the single slit diffraction pattern.

When the intensity of the electron beam is gradually reduced, the same pattern is observed after a sufficiently long time. If the intensity is such that essentially a single electron falls on the slit plate at a time, the same intensity pattern is observed if a sufficiently large number of electrons falls on the slits over a long period of time. It, therefore, appears that each electron interacts with both slits at a time, through according to the classical view a single electron would be expected to pass through only one of the slits at a time. Thus, if the electron is regarded as a particle, its motion at one of the slits is affected by the conditions at the other slit. This violates the *condition of locations* (valid strictly in classical physics) according to which the behaviour of a particle at a point should be influenced by the condition (e.g., fields) at that point only.

The above argument also valid in case of two-slit interference pattern produced by light when considered from the photon point of view, i.e., each photon would appear to interact with both the slits simultaneously.

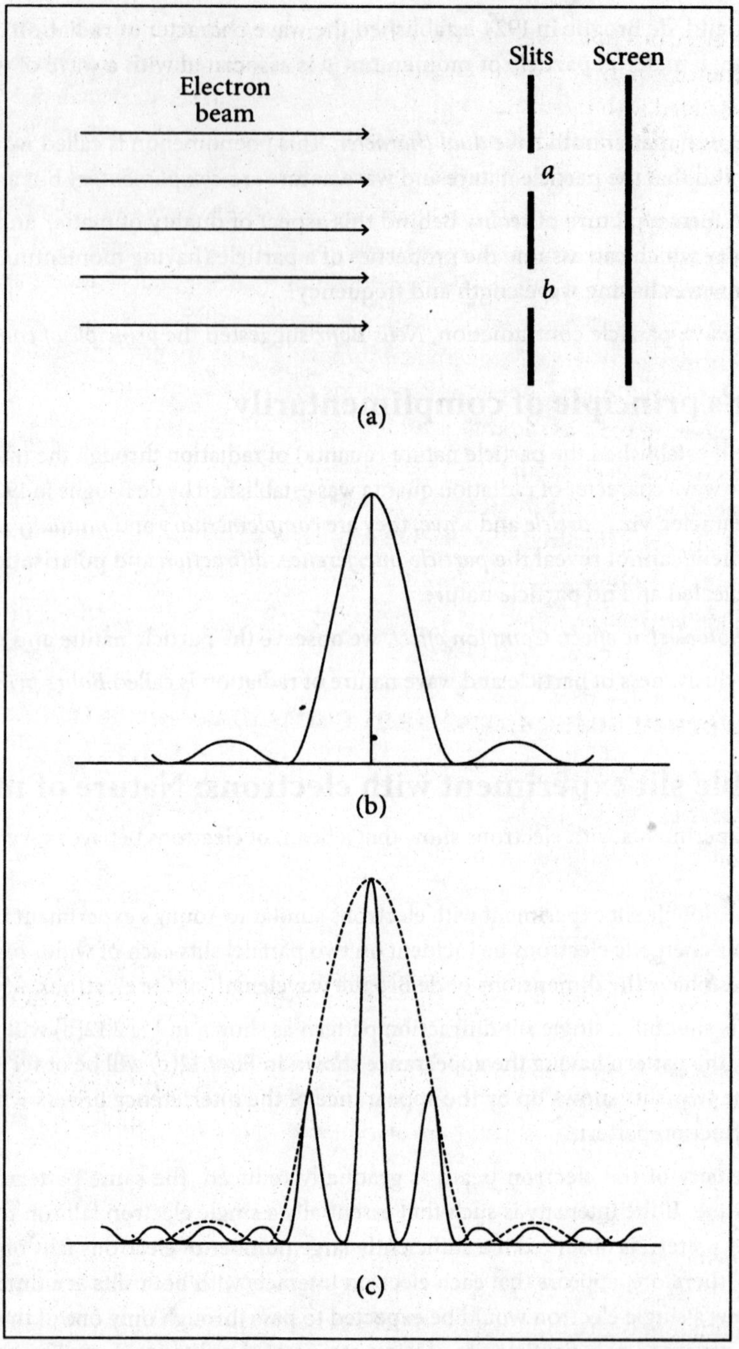

Fig 2.12: Double slit experiment with electrons: (a) Experimental arrangement, (b) Diffraction pattern with one slit closed, and (c) Diffraction pattern with both slits open.

But, neither the electrons nor the photons can be split into smaller fragments. Hence, the results of the double slit interference experiments with electrons or photons can only be interpreted by assuming wave characteristics associated with them. However, these waves have no classical analogue. Instead, we have to interpret *the modulus squared of the wave amplitude as the probability density and its integral over a finite region of space as the probability of finding the particle in that region.*

The amplitude squared of a classical wave, like the e.m. wave is a measure of a physically measurable quantity like the energy density or the charge density. For example, if it measured the charge density, then for a single incident electron in a diffraction experiment, the corresponding wave packet is so split by the diffracting slits that at the position of the maxima, the detectors detect a fraction of the (amplitude) each. On the average this may be true, if a large number of incident electrons are considered, but it is certainly not true if a single electron is incident on the diffracting plate at a time. This will be detected as a whole by either one or the other detector as the charge on the individual electrons does not split.

In quantum mechanics, even though we regarded the incident electron wave as divided into two parts in diffraction experiment, it is the intensity of the wave in a given direction which is proportional to the absolute square of the wave amplitude. So the intensity has a probabilistic interpretation. A quantity which depends quadricaly on the amplitude always represents the probability of a physical event to happen.

Hence, *the probability interpretation is a distinguishing feature in quantum mechanics in contrast to classical wave theory.*

2.19 Answers of some important conceptual questions

2.1. *What do you mean by wave-particle duality?*

Ans. See Art. 2.16.

2.2. *What is Bohr's principle of complementarity?*

Ans. See Art. 2.17.

2.3. *Justify that a pure sine wave is unable to represent a moving particle.*

Ans. If possible, let a moving particle be represented by a pure sine wave as

$$\psi(x,t) = A\sin(kx - \omega t).$$

The wave extends from $-\infty$ to $+\infty$. It is characterised by wavelength λ and momentum $p = h/\lambda$ and is completely *non-localised*. Since, a material particle (like electron) is completely localised, therefore, *a monochromatic sine wave cannot represent a moving particle*. It implies that de Broglie's waves are not *monochromatic*.

2.4. *A moving particle (like electron, etc.) can be represented by a wave packet.—Explain.*

Ans. A moving particle (like electron) is associated with *de Broglie wave* which cannot be a monochromatic one. A wave packet formed due to the superposition of several monochromatic waves with the same amplitude and phase but slightly differing in frequency can, however represent a particle. This is because the wave packet so formed is localised in a small region of space like the particle. The particle velocity (v) is equal to the group velocity (v_g) of the *wave packet* and as such a moving particle can be represented by a *wave packet*.

2.5. *Show that de Broglie wave packets are only theoretical concepts which are employed to explain the experimental findings.*

Ans. We know, that the wave representing a particle (like electron) cannot be a *monochromatic wave* instead it is a *wave packet* which is localised in a small region in space. Such wave packets are capable to represent a particle. The separation between two successive amplitudes in a wave packet λ_0 is characteristic of wave packet and has particle like location in space. But, a particle such as an electron or proton retains its particle nature like mass and charge which are *never divided in any diffraction (viz., double slit diffraction experiment with electron beam) experiment.* Because, if the intensity of the incident electron beam in a double slit be so reduced that the beam contains only one electron, then also the nature of the diffraction patterns remains unaltered. A single electron must fall in any one of the slits, it cannot fall on both the slits simultaneously as its mass and charge cannot be divided, i.e., the electron retains its particle nature even though the diffraction pattern remains the same. For this reason a particle in motion cannot be supposed to be just a wave packet.

2.6. *Why do we not observe de Broglie wave character of a cricket ball of mass* 0.5 kg *flying with velocity* 13.9 m/s.

Ans. The de Broglie wavelength of a particle of momentum $p = mv$ is

$$\lambda = \frac{h}{p} = \frac{h}{mv}$$

where m and v are respectively the mass and velocity of the particle and h is the Planck's constant.

For the cricket ball, $m = 0.5$ kg, $v = 13.9$ m/s and assuming $h = 6.62 \times 10^{-34}$ J-s, the de Broglie wavelength of the cricket ball is

$$\lambda = \frac{6.62 \times 10^{-34}}{0.5 \times 13.9} = 10^{-34} \text{ m} = 10^{-23} \text{ Å}$$

which is insignificant compared to the size of the cricket ball.

2.7. *de Broglie waves undergo diffraction like X-rays,—why?*

Ans. An essential property of wave is that it can undergo *diffraction,* for which the essential condition is that the wavelength of the wave must be of the order of grating element.

The wavelength of de Broglie wave associated with an electron accelerated through a PD of 100 V in vacuum is $\lambda = 1.227$ Å. This wavelength is comparable to the wavelength of X-rays which undergo diffraction through crystals. That is why de Broglie waves also undergo diffraction through crystals.

2.8. *An electron and a proton both are associated with de Broglie waves. Which one will have greater de Broglie wavelength if they move with the same velocity.*

Ans. The de Broglie wavelength of an electron of mass m_e moving with velocity $v_e (\ll c)$ is

$$\lambda_c = \frac{h}{m_e v_e}$$

Similarly, the de Broglie wavelength of a proton of mass m_p moving with velocity v_p is

$$\lambda_p = \frac{h}{m_p v_p}$$

$$\therefore \frac{\lambda_e}{\lambda_p} = \frac{m_p v_p}{m_e v_e} = \frac{m_p}{m_e} \quad [\because v_e = v_p]$$

Now, since $m_p > m_e$, $\frac{m_p}{m_e} > 1$.

$$\therefore \frac{\lambda_e}{\lambda_p} > 1 \text{ or, } \lambda_e > \lambda_p$$

That is *the de Broglie wavelength of the electron moving with same velocity is greater than that of the proton.*

2.9. *Show that a neutron beam can show diffraction interference phenomena like X-rays.*

Ans. The de Broglie wavelength associated with a thermal neutron at temperature T is given by

$$\lambda = \frac{h}{\sqrt{2mkT}}$$

where Planck's constant, $h = 6.63 \times 10^{-34}$ J-s, mass of the neutron, $m = 1.676 \times 10^{-27}$ kg, temperature, $T = 300$ K and Boltzmann constant, $k = 1.38 \times 10^{-23}$ J/K.

$$\therefore \lambda = \frac{6.63 \times 10^{-34}}{\sqrt{2 \times 1.676 \times 10^{-27} \times 1.38 \times 10^{-23} \times T}} = \frac{30.8 \times 10^{-10}}{\sqrt{T}} \text{ m or, } \lambda = \frac{30.8}{\sqrt{T}} \text{ Å}$$

For $T = 300$ K, $\lambda = \frac{30.8}{\sqrt{300}} = \frac{30.8}{17.32} = 1.77$ **Å.**

Hence, the de Broglie wavelength of a thermal neutron is of the order of interatomic distance in a crystal. This suggests that *the neutron beam can show diffraction and interference phenomena like X-rays of the same wavelength.*

2.10. *Show that de Broglie phase velocity exceeds the velocity of light.*

Ans. In a *non-relativistic case,* for a particle of mass m moving with a velocity v the total energy $E = mc^2$ and momentum $p = mv$.

$\therefore$ the phase velocity of the associated de Broglie wave is

$$V_p = \frac{E}{p} = \frac{mc^2}{mv} = \frac{c^2}{v}$$

As $v < c$, *the phase velocity of the associated wave v_p is always greater than c, the velocity of light in vacuum.*

In *relativistic case,* for the de Broglie waves associated with a moving particle of rest mass m_0 having momentum p the total energy is given by $E = \left[m_0^2c^4 + p^2c^2\right]^{1/2}$.

$$\therefore v_p = \frac{E}{p} = \left[\frac{m_0^2c^4 + p^2c^2}{p^2}\right]^{1/2} = c\left[\frac{m_0^2c^2}{p^2} + 1\right]^{1/2} = c\left[\frac{m_0^2c^2\lambda^2}{h^2} + 1\right]^{1/2}$$

$\because \frac{m_0^2c^2\lambda^2}{h^2}$ is always a positive quantity for non-zero values of m_0 or λ, *the phase velocity (v_p) of the associated wave is always greater than the velocity of light in vacuum.* If the velocity of the matter wave is equal to or greater than the velocity of light, then the associated wave with the particle will move faster than the particle itself, thus leaving the particle far behind.

Hence, *a moving material particle cannot be equivalent to a single wave train.* It is always equal to a group of waves and the particle velocity is always equal to that of a group of waves.

2.11. *de Broglie waves show dispersion in vacuum.—Why?*

Ans. The velocity of de Broglie wave depends upon wavelength. As a result, de Broglie waves of different wavelengths have different phase velocities. Thus, vacuum is a dispersive medium so far as de Broglie waves are concerned but is not a dispersive medium for electromagnetic waves as in their case the waves of all wavelengths travel with the same velocity in vacuum.

2.20 Solved problems

» ***Example* 2.1.** What voltage must be applied to an electron microscope to produce electrons of wavelength 0.40 Å?

Solution: We have the de Broglie wavelength of an electron of mass m, charge e and moving under a potential of V volts is $\lambda = \dfrac{h}{\sqrt{2meV}}$.

Here, $\lambda = 0.40$ Å $= 0.40 \times 10^{-10}$ m, $h = 6.62 \times 10^{34}$ J-s, $m = 9.1 \times 10^{-31}$ kg and $e = 1.60 \times 10^{-19}$ C.

$$\therefore \lambda = \frac{6.62 \times 10^{-34}}{\sqrt{2 \times 9.1 \times 10^{-31} \times 1.60 \times 10^{-19}}} \cdot \frac{1}{\sqrt{V}}$$

$$\text{or, } \sqrt{V} = \frac{6.62 \times 10^{-34}}{0.4 \times 10^{-10} \times \sqrt{2 \times 9.1 \times 10^{-31} \times 1.6 \times 10^{-19}}} = 30.6$$

$\therefore$ V **= 936.3 volts.**

» ***Example* 2.2.** What is the de Broglie wavelength of an electron moving with velocity, $v = \frac{3}{5}c$?

Solution: The de Broglie wavelength of a particle of mass m moving with velocity v is $\lambda = h/mv$.

Here, $v/c = 3/5$, so the electron mass undergoes relativistic variation according to the relation

$$m = \frac{m_0}{\sqrt{1 - v^2/c^2}}, \; m_0 \text{ is the rest mass of the electron}$$

$$\therefore \lambda = \frac{h}{m_0 v}\sqrt{1 - v^2/c^2} = \frac{6.63 \times 10^{-34}}{9.11 \times 10^{-31} \times \frac{3}{5}c} \times \sqrt{1 - \frac{9}{25}}$$

$$= \frac{6.63 \times 10^{-34} \times 4}{9.11 \times 10^{-31} \times 3 \times 3 \times 10^{8}} = 0.323 \times 10^{-11} \text{ m} = \mathbf{0.0323\ Å}$$

» ***Example* 2.3.** You have 10 eV photon and 10 eV electron. Which one has shorter wavelength? Explain your result.

Solution: Energy of a 10 eV photon, $E = 10 \times 1.6 \times 10^{-19}$ J [$\because$ 1 eV $= 1.6 \times 10^{-19}$ J]

Again,

$$E = h\nu = \frac{hc}{\lambda}$$

$$\therefore \lambda = \frac{h}{p} = \frac{hc}{E} = \frac{6.62 \times 10^{-34} \times 3 \times 10^{8}}{10 \times 1.6 \times 10^{-19}} = 1240 \times 10^{-10} \text{ m} = 1240 \text{ Å}$$

Now, for an electron, $E = \frac{1}{2}mv^2$.

$\therefore mv = \sqrt{2mE}$ and associated de Broglie wavelength is

$$\lambda = \frac{h}{mv} = \frac{h}{\sqrt{2mE}} = \frac{6.62 \times 10^{-34}}{(2 \times 9.1 \times 10^{-31} \times 10 \times 1.6 \times 10^{-19})^{1/2}}$$

$$= 3.88 \times 10^{-10} \text{ m} = \mathbf{3.88\ Å}$$

The de Broglie wavelength λ of a particle is $\lambda \propto \frac{1}{\sqrt{mE}}$ and for a photon $\lambda \propto \frac{1}{E}$.

This shows that de Broglie wavelength of an electron is much smaller as compared to that of a photon for the same energy E.

» ***Example* 2.4.** What would be the wavelength of a quantum of radiant energy emitted if an electron is transmitted into radiation and went into one quantum?

Solution: The energy of a quantum is $E_1 = h\nu = \frac{hc}{\lambda}$.

The energy of an electron transmitted into radiation is $E_2 = mc^2$.

Here

$$E_1 = E_2 \text{ or, } \frac{hc}{\lambda} = mc^2 \text{ or, } \lambda = \frac{h}{mc} = \frac{6.62 \times 10^{-34}}{9.1 \times 10^{-31} \times 3 \times 10^8} = \mathbf{0.0244\ Å}$$

» ***Example* 2.5.** An electron and a proton have the same de Broglie wavelength. Prove that the energy of the electron is greater than that of the proton.

Solution: Let the particle of mass m moves with velocity v. Then the associated wavelength is $\lambda = h/mv$.

For the electron $\lambda_e = h/mv$ [m, v are the mass and velocity of the electron] and for the proton, $\lambda_p = h/MV$ [M, V are the mass and velocity of the proton].

But, $\lambda_e = \lambda_p$.

$$\therefore \frac{h}{mv} = \frac{h}{MV} \text{ or, } mv = MV$$

Now, energy of the electron is

$$E_e = \frac{1}{2}mv^2 = \frac{1}{2}MV \cdot v = \frac{1}{2}mV \cdot \frac{MV}{m} = \frac{1}{2}MV^2 \cdot \frac{M}{m} = \frac{E_p M}{m}$$

$$\therefore \frac{E_e}{E_p} = \frac{M}{m} > 1, \therefore \mathbf{E_e > E_p}$$

i.e., the energy of the electron is greater than that of the proton if both have the same de Broglie wavelength.

» ***Example* 2.6.** Find out the energy of a photon having wavelength 1 Å [$h = 6.62 \times 10^{-27}$ erg].

Solution: The de Broglie wavelength of the photon is given by

$$\lambda = \frac{h}{p} \text{ or the momentum } p = \frac{h}{\lambda} \text{ or, } p = \frac{6.62 \times 10^{-27}}{10^{-8}}$$

Now, energy of the photon is

$$E = pc \,[c = \text{velocity of light}]$$

$$\text{or, } E = \frac{6.62 \times 10^{-27}}{10^{-8}} \times 3 \times 10^{10} \text{ erg} = \mathbf{1.24\ eV}\ [\because 1 \text{ eV} = 1.6 \times 10^{-19} \text{ J}]$$

» ***Example* 2.7.** Find the wavelength of the de Broglie wave associated with an electron having energy 1 MeV.

Solution: We have the rest mass energy of the electron as $m_0c^2 = 5.12 \times 10^5$ eV $= 5.12 \times 10^5 \times 1.6 \times 10^{-19}$ J $= 1.6 \times 0.512 \times 10^{-13}$ J.

The kinetic energy of the electron is given by $E_k = 1$ MeV $= 10^6$ eV $= 10^6 \times 1.6 \times 10^{-19}$ J $= 1.6 \times 10^{-13}$ J.

Since, kinetic energy E_k is comparable to rest mass energy m_0c^2 the relativistic expression for λ is to be used.

$$\therefore \lambda = \frac{hc}{\sqrt{E_k(E_k + 2m_0c^2)}} = \frac{6.63 \times 10^{-34} \times 3 \times 10^8}{\sqrt{(1.6 \times 10^{-13})(1.6 \times 10^{-13} + 2 \times 1.6 \times 0.512 \times 10^{-13})}}$$

$$= \frac{3 \times 6.63 \times 10^{-26}}{1.6 \times 10^{-13}\sqrt{(1 + 1.024)}} = \frac{19.89 \times 10^{-13}}{2.276} = 8.739 \times 10^{-13}$$

$$= 0.008739 \times 10^{-10} \text{ m} = \mathbf{0.008739\ Å}$$

» ***Example* 2.8.** An α-particle is accelerated through a potential difference of 2000 volts. What is the wavelength of the associated de Broglie wave?

Given mass of a proton = 1.67×10^{-24} gm, Plank's constant, $h = 6.62 \times 10^{-27}$ erg-sec.

Solution: We have the de Broglie wavelength of the α-particle of mass m moving with velocity v is

$$\lambda = \frac{h}{mv} = \frac{h}{\sqrt{2mE}}$$

But, $E = \frac{2eV}{300}$, where V is in volt and $2e$ is the charge on the α-particle.

$$\therefore \lambda = \frac{h}{\sqrt{2m \times \frac{2eV}{300}}} = h\sqrt{\frac{75}{meV}}$$

Here, $m = 4 \times$ mass of a proton $= 4 \times 1.67 \times 10^{-24}$ gm, $e = 4.8 \times 10^{-10}$ esu and $h = 6.62 \times 10^{-27}$ erg-sec.

$$\therefore \lambda = 6.62 \times 10^{-27} \times \sqrt{\frac{75}{4 \times 1.67 \times 10^{-24} \times 4.8 \times 10^{-10} \times 2000}}$$

$$= 2.3 \times 10^{-11} \text{ cm} = \mathbf{2.3 \times 10^{-3}\ Å}$$

» ***Example* 2.9.** Calculate the de Broglie wavelength associated with a 50 eV electron.

Solution: We have $\lambda = \frac{h}{\sqrt{2mE}}$.

Here, Planck constant, $h = 6.6 \times 10^{-27}$ erg-sec, mass of the electron, $m = 9.1 \times 10^{28}$ gm and KE, E = 50 eV = $50 \times 1.6 \times 10^{-12}$ erg.

$$\therefore \lambda = \frac{6.6 \times 10^{-27}}{\sqrt{2 \times 9.1 \times 10^{-28} \times (50 \times 1.6 \times 10^{-12})}} = \frac{6.6 \times 10^{-8}}{3.8} = 1.66 \times 10^{-8} \text{ cm} = \mathbf{1.66\ Å}$$

» ***Example* 2.10.** Energy of a particle at absolute temperature T is of the order of kT. Calculate the wavelength of thermal neutrons at 27 °C. [*Given*: the mass of the neutron = 1.67×10^{-24} gm. Planck's constant, h = 6.67×10^{-27} erg-sec and Boltzmann constant, $k = 8.6 \times 10^{-5}$ eV-K.]

Solution: We have de Broglie wavelength of the neutron is

$$\lambda = \frac{h}{\sqrt{2mE}}$$

where $E = kT$.

$$\therefore \lambda = \frac{h}{\sqrt{2mkT}}$$

Using $h = 6.6 \times 10^{27}$ erg-sec, $m = 1.67 \times 10^{-24}$ gm, $k = 8.6 \times 10^{-5}$ eV-K $= 8.6 \times 10^{-5} \times 1.6 \times 10^{-12}$ erg K $= 1.376 \times 10^{-16}$ erg-K and $T = 27 + 273 = 300$ K, we get

$$\lambda = \frac{6.6 \times 10^{-27}}{\sqrt{2 \times 1.67 \times 10^{-24} \times 1.376 \times 10^{-16} \times 300}} = 1.78 \times 10^{-8} \text{ cm} = \mathbf{1.78\ Å}$$

» ***Example* 2.11.** Find out the glancing angle at which electrons of 100 eV should fall on the lattice planes of a crystal to obtain a strong Bragg reflection in the first order [lattice spacing, $d = 2.15$ Å]. **[BU (H) 2003]**

Solution: The de Broglie wavelength of an electron accelerated through a potential of V volt is

$$\lambda = \frac{12.26}{\sqrt{V}} = \frac{12.26}{\sqrt{100}} = 1.226 \text{ Å}$$

Again, for the nth order Bragg diffraction maximum $2d \sin\theta = n\lambda$, θ is the glancing angle of incidence.

For 1st order $n = 1$.

$$\therefore 2d \sin\theta = \lambda \text{ or, } \sin\theta = \frac{\lambda}{2d} = \frac{1.226}{2 \times 2.15} = \mathbf{0.285}$$

$\therefore \theta = 16.5°$.

» ***Example* 2.12.** A beam of mono-energetic neutrons corresponding to 27 °C is allowed to fall on a crystal. A first order reflection is observed at a glancing angle 30°. Calculate the inter-planar spacing of the crystal.

Given Planck constant, $h = 6.62 \times 10^{-34}$ J-s, mass of neutron $m = 1.67 \times 10^{-27}$ kg, Boltzmann constant, $k = 1.38 \times 10^{-23}$ J/K.

Solution: From Bragg law of diffraction, we have $2d \sin\theta = n\lambda$, where the glancing angle, $\theta = 30°$, order no. of diffraction, $n = 1$.

$\therefore 2d \sin 30° = \lambda$ or, $2d \times \frac{1}{2} = \lambda$ or, $\lambda = d$.

Now, energy of the neutron, $E = kT$, where k is the Boltzmann constant and T is the absolute temperature.

Here, $T = 27\,°\text{C} = 300$ K, $k = 1.38 \times 10^{-23}$ J/K, $E = 1.38 \times 10^{-23} \times 300 = 4.14 \times 10^{-21}$ J.

Again, momentum, $p = \sqrt{2mE} = \sqrt{2 \times 1.67 \times 10^{-27} \times 4.14 \times 10^{-21}}$.

$$\therefore d = \lambda = \frac{h}{p} = \frac{6.62 \times 10^{-34}}{\sqrt{2 \times 1.67 \times 10^{-27} \times 4.14 \times 10^{-21}}} = 1.78 \times 10^{-10} \text{ m} = \mathbf{1.78\ Å}$$

» ***Example* 2.13.** A 0.083 eV neutron beam scatters from an unknown sample and a Bragg reflection peak is observed around 22°. What is the spacing of Bragg plane?

Solution: The de Broglie wavelength of neutron beam is given by

$$\lambda = \frac{h}{p} = \frac{h}{\sqrt{2m_0 E_k}} = \frac{hc}{\sqrt{2(m_0c^2)E_k}} = \frac{12.40 \times 10^3 \text{ eV Å}}{\sqrt{2(940 \times 10^6 \text{ eV})(0.083 \text{ eV})}} = 0.993 \text{ Å}$$

For the first order peak

$$d = \frac{\lambda}{2\sin\theta} = \frac{0.993\text{ Å}}{2\sin 22^\circ} = \mathbf{1.33\text{ Å}}$$

» ***Example* 2.14.** What is the de Broglie wavelength of electrons of energy 0.1 MeV using both relativistic and non-relativistic expressions for λ? What is the percentage error in using non-relativistic expression for λ?

***Solution*:** The non-relativistic de Broglie wavelength of an electron of mass m and energy eV is

$$\lambda_{nr} = \frac{h}{\sqrt{2meV}} = \frac{12.26}{\sqrt{V}} = \frac{12.26}{\sqrt{0.1\times10^6}} = 0.0388\text{ Å}$$

The relativistic expression for the de Broglie wavelength of the same electron is

$$\lambda_r = \frac{300ch}{e\sqrt{V(V+1.02\times10^6)}} = \frac{12.41\times10^3}{\sqrt{V(V+1.02\times10^6)}}$$

$$= \frac{12.41\times10^3}{\sqrt{0.1\times10^6(0.1+1.02)\times10^6}} = 0.0371\text{ Å}$$

Now, the % error in using non-relativistic expression is

$$\frac{\lambda_{nr}-\lambda_r}{\lambda_{nr}}\times100\,\% = \frac{(0.0388-0.0371)}{0.0388}\times100\,\% = \mathbf{4.38\,\%}$$

» ***Example* 2.15.** A crystalline material has a set of Bragg planes separated by 1.1 Å. For 2 eV neutrons what is the highest order Bragg reflection?

***Solution*:** The de Broglie wavelength of the neutron is given by

$$\lambda = \frac{h}{m_0 v} = \frac{h}{\sqrt{2m_0E_k}} = \frac{hc}{\sqrt{2(m_0c^2)E_k}} = \frac{12.40\times10^3\text{ eV Å}}{\sqrt{2(940\times10^6\text{ eV})2\text{ eV}}} = 0.202\text{ Å}$$

The maximum angle that can be attained is 90°. Then from Bragg's law

$$2d\sin\theta = n\lambda \text{ or, } 2(1.1)\sin 90^\circ = n(0.202) \text{ or, } n = 10.89$$

Since, n must be an integer, the highest order $\boldsymbol{n} = \mathbf{10}$.

» ***Example* 2.16.** Thermal neutrons incident on sodium chloride crystal (interatomic spacing 2.81 Å) undergo first order diffraction from the principal Bragg planes at an angle of 20°. Find out the energy of the thermal neutrons.

***Solution*:** The condition for the first order Bragg reflection

$$\lambda = 2d\sin\theta = 2(2.18)\sin 20^\circ = 1.922\text{ Å}$$

The associated de Broglie wavelength of neutron

$$\lambda = \frac{h}{p} = \frac{hc}{\sqrt{2(m_0c^2))E}} \text{ or, } 1.99 = \sqrt{\frac{12.40\times10^3}{2(940\times10^6)E}} \text{ or, } E = \mathbf{0.0221\text{ eV}}$$

» ***Example* 2.17.** Show that the de Broglie wavelength to the Compton wavelength of a particle is $\sqrt{\left(\frac{c}{v}\right)^2 - 1}$.

***Solution*:** We have the Compton wavelength of a particle is

$$\lambda_c = \frac{h}{m_0 c}$$

and the de Broglie wavelength of a particle is

$$\lambda_d = \frac{h}{mv}$$

where v = velocity, m_0 = rest mass of the particle and $m = \frac{m_0}{\sqrt{1-\frac{v^2}{c^2}}}$ = the mass of the particle when it is moving with velocity v.

$$\therefore \frac{\lambda_d}{\lambda_c} = \frac{h}{mv}\cdot\frac{m_0 c}{h} = \frac{m_0 e}{mV} = \frac{m_0 e}{v}\frac{\sqrt{1-\frac{v^2}{c^2}}}{m_0} \quad \text{or,} \quad \frac{\lambda_d}{\lambda_c} = \sqrt{\left(\frac{c}{v}\right)^2 - 1}$$

» ***Example* 2.18.** Calculate in MeV the energy of a proton having de Broglie wavelength of 1 fm = 10^{-15} m.

***Solution*:** The de Broglie wavelength

$$\lambda = \frac{hc}{\sqrt{T(T+2m_0c^2)}} = T^2 + 1876T - 1542564 = 0$$

when m_0c^2 = 938 MeV, hc = 1242 eV nm

$$\lambda = 1\text{ fm} = T = \mathbf{618\ MeV}$$

$$E = T + m_0c^2 = \mathbf{1556\ MeV}$$

» ***Example* 2.19.** Calculate the wavelength associated with proton moving with velocity 10^7 cms^{-1}? **[WBCS 1995]**

***Solution*:** Write here please.

» ***Example* 2.20.** Find the de Broglie wavelength of (i) 50 keV electrons and (ii) 50 keV protons. Compare the amount of diffraction of these particles in the first order by crystal with grating space of 9.5 Å. **[CU (H) 1999]**

***Solution*:**

(i) For 50 keV electrons. The de Broglie wavelength of electron is

$$\lambda_e = \frac{h}{p} = \frac{hc}{\sqrt{T(T+2m_0c^2)}} = \frac{1242\text{ eV nm}}{\sqrt{(50\text{ keV})\ (50\text{ keV} + 2\times 0.51\text{ MeV})}} \quad \text{[rest energy = 0.51 MeV]}$$

$$= \frac{1242\text{ nm}}{\sqrt{50\times10^3(50\times10^3 + 2\times0.51\times10^6)}} = 5.37\times10^{-3}\text{ nm} = 5.37\times10^{-3}\times10^{-9}\text{ m}$$

or, $\lambda_c = 5.37\times10^{-12}$ m

Now, by Bragg's law, $2d\sin\theta_e = n\lambda_e$, θ_e is the Bragg angle of diffraction for electron.

For d = 9.5 Å = 9.5×10^{-10} m, n = 1

$$\sin\theta_e = \frac{n\lambda_e}{2d} = \frac{1\times(5.37\times10^{-12}\text{ n})}{2\times(9.5\times10^{-10}\text{ m})} \quad \text{i.e., } \boldsymbol{\theta_e \simeq 0.16°}$$

(ii) For 50 keV proton, the de Broglie wavelength of proton is

$$\lambda_d = \frac{h}{p} = \frac{hc}{\sqrt{T(T+2m_0c^2)}} = \frac{1242 \text{ eV nm}}{\sqrt{50 \text{ keV } (50 \text{ keV} + 2 \times 938 \text{ MeV})}} \quad \text{[rest every = 938 MeV]}$$

$$= \frac{1242 \text{ nm}}{\sqrt{50(50 * 10^3 + 2 \times 938 \times 10^6)}} = 0.1282 \times 10^{-3} \text{ nm} = 0.1282 \times 10^{-3} \times 10^{-9}$$

$$\therefore \lambda_p = 0.1282 \times 10^{-12} \text{ m}$$

By Bragg's law, $2d \sin\theta_p = n\lambda_p$.

For $d = 0.4$ Å $= 0.5 \times 10^{-10}$ m, $n = 1$.

$$\therefore \sin\theta_p = \frac{n\lambda_p}{2d} = \frac{1 \times (0.1282 \times 10^{-12} \text{ m})}{2 \times (9.5 \times 10^{-10} \text{ m})} \quad \text{i.e., } \theta_p = \mathbf{0.004}$$

θ_p is the Bragg angle of diffraction for proton.

Now the amount of diffraction of electrons and protons in the first order by a crystal of grating space of 9.5 Å in a direction θ is given by $dQ \propto d \sin\theta$.

$\therefore$ for electrons $(dQ)_e \propto d \sin\theta_e$ and for proton $(dQ)_p \propto d \sin\theta_p$.

$\therefore (dQ)_e/(dQ)_p = \sin\theta_e/\sin\theta_p \because \theta_e > \theta_p$, $(dQ)_e > (dQ)_p$, i.e., the amount of diffraction of electrons will be more than the amount of diffraction of protons in the same direction.

» ***Example* 2.21. Photoelectrons are liberated by ultraviolet light of wavelength 3000 Å from a metallic surface for which the photoelectric threshold is 4000 Å. Calculate the de Broglie wavelength of electrons with maximum KE.**

***Solution*: The maximum KE of the electron**

$$T_{max} = h(\nu - \nu_0) = h\left(\frac{c}{\lambda} - \frac{c}{\lambda_0}\right) = hc\left(\frac{1}{\lambda} - \frac{1}{\lambda_0}\right) = (1242 \text{ eV nm})\left(\frac{1}{3000 \text{ Å}} - \frac{1}{4000 \text{ Å}}\right)$$

$$= \frac{1242 \times 10^{-9} \text{ m}}{10^{-10} \text{ m}}\left(\frac{1}{3000} - \frac{1}{4000}\right) = 1.035 \text{ eV}$$

The de Broglie wavelength of these photoelectrons emitted with maximum KE is

$$\lambda = \frac{h}{\sqrt{2mT_{max}}} = \frac{6.626 \times 10^{-34} \text{ J-s}}{2 \times (9.1 \times 10^{-31} \text{ kg}) \times (1.035 \times 1.6 \times 10^{-19} \text{ J})}$$

$$= 1.21 \times 10^{-9} \text{ m or, } \lambda = \mathbf{1.21 \text{ m}}$$

» ***Example* 2.22. An electron has a de Broglie wavelength of 0.15 Å. Compute the phase and group velocities of the de Broglie wave. Find the KE of the electron.** **[CU (H) 1995, 2015]**

***Solution*: Momentum of the proton**

$$p = \frac{h}{\lambda} = \frac{6.626 \text{ J-s}}{0.15 \times 10^{-10} \text{ m}} = 4.4 \times 10^{23} \text{ kg m s}^{-1}$$

Again,

$$p = mv = \frac{m_0}{\sqrt{1-\frac{v^2}{c^2}}}v \text{ or, } p = \frac{m_0^2 v^2}{1-\frac{v^2}{c^2}}$$

$$\text{i.e., } p^2 - p^2\frac{v^2}{c^2} = m_0^2 v^2 \text{ or, } v^2\left(m_0^2 + \frac{p^2}{c^2}\right) \text{ or, } p^2 = v = \frac{p}{\sqrt{m_0^2 + \frac{p^2}{c^2}}}, \text{ particle velocity}$$

$$\text{or, } v = \frac{4.4\times10^{-23}\,\text{kg m s}^{-1}}{\sqrt{(9.1\times10^{-31}\,\text{kg})^2 + \left(\frac{4.4\times10^{-23}\,\text{kg m s}^{-1}}{3\times10^8\,\text{ms}^{-1}}\right)^2}} = 4.77\times10^7\,\text{ms}^{-1} = \frac{4.77\times10^7}{3\times10^8}$$

or, $v = \mathbf{0.159c}$

Again, $v_g = v = 0.159c$.

$$v_{\text{ph}} = \frac{c^2}{v} = \frac{c^2}{0.159c} = \frac{c}{0.159} \text{ or, } v_{\text{ph}} = \mathbf{6.29 > c}$$

$$\therefore T = (m - m_0)c^2 = \left(\frac{m_0}{\sqrt{1-\frac{v^2}{c^2}}} - m_0\right)c^2 = m_0c^2\left[\frac{1}{\sqrt{1-\frac{v^2}{c^2}}} - 1\right]$$

$$= (0.51\times10^6\,\text{eV})\left[\frac{1}{\sqrt{\left(1-\frac{0.159c}{c}\right)^2}} - 1\right] = 6.57\times10\,\text{eV}.$$

$\therefore T = \mathbf{6.57\ keV}$

» ***Example* 2.23.** **Calculate the de Broglie wavelength of electrons of energy 10^4 eV and compare it with the wavelength of electromagnetic radiation for which the photon has the same energy.** **[CU (H) 1997]**

***Solution*: The de Broglie wavelength of electron is**

$$\lambda_e = \frac{hc}{\sqrt{T(T+2m_0c^2)}} = \frac{1242\,\text{eV nm}}{\sqrt{10^4\,\text{eV}(10^4\,\text{eV} + 2\times0.51\times10^6\,\text{eV})}}$$

$$= \frac{1242\,\text{nm}}{\sqrt{10^4(10^4 + 2\times0.51\times10^6)}} = 0.0122\,\text{nm}$$

The de Broglie wavelength of photon is

$$\lambda_{\text{ph}} = \frac{h}{p} = \frac{hc}{pc} = \frac{hc}{E} = \frac{1242\,\text{eV nm}}{10^4\,\text{eV}} = 0.1242\,\text{nm}$$

$$\therefore \frac{\lambda_e}{\lambda_{\text{ph}}} = \frac{0.0122\,\text{nm}}{0.1242\,\text{nm}} \simeq 0.1 \text{ or, } \frac{\lambda_e}{\lambda_{\text{ph}}} = \frac{1}{10}$$

» ***Example* 2.24.** **Electron of KE 100 eV are observed to be scattered from a crystal in Davision-Germer's diffraction experiment. The first maximum occurs at 10°. Obtain the lattice constant and number of peaks observed.**

***Solution*: The de Broglie wavelength of electrons of KE 100 eV is given by**

$$\lambda = \frac{h}{\sqrt{2mT}} = \frac{6.626\times10^{-34}\,\text{J-s}}{\sqrt{2\times(9.1\times10^{-31}\,\text{kg})\times(100\times1.6\times10^{-19}\,\text{J})}} = 1.228\times10^{-10}\,\text{m} = 1.228\,\text{Å}$$

If Bragg's condition $2d \sin\theta = n\lambda$ is obeyed, they produce constructive interference.

For the first maximum, $n = 1$

$$\therefore 2d \sin\theta = \lambda \text{ or, } d = \frac{\lambda}{2\sin\theta}$$

From Bragg's law

$$\sin\theta = \frac{n\lambda}{2d} \leqslant 1 \text{ or, } \frac{2d}{\lambda} = \frac{2 \times 3.54\ \text{Å}}{1.228\ \text{Å}}$$

$\therefore n \leqslant 5.77.$

Hence, the number of peaks is 5.

EXERCISES

2(A): Objective (or MCQ) Type Questions

Select the correct answer:

2.1. The de Broglie matter wave is a

(a) monochromatic wave (c) standing wave
(b) polychromatic wave (d) wave packet.

2.2. de Broglie proposed the hypothesis of matter wave by examining the parallelism of

(a) light and sound (c) electricity and magnetism
(b) heat and motion (d) mechanics and optics.

2.3. The de Broglie wavelength is associated with moving

(a) charged particles (c) sub-atomic particles
(b) macroscopic objects (d) electrically neutral particles.

2.4. The de Broglie wavelength of a particle of mass m moving with velocity v is

(a) $\lambda = \frac{m}{hv}$ (c) $\lambda = \frac{mv}{h}$
(b) $\lambda = \frac{h}{mv}$ (d) $\lambda = \frac{v}{mh}$.

2.5. Phase velocity and group velocity are equal when the medium is

(a) dispersive (c) isotropic
(b) non-dispersive (d) gas.

2.6. The relation between phase velocity and group velocity is

(a) $u = \frac{v_g}{c^2}$ (c) $u = c^2 v_g$
(b) $u = \frac{c^2}{v_g}$ (d) $u = c^2 + v_g$.

2.7. Which statement is correct?

(a) group velocity (v_g) = phase velocity (v_p)

(b) $v_g > v_p$

(c) $v_p > v_g$

(d) $v_p = \frac{1}{v_g}$.

2.8. Which statement is correct?

(a) phase velocity (v_p) of light wave is independent of λ in vacuum

(b) phase velocity of matter wave depends on λ in vacuum

(c) phase velocity of matter wave is independent on λ in vacuum

(d) phase velocity of light wave is dependent on λ in vacuum.

2.9. Davisson and Germer experiment is related to

(a) interference

(b) diffraction

(c) reflection

(d) polarisation.

2.10. Matter wave associated with a moving particle is

(a) a finite monochromatic wave train having velocity less than that of light

(b) an infinite monochromatic wave train having a phase velocity greater than that of light

(c) a wave packet having group velocity equal to that of the moving particle

(d) a wave packet having phase velocity equal to the group velocity.

2.11. Davisson and Germer experiment suggests that the electron is

(a) a particle

(b) a wave

(c) partly wave and partly particle

(d) none of them.

2.12. In Davisson and Germer experiment with electron beam and nickel crystal, the maximum diffraction appears at

(a) 45° with 50 volts

(b) 50° with 54 volts

(c) 54° with 50 volts

(d) 60° with 75 volts.

2.13. Davisson and Germer experiment proved that if λ be the de Broglie wavelength and V the accelerating voltage, then

(a) $\lambda \propto \sqrt{V}$

(b) $\lambda \propto \frac{1}{\sqrt{V}}$

(c) $\lambda \propto V$

(d) $\lambda \propto \frac{1}{V}$.

2.14. The series of concentric rings in G.P. Thomson experiment are similar to

(a) Newton's rings

(b) Debye-Scherrer rings

(c) interference rings

(d) diffraction rings.

2.15. If λ_r and λ_{nr} be the relativistic and non-relativistic wavelengths of the electron, then

(a) $\lambda_r > \lambda_{nr}$

(b) $\lambda_{nr} > \lambda_r$

(c) $\lambda_{nr} = \lambda_r$

(d) $\lambda_{nr} = \frac{1}{\lambda_r}$.

2.16. de Broglie wavelength of a body of mass m and kinetic energy E is

(a) $\frac{2mh}{\sqrt{E}}$ (c) $\frac{h}{2mp}$

(b) $\frac{h}{\sqrt{2mp}}$ (d) $\frac{\sqrt{2mE}}{E}$. **[WBUT 2005]**

2.17. An α-particle is 4 times heavier than a proton. If a proton and an α-particle are moving with the same velocity, how are their de Broglie wavelengths compared?

(a) $\lambda_p = \lambda_\alpha$ (c) $\lambda_p = \lambda_{\alpha/2}$

(b) $\lambda_p = 4\lambda_\alpha$ (d) $\lambda_p = \lambda_{\alpha/4}$. **[WBUT 2006]**

2.18. Consider the following statements about the de Broglie waves:

(i) The group velocity of these waves is equal to $\frac{dw}{dk}$.

(ii) The phase velocity of these waves is equal to $\frac{c^2}{v}$, where v is the velocity of the particle and c is the velocity of light.

(iii) Which of the statements given above is/are correct?

The face velocity of these waves is c^2/v, where c is the velocity of light and v is the velocity of the particle. Which of the statements is/are correct?

(a) 1 and 2 (c) 1 and 3

(b) 2 and 3 (d) 1, 2 and 3. **[IAS(Pre) 2004]**

2.19. The correct expression for the de Broglie wavelength of a particle (T is the KE) is

(a) $\lambda = \frac{hc}{\sqrt{T+2m_0c^2}}$ (c) $\lambda = \frac{hc}{\sqrt{T(T+2m_0c^2)}}$

(b) $\lambda = \frac{hc}{\sqrt{T(m_0c^2)}}$ (d) $\lambda = \frac{hc}{\sqrt{T+m_0c^2}}$. **[cf. IAS(Pre) 2004]**

2.20. The de Broglie wavelengths of a proton and an α-particle and equal. The ratio of their velocities is

(a) 1 : 3 (c) 1 : 2

(b) 4 : 1 (d) 1 : 6. **[IAS(Pre) 2003]**

2.21. The relationship between the angular frequency is and the wave number k (assuming de Broglie's hypothesis) of a particle associated with it is $\omega = ck^2$, where c is constant. The group velocity of the particle wave is

(a) $\frac{1}{2}ck$ (c) $2ck$

(b) $1ck$ (d) $4ck$. **[IAS(Pre) 2002]**

2.22. If a proton and an electron have same de Broglie wavelength then

(a) Both have same KE (c) Electron has more KE then proton

(b) Proton has more KE than electron (d) Both have same velocity. **[IAS(Pre) 2001]**

2.23. With symbols having the usual meaning, the wavelength of electron E is given by

(a) $2mhE$ (c) $\frac{h}{\sqrt{2mE}}$

(b) $\frac{2h}{mE}$ (d) $\frac{2\sqrt{2mE}}{\lambda}$. **[IAS(Pre) 2000]**

2.24. The de Broglie wavelength of a material particles which are in thermal equilibrium at temperature T is

(a) $\frac{h}{\sqrt{2mkT}}$ (c) $\frac{h}{\sqrt{mkT}}$

(b) $\frac{h}{2mkT}$ (d) $\frac{h}{\sqrt{2kT}}$. **[cf. IAS(Pre) 1999]**

2.25. The phase velocity (v_{ph}) and the group velocity (v_g) of a de Broglie wave in free space is (speed of light is c) are related as

(a) $\frac{v_{ph}}{v_g} = \sqrt{2}$ (c) $\frac{v_{ph}}{v_g} = 0.5$

(b) $v_{ph} v_g = c^2$ (d) $v_{ph} v_g = c\sqrt{2}$.

2.26. Which one of the following pairs of phenomena illustrates the aspect of wave particle duality?

(a) Compton effect and Bragg's law (d) Bragg's law and photo electric effect.

(b) Photo electric effect and Compton effect **[IAS(Pre) 1998]**

(c) Compton effect and Pauli's principle

2.27. An electron (mass = m_0) and a proton (mass = 1836 m) are moving with the same speed. The ratio of the de Broglie wavelength $\frac{\lambda_{electron}}{\lambda}$ will be

(a) 1 (c) $\frac{1}{1836}$

(b) 1836 (d) 918. **[IAS(Pre) 1997]**

2.28. The wavelength of 10 keV electron beam is 0.1227 Å. When these waves are diffracted from a metal foil having d = 0.55 Å, the first maxima occurs at an angle θ where θ is

(a) 0.1116 (c) 0.4464

(b) 0.2232 (d) 0.8928. **[IAS(Pre) 1996]**

2.29. The magnitude of de Broglie wavelength (λ) of electron proton (p), neutron (n) and α-particle are (α) all have the same energy of 1 MeV in the increasing order will follow the sequence.

(a) $\lambda_e, \lambda_p, \lambda_n, \lambda_\alpha$ (c) $\lambda_\alpha, \lambda_n, \lambda_p, \lambda_e$

(b) $\lambda_e, \lambda_n, \lambda_p, \lambda_\alpha$ (d) $\lambda_p, \lambda_e, \lambda_\alpha, \lambda_n$. **[IAS(Pre) 1995]**

2.30. The value of wavelength for a beam of 150 volt electron is

(a) 0.5 Å (c) 1.5 Å

(b) 1.0 Å (d) 15.0 Å. **[IAS(Pre) 1994]**

2.31. The wavelength of the yellow spectral emission of sodium is 5890 Å. At what KE would an electron have the same de Broglie wavelength?

(a) 2.4×10^{-17} J (c) 4.4×10^{-6} eV
(b) 2.4×10^{-17} eV (d) 4.4×10^{-6} J. **[cf. WBCS 2006]**

2.32. Through electrons and protons are emitted during β^- decay of nucleus. The cannot stay within the nucleus. This can be shown using.
(a) Pauli's exclusion principle (c) Heisenberg's uncertainty principle
(b) Born's interpretation (d) de Broglie hypothesis.

2.33. Which of the following statements are correct?
(i) Matter waves are electromagnetic waves
(ii) Matter waves are associated with particles which are microscopic
(iii) Matter waves are associated with all particles
(iv) A charged material particle gives rise to electromagnetic waves.
(a) 1 and 3 (c) 1 and 2
(b) 2 and 4 (d) 3 and 4.

2.34. Matter waves are
(a) electromagnetic waves (c) waves produced in a material medium
(b) longitudinal waves (d) neither longitudinal nor transverse waves.

2.35. A particle of mass m has KE T and associated wavelength λ. If KE is $R/4$, its de Broglie wavelength will be
(a) same (c) doubled
(b) halved (d) 4λ.

ANSWERS

2.1. (d).	**2.6.** (b).	**2.11.** (b).	**2.16.** (b).	**2.21.** (c).	**2.26.** (b).	**2.31.** (c).
2.2. (d).	**2.7.** (c).	**2.12.** (b).	**2.17.** (b).	**2.22.** (c).	**2.27.** (b).	**2.32.** (c).
2.3. (c).	**2.8.** (b).	**2.13.** (b).	**2.18.** (d).	**2.23.** (c).	**2.28.** (a).	**2.33.** (d).
2.4. (b).	**2.9.** (b).	**2.14.** (b).	**2.19.** (c).	**2.24.** (a).	**2.29.** (c).	**2.34.** (d).
2.5. (b).	**2.10.** (c).	**2.15.** (b).	**2.20.** (b).	**2.25.** (b).	**2.30.** (b).	**2.35.** (c).

2(B): Short Answer Type Questions

2.1. What is wave-particle duality?

2.2. What are matter waves? What is the wavelength of a particle of mass m moving with velocity v? **[WBUT 2002]**

2.3. Find an expression for de Broglie wavelength for a photon.

2.4. Show that the nth Bohr orbit is associated with n de Broglie waves where n is an integer.

2.5. Obtain Bohr's quantisation of angular momentum from de Broglie's hypothesis.

2.6. Explain that a monochromatic wave cannot represent a particle.

2.7. Discuss the formation of a wave packet.

2.8. What do you mean by phase velocity and group velocity? Which is greater?

2.9. Show that the de Broglie wavelength λ and the accelerating potential of an electron are related by $\lambda \propto 1/\sqrt{V}$.

2.10. Why do we not observe wave nature of household objects? [BU (H) 1993]

2.11. Show that the de Broglie wavelength λ associated with an electron of mass m and kinetic energy E is given by $\lambda = \frac{h}{\sqrt{2mE}}$.

2.12. Explain the principle of complementarity.

2.13. Justify that the experimental arrangement for the diffraction of X-rays can be employed for the diffraction of neutrons.

2.14. Show that the phase velocity of de Broglie waves depends on the wavelength even in vacuum.

2.15. Show that in a non-dispersive medium, group velocity is equal to wave velocity.

2.16. Show that the de Broglie wavelength λ of electrons of energy $E = eV$ is given by $\lambda = \frac{h}{\sqrt{2meV}}$. [WBUT 2005]

2.17. (a) What is the experimental evidence in favour of de Broglie's hypothesis of matter waves? [WBUT 2005]

(b) Why in case of moving electrons quantum mechanics is used, while for moving cars, we use Newtonian mechanics? Explain. [WBUT 2005]

2.18. Why unmodified lines appear in Compton scattering? [WBUT 2003]

2.19. Show that the circumference of the Bohr orbits are integral multiples of the de Broglie wavelength of the electron. [BU (H) 2009]

2.20. Analyse the intensity pattern that would be obtained on the screen in Young's double slit experiment with an electron beam. [BU (H) 2011]

2.21. Explain the existence of station any orbits in Bohr's theory on the basis of matter waves. [KU (H) 2014]

2.22. Show that xp_x, Et and $\hbar$ have the same dimension.

[*Hints:* $[xp_x] = [xmv_x] = [LM\frac{L}{T}] = [ML^2T^{-1}]$.

$[Et]$ = [force · distance ·t] $= [MLT^{-2} \cdot L \cdot T] = [ML^2T^{-1}]$.

$[\hbar] = \left[\frac{h}{2\pi}\right] = \left[\frac{E}{\nu}\right] = \left[\text{force} \cdot \text{distance} \cdot \frac{1}{\nu}\right] = [MLT^{-2} \cdot L \cdot T] = [ML^2T^{-1}]$.]

2(C): Broad Answer Type Questions

2.1. Discuss the basic results that led to the formulation of the wave concept of matter. What are de Broglie waves? Discuss their wavelengths and explain how they afford a correct description of the motion of a particle.

2.2. Derive a formula for the de Broglie wavelength of a particle in terms of kinetic energy E and the rest mass energy m_0c^2. [CU (H) 1992]

2.3. (a) State de Broglie's postulate on the wave-like property of particles. What is the energy of neutrons suitable for crystallographic studies?

(b) Briefly describe an experiment which demonstrates the wave-like property of matter. [BU (H) 2000]

2.4. (a) Show that the circumference of a circular Bohr orbit accommodates an integer number of de Broglie wavelengths of the orbiting electron. [BU (H) 2001]

(b) What is the de Broglie wavelength of a non-relativistic particle of mass m moving with kinetic energy W? [BU (H) 2003]

2.5. (a) Show that the wavelength λ associated with an electron of mass m and kinetic energy E is

$$\lambda = \frac{h}{\sqrt{2mE}}$$

h is the Planck's constant.

(b) Describe the Davisson and Germer experiment giving emphasis on the interpretation of the result of the experiment. [BU (H) 1998]

2.6. (a) Show that the de Broglie wavelength of a particle of mass m and kinetic energy E is given by

$$\lambda = \frac{hc}{\sqrt{E(E+2m_0c^2)}}$$

(b) Derive relativistically correct formula that gives the de Broglie wavelength of a charged particle in terms of the potential difference V through which it has been accelerated. What is the non-relativistic approximation of this formula valid for $eV \ll m_0c^2$?

2.7. Demonstrate that the group velocity of a wave packet is equal to the velocity of a freely moving particle. Consider the relativistic and non-relativistic cases.

2.8. Deduce the relation between the phase velocity and group velocity for the de Broglie waves. Which of these two is associated with the particle velocity?

In what way does the behaviour of the phase velocity of the de Broglie waves differ from that of the light wave?

2.9. Describe G.P. Thomson's experiment that demonstrates the wave-like behaviour of electrons.

Discuss how the dual nature of particles can be reconciled on the basis of de Broglie's theory. What are the practical applications of wave nature of electrons?

2.10. (a) Find out the ratio of Compton wavelength and de Broglie's wavelength.

(b) Describe two experiments—one to illustrate the wave nature of matter and the other to illustrate the particle nature of light.

(c) Show that the de Broglie waves show dispersion in vacuum.

2.11. Discuss briefly the experiment of Davisson and Germer pointing out clearly how it demonstrates the existence of de Broglie matter waves. What is the wave-particle duality? [BU (H) 2006]

2.12. (a) State and explain de Broglie's hypothesis.

(b) Show that the de Broglie wavelength λ of electrons of charge e and energy E (in eV) is given by $\lambda = \frac{h}{[\sqrt{2meV}]}$. [WBUT 2005]

2.13. Starting from de Broglie's hypothesis show that the group velocity associated with a particle is the same as the particle velocity. **[WBUT 2007]**

2.14. Describe briefly the Davisson-Germer experiment. What inferences may you draw from this experiment? Explain what do you mean by *wave packet*. **[WBUT 2006]**

2.15. What is the de Broglie wavelength associated with an electron having kinetic energy of 100 eV? **[CU (H) 2011]**

2.16. Find the ratio of de Broglie wavelength of an electron and an α-particle accelerated by the same electric potential difference. **[CU (H) 2011]**

2.17. Consider a photon and an electron each having energy $E = 20$ GeV. Whose de Broglie wavelength is smaller? Justify your answer. **[BU (H) 2015]**

2.18. Calculate the de Broglie wavelength of a ball of mass 0.01 kg moving with a speed of 100 m/s and comment on the wave nature of this object. **[BU (H) 2014]**

2.19. Show that the circumference of the Bohr orbits are integral multiples of the de Broglie wavelength of the electron. **[BU (H) 2009]**

2.20. Describe briefly Davision-Germer experiment. Does the result in the light of de Broglie hypothesis pointing out how it demonstrates the existence of de Broglie matter wave. **[IAS 1995]**

2.21. What are matter waves? Obtain an equation for the wavelength of such waves? How are such wavelengths measured (verified)? **[WBCS cf. 1998, 2001, cf. 1996]**

2.22. Define and explain the terms wave packet, phase velocity and group velocity. **[WBCS 2005]**

2.23. Derive the relation between the velocity of a group of waves and the wave velocity of the individual waves of the group. **[WBCS 2008]**

2.24. Define group velocity. Show that the group velocity of a wave packet representing a particle is equal to the classical velocity of the particle. **[CU (H) 2005; cf. 2001, 2014; KU (H) 2008, 2009; cf. WBCS 2005]**

2.25. State and explain Heisenberg's uncertainty principle using situations where one of the parameters is

(a) linear distance x

(b) angle ϕ

(c) energy E.

2.26. Show that the motion of a classical particle is an alleges to the motion of a wave packet which can be constructed by super position of a large number of plane waves. Construct such a wave packet and derive an expression for the group velocity. **[IAS 2002]**

2.27. Using Heisenberg's uncertainty principle calculate the mass being exchanged between nucleus inside the nucleus, taking the nuclear mass as 1.4 fm. **[IAS 1995]**

2.28. Using Heisenberg principle find ground state energy of Bohr radius of hydrogen atom. **[KU (H) 2013; IAS 1997]**

2.29. Using uncertainty principle $\Delta x\,\Delta p$, estimate the minimum energy of a particle in a simple harmonic potential $u = \frac{1}{2}kx^2$. **[IAS 1999]**

2.30. What is zero-point energy? Show that the existence of zero-point energy is in conformity with uncertainty principle. **[WBCS 2006]**

2.31. State and explain Heisenberg's uncertainty principle. Use it to prove that electrons cannot exist inside the nucleus.

2.32. How does then one account for the beta decay of a nucleus? **[IAS 2000]**

(a) Explain the concept of group velocity of waves.

(b) Derive a relation between phase velocity and group velocity. **[Kakatiya U. 2016]**

2.33. (a) What do you understand by wave function?

(b) Obtain operator value of energy $\hat{E}$.

(c) Find out relationship between group velocity (v_g) and wave velocity (v_p).

(d) What is uncertainty principle? Apply this to prove the non-existence of electron in the nucleus.

[Raj Gandhi U. 2016]

2(D): Numerical Problems

2.1. Calculate the de Broglie wavelength associated with 50 eV electron.
[$h = 6.6 \times 10^{-34}$ J-s, $m = 9.1 \times 10^{-31}$ kg, 1 eV = 1.6×10^{-19} J] [*Ans.* 1.66 Å]

2.2. Determine the group velocity of the wave corresponding to a de Broglie wavelength $\lambda = \frac{h}{p}$.
[*Ans.* $v_g = v$]

2.3. Show that the ratio of the de Broglie wavelength to the Compton wavelength of a particle is $\sqrt{\left(\frac{c}{p}\right)^2 - 1}$.
[$h = 6.6 \times 10^{-34}$ J-s, mass of the electron = 9.1×10^{-31} kg, 1 eV = 1.6×10^{-19} J]

2.4. A 0.083 eV electron beam scatters from an unknown sample and a Bragg reflection peak observed around 22°. What is the Bragg plane spacing? [*Ans.* 1.33 Å]

2.5. The non-relativistic calculation of the de Broglie wavelength of a proton is found to be in error by 5%. What is the kinetic energy of the electron? [*Ans.* 192 MeV]

2.6. The de Broglie wavelength of an electron calculated non-relativistically is found to be in error by 5%. What is its kinetic energy? [*Ans.* 0.015 MeV]

2.7. A particle of rest mass m_0 has a characteristic frequency v_0 given by the relation $hv_0 = m_0c^2$. The particle travels with the velocity v relative to a stationary observer. By using special theory of relativity, prove that the observer sees a progressive wave of phase velocity $\frac{c^2}{v}$ and of wavelength $\frac{h}{mv}$.

2.8. Find the de Broglie wavelength of 15 keV protons. (Rest mass of proton = 1.67×10^{-27} kg)
[*Ans.* 0.002343 Å]

2.9. If the kinetic energy of an electron in the first Bohr orbit is 13.6 eV, what is the associated de Broglie wavelength? [*Ans.* 3.33 Å]

2.10. A beam of neutrons with kinetic energy 0.02 eV is incident on KCl powder. The lattice spacing of KCl is 3.14 Å. What is the radius of the circle on a flat photographic plate placed 5 cm behind the target from the first order reflections from the Bragg planes with spacing 3.14 Å. [*Ans.* 3.85 cm]

2.11. A bullet of mass 40 gm travels at 1000 m/s

(a) What wavelength can be associated with it?

(b) Why does not the wave nature of the bullet reveal itself through diffraction effects?

2.12. A thermal neutron has a kinetic energy $\frac{3}{2}$kT, where T is the room temperature (300 K).

(a) What is the energy of the neutron in electron-volt?

(b) What is its de Broglie wavelength?

2.13. In Compton scattering a photon of frequency ν is scattered by an electron of rest mass m_0. The scattered photon of frequency ν' travels in a direction inclined at 90° with the initial direction. Show that the de Broglie wavelength of the recoil electron is $\lambda = \frac{c}{\sqrt{\nu^2+\nu'^2}}$.

2.14. Calculate the de Broglie wavelength of electrons of energy 10^4 eV and compare it with the wavelength of e.m. radiation for which the photon has the same energy. **[CU (H) 1997]**

2.15. Consider a γ-ray beam with $\lambda = 1.8 \times 10^{-2}$ Å. If the radiation scattered from free electrons is viewed at 90° to the incident beam what is the Compton shift? **[CU (H) 1996]**

[*Ans.* 0.0242 Å]

2.16. Find the de Broglie wavelength of a 15 keV electron. **[BU (H) 1998]**

2.17. What is the de Broglie wavelength of a neutron of kinetic energy 10 MeV? **[BU (H) 2002]**

2.18. If the uncertainty in the position of an electron is 4×10^{-10} m, calculate the uncertainty in its momentum. **[WBUT 2005]**

2.19. The speed of electron in first orbit is 2×10^6 ms^{-1}. Find the de Broglie wavelength. **[KU (H) 2015]**

[*Ans.* 3.6 Å]

2.20. A neutron takes 1.0 minute to traverse a distance of 20.0 m. Obtain its velocity, KE and de Broglie wavelength ($m_n = 1.675 \times 10^{-27}$ kg). **[CU (H) 1995]**

[*Ans.* $\frac{1}{3}$ ms^{-1}; 5.8×10^{-10} eV; 1.19×10^{-6} m]

2.21. What is the ratio of de Broglie wavelength of a proton and an α-particle if both are accelerated by same electric potential difference. **[KU (H) 2003]**

[*Ans.* $2\sqrt{2}$]

2.22. (a) Calculate the de Broglie wavelength of: (i) *a* 1 keV (b) *a* 1 MeV electron.

[cf. CU (H) 1999; cf. BU (U) 1998; cf. KU (H) 2007]

(b) Compare with those of protons having same energy.

2.23. Obtain the value of v/c and the KE T for which the relativistic calculation of de Broglie wavelength of electron will be error by 0.05 %, 1 % and 10 %. **[cf. BU (H) 1997]**

[*Ans.* $v/c = 0.6; 0.28, 0.7$]

2.24. Calculate the wavelength associated with proton moving with velocity 10^7 cm s^{-1}. **[WBCS 1995]**

[*Ans.* 3968 fm]

2.25. An electron of energy 200 eV is passing through a circular hold of radius 10^{-4} cm. What is the uncertainty introduced in the angle of emergence? **[CU (H) 2004, 2015]**

2.26. Photons are liberated by ultraviolet light of wavelength 3000 Å from a metallic surface for which the photoelectric threshold is at 4000 Å. Calculate the de Broglie wavelength of electrons with maximum KE. [*Ans.* 1.21 nm]

2.27. The momentum of an electron is 600 keV/c. Determine its de Broglie wavelength and phase and group velocity of its de Broglie waves. **[IAS 2000]**

[*Ans.* de Broglie wavelength $\lambda = 0.0207$ Å; group velocity $v_g = 0.76c$; phase velocity $v_{ph} = 1.3c > c$]

2.28. If an electron is confined within a region of 10^{-10} m, find the minimum uncertainty in its momentum and velocity. **[WBCS 2005; CU (H) 2001, 2012, 2014]**

2.29. Calculate the de Broglie wavelength of a proton and an electron each of energy 10 MeV. **[IAS 1999]**

Chapter 3

Heisenberg's Uncertainty Principle

3.1 The uncertainty principle

The uncertainty principle or the principle of indeterminacy was proposed by *Heisenberg in* 1927. This marvellous principle is a direct consequence of the dual nature of matter.

According to classical mechanics, a moving particle has a definite momentum and position in space. Both the position and momentum of such a particle can be definitely determined. Recently, the classical view has been proved to be an approximation and is inadequate for describing the dynamical behaviour of the macroscopic objects. It cannot describe satisfactorily the behaviour of sub-atomic particles.

In quantum mechanics, a particle is described by a wave packet of linear extension (Δx), which surrounds the position of the classical particle and moves with group velocity. According to *Max Born's* probability interpretation, the particle may be found everywhere within the wave packet. This means that the position of the particle is uncertain within the limits (Δx) of the wave packet. So we may, say, that the particle may be found anywhere within Δx.

Again, a wave packet is formed by the superposition of large number of waves having different wavelengths, the wavelength of the wave packet cannot be said definitely. What we can say, is that the wavelength of the wave packet lies within certain range λ to $\lambda + \Delta\lambda$. As the momentum of a particle is related to wavelength by

$$p = \frac{h}{\lambda} \text{ or, } \Delta p = -\frac{h}{\lambda^2}\Delta\lambda$$

the particle momentum is also within p to $p + \Delta p$. So we cannot definitely, say, what is the momentum of the particle.

Hence, because of the wave nature of particle there is an uncertainty (Δx) in position of the particle as well as an uncertainty in corresponding momentum (Δp).

If the number of superposed wave as well as their range of wavelength ($\Delta\lambda$) be increased, then the linear spread (Δx) of the wave packet decreases. But, simultaneously the uncertainty (Δp) in momentum increases as $\Delta p \propto \Delta\lambda$. When $\Delta\lambda = \infty$, $\Delta p = \infty$ and $\Delta x = 0$, i.e., the wave pocket reduces to a point. In this position, there is no uncertainty in the position of the particle.

Now, for a definite λ, $\Delta\lambda = 0$, then $\Delta p = 0$ ($\because \Delta p \propto \Delta\lambda$), and there is no uncertainty in the momentum of the particle, but in this case, $\Delta x = \infty$, i.e., the position uncertainty is infinite, since the associated wave with the particle may now extend up to ∞.

3.2 Statement of the uncertainty principle

The uncertainty principle states that it is fundamentally impossible to determine simultaneously the position and the momentum of a particle to an accuracy better than one quantum of action h.

The mathematical expression of the principle is $\Delta x\, \Delta p_x \geqslant \hbar$, where $\hbar = h/2\pi = 1.054 \times 10^{-31}$ J-s, h being the Planck's constant $= 6.62 \times 10^{-34}$ J-s. Here, Δx is the error (uncertainty) in the determination of position of the particle and Δp_x is the error (uncertainty) in the determination of x-component of momentum. If the position be determined accurately, then Δx becomes smaller and consequently larger is Δp_x and vice versa. It is to be kept in mind that *the uncertainties are not in the apparatus used for the determination, but in nature itself.*

In three dimensions the uncertainty relations are expressed as follows:

$$\Delta x\, \Delta p_x \geqslant \hbar$$

$$\Delta y\, \Delta p_y \geqslant \hbar$$

$$\Delta z\, \Delta p_z \geqslant \hbar$$

3.2.1 Elementary proof of uncertainty relation

The de Broglie wavelength λ of a particle moving along the x-direction is given by $\lambda = h/p_x$, where p_x is the linear momentum of the particle along x-direction.

$$\text{or, } p_x = \frac{h}{\lambda} = \frac{h}{2\pi}\frac{2\pi}{\lambda} = \hbar k \tag{3.1}$$

If the wave packet be considered as the superposition of two simple harmonic waves of propagation constants k and $k + \Delta k$, then linear extension of the wave packet is

$$\Delta x = \frac{2\pi}{\Delta k} \tag{3.2}$$

As the particle must be somewhere within the wave packet, Δx is the uncertainty in the position of the particle and Δk is the uncertainty in the corresponding propagation constant of the associated wave.

The uncertainty in the momentum Δp_x is obtained from eq. (3.1)

$$\Delta p_x = \hbar \Delta k \tag{3.3}$$

From eqs (3.2) and (3.3), we get the product of the uncertainties as

$$\Delta x\, \Delta p_x = \frac{2\pi}{\Delta k} \times \hbar \Delta k = 2\pi\hbar = h$$

Actually the above equality should be replaced by an inequality.

$$\therefore \Delta x\, \Delta p_x \geqslant h \tag{3.4}$$

Here, we have assumed that the wave packet is formed by the superposition of two simple harmonic waves which gives rise to a series of *wave packets*. A single wave packet is formed by the superposition of a large number of plane waves of k values slightly different from one another. By the use of *Fourier analysis* it can be shown that

$$\Delta x \geqslant \frac{1}{\Delta k} \tag{3.5}$$

From eqs (3.3) and (3.5), we get

$$\Delta x\, \Delta p_x \geqslant \frac{1}{\Delta k} \hbar \Delta k \text{ or, } \Delta x\, \Delta p_x \geqslant \hbar \tag{3.6}$$

3.2.2 Conclusions from uncertainty relations

From the uncertainty relation of Heisenberg, we conclude that

(a) In eq. (3.6), if $\Delta x = 0$, then

$$\Delta p_x = \frac{\hbar}{\Delta x} \text{ or, } \Delta p_x = \frac{\hbar}{0} = \infty$$

This means that if the uncertainty in the position measurement of a moving particle be zero, then at the same moment the uncertainty in the momentum measurement will be infinity.

(b) If the momentum of a moving particle at any instant be determined accurately (i.e., $\Delta p_x = 0$), then the uncertainty (error) in the determination of position of the particle will be infinite. So from eq. (3.6)

$$\Delta x = \frac{\hbar}{\Delta p_x} = \frac{h}{0} = \infty$$

(c) For a particle of mass m moving with velocity v_x along the x-direction, its momentum is $p_x = mv_x$. From eq. (3.6)

$$\Delta x\, \Delta p_x = \hbar \text{ or, } \Delta x\, \Delta v_x = \frac{\hbar}{m}$$

For a heavy particle $m \to \infty$ and in that case

$$\Delta x\, \Delta v_x = \frac{\hbar}{\infty} = 0$$

For such a particle both the position and momentum can be determined accurately. This is true for *macroscopic bodies* whose motions are discussed in classical mechanics. Thus, the *uncertainty relation has no meaning in classical mechanics*. But, for *microscopic objects* such as electron, proton, atom, etc., quantum mechanics is applicable and as such the uncertainty relation is a reality in describing the dynamical behaviour of such objects.

3.2.3 Time-energy uncertainty relation

The time-energy uncertainty principle is stated as follows:

In any simultaneous determination of time and energy of a moving particle, the product of the uncertainties is equal to the Planck's constant.

$$\text{i.e., } \Delta E \Delta t \geqslant \hbar \left[\hbar = \frac{h}{2\pi}\right]$$

Here, ΔE is the uncertainty in the measurement of energy and Δt the corresponding uncertainty in the measurement of time.

3.2.4 Elementary proof of time-energy uncertainty relation

The time-energy uncertainty relation can be derived by considering that a moving particle is represented by a wave packet whose group velocity is equal to the particle velocity.

Now, if the wave packet of width Δx moves along the x-axis with group velocity v_g, then

$$v_g = \frac{\Delta x}{\Delta t} \text{ or, } \Delta t = \frac{\Delta x}{v_g} \tag{3.7}$$

where Δt is the time in which the wave packet moves through Δx. Δx and Δt are the uncertainties in x-coordinate and time associated with the particle.

If m be the mass, p_x the momentum and E the kinetic energy of the particle, then

$$E = \frac{p_x^2}{2m} \text{ or, } \Delta E = \frac{2p_x\,\Delta p_x}{2m} = \frac{p_x}{m}\cdot\Delta p_x = \frac{mv_x}{m}\Delta p_x \text{ or, } \Delta E = v_x\,\Delta p_x \tag{3.8}$$

From eqs (3.7) and (3.8), we get

$$\Delta E\cdot\Delta t = v_x\,\Delta p_x\cdot\frac{\Delta x}{v_g} = \Delta x\,\Delta p_x\ [\because v_x = v_g]$$

But, $\Delta x\cdot\Delta p_x \geqslant \hbar$.

$$\therefore \Delta E \Delta t \geqslant \hbar \tag{3.9}$$

Equation (3.1) represents energy-time uncertainty relation.

3.2.5 Conclusion from eq. (3.9)

From eq. (3.1)

$$\Delta t = \frac{\hbar}{\Delta E} \tag{3.10}$$

If ΔE be the maximum uncertainty in the determination of energy of a system in a particular state, then from relation (3.10), Δt is the minimum, i.e., the system remains for a minimum time in the state.

Again,

$$\Delta E = \frac{\hbar}{\Delta t} \tag{3.11}$$

The relation shows that if a system remains in a particular state for a maximum interval of time, then the uncertainty in the measurement of energy is the minimum.

3.2.6 Angular position-angular momentum uncertainty relation

The angular position-angular momentum uncertainty principle is stated as below:

In any simultaneous determination of angular position and angular momentum of a particle, the product of the uncertainties in angular momentum ΔL and angular position $\Delta\phi$ is equal to or greater than Planck's constant h, i.e.,

$$\Delta L\cdot\Delta\phi \geqslant \hbar\left(\hbar = \frac{h}{2\pi}\right)$$

3.2.7 Proof of angular position-angular momentum uncertainty relation

We consider a particle of mass m moving in a circle of radius r with a velocity v. The uncertainty relation is applicable along the direction of the tangent to the circular path of motion of the particle.

If s be the linear displacement measured along the circle and p_s the linear momentum, then

$$\Delta p_s \Delta s \geqslant \hbar$$

If ϕ be the angular displacement related to the arc length s, then

$$\phi = \frac{s}{r} \text{ or, } \Delta\phi = \frac{\Delta s}{r} \text{ or, } \Delta s = r\Delta\phi \tag{3.12}$$

Again, $\because p_s = mv$ the angular momentum is $L = p_s r = mvr$.

$$\therefore \Delta p_s = \frac{\Delta L}{r} \tag{3.13}$$

From eqs (3.12) and (3.13), we get

$$\Delta p_s \Delta s = \frac{\Delta L}{r} r\Delta\phi = \Delta L \Delta\phi = \hbar \text{ or, } \Delta L \cdot \Delta\phi \geqslant \hbar \tag{3.14}$$

Equation (3.14) represents angular momentum-angular position uncertainty relation. Eq. (3.14) shows that

$$\Delta L = \frac{\hbar}{\Delta\phi} \to 0, \text{ when } \Delta\phi \to \infty$$

$$\Delta\phi = \frac{\hbar}{\Delta L} \to 0, \text{ when } \Delta L \to \infty$$

Alternative proof of equation (3.14): If I be the moment of inertia of a particle rotating about an axis with angular velocity ω, then its kinetic energy is given by

$$E = \tfrac{1}{2}I\omega^2 \text{ or, } \Delta E = \tfrac{1}{2}I2\omega\Delta\omega = \omega\Delta L \text{ or, } \omega\Delta L = \Delta E \text{ or, } \Delta L = \frac{\Delta E}{\omega} \tag{3.15}$$

where $L = I\omega$ the angular momentum of the rotating particle.

If ϕ be the angular position of the particle at any time t, then

$$\phi = \omega t \text{ or, } \Delta\phi = \omega\Delta t \tag{3.16}$$

Multiplying eqs (3.15) and (3.16), we get

$$\Delta L \cdot \Delta\phi = \frac{\Delta E}{\omega} \times \omega\Delta t = \Delta E \cdot \Delta t \geqslant \hbar$$

$$\therefore \Delta L \cdot \Delta\phi \geqslant \hbar \text{ [same as eq. (3.14)]} \tag{3.17}$$

3.2.8 Heisenberg's uncertainty principle from de Broglie wave concept

According to de Broglie's hypothesis, a particle in motion is associated with a wave packet moving with group velocity (v_g) equal to the particle velocity (v).

Let a wave group consist of two components of equal amplitude and frequencies differing by a small amount and represented as

$$\psi_1 = A\cos(\omega_1 t - k_1 x)$$

$$\text{and } \psi_2 = A\cos(\omega_2 t - k_2 x)$$

where $\dfrac{\omega_1}{k_1} = \dfrac{2\pi\nu_1}{2\pi/\lambda_1} = \nu_1\lambda_1 = \nu_{p_1}$.

Similarly, $\dfrac{\omega_2}{k_2} = \nu_{p_2}$; ν_{p_1} and ν_{p_2} being the phase velocities of the two waves.

The resultant displacement is given by

$$\psi = \psi_1 + \psi_2 = 2A\cos(\omega t - kx)\cos\left(\frac{\Delta x}{2}t - \frac{\Delta k}{2}x\right) \tag{3.18}$$

where $\omega = \dfrac{\omega_1 + \omega_2}{2}$, $k = \dfrac{k_1 + k_2}{2}$ and $\Delta\omega = \omega_1 - \omega_2$, $\Delta k = k_1 - k_2$.

The resultant wave represented by eq. (3.18) is a wave packet which travels with group velocity

$$\nu_g = \frac{\Delta\omega}{\Delta k}.$$

This group velocity is equal to the particle velocity. The wave packet so formed has a linear spread Δx and the particle represented by the wave packet may be anywhere within Δx. Thus, there is an uncertainty in the position of the particle by an amount Δx.

Now, the condition for the formation of the wave packet is

$$\cos\left(\frac{\Delta\omega}{2}t - \frac{\Delta k}{2}x\right) = 0 \text{ or, } \frac{\Delta\omega}{2}t - \frac{\Delta k}{2}x = (2n+1)\frac{\pi}{2} \text{ [where } n = 0, 1, 2, \ldots]$$

If $\Delta x = x_1 - x_2$, where x_1 and x_2 are the positions of two consecutive nodes of the wave packet, then

$$\frac{\Delta\omega}{2}t - \frac{\Delta k}{2}x_1 = (2n+1)\frac{\pi}{2}$$

$$\text{and } \frac{\Delta\omega}{2}t - \frac{\Delta k}{2}x_2 = (2n+3)\frac{\pi}{2}$$

$$\text{or, } \frac{\Delta k}{2}(x_1 - x_2) = \pi$$

$$\text{or, } \frac{\Delta k}{2}\Delta x = \pi$$

where Δx is the error in the measurement of position of the particle.

$$\therefore \Delta x = \frac{2\pi}{\Delta k} = \frac{2\pi}{\frac{2\pi}{h}\Delta p} = \frac{h}{\Delta p}\left[\because k = \frac{2\pi p}{h}\right]$$

$$\text{or, } \Delta x \cdot \Delta p = h$$

where Δp is the error in the measurement of the momentum p of the particle.

If we consider a group consisting of a large number of waves of continuously varying frequencies, we get

$$\Delta x\,\Delta p \geqslant \frac{\hbar}{2}, \left[\text{where } \hbar = \frac{h}{2\pi}\right]$$

3.2.9 Uncertainty relation from standard deviations of x and p_x

The mean squared deviation of x is given by

$$\begin{aligned}\langle\Delta x\rangle^2 &= \left\langle(x-\langle x\rangle)^2\right\rangle = \left\langle x^2 - 2x\langle x\rangle + \langle x\rangle^2\right\rangle = \langle x^2\rangle - 2\langle x\rangle\langle x\rangle + \langle x\rangle^2 \\ &= \langle x^2\rangle - 2\langle x\rangle^2 + \langle x\rangle^2 = \langle x^2\rangle - \langle x\rangle^2\end{aligned} \quad (3.19)$$

where $\langle\ \rangle$ denotes the expectation values in quantum mechanics.

Similarly, the mean squared deviation of momentum p_x is given by

$$\langle\Delta p_x\rangle^2 = \left\langle(p_x-\langle p_x\rangle)^2\right\rangle = \langle p_x^2\rangle - \langle p_x\rangle^2 \quad (3.20)$$

Now, the uncertainties in x and p_x are given by

$$\Delta x = \left[\langle x^2\rangle - \langle x\rangle^2\right]^{1/2} \quad (3.21)$$

$$\text{and } \Delta p_x = \left[\langle p_x^2\rangle - \langle p_x\rangle^2\right]^{1/2} \quad (3.22)$$

Making use of definitions of quantum mechanical expectation values, we get

$$\Delta x\cdot\Delta p_x = \left[\langle x^2\rangle - \langle x\rangle^2\right]^{1/2}\cdot\left[\langle p_x^2\rangle - \langle p_x\rangle^2\right]^{1/2} \geqslant \frac{\hbar}{2}$$

$$\text{or, } \Delta x\cdot\Delta p_x \geqslant \frac{\hbar}{2}\quad\left[\langle p_x\rangle = -i\hbar\frac{\partial}{\partial x}\right]$$

If in general q and p be two canonically conjugate variables, viz., the generalised position and generalised momentum coordinates, then according to the definition of uncertainty relation

$$\Delta q\Delta p \geqslant \frac{\hbar}{2} \quad (3.23)$$

The Heisenberg uncertainty relation points out the fundamental limitation of experimental capabilities in the measurement involving microscopic systems. If we accept the limitation predicted by this principle, then the laws of nature become consistent with experimental results. The entire subject matter of quantum mechanics is based on the uncertainty principle.

3.3 Causality in quantum mechanics

According to the *principle of causality*, knowledge of the behaviour of a system at one instant is sufficient for the prediction of its future behaviour. Thus, it is the quantum analogue of the classical principle that the future behaviour of a particle can be determined if its position and velocity are known at a given *instant*. The law of causality, however holds strictly in classical mechanics. As no observation can be made without disturbing a microsystem, it does not hold good for atomic systems.

According to *Max Born*, the principle of causality can be viewed from *two alternate angles*. In wave-particle picture the principle of causality breaks down. But, if the system be described by a wave function $\psi(\vec{r},t)$, which satisfies a differential equation of first order in time, then the behaviour of the system at a point $P(\vec{r})$ at any time t can be described uniquely if ψ is known at $t = 0$ at the same point $P(\vec{r})$. This shows that even for microsystems, the principle of causality is true. But, the measurements on a microsystem cannot be carried out without seriously disturbing it.

Hence, it is impossible to determine the state of the system unambiguously at $t = 0$. Thus, the principle of causality breaks down. Therefore, *the probabilistic interpretation of the observed events are to be adopted in accordance with the uncertainty principle.*

3.4 Concept of Bohr orbit in view of uncertainty principle

We have the energy-time uncertainty relation, $\Delta E \cdot \Delta t \geqslant \hbar$.

According to the concept of Bohr, each electron revolves around the nucleus in quantised stationary orbits with definite energy E. So the uncertainty is $\Delta E = 0$.

So

$$\Delta t = \frac{\hbar}{\Delta E} = \frac{\hbar}{0} = \infty$$

Hence, all energy states of the Bohr atom must have infinite lifetime ($\tau = \infty$). *This implied that radiation would not occur due to the transition of electrons.* But, actually $\Delta t \sim 10^{-8}$ sec leading to the occurrence of emission of radiation due to electronic transition. *Thus, the concept of Bohr orbit violates the Heisenberg's uncertainty principle.*

Again, if Δt is finite, then $\Delta E = \frac{\hbar}{\Delta t}$ is also finite. This means that the energy levels in the excited state must have a finite spread. As a result the radiation emitted due to the transition of an electron from the excited state to the ground state is not monochromatic. *This leads to the conclusion that the spectral lines can never be infinitely sharp, but must have a natural broadening. This is again in contradiction with Bohr's theory of spectral series.*

3.5 Illustration of Heisenberg's uncertainty principle

The truth of Heisenberg's uncertainty principle may be illustrated by the following thought experiments: A thought experiment is an imaginary experiment which is not possible to be performed in practice and it does not violate any fundamental law of nature.

3.5.1 Electron diffraction through a single slit

Let a mono-energetic parallel beam of electrons moving along the x-direction be incident on a single slit of width Δy. As the electron in motion is associated with de Broglie wave, the electron beam after passing through the slit undergoes diffraction producing a Fraunhofer diffraction pattern on the photo plate P, placed perpendicular to the x-axis [Fig 3.1].

The first minimum of the diffraction pattern satisfies the relation $\Delta y \sin\theta = \lambda$, where θ is the angle of diffraction for the first order and λ is the de Broglie wavelength. Now, the uncertainty in the position of an electron in the slit is Δy as it is very difficult to locate exactly the position of the electron in tne slit, i.e.,

$$\Delta y = \frac{\lambda}{\sin\theta} \tag{3.24}$$

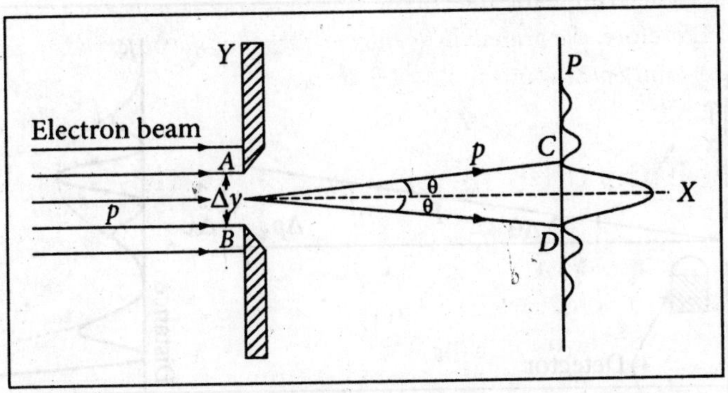

Fig 3.1 Diffraction of electron wave through a single slit

Before incidence, let the momentum of the electron along the x-direction be p. After diffraction if the electron moves along OC, then the component of momentum p along the y-direction is $p\sin\theta$. If the electron moves along OD after diffraction, then the y-component momentum is $-p\sin\theta$.

Since, the electron may be anywhere within $-\theta$ to $+\theta$, the y-component of momentum must be anywhere between $-p\sin\theta$ to $p\sin\theta$.

Thus, the uncertainty in the momentum measurement of the electron is

$$\Delta p_y = p\sin\theta - (-p\sin\theta)$$

$$\Delta p_y = 2p\sin\theta$$

$$\text{or, } \Delta p_y = 2\frac{h}{\lambda}\sin\theta \tag{3.25}$$

$$\because \lambda = \frac{h}{p} \tag{3.26}$$

Multiplying eqs (3.24) and (3.25), we get

$$\Delta y\,\Delta p_y = \frac{\lambda}{\sin\theta} \times \frac{2h}{\lambda}\sin\theta = 2h$$

$$\text{or, } \Delta y\,\Delta p_y = 2h \tag{3.27}$$

$$\text{or, } \Delta y\,\Delta p_y \geqslant \frac{\hbar}{2} \tag{3.28}$$

To decrease the uncertainty in the determination position of the electron, Δy must be decreased. But, eq. (3.24) shows that a decrease in Δy leads to the increase in θ and the diffraction pattern will be wider. This will cause a larger uncertainty in Δp. Thus, the electron diffraction experiment establishes the truth of uncertainty principle.

3.5.2 Electron diffraction through double slit

We consider a double slit AB with slit to slit distance d as shown in Fig 3.1(a). A parallel beam of electrons moving along the x-direction is incident from left of the double slit. As the electrons are associated with de Broglie wave of wavelength λ, they produce double slit diffraction pattern on the screen KL placed perpendicular to X-direction after passing through the double slit. The intensity pattern consists of interference pattern modulated by single slit diffraction pattern as shown in Fig 3.1(a).

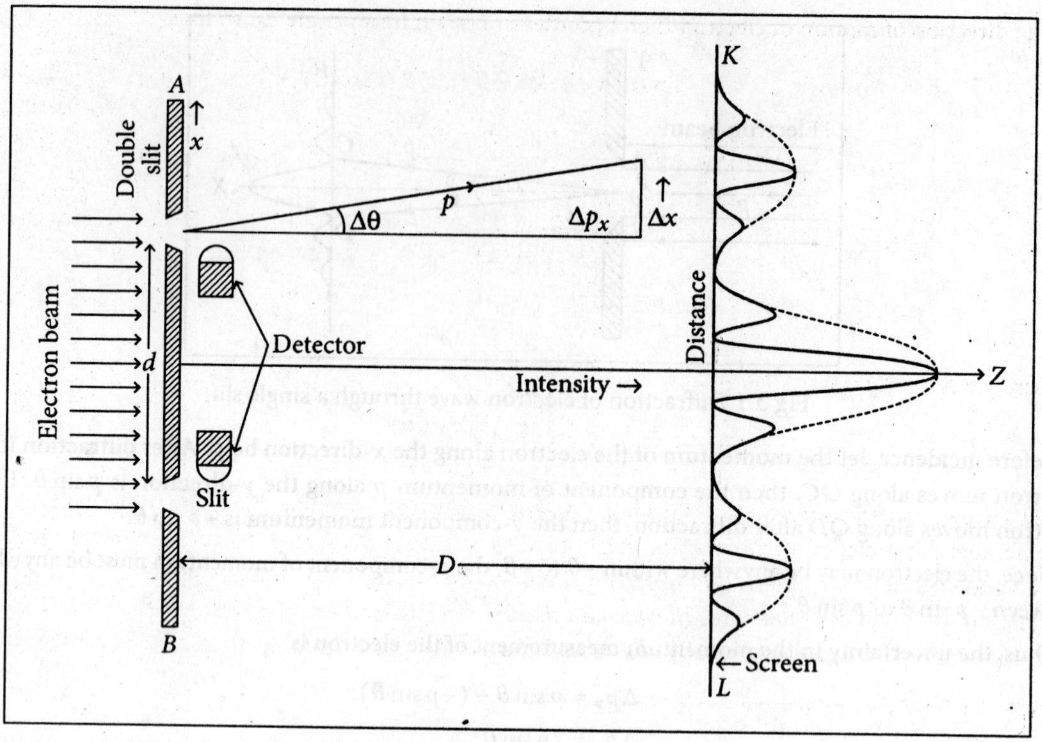

Fig 3.2 Electron diffraction through double slit.

If D be the distance from slit to screen, then the distance between successive maxima, i.e., fringe width

$$\beta = \frac{\lambda D}{d}$$

where $\lambda = \frac{h}{p}$ is the de Broglie wavelength associated with the electron and p, the momentum of it.

To detect electron passing through slits, a detector that will collide with the electron. For the detector to observe through which slit the electron passes, it should measure the x-component of the electron's potion to precision that is at least better than half the distance between the slits, i.e.,

$$\Delta x < \frac{d}{2}$$

If Δp_x be the uncertainty in the x-component of moment of the electron, then according to uncertainty relation.

$$\Delta x \, \Delta p_x \sim \hbar \text{ or, } \Delta p_x \sim \frac{\hbar}{\Delta x}$$

$$\text{or, } \Delta p_x > \frac{2\hbar}{d} \left[\because \Delta x < \frac{d}{2} \right]$$

The direction of motion of electron then becomes uncertain by an angle $\Delta\theta$ so that

$$\sin\Delta\theta = \frac{\Delta p_x}{p} \simeq \Delta\theta \ (\because \Delta\theta \text{ is small})$$

$$\text{or, } \Delta\theta = \frac{\Delta p_x}{p} > \frac{2\hbar/d}{p}$$

$$\text{or, } \Delta\theta > \frac{d\hbar/d}{h/\lambda} = \frac{2h\lambda}{2\pi\, dh}$$

$$\text{or, } \Delta\theta > \frac{\lambda}{\pi d}$$

Also $\sin\Delta\theta \simeq \Delta\theta \simeq \tan\Delta\theta = \dfrac{\Delta x}{D}$.

$$\therefore \Delta x = D\Delta\theta > \frac{\lambda}{\pi d} \text{ or, } \Delta x > \frac{\lambda DF}{\pi d} = \frac{\beta}{\pi} \sim 0.3\pi \text{ or, } \Delta x \sim \beta$$

This shows that, the uncertainty in position (Δx) of the electron is comparable to the fringe width β which implies that the stretches of the drak and bright regions on the screen are not defined. This means that there will be no well-defined pattern on the screen. *Then we can conclude that, uncertainty principle leads to the fact that if we try to determine through which slit the electron passes, then the interference pattern disappears. This is found to be true experimentally and as such the truth of the uncertainty relation is established.*

3.5.3 γ-ray microscope experiment

The hypothetical experiment was suggested by Bohr and was developed by Heisenberg to establish the validity of uncertainty principle. The use of γ-ray in the experiment can be justified as follows:

The atomic radius is ~ 10^{-11} m. So to determine the position of an electron with an uncertainty of about 10 % of the atomic radius, radiation of wavelength ~ 10^{-12} m = 0.01 Å is needed. The wavelength of γ-rays being ~ 0.01 Å, it is an efficient agent to conduct the experiment.

We suppose, O be the position of the electron and λ the wavelength of the γ-ray. If α be the semi-vertical angle of the cone of the γ-rays entering into the microscope objective from the electron, then the resolving of the microscope is given by

$$\Delta x = \frac{\lambda}{2\sin\alpha} \tag{3.29}$$

Here, Δx is the distance between two points which can just be resolved by the microscope. This suggests that for greater resolving power Δx should be small which can be achieved if λ be small. We know, the γ-rays emitted from the atomic nuclei have smallest wavelength available in nature. That is why γ-rays are used to locate the position of the electron with minimum uncertainty. In this connection, it must be remembered that we cannot observe an object with the help of a microscope which uses γ-ray beam. The microscope using lens, etc. cannot refract γ-rays. This is why the experiment is known as *a thought experiment.* [c.f. *Three elementary Examples of the uncertainty Principle* by D.S. Kothari, Phil, Mag., 27, 62(1939).]

The incident γ-ray photon is scattered by the electron according to Compton effect and enters into the microscope. During bouncing off into the microscope the photon transfers momentum to the electron. The amount of transferred momentum is uncertain as the aperture of the microscope is definite.

The uncertainty in momentum of the recoil electron can be calculated as follows:

We suppose that a γ-ray photon of momentum $\frac{h\nu}{c}$ strikes an electron of mass m at rest. After scattering the photon enters into the microscope and making an angle ϕ with the direction of incidence and the electron moves with momentum mv along a direction making an angle θ with the x-axis as shown in Fig 3.2.

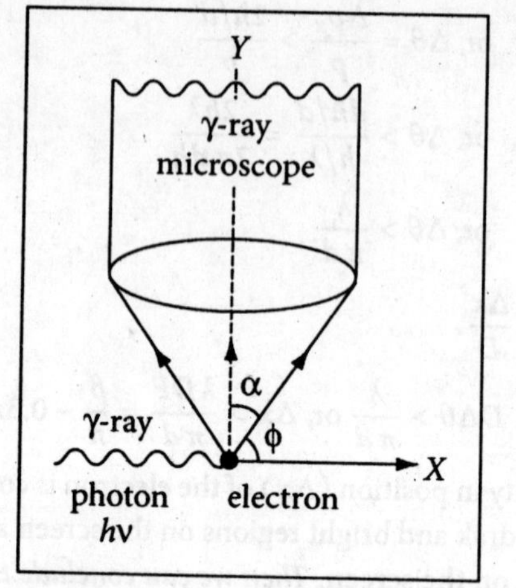

Fig 3.3 γ-ray microscope experiment

Now, from the momentum conservation principle along the x-direction [Fig 3.3]

$$\frac{h\nu}{c} = \frac{h\nu'}{c}\cos\phi + mv\cos\theta$$

$$\text{or, } p_x = mv\cos\theta = \frac{h\nu}{c} - \frac{h\nu'}{c}\cos\phi$$

$$\text{or, } p_x = \frac{h}{c}(\nu - \nu'\cos\phi)$$

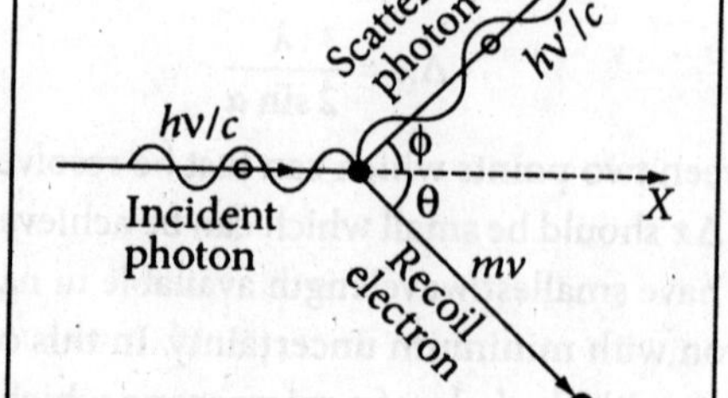

Fig 3.4 Scattering of photon by an electron

Here, p_x is the x-component of momentum of the electron.

Since, the limits of the angle ϕ through which the photon enters the microscope objective are from $\left(\frac{\pi}{2} - \alpha\right)$ to $\left(\frac{\pi}{2} + \alpha\right)$, the spread in the x-component of momentum is given by

$$\frac{h}{c}\left[\nu - \nu' \cos\left(\frac{\pi}{2} - \alpha\right)\right] \leqslant p_x \leqslant \frac{h}{c}\left[\nu - \nu' \cos\left(\frac{\pi}{2} + \alpha\right)\right]$$

$$\text{or,}\ \frac{h}{c}\left[\nu - \nu' \sin\alpha\right] \leqslant p_x \leqslant \frac{h}{c}\left[\nu + \nu' \sin\alpha\right]$$

The uncertainty in the momentum is given by

$$\Delta p_x = \frac{h}{c}[\nu + \nu' \sin\alpha] - \frac{h}{c}[\nu - \nu' \sin\alpha] = \frac{h\nu}{c} + \frac{h\nu'}{c}\sin\alpha - \frac{h\nu}{c} + \frac{h\nu'}{c}\sin\alpha$$

$$= \frac{2h\nu'}{c}\sin\alpha = \frac{2h}{\lambda}\sin\alpha \tag{3.30}$$

From eqs (3.29) and (3.30)

$$\Delta x\, \Delta p_x = \frac{\lambda}{2\sin\alpha} \times \frac{2h}{\lambda}\sin\alpha = h$$

which is consistent with $\Delta x\, \Delta p_x \geqslant \hbar$.

Hence, the experiment established the validity of the uncertainty principle.

3.6 Bohr's principle of complementarity

In 1928, Bohr introduced this principle in order to understand the implications of uncertainty principle in more physical terms. The principle states that *wave and particle properties of electromagnetic radiation and matter are merely complementary ways of regarding the same phenomenon.*

In any experiment in which physical entity exhibits its wave properties, it is impossible to attribute corpuscular characteristics to it. This means that particle and wave aspects of a physical entity are complementary and cannot be exhibited simultaneously.

According to uncertainty principle, both the position and momentum of a physical system cannot be simultaneously determined with arbitrary accuracy. If we try to improve the precision of measurement of position, the precision in the momentum measurement will be reduced. The uncertainties in the simultaneous measurement of position and momentum is an inherent limitation of nature and not due to any defect in the measurement method.

To illustrate the principle of complementarity, we consider the γ-ray thought experiment. The limit of resolution of the microscope is

$$\Delta x = \frac{\lambda'}{\sin\alpha}$$

where Δx is the uncertainty in the position of the electron. If the wavelength of radiation is very small, then

$$\Delta x = \frac{\lambda'}{\sin\alpha} \simeq 0$$

That is the position of the electron just after the scattering is accurately known. But, since from the uncertainty principle

$$\Delta p_x = \frac{\hbar}{\Delta x} = \frac{\hbar}{0} = \infty$$

the uncertainty in momentum is infinite, i.e., there will be no knowledge of p_x. The de Broglie wavelength

$$\lambda = \frac{h}{p_x} = \frac{h}{\infty} = 0$$

i.e., *the wave nature of the electron will disappear.*

Similarly, if λ' is large, then p_x just after the scattering will be found with good accuracy, but the position of the electron will be unknown. This means that under this condition wave nature will be prominent and particle nature will disappear. *Hence, in any experiment both the particle nature and wave nature cannot be detected simultaneously.* This is the principle of complementarity proposed by N. Bohr.

3.7 Applications of the uncertainty principle

3.7.1 Non-existence of free electrons in the nucleus

If we suppose that a free electron can exist in the nucleus, then the maximum uncertainty Δx in its position will be the same as the diameter of the nucleus. But, the diameter of the nucleus is 2×10^{-14} m.

$$\therefore \Delta x = 2 \times 10^{-14} \text{ m} \tag{3.31}$$

If Δp_x be the uncertainty in the corresponding momentum, then according to the uncertainty principle

$$\Delta x \, \Delta p_x \geqslant \hbar \text{ or, } \Delta p_x \geqslant \frac{\hbar}{\Delta x}$$

The minimum uncertainty in the momentum is given by

$$\Delta p_x = \frac{\hbar}{\Delta x} = \frac{6.63 \times 10^{-34}}{2\pi \times 2 \times 10^{-14}} = \frac{6.63 \times 10^{20}}{4\pi} = 0.5278 \times 10^{-20} \text{ kg-m/s} \tag{3.32}$$

So if the electron exists in the nucleus, then its minimum momentum must be

$$p_{\min} = \Delta p_x = 0.5278 \times 10^{-20} \text{ kg-m/s}$$

Now, total relativistic energy of the electron is

$$E_{\min}^2 = p_{\min}^2 c^2 + m_0^2 c^4 = \left(0.5278 \times 10^{-20} \times 3 \times 10^8\right)^2 + \left(9.11 \times 10^{-31}\right)^2 \left(3 \times 10^8\right)^4$$
$$= (0.5278)^2 \times 9 \times 10^{-24} + 9.11^2 \times 81 \times 10^{-30}$$
$$= 2.5 \times 10^{-24} + 6.72 \times 10^{-27} \text{ J}^2$$

Neglecting the 2nd term w.r.t. the first term due to smallness, we have

$$E_{\min} = \sqrt{2.5 \times 10^{-24}} = 1.58 \times 10^{-12} \text{ J} = \frac{1.58 \times 10^{-12}}{1.6 \times 10^{-19}} \text{ eV} \left[\because 1 \text{ eV} = 1.6 \times 10^{-19} \text{ J}\right]$$
$$= 0.9875 \times 10^7 = 9.875 \times 10^6 \text{ eV} = 9.875 \text{ MeV}$$

If a free electron would exists in the nucleus, it must have a minimum energy of about 9.0 MeV.

Again, the highest value of kinetic energy of a β-particle emitted spontaneously by a nucleus is found to be about 4 MeV. As this is much less than the calculated value of the minimum kinetic energy which the electron should possess (to exist in the nucleus), we conclude that *electrons cannot be constituent of the atomic nucleus; in other word, the nucleus cannot harbour an electron.* The electron of β-particles from the radioactive nucleus is due to the transformation of a neutron into a proton producing an electron.

3.7.2 Existence of protons, neutrons and α-particles in the nucleus

For a particle to be within the nucleus, the uncertainty in its position should be of the order of the radius of the nucleus. If the nuclear diameter is ~ 10^{-14} m, then the uncertainty in the position of the particle (proton, neutron or α-particle) $\Delta x \sim 10^{-14}$ m.

But, according to uncertainty relation, $\Delta x\,\Delta p_x \geqslant \hbar/2$, where Δp_x is the position uncertainty of the particle and $\hbar = h/2\pi$.

$$\therefore \Delta p_x \geqslant \frac{\hbar}{2\Delta x} \text{ or, } \Delta p_x \geqslant \frac{h}{4\pi\Delta x} = \frac{6.6\times10^{-34}}{4\pi\times10^{-14}} = 5.3\times10^{-21} \text{ kg-m/s}$$

Thus, the total momentum of the proton, neutron or α-particle in the nucleus must be at least ~ 5.3×10^{-21} kg-m/s.

Now, the rest mass of the proton = the rest mass of the neutron = $m_0 = 1.67\times10^{-27}$ kg.

The velocity of the particle (proton or neutron)

$$v = \frac{5.3\times10^{-21}}{1.67\times10^{-27}} = 3.1\times10^6 \text{ m/s}$$

which is in the non-relativistic region.

So the kinetic energy

$$E_K = \frac{p^2}{2m} = \frac{(\Delta p_x)^2}{2m} = \frac{(5.3\times10^{-2})^2}{2\times1.67\times10^{-27}} = 8.4\times10^{-15}\text{ J} = \frac{8.4\times10^{-15}}{1.6\times10^{-19}} = 52 \text{ keV} \tag{3.33}$$

Since, the rest mass of an α-particle is ~ 4 times the rest mass of proton, the α-particle should have a minimum kinetic energy of $\frac{52}{4}$ = **13 keV.**

It has been found experimentally that the energy carried by the proton or neutron is greater than **52 keV** and hence they can exist in the nucleus.

Similarly, the energy carried by an α-particle has been found to be more than **13 keV** and, therefore, it can also exist in the nucleus.

3.7.3 Existence of zero-point energy

We consider that a particle of rest mass m_0 is confined to move under the influence of a potential of linear dimension l. The uncertainty in the position of the particle $\Delta x = l/2$ as it can be found anywhere within the region.

According to uncertainty relation

$$\Delta x\,\Delta p_x = \frac{\hbar}{2}$$

Δp_x is the corresponding momentum uncertainty.

$$\therefore \Delta p_x = \frac{\hbar}{2\Delta x} = \frac{\hbar}{l}$$

If the uncertainty in momentum Δp_x is equal to the minimum value of momentum of the particle, then

$$p_x = \Delta p_x = \frac{\hbar}{l}$$

The minimum kinetic energy of the particle is

$$E = \frac{p_x^2}{2m_0} = \frac{\hbar^2}{2m_0 l^2} \tag{3.34}$$

Thus, we note that the minimum value of kinetic energy of a particle confined to move in a region of space under the action of a potential is finite and not zero.

We know, any physical system with lowest energy is close to absolute zero of temperature. This means that the kinetic energy of a system even at 0 K is finite and not zero. This energy is *zero-point energy.*

Hence, a particle at 0 K has finite momentum and motion which is called the *zero-point motion.*

3.7.4 Ground state energy of a linear harmonic oscillator (LHO) (zero-point energy)

Let a particle of mass m executes SHM along the x-axis and let at any time t its position is x from the mean position.

Total energy of the LHO is

$$E = E_K + E_p = \frac{p^2}{2m} + \frac{1}{2}m\omega^2x^2 = \frac{(\Delta p)^2}{2m} + \frac{1}{2}m\omega^2(\Delta x)^2 \tag{3.35}$$

Assuming that the maximum uncertainty in the position to be $\Delta x = x$, from the uncertainty principle the minimum uncertainty in momentum for simultaneous determination is

$$\Delta p = \frac{\hbar}{2\Delta x} = \frac{\hbar}{2x}$$

Since, the momentum (p) of the LHO cannot be less than Δp, the minimum value of momentum, $p = \hbar/x$.

$$\therefore E = \frac{1}{2m}\left(\frac{\hbar^2}{2x}\right)^2 + \frac{1}{2}m\omega^2x^2 = \frac{\hbar^2}{8mx^2} + \frac{1}{2}m\omega^2x^2 \tag{3.36}$$

Now, differentiating eq. (3.36) w.r.t. x, we get

$$\frac{dE}{dx} = -\frac{\hbar^2}{4mx^3} + m\omega^2x = 0$$

Since, the ground state energy of the LHO is the minimum

$$\therefore \frac{dE}{dx} = 0 \text{ for the ground state or, } -\frac{\hbar^2}{4mx^3} + m\omega^2x = 0 \text{ or, } m\omega^2x = \frac{\hbar^2}{4mx^3}$$

$$\text{or, } x^4 = \frac{\hbar^2}{4m^2x^2} \text{ or, } x^2 = \frac{\hbar^2}{2m\omega} \tag{3.37}$$

We can show that for this value of x^2, $\frac{d^2E}{dx^2}$ is positive.

$\therefore$ when $x^2 = \frac{\hbar}{2m\omega}$, E is minimum.

$$\therefore E_{\min} = \frac{\hbar^2}{8m}\frac{2m\omega}{\hbar} + \frac{1}{2}m\omega^2 \cdot \frac{\hbar}{2m\omega} = \frac{1}{4}\hbar\omega + \frac{1}{4}\hbar\omega = \frac{1}{2}\hbar\omega \tag{3.38}$$

Equation (3.38) represents the minimum energy of the LHO.

3.7.5 Size of the hydrogen atom

In a hydrogen atom the electron moves round the proton in a circular orbit of radius r (say).

Hence, the maximum uncertainty in the position of the electron w.r.t. the proton is $\Delta r = r$.

From the uncertainty principle the minimum uncertainty in the corresponding momentum in simultaneous measurement is

$$\Delta p = \frac{\hbar}{\Delta r} = \frac{\hbar}{r}$$

Since, the momentum of the electron cannot be less than Δp, $\therefore p = \frac{\hbar}{r}$.

The kinetic energy of the electron is

$$E_K = \frac{p^2}{2m} = \frac{\hbar^2}{2mr^2} \tag{3.39}$$

The potential energy

$$V = -\frac{e^2}{4\pi\varepsilon_0 r} \tag{3.40}$$

$\therefore$ total energy of the electron is

$$E = E_K + V = \frac{\hbar^2}{2mr^2} - \frac{e^2}{4\pi\varepsilon_2 r} \tag{3.41}$$

$$\therefore \frac{dE}{dr} = -\frac{\hbar^2}{mr^3} + \frac{e^2}{4\pi\varepsilon_0 r^2}$$

But, ground state E is $E_{\min}$. $\therefore \frac{dE}{dr} = 0$.

So

$$-\frac{\hbar^2}{mr^3} + \frac{e^2}{4\pi\varepsilon_0 r^2} = 0 \text{ or, } \frac{e^2}{4\pi\varepsilon_0} = \frac{\hbar^2}{mr}$$

$$\therefore r = \frac{4\pi\varepsilon_0\hbar^2}{me^2} = \frac{4\pi\varepsilon_0}{me^2}\left(\frac{h}{2\pi}\right)^2 = \frac{\varepsilon_0 h^2}{\pi me^2} \tag{3.42}$$

We can show that for $r = \frac{\varepsilon_0 h^2}{\pi me^2}$, $\frac{d^2E}{dr^2}$ = a positive quantity.

Hence, the radius of the electron path is $r = \frac{\varepsilon_0 h^2}{\pi me^2}$ in the ground state.

If $h = 6.63 \times 10^{-34}$ J, $m = 9.11 \times 10^{-31}$ kg, $e = 1.6 \times 10^{-19}$ C, $\varepsilon_0 = 8.85 \times 10^{-12}$ C^2/N-m^2, then

$$r = \frac{\left(8.85 \times 10^{-12}\right) \times \left(6.63 \times 10^{-34}\right)^2}{\pi \times 9.11 \times 10^{-31} \times \left(1.6 \times 10^{-19}\right)^2} = 5.31 \times 10^{-11} \text{ m}$$

3.7.6 Ground state energy of the electron in the hydrogen atom

We have the total energy of the electron in the hydrogen atom is

$$E = \frac{\hbar^2}{2mr^2} - \frac{e^2}{4\pi\varepsilon_0 r} \text{ [cf. Problem 5]}$$
$$= \frac{1}{r}\left[\frac{\hbar^2}{2m}\cdot\frac{1}{r} - \frac{e^2}{4\pi\varepsilon_0}\right]$$

Putting $r = \dfrac{me^2}{4\pi\varepsilon_0\hbar^2}$, the value of the radius of the electron in the ground state, we get

$$E = \frac{me^2}{4\pi\varepsilon_0\hbar^2}\left[\frac{\hbar^2}{2m}\frac{me^2}{4\pi\varepsilon_0\hbar^2} - \frac{e^2}{4\pi\varepsilon_0}\right] = \frac{me^2}{4\pi\varepsilon_0\hbar^2}\left[\frac{e^2}{8\pi\varepsilon_0} - \frac{e^2}{4\pi\varepsilon_0}\right]$$
$$= -\frac{me^4}{32\pi^2\varepsilon_0^2\hbar^2} = -\frac{me^4}{32\pi^2\varepsilon_0^2}\left(\frac{2\pi}{h}\right)^2 = -\frac{me^4}{8\varepsilon_0^2 h^2} \quad (3.43)$$

Using $m = 9.11 \times 10^{-31}$ kg, $e = 1.6 \times 10^{-19}$ C, $\varepsilon_0 = 8.85 \times 10^{-12}$ C^2/N-m^2 and $h = 6.63 \times 10^{-34}$ J, we get

$$E = -\frac{(9.11 \times 10^{-31}\text{ kg})(1.6 \times 10^{-19})^2}{8 \times (8.85 \times 10^{-12})^2 \times (6.63 \times 10^{-34})^2}$$

$$\therefore E = -\mathbf{13.6\ eV} \text{ [putting 1 eV} = 1.6 \times 10^{-19}\text{ J]}$$

3.7.7 Uncertainty in frequency of radiation emitted by an atom

If the lifetime of an excited state be Δt, and ΔE the uncertainty in energy of the state, then according to the uncertainty relation

$$\Delta E \cdot \Delta t \geqslant \frac{\hbar}{2} \text{ or, } \Delta E \geqslant \frac{\hbar}{2\Delta t} \text{ or, } \Delta E = \frac{h}{4\pi\Delta t}$$

But, $E = h\nu$. $\therefore \Delta E = h\Delta\nu$, where $\Delta\nu$ is the uncertainty in frequency of the emitted photon.

$$\therefore \Delta\nu = \frac{\Delta E}{h} = \frac{h}{4\pi\Delta t}\cdot\frac{1}{h} = \frac{1}{4\pi\Delta t}$$

Now, the average lifetime of the excited state of an atom is $\Delta t \sim 10^{-8}$ s.

$$\therefore \Delta\nu = \frac{1}{4\pi \times 10^{-8}} = \mathbf{0.8 \times 10^7\ Hz}$$

Thus, the radiation emitted by an atom cannot have a definite frequency (ν), but it must have a range from $\nu - \Delta\nu$ to $\nu + \Delta\nu$, i.e., *the radiation emitted must have a band of frequency.*

3.7.8 Binding energy of an electron in an atom

The electron revolving in a circular orbit of radius r around the nucleus, has the position uncertainty $\Delta x = r$. According to the uncertainty principle

$$\Delta x\,\Delta p = \frac{\hbar}{2} \text{ or, } \Delta p = \frac{\hbar}{2r}$$

Thus, the minimum uncertainty in the momentum p of the electron in the orbit is

$$p = \Delta p = \frac{\hbar}{2r}$$

Since, $r \simeq 10^{-10}$ m

$$p = \frac{\hbar}{2r} = \frac{6.6 \times 10^{-34}}{4\pi \times 10^{-10}} = 5.2 \times 10^{-25} \text{ kg-m/s}$$

The kinetic energy of the electron

$$E_K = \frac{p^2}{2m} = \frac{(5.2 \times 10^{-25})^2}{2 \times 9.1 \times 10^{-31}} = 1.5 \times 10^{-19} \text{ J} \simeq 1 \text{eV}$$

If Z be the atomic number of the nucleus, then the potential energy of the electron is

$$V = \frac{-Ze^2}{4\pi\varepsilon_0 r} = -Z \frac{(1.6 \times 10^{-19})^2 \times (9 \times 10^9)}{10^{-9}} = -Z \times 14.4 \text{ eV}$$

$\therefore$ total energy, $E = E_K + V = 1 - 14.4Z$ eV.

For hydrogen, $Z = 1$

$$E = 1 - 14.4 = -13.4 \text{ eV}$$

For helium, $Z = 2$

$$E = 1 - 2 \times 14.4 = 1 - 28.8 = -27.8 \text{ eV}$$

Actually the value of E for hydrogen is -13.6 eV and that for helium is -24.6 eV. The calculated E values by other methods agree closely with these values.

3.8 Answers of some important conceptual questions

3.1. *Show that for a classical (macroscopic) particles, unlike quantum mechanical particle, the position and movement can be measured accurately.*

Ans. If Δx and Δp_x be the uncertainties in position and corresponding momentum of a moving particle in simultaneous measurement, then according to uncertainty relation $\Delta x \, \Delta p_x = \hbar$.

But, $p_x = mv_x$.

$$\therefore \Delta x \, \Delta p_x = \frac{\hbar}{m} \quad [\because \Delta p_x = m\Delta v_x]$$

For a classical particle, the mass m is very large, i.e., $m \to \infty$.

$$\therefore \Delta x \, \Delta v_x = \frac{\hbar}{\infty} = 0$$

This shows that the position and momentum of classical particles can be determined accurately with $\Delta x = 0$ and $\Delta p_x = 0$.

So the uncertainty principle has no meaning in classical mechanics.

3.2. *A beam of short wavelength gives accurately the position of a particle.—Justify.* **[GNDU 1977, 96]**

Ans. If Δx and Δp_x be the uncertainties in the position and momentum of a particle, then according to the uncertainty relation, $\Delta x \, \Delta p_x = \hbar/2$.

Ans. If Δx and Δp_x be the uncertainties in the position and momentum of a particle, then according to the uncertainty relation $\Delta x \, \Delta p_x = \hbar/2$.

If Δx is small, then Δp_x must be large as $\Delta x \, \Delta p_x$ = constant.

But $p_x = h/\lambda$, where λ is the de Broglie wavelength.

Therefore, for large p_x (or Δp_x) for which Δx would be small, it is essential that λ should be small.

3.3. *If the average interval of time between the excitement and de-excitement of the atom is 10^{-8} s, then what is the uncertainty in the frequency of light emitted by the atom?*

Ans. If ΔE be the uncertainty in the energy of the state, then according to uncertainty principle

$$\Delta E \Delta t \geqslant \frac{\hbar}{2} \text{ or, } \Delta \geqslant \frac{\hbar}{2\Delta t} = \frac{h}{4\pi \Delta t} \quad [\because \hbar = h/2\pi]$$

But, $\Delta E = h\Delta \nu$.

$$\therefore \Delta \nu = \frac{\Delta E}{h} = \frac{h}{4\pi h \Delta t} = \frac{1}{4\pi \Delta t}$$

Given, the average lifetime of the excited atom is 10^{-8} s, i.e., $\Delta t = 10^{-8}$ s.

$$\therefore \Delta \nu = \frac{1}{4\pi \times 10^{-8}} = 0.8 \times 10^{7} \text{ Hz}$$

Thus, the radiation emitted from an excited atom cannot have a precise value on the other hand; there is an uncertainty in the frequency (0.8×10^7 Hz) of light emitted by the atom.

3.4. *Show that the size of an elementary phase cell is $(\hbar/2)^{3N}$.*

Ans. The state of a microsystem in *quantum statistics* is defined by three position coordinates, viz., x, y, z and three momentum coordinates p_x, p_y, p_z. Thus, the six-dimensional position-momentum space is necessary for a dynamical system which is known as *phase space*.

If the system is located in the position coordinate space between $x \to x + \Delta x, y \to y + \Delta y, z \to z + \Delta z$ and momentum space between $p_x \to p_x + \Delta p_x, p_y \to p_y + \Delta p_y, p_z \to p_z + \Delta p_z$, then the phase volume under the disposal of the system is $\Delta x\, \Delta y\, \Delta z\, \Delta p_x\, \Delta p_y\, \Delta p_z$.

This is six-dimensional space and is called the Γ-space.

Now, according to uncertainty relation, $\Delta x\, \Delta p_x = \hbar/2$, $\Delta y\, \Delta p_y = \hbar/2$ and $\Delta z\, \Delta p_z = \hbar/2$.

$\therefore \Delta x\, \Delta p_x\, \Delta y\, \Delta p_y\, \Delta z\, \Delta p_z = (\hbar/2)^3$.

A microscopic system consisting of N-microsystems is described by $6N$ variables. So in Γ-space, the size of the elementary cell $= (\hbar/2)^{3N}$.

3.5. *A particle is diffracted at a certain angle and the uncertainty in the measurement of the angle is one second of arc. What is the error in the measurement of its angular momentum?*

Ans. Uncertainty in the angle of diffraction measurement is

$$\Delta\theta = \text{one second} = \frac{\pi}{60 \times 60 \times 180} \text{ radian}$$

$\therefore$ uncertainly in the measurement of corresponding angular measurement is

$$\Delta L = \frac{h/2\pi}{\Delta\theta} = \frac{6.62 \times 10^{-34} \times 60 \times 60 \times 180}{2\pi \times \pi} = 2.168 \times 10^{-29} \text{ J-s}$$

3.6. *Show how one can arrive at Bohr's quantisation condition on the basis of de Broglie's hypothesis of matter waves.* **[CU (H) 1998]**

Ans. Let an electron of mass m rotate in a circular orbit of radius r around the nucleus with a speed v. Then the angular momentum, $L = mvr$.

The circumference of the electron orbit = $2\pi r$ and if it contains n de Broglie wavelength λ, then

$$2\pi r = n\lambda = \frac{nh}{mv} \quad \text{or,}\quad mvr = L = n\frac{h}{2\pi} = h\hbar$$

This is *Bohr's quantisation condition.*

3.7. *By applying uncertainty principle explain non-existence of electrons in the atomic nucleus.*

[PU 1997; HPU 1995; GNDU 1996]

Ans. See Art. 3.7.1.

3.8. *Using uncertainty principle show that protons, neutrons and α-particles can exist within the nucleus.*

Ans. See Art. 3.7.2.

3.9. *Apply Heisenberg's uncertainty principle to explain the following:*

(a) Existence of finite zero-point energy

(b) Binding energy of an electron in a hydrogen atom is of the order of 13.6 eV

(c) The minimum energy of harmonic oscillator

(d) Size of the hydrogen atom.

Ans. See (a) Art. 3.7.3; (b) Art. 3.7.8; (c) Art 3.7.4; (d) Art. 3.7.5.

3.10. *Resolving power of a microscope depends upon wavelength of light used. In what way electron microscope is superior?*

Ans. The resolving power (RP) of a microscope is given by RP = A/λ, where A is the aperture (diameter) of the lens used as objective and λ is the wavelength of light. Thus, smaller the wavelength, higher is the resolving power. If we use ultraviolet light, it has a wavelength 4000 Å = 4000×10^{-10} m = 4×10^{-7} m.

But, if we use an electron microscope and accelerate electrons through a potential difference of 64 kV = 64000 V, the velocity v attained by the electron is given by

$$eV = \frac{1}{2}mv^2 \quad \text{or,}\quad v = \sqrt{\frac{2eV}{m}} = \sqrt{\frac{2 \times 16 \times 10^{-19} \times 64 \times 10^3}{9.1 \times 10^{-31}}} = 1.5 \times 10^8 \text{ m/s}$$

$$\therefore \frac{v}{c} = \frac{1.5 \times 10^8}{3 \times 10^8} = \frac{1}{2}$$

As the velocity is only half of the velocity of light, we use relativity formula for the de Broglie wavelength of the electron.

$$\therefore \lambda = \frac{h}{m_0 c} \times \frac{(1 - v^2/c^2)^{1/2}}{v/c}$$

$$= \frac{6.62 \times 10^{-34}}{9.1 \times 10^{-38} \times 3 \times 10^8} \cdot \frac{(1-\frac{1}{4})^{1/2}}{1/2} \quad \text{[substituting } v/c = 1/2 \text{ and values of other constants]}$$

$$= \frac{6.62 \times 10^{-34} \times \sqrt{3}}{9.1 \times 10^{-31} \times 3 \times 10^8} = \frac{6.62 \times 10^{-34}}{9.1 \times 10^{-31} \times 1.732 \times 10^8} = \mathbf{4.2 \times 10^{-12}\ m}$$

This wavelength is 10^5 times smaller than the wavelength of ultraviolet/visible light. Thus, an electron microscope has resolving power 10^5 times the resolving power of an optical microscope. An electron microscope, therefore, can produce sharp images with higher magnification and is superior to the optical microscope.

3.11. *Clarify the difference between classical mechanics and quantum mechanics.*

Ans. The following table shows the difference between classical mechanics and quantum mechanics.

Classical mechanics	*Quantum mechanics*
1. Classical mechanics deals with macroscopic bodies.	1. Quantum mechanics deals with microscopic bodies
2. Classical mechanics (also called *Newtonian mechanics*) was developed on the basis of Newton's laws of mechanics, i.e., $P = mf$, where m and f are the mass and acceleration of the particle.	2. Quantum mechanics was developed on the basis of Schrödinger wave equation $-\frac{\hbar^2}{2m}\nabla^2\psi + V\psi = i\hbar\frac{\partial\psi}{\partial t}$ where ψ is the wave function representing the particle in a potential field V.
3. If the initial position and momentum of the particle as well as the forces acting on it are known, then the future position and momentum can be determined in classical mechanics.	3. There is an inherent uncertainty in position and momentum of the particle in quantum mechanics according to uncertainty principle. So the future position and momentum of the particle cannot be accurately determined.
4. Classical mechanics is deterministic.	4. Quantum mechanics is probabilistic
5. Energy and momentum in classical mechanics are continuous.	5. Energy and momentum in quantum mechanics are discrete.

3.9 Solved problems

» ***Example* 3.1.** An electron is confined in a box of length 10^{-9} m. Calculate the minimum uncertainty in the measurement of its velocity. **[CU 1992]**

***Solution*:** If Δx and Δp_x be the uncertainties in the position and corresponding momentum of the electron, then according to uncertainty relation, $\Delta x\,\Delta p_x = h$ or, $(\Delta x)_{\max}(\Delta p_x)_{\min} = h$.

Now, $(\Delta x)_{\max} = 10^{-9}$ m.

$$\therefore (\Delta p_x)_{\min} = \frac{h}{(\Delta x)_{\max}} = \frac{6.62\times10^{-34}}{10^{-9}} = 6.62\times10^{-25}\ \text{kg-m/s}$$

But,

$$m(\Delta v_x)_{\min} = (\Delta p_x)_{\min} \text{ or, } (\Delta v_x)_{\min} = \frac{(\Delta p_x)_{\min}}{m} = \frac{6.62\times10^{-25}}{9.1\times10^{-31}} = \mathbf{7.27\times10^5\ m/s}$$

» ***Example* 3.2.** Assume that the uncertainty in the location of a particle is equal to its de Broglie wavelength. Show that the uncertainty in the velocity is equal to its velocity.

Solution: According to the uncertainty principle

$$\Delta x\,\Delta p_x = h \quad \text{[where } \Delta x \text{ and } \Delta p_x \text{ are the uncertainties in position and momentum]}$$

$$\text{or, } \Delta x m \Delta v_x = h$$

$$\text{or, } \Delta v_x = \frac{h}{m\Delta x} = \frac{h}{m\lambda} \quad [\because \Delta x = \lambda]$$

$$= \frac{h}{m \times \frac{h}{mv}} = v \quad \left[\because \lambda = \frac{h}{mv}\right]$$

∴ **the uncertainty in velocity is equal to its velocity.**

» ***Example* 3.3.** Suppose, that the momentum of a certain particle can be measured to an accuracy of one part in 1000. Determine the minimum uncertainty in the position of the particle if the particle is (i) a 5×10^{-3} kg mass moving with a speed of 2 m/s (ii) an electron moving with a speed of 1.8×10^8 m/s.

Solution: Here, if Δp be the uncertainty in momentum p, then

$$\frac{\Delta p}{p} = 10^{-3} \text{ or, } \Delta p = 10^{-3}\,p = 10^{-3}\,mv$$

Then from the uncertainty relation, $\Delta x\,\Delta p_x \geqslant \hbar/2$.

∴ the uncertainty in position

$$\Delta x \geqslant \frac{h}{4\pi\Delta p} = \frac{h}{4\pi mv \times 10^{-3}}$$

$$\text{(i) } \Delta x = \frac{6.63 \times 10^{-34}}{4\pi \times 10^{-3}\,(5 \times 10^{-3}\text{ kg}\,(2\text{ m/s})} = 5.28 \times 10^{-30}\text{ m} = \mathbf{5.28 \times 10^{-20}\ \text{Å}}.$$

∴ the minimum uncertainty in position is 5.28×10^{-20} Å (unmeasurable).

(ii) If m be the mass of the electron, while moving with velocity v and m_0 the rest mass, then

$$m = \frac{m_0}{\sqrt{1 - \frac{v^2}{c^2}}}$$

$$\therefore \Delta x = \frac{h\sqrt{1 - v^2/c^2}}{4\pi\,m_0 v \times 10^{-3}} = \frac{(6.63 \times 10^{-34})\sqrt{1 - (0.6)^2}}{4\pi 10^{-3} \cdot (9.11 \times 10^{-31})\,(1.8 \times 10^8\text{ m/s})} = 2.57 \times 10^{-10}\text{ m} = \mathbf{2.57\ \text{Å}}$$

» ***Example* 3.4.** An electron has a speed of 1 km/s with an accuracy of 0.05 %. Calculate the uncertainty with which the position of the electron can be located. **[PU 1994]**

Solution: Uncertainty in the momentum measurement is

$$\Delta p = \Delta(mv) = m\Delta v = 9.1 \times 10^{-31} \times 1000 \times \frac{0.05}{100} = 4.55 \times 10^{-31}\text{ kg-m/s}$$

Error in the measurement of position is

$$\Delta x = \frac{h}{\Delta p} = \frac{6.62 \times 10^{-34}}{4.55 \times 10^{-31}} = 1.45 \times 10^{-3}\text{ m}$$

» ***Example* 3.5.** A proton is confined to a nucleus of radius 5×10^{-15} m. Calculate the minimum uncertainty in its momentum. What is the minimum uncertainty in the kinetic energy of the proton? $[m_p = 1.67 \times 10^{-27}\text{ kg}]$

Solution: Here, the nuclear diameter = $2 \times 5 \times 10^{-15} = 10^{-14}$ m.

For a proton to be confined in such a nucleus the minimum uncertainty in its position is

$$\Delta p = \frac{h/2\pi}{4x} = \frac{6.62 \times 10^{-34}}{2 \times 3.14 \times 10^{-14}} = 1.05 \times 10^{-20} \text{ kg-m/s}$$

The uncertainty in momentum of the proton must be at least comparable to its momentum.

$\therefore p = 1.05 \times 10^{-20}$ kg-m/s.

$\therefore$ minimum KE

$$(E_K)_{\min} = \frac{p^2}{2m} = \frac{(1.05 \times 10^{-20})^2}{2 \times 1.67 \times 10^{-27}} = 3.3 \times 10^{-14} \text{ J} = \mathbf{0.206 \text{ MeV}}$$

» ***Example* 3.6.** The average time interval that elapses between the excitation of an atom and the time it radiates energy is 10^{-8} sec. Calculate the uncertainty in the energy of emitted photons and the limit of accuracy with which the frequency of the emitted radiation may be determined.

Solution: If ΔE and Δt be the uncertainty in energy and time interval, then according to the uncertainty relation

$$\Delta E \cdot \Delta t = h \text{ or, } \Delta E = \frac{h}{\Delta t} = \frac{6.62 \times 10^{-34}}{10^{-8}} = 6.62 \times 10^{-26} \text{ J} = \frac{6.62 \times 10^{-26}}{1.6 \times 10^{-19}} = 4.14 \times 10^{-7} \text{ eV}$$

But, $\Delta E = \Delta(h\nu) = h\Delta\nu$.

$$\therefore \Delta\nu = \frac{\Delta E}{h} = \frac{6.62 \times 10^{-26}}{1.6 \times 10^{-34}} = \mathbf{10^8 \text{ cycle/sec}}$$

Hence, the limit of accuracy with which the frequency of the emitted radiation may be determined is $\mathbf{\Delta\nu = 10^8}$ **Hz.**

» ***Example* 3.7.** What is the minimum uncertainty in the energy state of an atom if an electron remains in this state for 10^{-8} s?

Solution: Since, the time available for energy measurement is 10^{-8}, the uncertainty in time $\Delta t = 10^{-8}$ sec.

Now, from the relation, $\Delta E \Delta t = \dfrac{h}{4\pi}$, we get

$$\Delta E = \frac{h}{4\pi \times \Delta t} \text{ or, } \Delta E = \frac{hC}{\Delta\pi C\Delta t} = \frac{12.4 \times 10^3}{4\pi(3 \times 10^8) \times 10^{-8} \times 10^{10}} = 0.329 \times 10^{-7} \text{ eV}$$

If τ be the mean lifetime of the excited state, then the *natural width* of the state is given by $\Gamma = h/(4\pi\tau)$.

Hence, the natural width of the state is 0.329×10^{-7} eV for this problem.

» ***Example* 3.8.** Find the average time that the atomic system remains in the energy state whose spectral line width is 10^{-4} Å for $\lambda = 4000$ Å.

Solution: Since, the natural width

$$\Gamma = \frac{h}{4\pi\tau} \quad \text{[where } \tau \text{ is the mean lifetime of the excited state]}$$

$$\text{or, } \tau = \frac{h}{4\pi\Gamma}$$

But, $\Gamma = \Delta E$ for $\Delta\lambda = 10^{-4}$ Å.

$$\because E = \frac{hc}{\lambda}, \therefore \Delta E = -\frac{hc}{\lambda^2}\Delta\lambda \text{ or, } |\Delta E| = \frac{hc}{\lambda^2}\Delta\lambda$$

$$\therefore \tau = \frac{h}{4\pi\Delta E} = \frac{h}{4\pi\left(\frac{hc}{\lambda^2}\Delta\lambda\right)} = \frac{\lambda^2}{4\pi c\Delta\lambda} = \frac{4\times10^{-7}}{4\pi(3\times10^8)\times10^{-14}} = \mathbf{4.24\times10^{-9}\ s}$$

***Example* 3.9.** In a Stern-Gerlach experiment, a collimated beam silver atom is produced by vapourising silver salt in a furnace at 1200 °C. If the beam travels 100 cm, use uncertainty principle to find the order of magnitude of the smallest spot that can be recorded on the detector. [*Given*: Planck's constant, $h = 6.6\times10^{-34}$ J-s; Boltzmann constant, $k = 1.4\times10^{-23}$ J/K. Mass of hydrogen atom = 1.66×10^{-27} kg.]

Solution: If the silver atom of mass m moves with velocity v along the x-axis from the furnace at temperature T, then its kinetic energy is $E_K = \frac{1}{2}mv^2 = \frac{3}{2}kT$.

$$\therefore v = \sqrt{\frac{3kT}{m}}$$

Time required by the silver atom to travel a distance of 100 cm is

$$t = \frac{100}{v} = \frac{100}{\sqrt{\frac{3kT}{m}}}$$

Now, time-energy uncertainty relation is $\Delta E\Delta t \sim \hbar$ or, $(\Delta E)_{min}(\Delta t)_{max} \simeq \hbar$.

$$\therefore (\Delta E)_{min} \simeq \frac{\hbar}{(\Delta t)_{max}} = \frac{\hbar}{t}\quad [\because (\Delta t)_{max} = t]$$

But, $(\Delta E)^2_{min} \dfrac{(\Delta p)^2_{min}}{2m}$.

$$\therefore \frac{(\Delta p)_{min}}{2m} = \frac{\hbar}{t} \text{ or, } (\Delta p)_{min} = \sqrt{\frac{2m\hbar}{t}}$$

From the uncertainty relation between position and momentum, we get $(\Delta x)_{max}(\Delta p)_{min} \simeq \hbar$

$$\text{or, } (\Delta x)_{max} \simeq \frac{\hbar}{(\Delta p)_{min}} = \frac{\hbar}{\sqrt{\frac{2m\hbar}{t}}} = \sqrt{\frac{\hbar t}{2m}} = \sqrt{\frac{\hbar}{2m}} \times \frac{\sqrt{100}}{(3kT/m)^{1/4}} = \frac{10\sqrt{\hbar}}{\sqrt{2}\,(3mkT)^{1/4}}$$

$$= \frac{10\times\left(\frac{6.6\times10^{-34}}{2\times3.14}\right)^{1/2}}{\sqrt{2}\,[3\times108\times1.66\times10^{-27}\times1.4\times10^{-23}\times1473]^{1/2}} = \mathbf{6.99\times10^{-7}\ m}$$

[Mass of silver atom is $m = 108\times1.66\times10^{-27}$ kg, Furnace temperature, $T = 1200 + 273 = 1473$ K]

The smallest spot on the detector corresponds to the maximum uncertainty in position of the silver atom. Hence, size of the smallest spot = 6.99×10^{-7} m.

***Example* 3.10.** A parallel beam of hydrogen atoms moving with velocity $v = 1.2\times10^3$ m/s falls normally on a diaphragm with a narrow slit, behind which a screen is placed at a distance of 1 m. Evaluate the width of the slit at which the width of the image on the screen is the minimum.

Solution: If the width of the slit be a and the additional broadening due to velocity uncertainty Δv_y caused by the slit be a', then $S(a) \simeq a + a'$.

If $\Delta x = a/2$ and the velocity spread of the train of waves is $\alpha \Delta v_y$, then $a' \simeq \left(\frac{2\hbar}{ma}\right)\frac{l}{v}$ [l = distance between the slit and the screen]

$$\therefore S(a) \simeq a + \left(\frac{2\hbar}{ma}\right)\frac{l}{v}. \qquad \therefore \frac{dS(a)}{da} = 1 - \frac{2\hbar}{ma^2}\frac{l}{v}$$

For $S(a)$ to be minimum, $\frac{dS(a)}{da} = 0. \therefore a^2 = \frac{2\hbar l}{mv}$.

Hence, the width of the image is minimum at

$$a = \sqrt{\frac{2\hbar l}{mv}} = \sqrt{\frac{2 \times 6.6 \times 10^{-34} \times 1}{2 \times 3.14 \times 9.1 \times 10^{-31} \times 10^8}} = \mathbf{10^{-5}\ m}$$

» ***Example* 3.11.** A nucleon is confined to a nucleus of diameter 5×10^{-14} m. Calculate the minimum uncertainty in the momentum of the nucleon.

Also calculate the minimum kinetic energy of the electron.

Solution: We have the uncertainty relation between position and momentum as

$$\Delta x\, \Delta p = h \text{ or, } (\Delta x)_{max} (\Delta p)_{min} = h$$

$$\therefore (\Delta p)_{min} = \frac{h}{(\Delta x)_{max}}$$

In this case, $(\Delta x)_{max} = 5 \times 10^{-14}$ m, $h = 6.626 \times 10^{-34}$ J-s.

$$(\Delta p)_{min} = \frac{6.626 \times 10^{-34}}{5 \times 10^{-14}} = 1.33 \times 10^{-20} \text{ kg-m/s}$$

Since, p cannot be less than $(\Delta p)_{min}$, we have $p_{min} = (\Delta p)_{min}$.

$$\therefore E_{min} = \frac{(\Delta p)^2_{min}}{2m} = \frac{(1.33 \times 10^{-20})^2}{2 \times 1.675 \times 10^{-27}} = 0.529 \times 10^{-13} \text{ J} = \frac{0.529 \times 10^{-13}}{1.6 \times 10^{-19}} \text{ eV} = \mathbf{0.33\ eV}$$

» ***Example* 3.12.** If a component of angular momentum of the electron in a hydrogen atom is known to be $2\hbar$ within 5 % error, show that its angular position in the plane perpendicular to that component cannot be specified at all.

Solution: According to uncertainty relation $\Delta L \Delta\theta \simeq \hbar$, where ΔL = uncertainty in angular momentum; $\Delta\theta$ = corresponding uncertainty in angular position.

$$\therefore \Delta L = \frac{5 \times 2\hbar}{100} \quad [\because \text{uncertainty in angular momentum is 5 \%}]$$

So

$$\Delta\theta = \frac{\hbar}{\Delta L} = \frac{\hbar \times 100}{5 \times 2\hbar} = \mathbf{10\ radians > 2\pi}$$

The angle in a plane perpendicular to the component of angular momentum cannot be greater than 2π. So the orbital angular position of the electron in the plane perpendicular to the given component of angular momentum cannot be specified at all.

» ***Example* 3.13.** The maximum uncertainty in the position of an electron in a nucleus is 2×10^{-14} m. Find the minimum uncertainty in its momentum. [$h = 6.63 \times 10^{-34}$ J-s] **[WBUT 2002]**

Solution: If Δx and Δp_x be the uncertainty in position and momentum of a particle, then according to Heisenberg's uncertainty relation, $\Delta x \, \Delta p_x \geqslant \hbar/2$.

Here, the maximum uncertainty in position, $\Delta x = 2 \times 10^{-14}$ m.

$\therefore$ the minimum uncertainty in position is

$$\Delta p_x = \frac{\hbar}{2\Delta x} = \frac{6.63 \times 10^{-34}}{(2 \times 3.14) \times 2 \times (2 \times 10^{-14})} = \mathbf{0.26 \times 10^{-20}\ kg\text{-}m/s}$$

***Example* 3.14.** On the average, an excited state of a system remains in that state for 10^{-11} s. What is the minimum uncertainty in the energy of an excited state? **[Question Bank, WBUT]**

Solution: If ΔE and Δt be the uncertainty in energy and time respectively, then according to uncertainty relation

$$\Delta E \cdot \Delta t \geqslant \frac{\hbar}{2}$$

$$\text{or, } \Delta E = \frac{\hbar}{2\Delta t} = \frac{1.054 \times 10^{-34}}{2 \times 10^{-11}} = \mathbf{0.527 \times 10^{-23}\ J}$$

***Example* 3.15.** If an electron is confined within a sphere of radius 1 Å, estimate the order of its minimum average kinetic energy using uncertainty principle. [Use non-relativistic relation.] **[BU (H) 2002]**

Solution: According to uncertainty relation, $\Delta x \, \Delta p_x \sim \hbar/2$.

Here, Δx and Δp_x are the uncertainties in position and x-component of momentum respectively.

$$\therefore \Delta p_x = \frac{\hbar}{2\Delta x} = \frac{1.054 \times 10^{-34}}{2 \times 2 \times 10^{-10}} \quad \left[\because \Delta x = 2 \times 1\,\text{Å} = 2 \times 10^{-10}\ \text{m and } \hbar = \frac{h}{2\pi} = 1.054 \times 10^{-34}\ \text{J}\right]$$

$$\text{or, } \Delta p_x = 0.2635 \times 10^{-24}\ \text{kg-m/s}$$

$\because p_x$ cannot be less than $(\Delta p_x)_{\text{min}}$, $p_{\text{min}} = (\Delta p_x)_{\text{min}}$.

$\therefore$ minimum energy of the electron is

$$E_{\text{min}} = \frac{(\Delta p_x)^2}{2m} = \frac{0.2635 \times 0.2635 \times 10^{-48}}{2 \times 9.1 \times 10^{-31}}\ \text{J} = \frac{0.2635 \times 0.2635}{18.2} \times 10^{-17}\ \text{J}$$

$$= \frac{0.2635 \times 0.2635 \times 10^{-17}}{18.2 \times 1.6 \times 10^{-19}}\ \text{eV} = \frac{0.2635 \times 0.2635}{18.2 \times 1.6} \times 10^{2}\ \text{eV} = \mathbf{0.24\ eV}$$

***Example* 3.16.** Estimate the maximum error in the determination of the velocity of an electron if it is constrained to move in the x-direction and if its x-coordinate is known with an uncertainty of 10^{-6} m.

[BU (H) 1999]

Solution: If Δx and Δp_x be the uncertainties in position along x-direction and the x-component of momentum respectively, then according to the uncertainty principle

$$\Delta x \, \Delta p_x = \frac{\hbar}{2}, \text{ or, } \Delta x m \Delta v_x = \frac{\hbar}{2}$$

$$\text{or, } \Delta v_x = \frac{\hbar}{2m\Delta x} \quad [\Delta v_x \text{ is the uncertainty in velocity}]$$

$$= \frac{1.054 \times 10^{-34}}{2 \times 9.1 \times 10^{-31} \times 10^{-6}}$$

where $\Delta x = 10^{-6}$ m and the mass of the electron, $m = 9.1 \times 10^{-31}$ kg.

$$\therefore \Delta v_x = \frac{1.054 \times 10^3}{18.2} = 0.057912 \times 10^3 = \mathbf{57.9\ m/s}$$

» ***Example* 3.17.** Find out the minimum momentum and the corresponding KE of an electron if it is to be confined within a nucleus having a diameter of 10^{-14} m. **[BU (H) 1996]**

Solution: Here, $\Delta x = 10^{-14}$ m, $\Delta p_x = ?$

From the uncertainty relation

$$\Delta p_x = \frac{\hbar}{2\Delta x} = \frac{1.054 \times 10^{-34}}{2 \times 10^{-14}} = 0.527 \times 10^{-20} \text{ kg-m/s } [\because \Delta x\, \Delta p_x \sim \hbar/2]$$

Now, the total relativistic energy of the electron is

$$E_{\min}^2 = (p_x c)^2 + m_0^2 c^4 = \left(0.527 \times 10^{-20} \times 3 \times 10^8\right)^2 + \left(9.11 \times 10^{-31}\right)^2 \times \left(3 \times 10^8\right)^4$$

$$= \left(0.527 \times 3 \times 10^{-12}\right)^2 + (9.11)^2 \times (81) \times 10^{-30}$$

$$[\because (p_x)_{\min} \text{ cannot be less than } \Delta p_x, \therefore p_{\min} = \Delta p_x]$$

$$\text{or, } E_{\min} \simeq (3 \times 0.527 \times 10^{-12}) = (1.571 \times 10^{-12})\text{ J} = \frac{1.571 \times 10^{-12}}{1.6 \times 10^{-19}} \text{ eV}$$

$$= 0.982 \times 10^7 \text{ eV} = 9.82 \times 10^6 \text{ eV} = \mathbf{9.82\ MeV}$$

Hence, the required momentum and kinetic energy of the electron are respectively 0.527×10^{-20} kgm/s and 9.82 MeV. As the maximum energy of the β-particles is ~ 4 MeV, the electron with 9.82 MeV energy cannot exist in the nucleus.

» ***Example* 3.18.** Apply the uncertainty relation to determine the ground state energy of a two-electron atom of atomic number Z.

Solution: If p_1 and p_2 be the momenta of two electrons, then from uncertainty relation

$$p_1 = \frac{\hbar}{r_1};\ p_2 = \frac{\hbar}{r_2} \tag{i}$$

where r_1 and r_2 are dimensions of the regions of the two electrons.

Kinetic energy

$$\text{KE} = \frac{p_1^2}{2m} + \frac{p_2^2}{2m} = \frac{\hbar^2}{2m}\left(\frac{1}{r_1^2} + \frac{1}{r_2^2}\right) \tag{ii}$$

Total energy

$$E = \frac{\hbar^2}{2m}\left(\frac{1}{r_1^2} + \frac{1}{r_2^2}\right) + \frac{e^2}{r_1 + r_2} - Ze\left(\frac{1}{r_1} + \frac{1}{r_2}\right) \tag{iii}$$

Here, the second term represents the interaction energy and the third term, the potential energy. For the ground state the energy should be minimum and hence

$$\frac{dE}{dr_1} = \frac{dE}{dr_2} = 0$$

This gives

$$r_1 = r_2 = \frac{\hbar^2}{me(Ze - \frac{e}{4})}$$

Using this result in (iii)

$$E = -\frac{me^4}{\hbar^2}\left(z - \frac{1}{4}\right)^2 \quad \text{(iv)}$$

Equation (iv) gives the desired ground state energy of two-electron atom of atomic number Z.

» ***Example* 3.19.** Calculate the uncertainty in the angle of emergence of 1 MeV electron passing through a hole of diameter of 20 micron.

Solution: The KE of the electron is $\frac{p^2}{2m} = 1.6 \times 10^{-3}$ J.

$$\therefore p = \left[2 \times 1.6 \times 10^{-13} \times 9.1 \times 10^{-31}\right] = 5.39 \times 10^{-22}$$

The order of uncertainty in the momentum is given by $\Delta p_x \, \Delta x = \hbar$.

$$\therefore \Delta p_x = \frac{\hbar}{\Delta x} = \frac{6.62 \times 10^{-34}}{2\pi \times 2 \times 10^{-5}} = 5.2 \times 10^{-30}$$

Hence, the uncertainty in the angle of emergence is given by

$$\Delta\theta = \frac{\Delta p_x}{p} = \frac{5.2 \times 10^{-30}}{5.39 \times 10^{-22}} = \mathbf{9.6 \times 10^{-9}\ rad}$$

» ***Example* 3.20.** A 20 gm rifle bullet takes 1 sec to reach its target. Find the order of magnitude of the spread of successive shots at the target considering that (i) the shots are fired accurately enough (ii) bullet is a point mass (iii) the effects of air and the earth motion can be neglected.

Solution: Under the conditions stated above, the minimum spread in aims will be observed. Now, according to uncertainty relation, the energy is uncertain by

$$(\Delta E)_{\min} = \frac{\hbar}{(\Delta t)_{\max}} = \frac{\hbar}{\text{time of travel of bullet}} = \frac{\hbar}{t}$$

∴ the uncertainty in position

$$(\Delta y)_{\max} = \frac{\hbar}{(\Delta p)_{\min}} = \frac{\hbar}{\sqrt{2m(\Delta E)_{\min}}} = \frac{\hbar}{\sqrt{2m(\hbar/t)}} = \sqrt{\frac{\hbar t}{2m}}$$

Here, $t = 1$ sec and $m = 20$ gm.

$$\therefore (\Delta y)_{\max} = \left[\frac{(6.625 \times 10^{-27}) \times 1}{(2 \times 3.14) \times 2 \times 20}\right]^{1/2} = \left[\frac{6.625 \times 10^{-27}}{3.14 \times 80}\right]^{1/2} = \mathbf{0.5 \times 10^{-14}\ cm}$$

Hence, the shot will miss the target by an amount of 0.5×10^{-14} cm which is negligibly small in this case, because the rings are on macroscopic scale.

» ***Example* 3.21.** If the average period that elapses between the excitation of an atom and the time it radiates energy is 10^{-8} sec, then calculate the uncertainty in the energy of emitted photon and limit of accuracy with which the frequency of emitted radiations can be determined.

Solution: From energy-time uncertainty principle, $\Delta E \Delta t = h$

$$\text{or, } \Delta E = \frac{h}{\Delta t} = \frac{6.62 \times 10^{-27}}{10^{-8}} \text{ ergs} = 6.62 \times 10^{-19} \text{ ergs} = \frac{6.62 \times 10^{-19}}{1.6 \times 10^{-12}} \text{ eV} = 4.14 \times 10^{-7} \text{ eV}$$

Again, $\Delta E = \Delta(h\nu) = h\Delta\nu$.

$$\therefore \Delta\nu = \frac{\Delta E}{h} = \frac{6.62 \times 10^{-19}}{6.62 \times 10^{-27}} \text{ cycles/sec} = 10^8 \text{ cycles/sec} = \mathbf{100\ MHz}$$

» ***Example* 3.22.** Using uncertainty principle estimate the ground state energy of the LHO.

[CU (H) 2004; cf. KU (H) 2000; BU (H) 2005]

Solution: The total energy of the LHO is

$$E = \frac{p^2}{2m} + \frac{1}{2} m\omega^2 x^2$$

Taking $p \sim \Delta p$ and $x \sim \Delta x$, where Δp and Δx are the uncertainties in momentum and the corresponding position of the LHO, we have

$$E \sim \frac{(\Delta p)^2}{2m} + \frac{1}{2} m\omega^2 (\Delta x)^2$$

From the uncertainty relation

$$\Delta x \, \Delta p \sim \hbar/2 \text{ or, } \Delta p \sim \frac{\hbar}{2\Delta x} = \frac{\hbar}{2a} \quad (\text{if } \Delta x = a)$$

$$\therefore E \sim \frac{\hbar^2}{8ma^2} + \frac{1}{2} m\omega^2 a^2. \quad \therefore \frac{dE}{da} = -\frac{\hbar^2}{4ma^3} + m\omega^2 a$$

But, E to be $E_{\min}$, $\dfrac{dE}{da} = 0$.

$$\therefore -\frac{\hbar^2}{8ma^3} + m\omega^2 a = 0 \text{ or, } -\frac{\hbar^2}{4ma^2} + m\omega^2 a^2 = 0 \text{ or, } a^2 = \frac{\hbar^2}{2m\omega}$$

$$\therefore E_{\min} = \frac{\hbar}{8m}[2m\omega] + \frac{1}{2} m\omega^2 \left(\frac{\hbar}{2m\omega}\right) = \frac{1}{2}\hbar\omega$$

» ***Example* 3.23.** In hydrogen atom an electron has angular momentum $2\hbar$ measured to an accuracy of 5 %. Show that its angular position in a perpendicular plane cannot be specified.

Solution: if $\Delta\theta$ and ΔJ_z be the uncertainties in angular position and angular momentum of the electron then from uncertainty relation

$$(\Delta J_z)(\Delta\theta) \sim \hbar \text{ or, } \Delta\theta \sim \frac{\hbar}{\Delta J_z} = \frac{h}{\frac{5}{100} 2\hbar} = 10 \text{ radians}$$

$$\text{or, } \Delta\theta = 10 \times \frac{180}{\pi} \text{ degrees} = 573°$$

Since, the angle in a plane cannot be greater than 360°, the angular position of the electron cannot be specified.

» ***Example* 3.24.** Calculate the smallest possible uncertainty in the position of an electron if (i) it is moving with speed 3×10^7 ms^{-1}, (ii) its speed is 360 ms^{-1} known to be accurate to 0.01 %.

Solution: If Δx and Δp_x be respectively the uncertainties in position and corresponding momentum of the electron, then

$$(\Delta x)(\Delta p_x) \sim \hbar \text{ or, } (\Delta x)(\Delta p_x) \sim \frac{h}{2\pi}$$

$$\text{or, } (\Delta x)\, m(\Delta v) \sim \frac{h}{2\pi}$$

$$\text{or, } \Delta x = \frac{h}{2\pi m \Delta v}$$

(i) When the electron speed is

$$v = 3\times 10^7\ \text{ms}^{-1} = \frac{3\times 10^7\ \text{ms}^{-1}}{3\times 10^8\ \text{ms}^{-1}} c = \frac{c}{10} \text{ (relativistic)}$$

Taking $v = 3\times 10^7\ \text{ms}^{-1} = \frac{c}{10}$

$$m = \frac{m_0}{\sqrt{1-\frac{v^2}{c^2}}} = \frac{9.1\times 10^{-31}\ \text{kg}}{\sqrt{1\left(\frac{c/10}{c}\right)^2}} = 9.1458\times 10^{-31}\ \text{kg}$$

$$\therefore \Delta x = \frac{6.626\times 10^{-34}\ \text{J-s}}{2(9.1458\times 10^{-31}\ \text{kg})(3\times 10^7\ \text{ms}^{-1})} = \mathbf{0.2384\ Å}$$

(ii) $v = 300\ \text{ms}^{-1}$ (non relativistic)

$$\Delta v = \frac{0.01}{100}\times 300\ \text{ms}^{-1} = 0.03\ \text{ms}^{-1} \text{ and } m \simeq m_0$$

$$\therefore \Delta x = \frac{h}{2\pi m_0 \Delta v} = \frac{6.626\times 10^{-34}\ \text{J-s}}{2\pi(9.1\times 10^{-31})(0.03\ \text{ms}^{-1})} = \mathbf{3.86\times 10^{-3}\ m}$$

» ***Example* 3.25.** The position and momentum of a 5 keV electron are simultaneously determined. If the position is located to within 0.5 Å, what is the percentage of uncertainty in its momentum? **[CU (H) 1998]**

Solution: The momentum p of the electron is $p = \sqrt{2mT} = \sqrt{2\times(9.1\times 10^{-3}\ \text{kg})\times 5\times 10^3\times 1.6\times 10^{-19}\ \text{J}}$

$$= 3.8\times 10^{-23}\ \text{kg m s}^{-1}$$

The position of the electron can be within 0.5 Å, i.e., uncertainty in position is $\Delta x = 0.5$ Å.

∴ from Heisenberg's uncertainty principle

$$(\Delta x)(\Delta p) \sim \frac{\hbar}{2} \text{ or, } \Delta p = \frac{\hbar}{2(\Delta x)} = \frac{6.626\times 10^{-24}\ \text{J-s}}{2\times 2\pi\times 0.5\times 10^{-10}\ \text{m}} = 1.05\times 10^{-24}\ \text{kg m s}^{-1}$$

∴ % uncertainty in momentum is

$$\frac{\Delta p}{p}\times 100 = \frac{1.05\times 10^{-24}\ \text{kg m s}^{-1}}{3.8\times 10^{-23}\ \text{kg m s}^{-1}}\times 100 = 2.76\,\%$$

» ***Example* 3.26.** An electron of energy 200 eV is passed through a circular hole of radius 10^{-4} cm. What is the uncertainty introduced in the angle of emergence? **[CU (H) 1997, 2004, 2014]**

Solution: Here, the momentum of the electron is

$$p = \sqrt{2mT} = \sqrt{2(9.1\times 10^{-31}\ \text{kg})(200\times 1.6\times 10^{-19}\ \text{J})} = 7.6\times 10^{-24}\ \text{kg m s}^{-1}$$

Now, Heisenberg's uncertainty relation is

$$\Delta x\, \Delta p_x \sim \hbar \text{ or, } \Delta p \sim \frac{\hbar}{\Delta x} = \frac{\frac{1}{2\pi}(6.62 \times 10^{-34})\text{ J-s}}{2 \times (10^{-4} \times 10^{-2}\text{ m})}$$

$\therefore \Delta p \sim 5.27 \times 10^{-29}\text{ kg m s}^{-1}$.

Now, the uncertainty in the angle of emergence is

$$\Delta\theta = \frac{\Delta p}{p} = \frac{5.27 \times 10^{-29}\text{ kg m s}^{-1}}{7.6 \times 10^{-24}\text{ kg m s}^{-1}}$$

$$= 6.9 \times 10^{6} \text{ radian} = 6.9 \times 10^{-6} \times \frac{180}{\pi} \times 60 \times 60 \text{ sec of arc}$$

$$= \mathbf{1.4\ seconds\ of\ arc}$$

» ***Example* 3.27.** Consider a nucleus of mass number $A = 64$ and radius $R = 4.8$ fm. Using uncertainty principle estimate uncertainty in the momentum of a nucleon and its KE.

Solution: Volume of the nucleus

$$V = \frac{4}{3}\pi R^3 \text{ or, } V = \frac{4}{3}\pi\left(4.8 \times 10^{-15}\text{ m}\right)^3 = 4.6 \times 10^{-23}\text{ m}^3$$

Let the uncertainty in position of nucleon (Δr) = radius of volume available to nucleon r, i.e., $\Delta r = r = 1.2 \times 10^{-15}$ m.

So, the uncertainty in momentum of nucleon is

$$\Delta p = \frac{\hbar}{\Delta r} = \frac{6.626 \times 10^{-34}\text{ J-s}}{2\pi \times (1.2 \times 10^{-15}\text{ m})} = \mathbf{8.79 \times 10^{-20}\ kg - m\, s^{-1}}$$

If the momentum of the nucleon is of the order of Δp (i.e., $p \sim \Delta p$), the uncertainty in KE of nucleon is

$$T = \frac{p^2}{2M} = \frac{(\Delta p)^2}{2M} = \frac{\left(8.79 \times 10^{-20}\text{ kg m s}^{-1}\right)^2}{2\pi \times (1.2 \times 10^{-15}\text{ m})} \quad [\text{mass } M = 1.87 \times 10^{-27}\text{ kg}]$$

or, $\mathbf{T = 14.375\ MeV}$

» ***Example* 3.28.** If an electron is confined within a region eta of 10^{-10} m, find the minimum uncertainty in its momentum and velocity. **[WBCS 2005; cf. KU (H) 2005; CU (H) 2001, 2012, 2014; BU (H) 1999]**

Solution: We have the uncertainty relation of position and momentum

$$(\Delta x)(\Delta p) \sim \frac{\hbar}{2} \text{ or, } \Delta p \sim \frac{\hbar}{2(\Delta x)}$$

As the electron can be found anywhere within a region 10^{-10} m, so $\Delta x = 10^{-10}$ m.

$$\therefore \Delta p \sim \frac{6.626 \times 10^{-34}\text{ J-s}}{2\pi \times 2 \times (10^{-10}\text{ m})} = \mathbf{5.27 \times 10^{-25}}\ kg - m\, s^{-1}$$

$$\therefore m\Delta v \sim 5.27 \times 10^{-25}\text{ kg m s}^{-1}$$

$$\text{or, } \Delta v \sim \frac{5.27 \times 10^{-25}\text{ kg m s}^{-1}}{9.1 \times 10^{-31}\text{ kg}} \quad [m = \text{mass of the electron} = 9.1 \times 10^{-31}\text{ kg}]$$

or, $\mathbf{\Delta v \sim 579 \times 10^{3}\ ms^{-1}}$

» ***Example* 3.29.** In an experiment an elementary particle is observed at intervals as short as 10^{-7} s. Find the minimum uncertainty in energy of the particle expressed in eV. **[KU (H) 2008]**

Solution: From Heisenberg's uncertainty relation for time-energy, we have

$$(\Delta E)(\Delta t) \sim \hbar \text{ or, } \Delta E \sim \frac{\hbar}{\Delta t} = \frac{6.626 \times 10^{-34}\text{ J-s}}{2\pi \times 10^{-7}\text{ s}} = 1.055 \times 10^{-27}\text{ J}$$

$$\text{or, } \Delta E \sim \frac{1.055 \times 10^{-27}}{1.6 \times 10^{-19}}\text{ eV} = \mathbf{6.59 \times 10^{-9}\ eV}$$

» ***Example* 3.30.** The life time of an atomic state is of the order of 10^{-8} s. Find the line width and the spread in frequency of radiation emitted by atom when it goes to ground state.

Solution: Given, lifetime of atomic state $\tau \sim 10^{-8}$ s.

In view of Heisenberg's uncertainty relation the minimum energy spread or line width is

$$\Delta E = \frac{\hbar}{r} = \frac{h}{2\pi r} = \frac{6.626 \times 10^{-34}\text{ J-s}}{2\pi \times 10^{-8}\text{ s}} = \mathbf{1.055 \times 10^{-26}\ J}$$

The minimum frequency spread when the atom goes to ground state is

$$\Delta \nu = \frac{\Delta E}{h} = \frac{1.055 \times 10^{-26}\text{ J}}{6.626 \times 10^{-34}} = \mathbf{15.9 \times 10^{6}\ s^{-1}}$$

» ***Example* 3.31.** For a particle of mass 10^{-6} gm, calculate the uncertainty if that in position is 8×10^{-11} cm. [Treat the case as non-relativistic] **[KU (H) 2003]**

Solution: We have the Heisenberg's uncertainty relation in position and momentum

$$\Delta x\, \Delta p = \frac{\hbar}{2} \text{ or, } x \cdot m(\Delta v) = \frac{\hbar}{2}$$

$$\text{or, } \Delta v = \frac{\hbar}{2m\Delta x} = \mathbf{6.59 \times 10^{-14}\ ms^{-1}}$$

» ***Example* 3.32.** 4 keV electrons from a source fall straight on a target 50 cm away. Find the radius of the electronic spot (order only) due to uncertainty principle, assuming focusing to be prefect.

Solution: If m and T be the energy of the electron, then its momentum

$$p = \sqrt{2mT} = \left(2 \times 9.1 \times 10^{-31}\text{ kg}\right) \times \left(4 \times 10^{3} \times 1.6 \times 10^{-19}\text{ J}\right)$$

$$= 3.4 \times 10^{-23}\text{ kg m s}^{-1} \left[\because \Delta x\, \Delta p_x \sim \hbar \therefore \Delta p = \frac{\hbar}{\Delta x}\right]$$

Again,

$$\Delta\theta = \frac{\Delta p}{p} = \frac{\Delta x}{50 \times 10^{-2}\text{ m}} \text{ or, } \frac{\hbar}{p\Delta x} = \frac{\Delta x}{50 \times 10^{-2}\text{ m}}$$

$$\text{or, } \Delta x = \sqrt{\frac{\hbar}{p} \times 50 \times 10^{-2}\text{ m}} = \frac{\left(6.626 \times 10^{-34}\text{ J-s}\right)\left(50 \times 10^{-2}\text{ m}\right)}{2\pi\left(3.4 \times 10^{-34}\text{ kg m s}^{-1}\right)} = 1.25 \times 10^{-6}\text{ m}$$

$\therefore$ radius of the electron spot is

$$\frac{\Delta x}{2} = \mathbf{6.25 \times 10^{-7}\ m}$$

» ***Example* 3.33.** Wavelength can be determined with accuracy of one part in 10^4. What is the uncertainty in the position of 1 Å X-ray photon when the wavelength is simultaneously measured? **[KU (H) 1997]**

Solution: From the principle of uncertainty

$$(\Delta x)(\Delta p_x) \sim \hbar \text{ or, } (\Delta x)\hbar(\Delta k) = \hbar \; [\because \Delta p_x = \hbar \Delta k]$$

$$\text{or, } (\Delta x)(\Delta k) = 1$$

» ***Example* 3.34.** An electron is confined within a region of 10^{-10} m, find the minimum uncertainty in its momentum and velocity. **[BU (H) 1999; WBCS 2005; KU (H) 2005; CU (H) 2001, 2012, 2014]**

Solution: We know that Heisenberg's uncertainty relation for position and momentum is given by

$$(\Delta x)(\Delta p_x) \sim \frac{\hbar}{2} \text{ or, } \Delta p_x \sim \frac{\hbar}{2(\Delta x)}$$

As the electron can be found anywhere within a region 10^{-10} n, so $\Delta x = 10^{-10}$ m.

$$\therefore \Delta p_x \sim \frac{6.626 \times 10^{-34} \text{ J-s}}{2\pi \times 2 \times (10^{-10} \text{ m})} = 5.27 \times 10^{-25} \text{ kg} - \text{m s}^{-1}$$

$$\text{or, } m\Delta v_x \sim 5.27 \times 10^{-25} \text{ kg m s}^{-1}$$

$$\therefore \Delta v_x \sim \frac{5.27 \times 10^{-25} \text{ kg m s}^{-1}}{9.1 \times 10^{-31} \text{ kg}} \quad [m = 9.1 \times 10^{-31} \text{ mass of electron}]$$

$$\text{or, } \Delta v_x \sim 5.79 \times 10^3 \text{ ms}^{-1}$$

EXERCISES

3(A): Objective (or MCQ) Type Questions

Select the correct answer:

3.1. If the number of superposed waves and the range of wavelength be increased, then the linear spread of the wave packet

(a) is increased (c) remains constant
(b) is decreased (d) is doubled.

3.2. Uncertainty principle is the consequence of

(a) wave nature of particle (c) wave-particle duality
(b) particle nature of wave (d) particle-particle interaction.

3.3. Uncertainties in canonically conjugated variables arise due to

(a) limitation of the apparatus (c) law of nature
(b) personal error (d) none of them.

3.4. For very heavy particle

(a) $\Delta x \, \Delta v_x \sim \hbar$ (c) $\Delta x \, \Delta v_x = \infty$
(b) $\Delta x \, \Delta v_x = 0$ (d) $\Delta x \, \Delta v_x = \lambda$.

3.5. If an atomic system remains in a particular state for a maximum interval of time, then the uncertainty in the measurement of energy is

(a) maximum
(b) zero
(c) minimum
(d) infinite.

3.6. Quantum mechanics is

(a) deterministic
(b) probabilistic
(c) partly probabilistic
(d) partly deterministic.

3.7. In view of uncertainty principle, the spectral lines are

(a) infinitely sharp
(b) broad
(c) intense
(d) faint.

3.8. In γ-ray thought experiment, γ-ray is used because γ-rays have

(a) high energy
(b) no visibility
(c) low wavelength
(d) pair production ability.

3.9. In the measurement of a physical entity an experiment can exhibit only

(a) wave aspect
(b) particle aspect
(c) wave and particle aspect simultaneously
(d) none of them.

3.10. A nucleus cannot harbour

(a) a proton
(b) a neutron
(c) an α-particle
(d) an electron.

3.11. The highest value of a β-particle emitted spontaneously from a nucleus has energy

(a) 10 Mev
(b) 4 Mev
(c) 0.5 MeV
(d) 15 keV.

3.12. The ground state energy of an LHO is

(a) 0
(b) $\frac{1}{2}\hbar\omega$
(c) $\frac{3}{2}\hbar\omega$
(d) ∞.

3.13. In view of uncertainty principle the radiation emitted by an atom can have

(a) a definite frequency
(b) a band of frequencies
(c) a definite phase
(d) a definite wavelength.

3.14. Uncertainty principle has significance only in

(a) macroscopic world
(b) microscopic world
(c) both macroscopic and microscopic world
(d) anti-world.

3.15. Uncertainty principle tells that

(a) A particle can have only position but no momentum
(b) A particle can have only momentum but no position
(c) One can determine simultaneously the position and momentum of a particle

(d) One cannot determine simultaneously the position and momentum of a particle. **[WBUT 2006]**

3.16. Which of (1) $\Delta x\, \Delta p_x \sim \hbar$, (2) $\Delta x\, \Delta p_x \ngtr \frac{\hbar}{2}$, (3) $\Delta x\, \Delta p_x \nless \frac{\hbar}{2}$, and (4) $\Delta x\, \Delta p_x \geqslant \frac{\hbar}{2}$ represent Heisenberg's uncertainty principle?

(a) 1 and 3 (c) 2 and 4

(b) 1 and 2 (d) 4 only.

ANSWERS

3.1. (b).	**3.4.** (b).	**3.7.** (b).	**3.10.** (d).	**3.13.** (b).	**3.16.** (a).
3.2. (c).	**3.5.** (c).	**3.8.** (c).	**3.11.** (b).	**3.14.** (b).	
3.3. (c).	**3.6.** (b).	**3.9.** (a) (b).	**3.12.** (b).	**3.15.** (d).	

3(B): Short Answer Type Questions

3.1. Is uncertainty principle the outcome of wave description of a particle? Justify.

3.2. State and explain uncertainty principle.

3.3. Why is uncertainty principle not applied for macroscopic objects in motion?

3.4. Obtain uncertainty relation from standard deviations of position coordinates (x) and the corresponding momentum coordinates (p_x).

3.5. Do you expect to observe wave and particle aspect of a physical entity simultaneously in the same experiment? Why?

3.6. What is about the Bohr's frequency condition if uncertainty principle is true?

3.7. Discuss the significance and importance of uncertainty principle.

3.8. Justify the use of γ-rays in γ-ray microscope (thought) experiment.

3.9. Using uncertainty relation show that a nucleus cannot harbour an electron. **[Question Bank, WBUT]**

3.10. *We cannot take measurement of physical property of a system without seriously disturbing it.*—Explain.

3.11. *Quantum mechanics is probabilistic whereas classical mechanics is deterministic.*—Discuss.

3.12. Estimate the minimum uncertainty in the momentum of an electron confined in a box of length 1 μm. **[BU (H) 2004]**

3.13. What is Heisenberg's uncertainty principle? **[WBUT 2005]**

3.14. Describe Heisenberg's thought experiment on the measurement of an electrons' position by photon and arrive at the position moment uncertainty relation. **[BU (H) 2013]**

3.15. State and explain Heisenberg's uncertainty principle. Explain why free electrons cannot exist in atomic nuclei. **[VU (H) 2009]**

3.16. State prove and comment on Heisenberg's uncertainty relation. **[cf. KU (H) 2015]**

3(C): Broad Answer Type Questions

3.1. (a) Starting from de Broglie wave concept obtain Heisenberg's uncertainty principle, $\Delta x\, \Delta p_x \geqslant \hbar$.

(b) Give one illustration of this principle.

3.2. (a) State and explain uncertainty principle. Discuss its significance. What is its importance?

(b) Describe single-slit electron diffraction experiment to prove the validity of the uncertainty principle.

3.3. (a) Give an elementary proof of the uncertainty relation between energy and time. What is its significance?

(b) Describe γ-ray microscope experiment to establish the truth of uncertainty principle. **[BU (H) 2004]**

3.4. (a) Show that for the rotational motion of a particle the uncertainty principle can be stated in the form, $\Delta L \cdot \Delta\phi \geqslant \hbar$, where ΔL is the uncertainty in the angular momentum of the particle and $\Delta\phi$ is the uncertainty in its angular position.

(b) Using uncertainty relation estimate the size of the hydrogen atom.

3.5. (a) Prove that proton, neutron and α-particle can exist within the nucleus. **[Question Bank, WBUT]**

(b) Using uncertainty relation prove that the lowest energy of an LHO is $\frac{1}{2}\hbar\omega$.

3.6. (a) What is the principle of complementarity? Explain.

(b) Discuss causality in quantum mechanics.

(c) Discuss the concept of Bohr orbit in view of uncertainty relation.

3.7. (a) State and explain Heisenberg uncertainty principle.

(b) Give an account of the γ-ray microscope experiment and discuss the results. **[WBUT 2003]**

3.8. (a) *Uncertainty principle is the consequence of wave packet description of the particles.*—Discuss.

(b) Evaluate the size and the ground state energy of hydrogen atom using uncertainty relation. **[BU (H) 2011]**

3.9. (a) Discuss the uncertainty principle in the context of observing an electron by a γ-ray microscope.

(b) What is the uncertainty relation as a consequence of a description of a particle by three-dimensional wave packet? **[BU (H) 2008]**

3.10. (a) Give an elementary proof of Heisenberg's uncertainty principle. **[cf. KU (H) 2011]**

(b) Explain from uncertainty relation if an α-particle can stay in nucleus.

(c) Estimate ground state energy of helium atom from Heisenberg's uncertainty relation.

3.11. (a) State and explain Heisenberg's uncertainty principle. Use it to prove that electrons cannot exist inside the nucleus. **[cf. CU (H) 1999; WBCS 2008; IAS 1997]**

(b) How does then one account for the beta decay of a nucleus? **[IAS 2000]**

3.12. (a) What is a zero-point energy? Show that the existence of zero point energy is inconformity with the uncertainty principle. **[CU (H) 1992, 1999; cf. KU (H) 2007; cf. WBCS 2006]**

(b) Briefly describe Heisenberg's thought experiment with γ-ray microscopy to illustrate uncertainty principle. **[BU (H) 2004]**

3.13. (a) Give an account of the γ-ray microscope experiment and discuss the results. Can such an experiment be actually performed? Explain your result. **[CU (H) 2006]**

(b) Using Heisenberg's uncertainty principle find the ground state energy of Bohr radius of hydrogen atom. **[KU (H) 2013; IAS 1997]**

3.14. (a) Using uncertainty principle $\Delta x\, \Delta p \sim \frac{\hbar}{2}$, estimate the minimum energy of a particle in simple harmonic potential $u = \frac{1}{2}kx^2$. **[IAS 1999]**

(b) Using Heisenberg's uncertainty principle calculate the mass being exchanged between nucleons inside the nucleus, taking the range of nuclear force as 1.4 fm. **[IAS 1995]**

3.15. (a) Discuss Heisenberg's uncertainty principle. Consider a linear harmonic oscillator for which the total energy is given by

$$E = \frac{p_x^2}{2m} + \frac{1}{2}m\omega^2 x^2$$

where the symbols have their usual meanings.

The particle is assumed to be confined to a region ~ a. Using the uncertainty principle to obtain the minimum energy (ground state energy) of the oscillator. **[BU (H) 2005; CU (H) 2004; KU (H) 2000]**

(b) Make a supposition of two harmonic waves the to construct a wave packet. Find the width of the wave packet and hence obtain uncertainty principle. **[KU (H) 1997]**

3.16. State and explain Heisenberg's uncertainty principle using situations where one of the parameters in (a) linear distance x, (b) angle ϕ, and (c) energy E.

3.17. Give an elementary proof of Heisenberg's uncertainty principle. **[cf. KU (H) 2011]**

3.18. (a) State and prove Heisenberg's uncertainty principle. Use it to prove that electrons cannot exist inside the nucleus. **[cf. CU (H) 1999; WBCS 2008]**

(b) How does then one account for the β-decay of a nucleus. **[IAS 2000]**

3.19. What is zero-point energy? Show that the existence of zero-point energy is in conformity with the uncertainty principle. **[WBCS 2006; KU (H) 2007; CU (H) 1992, 1999]**

3.20. State Heisenberg's uncertainty principle. Use it to estimate the ground state energy of a particle confined within a rigid one-dimensional box of length L. **[CU (H) 1999]**

3.21. Briefly describe Heisenberg's thought experiment with γ-ray microscope to illustrate the uncertainty principle. **[BU (H) 2004]**

3.22. Using Heisenberg's uncertainty principle. Calculate the mass being exchanged between nucleus inside the nucleus, taking the range of unclear force as 1.4 fm. **[IAS 1995]**

3.23. Using Heisenberg's uncertainty relation find the ground state energy and Bohr radius of hydrogen atom. **[IAS 1997; KU (H) 2013]**

3.24. State Heisenberg's uncertainty principle. Explain how the diffraction of electrons by a narrow slit together with de Broglie hypothesis illustrate this principle.

3(D): Numerical Problems

3.1. Calculate the indeterminacy of the velocity of an electron in a hydrogen atom assuming the size of the atom to be about 10^{-8} cm.

Compare the obtained value with the velocity of an electron in the first Bohr orbit.

[*Ans.* 10^6 m/s, 2.2×10^6 m/s]

3.2. A boy on the top of a vertical ladder of height h is dropping marbles of mass m to the floor. To aim well, the boy is using equipment of the highest possible precision. Estimate the typical distance by which he will miss the crack on the floor he is aiming at. [*Ans.* $S = \left(\frac{2\hbar^2 h}{m^2 g}\right)^{1/4}$]

3.3. Suppose, that the x-component of the velocity of a particle 2×10^{-4} kg mass is measured to an accuracy of $\pm 10^{-6}$ m/s. What, then is the limit of the accuracy with which we can locate the particle along the x-axis? [*Ans.* 1.32×10^{-25} m]

3.4. On the average, an excited state of a system exists in that state for 10^{-11} s. What is the minimum uncertainty in the energy of an excited state? **[Question Bank, WBUT]**

[*Ans.* 1.054×10^{-23} J]

3.5. β-particles of kinetic energy 1 MeV are emitted from a radioactive nucleus of mass number 64. Prove by the application of uncertainty principle, that these particles cannot exist inside the nucleus, given rest mass energy of the electron 0.5 MeV.

3.6. Estimate the maximum error in the determination of velocity of an electron if it is constrained to move in the x-direction and if its x-coordinate is known with an uncertainty 10^{-8} m. **[BU (H) 1999]**

3.7. A microscope using photons is employed to locate an electron in an atom to within a distance 0.25 Å. What is the uncertainty in the velocity of the electron located in this way? **[BU (H) 1993]**

3.8. Wavelengths can be determined with accuracies of 1 part in 10^6. What is the uncertainty in the position of a 1 Å X-ray photon when its wavelength is simultaneously measured? [*Ans.* 1.6×10^{-5} m]

3.9. An unstable elementary particle, called the *eta-meson,* has a rest mass of 549 MeV/C^2 and a mean lifetime of 7.00×10^{-19} s. What is the uncertainty in its rest mass?

3.10. Explain why spectral lines always have a width. If lifetime of an excited state is 10^{-8} s, determine the natural width in frequency scale.

3.11. An oil drop of mass 10^{-12} gm is floating on the top of a liquid. Its position at any instant can be determined with a probable error of 10^{-4} cm. Calculate the uncertainty in the value of its velocity.

3.12. A marine radar operating at a frequency of 9400 MHz emits groups of electromagnetic waves 0.08 μs in duration. The time needed for the reflections of these groups to return indicates the distance to a target.

(a) Find the length of each group and the number of waves it contains. [*Ans.* 24 m; 752 waves]

(b) What is the approximate maximum bandwidth the radar receiver must be able to process?

[*Ans.* 12.5 MHz]

3.13. Calculate the minimum energy of a proton for its existence within the nucleus of diameter 10^{-14} m.

[*Ans.* 4.9 MeV]

3.14. A particle of mass is confined to a one-dimensional line of length l. From uncertainty principle, determine the value of the smallest energy that the body can have. From this result what conclusion can you draw?

[*Ans.* $E_1 = \frac{h^2}{8ml^2}$; According to uncertainty principle, bound system cannot have zero energy.]

3.15. To explain the nuclear forces, Yukawa has suggested the existence of a massive meson. Applying the uncertainty, principlle derive a relation between the range of the force and the mass of the meson.

[*Hints:* The uncertainty in the position of a nucleon during transit, when it emits a meson absorbed by another nucleon, is of the order of the range of the force.

Thus, $\Delta x = x_0$, where x_0 = the range of the force.

According to uncertainty relation

$$\Delta p_x \simeq \frac{\hbar}{\Delta x} \simeq \frac{\hbar}{x_0}$$

Again, for a relativistic particle

$$\Delta p_x = mc \text{ or, } mc = \frac{\hbar}{x_0} \text{ or, } x_0 = \frac{\hbar}{mc}$$

Putting $x_0 = 1.5 \times 10^{-15}$ m, $\hbar = 1.054 \times 10^{-27}$ J-s, $C = 3 \times 10^8$ m/s, we get

$$m = \frac{\hbar}{x_0 c} = 0.234 \times 10^{-27} \text{ kg}$$

This gives an estimate of the mass m of the meson.]

3.16. Estimate the uncertainty in position of a neutron moving with velocity 5×10^6 m/s. **[BU (H) 2015]**

3.17. State Heisenberg's uncertainty principle. The uncertainty in the mass measurement of an elementary particle is 5 MeV. Estimate its lifetime. **[CU (H) 2008]**

3.18. Consider the one-dimensional wave packet

$$\psi(x) = \frac{1}{\sqrt{2\pi}} \oint_{-\infty}^{\infty} \phi(k)\, e^{ik}\, dk$$

with the amplitude at $k(= p/\hbar)$

$$\phi(k) = \left(\frac{a^2}{2\pi}\right)^{1/4} \exp\left[-\frac{a^2}{4}(k - k_0)^2\right]$$

(a) Show by direct integration that

$$\psi(x) = \left(\frac{2}{\pi a^2}\right)^{1/4} e^{-x^2/a^2}\, e^{ik_0 x}$$

(b) Obtain the minimum uncertainty relation $\Delta x \cdot \Delta p_x = \frac{\hbar}{2}$, where Δx and Δp_x are the half-widths defined by

$$\frac{|\psi(\pm\Delta x)|^2}{|\psi(0)|^2} = \frac{|\phi(k_0 \pm \Delta k)|^2}{|\phi(k_0)|^2} = e^{-1/2}$$

Chapter 4

The Schrödinger's Wave Equation

4.1 Classical mechanics versus quantum mechanics

Classical (or, Newtonian) mechanics deals with the motion of particle subjected to externally applied forces. It takes for granted that the properties like particle's mass, position, velocity and acceleration can be measured *correctly*. This is of course, valid in our every day experience. It provides the correct explanation for the dynamical behaviour of the bodies as the values it predicts for observable quantities agree well with the experimentally measured values of those quantities.

Quantum mechanics also deals with the observable magnitudes of dynamical systems, but the term *observable magnitude* bears different significance in view of uncertainty principle in the atomic realm. According to classical mechanics both the position and momentum of a moving particle are ascertainable at every instant of time, but in quantum mechanics the simultaneous measurement of position and momentum of a moving particle with precision is impossible. That is why classical mechanics is *deterministic* whereas quantum mechanics is *probabilistic.*

For example, classically the radius of the electron orbit in the ground state of hydrogen atom is exactly 5.3×10^{-11} m, but quantum mechanically this is the most probable value of the radius.

Classical mechanics is an approximate version of quantum mechanics. The certainties claimed in classical mechanics are only apparent. The agreement of the predicted value and the experimental value is due to the fact that macroscopic bodies consists of large number of individual atoms and the departures from the average behaviour are unnoticeable. Thus, quantum mechanics includes the two sets of physical principle, one for the macro-world and the other for the micro-world.

4.2 The wave function

The de Broglie relation between the momentum (p) and the associated wavelength of a moving particle, i.e., $\lambda = h/p$ has been experimentally verified for photons and material particles like electrons. This suggests that localised particles and photons are possible to be represented by a wave packet $\psi(\vec{r}, t)$. $\psi(\vec{r}, t)$ is the associated de Broglie wave function of a moving particle and is a function of space $(\vec{r})$ and time (t), which gives nearly complete information about the state of a physical system at a particular time. The value of the state function $\psi(\vec{r}, t)$ at a particular point $(\vec{r})$ at a particular time (t) represents the probability of finding the particle at that point $(\vec{r})$ in time (t).

To find out the nature of the wave function $\psi(\vec{r}, t)$ we proceed as follows:

Let us assume that

(a) the quantum mechanical particle is *free*, i.e., external forces are absent

(b) the motion is *non-relativistic* and one-dimensional, say, along the x-axis.

If p be the momentum and E the energy of the particle under consideration then the de Broglie wavelength of the associated wave is $\lambda = h/p$ and the frequency, $\nu = E/h$. The wave vector

$$k = \frac{2\pi}{\lambda} = \frac{p}{\hbar} \tag{4.1}$$

The angular frequency

$$\omega = 2\pi\nu = \frac{E}{\hbar} \tag{4.2}$$

Since, the particle of mass m is considered to be free

$$E = \frac{p^2}{2m}$$

$\therefore$ from eq. (4.1)

$$k = \frac{\sqrt{2mE}}{\hbar} \tag{4.3}$$

The possible harmonic waves representing a particle travelling along the positive x-direction are

$$\psi_1 = a\sin(\omega t - kx) \text{ or, } a\cos(\omega t - kx) \tag{4.4}$$

$$\text{and } \psi_2 = Ae^{-i(\omega t - kx)} \text{ or, } Ae^{i(\omega t - kx)} \tag{4.5}$$

The waves represented by eq. (4.4) are real, but those represented by eq. (4.5) are complex.

Since,

$$e^{-i(\omega t - kx)} = \cos(\omega t - kx) - i\sin(\omega t - kx) \tag{4.6}$$

$$\text{and } e^{i(\omega t - kx)} = \cos(\omega t - kx) + i\sin(\omega t - kx) \tag{4.7}$$

Equations (4.6) and (4.7) show that the complex waves are the linear combinations of real waves with a complex coefficient.

Now, let a particle travelling along the positive x-direction be represented by $\psi_1 = a\sin(\omega t - kx)$ and a particle travelling along negative x-direction be represented by $\psi_1' = a\sin(\omega t + kx)$.

Then according to *the principle of superposition* the resultant displacement at a distance x and at time t is

$$\psi = \psi_1 + \psi_1' = a\sin(\omega t - kx) + a\sin(\omega t + kx) \text{ or, } \psi = 2a\sin\omega t\cos kx \tag{4.8}$$

Now, at $t = 0$, $\psi = 0$, i.e., the wave function vanishes which means the probability of finding the particle is zero everywhere. As the particle cannot disappear everywhere we conclude that eq. (4.8) cannot represent a moving particle.

Hence, the wave represented by eq. (4.4) cannot represent a particle in motion.

But, the waves $\psi_2 = Ae^{-i(\omega t-kx)}$ and $\psi_2' = Ae^{-i(\omega t+kx)}$ representing particles travelling along the positive and negative x-directions respectively, then the resultant displacement at x and t is given by the superposition principle as

$$\psi = \psi_2 + \psi_2' = Ae^{-i(\omega t-kx)} + Ae^{-i(\omega t+kx)} \text{ or, } \psi = 2Ae^{-i\omega t}\cos kx \tag{4.9}$$

Since, at $t = 0$, $\psi = 2A\cos kx$, i.e., ψ does not vanish identically everywhere and is an acceptable wave function representing a moving particle. So ψ_2 and ψ_2' both are consistent with the principle of superposition. But, according to convention we accept $\psi_2 = Ae^{-i(\omega t-kx)}$ as a suitable wave function to represent a moving particle. This is due to the fact that the superposition $\psi = \psi_2 + \psi_2' = 2A\cdot\cos(\omega t - kx)$ is an unacceptable wave function. Thus, the acceptable wave function representing a quantum mechanical system is

$$\psi = Ae^{-i(\omega t-kx)} \tag{4.10}$$

where, we have omitted the subscript 2 of ψ for convenience. Eq. (4.10) can also be expressed as

$$\psi(x,t) = A\exp\left[-i\left(\frac{Et}{\hbar} - \frac{\sqrt{2mE}}{\hbar}x\right)\right] \tag{4.11}$$

$$\therefore \quad \frac{\partial\psi(x,t)}{\partial t} = -i\frac{E}{\hbar}\psi(x,t) \tag{4.12}$$

$$\text{and} \quad \frac{\partial^2\psi(x,t)}{\partial x^2} = -\frac{2mE}{\hbar^2}\psi(x,t) \tag{4.13}$$

Now, multiplying eq. (4.12) by $i\hbar$ and eq. (4.13) by $-\frac{\hbar^2}{2m}$, we get

$$i\hbar\frac{\partial\psi(x,t)}{\partial t} = \hbar\omega\psi(x,t) \quad [\because E = \hbar\omega] \tag{4.14}$$

$$\text{and} \quad -\frac{\hbar^2}{2m}\frac{\partial^2\psi(x,t)}{\partial x^2} = \hbar\omega\psi(x,t) \tag{4.15}$$

From eqs (4.14) and (4.15), we get

$$i\hbar\frac{\partial\psi(x,t)}{\partial t} = -\frac{\hbar^2}{2m}\frac{\partial^2\psi(x,t)}{\partial x^2} \tag{4.16}$$

Equation (4.16) is the *Schrödinger equation* for a free particle in one dimension. Schrödinger equation plays the same role in quantum mechanics as Newton's second law does in classical mechanics.

4.3 Matter waves are complex harmonic function of x and t

Equation (4.10) shows that the wave function $\psi(x,t)$ representing a moving particle is a complex function of x and t. This can also be shown as below:

Let a particle moving along the x-direction be represented by

$$\psi(x,t) = A\cos(kx - \omega t) + B\sin(kx - \omega t) \tag{4.17}$$

where A and B are arbitrary constants

$$\therefore \frac{\partial \psi}{\partial x} = k[-A\sin(kx-\omega t) + B\cos(kx-\omega t)] \tag{4.18}$$

$$\text{and } \frac{\partial^2 \psi}{\partial x^2} = -k^2[A\cos(kx-\omega t) + B\sin(kx-\omega t)] = -k^2\psi \tag{4.19}$$

$$\text{and } \frac{\partial \psi}{\partial t} = -\omega[-A\sin(kx-\omega t) + B\cos(kx-\omega t)] \tag{4.20}$$

$$\frac{\partial^2 \psi}{\partial t^2} = -\omega^2\psi \tag{4.21}$$

$$\therefore \frac{\partial \psi}{\partial t} = -\frac{\omega}{k}\frac{\partial \psi}{\partial x} \tag{4.22}$$

$$\text{and } \frac{\partial^2 \psi}{\partial t^2} = \frac{\omega^2}{k^2}\frac{\partial^2 \psi}{\partial x^2} \tag{4.23}$$

As $\omega/k = E/p = p/2m$ depends on the particular state of motion of the particle and has different values for different particle momentum (p) the eqs (4.22) as well as eq. (4.23) are unsuitable for describing a matter wave. We need a single equation which will be satisfied by matter waves of all momenta of the moving particle. But, $\omega/k^2 = \hbar/2m$ is dependent on the mass of the particle and not on its momentum p. Now, taking the ratio of eqs (4.22) and (4.19), we get

$$\frac{\left(\frac{\partial \psi}{\partial t}\right)}{\left(\frac{\partial^2 \psi}{\partial x^2}\right)} = \frac{-\omega[-A\sin(kx-\omega t) + B\cos(kx-\omega t)]}{-k^2[A\cos(kx-\omega t) + B\sin(kx-\omega t)]}$$

$$= \frac{\omega A\left[-\sin(kx-\omega t) + \frac{B}{A}\cos(kx-\omega t)\right]}{k^2 A\left[\cos(kx-\omega t) + \frac{B}{A}\sin(kx-\omega t)\right]}$$

$$= \frac{\omega}{k^2}\frac{\left[\frac{B}{A}\cos(kx-\omega t) - \sin(kx-\omega t)\right]}{\left[\cos(kx-\omega t) + \frac{B}{A}\sin(kx-\omega t)\right]} = \frac{\omega}{k^2} \tag{4.24}$$

$$\text{or, } \frac{\partial \psi}{\partial t} = \frac{\omega}{k^2}\frac{\partial^2 \psi}{\partial x^2} \tag{4.25}$$

$$\text{if } \frac{(B/A)}{1} = \frac{-1}{(B/A)} \text{ or, } \frac{A}{B} = -\frac{B}{A}$$

$$\text{or, } B^2 = -A^2 \text{ or, } B = \pm iA \tag{4.26}$$

Now, from eq. (4.17)

$$\psi(x,t) = A[\cos(kx-\omega t) + i\sin(kx-\omega t)]$$

$$\text{or, } \psi = Ae^{i(kx-\omega t)} \text{ or, } \psi(x,t) = Ae^{-i(\omega t - kx)}$$

The above function $\psi(x,t)$ can represent a quantum mechanical system. But, $\psi(x,t)$ is a complex function of x and t.

Hence, we conclude that the matter waves associated with a moving particle of momentum (p) are complex harmonic function of x and t.

4.4 Characteristics of the Schrödinger equation

The one-dimensional Schrödinger equation for a free particle of mass m is given by

$$i\hbar\frac{\partial\psi(x,t)}{\partial t} = -\frac{\hbar^2}{2m}\frac{\partial^2\psi(x,t)}{\partial x^2} \text{ [same as eq. (4.16)]} \tag{4.27}$$

1. A close look into eq. (4.27) reveals that

$$\frac{\partial\dot{\psi}(x,t)}{\partial t} \text{ and } \frac{\partial^2\psi(x,t)}{\partial x}$$

occur only in the first power, i.e., linearly. This implies that not only harmonic waves of the form: $\psi(x,t) = Ae^{-i(\omega t-kx)}$ are the solutions of eq. (4.27), but any linear combination or superposition of harmonic waves such as

$$\psi(x,t) = \int A(k)\,e^{-i(\omega t-kx)}\,dk \tag{4.28}$$

are also the solution of the same equation. This is an important property of Schrödinger equation. According to this principle if $\psi_1(x,t)$ and $\psi_2(x,t)$ be two-wave functions of the system, then $\psi(x,t) = C_1\psi_1(x,t) + C_2\psi_2(x,t)$ will also be a possible wave function of the system satisfying eq. (4.27). The eq. (4.28) represents a wave packet.

Hence, Schrödinger wave equation also has perfectly good wave packet solutions.

2. The wave eq. (4.27) is in the first order in time implying that if ψ is known at some point $\vec{r}$ at the initial time t_0, then at any subsequent time t, the wave function $\psi(\vec{r},t)$ at the same point will be uniquely determined.
3. All the partial differential equations of classical physics contains only real parameters. But, Schrödinger wave equation contains the imaginary unit i explicitly. This sets apart Schrödinger equation from all other classical partial differential equation.
4. The wave function $\psi(r,t)$ representing the matter wave must be a complex function of x and t.
5. The wave eq. (4.27) is consistent with de Broglie's hypothesis and correspondence principle.

4.5 One-dimensional time-dependent Schrödinger equation

4.5.1 When the particle is free

We consider a particle of mass m moving with velocity v along the positive x-direction. The velocity of the particle (v) is small compared to the velocity of light (c). The wave function associated with such a particle is represented as $\psi(x,t) = Ae^{-i(\omega t-kx)} = Ae^{i(kx-\omega t)}$, where A is a constant and $\omega/v = k$.

If λ be the wavelength, then we have from de Broglie's theory

$$p = \frac{h}{\lambda} = \hbar k,\ k = \frac{2\pi}{\lambda}$$

If p_x be the momentum of the particle, then its kinetic energy

$$E = \frac{p^2}{2m} = \frac{\hbar^2k^2}{2m} = h\nu = \hbar\omega$$

So the wave function in terms of p_x and E is

$$\psi(x,t) = Ae^{i/\hbar(p_x x-Et)} \tag{4.29}$$

The above equation is a plane continuous simple harmonic wave equation associated with a particle of accurately known momentum p_x and total energy E moving in the positive x-direction.

The desired differential equation should be such that the wave function $\psi(x,t)$ represented by eq. (4.29) must be a solution of it.

Differentiating eq. (4.29) w.r.t. x, we get

$$\frac{\partial\psi}{\partial x} = A\left(\frac{i}{\hbar}\right)p_x e^{\frac{i}{\hbar}(p_x x - Et)} = \left(\frac{i}{\hbar}\right)p_x\psi \tag{4.30}$$

$$\text{and } \frac{\partial^2\psi}{\partial x^2} = \left(\frac{i}{\hbar}\right)p_x\frac{\partial\psi}{\partial x} = \left(\frac{i}{\hbar}\right)p_x\left(\frac{i}{\hbar}\right)p_x\psi = -\frac{1}{\hbar^2}p_x^2\psi \tag{4.31}$$

$$\text{or, } p_x^2\psi = -\hbar^2\frac{\partial^2\psi}{\partial x^2} \tag{4.32}$$

Again, differentiating eq. (4.29) w.r.t. t

$$\frac{\partial\psi}{\partial t} = A\left(\frac{i}{\hbar}\right)(-E)\, e^{\frac{i}{\hbar}(p_x x - Et)} = -\frac{i}{\hbar}E\psi = \frac{E\psi}{i\hbar}$$

$$\text{or, } E\psi = i\hbar\frac{\partial\psi}{\partial t} \tag{4.33}$$

$$\text{or, } p_x^2\psi = i\hbar 2m\frac{\partial\psi}{\partial t}\quad\left[\because E = \frac{p_x^2}{2m}\right] \tag{4.34}$$

From eqs (4.32) and (4.34), we get

$$-\hbar^2\frac{\partial^2\psi}{\partial x^2} = i\hbar 2m\frac{\partial\psi}{\partial t} \text{ or, } -\frac{\hbar^2}{2m}\frac{\partial^2\psi}{\partial x^2} = i\hbar\frac{\partial\psi}{\partial t} \tag{4.35}$$

Equation (4.35) is the *one-dimensional time-dependent Schrödinger equation of a free particle of mass m*.

Since, this equation does not contain momentum (p_x), kinetic energy (E), wave number (k) and frequency (ν), it can represent any of the component wave of the wave packet representing the moving particle.

4.5.2 When the particle is acted upon by a force field derivable from a potential $V(x)$

Let a particle of mass m move along the x-direction with momentum p_x in a force field derivable from the potential $V(x)$, which does not explicitly depend on time and hence conservative. The total energy E of the particle is given by

$$\frac{p_x^2}{2m} + V = E \text{ or, } \frac{p_x^2}{2m}\psi + V\psi = E\psi \tag{4.36}$$

Now, the wave function representing the above particle is given by $\psi(x,t) = Ae^{\frac{i}{\hbar}(p_x x - Et)}$.

Differentiating the above equation w.r.t. x, we get

$$\frac{\partial\psi}{\partial x} = A\left(\frac{i}{\hbar}\right)p_x e^{\frac{i}{\hbar}(p_x x - Et)} = \left(\frac{i}{\hbar}\right)p_x\psi$$

$$\text{and } \frac{\partial^2\psi}{\partial x^2} = \left(\frac{i}{\hbar}\right)p_x\frac{\partial\psi}{\partial x} = \left(\frac{i}{\hbar}\right)p_x\left(\frac{i}{\hbar}\right)p_x\psi = -\frac{1}{\hbar^2}p_x^2\psi \text{ or, } p_x^2\psi = -\hbar^2\frac{\partial^2\psi}{\partial x^2}$$

Again, differentiating $\psi(x,t)$ w.r.t. time t, we get

$$\frac{\partial\psi}{\partial t} = A\left(\frac{i}{\hbar}\right)(-E)\, e^{\frac{i}{\hbar}(p_x x - Et)} = -\frac{i}{\hbar}E\psi = \frac{E\psi}{i\hbar} \text{ or, } E\psi = i\hbar\frac{\partial\psi}{\partial t} \tag{4.37}$$

Substituting the values of $p_x^2\,\psi$ and $E\psi$ in eq. (4.36), we get

$$-\frac{\hbar^2}{2m}\frac{\partial^2\psi}{\partial x^2} + V\psi = i\hbar\frac{\partial\psi}{\partial t} \tag{4.38}$$

This is the *time-dependent one-dimensional Schrödinger wave equation for a particle moving under the action of a conservative force field derivable from $V(x)$.*

4.6 Time-dependent and time-independent part of the one-dimensional wave equation

The Schrödinger time-dependent equation for a particle moving along the x-direction in a potential field V, which is independent of t is given by

$$-\frac{\hbar^2}{2m}\frac{\partial^2\psi}{\partial x^2} + V\psi = i\hbar\frac{\partial\psi}{\partial t} \text{ [same as eq. (4.38)]} \tag{4.39}$$

The equation being a partial differential equation of two variables x and t we can write:

$$\psi(x,t) = u(x)\,f(t) \tag{4.40}$$

Using this value of $\psi(x,t)$ in eq. (4.37), we get

$$-\frac{\hbar^2}{2m}f\frac{d^2u}{dx^2} + Vuf = i\hbar u\frac{df}{dt}$$

Dividing both sides of the above equation by uf, we get

$$-\frac{\hbar^2}{2m}\frac{1}{u}\frac{d^2u}{dx^2} + V = i\hbar\frac{1}{f}\frac{df}{dt} \tag{4.41}$$

Since, the left-hand side is a function of x and the right-hand side is a function of t only, the left-hand side of eq. (4.40) is the time-independent part and the right-hand side is the time-dependent part of the Schrödinger wave eq. (4.37). As x and t are independent variable, each side of eq. (4.40) must be equal to a constant. So

$$-\frac{\hbar^2}{2m}\frac{1}{u}\frac{d^2u}{dx^2} + V = i\hbar\frac{1}{f}\frac{df}{dt} = E \text{ [}E = \text{total energy which is constant]} \tag{4.42}$$

4.6.1 Time-dependent part

The eq. (4.37) is

$$E\psi = i\hbar\frac{\partial\psi}{\partial t}$$

Putting $\psi = u(x)\,f(t)$, we get

$$i\hbar u\frac{df}{dt} = Euf$$

Dividing both sides of the above equation by uf, we get

$$i\hbar\frac{1}{f}\frac{df}{dt} = E \tag{4.43}$$

Thus, eq. (4.43) shows that the time-dependent part of eq. (4.40) is equal to the total energy E of the particle.

$$\text{or, } \frac{df}{f} = \frac{E}{i\hbar}dt = -\frac{iE}{\hbar}dt \tag{4.44}$$

Integrating eq. (4.44), we get

$$\ln f = -\frac{iE}{\hbar}t + \ln C, \text{ where } C \text{ is a constant}$$

$$\text{or, } \ln\left(\frac{f}{C}\right) = -\frac{iE}{\hbar}t \text{ or, } \frac{f}{C} = e^{-\left(\frac{iE}{\hbar}\right)t} \text{ or, } f(t) = Ce^{-\left(\frac{iE}{\hbar}\right)t} \tag{4.45}$$

Equation (4.45) represents the *time-dependent part* of the solution $\psi(x,t)$ of the Schrödinger equation.

4.6.2 Time-Independent Part

We have from eq. (4.42)

$$-\frac{\hbar^2}{2m}\frac{1}{u}\frac{d^2u}{dx^2} + V = E \text{ or, } -\frac{\hbar^2}{2m}\frac{d^2u}{dx^2} + Vu = Eu \tag{4.46}$$

Equation (4.45) is the *one-dimensional time-independent Schrödinger equation* for a particle moving along x-direction in a potential field $V(x)$.

Equation (4.46) is called one-dimensional *steady state Schrödinger equation* and its solution $u(x)$ is known as the *time-independent wave function* or the *steady state wave function.*

Now, eq. (4.46) can be expressed as

$$\frac{d^2u}{dx^2} + \frac{2m}{\hbar^2}(E - V)u = 0 \tag{4.47}$$

The solution of the time-dependent Schrödinger equation is

$$\psi(x,t) = u(x)f(t) = Cu(x)e^{-\frac{iEt}{\hbar}} \text{ or, } \psi(x,t) = u(x)e^{-i\omega t} \tag{4.48}$$

Since, $E = h\nu = \frac{h}{2\pi}2\pi\nu = \hbar\omega$ and taking $C = 1$. Here, $\psi(x,t)$ corresponds to a particular energy E and as such it is a particular solution of the time-dependent Schrödinger equation.

General solution of time-dependent Schrödinger equation: If $\psi_1(x,t), \psi_2(x,t), \ldots, \psi_n(x,t)$ be a set of particular solutions of a system corresponding to energy $E_1, E_2, \ldots, E_n$, then the general solution of the wave equation is given by

$$\psi(x,t) = \psi_1(x,t) + \psi_2(x,t) + \cdots + \psi_n(x,t)$$

$$= C_1u_1(x)e^{-\frac{iE_1t}{\hbar}} + C_2u_2(x)e^{-\frac{iE_2t}{\hbar}} + \cdots + C_nu_n(x)e^{-\frac{iE_nt}{\hbar}}$$

$$\text{or, } \psi(x,t) = \sum_{n=1}^{n} C_nu_n(x)e^{-\frac{iE_nt}{\hbar}}$$

$$\text{or, } \psi(x,t) = \sum_{n=1}^{n} C_nu_n(x)e^{-i\omega_nt} \tag{4.49}$$

The general solution given by eq. (4.49) does not satisfy Schrödinger time-independent eq. (4.46).

4.7 Three-dimensional Schrödinger equation

We consider a particle of mass m moving along any direction with reference to the origin of a Cartesian coordinate system in a force field, $\vec{F} = -\nabla V$, where V is the potential and is a function of x, y, z. As V is independent of time, the force field is conservative in nature. According to de Broglie's hypothesis the particle is associated with a wave packet represented by the wave function

$$\psi(\vec{r}, t) = Ae^{-\frac{i}{\hbar}(\vec{p}\cdot\vec{r} - Et)} \tag{4.50}$$

where p is the momentum and E the energy of the particle.

Since, the motion of the particle is non-relativistic, its total energy E is given by

$$E = \frac{p^2}{2m} + V(\vec{r}) \text{ or, } \frac{1}{2m}(p_x^2 + p_y^2 + p_z^2) + V(\vec{r}) = E$$

where $\vec{p} = \hat{i}p_x + \hat{j}p_y + \hat{k}p_z$, $\hat{i}, \hat{j}, \hat{k}$ being triad of unit vectors.

Multiplying both sides of the above equation by ψ, we get

$$\frac{1}{2m}(p_x^2 + p_y^2 + p_z^2)\,\psi + V\psi = E\psi \tag{4.51}$$

Now, eq. (4.50) may be expressed as

$$\psi(\vec{r}, t) = Ae^{\frac{i}{\hbar}(p_x x + p_y y + p_z z - Et)} \quad [\because \vec{p}\cdot\vec{r} = p_x x + p_y y + p_z z] \tag{4.52}$$

Differentiating eq. (4.52) w.r.t. x, we get

$$\frac{\partial\psi}{\partial x} = A\left(\frac{i}{\hbar}\right)p_x e^{\frac{i}{\hbar}(p_x x + p_y y + p_z z - Et)} = \left(\frac{i}{\hbar}\right)p_x\psi \tag{4.53}$$

$$\text{and } \frac{\partial^2\psi}{\partial x^2} = \left(\frac{i}{\hbar}\right)p_x\frac{\partial\psi}{\partial x} = \left(\frac{i}{\hbar}\right)p_x\left(\frac{i}{\hbar}\right)p_x\psi = -\frac{1}{\hbar^2}p_x^2\psi \text{ or, } p_x^2\psi = -\hbar^2\frac{\partial^2\psi}{\partial x^2} \tag{4.54}$$

Similarly,

$$p_y^2\psi = -\hbar^2\frac{\partial^2\psi}{\partial y^2} \tag{4.55}$$

$$\text{and } p_z^2\psi = -\hbar^2\frac{\partial^2\psi}{\partial z^2} \tag{4.56}$$

Differentiating eq. (4.52) w.r.t. t, we get

$$\frac{\partial\psi}{\partial t} = A\left(\frac{i}{\hbar}\right)(-E)\,e^{\frac{i}{\hbar}(p_x x + p_y y + p_z z - Et)} = -\left(\frac{i}{\hbar}\right)E\psi \text{ or, } E\psi = -\frac{\hbar}{i}\frac{\partial\psi}{\partial t} = i\hbar\frac{\partial\psi}{\partial t} \tag{4.57}$$

Now, using eqs (4.54), (4.55), (4.56) and (4.57) in eq. (4.51), we get

$$-\frac{\hbar^2}{2m}\left(\frac{\partial^2\psi}{\partial x^2} + \frac{\partial^2\psi}{\partial y^2} + \frac{\partial^2\psi}{\partial z^2}\right) + V\psi = i\hbar\frac{\partial\psi}{\partial t} \tag{4.58}$$

$$\text{or, } -\frac{\hbar^2}{2m}\nabla^2\psi + V\psi = i\hbar\frac{\partial\psi}{\partial t} \tag{4.59}$$

where $\nabla^2 = \frac{\partial^2}{\partial x^2} + \frac{\partial^2}{\partial y^2} + \frac{\partial^2}{\partial z^2}$.

Equation (4.59) is the three-dimensional wave equation of a particle moving in a conservative force field $F = -\nabla V$ and was deduced by Schrödinger in 1926. It is interesting to note that the equation is of *second order in space coordinates* and *first order in time.* The solution $\psi(\vec{r}, t)$ of the equation gives detailed information regarding the dynamical behaviour of the particle in motion.

Equation (4.59) can be expressed as

$$\left[-\frac{\hbar^2}{2m}\nabla^2 + V(\vec{r})\right]\psi = i\hbar\frac{\partial\psi}{\partial t} \text{ or, } \hat{H}\psi = i\hbar\frac{\partial\psi}{\partial t} \tag{4.60}$$

where $\hat{H} = -\frac{\hbar^2}{2m}\nabla^2 + V(\vec{r})$ is the *three-dimensional Hamiltonian* operator of the moving particle.

4.8 Three-dimensional Schrödinger wave equation for a free particle

If $V(\vec{r}, t) = 0$, then the motion of the particle is not influenced by any force and the particle is called a *free particle.* Putting $V = 0$ in eq. (4.59), we get

$$-\frac{\hbar^2}{2m}\nabla^2\psi = i\hbar\frac{\partial\psi}{\partial t} \tag{4.61}$$

This is the *three-dimensional wave equation for a free particle moving in three-dimension.*

4.9 Time-dependent and time-independent part of three-dimensional Schrödinger wave equation

If $u(\vec{r})$ and $f(t)$ be the time-independent and time-dependent part of $\psi(\vec{r}, t)$, then

$$\psi(\vec{r}, t) = u(\vec{r})\, f(t) \tag{4.62}$$

Substituting eq. (4.62) in eq. (4.59), we get

$$-\frac{\hbar^2}{2m}f\left[\frac{\partial^2 u}{\partial x^2} + \frac{\partial^2 u}{\partial y^2} + \frac{\partial^2 u}{\partial z^2}\right] + Vuf = i\hbar u\frac{\partial f}{\partial t}$$

Dividing both sides of the above equation by uf, we get

$$-\frac{\hbar^2}{2m}\frac{1}{u}\left[\frac{\partial^2 u}{\partial x^2} + \frac{\partial^2 u}{\partial y^2} + \frac{\partial^2 u}{\partial z^2}\right] + V = i\hbar\frac{1}{f}\frac{\partial f}{\partial t} \tag{4.63}$$

Since, the left side of the above equation is a function of space coordinates and the right side is a function of time, each side is equal to a constant E, where E is the total energy of the system.

$$\therefore -\frac{\hbar^2}{2m}\frac{1}{u}\left[\frac{\partial^2 u}{\partial x^2} + \frac{\partial^2 u}{\partial y^2} + \frac{\partial^2 u}{\partial z^2}\right] + V = i\hbar\frac{1}{f}\frac{\partial f}{\partial t} = E \tag{4.64}$$

(a) Time-dependent part

From eq. (4.64), we write:

$$i\hbar\frac{1}{f}\frac{df}{dt} = E \text{ or, } \frac{df}{f} = \frac{1}{i\hbar}Edt = -\frac{i}{\hbar}Edt \tag{4.65}$$

Solving eq. (4.65), we get

$$f = Ce^{-\frac{iEt}{\hbar}} \tag{4.66}$$

where C is a constant.

(b) Time-independent part

From eq. (4.64), we write:

$$-\frac{\hbar^2}{2m}\frac{1}{u}\left[\frac{\partial^2 u}{\partial x^2}+\frac{\partial^2 u}{\partial y^2}+\frac{\partial^2 u}{\partial z^2}\right]+V=E$$

$$\text{or, } -\frac{\hbar^2}{2m}\left[\frac{\partial^2 u}{\partial x^2}+\frac{\partial^2 u}{\partial y^2}+\frac{\partial^2 u}{\partial z^2}\right]+Vu=Eu \tag{4.67}$$

$$\text{or, } -\frac{\hbar^2}{2m}\left[\frac{\partial^2}{\partial x^2}+\frac{\partial^2}{\partial y^2}+\frac{\partial^2}{\partial z^2}\right]u+Vu=Eu$$

$$\text{or, } -\frac{\hbar^2}{2m}\nabla^2 u+Vu=Eu$$

$$\text{or, } \left[\frac{\hbar^2}{2m}\nabla^2+V\right]u=Eu \tag{4.68}$$

Equation (4.68) may also be expressed as

$$\nabla^2 u+\frac{2m}{\hbar^2}(E-V)\,u=0 \tag{4.69}$$

Equations (4.67), (4.68) and (4.69) are the three different forms of *three-dimensional time-independent Schrödinger equation for a particle moving in a force field,* $\vec{F}=-\nabla V$. For a free particle (when $V=0$) the corresponding equation is

$$\nabla^2 u+\frac{2mE}{\hbar^2}u=0 \tag{4.70}$$

(c) Three-dimensional time-dependent wave function

The time-dependent and time-independent solutions of the Schrödinger equation are respectively

$$f(t)=Ce^{-\frac{iEt}{\hbar}}=Ce^{-i\omega t} \text{ and } u(\vec{r}).\ [\because E=\hbar\omega]$$

The three-dimensional wave function is, therefore,

$$\psi(\vec{r},t)=u(\vec{r})\,f(t) \text{ or, } \psi(\vec{r},t)=Cu(\vec{r})\,e^{-i\omega t} \tag{4.71}$$

Usually the value of the constant C is taken as 1. The wave eq. (4.71) represents a monochromatic wave of particular energy E.

4.10 Physical interpretation of the wave function $\psi(\vec{r},t)$

We have seen that the waves associated with a particle are represented by a function $\psi(\vec{r},t)$, called the *wave function*. Regarding $\psi(\vec{r},t)$ several questions arise:

(i) Is $\psi(\vec{r},t)$ a measurable quantity?

(ii) What precisely does $\psi(\vec{r},t)$ describe?

(iii) What feature of the particle is related to $\psi(\vec{r},t)$?

Unless the properties of the waves represented by $\psi(\vec{r},t)$ and the way in which $\psi(\vec{r},t)$ used in the calculations be known, satisfactory answers to the above questions cannot be expected. Still the wave function $\psi(\vec{r}, t)$, which describes the space-time behaviour of the particle is such that its magnitude is large in the region where the probability of finding the particle is large. In other region, where the probability of finding the particle is small, the magnitude of $\psi(\vec{r}, t)$ is small. In other word, $\psi(\vec{r}, t)$ may be regarded as the probability of finding the particle around a particular position in space.

Again, the probability (P) is a real and non-negative quantity whereas $\psi(\vec{r}, t)$ is a complex function; we shall assume that the quantity

$$|\psi(\vec{r}, t)|^2 \, d^3r = \psi^*(\vec{r}, t)\, \psi(\vec{r}, t)\, d^3r \tag{4.72}$$

is proportional to the probability of finding the particle in the volume element d^3r around the point r at time t. Here, $\psi^*(\vec{r}, t)$ is the complex conjugate of $\psi(\vec{r}, t)$.

Hence, the total probability of finding the particle anywhere in space is

$$P(\vec{r}, t) \propto \int_{-\infty}^{+\infty} |\psi(\vec{r}, t)|^2 \, d^3r$$

The position probability density is defined as

$$\rho(\vec{r}, t) = \frac{|\psi(\vec{r}, t)|^2}{\int_{-\infty}^{+\infty} |\psi(\vec{r}, t)|^2 \, d^3r} \tag{4.73}$$

Thus, according to Max Born, the most direct physical interpretation of the wave function $\psi(\vec{r}, t)$ is that its modulus square determines the probability density of the particle in space.

Since, $|\psi(\vec{r}, t)|^2$ is a positive quantity, $P(\vec{r}, t)$ is always positive. So the total probability of finding the particle anywhere in space is

$$P(\vec{r}, t) = \int_{-\infty}^{+\infty} \rho(\vec{r}, t)\, d^3r = \frac{\int_{-\infty}^{+\infty} |\psi(\vec{r}, t)|^2 \, d^3r}{\int_{-\infty}^{+\infty} |\psi(\vec{r}, t)|^2 \, d^3r} = 1 \tag{4.74}$$

This is in accordance with the expectation.

4.11 Normalisation of the wave function

The Schrödinger wave equation is linear and homogeneous in ψ and its derivatives $\left(\text{viz., } \frac{\partial^2 \psi}{\partial x^2} \text{ and } \frac{\partial \psi}{\partial t}\right)$. That is why if any solution of it be multiplied by a constant, then the resulting wave function will also be a solution. To avoid this arbitrariness, it is necessary to impose normalisation condition.

Let us suppose that $\psi_1(\vec{r}, t)$ is a solution of the Schrödinger equation (4.68) and

$$\int_{-\infty}^{+\infty} |\psi_1(\vec{r}, t)|^2 \, d^3r = N^2 \tag{4.75}$$

where $|\psi_1(\vec{r}, t)|^2 = \psi_1^*(\vec{r}, t)\, \psi(\vec{r}, t)$ is a positive number and so also the integral. N^2 is, therefore, a real positive number and is called the *norm* of $\psi(\vec{r}, t)$.

Now, we consider another wave function $\psi(\vec{r}, t)$, which is related to $\psi_1(\vec{r}, t)$ by the relation

$$\psi(\vec{r}, t) = \frac{1}{N}\, \psi_1(\vec{r}, t) \tag{4.76}$$

Here, $\psi(\vec{r},t)$ is also a solution of eq. (4.68), since it differs from the solution $\psi_1(\vec{r},t)$ by the factor $1/N$, which is a constant.

Substituting this value of $\psi(\vec{r},t)$ in eq. (4.75), we get

$$\int_{-\infty}^{+\infty} N^2 |\psi(\vec{r},t)|^2 \, d^3r = N^2 \text{ or, } \int_{-\infty}^{+\infty} |\psi(\vec{r},t)|^2 \, d^3r = 1 \tag{4.77}$$

The wave function $\psi(\vec{r},t)$, which satisfies eq. (4.77) is said to be *normalised. Hence, a normalised wave function must have unit norm.*

Again, N in eq. (4.76) must be finite, since $|\psi(\vec{r},t)|^2$ is finite, when $|\vec{r}|$ is finite and $|\psi(\vec{r},t)|^2 \to 0$ as $|\vec{r}| \to \infty$. Eq. (4.77) is known as the *condition of normalisation.*

Now, for one-dimensional motion of the particle, the normalisation condition is given by

$$\int_{-\infty}^{+\infty} \psi^*(x,t)\,\psi(x,t)\,dx = \int_{-\infty}^{+\infty} |\psi(x,t)|^2 \, dx = 1$$

Since, for any solution of Schrödinger equation

$$\int_{-\infty}^{+\infty} \psi^*(\vec{r},t)\,\psi(\vec{r},t)\,d^3r = \int_{-\infty}^{+\infty} |\psi(\vec{r},t)|^2 \, d^3r$$

is independent of time.

$$\therefore \int_{-\infty}^{+\infty} \psi^*(\vec{r},t)\,\psi(\vec{r},t)\,d^3r = \int_{-\infty}^{+\infty} \psi^*(\vec{r})\,\psi(\vec{r})\,d^3r$$

Hence, the normalisation condition is

$$\int_{-\infty}^{+\infty} \psi^*(\vec{r})\,\psi(\vec{r})\,d^3r = 1 \tag{4.78}$$

For one-dimensional motion the condition of normalisation is

$$\int_{-\infty}^{+\infty} \psi^*(x)\,\psi(x)\,dx = 1 \tag{4.79}$$

Note: The wave function $\psi(\vec{r},t)$ remains undetermined even after normalisation. The extent of unnormalisation of $\psi(\vec{r},t)$ is a phase factor of the form: $e^{i\alpha}$, where α is a constant. In such case, ψ is to be multiplied by $e^{i\alpha}$ without affecting the condition eq. (4.77).

4.12 Orthogonality of wave functions

Consider a set of wave functions satisfying Schrödinger wave equation, viz., $\psi_1(\vec{r})$, $\psi_2(\vec{r})$, $\psi_3(\vec{r})$, etc. such that any two functions $\psi_m(\vec{r})$ and $\psi_n(\vec{r})$ in the set satisfy the condition:

$$\int \psi_m^*(\vec{r})\,\psi_n(\vec{r})\,d\tau = 0 \tag{4.80}$$

then the functions in the set are said to be *orthogonal* in analogy with the condition of orthogonality of two vectors.

If $\int \psi_m^*(\vec{r})\,\psi_m(\vec{r})\,d\tau = 1$, then each function is *normalised.* Then the set of functions under consideration is called *orthonormal set* of wave functions.

The condition of orthogonality and normalisation of the wave function can be expressed as

$$\int \psi_m^*(\vec{r})\,\psi_n(\vec{r})\,d\tau = \delta_{mn} \tag{4.81}$$

δ_{mn} is the Kronecker delta, where

$$\delta_{mn} = \begin{cases} 1, & \text{when } m = n \\ 0, & \text{when } m \neq n \end{cases}$$

Completeness: If the normalised orthogonal wave functions are single-valued, continuous, finite throughout the entire space of the system and satisfy the Schrödinger wave equation, then they can be expanded as a series of other orthogonal wave functions of the same variable.

Thus, if $\psi(\vec{r})$ has the above properties and $\phi_1(\vec{r}), \phi_2(\vec{r}), \ldots, \phi_n(\vec{r})$ are a set of mutually orthogonal wave functions, then

$$\psi(\vec{r}) = C_1\phi_1 + C_2\phi_2 + \cdots + C_m\phi_m \tag{4.82}$$

where $C_1, C_2, \ldots, C_m$ are constants.

Now, multiplying both sides of eq. (4.82) by ϕ_m^* and, then integrating over all space, we get

$$\int \phi_m^*\psi\, d\tau = C_1 \int \phi_m^*\phi_1\, d\tau + C_2 \int \phi_m^*\phi_2\, d\tau + \cdots + C_m \int \phi_m^*\phi_m d\tau + \ldots \tag{4.83}$$

Again, since

$$\int \phi_m^*\phi_n\, d\tau = \begin{cases} 0, & \text{for } m \neq n \\ 1, & \text{for } m = n \end{cases}$$

for mutually orthonormal wave functions, the equation eq. (4.83) reduces to

$$\int \phi_m^*\psi\, d\tau = C_m \tag{4.84}$$

The values of the expansion coefficients $C_1, C_2, \ldots, C_m$ can be found out using eq. (4.84).

The complex conjugate of the wave function ψ in eq. (4.82) can be expressed as

$$\psi^* = C_1^*\phi_1^* + C_2^*\phi_2^* + \cdots + C_m^*\phi_m^* + \ldots \tag{4.85}$$

So $\psi^*\psi = [C_1^*\phi_1^* + C_2^*\phi_2^* + \cdots + C_m^*\phi_m^* + \ldots] \times [C_1\phi_1 + C_2\phi_2 + \cdots + C_m\phi_m + \ldots]$.

Again, since ψ is normalised $\int \psi^*\psi\, d\tau = 1$.

$$\therefore \int [C_1^*\phi_1^* + C_2^*\phi_2^* + \cdots + C_m^*\phi_m^*] \times [C_1\phi_1 + C_2\phi_2 + \cdots + C_m\phi_m + \ldots] d\tau = 1 \tag{4.86}$$

$$\text{or, } C_1^*C_1 + C_2^*C_2 + \cdots + C_m^*C_m = 1 \quad \left[\because \int \phi_m\phi_n\, d\tau = \begin{cases} 0, & \text{for } m \neq n \\ 1, & \text{for } m = n \end{cases}\right]$$

$$\text{or, } \sum_m C_m^*C_m = 1 \tag{4.87}$$

Thus, the wave function

$$\psi(\vec{r}, t) = \sum_m C_m\phi_m(\vec{r}, t)$$

satisfying Schrödinger's wave equation is normalised if the coefficients satisfy relation (4.87). Here, $\phi_m(\vec{r}, t)$ is a *complete set* of eigenfunction.

4.13 Stationary states

If the probability density, $\rho = \psi^*\psi$ is independent of time for a particular state, then the state under consideration is called a *stationary state.*

Now, corresponding to various values of the constant energy E of a particle, the general solution of three-dimensional time-dependent Schrödinger equation is given by

$$\psi(\vec{r},t) = \sum_n C_n \psi_n(\vec{r})\, e^{-iE_n t/\hbar} \tag{4.88}$$

The probability density

$$\begin{aligned}\rho(\vec{r},t) &= \psi^*(\vec{r},t)\,\psi(\vec{r},t)\\ &= \sum_n C_n^* C_n\, \psi_n^*(\vec{r})\,\psi(\vec{r}) + \sum_n\sum_m C_m^* C_n\, \psi_m^*(\vec{r})\psi_n(\vec{r}) \exp\left[\frac{i(E_m - E_n)^t}{\hbar}\right]\end{aligned} \tag{4.89}$$

Equation (4.89) shows that $\rho(\vec{r},t)$ depends on time t and the state expressed by eq. (4.88) is not a stationary state. $\rho(\vec{r},t)$ will be independent of t if C_n's are zero for all values except for one value E_k. In such case, the wave function will contain only single term and will be represented by

$$\psi(\vec{r},t) = C_k \psi_k(\vec{r})\, e^{-iE_k t/\hbar} \tag{4.90}$$

Now, for this energy state E_k

$$\psi^*(\vec{r},t)\,\psi(\vec{r},t) = C_k^* C_k \psi_k^*(\vec{r})\,\psi_k(\vec{r}) = \rho(\vec{r}) \tag{4.91}$$

$\rho(\vec{r})$ is thus independent of time t and the energy state represented by E_k is called the *stationary state.* The solution represented by eq. (4.91) is the *stationary state solution of the Schrödinger equation,* since ρ is independent of time.

4.14 Conservation of probability

We know, according to the probability interpretation of the normalised wave function, the probability of finding the particle over all space is unity. This implies that the particle must be found somewhere in space. If the particle does not decay or disappear in some other way, then the above statement must be true at all values of time.

In other words, the total probability must be conserved, i.e., $|\psi(\vec{r},t)|^2$ must be independent of time.

So

$$\frac{\partial}{\partial t}\int |\psi(\vec{r},t)|^2\, d\tau = \int \frac{\partial}{\partial t}|\psi(\vec{r},t)|^2\, d\tau = \int \frac{\partial \rho(\vec{r},t)}{\partial t}\, d\tau = 0$$

where $\rho(\vec{r},t) = |\psi(\vec{r},t)|^2$.

Now, we have the time-dependent three-dimensional Schrödinger equation for a particle moving in a force field, $\vec{F} = -\nabla V(\vec{r},t)$ is given by

$$i\hbar\frac{\partial \psi(\vec{r},t)}{\partial t} = -\frac{\hbar^2}{2m}\nabla^2 \psi(\vec{r},t) + V(\vec{r},t)\psi(\vec{r},t) \tag{4.92}$$

Replacing $\psi(\vec{r},t)$ by its complex conjugate, we get

$$-i\hbar\frac{\partial \psi^*(\vec{r},t)}{\partial t} = -\frac{\hbar^2}{2m}\nabla^2 \psi^*(\vec{r},t) + V(\vec{r},t)\,\psi^*(\vec{r},t) \tag{4.93}$$

Multiplying eq. (4.92) on the left by $\psi^*(\vec{r}, t)$ and eq. (4.93) by $\psi(\vec{r}, t)$, we get

$$i\hbar\psi^*\frac{\partial\psi}{\partial t} = -\frac{\hbar^2}{2m}\psi^*\nabla\psi + V(\vec{r}, t)\psi^*\psi \tag{4.94}$$

$$\text{and } -i\hbar\psi\frac{\partial\psi^*}{\partial t} = -\frac{\hbar^2}{2m}\psi\nabla\psi^* + V(\vec{r}, t)\psi\psi^* \tag{4.95}$$

Subtracting eq. (4.95) from eq. (4.94), we get

$$i\hbar\left(\psi^*\frac{\partial\psi}{\partial t} + \psi\frac{\partial\psi^*}{\partial t}\right) = -\frac{\hbar^2}{2m}[\psi^*\nabla^2\psi - \psi\nabla^2\psi^*] \tag{4.96}$$

$$\text{or, } \frac{\partial}{\partial t}(\psi^*\psi) + \frac{i\hbar}{2m}[\psi\nabla^2\psi^* - \psi^*\nabla^2\psi] = 0$$

$$\text{or, } \frac{\partial}{\partial t}(\psi^*\psi) + \frac{i\hbar}{2m}\left[\frac{\partial}{\partial x}\left(\psi\frac{\partial\psi^*}{\partial x} - \psi^*\frac{\partial\psi}{\partial x}\right) + \frac{\partial}{\partial y}\left(\psi\frac{\partial\psi^*}{\partial y} - \psi^*\frac{\partial\psi}{\partial y}\right) + \frac{\partial}{\partial z}\left(\psi\frac{\partial\psi^*}{\partial z} - \psi^*\frac{\partial\psi}{\partial z}\right)\right] = 0$$

$$\text{or, } \frac{\partial\rho(\vec{r}, t)}{\partial t} + \nabla\cdot\vec{J}(\vec{r}, t) = 0 \tag{4.97}$$

where

$$\rho(\vec{r}, t) = \psi^*\psi \tag{4.98}$$

$$\text{and } \nabla\cdot\vec{J} = \frac{\partial J_x}{\partial x} + \frac{\partial J_y}{\partial y} + \frac{\partial J_z}{\partial z} \tag{4.99}$$

$$\text{with } J_x = \frac{i\hbar}{2m}\left[\psi\frac{\partial\psi^*}{\partial x} - \psi^*\frac{\partial\psi}{\partial x}\right] \tag{4.100}$$

$$J_y = \frac{i\hbar}{2m}\left[\psi\frac{\partial\psi^*}{\partial y} - \psi^*\frac{\partial\psi}{\partial y}\right] \tag{4.101}$$

$$J_z = \frac{i\hbar}{2m}\left[\psi\frac{\partial\psi^*}{\partial z} - \psi^*\frac{\partial\psi}{\partial z}\right] \tag{4.102}$$

Therefore,

$$\vec{J}(\vec{r}, t) = \frac{i\hbar}{2m}[\psi\nabla\psi^* - \psi^*\nabla\psi] \tag{4.103}$$

Now, integrating $\int\rho(\vec{r}, t)\,d\tau$ over all space, we get

$$\frac{\partial}{\partial t}\int_\tau\rho(\vec{r}, t)\,d\tau = \int_\tau\frac{\partial}{\partial t}(\psi^*\psi)\,d\tau = -\int_\tau\nabla\cdot\vec{J}d\tau$$

$$\text{or, } \frac{\partial}{\partial t}\int_\tau\rho(\vec{r}, t)\,d\tau = -\oint\vec{J}\cdot\hat{n}dS \tag{4.104}$$

where S is the closed surface enclosing volume τ.

Since, the volume integral is over all space, the surface S is at infinity, i.e., $|\vec{r}| \to \infty$. This implies that ψ and $\nabla\psi$ vanish at the surface S and $\vec{J}$, also vanishes sufficiently fast as $|\vec{r}| \to \infty$.

So

$$\oint_S\vec{J}\cdot\hat{n}dS = 0$$

Hence,

$$\frac{\partial}{\partial t}\int_\tau\rho(\vec{r}, t)\,d\tau = 0 \text{ or, } \int_\tau\rho(\vec{r}, t)\,d\tau \tag{4.105}$$

is independent of time and satisfies the *conservation of total probability*. This is analogous to the classical conservation law of matter or charge in absence of source or sink.

4.15 Probability current density

The equation

$$\frac{\partial \rho(\vec{r},t)}{\partial t} + \nabla \cdot \vec{J}(\vec{r},t) = 0$$

is the familiar equation of continuity in hydrodynamics where $\rho(\vec{r},t)$ and $\vec{J}(\vec{r},t)$ represent respectively the density of matter and current density. In the present case, $\rho(\vec{r},t) = \psi^* \psi$ of eq. (4.98) represents the probability density. The probability current density

$$\vec{J}(\vec{r},t) = -\frac{i\hbar}{2m}[\psi^* \nabla\psi - (\nabla\psi^*)\,\psi] \tag{4.106}$$

Equation (4.97) is valid, if there is no creation or destruction of probability. An increase or decease of $\frac{\partial \rho(\vec{r},t)}{\partial t} d\tau\, dt$ in the probability is compensated by the corresponding decrease or increase elsewhere through an inflow or outflow of probability $[\nabla \cdot \vec{J}(\vec{r},t)]\, d\tau\, dt$ across the boundaries of volume element $d\tau$.

A state for which $\nabla \cdot \vec{J}(\vec{r},t) = 0$, $\rho(\vec{r},t)$ = constant in time, then the corresponding state is a *stationary state*.

4.16 Conditions of admissibility of the wave function

The three-dimensional time-dependent Schrödinger wave equation for a particle moving in a potential field $V(\vec{r},t)$ is

$$-\frac{\hbar^2}{2m}\nabla^2 \psi(\vec{r},t) = i\hbar\frac{\partial \psi(\vec{r},t)}{\partial t}$$

The above equation being a second order linear differential equation may give rise to several solutions. But, every solution of the equation cannot represent a matter wave associated with the particle. Only those solutions satisfying certain conditions may represent a particle. For those wave functions $|\psi(\vec{r},t)|^2\, d\tau$ should represent the probability of finding the particle in the volume element $d\tau$.

The admissibility conditions on the wave functions are as follows:

(i) $\psi(\vec{r},t)$ *must be finite everywhere in space.* This is a must because $|\psi(\vec{r},t)|^2\, d\tau$ for an infinitesimal volume element $d\tau$ should be between 0 and 1.

(ii) $\psi(\vec{r},t)$ *must be single-valued everywhere.* This is necessary because at any given physical point $\vec{r}$, $\psi(\vec{r})$ should have a unique value and, then only the probability density $\rho = |\psi|^2$ may be uniquely defined. Accordingly, if α be an angle and any one of the variables x, y, z, then ψ must have only one value at (x, y, z), i.e., $\psi(\alpha) = \psi(\alpha + 2\pi n)$, where n is an integer.

(iii) ψ *and its first partial derivatives w.r.t. its variables are continuous.* This implies, there must not exist any sudden change in ψ, when its variables undergoes a change.

(iv) $\psi(x)$ *must be square integrable so that the integral of its modulus squared is finite.* This is necessary as the probability

$$\rho = \int_\tau |\psi|^2\, d\tau \text{ is finite or zero}$$

(v) *ψ is zero at any surface at which there is an infinite potential step; the component of $\nabla\psi$ normal to the surface is undetermined.*

A wave function which satisfies all the above conditions is an *admissible wave function* and is also called a *well behaved wave function* in wave mechanics.

4.17 Solved problems

» ***Example* 4.1.** The wave function of a particle is given by $\psi(x) = Ce^{-\alpha^2x^2}$, $-\infty < x < +\infty$, where C and α are constants. Calculate the probability of finding the particle in the region $0 < x < \infty$. **[BU (H) 2002]**

Solution: The probability of finding the particle in the region $0 < x < \infty$ is given by

$$P = \int_0^\infty \psi^*(x)\,\psi(x)\,dx = \int_0^\infty C^* e^{-\alpha^2x^2}\cdot Ce^{-\alpha^2x^2}\,dx = C^*C\int_0^\infty e^{-2\alpha^2x^2}\,dx = C^2\,\frac{1}{2}\sqrt{\frac{\pi}{2\alpha^2}} \tag{i}$$

[Since, from standard integral, $\int_0^\infty e^{-\alpha x^2}\,dx = \frac{1}{2}\sqrt{\frac{\pi}{\alpha}}$]

Now, to normalise the wave function

$$\int_{-\infty}^{+\infty} |\psi(x)|^2\,dx = 1 \text{ or, } \int_{-\infty}^{+\infty} |C|^2\,e^{-2\alpha^2x^2}\,dx = 1 \text{ or, } |C|^2\sqrt{\frac{\pi}{2\alpha^2}} = 1 \text{ or, } |C|^2 = \sqrt{\frac{2\alpha^2}{\pi}}$$

From (i)

$$P = \sqrt{\frac{2\alpha^2}{\pi}}\cdot\frac{1}{2}\sqrt{\frac{\pi}{2\alpha^2}} = \frac{1}{2}$$

Hence, the probability of finding the particle in $-0 < x < +\infty$ is $\frac{1}{2}$.

» ***Example* 4.2.** Normalise the one-dimensional wave function given by

$$\psi(x) = \begin{cases} A\sin\left(\frac{\pi x}{l}\right), & 0 < x < l \\ 0, & \text{outside} \end{cases}$$ **[BU (H) 1998, 2001; HPU 1995]**

Solution: The wave function $\psi(x)$ is said to be normalised if it satisfies the relation

$$\int_{-\infty}^{+\infty} \psi^*(x)\,\psi(x)\,dx = 1 \tag{i}$$

As the wave function $\psi(x)$ exists only in the region $0 < x < l$ and is zero outside, we have from (i)

$$\int_0^l \psi^*(x)\,\psi(x)\,dx = 1 \text{ or, } \int_0^l A^*\sin\left(\frac{\pi x}{l}\right)A\sin\left(\frac{\pi x}{l}\right)dx = 1 \text{ or, } A^*A\int_0^l \sin^2\left(\frac{\pi x}{l}\right)dx = 1$$

$$\text{or, } A^*A\int_0^l \frac{1}{2}\left[1-\cos\left(\frac{2\pi x}{l}\right)\right]dx = 1 \text{ or, } A^*A\left[\int_0^l dx - \int_0^l \cos\frac{2\pi x}{l}\,dx\right] = 2$$

$$\text{or, } |A|^2\left[x - \frac{l}{2\pi}\left\{\sin\left(\frac{2\pi x}{l}\right)\right\}\right]_0^l = 2 \text{ or, } |A|^2\,[l - 0 + 0] = 2 \text{ or, } |A|^2 = \frac{2}{l}$$

$$\text{or, } |A| = A = \sqrt{\frac{2}{l}} \text{ [assuming } A \text{ to be real]}$$

Hence, the normalised wave function is

$$\psi(x) = \sqrt{\frac{2}{l}}\sin\left(\frac{\pi x}{l}\right) \tag{ii}$$

***Example* 4.3.** A particle is represented by the wave function $\psi(x) = e^{-|x|}\sin\alpha x$. What is the probability that its position to the right of the point $x = 1$?

Solution: Here, let $\psi_1(x) = e^{x}\sin\alpha x$, for $x < 0$

and $\psi_2(x) = e^{-x}\sin\alpha x$, for $x > 0$

$$\therefore \int_{-\infty}^{+\infty} |\psi(x)|^2\, dx = 1/C^2, \text{ where C is the normalisation constant}$$

$$\text{or, } \int_{-\infty}^{0} |\psi_1(x)|^2\, dx + \int_{0}^{\infty} |\psi_2(x)|^2\, dx = \int_{-\infty}^{0} e^{2x}\sin^2\alpha x\, dx + \int_{0}^{\infty} e^{-2x}\sin^2\alpha x\, dx$$

$$= \int_{-\infty}^{0} e^{2x}\left[\frac{1-\cos 2\alpha x}{2}\right] dx + \int_{0}^{\infty} e^{-2x}\left[\frac{1-\cos 2\alpha x}{2}\right] dx$$

$$= \frac{1}{2} - \frac{1}{2}\int_{-\infty}^{0} e^{2x}\cos 2\alpha x\, dx - \frac{1}{2}\int_{0}^{\infty} e^{-2x}\cos 2\alpha x\, dx$$

$$= \frac{1}{2} - \frac{1}{4(1+\alpha^2)} - \frac{1}{4(1+\alpha^2)} = \frac{\alpha^2}{2(1+\alpha^2)} = 1/C^2$$

$$\therefore C = \sqrt{\frac{2(1+\alpha^2)}{\alpha^2}}$$

So the normalised wave function is

$$\psi(x) = \sqrt{\frac{2(1+\alpha^2)}{\alpha^2}}\, e^{-|x|}\sin\alpha x$$

The required probability is given by

$$P = \int_{1}^{\infty} |\psi|^2\, dx = \frac{1}{C^2}\int_{1}^{\infty} e^{-2x}\sin^2\alpha x = \frac{e^{-2}}{4(1+\alpha^2)}\left[1+\alpha^2 - \cos 2\alpha + \alpha\sin 2\alpha\right]$$

***Example* 4.4.**

(i) Calculate the normalisation constant for a wave function (at $t = 0$) given by $\psi(x) = ae^{-\frac{(\alpha^2x^2)}{2}}e^{ikx}$ known as a Gaussian wave packet. Determine

(ii) the probability density and

(iii) probability current density of the wave packet. **[NBU (H) 2003]**

Solution:

(i) According to the condition of normalisation

$$\int_{-\infty}^{+\infty} |\psi(x)|^2\, dx = |a|^2\int_{-\infty}^{+\infty} e^{-\alpha^2x^2}\, dx = 1$$

So

$$|a|^2\frac{\sqrt{\pi}}{\alpha} = 1 \text{ or, } |a| = \left(\frac{\alpha}{\sqrt{\pi}}\right)^{1/2}$$

This gives the normalisation constant of the wave function.

$\therefore$ probability density

$$\rho = \psi^*\psi = a^* e^{-a^2x^2/2}e^{-ikx} \times ae^{-a^2x^2/2}e^{ikx} = \frac{a}{\sqrt{\pi}}e^{-a^2x^2}$$

(ii) We have the wave function

$$\psi(x) = ae^{-\frac{\alpha^2x^2}{2}}e^{ikx}$$

$$\therefore \psi^*(x) = a^* e^{-\alpha^2x^2/2} e^{-ikx}$$

(iii) We have

$$\frac{\partial\psi(x)}{\partial x} = a(ik - \alpha^2 x)\, e^{-\alpha^2x^2/2} e^{ikx}\;;$$

$$\frac{\partial\psi^*(x)}{\partial x} = a^*(-ik - \alpha^2 x)\, e^{-\alpha^2x^2/2} e^{-ikx}$$

$\therefore$ the probability current density is

$$J_x = \frac{i\hbar}{2m}\left[\psi\frac{\partial\psi^*}{\partial x} - \psi^*\frac{\partial\psi}{\partial x}\right] = \frac{i\hbar}{2m}|a|^2\,(-ik - \alpha^2 x - ik + \alpha^2 x)\, e^{-\alpha^2x^2} = \frac{\hbar k}{m}|a|^2\, e^{-\alpha^2x^2} = \frac{\hbar k}{m}p$$

» ***Example* 4.5.** A plane wave is given by the wave function $\psi(x) = Ae^{ikx}$ in one-dimension. Find the probability current density.

Solution: The probability current density is given by

$$J = \frac{\hbar}{2mi}[\psi^*\nabla\psi - \psi\nabla\psi^*] = \frac{\hbar}{m}\left[\frac{\psi^*\nabla\psi - \psi\nabla\psi^*}{2i}\right]$$

$$\therefore J = \frac{\hbar}{m}\,\mathrm{Im}(\psi^*\nabla\psi) \text{ in one-dimension}$$

Since, $\psi(x) = Ae^{ikx}$

$$\therefore \frac{\partial\psi}{\partial x} = A(ik)\, e^{ikx}; \qquad \psi^* = A^* e^{-ikx}$$

$$\therefore J = \frac{\hbar}{m}\,\mathrm{Im}\left(A^* e^{-ikx} Ae^{ikx}\cdot ik\right) = \frac{\hbar}{m}\,\mathrm{Im}(|A|^2\, ik) = \frac{\hbar k}{m}|A|^2$$

But, $\frac{\hbar k}{m} = v$, velocity. $\therefore J = v|A|^2$.

Now, the probability density of the particle is $\rho = \psi^*\psi = A^* e^{-ikx} Ae^{ikx} = |A|^2$.

$\therefore$ the probability current density is J = velocity × probability density

$= v\rho =$ **flow of probability per second through unit area**

» ***Example* 4.6.** A one-dimensional wave function is given by $\psi(x) = \sqrt{a}e^{-ax}$, find the probability of finding the particle between $x = 1/a$ and $x = 2/a$.

Solution: Probability of finding the particle between x and $x + dx$ is

$$P(x)\,dx = \psi^*\psi\,dx = \sqrt{a}e^{-ax}\cdot\sqrt{a}e^{-ax} = ae^{-2ax}\,dx$$

Now, the probability of finding the particle between $x = 1/a$ and $x = 2/a$ is

$$P = \int_{1/a}^{2/a} ae^{-2ax}\,dx = a\left[\frac{e^{-2ax}}{-2a}\right]_{1/a}^{2/a} = -\frac{1}{2}\left[e^{-2a\times\frac{2}{a}} - e^{-2a\times\frac{1}{a}}\right] = \frac{e^{-2} - e^{-4}}{2}$$

***Example* 4.7.** If $\psi_1(x,t)$ and $\psi_2(x,t)$ are both the solutions of Schrödinger's wave equation for a given potential $V(x,t)$, then show that $\psi = a_1\psi_1 + a_2\psi_2$, in which a_1 and a_2 are arbitrary constants, is also a solution. **[PU 1991; BU (H) 2006]**

Solution: One-dimensional time-dependent Schrödinger equation is

$$-\frac{\hbar^2}{2m}\frac{\partial^2}{\partial x^2}\psi(x,t) + V(x,t)\,\psi(x,t) = i\hbar\frac{\partial}{\partial t}\psi(x,t) \qquad \text{(i)}$$

Since, $\psi_1(x,t)$ and $\psi_2(x,t)$ are both solutions of the above equation, we have

$$-\frac{\hbar^2}{2m}\frac{\partial^2}{\partial x^2}\,\psi_1(x,t) + V(x,t)\psi_1(x,t) = i\hbar\psi_1\psi_1(x,t) \qquad \text{(ii)}$$

$$\text{and } -\frac{\hbar^2}{2m}\frac{\partial^2}{\partial x^2}\,\psi_2(x,t) + V(x,t)\,\psi_2(x,t) = i\hbar\frac{\partial}{\partial t}\,\psi_2(x,t) \qquad \text{(iii)}$$

Multiplying (ii) by a_1 (iii) by a_2 and, then adding, we get

$$-\frac{\hbar^2}{2m}\frac{\partial^2}{\partial x^2}(a_1\psi_1 + a_2\psi_2) + V(a_1\psi_1 + a_2\psi_2) = i\hbar\frac{\partial}{\partial t}(a_1\psi_1 + a_2\psi_2) \qquad \text{(iv)}$$

Comparing (iv) with (i), we get $\psi(x,t) = a_1\psi_1 + a_2\psi_2$.

$\therefore \psi = a_1\psi_1 + a_2\psi_2$ is also a solution of Schrödinger time-dependent wave equation.

***Example* 4.8.** The wave function of a particle constraint to move along $x(-\infty < x < \infty)$ at a certain instant is given by $\psi(x) = A\exp\left[-\left(\frac{x^2}{a^2} + ibx\right)\right]$ [a, b are real constants]. Find the normalisation constant A. **[BU (H) 1993]**

Solution: The condition of normalisation of the wave function of $\psi(x)$ is

$$\int_{-\infty}^{+\infty} \psi^*(x)\,\psi(x)\,dx = 1$$

But, $\psi(x) = Ae^{-x^2/a^2 + ibx}$.

$\therefore \psi^*(x) = A^*e^{-x^2/a^2 - ibx}$.

Now,

$$\int_{-\infty}^{+\infty} \psi^*(x)\psi(x)\,dx = \int_{-\infty}^{\infty} |A|^2 e^{-x^2/a^2 - ibx} \times e^{-x^2/a^2 + ibx}\,dx = A^2\int_{-\infty}^{+\infty} e^{-\frac{2x^2}{a^2}}\,dx = 1 \qquad \text{(i)}$$

Putting $\frac{2x^2}{a^2} = z$, $x^2 = \frac{a^2 z}{2}$.

$$\therefore x = \frac{a\sqrt{z}}{\sqrt{2}} \text{ or, } 2x\,dx = \frac{a^2}{2}\,dz \text{ or, } dx = \frac{a^2}{4x}\,dz = \frac{a^2 dz}{4 \times \frac{az^{1/2}}{\sqrt{2}}} = \frac{az^{-1/2}\,dz}{2\sqrt{2}}$$

$\therefore$ from (i)

$$A^2\int_{-\infty}^{+\infty} e^{-z}\frac{a}{2\sqrt{2}}z^{-\frac{1}{2}}\,dz = 1 \text{ or, } A^2\frac{a}{2\sqrt{2}}\int_{-\infty}^{+\infty} e^{-z}z^{+\frac{1}{2}-1}\,dz = 1 \text{ or, } A^2\frac{a}{2\sqrt{2}}\Gamma\left(\frac{1}{2}\right) = 1$$

$$\text{or, } A^2\frac{a}{2\sqrt{2}}\sqrt{\frac{\pi}{2}} = 1 \text{ or, } A^2 = \frac{4}{\sqrt{\pi}}\cdot\frac{1}{a} \text{ or, } A = \left(\frac{4}{\sqrt{\pi}}\frac{1}{a}\right)^{1/2}$$

» ***Example* 4.9.** Consider an infinite square well potential in one dimension

$$V = \begin{cases} 0, & \text{for } -a < x < a \\ \infty, & \text{for } x < -a \text{ and } x > a \end{cases}$$

The wave function of a particle trapped in this well is found to be

$$\psi(x) = C\left[\cos\left(\frac{\pi x}{2a}\right) + \sin\left(\frac{3\pi x}{a}\right) + \frac{1}{4}\cos\left(\frac{3\pi x}{2a}\right)\right]$$

inside the well and $\psi(x) = 0$ outside the well. Calculate C. **[BU (H) 1996]**

Solution: According to the normalisation condition

$$\int_{-\infty}^{+\infty} \psi^*(x)\,\psi(x)\,dx = 1$$

$$\text{or, } 2\int_0^a |C|^2\left[\cos\left(\frac{\pi x}{2a}\right) + \sin\left(\frac{3\pi x}{a}\right) + \frac{1}{4}\cos\left(\frac{3\pi x}{2a}\right)\right]^2 dx = 1$$

$$\text{or, } 2|C|^2\left[\int_0^a \cos^2\left(\frac{\pi x}{2a}\right)dx + \int_0^a \sin^2\left(\frac{3\pi x}{2a}\right)dx + \frac{1}{16}\int_0^a \cos^2\left(\frac{3\pi x}{2a}\right)dx + \int_0^a 2\cos\left(\frac{\pi x}{2a}\right)\sin\left(\frac{3\pi x}{a}\right)dx\right.$$

$$\left. + \int_0^a \frac{1}{2}\sin\left(\frac{3\pi x}{a}\right)\cos\left(\frac{3\pi x}{2a}\right)dx + \int_0^a \frac{1}{2}\cos\left(\frac{\pi x}{2a}\right)\cos\left(\frac{3\pi x}{2a}\right)dx\right] = 1 \qquad \text{(i)}$$

But,

$$\int_0^a 2\cos\left(\frac{\pi x}{2a}\right)\sin\left(\frac{3\pi x}{a}\right)dx = 0$$

$$\text{and } \frac{1}{2}\int_0^a \cos\left(\frac{\pi x}{2a}\right)\cos\left(\frac{3\pi x}{a}\right)dx = \frac{1}{4}\int_0^a 2\cos\left(\frac{\pi x}{2a}\right)\cos\left(\frac{3\pi x}{a}\right)dx$$

$$= \frac{1}{4}\left[\int_0^a \sin\left(\frac{\pi x}{2a} + \frac{3\pi x}{2a}\right)dx + \int_0^a \sin\left(\frac{3\pi x}{2a} - \frac{\pi x}{2a}\right)dx\right] = 0$$

∴ from (i)

$$2|C|^2\left[\int_0^a \cos^2\frac{\pi x}{2a}dx + \int_0^a \sin^2\frac{3\pi x}{2a}dx + \frac{1}{16}\int_0^a \cos^2\frac{3\pi x}{2a}dx\right] = 1$$

$$\text{or, } 2|C|^2\left[\frac{a}{2} + \frac{a}{2} + \frac{1}{16}\frac{a}{2}\right] = 1 \text{ or, } 2|C|^2 \times \frac{33a}{32} = 1 \text{ or, } |C|^2 = \frac{16}{33a}$$

$$\therefore C = \sqrt{\frac{16}{33a}} = \frac{4}{\sqrt{33a}}$$

» ***Example* 4.10.** The normalised ground state wave function of an electron in a hydrogen atom is given by $\psi(\vec{r}) = \left[\frac{1}{(\pi a_0^3)}\right]^{1/2} e^{-r/a_0}$, where a_0 is the first Bohr radius.

(i) Verify that $\psi(\vec{r})$ is normalised.

(ii) Obtain the probability of finding the ground state electron at a distance greater than a_0 from the nucleus. **[BU (H) 1994, 1997, 2002]**

Solution:

(i) According to the condition of normalisation

$$\int_{r=0}^{\infty} \psi^*(r)\,\psi(r) 4\pi r^2\,dr = 1$$

Now, let

$$I = 4\pi\left(\frac{1}{\pi a_0^3}\right)\int_0^{\infty} e^{-2r/a_0} r^2\,dr \qquad \text{(i)}$$

Putting

$$x = \frac{2r}{a_0};\; r = \frac{a_0 x}{2},\; dr = \frac{a_0}{2}\,dx \text{ and } r^2\,dr = \left(\frac{a_0}{2}\right)^3 x^2\,dx$$

$\therefore$ from (i)

$$I = 4\pi\left(\frac{1}{\pi a_0^3}\right)\left(\frac{a_0}{2}\right)^3\int_0^{\infty} e^{-x} x^2\,dx = \frac{1}{2}\Gamma(3) = \frac{1}{2}2! = 1 \qquad \text{(ii)}$$

Hence, the given wave function $\psi(r)$ is normalised.

(ii) Now, the probability of finding the ground state electron is

$$P = \int_{a_0}^{\infty} \psi^*(r)\,\psi(r) 4\pi r^2\,dr = 4\pi\left(\frac{1}{\pi a_0^3}\right)\int_{a_0}^{\infty} e^{-2r/a_0} r^2\,dr = \frac{4}{a_0^3}\int_{a_0}^{\infty} e^{-2r/a_0} r^2\,dr \qquad \text{(iii)}$$

Put $x = r - a_0$, $\therefore r = x + a_0$ or, $r^2 = x^2 + 2a_0 x + a_0^2$ and $dr = dx$.

$$\therefore \int_{a_0}^{\infty} e^{-2r/a_0} r^2\,dr = \int_0^{\infty} e^{-2\left(\frac{x}{a_0}+1\right)}(x^2 + 2a_0 x + a_0^2)\,dx$$

$$= e^{-2}\left[\int_0^{\infty} e^{-\frac{2x}{a_0}} x^2\,dx + 2a_0\int_0^{\infty} e^{-\frac{2x}{a_0}} x\,dx + a_0^2\int_0^{\infty} e^{-\frac{2x}{a_0}}\,dx\right]$$

Now, putting $\dfrac{2x}{a_0} = z$, $dx = \dfrac{a_0}{2}\,dz$, $x^2 = \dfrac{a_0^2}{4}z^2$ and $x = \dfrac{a_0}{2}z$

$$\therefore \int_{a_0}^{\infty} e^{-2r/a_0} r^2\,dr = e^{-2}\left[\frac{a_0^3}{8}\int_0^{\infty} e^{-z} z^2\,dz + 2a_0\frac{a_0^3}{4}\int_0^{\infty} e^{-z} z\,dz + a_0^2 \times \frac{a_0}{2}\int_0^{\infty} e^{-z}\,dz\right]$$

$$= e^{-2}a_0^3\left[\frac{1}{8}\Gamma(3) + \frac{1}{2}\Gamma(2) + \frac{1}{2}\right] = \frac{5}{4}e^{-2}a_0^3$$

$$\therefore P = \frac{4}{a_0^3} \times \frac{5}{4}e^{-2}a_0^3 = 5e^{-2}$$

$\therefore$ the probability of finding the ground state electron at a distance greater than a_0 from the nucleus is $5e^{-2}$.

» ***Example* 4.11.** The wave function of a particle is given by $\psi(x) = Ae^{-\alpha x^2}$, $-\infty < x < \infty$, where A and α are constants. Calculate the probability of finding the particle in the region $0 < x < \infty$. **[BU (H) 1995]**

Solution: According to the condition of normalisation

$$\int_{-\infty}^{+\infty} |\psi(x)|^2\,dx = 1 \text{ or, } \int_{-\infty}^{+\infty} |A|^2 e^{-2\alpha x^2}\,dx = 1 \text{ or, } 2\int_0^{\infty} |A|^2 e^{-2\alpha x^2}\,dx = 1$$

$$\text{or, } 2|A|^2 \cdot \frac{1}{4}\sqrt{\frac{2\pi}{\alpha}} = 1 \text{ or, } |A|^2 = 2\sqrt{\frac{\alpha}{2\pi}}$$

Again, the probability of finding the particle in the region $0 < x < \infty$ is

$$P = \int_0^\infty |\psi(x)|^2\, dx = |A|^2 \int_0^\infty e^{-2\alpha x^2}\, dx = |A|^2 \int_0^\infty e^{-2\alpha^2} x\, dx = |A|^2 \frac{\sqrt{2\alpha}}{4\pi} \int_0^\infty e^{-z} z^{-1/2}\, dz$$

[Putting $2\alpha x^2 = z$, $\therefore 4\alpha x\, dx = dz$, $\therefore dx = \frac{\sqrt{2\alpha}}{4\alpha} z^{-1/2}\, dz$]

$$\therefore P = |A|^2 \frac{1}{4}\sqrt{\frac{2}{\alpha}}\Gamma\left(\frac{1}{2}\right) = |A|^2 \frac{1}{4}\sqrt{\frac{2}{\alpha}}\sqrt{\pi} = |A|^2 \frac{1}{4}\sqrt{\frac{2\pi}{\alpha}} = 2\sqrt{\frac{\alpha}{2\pi}} \times \frac{1}{4}\sqrt{\frac{2\pi}{\alpha}} = \frac{1}{2}$$

$\therefore$ the probability of finding the particle in the region $0 < x < \alpha$ is $\frac{1}{2}$.

» ***Example* 4.12.** The normalised ground state wave function of a one-electron atom is given by

$$\psi(r) = \pi^{-1/2}\left(\frac{z}{a_0}\right)^{3/2} e^{-\frac{zr}{a_0}}$$

where the notations have usual meaning. Evaluate the probability that the electron in this state will be found at a distance greater than $2a_0/z$. **[BU (H) 2000]**

***Solution*:** The probability of finding the electron at a distance greater than $2a_0/z$ is

$$P = \int_{2a_0/z}^\infty |\psi(r)|^2\, 4\pi r^2\, dr = 4\pi \int_{2a_0/z}^\infty |\psi(r)|^2 r^2\, dr = 4\pi \int_0^\infty |\psi|^2 r^2\, dr - 4\pi \int_0^{2a_0/z} |\psi|^2 r^2\, dr$$

$$= 1 - 4\left(\frac{z}{a_0}\right)^3 \int_0^{2a_0/z} r^2 e^{-\frac{2zr}{a_0}}\, dr = 1 - 4\left(\frac{z}{a_0}\right)^3 \left(\frac{a}{2z}\right)^4 \int_0^4 x^2 e^{-x}\, dx \left[x = \frac{2z}{a_0} r\right]$$

$$= 1 - \tfrac{1}{2}(2 - 26e^{-4}) = 13e^{-4} \simeq \mathbf{0.15}$$

» ***Example* 4.13.** Prove that in the case of the wave function, $\psi(r) = \frac{e^{ikr}}{r}$, the probability current density is given by $J = v/r^2$, where v is the velocity of the particle. **[CU (H) 2006, 2009; KU (H) 2011]**

***Solution*:** The probability current density is

$$J = \frac{\hbar}{2mi}[\psi^* \nabla\psi - \psi \nabla\psi^*] = \frac{\hbar}{m}\left[\frac{\psi^* \nabla\psi - \psi \nabla\psi^*}{2i}\right] = \frac{\hbar}{m}\,\mathrm{Im}(\psi^* \nabla\psi)$$

The given wave function $\psi(r)$ is a function of r only we take the r-component of $\nabla\psi = \frac{\partial\psi}{\partial r}$

$$\therefore J_r = \frac{\hbar}{m}\,\mathrm{Im}\left(\psi^* \frac{\partial\psi}{\partial r}\right) = \frac{\hbar}{m}\,\mathrm{Im}\left[\frac{e^{-ikr}}{r}\frac{1}{r}\left(e^{ikr} ik - \frac{e^{ikr}}{r}\right)\right] = \frac{\hbar}{m}\,\mathrm{Im}\left[\frac{1}{r^2}\left(ik - \frac{1}{r}\right)\right]$$

$$= \frac{\hbar}{m} \cdot \frac{k^2}{r^2} = \frac{v}{r^2} \left[\because v = \frac{\hbar k}{m} = \text{velocity of the particle}\right]$$

» ***Example* 4.14.** Find the value of the normalisation constant A for the wave function, representing a particle moving in a box of dimension $a \times b \times c$, where

$$\psi(x, y, z) = A \sin\left(\frac{n_x \pi x}{a}\right) \sin\left(\frac{n_y \pi y}{b}\right) \sin\left(\frac{n_z \pi y}{c}\right).$$ **[WBUT 2003]**

***Solution*:** The given wave function can be expressed as $\psi(x, y, z) = \psi(x)\,\psi(y)\,\psi(z)$, where

$$\psi(x) = B_x \sin\frac{n_x \pi x}{a}$$

Similarly,

$$\psi(y) = B_y \sin\frac{n_y \pi x}{a} \text{ and } \psi(z) = B_z \sin\frac{n_y \pi x}{a}$$

Since, the free particle is moving inside a box of length a the probability of finding the particle inside the box is 1.

$$\therefore \int_0^a \psi^*(x)\,\psi(x)\,dx = 1 \text{ or, } \int_0^a B_x \sin\frac{n_x \pi x}{a} B_x^* \sin\frac{n_x \pi x}{a}\,dx = 1$$

$$\text{or, } |B_x|^2 \times \int_0^a \sin^2\frac{n_x \pi x}{a}\,dx = 1 \text{ or, } \frac{|B_x|^2}{2} \times \int_0^a \left(1 - \cos\frac{2n_x \pi x}{a}\right) dx = 1$$

$$\text{or, } \frac{|B_x|^2}{2} a = 1 \text{ or, } |B_x| = \sqrt{\frac{2}{a}}$$

Similarly,

$$|B_y| = \sqrt{\frac{2}{b}} \text{ and } |B_z| = \sqrt{\frac{2}{c}}$$

$\therefore$ the normalised wave function is

$$\psi(x, y, z) = \frac{2\sqrt{2}}{\sqrt{abc}} \sin\frac{n_x \pi x}{a} \sin\frac{n_y \pi y}{b} \sin\frac{n_z \pi z}{c}$$

and the normalisation constant $A = \dfrac{2\sqrt{2}}{\sqrt{abc}}$.

» ***Example* 4.15.** Normalise the function $\sin\frac{n\pi x}{L}$ of a particle moving in a region $0 \leqslant x \leqslant L$. Calculate the expectation value of the momentum of the particle.

Solution: Let the wave function be

$$\psi(x) = N\frac{\sin n\pi x}{L}$$

N is the normalisation constant.

Now, the normalisation condition is

$$\int_0^L \psi^*(x)\psi(x)\,dx = 1 \text{ or, } \int_0^L \left(N\sin\frac{n\pi x}{L}\right)^* \left(\sin\frac{n\pi x}{L}\right) dx = 1$$

$$\text{or, } |N|^2 \int_0^L \sin^2\frac{n\pi x}{L}\,dx = 1$$

$$\text{or, } |N|^2 \left[\int_0^L \frac{1}{2}\left(1 - \cos\frac{2n\pi x}{L}\,dx\right)\right] = 1$$

$$\text{or, } \frac{|N|^2}{2}(L - 0) = 1 \text{ or, } N = \sqrt{\frac{2}{L}}$$

The normalised wave function is

$$\psi(x) = \sqrt{\frac{2}{L}} \sin\frac{n\pi x}{L}$$

Now, the expectation value of momentum is

$$\langle \hat{p} \rangle = \int_0^L \psi^* \hat{p} \psi \, dx$$

$$= \int_0^L \left(\sqrt{\frac{2}{L}} \sin \frac{n\pi x}{L} \right)^* \left(-i\hbar \frac{\partial}{\partial x} \right) \left(\sqrt{\frac{2}{L}} \sin \frac{n\pi x}{L} \right) dx$$

$$= -i\hbar \frac{2}{L} \int_0^L \sin \frac{n\pi x}{L} \cos \frac{n\pi x}{L} \left(\frac{n\pi}{L} \right) dx$$

$$= -i\hbar \frac{n\pi}{L^2} \int_0^L \sin \frac{2n\pi x}{L} \, dx = -\frac{i\Gamma n\pi}{L^2} \cdot 0$$

$$\therefore \langle \hat{p} \rangle = 0$$

» ***Example* 4.16.** Given $\psi(\vec{r}, t) = Ae^{i(\vec{k}\cdot\vec{r}-\omega t)}$. Find the probability current density. **[CU (H) 2012; KU (H) 2009]**

***Solution*:** The expression for probability current density is

$$\vec{S} = \frac{\hbar}{2m} [\psi^* \nabla \psi - (\nabla \psi^*) \psi]$$

Now, $\nabla \psi = \nabla A e^{i(\vec{k}\cdot\vec{r}-\omega t)} = A(\nabla e^{i\vec{k}\cdot\vec{r}}) = i\vec{k} A e^{i(\vec{k}\cdot\vec{r})} = i\vec{k}\psi$

and $\nabla \psi^* = \nabla A^* e^{-i(\vec{k}\cdot\vec{r})} = A^* (\nabla e^{-i\vec{k}\cdot\vec{r}}) e^{i\omega t} = -i\vec{k} A^* e^{-i(\vec{k}\cdot\vec{r}-\omega t)} = -i\vec{k}\psi^*$

$$\therefore \vec{S} = \frac{\hbar}{2m} [\psi^* i\vec{k}\psi - (-i\vec{k}\psi^*)\psi] = \frac{\hbar}{2im} 2i\vec{k}\psi^*\psi = \frac{\hbar\vec{k}}{m}\psi^* Y = \frac{\vec{p}}{m}\psi^*\psi = \vec{v}\psi^*\psi$$

where $\vec{v} = \frac{\vec{p}}{m}$ is the velocity of the particle.

The probability density

$$\rho = \psi^* \psi = \left[A^* e^{i(\vec{k}\cdot\vec{r})} \right] \left[A e^{i(\vec{k}\cdot\vec{r}-\omega t)} \right] = |A|^2$$

$$\therefore \vec{S} = \vec{v}\rho = \frac{\hbar\vec{k}}{m} |A|^2$$

» ***Example* 4.17.** A wave function is given by

$$\psi(x) = N \left(\frac{x}{x_0} \right)^n e^{-x/x_0}$$

where N, n, x_0 are constants.

From Schrödinger equation find potential $V(x)$ and energy E for which $\psi(x)$ is an eigenfunction.

***Solution*:** The time independent Schrödinger equation is given by

$$\frac{d^2}{dx^2}\psi(x) + \frac{2m}{\hbar^2}[V - V(x)]\psi(x) = 0$$

$$\text{or,} \quad \frac{d^2}{dx^2} N \left(\frac{x}{x_0} \right)^2 e^{-x/x_0} + \frac{2m}{\hbar^2}[E - V(x)]\psi(x) = 0$$

Now,

$$\frac{d^2}{dx^2}N\left(\frac{x}{x_0}\right)^n e^{-x/x_0} = \frac{N}{x_0^n}\frac{d}{dx}\frac{d}{dx}\left(x^n e^{-x/x_0}\right) = \frac{N}{x_0^n}\frac{d}{dx}\left[nx^{n-1}e^{-x/x_0} + x^n\left(-\frac{1}{x_0}\right)e^{-x/x_0}\right]$$

$$= \frac{N}{x_0^n}\frac{d}{dx}\left(nx^{n-1} - \frac{x^n}{x_0}\right)e^{-x/x_0}$$

$$= \frac{N}{x_0^n}\left[\left\{n(n-1)x^{n-2} - \frac{n}{x_0}x^{n-1}\right\}e^{-x/x_0} + \left(nx^{n-1} - \frac{x^n}{x_0}\right)\left(-\frac{1}{x_0}\right)e^{-x/x_0}\right]$$

$$= \frac{N}{x_0^n}\left[n(n-1)x^{n-2} - \frac{n}{x_0}x^{n-1} - \frac{1}{x_0}\left(nx^{n-1} - \frac{x^n}{x_0}\right)\right]$$

$$= \left[\frac{n(n-1)}{x^2} - \frac{2n}{x_0 x} + \frac{1}{x_0^2}\right]N\frac{x^2}{x_0^n}e^{-x/x_0} = \left[\frac{n(n-1)}{x^2} - \frac{2n}{x_0 x} + \frac{1}{x_0^2}\right]\psi(x)$$

$$\therefore \left[\frac{n(n-1)}{x^2} - \frac{2n}{x x_0} + \frac{1}{x_0^2}\right]\psi(x) + \frac{2m}{\hbar^2}[E - V(x)]\psi(x) = 0$$

$$\text{or, } E - V(x) = -\frac{\hbar^2}{2m}\left[\frac{n(n-1)}{x^2} - \frac{2n}{x_0 x} + \frac{1}{x_0^2}\right]$$

When $x \to \infty$, $V(x) = 0$.

$$\therefore E = -\frac{\hbar^2}{2mx_0^2}$$

Also

$$-\frac{\hbar^2}{2mx_0^2} - V(x) = -\frac{\hbar^2}{2m}\left[\frac{n(n-1)}{x^2} - \frac{2m}{x_0 x} + \frac{1}{x_0^2}\right]$$

$$\text{or, } V(x) = \frac{\hbar^2}{2m}\left[\frac{n(n-1)}{x^2} - \frac{2m}{x_0 x}\right]$$

» ***Example*** **4.18. Find the probability that a particle can the particle is limited to the x-axis and has the wave function**

$$\psi = \begin{cases} ax, & \text{for } 0 \leqslant x \leqslant 1 \\ 0, & \text{for } x < 0 \text{ and } x > 0 \end{cases}$$

Also find the average value of x. **[cf. BU (H) 2006]**

Solution: From the condition of normalisation of the wave function we have

$$\int_{-\infty}^{+\infty} \psi^*\psi\, dx = 1 \text{ or, } \int_0^1 (ax)^*(ax)\, dx = 1 \text{ or, } a^2\int_0^1 x^* x\, dx = 1 \text{ or, } a^2\int^1 x^2\, dx = 1 \text{ or, } a^2 \cdot \frac{1}{3} = 1$$

$\therefore a = \sqrt{3}$.

$$\therefore \psi = \begin{cases} \sqrt{3}x, & \text{for } x \leqslant x \leqslant 1 \\ 0, & \text{for } x < 0, x > 0 \end{cases}$$

Now, the probability of the particle that can be found in between $x = 0.45$ and $x = 0.55$ is

$$\int_{0.45}^{0.55} \psi^*\psi\, dx = \int_{0.45}^{0.55} (\sqrt{3}x)^*(\sqrt{3}x)\, dx = 3\int_{0.45}^{0.55} x^2\, dx$$

$$= \left[3\cdot\frac{x^3}{3}\right]_{0.45}^{0.55} = \left[x^3\right]_{0.45}^{0.55} = (0.55)^3 - (0.45)^3 = \mathbf{0.08}$$

$$\text{and } \langle\hat{x}\rangle = \int_{-\infty}^{+\infty} \psi^*\hat{x}\psi\, dx = \int_{-\infty}^{+\infty} |\psi|^2 x\, dx = \int_0^1 |\sqrt{3}x|^2 x\, dx$$

$$= 3\int_0^1 x^3\, dx = 3\cdot\left[\frac{x^4}{4}\right]_0^1 = \frac{3}{4}(1-0) = \mathbf{0.75}$$

» ***Example* 4.19.** Normalise the following wave function

$$\psi(x) = \begin{cases} Ne^{-\alpha x}, & (x > 0) \\ Ne^{\alpha x}, & (x < 0) \end{cases}$$

where α is a positive constant.

***Solution*:** The condition of normalisation is

$$\int_{-\infty}^{+\infty} \psi^*\psi\, dx = 1 \text{ or, } \int_{-\alpha}^{0} (Ne^{\alpha x})^*(Ne^{\alpha x})\, dx + \int_0^{\infty} (Ne^{-\alpha x})^*(Ne^{-\alpha x})\, dx = 1$$

$$\text{or, } |N|^2 \int_{\alpha}^{0} e^{2\alpha x}\, dx + |N|^2 \int_{0 e^{-2\alpha x}}^{\infty} dx = 1$$

$$\text{or, } |N|^2 \left[\frac{e^{2\alpha x}}{2\alpha}\Bigg|_{-\infty}^{0} + \frac{e^{-2\alpha x}}{-2\alpha}\Bigg|_{-\infty}^{0}\right] = 1$$

$$\text{or, } |N|^2 \left[\frac{1}{2\alpha} + \frac{1}{2\alpha}\right] = 1 \quad (\because e^{-\alpha} = 0)$$

$$\text{or, } |N|^2 = \alpha \text{ or, } N = \sqrt{\alpha}$$

$$\therefore \psi(x) = \begin{cases} \sqrt{\alpha}e^{-\alpha x}, & \text{for } x > 0 \\ \sqrt{\alpha}e^{\alpha x}, & \text{for } x < 0 \end{cases}$$

» ***Example* 4.20.** Prove that the expectation value of the square of an observable quantity is always positive.

[cf. CU (H) 2008; KU (H) 2013, 2015]

***Solution*:** We have

$$\langle\hat{A}^2\rangle = \int_{-\infty}^{\infty} \hat{A}^2\psi\, dx = \int_{-\infty}^{+\infty} \psi^*\hat{A}\hat{A}\psi\, dx$$

Let $\hat{A}\psi = \phi$, then

$$\langle\hat{A}^2\rangle = \int_{-\infty}^{+\infty} \psi^*\hat{A}\phi\, dx$$

$\hat{A}$ being Hermitian

$$\langle\hat{A}^2\rangle = \int_{-\infty}^{+\infty} \phi\hat{A}^*\psi^*\, dx = \int_{-\infty}^{+\infty} (\hat{A}\psi)\hat{A}^*\psi^*\, dx$$

$$= \int_{-\infty}^{+\infty} (\hat{A}\psi)(\hat{A}\psi)^*\, dx = \int_{-\infty}^{+\infty} |\hat{A}\psi|^2\, dx$$

$\therefore \langle \hat{A}^2 \rangle \geqslant 0$. Hence, the result follows.

» ***Example* 4.21.** Consider the Gaussian wave packet $\exp\left[-\frac{(x-x_0)^2}{2\sigma^2}\right]\exp\left[\frac{ip_0x}{\hbar}\right]$.

(i) Normalise the wave function.

(ii) Find position probability density and position probability current density.

(iii) Show that it corresponds minimum uncertainty product. **[cf. CU (H) 2008, 2015; cf. KU (H) 2013]**

***Solution*:**

(i) The wave function for the Gaussian wave packet is $\psi(x) = \exp\left[-\frac{(x-x_0)^2}{2\sigma^2}\right]\exp\left[\frac{ip_0x}{\hbar}\right]$.

The normalised Gaussian wave function is

$$\psi(x) = N\exp\left[-\frac{(x-x_0)^2}{2\sigma^2}\right]\exp\left[\frac{ip_0x}{\hbar}\right]$$

$$\text{and } \psi(x)^* = N\exp\left[-\frac{(x-x_0)^2}{2\sigma^2}\right]\exp\left[\frac{ip_0x}{\hbar}\right]$$

where N is the normalisation constant.

From the condition of normalisation

$$\int_{-\infty}^{\infty} \psi^*\psi\, dx = 1$$

$$\text{or, } \int_{-\infty}^{+\infty} (N^*)\exp\left[-\frac{(x-x_0)^2}{2\sigma^2}\right]\exp\left[\frac{ip_0x}{\hbar}\right]\left(N\exp\left[-\frac{(x-x_0)^2}{2\sigma^2}\right]\exp\left[\frac{ip_0x}{\hbar}\right]\right) = 1$$

$$\text{or, } |N|^2\int_{-\infty}^{+\infty}\exp\left[-\frac{(x-x_0)^2}{\sigma^2}\right]dx = 1 \text{ or, } |N|^2\int_{-\infty}^{+\infty} e^{-x'^2/\sigma^2}dx' = 1\, [x' = (x-x_0)]$$

$$\text{or, } |N|^2\frac{\sqrt{\pi}}{\sqrt{1/\sigma^2}} = 1 \text{ or, } |N|^2 = \frac{1}{\sigma\sqrt{\pi}} \text{ or, } N = \frac{1}{\sqrt{\sigma\sqrt{\pi}}} \left[\because e^{-\beta x^2\,dx = \sqrt{\pi/\beta}}\right]$$

$\therefore$ the normalised wave function for the Gaussian wave packet is

$$\tau(x) = \frac{1}{\sqrt{\sigma\sqrt{\pi}}}\exp\left[-\frac{(x-x_0)^2}{2\sigma^2}\right]\exp\left[\frac{ip_0x}{\hbar}\right]$$

(ii) Position probability density

$$\rho = \psi^*\psi = |\psi|^2 = \left|\frac{1}{\sqrt{\sigma\sqrt{\pi}}}\exp-\left[\frac{(x-x_0)^2}{2\sigma^2}\right]\exp\left[\frac{ip_0x}{\hbar}\right]\right|^2$$

$$\rho = \frac{1}{\sqrt{\sigma\sqrt{\pi}}}\exp\left[-\frac{(x-x_0)^2}{\sigma^2}\right]$$

and the position probability current density is

$$S = \frac{\hbar}{2im}\left(\psi^*\frac{\partial\psi}{\partial x} - \frac{\partial\psi^*}{\partial x}\psi\right)$$

Now

$$\frac{\partial \psi}{\partial x} = -\frac{1}{2\sigma^2}2(x-x_0)\,\psi + \frac{ip_0}{\hbar} = \left(-\frac{x-x_0}{\sigma^2} + \frac{ip_0}{\hbar}\right)\psi$$

$$\text{and } \frac{\partial \psi^*}{\partial x} = -\frac{1}{2\sigma^2}2(x-x_0)\,\psi^* - \frac{ip_0\,\psi^*}{\hbar} = \left(-\frac{x-x_0}{\sigma^2} - \frac{ip_0}{\hbar}\right)\psi^*$$

$$\therefore S = \frac{\hbar}{2im}\left[\psi^*\left(-\frac{x-x_0}{\sigma^2} + \frac{ip_0}{\hbar}\right)\psi - \left(-\frac{x-x_0}{\sigma^2} - \frac{ip_0}{\hbar}\,\psi^*\psi\right)\right]$$

$$= \frac{\hbar}{2im}\frac{2ip_0}{\hbar}\psi^*\psi = \frac{p_0}{m}\psi^*\psi = v_0\rho$$

$$\text{or, } S = \frac{p_0}{m}\frac{1}{\sigma\sqrt{\pi}}\exp\left[-\frac{(x-x_0)^2}{\sigma^2}\right]\left[\because v_0 = \frac{p_0}{m} = \text{velocity}\right]$$

(iii) We have

$$\langle \hat{x} \rangle = \int_{-\infty}^{+\infty} \psi^* \hat{x}\,\psi\, dx = \int_{-\infty}^{+\infty} |\psi|^2 x\, dx = \int_{-\infty}^{+\infty} \frac{1}{\sigma\sqrt{\pi}} e^{\frac{-(x-x_0)^2}{\sigma^2}} x\, dx$$

Putting

$$\frac{x-x_0}{\sigma} = t \text{ or, } x - x_0 = \sigma t,\; dx = \sigma\, dt$$

we get

$$\langle \hat{x} \rangle = \frac{1}{\sigma\sqrt{\pi}}\int_{-\infty}^{+\infty} e^{-t^2}(x_0+\sigma t)(\sigma\, dt) = \frac{x_0}{\sqrt{\pi}}\int_{-\infty}^{+\infty} e^{-t^2} dt + \frac{\sigma}{\sqrt{\pi}}\int_{-\infty}^{+\infty} e^{-t^2} t\, dt$$

$$= \frac{x_0}{\sqrt{\pi}}\sqrt{\pi} + 0 \left[\because \int_{-\infty}^{+\infty} e^{-t^2} dt = \sqrt{\pi} \text{ and the second integral is odd}\right]$$

$\therefore \langle \hat{x} \rangle = x_0.$

Again

$$\langle \hat{x}^2 \rangle = \int_{-\infty}^{+\infty} \psi^* \hat{x}^2 \psi\, dx = \int_{-\infty}^{+\infty} \psi^* x^2 \psi\, dx = \int_{-\infty}^{+\infty} |\psi|^2 x^2\, dx = \frac{1}{\sigma\sqrt{\pi}}\int_{-\infty}^{+\infty} e^{\frac{-(x-x_0)^2}{\sigma^2}} x^2\, dx$$

$$= \frac{1}{\sigma\sqrt{\pi}}\int_{-\infty}^{+\infty} e^{-t^2}(x_0+\sigma t)^2(\sigma\, dt) = \frac{1}{\sigma\sqrt{\pi}}\int_{-\infty}^{+\infty} e^{-t^2}(x_0^2 + 2x_0\sigma t + \sigma^2 t^2)\,\sigma\, dt$$

$$= \frac{x_0^2}{\sqrt{\pi}}\int_{-\infty}^{+\infty} e^{-t^2} dt + \frac{2x_0\sigma}{\sqrt{\pi}}\int_{-\infty}^{+\infty} e^{-t^2} t\, dt + \frac{\sigma^2}{\sqrt{\pi}}\int_{-\infty}^{+\infty} e^{-t^2} t^2\, dx$$

$$= \frac{x_0^2}{\sqrt{\pi}}\cdot\sqrt{\pi} + 0 + \frac{\sigma^2}{\sqrt{\pi}}\cdot\frac{\sqrt{\pi}}{2}$$

$$\left[\because \text{the second integral is odd and } \int_{-\infty}^{+\infty} e^{-\beta x^2} x^2\, dx = \frac{1}{2}\sqrt{\frac{\pi}{\beta^2}}\right]$$

$\therefore \langle \hat{x}^2 \rangle = x_0^2 + \dfrac{\sigma^2}{2}.$

Now, the average value of momentum p is

$$\langle\hat{p}\rangle = \int_{-\infty}^{+\infty} \psi^* \hat{p}\psi\, dx = \int_{-\infty}^{+\infty} \psi^* \left(-i\hbar\frac{\partial}{\partial x}\psi\right) dx = -i\hbar \int_{-\infty}^{+\infty} \psi^* \frac{\partial\psi}{\partial x}\, dx$$

$$= -i\hbar \int_{-\infty}^{+\infty} \psi^* \left(-\frac{x - x_0}{\sigma^2} + \frac{ip_0}{\hbar}\right)\psi\, dx = -i\hbar\left[\left(\frac{ip_0}{\hbar} + \frac{x_0}{\sigma^2}\right)(1) - \frac{1}{\sigma^2}\langle\hat{x}\rangle\right]$$

$$= p_0 - i\hbar\frac{x_0}{\sigma^2} + i\hbar\frac{x_0}{\sigma^2}$$

$\therefore$ $\langle\hat{p}\rangle = p_0$.

Again,

$$\langle\hat{p}^2\rangle = \int_{-\infty}^{+\infty} \psi^* \hat{p}^2 \psi\, dx = \int_{-\infty}^{+\infty} \psi^* \left(-\hbar^2\frac{\partial^2\psi}{\partial x^2}\right) dx = -\hbar^2 \int_{-\infty}^{+\infty} \psi^* \frac{\partial}{\partial x}\left(\frac{\partial\psi}{\partial x}\right) dx$$

$$= -\hbar^2 \int_{-\infty}^{+\infty} \psi^* \frac{\partial}{\partial x}\left(-\frac{x - x_0}{\sigma^2} + \frac{ip_0}{\hbar}\right)\psi\, dx$$

$$= -\hbar^2 \int_{-\infty}^{+\infty} \psi^* \frac{\partial}{\partial x}\left[-\frac{x\psi}{\sigma^2} + \left(\frac{ip_0}{\hbar^2} + \frac{x_0}{\sigma^2}\right)\psi\right] dx$$

$$= -\hbar^2 \int_{-\infty}^{+\infty} \psi^* \left[-\frac{\psi}{\sigma^2} - \frac{x}{\sigma^2}\frac{\partial\psi}{\partial x} + \left(\frac{ip_0}{\hbar} + \frac{x_0}{\sigma^2}\right)\right]\frac{\partial\psi}{\partial x}\, dx$$

$$= \frac{\hbar^2}{\sigma^2}\int_{-\infty}^{+\infty} \psi^*\psi\, dx - \hbar^2 \int_{-\infty}^{+\infty} \psi^* \left[\left(\frac{ip_0}{\hbar} + \frac{x_0}{\sigma^2}\right) - \frac{x}{\sigma^2}\right]$$

$$= \frac{\hbar^2}{\sigma^2} - \hbar^2 \int_{-\infty}^{+\infty} \psi^* \left[\left(\frac{ip_0}{\hbar} + \frac{x_0}{\sigma^2}\right)^2 - \frac{2}{\sigma^2}\left(\frac{ip_0}{\hbar} + \frac{x_0}{\sigma^2}\right)\right]$$

$$= \frac{\hbar^2}{\sigma^2} - \hbar^2\left(\frac{ip_0}{\hbar} + \frac{x_0}{\sigma^2}\right)^2 \int_{-\infty}^{+\infty} \psi^*\psi\, dx - \frac{\hbar^2}{\sigma^4}\int_{-\infty}^{+\infty} \psi^* x^2 \psi\, dx$$

$$= \frac{\hbar^2}{\sigma^2} - \hbar^2\left(\frac{ip_0}{\hbar} + \frac{x_0}{\sigma^2}\right)^2 (1) + \frac{2\hbar^2}{\sigma^2}\left(\frac{ip_0}{\hbar} + \frac{x_0}{\sigma^2}\right)\langle\hat{x}\rangle - \frac{\hbar^2}{\sigma^4}\langle\hat{x}\rangle$$

$$= \frac{\hbar^2}{\sigma^2} - \hbar^2\left(\frac{ip_0}{\hbar} + \frac{x_0}{\sigma^2}\right) + \frac{2\hbar^2}{\sigma^2}\left(\frac{ip_0}{\hbar} + \frac{x_0}{\sigma^2}\right)x_0 - \frac{\hbar^2}{\sigma^4}\left(x_0^2 + \frac{\sigma^2}{2}\right)$$

$$\therefore \langle\hat{p}^2\rangle = \frac{\hbar^2}{\sigma^2} - \frac{\hbar^2}{\sigma^4}\left(x_0^2 + \frac{\sigma^2}{2}\right) - \hbar^2\left(\frac{ip_0}{\hbar} + \frac{x_0}{\sigma^2}\right)\left(\frac{ip_0}{\hbar} + \frac{x_0}{\sigma^2} - \frac{2x_0}{\sigma^2}\right)$$

$$= \frac{\hbar^2}{2\sigma^2} - \frac{\hbar^2 x_0^2}{\sigma^4} - \hbar^2\left(\frac{ip_0}{\hbar} + \frac{x_0}{\sigma^2}\right)\left(\frac{ip_0}{\hbar} - \frac{x_0}{\sigma^2}\right)$$

$$= \frac{\hbar^2}{2\sigma^2} - \frac{\hbar^2 x_0^2}{\sigma^4} - \hbar^2\left[\left(\frac{ip_0}{\hbar}\right)^2 - \left(\frac{x_0}{\sigma^2}\right)^2\right]$$

$$= \frac{\hbar^2}{2\sigma^2} - \frac{\hbar^2 x_0^2}{\sigma^4} - \hbar^2\left(-\frac{p_0^2}{\hbar^2} - \frac{x_0^2}{\sigma^4}\right)$$

$$\therefore \langle\hat{p}^2\rangle = p_0^2 + \frac{\hbar^2}{2\sigma^2}$$

Uncertainty in

$$x = 4x = \sqrt{\langle\hat{x}^2\rangle - \langle\hat{x}\rangle^2} = \sqrt{\left(x_0^2 + \frac{\sigma^2}{2}\right) - x_0^2} = \frac{\sigma}{\sqrt{2}}$$

Uncertainty in

$$p = \Delta p = \langle \hat{p}^2 \rangle - \langle \hat{p} \rangle^2 = \sqrt{\left(p_0^2 + \frac{\hbar^2}{2\sigma^2}\right) - p_0^2} = \frac{\hbar}{\sigma\sqrt{2}}$$

Uncertainty product

$$(\Delta x)(\Delta p) = \frac{\sigma}{\sqrt{2}} \cdot \frac{\hbar}{\sigma\sqrt{2}} \text{ or, } (\Delta x)(\Delta p) = \frac{\hbar}{2} \text{ (minimum)}$$

Hence, Gaussian wave packet corresponds to minimum uncertainty product.

» ***Example* 4.22.** The quantum state of a two levels system with energy in eV is given as $\psi(\vec{r}, t) = 0.80$ $\psi_1(\vec{r}) + 0.60\frac{e^{-it}}{\hbar}\psi_2(\vec{r})$. What is the probability of finding the system in the upper level? What is the outcome of energy measurement of the system? **[cf. BU (H) 2006]**

***Solution*:** $\psi_1(\vec{r}) \cdot \psi_2(\vec{r})$ are stationary states with energy values E_1 and E_2 respectively. $\psi(\vec{r}, t)$ is the quantum state obtained from a linear combination of $\psi_1(\vec{r})$ and $\psi_2(\vec{r})$ with C_1. C_2 being the expansion coefficients.

$$\begin{aligned}\psi_1(\vec{r}, t) &= C_1\,\psi_1(\vec{r})\,e^{-\frac{iEt}{\hbar}} + C_2\,\psi_2(\vec{r})\,e^{-\frac{iE_2t}{\hbar}} \\ &= 0.80\psi_1(\vec{r}) + 0.60e^{-it/\hbar}\,\psi_2(\vec{r})\end{aligned}$$

Comparing $C_1 = 0.80$, $C_2 = 060$, $E_1 = 0$ eV, $E_2 = 1$ eV .

$\therefore |C_1|^2 = 0.64 \rightarrow$ probability of finding the system in the upper level.

$|C_2|^2 = 0.36 \rightarrow$ probability of finding the system in the lower level.

Hence, the outcome of energy measurement and $E_1 = 0$ eV (with probability 64 %) and $E \simeq 1$ eV (with probability 36 %).

» ***Example* 4.23.** A three-level quantum system has energy eigenvalue 0, 1, 2 MeV. If the probabilities of the system at time t, to be in the first two eigenstates are 49 % and 36 % respectively, write down the wave function for the system. **[BH (U) 2005]**

***Solution*:** Let $\hat{H}$ be the Hamiltonian of the system with eigenstates $\psi_1(x)$, $\psi_2(x)$, $\psi_3(x)$ and energies $E_1 = 0$ MeV, $E_2 = 1$ MeV, $E_3 = 2$ MeV respectively. Then

$$\begin{aligned}\hat{H}Y_1 &= E_1\,\psi_1 = 0 \\ \hat{H}Y_2 &= E_2\,\psi_2 = \psi_2 \\ \hat{H}Y_3 &= E_3\,\psi_3 = 2\psi_3\end{aligned}$$

The linear combination of these stationary with expansion coefficients C_1, C_2, C_3 is

$$\psi(x,0) = C_1\,\psi_1(x) + C_2\,\psi_2(x) + C_3\,\psi_3(x)$$

$$\begin{aligned}\text{and } \psi(x,t) &= C_1\,\psi_1(x)\,e^{-\frac{iE_1t}{\hbar}} + C_2\,\psi_2(x)\,e^{-\frac{iE_2t}{\hbar}} + C_3\,\psi_3(x)\,e^{-\frac{iE_3t}{\hbar}} \\ &= C_1\,\psi_1(x) + C_2\,\psi_2(x)\,e^{-\frac{it}{\hbar}} + C_3\,\psi_3(x) + e^{-\frac{2it}{\hbar}}\end{aligned}$$

where $|C_1|^2$ = probability of first eigenstate = $0.49 = |0.7|^2$

$|C_2|^2$ = probability of second eigenstate = $0.36 = |0.6|^2$

$|C_3|^2$ = probability of third eigenstate = $1 - |C_1|^2 - |C_2|^2 = 1 - 0.49 - 0.36 = 0.15$.

$$\therefore \boldsymbol{\psi(x,t) = 0.7\psi_1(x) + 0.6\psi_2(x)\,e^{-it/\hbar} + \sqrt{0.15}\,\psi_3 e^{-i2t/\hbar}}.$$

This is the wave function for the system.

Example 4.24. A system has two possible energy values E_0 and $2E_0$ and at a certain instant it is in a state in which the expectation value of energy is $\frac{3}{2}E_0$. Calculate the wave function in the state given that ψ_0 and ψ_1 are the wave functions corresponding to the two possible stationary states. What is the wave function after a time elapsed? [CU (H) 2002]

Solution: If $\hat{H}$ be the Hamiltonian, then $\hat{H}\psi_0 = E_0\,\psi_0$; $\hat{H}\psi_1 = E_1\,\psi_1 = 2E_0\,\psi_1$.

We have $\psi = C_0\,\psi_0 + C_1\,\psi_1$, C_0, C_1 are constants where ψ_0 and ψ_1 are ortho normalised and represent the base state.

Now, the expectation value of $\langle \hat{H} \rangle$ is

$$\begin{aligned}\langle \hat{H} \rangle &= \psi^{+}\hat{H}\psi_{*} = (C_0\,\psi_0 + C_1\,\psi_1)^{+}\hat{H}(C_0\,\psi_0 + C_1\,\psi_1) = \tfrac{3}{2}E_0\\ &= (C_0^*\,\psi_0^+ + C_1^*\,\psi_1^+)(C_0E_0\,\psi_0 + C_1 2E_0\,\psi_1) = \tfrac{3}{2}E_0\\ &= |C_0|^2 E_0\,\psi_0^+\,\psi_0 + 2C_0^* C_1 E_0\,\psi_0^+\,\psi_1 + C_1^* C_0 E_0\,\psi_1^+\,\psi_0 + 2|C_1|^2 E_0\,\psi_1^+\,\psi_1^1 = \tfrac{3}{2}E_0\\ &= |C_0|^2 E_0(1) + 2C_0^* C_1 E_0(0) + C_1^* C_0 E_0(0) + 2|C_1|^2 E_0(1) = \tfrac{3}{2}E_0\\ &= |C_0|^2 E_0 + 2|C_1|^2 E_0 = \tfrac{3}{2}E_0\end{aligned}$$

or, $|C_0|^2 + 2|C_1|^2 = \frac{3}{2}$ (i)

Probability of conservation gives $\psi^{+}\psi = 1$

$$\begin{aligned}&\text{or, } (C_0\,\psi_0 + C_1\,\psi_1)^{+}(C_0\,\psi_0 + C_1\,\psi_1) = 1\\ &\text{or, } |C_0|^2\,\psi_0^+\,\psi_0 + C_0^* C_1\,\psi_0^+\,\psi_1 + C_1^* C_0\,\psi_1^+\,\psi_0 + |C_1|^2\,\psi_1^+\psi = 1\\ &\text{or, } |C_0|^2 + |C_1|^2 = 1 \qquad \text{(ii)}\end{aligned}$$

From (i) and (ii), we get $C_0 = 1/\sqrt{2}$, $C_1 = 1/\sqrt{2}$.

$\therefore \psi = \frac{1}{\sqrt{2}}\psi_0 + \frac{1}{\sqrt{2}}\psi_1$.

The wave function after time t is

$$\psi(t) = \frac{1}{\sqrt{2}}e^{-\frac{iEt}{\hbar}}\,\psi_0 + \frac{1}{\sqrt{2}}e^{-\frac{ieE_0 t}{\hbar}}\,\psi$$

EXERCISES

4(A): Objective (or MCQ) Type Questions

Select the correct answer:

4.1. Matter wave functions are

(a) periodic functions
(b) imaginary functions of x and t
(c) real functions of x and t
(d) complex functions of x and t.

4.2. Shrödinger wave equation is the foundation of

(a) relativistic quantum mechanics
(b) non-relativistic quantum mechanics
(c) classical mechanics
(d) hydrodynamics.

4.3. Schrödinger one-dimensional wave equation is

(a) first order in x and second order in t
(b) second order in x and first order in t
(c) first order in both x and t
(d) second order in both x and t.

4.4. The probability of finding a particle represented by $\psi(\vec{r},t)$ in unit volume is

(a) $|\psi(r,t)|$
(b) $|\psi(r,t)|^2$
(c) $|\psi(r,t)|^3$
(d) $\psi(r,t)$.

4.5. The wave function $\psi(\vec{r},t)$ is said to be normalised, if

(a) $\int_{\text{all space}} |\psi(r,t)|\, d\tau = 1$
(b) $\int_{\text{all space}} |\psi(r,t)|^2\, d\tau = 1/N^2$
(c) $\int_{\text{all space}} |\psi(r,t)|^2\, d\tau = 1$
(d) $\int_{\text{all space}} |\psi^*(\vec{r},t)|\, d\tau = 1$.

4.6. Two-wave functions $\psi_m(\vec{r},t)$ and $\psi_n(\vec{r},t)$ are said to be orthogonal, if

(a) $\int \psi_m(\vec{r},t)\,\psi_n(\vec{r},t)\, d\tau = 1$
(b) $\int \psi_m(\vec{r},t)\,\psi_n(\vec{r},t)\, d\tau = 0$
(c) $\int |\psi_m(\vec{r},t)|^2\, d\tau = 1$
(d) $\int |\psi_n(\vec{r},t)|^2\, d\tau = 0$.

4.7. A stationary state is that for which the probability of finding the particle at a point in space is

(a) dependent of time
(b) independent of time
(c) dependent on x
(d) independent of x.

4.8. A wave function $\psi(\vec{r},t)$ is admissible, if

(a) ψ is single-valued and finite
(b) ψ is single-valued
(c) ψ is finite
(d) ψ is finite and multivalued.

4.9. Which of the wave function is the solution of Schrödinger equation?

(a) $A\sec(x)$
(b) $A\tan(x)$
(c) $A\exp(-x^2)$
(d) $A\exp(x^2)$.

4.10. A particle is moving in one dimension and is described by the wave function

$$\psi(x,t) = \begin{cases} 0, & x < 0 \\ Axe^{-\beta x}e^{ict/\hbar}, & x > 0 \end{cases}$$

where A, B, C are positive constants. Which of the following holds?

(a) particle is bound
(b) particle is unbound
(c) particle oscillates
(d) particle executes damped oscillator.

4.11. What is the probability current density corresponding to the wave function $\psi = Ae^{-\alpha x}$, α = real constant, A = complex constant.

(a) $|A|^2$
(b) $|\alpha|^2$
(c) 0.
(d) $|A|^2 e^{-2\alpha x}$.

ANSWERS

4.1. (d).	4.3. (b).	4.5. (c).	4.7. (b).	4.9. (c).	4.11. (c).
4.2. (b).	4.4. (b).	4.6. (b).	4.8. (a).	4.10. (a).	

4(B): Short Answer Type Questions

4.1. Why the wave function representing a moving particle must be complex?

4.2. Write down Schrödinger time-dependent and time-independent equations for a particle.

4.3. Write down the time-dependent Schrödinger equation for a particle of mass m moving in a real potential $V(x, y, z, t)$. **[BU (H) 2000]**

4.4. In what respect does the Schrödinger equation differ from classical wave equation? **[BU (H) 1995]**

4.5. What solutions of the Schrödinger time-dependent equation are called *stationary*? **[BU (H) 1999]**

4.6. Write down the classical Hamiltonian and the time-independent Schrödinger equation for a particle. **[BU (H) 2002]**

4.7. Why the Schrödinger equation is not valid for a relativistic particle? **[HPU 1995]**

4.8. What is the physical interpretation of the wave function ψ? Explain Born's interpretation. **[HPU 1996, LU 1996, PU 1995]**

4.9. What is the need for normalising a wave function? Write the normalisation condition of two-wave functions.

4.10. State the orthonormal condition of two-wave functions.

4.11. Prove that normalisation is independent of time. **[PU 1991; KU 1991]**

4.12. Justify the statement that the probability current density cannot be directly measured.

[*Hints:* $\vec{J}(\vec{r}, t)$ is not possible to be directly measured in the sense $P(\vec{r}, t)$ is. This is because measurement of flux would require simultaneous measurement of position and velocity which is not consistent with uncertainty principle.]

4.13. Under what conditions the solution of Schrödinger equation is said to be well behaved?

4.14. If ψ_m and ψ_n are two solutions of the Schrödinger equation, then show that $A\psi_m + B\psi_n$ is also a solution of the same equation.

4.15. Does the function $f(x) = 4 - |x|$ represent a physically acceptability of wave function? Justify your answer. **[BU (H) 2015]**

4.16. What is the significance of the continuity equation for matter waves?

4.17. Normalise the wave function, $\psi(x) = e^{-|x|} \sin \alpha x$.

4.18. What is the physical meaning of normalisation of a wave function. Find the dimension of ϕ. **[BU (H) 2012]**

4.19. (a) What is meant by the statement, *the Schrödinger equation is linear in the wave function,* ψ?

(b) How can you relate the wave nature of a material particle like electron to the linearity of the Schrödinger equation?

4.20. (a) Obtain continuity equation starting from time-dependent Schrödinger equation and identify the probability density.

(b) Calculate the probability current density for the wave function $\phi = e^{ikr}/r$. **[BU (H) 2008]**

4.21. Show that the probability current density vanishes if the wave function is real. **[CU (H) 1999]**

[*Hints*: Since, $\psi^* = \psi$ for real wave function, therefore,

$$\vec{S} = \frac{\hbar}{2im}[\psi^* \nabla\psi - (\nabla\psi)^*\psi] = 0]$$

4.22. Which of $\psi_1 = Ae^{x^2}$ and $\psi_2 = Ae^{-x^2}$ is well-behaved quantum mechanical function in the range $-\infty < x < +\infty$. **[CU (H) 2003]**

[*Hints*: At $x \to \pm\infty$, $\psi_1 \to \infty$. $\therefore$ ψ_1 is not diverging at $x = \pm\infty$. So, ψ_1 is not well-behaved. But at $x \to \pm\infty$, $\psi_2 \to 0$. $\therefore$ ψ_2 is bounded and is well-behaved wave function.]

4.23. Normalise the wave function $\psi(x) = Ne^{-x^2/2}$.

4.24. How many angular momentum components can be measured simultaneously? **[CU (H) 1999; cf. BU (H) 2006]**

[*Ans.* Only use]

4.25. Can kinetic energy and linear momentum of a quantum mechanical system be specified simultaneously? **[KU (H) 2002]**

[*Ans.* Yes. As $[\hat{T}, \hat{p}] = \left[\frac{\hat{p}^2}{2m}, \hat{p}\right] = 0$ commutes]

4.26. Is it possible measure accurately all the three components of angular momentum of a quantum system at the same time? Give reasons. **[BU (H) 2001]**

4.27. Derive a conservation equation from Schrödinger time-dependent equation. **[CU (H) 2006]**

4.28. What are the basic postulates of quantum mechanics? **[CU (H) 2004]**

4.29. If $\psi(r, t)$ is a Schrödinger wave function prove that $y^*(x, t)\psi(x, t)$ is necessarily real and either positive or zero. **[CU (H) 2006]**

4.30. What is probability density in terms of wave function? **[KU (H) 2009]**

4.31. Comment on the normalisability of the wave function $\psi(x) = Ae^{ikx}$. **[KU (H) 2014]**

4(C): Broad Answer Type Questions

4.1. In what respect does the Schrödinger equation differ from classical wave equation? Explain the term *stationary state* of a quantum mechanics system. In a stationary state E, what is the form of the time-dependent part of the wave function? **[BU (H) 1995]**

4.2. Explain the need for differential wave equation. Starting from the wave equation and introducing energy and momentum of the particle obtain three-dimensional Schrödinger equation in time-dependent form. **[HPU 1991, 1993, 1996; PU 1991; LU 1995]**

4.3. Obtain three-dimensional Schrödinger equation from time-dependent Schrödinger equation. **[PU 1997; LU 1996; HPU 1996]**

4.4. What is normalisation of a wave function? Find the normalisation factor of Schrödinger wave function. Prove that normalisation is independent of time. **[KU 1991; PU 1991]**

4.5. Derive Schrödinger time-dependent wave equation for a particle of mass m moving along x-axis in a potential $V(\vec{r}, t)$.

Hence, obtain Schrödinger time-independent equation for the same particle. State the equations for a free particle.

4.6. What do you mean by probability and probability current density for a quantum mechanical system? Deduce an expression for the latter for three-dimensional motion and the law of conservation of probability density.

What are stationary states? Why are they so called?

4.7. (a) What are the characteristics of Schrödinger wave equation?

(b) Discuss the admissibility conditions on the wave equation.

4.8. What are stationary states? Starting from the time-dependent Schrödinger equation find the time-independent equation satisfied by the wave function for stationary states. **[BU (H) 1993]**

4.9. What solutions of Schrödinger time-dependent equation are called stationary? When are such solutions obtained? What is the form of the time-dependent part of the wave function for a stationary state? **[BU (H) 1999]**

4.10. Verify that Schrödinger equation is linear. **[CU (H) 1997, 1993]**

4.11. If $\psi(x, t)$ is a Schrödinger wave function, prove that $\psi^*(x, t)\,\psi(x, t)$ is necessarily real and either positive or zero.

4.12. Given that a general solution of the time-independent one-dimensional Schrödinger equation can be written as

$$\phi(x, t) = \sum_{n=1}^{\infty} a_n \psi_n e^{iE_n t/\hbar}$$

Check whether the probability density is independent of time or not. The symbols have their usual meanings. **[NBU (H) 2003; BU (H) 2004]**

4.13. Calculate the normalisation constant and probability density of a wave function given by $\psi(x) = Ae^{-\sigma^2x^2/2}e^{ikx}$. **[NBU (H) 2003]**

4.14. Discuss a method to arrive at the Schrödinger equation for the wave function ψ representing de Broglie wave associated with a particle. **[BU (H) 2006]**

4.15. (a) What are the basic postulates of Quantum Mechanics? **[BU (H) 2011]**

(b) Deduce time-dependent Schrödinger equation for a free particle of mass m.

(c) Prove that any coordinate and its conjugate momentum do not commute. **[WBUT 2005]**

4.16. $\psi(x, t)$ is a wave function in one space dimension, satisfying the Schrödinger equation.

(a) Write the statistical interpretation of this wave function.

(b) Show with the help of Schrödinger equation, that the space integrated probability is independent of time. What is the significance of this result so far as the normalisation of the wave function is conserved? **[BU (H) 2014]**

4.17. (a) From the definition of probability density and current density establish the equation of continuity.

(b) Find the normalisation constant of the unnormalised wave function, $\psi = A\exp(-r/a_0)$, a_0 is a constant. **[BU (H) 2012]**

4.18. Consider the normalised wave function ψ :

$$\psi = \sqrt{\frac{1}{19}}\phi, \sqrt{\frac{2}{19}}\phi_2, \sqrt{\frac{3}{19}}\phi_3$$

where $\phi_n (n = 1, 2, 3)$ is a complete set of orthogonal energy eigenfunction $H\phi_n = nE_0\phi_n$.

(a) Find the normalisation constant for ϕ.

(b) If the energy is measured on large number of identical systems, each initially having ϕ as the wave function. What are the possible results and what are their probabilities?

(c) Find the average value of measured energy? **[BU (H) 2014]**

4.19. What solution of the Schrödinger time-dependent equation are called *stationary*? When are such solutions obtained? **[BU (H) 2009]**

4.20. Write down time-dependent Schrödinger equation for a particle of mass m moving in a real potential. From here obtained the equation of continuity. Give its physical interpretation. Show that conservation of probability follows. **[KU (H) 2014]**

4(D): Numerical Problems

4.1. The wave function of a particle at time t is given as

$$\psi(\vec{r}, t) = a_1\phi_1(\vec{r})\, e^{-iE_1t/\hbar} + a_2\phi_2(\vec{r})\, e^{-iE_2t/\hbar}$$

where ϕ_1 and ϕ_2 are two normalised wave functions with energies E_1 and E_2 $(E_1 \neq E_2)$; a_1, a_2 are constants.

(a) Calculate the probability density.

(b) Does ψ represent a stationary state? Justify your answer. **[BU (H) 2004]**

4.2. Normalise the momentum wave function $a(\vec{P}) = Ne^{-\frac{\alpha}{\hbar}|\vec{P}|}$. [*Ans.* $a(\vec{P}) = \left(\frac{\alpha^3}{\pi\hbar^3}\right)^{1/2} e^{-\frac{\alpha}{\hbar}|\vec{P}|}$]

4.3. Consider two stationary state solutions $\psi_1(\vec{r}, E_1)$ and $\psi_2(\vec{r}, \vec{E}_2)$ for the time-independent Schrödinger equation corresponding to the two different energies E_1 and E_2 respectively. Prove that

$$\int_\tau \psi_1^*(\vec{r}, E_1)\, \psi_2(\vec{r}, E_2)\, d\tau = 0$$

4.4. The following wave packet describes the one-dimensional motion of a free particle of mass m :

$$\psi(x, t) = (2\pi)^{-1/4}\left(a + \frac{i\hbar t}{2ma}\right)\exp\left[\frac{-x^2}{4a^2 + \left(\frac{2i\hbar t}{m}\right)}\right].$$

Show that the above wave function is properly normalised.

4.5. Normalise the wave function $\psi(x) = e^{-|x|}\cos\alpha x$.

4.6. Find the normalisation constant for a wave function given by (at $t = 0$)

$$\psi(x) = Ae^{-\alpha^2x^2/2}e^{ikx}.$$

[*Ans.* $A = \left(\alpha/\sqrt{\pi}\right)^{1/2}$]

4.7. The Gaussian wave packet at $t = 0$ is given by $\psi(x, 0) = Ne^{-\frac{(x-x_0)^2}{2\sigma^2}} e^{\frac{i}{\hbar}p_0x}$. Calculate the normalisation constant. [*Ans.* $\frac{1}{\sqrt{\sigma\sqrt{\pi}}}$]

4.8. A particle moving along the positive direction of the x-axis in a region of potential energy $V(x)$ is represented by a wave packet given by

$$\psi(x,t) = \frac{1}{\sqrt{2\pi\hbar}} \int_{-\infty}^{+\infty} A(p)\, e^{\frac{i}{\hbar}(px - Et)}\, dp$$

Obtain the one-dimensional time-dependent Schrödinger wave equation using the expression of non-relativistic total energy.

4.9. A particle on the x-axis has the wave function $\psi = cx^2$ between $x = 0$ and $x = 2$.

(a) Normalise the wave function over the interval.

(b) Find the probability that the particle can be found between $x = 0.50$ and $x = 0.60$. **[BU (H) 2005]**

[*Ans.* $\psi = \sqrt{\frac{5}{32}}\, x^2$, $P = 1.5 \times 10^{-3}$]

4.10. The stationary state of a two-level system with energy eV is given by

$$\psi(r,t) = 0.80 u_0(r) + 0.60 e^{-it/\hbar} u_1(r)$$

What is the probability of finding the system in the upper level? What is the outcome of energy measurement of the system? **[BU (H) 2006]**

4.11. A system has two energy states ε_0 and $3\varepsilon_0$. ψ_1 and ψ_2 are the corresponding normalised wave functions. At an instant the system is in a superposed state $\psi = C_1\psi_1 + C_2\psi_2$ and $\psi_1 = 1/\sqrt{2}$.

(a) Find the value of C_2, if ψ is normalised.

(b) What is the probability that an energy measurement would yield a value $3\varepsilon_0$?

(c) Find the expectation value of energy. **[WBUT 2007]**

4.12. A particle wave function can be written as

$$\psi(x,t) = \frac{1}{\sqrt{2\pi}} e^{-\frac{1}{(kx-\omega t)}}$$

Find the energy of the particle using wave function and the time-independent Schrödinger equation. **[BU (H) 2010]**

4.13. A one-dimensional wave function is given by $\psi(x) = \sqrt{a}e^{-ax}$; find the probability of finding the particle between $x = \frac{1}{a}$ and $x = \frac{2}{a}$. **[CU (H) 2013]**

4.14. For the wave function $\psi = A \exp i(ax - \omega t)$, find the probability current density (A is constant). **[CU (H) 2012]**

4.15. Obtain the probability current density corresponding to the following wave function:

$$\psi(x) = \frac{1}{(\pi)^{1/4}} \exp\left[-\frac{x^2}{2a^2} + \frac{ip_0 x}{\hbar}\right]$$

where a and p_0 are constants. **[CU (H) 2008]**

4.16. (a) What is meant by the statement, *the Schrödinger equation is linear in the wave function* ψ?

(b) How can you relate the wave nature of a material particle like electron to the linearity of the Schrödinger equation? **[BU (H) 2010]**

4.17. (a) Obtain continuity equation starting from time-dependent Schrödinger equation and identify the probability density.

(b) Calculate the probability current density for the wave function $\psi = e^{ikr}/r$. **[BU (H) 2008]**

4.18. A free particle wave function can be written as $\psi(x, y) = \frac{1}{\sqrt{2\pi}} e^{-i(kx-\omega t)}$.

Find the energy of the particle using this wave function and the time-independent Schrödinger equation. **[BU (H) 2010]**

4.19. Find the probability that a particle can be found between $x = 0.45$ and $x = 0.55$, when the wave function:

$$\psi(x) = \begin{cases} ax, & \text{for } x < 0 < 1 \\ 0, & \text{for } x < 0 \text{ and } x > 1. \end{cases}$$

4.20. Write down Schrödinger time-dependent equation and from it obtain the time-independent equation for the case when the potential energy is independent of time. **[NBU (H) 2009]**

4.21. Starting from Schrödinger equation deduce law of conservation of probability current density. Calculate the probability current density for the wave function, $\psi = e^{ikr}/r$. **[VU (H) 2009]**

4.22. A particle on the x-axis has the wave function $\psi(x) = cx^2$ between $x = 0$ and $x = 2$.

(a) Normalise the wave function over the interval.

(b) Find the probability that the particle can be found between $x = 0$ and $x = 0.6$.

(c) Find the expectation value of the particles position x. **[*Ans.* $\sqrt{5/32}, 1.45 \times 10^{-3}, 1.667$]**

4.23. The wave function of a particle in one dimension is

$$\psi(x) = \begin{cases} \sqrt{15/a}A(a^2 - x^2), & \text{for } -a \leqslant x \leqslant a \\ 0, & \text{for } |x| > a \end{cases}$$

Find the value of A that will normalise $\psi(x)$. Calculate the expectation values of $\hat{x}, \hat{p}\hat{x}^2, \hat{p}^2$ and uncertainty product. **[BU (H) 2003]**

[*Ans.* $\frac{1}{4a^2}, 0, 0, \frac{a^2}{7}, \frac{5\hbar^2}{2a^2}, 0.598\hbar > \frac{\hbar}{2}$]

4.24. Can KE and linear momentum of a quantum mechanical system be specified simultaneously? **[KU (H) 2002]**

[*Ans.* Yes. As $[\hat{T}, \hat{p}] = \left[\frac{\hat{p}^2}{2m}, \hat{p}\right] = 0$ commutes]

4.25. Is it possible to measure accurately all the three components of angular momentum of a quantum system at the same time? Give reasons. **[CU (H) 2001]**

4.26. Derive a conservation equation from Schrödinger time-dependent equation. **[CU (H) 2006]**

4.27. What are the basic postulates of quantum mechanics? **[CU (H) 2004]**

4.28. If $\psi(r, t)$ is a Schrödinger wave function prove that $\psi^*(x, t)\,\psi(x, t)$ is necessarily real and either positive or zero. **[CU (H) 2006]**

4.29. What is probability density in terms of wave function? **[KU (H) 2009]**

4.30. Comment on the normalisability of the wave function $\psi(x) = Ae^{ikx}$. **[KU (H) 2014]**

Chapter 5

The Operator Formalism in Quantum Mechanics

5.1 Operator

An operator is a mathematical rule (or, procedure), which operating on one function transforms it into another function.

For example, if A is an operator (represented by $\hat{A}$) and $\psi(x)$ is a function of x, then $\hat{A}\,\psi(x) = \phi(x)$, where $\phi(x)$ is the transformed function.

The following table shows the mathematical operations on the function $f(x) = x^m$ and the results of operations.

Table 5.1: Mathematical operators and the results of operations

Operations	***Symbols of the operators***	***Results of operation $f(x) = x^m$***
1. Taking the square root	$\sqrt{\ }$	$\sqrt{x^m} = x^{\frac{m}{2}}$
2. Taking the square	$(\ )^2$	$(x^m)^2 = x^{2m}$
3. Addition	$+$	$x^m + x^n$
4. Subtraction	$-$	$x^m - x^n$
5. Multiplication by a constant	$\times a$	ax^m
6. Differentiation w.r.t. x	$\frac{d}{dx}$	$\frac{d}{dx}(x^m) = mx^{m-1}$
7. Integration w.r.t. x	$\int (\ )\,dx$	$\int x^m\,dx = \frac{x^{m+1}}{m+1} + k$

An operator has no meaning when it stands alone. When an operator operates on a function as

$$\text{(Operator) (function)} = \text{Another function} \tag{5.1}$$

the function here is called an *operand*.

Each dynamical variable in quantum mechanics is represented by an operator which provides a link with the latter through the correspondence principle.

5.2 Operator algebra

(i) ***Addition and subtraction of operators***: If $\hat{A}$ and $\hat{B}$ are two different operators, then

$$(\hat{A}+\hat{B})\,\psi(x) = \hat{A}\,\psi(x) + \hat{B}\,\psi(x) \tag{5.2}$$

$$\text{and } (\hat{A}-\hat{B})\,\psi(x) = \hat{A}\,\psi(x) - \hat{B}\,\psi(x) \tag{5.3}$$

where $\psi(x)$ is an operand.

From eqs (5.2) and (5.3), we obtain

$$\hat{A}+\hat{B} = \hat{B}+\hat{A} \tag{5.4}$$

$$\text{and } \hat{A}-\hat{B} = -\hat{B}+\hat{A} \tag{5.5}$$

Thus, by adding and subtracting operators new operators can be obtained.

(ii) ***Multiplication of operators***: If $\hat{A}$ and $\hat{B}$ be two different operators and $\psi(x)$ an operand, then $\hat{A}\,\psi(x)$ means that $\psi(x)$ is to be first operated on by $\hat{B}$ to obtain another function $\phi(x)$. Then $\phi(x)$ is operated on by $\hat{A}$ to obtain the final function $g(x)$ as follows:

$$\hat{B}\,\psi(x) = \phi(x)$$

$$\hat{A}\phi(x) = g(x)$$

so that

$$\hat{A}\hat{B}\,\psi(x) = g(x) \tag{5.6}$$

The order of application of operators on the operand is to be from right to left.

Now, if $\hat{B} = \hat{A}$, then

$$\hat{A}\hat{B}\,\psi(x) = \hat{A}\hat{A}\,\psi(x) = \hat{A}^2\,\psi(x) \tag{5.7}$$

Hence, if the same operator is to be applied several times in succession, then the operator is to be written with a power.

5.3 Commutators

The two operators $\hat{A}$ and $\hat{B}$ are said to commute if $\hat{A}\hat{B} = \hat{B}\hat{A}$. The commutator or the commutator bracket of the operators $\hat{A}$ and $\hat{B}$ is defined as

$$[\hat{A},\hat{B}] = \hat{A}\hat{B} - \hat{B}\hat{A} \tag{5.8}$$

Hence, when $\hat{A}$ and $\hat{B}$ commute, the commutator

$$[\hat{A}, \hat{B}] = 0 \tag{5.9}$$

$$\text{or, } \hat{A}\hat{B} - \hat{B}\hat{A} = 0$$

so that

$$\hat{A}\hat{B} = \hat{B}\hat{A} \tag{5.10}$$

On the other hand, if $\hat{A}$ and $\hat{B}$ do not commute, then $[\hat{A}, \hat{B}] \neq 0$ so that

$$[\hat{A}, \hat{B}] \neq [\hat{B}\hat{A}] \tag{5.11}$$

For example,

(i) Let

$$\hat{A} = \frac{\partial}{\partial x} \text{ and } \hat{B} = \frac{\partial^2}{\partial x^2}$$

and $\psi(x)$ be an operand, then

$$\hat{A}\hat{B}\,\psi(x) = \frac{\partial}{\partial x}\frac{\partial^2}{\partial x^2}\psi(x) = \frac{\partial}{\partial x}\frac{\partial^2 \psi}{\partial x^2} = \frac{\partial^3 \psi}{\partial x^3}$$

$$\text{and } \hat{B}\hat{A}\,\psi(x) = \frac{\partial^2}{\partial x^2}\frac{\partial}{\partial x}\psi(x) = \frac{\partial^2}{\partial x^2}\frac{\partial \psi}{\partial x} = \frac{\partial^3 \psi}{\partial x^3}$$

Therefore,

$$\hat{A}\hat{B}\,\psi(x) - \hat{B}\hat{A}\,\psi(x) = \frac{\partial}{\partial x}\frac{\partial^2}{\partial x^2}\psi(x) - \frac{\partial^2}{\partial x^2}\frac{\partial}{\partial x}\psi(x)$$

$$\text{or, } \left(\hat{A}\hat{B} - \hat{B}\hat{A}\right)\psi(x) = \left(\frac{\partial}{\partial x}\frac{\partial^2}{\partial x^2} - \frac{\partial^2}{\partial x^2}\frac{\partial}{\partial x}\right)\psi(x) = \frac{\partial^3 \psi}{\partial x^3} - \frac{\partial^3 \psi}{\partial x^3} = 0$$

$$\text{or, } \left(\hat{A}\hat{B} - \hat{B}\hat{A}\right)\psi(x) = 0 \text{ or, } \hat{A}\hat{B} - \hat{B}\hat{A} = 0\ [\because \psi(x) \neq 0]$$

$$\text{i.e., } [\hat{A}, \hat{B}] = \left[\frac{\partial}{\partial x}, \frac{\partial^2}{\partial x^2}\right] = \left[\frac{\partial}{\partial x}\frac{\partial^2}{\partial x^2} - \frac{\partial^2}{\partial x^2}\frac{\partial}{\partial x}\right] = 0 \tag{5.12}$$

Hence, if $\psi(x)$ is a differentiable function, then $\frac{\partial}{\partial x}$ and $\frac{\partial^2}{\partial x^2}$ commute.

(ii) Now, we consider another two operators $\hat{A}$ and $\hat{B}$ such that $\hat{A} = x$ and $\hat{B} = \frac{d}{dx}$.

If $\psi(x)$ be a differentiable function, then

$$\hat{A}\hat{B}\,\psi(x) = \hat{x}\frac{d}{dx}\psi(x) = x\frac{d\psi}{dx}$$

$$\text{and } \hat{B}\hat{A}\,\psi(x) = \frac{d}{dx}\hat{x}\,\psi(x) = \frac{d}{dx}x\psi(x) = x\frac{d\psi}{dx} + \psi(x)$$

Hence,

$$(\hat{A}\hat{B} - \hat{B}\hat{A})\,\psi(x) = \left(\hat{x}\frac{d}{dx} - \frac{d}{dx}\hat{x}\right)\psi(x) = x\frac{d\psi}{dx} - x\frac{d\psi}{dx} - \psi(x) = (-1)\,\psi(x)$$

or, $\hat{A}\hat{B} - \hat{B}\hat{A} = -1$

$$\text{or, } \left(x, \frac{d}{dx}\right) = x\frac{d}{dx} - \frac{d}{dx}x = -1 \qquad (5.13)$$

i.e., the operators x and $\frac{d}{dx}$ do not commute.

Equation (5.13) is an operator equation. The right-hand side of this equation is also an operator, called the *negative of a unit operator* or *identity operator*.

It can be shown that any operator commutes with itself or any power of it. Thus, for an operator $\hat{A}$

$$\hat{A}\hat{A}^m - \hat{A}^m\hat{A} = 0 \qquad (5.14)$$

5.4 Linear operator

Operators used in quantum mechanics are linear operators having the following properties:

(i) The operator $\hat{A}$ is said to be linear, if

$$\hat{A}\,[\,C\psi(x)\,] = C\hat{A}\,\psi(x) \qquad (5.15)$$

where C is a constant and $\psi(x)$ is a function of x

$$\text{or, } \hat{A}C\,\psi(x) = C\hat{A}\,\psi(x)$$

Omitting $\psi(x)$ from both sides, we get

$$\hat{A}C = C\hat{A}$$

i.e., $\hat{A}$ commutes with an arbitrary constant.

Example: If $\hat{A} = \frac{d}{dx}$, then

$$\hat{A}\,[\,C\psi(x)\,] = \frac{d}{dx}[\,C\psi(x)\,] = C\frac{d}{dx}\psi(x)$$

Note: Since, $\sqrt{\psi_1(x) + \psi_2(x)} \neq \sqrt{\psi_1(x)} + \sqrt{\psi_2(x)}$ the operator '$\sqrt{\ }$' is nonlinear.

(ii) Distributive law is obeyed by linear operator.

If $\psi_1(x)$ and $\psi_2(x)$ be two arbitrary functions, then

$$\hat{A}\,[\,\psi_1(x) + \psi_2(x)\,] = \hat{A}\,\psi_1(x) + \hat{A}\,\psi_2(x) \qquad (5.16)$$

Example: If $\hat{A} = \frac{d}{dx}$, then for $\hat{A}$ to be linear operator

$$\frac{d}{dx}[\,\psi_1(x) + \psi_2(x)\,] = \frac{d}{dx}\psi_1(x) + \frac{d}{dx}\psi_2(x)$$

The linear operators obey the following laws:

(a) $(\hat{A} \pm \hat{B})\,\hat{C} = \hat{A}\hat{C} \pm \hat{B}\hat{A}$ (b) $(\hat{A}\hat{B})\,\hat{C} = \hat{A}\,(\hat{B}\hat{C})$ (c) $\hat{A}\hat{B} \neq \hat{B}\hat{A}$.

Addition is commutative, but in general, multiplication is not-commutative. In special case $\hat{A}\hat{B} = \hat{B}\hat{A}$, then $\hat{A}$ and $\hat{B}$ are said to be commuting.

5.4.1 Non-linear opeator

An operator $\hat{a}$ is said to be non-linear if

$$\hat{a}(C_1\psi_1 + C_2\psi_2) \neq C_1\hat{a}\psi_1 + C_2\hat{a}\psi_2$$

where C_1 and C_2 are constants and ψ_1 and ψ_2 are functions.

5.4.2 Antilinear operator

An operator $\hat{a}$ is said to be anti linear if

$$\hat{a}(C_1\psi + C_2\psi_2) = C_1^*(\hat{a}\psi_1) + C_2^*(\hat{a}\psi_2)$$

* denotes complex conjugate, i.e., replacement of i by $i = \sqrt{-1}$.

The linear operators which we generally use in quantum mechanics are

(i) ***Multiplying functions***: These functions when operate from the left of a function $\psi(x)$ give rise to a new function.

For example, $\hat{x}\,\psi(x) = xf(x)$.

These operators are known as *c numbers* in quantum mechanics.

(ii) ***Differential operators***: These operators operating from the left of a function $\psi(x)$ generates a new function $\phi(x)$.

For example,

$$\frac{d}{dx}\psi(x) = \frac{d}{dx}(\sin x) = \cos x = g(x) \text{ (say)},$$

where $f(x) = \sin x$ and $g(x) = \cos x$. These operators are known as *q numbers* in quantum mechanics.

(iii) ***Mixture of multiplying function with differential operators***: If $\hat{x}$ and d/dx be a multiplying and a differential operator respectively and $f(x) = e^{ax}$ an operand, then

$$\left(\hat{x} + \frac{d}{dx}\right) f(x) = \left(x + \frac{d}{dx}\right) e^{ax} = xe^{ax} + ae^{ax} = (x + a)\, e^{ax}$$

There are many more types of linear operator.

5.5 Vector operator

A vector operator ∇ is defined by

$$\nabla = \hat{i}\frac{\partial}{\partial x} + \hat{j}\frac{\partial}{\partial y} + \hat{k}\frac{\partial}{\partial z} \tag{5.17}$$

in Cartesian coordinates where $\hat{i}$, $\hat{j}$, $\hat{k}$ are the triad of unit vectors.

If $\psi(x)$ be a scalar function of x, y, z, then we get a vector called the *gradient* of $\psi(x)$

$$\nabla\psi = \hat{i}\frac{\partial\psi}{\partial x} + \hat{j}\frac{\partial\psi}{\partial y} + \hat{k}\frac{\partial\psi}{\partial z} \tag{5.18}$$

5.6 Laplacian operator

The Laplacian operator plays an important role in quantum mechanics and is defined by

$$\nabla^2 = \frac{\partial^2}{\partial x^2} + \frac{\partial^2}{\partial y^2} + \frac{\partial^2}{\partial z^2} \tag{5.19}$$

5.7 Null operator

A null or zero operator is that which operating on a function reduces it to zero. Thus, if $\hat{A}\,\psi(x) = 0$, $\hat{A}$ is, then a null operator.

5.8 Inverse operator

If A and B be any two different linear spaces and $\hat{T}$ is any linear operator, then

$$B = \hat{T}\,A \text{ and } A = \hat{T}^{-1}B \tag{5.20}$$

Here, $\hat{T}$ is a linear operator and $\hat{T}^{-1}$ is an inverse operator.

5.9 Singular and nonsingular operators

If two operators $\hat{A}$ and $\hat{B}$ have the relation $\hat{A}\hat{B} = \hat{B}\hat{A} = I$, where I is the identity (or, unit) operator, then $\hat{A}$ and $\hat{B}$ are reciprocal to each other, i.e.,

$$\hat{A}^{-1} = \hat{B},\ \hat{B}^{-1} = A \tag{5.21}$$

So an operator for which the reciprocal exists is called *nonsingular*.

If $\phi(x) = A\psi(x)$ holds good, then $\hat{A}$ has reciprocal and therefore

$$\psi(x) = A^{-1}\phi(x) \tag{5.22}$$

If for non-zero $\psi(x)$, $A\psi(x) = 0$, then $\hat{A}$ has no reciprocal and $\hat{A}$ is a singular.

Hence, an operator which has no reciprocal, is called a *singular operator*.

5.10 Hermitian operator

An operator $\hat{A}$ is Hermitian, if it satisfies the condition

$$\int \psi_1^* \cdot (\hat{A}\psi_2)\, d\tau = \int (\hat{A}\psi_1)^* \cdot \psi_2\, d\tau \tag{5.23}$$

Here, ψ_1^* is the complex conjugate of ψ_1. ψ_1 and ψ_2 are two eigenfunctions of the operator $\hat{A}$. A Hermitian operator is *linear and has real eigenvalues.*

5.11 Adjoint (or Hermitian conjugate) of an operator

If the operator $\hat{A}$ and a linear operator $\hat{A}^\dagger$ satisfy the condition

$$\int \psi_1^* \hat{A}^\dagger\, \psi_2\, d\tau = \int (\hat{A}\psi_1)^* \psi_2\, d\tau \tag{5.24}$$

for any two arbitrary state functions ψ_1 and ψ_2, then $\hat{A}^\dagger$ is called the *adjoint* or *Hermitian conjugate* of the operator $\hat{A}$. If $\hat{A}$ is Hermitian, then $\hat{A}$ would satisfy eq. (5.23). Now, comparing eqs (5.23) and (5.24), we get $\hat{A}^\dagger = \hat{A}$, then the operator $\hat{A}$ is said to be *self-adjoint*.

Similarly, if $\hat{A} = \hat{A}^\dagger$, then $\hat{A}$ is Hermitian.

5.12 Parity operator

If $\psi(x)$ be a function of x and $\hat{\pi}$ be the parity operator, then

$$\hat{\pi}\psi(x) = \psi(-x) \tag{5.25}$$

i.e., when $\psi(x)$ is operated by the parity operator, it gets reflected in its coordinates.

If $\psi_1(x)$ and $\psi_2(x)$ be two different functions of x, then $\hat{\pi}[\psi_1(x) + \psi_2(x)] = \psi_1(-x) + \psi_2(-x) = \hat{\pi}\psi_1(x) + \hat{\pi}\psi_2(x)$, also $\hat{\pi}c\psi(x) = c\psi(-x) = c\hat{\pi}\psi(x)$, $\hat{\pi}$ operator is a linear operator.

5.13 Identity operator

An operator which leaves a function $\psi(x)$ unchanged in space after operation is an identity operator $\hat{I}$. Mathematically,

$$\hat{I}\psi = \psi = \psi\hat{I} \tag{5.26}$$

It can be shown that the product of $\hat{A}\hat{B}$ of two unitary operators $\hat{A}$ and $\hat{B}$ is also a unitary operator.

Let us consider the product $\hat{A}\hat{B}$ of two unitary operators $\hat{A}$ and $\hat{B}$, then

$$(\hat{A}\hat{B})(\hat{A}\hat{B})^\dagger = (\hat{A}\hat{B})(\hat{B}^\dagger\hat{A}^\dagger) = \hat{A}(\hat{B}\hat{B}^\dagger)\hat{A}^\dagger(\hat{A}\hat{B})^\dagger = \hat{A}\hat{A}^\dagger = \hat{I} \tag{5.27}$$

where $\hat{B}\hat{B}^\dagger = \hat{I}$, $\hat{A}\hat{A}^\dagger = \hat{I}$.

Similarly,

$$(\hat{A}\hat{B})^\dagger(\hat{A}\hat{B}) = (\hat{B}^\dagger\hat{A}^\dagger)(\hat{A}\hat{B}) = \hat{B}^\dagger(\hat{A}^\dagger\hat{A})\hat{B} = \hat{B}^\dagger\hat{B} = \hat{I} \tag{5.28}$$

$$\therefore (\hat{A}\hat{B})(\hat{A}\hat{B})^\dagger = (\hat{A}\hat{B})^\dagger(\hat{A}\hat{B}) = \hat{I} \tag{5.29}$$

5.14 Unitary operator

An operator $\hat{U}$ is said to be unitary if its inverse and adjoint are identical, i.e.,

$$\hat{U}^\dagger = \hat{U}^{-1} \tag{5.30}$$

Again, $\hat{U}\hat{U}^{-1} = \hat{I} = \hat{U}^{-1}U$.

$$\therefore \hat{U}\hat{U}^\dagger = \hat{I} \text{ or, } \hat{U}^\dagger\hat{U} = \hat{I} \tag{5.31}$$

where $\hat{I}$ is the identity operator.

5.15 Operators corresponding to different dynamical variables (observables)

In quantum mechanics every dynamical variable is assigned to a linear operator.

(a) ***Cartesian coordinate operator***: The Cartesian coordinate operators are represented by $\hat{x}, \hat{y}, \hat{z}$ corresponding to the x, y, z coordinates of a particle in Cartesian coordinate space. The result of operation of any one of the operator on a wave function is the multiplication of the wave function by the corresponding coordinate, i.e.,

$$\hat{x}\,\psi(x) = x\psi(x) \tag{5.32}$$

(b) ***Linear momentum operator***: Operators corresponding to the linear momentum components, viz., p_x, p_y and p_z can be found out in the following way:

Let $\psi(x,t) = Ae^{i(kx-\omega t)}$ represent a free particle, then

$$\frac{\partial}{\partial x}\psi(x,t) = \frac{\partial}{\partial x}\left[Ae^{i(kx-\omega t)}\right] = A(ik)\,e^{i(kx-\omega t)} = ik\psi(x,t)$$

Multiplying both sides of the above equation by $\frac{\hbar}{i}$ from left, we get

$$\frac{\hbar}{i}\frac{\partial}{\partial x}\psi(x,t) = \frac{\hbar}{i}ik\psi(x,t) = \hbar k\psi(x,t)$$

$$\text{or, } \frac{\hbar}{i}\frac{\partial}{\partial x}\psi(x,t) = p_x\psi(x,t) \tag{5.33}$$

where $\hbar k = p_x$ the x-component of momentum of the particle.

Thus, the dynamical variable momentum can be represented by the mathematical operator $\frac{\hbar}{i}\frac{\partial}{\partial x}$.

So

$$\hat{p}_x = \frac{\hbar}{i}\frac{\partial}{\partial x} = -i\hbar\frac{\partial}{\partial x} \text{ and } \hat{p}_x^* = i\hbar\frac{\partial}{\partial x} \tag{5.34}$$

Similarly,

$$\hat{p}_y = -i\hbar\frac{\partial}{\partial y} \text{ and } \hat{p}_z = -i\hbar\frac{\partial}{\partial z} \tag{5.35}$$

Also the operator for the total momentum p is

$$\hat{p} = \hat{i}p_x + \hat{j}p_y + \hat{k}p_z = \left[\hat{i}\left(-i\hbar\frac{\partial}{\partial x}\right) + \hat{j}\left(-i\hbar\frac{\partial}{\partial y}\right) + \hat{k}\left(-i\hbar\frac{\partial}{\partial z}\right)\right]$$

$$= -i\hbar\left[\hat{i}\frac{\partial}{\partial x} + \hat{j}\frac{\partial}{\partial y} + \hat{k}\frac{\partial}{\partial z}\right]$$

$$\text{or, } \hat{p} = -i\hbar\nabla \tag{5.36}$$

That the operator $\hat{p}_x$ is Hermitian can be shown as below:

If $\psi_1(x)$ and $\psi_2(x)$ be any two-wave functions, then

$$\int_{-\infty}^{+\infty} \psi_1^* \hat{p}_x \psi_2 \, dx = \int_{-\infty}^{+\infty} \psi_1^*\left(-i\hbar\frac{\partial}{\partial x}\right)\psi_2 \, dx = -i\hbar\left[\psi_1^*\psi_2\right]_{-\infty}^{+\infty} + i\hbar\int_{-\infty}^{+\infty} \psi_2\frac{\partial \psi_1^*}{\partial x}\,dx$$

But, $\left[\psi_1^*\psi_2\right]_{-\infty}^{+\infty} = 0$, since both ψ_1 and ψ_2 vanish at infinity.

$$\therefore \int_{-\infty}^{+\infty} \psi_1^* \hat{p}_x \psi_2\, dx = \int_{-\infty}^{+\infty} \psi_2 \left(i\hbar\frac{\partial}{\partial x}\right)\psi_1^*\, dx = \int_{-\infty}^{+\infty} \psi_2\left(-i\hbar\frac{\partial}{\partial x}\right)^* \psi_1^*\, dx$$

$$= \int_{-\infty}^{+\infty} \psi_2 \hat{p}_x^* \psi_1^*\, dx \tag{5.37}$$

Thus, $\hat{p}_x$ is Hermitian and as such linear momentum of a dynamical system is observable.

(c) ***Potential energy operator*:** The operator corresponding to the potential energy $V(\vec{r})$ is represented by $\hat{V}(\vec{r})$. The result of operation of the operator on a function $\psi(x)$ is simply to multiply $\psi(x)$ by $V(\vec{r})$. Thus,

$$\hat{V}(\vec{r})\,\psi = V(\vec{r})\,\psi \tag{5.38}$$

(d) ***Kinetic energy operator*:** We have from the x-component of linear momentum operator $(\hat{p}_x)$

$$\frac{\hbar}{i}\frac{\partial\psi}{\partial x} = p_x\psi$$

Differentiating w.r.t. x

$$\frac{\hbar}{i}\frac{\partial^2\psi}{\partial x^2} = p_x\frac{\partial\psi}{\partial x}$$

But,

$$\frac{\partial\psi}{\partial x} = \frac{i}{\hbar}p_x\psi$$

$$\therefore \frac{\hbar}{i}\frac{\partial^2\psi}{\partial x^2} = p_x\frac{i}{\hbar}p_x\psi = p_x^2\frac{i}{\hbar}\psi$$

$$\text{or, } \frac{\hbar^2}{i^2}\frac{\partial^2\psi}{\partial x^2} = p_x^2\psi \text{ or, } -\hbar^2\frac{\partial^2\psi}{\partial x^2} = p_x^2\psi. \tag{5.39}$$

Dividing both sides by $2m$, we get

$$-\frac{\hbar^2}{2m}\frac{\partial^2\psi}{\partial x^2} = \frac{p_x^2}{2m}\psi \tag{5.40}$$

$$\text{or, } -\frac{\hbar^2}{2m}\frac{\partial^2\psi}{\partial x^2} = K\psi \tag{5.41}$$

where $K = \dfrac{p_x^2}{2m}$, the kinetic energy of the particle.

Hence, $-\dfrac{\hbar^2}{2m}\dfrac{\partial^2}{\partial x^2}$ is the operator corresponding to kinetic energy of the particle for motion along the x-axis and is denoted by $\hat{K}$.

So

$$\hat{K} = -\frac{\hbar^2}{2m}\frac{\partial^2}{\partial x^2} \tag{5.42}$$

For three dimensions

$$\hat{K} = -\frac{\hbar^2}{2m}\nabla^2 \tag{5.43}$$

(e) ***Total energy operator***: **If $\psi(x,t) = Ae^{i(kx-\omega t)}$ be the wave function representing a moving particle then**

$$\frac{\partial}{\partial t}\psi(x,t) = \frac{\partial}{\partial t}\left[Ae^{i(kw-\omega t)}\right] = A(-i\omega)\,e^{i(kx-\omega t)} = (-i\omega)\,\psi(x,t)$$

Multiplying both sides by $i\hbar$ from left, we get

$$i\hbar\frac{\partial}{\partial t}\psi(x,t) = i\hbar(-i\omega)\,\psi(x,t) = \hbar\omega\psi(x,t) = E\psi(x,t) \tag{5.44}$$

where the total energy of the particle is $E = \hbar\omega$. Eq. (5.37) shows that the operator corresponding to E is

$$\hat{E} = i\hbar\frac{\partial}{\partial t} \tag{5.45}$$

(f) ***Hamiltonian operator***: **We have for the motion along the x-direction of a particle in a potential $V(x)$ the total energy**

$$E = \frac{p_x^2}{2m} + V(x) \text{ or, } E\psi = \frac{p_x^2}{2m}\psi + V(x)\,\psi \text{ or, } \frac{p_x^2}{2m}\psi = (E - V)\,\psi$$

$$\text{or, } -\frac{\hbar^2}{2m}\frac{\partial^2\psi}{\partial x^2} = (E - V)\,\psi \left[\because \frac{p_x^2}{2m} = -\frac{\hbar^2}{2m}\frac{\partial^2}{\partial x^2}\right]$$

$$\text{or, } -\frac{\hbar^2}{2m}\frac{\partial^2\psi}{\partial x^2} + V\psi = E\psi \text{ or, } \left[-\frac{\hbar^2}{2m}\frac{\partial^2}{\partial x^2} + V\right]\psi = E\psi \tag{5.46}$$

Thus, $\left[-\frac{\hbar^2}{2m}\frac{\partial^2}{\partial x^2} + V\right]$ is the total energy operator for the motion of the particle of mass m along the x-direction.

Thus,

$$\hat{H}_x = -\frac{\hbar^2}{2m}\frac{\partial^2}{\partial x^2} + V$$

$\hat{H}_x$ is the Hamiltonian operator.

For three-dimensional motion

$$-\frac{\hbar^2}{2m}\left(\frac{\partial^2}{\partial x^2} + \frac{\partial^2}{\partial y^2} + \frac{\partial^2}{\partial z^2}\right) + V = -\frac{\hbar^2}{2m}\nabla^2 + V$$

Thus, the Hamiltonian operator for the three-dimensional motion is

$$\hat{H} = -\frac{\hbar^2}{2m}\nabla^2 + V \tag{5.47}$$

If there are several particles, then

$$\hat{H} = -\frac{\hbar^2}{2}\sum_i \frac{\nabla_i^2}{m_i} + V \tag{5.48}$$

(g) ***Angular momentum operators:***

(i) ***Cartesian coordinates***: **For a particle of linear momentum $\vec{p}$ and radius vector $\vec{r}$ from the fixed point, the angular momentum of the particle about the fixed point O [Fig 5.1] is $\vec{L} = \vec{r} \times \vec{p}$.**

In operator notation $\hat{L} = \hat{r} \times \hat{p} = \hat{r} \times -i\hbar\nabla = -i\hbar\hat{r} \times \nabla$.

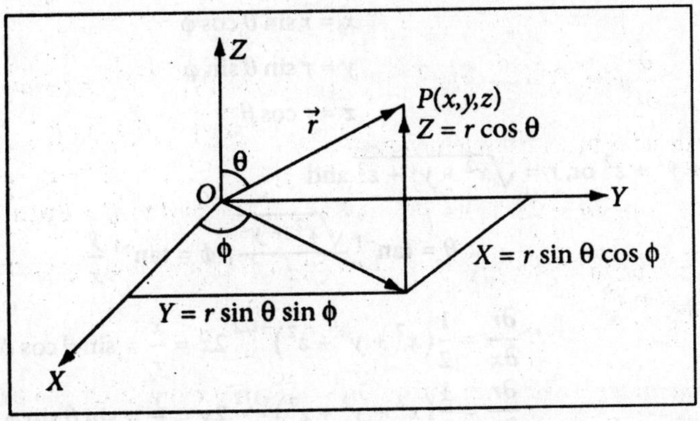

Fig 5.1 Position vector r of the point P in spherical polar coordinates

Now,

$$\hat{r} = \hat{i}\hat{x} + \hat{j}\hat{y} + \hat{k}\hat{z} \text{ and } \nabla = \hat{i}\frac{\partial}{\partial x} + \hat{j}\frac{\partial}{\partial y} + \hat{k}\frac{\partial}{\partial z}$$

$$\therefore \hat{L} = -i\hbar \begin{vmatrix} \hat{i} & \hat{j} & \hat{k} \\ \hat{x} & \hat{y} & \hat{z} \\ \frac{\partial}{\partial x} & \frac{\partial}{\partial y} & \frac{\partial}{\partial z} \end{vmatrix}$$

$$\text{or, } \hat{L} = -i\hbar\left[\hat{i}\left(\hat{y}\frac{\partial}{\partial z} - \hat{z}\frac{\partial}{\partial y}\right) + \hat{j}\left(\hat{z}\frac{\partial}{\partial x} - \hat{x}\frac{\partial}{\partial z}\right) + \hat{k}\left(\hat{x}\frac{\partial}{\partial y} - \hat{y}\frac{\partial}{\partial z}\right)\right]$$

$$= \hat{i}\hat{L}_x + \hat{j}\hat{L}_y + \hat{k}\hat{L}_z$$

where

$$\left.\begin{aligned} \hat{L}_x &= -i\hbar\left(\hat{y}\frac{\partial}{\partial z} - \hat{z}\frac{\partial}{\partial y}\right) \\ \hat{L}_y &= -i\hbar\left(\hat{z}\frac{\partial}{\partial x} - \hat{x}\frac{\partial}{\partial z}\right) \\ \hat{L}_z &= -i\hbar\left(\hat{x}\frac{\partial}{\partial y} - \hat{y}\frac{\partial}{\partial x}\right) \end{aligned}\right\} \tag{5.49}$$

where $\hat{L}_x$, $\hat{L}_y$ and $\hat{L}_z$ are the x-, y- and z-component angular momentum operators respectively.

The total angular momentum operator is

$$\hat{L} = \hat{i}\hat{L}_x + \hat{j}\hat{L}_y + \hat{k}\hat{L}_z.$$

$$\therefore \hat{L}\cdot\hat{L} = \hat{L}^2 = \hat{L}_x^2 + \hat{L}_y^2 + \hat{L}_z^2 \tag{5.50}$$

(ii) ***Spherical polar coordinates:*** If (x, y, z) be the coordinates of the point P in Cartesian coordinates and (r, θ, ϕ) be those in spherical polar coordinates [Fig 5.1], then

$$x = r\sin\theta\cos\phi$$
$$y = r\sin\theta\sin\phi$$
$$z = r\cos\theta$$

$\therefore r^2 = x^2 + y^2 + z^2$ or, $r = \sqrt{x^2+y^2+z^2}$ and

$$\theta = \tan^{-1}\frac{\sqrt{x^2+y^2}}{z},\ \phi = \tan^{-1}\frac{y}{x}$$

$$\therefore \frac{\partial r}{\partial x} = \frac{1}{2}\left(x^2+y^2+z^2\right)^{-1/2} 2x = \frac{x}{r} = \sin\theta\cos\phi \tag{5.51}$$

$$\frac{\partial r}{\partial y} = \frac{1}{2}\left(x^2+y^2+z^2\right)^{-1/2} 2y = \frac{y}{r} = \sin\theta\sin\phi \tag{5.52}$$

$$\frac{\partial r}{\partial z} = \frac{1}{2}\left(x^2+y^2+z^2\right)^{-1/2} 2z = \frac{z}{r} = \cos\theta \tag{5.53}$$

Again, $z = r\cos\theta$; $\therefore \cos\theta = z/r$.

$$\therefore -\sin\theta\frac{\partial\theta}{\partial x} = -\frac{z}{r^2}\frac{\partial r}{\partial x} = -\frac{1}{r^2}r\cos\theta\sin\theta\cos\phi$$

$$\therefore \frac{\partial\theta}{\partial x} = \frac{1}{r}\cos\theta\cos\phi \tag{5.54}$$

Now,

$$-\sin\theta\frac{\partial\theta}{\partial y} = -\frac{z}{r^2}\frac{\partial r}{\partial y} = -\frac{1}{r^2}r\cos\theta\sin\theta\sin\phi$$

$$\therefore \frac{\partial\theta}{\partial y} = \frac{1}{r}\cos\theta\sin\phi \tag{5.55}$$

and

$$-\sin\theta\frac{\partial\theta}{\partial z} = -\frac{z}{r^2}\frac{\partial r}{\partial z} + \frac{1}{r} = \frac{1}{r}\left(1-\cos^2\theta\right) = \frac{\sin^2\theta}{r}$$

$$\therefore \frac{\partial\theta}{\partial z} = -\frac{1}{r}\sin\theta \tag{5.56}$$

Again,

$$\tan\phi = \frac{y}{x}$$

Hence,

$$\sec^2\phi\frac{\partial\phi}{\partial x} = -\frac{y}{x^2} = -\frac{y}{x}\cdot\frac{1}{x} = -\frac{\sin\phi}{\cos\phi}\cdot\frac{1}{r\sin\theta\cos\phi}$$

$$\text{or, } \frac{\partial\phi}{\partial x} = -\frac{\sin\phi}{r\sin\theta} \text{ and } \sec^2\phi\frac{\partial\phi}{\partial y} = \frac{1}{x} = \frac{1}{r\sin\theta\cos\phi}$$

$$\therefore \frac{\partial\phi}{\partial y} = \frac{\cos\phi}{r\sin\theta} \tag{5.57}$$

Also

$$\frac{\partial \phi}{\partial z} = 0 \tag{5.58}$$

Now,

$$\frac{\partial}{\partial x} = \frac{\partial r}{\partial x}\frac{\partial}{\partial r} + \frac{\partial \theta}{\partial x}\frac{\partial}{\partial \theta} + \frac{\partial \phi}{\partial x}\frac{\partial}{\partial \phi} = \sin\theta\cos\phi\frac{\partial}{\partial r} + \frac{1}{r}\cos\theta\cos\phi\frac{\partial}{\partial \theta} - \frac{\sin\phi}{r\sin\theta}\frac{\partial}{\partial \phi} \tag{5.59}$$

$$\frac{\partial}{\partial y} = \frac{\partial r}{\partial y}\frac{\partial}{\partial r} + \frac{\partial \theta}{\partial y}\frac{\partial}{\partial \theta} + \frac{\partial \phi}{\partial y}\frac{\partial}{\partial \phi} = \sin\theta\sin\phi\frac{\partial}{\partial r} + \frac{1}{r}\cos\theta\cos\phi\frac{\partial}{\partial \theta} + \frac{\cos\phi}{r\sin\theta}\frac{\partial}{\partial \phi} \tag{5.60}$$

$$\frac{\partial}{\partial z} = \frac{\partial r}{\partial z}\frac{\partial}{\partial r} + \frac{\partial \theta}{\partial z}\frac{\partial}{\partial \theta} + \frac{\partial \phi}{\partial z}\frac{\partial}{\partial \phi} = \cos\theta\frac{\partial}{\partial r} - \frac{\sin\theta}{r}\frac{\partial}{\partial \theta} \tag{5.61}$$

$$\therefore \left(y\frac{\partial}{\partial z} - z\frac{\partial}{\partial y}\right) = r\sin\theta\sin\phi\cos\phi\frac{\partial}{\partial r} - \sin^2\theta\sin\phi\frac{\partial}{\partial \theta}$$

$$- r\cos\theta\sin\theta\sin\phi\frac{\partial}{\partial r} - \cos^2\theta\sin\phi\frac{\partial}{\partial \theta} - \frac{\cos\theta}{\sin\theta}\cos\phi\frac{\partial}{\partial \phi}$$

$$= -\left(\sin\phi\frac{\partial}{\partial \theta} + \cot\theta\cos\phi\frac{\partial}{\partial \phi}\right)$$

$$\left(z\frac{\partial}{\partial x} - x\frac{\partial}{\partial z}\right) = r\cos\theta\sin\theta\cos\phi\frac{\partial}{\partial r} + \cos^2\theta\cos\phi\frac{\partial}{\partial \theta}$$

$$- \cot\theta\sin\phi\frac{\partial}{\partial \phi} - r\sin\theta\cos\phi\cos\theta\frac{\partial}{\partial r} + \sin^2\theta\cos\phi\frac{\partial}{\partial \theta}$$

$$= \left(\cos\phi\frac{\partial}{\partial \theta} - \cot\theta\sin\phi\frac{\partial}{\partial \phi}\right)$$

$$\left(x\frac{\partial}{\partial y} - y\frac{\partial}{\partial x}\right) = r\sin^2\theta\cos\phi\sin\phi\frac{\partial}{\partial r} + \sin\theta\cos\theta\sin\phi\cos\phi\frac{\partial}{\partial \theta}$$

$$+ \cos^2\phi\frac{\partial}{\partial \phi} - r\sin^2\theta\sin\phi\cos\phi\frac{\partial}{\partial r} - \sin\theta\sin\phi\cos\theta\cos\phi\frac{\partial}{\partial \theta} + \sin^2\phi\frac{\partial}{\partial \phi}$$

$$= \frac{\partial}{\partial \phi}$$

$$\therefore \hat{L}_x = -i\hbar\left(y\frac{\partial}{\partial z} - z\frac{\partial}{\partial y}\right) = -i\hbar\left(-\sin\phi\frac{\partial}{\partial \theta} - \cot\theta\cos\phi\frac{\partial}{\partial \phi}\right) \tag{5.62}$$

$$\hat{L}_y = -i\hbar\left(z\frac{\partial}{\partial x} - x\frac{\partial}{\partial z}\right) = -i\hbar\left(\cos\phi\frac{\partial}{\partial \theta} - \cot\theta\sin\phi\frac{\partial}{\partial \phi}\right) \tag{5.63}$$

$$\hat{L}_z = -i\hbar\left(x\frac{\partial}{\partial y} - y\frac{\partial}{\partial x}\right) = -i\hbar\frac{\partial}{\partial \phi} \tag{5.64}$$

Now, substituting the values of $\hat{L}_x$, $\hat{L}_y$ and $\hat{L}_z$ in eq. (5.50), we get

$$\hat{L}^2 = \hat{L}_x^2 + \hat{L}_y^2 + \hat{L}_z^2 = -\hbar^2\left[\frac{1}{\sin\theta}\frac{\partial}{\partial \theta}\left(\sin\theta\frac{\partial}{\partial \theta}\right) + \frac{1}{\sin^2\theta}\frac{\partial^2}{\partial \phi^2}\right]$$

(iii) ***Commutation relation of orbital angular momentum with position:*** We have

$$[\hat{L}_x, \hat{x}] = [\hat{L}_x\hat{x} - \hat{x}\hat{L}_x] = \left[y\left(\frac{\hbar}{i}\frac{\partial}{\partial z}\right) - z\left(\frac{\hbar}{i}\frac{\partial}{\partial y}\right)\right]x - x\left[y\left(\frac{\hbar}{i}\frac{\partial}{\partial z}\right) - z\left(\frac{\hbar}{i}\frac{\partial}{\partial y}\right)\right]$$

Now,

$$(\hat{L}_x\hat{x} - \hat{x}\hat{L}_x)\,\psi(x) = \left[y\left\{\frac{\hbar}{i}\frac{\partial}{\partial z}(x\psi)\right\} - z\left\{\frac{\hbar}{i}\frac{\partial}{\partial y}(x\psi)\right\}\right] - x\left\{y\left(\frac{\hbar}{i}\frac{\partial\psi}{\partial z}\right) - z\left(\frac{\hbar}{i}\frac{\partial\psi}{\partial y}\right)\right\} = 0$$

$$\therefore [\hat{L}_x, \hat{x}] = 0 \tag{5.65}$$

Similarly, we can show that

$$[\hat{L}_x, \hat{y}] = i\hbar z \qquad [\hat{L}_y, \hat{y}] = 0 \qquad [\hat{L}_z, \hat{y}] = -i\hbar x$$
$$[\hat{L}_x, \hat{z}] = -i\hbar y \qquad [\hat{L}_y, \hat{z}] = i\hbar z \qquad [\hat{L}_z, \hat{z}] = 0$$
$$[\hat{L}_y, \hat{z}] = -i\hbar z \qquad [\hat{L}_z, \hat{x}] = i\hbar y$$

Also $[\hat{L}_x, p_x] = 0$, $[\hat{L}_x, p_y] = i\hbar p_z$, $[\hat{L}_x, p_z] = -i\hbar p_y$.

(iv) ***Commutation relations for the various components of orbital angular momentum:*** Let us consider the commutation relation between $\hat{L}_x$ and $\hat{L}_y$.

$$\therefore [\hat{L}_x, \hat{L}_y] = \hat{L}_x\hat{L}_y - \hat{L}_y\hat{L}_x$$

Now,

$$\hat{L}_x\hat{L}_y = (yp_z - zp_y)(zp_x - xp_z) = yp_zzp_x - yp_zxp_z - zp_yzp_x + zp_yxp_z$$
$$= yp_xp_zz + zp_zp_yx - yp_zxp_z - zp_yzp_x \left[\because \hat{L}_x = -i\hbar\left(\hat{y}\frac{\partial}{\partial z} - \hat{z}\frac{\partial}{\partial y}\right) = yp_z - zp_y, \text{ etc.}\right] \tag{5.66}$$

Similarly,

$$L_yL_x = (zp_x - xp_z)(yp_z - zp_y) = zp_xyp_z - zp_xzp_y - xp_zyp_z + xp_zzp_y$$
$$= yp_xzp_z + p_zzp_yx - yp_zxp_z - zp_yzp_x \tag{5.67}$$

From eqs (5.66) and (5.67) substituting the values of L_xL_y and L_yL_x in equation $[\hat{L}_x, \hat{L}_y] = \hat{L}_x\hat{L}_y - \hat{L}_y\hat{L}_x$, we obtain

$$[\hat{L}_x, \hat{L}_y] = yp_x[p_z, z] + [z, p_z]p_yx = xp_y[z, p_z] - yp_x[z, p_z]$$
$$= -i\hbar(xp_y - yp_x) = i\hbar\,\hat{L}_z \tag{5.68}$$

Similarly, we get

$$[\hat{L}_y, \hat{L}_z] = i\hbar\,\hat{L}_x \tag{5.69}$$

$$\text{and } [\hat{L}_z, \hat{L}_x] = i\hbar\,\hat{L}_y \tag{5.70}$$

(v) ***Commutation relation of $\hat{L}^2$ with the components $\hat{L}_x, \hat{L}_y$ and $\hat{L}_z$*:** We consider

$$[\hat{L}^2, \hat{L}_x] = [\hat{L}_x^2 + \hat{L}_y^2 + \hat{L}_z^2, \hat{L}_x] = [\hat{L}_x^2, \hat{L}_x] + [\hat{L}_y^2, \hat{L}_x] + [\hat{L}_z^2, \hat{L}_x]$$

$$\text{or, } [\hat{L}^2, \hat{L}_x] = [\hat{L}_y\hat{L}_y, \hat{L}_x] + [\hat{L}_z\hat{L}_z, \hat{L}_x] \; [\because \hat{L}_x^2, \hat{L}_x = 0]$$

$$= \hat{L}_z[\hat{L}_z, \hat{L}_x] + [\hat{L}_z, \hat{L}_x]\,\hat{L}_z + \hat{L}_y[\hat{L}_y, \hat{L}_x] + [\hat{L}_y, \hat{L}_x]\,\hat{L}_y$$

$$= \hat{L}_z(i\hbar\,\hat{L}_y) + (i\hbar\,\hat{L}_y)\,\hat{L}_z + \hat{L}_y(-i\hbar\,\hat{L}_z) + (-i\hbar\,\hat{L}_z)\,\hat{L}_y$$

$$= i\hbar[\hat{L}_z\hat{L}_y + \hat{L}_y\hat{L}_z - \hat{L}_y\hat{L}_z - \hat{L}_z\hat{L}_y] = 0 \tag{5.71}$$

Similarly, we can show that

$$[\hat{L}^2, \hat{L}_y] = 0 \text{ and } [\hat{L}^2, \hat{L}_z] = 0 \tag{5.72}$$

(vi) ***Commutation relation of $\hat{L}_z$ with Ladder operators $\hat{L}_+$ and $\hat{L}_-$*:** The ladder operators are defined by

$$\hat{L}_+ = \hat{L}_x + i\hat{L}_y \tag{5.73}$$

$$\text{and } \hat{L}_- = \hat{L}_x - i\hat{L}_y \tag{5.74}$$

Now,

$$[\hat{L}_z, \hat{L}_+] = [\hat{L}_z, \hat{L}_x + i\hat{L}_y] = [\hat{L}_z, \hat{L}_x] + i[\hat{L}_z, \hat{L}_y] = i\hbar\,\hat{L}_y + i(-i\hbar\,\hat{L}_x)$$

$$= i\hbar\,\hat{L}_y + \hbar\,\hat{L}_x = \hbar(\hat{L}_x + i\hat{L}_y) = \hbar\,\hat{L}_+ \tag{5.75}$$

$$\text{and } [\hat{L}_z, \hat{L}_-] = [\hat{L}_z, \hat{L}_x - i\hat{L}_y] = [\hat{L}_z, \hat{L}_x] - i[\hat{L}_z, \hat{L}_y] = i\hbar\,\hat{L}_y - i(-i\hbar\,\hat{L}_x)$$

$$= i\hbar\,\hat{L}_y - \hbar\,\hat{L}_x = -\hbar(\hat{L}_x - i\hat{L}_y) = -\hbar\,\hat{L}_- \tag{5.76}$$

(vii) ***Commutation relation of $\hat{L}_+$ and $\hat{L}_-$*:**

$$[\hat{L}_+, \hat{L}_-] = [(\hat{L}_x + i\hat{L}_y), (\hat{L}_x - i\hat{L}_y)] = [\hat{L}_x, \hat{L}_x] - i[\hat{L}_x, \hat{L}_y] + i[\hat{L}_y, \hat{L}_x] + [\hat{L}_y, \hat{L}_y]$$

$$= 0 - i(i\hbar\,\hat{L}_z) + i(-i\hbar\,\hat{L}_z) + 0 = \hbar\,\hat{L}_z + \hbar\,\hat{L}_z = 2\hbar\,\hat{L}_z \tag{5.77}$$

Table 5.2: Different dynamical variables and their corresponding operators

Dynamical variables	***Symbol***	***Operator***
1. Position vector	$\vec{r}$	$\hat{r}$
2. Cartesian components of r	x, y, z	$\hat{x}\,\hat{y}\,\hat{z}$
3. Linear momentum	p	$\hat{p} = -i\hbar\nabla$
4. Cartesian components of p	$p_x\ p_y\ p_z$	$\hat{p}_x = -i\hbar\frac{\partial}{\partial x}$ $\hat{p}_y = -i\hbar\frac{\partial}{\partial y}$ $\hat{p}_z = -i\hbar\frac{\partial}{\partial z}$

Dynamical variables	Symbol	Operator
5. Angular momentum	$\vec{L} = \vec{r} \times \vec{p}$	$\hat{L} = -i\hbar(\vec{r} \times \nabla)$
6. Cartesian components of L	$L_x = yp_z - zp_y$ $L_y = zp_x - xp_z$ $L_z = xp_y - yp_x$	$\hat{L}_x = -i\hbar\left(y\frac{\partial}{\partial z} - z\frac{\partial}{\partial y}\right)$ $\hat{L}_y = -i\hbar\left(z\frac{\partial}{\partial x} - x\frac{\partial}{\partial z}\right)$ $\hat{L}_z = -i\hbar\left(x\frac{\partial}{\partial y} - y\frac{\partial}{\partial x}\right)$
7. Hamiltonian	H	$\hat{H} = -\frac{\hbar^2}{2m}\nabla^2 + V(r)$
8. Total energy	$\frac{1}{2}\sum_n\left[\frac{1}{m_n}(p_{xn}^2 + p_{yn}^2 + p_{zn}^2) + V(x_n, y_n, z_n)\right]$	$-\frac{\hbar^2}{2}\sum_n\left[\frac{1}{m_n}\left(\frac{\partial^2}{\partial x_n^2} + \frac{\partial^2}{\partial y_n^2} + \frac{\partial^2}{\partial z_n^2}\right) + V(x_n, y_n, z_n)\right]$

5.16 Eigenfunctions and eigenvalues

If an operator $\hat{A}$ operating on a function $\psi(x)$ multiplies the latter by a constant λ, then $\psi(x)$ is called the *eigenfunction* of $\hat{A}$ belonging to the eigenvalue λ. A set of eigenvalues λ_n and a set of eigenfunctions ψ_n each belong to operator $\hat{A}$, i.e.,

$$\hat{A}\,\psi_n(x) = \lambda_n \psi_n(x) \tag{5.78}$$

Equation (5.78) is called the *eigenvalue equation.*

Hence, eigenfunctions $\psi_n(x)$ are a set of functions which when operated by the operator $\hat{A}$ remains unchanged and are multiplied by the corresponding eigenvalues λ_n. All operators of quantum mechanics have eigenvalues and eigenfunctions. [The German word *eigen* means proper or characteristics.]

For example, if

$$\hat{A} = \frac{d^2}{dx^2} \text{ and } \psi(x) = ae^{-2x}$$

then

$$\hat{A}\,\psi(x) = \frac{d^2}{dx^2}\left(ae^{-2x}\right) = 4ae^{-2x} = 4\psi(x) \tag{5.79}$$

Here, $\psi(x)$ is the eigenfunction of the operator $\frac{d^2}{dx^2}$ and 4 is the eigenvalue belonging to $\psi(x)$.

An eigenfunction $\psi(x)$ must be *well-behaved.* A well-behaved function $\psi(x)$ must satisfy the following conditions, viz.,

(i) $\psi(x)$ must be everywhere single-valued and continuous.

(ii) $\int |\psi(x)|^2 \, d\tau$ must be finite, i.e., it must be square-integrable.

(iii) $\psi(x) \to$ finite as $x \to \pm\infty$.

(iv) ψ and $\frac{d\psi}{dx}$ must be continuous everywhere.

5.17 Well-behaved and the admissible solutions

If $\hat{A} = \dfrac{d^2}{dx^2}$ and $\psi(x)$ be a function of (x), then let

$$\hat{A}\,\psi(x) = \frac{d^2\psi}{dx^2} = -\lambda\psi \text{ or, } \frac{d^2\psi}{dx^2} = -\lambda\psi \tag{5.80}$$

The general solution of eq. (5.80) is

$$\psi(x) = Ae^{i\sqrt{\lambda}x} + Be^{-i\sqrt{\lambda}x} \tag{5.81}$$

where A and B are arbitrary constants. The above equation may also be expressed as

$$\psi(x) = C\cos\sqrt{\lambda}\,x + D\sin\sqrt{\lambda}\,x \tag{5.82}$$

The actual form of the solution depends on the nature of λ as follows:

(i) If $\lambda > 0$, then $\sqrt{\lambda}$ is a non-vanishing positive number. Under this condition $\psi(x)$ is single-valued and finite as

$$-1 \leqslant \sin\theta \leqslant 1$$

$$\text{and } -1 \leqslant \cos\theta \leqslant 1.$$

$\psi(x)$ is, then a well-behaved wave function.

(ii) If $\lambda = 0$, then from eq. (5.80)

$$\frac{d^2\psi}{dx^2} = 0 \tag{5.83}$$

The solution of this equation is

$$\psi(x) = Fx + G \tag{5.84}$$

where F and G are two arbitrary constants. As $|\psi(x)| \to \infty$, when $x \to \pm\infty$, the function $\psi(x)$ is not well-behaved. If $F = 0$, then $\psi(x) = G$ and $\psi(x)$ is again well-behaved as it is now finite and constant.

(iii) If $\lambda < 0$, then $\sqrt{\lambda}$ is non-vanishing imaginary number. If $\sqrt{\lambda} = ib$, where b is a real and positive number, then from eq. (5.81)

$$\psi(x) = Ae^{-bx} + Be^{bx} \tag{5.85}$$

Here, $\psi(x)$ is single-valued and continuous, but $|\psi(x)| \to \infty$ as $x \to \pm\infty$. So, for $\lambda < 0$, $\psi(x)$ cannot be well-behaved.

If for a particular value of λ, $\psi(x)$ is well-behaved, then λ is called the *eigenvalue* and $\psi(x)$ is called the *eigenfunction* of the operator

$$\hat{A} = \frac{d^2}{dx^2}$$

If a string lying along the x-direction and vibrating along the y-direction be fixed at $x = 0$ and $x = l$, then the general expression for the displacement of any point of the string at a distance x and time t is given by

$$y_n(x,t) = [A_n\cos\omega_n t + B_n\sin\omega_n t]\sin\frac{n\pi x}{l} \tag{5.86}$$

under the condition $y = 0$ at $x = 0$ and $x = l$. Here

$$\omega_n = \frac{n\pi c}{l}$$

and A_n and B_n are two arbitrary constants, c is the velocity of the wave and $n = 0, 1, 2, \ldots$ Equation (5.86) represents an eigenfunction of the problem.

Using superposition principle the general solution is expressed as

$$y(x,t) = \sum_{n=1}^{\infty} [A_n \cos \omega_n t + B_n \sin \omega_n t] \sin \frac{n\pi x}{l} \tag{5.87}$$

Equation (5.87) represents a set of eigenfunctions which are orthogonal and complete.

Hence, the stationary state solutions of the Schrödinger time-independent equation may be superposed to give a general solution as discussed above.

5.18 Simultaneous eigenfunctions and commutativity of operators

Let $\hat{A}$ and $\hat{B}$ be two quantum mechanical operators representing two dynamical variables A and B and operate on a wave function $\psi(x)$ to give

$$\hat{A}\,\psi(x) = \lambda_1\,\psi(x) \tag{5.88}$$

$$\hat{B}\,\psi(x) = \lambda_2\,\psi(x) \tag{5.89}$$

where λ_1 and λ_2 are two constants. $\psi(x)$ is, then said to be the simultaneous eigenfunction of $\hat{A}$ and $\hat{B}$ belonging to the eigenvalues λ_1 and λ_2 respectively. This means that measurement of A and B in the state $\psi(x)$ simultaneously gives the two definite eigenvalues λ_1 and λ_2 respectively.

For example, the operators $\hat{p}_x, \hat{p}_y, \hat{p}_z$ corresponding to the three components p_x, p_y and p_z of linear momentum p have the simultaneous eigenfunction

$$\psi(\vec{r}) = Ae^{\frac{i}{\hbar}(p_x x + p_y y + p_z z)}$$

Now, we can show that *if the simultaneous wave function of two operators $\hat{A}$ and $\hat{B}$ constitute a complete set, then the operators must commute, i.e., $[\hat{A}, \hat{B}] = 0$.*

From eqs (5.88) and (5.89)

$$\hat{B}\hat{A}\,\psi(x) = \hat{B}\lambda_1\,\psi(x) = \lambda_1\hat{B}\,\psi(x) = \lambda_1\lambda_2\,\psi(x) \tag{5.90}$$

$$\text{and } \hat{A}\hat{B}\,\psi(x) = \hat{A}\lambda_2\,\psi(x) = \lambda_2\hat{A}\,\psi(x) = \lambda_2\lambda_1\,\psi(x) \tag{5.91}$$

$$\therefore \hat{A}\hat{B}\,\psi(x) - \hat{B}\hat{A}\,\psi(x) = \left(\hat{A}\hat{B} - \hat{B}\hat{A}\right)\psi(x) = (\lambda_2\lambda_1 - \lambda_1\lambda_2)\,\psi(x) = 0 \tag{5.92}$$

If $\psi(x)$ be replaced by $\psi_n(x)$, the eigenfunction for nth quantum state, then

$$\hat{A}\hat{B}\,\psi_n(x) = \hat{B}\hat{A}\,\psi_n(x) \tag{5.93}$$

If all the eigenfunctions $\psi_1(x), \psi_2(x), \ldots, \psi_n(x), \ldots$ constitute a complete set, then

$$\psi(x) = \sum_n C_n \psi_n(x) \tag{5.94}$$

From eq. (5.92)

$$\hat{A}\hat{B}\,\psi(x) = \hat{B}\hat{A}\,\psi(x) \text{ or, } \hat{A}\hat{B} - \hat{B}\hat{A} = 0 \text{ or, } \left[\hat{A}, \hat{B}\right] = 0 \tag{5.95}$$

In this connection, we get the following theorems:

► ***Theorem* 1.** If $\psi_n(x)$ be a set of orthogonal eigenfunctions of two operators $\hat{A}$ and $\hat{B}$, then they must commute, i.e., $[\hat{A}, \hat{B}] = 0$.

Proof: We consider

$$[\hat{A}, \hat{B}]\psi(x) = [\hat{A}\hat{B} - \hat{B}\hat{A}]\psi(x) = [\hat{A}\hat{B} - \hat{B}\hat{A}]\sum_n C_n\psi_n$$

$$= \hat{A}\hat{B}\sum_n C_n\psi_n - \hat{B}\hat{A}\sum_n C_n\psi_n = \sum_n C_n\hat{A}(\hat{B}\psi_n) - \sum_n C_n\hat{B}(\hat{A}\psi_n) \quad \text{(i)}$$

But,

$$\hat{A}\psi_n = \lambda_{1n}\psi_n$$

$$\text{and } \hat{B}\psi_n = \lambda_{2n}\psi_n$$

$$\therefore [\hat{A}\hat{B} - \hat{B}\hat{A}]\psi = \sum_n C_n\hat{A}(\lambda_{2n}\psi_n) - \sum_n C_n\hat{B}(\lambda_{1n}\psi_n) = \sum_n C_n\lambda_{2n}(\hat{A}\psi_n) - \sum_n C_n\lambda_{1n}(\hat{B}\psi_n)$$

$$= \sum_n C_n(\lambda_{2n}\lambda_{1n} - \lambda_{1n}\lambda_{2n})\psi_n = 0 \quad \text{(ii)}$$

Since, $\psi(x)$ is an arbitrary function, $\psi(x) \neq 0$.

$$\therefore \hat{A}\hat{B} - \hat{B}\hat{A} = 0 \text{ or, } [\hat{A}, \hat{B}] = 0$$

► ***Theorem* 2.** If the operators $\hat{A}$ and $\hat{B}$ commute, then they have the same set of eigenvalues.

Proof: If $\psi_n(x)$ be an eigenfunction of the operator $\hat{A}$, then $\hat{A}\psi_n = \lambda_{1n}\psi_n$, where λ_{1n} is the eigenvalue of the state ψ_n.

Now,

$$\hat{A}\hat{B}\psi_n = \hat{B}\hat{A}\psi_n = \lambda_{1n}\hat{B}\psi_n \quad \text{(i)}$$

since $\hat{A}\hat{B} = \hat{B}\hat{A}$.

Equation (5.96) shows that $\hat{B}\psi_n$ is an eigenfunction of $\hat{A}$ having eigenvalue λ_{1n}. This is possible if $\hat{B}\psi_n$ is a multiple of ψ_n, i.e.,

$$\hat{B}\psi_n = \lambda_{2n}\psi_n \quad \text{(ii)}$$

Equation (5.97) shows that ψ_n is also an eigenfunction of $\hat{B}$.

5.19 Expansion of an arbitrary function in terms of eigenfunctions

If the operator $\hat{A}$ operating on eigenfunction $\psi_n(x)$ gives the eigenvalue λ_n, then

$$\hat{A}\psi_n(x) = \lambda_n\psi_n(x) \quad (5.96)$$

Generally, the solution of the eigenvalue equation as above yields a whole set of solutions with different eigenvalues. The total set of possible solutions of the eigenvalue equation constitute a complete set of eigenfunctions having different eigenvalues.

If $\psi_1(x), \psi_2(x), \ldots, \psi_n(x)$ be the set of eigenfunctions of $\hat{A}$ with eigenvalues $\lambda_1, \lambda_2, \ldots, \lambda_n$, then in accordance with the principle of superposition

$$\psi(x) = C_1\psi_1(x) + C_2\psi_2(x) + \cdots + C_n\psi_n(x) = \sum_n C_n\psi_n(x) \tag{5.97}$$

$\psi(x)$ obeys the same boundary conditions as ψ_n's.

The modulus squared $|C_n|^2$ of the amplitudes of ψ_n gives the probability of $\psi_n(x)$ in state $\psi(x)$. Using the orthogonality property of $\psi_n(x)$, C_n can be determined as follows:

At $t = 0$, we have

$$\psi(x,0) = \psi(x) = \sum_n C_n\psi_n(x)$$

Multiplying both sides by ψ_m^* and, then integrating, we get

$$\int \psi(x)\,\psi_m^*(x)\,dx = \sum_n C_n \int \psi_m^*(x)\,\psi_n(x)\,dx = C_m \int \psi_m^*(x)\,\psi_m(x)\,dx = C_m \int |\psi_m(x)|^2\,dx$$

Since,

$$\int \psi_m^*(x)\,\psi_n(x)\,dx = \begin{cases} 1, & \text{for } m = n \\ 0, & \text{for } m \neq n \end{cases}$$

$$\therefore C_m = \frac{\int \psi(x)\,\psi_m^*(x)\,dx}{\int |\psi_m(x)|^2\,dx} \tag{5.98}$$

If the eigenfunctions are orthogonal, then

$$\int |\psi_m(x)|^2\,dx = 1$$

Hence,

$$C_m = \int \psi(x)\,\psi_m^*(x)\,dx \tag{5.99}$$

5.20 Completeness relation

We know, an eigenfunction $\psi(x)$ of a state can be expanded in the form:

$$\psi(x) = \sum_n C_n\psi_n(x)$$

which forms a complete orthogonal set of functions.

If we include only m terms $(m < n)$, then the above relation can be written as

$$\psi(x) = \sum_{n=0}^{m} C_n\phi_n(x) + R_m(x) \tag{5.100}$$

$\phi_n(x)$ are orthonormal set of eigenfunctions.

If

$$\lim_{m\to\infty} \int |R_m(x)|^2\,dx = 0$$

then $\phi_n(x)$ is said to be a complete set of eigenfunctions.

5.21 Linearly dependent and independent wave functions

Let $\psi_1(x)$ and $\psi_2(x)$ be two functions and their linear combination is

$$\psi(x) = C_1\psi_1(x) + C_2\psi_2(x) = 0 \tag{5.101}$$

Equation (5.101) is satisfied if $C_1 = C_2 = 0$. Under this condition $\psi_1(x)$ and $\psi_2(x)$ are said to be *linearly independent*.

On the other hand, if either, $C_1 \neq 0$ or $C_2 \neq 0$ or both C_1 and C_2 not equal to zero, then $\psi_1(x)$ and $\psi_2(x)$ are said to be *linearly dependent*.

Differentiating eq. (5.101), we get

$$\psi'(x) = C_1\psi_1'(x) + C_2\psi_2'(x) \tag{5.102}$$

Equations (5.101) and (5.102) are two linear equations with two unknowns C_1 and C_2. These equations can have non-zero solutions if the *Wronskians* of $\psi_1(x)$ and $\psi_2(x)$ is zero, i.e.,

$$W = \begin{vmatrix} \psi_1(x) & \psi_2(x) \\ \psi_1'(x) & \psi_2'(x) \end{vmatrix} = \psi_1(x)\,\psi_2'(x) - \psi_2(x)\,\psi_1'(x) = 0 \tag{5.103}$$

So the condition for $\psi_1(x)$ and $\psi_2(x)$ to be linearly independent is $W \neq 0$. $\psi_1(x)$ and $\psi_2(x)$ are, however linearly dependent, if $W = 0$. This definition and test hold good for more than two-wave functions.

5.22 Degeneracy

If the Schrödinger equation has more than one linearly independent solutions for the same value of energy E, then E is said to be *degenerate*.

We consider the eigenvalue equation

$$\hat{A}\,\psi(x) = \lambda\psi(x)$$

If $\hat{A} = \frac{d^2}{dx^2}$, then

$$\frac{d^2\,\psi(x)}{dx^2} = \lambda\psi(x) \tag{5.104}$$

The solution of the above equation for $\lambda > 0$ is

$$\psi(x) = C_1 e^{\sqrt{\lambda}\,x} + C_2\, e^{-\sqrt{\lambda}\,x} \tag{5.105}$$

$$= C_1\psi_1(x) + C_2\psi_2(x) \tag{5.106}$$

where $\psi_1(x) = e^{\sqrt{\lambda}\,x}$ and $\psi_2(x) = e^{-\sqrt{\lambda}\,x}$.

Equation (5.104) is satisfied by each of $\psi_1(x)$ and $\psi_2(x)$ for the same eigenvalue λ. Here, $\psi_1(x)$ and $\psi_2(x)$ are linearly independent.

If $\psi_{n1}, \psi_{n2}, \ldots, \psi_{nk}$ be a set of k linearly independent solutions of eq. (5.104) for the particular eigenvalue λ_n, then λ_n is called *k-fold degenerate*. The linear combination of the above set of wave functions is expressed as

$$\psi_n = C_{n1}'\,\psi_{n1} + C_{n2}'\,\psi_{n2} + \cdots + C_{nk}'\,\psi_{nk} \tag{5.107}$$

ψ_n is also an eigenfunction of eq. (5.104).

5.23 Reality of energy eigenvalues

Let the wave function

$$\psi_n(\vec{r},t) = \psi_n(\vec{r})\, e^{-iE_n t/\hbar} \tag{5.108}$$

represent an eigenstate of the Hamiltonian operator $\hat{H}$ of energy eigenvalue E_n. We have to prove that E_n is real.

The probability density of the state is given by

$$\rho(\vec{r},t) = \psi_n^*(\vec{r},t)\,\psi_n(\vec{r},t) = \psi_n^*(\vec{r})\, e^{iE_n t/\hbar}\psi_n(\vec{r})\, e^{-iE_n t/\hbar} = \psi_n^*(\vec{r})\,\psi_n(\vec{r})\, e^{-i(E_n-E_n^*)t/\hbar} \tag{5.109}$$

Therefore, the probability current density is

$$\begin{aligned}\vec{J}(\vec{r},t) &= \frac{i\hbar}{2m}\left[\psi_n(\vec{r},t)\,\nabla\psi_n^*(\vec{r},t) - \psi_n^*(\vec{r},t)\,\nabla\psi_n(\vec{r},t)\right]\\ &= \frac{i\hbar}{2m}\left[\psi_n(\vec{r})\,\nabla\psi_n^*(\vec{r}) - \psi_n^*(\vec{r})\,\nabla\psi_n(\vec{r})\right]\times e^{-i(E_n-E_n^*)t/\hbar}\\ &= \vec{J}(\vec{r})\, e^{-i(E_n-E_n^*)t/\hbar}\end{aligned}$$

$$\therefore \nabla\cdot\vec{J}(\vec{r},t) = \nabla\cdot\vec{J}(\vec{r})\, e^{-i(E_n-E_n^*)t/\hbar} \tag{5.110}$$

Now, the equation of continuity is

$$\nabla\cdot\vec{J}(\vec{r},t) + \frac{\partial\rho}{\partial t} = 0 \text{ or, } \frac{\partial\rho}{\partial t} = -\nabla\cdot\vec{J}(\vec{r},t)$$

$$\text{or, } \frac{\partial}{\partial t}\int_\tau \rho\, d\tau = -\int_\tau \nabla\cdot\vec{J}(\vec{r},t)\, d\tau = -\oint_S \vec{J}(\vec{r},t)\cdot d\vec{S} \tag{5.111}$$

From eqs (5.109), (5.110) and (5.111), we get

$$-\frac{i}{\hbar}(E_n - E_n^*)\, e^{-i(E_n-E_n^*)t/\hbar}\int_\tau \psi_n^*(\vec{r})\,\psi_n(\vec{r})\, d\tau = e^{-i(E_n-E_n^*)t/\hbar}\oint_S \vec{J}(\vec{r})\cdot d\vec{S} \tag{5.112}$$

Now,

$$\begin{aligned}\oint_S \vec{J}(\vec{r})\cdot d\vec{S} &= \frac{i\hbar}{2m}\oint_S \left[\psi_n(\vec{r})\,\nabla\psi_n^*(\vec{r}) - \psi_n^*(\vec{r})\,\nabla\psi_n(\vec{r})\right]\cdot d\vec{S}\\ &= 0 \text{ as } \psi_n(\vec{r}) \to 0 \text{ as } |\vec{r}| \to \infty\end{aligned} \tag{5.113}$$

So

$$-\frac{i}{\hbar}(E_n - E_n^*)\, e^{-i(E_n-E_n^*)t/\hbar}\int_\tau \psi_n^*(\vec{r})\,\psi_n(\vec{r})\, d\tau = 0 \tag{5.114}$$

But,

$$-\frac{i}{\hbar}\int_\tau \psi_n^*(\vec{r})\,\psi_n(\vec{r})\, d\tau \neq 0$$

$$\therefore (E_n - E_n^*) = 0 \text{ or, } E_n = E^* \tag{5.115}$$

This shows that E_n is real.

5.24 Expectation values of dynamical quantities

The wave function $\psi(\vec{r}, t)$, according to Born, has probabilistic interpretation and as such it is essential to calculate the average or expectation value of any dynamical quantity defined by $\psi(\vec{r}, t)$. The space coordinates, momenta and energy of a system are the dynamical quantities in Physics.

The average or expectation value of a dynamical variable is defined as the mathematical expectation for the result of a single measurement on each of the large number of identical systems.

Or

It is defined as the average of the result of a large number of independent measurements on the same system.

Suppose, we have to find the expectation value $\langle x \rangle$ of position coordinate at time t of an electron moving along the x-direction in electron diffraction experiment. Let the number of electrons be N and each of them is described the wave function $\psi(x, t)$. At time t we measure the position of each electron and find the number of electrons in between x and $x + dx$.

Then the probability of an electron to be found in the position between x and $x + dx$ is

$$P(dx) = \frac{\text{number of electrons between } x \text{ and } x + dx}{\text{total number of electrons } (N)}$$

Now, according to wave mechanics, the probability of finding an electron between x and $x + dx$ is

$$P(dx) = \psi^*(x, t)\,\psi(x, t)\,dx$$

if $\psi(x, t)$ is normalised.

So the number of electrons between x and $x + dx$ is

$$dN = N\psi^*(x, t)\,\psi(x, t)\,dx$$

Now, if $x_1, x_2, x_3, \ldots, x_N$ be the measured values of position for all the N electrons, then

$$x_1 + x_2 + x_3 + \cdots + x_N = \int_{-\infty}^{+\infty} xN\psi^*(x, t)\,\psi(x, t)\,dx$$

$$\text{or,} \quad \frac{x_1 + x_2 + x_3 + \cdots + x_N}{N} = \int_{-\infty}^{+\infty} x\psi^*(x, t)\,\psi(x, t)\,dx$$

But,

$$\frac{x_1 + x_2 + x_3 + \cdots + x_N}{N} = \langle x \rangle$$

the average or expectation value of position.

$$\therefore \langle x \rangle = \int_{-\infty}^{+\infty} \psi^*(x, t)\,x\psi(x, t)\,dx \tag{5.116}$$

if $\psi(x, t)$ is normalised. If, however, $\psi(x, t)$ is not normalised, then

$$\langle x \rangle = \frac{\int_{-\infty}^{+\infty} \psi^*(x, t)\,\hat{x}\,\psi(x, t)\,dx}{\int_{-\infty}^{+\infty} \psi^*(x, t)\,\psi(x, t)\,dx} \tag{5.117}$$

In general, for any dynamical variable represented by the operator $\hat{A}$, the expectation value is given by

$$\langle A\rangle = \frac{\int_{-\infty}^{+\infty} \psi^*(\vec{r},t)\,\hat{A}\psi(\vec{r},t)\,d\tau}{\int_{-\infty}^{+\infty} \psi^*(\vec{r},t)\,\psi(\vec{r},t)\,d\tau} \tag{5.118}$$

$\psi(\vec{r},t)$ being the *unnormalised* wave function in three dimensions. If $\psi(\vec{r},t)$ is *normalised*, then

$$\langle A\rangle = \int_{-\infty}^{+\infty} \psi^*(\vec{r},t)\,\hat{A}\,\psi(\vec{r},t)\,d\tau \tag{5.119}$$

In the similar way the expectation value of $\vec{r}$ is

$$\langle r\rangle = \int_{-\infty}^{+\infty} \psi^*(\vec{r},t)\,\hat{r}\psi(\vec{r},t)\,d\tau \tag{5.120}$$

The expectation value of $f(\vec{r})$ is

$$\langle f(\vec{r})\rangle = \int_{-\infty}^{+\infty} \psi^*(\vec{r},t)\,f(\vec{r})\,\psi(\vec{r},t)\,d\tau \tag{5.121}$$

The expectation value of the potential energy $V(\vec{r},t)$ is

$$\langle V(\vec{r},t)\rangle = \int_{-\infty}^{+\infty} \psi^*(\vec{r},t)\,V(\vec{r},t)\,\psi(\vec{r},t)\,d\tau \tag{5.122}$$

(a) ***Expectation value of*** $\boldsymbol{p_x}$**:** The expectation value of the x-component of linear momentum (p_x) is given by

$$\langle p_x\rangle = \int_{-\infty}^{+\infty} \psi^*(x,t)\,\hat{p}_x\,\psi(x,t)\,dx = \int_{-\infty}^{+\infty} \psi^*(x,t)\left(\frac{\hbar}{i}\frac{\partial}{\partial x}\right)\psi(x,t)\,dx \tag{5.123}$$

(b) ***Expectation value of kinetic energy*** **(*K*):** The expectation value of the kinetic energy is given by

$$\langle K\rangle = \int_{-\infty}^{+\infty} \psi^*(x,t)\hat{K}\psi(x,t)\,dx = \int_{-\infty}^{+\infty} \psi^*(x,t)\left(-\frac{\hbar^2}{2m}\frac{\partial^2}{\partial x^2}\right)\psi(x,t)\,dx \quad \left[\because \hat{K} = -\frac{\hbar^2}{2m}\frac{\partial^2}{\partial x^2}\right]$$

$$\text{or, } \langle K\rangle = -\frac{\hbar^2}{2m}\int_{-\infty}^{+\infty} \psi^*(x,t)\frac{\partial}{\partial x}\left(\frac{\partial\psi}{\partial x}\right)dx = -\frac{\hbar^2}{2m}\left[\left\{\psi^*\frac{\partial\psi}{\partial x}\right\}_{-\infty}^{+\infty} - \int_{-\infty}^{+\infty}\frac{\partial\psi^*}{\partial x}\cdot\frac{\partial\psi}{\partial x}\,dx\right]$$

Since, $\psi^*(x,t) \to 0$ as $x \to \pm\infty$

$$\therefore \left\{\psi^*\frac{\partial\psi}{\partial x}\right\}_{-\infty}^{+\infty} = 0 \text{ or, } \langle K\rangle = \frac{\hbar^2}{2m}\int_{-\infty}^{+\infty}\frac{\partial\psi^*}{\partial x}\cdot\frac{\partial\psi}{\partial x}\,dx \tag{5.124}$$

$$\therefore \langle K\rangle = \frac{\hbar^2}{2m}\int_{-\infty}^{+\infty}\left|\frac{\partial\psi}{\partial x}\right|^2 dx \tag{5.125}$$

(c) ***Expectation value of energy*** $\boldsymbol{(E_n)}$**:** If E_n be the energy of an eigenstate $\psi_n(\vec{r},t)$, then since E_n is real

$$\langle E\rangle = \int_{-\infty}^{+\infty} \psi_n^*(\vec{r},t)\,\hat{E}\psi(\vec{r},t)\,d\tau$$

If $\psi_n(\vec{r},t) = \psi_n(\vec{r})\,e^{-iE_n\hbar/t}$, then

$$\langle E\rangle = \int_{-\infty}^{+\infty} \psi_n^*(\vec{r})\,e^{iE_nt/\hbar}\left(i\hbar\frac{\partial}{\partial t}\right)\psi_n(r)\,e^{-iE_nt/\hbar}\,d\tau \quad \left[\because \hat{E} = i\hbar\frac{\partial}{\partial t}\right]$$

$$= i\hbar\int_{-\infty}^{+\infty} \psi_n^*(\vec{r})\,e^{iE_nt/\hbar}(-iE_n/\hbar)\psi_n(\vec{r})\,e^{iE_nt/\hbar}\,d\tau = E_n\int_{-\infty}^{+\infty}\psi_n^*(\vec{r})\,\psi_n(\vec{r})\,d\tau$$

If ψ_n is normalised, then

$$\int_{-\infty}^{+\infty} \psi_n^* (\vec{r})\, \psi_n (\vec{r})\, d\tau = 1$$

$$\therefore \langle E \rangle = E_n \tag{5.126}$$

Equation (5.128) shows that $\langle E \rangle$ is constant w.r.t. time.

5.25 Ehrenfest theorem

The theorem states that the Schrödinger wave equation leads to the satisfaction of Newton's laws of motion on the average.

Alternatively, the average motion of a wave packet agrees with the motion of the corresponding classical particle.

If $\vec{p}$ be the momentum of a particle moving in a potential field V, then according to Newton's laws of motion

$$\frac{d\vec{p}}{dt} = -\nabla V \tag{5.127}$$

$$\text{and } \frac{d\vec{r}}{dt} = \frac{\vec{p}}{m} \tag{5.128}$$

In component form:

$$\left.\begin{aligned} \frac{dp_x}{dt} &= -\frac{\partial V}{\partial x}, & \frac{dp_y}{dt} &= -\frac{\partial V}{\partial y}, & \frac{dp_z}{dt} &= -\frac{\partial V}{\partial z} \\ \text{and } \frac{dx}{dt} &= \frac{p_x}{m}, & \frac{dy}{dt} &= \frac{p_y}{m}, & \frac{dz}{dt} &= \frac{p_z}{m} \end{aligned}\right\} \tag{5.129}$$

Now,

$$\frac{dx}{dt}, \frac{dy}{dt}, \frac{dz}{dt}; \frac{dp_x}{dt}, \frac{dp_y}{dt}, \frac{dp_z}{dt}$$

are not possible to define in quantum mechanics in the classical sense. The approximate values of the above quantities can, however be found by considering the time rate of change of $\langle x \rangle, \langle y \rangle, \langle z \rangle; \langle p_x \rangle, \langle p_y \rangle, \langle p_z \rangle$. In case of width of quantum mechanical wave packet tending to zero

$$\left.\begin{aligned} \frac{d\langle x \rangle}{dt} &= \frac{dx}{dt}, & \frac{d\langle y \rangle}{dt} &= \frac{dy}{dt}, & \frac{d\langle z \rangle}{dt} &= \frac{dz}{dt} \\ \text{and } \frac{d}{dt}\langle p_x \rangle &= \frac{dp_x}{dt}, & \frac{d}{dt}\langle p_y \rangle &= \frac{dp_y}{dt}, & \frac{d}{dt}\langle p_z \rangle &= \frac{dp_z}{dt} \end{aligned}\right\} \tag{5.130}$$

To prove Ehrenfest's theorem we proceed as follows:

The average or expectation value of position $(\vec{r})$ of a particle represented by $\psi(\vec{r}, t)$ is

$$\langle \vec{r} \rangle = \int \psi^* \hat{r} \psi \, d\tau$$

where $d\tau$ = the volume element.

$\therefore$ the average value of velocity of the particle is

$$\frac{d\langle \vec{r} \rangle}{dt} = \int \left(\frac{\partial \psi^*}{\partial t} r \psi + \psi^* r \frac{\partial \psi}{\partial t} \right) d\tau \tag{5.131}$$

Now, the Schrödinger equation is

$$i\hbar\frac{\partial\psi(\vec{r},t)}{\partial t}=-\frac{\hbar^2}{2m}\nabla^2\psi(\vec{r},t)+V\psi(\vec{r},t) \tag{5.132}$$

where m is the mass of the particle.

The complex conjugate of the above equation is

$$-i\hbar\frac{\partial\psi^*(\vec{r},t)}{\partial t}=-\frac{\hbar^2}{2m}\nabla^2\psi^*(\vec{r},t)+V\psi^*(\vec{r},t) \tag{5.133}$$

Substituting the values of $\frac{\partial\psi(\vec{r},t)}{\partial t}$ and $\frac{\partial\psi^*(\vec{r},t)}{\partial t}$ from eqs (5.132) and (5.133), we get from eq. (5.131)

$$\frac{d\langle\vec{r}\rangle}{dt}=\frac{1}{i\hbar}\left[\int\psi^* r\left(-\frac{\hbar^2}{2m}\nabla^2\psi+V\psi\right)d\tau-\int\left(-\frac{\hbar^2}{2m}\nabla^2\psi^*+V\psi^*\right)r\psi\, d\tau\right]$$

$$=\frac{i\hbar}{2m}\int\left[\psi^* r\nabla^2\psi-(\nabla^2\psi^*)\,r\psi\right]d\tau \tag{5.134}$$

But,

$$\int(\nabla^2\psi^*)\,r\psi\, d\tau=\int[\nabla\cdot(r\psi\,\nabla\psi^*)]\,d\tau-\int(\nabla\psi^*)\cdot\nabla(r\psi)\,d\tau \tag{5.135}$$

Again, by *Gauss' divergence theorem*,

$$\int[\nabla\cdot(r\psi\,\nabla\psi^*)\,d\tau]=\int(r\psi\,\nabla\psi^*)_n\cdot d\vec{S} \tag{5.136}$$

where $(r\psi\,\nabla\psi^*)_n$ represents the normal component of the vector $(r\psi\,\nabla\psi^*)$.

Since, $\psi(\vec{r},t)\to 0$ as $\vec{r}\to\pm\infty$

$$\int(r\psi\,\nabla\psi^*)_n\cdot d\vec{S}=0 \tag{5.137}$$

$$\therefore\int(\nabla^2\psi^*)\,r\psi\, d\tau=-\int\nabla\psi^*\cdot\nabla(r\psi)\,d\tau$$

$$=-\left[\int\nabla\cdot(\psi^*\nabla(r\psi))\,d\tau-\int\psi^*\nabla^2(r\psi)\,d\tau\right] \tag{5.138}$$

But, as before

$$\int\nabla\cdot(\psi^*\nabla(r\psi))\,d\tau=0 \tag{5.139}$$

So

$$\int(\nabla^2\psi^*)\,r\psi\, d\tau=\int\psi^*\nabla^2(r\psi)\,d\tau \tag{5.140}$$

Now, substituting eq. (5.140) in eq. (5.134), we get

$$\frac{d\langle\vec{r}\rangle}{dt}=\frac{i\hbar}{2m}\int\left[\psi^*\left\{r\nabla^2\psi-\nabla^2(r\psi)\right\}\right]d\tau=-\frac{i\hbar}{m}\int\psi^*\nabla\psi\, d\tau$$

$$=\frac{1}{m}\int\psi^*(-i\hbar\nabla)\,\psi\, d\tau=\frac{1}{m}\int\psi^*\hat{p}\psi\, d\tau\quad[\because\hat{p}=-i\hbar\nabla]$$

$$=\frac{\langle\vec{p}\rangle}{m} \tag{5.141}$$

Since, $\langle\vec{r}\rangle$ is a real number, $\langle\vec{p}\rangle$ is also real.

Differentiating $\langle \vec{p} \rangle$ as in case of $\langle \vec{r} \rangle$ above we can show that

$$\frac{d\langle \vec{p} \rangle}{dt} = -\langle \nabla V(\vec{r}, t) \rangle \tag{5.142}$$

But, $\langle \nabla V(\vec{r}, t) \rangle$ is the average value of force $\langle \vec{F} \rangle$ acting on the particle moving in potential field $V(\vec{r}, t)$.

Now, according to Newton's second law

$$\langle \vec{F} \rangle = \frac{d}{dt} \langle \vec{p} \rangle \tag{5.143}$$

∴ eq. (5.142) can be written as

$$m \frac{d\langle \vec{r} \rangle}{dt} = -\langle \nabla V(\vec{r}, t) \rangle \tag{5.144}$$

where $\frac{d}{dt} \langle \vec{p} \rangle = m \frac{d}{dt} \langle \vec{r} \rangle$. Eq. (5.144) represents the *Newton's second law of motion*. This proves Ehrenfest's theorem.

This theorem shows that *a wave packet moves like a classical particle whenever the expectation values are considered in the macroscopic limit*. Thus, Ehrenfest theorem establishes the truth of Bohr's correspondence principle because *if the expectation values of dynamical quantities are considered, quantum mechanics gives the values of classical mechanics*.

5.26 Schwarz inequality

The inequality states that

$$\int f^* f d^3\vec{r} \int \phi^* \phi \, d^3\vec{r} \geqslant \tfrac{1}{4} \left[\int (f^* \phi + \phi^* f) \, d^3\vec{r} \right]^2$$

where f and ϕ are *well-behaved* functions so that the above integrals exist.

Proof: Let

$$A = \int f^* f d^3 r; \; B = \int f^* \phi \, d^3 r; \; C = \int \phi^* \phi \, d^3 r$$

Now, we consider the integral

$$\int (\lambda f^* + \phi^*)(\lambda f + \phi) \, d^3 r = A\lambda^2 + (B + B^*)\lambda + C$$

where λ is a real parameter.

Now,

$$\int (\lambda f^* + \phi^*)(\lambda f + \phi) \, d^3 r$$

is a positive definite so that $A\lambda^2 + (B + B^*)\lambda + C > 0$ for all values of λ.

∴ the equation

$$A\lambda^2 + (B + B^*)\lambda + C = 0$$

cannot have any real root and as such

$$4AC \geqslant (B + B^*)^2 \text{ or, } AC \geqslant \tfrac{1}{4}(B + B^*)^2$$

$$\text{or, } \int f^* f d^3 r \int \psi^* \phi \, d^3 r \geqslant \tfrac{1}{4} \left[\int (f^* \phi + \phi^* f) \, d^3 r \right]^2$$

This is **Schwarz Inequality.**

This can also be expressed as

$$\left|f^{*}\phi\, d^{3}r\right|^{2} \leqslant \int f^{*} f d^{3}r \int \phi^{*}\phi\, d^{3}r$$

5.27 Exact statement and proof of the uncertainty principle for wave packets

By analogy with the standard deviation in statistical mechanics the uncertainty in the position of a particle moving in the x-direction is defined by

$$\Delta x = \langle x - \langle x \rangle \rangle^{1/2} \tag{5.145}$$

where $\langle x \rangle$ is the expectation value of x and is given by

$$\langle x \rangle = \int \psi^{*}(x)\, x\psi(x)\, dx \tag{5.146}$$

where $\psi(x)$ is normalised.

In a similar way the uncertainty in momentum is given by

$$\langle p_x \rangle = \langle p_x - \langle p_x \rangle \rangle^{1/2} \tag{5.147}$$

where $\langle p_x \rangle$ is the expectation value of p_x and is given by

$$\langle p_x \rangle = \int \psi^{*}\left(-i\hbar \frac{\partial}{\partial x}\right)\psi\, dx = -i\hbar \int \psi^{*}\frac{\partial \psi}{\partial x}\, dx \tag{5.148}$$

Let the wave function $\psi(x)$ is normalised and initially centred at $x = 0$, then $\langle x \rangle = 0$ and $\langle p_x \rangle = 0$.

Now, consider the integral

$$i\hbar \int_{-\infty}^{+\infty} \psi^{*}\frac{d}{dx}(x\psi)\, dx = i\hbar\left[\psi x \psi^{*}\right]_{-\infty}^{+\infty} - i\hbar \int_{-\infty}^{+\infty} x\psi \frac{d\psi^{*}}{dx}\, dx$$

$$= 0 - i\hbar \int_{-\infty}^{+\infty} x\psi \frac{d\psi^{*}}{dx}\, dx \; [\because \psi \to 0 \text{ as } x \to \pm\infty]$$

$$\text{or, } i\hbar \int \frac{d\psi^{*}}{dx} x\psi\, dx = -i\hbar \int \psi^{*}\frac{d}{dx}(x\psi)\, dx = -i\hbar\left[\int \psi^{*}\frac{d\psi}{dx} x\, dx + \int \psi^{*}\psi\, dx\right]$$

$$= -i\hbar\left[\int \psi^{*}\frac{d\psi}{dx} x\, d\psi + 1\right]\left[\because \int \psi^{*}\psi\, dx = 1 \text{ as } \psi(x) \text{ is normalised}\right]$$

$$\therefore i\hbar \int \frac{d\psi^{*}}{dx} x\psi\, dx = -i\hbar \int \psi^{*}\frac{d\psi}{dx} x\, dx - i\hbar$$

$$\text{or, } i\hbar \int \frac{d\psi^{*}}{dx} x\psi\, dx + i\hbar \int \psi^{*}\frac{d\psi}{dx} x\, dx = -i\hbar \tag{5.149}$$

$$\text{or, } 2i \times \text{imaginary part of } \int i\hbar \frac{d\psi^{*}}{dx} x\psi\, dx = -i\hbar \tag{5.150}$$

or, squaring both sides

$$4\left|\text{imaginary part of } \int i\hbar \frac{d\psi^{*}}{dx} x\psi\, dx\right|^{2} = \hbar^{2} \tag{5.151}$$

But,

$$\left|\int i\hbar\frac{d\psi^*}{dx}x\psi\,dx\right| \geqslant \left|\text{imaginary part of} \int i\hbar\frac{\psi^*}{x}x\psi\,dx\right|.$$

$$\therefore 4\left|\int i\hbar\frac{d\psi^*}{dx}x\psi\,dx\right|^2 \geqslant \hbar^2 \tag{5.152}$$

Now, Schwarz inequality is

$$|f^*\phi\,dx|^2 \leqslant \int f^*f\,dx \int \phi\phi^*dx \tag{5.153}$$

Putting $i\hbar\frac{d\phi^*}{dx} = f^*$ and $\phi = x\psi$ in eq. (5.151), we get

$$4\left|\int i\hbar\frac{d\psi^*}{dx}x\psi\,dx\right|^2 \leqslant \int i\hbar\frac{d\psi^*}{dx}\left(-i\hbar\frac{d\psi}{dx}\right)dx \int x\psi\,x\psi^*\,dx$$

∴ eq. (5.151) can be written as

$$4\int i\hbar\frac{d\psi^*}{dx}\left(-i\hbar\frac{d\psi}{dx}\right)dx \int \psi^*x^2\psi\,dx \geqslant \hbar^2$$

$$\text{or, } \int\left|-i\hbar\frac{d\psi}{dx}\right|^2 \int \psi^*x^2\psi\,dx \geqslant \frac{\hbar^2}{4}$$

$$\text{or, } \langle p_x^2\rangle\langle x^2\rangle \geqslant \frac{\hbar^2}{4} \tag{5.154}$$

$$\because \langle p_x^2\rangle = \int\left|-i\hbar\frac{d\psi}{dx}\right|^2 dx \text{ and } \langle x^2\rangle = \int \psi^*x^2\psi\,dx \tag{5.155}$$

Now,

$$(\Delta x)^2 = \langle x - \langle x\rangle\rangle^2 = \langle x^2\rangle \;[\because \langle x\rangle = 0]$$

$$\text{and } (\Delta p_x)^2 = \langle p_x - \langle p_x\rangle\rangle^2 = \langle p_x^2\rangle \;[\because \langle p_x\rangle = 0]$$

$$\therefore (\Delta x)^2(\Delta p_x)^2 \geqslant \frac{\hbar^2}{4} \text{ or, } \Delta x\,\Delta p_x \geqslant \frac{\hbar}{2} \tag{5.156}$$

This is the *exact statement of the Heisenberg's uncertainty relation.*

5.28 Fundamental postulates of quantum mechanics

In the Schrödinger formulation of quantum mechanics, a complex wave function $\psi(\vec{r}, t)$ has been introduced to represent the de Broglie wave associated with a moving particle. On the basis of the earlier discussion the following postulates are proposed to describe a quantum mechanical system consisting of a single particle:

***Postulate* I**: *A system consisting of a single particle moving in a conservative force field is associated with a complex wave function $\psi(\vec{r}, t)$, which completely describes the space-time behaviour of the particle consistent with the uncertainty principle.*

***Postulate* II**: *Every dynamical variable (e.g., position, velocity, momentum, angular momentum, energy, etc.) of the system is represented by an operator in quantum mechanics. These operators are linear and Hermitian.*

Table 5.2 shows the list of the dynamical variables and their associated operators.

***Postulate* III**: *The only possible values of dynamical variables A that may be experimentally measured are the eigenvalues λ of the operator $\hat{A}$ corresponding to the eigenvalue equation $\hat{A}\psi_n = \lambda_n\psi_n$, where ψ_n is the eigenfunction of $\hat{A}$ belonging to the eigenvalue λ_n.*

The following points must be kept in mind:

(a) The eigenvalue λ_n must be real as the observable physical quantities have real magnitude.

(b) The eigenfunctions ψ_n are well-behaved and form a complete set which is orthogonal.

(c) An observation on a dynamical variable represented by $\hat{A}$ on an eigenstate ψ_n of a system definitely leads to the result λ_n.

***Postulate* IV**: *The probability density of finding a particle at x, y, z at time t is $\rho = \psi^*\psi$, where $\psi^*\psi$ is always real as ρ is a real quantity. The total probability of finding the particle over the entire volume is*

$$P = \int_{\text{all space}} \rho\, d\tau = \int_{\text{all space}} \psi^*\psi\, d\tau = 1$$

since the particle must be somewhere in space.

In this case, the wave function ψ is said to be normalised.

***Postulate* V**: *The average or expectation value of a dynamical variable A associated with the operator $\hat{A}$ is given by*

$$\langle A\rangle = \int_\tau \psi^*\hat{A}\psi\, d\tau$$

if ψ is normalised. If ψ is not normalised, then

$$\langle A\rangle = \frac{\int_\tau \psi^*\hat{A}\psi\, d\tau}{\int_\tau \psi^*\psi\, d\tau}$$

5.29 Schmidt orthogonalisation process

If the wave functions ψ_1 and ψ_2 have the same eigenvalue λ (i.e., if ψ_1 and ψ_2 are degenerated), then they may not be orthogonal.

However, ψ_1 and ψ_2 can be orthogonalised as below:

Let

$$\hat{A}\psi_1 = \lambda\psi_1 \tag{5.157}$$

$$\text{and } \hat{A}\psi_2 = \lambda\psi_2; \tag{5.158}$$

then the linear combination of ψ_1 and ψ_2, i.e., $\psi = \psi_1 + C\psi_2$ obeys the eigenvalue equation

$$\hat{A}(\psi_1 + C\psi_2) = \lambda(\psi_1 + C\psi_2)$$

where C is a constant.

Now, if ψ is orthogonal to ψ_1, then

$$\int \psi_1^* \psi \, d\tau = 0 \text{ or, } \int \psi_1^* (\psi_1 + C\psi_2)\, d\tau = 0 \text{ or, } \int \psi_1^* \psi_1 \, d\tau + C \int \psi_1^* \psi_2 \, d\tau = 0$$

$$\text{or, } C = -\frac{\int \psi_1^* \psi_1 \, d\tau}{\int \psi_1^* \psi_2 \, d\tau} \tag{5.159}$$

Thus, if the value of C is given by the eq. (5.159), then $\psi = (\psi_1 + C\psi_2)$ and ψ_1 are orthogonal.

5.30 Momentum wave function for a free particle

We consider a non-relativistic free particle of mass m moving along the positive x-direction with a velocity v_x. Then its kinetic energy is $E = \frac{1}{2}mv_x^2$ and momentum $p = mv_x$.

The de Broglie wavelength λ and frequency ν associated with the particle are

$$\lambda = \frac{h}{p_x} \text{ and } \nu = \frac{E}{h}$$

The wave propagation vector

$$k_x = \frac{2\pi}{\lambda} = \frac{2\pi}{h/p_x} = \frac{p_x}{h/2\pi} = \frac{p_x}{\hbar}$$

The angular frequency

$$\omega = 2\pi\nu = \frac{2\pi E}{h} = \frac{E}{\hbar}$$

Now, a plane wave travelling along the positive x-direction is given by

$$\psi(x,t) = Ae^{i(k_x x - \omega t)} = Ae^{i(\frac{p_x}{\hbar}x - \frac{E}{\hbar}t)} = Ae^{\frac{i}{\hbar}(p_x x - Et)}$$

If a large number of such waves travel along the positive x-direction simultaneously, then a wave packet is formed. This wave packet of small extension is given by Fourier's theorem

$$\psi(x,t) = \frac{1}{\sqrt{2\pi\hbar}} \int_{-\infty}^{+\infty} A(p_x)\, e^{\frac{i}{\hbar}(p_x x - Et)}\, dp_x \tag{5.160}$$

The function $\psi(x,t)$ is the momentum wave function for the free particle moving along the x-direction. The amplitude $A(p_x)$ of the wave function is given by the Fourier transform

$$A(p_x) = \frac{1}{\sqrt{2\pi\hbar}} \int_{-\infty}^{+\infty} \psi(x,t) e^{-\frac{i}{\hbar}(p_x x - Et)}\, dx \tag{5.161}$$

The momentum wave function for the above particle in three dimensions is

$$\psi(\vec{r},t) = \frac{1}{(2\pi\hbar)^{3/2}} \int_{-\infty}^{+\infty} A(\vec{p}) e^{\frac{i}{\hbar}(\vec{p}\cdot\vec{r} - Et)}\, d^3\vec{p} \tag{5.162}$$

where

$$A(\vec{p}) = \frac{1}{(2\pi\hbar)^{3/2}} \int_{-\infty}^{+\infty} \psi(\vec{r},t) e^{-\frac{i}{\hbar}(\vec{p}\cdot\vec{r} - Et)}\, d^3\vec{r} \tag{5.163}$$

Here, $d^3\vec{p} = dp_x\, dp_y\, dp_z$ is the volume element in momentum space.

5.31 Momentum eigenfunctions

The momentum eigenvalue equation for a particle of momentum p_x moving along the positive x-direction is given by

$$-i\hbar\frac{\partial\psi}{\partial x} = p_x\psi \tag{5.164}$$

where $\psi(x)$ is the momentum eigenfunction for momentum p_x.

The solution of eq. (5.164) is

$$\psi_{p_x}(x) = \frac{1}{\sqrt{2\pi\hbar}}e^{\frac{i}{\hbar}p_x x} \tag{5.165}$$

p_x has a continuous spectrum of real eigenvalues ranging from $-\infty < p_x < +\infty$.

Now, if $\psi(x)$ is orthonormal, then

$$\int_{-\infty}^{+\infty}\psi_{p_x}^*(x)\,\psi_{p_x'}^*(x)\,dx = \delta(p_x - p_x') \tag{5.166}$$

where $\delta(p_x - p_x')$ is the *Dirac delta function.*

The complete set of momentum eigenfunction is given by Fourier's theorem as below:

$$\psi(x) = \frac{1}{(2\pi\hbar)^{3/2}}\int_{-\infty}^{+\infty}A(p_x)\,e^{\frac{i}{\hbar}p_x x}\,dp_x \tag{5.167}$$

For the motion of the particle in three dimensions the eigenvalue equation is

$$-i\hbar\nabla^2\psi(\vec{r}) = p\psi(\vec{r}) \tag{5.168}$$

The normalised solutions which are the momentum eigenfunctions are

$$\psi_p(\vec{r}) = \frac{1}{(2\pi\hbar)^{3/2}}e^{\frac{i}{\hbar}(d^3\cdot\vec{p})} \tag{5.169}$$

The condition for orthonormality of $\psi_p(\vec{r})$ is

$$\int\psi_p^*(\vec{r})\,\psi_{p'}(\vec{r})\,d^3\vec{r} = \delta(\vec{p} - \vec{p}\,') \tag{5.170}$$

$\psi_p(\vec{r})$ forms a complete orthonormal set of momentum eigenfunctions and by Fourier's theorem the momentum eigenfunction for the particle of moment $\vec{p}$ moving in three dimensions is

$$\psi(\vec{r}) = \frac{1}{(2\pi\hbar)^{3/2}}\int A(\vec{p})\,e^{\frac{i}{\hbar}\vec{p}\cdot\vec{r}}\,d^3\vec{p} \tag{5.171}$$

where $A(\vec{p})$ is the amplitude of the momentum wave function $\psi(\vec{r})$ and is given by Fourier transform

$$A(\vec{p}) = \frac{1}{(2\pi\hbar)^{3/2}}\int\psi(\vec{r})\,e^{-\frac{i}{\hbar}\vec{p}\cdot\vec{r}}\,d^2\vec{r} \tag{5.172}$$

5.32 Schrödinger equation in momentum representation

The three-dimensional time-dependent Schrödinger equation for a particle of mass m and momentum $\vec{p}$ moving in a potential $V(\vec{r})$ is given by

$$-\frac{\hbar^2}{2m}\nabla^2\psi(\vec{p}) + V\psi(\vec{p}) = i\hbar\frac{\partial\psi(\vec{p})}{\partial t} \tag{5.173}$$

The plane wave solution of the above equation is

$$\psi_p(\vec{p}) = Ae^{\frac{i}{\hbar}(\vec{p}\cdot\vec{r}-Et)}$$

where E is the total energy of the particle.

The wave packet formed due to the superposition of the above wave function is given by Fourier theorem:

$$\psi(\vec{p}) = \frac{1}{(2\pi\hbar)^{3/2}}\int A(\vec{p})e^{\frac{i}{\hbar}(\vec{p}\cdot\vec{r}-Et)}\,d\vec{p}.$$

$$\therefore \frac{\partial\psi}{\partial t} = \frac{1}{(2\pi\hbar)^{3/2}}\int\left(-\frac{iE}{\hbar}\right)A(\vec{p})e^{\frac{i}{\hbar}(\vec{p}\cdot\vec{r}-Et)}\,d\vec{p} \tag{5.174}$$

$$\text{and } \frac{\partial^2\psi}{\partial x^2} = \frac{1}{(2\pi\hbar)^{3/2}}\int A(\vec{p})\left(-\frac{p_x^2}{\hbar^2}\right)e^{\frac{i}{\hbar}(\vec{p}\cdot\vec{r}-Et)}\,d\vec{p}$$

$$\frac{\partial^2\psi}{\partial y^2} = \frac{1}{(2\pi\hbar)^{3/2}}\int A(\vec{p})\left(-\frac{p_y^2}{\hbar^2}\right)e^{\frac{i}{\hbar}(\vec{p}\cdot\vec{r}-Et)}\,d\vec{p}$$

$$\text{and } \frac{\partial^2\psi}{\partial z^2} = \frac{1}{(2\pi\hbar)^{3/2}}\int A(\vec{p})\left(-\frac{p_z^2}{\hbar^2}\right)e^{\frac{i}{\hbar}(\vec{p}\cdot\vec{r}-Et)}\,d\vec{p}$$

where $p^2 = p_x^2 + p_y^2 + p_z^2$.

$$\therefore \nabla^2\psi(\vec{p}) = \frac{\partial^2\psi}{\partial x^2} + \frac{\partial^2\psi}{\partial y^2} + \frac{\partial^2\psi}{\partial z^2} = \frac{1}{(2\pi\hbar)^{3/2}}\int\left(-\frac{p^2}{\hbar^2}\right)A(\vec{p})e^{\frac{i}{\hbar}(\vec{p}\cdot\vec{r}-Et)}\,d\vec{p} \tag{5.175}$$

Now, from eqs (5.172), (5.173) and (5.174)

$$\frac{1}{(2\pi\hbar)^{3/2}}\int\left(\frac{p^2}{2m} - E\right)A(\vec{p})e^{\frac{i}{\hbar}(\vec{p}\cdot\vec{r}-Et)}\,d\vec{p} + V\psi = 0 \tag{5.176}$$

This is the Schrödinger equation in momentum representation for the three-dimensional motion of a particle moving in potential $V(\vec{r})$.

If the particle is free, i.e., $V(\vec{r}) = 0$, then eq. (5.175) reduces to

$$\frac{1}{(2\pi\hbar)^{3/2}}\int\left(\frac{p^2}{2m} - E\right)A(\vec{p})e^{\frac{i}{\hbar}(\vec{p}\cdot\vec{r}-Et)}\,d\vec{p} = 0 \tag{5.177}$$

5.33 Non-normalisable wave functions

We have the condition of normalisation of the wave function $\psi_1(\vec{r}, t)$, which is also a solution of Schrödinger wave equation as

$$\int\psi_1^*(\vec{r}, t)\psi_1(\vec{r}, t)\,d^3\vec{r} = N^2 \text{ or, } \int|\psi_1|^2\,d^3r = N^2$$

where the value of the integral

$$\int|\psi_1|^2\,d^3r$$

is finite, i.e., N is finite. There are physical problems in which N^2 is infinite. As *for example*, for the wave function of a particle of certain momentum, i.e., $\psi(\vec{r}, t) = Ae^{+i(\vec{k}\cdot\vec{r}-\omega t)}$, we have

$$\int|\psi(\vec{r}, t)|^2\,d^3r = |A|^2\int d^3\vec{r} = \infty$$

since A is a constant and

$$\left|e^{i[(\vec{k}\cdot\vec{r})-\omega t]}\right|^2 = 1$$

This shows that the probability $|A|^2$ of finding the particle everywhere in space is the same. But, this is absurd.

Hence, it is not possible for a quantum mechanical system to be unnormalised. Still due to mathematical advantages such wave functions have found place in quantum mechanics.

5.34 Box normalisation

Let a particle be confined in a large box so that integral

$$\int |\psi_1(\vec{r},t)|\, d^3r = N^2$$

taken over the interior of the box is finite. By multiplying with a suitable constant we can make the norm N^2 of the wave function to be unity. If we normalise $\psi(\vec{r},t)$ and, then integrate over the limit of infinite volume of the box, we get the correct result. This is called the *box normalisation*.

Now, the momentum eigenfunction $\psi_k(\vec{r})$ is expressed as $\psi_k(\vec{r}) = \psi(k)\, e^{i\vec{k}\cdot\vec{r}}$, $\psi_k(\vec{r})$ may have both either continuous range of eigenvalues or a set of discrete eigenvalues. In case of discrete set of eigenvalues, $\psi_k(\vec{r})$ is normalised on an arbitrarily large, but finite cubical box of side L centred at the origin. So

$$\int \psi_k^* \psi_k d^3r = \psi^*(k)\,\psi(k) \int_{-L/2}^{+L/2} e^{-i\vec{k}\cdot\vec{r}} \cdot e^{+i\vec{k}\cdot\vec{r}}\, d^3r = |\psi_k|^2 \int_{-L/2}^{+L/2} d^3r = |\psi_k|^2 L^3 = 1 \text{ or, } \psi(k) = L^{-3/2}$$

$\therefore$ the normalised wave function is $\psi_k(\vec{r}) = L^{-3/2} e^{i\vec{k}\cdot\vec{r}}$.

If a charged particle passes through a periodic potential in a crystal lattice, then the wave function assumes the preassigned value after a translation L on either direction. Therefore, it can be assumed that there is a box of side L whose centre is at the origin. In such a case, the wave function remains unchanged if x, y or z is increased by L.

$\therefore e^{i(\vec{k}\cdot\vec{r})} = e^{i(k_x x + k_y y + k_z z)} = e^{ik_x(x+L)} e^{ik_y(y+L)} e^{ik_z(z+L)}$.

$$\therefore k_x L = 2\pi n_x,\ k_y L = 2\pi n_y,\ k_z L = 2\pi n_z$$

$$\text{or, } k_x = \frac{2\pi n_x}{L},\ k_y = \frac{2\pi n_y}{L},\ k_z = \frac{2\pi n_z}{L}$$

where n_x, n_y, n_z are integers including zero

$$\text{or, } |\vec{k}| = \frac{2\pi}{L} n$$

i.e., the admissible momentum vectors get normalised as a result of box normalisation.

The orthogonality condition in case of box normalisation is, then given by

$$\int \psi_{k'}^*(\vec{r})\, \psi_k(\vec{r})\, dr = \int_{-L/2}^{+L/2} L^{-3} e^{-i\vec{k}'\cdot\vec{r}} e^{i\vec{k}\cdot\vec{r}}\, dr$$

$$= L^{-3} \int_{-L/2}^{+L/2} e^{i(k_x - k_x')x}\, dx \int_{-L/2}^{+L/2} e^{i(k_y - k_y')y}\, dy \int_{-L/2}^{+L/2} e^{i(k_z - k_z')z}\, dz$$

where $dr = dx\, dy\, dz$ in the Cartesian space and $\vec{k}\cdot\vec{r} = k_x x + k_y y + k_z z$.

$$\therefore \int \psi_{k'}^{*}(\vec{r})\,\psi_k(\vec{r})\,dr = \delta_{k_x k_x'}\delta_{k_y k_y'}\delta_{k_z k_z'} = \delta_{\vec{k},\vec{k}'}$$

where

$$\delta_{\vec{k},\vec{k}'} = \begin{cases} 0, & \text{when } k \neq k' \\ 1, & \text{when } k = k' \end{cases}$$

5.35 Linear vector space: Hilbert space and basis vector

Any vector in 3D space can be represented in terms of 3 basis vectors (3 building blocks) $\hat{x} \equiv \hat{e}_1$; $\hat{y} \equiv \hat{e}_2$; $\hat{z} \equiv \hat{e}_3$ which are orthogonal ($\hat{x}\cdot\hat{y} = 0$; $\hat{y}\cdot\hat{z} = 0$; $\hat{z}\cdot\hat{x} = 0$, i.e., $\hat{e}_i \cdot \hat{e}_j = 0$ for $i \neq j$) are normalised ($\hat{x}\cdot\hat{x} = 1$; $\hat{y}\cdot\hat{y} = 1$; $\hat{z}\cdot\hat{z} = 1$ i.e., $\hat{e}_i \cdot \hat{e}_i = 1$). The basis vector $\hat{e}_i$s are orthogonal, i.e., $\hat{e}_i \cdot \hat{e}_j = \delta_{ij}$. Here, the vector space is linear. Any arbitrary vector $\vec{V}$ can be explained as a linear combination of these basis vector as follows:

$$\vec{V} = V_x\hat{x} + V_y\hat{y} + V_z\hat{z} \equiv V_1\hat{e}_1 + V_2\hat{e}_2 + V_3\hat{e}_3 = \sum_{i=1}^{3} V_i\hat{e}_i$$

where $V_x = V_1$, $V_y = V_2$, $V_z = V_3$ are the expansion coefficients and are identified as the projections of $\vec{V}$ along $\hat{x} = \hat{e}_1$, $\hat{y} = \hat{e}_2$, $\hat{z} = \hat{e}_3$. That is the expansion coefficient V_i is the projection of $\vec{V}$ along $\hat{e}_i$.

Taking dot product of $\vec{V}$ with $\hat{e}_j$, we get

$$\hat{e}_j \cdot \vec{V} = \hat{e}_j \cdot \sum_{i=1}^{3} V_i\hat{e}_i = \sum_{i=1}^{3} V_i\hat{e}_i \cdot \hat{e}_j = \sum_{i=1}^{3} V_i\delta_{ij} = V_j$$

$$\therefore \vec{V} = \sum_{i=1}^{3} V_i\hat{e}_i = \sum_{i=1}^{3}(\hat{e}_i \cdot \vec{V})\hat{e}_i = \sum_{i=1}^{3} \hat{e}_i(\hat{e}_i \cdot \vec{V})$$

The basis vectors are represented by column matrices having three rows (in 3D space) as

$$\hat{e}_i = \begin{pmatrix} 1 \\ 0 \\ 0 \end{pmatrix}, \hat{e}_2 = \begin{pmatrix} 0 \\ 1 \\ 0 \end{pmatrix}, \hat{e}_3 = \begin{pmatrix} 0 \\ 0 \\ 0 \end{pmatrix}; \quad \vec{V}_1 = \begin{pmatrix} V_x \\ V_y \\ V_z \end{pmatrix} \equiv \begin{pmatrix} V_1 \\ V_2 \\ V_3 \end{pmatrix}$$

The dot product (or scalar product) is defined as the matrix product of a row matrix and a column matrix. The vector occurring before the dot is a row matrix and the vector occurring after the dot is a column matrix, i.e.,

$$\vec{A}\cdot\vec{B} = (A_1\ A_2\ A_3)\begin{pmatrix} B_1 \\ B_2 \\ B_3 \end{pmatrix} = A_1B_1 + A_2B_2 + A_3B_3 = \text{a number}$$

This can be generalised to vector belonging to N-dimensional space where basis vectors are represented by column matrices having N rows as

$$\hat{e}_1 = \begin{pmatrix} 1 \\ 0 \\ \vdots \\ 0 \end{pmatrix}, \quad \hat{e}_2 = \begin{pmatrix} 0 \\ 1 \\ 0 \\ \vdots \\ 0 \end{pmatrix}, \quad \ldots, \quad \hat{e}_N = \begin{pmatrix} 0 \\ 0 \\ \vdots \\ 1 \end{pmatrix}$$

and any arbitrary vector is $\vec{V} = \begin{pmatrix} V_1 \\ V_2 \\ \vdots \\ V_n \end{pmatrix}$.

An infinite dimensional linear vector space can be defined using a set of vector functions $\psi_1, \psi_2, \psi_3, \ldots$ as orthogonal basis $(\psi_i^+ \psi_j)$ which are in general complex and represent eigenfunction or eigenvector of an operator $\hat{a}$ corresponding to an observable A. This infinite-dimensional linear vector space is called the *Hilbert space.*

The dimension of Hilbert space has nothing to do with the physical dimension of the system.

State vector and wave function

Let us define a state vector in infinite-dimensional linear vector space (i.e., Hilbert space) which contains complete information about the system and is denoted by the symbol $|\psi\rangle$ (called ket ψ) which is a column matrix. It is such that its projection in coordinate space is $C(p)$ called the *momentum space* wave function.

In 3D space complete information is contained in the vector $\vec{A}$. It is such that its projection along X-axis ($\hat{x}$ unit vector) is A_x; $\vec{A}$ and A_y are interrelated as $A_x = \hat{x} \cdot \vec{A}$; $A_y = \hat{y} \cdot \vec{A}$ and so on.

ket ψ or $|\psi\rangle$ is like $\vec{A}$; $\psi(x), C(p)$ are like A_x, A_y.

We define $|\psi\rangle^+ = \langle\psi|$ called bra ψ which is a row matrix

$$|\phi\rangle^+ \langle\psi| = \langle\phi||\psi\rangle = \langle\phi|\psi\rangle = \int \phi^*(x)\,\psi(x)\,dx$$

is called the *scalar product* of the 2 kets $|\psi\rangle$ and $\langle\psi|$.

$$\therefore \psi(x) = \langle x \,|\, \psi\rangle \rightarrow \text{projection of } |\tau\rangle \text{ on } |x\rangle$$

$$\text{and } c(p) = \langle p \,|\, \psi\rangle \rightarrow \text{projection of } |\tau\rangle \text{ on } |p\rangle$$

The ket $|\ \rangle$ and bra $\langle\ |$ notations are elegant and compact introduced by *Dirac.*

5.36 Answers of some important conceptual questions

5.1. *What is an observable (or dynamical variable)?—Explain.*

Ans. A quantity obtained by the process of observation or measurement on a physical system is called an observable. The result of the measurement of an observable in a system is expressed by a number. An observable is always a real entity as it is the result of actual measurement.

The act of measurement disturbs a physical system. Therefore, the value of an observable depends upon the interaction between the system and the measuring device and as such spread around some mean value. The value of an observable generally depends upon time. So time is a universal parameter but not itself observable. The value of an observable is a *constant* of the system if it does not change with time, e.g., the physical constants like the rest mass and charge on the electron, photon, etc. belong to this class of observables.

Classical mechanics assumed that

(i) the process of measurement does not disturb the system

(ii) all observables of a system are characteristic of the system irrespective of the measuring device

(iii) various observables of the system can be measured with utmost accuracy.

On the other hand, in quantum mechanics

(i) the value of an observable, though characteristic of the system, depends upon the measuring device

(ii) in view of Heisenberg's uncertainty principle, simultaneous measurement of conjugate variables like position and momentum, time and energy have a fundamental error of measurement

(iii) observables like position coordinates components of linear momentum, the total energy though have the same physical meaning as in classical mechanics, observables like spin, symmetry, etc. hence no equivalence in classical mechanics.

5.2. *Explain the term operator. What is a Hermitian operator?*

Ans. An operator is a mathematical rule (or procedure) which operating on one function transforms it into another function.

For example, if A is an operator (represented by $\hat{A}$) and $\psi(x)$ is a function of x, then $\hat{A}\,\psi(x) = \phi(x)$, where $\phi(x)$ is the transformed function.

Mathematical operations in algebra and calculus like adding, subtracting, multiplying, dividing finding the square root, differentiating w.r.t. a variable or integration are respectively represented by characteristic symbols like $+, -, \times, \div, \sqrt{\ }, \frac{\partial}{\partial x}, \frac{d}{dt}$ and $\int f\,dx$ can be consider as operators.

Hermitian operator: An operator $\hat{A}$ is called Hermitian if it satisfies the equation

$$\int_{-\infty}^{+\infty} f^*(\hat{A}g)\,dx = \int_{-\infty}^{+\infty} (\hat{A}f)^* g\,dx$$

where $f(x)$ and $g(x)$ are any two well-behaved functions of x.

5.3. *Why the Schrödinger's wave equation is not valid for relativistic particle?*

Ans. Schrödinger's wave equation is not valid for classical particles because in deriving this equation, we use classical (not-relativistic) expression for total energy, $E = \frac{p^2}{2m} + V$. We also take the momentum of the particle as non-relativistic and equal to mv. The kinetic energy is taken as $\frac{1}{2}mv^2$, which is also a non-relativistic expression.

5.4. *What are eigenfunctions and eigenvalues?*

Ans. See Art. 5.16.

5.5. *What is a well-behaved function?*

A wave function $\phi(x)$ is called well-behaved if it satisfies the following conditions:

(i) $\psi(x)$ must be everywhere single valued and continuous

(ii) $\int |\psi(x)|^2 d\tau$ must be finite, i.e., it must be square integrable

(iii) $\psi(x) \to$ finite as $x \to \pm\infty$

(iv) ψ and $\frac{d\psi}{dx}$ must be continuous everywhere.

5.6. *What do you mean by expectation value of a dynamical variable?*

Ans. See Art. 5.24.

5.7. *Why cannot we represent matter waves associated with a free particle by a wave function $\psi(x,t) = A\sin(\omega t - kx)$?*

Ans. The wave function $\psi(x,t) = A\sin(\omega t - ix)$ represents a wave of wavelength λ or wavenumber $k = 2\pi/\lambda$ and frequency ν (or angular frequency $\omega = 2\pi\nu$) associated with a particle travelling along positive X-direction. Hence, a wave travelling in the negative will be associated with a wave $\psi(x,t) = A\sin(\omega t + kx)$.

According to the principle of superposition, the wave function can also be represented by

$$\psi(x,t) = A\sin(\omega t - kx) + A\sin(\omega t + kx) \quad \text{(i)}$$

$$= 2A\sin\omega t\cos kx \quad \text{(ii)}$$

Hence, the wave function $\psi(x,t) = 2A\sin\omega t\cos kx$ should also be a positive quantum mechanical wave function. But, the wave function represented by (i) is not a stationary wave function because at $t = 0$, *it vanishes identically everywhere, i.e., the probability of finding the particle is zero everywhere.* Since, the particle cannot disappear, the wave function (i) is not acceptable.

Thus, we conclude that $\psi(x,t) = A\sin(\omega t - ix)$ does not represent a matter wave associated with a free particle.

5.8. *The operator for z-component of angular momentum is $L_z = -i\hbar\frac{\partial}{\partial\phi}$, determine whether or not $\sin m\phi$ is its eigenfunction.*

Ans. Let $f = \sin m\phi$ and $\hat{L}_z = -i\hbar\frac{\partial}{\partial\phi}$.

$$\therefore \hat{A}f = \hat{L}\sin m\phi = -i\hbar\frac{\partial}{\partial\phi}\sin m\phi = -i\hbar m\cos m\phi$$

As $L_z \sin m\phi \neq$ *a* constant $\times \sin m\phi$, $\sin m\phi$ is not an eigenfunction of angular momentum operator, $\hat{L}_z = -i\hbar\frac{\partial}{\partial\phi}$.

5.9. *Show that normalisation is independent of time.*

Ans. The wave function can be written as

$$\psi(x,t) = \psi(x)f(t) = \psi(x)e^{-iEt/\hbar} \; [\because f(t) = e^{-iEt/\hbar}]$$

Now, according to the condition of normalisation

$$\int_{-\infty}^{+\infty}\psi(x)e^{-iEt/\hbar}\psi^* e^{iEt/\hbar} = 1 \text{ or, } \int_{-\infty}^{+\infty}\psi(x)\psi^*(x) = 1$$

As $\psi(x)$ and $\psi^*(x)$ are functions of x only and are independent of time, the probability of locating a system in the region $-\infty$ to $+\infty$ continues to be one for all times, i.e., normalisation is independent of time.

5.37 Solved problems

» ***Example* 5.1.** Show that $\left[x, \frac{\partial^2}{\partial x^2}\right] = -2\frac{\partial}{\partial x}$.

Solution: We have

$$\left[x, \frac{\partial^2}{\partial x^2}\right] = \left[x, \frac{\partial}{\partial x}\cdot\frac{\partial}{\partial x}\right] = \left[x, \frac{\partial}{\partial x}\right]\frac{\partial}{\partial x} + \frac{\partial}{\partial x}\left[x, \frac{\partial}{\partial x}\right] = (-1)\frac{\partial}{\partial x} + \frac{\partial}{\partial x}(-1) \quad \left[\because x, \frac{\partial}{\partial x} = -1\right]$$

$$= -2\frac{\partial}{\partial x}$$

» ***Example* 5.2.** If A and B are two operators, then show that $[A, B^{-1}] = -B^{-1}[A, B]B^{-1}$.

Solution: We have the right-hand side

$$-B^{-1}[A, B]B^{-1} = -B^{-1}(AB - BA)B^{-1} = -B^{-1}\left(ABB^{-1} - BAB^{-1}\right) = -B^{-1}ABB^{-1} + B^{-1}BAB^{-1}$$

$$= -B^{-1}A + AB^{-1} = AB^{-1} - B^{-1}A = [A, B^{-1}] = \text{left-hand side}$$

» ***Example* 5.3.** Show that $\left(\frac{\partial}{\partial x} + \frac{\partial}{\partial y}\right)^2 = \frac{\partial^2}{\partial x^2} + 2\frac{\partial^2}{\partial x \partial y} + \frac{\partial^2}{\partial y^2}$

Solution: Let ψ be an arbitrary function. Then

$$\left(\frac{\partial}{\partial x} + \frac{\partial}{\partial y}\right)^2 \psi = \left(\frac{\partial}{\partial x} + \frac{\partial}{\partial y}\right)\left(\frac{\partial}{\partial x} + \frac{\partial}{\partial y}\right)\psi = \left(\frac{\partial}{\partial x} + \frac{\partial}{\partial y}\right)\left(\frac{\partial \psi}{\partial x} + \frac{\partial \psi}{\partial y}\right) = \frac{\partial^2 \psi}{\partial x^2} + \frac{\partial^2 \psi}{\partial x \partial y} + \frac{\partial^2 \psi}{\partial y \partial x} + \frac{\partial^2 \psi}{\partial y^2}$$

$$= \frac{\partial^2 \psi}{\partial x^2} + 2\frac{\partial^2 \psi}{\partial x \partial y} + \frac{\partial^2 \psi}{\partial y^2} = \left(\frac{\partial^2}{\partial x^2} + 2\frac{\partial^2}{\partial x \partial y} + \frac{\partial^2}{\partial y^2}\right)\psi$$

$\because \psi$ is arbitrary

$$\therefore \left(\frac{\partial}{\partial x} + \frac{\partial}{\partial y}\right)^2 = \frac{\partial^2}{\partial x^2} + 2\frac{\partial^2}{\partial x \partial y} + \frac{\partial^2}{\partial y}$$

» ***Example* 5.4.** If $\hat{A}$ and $\hat{B}$ are two operators which commute with their commutator $[\hat{A}, \hat{B}]$, prove that $[\hat{A}, \hat{B}^n] = n\hat{B}^{n-1}[\hat{A}, \hat{B}]$.

Solution: We have $\hat{A}[\hat{A}, \hat{B}] = [\hat{A}, \hat{B}]\hat{A}$, $\hat{B}[\hat{A}, \hat{B}] = [\hat{A}, \hat{B}]\hat{B}$.

The result is true for $n = 1$.

Also $[\hat{A}, \hat{B}^2] = [\hat{A}\hat{B}\hat{B}] = \hat{B}[\hat{A}, \hat{B}] + [\hat{B}, \hat{B}]\hat{B} = \hat{B}[\hat{A}, \hat{B}] + \hat{B}[\hat{A}, \hat{B}] = 2B^{2-1}[\hat{A}, \hat{B}]$.

The result is true for $n = 2$.

Now, let this be true for $n = k$, then $[\hat{A}, \hat{B}^k] = k\hat{B}^{k-1}[\hat{A}, \hat{B}]$

$$\text{and } [\hat{A}, \hat{B}^{k+1}] = [\hat{A}, \hat{B}^k]\hat{B} + \hat{B}^k[\hat{A}, \hat{B}] = k\hat{B}^{k-1}[\hat{A}, \hat{B}] + \hat{B}^k[\hat{A}, \hat{B}]$$

$$= k\hat{B}^k[\hat{A}, \hat{B}] + \hat{B}^k[\hat{A}, \hat{B}] = (k+1)\hat{B}^{k+1-1}[\hat{A}, \hat{B}]$$

Thus, the result is true for $n = k + 1$.

Hence, for any positive number n, $[\hat{A}, \hat{B}^n] = n\hat{B}^{n-1}[\hat{A}, \hat{B}]$.

» ***Example* 5.5.** Show that $[x^n, \hat{p}_x] = -i\hbar n x^{n-1}$, n is a positive integer. **[CU (H) 2011; KU (H) 2014]**

Solution: If ψ is a state function, then

$$\hat{p}_x x^n \psi = (-i\hbar)\frac{\partial}{\partial x}(x^n \psi) = (-i\hbar)\, n x^{n-1}\psi + (-i\hbar)\, x^n \frac{\partial \psi}{\partial x}$$

$$\therefore [x^n, \hat{p}_x] = [-i\hbar]nx^{n-1}$$

» ***Example* 5.6.** If $\hat{H} = \frac{p^2}{2m} + V(x)$, then show that $\left[x, \left[x, \hat{H}\right]\right] = -\frac{\hbar^2}{m}$.

***Solution*:** We know,

$$[x, \hat{H}] = \left[x, \frac{p^2}{2m}\right] + [x, V(x)] = \frac{1}{2m}[x, p^2] + 0 = \frac{1}{2m} \cdot 2i\hbar p = \frac{i\hbar}{m}p$$

$$\therefore \left[x, \left[x, \hat{H}\right]\right] = \left[x, \frac{i\hbar}{m}p\right] = \frac{i\hbar}{m}[x, p] = \frac{i\hbar}{m}i\hbar = -\frac{\hbar^2}{m}$$

» ***Example* 5.7.** A particle is in the state described by the eigenfunction ψ of the operator $\hat{\alpha}$ that does not depend explicitly upon time. Show that the corresponding eigenvalue λ does not vary with time if $\hat{\alpha}$ commutes with the Hamiltonian operator.

***Solution*:** Here, the eigenvalue equation is $\hat{\alpha}\psi = \lambda\psi$.

$$\therefore \frac{\partial \hat{\alpha}}{\partial t}\psi + \hat{\alpha}\frac{\partial \psi}{\partial t} = \frac{\partial}{\partial t}(\lambda\psi) = \frac{\partial \lambda}{\partial t}\psi + \lambda\frac{\partial \psi}{\partial t}$$

If $\frac{\partial \hat{\alpha}}{\partial t} = 0$, then $\hat{\alpha}\frac{\partial \psi}{\partial t} = \frac{\partial \lambda}{\partial t}\psi + \lambda\frac{\partial \psi}{\partial t}$.

Again,

$$-i\hbar\frac{\partial \psi}{\partial t} = \hat{H}\psi \text{ or, } \frac{\partial \psi}{\partial t} = -\frac{1}{i\hbar}\hat{H}\psi = \frac{i}{\hbar}\hat{H}\psi$$

$$\therefore \frac{\partial \lambda}{\partial t}\psi = \frac{i}{\hbar}\left(\hat{\alpha}\hat{H} - \lambda\hat{H}\right)\psi$$

If $\hat{\alpha}$ and $\hat{H}$ commute, then $\hat{\alpha}\hat{H} = \hat{H}\hat{\alpha}$ or, $\hat{\alpha}\hat{H}\psi = \hat{H}\hat{\alpha}\psi = \lambda\hat{H}\psi$.

$$\therefore \frac{\partial \lambda}{\partial t}\psi = \frac{i}{\hbar}\left(\lambda\hat{H}\psi - \lambda\hat{H}\psi\right) = \frac{i}{\hbar}\left(\lambda\hat{H} - \lambda\hat{H}\right)\psi = 0$$

$\because \psi \neq 0$, $\therefore \frac{\partial \lambda}{\partial t} = 0$, i.e., the eigenvalue λ does not vary with time.

» ***Example* 5.8.** If $\hat{\alpha}$ and $\hat{\beta}$ are two operators which commute with their commutator $\left[\hat{\alpha}, \hat{\beta}\right]$, then prove that $e^{\hat{\alpha}}e^{\hat{\beta}} = e^{\hat{\alpha}+\hat{\beta}+\frac{1}{2}[\hat{\alpha},\hat{\beta}]}$.

***Solution*:** We assume that $f(\lambda) = e^{\lambda\hat{\alpha}}e^{\lambda\hat{\beta}}$.

Then

$$\frac{df}{d\lambda} = \hat{\alpha}e^{\lambda\hat{\alpha}}e^{\lambda\hat{\beta}} + e^{\lambda\hat{\alpha}}\hat{\beta}e^{\lambda\hat{\beta}} = \hat{\alpha}e^{\lambda\hat{\alpha}}e^{\lambda\hat{\beta}} + e^{\lambda\hat{\alpha}}\hat{\beta}e^{-\lambda\hat{\alpha}}e^{\lambda\hat{\alpha}}\hat{\beta}e^{\lambda\hat{\beta}} \tag{i}$$

Again,

$$e^{\lambda\hat{\alpha}}\hat{\beta} = \left(1 + \lambda\hat{\alpha} + \frac{\lambda^2}{2!}\hat{\alpha}^2 + \frac{\lambda^3}{3!}\hat{\alpha}^3 + \ldots\right)\hat{\beta} = \hat{\beta} + \lambda\hat{\alpha}\hat{\beta} + \frac{\lambda^2}{2!}\hat{\alpha}^2\hat{\beta} + \frac{\lambda^3}{3!}\hat{\alpha}^3\hat{\beta} + \ldots \tag{ii}$$

But,

$$[\hat{\alpha}^n, \hat{\beta}] = n\hat{\alpha}^{n-1}[\hat{\alpha}, \hat{\beta}] \text{ or, } \hat{\alpha}^n\hat{\beta} = \hat{\beta}\hat{\alpha} + n\hat{\alpha}^{n-1}\left[\hat{\alpha}, \hat{\beta}\right] \tag{iii}$$

Using eq. (iii) in eq. (ii), we get

$$e^{\lambda\hat{\alpha}}\hat{\beta} = \hat{\beta} + \lambda\left(\hat{\alpha}\hat{\beta} + [\hat{\alpha},\hat{\beta}]\right) + \frac{\lambda^2}{2!}\left(\hat{\beta}\hat{\alpha}^2 + 2\hat{\alpha}[\hat{\alpha},\hat{\beta}]\right)$$

$$= \hat{\beta} + \lambda\hat{\beta}\hat{\alpha} + \frac{\lambda^2}{2!}\hat{\beta}\hat{\alpha}^2 + \cdots + \lambda[\hat{\alpha},\hat{\beta}] + \lambda^2\hat{\alpha}[\hat{\alpha},\hat{\beta}] + \ldots$$

$$= \hat{\beta}\left(1 + \lambda\hat{\alpha} + \frac{\lambda^2}{2!}\hat{\alpha}^2 + \ldots\right) + \lambda[\hat{\alpha},\hat{\beta}]\left(1 + \lambda\hat{\alpha} + \frac{\lambda^2}{2!}\hat{\alpha}^2 + \ldots\right)$$

$$= \hat{\beta}e^{\lambda\hat{\alpha}} + \lambda[\hat{\alpha},\hat{\beta}]e^{\lambda\hat{\alpha}} \tag{iv}$$

But, $e^{\lambda\hat{\beta}}\hat{\beta}e^{-\lambda\hat{\alpha}} = \hat{\beta} + \lambda[\hat{\alpha},\hat{\beta}]$.

Now, using the above relation in eq. (i), we get

$$\frac{df}{d\lambda} = \hat{\alpha} + \hat{\beta} + \lambda[\hat{\alpha},\hat{\beta}]f(\lambda), \quad \frac{df}{f(\lambda)} = \left(\hat{\alpha} + \hat{\beta} + \lambda[\hat{\alpha},\hat{\beta}]\,d\lambda\right)$$

Integrating w.r.t. λ

$$\ln f(\lambda) = (\hat{\alpha}+\hat{\beta})\lambda + [\hat{\alpha},\hat{\beta}]\frac{\lambda^2}{2} \text{ or, } f(\lambda) = e^{\left\{(\hat{\alpha}+\hat{\beta})\lambda + [\hat{\alpha},\hat{\beta}]\frac{\lambda^2}{2}\right\}} \text{ or, } e^{\lambda\hat{\alpha}}e^{\lambda\hat{\beta}} = e^{\left\{(\hat{\alpha}+\hat{\beta})\lambda + [\hat{\alpha},\hat{\beta}]\frac{\lambda^2}{2}\right\}}$$

Putting $\lambda = 1$, we get

$$e^{\hat{\alpha}}e^{\hat{\beta}} = e^{\left[\hat{\alpha}+\hat{\beta}+\frac{1}{2}(\hat{\alpha},\hat{\beta})\right]}$$

» ***Example* 5.9.** Show that (i) $\hat{L}_z$ (ii) $\hat{p}_x$ (iii) $\hat{H}$ is Hermitian.

Solution: If $\hat{A}$ is an Hermitian operator, then we know that

$$\int \psi^*(x)\hat{A}\phi\,d\tau = \int \phi\hat{A}^*\psi^*(x)\,d\tau$$

(i) Now,

$$\hat{L}_z = -i\hbar\frac{\partial}{\partial\phi}$$

where ϕ ranges from 0 to 2π.

If $\psi(x)$ and $g(x)$ be two functions, then

$$\int_0^{2\pi}\psi^*\hat{L}_z g\,d\phi = \int_0^{2\pi}\psi^*\left(-i\hbar\frac{\partial}{\partial\phi}\right)g\,d\phi = -i\hbar\int_0^{2\pi}\psi^*\frac{\partial g}{\partial\phi}\,d\phi$$

Integrating by parts, we get

$$\int_0^{2\pi}\psi^*\hat{L}_z g\,d\phi = -i\hbar\left[\psi^* g\right]_0^{2\pi} + i\hbar\int_0^{2\pi}g\frac{\partial\psi^*}{\partial\phi}\,d\phi = 0 + i\hbar\int_0^{2\pi}g\frac{\partial\psi^*}{\partial\phi}\,d\phi$$

$\because \psi^*(2\pi) = \psi^*(0)$ and $g(2\pi) = g(0)$ { Because $\psi(x)$ and $g(x)$ are single-valued.

So

$$\int_0^{2\pi}\psi^*\hat{L}_z g\,d\phi = \int_0^{2\pi}g\left(i\hbar\frac{\partial}{\partial\phi}\right)\psi^*\,d\phi = \int_0^{2\pi}g\hat{L}_z\psi^*\,d\phi$$

$\therefore \hat{L}_z$ is an Hermitian operator.

(ii) We know,

$$\hat{p}_z = -i\hbar\frac{\partial}{\partial x}$$

$$\therefore \int_{-\infty}^{+\infty} \phi^* \hat{p}_x g\, dx = \int_{-\infty}^{+\infty} \psi^* \left(-i\hbar\frac{\partial}{\partial x}\right) g\, dx = -i\hbar \int_{-\infty}^{+\infty} \psi^* \frac{\partial g}{\partial x} dx$$

$$= -i\hbar\,[\psi^* g]_{-\infty}^{+\infty} + i\hbar \int_{-\infty}^{+\infty} g\frac{\partial \psi^*}{\partial x} dx = 0 + i\hbar \int_{-\infty}^{+\infty} g\frac{\partial \phi^*}{\partial x} dx$$

$\because \psi \to 0$ as $x \to \pm\infty$ and $g \to 0$ as $x \to \pm\infty$

$$\therefore \int_{-\infty}^{+\infty} \psi^* \hat{p}_x g\, dx = \int_{-\infty}^{+\infty} g\left(i\hbar\frac{\partial}{\partial x}\right)\psi^* dx = \int_{-\infty}^{+\infty} g\hat{p}_x^* \psi^* dx$$

$\therefore \hat{p}_x$ is Hermitian.

(iii) We have the Hamiltonian operator

$$\hat{H} = -\frac{\hbar^2}{2m}\nabla^2 + V(\vec{r})$$

where $V(\vec{r})$ is real.

If $\psi(\vec{r})$ is normalised, then

$$\int \psi^*(r)\hat{H}\psi(r)\, d\tau = \int \psi^*\left(-\frac{\hbar^2}{2m}\nabla^2 + V\right)\psi\, d\tau = -\frac{\hbar^2}{2m}\int \psi^*\nabla^2\psi\, d\tau + \int \psi^* V\psi\, d\tau \quad \text{(i)}$$

Now,

$$\int_\tau \psi^*\nabla^2\psi\, d\tau = \int_\tau \nabla\cdot(\psi^*\nabla\psi)\, d\tau - \int_\tau \nabla\psi^* \cdot \nabla\psi\, d\tau = \oint_S \psi^*\nabla\psi\cdot d\vec{S} - \int_\tau \nabla\psi^*\cdot\nabla\psi\, d\tau \quad \text{(ii)}$$

If the surface S of volume τ is ∞, at $r \to \pm\infty$, then ψ and ψ^* vanish.

$$\therefore \int_\tau \psi^*\nabla^2\psi\, d\tau = -\int \nabla\psi^*\cdot\nabla\psi\, d\tau = \int \nabla\cdot(\psi\nabla\psi^*)\, d\tau + \int \psi\nabla^2\psi^*\, d\tau$$

$$= -\oint_S \psi\nabla\psi^*\cdot d\vec{S} + \int_\tau \psi\nabla^2\psi^*\, d\tau = 0 + \int_\tau \psi\nabla^2\psi^*\, d\tau \quad \text{(iii)}$$

If V is real, then

$$\int \psi^* V\psi\, d\tau = \int \psi V\psi^*\, d\tau \quad \text{(iv)}$$

From eqs (i), (iii) and (iv), we get

$$\int \psi^*\hat{H}\psi\, d\tau = -\frac{\hbar^2}{2m}\int \psi\nabla^2\psi^*\, d\tau + \int \psi V^*\psi^*\, d\tau$$

$$\therefore \int \psi^*\left(-\frac{\hbar^2}{2m}\nabla^2 + V\right)\psi\, d\tau = \int \psi\left(-\frac{\hbar^2}{2m}\nabla^2 + V\right)^*\psi^*\, d\tau \quad \text{(v)}$$

$$\therefore \hat{H} = \left[-\frac{\hbar^2}{2m}\nabla^2 + \hat{V}\right] \text{ is an Hermitian operator}$$

» ***Example* 5.10.** Show that the expectation value of H is the total energy E of the system.

Solution: We have the expectation value of H is

$$\langle \hat{H} \rangle = \int \psi^* \hat{H} \psi \, d\tau$$

where ψ is a wave function and $d\tau$ is the volume element.

$$\therefore \langle \hat{H} \rangle = \int \psi^* E \psi \, d\tau = E \int \psi^* \psi \, d\tau = E$$

$\because \hat{H}\psi = E\psi$ and $\int \psi^* \psi \, d\tau = 1$ as ψ is normalised.

$\therefore \langle \hat{H} \rangle$ is the total energy E of the system which is real.

» ***Example* 5.11.** Show that the eigenvalue of an Hermitian operator is real.

Solution: Let $\psi(x)$ be an eigenfunction of the Hermitian operator $\hat{A}$ related to eigenvalue λ.

Then $\hat{A}\psi = \lambda\psi$ and $\hat{A}^* \psi^* = \lambda^* \psi^*$.

$$\therefore \int \psi^* \hat{A} \psi \, dx = \int \psi^* \lambda \psi \, dx = \lambda \int \psi^* \psi \, dx$$

$$\text{and} \int \psi \hat{A}^* \psi^* \, dx = \int \psi \lambda^* \psi^* \, dx = \lambda^* \int \psi \psi^* \, dx$$

$$\because \int \psi^* \hat{A} \psi \, dx = \int \psi \hat{A}^* \psi^* \, dx \text{ as } \hat{A} \text{ is Hermitian}$$

$$\therefore \lambda \int \psi^* \psi \, dx = \lambda^* \int \psi \psi^* \, dx \text{ or, } \lambda = \lambda^*$$

$\therefore \lambda$ is real, i.e., the eigenvalue of an Hermitian operator is real.

» ***Example* 5.12.** Show that the component of angular momentum about any axis forms a discrete eigenspectrum.

Solution: If $\hat{L}_z$ be the z-component of the angular momentum operator and λ the corresponding eigenvalue, then according to quantum mechanics

$$\hat{L}_z \psi = \lambda \psi \text{ or, } -i\hbar \frac{\partial \psi}{\partial \phi} = \lambda \psi \text{ or, } \frac{\partial \psi}{\partial \phi} = -\frac{1}{i\hbar} \lambda \psi = \frac{i}{\hbar} \lambda \psi \text{ or, } \psi = e^{\frac{i}{\hbar}(\lambda \phi)} = \cos\left(\frac{\lambda}{\hbar}\phi\right) + i \sin\left(\frac{\lambda}{\hbar}\phi\right) \quad \text{(i)}$$

As ψ is single-valued function

$$\therefore e^{\frac{i\lambda}{\hbar}\phi} = e^{\frac{i\lambda}{\hbar}(\psi + 2\pi)}.$$

$$\therefore e^{ik\phi} = e^{ik(\phi+2\pi)} \quad \left[k = \frac{\lambda}{\hbar}\right]$$

$$\text{i.e., } e^{2\pi i k} = 1 \text{ or, } \cos 2\pi k + i \sin 2\pi k = 1 \quad \text{(ii)}$$

Equation (ii) is valid, if

$$k = 0, \pm 1, \pm 2, \ldots, \pm n$$

$$\text{or, } \lambda = k\hbar = 0, \pm\hbar, \pm 2\hbar, + \cdots + n\hbar \quad \text{(iii)}$$

The above eq. (iii) shows that λ must be zero or has a discrete set of values.

» ***Example* 5.13.** Show that in absence of any restriction, the component of linear momentum along any axis forms a continuous eigenspectrum.

Solution: We have the eigenvalue equation

$$\hat{p}_x \psi = \lambda \psi \quad \text{(i)}$$

where $\hat{p}_x$ is the x-component of linear momentum operator associated with eigenvalue λ and ψ is the eigenfunction relating to $\hat{p}_x$.

$$\therefore -i\hbar\frac{d\psi}{dx} = \lambda\psi \qquad \text{(ii)}$$

Solving the above equation, we get the eigenfunction

$$\psi = Ce^{\frac{i}{\hbar}\lambda x} \qquad \text{(iii)}$$

For ψ to be a quantum mechanical wave function λ must be real. So $\lambda = p$, where p is any real number.

Hence, the x-component of linear momentum operator has a continuous infinite spectrum of eigenvalues.

» ***Example* 5.14.** Deduce the equation of motion in quantum mechanics.

Hence, show that for a conservative system, the energy is a constant of motion in quantum mechanics.

***Solution*:** We consider a dynamical variable A whose associated operator is $\hat{A}$. The expectation value of A is

$$\langle A\rangle = \int \psi^*(x,t)\hat{A}\psi(x,t)\,dx$$

$$\therefore \frac{d}{dt}\langle A\rangle = \frac{d}{dt}\int \psi^*(x,t)\hat{A}\psi(x,t)\,dx = \int\left[\frac{\partial\psi^*}{\partial t}\hat{A}\psi + \psi^*\frac{\partial\hat{A}}{\partial t}\psi + \psi^*\hat{A}\frac{\partial\psi}{\partial t}\right]dx \qquad \text{(i)}$$

Now, according to Schrödinger's equation

$$i\hbar\frac{\partial\psi}{\partial t} = \hat{H}\psi \text{ or, } \frac{\partial\psi}{\partial t} = -\frac{i}{\hbar}\hat{H}\psi \text{ and } \frac{\partial\psi^*}{\partial t} = \frac{i}{\hbar}\hat{H}^*\psi^*$$

$$\therefore \frac{d}{dt}\langle A\rangle = \frac{i}{\hbar}\int\{(\hat{H}^*\psi^*)\hat{A}\psi - \psi^*\hat{A}(\hat{H}\psi)\}\,dx + \int\psi^*\frac{\partial\hat{A}}{\partial t}\psi\,dx$$

But, $\hat{H}$ is an Hermitian operator.

$$\therefore \int(\hat{H}^*\psi^*)\hat{A}\psi\,dx = \int(\hat{A}\psi)\hat{H}^*\psi^*\,dx = \int\psi^*\hat{H}\hat{A}\psi\,dx$$

Hence,

$$\frac{d}{dt}\langle A\rangle = \frac{i}{\hbar}\int\psi^*(\hat{H}\hat{A}-\hat{A}\hat{H})\psi\,dx + \int\psi^*\frac{\partial\hat{A}}{\partial t}\psi\,dx \qquad \text{(ii)}$$

If $\hat{A}$ does not depend on time, i.e., $\frac{\partial\hat{A}}{\partial t} = 0$, then

$$\frac{d}{dt}\langle A\rangle = \frac{i}{\hbar}\int\psi^*(\hat{H}\hat{A}-\hat{A}\hat{H})\psi\,dx = \frac{i}{\hbar}\langle\hat{H}\hat{A}-\hat{A}\hat{H}\rangle \qquad \text{(iii)}$$

where $\langle\hat{H}\hat{A}-\hat{A}\hat{H}\rangle$ represents the expectation value of $[\hat{H},\hat{A}]$.

$$\therefore \frac{dA}{dt} = \frac{i}{\hbar}[\hat{H},\hat{A}] \qquad \text{(iv)}$$

Equation (iv) is known as the *equation of motion* in quantum mechanics.

If $[\hat{H},\hat{A}] = 0$, i.e., if $\hat{H}$ and $\hat{A}$ commute, then

$$\frac{dA}{dt} = 0 \qquad \text{(v)}$$

Hence, the dynamical variable A is constant of motion.

Again, in classical mechanics for a conservative system the Hamiltonian is a constant of motion. So $\frac{dH}{dt} = 0$.

$$\therefore \frac{dH}{dt} = \frac{i}{\hbar}\left[\hat{H}, \hat{H}\right] = 0 \text{ or, } H = \text{constant}$$

Thus, for a conservative system the total energy of the dynamical system is a constant of motion.

» **Example 5.15.** Calculate the value of $\langle r \rangle$ related to the wave function $\psi(r) = \sqrt{\frac{1}{\pi a^3}}\, e^{-\frac{r}{a}}$, where a is a constant.

Solution: We have the expectation value of r as

$$\langle r \rangle = \int \psi^* r \psi \, d\tau = \int_0^\infty r^2\, dr \int_0^\pi \sin\theta\, d\theta \int_0^{2\pi} d\phi\, (\psi^* r \psi) \quad [\because d\tau = r^2 \sin\theta\, dr\, d\theta\, d\phi]$$

$$= \left(\sqrt{\pi a^3}\right)^2 \times 4\pi \int_0^\infty r^3 e^{-\frac{2r}{a}}\, dr = \pi a^3 \times 4\pi \int_0^\infty r^3 e^{-\frac{2r}{a}}\, dr$$

$$= \pi a^3 \times 4\pi \times \frac{3}{8\pi^2 a^2} = \frac{3a}{2}$$

» **Example 5.16.** Calculate $\langle p \rangle$ and $\langle p^2 \rangle$ for the wave function:

$$\psi(x) = \begin{cases} \left(\frac{2}{L}\right)^{1/2} \sin\frac{\pi x}{L}, & 0 < |x| < L \\ 0, & |x| > L \end{cases}$$

Solution: Now,

$$\langle p \rangle = \frac{\int_{-\infty}^{+\infty} \psi^*(x)\, \hat{p}\psi(x)\, dx}{\int_{-\infty}^{+\infty} \psi^*(x)\, \psi(x)\, dx} = \frac{\frac{1}{2}\int_{-L}^{+L} \left(\frac{2}{L}\right) \sin\frac{\pi x}{L} \left(-i\hbar \frac{d}{dx}\right) \sin\frac{\pi x}{L}\, dx}{\int_{-\infty}^{+\infty} \psi^*(x)\, \psi(x)\, dx}$$

$$= \frac{1}{2}\int_{-L}^{+L} \left(\frac{2}{L}\right) \sin\frac{\pi x}{L} (-i\hbar) \frac{\pi}{L} \cos\frac{\pi x}{L}\, dx = \frac{\pi\hbar}{2iL^2}\int_{-L}^{+L} 2\sin\frac{\pi x}{L}\cos\frac{\pi x}{L}\, dx$$

$$= \frac{\pi\hbar}{2iL^2}\int_{-L}^{+L} \sin\frac{2\pi x}{L}\, dx = 0$$

$\because \sin\left(\frac{2\pi x}{L}\right)$ is an odd function

$$\therefore \int_{-L}^{+L} \sin\left(\frac{2\pi x}{L}\right) dx = 0$$

Again,

$$\langle p^2 \rangle = \frac{\int_{-\infty}^{+\infty} \psi^*(x) \left(-\hbar^2 \frac{d^2}{dx^2}\right) \psi(x)\, dx}{\int_{-\infty}^{+\infty} \psi^*(x)\, \psi(x)\, dx} = \frac{\hbar^2}{2}\frac{2}{L}\left(\frac{\pi}{L}\right)^2 \int_{-L}^{+L} \sin\frac{\pi x}{L} \sin\frac{\pi x}{L}\, dx$$

$$= \frac{\pi^2\hbar^2}{2L^3}\int_{-L}^{+L} 2\sin^2\frac{\pi x}{L}\, dx = \frac{\pi^2\hbar^2}{2L^3}\int_{-L}^{+L}\left(1 - \cos\frac{2\pi x}{L}\right) dx$$

$$= \frac{\pi^2\hbar^2}{2L^3}\left[x - \frac{\sin\frac{2\pi x}{L}}{\frac{2\pi}{L}}\right]_{-L}^{+L} = \frac{\pi^2\hbar^2}{2L^3}\left[2L - \frac{L}{2\pi}(0-0)\right] = \frac{\pi^2\hbar^2}{L^2}$$

» ***Example* 5.17.** If

$$\psi(x) = \left[\frac{2(1+\alpha^2)}{\alpha^2}\right]^{1/2} e^{-|x|} \sin \alpha x$$

be the wave function for a particle, then calculate $\langle x\rangle$ and $\langle p^2\rangle$.

***Solution*: We have**

$$\psi(x) = \left[\frac{2(1+\alpha^2)}{\alpha^2}\right]^{1/2} e^{x} \sin \alpha x, \text{ for } x < 0$$

$$\psi(x) = \left[\frac{2(1+\alpha^2)}{\alpha^2}\right]^{1/2} e^{-x} \sin \alpha x, \text{ for } x > 0$$

Now,

$$\langle x\rangle = \int_{-\infty}^{+\infty} \psi^*(x)\hat{x}\,\psi(x)\,dx = \frac{2(1+\alpha^2)}{\alpha^2}\left[\int_{-\infty}^{0} xe^{2x}\sin^2\alpha x\,dx + \int_{0}^{\infty} xe^{-2x}\sin^2\alpha x\,dx\right] = 0$$

Also

$$\langle p^2\rangle = -\hbar^2\int_{-\infty}^{+\infty}\psi^*\frac{d^2\psi}{dx^2}\psi$$

$$= -\hbar^2\left[\int_{-\infty}^{0} e^{x}\sin\alpha x\frac{d^2}{dx^2}(e^{x}\sin\alpha x)\,dx + \int_{0}^{\infty} e^{-x}\sin\alpha x\frac{d^2}{dx^2}(e^{-x}\sin\alpha x)\,dx\right]$$

$$= \hbar^2(1+\alpha^2)$$

» ***Example* 5.18. Show that for a three-dimensional wave packet**

$$\frac{d}{dx}\langle x^2\rangle = \frac{1}{m}\left[\langle xp_x\rangle + \langle p_x x\rangle\right]$$

***Solution*: We have**

$$\frac{d}{dt}\langle x\rangle = \langle \hat{H},\hat{x}^2\rangle, \because \frac{\partial\langle x\rangle}{\partial t} = 0$$

Now, the Hamiltonian in three dimensions is

$$\hat{H} = \frac{\hat{p}_x^2+\hat{p}_y^2+\hat{p}_z^2}{2m} + \hat{V}(x,y,z)$$

$$\therefore \left[\hat{H},\hat{x}^2\right] = \frac{1}{2m}\left[(\hat{p}_x^2+\hat{p}_y^2+\hat{p}_z^2),\hat{x}^2\right] + \hat{V}(x,y,z)$$

Again,

$$\frac{1}{2m}\left[\hat{p}_x^2,\hat{x}^2\right] = \frac{1}{2m}\left\{\hat{p}_x\left[\hat{p}_x,\hat{x}^2\right] + \left[\hat{p}_x,\hat{x}^2\right]\hat{p}_x^2\right\}$$

$$= \frac{1}{2m}\left\{\hat{p}_x\left(\hat{x}\left[\hat{p}_x,\hat{x}\right] + \left[\hat{p}_x,\hat{x}\right]\hat{x}\right) + \left(\hat{x}\left[\hat{p}_x,\hat{x}\right] + \left[\hat{p}_x,\hat{x}\right]\hat{x}\right)\hat{p}_x\right\}$$

$$= \frac{1}{2m}\left\{\hat{p}_x\hat{x}\frac{\hbar}{i} + \frac{\hbar}{i}\hat{p}_x\hat{x} + \hat{x}\frac{\hbar}{i}\hat{p}_x + \frac{\hbar}{i}\hat{x}\hat{p}_x\right\} = \frac{1}{2m}\frac{2\hbar}{i}(\hat{p}_x\hat{x}+\hat{x}\hat{p}_x)$$

We can also prove that $\left[p_y^2,\hat{x}\right] = \left[\hat{p}_z^2,\hat{x}^2\right] = 0$ and $\left[\hat{V},\hat{x}^2\right] = 0$.

Hence,

$$\frac{d}{dt}\langle x^2\rangle = \left\langle \frac{i}{\hbar}\frac{1}{2m}\left[\hat{p}_x^2, \hat{x}^2\right]\right\rangle = \left\langle \frac{i}{\hbar}\frac{1}{2m}\frac{2\hbar}{i}\left(p_x x + x p_x\right)\right\rangle = \frac{1}{m}\left(\langle x p_x\rangle + \langle p_x x\rangle\right)$$

» ***Example* 5.19.** Write down the Hamiltonian operator H for a linear harmonic oscillator, verify that the function, $\psi(x) = \left(\frac{2\alpha}{\sqrt{\pi}}\right)^{1/2} \alpha x e^{-\frac{(\alpha^2 x^2)}{2}}$ is an eigenfunction of H. What is the corresponding eigenvalue?

[BU (H) 1999]

Solution: The Hamiltonian operator for the LHO is

$$H = -\frac{\hbar^2}{2m}\nabla^2 + V(r)$$

where $\hbar = \frac{h}{2\pi}$, h is the Planck's constant and

$$\nabla^2 = \frac{\partial^2}{\partial x^2} + \frac{\partial^2}{\partial y^2} + \frac{\partial^2}{\partial z^2}$$

$V(r)$ is the potential energy of the oscillator.

For an LHO, $\nabla^2 = \frac{\partial^2}{\partial x^2}$ and $V(x) = \frac{1}{2}kx^2$.

Considering the motion along the x-direction only

$$\hat{H} = -\frac{\hbar^2}{2m}\frac{\partial^2}{\partial x^2} + \frac{1}{2}kx^2$$

Now,

$$\psi(x) = \left(\frac{2\alpha}{\sqrt{\pi}}\right)^{1/2} \alpha x e^{-\frac{(\alpha^2 x^2)}{2}}$$

$$\therefore \frac{\partial \psi}{\partial x} = \left(\frac{2\alpha}{\sqrt{\pi}}\right)^{1/2}\left[\alpha x e^{-\frac{(\alpha^2 x^2)}{2}}\left(-\frac{2\alpha^2 x}{2}\right) + \alpha\cdot 1\cdot e^{-\frac{(\alpha^2 x^2)}{2}}\right]$$

$$= \left(\frac{2\alpha}{\sqrt{\pi}}\right)^{1/2}\left[-\alpha^3 x^2 e^{-\frac{(\alpha^2 x^2)}{2}} + \alpha e^{-\frac{(\alpha^2 x^2)}{2}}\right] = \left(\frac{2\alpha}{\sqrt{\pi}}\right)^{1/2} e^{-\frac{(\alpha^2 x^2)}{2}}\left(\alpha - \alpha^3 x^2\right)$$

$$\text{and } \frac{\partial^2 \psi}{\partial x^2} = \left(\frac{2\alpha}{\sqrt{\pi}}\right)^{1/2}\left[\left(\alpha - \alpha^3 x^2\right) e^{-\frac{(\alpha^2 x^2)}{2}}\left(-\alpha^2 x\right) + e^{-\frac{(\alpha^2 x^2)}{2}}\left(0 - 2\alpha^3 x\right)\right]$$

$$= \left(\frac{2\alpha}{\sqrt{\pi}}\right)^{1/2}\left[-\alpha^2 x^2\left(\alpha - \alpha^3 x\right) e^{-\frac{(\alpha^2 x^2)}{2}} - 2\alpha^3 x e^{-\frac{(\alpha^2 x^2}{2}}\right]$$

$$= \left(\frac{2\alpha}{\sqrt{\pi}}\right)^{1/2} e^{-\frac{(\alpha^2 x^2)}{2}}\left[-\alpha^3 x + \alpha^5 x^3 - 2\alpha^3 x\right]$$

$$= \left(\frac{2\alpha}{\sqrt{\pi}}\right)^{1/2} e^{-\frac{(\alpha^2 x^2)}{2}}\left(\alpha^5 x^3 - 3\alpha^3 x\right) = \left(\frac{2\alpha}{\sqrt{\pi}}\right)^{1/2} e^{-\frac{(\alpha^2 x^2)}{2}} \alpha^3 x\left(\alpha^2 x^2 - 3\right)$$

$$\therefore \hat{H}\psi = -\frac{\hbar^2}{2m}\frac{\partial^2 \psi}{\partial x^2} + \frac{1}{2}kx^2\psi$$

$$= -\frac{\hbar^2}{2m}\left[\left(\frac{2\alpha}{\sqrt{\pi}}\right)^{1/2} \alpha x e^{-\frac{(\alpha^2x^2)}{2}} \alpha^2(\alpha^2x^2-3)\right] + \frac{1}{2}k^2x^2\left(\frac{2\alpha}{\sqrt{\pi}}\right)^{1/2} \alpha x e^{-\frac{(\alpha^2x^2)}{2}}$$

$$= \left(\frac{2\alpha}{\sqrt{\pi}}\right)^{1/2} \alpha x e^{-\frac{(\alpha^2x^2)}{2}}\left[(\alpha^4x^2-3\alpha^2)\left(\frac{\hbar^2}{2m}\right)+\frac{1}{2}kx^2\right]$$

$$= \left(\frac{2\alpha}{\sqrt{\pi}}\right)^{1/2} \alpha x e^{-\frac{(\alpha^2x^2)}{2}}\left[\frac{km}{\hbar^2}x^2\times\left(-\frac{\hbar^2}{2m}\right)+3\alpha^2\frac{\hbar^2}{2m}+\frac{1}{2}kx^2\right]\left[\text{Putting } \alpha^4=\frac{km}{\hbar^2}\right]$$

$$= \left(\frac{2\alpha}{\sqrt{\pi}}\right)^{1/2} \alpha x e^{-\alpha^2x^2}\left[-\frac{1}{2}kx^2+\frac{1}{2}kx^2+3\frac{\alpha^2\hbar^2}{2m}\right] = \frac{3\alpha^2\hbar^2}{2m}\psi$$

$$\therefore \hat{H}\psi = \frac{3\alpha^2\hbar^2}{2m}\psi = E\psi$$

So the eigenvalue corresponding to $\hat{H}$ is

$$E = \frac{3\alpha^2\hbar^2}{2m}$$

$\psi(x) = \left(\frac{2\alpha}{\sqrt{\pi}}\right)^{1/2} \alpha x e^{-\frac{(\alpha^2x^2)}{2}}$ is an eigenfunction of $\hat{H}$.

And

$$E = \frac{3\alpha^2\hbar^2}{2m} = 3\left(\frac{km}{\hbar^2}\right)^{1/2}\frac{\hbar^2}{2m} = 3\sqrt{\frac{km}{\hbar^2}\frac{\hbar^2}{2m}} = \frac{3}{2}\sqrt{\frac{k\hbar^2}{m}} = \frac{3\hbar}{2}\sqrt{\frac{k}{m}}$$

» ***Example* 5.20.** Prove that if there are two solutions of the one-dimensional time-independent Schrödinger wave equation for different values of energy E_n and E_m

$$\int_{-\infty}^{+\infty} \psi_m^* \psi_n \, dx = 0.$$

Solution: The one-dimensional time-independent Schrödinger equations for ψ_m and ψ_n are

$$-\frac{\hbar^2}{2m}\frac{d^2\psi_m}{dx^2} + V\psi_m = E_m\psi_m \quad \text{(i)}$$

$$\text{and } -\frac{\hbar^2}{2m}\frac{d^2\psi_n}{dx^2} + V\psi_n = E_n\psi_n \quad \text{(ii)}$$

Since, V, the potential energy, is real, $V^* = V$.

$$\therefore -\frac{\hbar^2}{2m}\frac{d^2\psi_m^*}{dx^2} + V\psi_m^* = E_m\psi_m^* \quad \text{(iii)}$$

Multiplying eq. (i) by ψ_m^* and eq. (iii) by ψ_n, we get

$$-\frac{\hbar^2}{2m}\psi_m^*\frac{d^2\psi_n}{dx^2} + V\psi_m^*\psi_n = E_n\psi_m^*\psi_n \quad \text{(iv)}$$

$$-\frac{\hbar^2}{2m}\psi_n\frac{d^2\psi_m^*}{dx^2} + V\psi_m^*\psi_n = E_m\psi_m^*\psi_n \quad \text{(v)}$$

From eqs (ii) and (iii)

$$\frac{\hbar^2}{2m}\left[\psi_m^*\frac{d^2\psi_n}{dx^2}-\psi_n\frac{d^2\psi_m^*}{dx^2}\right]=(E_m-E_n)\,\psi_m^*\,\psi_n$$

$$\text{or, }\psi_m^*\frac{d^2\psi_n}{dx^2}-\psi_n\frac{d^2\psi_m^*}{dx^2}=\frac{2m}{\hbar^2}(E_m-E_n)\,\psi_m^*\,\psi_n$$

$$\text{or, }\left[\psi_m^*\frac{d}{dx}\left(\frac{d\psi_n}{dx}\right)+\frac{d\psi_n}{dx}\cdot\frac{d\psi_m^*}{dx}\right]-\left[\psi_n\frac{d}{dx}\left(\frac{d\psi_m^*}{dx}\right)+\frac{d\psi_n}{dx}\cdot\frac{d\psi_m^*}{dx}\right]=\frac{2m}{\hbar^2}(E_m-E_n)\,\psi_m^*\,\psi_n$$

$$\text{or, }\frac{d}{dx}\left(\psi_m^*\frac{d\psi_n}{dx}\right)-\frac{d}{dx}\left(\psi_n\frac{d\psi_m^*}{dx}\right)=\frac{2m}{\hbar^2}(E_m-E_n)\,\psi_m^*\,\psi_n$$

$$\text{or, }\frac{d}{dx}\left[\psi_m^*\frac{d\psi_n}{dx}-\psi_n\frac{d\psi_m^*}{dx}\right]=\frac{2m}{\hbar^2}(E_m-E_n)\,\psi_m^*\,\psi_n$$

Integrating from $x = -\infty$ to $x = +\infty$

$$\int_{-\infty}^{+\infty}\frac{d}{dx}\left(\psi_m^*\frac{d\psi_n}{dx}-\psi_n\frac{d\psi_m^*}{dx}\right)dx=\frac{2m}{\hbar^2}(E_m-E_n)\int_{-\infty}^{+\infty}\psi_m^*\,\psi_n\,dx$$

$$\text{or, }\left[\psi_m^*\frac{d\psi_n}{dx}-\psi_n\frac{d\psi_m^*}{dx}\right]_{-\infty}^{+\infty}=\frac{2m}{\hbar^2}(E_m-E_n)\int_{-\infty}^{+\infty}\psi_m^*\,\psi_n\,dx \qquad \text{(vi)}$$

But, ψ and $\frac{d\psi}{dx}\to 0$ as $x\to\pm\infty$, as ψ is well-behaved

$$\therefore\ \frac{2m}{\hbar^2}(E_m-E_n)\int_{-\infty}^{+\infty}\psi_m^*\,\psi_n\,dx=0 \qquad \text{(vii)}$$

Since, $E_m \neq E_n$

$$\int_{-\infty}^{+\infty}\psi_m^*\,\psi_n\,dx=0 \text{ (proved)} \qquad \text{(viii)}$$

» **Example 5.21.** Show that $\left[x,\frac{d}{dx}\right]=-1$ and $\left[\frac{d}{dx},x\right]=1.$ **[CU (H) 2007]**

Solution: We consider

$$\left[x,\frac{d}{dx}\right]\psi=x\frac{d\psi}{dx}-\frac{d}{dx}(x\psi)=x\frac{d\psi}{dx}-\left(x\frac{d\psi}{dx}+\psi\right)=-\psi$$

$$\text{or, }\left[x,\frac{d}{dx}\right]=-1\ (\because \psi \text{ is arbitrary})$$

Again, consider

$$\left[\frac{d}{dx},x\right]\psi=\frac{d}{dx}(x\psi)-x\frac{d\psi}{dx}=\psi+x\frac{d\psi}{dx}-x\frac{d\psi}{dx}=\psi$$

$$\text{or, }\left[\frac{d}{dx},x\right]=1\ (\because \psi \text{ is arbitrary})$$

» **Example 5.22.** Check whether $[\hat{A},\hat{B}]$ is Hermitian if $\hat{A}$ and $\hat{B}$ are Hermitian. **[CU (H) 2002, 2005, 2013]**

Solution: Suppose, $\hat{p}=[\hat{A},\hat{B}]=\hat{A}\hat{B}-\hat{B}\hat{A}.$

$$\therefore \hat{P}^\dagger=(\hat{A}\hat{B}-\hat{B}\hat{A})^\dagger=(\hat{A}\hat{B})^\dagger-(\hat{B}\hat{A})^\dagger=\hat{B}\hat{A}-\hat{A}^\dagger\hat{B}^\dagger=\hat{B}\hat{A}-\hat{A}\hat{B}$$

$$\therefore\ (\hat{A}^\dagger=\hat{A},\ \hat{B}^\dagger=\hat{B} \text{ as } \hat{A},\hat{B} \text{ are Hermitian})$$

$$=-(\hat{A}\hat{B}-\hat{B}\hat{A})=-\hat{P} \text{ or, } \hat{P}=[\hat{A},\hat{B}]$$

$\because \hat{p}^+$ is not Hermitian but anti-Hermitian or skew Hermitian.

» **Example 5.23.** Show that if $\hat{A}, \hat{B}$ are Hermitian, then $i[\hat{A}, \hat{B}]$ is Hermitian. **[KU (H) 2010]**

Solution: Let $\hat{P} = i[\hat{A}, \hat{B}] = i(\hat{A}\hat{B} - \hat{B}\hat{A})$.

$$\therefore \hat{P}^\dagger = -i[(\hat{A}\hat{B})^\dagger - (\hat{B}\hat{A})^\dagger] = -i[\hat{B}^\dagger \hat{A}^\dagger - \hat{A}^\dagger \hat{B}^\dagger]$$
$$= -i(\hat{B}\hat{A} - \hat{A}\hat{B}) \; (\because \hat{A} = \hat{A}^\dagger, \hat{B} = \hat{B}^\dagger)$$
$$= i(\hat{A}\hat{B} - \hat{B}\hat{A}) = i[\hat{A}, \hat{B}] = \hat{P}$$

$\therefore \hat{P}$ is Hermitian.

» **Example 5.24.** Show that $\left[\dfrac{d}{dx}, x^n\right] = nx^{n-1}$.

Solution: We have

$$\left[\frac{d}{dx}, x^n\right]\psi = \left(\frac{d}{dx}x^n - x^n\right)\psi = \frac{d}{dx}(x^n\psi) - x^n\frac{d\psi}{dx} = x^n\frac{d\psi}{dx} + nx^{n-1}\psi - x^n\frac{d\psi}{dx} = nx^{n-1}\psi$$

$$\therefore \left[\frac{d}{dx}, x^n\right] = nx^{n-1} \text{ (as } \psi \text{ is arbitrary)}$$

» **Example 5.25.** Prove that

(a) $[\hat{A}, \hat{B}\hat{C}] = [\hat{A}, \hat{B}]\hat{C} + \hat{B}[\hat{A}, \hat{C}]$. **[CU (H) 2012, 2014]**

(b) $[\hat{A}\hat{B}, \hat{C}] = [\hat{A}, \hat{C}]\hat{B} + \hat{A}[\hat{B}, \hat{C}]$. **[CU (H) 2002]**

(c) $[\hat{A}, [\hat{B}, \hat{C}]] + [\hat{B}, [\hat{C}, \hat{A}]] + [\hat{C}, [\hat{A}, \hat{B}]] = 0$. **[CU (H) 2006]**

Solution:

(a) $[\hat{A}, \hat{B}\hat{C}] = \hat{A}\hat{B}\hat{C} - \hat{B}\hat{C}\hat{A} = \hat{A}\hat{B}\hat{C} - \hat{B}\hat{A}\hat{C} + \hat{B}\hat{A}\hat{C} - \hat{B}\hat{C}\hat{A} = (\hat{A}\hat{B} - \hat{B}\hat{A})\hat{C} + \hat{B}(\hat{A}\hat{C} - \hat{C}\hat{A})$
$= [\hat{A}, \hat{B}]\hat{C} + \hat{B}[\hat{A}, \hat{C}]$ (proved)

(b) $[\hat{A}\hat{B}, \hat{C}] = \hat{A}\hat{B}\hat{C} - \hat{C}\hat{A}\hat{B} = \hat{A}\hat{C}\hat{B} - \hat{C}\hat{A}\hat{B} + \hat{A}\hat{B}\hat{C} - \hat{A}\hat{C}\hat{B} = (\hat{A}\hat{C} - \hat{C}\hat{A})\hat{B} + \hat{A}(\hat{B}\hat{C} - \hat{C}\hat{B})$
$= [\hat{A}, \hat{C}]\hat{B} + \hat{A}[\hat{B}, \hat{C}]$ (proved)

(c) $[\hat{A}, [\hat{B}, \hat{C}]] + [\hat{B}[\hat{C}, \hat{A}]] + [\hat{C}, [\hat{A}, \hat{B}]] = [\hat{A}, \hat{B}\hat{C} - \hat{C}\hat{B}] + [\hat{B}, \hat{C}\hat{A} - \hat{A}\hat{C}] + [\hat{C}, \hat{A}\hat{B} - \hat{B}\hat{A}]$
$= \hat{A}(\hat{B}\hat{C} - \hat{C}\hat{B}) - (\hat{B}\hat{C} - \hat{C}\hat{B}) + \hat{B}(\hat{C}\hat{A} - \hat{A}\hat{C})$
$-(\hat{C}\hat{A} - \hat{A}\hat{C})\hat{B} + \hat{C}(\hat{A}\hat{B} - \hat{B}\hat{A}) - (\hat{A}\hat{B} - \hat{B}\hat{A})\hat{C} = 0$ (proved)

» **Example 5.26.** Show that $[\hat{x}^n, \hat{p}] = i\hbar n\hat{x}^{n-1}$. **[CU (H) 2006, 2009; KU (H) 2011, 2015]**

Solution: We have

$$[\hat{x}^n, \hat{p}]\psi = (\hat{x}^n\hat{p} - \hat{p}\hat{x}^n)\psi = x^n\left(-i\hbar\frac{\partial}{\partial x}\right)\psi - \left(-i\hbar\frac{\partial}{\partial x}\right)x^n\psi = -i\hbar\left[x^n\frac{\partial\psi}{\partial x} - \frac{\partial}{\partial x}(x^n\psi)\right]$$
$$= -i\hbar\left[x^n\frac{\partial\psi}{\partial x} - \left(x^n\frac{\partial\psi}{\partial x} + \psi nx^{n-1}\right)\right] = i\hbar nx^{n-1} = i\hbar n\hat{x}^n\psi$$

$\because \psi$ is arbitrary, $[\hat{x}^n, \hat{p}] = i\hbar n\hat{x}^{n-1}$ (proved).

» **Example 5.27.** Find $[\hat{x}, \hat{p}^2]$. **[KU (H) 1998]**

Solution: We consider

$$[\hat{x},\hat{p}^2]\psi = (\hat{x}\hat{p}^2 - \hat{p}^2\hat{x})\,\psi = x^2\left(-\hbar^2\frac{\partial^2\psi}{\partial x^2}\right) - \left(-\hbar^2\frac{\partial^2}{\partial x^2}\right)x^2\,\psi = -\hbar^2\left[x^2\frac{\partial^2\psi}{\partial x^2} - \frac{\partial^2}{\partial x^2}(x^2\,\psi)\right]$$

$$= -\hbar^2\left[x^2\frac{\partial^2\psi}{\partial x^2} - \left(x^2\frac{\partial\psi}{\partial x^2} + 2x\frac{\partial\psi}{\partial x} + 2\psi + 2x\frac{\partial\psi}{\phi x}\right)\right] = \hbar^2\left(4x\frac{\partial\psi}{\partial x} + 2\psi\right)$$

$$= -4i^2\hbar^2 x\frac{\partial\psi}{\partial x} + 2\hbar^2\,\psi = 4i\hbar x\left(-i\hbar\frac{\partial}{\partial x}\right)\psi + 2\hbar^2\,\psi = 4i\hbar\hat{x}\hat{p}\psi + 2\hbar^2\,\psi$$

or, $[\hat{x}^2\hat{p}^2] = 4i\hbar^2\hat{x}\hat{p} + 2\hbar^2$ ($\because \psi$ is arbitrary)

» ***Example* 5.28.** Find $[\hat{x}^2,\hat{p}]$. **[KU (H) 2003]**

Solution: We have

$$[\hat{x}^2,\hat{p}]\psi = (\hat{x}^2\hat{p} - \hat{p}\hat{x}^2)\,\psi = x^2\left(-i\hbar\frac{\partial}{\partial x}\right)\psi - \left(-i\hbar\frac{\partial}{\partial x}\right)x^2\,\psi = -i\hbar\left[x^2\frac{\partial\psi}{\partial x} - \frac{\partial}{\partial x}(x^2\,\psi)\right]$$

$$= -i\hbar\left[x^2\frac{\partial\psi}{\partial x} - \left(x^2\frac{\partial\psi}{\partial x} + \psi 2x\right)\right] = 2i\hbar x\psi = 2i\hbar\hat{x}\,\psi$$

$\because \psi$ is arbitrary, $[\hat{x}^2,\hat{p}] = 2i\hbar\hat{x}$.

» ***Example* 5.29.** Find $[x^2,\hat{p}_y]$. **[KU (H) 2003, 2015]**

Solution: We have $[\hat{x}^2,\hat{p}_y]\left[[\hat{x}\hat{x},\hat{p}_y]\right] = \hat{x}\left[\hat{x},\hat{p}_y\right] + \left[\hat{x},\hat{p}_y\right]\hat{x} = \hat{x}(0) + (0)\hat{x} = 0.$

» ***Example* 5.30.** Evaluate $[\hat{x},\sin\hat{p}]$. **[CU (H) 2000]**

Solution:

$$[\hat{x},\sin\hat{p}] = \left[\hat{x}\hat{p} - \frac{\hat{p}^2}{3!} + \frac{\hat{p}^5}{5!} - \ldots\right]\left[\because \sin\theta = \sum_{n=0}^{\infty}(-1)^n\frac{\theta^{2n+1}}{(2n+1)!}\right]$$

$$= [\hat{x},\hat{p}] - \frac{1}{3!}[\hat{x},\hat{p}^3] + \frac{1}{5!}[\hat{x},\hat{p}^5] - \ldots$$

$$= i\hbar - \frac{1}{3!}3i\hbar\hat{p}^{3-1} + \frac{1}{5!}\hbar\hat{p}^{5-1} - \ldots\left[\because [\hat{x},\hat{p} = i\hbar] \text{ and } [\hat{x},\hat{p}^n] = ni\hbar\hat{p}^{n-1}\right]$$

$$= i\hbar\left[1 - \frac{p^2}{2!} + \frac{p^4}{4!} - \ldots\right] = i\hbar\cos p\left[\because \cos\theta\sum_{n=0}^{\infty}(-1)^n\frac{\theta^{2n}}{(2n)!}\right]$$

» ***Example* 5.31.** Prove that if $\hat{A}$ and $\hat{B}$ are constants of motion and $\hat{H}$ is the Hamiltonian, then $[\hat{A},\hat{B}]$ is a constant of motion.

Solution: Let $\hat{A},\hat{B},\hat{H}$ are all explicitly time-independent.

We consider

$$[[\hat{A},\hat{B}],\hat{H}] = [(\hat{A}\hat{B} - \hat{B}\hat{A}),\hat{H}]$$
$$= [\hat{A}\hat{B},\hat{H}] - [\hat{B}\hat{A},\hat{H}]$$
$$= \hat{A}[\hat{B},\hat{H}] + [\hat{A},\hat{H}]\hat{B} - \hat{B}[\hat{A},\hat{H}] - [\hat{B},\hat{H}] = 0$$

$\because [\hat{A},\hat{H}] = 0, [\hat{B},\hat{H}] = 0$ (as $\hat{A},\hat{B}$ are constants of motion).

So, $[\hat{A},\hat{B}]$ is a constant of motion.

» ***Example* 5.32.** Show that $[\hat{A},\hat{B}^{-1}] = -\hat{B}^{-1}[\hat{A},\hat{B}]\hat{B}^{-1}$.

Solution: RHS $= -\hat{B}^{-1}[\hat{A},\hat{B}]\hat{B}^{-1} = -\hat{B}^{-1}(\hat{A}\hat{B}-\hat{B}\hat{A})\hat{B}^{-1} = -\hat{B}^{-1}\hat{A}\hat{B}\hat{B}^{-1} + \hat{B}^{-1}\hat{A}\hat{B}^{-1} = -\hat{B}^{-1}\hat{A}\hat{I} + \hat{I}\hat{A}\hat{B}^{-1}$
$= -(\hat{B}^{-1}\hat{A} - \hat{A}\hat{B}^{-1}) = -[\hat{B}^{-1},\hat{A}] = [\hat{A},\hat{B}^{-1}] =$ LHS

» ***Example* 5.33.** Show that an operator always commutes with its inverse.

Solution: We have $[\hat{A},\hat{A}] = \hat{A}^{-1}\hat{A} - \hat{A}^{-1} = \hat{I} - \hat{I} = 0$ (proved).

» ***Example* 5.34.** If $\hat{A}$ and $\hat{B}$ are Hermitian operators show that $\hat{A}\hat{B}+\hat{B}\hat{A}$ is Hermitian while $\hat{A}\hat{B}-\hat{B}\hat{A}$ is anti-Hermitian. Is the same true for $(1+i)\hat{A}\hat{B}+(1-i)\hat{B}\hat{A}$? **[cf. KU (H) 2012, 2014]**

Solution: We consider $(\hat{A}\hat{B}+\hat{B}\hat{A})^\dagger = (\hat{A}\hat{B})^\dagger + (\hat{B}\hat{A})^\dagger = \hat{B}^\dagger\hat{A}^\dagger + \hat{A}^\dagger\hat{B}^\dagger = \hat{B}\hat{A}+\hat{A}\hat{B}$ [$\because \hat{A}=\hat{A}^\dagger \hat{B}=\hat{B}^\dagger$ they are Hermitian].

$\therefore (\hat{A}\hat{B}+\hat{B}\hat{A})^\dagger = (\hat{A}\hat{B}+\hat{B}\hat{A})$.

So, $(\hat{A}\hat{B}+\hat{B}\hat{A})$ is Hermitian.

Now $(\hat{A}\hat{B}-\hat{B}\hat{A})^\dagger = (\hat{A}\hat{B})^\dagger - (\hat{B}\hat{A})^\dagger = \hat{B}^\dagger\hat{A}^\dagger - \hat{A}^\dagger\hat{B}^\dagger = \hat{B}\hat{A}-\hat{A}\hat{B}$ [$\because \hat{A}=\hat{A}^\dagger, \hat{B}=\hat{B}^\dagger$ are Hermitian]
$= -(\hat{A}\hat{B}-\hat{B}\hat{A})$

or $\hat{A}\hat{B}-\hat{B}\hat{A}$ is not Hermitian but anti-Hermitian (or, skew Hermitian).

» ***Example* 5.35.** Show that the product of two unitary operator is a unitary operator. **[BU (H) 2003]**

Solution: Suppose $\hat{A}$ and $\hat{B}$ be two unitary operators.

$$\therefore \hat{A}\hat{A}^\dagger = \hat{A}^\dagger\hat{A} = \hat{I}$$
$$\hat{B}\hat{B}^\dagger = \hat{B}^\dagger\hat{B} = \hat{I}$$

Now, we consider $(\hat{A}\hat{B})(\hat{A}\hat{B})^\dagger = \hat{A}\hat{B}\hat{B}^\dagger\hat{A}^\dagger = \hat{A}\hat{I}\hat{A}^\dagger = \hat{A}\hat{A}^\dagger = \hat{I}$
and $(\hat{A}\hat{B})^\dagger(\hat{A}\hat{B}) = \hat{B}^\dagger\hat{A}^\dagger\hat{A}\hat{B} = \hat{B}^\dagger\hat{I}\hat{B} = \hat{B}^\dagger B = \hat{I}$.

$\therefore (\hat{A}\hat{B})(\hat{A}\hat{B})^\dagger = (\hat{A}\hat{B})^\dagger(\hat{A}\hat{B}) = \hat{I}$.

So, $\hat{A}\hat{B}$ is a unitary operator.

» ***Example* 5.36.** Show that the inverse of a unitary matrix is unitary. **[BU (H) 2003]**

Solution: If $\hat{A}$ be a unitary matrix, then $\hat{A}\hat{A}^\dagger = \hat{A}^\dagger\hat{A} = \hat{I}$ or, $\hat{A}^\dagger = \hat{A}^{-1}$.

Now $(\hat{A}^{-1})(\hat{A}^{-1}) = \hat{A}+(\hat{A}+)+ = \hat{A}+\hat{A} = \hat{I}$
$(\hat{A}^{-1})(\hat{A}^{-1}) = (\hat{A}+)\hat{A}+ = \hat{A}\hat{A}+ = \hat{I}$.
$\therefore (\hat{A}^{-1})(\hat{A}^{-1}) = (\hat{A}^{-1})+(\hat{A}^{-1}) = \hat{I}$.

So, $\hat{A}^{-1}$ is a unitary matrix.

» ***Example* 5.37.** If $\hat{A}$ is non-singular show that the eigenvalues of $\hat{A}^{-1}$ are reciprocals of those of $\hat{A}$ and every eigenvectors of $\hat{A}$ is also an eigenvector of A^{-1}. **[BU (H) 1998]**

Solution: The eigenvalue equation for $\hat{A}$ is $\hat{A}\psi = a\psi$.

Multiplying by $\hat{A}^{-1}$ from left ($\because \hat{A}$ is non-singular $|\hat{A}| \neq 0$).

$\therefore \hat{A}^{-1}\hat{A}\psi = \hat{A}^{-1}a\psi$ or, $\hat{A}^{-1}\hat{A}\psi = \hat{I}\psi = aA^{-1}\psi$ or, $A^{-1}\psi = \frac{1}{a}\psi$.

$\therefore \hat{A}^{-1}$ has eigenvalue $\frac{1}{a}$ (= reciprocal of the eigenvalue of $\hat{A}$) with the eigenfunction ψ (= same that of $\hat{A}$).

» ***Example* 5.38.** The modulus of eigenvalue of unitary matrix is unity. Show. **[BU (H) 1999]**

Solution: If $\hat{A}$ be a unitary matrix, then $AA^\dagger = \hat{A}^\dagger\hat{A} = I$.

Eigenvalue equation for $\hat{A}$ is

$$\hat{A}\psi = a\psi \tag{i}$$

where ψ is the eigenfunction and is the eigenvalue.

The Hermitical conjugate

$$(\hat{A}\psi)^\dagger = (a\psi)^4 \text{ or, } \psi^\dagger A^\dagger = \psi^\dagger a^\dagger \tag{ii}$$

Operating (ii) by (i), from right

$$(\psi^\dagger \hat{A}^\dagger)(\hat{A}\psi) = (\psi^\dagger a^\dagger)(a\psi)$$
$$\text{or, } \psi^\dagger \hat{A}^\dagger \hat{A}\psi = \psi^\dagger a^\dagger a\psi$$

As $\hat{A}^\dagger \hat{A} = I \Rightarrow \psi^\dagger\psi = \psi^\dagger a^\dagger a\psi \Rightarrow a^\dagger a = 1$ or, $a^* a = 1 \Rightarrow |a|^2 = 1$ or, $a = \pm 1$.

$\therefore |a| = 1$. Hence, the modulus of unitary matrix is unity.

» ***Example* 5.39.** Find the eigenfunction and eigenvalue of the operator $-i\frac{d}{dx}$ assuming eigenfunction to have a period L. **[KU (H) 2015; BU (H) 2012]**

Solution: We have

$$-i\frac{d\psi}{dx} = \lambda\psi \text{ or, } \int \frac{d\psi}{\psi} = i\lambda \int dx + \ln N \text{ or, } \psi = Ne^{i\lambda x}$$

Now,

$$\psi(x) = \psi(x+1) \text{ or, } Ne^{i\lambda x} = Ne^{i\lambda(x+L)}$$
$$\text{or, } e^{i\lambda L} = 1 = e^{i2\pi n} \ (n = 0, \pm 1, \pm 2, \dots)$$
$$\text{or, } \lambda = \frac{2n\pi}{L} \rightarrow \text{eigenvalue}$$

From the condition of normalisation

$$\int_0^L \psi^*\psi\, dx = 1 \text{ i.e., } N = \frac{1}{\sqrt{L}}$$

$$\therefore \psi = \frac{e^{i\lambda x}}{\sqrt{L}}$$

is the eigenfunction.

» ***Example* 5.40.** If a system has two eigenstates ψ_1 and ψ_2 with eigenvalues E_1 and E_2 under what condition a linear combination $(a_1\psi_1 + a_2\psi_2)$ is also an eigenstate? **[KU (H) 2004]**

Solution: We have $\hat{H}\psi_1 = E_1\psi_1$; $\hat{H}\psi_2 = E_2\psi_2$; $\psi = a_1\psi_1 + a_2\psi_2$.

$\therefore \hat{H}\psi = \hat{H}(a_1\psi_1 + a_2\psi_2) = a_1\hat{H}\psi_1 + a_2\hat{H}\psi_2 = a_1E_1\psi_1 + a_2E_2\psi_2 \neq \psi$.

$\therefore \psi$ is not an eigenstate of $\hat{H}$.

If $E_1 = E_2 = E$ (say) (doubly degenerate level), then $\hat{H}\psi = a_1E_1\psi_1 + a_2E_2\psi_2 = E_1(a_1\psi_1 + a_2\psi_2) = E\psi$, i.e., $\psi = a_1\psi + a_2\psi_2$ is a eigenstate of $\hat{H}$ if $E_1 = E_2 = E$.

» ***Example* 5.41.** Show that $u(x) = e^{-x^2/2}$ is an eigenfunction of the operator $\left(\frac{d^2}{dx^2} - x^2\right)$. **[KU (H) 2000, 2010]**

Solution: We have

$$\left(\frac{d^2}{dx^2}-x^2\right)u(x)=\frac{d^2}{dx^2}e^{-x^2/2}-x^2e^{-x^2/2}=\frac{d}{dx}\left[e^{-x^2/2}\left(-\frac{1}{2}\right)(2x)\right]-x^2e^{-x^2/2}$$

$$=-\frac{d}{dx}\left(xe^{-x^2/2}\right)-x^2e^{-x^2/2}=-\left[e^{-x^2/2}+xe^{-x^2/2}\left(-\frac{1}{2}\right)(2x)\right]-x^2e^{-x^2/2}$$

$$=-e^{-x^2/2}+x^2e^{-x^2/2}-x^2e^{-x^2/2}=-u(x)$$

Hence, $u(x)$ is an eigenfunction of the operator $\left(\frac{d^2}{dx^2}-x^2\right)$ with eigenvalue –1.

» ***Example* 5.42.** Find the eigenfunction of the operator given by $x+\frac{d}{dx}$. [CU (H) 2008; KU (H) 2008]

Solution: Let ψ be the eigenfunction corresponding to the eigenvalue λ, then the eigenvalue equation is

$$\left(x+\frac{d}{dx}\right)\psi=\lambda\psi \text{ or, } x\psi+\frac{d\psi}{dx}=\lambda\psi \text{ or, } \frac{d\psi}{dx}=(\lambda-x)\,\psi \text{ or, } \frac{d\psi}{\psi}=(\lambda-x)\,dx$$

Integrating,

$$\int\frac{d\psi}{\psi}=\int(\lambda-x)\,dx+C \text{ or, } \ln\psi=\lambda x-\frac{x^2}{2}+C$$

Let $x=0,\psi=\psi_0$, then $C=\ln\psi_0$.

$$\therefore \ln\psi=\lambda x-\frac{x^2}{2}+\ln\psi_0 \text{ or, } \ln\frac{\psi}{\psi_0}=\lambda x-\frac{x^2}{2}$$

$$\therefore \psi=\psi_0e^{(x^2/2-\lambda x)}$$

» ***Example* 5.43.** Show that (i) $[f(\hat{x}),\hat{p}]=i\hbar\frac{\partial f}{\partial x}$ (ii) $[\hat{x},f(\hat{p})]=i\hbar\frac{\partial f}{\partial p}$, where f denotes a polynomial. Write down corresponding 3D relations. [KU (H) 2011]

Solution:

(i) Suppose $f(x)=\sum a_nx^n$, then

$$[f(\hat{x}),\hat{p}]=[f(x),\hat{p}]=\left[\sum_n a_nx^n,\hat{p}\right]=\sum_n a_n\,[x^n,\hat{p}]\ \{[\hat{x}^n,\hat{p}]=i\hbar n x^{n-1}\}$$

$$=\sum_n a_n\left(i\hbar n\hat{x}^{n-1}\right)=i\hbar\sum a_nnx^{n-1}=i\hbar\sum a_nx^n=i\hbar\frac{\partial}{\partial x}f(x)$$

(ii) Let $f(\hat{p})\sum_n a_n\hat{p}^n$, then

$$[\hat{x},f(\hat{p})]=\left[\hat{x}\sum_n a_n\hat{p}^n\right]=\sum_n a_n\,[\hat{x},\hat{p}^n]\ \left[\because\ [\hat{x},\hat{p}^n]=ni\hbar\hat{p}^{n-1}\right]$$

$$=i\hbar\sum_n a_nnp^{n-1}=i\hbar\frac{\partial}{\partial\hat{p}}\sum_n a_np^n=i\hbar\frac{\partial\hat{f}(\hat{p})}{\partial p}$$

The 3D relations corresponding to (i) is $[f*(\vec{r}),\hat{p}]=i\hbar\nabla f(\vec{r})$ is corresponding to (ii) is $[\hat{r},f(\vec{p})]=i\hbar\nabla_{\vec{p}}f(\vec{p})$, where

$$\nabla\equiv\left(\frac{\partial}{\partial x},\frac{\partial}{\partial y},\frac{\partial}{\partial z}\right) \text{ and } \nabla_{\vec{p}}\equiv\left(\frac{\partial}{\partial p_x},\frac{\partial}{\partial p_y},\frac{\partial}{\partial p_z}\right)$$

» ***Example* 5.44.** Find the eigenfunction of the operator given by $x + \frac{d}{dx}$. **[CU (H) 2008; KU (H) 2008]**

***Solution*:** Let λ be the eigenvalue belonging to the operator $x + \frac{d}{dx}$ and ψ; the corresponding eigenfunction. Then the eigenvalue equation is

$$\left(x + \frac{\phi}{dx}\right)\psi = \lambda x \text{ or, } x\psi + \frac{d\psi}{dx} = \lambda\psi \text{ or, } \frac{d\psi}{dx} = (\lambda - x)\,\psi \text{ or, } \frac{d\psi}{\psi}(\lambda - x)\,dx$$

Integrating,

$$\int \frac{d\psi}{\psi} = \int (\lambda - x)\,dx + C \text{ or, } \ln\psi = \lambda x - \frac{x^2}{c} + C$$

Let $x = 0$, $\psi = \psi_c$, then $C = \ln\psi_0$.

$$\therefore \ln\psi = \lambda x - \frac{x^2}{2} + \ln\psi_0 \text{ or, } \ln\frac{\psi}{\psi_0} = \lambda x - \frac{x^2}{2} \text{ or, } \psi = \psi_0 e^{-(x^2/2)-\lambda x}$$

» ***Example* 5.45.** Show that $\hat{p}_x = -i\hbar\frac{\partial}{\partial x}$ is Hermitian. **[CU (H) 2013]**

***Solution*:** We consider the integral

$$I = \int \psi_1^* \hat{p}_x \psi_2\,dx = \int \psi_1^*\left(-i\hbar\frac{\partial}{\partial x}\right)\psi_2\,dx = -i\hbar\int_{-\infty}^{\infty}\psi_1\frac{\partial}{\partial x}\psi_2\,dx$$

$$\text{or, } I = -i\hbar\left[\psi_1^*\int\frac{\partial\psi_2}{\partial x}dx\Big|_{-\infty}^{\infty} - \int\frac{\partial}{\partial x}\psi_1^*\left[\int\frac{\partial\psi_2}{\partial x}dx\right]dx\right] \text{ [integrating by parts]}$$

$$= -i\hbar\left[\psi_1^*\,\psi_2\Big|_{-\infty}^{\infty} - \int\frac{\partial\psi_1^*}{\partial x}\psi_2\,dx\right] = -i\hbar\left[0 - \int\frac{\partial\psi_1^*}{\partial x}\psi_2\,dx\right] = i\hbar\int\frac{\partial\psi_1^*}{\partial x}\psi_2\,dx$$

$$= \int\left(-i\hbar\frac{\partial\psi_1}{\partial x}\right)^*\psi_2\,dx = \int(\hat{p}_x\psi_1)\,\psi_2\,dx = \int\psi_2\hat{p}_x^*\psi_1^*\,dx$$

$$\text{or, } I = \int\psi_1^*\hat{p}_x\psi_2\,dx = \int\psi_2\hat{p}_x^*\psi_1^*\,dx$$

i.e., p_x is Hermitian.

» ***Example* 5.46.** Show that $[\hat{l}_x, \hat{l}_\pm] = \pm i\hbar\hat{l}_\pm$. Hence, show that the action of $L_\pm$ on an eigenstate of $\hat{L}_2$ is to increase or decrease respectively the eigenvalue by $\hbar$. **[CU (H) 2009]**

***Solution*:** We have

$$[\hat{l}_z, \hat{l}_\pm] = [\hat{l}_z, \hat{l}_x + i\hat{l}_y] = [\hat{l}_z, \hat{l}_x] \pm i[\hat{l}_z, \hat{l}_y] = i\hbar\hat{l}_y \pm i(i\hbar\hat{l}_x)$$
$$= i\hbar\hat{l}_y \pm \hbar\hat{l}_x = \pm\hbar\hat{l}_x + i\hbar\hat{l}_y \; (\because -i^2 = 1)$$
$$= \pm\hbar(\hat{l}_x \pm i\hat{l}_y) = \pm\hbar\hat{l}_\pm$$

Let ψ_{lm_l} be the eigenstate of $\hat{l}_z$ with eigenvalue $m_l\hbar$.

$\therefore \hat{l}_z\psi_{lm_l} = m_l\hbar\psi_{lm_k}$.

Operating from left by $\hat{l}^l_\pm$, $\hat{l}_\pm\hat{l}_z\psi_{lm_l} = m_l\hbar\hat{l}_\pm\psi_{lm_l}$.

Using $[\hat{l}_z, \hat{l}_\pm] = \pm\hbar\hat{l}_\pm$ or, $\hat{l}_z\hat{l}_\pm - \hat{l}_\pm\hat{l}_z = \pm\hbar\hat{l}_\pm$ or, $\hat{l}_\pm\hat{l}_z = \hat{l}_z\hat{l}_\pm \mp \hbar\hat{l}_\pm$.

$\therefore (\hat{l}_z\hat{l}_\pm \mp \hbar\hat{l}_\pm)\,\psi_{lm_l} = m_l\hbar\hat{l}_\pm\psi_{lm_l} = m_l\hbar\hat{l}_\pm\psi_{lm_l}$ or, $\hat{l}_z\hat{l}_\pm\psi_{lm_l} = m_l\hbar\hat{l}_\pm\psi_{lm_l} \pm \hbar\hat{l}_z\psi_{lm_l} = (m_l \pm 1)\hbar\hat{l}_\pm\psi_{lm_l}$.

Operating by $\psi^\dagger_{lm_l}$ from left, $\psi^\dagger_{lm_l}\hat{l}_z\hat{l}_\pm\psi_{lm_l} = (m_l \pm 1)\hbar\psi^\dagger_{lm_l'}\hat{l}_\pm\psi_{lm_l}$.

As $\hat{l}_z\psi_{lm_l'} = m_{l'}\hbar\psi_{lm_{l'}}$ or, $(\hat{l}_z\psi_{lm_l'})^\dagger = (m_l'\hbar\psi_{lm_l'})^\dagger$ or, $\psi^\dagger_{lm_l'}\hat{l}_z = \psi^\dagger_{lm'}m_{l'}\hbar$ ($\because \hat{l}_z$ = Hermitian = $\hat{l}_z^\dagger$).

$\therefore \psi^\dagger_{lm_l'} m_l'\hbar \hat{l}_\pm \psi_{lm_l} = (m_l + 1)\hbar \psi^\dagger_{lm_l'} \hat{l}_\pm \psi_{lm_l}$ or, $m_l'\hbar(\psi^\dagger_{lm_l'} \hat{l}_\pm \psi_{lm_l}) = (m_l \pm 1)\hbar \left(\psi^\dagger_{lm_l'} \hat{l}_\pm \psi_{lm_l}\right)$.

$\therefore m_l' = m_l \pm 1$ (here, + for $\hat{l}_+$ and − for $\hat{l}_-$).

Hence, $\hat{l}_+$ acts as a step up or raising operator and increases the $\hat{l}_z$ eigenvalue $m_l\hbar$ of a state ψ_{lm_l} by $\hbar$ to $(m_l + 1)\hbar$. And $\hat{l}_-$ acts as a step down or lowering operator and decreases to $\hat{l}_z$ eigenvalue $m_l\hbar$ of a state ψ_{lm_l} to $(m_l - 1)\hbar$.

» ***Example* 5.47.** Find $[\hat{l}_x^2 + \hat{l}_y^2, \hat{l}_z^2]$, $[\hat{l}^2, \hat{l}_+]$, $[\hat{l}^2, \hat{l}_{-1}]$, $[\hat{l}_x, \hat{l}_+]$, $[\hat{l}_x, \hat{l}_-]$, $[\hat{l}_y, \hat{l}_+]$, $[\hat{l}_y, \hat{l}_-]$, $[\hat{l}_z, \hat{l}_+]$, $[\hat{l}_z, \hat{l}_-]$, $[\hat{l}_+, \hat{l}_-]$. **[cf. CU (H) 2012, 2014]**

Solution: $[\hat{l}_x^2 + \hat{l}_y^2, \hat{l}_z^2] = [\hat{l}^2 - \hat{l}_z^2, \hat{l}_z^2] = [\hat{l}^2, \hat{l}_z^2] - [\hat{l}_z^2, \hat{l}_z^2] = \hat{l}_z[\hat{l}^2, \hat{l}_z] + [\hat{l}^2, \hat{l}_z]\hat{l}_z = 0.$

$$[\hat{l}^2, \hat{l}_\pm] = [\hat{l}^2, \hat{l}_x + i\hat{l}_y] = [\hat{l}^2, \hat{l}_z] \pm i[\hat{l}^2, \hat{l}_y] = 0 + i(0) = 0.$$

$$[\hat{l}_x, \hat{l}_\pm] = [\hat{l}_x, \hat{l}_x \pm i\hat{l}_y] = [\hat{l}_x, \hat{l}_x] \pm i[\hat{l}_x, \hat{l}_y] = 0 \pm i(i\hbar l_z) = \mp\hbar l_z.$$

$$[\hat{l}_y, \hat{l}_\pm] = [\hat{l}_y, \hat{l}_x \pm i\hat{l}_y] = [\hat{l}_y, \hat{l}_x] \pm [\hat{l}_y, \hat{l}_y] = -i\hbar\hat{l}_z \pm i(0) = -i\hbar\hat{l}_z.$$

$$[\hat{l}_z, \hat{l}_\pm] = [\hat{l}_x, \hat{l}_x \pm i\hat{l}_y] = [\hat{l}_z, \hat{l}_x] \pm i[\hat{l}_z, \hat{l}_y] = i\hbar\hat{l}_y \pm i(-i\hbar\hat{l}_z) = i\hbar\hat{l}_y \pm (-i^2)\hbar\hat{l}_x$$
$$= i\hbar\hat{l}_y \pm \hbar\hat{l}_x = \pm\hbar\hat{l}_x + i\hbar\hat{l}_y = \pm\hbar(\hat{l}_x \pm i\hat{l}_y) = \pm\hbar_\pm$$

$$[\hat{l}_+, \hat{l}_-] = [\hat{l}_x + i\hat{l}_y, \hat{l}_x - i\hat{l}_y] = [\hat{l}_x, \hat{l}_x] + i[\hat{l}_y, \hat{l}_x] - i[\hat{l}_x, \hat{l}_y] + [\hat{l}_y, \hat{l}_y]$$
$$= 0 + i(-i\hbar\hat{l}_z) - i(i\hbar\hat{l}_z) + 0 = 2\hbar\hat{l}_z.$$

» ***Example* 5.48.** (a) Find $\hat{l}_+\hat{l}_-$, $\hat{l}_-\hat{l}_+$ (b) Prove that $\hat{l}^2 = \frac{1}{2}(\hat{l}_+\hat{l}_- + \hat{l}_-\hat{l}_+) + \hat{l}_z^2$. **[cf. CU (H) 2015]**

Solution:

(a) $\hat{l}_+\hat{l}_- = (\hat{l}_x + i\hat{l}_y)(\hat{l}_x - i\hat{l}_y) = \hat{l}_x^2 + \hat{l}_y^2 - i(\hat{l}_x\hat{l}_y - \hat{l}_y\hat{l}_x) = \hat{l}_x^2 + \hat{l}_y^2 + i\hat{l}_z^2 - i[\hat{l}_x, \hat{l}_y] = \hat{l}^2 - \hat{l}_z^2 - i(i\hbar\hat{l}_z)$
$= \hat{l}^2 - \hat{l}_z^2 + \hbar\hat{l}_z^2$

$\hat{l}_-\hat{l}_+ = (\hat{l}_x - i\hat{l}_y)(\hat{l}_x + i\hat{l}_y) = \hat{l}_x^2 + \hat{l}_y^2 + i(\hat{l}_x\hat{l}_y - \hat{l}_y\hat{l}_x) = \hat{l}_x^2 + \hat{l}_y^2 + \hat{l}_z^2 - \hat{l}_z^2 + i[\hat{l}_x, \hat{l}_y] = \hat{l}^2 - \hat{l}_z^2 + i(i\hbar\hat{l}_z^2)$
$= \hat{l}^2 - \hat{l}_z^2 - \hbar\hat{l}_z$

(b) $\hat{l}_+\hat{l}_- + \hat{l}_-\hat{l}_+ = 2\hat{l}^2 - 2\hat{l}_z^2$.

$\hat{l}^2 = \frac{1}{2}(\hat{l}_+\hat{l}_- + \hat{l}_-\hat{l}_+) + \hat{l}_z^2$ (proved).

» ***Example* 5.49.** Show that $\hat{x}\hat{p}_x + \hat{p}_x\hat{x}$ is Hermitian. **[CU (H) 2003, 2008, 2015]**

Solution: We have $(\hat{x}\hat{p}_x + \hat{p}_x\hat{x})^\dagger = (\hat{x}\hat{p}_x)^\dagger + (\hat{p}_x\hat{x})^\dagger = \hat{p}_x^\dagger\hat{x}^\dagger + \hat{x}^\dagger\hat{p}_x^\dagger$.

Since, $\hat{x}$, $\hat{p}_x$ are Hermitian operators $\hat{x}^\dagger = \hat{x}$; $\hat{p}_x^\dagger = \hat{p}_x$.

$\therefore (\hat{x}\hat{p}_x + \hat{p}_x\hat{x})^\dagger = \hat{p}_x\hat{x} + \hat{x}\hat{p}_x = \hat{x}\hat{p}_x + \hat{p}_x\hat{x}$.

So, $\hat{x}\hat{p}_x + \hat{p}_x\hat{x}$ is Hermitian.

» ***Example* 5.50.** Prove that $(\vec{\sigma} \cdot \vec{A})(\vec{\sigma} \cdot \vec{B}) = \vec{A} \cdot \vec{B} + i\vec{\sigma} \cdot (\vec{A} \times \vec{B})$.

***Solution*:**

$$\text{LHS} = (\dot{\vec{\sigma}} \cdot \vec{A})(\dot{\vec{\sigma}} \cdot \vec{B}) = (\dot{\sigma}_x A_x + \dot{\sigma}_y A_y + \dot{\sigma}_z A_z)(\dot{\sigma}_x B_x + \dot{\sigma}_y B_y + \dot{\sigma}_z B_z)$$

$$= \dot{\sigma}_x^2 A_x B_x + \dot{\sigma}_x^2 A_y B_y + \dot{\sigma}_x^2 A_z B_z + \dot{\sigma}_x \dot{\sigma}_y A_x B_x + \dot{\sigma}_x \dot{\sigma}_z A_x B_z + \dot{\sigma}_y \dot{\sigma}_z A_y B_z + \dot{\sigma}_y \dot{\sigma}_z A_y B_z + \dot{\sigma}_z \dot{\sigma}_x A_z B_x + \dot{\sigma}_z \dot{\sigma}_y A_z B_y$$

$$= (A_x B_x + A_y B_y + A_z B_z) + \dot{\sigma}_x \dot{\sigma}_y (A_x B_y - A_y B_x) + \dot{\sigma}_y \dot{\sigma}_z (A_y B_z - A_z B_y) + \dot{\sigma}_z \dot{\sigma}_x (A_z B_x - A_x B_z)$$

$$(\text{using } \hat{\sigma}_x^2 = \hat{\sigma}_y^2 = \hat{\sigma}_z^2 = \hat{I};\ \hat{\sigma}_y \hat{\sigma}_x = -\hat{\sigma}_x \hat{\sigma}_y;\ \hat{\sigma}_z \hat{\sigma}_y = -\hat{\sigma}_y \hat{\sigma}_z;\ \hat{\sigma}_x \hat{\sigma}_z = -\hat{\sigma}_z \hat{\sigma}_y)$$

$$= (\vec{A} \cdot \vec{B}) + i\hat{\sigma}_z (\vec{A} \times \vec{B})_z + i\hat{\sigma}_x (\vec{A} \times \vec{B})_x + i\hat{\sigma}_y (\vec{A} \times \vec{B})_y$$

$$[\because \hat{\sigma}_x \hat{\sigma}_y = i\hat{\sigma}_z;\ \hat{\sigma}_y \hat{\sigma}_z = i\hat{\sigma}_x;\ \hat{\sigma}_z \hat{\sigma}_x = i\hat{\sigma}_y]$$

$$= (\vec{A} \cdot \vec{B}) + i\hat{\sigma} \cdot (\vec{A} \times \vec{B}) = \text{RHS (proved)}$$

» ***Example* 5.51.** What is the result of measurement of $\hat{\sigma}_z$ on a spin $\frac{1}{2}$ particle beam in a state $\frac{1}{\sqrt{2}}\begin{pmatrix} 1 \\ -1 \end{pmatrix}$?

***Solution*:** The eigenvectors of $\hat{\sigma}_z$ are $\chi_{\text{up}} = \begin{pmatrix} 1 \\ 0 \end{pmatrix}$ and $\chi_{\text{down}} = \begin{pmatrix} 0 \\ 1 \end{pmatrix}$.

The eigenvectors χ_{up} and χ_{down} are orthonormalised and form a complete set. We choose them as basis vectors.

So, $\chi = \frac{1}{\sqrt{2}}\begin{pmatrix} 1 \\ -1 \end{pmatrix}$ in their terms as

$$\chi = \frac{1}{\sqrt{2}}\begin{pmatrix} 1 \\ -1 \end{pmatrix} = \frac{1}{\sqrt{2}}\begin{pmatrix} 1 \\ 0 \end{pmatrix} - \frac{1}{\sqrt{2}}\begin{pmatrix} 0 \\ 1 \end{pmatrix} = \frac{1}{\sqrt{2}}\xi_{\text{up}} - \frac{1}{\sqrt{2}}\xi_{\text{down}}$$

So, in a measurement the probability of finding the particle beam in spin up state is

$$\left(\frac{1}{\sqrt{2}}\right)^2 = \frac{1}{2} = 50\,\%$$

In spin down state the probability of finding the particle is $\left[-\frac{1}{\sqrt{2}}\right]^2 = \frac{1}{2}$.

EXERCISES

5(A): Objective (or MCQ) Type Questions

Select the correct answer:

5.1. If $\hat{A}$ and $\hat{B}$ be two commutating operators, then

(a) $[\hat{A}, \hat{B}] = -i\hbar$
(b) $[\hat{A}, \hat{B}] = 0$
(c) $[\hat{A}, \hat{B}] = -1$
(d) $[\hat{A}, \hat{B}] = +1$.

5.2. If the operator $\hat{A}$ commutes with itself, then

(a) $\hat{A}\hat{A}^m - \hat{A}^m\hat{A} = -1$
(b) $\hat{A}\hat{A}^m - \hat{A}^m\hat{A} = 0$
(c) $\hat{A}\hat{A}^m = \hat{A}^m\hat{A} + 1$
(d) $\hat{A}\hat{A}^m + \hat{A}^m\hat{A} = 0$.

5.3. If $\hat{A}$ is a linear operator and C is a constant, then

(a) $\hat{A}C = C\hat{A}$

(b) $\hat{A} = \hat{A} + C$

(c) $\hat{A}C = \hat{A}$

(d) $\hat{A}^2 = C\hat{A}$.

5.4. If $\frac{d}{dx}$ is a linear operator, then

(a) $\frac{d}{dx}[\psi_1(x) + \psi_2(x)] = \frac{d}{dx}\psi_1(x) + \frac{d}{dx}\psi_2(x)$

(b) $\frac{d}{dx}[\psi_1(x) + \psi_2(x)] \neq \frac{d}{dx}\psi_1(x) + \frac{d}{dx}\psi_2(x)$

(c) $\frac{d}{dx}[\psi_1(x) + \psi_2(x)] = \frac{d}{dx}\psi_1(x) - \frac{d}{dx}\psi_2(x)$

(d) $\frac{d}{dx}[\psi_1(x) + \psi_2(x)] = \frac{d}{dx}\psi_1(x) \times \frac{d}{dx}\psi_2(x)$.

5.5. If $\int \psi_1^* \cdot (\hat{A}\psi_2)\, d\tau = \int (\hat{A}\psi_1)^* \cdot \psi_2\, d\tau$, then $\hat{A}$ is

(a) linear

(b) Hermitian

(c) multiplicative

(d) additive.

5.6. The parity operator $\hat{\pi}$ is

(a) Hermitian

(b) linear

(c) multiplicative

(d) unitary.

5.7. If $\hat{A} = \frac{d^2}{dx^2}$ and $\psi(x) = ae^{-2x}$, then the eigenvalue of $\psi(x)$ is

(a) 2

(b) 4

(c) 1

(d) 3.

5.8. $\psi(x)$ is well-behaved if $\psi(x)$ is single-valued and

(a) $|\psi(x)| \to 0$ as $x \to \pm\infty$

(b) $|\psi(x)| \to \infty$ as $x \to \pm\infty$

(c) $|\psi(x)| \to 0$ as $x \to 0$

(d) $|\psi(x)| \to$ finite value as $x \to \pm\infty$.

5.9. If the simultaneous wave function of two operators $\hat{A}$ and $\hat{B}$ form a complete set, then

(a) $[\hat{A}, \hat{B}] = -1$

(b) $[\hat{A}, \hat{B}] = -i\hbar$

(c) $[\hat{A}, \hat{B}] = 0$

(d) $[\hat{A}, \hat{B}] = \infty$.

5.10. If the eigenfunction $\psi_m(x)$ is orthonormal, then $\int |\psi_m(x)|^2\, dx$ is

(a) 0

(b) −1

(c) 1

(d) ∞.

5.11. If the Wronskian

$$W = \begin{vmatrix} \psi_1(x) & \psi_2(x) \\ \psi_1'(x) & \psi_2'(x) \end{vmatrix} = 0$$

then the wave functions $\psi_1(x)$ and $\psi_2(x)$ are

(a) normalised

(b) orthonormal

(c) linearly independent

(d) linearly dependent.

5.12. Unlike classical mechanics, quantum mechanics deals with expectation or average value of a dynamical quantity because in quantum mechanics the position of the quantity is

(a) definitely known
(b) definitely not known
(c) smeared in the wave packet
(d) uncertain in accordance with the uncertainty principle.

5.13. The wave function $\psi(x)$ is normalised, if $\int_\tau |\psi(x)|^2\, dx$ is equal to

(a) 0 (c) $-i\hbar$
(b) 1 (d) -1.

5.14. If $\hat{p}$ be the operator associated with the linear momentum of a particle, then $\hat{p}$ is

(a) $-i\hbar\nabla$ (c) $-\frac{\hbar}{i}\nabla$
(b) $i\hbar\nabla^2$ (d) $-\frac{\hbar}{i}\nabla^2$.

5.15. Quantum mechanics gives the values of classical mechanics, if

(a) the expectation values are considered (c) $\psi(x) = Ae^{+i(k_x x - \omega t)}$
(b) $|\psi|^2$ is considered (d) $\psi(x) = \text{constant}$.

5.16. Stationary states are those for which the probability density ρ is

(a) time-dependent (c) space-dependent
(b) time-independent (d) space-independent.

5.17. Though probability density $\rho(\vec{r}, t)$ is directly measurable, probability current density $J(\vec{r}, t)$ is not, because

(a) the position cannot be accurately measured
(b) the velocity (momentum) cannot be accurately measured
(c) the measurement of flux, i.e., simultaneous measurement of position and momentum cannot be accurately measured
(d) the energy cannot be accurately measured.

5.18. The value of $\left[\hat{x}, \frac{\hat{\partial}}{\partial x}\right]$ is

(a) 1 (c) none of the above.
(b) -1 **[WBUT 2005]**

5.19. What is the value of $\left[\hat{x}, \frac{\hat{\partial}}{\partial x}\right]$?

(a) -1 (c) -2
(b) 0 (d) 1. **[WBUT 2007]**

5.20. Consider the following statements regarding eigenvalues and eigenfunctions in quantum mechanics:

(a) Each eigenfunction belongs to only one eigenvalue.
(b) One or more eigenfunctions may belong to one eigenvalue.
(c) Eigenfunctions belonging to different eigenvalues are orthogonal. Which of the statements are correct?

(a) 1 and 2
(b) 2 and 3
(c) 1 and 3
(d) 1, 2 and 3. **[cf. IAS(Pre) 2003]**

5.21. Identify the linear operator

(a) $\hat{A}\psi(x) = \psi(-x)$ [Parity operator]
(b) $\hat{A}\psi(x) = \psi^*(x)$ [Complex operator]
(c) $\hat{A}\psi(x) = \psi^2(x)$ [Squaring operator]
(d) $\hat{A}\psi(x) = \psi(x) + kx$.

[cf. CU (H) 2005, 2009; KU (H) 1998]

In quantum mechanics a classical dynamical quantity is replaced by the corresponding quantum mechanical operator by

(a) $i\hbar\dfrac{\partial}{\partial x}$
(b) $-i\hbar\dfrac{\partial}{\partial x}$
(c) $\dfrac{2\pi\hbar}{i}\dfrac{\partial}{\partial x}$
(d) $\dfrac{\hbar}{i2\pi}\dfrac{\partial}{\partial x}$. **[IAS(Pre)2002]**

5.22. If $[x, p] = i\hbar$, the value of $[x^3, p]$ is

(a) $2i\hbar x^2$
(b) $-2i\hbar x^2$
(c) $3i\hbar x^2$
(d) $-3i\hbar x^2$

5.23. A particle is moving in one dimension and is described by the wave function

$$\psi(x) = \begin{cases} 0, & x < 0 \\ Axe^{-\beta x}e^{ict/\hbar}, & x > 0 \end{cases}$$

where *A, B, C* are constants (positive). Which of the following holds?

(a) particle is bound
(b) particle is unbound
(c). particle oscillates
(d) particle executes damped oscillation.

5.24. Consider the following statements regarding eigenvalues and eigenfunctions in quantum mechanics.

1. Each eigenfunction belongs to only one eigenvalue.

2. One or more eigenfunctions may belong to one eigenvalue.

3. Eigenfunctions belongs to different eigenvalues are orthogonal.

Which of the statements are correct?

(a) 1 and 2
(b) 2 and 3
(c) 1 and 3
(d) 1, 2 and 3. **[cf. IAS Pre. 2003]**

ANSWERS

5.1. (b).	**5.5.** (b).	**5.9.** (c).	**5.13.** (b).	**5.17.** (c).	**5.21.** (a).
5.2. (b).	**5.6.** (b).	**5.10.** (c).	**5.14.** (a).	**5.18.** (b).	**5.22.** (d).
5.3. (a).	**5.7.** (b).	**5.11.** (c).	**5.15.** (a).	**5.19.** (a).	**5.23.** (a).
5.4. (a).	**5.8.** (a).	**5.12.** (d).	**5.16.** (b).	**5.20.** (d).	**5.24.** (a).

5(B): Short Answer Type Questions

5.1. What is an operator and how is it represented?

5.2. What is an identity operator?

5.3. What is an Hermitian operator?

5.4. What is the probability of finding a particle in the small region Δx?

5.5. What is the necessity of normalising a wave function?

5.6. State the condition of normalisation of a wave function.

5.7. What is the condition of orthonormality of two-wave functions?

5.8. State the eigenvalue equation. What are eigenfunctions and eigenvalues?

5.9. Write down operators for position, momentum, kinetic energy, Hamiltonian and angular momentum.

5.10. Show that the eigenvalue corresponding to Hermitian operator is real.

5.11. If $\hat{\alpha}$ and $\hat{\beta}$ commute, then they have the same eigenfunction. Prove this.

5.12. If ψ is an eigenfunction of both $\hat{\alpha}$ and $\hat{\beta}$, then prove that $\left[\hat{\alpha}\hat{\beta} - \hat{\beta}\hat{\alpha}\right] = 0$.

5.13. State the significance of the equation

$$\frac{d}{dt}\langle p_x \rangle = \langle F_x \rangle$$

5.14. Show that $\hat{L}_z$ and $\hat{p}_z$ are Hermitian operators.

5.15. What are stationary states?

5.16. Verify that

(a) $\left(\frac{\partial}{\partial x} + x\right)\left(\frac{\partial}{\partial x} - x\right) = \frac{\partial^2}{\partial x^2} - x^2 - 1$ (b) $\left(x + \frac{d}{dx}\right)^2 = 1 + x^2 + 2x\frac{d}{dx} + \frac{d^2}{dx^2}$.

5.17. Prove that if two eigenfunctions operated by the same Hamiltonian operator give two eigenvalues, then they will be orthogonal functions.

5.18. Show that $\psi(x) = 4x$ is an eigenfunction of the momentum operator $\frac{d^2}{dx^2}$ belonging to the eigenvalue −16.

5.19. Show that the operator for simultaneous measurable variables commute.

5.20. Prove that $\left[\hat{A}, (\hat{B} + \hat{C})\right] + \left[\hat{B}, (\hat{C} + \hat{A})\right] + \left[\hat{C}, (\hat{A} + \hat{B})\right] = 0$.

5.21. State and explain the significance of the Ehrenfest's theorem.

5.22. Show that (a) $\frac{\hat{x}\hat{p} + \hat{p}\hat{x}}{2}$ (b) $i\left(\hat{p}^2\hat{x} - \hat{x}\hat{p}_x^2\right)$ are Hermitian operators.

5.23. Show that $\hat{L}^2$ commutes with L_x, L_y and L_z.

5.24. A particle moving along the x-axis is described by the function

$$\psi(x) = \frac{x + ix}{(1 + ix)^2}$$

Normalise $\psi(x)$.

5.25. Discuss the uncertainty principle and write down in terms of expectation values of dynamical variables.

5.26. Show that $[\hat{H}, \hat{p}] = 0$.

5.27. **(a) Explain in words the meaning of (i) the normalisation of a wave function (ii) the expectation value of an observable.**

(b) Prove that $\langle L_x \rangle = \langle L_y \rangle = 0$ in a state which is simultaneous eigenstate of L^2 and L_z. **[BU (H) 1997]**

5.28. Write down the linear momentum in operator form. **[BU (H) 2007]**

5.29. Find the eigenvalue of the operator $\frac{\partial^2}{\partial x^2}$ for one of its eigenfunction $\psi = e^x \cdot \sin x$. What are eigenvalues of parity operator? **[BU (H) 2007; CU (H) 2013; KU (H) 2013]**

[*Ans.* 1, −1; ±1]

5.30. Examine if $\hat{A}$ is linear if (a) $\hat{A}\,\psi(x) = \psi(-x)$ (b) $\hat{A}\,\psi(x) = \psi^*(x)$. **[CU (H) 2015]**

[*Ans.* **(a) Linear (b) Antilinear]**

5.31. Write down the angular momentum commutation relations. What is the implication of those results? **[CU (H) 2007, 2008]**

[*Ans.* $l_i l_j = i\hbar E_{ijk}; [\hat{l}^2, \hat{l}_i = 0]$]

5.32. Show that $[\hat{l}_x^2 + \hat{l}_y^2 \hat{l}_x] = 0$. **[CU (H) 2006]**

5.33. What is the implication of the result $[\hat{H}, \dot{\vec{l}}] = 0$? **[CU (H) 2004, 2013, 2015]**

[*Hints*: **$[\hat{H}, \vec{l}] = 0$ implies $\vec{l}$ is conserved.]**

5.34. How many angular momentum components can be measured simultaneously? **[BU (H) 2003; CU (H) 1999]**

[*Ans.* **Only one]**

5.35. Can the x-component of angular momentum and x-component of linear momentum of a quantum particle be specified simultaneously? **[KU (H) 2003]**

[*Hints*: Since, $[\hat{l}_x, \hat{p}_x] = 0$, $\hat{l}_x, \hat{p}_x$ **commutes and as such they have simultaneous eigenstate and can be measured simultaneously with infinite accuracy.]**

5.36. Define linear operator. Define Hermitian operator. **[CU (H) 2013; KU (H) 2012]**

5.37. If two operators commute what does they physically mean? **[CU (H) 2012, 2013; KU (H) 2009]**

5.38. Show that the eigenvalues of Hermitian operator are real. **[CU (H) 1996]**

5.39. Why is it useful to find observables commuting with the Hamiltonian of a quantum system? **[BU (H) 2006]**

5.40. When are two eigenfunctions of an operator said to be degenerate? **[BU (H) 2003]**

5.41. Can KE and linear momentum of a quantum mechanical system be specified simultaneously? **[KU (H) 2002]**

[*Hints*: $\because [\hat{T}, \hat{p}] = \left[\frac{\hat{p}}{2m}, \hat{p}\right] = 0$. $\therefore$ $\hat{T}, \hat{p}$ **commutes and can be specified simultaneously.]**

5.42. Is it possible to measure both x and z-components of the electron spin simultaneously? **[KU (H) 2016]**

[*Hints*: $[\hat{s}_x, \hat{s}_z] \neq 0$, $\therefore \hat{s}_x, \hat{s}_z$ **do not commute and as such they cannot be measured simultaneously with infinite accuracy?]**

5(C): Broad Answer Type Questions

5.1. **(a) Discuss the role of operators in quantum mechanics. In Schrödinger representation what are the basic operators relating to dynamical variables.**

(b) What are the other forms of operator representation of those dynamical variables?

5.2. **(a) What are linear operators? Discuss the properties of linear operators. What are the different forms of linear operator which we use in quantum mechanics?**

(b) What are Hermitian operators? Show that $\hat{p}_x$ is Hermitian and has real eigenvalues.

5.3. **(a) Write down eigenvalue equation. What are eigenvalues and eigenfunctions? State the conditions to be satisfied by a well-behaved function.**

(b) State the conditions of normalisation and orthogonality of two-wave functions.

5.4. **(a) What is an *observable* corresponding to a quantum mechanical system? Elaborate on the results of measurement of an observable.**

(b) Establish the relations: $[L_x, L_y] = i\hbar L_z$; $[L^2, L_z] = 0$, where the notations have usual meanings. What conclusions about the eigenfunctions of the operators involved can be drawn from these relations? **[BU (H) 2000]**

5.5. **(a) Write down the time-dependent Schrödinger equation for a particle of mass m moving in a real potential $V(x, y, z, t)$.**

From this equation obtain the continuity equation involving the probability density and the probability current density.

(b) For the above particle establish the relation:

$$\frac{d}{dt}\langle x\rangle = \frac{\langle p_x\rangle}{m}$$

where $\langle x\rangle$ and $\langle p_x\rangle$ are the expectation values of position and momentum respectively.

[BU (H) 2000]

5.6. **(a) What is meant by an eigenfunction of an operator? When are two eigenfunctions of an operator said to be degenerate?**

(b) Derive a continuity equation from the time-dependent Schrödinger equation of a particle moving in a real potential.

Give a physical interpretation of the continuity equation you derive. **[BU (H) 2003]**

5.7. **(a) Define expectation value of a dynamical variable. Find out expectation value of energy of an eigenstate $\psi_n(\vec{r}, t)$.**

(b) State and prove Ehrenfest's theorem.

5.8. **(a) Give the exact statement and proof of Heisenberg uncertainty relation for wave packets.**

(b) Obtain Schrödinger equation in momentum representation.

5.9. **(a) State the postulates of quantum mechanics.** **[BU (H) 2012]**

(b) Find out momentum wave function and momentum eigenfunction for a free particle.

5.10. **(a) Define eigenvalues and eigenfunctions of an operator. Show that the eigenfunctions corresponding to different eigenvalues of a Hermitian operator are orthogonal.**

(b) Is it possible to measure accurately all the three components of angular momentum of a quantum system at the same time? Give reasons for your answer.
What are the permitted values of the square of the orbital angular momentum and one of its components? **[BU (H) 2001]**

5.11. (a) What does the wave function $\psi(x)$ of a particle in quantum mechanics represent? What are the conditions it must satisfy and why?
(b) State in words the meaning of the expectation value of the position variable x in quantum mechanical context. **[BU (H) 1994, 2012]**

5.12. (a) Why is the wave nature of matter not apparent in our daily observations?
(b) What is the physical meaning of normalisation of a wave function? **[BU (H) 1994]**

5.13. (a) Why is an observable in quantum mechanics is represented by Hermitian operator? Examine whether the operator $i\frac{\partial}{\partial x}$ is Hermitian or not.
(b) What are stationary states? Starting from the time-dependent Schrödinger equation find the time-independent equation satisfied by the wave function for stationary states. **[BU (H) 1994]**

5.14. Starting from De-Broglie's hypothesis show that the group velocity associated with a particle is the same as the particle velocity? **[WBUT 2007]**

5.15. (a) What is the function in the spherical polar coordinate system which is the eigenfunction of the operators $\hat{L}^2$ and $\hat{L}_z^2$?
(b) Why is that only one component of $\vec{L}$ is quantised?
(c) Show that $\hat{L}_x\hat{L}_y - \hat{L}_y\hat{L}_x = i\hbar\,\hat{L}_z$, where $\hat{L}_x, \hat{L}_y$ and $\hat{L}_z$ are the x, y and z components of $\vec{L}$ respectively. **[BU (H) 2009]**

5.16. Answer the following questions:
(a) What is an observable?
(b) What kind of operator can represent an observable?
(c) What are the results of measurement of an observable?
(d) State the *expansion postulate* in language and by equation. **[BU (H) 2009]**

5.17. (a) What do you understand by *expectation value* and eigenvalue of dynamical variable? Under what conditions the eigenvalue of a dynamical variable is equal to its expectation value?
(b) Explain the significance of wave function. What do you mean by stationary state?
(c) Find the eigenvalues and normalised eigenfunction of a particle confined in a one-dimensional box with perfectly rigid walls. Can the possible energy of such a particle be zero? Explain your answer. **[NB(H) 2009]**

5.18. (a) Prove that the expectation value of the square of a Hermitian operator is always positive.
(b) Prove that $[\hat{L}_x, \hat{L}_y] = i\hbar\,\hat{L}_z$, where $\hat{L}_x, \hat{L}_y, \hat{L}_z$ are the components of the orbital angular momentum. **[KU (H) 2015, 2014]**

5.19. (a) From the equation of motion $\dot{\vec{A}} = \frac{i}{\hbar}[\vec{H}, \vec{A}]$ deduce the relation $\dot{p} = -\frac{\partial V}{\partial x}$ and $\dot{x} = p/m$ for an one-dimensional system.

(b) Show that the commutation relation is satisfied by the operators $p_{op} = P$ and $X_{op} = i\hbar\frac{\partial}{\partial p}$.

[KU (H) 2011]

5.20. (a) Prove that the eigenvalues of Hermitian operator are real.

(b) Prove that $\frac{d}{dt}\langle x\rangle = \frac{1}{m}\langle p_x\rangle$.

(c) Using the operator representation, prove that $[x, p_x] = i\hbar$. **[KU (H) 2010]**

5.21. (a) Define parity operator. Hence, show that parity operator commutes with $H = p^2/2m + V(x)$ if $V(-x) = V(x)$.

(b) Prove that (i) $[L^2, Lx] = 0$ (ii) $[L_z, x] = i\hbar y$ (iii) $[L_z, P_z] = 0$ (iv) $\hat{A} = \frac{i}{\hbar}[H, A]$. **[KU (H) 2010]**

5.22. (a) What is the physical significance of normalisation? Discuss the methods. **[CU (H) 2012]**

[*Hints*: (a) To fix up probability scale. (b) To present position probability amplitude. (c) To conserve probability.]

(b) Establish $[\hat{l}_x, \hat{l}_y] = i\hbar\hat{l}_z; [\hat{l}^2, \hat{l}_z] = 0$. What conclusions about the eigenfunction of the operators involved can be drawn from the relations? **[BU (H) 2000; KU (H) 2012, 2014, 2015]**

5.23. A particle on the x-axis has the wave function $\psi(x) = cx^2$ between $x = 0$ and $x = 2$. (a) Normalise the wave function over the interval. (b) Find the probability that the particle can be found between $x = 0.5$ and $x = 0.6$. (c) Find the expectation value of the particles position x. **[BU (H) 2005]**

[*Ans.* $\sqrt{5/32}, 1.45\times10^{-3}, 1.667$]

5(D): Numerical Problems

5.1. Find the common eigenfunctions of the operators (a) $\hat{x}$ and $\hat{p}_y$ (b) $\hat{p}_x$ and $\hat{p}_x^2$.

5.2. Show that the function $\psi(x) = xe^{-\frac{x^2}{2}}$ is an eigenfunction of the operator $-\frac{d^2}{dx^2} + x^2$ belonging to the eigenvalue 3.

5.3. Show that if a wave function $\psi(x)$ can be expanded as a series of orthonormal functions $\psi_1(x), \psi_2(x), \ldots, \psi_n(x)$ in the form: $\psi(x) = C_1\psi_1(x) + C_2\psi_2(x) + \cdots + C_n\psi_n(x) + \ldots$, where $C_1, C_2, \ldots, C_n$ are constants, then the condition of normalisation of $\psi(x)$ is

$$\sum|C_i|^2 = |C_1|^2 + |C_2|^2 + \cdots + |C_n|^2 = 1$$

5.4. Normalise $\psi_1(x) = A_1e^{-ax^2}$ and $\psi_2(x) = A_2xe^{-ax^2}$ over the interval $-\infty \leqslant x \leqslant +\infty$. Are ψ_1 and ψ_2 orthogonal over the same interval? [*Ans.* $A_1 = \left(\frac{2a}{\pi}\right)^{1/4}, A_2 = \sqrt{4a}\left(\frac{2a}{\pi}\right)^{1/4}$, yes]

5.5. The wave function of a particle is represented by $\psi(x, 0) = Ne^{-x^2/2a^2} + ik_0x$.

(a) Find the value of N.

(b) In what region of space the particle is most likely to be found?

[*Ans.* $|N| = \frac{1}{\sqrt{a}\sqrt{\sqrt{\pi}}}$, in a width a on either side of $x = 0$]

5.6. Show that in quantum mechanics, the time-dependence of the expectation value of a dynamical variable q can be expressed as

$$\frac{d}{dt}\langle q\rangle = \left\langle \frac{i}{\hbar}\left[\hat{H},\hat{Q}\right] + \frac{\partial \hat{Q}}{\partial t}\right\rangle$$

where $\hat{Q}$ is the operator corresponding to q and $\left[\hat{H},\hat{Q}\right]$ is the commutator bracket of Hamiltonian operator with $\hat{Q}$.

5.7. Find the current density carried by a plane wave Ae^{ikx} in one dimension and verify that it satisfies the equation of continuity in one dimension.

Can a real wave function have a current? [*Ans.* A real wave function has no current]

5.8. Find the values of $\langle x\rangle, \langle x^2\rangle, \langle p_x\rangle$ and $\langle p_x^2\rangle$ for the Gaussian wave packet

$$\psi(x) = \left(\frac{1}{\sigma\sqrt{\pi}}\right)^{1/2} e^{-\frac{x^2}{2\sigma^2}} e^{ik_0 r}$$

[*Ans.* $\langle x\rangle = 0, \langle x^2\rangle = \frac{\sigma^2}{2}, \langle p_x\rangle = \hbar k_0, \langle p_x^2\rangle = \hbar^2 k_0^2 + \frac{\hbar^2}{2\sigma^2}$]

5.9. Determine probability density and probability current density for the wave function in problem 8.

[*Ans.* $\rho = \frac{\sigma}{\sqrt{\pi}} e^{-\sigma^2 x^2}, J_x = \frac{\hbar k}{m}\rho$]

5.10. Generalise the uncertainty relation $\Delta x\,\Delta p_x \geqslant \frac{\hbar}{2}$ for the case when $\langle x\rangle \neq 0$ and $\langle p_x\rangle \neq 0$.

5.11. If $\psi(x) = Ae^{-m\omega x^2/\hbar}$ find the expectation values of momentum and the position.

5.12. Find the probability current density to the wave function $f(r) = \frac{e^{ikr}}{r}$. [*Ans.* $J_r = \frac{v}{r^2}$]

5.13. Show that if $\psi_1(\vec{r})$ and $\psi_2(\vec{r})$ are two independent solutions of the Schrödinger equation, then $\psi(\vec{r}) = a_1\psi_1(\vec{r}) + a_2\psi_2(\vec{r})$ is also a solution. What does it imply? **[BU (H) 2006]**

[*Hints*: The Schrödinger equation for a free particle of mass m is

$$\nabla^2\psi(\vec{r}) + \frac{2m}{\hbar^2}\psi(\vec{r}) = 0 \quad \text{(i)}$$

Since, $\psi_1(\vec{r}_1)$ and $\psi_2(\vec{r}_2)$ satisfy Schrödinger equation

$$\therefore \nabla^2\psi_1(\vec{r}) + \frac{2m}{\hbar^2}\psi_1(\vec{r}) = 0 \quad \text{(ii)}$$

$$\text{and } \nabla^2\psi_2(\vec{r}) + \frac{2m}{\hbar^2}\psi_2(\vec{r}) = 0 \quad \text{(iii)}$$

Multiplying (2) by a_1 and (3) by a_2 and, then adding, we get

$$\nabla^2\left[a_1\psi_1(\vec{r}) + a_2\psi_2(\vec{r})\right] + \frac{2m}{\hbar^2}\left[a_1\psi_1(\vec{r}) + a_2\psi_2(\vec{r})\right] = 0$$

$$\text{or, } \nabla^2\psi(\vec{r}) + \frac{2m}{\hbar^2}\psi(\vec{r}) = 0 \quad \text{(iv)}$$

where

$$\psi(\vec{r}) = a_1\psi_1(\vec{r}) + a_2\psi_2(\vec{r}) \quad \text{(v)}$$

$\neq 0$ as $\psi_1(\vec{r})$ and $\psi_2(\vec{r})$ are linearly independent.

Hence, $\psi(\vec{r})$ given by (v) is also a solution of the Schrödinger equation. But, $\psi(\vec{r})$ expressed by (v) represents the superposition of two linearly independent solutions of (i).

Hence, we may, say, that *the solution of Schrödinger equation obeys the superposition principle of a set of linearly independent solutions.*]

5.14. The wave function of a particle moving in one dimension is given by

$$\psi(x) = \begin{cases} \sqrt{\frac{15}{a}} A(a^2 - x^2), & \text{for } -a \leqslant x \leqslant a \\ 0, & \text{for } |x| > a \end{cases}$$

(a) Find the value of A that will normalise $\psi(x)$.

(b) Calculate the expectation value of x and p. **[BU (H) 2003]**

5.15. (a) The linear momentum operator is $\hat{p} = -i\hbar\nabla$. Find the commutator $[x, p_x]$, i.e., $(xp_x - p_x x)$.

(b) If $\vec{L}$ is the orbital angular momentum operator find out $[L_x, L_y]$.

(c) If ψ_{lm} is an eigenstate of L^2 and L_z with eigenvalues $l(l+1)\hbar^2$ and $m\hbar$ respectively, show that $\phi = (L_x + iL_y)\psi_{lm}$ is also an eigenstate of L^2 and L_z and determine the corresponding eigenvalues. **[BU (H) 1996]**

5.16. Show that $\hat{x} = i\hbar\frac{\partial}{\partial p}$ and $\hat{p} = p$ is consistent with the commutation relation $[\hat{x}, \hat{p}] = i\hbar$. **[WBUT 2007]**

5.17. Show that $L_xL_y - L_yL_x = i\hbar L_z$, where L_x, L_y, L_z are the components of the angular momentum operator. **[BU (H) 2015]**

5.18. Show that the one-dimensional momentum operator $\hat{p}$ is self-adjoint. **[BU (H) 2014]**

5.19. For any operator which does not have any explicit time-dependence, show that $\frac{\partial}{\partial t}\langle\hat{O}\rangle = \frac{i}{\hbar}\langle[\hat{H}, \hat{O}]\rangle$, where a $\langle\quad\rangle$ means expectation value and $[\hat{H}, \hat{O}]$ is the commutator of $\hat{O}$ with the Hamiltonian H. **[BU (H) 2014]**

5.20. (a) Write down the expression for the expectation value of x, i.e., $\langle x\rangle$ in terms of $\psi(x, t)$. What is the physical meaning of $\langle x\rangle$?

(b) Calculate $\frac{d\langle x\rangle}{dt}$ assuming $\frac{d\langle x\rangle}{dt} = \frac{\langle p\rangle}{m}$, where $\langle p\rangle$ is the expectation value of momentum p, identify the momentum operator. **[BU (H) 2014]**

5.21. What is the physical meaning of normalisation of a wave function. Find the dimension of ψ. **[BU (H) 2012]**

5.22. Find the eigenfunction ψ and the eigenvalue of the operator, $-\left(i\frac{d}{dx}\right)$, If $\psi(x) = \psi(x + a)$; a is a constant.

5.23. Find the commutator $[\hat{A}, \hat{B}]$, where $\hat{A} = x^3$ and $\hat{B} = x\frac{d}{dx}$. **[BU (H) 2012]**

5.24. Find the eigenvalue of the operator $\hat{A} = -\frac{d^2}{dx^2} + x^2$, corresponding to the eigenfunction $\psi = e^{-x^2/2}$. **[BU (H) 2011]**

5.25. Show that for the particle moving along the x-axis with position coordinate x and momentum p, $\langle xp\rangle - \langle px\rangle = i\hbar$. **[BU (H) 2011]**

5.26. If $\hat{A}$ and $\hat{B}$ are two Hermitian operators, then the commutator $[\hat{A}, \hat{B}]$ is an anti-Hermitian operator. **[CU (H) 2014]**

5.27. If two operators $\hat{A}$ and $\hat{B}$ commute, show that they will have simultaneous eigenfunctions. [CU (H) 2014]

5.28. Show that the eigenvalues of an Hermitian operator are real. [CU (H) 2014]

5.29. If $\hat{H} = \frac{p^2}{2m} + V(x)$, show that $[x, [x, \hat{H}]] = -\frac{\hbar^2}{2m}$. [CU (H) 2013]

5.30. Prove that a Hermitian operator has real eigenvalues and orthogonal eigenvectors. [CU (H) 2014]

5.31. (a) Find the constant B, which makes $e^{-\alpha x^2}$ an eigenfunction of the operator

$$\left(\frac{d^2}{dx^2} - Bx^2\right).$$

(b) Evaluate $\left[\hat{L}_x^2 + \hat{L}_y^2, \hat{L}_z^2\right]$, where $\hat{L}_x, \hat{L}_y, \hat{L}_z$ are the component of angular momentum operator. [CU (H) 2012]

5.32. Evaluate the commutation $[x^n, p_x]$. [KU (H) 2014; CU (H) 2011]

5.33. Find the eigenvalues and eigenfunctions of the angular momentum operator $\hat{L}_z = i\hbar\frac{\partial}{\partial\phi}$.

5.34. Show that if ψ be an eigenfunction of the operator $\hat{A}$ with eigenvalue λ, then it is also eigenfunction of with eigenvalue e^{λ}. [CU (H) 2009]

5.35. Show that $[\hat{A}, \hat{B}] + [\hat{B}, \hat{A}] = 0$, where $\hat{A}$ and $\hat{B}$ are Hermitian operator. [CU (H) 2009]

5.36. Examine if the operator $\tilde{A}$ is linear in the examples (a) $\tilde{A}\psi(x) = \psi(x)$ (b) $\tilde{A}\psi(x) = \psi^*(x)$. [CU (H) 2009]

5.37. Establish the following relation of commutation algebra: $[\hat{A}, \hat{B}\hat{C}] = [\hat{A}, \hat{B}]\hat{C} + \hat{B}[\hat{A}, \hat{C}]$ and hence prove that $[\hat{x}, \hat{p}_x^n] = in\hbar\hat{p}_x^{n-1}$. [NBU (H) 2009]

5.38. (a) Obtain the eigenfunction of the operator $\left(-i\frac{d}{dx}\right)$ assuming the eigenfunction to have a period L.

(b) Show that $(1+i)\hat{A}\hat{B} + (1-i)\hat{B}\hat{A}$ is Hermitian provided $\hat{A}$ and $\hat{B}$ are Hermitian operators.

5.39. (a) For an operator $\hat{A}$ corresponding to an observable A, prove that

$$\frac{d}{dt}\langle A\rangle = \left\langle\frac{\partial\hat{A}}{\partial t}\right\rangle + \frac{\langle[\hat{A}, \hat{H}]\rangle}{i\hbar}$$ [KU (H) 2013]

(b) Find the commutation relation of $[x, \sin p_x]$. [KU (H) 2014, 2012]

5.40. Show that $\frac{d}{dt}\langle x\rangle = \langle p_x\rangle/m$, for a particle moving along X-axis. [KU (H) 2012]

5.41. Prove that the eigenvalues of a Hermitian operator are real, and momentum operator $(\hat{p}_x)$ is Hermitian. [VU (H) 2009]

5.42. Show that $[\hat{L}_y, \hat{L}_x] = i\hbar\,\hat{L}_x, [\hat{L}_z, \hat{L}] = -i\hbar\,\hat{L}$. [VU (H) 2009]

5.43. What is Hermitian operator? Prove that the eigenfunctions of a Hermitian operator belonging to different eigenvalues are orthogonal. [VU (H) 2009]

Chapter 6

Applications of Schrödinger's Equation to One-dimensional Problems—Free States

6.1 Introduction

Schrödinger wave equation can be applied to many simple problems in quantum mechanics and it is possible to find out the exact solutions under appropriate boundary conditions. These exact solution provides detailed information regarding the physical properties like *quantisation of energy, quantum numbers, zero point energy*, etc., of the relevant problems.

In this chapter we shall apply Schrödinger equation to two types of simple cases, viz.,

1. one-dimensional motion of a particle under *no force*, i.e., potential $V = 0$ (free particle).
2. one-dimensional motion of a stream of particle through a *force field*, i.e., $V \neq 0$ (potential barrier problems).

6.2 Boundary conditions at the surface of finite potential

To solve Schrödinger wave equation relating to any specific problem we require some conditions to be fulfilled by the wave function $\psi(x)$ and its first derivative $\frac{d\psi}{dx}$ at the surface of finite potential.

If a particle moves in a finite potential $V(x)$ along the x-direction (Fig 6.1), then the time-independent Schrödinger equation is

$$-\frac{\hbar^2}{2m}\frac{d^2\psi}{dx^2} + V(x)\,\psi = E\psi \text{ or, } \frac{d^2\psi}{dx^2} = \frac{2m}{\hbar^2}(V-E)\,\psi \tag{6.1}$$

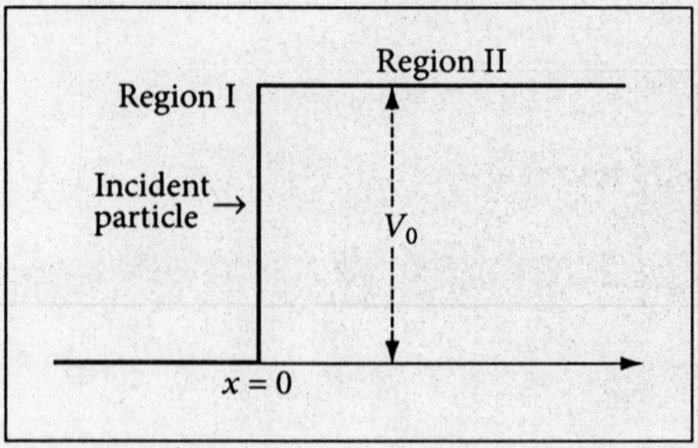

Fig 6.1 One-dimensional step potential

We suppose that potential $V(x)$ is discontinuous at $x = a$. Integrating eq. (6.1) from $x = a - \varepsilon$ to $x = a + \varepsilon$, where $\varepsilon \to 0$, we get

$$\int_{a-\varepsilon}^{a+\varepsilon} \frac{d^2\psi}{dx^2}\,dx = \frac{2m}{\hbar^2}\int_{a-\varepsilon}^{a+\varepsilon}(V - E)\,\psi\,dx$$

$$\text{or,}\quad \int_{a-\varepsilon}^{a+\varepsilon} \frac{d}{dx}\left(\frac{d\psi}{dx}\right)dx = \frac{2m}{\hbar^2}\int_{a-\varepsilon}^{a+\varepsilon}(V - E)\,\psi\,dx$$

$$\text{or,}\quad \left(\frac{d\psi}{dx}\right)_{a+\varepsilon} - \left(\frac{d\psi}{dx}\right)_{a-\varepsilon} = \frac{2m}{\hbar^2}\int_{a-\varepsilon}^{a+\varepsilon}(V - E)\,\psi\,dx \qquad (6.2)$$

In the limit $\varepsilon \to 0$

$$\int_{a-\varepsilon}^{a+\varepsilon}(V - E)\,\psi\,dx = \int_{a}^{a}(V - E)\,\psi\,dx = 0$$

$$\therefore \lim_{\varepsilon\to 0}\left(\frac{d\psi}{dx}\right)_{a-\varepsilon} = \lim_{\varepsilon\to 0}\left(\frac{d\psi}{dx}\right)_{a+\varepsilon}$$

Hence, $\frac{d\psi}{dx}$ is continuous everywhere even at the point $x = a$ the point where $V(x)$ is discontinuous.

Hence, the required boundary conditions are

1. $\psi(x)$ must be finite and single-valued **2.** $\frac{d\psi}{dx}$ must be continuous and finite.

6.3 Boundary conditions at the surface of infinite potential

Let a particle of mass m moves along the x-axis from left to right in a region, where $V = 0$. After sometime the particle reaches a region where $V \to \infty$ [Fig 6.2].

The one-dimensional time-independent Schrödinger equation for a particle of mass m moving in a potential V is given by

$$\frac{d^2\psi}{dx^2} + \frac{2m}{\hbar^2}(E - V)\,\psi = 0$$

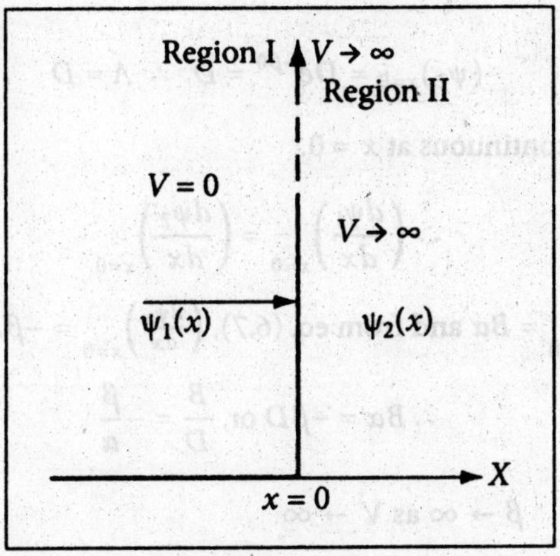

Fig 6.2 Infinite potential

For region I

$$\frac{d^2\psi_1}{dx^2} + \frac{2m}{\hbar^2}(E)\,\psi_1 = 0\ [\because V = 0] \text{ or, } \frac{d^2\psi_1}{dx^2} + \alpha^2\psi_1 = 0 \tag{6.3}$$

where $\alpha^2 = \dfrac{2mE}{\hbar^2}$.

The solution of eq. (6.3) for $x < 0$, where $\psi = \psi_1$ and

$$\psi_1 = A\cos\alpha x + B\sin\alpha x \tag{6.4}$$

For region II

$$\frac{d^2\psi_2}{dx^2} + \frac{2m}{\hbar^2}(E - V)\,\psi_2 = 0 \text{ or, } \frac{d^2\psi_2}{dx^2} - \frac{2m}{\hbar^2}(V - E)\psi_2 = 0\ [\because V > E]$$

$$\text{or, } \frac{d^2\psi_2}{dx^2} - \beta^2\psi_2 = 0 \tag{6.5}$$

where $\beta = \dfrac{2m(V - E)}{\hbar^2}$.

The solution of eq. (6.5) is

$$\psi_2 = Ce^{\beta x} + De^{-\beta x} \tag{6.6}$$

Since, $Ce^{\beta x} \to \infty$ as $x \to \infty$, it is not acceptable.

We write:

$$\psi_2 = De^{-\beta x} \tag{6.7}$$

Now, ψ_1 and ψ_2 must be continuous at $x = 0$.

$$\therefore (\psi_1)_{x=0} = (\psi_2)_{x=0}$$

From eq. (6.4)

$$(\psi_1)_{x=0} = A\cos\alpha 0 + B\sin\alpha 0 = A$$

From eq. (6.7)

$$(\psi_2)_{x=0} = De^{-\beta 0} = D. \therefore A = D \tag{6.8}$$

Again, $\left(\frac{d\psi}{dx}\right)$ should also be continuous at $x = 0$.

$$\therefore \left(\frac{d\psi_1}{dx}\right)_{x=0} = \left(\frac{d\psi_2}{dx}\right)_{x=0}$$

Now, from eq. (6.4), $\left(\frac{d\psi_1}{dx}\right)_{x=0} = B\alpha$ and from eq. (6.7), $\left(\frac{d\psi_2}{dx}\right)_{x=0} = -\beta D$.

$$\therefore B\alpha = -\beta D \text{ or, } \frac{B}{D} = -\frac{\beta}{\alpha} \tag{6.9}$$

Now, since $\beta = \frac{2m}{\hbar}(V - E)$, $\therefore \beta \to \infty$ as $V \to \infty$.

Also as $B \neq \infty, D = 0$.

$$\therefore A = D = 0; \text{ hence, } \psi_2 = De^{-\beta x} = 0 \tag{6.10}$$

This shows that at the surface of infinite potential the wave function $\psi_2 = 0$.

For a physically meaningful solution, $\psi(x)$ must satisfy the conditions given below:

1. $\int_{-\infty}^{+\infty} |\psi|^2\, dx$ must be finite.
2. ψ must be continuous and single-valued.
3. $\frac{d\psi}{dx}$ must be continuous and finite.

6.4 A single step potential

The potential function $V(x)$ in a one-dimensional single step potential undergoes a discontinuous change as shown in Fig 6.1. The potential function $V(x)$ of the potential step may be expressed as

$$V(x) = \begin{cases} 0, & \text{for } x < 0 \\ V_0, & \text{for } x > 0 \end{cases} \tag{6.11}$$

The potential difference between the dees of a cyclotron or the potential barrier at the surface of a metal is approximately similar to this step potential.

Let a particle of mass m and kinetic energy E be travelling parallel to the x-axis and incident on the potential step at $x = 0$.

Since, $V(x) = 0$ for $x < 0$ and $V(x) = V_0$ for $x > 0$, the energy E of the particle is wholly kinetic for $x < 0$, and partly kinetic and partly potential for $x > 0$.

From classical point of view in region I, the particle can move freely as $V(x) = 0$. But, in region II, if $E < V_0$, the particle will remain in region I for ever and sometimes will be reflected back at $x = 0$. But, for $E > V_0$ the particle will not be reflected back at $x = 0$, instead will enter the region II.

Quantum mechanically, it will be shown that the electrons will be partly reflected and partly transmitted at the discontinuity.

Case A: $(E > V_0)$

The time-independent Schrödinger equation for region I is

$$-\frac{\hbar^2}{2m}\frac{d^2\psi_1}{dx^2} = E\psi_1 \text{ or, } \frac{d^2\psi_1}{dx^2} + \frac{2mE}{\hbar^2}\psi_1 = 0 \text{ or, } \frac{d^2\psi_1}{dx^2} + \alpha^2\psi_1 = 0 \tag{6.12}$$

where $\alpha = \sqrt{\frac{2mE}{\hbar^2}}$ and ψ_1 is the wave function for the particle moving in region I.

For region II, the time-independent Schrödinger equation is

$$\frac{d^2\psi_2}{dx^2} + \frac{2m}{\hbar^2}(E - V_0)\psi_2 = 0 \text{ or, } \frac{d^2\psi_2}{dx^2} + \beta^2\psi_2 = 0 \tag{6.13}$$

where $\beta = \sqrt{\frac{2m(E-V_0)}{\hbar^2}}$ and ψ_2 is the wave function for the particle in region II.

The general solutions of eqs (6.12) and (6.13) are

$$\psi_1 = Ae^{i\alpha x} + Be^{-i\alpha x} \tag{6.14}$$

$$\psi_2 = Ce^{i\beta x} + De^{-i\beta x} \tag{6.15}$$

The terms $e^{i\alpha x}$ and $e^{-i\alpha x}$ represent respectively a wave travelling along the positive x-direction and a wave travelling along the negative x-direction, i.e., a reflected wave.

Again, the terms $e^{i\beta x}$ and $e^{-i\beta x}$ represent respectively a wave travelling along the positive x-direction and a wave travelling along the negative x-direction (a reflected wave). But, due to the presence of only one discontinuity at $x = 0$ there cannot be any reflected wave in region II.

Hence, the term $De^{-i\beta x}$ is discarded.

$$\therefore \psi_2 = Ce^{i\beta x} \tag{6.16}$$

The values of the constants A, B and C can be determined by applying the following boundary conditions:

1. $(\psi_1)_{x=0} = (\psi_2)_{x=0}$. 2. $\left(\frac{d\psi_1}{dx}\right)_{x=0} = \left(\frac{d\psi_2}{dx}\right)_{x=0}$.

Using the above boundary condition in eqs (6.14) and (6.16), we get

$$A + B = C \tag{6.17}$$

$$\text{and } \alpha A - \alpha B = \beta C \tag{6.18}$$

$$\text{or, } \alpha A - \alpha B = \beta(A + B)$$

$$\text{or, } (\alpha - \beta)A = (\alpha + \beta)B \text{ or, } B = \frac{\alpha - \beta}{\alpha + \beta}A \tag{6.19}$$

Now, from eqs (6.17) and (6.19), we get

$$A + \frac{\alpha - \beta}{\alpha + \beta}A = C \text{ or, } C = \left(\frac{2\alpha}{\alpha + \beta}\right)A \tag{6.20}$$

$$\therefore \psi_1 = A\left[e^{i\alpha x} + \left(\frac{\alpha - \beta}{\alpha + \beta}\right)e^{-i\alpha x}\right] \tag{6.21}$$

$$\text{and } \psi_2 = \left(\frac{2\alpha}{\alpha + \beta}\right)Ae^{i\alpha\beta_2 x} \tag{6.22}$$

The probability current density for the incident beam of particles in region I is

$$J_i = \frac{\hbar\alpha}{m}(Ae^{i\alpha x})(Ae^{i\alpha x})^* = \frac{\hbar\alpha}{m}(AA^* e^{i\alpha x}\cdot e^{-i\alpha x}) \text{ or, } J_i = \frac{\hbar\alpha}{m}|A|^2 \quad (6.23)$$

The probability current density for the reflected beam is

$$J_r = \frac{\hbar\alpha}{m}(Be^{-i\alpha x})(Be^{-i\alpha x})^* = \frac{\hbar\alpha}{m}|B|^2 \quad (6.24)$$

The net probability current density in the direction from left to right is

$$J = J_i - J_r = \frac{\hbar\alpha}{m}(|A|^2 - |B|^2)$$

But, $|B|^2 = \dfrac{(\alpha-\beta)^2}{(\alpha+\beta)^2}|A|^2$ [from eq. (6.19)]

$$\therefore J = \frac{\hbar\alpha}{m}|A|^2\left[1 - \frac{(\alpha-\beta)^2}{(\alpha+\beta)^2}\right] = \frac{\hbar\alpha}{m}|A|^2\left[\frac{4\alpha\beta}{(\alpha+\beta)^2}\right] \text{ or, } J = \frac{4\alpha\beta}{(\alpha+\beta)^2}\frac{\hbar\alpha}{m}|A|^2 \quad (6.25)$$

Again, the probability current density for the transmitted beam of particles in region II is

$$J_t = \frac{\hbar\beta}{m}\psi_2\psi_2^* = \frac{\hbar\beta}{m}(Ce^{i\beta x})(Ce^{i\beta x})^* = \frac{\hbar\beta}{m}|C|^2$$

But, $C = \dfrac{2\alpha}{\alpha+\beta}A$.

$$\therefore J_t = \frac{\hbar\beta}{m}\left(\frac{2\alpha}{\alpha+\beta}\right)^2|A|^2 = \frac{4\alpha^2}{(\alpha+\beta)^2}\frac{\hbar\beta}{m}|A|^2 \quad (6.26)$$

From eqs (6.25) and (6.26), $J = J_t$, i.e., the net probability current density from left to right in region I is equal to the probability current density in region II.

Again,

$$J_r + J_t = \frac{\hbar\alpha}{m}|A|^2\frac{(\alpha-\beta)^2}{(\alpha+\beta)^2} + \frac{4\alpha}{(\alpha+\beta)^2}\frac{\hbar\beta}{m}|A|^2 = \left[\frac{(\alpha-\beta)^2}{(\alpha+\beta)^2} + \frac{4\alpha\beta}{(\alpha+\beta)^2}\right]\frac{\hbar\alpha}{m}|A|^2$$

$$= \left[\frac{(\alpha-\beta)^2}{(\alpha+\beta)^2} + \frac{4\alpha\beta}{(\alpha+\beta)^2}\right]J_i$$

$$\text{or, } J_r + J_t = J_i \quad (6.27)$$

This shows that *the total number of particles is conserved in the process.*

Reflection and transmission coefficients

The reflection coefficient is defined as

$$R = \frac{\text{Reflected flux } (J_r)}{\text{Incident flux } (J_i)} = \frac{\frac{\hbar\alpha}{m}|B|^2}{\frac{\hbar\alpha}{m}|A|^2} = \frac{|B|^2}{|A|^2} = \left(\frac{\alpha-\beta}{\alpha+\beta}\right)^2 \quad \left[\because B = \frac{\alpha-\beta}{\alpha+\beta}A\right]$$

The transmission coefficient is

$$T = \frac{\text{Transmitted flux } (J_t)}{\text{Incident flux } (J_i)} = \frac{\frac{\hbar\beta}{m}|C|^2}{\frac{\hbar\alpha}{m}|A|^2} = \frac{\beta}{\alpha}\frac{|C|^2}{|A|^2} = \frac{\beta}{\alpha}\frac{4\alpha^2}{(\alpha+\beta)^2} = \frac{4\alpha\beta}{(\alpha+\beta)^2} \quad \left[\because C = \frac{2\alpha}{\alpha+\beta}A\right]$$

$$\therefore T + R = \frac{4\alpha\beta}{(\alpha+\beta)^2} + \frac{(\alpha-\beta)^2}{(\alpha+\beta)^2} = 1 \tag{6.28}$$

Equation (6.28) indicates that at the boundary between regions I and II, the probability of finding the particle is conserved.

Equation (6.20) shows that $|C| > |A|$ as $\alpha > \beta$.

In other words, the amplitude (C) of the transmitted wave is greater than that of the incident wave (i.e., A). The de Broglie wavelength in region I will be shorter than that in region II. This is due to the fact that the kinetic energy of the particle in region I is greater than the kinetic energy in region II. The nature of the waveform is shown in Fig 6.3.

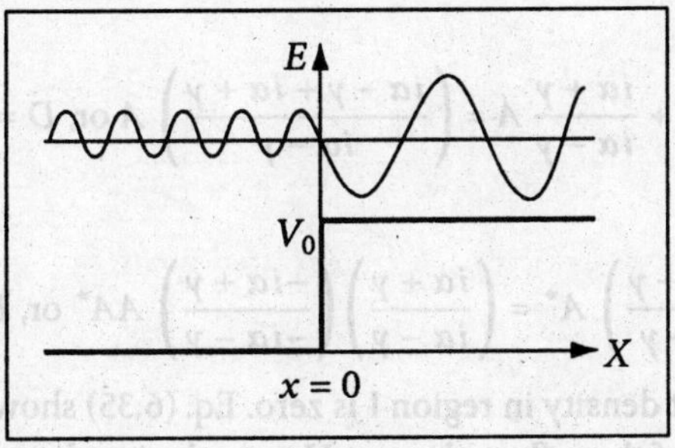

Fig 6.3 A single step potential with $E > V_0$

The analysis thus shows that for $E > V_0$, *the incident particle has certain probability of being reflected and certain probability of being transmitted.*

Case B: ($E < V_0$)

For region I ($x < 0$) the Schrödinger wave equation is

$$\frac{d^2\psi_1}{dx^2} + \alpha^2\psi_1 = 0, \text{ where } \alpha^2 = \frac{2mE}{\hbar^2}$$

The general solution of the above equation is as above

$$\psi_1 = Ae^{i\alpha x} + Be^{-i\alpha x}$$

For region II ($x > 0$) the wave equation is

$$\frac{d^2\psi_2}{dx^2} - \frac{2m}{\hbar^2}(V_0 - E)\psi_2 = 0 \; (\because V_0 > E)$$

$$\text{or, } \frac{d^2\psi_2}{dx^2} - \gamma^2\psi_2 = 0 \; \left[\gamma^2 = \frac{2m}{\hbar^2}(V_0 - E)\right]$$

The general solution of the above equation is

$$\psi_2 = Ce^{\gamma x} + De^{-\gamma x} \tag{6.29}$$

But, the term, $e^{\gamma x} \to \infty$ as $x \to \infty$, which contradicts the condition of well-behaved wave function.

$$\therefore \psi_2 = De^{-\gamma x} \tag{6.30}$$

The values of the constants A, B and D can be calculated by applying the following boundary conditions:

1. $(\psi_1)_{x=0} = (\psi_2)_{x=0}$. 2. $\left(\frac{d\psi_1}{dx}\right)_{x=0} = \left(\frac{d\psi_2}{dx}\right)_{x=0}$.

Thus, we get

$$A + B = D \tag{6.31}$$

$$\text{and } i\alpha A - i\alpha B = -\gamma D \tag{6.32}$$

Solving, we get

$$i\alpha(A-B) = -\gamma(A+B) \text{ or, } (i\alpha+\gamma)A = (i\alpha-\gamma)B \text{ or, } B = \frac{i\alpha+\gamma}{i\alpha-\gamma}A \tag{6.33}$$

and

$$D = A + \frac{i\alpha+\gamma}{i\alpha-\gamma}A = \left(\frac{i\alpha-\gamma+i\alpha+\gamma}{i\alpha-\gamma}\right)A \text{ or, } D = \frac{2i\alpha}{i\alpha-\gamma}A \tag{6.34}$$

Now, from eq. (6.33)

$$BB^* = \left(\frac{i\alpha+\gamma}{i\alpha-\gamma}\alpha\right)A\left(\frac{-i\alpha+\gamma}{i\alpha-\gamma}\right)A^* = \left(\frac{i\alpha+\gamma}{i\alpha-\gamma}\right)\left(\frac{-i\alpha+\gamma}{-i\alpha-\gamma}\right)AA^* \text{ or, } BB^* = AA^* \text{ or, } |B|^2 = |A|^2 \tag{6.35}$$

Thus, the probability current density in region I is zero. Eq. (6.35) shows that the intensity of the incident wave is equal to the intensity of the reflected wave. Hence, the incident and reflected probability currents cancel each other. This can be proved as follows:

We have the probability current density in region II as

$$J_t = -\frac{i\hbar}{2m}\left[\psi_2^*\frac{d\psi_2}{dx} - \psi_2\frac{d\psi_2^*}{dx}\right] = -\frac{i\hbar}{2m}\left[D^*e^{-\gamma x}D(-\gamma)\,e^{-\gamma x} - De^{-\gamma x}\cdot D^*(-\gamma)\,e^{-\gamma x}\right]$$

$$\text{or, } J_t = \frac{i\hbar}{2m}DD^*\gamma\left[e^{-2\gamma x} - e^{-2\gamma x}\right] = 0 \tag{6.36}$$

Reflection and transmission coefficients

The reflection coefficient

$$R = \frac{J_r}{J_i} = \frac{(Be^{-i\alpha x})\,(Be^{-i\alpha x})^*\,\frac{\hbar\alpha}{m}}{(Ae^{i\alpha x})\,(Ae^{i\alpha x})^*\,\frac{\hbar\alpha}{m}} = \frac{BB^*}{AA^*} = \frac{|B|^2}{|A|^2}$$

Since, $|B|^2 = |A|^2$ for $V_0 > E$

$$\therefore R = 1 \tag{6.37}$$

Hence, the incident wave is totally reflected at the potential step at $x = 0$.

The transmission coefficient

$$T = \frac{J_t}{J_i} \tag{6.38}$$

But, for region II, $J_t = 0$. $\therefore$ $T = 0$.

The above analysis leads to the following conclusions:

1. The incident wave is totally reflected at the potential step.
2. In region II, the probability current is zero.

3. In region II, the wave function, $\psi_2 = De^{-\gamma x}$, i.e., ψ_2 is exponentially damped [Fig 6.4]. This leads to a finite probability of finding the particle in region II.

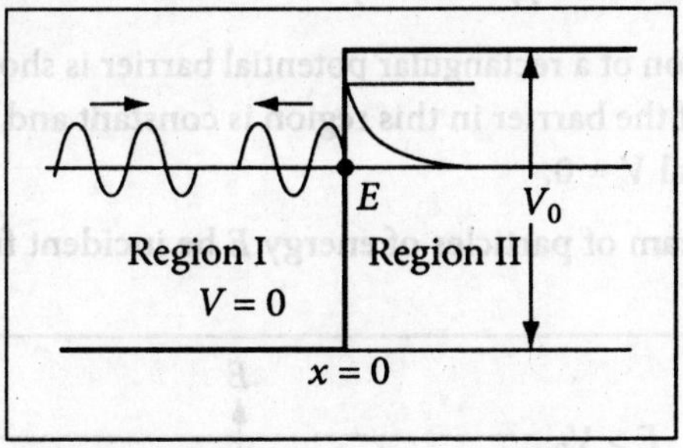

Fig 6.4 Step potential with $E < V_0$

The phenomenon is very much similar to the case of *total internal reflection of electromagnetic waves,* while travelling from denser to a rarer medium.

If for finite energy E the potential $V_0 \to \infty$, then $\gamma = \sqrt{\dfrac{2m(V_0 - E)}{\hbar^2}} \to \infty$.

The wave function ψ_2 in this case, is $\psi_2 = De^{-\gamma x} \to De^{-\alpha x} \to 0$.

Again,

$$B = \left(\frac{i\alpha + \gamma}{i\alpha - \gamma}\right) A \text{ and } D = \frac{2i\alpha}{i\alpha - \gamma} A$$

$$\therefore \lim_{\gamma \to 0} \frac{B}{A} = \lim_{\gamma \to 0} \frac{\alpha - i\gamma}{\alpha + i\gamma} = \lim_{\gamma \to 0} \frac{\frac{\alpha}{\gamma} - i}{\frac{\alpha}{\gamma} + i} = \frac{-i}{i} = -1 \text{ or, } \frac{B}{A} = -1 \text{ or, } A + B = 0 \tag{6.39}$$

Also

$$\lim_{\gamma \to \infty} \frac{D}{A} = \frac{2i\alpha}{i\alpha - \gamma} = \frac{2\alpha}{\alpha + i\gamma} = 0$$

$$\therefore D = 0 \tag{6.40}$$

The wave function in region I is $\psi_1(x) = Ae^{i\alpha x} + Be^{-i\alpha x}$.

When $x = 0$

$$\psi_1(0) = A + B = 0 \tag{6.41}$$

and the wave function in region II is $\psi_2(x) = De^{-\gamma x}$.

At $x = 0$, $\psi_2(0) = D = 0$.

The slope of ψ_1 at $x = 0$ is

$$\left(\frac{d\psi_1}{dx}\right)_{x=0} = A(i\alpha)\, e^{i\alpha 0} + B(-i\alpha)\, e^{-i\alpha 0} = i\alpha(A - B) = i\alpha[A - (-A)] = 2i\alpha A \; [\because A + B = 0] \tag{6.42}$$

Hence, at $x = 0$, where the value of $V(x)$ suddenly increases from 0 to ∞, the wave function $\psi_1(0)$ becomes zero.

Also the slope of ψ_1 at $x = 0$, i.e., $\left(\dfrac{d\psi}{dx}\right)_{x=0}$ suddenly decreases from $2i\alpha A$ to zero.

6.5 One-dimensional rectangular potential barrier (quantum mechanical tunneling effect)

The diagrammatic representation of a rectangular potential barrier is shown in Fig 6.5 which extends from $x = 0$ to $x = a$. The potential of the barrier in this region is constant and equal to V_0. On the left and right sides of the barrier the potential $V = 0$.

Now, we suppose that a stream of particles of energy E be incident from left on the barrier surface at $x = 0$.

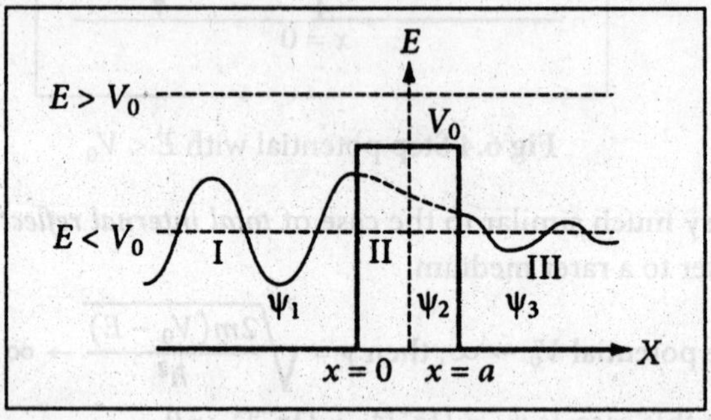

Fig 6.5 One-dimensional rectangular potential barrier

The following two cases arise:

1. If $E > V_0$, then according to classical mechanics the particles will be wholly transmitted and no reflection is possible. But, quantum mechanically there is always some probability at $x = 0$ and $x = a$.
2. If $E < V_0$, then classically the particles will be wholly reflected and hence penetration through the barrier is *impossible*. But, quantum mechanically there is always some probability of penetration into the barrier and appearance of the particles in region III. This finite probability of transmission through the barrier even $E < V_0$ is called *the quantum mechanical tunneling effect*.

Case A: When $E > V_0$

For region I, $(x < 0)$, where $V(x) = 0$, the time-independent Schrödinger equation is

$$\frac{d^2\psi_1}{dx^2} + \frac{2m}{\hbar^2}E\psi_1 = 0 \text{ or, } \frac{d^2\psi_1}{dx^2} + \alpha^2\psi_1 = 0 \tag{6.43}$$

where $\alpha^2 = \dfrac{2mE}{\hbar^2}$ and ψ_1 is the wave function in region I.

The general solution of eq. (6.45) is

$$\psi_1(x) = Ae^{i\alpha x} + Be^{-i\alpha x} \tag{6.44}$$

For region II, $(0 < x < a)$, where $V(x) = V_0$ the Schrödinger equation is

$$\frac{d^2\psi_2}{dx^2} + \frac{2m}{\hbar^2}(E - V_0)\psi_2 = 0 \text{ or, } \frac{d^2\psi_2}{dx^2} + \beta^2\psi_2 = 0 \tag{6.45}$$

where $\beta^2 = \dfrac{2m(E - V_0)}{\hbar^2}$ and ψ_2 is the wave function in region II.

The general solution of eq. (6.45) is

$$\psi_2(x) = Ce^{i\beta x} + De^{-i\beta x} \tag{6.46}$$

Here, $De^{-i\beta x}$ represents the wave which is reflected at the boundary at $x = a$.

For region III, $(x > a)$, where $V(x) = 0$, again the Schrödinger equation is

$$\frac{d^2 \psi_3}{dx^2} + \alpha^2 \psi_3 = 0 \tag{6.47}$$

and the general solution is

$$\psi_3(x) = Fe^{i\alpha x} \tag{6.48}$$

Since, in region III, the wave propagates to the right only, there is no question of reflection and as such the $e^{-i\alpha x}$ term is absent from eq. (6.48).

The five constants A, B, C, D and F can be obtained by applying the boundary conditions at $x = 0$ and $x = a$.

The boundary conditions are

1. ψ is continuous at $x = 0$ and $x = a$ as the probability density must be continuous.
2. $\frac{d\psi}{dx}$ at $x = 0$ and $x = a$ is continuous, since probability current must be continuous.

Applying the boundary conditions at $x = 0$ and $x = a$, we get

$$A + B = C + D \tag{6.49}$$

$$A - B = \frac{\beta}{\alpha}(C - D) \tag{6.50}$$

$$Ce^{i\beta a} + De^{-i\beta a} = Fe^{i\alpha a} \tag{6.51}$$

$$Ce^{i\beta a} - De^{-i\beta a} = \frac{\alpha}{\beta}e^{i\alpha a} \tag{6.52}$$

Now, solving eqs (6.49) and (6.50), we get

$$A = \frac{C}{2}\left(1 + \frac{\beta}{\alpha}\right) + \frac{D}{2}\left(1 - \frac{\beta}{\alpha}\right) \tag{6.53}$$

$$B = \frac{C}{2}\left(1 - \frac{\beta}{\alpha}\right) + \frac{D}{2}\left(1 + \frac{\beta}{\alpha}\right) \tag{6.54}$$

Solving eqs (6.49) and (6.50), we get

$$C = \frac{F}{2}\left(1 + \frac{\alpha}{\beta}\right) e^{i(\alpha-\beta)a} \tag{6.55}$$

$$D = \frac{F}{2}\left(1 - \frac{\alpha}{\beta}\right) e^{i(\alpha+\beta)a} \tag{6.56}$$

Substituting the vålues of C and D from eqs (6.55) and (6.56) in eqs (6.53) and (6.54), we get

$$A = \frac{F}{4}e^{i\alpha a}\left[\left(1 + \frac{\beta}{a}\right)\left(1 + \frac{\alpha}{\beta}\right)e^{-i\beta a} + \left(1 - \frac{\beta}{a}\right)\left(1 - \frac{\alpha}{\beta}\right)e^{i\beta a}\right] \tag{6.57}$$

$$B = \frac{F}{4}e^{i\alpha a}\left[\left(1 - \frac{\beta}{\alpha}\right)\left(1 + \frac{\alpha}{\beta}\right)e^{-i\beta a} + \left(1 + \frac{\beta}{\alpha}\right)\left(1 - \frac{\alpha}{\beta}\right)e^{i\beta a}\right] \tag{6.58}$$

From eq. (6.57)

$$\frac{F}{A} = \frac{4e^{-i\alpha a}}{\left(1+\frac{\beta}{\alpha}\right)\left(1+\frac{\alpha}{\beta}\right)e^{-i\beta a} + \left(1-\frac{\beta}{\alpha}\right)\left(1-\frac{\alpha}{\beta}\right)e^{i\beta a}}$$

$$= \frac{2\alpha\beta e^{-i\alpha a}}{(\alpha^2+\beta^2)\left(\frac{e^{-i\beta a}-e^{i\beta a}}{2}\right) + 2\alpha\beta\left(\frac{e^{-i\beta a}+e^{i\beta a}}{2}\right)}$$

$$= \frac{2\alpha\beta e^{-i\alpha a}}{2\alpha\beta\cos\beta a - i(\alpha^2+\beta^2)\sin\beta a} \tag{6.59}$$

The transmission coefficient (T) and reflection coefficient (R).

The transmission coefficient is given by

$$T = \frac{J_t}{J_i} = \frac{\frac{\hbar k}{m}(Fe^{iax})(Fe^{-iax})^*}{\frac{\hbar k}{m}(Ae^{iax})(Ae^{iax})^*} = \frac{FF^*}{AA^*} = \frac{|F|^2}{|A|^2} = \left(\frac{F}{A}\right)\left(\frac{F}{A}\right)^*$$

$$= \frac{4\alpha^2\beta^2}{4\alpha^2\beta^2\cos^2\beta a + (\alpha^2+\beta^2)^2\sin^2\beta a} = \frac{1}{1+\frac{(\alpha^2-\beta^2)^2}{4\alpha^2\beta^2}\sin^2\beta a}$$

But,

$$\alpha^2 - \beta^2 = \frac{2mE}{\hbar^2} - \frac{2m}{\hbar^2}(E-V_0) \text{ or, } (\alpha^2-\beta^2)^2 = \left(\frac{2m}{\hbar^2}\right)^2 V_0^2$$

$$\text{and } \alpha^2\beta^2 = \left(\frac{2mE}{\hbar^2}\right)\left[\frac{2m}{\hbar^2}(E-V_0)\right] = \left(\frac{2m}{\hbar^2}\right)^2 E(E-V_0)$$

$$\therefore T = \frac{1}{1+\frac{V_0^2}{4E(E-V_0)}\sin^2\beta a} \tag{6.60}$$

From eq. (6.60) we note that $T = 1$, if $\sin^2\beta a = 0$ or $\beta a = n\pi$, $n = 1, 2, 3, \ldots$, then

$$a = \frac{n\pi}{\beta} = n\frac{\pi\hbar}{\beta\hbar} = n\frac{h/2}{p} = n\frac{\lambda_2}{2}$$

where $\hbar = h/2\pi$ and $\beta\hbar = p$, the momentum of the particle in region II. $\lambda_2 = h/p$, the de Broglie wavelength of the particle.

Hence, *if the barrier region contains an integral multiple of half de Broglie wavelengths, then the barrier transmission is complete. This is Ramsauer-Townsend Effect.* It explains the phenomenon of scattering of low energy electrons from the atoms of noble gases.

Equation (6.60) shows that T oscillates for $E/V_0 > 1$ and approaches to unity for $E\gamma V_0$ [Fig 6.6].

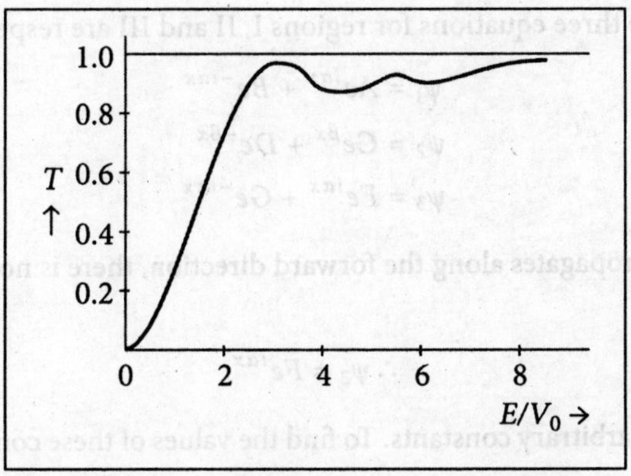

Fig 6.6 Transmission through a rectangular potential barrier as a function of energy

The reflection coefficient is given by

$$R = \left(\frac{B}{A}\right)\left(\frac{B}{A}\right)^* = \frac{\dfrac{(\alpha^2-\beta^2)^2}{4\alpha^2\beta^2}\sin^2\beta a}{1+\left(\dfrac{\alpha^2-\beta^2}{4\alpha^2\beta^2}\right)\sin^2\beta a} = \frac{\dfrac{V_0^2}{4E(E-V_0)}\sin^2\beta a}{1+\dfrac{V_0^2}{4E(E-V_0)}\sin^2\beta a} \tag{6.61}$$

From eqs (6.60) and (6.61)

$$R + T = 1$$

The natures of the wave functions ψ_1, ψ_2 and ψ_3 in regions I, II and III respectively are shown in Fig 6.7.

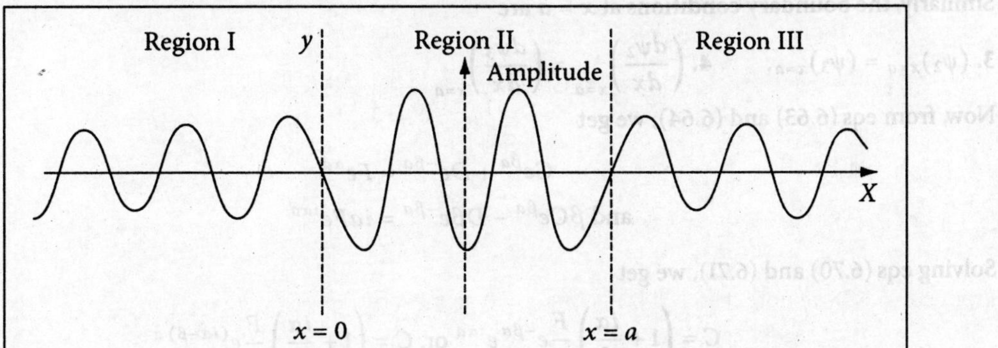

Fig 6.7 Wave function of a particle in a rectangular potential barrier

Case B: When $E < V_0$

The time-independent Schrödinger equations for regions I, II and III are given by

$$\frac{d^2\psi_1}{dx^2} + \alpha^2\psi_1 = 0, \qquad \frac{d^2\psi_2}{dx^2} - \beta^2\psi_2 = 0, \qquad \frac{d^2\psi_3}{dx^2} + \alpha^2\psi_3 = 0$$

where $\alpha^2 = \dfrac{2mE}{\hbar^2}$ and $\beta^2 = \dfrac{2m}{\hbar^2}(V_0 - E)$.

The solutions of the above three equations for regions I, II and III are respectively

$$\psi_1 = Ae^{i\alpha x} + Be^{-i\alpha x} \tag{6.62}$$

$$\psi_2 = Ce^{\beta x} + De^{-\beta x} \tag{6.63}$$

$$\psi_3 = Fe^{i\alpha x} + Ge^{-i\alpha x} \tag{6.64}$$

In region III as the wave propagates along the forward direction, there is no reflection and hence $Ge^{-i\alpha x}$ is discarded.

$$\therefore \psi_3 = Fe^{i\alpha x} \tag{6.65}$$

Here, A, B, C, D and F are arbitrary constants. To find the values of these constants we apply the following boundary conditions:

1. $(\psi_1)_{x=0} = (\psi_2)_{x=0}$. 2. $\left(\frac{d\psi_1}{dx}\right)_{x=0} = \left(\frac{d\psi_2}{dx}\right)_{x=0}$.

Applying these boundary conditions to eqs (6.62) and (6.63), we get

$$A + B = C + D \tag{6.66}$$

$$\text{and } i\alpha A - i\alpha B = \beta C - \beta D \tag{6.67}$$

Solving the above two equations, we get

$$A = \left(1 - \frac{i\beta}{\alpha}\right)\frac{C}{2} + \left(1 + \frac{i\beta}{\alpha}\right)\frac{D}{2} \tag{6.68}$$

$$\text{and } B = \left(1 + \frac{i\beta}{\alpha}\right)\frac{C}{2} + \left(1 - \frac{i\beta}{\alpha}\right)\frac{D}{2} \tag{6.69}$$

Similarly, the boundary conditions at $x = a$ are

3. $(\psi_2)_{x=a} = (\psi_3)_{x=a}$. 4. $\left(\frac{d\psi_2}{dx}\right)_{x=a} = \left(\frac{d\psi_3}{dx}\right)_{x=a}$.

Now, from eqs (6.63) and (6.64), we get

$$Ce^{\beta a} + De^{-\beta a} = Fe^{\alpha a} \tag{6.70}$$

$$\text{and } \beta Ce^{\beta a} - D\beta e^{-\beta a} = i\alpha Fe^{i\alpha a} \tag{6.71}$$

Solving eqs (6.70) and (6.71), we get

$$C = \left(1 + \frac{i\alpha}{\beta}\right)\frac{F}{2}e^{-\beta a}e^{i\alpha a} \text{ or, } C = \left(1 + \frac{i\alpha}{\beta}\right)\frac{F}{2}e^{(i\alpha - \beta)a} \tag{6.72}$$

$$\text{and } D = \left(1 - \frac{i\alpha}{\beta}\right)\frac{F}{2}e^{(i\alpha + \beta)a} \tag{6.73}$$

Using these values of C and D in eqs (6.68) and (6.69), we get

$$A = \frac{F}{4}e^{i\alpha a}\left[\left(1 + \frac{i\beta}{a}\right)\left(1 + \frac{\alpha}{i\beta}\right)e^{-\beta a} + \left(1 - \frac{i\beta}{\alpha}\right)\left(1 - \frac{\alpha}{i\beta}\right)e^{\beta a}\right] \tag{6.74}$$

$$B = \frac{F}{4}e^{i\alpha a}\left[\left(1 - \frac{i\beta}{\alpha}\right)\left(1 + \frac{\alpha}{i\beta}\right)e^{-\beta a} + \left(1 + \frac{i\beta}{\alpha}\right)\left(1 - \frac{\alpha}{i\beta}\right)e^{\beta a}\right] \tag{6.75}$$

Transmission Coefficient and Reflection Coefficient

The transmission coefficient is given by

$$T = \frac{J_t}{J_i} = \frac{\frac{\hbar k}{m}(Fe^{i\alpha x})(Fe^{i\alpha x})^*}{\frac{\hbar k}{m}(Ae^{i\alpha x})(Ae^{i\alpha x})} = \left(\frac{F}{A}\right)\left(\frac{F}{A}\right)^* \tag{6.76}$$

Now, from eq. (6.74), we get

$$\frac{F}{A} = \frac{4}{e^{i\alpha a}\left[\left(1+\frac{i\beta}{\alpha}\right)\left(1+\frac{\alpha}{i\beta}\right)e^{-\beta a} + \left(1-\frac{i\beta}{\alpha}\right)\left(1-\frac{\alpha}{i\beta}\right)e^{\beta a}\right]}$$

$$\therefore T = \left(\frac{F}{A}\right)\left(\frac{F}{A}\right)^* = \frac{1}{\cosh^2(\beta a) + \frac{1}{4}\left(\frac{\beta}{\alpha} - \frac{\alpha}{\beta}\right)^2 \sinh^2(\beta a)} = \frac{1}{1+\sinh^2(\beta a) + \left(\frac{\beta^2-\alpha^2}{2\beta\alpha}\right)^2 \sinh^2(\beta a)}$$

$$= \frac{1}{1+\sinh^2(\beta a)\left[1+\left(\frac{\beta^2-\alpha^2}{2\alpha\beta}\right)^2\right]} = \frac{1}{1+\left(\frac{\beta^2+\alpha^2}{4\alpha^2\beta^2}\right)\sinh^2\beta a} \tag{6.77}$$

Substituting $\alpha^2 = \frac{2mE}{\hbar^2}$ and $\beta^2 = \frac{2m(V_0-E)}{\hbar^2}$, we get from eq. (6.77)

$$T = \frac{1}{1+\frac{V_0^2}{4E(V_0-E)}\sinh^2(\beta a)} \tag{6.78}$$

The reflection coefficient is given by

$$R = \frac{J_r}{J_i} = \frac{\frac{\hbar k}{m}(Be^{-i\alpha x})(Be^{-i\alpha x})^*}{\frac{\hbar k}{m}(Ae^{i\alpha x})(Ae^{i\alpha x})^*} = \left(\frac{B}{A}\right)\left(\frac{B}{A}\right)^* = \frac{|B|^2}{|A|^2} \tag{6.79}$$

Now, from eqs (6.74) and (6.75)

$$\frac{B}{A} = \frac{\left(1-\frac{i\beta}{\alpha}\right)\left(1+\frac{\alpha}{i\beta}\right)e^{-\beta a} + \left(1+\frac{i\beta}{\alpha}\right)\left(1-\frac{\alpha}{i\beta}\right)e^{\beta a}}{\left(1+\frac{i\beta}{\alpha}\right)\left(1+\frac{\alpha}{i\beta}\right)e^{-\beta a} + \left(1-\frac{i\beta}{\alpha}\right)\left(1-\frac{\alpha}{i\beta}\right)e^{\beta a}}$$

$$\therefore R = \left(\frac{B}{A}\right)\left(\frac{B}{A}\right)^* \quad \left[\left(\frac{B}{A}\right)^* \text{ is the complex conjugate of } (B/A)\right]$$

$$= \frac{\frac{(\beta^2+\alpha^2)^2}{4\alpha^2\beta^2}\sinh^2\beta a}{\cosh^2\beta a + \frac{1}{4}\frac{(\beta^2-\alpha^2)^2}{4\alpha^2\beta^2}\sinh^2\beta a} = \frac{\left(\frac{\beta^2+\alpha^2}{2\beta\alpha}\right)^2\sinh^2\beta a}{\cosh^2\beta a + \left(\frac{\beta^2-\alpha^2}{2\beta\alpha}\right)^2\sinh^2\beta a}$$

$$\text{or, } R = \frac{\left(\frac{\beta^2+\alpha^2}{2\beta\alpha}\right)^2 \sinh^2 \beta a}{1+\left(\frac{\beta^2+\alpha^2}{2\beta\alpha}\right)^2 \sinh^2 \beta a} \tag{6.80}$$

Quantum mechanical tunneling

From eq. (6.77) we see that the transmission coefficient $T \neq 0$, which means that there is a finite probability of transmission of the particles through the potential barrier of height V_0 and width a, even if $E < V_0$. This cannot be accounted for in view of classical theory.

Thus, quantum mechanical tunneling is the phenomenon of transmission of a particle through a potential barrier of finite width and height when its energy is less than the barrier height.

The transmission probability depends on V_0 and a such that T is small if V_0 and a are large.

Again, when $a \to$ large value $\beta a \to \infty$.

$$\therefore \sinh^2(\beta a) = \left(\frac{e^{\beta a} - e^{-\beta a}}{2}\right)^2 = \frac{e^{2\beta a}}{4} \gg 1$$

$$\therefore T = \frac{16\alpha^2\beta^2}{(\beta^2+\alpha^2)} e^{-2\beta a} = \frac{16(V_0 - E)}{V_0^2} e^{-2\beta a} \tag{6.81}$$

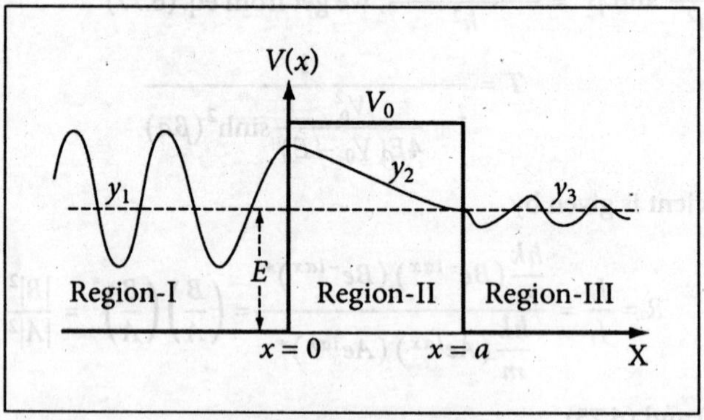

Fig 6.8 y_1, y_2 and y_3 for $E < V_0$

The waveforms in region I, II and III are shown in Fig 6.8. The appearance of the incident and transmitted waves are the same as in case $E > V_0$. But, for $E < V_0$ there is no wave inside the barrier except an exponentially decaying function as is evident from eq. (6.81).

From eqs (6.77) and (6.80) it is easily seen that $R + T = 1$.

From the foregoing discussion the following conclusion are drawn:

Case A: When $E > V_0$

1. When $E \to V_0$

$$\beta a = a\sqrt{\frac{2m}{\hbar^2}(E - V_0)} \ll 1$$

$$\therefore \sin \beta a = a\sqrt{\frac{2m}{\hbar^2}(E - V_0)} \text{ and } \sin^2 \beta a = \frac{2m(E - V_0)\, a^2}{\hbar^2}$$

From eq. (6.60)

$$T = \frac{1}{1 + \frac{V_0^2}{4E(E - V_0)} \sin^2 \beta a} = \frac{1}{1 + \frac{V_0^2}{4E(E - V_0)} \times \frac{2m(E - V_0)\, a^2}{\hbar^2}} = \frac{1}{1 + \frac{2mV_0^2 a^2}{4E\hbar^2}} \simeq \frac{1}{1 + \frac{mV_0^2 a^2}{2E\hbar^2}}$$

Putting $E \simeq V_0$

$$T = \frac{1}{1 + \frac{mV_0 a^2}{2\hbar^2}} \tag{6.82}$$

2. If $E > V_0$, then $(E - V_0)$ increases, $\beta = \sqrt{\frac{2m(E - V_0)}{\hbar^2}}$ also increases. But, $(E - V_0)$ increases more rapidly than $\sin^2 \beta a$ for given values of a and V_0.
   Hence, T increases with the increase of energy E of the incident particle.
3. In this case, as shown before that if $a = n(\lambda/2)$, then there is a perfect transmission of the incident beam of particles.

Case B: When $E < V_0$

1. There is a finite probability of transmission of a particle through a potential barrier of width a and height V_0, even if $E < V_0$.
2. The factor $(V_0 - E)$ decreases with the increase of E.

$$\therefore \beta = \sqrt{\frac{2m(V_0 - E)}{\hbar^2}}$$

also decreases. The rate of decrease of $\sin^2 \beta a$ for a fixed value of a and V_0 is more than that of $(V_0 - E)$. So T increases steadily with E.

3. When $E \to V_0$

$$\beta a = a\sqrt{2m(V_0 - E)} \ll 1$$

$$\therefore \sinh \beta a = \sqrt{\frac{2m(V_0 - E)}{\hbar^2}}\, a \text{ and } \sinh^2 \beta a = \frac{2m(V_0 - E)\, a}{\hbar^2}$$

Then from eq. (6.77)

$$T = \frac{1}{1 + \frac{V_0^2}{4E(V_0 - E)} \times \frac{2m(V_0 - E)\, a}{\hbar^2}} \simeq \frac{1}{1 + \frac{2mV_0^2 a^2}{4E\hbar^2}} = \frac{1}{1 + \frac{mV_0 a^2}{2\hbar^2}} \text{ (as } E \to V_0) \tag{6.83}$$

This value T, in case $E > V_0$, is the same as the value of T for $E < V_0$. T is small, if V_0 and a are large.

Application of tunneling effect

Tunneling effect provides satisfactory explanation of the following phenomena:

1. α-particle emission from radioactive elements
2. Field emission of electrons from a cold metallic surface

3. The reverse breakdown of semiconductor diodes
4. Switching action of a tunnel diode
5. Electrical breakdown of insulators.

6.6 Application of barrier penetration (α-decay)

Heavy elements ($Z \geqslant 81$) are generally unstable. In an attempt to acquire stability, they undergo spontaneous disintegration with the emission of α-particles, β-particles and γ-rays. α-decay is the process in which a nucleus loses an α-particle and is converted into another nucleus with charge number less by two and mass number less by four. The decay mode is expressed as ${}_Z X^A \rightarrow {}_{Z-2} Y^{A-4} + {}_2 He^4$ (α-particle).

The potential energy of an α-particle of charge $2e$ at a large distance r from a daughter nucleus of $(Z-2)$ is

$$V(r) = \frac{(Z-2)\,e}{4\pi\varepsilon_0 r} \times 2e = \frac{2(Z-2)\,e^2}{4\pi\varepsilon_0 r}$$

If r is decreased, $V(r)$ is increased and becomes the maximum (V_0) at $r = R$, where R is the effective nuclear radius. This maximum potential energy is equal to the work done against the Coulombian repulsive force from $r = \infty$ to $r = R$. Inside the nucleus the α-particle is acted on by strong short range nuclear force over a short distance R_0 from the centre of the nucleus.

Since, the force on the α-particle is attractive, the related potential energy is negative. The graph of $V(r)$ versus r is of the form as shown in Fig 6.9.

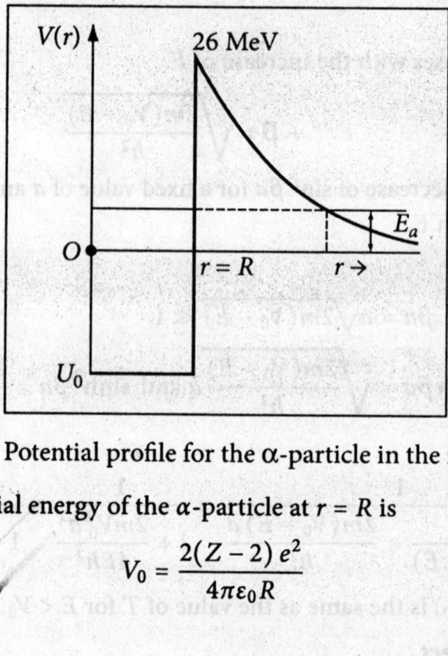

Fig 6.9 Potential profile for the α-particle in the nucleus

The maximum positive potential energy of the α-particle at $r = R$ is

$$V_0 = \frac{2(Z-2)\,e^2}{4\pi\varepsilon_0 R}$$

For ${}_{92}U^{238}$

$$V_0 = \frac{2(92-2) \times (1.6 \times 10^{-19})^2 \times 8.99 \times 10^9}{10^{-14}} = \textbf{26 MeV} \text{ (using standard values of } Z, e, 4\pi\varepsilon_0 \text{ and } R).$$

The α-particles emitted by most of the radioactive nuclei have energies from $E_\alpha = 5$ to 6 MeV and in some cases about 10 MeV. Thus, E_α is much lower than V_0. Thus, according to classical mechanics it is very difficult to understand how the α-particles contained inside the nucleus can go over a potential barrier which is more than twice as high as their total energy.

Quantum mechanically, however there is always certain probability of the α-particle of penetrating through the barrier. The transmission coefficient is given by

$$T \sim e^{-2\beta a} = e^{\left[-2\sqrt{\frac{2m}{\hbar^2}(V_0-E)}\cdot\alpha\right]}, \text{ where } 2\beta a \gg 1$$

We here assume that the α-particle moves back and forth almost freely inside the nucleus of diameter 10^{-12} cm with a velocity ~ 10^7 m/s. Then the probability per second to come out of the nucleus is $P = 10^{21} T$.

The mean lifetime of the nucleus is

$$\tau = \frac{1}{P} = \frac{10^{-21}}{T} \text{ sec}$$

Now, we have

$$T = \frac{e^{-2\beta a}}{\left[1+\left(\frac{\beta}{\alpha}\right)^2\right]\left[1+\left(\frac{\alpha}{\beta}\right)^2\right]} \simeq e^{-2\beta a}$$

Since,

$$\left[1+\left(\frac{\beta}{\alpha}\right)^2\right]\left[1+\left(\frac{\alpha}{\beta}\right)^2\right] \simeq 1$$

Putting the mass of the α-particle, $m = 6.4 \times 10^{-2.7}$ kg, 1 eV = 1.6×10^{-19} erg and $a \simeq 3 \times R = 3 \times 10^{-14}$ cm, we get

$$2\beta a = 2\left[\sqrt{\frac{2m}{\hbar^2}(V_0 - E)}\right] a = 2\left[\sqrt{\frac{12.8 \times 10^{-27} \times 12 \times 10^6 \times 1.6 \times 10^{-19}}{1.07 \times 10^{-34}}}\right] \times 3 \times 10^{-14} = 90$$

$\therefore T \sim 10^{-19}$.

So $P = 10^{21} e^{-19} = 10^{-18}/\text{s} = 10^{-11}/\text{year}$.

$\therefore$ mean life $\tau = 1/P =$ **10^{11} years.**

A more accurate treatment of α-decay is provided by WKB approximation method.

6.7 One-dimensional square well potential (free states)

A diagrammatic representation of a one-dimensional square well potential extending from $x = 0$ to $x = a$ is shown in Fig 6.10. The three regions of the potential well are as below:

Region I $(x < 0)$: $V = V_0$
Region II $(0 < x < a)$: $V = 0$
Region III $(x > a)$: $V = V_0$

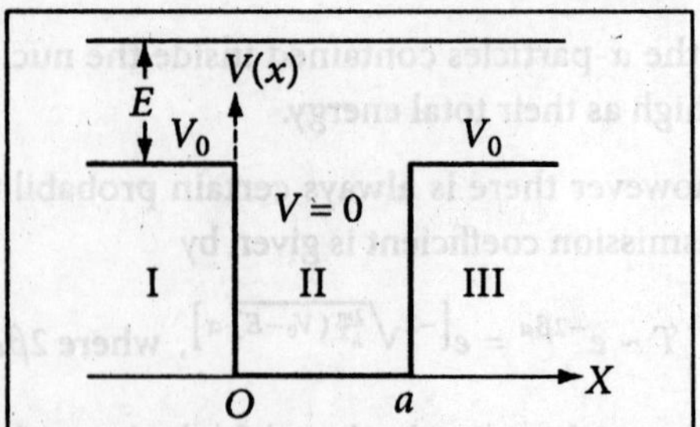

Fig 6.10 Square well potential (one-dimensional) with $E > V_0$

Case I: $E > V_0$

Let a stream of particles each of mass m and energy $E > V_0$ be travelling along the positive x-direction from left to the right. The time-independent Schrödinger equations for region I, II and III are as below:

$$\frac{d^2\psi_1}{dx^2} + \alpha^2\psi_1 = 0 \tag{6.84}$$

$$\frac{d^2\psi_2}{dx^2} + \beta^2\psi_2 = 0 \tag{6.85}$$

$$\text{and } \frac{d^2\psi_3}{dx^2} + \gamma^2\psi_3 = 0, \tag{6.86}$$

$$\text{where } \alpha^2 = \gamma^2 = \frac{2m}{\hbar^2}(E - V_0) \tag{6.87}$$

$$\text{and } \beta^2 = \frac{2mE}{\hbar^2} \tag{6.88}$$

The general solutions of eqs (6.84), (6.85) and (6.85) are given by

$$\psi_1 = Ae^{i\alpha x} + Be^{-i\alpha x} \tag{6.89}$$

$$\psi_2 = Ce^{i\beta x} + De^{-i\beta x} \tag{6.90}$$

$$\psi_3 = Fe^{i\alpha x} + Ge^{-i\beta x} \text{ or, } \psi_3 = Fe^{i\alpha x} \tag{6.91}$$

The term $Ge^{-i\alpha x}$ is discarded as there is no case of reflection of the wave from region III and the wave travels only along the positive x-axis.

To find out the unknown constants A, B, C, D and F we apply the following boundary conditions:

1. $\psi_1(0) = \psi_2(0)$

$$\left(\frac{d\psi_1}{dx}\right)_{x=0} = \left(\frac{d\psi_2}{dx}\right)_{x=0} \tag{6.92}$$

2. $\psi_2(a) = \psi_3(a)$

$$\left(\frac{d\psi_2}{dx}\right)_{x=a} = \left(\frac{d\psi_3}{dx}\right)_{x=a} \tag{6.93}$$

Applying boundary condition (i) at $x = 0$, we get

$$A + B = C + D \tag{6.94}$$

$$\text{and } i\alpha(A - B) = i\beta(C - D) \tag{6.95}$$

At $x = a$

$$Ce^{i\beta a} + De^{-i\beta a} = Fe^{i\alpha a} \tag{6.96}$$

$$\text{and } Ce^{i\beta a} - De^{-i\beta a} = \frac{\alpha}{\beta}Fe^{i\alpha a} \tag{6.97}$$

Applying eqs (6.94) and (6.95), we get

$$2A = \left(1+\frac{\beta}{\alpha}\right)C + \left(1-\frac{\beta}{\alpha}\right)D \tag{6.98}$$

Again, adding eqs (6.96) and (6.97) and, then dividing both sides by $2e^{i\beta a}$ we get

$$C = \frac{1}{2}\left(1+\frac{\alpha}{\beta}\right)Fe^{i(\alpha-\beta)a} \tag{6.99}$$

Subtracting eq. (6.97) from eq. (6.96) and, then dividing both sides by $2e^{-ik_2a}$, we get

$$D = \frac{1}{2}\left(1-\frac{\alpha}{\beta}\right)Fe^{i(\alpha+\beta)a} \tag{6.100}$$

Using these values of C and D in eq. (6.98), we get

$$2A = \left[\left(1+\frac{\beta}{\alpha}\right)\frac{1}{2}\left(1+\frac{\alpha}{\beta}\right)e^{-i\beta a} + \left(1-\frac{\beta}{\alpha}\right)\frac{1}{2}\left(1-\frac{\alpha}{\beta}\right)e^{ika}\right]Fe^{i\alpha a}$$

$$= \left[\frac{1}{2}\left\{2+\left(\frac{\alpha}{\beta}+\frac{\beta}{\alpha}\right)\right\}e^{-i\beta a} + \frac{1}{2}\left\{2-\left(\frac{\alpha}{\beta}+\frac{\beta}{\alpha}\right)\right\}e^{-i\beta a}\right]Fe^{i\alpha a}$$

$$\therefore A = \left[\frac{e^{i\beta a}+e^{-i\beta a}}{2} - \frac{i}{2}\left(\frac{\alpha}{\beta}+\frac{\beta}{\alpha}\right)\left(\frac{e^{i\beta a}-e^{-i\beta a}}{2i}\right)\right]Fe^{i\alpha a} = \left[\cos\beta a - \frac{i}{2}\left(\frac{\alpha}{\beta}+\frac{\beta}{\alpha}\right)\sin\beta a\right]Fe^{i\alpha a}$$

$$\therefore \left(\frac{A}{F}\right) = \left[\cos\beta a - \frac{i}{2}\left(\frac{\alpha}{\beta}+\frac{\beta}{\alpha}\right)\sin\beta a\right]e^{i\alpha a} \tag{6.101}$$

The complex conjugate of (A/F) is

$$\left(\frac{A}{F}\right)^* = \left[\cos\beta a + \frac{i}{2}\left(\frac{\alpha}{\beta}+\frac{\beta}{\alpha}\right)\sin\beta a\right]e^{-i\alpha a} \tag{6.102}$$

Now, the incident and reflected waves in regions I and II are represented as

$$\psi_{1i} = Ae^{i\alpha x};\ \psi_{1r} = Be^{-i\alpha x} \tag{6.103}$$

$$\psi_{2i} = Ce^{i\beta x};\ \psi_{2r} = De^{-i\beta x} \tag{6.104}$$

The transmitted wave in region III is

$$\psi_{3t} = Fe^{i\alpha x} \tag{6.105}$$

Now, the transmission coefficient T is given by

$$T = \frac{J_{3t}}{J_{1i}} \tag{6.106}$$

where J_{3t} and J_{1i} are the probability current density of the transmitted and incident wave respectively.

But,

$$J_{1i} = \frac{i\hbar}{2m}\left[\psi_{1i}\psi_{1i}^{*\prime} - \psi_{1i}^{*}\psi_{1i}^{\prime}\right] = \frac{\hbar}{m}\alpha|A|^2 \tag{6.107}$$

$$\text{and } J_{3t} = \frac{i\hbar}{2m}\left[\psi_{3t}\psi_{3t}^{*\prime} - \psi_{3t}^{*}\psi_{3t}^{\prime}\right] = \frac{\hbar\alpha}{m}|F|^2 \tag{6.108}$$

Now, from eqs (6.105), (6.106) and (6.108), we get

$$T = \frac{\left(\frac{\hbar\alpha}{m}\right)|F|^2}{\left(\frac{\hbar\alpha}{m}\right)|A|^2} = \frac{|F|^2}{|A|^2} = \left(\frac{F}{A}\right)\left(\frac{F}{A}\right)^* \tag{6.109}$$

Now, substituting the values of $\left(\frac{F}{A}\right)$ and $\left(\frac{F}{A}\right)^*$ in eq. (6.109) from eqs (6.101) and (6.102), we get

$$T = \frac{1}{\cos^2\beta a + \frac{1}{4}\left(\frac{\alpha}{\beta}+\frac{\beta}{\alpha}\right)^2\sin^2\beta a} = \frac{1}{1-\sin^2\beta a + \frac{1}{4}\left(\frac{\alpha}{\beta}+\frac{\beta}{\alpha}\right)^2\sin^2\beta a}$$

$$= \frac{1}{1+\frac{1}{4}\left[\left(\frac{\alpha}{\beta}+\frac{\beta}{\alpha}\right)^2-4\right]\sin^2\beta a} = \frac{1}{1+\frac{1}{4}\left(\frac{\alpha}{\beta}-\frac{\beta}{\alpha}\right)^2\sin^2\beta a} \tag{6.110}$$

$$\because \alpha = \sqrt{\frac{2m(E-V_0)}{\hbar^2}},\ \beta = \sqrt{\frac{2m}{\hbar^2}E}.\ \therefore\ \frac{\alpha}{\beta} = \sqrt{\frac{(E-V_0)}{E}} \text{ and } \frac{\beta}{\alpha} = \sqrt{\frac{E}{(E-V_0)}}$$

$$\therefore \left(\frac{\alpha}{\beta}-\frac{\beta}{\alpha}\right) = \sqrt{\frac{E-V_0}{E}} - \sqrt{\frac{E}{E-V_0}} = \frac{E-V_0-E}{\sqrt{E(E-V_0)}} \text{ or, } \left(\frac{\alpha}{\beta}-\frac{\beta}{\alpha}\right)^2 = \frac{+V_0^2}{E(E-V_0)} \tag{6.111}$$

$\therefore$ substituting this value of $\left(\frac{\alpha}{\beta}-\frac{\beta}{\alpha}\right)^2$ in eq. (6.110), we get

$$T = \frac{1}{1+\frac{V_0^2}{4E(E-V_0)}\sin^2\beta a} \tag{6.112}$$

Equation (6.112) shows that the transmission coefficient (T) undergoes a fluctuation as E increases.

Ramsauer effect

If in region II, the de Broglie wavelength $\lambda_2 = 2a$, then $a = \lambda_2/2 = \pi/\beta$ or, $\beta a = \pi$.

$\therefore$ $\sin\beta a = 0$. This gives $T = 1$ [from eq. (6.112)].

In other words, *the transmission of the incident wave is* 100 % through the potential well when $a = \lambda/2$.

From the boundary conditions (1) and (2)

$$A + B = C + D,\ A - B = \frac{\beta}{\alpha}(C-D)\ [\because e^{i\beta a} = -1]$$

$$\text{and } C + D = -Fe^{i\alpha a},\ C - D = -\frac{\alpha}{\beta}Fe^{i\alpha a}$$

Solving the above equations, we get $B = 0$.

$$\therefore A = -Fe^{i\alpha a} \left[\begin{array}{l} \because C + D = A + 0 = A \\ \text{and } C + D = -Fe^{i\alpha a} \end{array}\right] \tag{6.113}$$

$$\text{or, } |A|^2 = |F|^2 \tag{6.114}$$

Equation (6.114) shows that although the amplitudes of the transmitted and reflected waves are not in phase, the transmission is cent percent through the potential well.

Now, *heavier noble gases* like Ar, Kr, Xe are *almost transparent to electrons* having certain value of energy $E = E_c$, the *critical energy* is given by

$$\beta^2 a^2 = \frac{2m}{\hbar^2} E_c a^2 = \pi^2 \text{ or, } E_c = \frac{\pi^2 \hbar^2}{2ma} \tag{6.115}$$

For the above noble gases, the atoms are spherically symmetrical with sharp boundaries which provides the possibility of the rectangular potential well approximation.

Since, for lighter noble gases like He and Ne, $V_0 a^2$ is not enough to make $\beta a = \pi$, *Ramsauer effect* is not observed. Physically, *the Ramsauer effect may be thought of as a diffraction of the electron around the rare gas atoms* (cf. *Quantum Mechanics* by L.I. Schiff).

6.8 Answer of some important conceptual questions

6.1. *What do you mean by free particle and bound particle?*

Ans. If a charged particle moves in a region of constant potential, then the force acting on the particle is zero. Such a particle is known as a *free particle*.

On the other hand, if the particle is confined in a limited region of space by potential energy so that the particle can move back and forth in this region, the particle is called a *bound particle*.

6.2. *What are quasi-quantised states?*

Ans. For a particle confined in a one-dimensional potential box, the energy of the particle in the nth quantum state is

$$E_n = \frac{n^2 \pi^2 \hbar^2}{2mL^2}, \quad n = 1, 2, 3, \ldots$$

where m is the mass of the particle, L is the width of the potential box and $\hbar = h/2\pi$.

Since, n is an integer, $E_n = E_1, E_2, E_3, \ldots$ for $n = 1, 2, 3, \ldots$.

Since, $p_n \propto 1/L$ and $E_n \propto 1/L^2$, for large values of L, the separation between the discrete momentum and energy values become small. For such case, the energy states are called *quasi-quantised states*.

6.3. *For a particle confined in a one-dimensional box, the quantum mechanical probability density $|\psi_n|^2$ varies with distance (x) but classical probability density is constant. Why?*

Ans. The quantum mechanical probability density for a particle confined in a one-dimensional box is

$$|\psi_n|^2 = \frac{2}{L} \sin^2 \frac{n\pi x}{L}$$

the different symbols have their usual significances.

$$\therefore |\psi_n|^2 = \text{maximum for } \frac{n\pi x}{L} = \frac{\pi}{2}, \frac{3\pi}{2}, \frac{5\pi}{2}, \ldots \text{ or, } x = \frac{L}{2n}, \frac{3L}{2n}, \frac{5L}{2n}, \text{etc.}$$

Since, $n = 1, 2, 3, \ldots, |\psi_n|^2$ is the maximum at different values of x in the box.

But, classically, the probability of finding the particle in a small region between x and $x + dx$ is $\frac{dx}{L}$, which is the same at every values of x. The probability density $1/L$ is a constant. This constancy in classical probability density is due to the uniform velocity of the particle in the box and perfect reflection at the walls.

6.4. ***What is the barrier penetration depth?***

Ans. **The penetration depth is defined as the distance in which the wave function ψ_2 reduces to $1/e$ of its value at $x = 0$.**

If a particle of energy E be incident from the left on a potential barrier of height $V_0 > E$, then the wave function in region II at $x = 0$ is $\phi_2 = De^{-\beta x} = De^{-\beta \times 0} = D$, where the different symbols have their usual meaning.

If Δx be the distance from $x = 0$, where the wave function reduces to $1/e$ of its value at $x = 0$. Then

$$(\psi_2)_{x+\Delta x}\,\Delta x = \frac{1}{e}(\psi_2)_{x=0} \text{ or, } De^{-\beta\Delta x} = \frac{1}{e}D \text{ or, } e^{-\beta\Delta x} = \frac{1}{e}$$

Taking log on both sides

$$-\beta\Delta x = -1 \text{ or, } \Delta x = \frac{1}{\beta} = \frac{\hbar}{\sqrt{2m(V_0 - E)}}$$

Δx is known as the *barrier penetration depth.*

6.5. ***An electron and a proton of same energy E incident on a potential barrier of height $V_0 > E$. Which one will have greater penetration depth?***

Ans. **The penetration depth d of a particle of mass m approaching a potential barrier of height $V_0 > E$ (E is the energy of the particle) is**

$$d = \frac{\hbar}{\sqrt{2m(V_0 - E)}}$$

For an electron

$$d_e = \frac{\hbar}{\sqrt{2m_e(V_0 - E)}}$$

and for a proton

$$d_p = \frac{\hbar}{\sqrt{2m_p(V_0 - E)}}$$

$$\therefore \frac{d_e}{d_p} = \sqrt{\frac{m_p}{m_e}} = \sqrt{\frac{1840 m_e}{m_e}} = \sqrt{1840}$$

$$\therefore d_e = \sqrt{1840}\, d_p \text{ i.e., } d_e > d_p$$

Hence, *the penetration depth (d_e) of an electron is greater than that of proton if both have the same energy.*

6.6. ***What do you mean by degeneracy? Explain with example.***

Ans. Definition: The property that two or more quantum states of a particle with different sets of quantum numbers and different eigenfunctions having the same value of momentum and energy is called degeneracy.

Example: For a particle confined in a three-dimensional potential box in the first excited state having quantum numbers as below

n_x	n_y	n_z
2	1	1
1	2	1
1	1	2

the value of momentum is $p_{2,1,1} = p_{1,2,1} = p_{1,1,2} = \frac{\pi\hbar}{a}\sqrt{6}$ and energy $E_{2,1,1} = E_{1,2,1} = E_{1,1,2} = \frac{\pi^2\hbar^2}{2ma^2}6$.

This shows that the states represented by wave functions $\psi_{2,1,1}$, $\psi_{1,2,1}$ and $\psi_{1,1,2}$ the momentum and energy values are the same, viz., $\frac{\pi\hbar}{a}\sqrt{6}$ and $\frac{\pi^2\hbar^2}{2ma^2}6$ respectively. Hence, the degeneracy of the first excited state of a particle in a three-dimensional rectangular box.

6.7. *Can a particle with zero energy exist in the box? What is zero-point energy?*

Ans. The energy of a particle of mass m is given by

$$E_n = \frac{n^2\pi^2\hbar^2}{2mL^2}$$

where L is the width of the potential box and the wave function is

$$\psi_n = \sqrt{\frac{2}{L}}\sin\frac{n\pi x}{L}$$

For $n = 0$, both $E_n = 0$ and $\psi_n = 0$. This shows that *a particle with zero energy cannot exist in the box.*

Again, for the first excited state with $n = 1$.

$$E_1 = \frac{\pi^2\hbar^2}{2mL^2} \text{ and } \psi_n = \sqrt{\frac{2}{L}}\sin\frac{\pi x}{L}$$

Thus, the minimum energy with which a particle can exist in the box is E_1. This is called *zero-point energy* because even if $T = 0$, then also E_1 and ψ_1 will exist in the box.

If $m \to \infty$ or, $L \to \infty$, then $E_1 \to 0$, the particle, then behaves as a *free particle.*

6.8. *A particle of energy $E < V_0$ approaching a potential barrier of height V_0, then classically, the particle cannot enter region II of the barrier, where $V = V_0$. Justify.*

Ans. If p be the momentum and m the mass of the particle of energy $E < V_0$, approaching the barrier, then

$$E = \frac{p^2}{2m} + V_0, \text{ at } x = 0$$

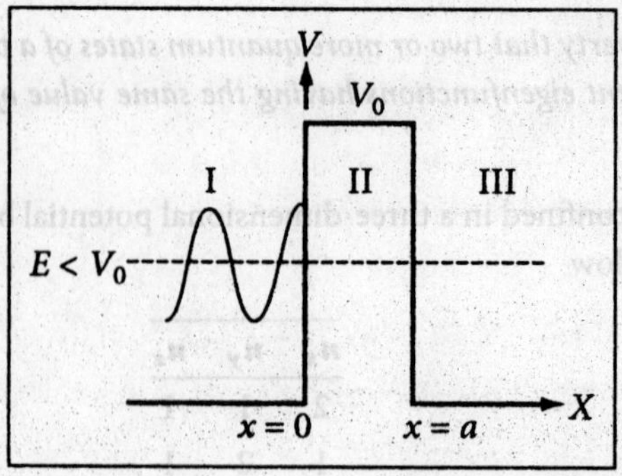

Fig 6.11 Particle of energy $E < V_0$ incident on a potential barrier

But, $E < V_0$

$$\therefore \frac{p^2}{2m} + V_0 < V_0 \text{ or, } \frac{p^2}{2m} < 0$$

This shows that $p^2 < 0$ or p is imaginary.

$\therefore$ classically, the particle cannot enter region II.

6.9. *Discuss how the width of a one-dimensional box influences colour of the system.*

Ans. By absorbing appropriate energy a particle in a one-dimensional box can undergo a transition from mth to nth quantum states.

The energies of the mth and nth quantum states are

$$E_m = \frac{m^2\pi^2\hbar^2}{2ma^2} \text{ and } E_n = \frac{n^2\pi^2\hbar^2}{2ma^2}$$

where a = width of the box.

$$\therefore \Delta E = E_m - E_n = \frac{\pi^2(m^2 - n^2)\hbar^2}{2ma^2}$$

According to Bohr's frequency condition of the emitted radiation

$$\nu = \frac{\Delta E}{\hbar} = \frac{\pi^2(m^2 - n^2)\,h^2}{8ma^2\pi^2 h} = \left(\frac{m^2 - n^2}{8ma^2}\right) h$$

This shows that $\lambda \propto a^2$. Colour of the emitted radiation is related to λ. By adjusting the width a of the box, the radiation may be made to appear beyond the visible region, when the system becomes coloured. Thus, the width a of the box influences the colour of the system.

6.10. *For* $_{84}Po^{210}$ *nucleus, the α-particle emitted has energy* 5.3 MeV *whereas the potential barrier existing at the surface of the nucleus is* V_0 = 23.58 MeV. *Then how an α-particle can come out of the nucleus during α-decay of* $_{84}Po^{210}$?

Ans. In view of classical physics, an α-particle of 5.3 MeV cannot come out of the nucleus by crossing a potential barrier of height 23.58 MeV. But, quantum mechanically it can be shown that there is a probability of the magnitude 1.77×10^{-56} of coming out of the nucleus due to *quantum mechanical tunneling*.

6.9 Solved problems

» ***Example* 6.1.** Calculate the percentage transmission of an electron beam of energy 3 eV when incident on a potential barrier of height 4 eV and width 20 Å.

Solution: Here, energy of each electron, $E = 3\,\text{eV} = 3 \times 1.6 \times 10^{-19}$ J; height of the barrier $V_0 = 4\,\text{eV} = 4 \times 1.6 \times 10^{-19}$ J; width of the barrier $a = 20\,\text{Å} = 20 \times 10^{-10} = 2 \times 10^{-9}$ m.

Now, the transmission coefficient is given by

$$T = \frac{16E}{V_0}\left(1 - \frac{E}{V_0}\right)e^{-2\beta a}$$

But,

$$2\beta a = 2a \times \sqrt{\frac{2m}{\hbar^2}(V_0 - h)} = \frac{2a}{\hbar}\sqrt{2m(V_0 - E)}$$

$$= \frac{2 \times (2 \times 10^{-19}}{1.054 \times 10^{-34}}\sqrt{2 \times 9.11 \times 10^{-31} \times 1.6 \times 10^{-19}} = 20.49$$

$$\therefore T = \frac{16E}{V_0}\left(1 - \frac{E}{V}\right)e^{-2\beta a} = \frac{16 \times 3}{4}\left(1 - \frac{3}{4}\right)e^{-2\beta a} = \frac{16 \times 3}{4} \times \frac{1}{4} \times (2.718)^{-20.49}$$

$$= \frac{3}{7.9 \times 10^8} = 3.797 \times 10^{-9}$$

Hence, the % transmission = $3.797 \times 10^{-9} \times 100 = \mathbf{3.797 \times 10^{-7}}$.

» ***Example* 6.2.** An electron beam is incident on a potential barrier of height 0.03 eV and of infinite width. Find the reflection coefficient and transmission coefficient at the barrier if the energy of the incident electron beam be (a) 0.04 eV (b) 0.025 eV (c) 0.03 eV.

Solution:

(a) in this case $E > V_0$

$$E = 0.04\,\text{eV} = 0.04 \times 1.6 \times 10^{-19}\,\text{J} = 0.064 \times 10^{-19}\,\text{J}$$

$$V_0 = 0.03\,\text{eV} = 0.03 \times 1.6 \times 10^{-19}\,\text{J} = 0.048 \times 10^{-19}\,\text{J}$$

The reflection coefficient

$$R = \left(\frac{\alpha - \beta}{\alpha + \beta}\right)^2, \text{ where } \alpha = \sqrt{\frac{2mE}{\hbar^2}} \text{ and } \beta = \sqrt{\frac{2m(E - V_0)}{\hbar^2}}$$

$$\therefore R = \frac{\left(\sqrt{\frac{2m}{\hbar^2}E} - \sqrt{\frac{2m}{\hbar^2}(E - V_0)}\right)^2}{\left(\sqrt{\frac{2m}{\hbar^2}E} + \sqrt{\frac{2m}{\hbar^2}(E - V_0)}\right)^2} = \left(\frac{\sqrt{E} - \sqrt{(E - V_0)}}{\sqrt{E} + \sqrt{(E - V_0)}}\right)^2$$

$$= \left(\frac{\sqrt{0.064 \times 10^{-19}} - \sqrt{(0.016 \times 10^{-19})}}{\sqrt{0.064 \times 10^{-19}} + \sqrt{(0.016 \times 10^{-19})}}\right)^2 \simeq 0.1$$

The transmission coefficient

$$T = \frac{4\alpha\beta}{(\alpha+\beta)^2}$$

T can be easily calculated from the relation $R + T = 1$. $\therefore$ $T = 1 - R = 1 - 0.1 = \mathbf{0.9}$.

(b) In this case, $E = 0.025\,\text{eV}$ and $V_0 = 0.03\,\text{eV}$. $\therefore$ $E < V_0$.

Here, $T = 0$ and $R = 1$, i.e., all the electrons that are incident on the barrier are reflected back and no electron will be transmitted through the barrier.

(c) In this case, $E = 0.03\,\text{eV}$ and $V_0 = 0.03\,\text{eV}$. $\therefore$ $E = V_0$. So $\beta = 0$.

$$\therefore R = \left(\frac{\alpha-\beta}{\alpha+\beta}\right)^2 = 1 \text{ and } T = \frac{4\alpha\beta}{(\alpha+\beta)^2} = 0$$

Here, also all the electrons are reflected back.

» ***Example* 6.3.** Electrons of energy 2 eV are incident on a barrier of 3.0 eV high and 0.4 nm wide. What is the value of transmission probability? **[PU 1991]**

Solution: Here, $E = 2\,\text{eV} = 2 \times 1.6 \times 10^{-19}$ J; $V_0 = 3\,\text{eV} = 3 \times 1.6 \times 10^{-19}$ J; $a = 0.4\,\text{nm} = 4 \times 10^{-10}$ m.

Transmission probability

$$T = \frac{1}{1 + \dfrac{V_0^2}{4E(V_0 - E)} \sinh^2 \sqrt{\dfrac{2m}{\hbar^2}(V_0 - E)} \cdot a}$$

But,

$$(V_0 - E) = (3 - 2) = 1\,\text{eV} = 1.6 \times 10^{-19}\,\text{J}$$

$$\text{and } \frac{V_0^2}{4E(V_0 - E)} = \frac{3 \times 1.6 \times 10^{-19} \times 3 \times 1.6 \times 10^{-19}}{4 \times 2 \times 1.6 \times 10^{-19} \times 1.6 \times 10^{-19}} = \frac{9}{8} = 1.125$$

$$\text{and } \sqrt{\frac{2m}{\hbar^2}(V_0 - E)} = \frac{2 \times 9 \times 10^{-31} \times 1.6 \times 10^{-19}}{1.05 \times 10^{-34}} \times 4 \times 10^{-10} = 2.045$$

$$\sinh(2.045) = 3.805$$

$$\therefore T = \frac{1}{1 + 1.125 \times (3.805)^2} = \mathbf{0.058}$$

» ***Example* 6.4.** What fraction of the incident electron beam of energy 36 eV will be reflected at the boundary of a step potential of height 20 eV? What would be the fraction of the proton beam of the same energy?

Solution: The reflection coefficient as shown in Art. 6.4 is

$$R = \frac{(\alpha-\beta)^2}{(\alpha+\beta)^2} = \frac{\left(\sqrt{E} - \sqrt{E - V_0}\right)^2}{\left(\sqrt{E} + \sqrt{E - V_0}\right)^2} = \frac{\left(\sqrt{36} - \sqrt{16}\right)^2}{\left(\sqrt{36} + \sqrt{16}\right)^2}$$

$$= 0.040 = \mathbf{4.0\,\%}\ [\because E - V_0 = 36 - 20 = 16]$$

For the proton beam of same energy, $R = 4.0\,\%$.

» ***Example* 6.5.** What is the width of the potential barrier for an α-particle emitted with energy 5.5 MeV from ${}_{86}\text{Rn}^{22}$ nucleus?

Solution: If R be the nuclear radius and r_0 the distance from the centre of the nucleus at which the potential energy of the α-particle is equal to the kinetic energy of the emitted α-particles, then the width of the potential barrier is $a = r - R_0$.

But,

$$R_0 = 1.5 \times 10^{-15} \times A^{1/3} = 1.5 \times 10^{-15} \times (222)^{1/3} = 1.5 \times 10^{-15} \times 6.055 = 9.08 \times 10^{-15}\,\text{m}$$

$$r = \frac{2(Z-2)\,e^2}{4\pi\varepsilon_0 E} = \frac{2 \times (86-2) \times (1.6 \times 10^{-19})^2}{4\pi \times 8.85 \times 10^{-12} \times 5.5 \times 1.6 \times 10^{-13}} = \frac{2 \times 84 \times (1.6)^2 \times 10^{-13}}{4\pi \times 8.85 \times 5.5 \times 1.6} = 43.97 \times 10^{-15}\,\text{m}$$

$$\therefore a = r - R = 43.97 \times 10^{-15} - 9.08 \times 10^{-15} = \mathbf{34.89 \times 10^{-15}\,m}$$

» ***Example* 6.6.** **A particle of energy 1 eV is incident on a rectangular potential barrier of height 2 eV and width 1 Å. Calculate the barrier transmission probability.**

Solution: The barrier transmission probability of the particle is given by

$$T = \frac{16E(V_0 - E)}{V_0^2} \times e^{-2\beta a} = \frac{16E(V_0 - E)}{V_0^2} e^{-2\left(\sqrt{2m(V_0-E)/\hbar^2}\right)a} \quad (\because \text{here } V_0 > E)$$

In the present problem, $V_0 = 2\,\text{eV} = 2 \times 1.6 \times 10^{-19}$ J; $m = 1.6 \times 10^{-27}$ kg; $a = 1\,\text{Å} = 10^{-10}$ m; $E = 1\,\text{eV} = 1.6 \times 10^{-19}$ J.

$$\therefore T = \frac{16 \times 1.6 \times 10^{-19} \times 1 \times 1.6 \times 10^{-19}}{(2 \times 1.6 \times 10^{-19})^2} \times e^{\left[\frac{2 \times 1.6 \times 10^{-27} \times 1 \times 1.6 \times 10^{-19}}{1.05 \times 10^{-34}}\right]^{1/2} \times 10^{-10}} = \mathbf{4e^{-43}}$$

» ***Example* 6.7.** **A potential barrier has a height of 5 eV and width 0.5 nm. If electrons of energies 1 eV and 2 eV are incident on the barrier, then find out their respective transmission probabilities. How are the transmission probabilities affected if the barrier be doubled in width?**

Solution: We have the transmission probability, $T \approx e^{-2\beta a}$, where $\beta = \sqrt{\dfrac{2m(V_0 - E)}{\hbar^2}}$ $[\because V_0 < E]$.

For 1 eV electron, $m = 9.1 \times 10^{-31}$ kg; $V_0 = 5(1 \times 10^{-19}\,\text{J})$; $E = 1 \times 10^{-19}$ J; $\hbar = 1.054 \times 10^{-34}$ J-s.

$$\therefore \beta = \frac{\sqrt{2 \times (9.1 \times 10^{-31})\,(5-1) \times 10^{-34}}}{1.054 \times 10^{-34}} = 1.0 \times 10^{10}/\text{m}$$

Since, the width, $a = 0.5\,\text{nm} = 5 \times 10^{-10}$ m, we have $T_1 \simeq e^{-2 \times (1.0 \times 10^{10}) \times 5 \times 10^{-10}} = e^{-10} = 4.5 \times 10^{-5}$.

For the 2 eV electron the transmission probability $T_2 \simeq e^{-2\beta a} = e^{-2\left(\sqrt{\frac{546}{1.054}}\right) \times 5 \times 10^{-10}} = e^{-8.9} = 1.4 \times 10^{-4}$.

$$\therefore \frac{T_2}{T_1} = \frac{1.4 \times 10^{-4}}{4.5 \times 10^{-5}} = \frac{14}{4.5} = 3.99$$

$$\because \beta = \frac{\sqrt{2 \times 9.1 \times 10^{-31} \times (5-2) \times 10^{-34}}}{1.054 \times 10^{-34}} = \frac{\sqrt{2 \times 9.1 \times 3 \times 10^{-65}}}{1.054 \times 10^{-34}} = \frac{\sqrt{6 \times 9.1 \times 10^3}}{1.054}$$

$$= \frac{\sqrt{54.6 \times 10^3}}{1.059} = \frac{\sqrt{54600}}{1.054} = \frac{\sqrt{546}}{\mathbf{1.054}}$$

$\therefore$ **2 eV electrons are three times more likely to tunnel through the barrier.**

If $a = 2 \times 5 \times 10^{-10} = 10^{-9}$ m, then it can be shown that the 1 eV and 2 eV electrons have transmission probabilities $T_1' = \mathbf{2.1 \times 10^{-9}}$ and $T_2' = \mathbf{1.9 \times 10^{-8}}$.

Now,

$$\frac{T_2'}{T_1'} = \frac{1.9 \times 10^{-8}}{2.1 \times 10^{-9}} = \frac{19}{2.1} \simeq 9$$

∴ we note that the transmission probability depends far more strongly on the width of the barrier than the energy of the incident electron.

» ***Example* 6.8.** A step potential in one dimension is expressed as

$$V(x) = \begin{cases} 0, & \text{for } x < 0 \\ V_0, & \text{for } x > 0 \end{cases}$$

Applying one-dimensional Schrödinger equation show that the wave function vanishes in the region $x < 0$, if $V_0 \to \infty$.

Solution: The one-dimensional time-independent Schrödinger equations for $x < 0$ and $x > 0$ respectively are

$$\frac{d^2\psi_1}{dx^2} + \frac{2mE}{\hbar^2}\psi_1 = 0 \text{ and } \frac{d^2\psi_2}{dx^2} - \frac{2m}{\hbar^2}(V_0 - E)\,\psi_2 = 0 \text{ [if } 0 < E < V_0]$$

The general solutions of the above equations are as below:

$$\psi_1 = A\cos\alpha x + B\sin\alpha x, \text{ for } x < 0$$

$$\psi_2 = Ce^{+\beta x} + De^{-\beta x}, \quad \text{for } > 0$$

where $\alpha^2 = \dfrac{2mE}{\hbar^2}$, $\beta^2 = \dfrac{2m}{\hbar^2}(V_0 - E)$.

Neglecting the diverging solution $Ce^{\beta x}$, we get $\psi_2 = De^{-\beta x}$.

The boundary conditions for evaluating A, B and D are

1. $(\psi_1)_0 = (\psi_2)_0$ **2.** $\left(\dfrac{d\psi_1}{dx}\right)_0 = \left(\dfrac{d\psi_2}{dx}\right)_0$.

Applying the boundary conditions, we get $A = D$ and $\alpha B = -\beta D$.

Since, $\beta^2 = 2m/\hbar^2(V_0 - E) \to \infty$ as $V_0 \to \infty$ and ψ_1 must be finite, so $(\alpha B = -\beta D)$ shows that $D \to 0$ as $V_0 \to \infty$. A is also zero under this condition.

Hence, $\psi_1 t$ vanishes for $x < 0$.

» ***Example* 6.9.** Consider an electron of energy E is incident on a step potential of height $V_0 = 10$ eV. Find the reflection coefficient R and transmission coefficient T when (i) $E = 5$ eV (ii) $E = 10$ eV (iii) $E = 15$ eV.

Solution:

(i) When $E = 5$ eV, $V_0 = 10$ eV; $E < V_0$.

∴ reflection coefficient $\boldsymbol{R = 1}$ and $\boldsymbol{T = 0}$. This means the electrons will be totally reflected and there will be no transmission through the barrier.

(ii) When $E = 10$ eV, $V_0 = 10$ eV; $E = V_0$.

Here, also $\boldsymbol{R = 1}$ and $\boldsymbol{T = 0}$. Here, also reflection 100 % and no transmission at all.

(iii) $E = 15$ eV, $V_0 = 10$ eV; $E > V_0$.

Then reflection coefficient

$$R = \left|\frac{1-\sqrt{1-V_0/E}}{1+\sqrt{1-V_0/E}}\right|^2$$

Here, $V_0/E = \frac{10}{15} = \frac{2}{5}$.

$$\therefore R = \left|\frac{1-\sqrt{1-\frac{2}{3}}}{1+\sqrt{1-\frac{2}{3}}}\right|^2 = \left|\frac{0.423}{1.577}\right|^2 = 0.072 = \mathbf{7.2\,\%}$$

» ***Example* 6.10.** A beam of electrons of energy 1 eV moves through a one-dimensional rectangular potential barrier of height 5 eV and width 1Å. Find the transmission coefficient.

Solution: Here, the barrier height $V_0 = 5\,\text{eV}$. Electron energy $E = 1\,\text{eV}$. So, $E < V_0$. Barrier width $a = 1\text{Å}$.

The barrier transmission coefficient is given by

$$T = \frac{1}{1 + \dfrac{V_0^2}{4E(V_0 - E)}\sinh^2 \beta L}$$

where $\beta = \sqrt{\dfrac{2m(V_0 - E)}{\hbar^2}}$.

$$\therefore \beta = \frac{(2\times 9.1\times 10^{-31}\,\text{kg})\,(5\,\text{eV} - 1\,\text{eV})}{\left(\dfrac{6.626\times 10^{-34}\,\text{Js}}{2\pi}\right)^2} = \sqrt{\frac{2(9.1\times 10^{-31}\,\text{kg})\,(2\pi)^2\,(4\times 1.6\times 10^{-19}\,\text{J})}{(6.626\times 10^{-34}\,\text{Js})^2}}$$

$$= 10^{10}\,\text{m}^{-1}$$

$$\therefore T = \frac{1}{1 + \dfrac{(5\,\text{eV})^2}{4(1\,\text{eV})\,(5\,\text{eV} - 1\,\text{eV})}\sinh^2[10^{10}\,\text{m}^{-1}\cdot 10^{-10}\,\text{m}]} = \frac{1}{1 + \dfrac{5^2}{4^2}(1.175)^2} \quad [\because \sin 1 = 1.175]$$

$$= 0.32 = \mathbf{32\,\%}$$

» ***Example* 6.11.** An electron beam and proton beam both having energy 25 eV are separately incident on a step potential of height 10 eV. Calculate the fraction reflected in the two cases.

Solution: Here, the energy of both electron beam and proton beam is $E = 25\,\text{eV}$ and the potential barrier height $V_0 = 10\,\text{eV}$. $\therefore E > V_0$.

The reflection coefficient

$$R = \left|\frac{1-\sqrt{1-\frac{V_0}{E}}}{1-\sqrt{1+\frac{V_0}{E}}}\right|^2 = \left|\frac{1-\sqrt{1-\frac{20}{25}}}{1-\sqrt{1-\frac{10}{25}}}\right|^2 = \frac{0.05}{3.15} = 0.016 = \mathbf{1.6\,\%}$$

Thus, the reflection coefficient for both the electron beam the proton beam are the same, viz., 0.016 = 1.6 %.

» ***Example* 6.12.** A particular trapped in a one-dimensional infinite square well within limits $0 \leqslant x \leqslant L$ is described by the wave function

$$\psi(x, 0) = A\sin\frac{5\pi x}{L}\cos\frac{2\pi x}{L}$$

(i) Normalise $\psi(x,0)$; (ii) Expand $\psi(x,0)$ in terms of energy eigenfunctions; (iii) Find $\psi(x,t)$; (iv) Is $\psi(x,t)$ a stationary state? **[KU (H) 2015]**

Solution: We have the wave function

$$\psi(x,0) = A\sin\frac{5\pi x}{L}\cos\frac{2\pi x}{L} = A\left[\sin\frac{(5+2)\pi x}{L} + \sin\frac{(5-2)\pi x}{L}\right]$$

$$= \frac{A}{2}\left(\sin\frac{7\pi x}{L} + \sin\frac{3\pi x}{L}\right)$$

(i) The condition of normalisation of $\psi(x,0)$ is

$$\int_{-\alpha}^{+\alpha} \psi^*(x,0)\,\psi(x,0)\,dx = 1 \text{ or, } \int_0^L |\psi(x,0)|^2\,dx = 1$$

$$\text{or, } \int_0^L \left[\frac{A}{2}\sin\frac{7\pi x}{L} + \sin\frac{3\pi x}{L}\right]^2 dx = 1$$

$$\text{or, } \frac{A^2}{4}\int_0^L \left(\sin^2\frac{7\pi x}{L} + \sin^2\frac{3\pi x}{L} + 2\sin\frac{7\pi x}{L}\sin\frac{2\pi x}{L}\right) dx = 1$$

$$\text{or, } \frac{A^2}{4}\left[\int_0^L \sin^2\frac{7\pi x}{L}\,dx + \int_0^L \sin^2\frac{3\pi x}{L}\,dx + 2\int_0^L \sin\frac{7\pi x}{L}\sin\frac{3\pi x}{L}\,dx = 1\right]$$

Now, since

$$\int_0^L \sin\frac{n\pi x}{L}\sin\frac{m\pi x}{L}\,dx = \frac{L}{2}\delta_{mn}$$

$$\therefore \frac{A^2}{4}\left[\frac{L}{2} + \frac{L}{2} + 2\times 0\right] = 1 \text{ or, } \frac{A^2}{4}L = 1$$

$$\therefore A = \frac{2}{\sqrt{L}}$$

$\therefore$ the normalised wave function is

$$\psi(x,0) = \frac{1}{\sqrt{L}}\left(\sin\frac{7\pi x}{L} + \sin\frac{3\pi x}{L}\right)$$

(ii) The energy eigenfunction and the energy eigenvalues are

$$\psi_n = \sqrt{\frac{2}{L}}\sin\frac{n\pi x}{L};\ E_n = \frac{n^2\pi^2\hbar^2}{2mL^2},\ n = 1, 2, 3, \ldots$$

$$\therefore \psi(x,0) = \frac{1}{\sqrt{2}}\left(\sqrt{\frac{2}{L}}\sin\frac{7\pi x}{L} + \sqrt{\frac{2}{L}}\sin\frac{3\pi x}{L}\right) \text{ or, } \psi(x,0) = \frac{1}{\sqrt{2}}(\psi_7 + \psi_3)$$

(iii) $\psi(x,t) = \frac{1}{\sqrt{2}}\psi_y e^{-i\frac{E_7 t}{\hbar}} + \frac{1}{\sqrt{2}}\psi_3 e^{-i\frac{E_3 t}{\hbar}}$.

This represents the time-dependence of $\psi(x,t)$.

(iv) $\hat{H}\psi(x,t) = \frac{1}{\sqrt{2}}(\hat{H}\psi_7)\,e^{-\frac{iE_7 t}{\hbar}} + \frac{1}{\sqrt{2}}(\hat{H}\psi_3)\,e^{-\frac{iE_3 t}{\hbar}} = \frac{E_7}{\sqrt{2}}\psi_7 e^{-\frac{iE_7 t}{\hbar}} + \frac{E_3}{\sqrt{2}}(\hat{H}\psi_3)\,e^{-\frac{iE_3 t}{\hbar}}$.

Thus, $\psi(x,t)$ is not a stationary state.

» ***Example* 6.13.** A particle of mass m is confined to a one-dimensional box of length L extending from $x = 0$ to $x = L$. Find the probability of finding the particle between $0 \leqslant x \leqslant \frac{1}{4}$ for an arbitrary value of quantum number n. **[KU (H) 2002]**

Solution: The probability of finding the particle between $0 \leqslant x \leqslant \frac{1}{4}$ is given by

$$P = \int_0^{L/4} \psi_n^* \psi_n \, dx = \int_0^{L/2} \left[\sqrt{\frac{2}{L}} \sin \frac{n\pi x}{L}\right]^* \left[\sqrt{\frac{2}{L}} \sin \frac{n\pi x}{L}\right] dx = \frac{2}{L} \int_0^{L/4} \sin^2 \frac{n\pi x}{L} dx$$

$$= \frac{1}{L} \int_0^{L/4} \left(1 - \cos \frac{2n\pi x}{L}\right) dx = \frac{1}{L} \left[\int_0^{L/4} dx - \int_0^{L/4} \cos \frac{2n\pi x}{L} dx\right]$$

$$= \frac{1}{L} \left[\frac{1}{4} - \frac{\sin \frac{n\pi x}{L}}{\frac{2n\pi}{L}}\right]_0^{L/4} = \frac{1}{L} \left[\frac{L}{4} - \frac{L}{2n\pi} \left(\sin \frac{n\pi}{2} - \sin 0\right)\right]$$

$$\therefore P = \frac{1}{4} - \frac{\sin \frac{n\pi}{2}}{2n\pi}$$

» ***Example* 6.14.** Suppose an electron is trapped in a one-dimensional box. Find the length of the box in units of Compton wavelength, so that the zero point energy is equal to rest mass energy of the electron. **[KU (H) 2002]**

Solution: If the electron is considered non-relativistic, then the speed v of the electron is very small compared speed c of light in free space.

$$\therefore \text{mass of the electron is } m = \frac{m_0}{\sqrt{1 - \frac{v^2}{c^2}}} \simeq m_0.$$

The zero point energy = rest mass energy of the electron

$$\text{or, } E_1 = \frac{\pi^2 \hbar^2}{2mL^2} = m_0 c^2,$$

where L is the length of the one-dimensional box containing the electron and m_0 = the rest mass of the electron.

$$\therefore \frac{\pi^2 \hbar^2}{2m_0 L^2} = m_0 c^2 \text{ or, } L^2 = \frac{\pi^2 \hbar^2}{2m_0^2 c^2} = \frac{\pi^2}{2} \left(\frac{h/2\pi}{m_0 C}\right)^2 \text{ or, } L = \frac{\lambda_c}{\sqrt{8}},$$

where $\lambda_c = \dfrac{h}{m_0 c}$ is the Compton wavelength of the electron.

∴ length of the box is

$$L = \frac{\lambda_c}{\sqrt{8}} = 0.35 \text{Å}$$

» ***Example* 6.15.** A particle is confined in a one-dimensional box. Show that the de Broglie wavelength of the particle in the first excited state is equal to the length of the box.

Solution: We know that the energy of the particle in the first excited state ($n = 2$) is

$$E_2 = \frac{2^2 \pi^2 \hbar^2}{2mL^2} = \frac{4\pi^2 \hbar^2}{2mL^2} \text{ or, } \frac{p^2}{2m} = E_2 = \frac{4\pi^2 \hbar^2}{2mL^2}$$

where p is the momentum of the particle of mass m.

$$\therefore p = \sqrt{\frac{4\pi^2\hbar^2}{L^2}} = \frac{2\pi\hbar}{2mL^2}$$

$\therefore$ de Broglie wavelength is

$$L = \frac{h}{p} = \frac{h}{\frac{2\pi\hbar^2}{L}} = \frac{Lh}{2\pi\frac{h}{2\pi}} = L \text{ (proved)}$$

» ***Example* 6.16.** Calculate the energy of an electron in ground state and first excited state when it is confined in an infinite well of width 0.5Å.

***Solution*:** If L be the width and m the mass of the electron in the infinite well, then its energy in the nth state is

$$E_n = \frac{n^2\pi^2\hbar^2}{2mL^2}$$

For the ground state the energy is

$$E_1 = \frac{\pi^2\hbar^2}{2mL^2} = \frac{\pi^2\left(\frac{6.626 \times 10^{-34}\text{ Js}}{2\pi}\right)^2}{2(9.1 \times 10^{-31}\text{ kg})\,(0.5 \times 10^{-10}\text{ m})^2} = 2.4 \times 10^{-17}\text{ J} = \frac{2.4 \times 10^{-17}}{1.6 \times 10^{-19}} = \mathbf{150\ eV}$$

For the first excited state the energy of the electron is

$$E_2 = \frac{2^2\pi^2\hbar^2}{2mL^2} = 4 \times E_1 = 4 \times 150\text{ eV} = \mathbf{600\ eV}$$

» ***Example* 6.17.** A beam photon in wavelength range (9.0 ± 1.0) nm strikes an electron trapped in an infinite one-dimensional potential well of width 1 nm staying in the ground state. To what higher states can be electron be excited?

***Solution*:** According to Bohr's frequency rule, the frequency of transition from nth state to pth state ($n > p$) is given by $\nu = (n^2 - p^2)\frac{h}{8mL^2}$.

The corresponding wavelength is

$$\lambda = \frac{c}{\nu} = \frac{8cmL^2}{(n^2 - p^2)\,h} \text{ or, } n^2 - p^2 = \frac{8cmL^2}{\lambda h}$$

Here, $p = 1$ (ground state) $L = 1\text{ nm} = 10^{-9}$ m.

$$\therefore n^2 - 1 = \frac{8(3 \times 10^8\text{ ms}^{-1})\,(9.1 \times 10^{-31}\text{ kg})\,(10^{-9}\text{ m})}{(9.0 \pm 1) \times 10^{-9}\text{ m} \times (6.626 \times 10^{-34}\text{ Js})} = \frac{3296}{9 \pm 1} = \frac{3296}{10}, \frac{3296}{8} = 329.6;\ 412$$

$$\text{or, } n^2 = 329.6 + 1;\ 412 + 1 = 330.6;\ 413.\ \therefore\ \mathbf{n = 20}$$

Thus, the possivalues of n to which the electron can be excited are $\boldsymbol{n = 18, n = 19, n = 20}$.

» ***Example* 6.18.** Can you measure the energy levels of a ball of mass 10 gm moving in a one-dimensional box of length of 10 cm?

Solution: We know that the energy level of a particle of mass in a one-dimensional box of length L is

$$E_n = \frac{n^2\pi^2\hbar^2}{2mL^2},\ n = 1, 2, 3, \ldots$$

$$= \frac{n^2\pi^2\left(\dfrac{6.626\times10^{-34}\text{ Js}}{2\pi}\right)}{2\times(10\times10^{-3}\text{ kg})(10\times10^{-2}\text{ m})^2} = (5.5\times10^{-64}\text{ J})^2 n^2$$

where $m = 10 \times 10^{-3}$ kg and $L = 10 \times 10^{-2}$ m.

For $n = 1$, $E_1 = \mathbf{5.5 \times 10^{-64}}$ **J.**

For $n = 2$, $E_2 = \mathbf{22 \times 10^{-64}}$ **J.**

For $n = 3$, $E_3 = \mathbf{49.5 \times 10^{-64}}$ **J.**

Since, 10^{-64} very small, E_1, E_2 and E_3 levels are very closely spaced and as such cannot be measured.

» ***Example* 6.19.** A particle is confined to move freely along a line freely between $x = 0$ and $x = L$. (a) Show that $\langle x\rangle = L/2$ regardless of the state of the particle. (b) Show that this agrees with the classical result.

[KU (H) 2004]

Solution: We know that the wave function for the particle moving in a quantum well between $x = 0$ and $x = L$ is given by

$$\psi_n = \sqrt{\frac{2}{L}}\sin\frac{n\pi x}{L},\ n = 1, 2, 3, \ldots$$

The quantum mechanical average value is

$$\langle\hat{x}\rangle_Q = \int_0^L \psi_n^*\hat{x}\,\psi_n\,dx = \int_0^L\left[\sqrt{\frac{2}{L}}\sin\frac{n\pi x}{L}\right]^* x\left[\sqrt{\frac{2}{L}}\sin\frac{n\pi x}{L}\right]dx = \frac{2}{L}\int_0^L x\sin^2\frac{n\pi x}{L}\,dx$$

$$= \frac{1}{L}\int_0^L x\left(1-\cos\frac{2n\pi x}{L}\right)dx = \frac{1}{L}\left[\int_0^L x\,dx - \int_0^L x\cos\frac{2n\pi x}{L}\,dx\right]$$

$$= \frac{1}{L}\left[\frac{x^2}{2}\right]_0^L - \left\{x\int\cos\frac{2n\pi x}{L}\,dx\Big|_0^L - \int_0^L\frac{d}{dx}(x)\int\cos\frac{2n\pi x}{L}\,dx\right\}dx$$

$$= \frac{1}{L}\left[\frac{L^2}{2} - \left\{x\frac{\sin\dfrac{2n\pi x}{L}}{\dfrac{2n\pi}{L}}\Bigg|_0^L - \int_0^L\frac{\sin\dfrac{2n\pi x}{L}}{\dfrac{2n\pi}{L}}\,dx\right\}\right]$$

$$= \frac{L}{2} - \frac{1}{2n\pi}\left[L\sin 2n\pi - 0 - \int_0^L\sin\frac{2n\pi x}{L}\,dx\right] = \frac{L}{2} + \frac{1}{2n\pi}\,\frac{(-)\cos\dfrac{2n\pi x}{L}}{\dfrac{2n\pi}{L}}\Bigg|_0^L$$

$$= \frac{L}{2} - \frac{L}{(2n\pi)^2}[\cos 2n\pi - \cos 0] = \frac{L}{2}.$$

$$\therefore\ \langle\hat{x}\rangle_Q = \frac{L}{2}$$

This holds for all states as it does not contain n.

Now, the classical average is

$$\langle \hat{x} \rangle_{Cl} = \frac{1}{L}\int_0^L x\,dx = \frac{1}{L}\frac{x^2}{2}\bigg|_0^L = \frac{L}{2} \text{ or, } \langle \hat{x} \rangle_{Cl} = \frac{L}{2}$$

Hence, $\langle \hat{x} \rangle_Q = \langle x \rangle_{Cl}$.

» ***Example* 6.20.** What fraction of a beam of electrons each of energy $E = 0.08$ eV is reflected from a step barrier of height $V_0 = 0.06$ eV and infinite width. **[IAS(Pre) 2007]**

Solution: The reflection coefficient of the electron beam from the step barrier is given by

$$R = \left|\frac{1-\sqrt{1-\dfrac{V_0}{E}}}{1+\sqrt{1-\dfrac{V_0}{E}}}\right|^2$$

Here, $V_0 = 0.06$ eV and $E = 0.08$ eV.

$$\therefore R = \left|\frac{1-\sqrt{1-\dfrac{0.06\text{ eV}}{0.08\text{ eV}}}}{1+\sqrt{1+\dfrac{0.06\text{ eV}}{0.08\text{ eV}}}}\right|^2 = \left|\frac{1-\sqrt{1-\dfrac{3}{4}}}{1+\sqrt{1+\dfrac{3}{4}}}\right|^2 = \left|\frac{1-\dfrac{1}{2}}{1+\sqrt{\dfrac{7}{2}}}\right|^2 = \left|\frac{\dfrac{1}{2}}{1+\dfrac{\sqrt{7}}{2}}\right|^2 \text{ or, } R = 0.11$$

» ***Example* 6.21.** For a particle of mass m in the ground state in a one-dimensional box which extends from $x = 0$ to $x = L$. What is the ratio of probability densities $\dfrac{\rho\left(x = \dfrac{L}{4}\right)}{\rho\left(x = \dfrac{L}{2}\right)}$?

Solution: The probability density in the ground state ($n = 1$) in a one-dimensional box extending from $x = 0$ to $x = L$ is

$$\rho_n = \frac{2}{L}\sin^2\frac{n\pi x}{L}, \text{ for } 0 < x < L$$

∴ the ratio of the probabilities is

$$\frac{\rho_{n=1}(L/4)}{\rho_n(L/2)} = \frac{\dfrac{2}{L}\sin^2\dfrac{\pi}{L}\dfrac{L}{4}}{\dfrac{2}{L}\sin^2\dfrac{\pi}{L}\dfrac{L}{2}} = \frac{\sin^2\dfrac{\pi}{4}}{\sin^2\dfrac{\pi}{2}} = \frac{1}{2}$$

EXERCISES

6(A): Objective (or MCQ) Type Questions

Select the correct answer:

6.1. The conditions to be obeyed by the solution of Schrödinger equation at the potential boundary at $x = 0$ are

(a) ψ must be continuous

(b) ψ must be single-valued

(c) $\frac{d\psi}{dx}$ must be continuous and finite

(d) ψ and $\frac{d\psi}{dx}$ must be continuous at $x = 0$.

6.2. When a beam of electron of energy E is incident on the boundary of a potential step of height V_0 ($V_0 > E$) the probability current density in the region $x < 0$ is

(a) finite

(b) zero

(c) infinity

(d) none of these.

6.3. The transmission coefficient of a particle of energy E through a potential step of height V_0 ($V_0 > E$) is

(a) $\frac{4\alpha\beta}{(\alpha+\beta)^2}$

(b) $\left(\frac{\alpha-\beta}{\alpha+\beta}\right)^2$

(c) $\left(\frac{\alpha+\beta}{\alpha-\beta}\right)^2$

(d) $\frac{(\alpha-\beta)^2}{4\alpha\beta}$.

6.4. If R and T be the reflection and transmission probabilities of a particle through a potential step, then

(a) $R - T = \text{constant}$

(b) $R + T = 1$

(c) $R + T = 0$

(d) $R + T = \infty$.

6.5. When the energy E of an incident particle approaches the barrier height V_0 the transmission probability is

(a) $T \to \infty$ as $a \to 0$

(b) $T \to 0$ as $a \to 0$

(c) $T \to \infty$ as $a \to \infty$

(d) $T \to 1$ as $a \to 0$.

6.6. When $E > V_0$ the width of the barrier contains

(a) integral multiple of de Broglie wavelength

(b) half-integral multiple of de Broglie wavelength

(c) odd multiple of de Broglie wavelength

(d) fraction of de Broglie wavelength.

6.7. Quantum mechanical tunneling is due to the

(a) particle nature of matter

(b) wave nature of matter

(c) dual nature of matter

(d) none of these.

6.8. The phenomena of α-decay could be understood on the basis of

(a) quantum mechanical tunneling

(b) classical mechanics

(c) electromagnetic theory

(d) semiclassical theory.

6.9. When a particle of mass m having an energy E meets a potential barrier of height $V_0 (E < V_0)$, the penetration depth Δx is

(a) $\frac{2m(V_0 - E)}{\hbar}$

(b) $\sqrt{\frac{2m(V_0 - E)}{\hbar^2}}$

(c) $\sqrt{\frac{\hbar^2}{2m(V_0 - E)}}$.

(d) $\frac{V_0 - E}{\hbar}$.

6.10. If an electron and a proton have the same energy, then the penetration depth of electron is

(a) greater than proton (c) equal

(b) smaller than proton (d) none of them.

6.11. If a 5 MeV α as well as β particles are incident on a potential barrier of height 10 MeV and width 2 fm, then the probability of transmission is

(a) higher for α particle (c) equal for both

(b) higher for β particle (d) zero for both.

6.12. A particle of energy E is incident from the left on a potential step of height V_0 at $x = 0$.

(a) If $E < V_0$ wave function for the particles zero for $x > 0$.

(b) If $E < V_0$ wave function is not zero for $x > 0$

(c) If $E > V_0$ reflection coefficient is not zero.

Which of the statements is/are correct?

(a) (i) only (b) (ii) only (c) (i) and (iii) (d) (ii) and (iii). **[IAS(Pre) 2008]**

6.13. A particle of mass m coming in from left with energy $E > 0$ encounters a barrier potential

$$\psi(x) = \begin{cases} V_0, & \text{for } |x| < L \\ 0, & \text{for } |x| > L \end{cases}$$

The wave function is given by

$$\psi(x) = \begin{cases} e^{ikx} + Be^{-ikx}, & \text{for } x < -1 \\ Ae^{ikx}, & \text{for } x > L \end{cases}$$

The constants A and B satisfy which one of the following relations?

(a) $A + B = 1$ (c) $|A| = |B|$

(b) $|A| + |B| = 1$ (d) $|A|^2 + |B|^2 = 1$. **[IAS(Pre) 2007]**

6.14. A particle of mass m is incident on a step of height V_0 (at $x = 0$). If $E < V_0$, the wave function of the particle for $x > 0$ will be of the form:

(a) e^{ikx} (c) e^{-kx}

(b) zero (d) $Ae^{ikx} + Be^{-ikx}$. **[IAS(Pre) 2005; 2006]**

6.15. The energy of a particle in a potential box and that of the hydrogen atom varies with the quantum number respectively as

(a) n^2 and $\frac{1}{n^2}$ (c) n^2 both

(b) $\frac{1}{n^2}$ and n^2 (d) $\frac{2}{n^2}$ and $\frac{1}{n^2}$. **[IAS(Pre) 2001]**

6.16. For a particle of mass m in a one-dimensional box what is the average of momentum p_x for the ground state?

(a) 0

(b) $\frac{h}{2L}$

(c) $\frac{L}{L}$

(d) $\frac{h}{2\pi L}$ [IAS(Pre) 2005]

ANSWERS

6.1. (d).	6.4. (b).	6.7. (b).	6.10. (a).	6.13. (d).	6.16. (a).
6.2. (b).	6.5. (d).	6.8. (a).	6.11. (d).	6.14. (c).	
6.3. (a).	6.6. (b).	6.9. (c).	6.12. (d).	6.15. (a).	

6(B): Short Answer Type Questions

6.1. State boundary conditions at the surface of a finite potential step.

6.2. What do you mean by *free state* and *bound state* of a particle.

6.3. What do you mean by *reflection coefficient* and *transmission of a particle incident* at the surface of a step potential?

6.4. If R and T be the reflection coefficient and transmission coefficient of a particle incident at the boundary of a potential barrier of height V_0, then show that $R + T = 1$, for $E < V_0$.

6.5. Show that the reflection coefficient R for a potential step for $E < V_0$ is given by

$$R = \left[\frac{1 - \sqrt{1 - V_0/E}}{1 + \sqrt{1 - V_0/E}}\right]^{1/2}$$

6.6. Explain the cause of failure of classical mechanics to explain the α-particle emission from radioactive elements.

6.7. What do you mean by the width of the potential barrier of a nucleus? Is the potential barrier of a nucleus is truly rectangular in shape? Explain.

6.8. What is *quantum mechanical tunneling*? Under what condition the transmission coefficient $T = 1$?

6.9. Show that for $T = 1$ the width of the potential well is exactly equal to half the de Broglie wavelength.

6.10. What is Ramsauer effect? Why this effect occurs for noble gases like Ar, Kr and Xe?

6.11. Mention some applications of *tunneling effect* in modern physics.

6.12. Show that for a potential $V(-\vec{r}) = -V(\vec{r})$ the wave function must be of even or odd parity. [CU (H) 2015]

6.13. Write down Hamiltonian for a freely moving particle and hence the corresponding Schrödinger equation. [KU (H) 2013]

6.14. Draw the energy level diagram of a particle in an infinite square well given by

$$V = \begin{cases} 0, & \text{for } 0 < x < L \\ \infty, & \text{for } x < 0 \text{ and } x > L \end{cases}$$

[KU (H) 2008]

6.15. Evaluate the commutator $[\hat{p}_x, \hat{H}]$ for a free particle of mass m moving along the positive x-axis, where p_x is the linear momentum and H is the Hamiltonian of the particle. Comment on the result. [KU (H) 2008]

6(C): Broad Answer Type Questions

6.1. (a) What is a one-dimensional potential step? **[PU 1997]**

(b) A particle of mass m is moving in one-dimensional potential given by

$$V = \begin{cases} 0, & \text{for } x < 0 \\ V_0, & \text{for } x \geqslant 0 \end{cases}$$

If the energy E of the incident particle is greater than V_0, then calculate the coefficient of reflection and transmission. **[LU 1992]**

6.2. What is *tunneling effect*? Use it to explain the phenomenon of α-decay.

6.3. A rectangular potential barrier is represented as

$$V = \begin{cases} 0, & \text{for } x < 0 \\ V_0, & \text{for } 0 \leqslant x \leqslant a \\ 0, & \text{for } x \geqslant a \end{cases}$$

Prove that for a particle of energy $E < V_0$, the transmission coefficient through the barrier is given by

$$T = \left[1 + \frac{V_0^2}{4E(V_0 - E)} \sinh^2 \alpha a\right]^{-1}, \text{ where } \alpha^2 = \frac{2m(V_0 - E)}{\hbar^2}$$

How does T change if (i) V_0 is increased (ii) barrier thickness a is increased?

6.4. A beam of electron of mass m and energy E moving in a region of zero potential energy, approaches a rectangular potential barrier of width a and height V_0, where $V_0 > E$. If $\beta a \gg 1$, where $\beta = \sqrt{\frac{2m(V_0-E)}{\hbar^2}}$, prove that the transmission coefficient is given by

$$T = \frac{16E}{V_0}\left(1 - \frac{E}{V_0}\right) e^{-2\beta a}$$

6.5. What is a fast-moving automobile's chance of being rebounded at a river bank instead of plunging into the river?

[*Hints*: This problem may be approximated to a potential well problem where $E > V_0$.]

6.6. Show that the coefficient of reflection of particles incident on a barrier is the same whether they approach from left or right.

6.7. A particle of total energy E is incident on a potential barrier described by

$$V(x) = \begin{cases} 0, & \text{for } x < 0 \\ V_0, & \text{for } 0 < x < a \\ 0, & \text{for } a < x \end{cases}$$

Obtain expressions for reflection and transmission coefficients and plot $|T|^2$ against E.

6.8. A particle travelling with energy E along x-axis has in its path a rectangular potential barrier of height $V_0 > E$ and width a.

(a) Show that there is a finite probability of transmission, even if $E < V_0$.

(b) Apply the result to explain the observed phenomenon of α-decay in nuclei.

6.9. Evaluate in the semiclassical approximation the transmission coefficient for a parabolic potential barrier of the form:

$$V(x) = \begin{cases} V_0\left(1 - \dfrac{x^2}{a^2}\right), & \text{for } -a \leqslant x \leqslant a \\ 0, & \text{for } |x| \geqslant 0 \end{cases}$$

[*Ans.* $T \simeq -\pi\sqrt{\frac{2m}{V_0}} \times \frac{a(V_0-E)}{\hbar}$]

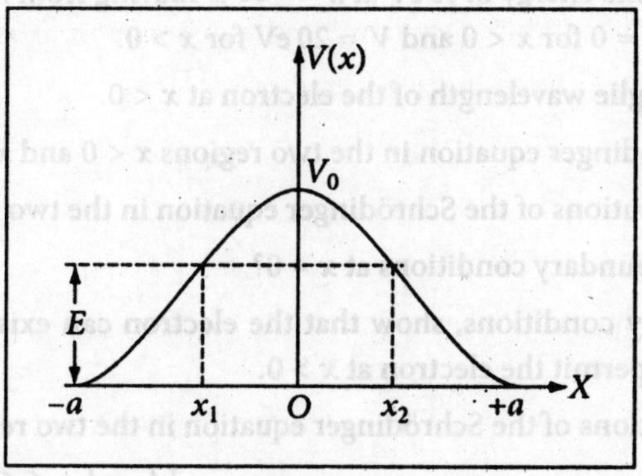

Fig 6.12

6.10. A particle is confined in a one-dimensional box with infinitely hard walls. Find the energy eigenvalues and normalised wave functions.

What is the need for normalisation? What is the expectation value of x of the ground state.

[BU (H) 1995, 2006]

6.11. A particle of mass m moving along a line with energy E is incident from left on a step potential

$$V(x) = \begin{cases} 0, & \text{for } x < 0 \\ V_0, & \text{for } x > 0 \end{cases}$$

V_0 being a constant.

(a) Write down the time-independent Schrödinger equation for the system.

(b) Obtain the solution of the wave equation for (i) $E > 0$ (ii) $0 < E < V_0$, illustrating the nature of the solutions with rough sketches.

(c) Calculate the transmission coefficient for the latter case and compare it with the classical result. Show that the sum of transmission and reflection coefficient is unity. **[BU (H) 1993, 1996, 1999, 2004]**

6.12. Consider the step potential shown in figure below:

A particle of energy $E < V_0$ is incident on the potential from left at $x = 0$. Obtain the expression for the transmission coefficient. **[BU (H) 1999]**

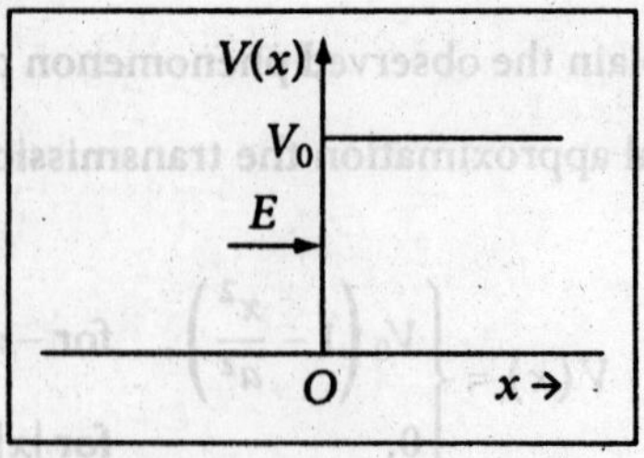

Fig 6.13

6.13. An electron with kinetic energy of 10 eV at $x = -\infty$ is moving from left to right along the x-axis. The potential energy is $V = 0$ for $x < 0$ and $V = 20$ eV for $x > 0$.

(a) Find the de Broglie wavelength of the electron at $x < 0$.

(b) Write the Schrödinger equation in the two regions $x < 0$ and $x > 0$.

(c) Find out the solutions of the Schrödinger equation in the two regions and interpret them.

(d) What are the boundary conditions at $x = 0$?

(e) Using boundary conditions, show that the electron can exist at $x > 0$. Why does classical mechanics not permit the electron at $x > 0$.

(f) Sketch the solutions of the Schrödinger equation in the two regions. [BU (H) 1999]

[*Ans.* $h = 6.63 \times 10^{-34}$ J-s $= 6.63 \times 10^{-27}$ ergs]

6.14. Consider a potential

$$V(x) = \begin{cases} 0, & \text{for } -a < x < a \\ +\infty, & \text{elsewhere} \end{cases}$$

(a) Find the bound state energy eigenvalues and (normalised) eigenfunctions. Show that the ground state energy is consistent with Heisenberg's uncertainty principle.

(b) Find the energy eigenvalues and wave functions of a particle moving in a two-dimensional box of size $L \times L$. [CU (H) 2014]

6.15. The state of a free particle is described by the following wave functions:

$$\psi(x) = \begin{cases} 0, & \text{for } x < -3a \\ c, & \text{for } -3a < x < a \\ 0, & \text{for } x > a \end{cases}$$

Determine c using the normalisation condition. Find the probability of finding the particle in the interval $[0, a]$. [CU (H) 2015]

6.16. (a) Find the energy eigenvalues and the normalised wave functions of particle confined in a one-dimensional box.

(b) Consider a potential barrier

$$V(x) = \begin{cases} 0, & \text{for } x < 0 \text{ and } x > a \\ V_0, & \text{for } 0 < x < a \end{cases}$$

where V_0 is positive. When a particle of mass m and energy $E(0 < E < V_0)$, approaches the barrier from the left, show that the transmission coefficient is

$$\left[1+\frac{V_0^2 \sinh^{2(\beta a)}}{4E(V_0-E)}\right]^{-1}$$

where $\beta = \sqrt{2m(V_0 - E)}/\hbar$. **[CU (H) 2011]**

6.17. **A particle of energy E is incident on the following step potential from left:**

$$V(x) = \begin{cases} 0, & x < 0, \text{ (Region I)} \\ V_0, & x \geqslant 0, \text{ (Region II)} \end{cases}$$

6.18. **A particle is confined in a one-dimensional potential box given by**

$$V(x) = \begin{cases} 0, & \text{for } 0 < x < a \\ \infty, & \text{otherwise} \end{cases}$$

Obtain the normalised wave functions and the allowed energy values for the system. **[CU (H) 2007]**

(a) Write down the time-independent Schrödinger equation in the two regions and obtain the solution for the cases: (i) $E > V_0$ (ii) $E < V_0$ up to one arbitrary constant.

(b) Sketch the probability density $|\psi(x)|^2$ in the two regions for: (i) $E > V_0$ (ii) $E < V_0$. **[BU (H) 2014]**

6.19. **(a) An otherwise free electron is trapped in a one-dimensional box of length L, having infinitely hard walls. Obtain an expression of quantised energy levels of this electron by considering its wave nature.**

(b) Solve Schrödinger equation for this electron with appropriate boundary conditions to obtain the energy eigenvalues and eigenfunctions.

(c) Show that the average position of the electron is the middle of the box for all quantum states.

[BU (H) 2013]

6.20. **(a) A particle of energy E is moving from left to right along X-axis and encounters potential barrier given by**

$$V(x) = \begin{cases} 0, & \text{for } x < 0 \\ V_0, & \text{for } 0 < x < a \\ 0, & \text{for } x > a \end{cases}$$

Write down Schrödinger equations and find the corresponding wave functions in the regions described above.

Show that the particle may be turned back by the potential barrier even if its energy $E > V_0$. Hence, evaluate the sum of transmission coefficient and reflection coefficient.

(b) Plot and discuss the variations of transmission coefficient and reflection coefficient with the energy of the incident particle. **[BU (H) 2012]**

6.21. **Consider a one-dimensional box with infinitely hard walls at $x = 0$ and $x = L$. A particle of mass m can move freely within the box and it does not loose any energy by collision with the walls.**

(a) Write down the Schrödinger equation for this particle in the region inside the box.

(b) Write down general solution of the Schrödinger equation and solution consistent with the boundary condition at $x = 0$.

(c) Obtain possible values of energy of the particle.

(d) Can the energy of the particle be zero? **[BU (H) 2009]**

6.22. (a) An otherwise free particle of mass m is confined in a one-dimensional box of length L. The walls of the box are infinitely hard and the particle does not loose energy while colliding with the walls. Find the minimum energy of such a particle from the uncertainty principle. **[BU (H) 2008]**

(b) Find the probability of finding the particle at x.

6.23. (a) A particle is confined to move in one-dimensional box with perfectly rigid walls at $x = 0$ and $x = l$. Get an expression for the normalised wave function of the particle.

(b) Assuming the wave function for one-dimensional box of length l find the expectation value of x.

(c) State the standard conditions the wave function must satisfy in quantum mechanics. **[NBU (H) 2008]**

6.24. A particle with mass m and energy E is confined in a one-dimensional potential barrier with potential function

$$V(x) = \begin{cases} 0, & \text{for } x < 0 \\ V_0, & \text{for } 0 < x < a \\ 0, & \text{for } a < x \end{cases}$$

If $E \geqslant V_0$ and $E < V_0$, calculate the transmission coefficient. **[KU (H) 2011]**

6.25. A particle of energy E moving in one dimension is incident from left (negative x region) on the step potential

$$V(x) = 0, \text{ for } x < 0$$

$$V(x) = V_0, \text{ for } x > 0$$

$$V_0 > 0$$

Calculate the reflection coefficient R. **[KU (H) 2010]**

6.26. (a) Write down the time-independent Schrödinger equation for a particle moving in one-dimensional if there is no force acting on it and find out the solution.

(b) If the particle is constrained to move such that $0 < x < L$, what are the boundary conditions on the wave function? **[CU (H) 2013]**

6.27. For particle trapped in infinite square well of width L

(a) Find the ground state wave function, $\langle x \rangle$, $\langle p_x \rangle$ and interpret. **[KU (H) 2012]**

(b) Sketch ground state and first excited state wave function if end positions are at (i) $x = 0, L$ (ii) $x = \pm L/2$. **[cf. KU (H) 2014]**

6.28. So far as α-decay is concerned how does the potential barrier appear around the nucleus? Explain clearly. Estimate the minimum potential energy by an α-particle in a nucleus of mass number $A = 210$ and $Z = 82$. Assume the radius of the nucleus to be $R = R_0 A^{1/3}$, $R_0 = 1.4$ fm. **[KU (H) 2013]**

6.29. For a particle in one-dimensional rectangular potential well-defined by $V(x) = 0$ for $x < 0$, $L > x$.

(a) Write down Schrödinger wave equation for the particle and solve it.

(b) Find reflection and transmission coefficients for $E > V_0$ case. [cf. KU (H) 2008, 2013]

6.30. A particle of energy E moving in one dimension is incident from left (negative x region) on a step potential

$$V(x) = \begin{cases} 0, & \text{for } x < 0 \\ V_0, & \text{for } x \geqslant 0 \; [V_0 > 0] \end{cases}$$

(a) Write down Schrödinger wave equation for the particle and solve it.

(b) Find reflection and transmission coefficient for $E > V_0$ case. [cf. KU (H) 2008, 2013]

6(D): Numerical Problems

6.1. A stream of electrons of kinetic energy 2.5 eV is incident on a potential barrier (rectangular in shape) of width 10^{-10} m and height 5 eV. Calculate the transmission coefficient. [*Ans.* $T = 0.6$]

6.2. An α-particle of energy 10 MeV approaches a potential barrier of height equal to 30 MeV. Determine the width of the potential barrier if the transmission coefficient is 2×10^{-3}.

$[m_\alpha = 4 \times 1.67 \times 10^{-27}$ kg, $\hbar = 1.054 \times 10^{-34}$ J-s] [*Ans.* 1.91×10^{-15} m]

6.3. What ratio of E/V_0 is necessary for scattering for a one-dimensional step potential, so that the transmission probability is 50 %? [*Ans.* 1.03]

6.4. A particle of mass 1 gm travelling with a velocity of 1 cm/sec encounters a potential barrier of height 10^{-7} J. Calculate the penetration depth of this particle.

[*Hints*: Penetration depth, $\alpha = \sqrt{\frac{2m(V_0 - E)}{\hbar^2}}$]

6.5. Calculate the transmission coefficient for a beam of electrons of energy 20 eV when incident on a potential barrier of 4.0 eV height and 0.4 nm wide. [*Ans.* 0.0122]

6.6. A beam of electrons is incident on a potential barrier 5 eV high and 0.5 nm wide. What energy should they possess so that half of them are to get through the barrier? [*Ans.* 4.9816 eV]

6.7. An electron streams each of energy 4 eV is incident on a potential barrier of height 5 eV. If the width of the barrier is 2×10^{-7} cm, what is the percentage transmission of this beam through the barrier? [*Ans.* 3.22×10^{-7} %]

6.8. An α-particle of energy 10 MeV is incident on a potential barrier of height 30 MeV. If the transmission coefficient is 0.002, then what is the width of the barrier? [*Ans.* 1.94×10^{-13} cm]

6.9. An electron and a proton of same energy (E) approach a potential barrier of height $V_0 > E$. Whose probability of getting through the barrier will be more? [*Ans.* $P_{\text{electron}} > P_{\text{proton}}$]

6.10. Find the eigenfunctions and the energy eigenvalues for a particle in one-dimensional infinite deep well $V(x) = 0$ for $-L/2 < x < L/2$ and $V(x) = \infty$ for $|x| \geqslant L/2$. Compare the solutions with the eigenfunctions and the energy eigenvalues if $V(x) = 0$ only for $0 < x < L$. [cf. KU (H) 2006, 2008]

6.11. (a) Find $\langle x \rangle$, $\langle x^2 \rangle$ and uncertainty in x in the region $x > 0$ for a particle with energy E incident from left to right on a step potential barrier of height $V_0 > E$ at $x = 0$.

(b) Hence, discuss whether the particle which penetrates the barrier in the region $x > 0$ can be observed here?

Chapter 7

Applications of Schrödinger's Equation to One-dimensional Problems—Bound States

7.1 What is a bound state?

If the motion of a particle is confined to a limited region of space by potential energy so that the particle can move back and forth in the region, then the particle is in a *bound state*. The simplest example of all motions in bound state is the motion of a particle in a one-dimensional box with zero potential energy and the potential energy is assumed to be *infinite* at the walls of the box.

The motion of a particle in three-dimensional box, one-dimensional box, motion of a simple harmonic oscillator, motion of a particle in a one-dimensional square well potential with $(E < V_0)$, motion of a particle in one-dimensional (also in three-dimensional) potential well, motion of a particle in spherically symmetric potential, etc. are the common examples of bound state. Application of *Schrödinger time-independent equation* to such problems leads to *discrete energy values* of the particle.

7.2 Particle in a one-dimensional box

We assume a particle of mass m and energy E moving in a region from $x = 0$ to $x = a$ with absolutely rigid, impenetrable potential walls of infinite height (Fig 7.1). So

$$V(x) = \begin{cases} \infty, & \text{for } x = 0 \text{ and } x = a \\ 0, & \text{for } 0 < x < a \end{cases}$$

The particle experiences no force in the box as $V(x) = 0$.

Also the particle during its motion collides with the rigid walls with no loss of energy and as such the total energy E of the particle remains constant. As $V(x) = \infty$ at both $x = 0$ and $x = a$, the particle needs infinite work to leave the region.

Since, this is impossible, the particle remains confined within the box and the probability of the particle to be at $x = 0$ and $x = a$ must be zero.

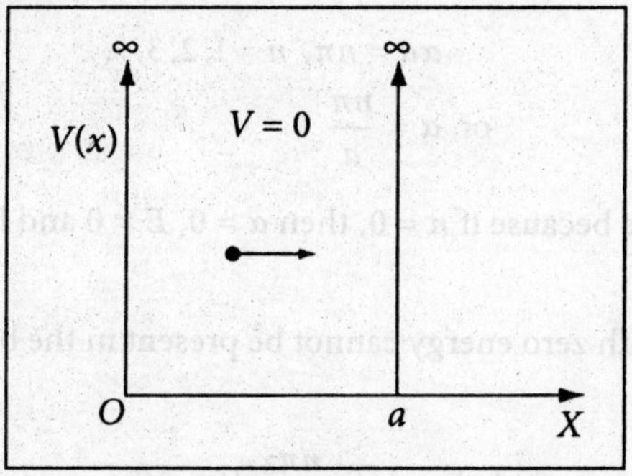

Fig 7.1 One-dimensional potential box

Since, probability is given by $|\psi|^2$, the wave function $\psi(x)$ at $x = 0$ and $x = a$ must vanish. That is

$$\psi(x)\Big|_{x=0} = \psi(x)\Big|_{x=a} = 0 \tag{7.1}$$

The potential energy curve of the appearance as shown in Fig 7.1 is called a *square well potential of infinite depth.*

7.2.1 The wave equation for the particle

Since, $V(x)$ is independent of time we apply time-independent Schrödinger wave equation in the region $0 \leqslant x \leqslant a$.

The *one-dimensional time-independent Schrödinger equation is*

$$-\frac{\hbar^2}{2m}\frac{d^2\psi}{dx^2} = E\psi$$

$$\text{or,}\ \frac{d^2\psi}{dx^2} + \frac{2m}{\hbar^2}E\psi = 0 \tag{7.2}$$

$$\text{or,}\ \frac{d^2\psi}{dx^2} + \alpha^2\psi = 0 \tag{7.3}$$

$$\text{where}\ \alpha = \sqrt{\frac{2mE}{\hbar^2}} \tag{7.4}$$

7.2.2 The solution of eq. (7.3)

- The general solution of eq. (7.3) is given by

$$\psi(x) = A\sin\alpha x + B\cos\alpha x \tag{7.5}$$

where A and B are constants of integration whose values may be found out by applying the following boundary conditions:

1. $\psi(x) = 0$ at $x = 0$ **2.** $\psi(x) = 0$ at $x = a$.

Applying condition (i) to eq. (7.5), we get

$$0 = A\sin 0 + B\cos 0 \text{ or, } B = 0$$

$$\therefore \psi(x) = A\sin\alpha x \tag{7.6}$$

Applying the condition (2) to eq. (7.6), we get $0 = A\sin\alpha a$.

Since, $A \neq 0$, $\sin\alpha a = 0$, then

$$\alpha a = n\pi,\ n = 1, 2, 3, \ldots$$

$$\text{or, } \alpha = \frac{n\pi}{a} \tag{7.7}$$

Here, $n = 0$ is not admissible because if $n = 0$, then $\alpha = 0$, $E = 0$ and hence $\psi(x) = 0$ everywhere inside of the box.

In other words, a particle with zero energy cannot bè present in the box.

Hence,

$$\psi_n(x) = A\sin\frac{n\pi x}{a} \quad [\text{for } 0 < x < a] \tag{7.8}$$

7.2.3 Energy eigenvalues

From eq. (7.7)

$$\sqrt{\frac{2m}{\hbar^2}E} = \frac{n\pi}{a} \text{ or, } E = \frac{n^2\pi^2\hbar^2}{2ma^2} \text{ or, } E_n = \frac{n^2\pi^2\hbar^2}{2ma^2} \tag{7.9}$$

So, we get an infinite sequence of discrete energy levels that corresponds to all integral values of n, where n is called the quantum number. As $a \to \infty$, $\Delta E_n \to 0$, i.e., the energy values are closer if the size of the box is larger forming a continuum.

Putting $n = 1$ in eq. (7.9), we get

$$\boxed{E_1 = \frac{\pi^2\hbar^2}{2ma^2}} \tag{7.10}$$

(i) E_1 is the lowest energy of the particle in the box and is called the *ground state energy level.* In terms of E_1, E_n can be expressed as

$$E_n = n^2 E_n \tag{7.11}$$

(ii) The possible energy values of the particle are $E_1, E_2, E_3, \ldots, E_n$ for $n = 1, 2, 3, \ldots, n$, which are discrete.

(iii) Since,

$$\Delta E = E_{n+1} - E_n = (n+1)^2 E_1 - n^2 E_1$$

$$\text{or, } \Delta E = (2n+1)E_1 \tag{7.12}$$

$$\text{or, } \Delta E = \frac{(2n+1)}{a^2}\frac{\pi^2\hbar^2}{2m}\left[\because E_1 = \frac{\pi^2\hbar^2}{2ma^2}\right] \tag{7.13}$$

Equation (7.13) shows that as $a \to \infty$, $\Delta E \to 0$, the energy levels lead to a continuum corresponding to a *free particle.*

The spacing between the energy levels increases with the increase of n, i.e., for higher energy levels. The energy level diagram for a particle confined to move in a one-dimensional box is shown in Fig 7.2.

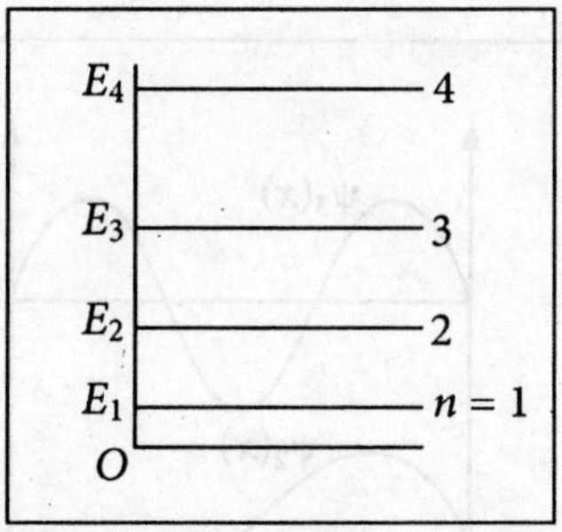

Fig 7.2 Energy level diagram for a particle confined to move in a one-dimensional box

7.2.4 Normalisation of the wave function

The wave function ψ_n corresponding to the energy E_n of the nth quantum state is given by

$$\psi_n(x) = \begin{cases} A \sin \frac{n\pi x}{a}, & \text{for } 0 < x < a \\ 0, & \text{for } 0 < x, x > a \end{cases} \tag{7.14}$$

Since, the particle must be somewhere within the box, the total probability of finding it there is unity.

$$\therefore \int_0^a \psi_n^*(x)\,\psi_n(x)\,dx = 1 \text{ or, } \int_0^a |\psi_n(x)|^2\,dx = 1 \text{ or, } A^2 \int_0^a \sin^2 \frac{n\pi x}{a}\,dx = 1$$

$$\text{or, } A^2 \int_0^a \frac{1}{2}\left[1 - \cos \frac{2\pi n x}{a}\right] dx = 1 \text{ or, } \frac{A^2}{2}\left[x - \frac{a}{2\pi n} \sin \frac{2\pi n x}{a}\right]_0^1 = 1$$

$$\text{or, } \frac{A^2 a}{2} = 1 \left[\because \int_0^a \sin \frac{2\pi n x}{a}\,dx = 0\right]$$

$$\text{or, } A = \sqrt{\frac{2}{a}} \tag{7.15}$$

Hence, the normalised wave function of the particle for the nth quantum state is

$$\boxed{\psi_n(x) = \sqrt{\frac{2}{a}} \sin \frac{n\pi x}{a}} \tag{7.16}$$

From eq. (7.16), it is evident that for $n = 1$ at $x = 0$ and $x = a$, $\psi_1(x) = 0$, i.e., the wave function on $\psi_1(x)$ has two nodes at $x = 0$ and $x = a$.

For $n = 2$, $\psi_2(x) = 0$ at $x = 0, \frac{a}{2}, a$.

Hence, $\psi_2(x)$ has three nodes at $x = 0, \frac{a}{2}, a$.

For $n = 3$, $\psi_3(x) = 0$ at $x = 0, \frac{a}{3}, \frac{2a}{3}, a$.

Hence, $\psi_3(x)$ has four nodes at $x = 0, \frac{a}{3}, \frac{2a}{3}, a$. In this way, the wave function $\psi_n(x)$ has $(n+1)$ nodes. The shapes of ψ_1, ψ_2 are depicted in Fig 7.3 with the positions of the nodes.

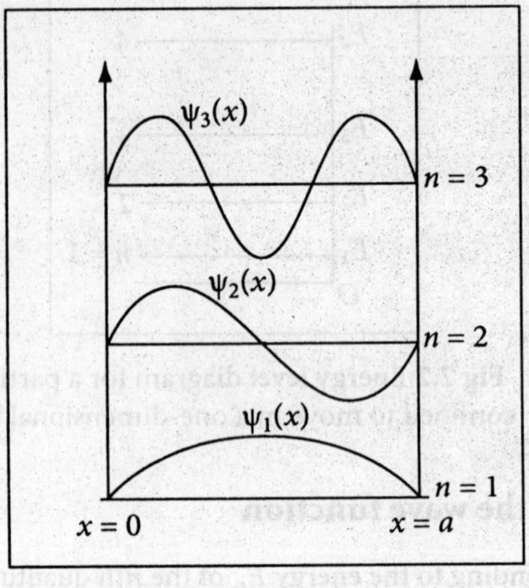

Fig 7.3 Shapes of ψ_1, ψ_2 and ψ_3.

7.2.5 Orthogonality of the wave functions

If $\psi_n(x)$ and $\psi_m(x)$ be two normalised wave functions corresponding to the quantum states n and m respectively in the interval $(0, a)$, then

$$\psi_n(x) = \sqrt{\frac{2}{a}} \sin\frac{n\pi x}{a}, \qquad \psi_m(x) = \sqrt{\frac{2}{a}} \sin\frac{m\pi x}{a}$$

Now,

$$\int_0^a \psi_m^* \psi_n dx = \frac{2}{a}\int_0^a \sin\frac{m\pi x}{a} \sin\frac{n\pi x}{a} dx = \frac{1}{a}\int_0^a \left[\cos\frac{(m-n)\pi x}{a} - \cos\frac{(m+n)\pi x}{a}\right] dx$$

$$= \frac{1}{a}\left[\frac{a}{\pi(m-n)} \sin\frac{(m-n)\pi x}{a} - \frac{a}{\pi(m+n)} \sin\frac{(m+n)\pi x}{a}\right]_0^a = 0 \text{ [when } m \neq n]$$

Hence, the wave functions $\psi_m(x)$ and $\psi_n(x)$ are mutually orthogonal in $(0, a)$. These are also normalised in $(0, a)$. So $\psi_m(x)$ and $\psi_n(x)$ are orthonormal in the interval $(0, a)$. $\psi_m(x)$ and $\psi_n(x)$ are also orthonormal in the interval $(-\infty, +\infty)$ as these functions are zero in $-\infty < x < 0$ and $a < x < \infty$.

7.2.6 Eigenvalues of the momentum

The energy E_n for the nth quantum state is related to the corresponding relation by

$$E_n = \frac{p_n^2}{2m} \text{ or, } p_n^2 = 2mE_n \text{ or, } p_n^2 = 2m\frac{n^2\pi^2\hbar^2}{2ma^2} = \frac{n^2\pi^2\hbar^2}{a^2} \text{ or, } p_n = \pm\frac{n\pi\hbar}{a} \tag{7.17}$$

The ± sign appearing in the right-hand side due to the fact that the particle moves back and forth in the box.

From eq. (7.16), it is evident that the eigenvalues of momentum are discrete, i.e.,

$$\Delta p_n = p_{n+1} - p_n = \pm\left[\frac{(n+1)\pi\hbar}{a} - \frac{n\pi\hbar}{a}\right]$$

$$\Delta p_n = \pm\frac{\pi\hbar}{a} \tag{7.18}$$

Equation (7.17) shows that the difference of momenta between two consecutive energy levels is constant and is equal to $\frac{\pi\hbar}{a}$, unlike the difference of energy between the same two energy levels.

7.2.7 Probability of finding the particle in a small range between x and $x + dx$

The probability of finding the particle between x and $x + dx$ is given by

$$P(x)\,dx = |\psi_n(x)|^2\,dx = \frac{2}{a}\sin^2\frac{n\pi x}{a}\,dx \tag{7.19}$$

The probability density

$$\rho(x) = \frac{2}{a}\sin^2\frac{n\pi x}{a} \tag{7.20}$$

But, $\sin^2\frac{n\pi x}{a}$ is the maximum for

$$\frac{n\pi x}{a} = \frac{\pi}{2}, \frac{3\pi}{2}, \frac{5\pi}{2} \text{ or, } x = \frac{a}{2n}, \frac{3a}{2n}, \frac{5a}{2n} \tag{7.21}$$

We know, the probability of finding the particle is the maximum at the position, where the amplitude of the wave function is the maximum. In that view the most probable position of finding the particle in state $n = 1$ is at $x = a/2$; $n = 2$ are at $x = a/4$ and $x = 3a/4$; $n = 3$ are at $x = a/6$, $x = 3a/6$, $x = 5a/6$.

The probability dependence of the particle in state $n = 0, 1$ and 2 is represented in Fig 7.4.

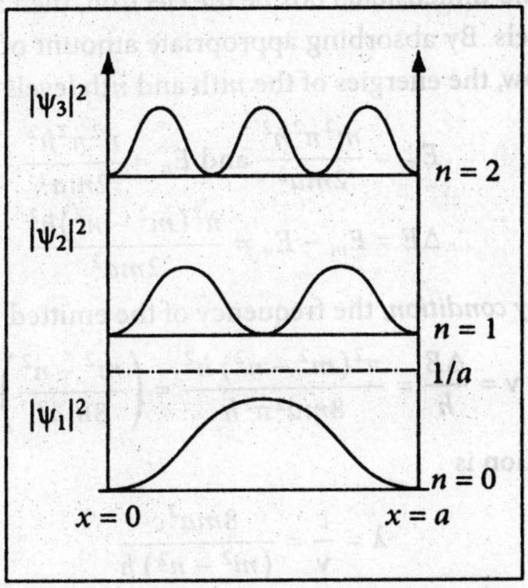

Fig 7.4 The probability dependence of the particle in state $n = 0, 1, 2$

According to classical mechanics the particle in such a potential box would move with constant velocity and suffer reflection at the rigid walls. The probability of finding the particle anywhere in the box is the same and $= dx/a$. The classical probability density $(1/a)$ in this case, does not depend on x and as such is represented by a straight line [Fig 7.4].

7.2.8 Zero-point energy

We have from eq. (7.10) the minimum kinetic energy of the particle in the box

$$E_1 = \frac{\pi^2\hbar^2}{2ma^2}$$

Since, the particle cannot have energy less than E_1, the energy of the particle in the box even if its temperature be reduced to 0 K will still be E_1 and will not be zero.

Hence,

$$E_1 = \frac{\pi^2 \hbar^2}{2ma^2}$$

is called the *zero-point energy*. This implies that the particle inside the box is not at rest even at 0 K. The result is that the position of the particle is not precisely known in the box.

Also as the mean kinetic energy is known, the momentum is not precisely known. Thus, the occurrence of zero-point energy is consistent with the *Heisenberg uncertainty principle*.

If the uncertainty in position of the particle is $\Delta x = a$, then from the uncertainty relation

$$\Delta x \, \Delta p_x = \hbar \text{ or, } a \Delta p_x = \hbar$$

We have $\Delta p_x = \hbar/a$, the uncertainty in the corresponding momentum. The minimum kinetic energy in this case, is

$$(\text{KE})_{\text{min}} = \frac{(\Delta p_x)^2}{2m} = \frac{\hbar^2}{2ma^2}$$

7.2.9 Factors influencing colour

If the particle confined in the one-dimensional box be the electron, then as, we have seen, it can have a large number of discrete energy levels. By absorbing appropriate amount of energy it can undergo transition from mth level to nth level. Now, the energies of the mth and nth levels are

$$E_m = \frac{m^2 \pi^2 \hbar^2}{2ma^2} \text{ and } E_n = \frac{n^2 \pi^2 \hbar^2}{2ma^2}$$

$$\therefore \Delta E = E_m - E_n = \frac{\pi^2 (m^2 - n^2) \hbar^2}{2ma^2} \tag{7.22}$$

According to *Bohr's frequency condition*, the frequency of the emitted radiation is

$$\nu = \frac{\Delta E}{h} = \frac{\pi^2 (m^2 - n^2) \, h^2}{8ma^2 \pi^2 h} = \left(\frac{m^2 - n^2}{8ma^2} \right) h \tag{7.23}$$

The wavelength of the transition is

$$\lambda = \frac{c}{\nu} = \frac{8ma^2 c}{(m^2 - n^2) \, h} \tag{7.24}$$

Equation (7.24) shows that $\lambda \propto a^2$. Thus, an increase in width a of the box leads to an increase in λ. By adjusting the value of a the radiation may be made to appear beyond the visible region where the system becomes coloured. Thus, *the width of the box a influences the colour of the system.*

7.2.10 Electronic transition possibility between a pair of states

Here, we shall show that *the transition between a pair of energy states is possible if the sum or difference of quantum numbers is an odd number. In case both are even numbers, then the transition is strictly forbidden.*

We know, the magnitude of the transition dipole moment integral is expressed as

$$e \int \psi_m \hat{x} \, \psi_n \, dx$$

e is the electronic charge and ψ_m and ψ_n are the eigenfunction corresponding to quantum states m and n respectively.

$$\because \psi_m = \sqrt{\frac{2}{a}} \sin \frac{m\pi x}{a} \text{ and } \psi_n = \sqrt{\frac{2}{a}} \sin \frac{n\pi x}{a}$$

$$\therefore \int_0^a \psi_m \hat{x} \psi_n \, dx = \frac{2}{a} \int_0^a \sin \frac{m\pi x}{a} \cdot x \cdot \sin \frac{n\pi x}{a} \, dx \tag{7.25}$$

Let $\pi x/a = \alpha$, then the above equation becomes

$$\int_0^a \psi_m \hat{x} \psi_n \, dx = \frac{2a}{\pi^2} \int_0^\pi \sin m\alpha \sin n\alpha \cdot \alpha \cdot d\alpha$$

$$= \frac{a}{\pi^2} \left[\int_0^\pi \cos(n-m)\alpha \cdot \alpha \cdot d\alpha - \int_0^\pi \cos(n+m)\alpha \cdot \alpha \cdot d\alpha \right]$$

$$= \frac{a}{\pi^2} [I_1 - I_2] \tag{7.26}$$

where

$$I_1 = \int_0^\pi \cos(n-m)\alpha \cdot \alpha \cdot d\alpha = \left[\alpha \cdot \frac{\sin(n-m)\alpha}{(n-m)} \right]_0^\pi - \int_0^\pi \frac{\sin(n-m)\alpha}{(n-m)} d\alpha$$

$$= -\frac{2}{(n-m)^2} \text{ if } (n-m) \text{ is odd}$$

$$= 0, \text{ if } (n-m) \text{ is even}$$

In a similar way we can show that

$$I_2 = \begin{cases} -\frac{2}{(n+m)^2}, & \text{if } (n+m) \text{ is odd} \\ 0, & \text{if } (n+m) \text{ is even.} \end{cases}$$

Hence, *the transition between a pair of states is possible if $(n \pm m)$ is odd.*

7.3 One-dimensional square well potential of infinite depth

If a particle be constrained to move in a potential as below

$$V(x) = \begin{cases} 0, & \text{for } -a/2 < x < a/2 \\ +\infty, & \text{for } |x| > a/2 \end{cases}$$

having perfectly rigid and impenetrable walls at $x = -a/2$ and $x = +a/2$, then also we can show that the energy eigenvalues are *discrete* similar to the previous case.

The Schrödinger time-independent equation for the particle in the region $-a/2 < x < a/2$ is

$$-\frac{\hbar^2}{2m}\frac{d^2\psi}{dx^2} = E\psi \text{ or, } \frac{d^2\psi}{dx^2} + \frac{2m}{\hbar^2}E\psi = 0 \tag{7.27}$$

where m is the mass of the particle. The general solution of eq. (7.27) is

$$\psi(x) = A\cos\alpha x + B\sin\alpha x, \tag{7.28}$$

$$\text{where } \alpha = \left(\frac{2mE}{\hbar^2}\right)^{1/2} \tag{7.29}$$

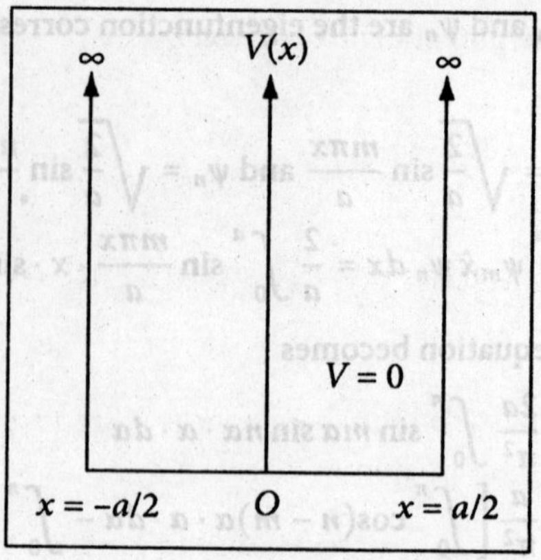

Fig 7.5 One-dimensional square well potential of finite depth

Here, A and B are two constants of integration which can be found out by applying the following boundary conditions:

1. $\psi = 0$ at $x = a/2$ **2.** $\psi = 0$ at $x = -a/2$.

Applying these boundary conditions to eq. (7.28), we obtain

$$A\cos\frac{\alpha a}{2} + B\sin\frac{\alpha a}{2} = 0,\ A\cos\frac{\alpha a}{2} - B\sin\frac{\alpha a}{2} = 0.\ \therefore\ A\cos\frac{\alpha a}{2} = 0 \text{ and } B\sin\frac{\alpha a}{2} = 0$$

Now, since both A and B cannot be zero (as in that case $\psi(x) = 0$ everywhere).

Also for given value of α or E, $\cos\frac{\alpha a}{2}$ and $\sin\frac{\alpha a}{2}$ cannot be zero.

There are two possible solutions, viz., either $B = 0$ and $\cos\frac{\alpha a}{2} = 0$

$$\therefore \frac{\alpha a}{2} = \frac{n\pi}{2} \text{ (} n \text{ being an odd integer)}$$

or, $A = 0$ and $\sin\frac{\alpha a}{2} = 0$.

$$\therefore \frac{\alpha a}{2} = \frac{n\pi}{2} \text{ (} n \text{ being an even integer)}$$

So

$$\psi(x) = A\cos\frac{n\pi x}{a} \text{ (when } n \text{ is odd)} \tag{7.30}$$

$$\text{and } \psi(x) = B\sin\frac{n\pi x}{a} \text{ (when } n \text{ is even)} \tag{7.31}$$

The energy eigenvalue for the nth quantum state is given by

$$\frac{\alpha a}{2} = \frac{n\pi}{2} \text{ or, } \sqrt{\frac{2mE_n}{\hbar^2}} = \frac{n\pi}{a} \text{ or, } \frac{2mE_n}{\hbar^2} = \frac{n^2\pi^2}{a^2} \text{ or, } E_n = \frac{n^2\pi^2\hbar^2}{2ma^2} \tag{7.32}$$

This shows that $E_n \propto 1/a^2$ for a particular value of n and m. *There is thus an infinite sequence of discrete energy levels.*

Similar results are obtained if the potential box is given by

$$V(x) = \begin{cases} 0, & \text{for } -a < x < a \\ +\infty, & \text{for } x < -a \text{ and } x > a. \end{cases}$$

7.4 One-dimensional square well potential of finite depth: Bound state

We now consider the motion of a particle of mass m approaching a square well potential of finite depth as shown in Fig 7.6. The potential well is represented as

$$V = \begin{cases} V_0, & \text{for region I } (x < -a) \\ 0, & \text{for region II } (-a < x < a) \\ V_0, & \text{for region III } (x > a) \end{cases}$$

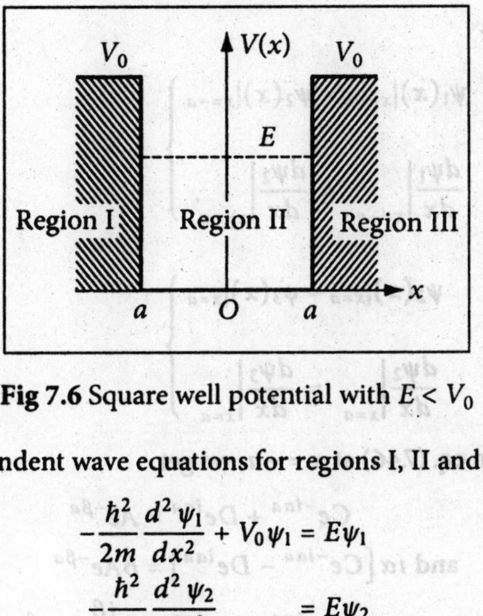

Fig 7.6 Square well potential with $E < V_0$

The Schrödinger's time-independent wave equations for regions I, II and III respectively are

$$-\frac{\hbar^2}{2m}\frac{d^2\psi_1}{dx^2} + V_0\psi_1 = E\psi_1 \tag{7.33}$$

$$-\frac{\hbar^2}{2m}\frac{d^2\psi_2}{dx^2} = E\psi_2 \tag{7.34}$$

$$-\frac{\hbar^2}{2m}\frac{d^2\psi_3}{dx^2} + V_0\psi = E\psi_3 \tag{7.35}$$

with energy of the particle $E < V_0$.

$$\text{or, } \frac{d^2\psi_1}{dx^2} - \frac{2m}{\hbar^2}(V_0 - E)\,\psi_1 = 0 \tag{7.36}$$

$$\frac{d^2\psi_2}{dx^2} + \alpha^2\psi_1 = 0 \tag{7.37}$$

$$\text{and } \frac{d^2\psi_3}{dx^2} - \beta^2\psi_3 = 0, \tag{7.38}$$

$$\text{where } \alpha^2 = \frac{2mE}{\hbar^2} \text{ and } \beta^2 = \frac{2m(V_0 - E)}{\hbar^2} \tag{7.39}$$

The general solutions of eqs (7.36), (7.37) and (7.38) are as follows:

$$\psi_1(x) = Ae^{\beta x} + Be^{-\beta x} \tag{7.40}$$

$$\psi_2(x) = Ce^{i\alpha x} + De^{-i\alpha x} \tag{7.41}$$

$$\psi_3(x) = Fe^{-\beta x} + Ge^{\beta x} \tag{7.42}$$

As $\psi_1(x)$ must remain well-behaved in region I, we must have $\psi_1(x)$ at $x = -\infty$ must be zero. For this to be true $B = 0$ in eq. (7.40).

Similarly, for $\psi_3(x)$ to remain well-behaved in region III, $G = 0$.

Thus, we get

$$\psi_1(x) = Ae^{\beta x} \tag{7.43}$$

$$\psi_2(x) = Ce^{i\alpha x} + De^{-i\alpha x} \tag{7.44}$$

$$\text{and } \psi_3(x) = Fe^{-\beta x} \tag{7.45}$$

The boundary conditions are

$$\left.\begin{array}{ll} \text{(i)} & \psi_1(x)|_{x=-a} = \psi_2(x)|_{x=-a} \\ \text{(ii)} & \left.\dfrac{d\psi_1}{dx}\right|_{x=-a} = \left.\dfrac{d\psi_2}{dx}\right|_{x=-a} \end{array}\right\} \tag{7.46}$$

$$\left.\begin{array}{ll} \text{(iii)} & \psi_2(x)|_{x=a} = \psi_3(x)|_{x=a} \\ \text{(iv)} & \left.\dfrac{d\psi_2}{dx}\right|_{x=a} = \left.\dfrac{d\psi_3}{dx}\right|_{x=a} \end{array}\right\} \tag{7.47}$$

Applying boundary condition eq. (7.46) at $x = -a$, we get

$$Ce^{-i\alpha a} + De^{i\alpha a} = Ae^{-\beta a} \tag{7.48}$$

$$\text{and } i\alpha\left[Ce^{-i\alpha a} - De^{i\alpha a}\right] = \beta Ae^{-\beta a} \tag{7.49}$$

$$\text{or, } Ce^{-i\alpha a} - De^{i\alpha a} = -\frac{i\beta}{a} Ae^{-\beta a} \tag{7.50}$$

Adding eqs (7.48) and (7.50), we get

$$2Ce^{-i\alpha a} = \left(1 - \frac{i\beta}{\alpha}\right) Ae^{-\beta a} \tag{7.51}$$

Subtracting eq. (7.50) from eq. (7.48), we get

$$2De^{i\alpha a} = \left(1 + \frac{i\beta}{a}\right) Ae^{-\beta a} \tag{7.52}$$

Now, taking the ratio of eqs (7.51) and (7.52), we get

$$\frac{C}{D}e^{-2i\alpha a} = \frac{\alpha - i\beta}{\alpha + i\beta} \text{ or, } \frac{C}{D} = \left(\frac{\alpha - i\beta}{\alpha + i\beta}\right) e^{2i\alpha a} \tag{7.53}$$

Also from the boundary conditions represented by eq. (7.47), we obtain by proceeding exactly in the same way as above

$$Ce^{i\alpha a} + De^{-i\alpha a} = Fe^{-\beta a} \tag{7.54}$$

$$i\alpha\left[Ce^{i\alpha a} - De^{-i\alpha a}\right] = \beta Fe^{-\beta a} \tag{7.55}$$

$$\text{or, } Ce^{i\alpha a} - De^{-i\alpha a} = \frac{i\beta}{a}Fe^{-\beta a} \tag{7.56}$$

Adding eqs (7.54) and (7.56), we get

$$2Ce^{i\alpha a} = \left(1 + \frac{i\beta}{\alpha}\right)Fe^{-\beta a} \tag{7.57}$$

Subtracting eq. (7.56) from eq. (7.54), we get

$$2De^{-i\alpha a} = \left(1 - \frac{i\beta}{\alpha}\right)Fe^{-\beta a} \tag{7.58}$$

$$\therefore \frac{C}{D}e^{2i\alpha a} = \frac{\alpha + i\beta}{\alpha - i\beta} \text{ or, } \frac{C}{D} = \frac{\alpha + i\beta}{\alpha - i\beta}e^{-2i\alpha a} \tag{7.59}$$

Multiplying eqs (7.53) and (7.59), we get

$$\frac{C}{D} \times \frac{C}{D} = \frac{\alpha - i\beta}{\alpha + i\beta} \times \frac{\alpha + i\beta}{\alpha - i\beta} \text{ or, } \frac{C^2}{D^2} = 1 \text{ or, } \frac{C}{D} = \pm 1 \tag{7.60}$$

Now, taking $C/D = +1$ or $C = D$, we get from eqs (7.48) and (7.54), we get

$$Ae^{-\beta a} = Fe^{-\beta a} \text{ or, } A = F \tag{7.61}$$

Again, since $C = D$, we get from eq. (7.44)

$$\psi_2(x) = 2C(e^{i\alpha x} + e^{-i\alpha x})$$

$$\text{or, } \psi_2(x) = 2C\cos\alpha x \tag{7.62}$$

From eq. (7.62) we note that

$$\psi_2(-x) = \psi_2(+x)$$

i.e., $\psi_2(x)$ has *even parity*. $\psi_2(x)$ for E_0 and E_2 are *symmetric* about $x = 0$ as shown in Fig 7.7.

Also for $C = -D$, $A = -F$ and

$$\psi_2(x) = 2C(e^{i\alpha x} - e^{-i\alpha x}) \text{ or, } \psi_2(x) = 2Ci\sin\alpha x \tag{7.63}$$

Here, $\psi_2(-x) = -\psi(+x)$, i.e., $\psi_2(x)$ is *antisymmetric*. $\psi_2(x)$ for E_1 is depicted in Fig 7.7.

It is evident from Fig 7.7 that $|\psi_2(x)|^2 \neq 0$ in regions I and III which is a new result not expected classically for $E < V_0$. $\psi_2(x)$ in the present analysis is seen to penetrate the walls at $x = \pm a$ and the amplitude of $\psi_2(x)$ falls off exponentially as $|x|$ increases beyond the walls at $x = \pm a$, i.e., in the classically forbidden region. *The form of the wave function $\psi_2(x)$ shows that it will remain appreciable beyond the well if $(V_0 - E)$ is small, i.e., E is large.*

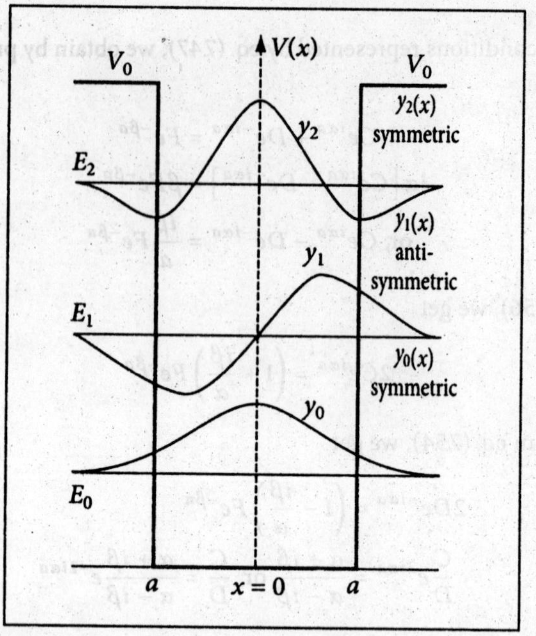

Fig 7.7 Eigenfunctions for the finite potential well

7.4.1 Energy eigenvalues

From eqs (7.53) and (7.59), we have

$$\frac{\alpha - i\beta}{\alpha + i\beta}e^{2i\alpha a} = \frac{\alpha + i\beta}{\alpha - i\beta}e^{-2i\alpha a} \text{ or, } (\alpha + i\beta)^2 e^{-2i\alpha a} = (\alpha + i\beta)\, e^{2i\alpha a}$$

$$\text{or, } (\alpha + i\beta)\, e^{-i\alpha a} = \pm(\alpha - i\beta)\, e^{i\alpha a} \tag{7.64}$$

The positive sign on the right-hand side of eq. (7.64) is related to *symmetric solution* and the negative sign to *antisymmetric solution*.

For *symmetric solution*

$$(\alpha + i\beta)\, e^{-i\alpha a} = (\alpha - i\beta)\, e^{i\alpha a}$$

$$\text{or, } \alpha[e^{i\alpha a} - e^{-i\alpha a}] = i\beta[e^{i\alpha a} - e^{-i\alpha a}]$$

$$\text{or, } \alpha[2i \sin \alpha a] = i\beta[2 \cos \alpha a]$$

$$\text{or, } \alpha \sin \alpha a = \beta \cos \alpha a \tag{7.65}$$

$$\text{or, } \alpha \frac{\sin \alpha a}{\cos \alpha a} = \beta \text{ or, } \alpha a \tan \alpha a = \beta a \tag{7.66}$$

For antisymmetric solution

$$(\alpha + i\beta)\, e^{-i\alpha a} = -(\alpha - i\beta)\, e^{i\alpha a}$$

$$\text{or, } \alpha \cos \alpha a = -\beta \sin \alpha a \tag{7.67}$$

$$\text{or, } \alpha a \cot \alpha a = -\beta a \tag{7.68}$$

Again,

$$\alpha^2 + \beta^2 = \frac{2mE}{\hbar^2} + \frac{2m(V_0 - E)}{\hbar^2} \text{ or, } (\alpha^2 + \beta^2) = \frac{2mV_0}{\hbar^2} \text{ or, } (\alpha^2 + \beta^2)\, a^2 = \frac{2mV_0 a^2}{\hbar^2} \tag{7.69}$$

Putting $p = \alpha a$ and $q = \beta a$, we have from eqs (7.66) and (7.69)

$$q = p \tan p \text{ and } p^2 + q^2 = \frac{2mV_0 a^2}{\hbar^2} \tag{7.70}$$

From eqs (7.68) and (7.69)

$$q = -p \cot p \text{ and } p^2 + q^2 = \frac{2mV_0 a^2}{\hbar^2} \tag{7.71}$$

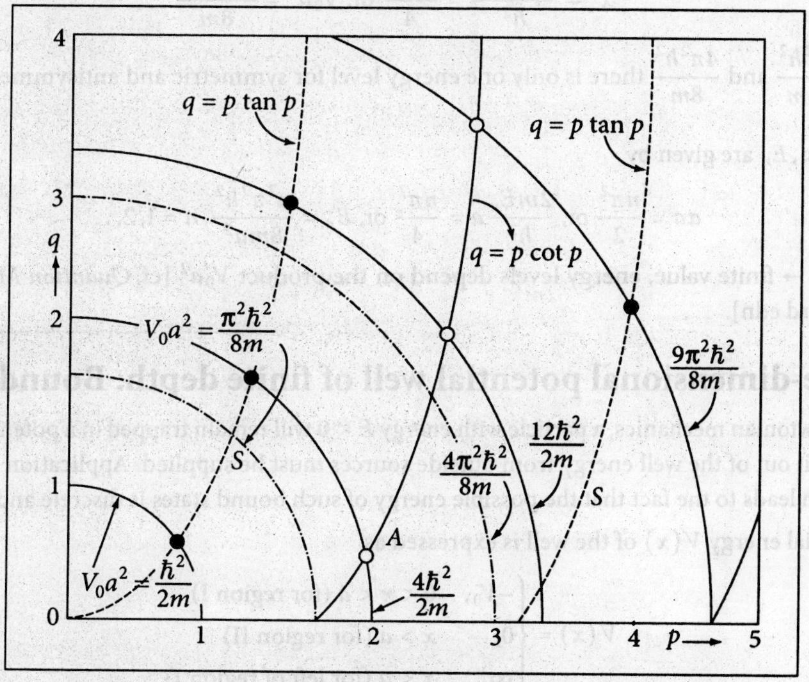

Fig 7.8 Graphical solution for a particle in square well-potential

Energy eigenvalues of the particle can be obtained by solving eqs (7.70) and (7.71) for *symmetric* and *antisymmetric* cases respectively.

Since, analytical solution is not possible, graphical solution is the only help. Since, α and β are restricted to positive values, the required energy values will be given by the intersections of $q = p \tan p$ and $q = -p \cot p$ in the first quadrant with the circle of known radius $\left(\frac{2mV_0 a^2}{\hbar^2}\right)^{1/2}$, [Fig 7.8].

Fig 7.8 shows the curves for three values of

$$V_0 a^2 = \frac{\hbar^2}{2m}, \frac{4\hbar^2}{2m}, \frac{12\hbar^2}{2m}$$

The shaded and unshaded circles represent respectively even and odd solutions. There is no odd solution for

$$V_0 a^2 < \frac{\pi^2 \hbar^2}{8m}$$

For the odd solution the first bound energy occurs for

$$V_0 a^2 = \frac{\pi^2 \hbar^2}{8m}$$

It may be noted that the limiting value $V_0 = E$ gives $p = \dfrac{n\pi}{2}(n = 0, 1, 2, \dots)$ and hence

$$x^2 = \frac{2mV_0 a^2}{\hbar^2} = \frac{n^2\pi^2}{4} \text{ or, } V_0 a^2 = \frac{n^2\pi^2\hbar^2}{8m}$$

Between $\dfrac{\pi^2\hbar^2}{8m}$ and $\dfrac{4\pi^2\hbar^2}{8m}$ there is only one energy level for symmetric and antisymmetric class and so on.

For $V_0 \to \infty$, E_n are given by

$$\alpha a = \frac{n\pi^2}{2} \text{ or, } \frac{2mE_n}{\hbar^2} a = \frac{n\pi^2}{4} \text{ or, } E_n = \frac{n^2\pi^2\hbar^2}{8ma^2},\ n = 1, 2, \dots \tag{7.72}$$

But, for $V_0 \to$ finite value, energy levels depend on the product $V_0 a^2$ [cf. *Quantum Mechanics* by L.I. Schiff, p. 37, 3rd edn].

7.5 One-dimensional potential well of finite depth: Bound states

In view of Newtonian mechanics, a particle with energy $E < 0$ will remain trapped in a potential well of depth $-V_0$. To take it out of the well energy from outside sources must be supplied. Application of Schrödinger's wave equation leads to the fact that the possible energy of such bound states is discrete and not continuous.

The potential energy $V(x)$ of the well is expressed as

$$V(x) = \begin{cases} -V_0, & 0 < x < a \text{ (for region I)} \\ 0, & x > a \text{ (for region II)} \\ \infty, & x < a \text{ (for left of region I)} \end{cases}$$

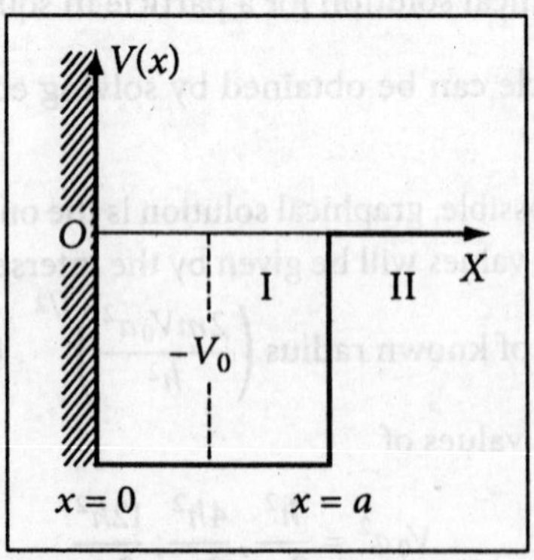

Fig 7.9 One-dimensional potential well

The region $0 < x < a$, where $V(x) = -V_0$ everywhere is the potential well of width a. For $x < 0$ the potential energy is infinity.

Therefore, the particle is reflected at $x = 0$ and as such the probability of finding the particle at $x = 0$ is zero. So at $x = 0$, $\psi(x) = 0$.

Now, the Schrödinger's time-independent equations for regions I and II are

$$\frac{d^2\psi_1}{dx^2} + \frac{2m}{\hbar^2}(V_0 - E)\,\psi_1 = 0 \tag{7.73}$$

[Since, here $E < 0$ and $V < 0$]

$$\frac{d^2\psi_2}{dx^2} - \frac{2m}{\hbar^2}E\psi_2 = 0 \tag{7.74}$$

$$\text{or, } \frac{d^2\psi_1}{dx^2} + \alpha^2\psi_1 = 0 \tag{7.75}$$

$$\text{and } \frac{d^2\psi_2}{dx^2} - \beta^2\psi_2 = 0 \;[\because E < 0 \text{ inside the well}] \tag{7.76}$$

$$\text{where } \alpha^2 = \frac{2m(V_0 - E)}{\hbar^2} \text{ and } \beta^2 = \frac{2mE}{\hbar^2} \tag{7.77}$$

The general solution of eq. (7.75) is

$$\psi_1(x) = A\sin\alpha x + B\cos\alpha x \tag{7.78}$$

But, since $\psi_1(x) = 0$ at $x = 0$

$$\therefore 0 = A\cdot 0 + B, \;\therefore B = 0$$

$$\therefore \psi_1(x) = A\sin\alpha x \tag{7.79}$$

The solution of eq. (7.76) is of the hyperbolic form and may be expressed as

$$\psi_2(x) = Ce^{-\beta x} + De^{\beta x} \tag{7.80}$$

Here, the term $De^{\beta x}$ is inadmissible as for $x \to \infty$, $\psi_2 \to$ finite or 0.

$$\therefore \psi_2(x) = Ce^{-\beta x} \tag{7.81}$$

Now, applying the boundary conditions

$$\text{(i) } \psi_1(x)|_{x=a} = \psi_2(x)|_{x=a} \text{ and (ii) } \left.\frac{d\psi_1}{dx}\right|_{x=a} = \left.\frac{d\psi_2}{dx}\right|_{x=a} \tag{7.82}$$

to eqs (7.80) and (7.81), we get

$$A\sin\alpha a = Ce^{-\beta a} \tag{7.83}$$

$$\alpha A\cos\alpha a = (-\beta)Ce^{-\beta a} \tag{7.84}$$

Dividing eq. (7.84) by eq. (7.83), we get

$$\alpha\cot\alpha a = -\beta \text{ or, } \cot\alpha a = -\frac{\beta}{\alpha} \tag{7.85}$$

Again, $\sin \alpha a = \pm \dfrac{1}{\sqrt{1+\cot^2 \alpha a}}$.

$$\therefore \sin \alpha a = \pm \frac{1}{\sqrt{1+\dfrac{\beta^2}{\alpha^2}}} = \frac{1}{\sqrt{1+\left(\dfrac{E}{V_0 - E}\right)}} = \pm\sqrt{\frac{V_0 - E}{V_0}} = \pm\sqrt{\frac{T}{V_0}} \tag{7.86}$$

where $T = V_0 - E$ is the kinetic energy of the particle in the well.

Now, from the equation

$$\alpha = \sqrt{\frac{2m(V_0 - E)}{\hbar^2}}$$

we get

$$\sqrt{V_0 - E} = \left(\sqrt{\frac{\hbar^2}{2m}}\right)\alpha \tag{7.87}$$

From eqs (7.86) and (7.87)

$$\sin \alpha a = \pm \frac{\hbar}{a\sqrt{2mV_0}} \alpha a \tag{7.88}$$

Equation (7.85) shows that $\cot \alpha a$ is negative. Therefore, only those solutions are to be taken into account for which $\cot(\alpha a)$ is negative. For $\cot(\alpha a)$ to be negative, αa must lie in 2nd, 4th, 6th, ... quadrants. Then the eq. (7.88) gives various bound state energies of the particle in the potential well.

To solve eq. (7.88) we plot $\sin \alpha a$ against αa when, we get a pair of straight lines having slopes

$$\frac{\hbar}{a\sqrt{2mV_0}} \text{ and } -\frac{\hbar}{a\sqrt{2mV_0}}$$

If $\left|\frac{\sin \alpha a}{\alpha a}\right| < \frac{2}{\pi}$, there will be the points of intersection of the straight lines and the sine curve in 2nd, 4th quadrants. The coordinates of the point of intersection will give energy of the bound states in the potential well. The point of intersection of αa and $\sin \alpha a$ for $x = 0$ does not contribute as $\psi(x) = 0$ for $x = 0$ everywhere.

The general form of the wave function inside the well is given by

$$\psi_n(x) = A \sin\left(\alpha x + \frac{\pi}{2}\right) \tag{7.89}$$

where n is the number of eigenstate.

Hence, *for the potential well of finite depth there are several discrete eigenvalues.*

7.6 A particle in a square well potential: Bound states

We consider a particle of mass m and total energy E confined in a potential well, defined by

$$V(x) = \begin{cases} 0, & x < -a \quad \text{(Region I)} \\ -V_0, & -a < x < a \quad \text{(Region II)} \\ 0, & a > x \quad \text{(Region III)} \end{cases} \tag{7.90}$$

In view of classical mechanics, if the total energy of the particle E is negative (i.e., $E < 0$), then the particle will remain confined in the region $x = -a$ and $x = +a$. But, for $E > 0$, the particle will experience a momentary force at $x = -a$ and $x = +a$, but it has freedom to go anywhere outside the well.

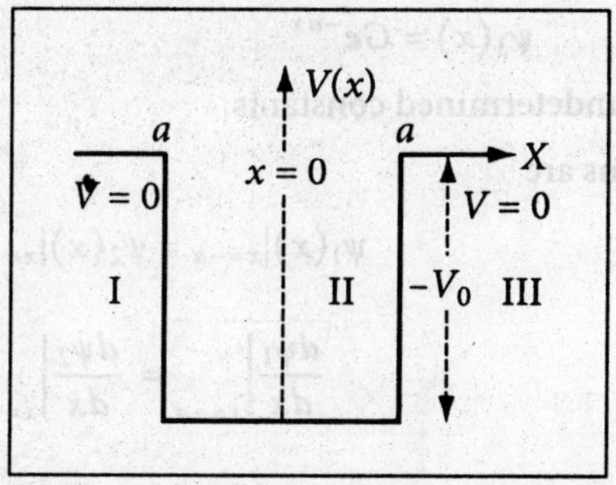

Fig 7.10 Square well potential of depth $-V_0$

Quantum mechanical result will show that there are discrete energy levels in the well and that the wave functions $\psi(x)$ have non-zero values in the classically forbidden regions I and III, when $E < 0$. The state with $E < 0$ is called the *bound state.*

The time-independent Schrödinger equation in regions I, II and III are respectively

$$-\frac{\hbar^2}{2m}\frac{d^2\psi_1}{dx^2} = E\psi, \; x < -a \tag{7.91}$$

$$-\frac{\hbar^2}{2m}\frac{d^2\psi_2}{dx^2} - V_0\psi = E\psi, \; -a < x < a \tag{7.92}$$

$$-\frac{\hbar^2}{2m}\frac{d^2\psi_3}{dx^2} = E\psi, \; x > a \tag{7.93}$$

The above equations can be written as

$$\frac{d^2\psi_1}{dx^2} - \alpha^2\psi_1 = 0 \tag{7.94}$$

$$\frac{d^2\psi_2}{dx^2} + \beta^2\psi_2 = 0 \tag{7.95}$$

$$\frac{d^2\psi_3}{dx^2} - \alpha^2\psi_3 = 0 \tag{7.96}$$

where $\alpha^2 = -\frac{2mE}{\hbar^2}$ and $\beta^2 = \frac{2m(E+V_0)}{\hbar^2}$.

The general solutions of eqs (7.94), (7.95) and (7.96) are

$$\psi_1(x) = Ae^{\alpha x} + Be^{-\alpha x} \tag{7.97}$$

$$\psi_2(x) = C\cos\beta x + D\sin\beta x \tag{7.98}$$

$$\psi_3(x) = Fe^{\alpha x} + Ge^{-\alpha x} \tag{7.99}$$

The term $Be^{-\alpha x} \to \infty$ as $x \to -\infty$, which is not admissible in eq. (7.97).

Similarly, the term $Fe^{+\alpha x}$ is to be excluded as $Fe^{+\alpha x} \to \infty$ as $x \to \infty$.

Hence,

$$\psi_1(x) = Ae^{\alpha x} \tag{7.100}$$

$$\psi_2(x) = C\cos\beta x + D\sin\beta x \tag{7.101}$$

$$\psi_3(x) = Ge^{-\alpha x} \tag{7.102}$$

The constants A, C, D and G are undetermined constants.

The general boundary conditions are

(i) $$\psi_1(x)|_{x=-a} = \psi_2(x)|_{x=-a} \tag{7.103}$$

(ii) $$\left.\frac{d\psi_1}{dx}\right|_{x=-a} = \left.\frac{d\psi_2}{dx}\right|_{x=-a} \tag{7.104}$$

Also

(iii) $$\psi_2(x)|_{x=a} = \psi_3(x)|_{x=a} \tag{7.105}$$

(iv) $$\left.\frac{d\psi_2}{dx}\right|_{x=a} = \left.\frac{d\psi_3}{dx}\right|_{x=a} \tag{7.106}$$

From boundary conditions (i) and (ii), we get

$$Ae^{-\alpha a} = C\cos\beta a - D\sin\beta a \tag{7.107}$$

$$\text{and } A(+\alpha)\,e^{-\alpha a} = +C\beta\sin\beta a + D\beta\cos\beta a \tag{7.108}$$

Also from conditions (iii) and (iv)

$$Ge^{-\alpha a} = C\cos\beta a + D\sin\beta a \tag{7.109}$$

$$\text{and } -\alpha Ge^{-\alpha a} = -\beta C\sin\beta a + \beta D\cos\beta a \tag{7.110}$$

Now, adding eqs (7.107) and (7.109)

$$2C\cos\beta a = (A+G)\,e^{-\alpha a} \tag{7.111}$$

Subtracting eq. (7.110) from eq. (7.108)

$$2\beta C\sin\beta a = (A+G)\alpha e^{-\alpha a} \tag{7.112}$$

Subtracting eq. (7.110) from eq. (7.108), we get

$$2D\sin\beta a = -(A-G)\,e^{-\alpha a} \tag{7.113}$$

Adding eqs (7.108) and (7.110)

$$2D\beta\cos\beta a = (A-G)\alpha e^{-\alpha a} \tag{7.114}$$

Now, from eq. (7.111) we see that if $(A+G) \neq 0$, then $C \neq 0$; also dividing eq. (7.112) by eq. (7.111), we get

$$\alpha = \beta\tan\beta a \tag{7.115}$$

Using this value of α in eq. (7.114), we get

$$2D\beta\cos\beta a = (A-G)\beta\tan\beta a \times e^{-\beta a\tan\beta a} \tag{7.116}$$

Multiplying eq. (7.116) by $\sin\beta a$, we get

$$2D\beta\cos\beta a\sin\beta a = (A-G)\beta\tan\beta a\sin\beta a e^{-\beta a\tan\beta a} \tag{7.117}$$

Using eq. (7.113) in eq. (7.117), we get

$$-(A-G)\,e^{-\beta a\tan\beta a}\times\beta\cos\beta a = (A-G)\beta\sin\beta a\tan\beta a e^{-\beta a\tan\beta a}$$

$$\text{or, } -(A-G)\cos^2\beta a = (A-G)\sin^2\beta a$$

$$\text{or, } (A-G)[\cos^2\beta a+\sin^2\beta a] = 0$$

$$\therefore A = G\ [\because \cos^2\beta a+\sin^2\beta a \neq 0]$$

$\therefore$ from eq. (7.114)

$$D = 0 \tag{7.118}$$

Putting $A = G$ in eq. (7.111), we get

$$2C\cos\beta a\cdot e^{\alpha a} = 2A$$

$$\therefore A = Ce^{\alpha a}\cos\beta a \tag{7.119}$$

Equations (7.115) and (7.119) provide the *first class solution* to the problem at hand.

Again, for $A \neq G$ and $D \neq 0$, we get from eqs (7.110) and (7.111) that

$$\alpha = -\beta\cot\beta a \tag{7.120}$$

We can also show that

$$A = -G \text{ and } C = 0 \tag{7.121}$$

$$\text{and } G = De^{\alpha a}\sin\beta a \tag{7.122}$$

This gives the *second class solution*.

Energy eigenvalues

We have

$$\alpha^2 = -\frac{2mE}{\hbar^2} \text{ and } \beta^2 = \frac{2m(E+V)}{\hbar^2}$$

$$\therefore \alpha^2+\beta^2 = \left(-\frac{2mE}{\hbar^2}\right)+\left(\frac{2m(E+V)}{\hbar^2}\right) = \frac{2mV}{\hbar^2} \text{ or, } (\alpha^2+\beta^2)\,a^2 = \frac{2mVa^2}{\hbar^2} = \frac{V}{\delta}, \tag{7.123}$$

$$\text{where } \delta = \frac{\hbar^2}{2ma^2} \tag{7.124}$$

The uncertainty in position of the particle is given by the half-width (a) of the potential well. Then the uncertainty in momentum is $(\hbar/a)$ and the minimum energy of the particle in the well is

$$\frac{(\hbar/a)^2}{2m} = \frac{\hbar^2}{2ma^2} = \delta \tag{7.125}$$

Thus, δ represents the energy of the particle in the well.

Now, for the first class solution, $\alpha/\beta = \tan\beta a$ is positive as α, β are positive. So for the admissible solution

$$2r\tfrac{\pi}{2} \leqslant \beta a \leq (2r+1)\tfrac{\pi}{2},\ r = 0, 1, 2, \ldots \tag{7.126}$$

Now,

$$\frac{V}{\delta} = \beta^2 a^2 \sec^2 \beta a \;[\because \alpha = \beta \tan \beta a]$$

$$\therefore \left(\frac{\delta}{V}\right)^{1/2} \beta a = |\cos \beta a| \tag{7.127}$$

For the second class solution of the Schrödinger equation, we have

$$(2r-1)\tfrac{\pi}{2} \leqslant \beta a \leqslant 2r\tfrac{\pi}{2},\; r = 1, 2, 3, \ldots \tag{7.128}$$

$$\left(\frac{V}{\delta}\right) = \beta^2 a^2 \operatorname{cosec}^2 \beta a \text{ or, } \left(\frac{\delta}{V}\right)^{1/2} \beta a = |\beta a \sin \beta a| \tag{7.129}$$

We note from the above analysis that the eqs (7.126), (7.127), (7.128) and (7.129) are satisfied by a set of discrete values of β. The β-values may be found out from the coordinates of the point of intersection of the straight lines $(\delta/V)^{1/2}\,\beta a$ vs βa with the curves obtained by plotting the values of $|\cos \beta a|$ and $|\sin \beta a|$ against βa on the same graph (Fig 7.11). The allowed portions of curves represented by eqs (7.127) and (7.129) are shown by solid lines and dotted lines respectively.

If the intersections occur at $\beta = \beta'$ (for $n = 0, 1, 2, \ldots$), then since $\beta^n = 2m/\hbar^2(E_n + V)$, we have

$$E_n = \frac{\hbar^2}{2m}\beta^n - V = \left[(\beta^n a)^2\left(\frac{\delta}{V}\right) - 1\right]V \tag{7.130}$$

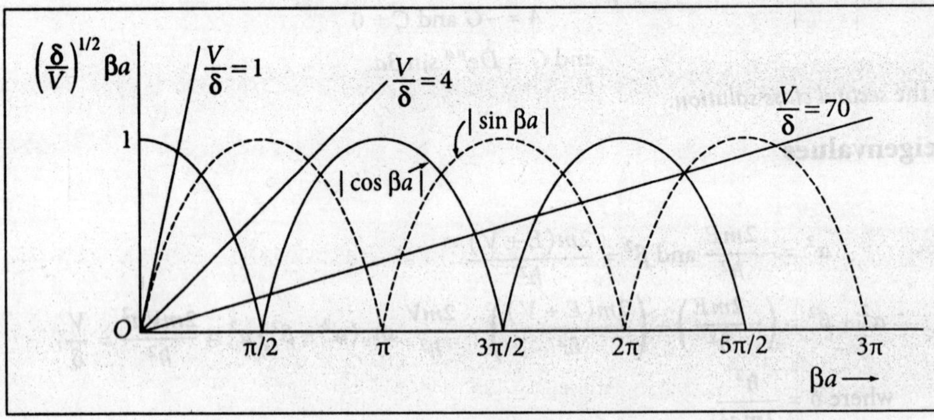

Fig 7.11 Plot of $\left(\frac{\delta}{V}\right)^{1/2} \beta a$ against βa

From Fig 7.11, it is evident that if $(\delta/V)^{1/2}\,\beta a = 1$ for $\frac{1}{2}\pi S \leqslant \beta a \leqslant \frac{1}{2}\pi(S+1)$, then there are $(S+1)$ intersections, i.e., the number of discrete energy levels is $(S+1)$, if

$$\tfrac{1}{2}\pi S\left(\frac{\delta}{V}\right)^{1/2} \leqslant 1 \leqslant \pi(S+1)\left(\frac{\delta}{V}\right)^{1/2} \text{ or, } S \leqslant \tfrac{2}{\pi}\left(\frac{V}{\delta}\right)^{1/2} < (S+1) \tag{7.131}$$

Equation (7.131) shows that the number of energy levels is more if V is large and there is at least one bound state energy of the particle in the well even for the weakest potential (V).

Eigenfunction

The wave functions in regions I, II and III are given by eqs (7.100), (7.101) and (7.102) respectively. Putting $A = G, D = 0$ and $A = Ce^{\alpha a}\cos\beta a$ in eqs (7.100), (7.101) and (7.102), we get the first class solutions in the above three regions as follows:

$$\left.\begin{aligned} \psi_1^n(x) &= (Ce^{\beta^n a}\cos\beta^n a)\,e^{\alpha^n x}, && \text{for } x < -a \\ \psi_2^n(x) &= C\cos\beta^n x, && \text{for } -a < x < a \\ \psi_3^n(x) &= (Ce^{\alpha^n a}\cos\beta^n a)\,e^{-\alpha^n x}, && \text{for } x > a \end{aligned}\right\} \quad (7.132)$$

where $n = 0, 2, 4, \ldots$

Since, $\psi^n(x) = \psi^n(-x)$, $\psi^n(x)$ is *symmetric* about $x = 0$. The wave function which has this property is said to have *even parity.*

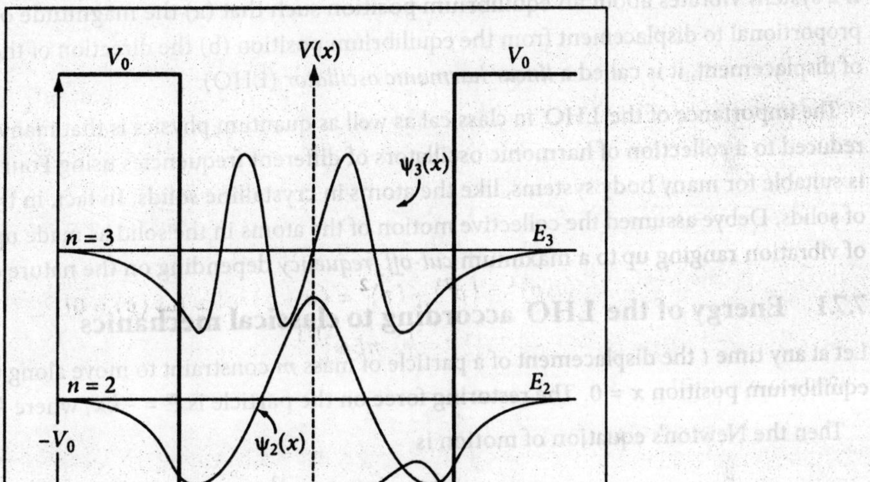

Fig 7.12 Symmetric and antisymmetric wave functions in the potential well

Again, for the second class solution, the eigenfunctions in regions I, II and III respectively are given by eqs (7.100), (7.101) and (7.102) with $A = -G$, $C = 0$ and $G = De^{\alpha a}\sin\beta a$ as

$$\left.\begin{aligned} \psi_1^n(x) &= -(De^{\alpha^n a}\sin\beta^n a)\, e^{\alpha^n x}, x < -a \\ \psi_2^n(x) &= D\sin\beta^n x, \qquad\qquad -a < x < a \\ \psi_3^n(x) &= (De^{\alpha^n} a\sin\beta^n a)\, e^{-\alpha^n x}, x > a \end{aligned}\right\} \tag{7.133}$$

for $n = 1, 3, 5, \ldots$

Here, since $\psi^n(x) = -\psi^n(-x)$, the wave functions are *antisymmetric* with *odd parity*.

The wave functions $\psi^n(x)$ are alternately of *even* and *odd parity* as $V(x)$ is symmetric w.r.t. 0 [Fig 7.12].

It is important to note the interesting property of $\psi^n(x)$, which has non-vanishing components outside the well, i.e., in region I and III. This result is quite unexpected according to classical mechanics. Moreover, in the classically forbidden region the probability density $|\psi(x)|^2 \propto e^{-2\alpha|x|}$ decreases exponentially suggesting that the particle is bound to the potential well and unable to go to infinitely large distances.

Hence, all the states with $(E < 0)$ are *bound states within the potential well.*

7.7 Linear harmonic oscillator

If a system vibrates about an equilibrium position such that (a) the magnitude of the force acting on it is proportional to displacement from the equilibrium position (b) the direction of the force is opposite to that of displacement, it is called a *linear harmonic oscillator* (LHO).

The importance of the LHO in classical as well as quantum physics is that many complex systems can be reduced to a collection of harmonic oscillators of different frequencies using Fourier analysis. The method is suitable for many body systems, like the atoms in crystalline solids. In fact, in the theory of specific heat of solids, Debye assumed the collective motion of the atoms in the solid as made up of LHO, the frequency of vibration ranging up to a maximum *cut-off frequency* depending on the nature of the system.

7.7.1 Energy of the LHO according to classical mechanics

Let at any time t the displacement of a particle of mass m constraint to move along the X-axis be x from the equilibrium position $x = 0$. The restoring force on the particle is $F = -kx$, where k is a positive constant.

Then the Newton's equation of motion is

$$m\frac{d^2x}{dt^2} = -kx \text{ or, } \frac{d^2x}{dt^2} + \omega^2 x = 0 \tag{7.134}$$

[where $\omega^2 = k/m$].

The general solution of eq. (7.134) is

$$x = a\sin(\omega t + \phi) \tag{7.135}$$

where a and ψ are constants which may be determined from the initial conditions. Eq. (7.135) represents the simple harmonic motion (SHM) of the particle with frequency

$$\nu = \frac{\omega}{2\pi} = \frac{1}{2\pi}\sqrt{\frac{k}{m}} \tag{7.136}$$

The potential energy $V(x)$ at displacement x is

$$V = \int_0^x \frac{dV}{dx}dx = \int_0^x -Fdx = -\int_0^x (-kx)\,dx = \tfrac{1}{2}kx^2 = \tfrac{1}{2}m\omega^2x^2 \left[\because \omega^2 = \frac{k}{m}\right]$$

The kinetic energy T at x is given by

$$T = \tfrac{1}{2}m\left(\frac{dx}{dt}\right)^2 = \tfrac{1}{2}ma^2\omega^2\cos^2(\omega t+\phi) = \tfrac{1}{2}m\omega^2a^2[1-\sin^2(\omega t+\phi)] = \tfrac{1}{2}m\omega^2(a^2-x^2)$$

∴ the total energy of the particle at x is

$$E = T + V = \tfrac{1}{2}m\omega^2(a^2-x^2) + \tfrac{1}{2}m\omega^2x^2$$

$$E = \tfrac{1}{2}m\omega^2a^2. \tag{7.137}$$

The total energy E is proportional to a^2. Classically, the particle can oscillate with any amplitude a and as such the total energy increases with the increase of a.

The potential energy, $V = \frac{1}{2}kx^2$ has a parabolic form (Fig 7.13) signifying that the particle executes motion in a potential well of parabolic form.

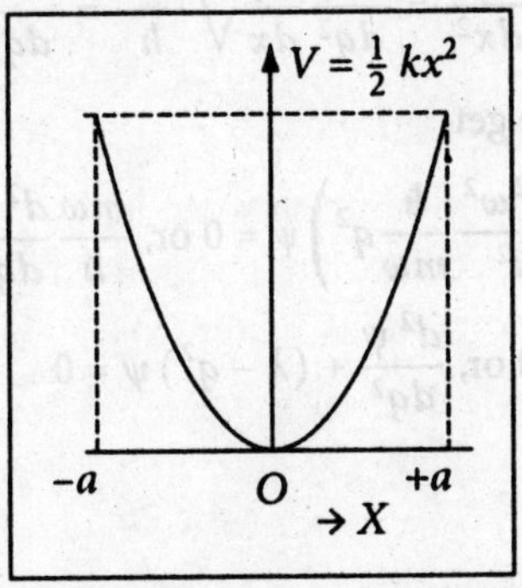

Fig 7.13 Potential energy curve of a LHO

7.7.2 Classical probability of finding the LHO within a small distance dx

The probability of finding the LHO within x and $x + dx$ from its equilibrium position is

$$= \frac{\text{time taken by the LHO to pass over } dx}{\text{period of oscillation}} \quad \text{or, } P(x)\,dx = \frac{2dt}{T} = \frac{2dt}{\left(\frac{2\pi}{\omega}\right)} = \frac{\omega dt}{\pi}$$

But, ∵ $x = a\sin(\omega t + \phi)$

$$\therefore dx = a\omega\cos(\omega t+\phi)\,dt$$

$$\text{or, } dt = \frac{dx}{a\omega\sqrt{1-x^2/a^2}} = \frac{dx}{\omega\sqrt{a^2-x^2}} = \frac{dx}{\omega\sqrt{(2E/m\omega^2)-x^2}} \quad \left[\because a^2 = \frac{2E}{m\omega^2}\right]$$

$$\therefore P(x)\,dx = \frac{dx}{\pi\sqrt{(2E/m\omega^2)-x^2}} \tag{7.138}$$

7.7.3 Quantum mechanical linear harmonic oscillator

The Hamiltonian operator for the one-dimensional linear harmonic oscillator is given by

$$\hat{H} = \frac{\hat{p}_x^2}{2m} + V(x) = \frac{\hbar^2}{2m}\frac{d^2}{dx^2} + \tfrac{1}{2}m\omega^2x^2 \tag{7.139}$$

$$\left[\because V(x) = \text{the potential energy at } x = \tfrac{1}{2}m\omega^2x^2\right]$$

The time-independent Schrödinger wave equation for the LHO is

$$-\frac{\hbar^2}{2m}\frac{d^2\psi}{dx^2}+V(x)\psi = E\psi \text{ or, } \frac{d^2\psi}{dx^2}+\frac{2mE}{\hbar^2}\psi-\frac{2m}{\hbar^2}\left(\frac{1}{2}m\omega^2x^2\right)\psi = 0$$

$$\text{or, } \frac{d^2\psi}{dx^2}+\left(\frac{2mE}{\hbar^2}\right)\psi-\left(\frac{m^2\omega^2}{\hbar^2}\right)x^2\psi = 0 \tag{7.140}$$

Equation (7.140) may be expressed in a simpler form by introducing a dimensional independent variable q such that

$$q=\sqrt{\frac{m\omega}{\hbar}}x=\sqrt{\alpha}x, \left[\text{where } \alpha=\frac{m\omega}{\hbar}\right] \text{ or, } x=\sqrt{\frac{\hbar}{m\omega}}q$$

$$\therefore \frac{d\psi}{dx}=\frac{d\psi}{dq}\frac{dq}{dx}=\frac{d\psi}{dq}\sqrt{\frac{m\omega}{\hbar}} \text{ or, } \frac{d^2\psi}{dx^2}=\frac{d^2\psi}{dq^2}\frac{dq}{dx}\sqrt{\frac{m\omega}{\hbar}}=\frac{d^2\psi}{dq^2}\frac{m\omega}{\hbar} \text{ or, } \frac{d^2\psi}{dx^2}=\frac{m\omega}{\hbar}\frac{d^2\psi}{dq^2} \tag{7.141}$$

Now, from eqs (7.140) and (7.141), we get

$$\frac{m\omega}{\hbar}\frac{d^2\psi}{dq^2}+\left(\frac{2mE}{\hbar^2}-\frac{m^2\omega^2}{\hbar^2}\frac{\hbar}{m\omega}q^2\right)\psi = 0 \text{ or, } \frac{m\omega}{\hbar}\frac{d^2\psi}{dq^2}+\left(\frac{2mE}{\hbar^2}-\frac{m\omega}{\hbar}q^2\right)\psi=0$$

$$\text{or, } \frac{d^2\psi}{dq^2}+\left(\frac{2E}{\hbar\omega}-q^2\right)\psi = 0 \text{ or, } \frac{d^2\psi}{dq^2}+(\lambda-q^2)\psi=0 \tag{7.142}$$

where $\lambda = 2E/\hbar\omega$.

7.7.4 Asymptotic solution

For large value of q such that $q^2\gamma\lambda$ we may neglect λ in eq. (7.142), then it reduces to the form:

$$\frac{d^2\psi}{dq^2}-q^2\psi=0 \tag{7.143}$$

The solution of the above equation is

$$\psi(q)=e^{\pm q^2/2} \tag{7.144}$$

which may be verified as follows:

Differentiating eq. (7.144) twice w.r.t. q, we get

$$\frac{d^2\psi}{dq^2}=q^2e^{\pm q^2/2}\pm e^{\pm q^2/2}=(q^2\pm 1)\,e^{\pm q^2/2} \tag{7.145}$$

Since, q is very large, $q^2 \pm 1 \simeq q^2$.

$$\therefore \frac{d^2\psi}{dq^2}=q^2e^{\pm q^2/2}=q^2\psi \text{ or, } \frac{d^2\psi}{dq^2}-q^2\psi=0$$

which is the same as eq. (7.143).

Now, the probability of finding the LHO along the x-axis is $|\psi(q)|^2$ and for $\psi = e^{q^2/2}$, $|\psi(q)|^2 \to \infty$ as $q \to e^{q^2/2}$.

Hence, $\psi(q) = \infty$ is not *well-behaved* and as such this solution is rejected. On the other hand, $|\psi(q)|^2 \to 0$ as $q \to \infty$ in $\psi(q) = e^{-q^2/2}$. So $\psi(q) = e^{-q^2/2}$ is an asymptotic solution of the wave eq. (7.143).

Let us now try with a possible solution

$$\psi(q) = e^{-q^2/2}\phi(q) \tag{7.146}$$

where $\phi(q)$ is a function of q.

Differentiating eq. (7.145) twice w.r.t. q, we get

$$\frac{d^2\psi}{dq^2} = e^{-q^2/2}\left[\frac{d^2\phi}{dq^2} - 2q\frac{d\phi}{dq} + (q^2-1)\phi\right] \tag{7.147}$$

Now, from eqs (7.146), (7.146) and (7.142), we get

$$e^{-q^2/2}\left[\frac{d^2\phi}{dq^2} - 2q\frac{d\phi}{dq} + (q^2-1)\phi\right] + e^{-q^2/2}(\lambda - q^2)\,\psi = 0 \text{ or, } e^{-q^2/2}\left[\frac{d^2\phi}{dq^2} - 2q\frac{d\phi}{dq} + (\lambda-1)\phi\right] = 0$$

But, $e^{-q^2/2} \neq 0$, except for $q \to \pm\infty$.

$$\therefore \frac{d^2\phi}{dq^2} - 2q\frac{d\phi}{dq} + (\lambda-1)\phi = 0 \tag{7.148}$$

7.7.5 Power series solution of eq. (7.148)

The power series solution of eq. (7.148) may be assumed as

$$\phi(q) = q^s\sum_{r=0}^{\infty} a_r q^r = \sum_r a_r q^{s+r} \tag{7.149}$$

Differentiating eq. (7.147) w.r.t. q, we get

$$\frac{d\phi}{dq} = \sum_r a_r(s+r)\,q^{s+r-1} \tag{7.150}$$

$$\text{and } \frac{d^2\phi}{dq^2} = \sum_r a_r(s+r)(s+r-1)\,q^{s+r-2} \tag{7.151}$$

Using these values of $\frac{d\phi}{dq}$ and $\frac{d^2\phi}{dq^2}$ in eq. (7.146) we get

$$\sum_r a_r(s+r)(s+r-1)\,q^{s+r-2} - 2q\sum_r a_r(s+r)\,q^{s+r-1} + (\lambda-1)\sum_r a_r q^{s+r} = 0$$

$$\text{or, } \sum_r a_r(s+r)(s+r-1)\,q^{s+r-2} - \sum_r a_r\{2(s+r) - (\lambda-1)\}q^{s+r} = 0 \tag{7.152}$$

The power series expressed by eq. (7.152) will be satisfied only if the coefficient of each power of q vanishes separately.

Equating the coefficient of q^{s-2} to zero, we get

$$a_0 s(s-1) = 0;\ \because a_0 \neq 0,\ s = 0 \text{ or, } (s-1) = 0 \tag{7.153}$$

Equating the coefficient of q^{s+1} to zero, $a_1(s+1)s = 0$. $\therefore$ either $a_1 = 0$ or, $s = 0$ or, $s = -1$.

But, as $s \neq -1$.

$$\therefore \text{either } a_1 = 0 \text{ or, } s = 0 \text{ or both} \tag{7.154}$$

Similarly, equating the coefficient of q^{s-r} to zero

$$a_{r+2}(s+r+2)(s+r+1) - a_r[2(s+r) - (\lambda - 1)] = 0 \text{ or, } a_{r+2} = \frac{2s+2r+1-\lambda}{(s+r+2)(s+r+1)} a_r \quad (7.155)$$

Equation (7.155) is called the *recursion formula*. From this equation the coefficients of each even power of q, i.e., $q^2, q^4, q^6, \ldots$ can be calculated if q_0 is known.

Similarly, the coefficients of $q^3, q^5, q^7, \ldots$ can be found out if q_1 is known.

7.7.6 Examining the validity of the series solution

From eq. (7.155) we note that for a large value of r

$$\lim_{r\to\infty} \frac{a_{r+2}}{a_r} = \frac{2}{r} \quad (7.156)$$

Again, considering the series expansion

$$e^{q^2} = 1 + q^2 + \frac{q^4}{2!} + \cdots + \frac{q^n}{\left(\frac{r}{2}\right)!} + \frac{q^{r+2}}{\left(\frac{r}{2}+1\right)!} + \cdots \quad (7.157)$$

$$= 1 + b_2 q^2 + b_4 q^4 + \cdots + b_r q^r + b_{r+2} r^{r+2} + \cdots \quad (7.158)$$

The recursion formula for the expansion of e^{q^2} is

$$\frac{b_{r+2}}{b_r} = \frac{\left(\frac{r}{2}\right)!}{\left(\frac{r}{2}+1\right)!} = \frac{1}{\frac{r}{2}+1} = \frac{2}{r+2}.$$

$$\therefore \lim_{r\to\infty} \frac{b_{r+2}}{b_r} = \frac{2}{r} \quad (7.159)$$

It is, therefore, clear that for large value of r the series for $\phi(q)$ and e^{q^2} behave similarly. Then the corresponding eigenfunction is

$$\psi(q) = e^{-q^2/2}\phi(q) = e^{-q^2/2} \cdot e^{q^2} = e^{q^2/2} \quad (7.160)$$

Equation (7.160) shows that for $q \to \infty$, $\psi(q) \to \infty$ and hence the wave function $\psi(q)$ is inadmissible as it does not satisfy the well-behaved character of the wave function.

7.7.7 Eigenvalues of the LHO

The requirement of the well-behaved solution imposes the condition that the series represented by eq. (7.149) must terminate after the rth term. The condition for this is

$$2s + 2r + 1 - \lambda = 0 \text{ or, } \lambda = 2s + 2r + 1 \quad (7.161)$$

$$\text{or, } \lambda = 2(s+r) + 1 = 2n + 1 \text{ [where } n = s + r \text{ and } n = 0, 1, 2, \ldots] \quad (7.162)$$

$$\text{or, } \frac{2E_n}{\hbar\omega} = 2n + 1 \left[\because \lambda = \frac{2E}{\hbar\omega}\right]$$

$$\text{or, } E_n = (2n+1)\frac{\hbar\omega}{2} \text{ or, } \boxed{E_n = \left(n + \tfrac{1}{2}\right)\hbar\omega,} \quad (7.163)$$

where $\omega = 2\pi\nu$, the angular frequency of the LHO and $n = 0, 1, 2, \ldots$.

The eq. (7.163) leads to the fact that there is a discrete set of energy eigenvalues which are equispaced in the interval of $\hbar\omega$.

For $n = 0$, we get

$$E_0 = \frac{\hbar\omega}{2} = \frac{h\nu}{2} \tag{7.164}$$

E_0 is called the *zero-point* (or, *null point*) energy of the oscillator.

7.7.8 The significance of the *zero-point* energy

According to old quantum theory, the energy of the oscillator is given by $E_n = nh\nu$. So for $n = 0$, i.e., in the ground state of the oscillator, $E_0 = 0$. But, the new quantum mechanical theory gives $E_0 = \frac{1}{2}\hbar\omega$. There is, therefore, a clear contradiction between old quantum theory and new quantum theory.

The existence of the zero-point energy is, however, consistent with the uncertainty principle of Heisenberg. That the old quantum theory is not in consistent with the uncertainty theory can be shown as below:

If both the kinetic energy and potential energy of the oscillator are zero, then in the ground state the total energy of the oscillator is zero. The oscillator in the ground state is at $x = 0$, i.e., at the bottom of the potential well and hence the position uncertainty $\Delta x = 0$. At the same time the kinetic energy of the oscillator is zero and hence the momentum, and momentum uncertainty of the oscillator $\Delta p = 0$. So $\Delta x \, \Delta p_x = 0$, which is a clear contradiction to uncertainty relation.

Since, uncertainty relation is an established principle and zero-point energy is consistent with it, the existence of zero-point energy of the LHO is quite justified.

The energy level diagrams of the LHO according to wave mechanics is shown in Fig 7.14.

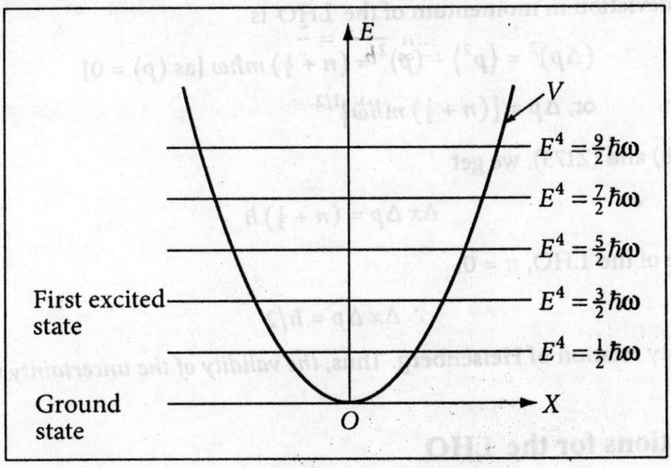

Fig 7.14 Energy levels for LHO

7.7.9 Uncertainty Relation and the LHO

Assuming the uncertainty in position of the oscillator to be of the order of its amplitude and the uncertainty in momentum to be the fluctuations in the value of its momentum between the limiting positions, the validity of the uncertainty relation in case of the LHO can be established.

At the limiting value of the displacement the total energy E of the oscillator is equal to its potential energy V as the kinetic energy is zero there.

But,

$$E_n = (n + \tfrac{1}{2})\,\hbar\omega \text{ and } V = \tfrac{1}{2}m\omega^2 x_n^2 \text{ [for the } n\text{th state of the oscillator]}$$

$$\therefore \tfrac{1}{2}m\omega^2 x_n^2 = (n + \tfrac{1}{2})\,\hbar\omega \text{ or, } x_n^2 = 2\,(n + \tfrac{1}{2})\,\hbar\omega \tag{7.165}$$

The mean squared displacement of the LHO is

$$\langle x^2 \rangle = \frac{x_n^2}{2} = (n + \tfrac{1}{2})\frac{\hbar}{m\omega} \tag{7.166}$$

$$\therefore (\Delta x)^2 = \langle x^2 \rangle - \langle x \rangle^2 = (n + \tfrac{1}{2})\frac{\hbar}{m\omega} \tag{7.167}$$

as the mean displacement of the oscillator is $\langle x \rangle = 0$.

$$\text{or, } \Delta x = \left[(n + \tfrac{1}{2})\frac{\hbar}{m\omega}\right]^{1/2} \tag{7.168}$$

The maximum KE $(T)_{\max}$ of the oscillator is

$$(T)_{\max} = \frac{p_{\max}^2}{2m} = E_n = (n + \tfrac{1}{2})\,\hbar\omega \tag{7.169}$$

$$\therefore p_{\max}^2 = 2\,(n + \tfrac{1}{2})\,m\hbar\omega \tag{7.170}$$

The mean squared momentum of the LHO is

$$\langle p^2 \rangle = \frac{p_{\max}^2}{2} = (n + \tfrac{1}{2})\,m\hbar\omega \tag{7.171}$$

The mean squared deviation in momentum of the LHO is

$$(\Delta p)^2 = \langle p^2 \rangle - \langle p \rangle^2 = (n + \tfrac{1}{2})\,m\hbar\omega \text{ [as } \langle p \rangle = 0] \tag{7.172}$$

$$\text{or, } \Delta p = [(n + \tfrac{1}{2})\,m\hbar\omega]^{1/2} \tag{7.173}$$

Now, from eqs (7.168) and (7.173), we get

$$\Delta x\,\Delta p = (n + \tfrac{1}{2})\,\hbar \tag{7.174}$$

For the ground state of the LHO, $n = 0$.

$$\therefore \Delta x\,\Delta p = \hbar/2 \tag{7.175}$$

which is the uncertainty relation of Heisenberg. Thus, *the validity of the uncertainty relation is established in case of the* LHO.

7.7.10 Wave functions for the LHO

Putting $\lambda = 2n + 1$ in eq. (7.148), we get

$$\frac{d^2\phi}{dq^2} - 2q\frac{d\phi}{dq} + 2nq = 0 \tag{7.176}$$

This is known as the *Hermite differential equation*. We can prove that the solution of eq. (7.176) is of the form:

$$\phi(q) = e^{q^2}\left(\frac{d}{dq}\right)^n e^{-q^2} \tag{7.177}$$

Multiplying eq. (7.177) by $(-1)^n$, we get Hermite polynomial of order n

$$H_n(q) = (-1)^n \phi(q) = (-1)^n e^{q^2} \left(\frac{d}{dq}\right)^n e^{-q^2} \tag{7.178}$$

The following are a few Hermite polynomials for even and odd values of n :

$n = 0,\ H_0(q) = 1$ $\quad$ $n = 1,\ H_1(q) = 2q$

$n = 2,\ H_2(q) = 4q^2 - 2$ $\quad$ $n = 3,\ H_3(q) = 8q^3 - 12q$

$n = 4,\ H_4(q) = 16q^4 - 48q^2 + 12$ $\quad$ $n = 5,\ H_5(q) = 32q^5 - 160q^3 + 120q$

Now, the eigenfunction for the LHO is given by

$$\psi_n(q) = N_n e^{-q^2/2} H_n(q) \tag{7.179}$$

Here, N_n is the normalisation constant. N_n can be found out from the condition of normalisation as follows:

$$\int_{-\infty}^{+\infty} |\psi_n(x)|^2\, dx = 1 \text{ or, } \frac{|N_n|^2}{\sqrt{\alpha}} \int_{-\infty}^{+\infty} H_n^2(q)\, e^{-q^2}\, dq = 1$$

But, $\int_{-\infty}^{+\infty} H_n^2(q)\, e^{-q^2/2}\, dq = 2^n(n!)\sqrt{\pi}.$

$$\therefore \frac{N_n^2}{\sqrt{\alpha}} \cdot 2^n(n!)\sqrt{\pi} = 1 \text{ or, } N_n = \left[\frac{\sqrt{\alpha}}{2^n n!\sqrt{\pi}}\right]^{1/2} \tag{7.180}$$

Thus, the normalised wave function for the LHO is

$$\psi_n(x) = \left[\frac{\sqrt{\alpha}}{2^n n!\sqrt{\pi}}\right]^{1/2} e^{-\alpha x^2/2} H_n\left(\sqrt{\alpha}x\right) \tag{7.181}$$

where α is equal to $m\omega/\hbar$

$$\text{or, } \psi_n(x) = \left(\frac{\alpha}{\pi}\right)^{1/4} (2^n n!)^{-1/2} e^{-\alpha x^2/2} H_n\left(\sqrt{\alpha}x\right) \tag{7.182}$$

This is the normalised eigenfunction of the oscillator in the nth state. The wave function corresponding to the lowest energy state, i.e., $n = 0$ is

$$\psi_0(x) = \left(\frac{\alpha}{\pi}\right)^{1/4} e^{-\alpha x^2/2} \left[\because H_0\left(\sqrt{\alpha}x\right) = 1\right] \tag{7.183}$$

The table below shows first few wave functions of the *oscillator*:

Table 7.1: First few wave function of the LHO

n	$\psi_n(x)$
0	$\psi_0(x) = \left(\frac{\alpha}{\pi}\right)^{1/4} e^{-\alpha x^2/2}$
1	$\psi_1(x) = \left(\frac{\alpha}{\pi}\right)^{1/4} \sqrt{2}\sqrt{\alpha}x e^{-\alpha x^2/2}$
2	$\psi_2(x) = \left(\frac{\alpha}{\pi}\right)^{1/4} \left(\frac{1}{\sqrt{2}}\right)\left(2\alpha x^2 - 1\right) e^{-\alpha x^2/2}$
3	$\psi_3(x) = \left(\frac{\alpha}{\pi}\right)^{1/4} \left(\frac{1}{\sqrt{3}}\right)\left[2\left(\alpha\sqrt{x}\right)^3 - 2\sqrt{\alpha}x\right] e^{-\alpha x^2/2}$
4	$\psi_4(x) = \frac{1}{2}\left(\frac{\alpha}{\pi}\right)^{1/4} \left(\frac{1}{\sqrt{6}}\right)\left[4\left(\sqrt{\alpha}x\right)^4 - 12\left(\sqrt{\alpha}x\right)^2 + 3\right] e^{-\alpha x^2/2}$
5	$\psi_5(x) = \frac{1}{2}\left(\frac{\alpha}{\pi}\right)^{1/4} \left(\frac{1}{\sqrt{15}}\right)\left[4\left(\sqrt{\alpha}x\right)^5 - 20\left(\sqrt{\alpha}x\right)^3 + 15\left(\sqrt{\alpha}x\right)\right] e^{-\alpha x^2/2}$

It is to be noted that $\psi_0(x)$ has no node for finite q. As n increases, the number of nodes progressively increases with n.

For example, $\psi_1(x)$ has a single node and $\psi_2(x)$ has two nodes as shown in Fig 7.15.

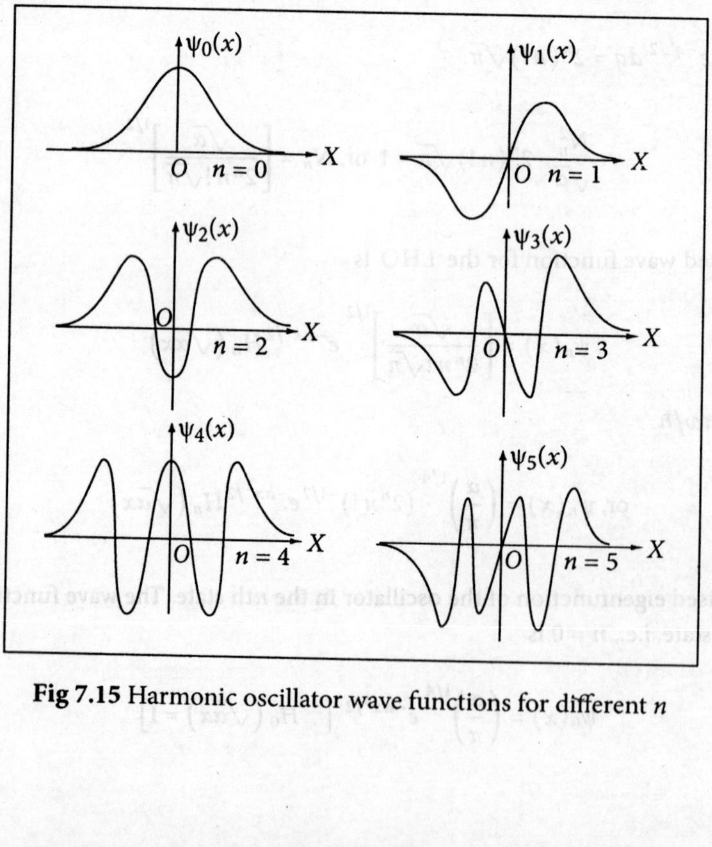

Fig 7.15 Harmonic oscillator wave functions for different n

7.7.11 Classical Probability Density and Quantum Mechanical Probability Density of an LHO

We know, the classical probability density of an LHO having amplitude a and energy $E = \frac{1}{2}m\omega^2 a^2$ is given by

$$P(x) = \frac{1}{\pi\sqrt{a^2 - x^2}} = \frac{1}{\pi\sqrt{\frac{2E}{m\omega^2} - x^2}}$$

This equation shows that the probability density of finding the oscillator is maximum at $x = \pm a$ and minimum at $x = 0$, i.e., at the equilibrium position. The classical probability densities for $n = 0$ and $n = 10$ are shown in Figs 7.16 and 7.17 respectively.

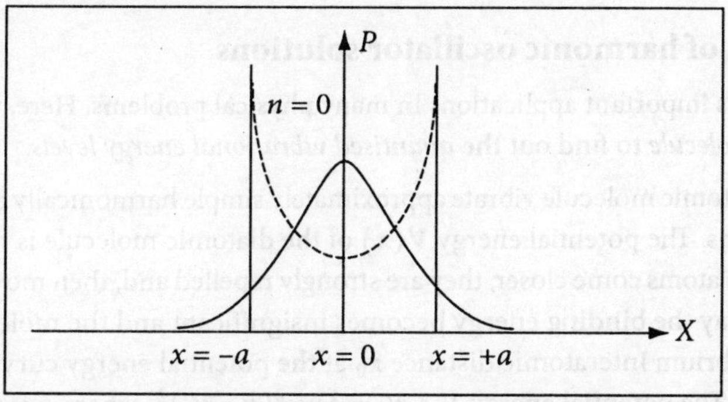

Fig 7.16 The classical and quantum mechanical probabilities are shown by dotted and solid lines respectively for $n = 0$.

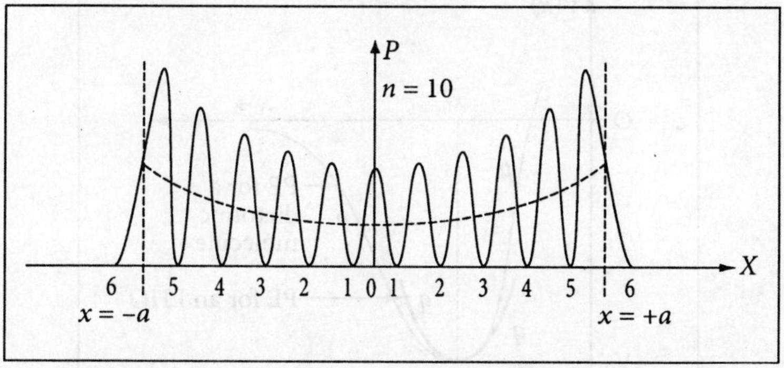

Fig 7.17 The classical and quantum mechanical probabilities are shown by dotted and solid lines respectively for $n = 10$.

The quantum mechanical probability of the LHO in the nth quantum state is $P(x) = |\psi_n(x)|^2$.

It can be shown that for $n = 0$, $|\psi_0(x)|^2$ assumes maximum value at $x = 0$ and decreases on either side of x in contradiction with the classical probability. However, for large value of n,, say, $n = 10$, $\psi_{10}(x)$ has five peaks on either side of x (Fig 7.17). The classical probability is shown by dotted line. It is clear from the figure that for large values of n, both the classical and quantum mechanical average probabilities agree with each other. With increase of n, this correspondence improves rapidly.

7.7.12 Selection Rules for the LHO

We have seen that the number of discrete energy levels of a linear harmonic oscillator is infinite and as such the spectrum might be very complex due to transition between different energy levels. The selection rule in the present case may, however be possible to derive by examining the transition of the dipole moment.

If $\psi_m(x)$ and $\psi_n(x)$ be the wave functions corresponding to vibrational quantum numbers m and n, then a transition is possible if the dipole moment of transition is not zero. That is

$$e\int_{-\infty}^{+\infty} \psi_m^*(x)\hat{x}\,\psi_n(x)\,dx \neq 0$$

It can be shown that the above integral is zero, if $m \neq (n \pm 1)$. So the selection rule is $\Delta n = \pm 1$ for the LHO. This shows that energy corresponding to fundamental frequency is emitted or absorbed.

7.7.13 Application of harmonic oscillator solutions

The solution of LHO has important applications in many physical problems. Here, we apply the solution to the case of a *diatomic molecule* to find out the *quantised vibrational energy levels.*

The two atoms of a diatomic molecule vibrate approximately simple harmonically at low energies along the line joining the two atoms. The potential energy $V(x)$ of the diatomic molecule is represented by curve pp in Fig 7.18. When the two atoms come closer, they are strongly repelled and, then move away from each other. After moving too far away the binding energy becomes insignificant and the molecule dissociates. When the molecule is in equilibrium interatomic distance x_0 at the potential energy curve (pp) approximates to that for the LHO (qq). The potential energy $V + V_0 = \frac{1}{2}\mu\omega^2(x - x_0)^2$, where μ is the reduced mass of the diatomic molecule and ω is the angular frequency of vibration.

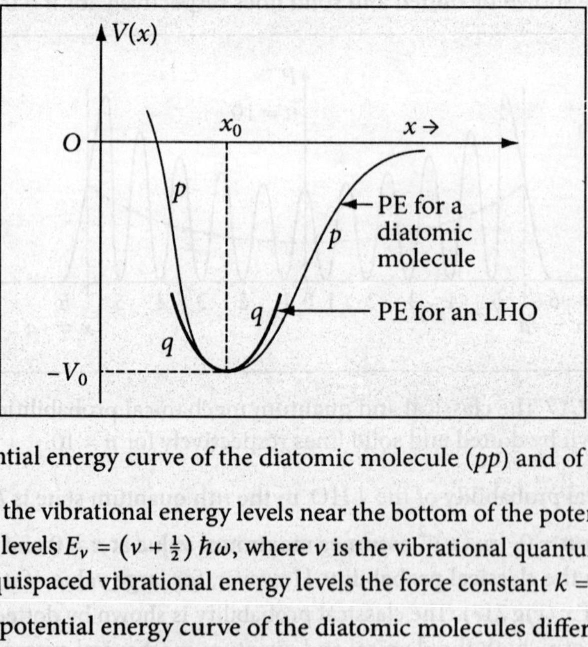

Fig 7.18 Potential energy curve of the diatomic molecule (pp) and of the LHO (qq)

It is, therefore, clear that the vibrational energy levels near the bottom of the potential energy curve follow the behaviour of the LHO levels $E_v = (v + \frac{1}{2})\,\hbar\omega$, where v is the vibrational quantum number having values $v = 0, 1, 2, \ldots$. From the equispaced vibrational energy levels the force constant $k = \mu\omega^2$ can be determined.

For higher energies the potential energy curve of the diatomic molecules differs widely from the LHO potential and the oscillator tends to be enharmonic. The energy levels at this state are crowded together and merges with the continuum leading to the dissociation of the diatomic molecule.

7.8 Answers of some important conceptual questions

7.1. *What is a potential well?*

Ans. If the potential distribution in a region of space is such that

$$V(x) = \begin{cases} 0, & x < -a \quad \text{(Region I)} \\ -V_0, & -a < x < a \quad \text{(Region II)} \\ 0, & a > x \quad \text{(Region III)} \end{cases}$$

Then it is a square are well potential, classically total energy E of the particle in the well is negative (i.e., $E < 0$) and as such the particle will remain confined in the region $x = -a$ and $x = +a$. But, for $E > 0$ the particle will experience a monetary force at $x = -a$ and $x = +a$, but it has freedom to go anywhere outside the well.

Quantum mechanically there are discrete energy levels in the well and that the wave function $\psi(x)$ have non-zero values in the classically forbidden regions I and III, when $E < 0$. The state with $E < 0$ is called the *bound state* in the potential.

7.2. *What is the probability density of a particle in a potential well? What is its implication?*

Ans. The interesting property associated with the wave functions $\psi^n(x)$ is that it has non-vanishing components outside the well, i.e., in regions I and III. This result is quite unexpected according to classical mechanics. Moreover, in the classically forbidden region the probability density $|\psi(x)|^2 \propto e^{-2a|x|}$ decreases exponentially implying that the particle is bound to the potential well and unable to go to infinitely large distances. Hence, all states with $E < 0$ are *bound states* within the potential well.

7.3. *Comment on the nature and parity of the wave function associated with a particle in a potential well.*

Ans. In case of a particle confined in a one-dimensional potential well, it can be shown that the associated wave functions are $\phi^n(x) = -\psi^n(-x)$ and hence, the wave functions are *antisymmetric with odd parity*.

7.4. *Discuss the characteristics of energy levels of a particle confined in a one-dimensional potential box.*

Ans. (a) The energy levels of the particle confined in a one-dimensional potential box are *discrete* since the energy of the nth quantum state is

$$E_n = \frac{n^2\pi^2\hbar^2}{2ma^2}, \quad n = 1, 2, 3, \ldots$$

$$\therefore E_1 = \frac{\pi^2\hbar^2}{2ma^2}, \quad E_2 = 4E_1, \quad E_3 = 9E_1, \quad E_n = n^2E_n$$

i.e., $E_1, E_2, \ldots, E_n$ are set of discrete energy values of the particle. For $n = 0$, $E_n \neq 0$. This means the particle in one-dimensional box cannot sit idle.

(b) The difference in energy between two consecutive levels is

$$\Delta E = (n+1)\frac{\pi^2\hbar^2}{2ma^2}$$

So, the spacing between the energy levels increases with the increase of n, i.e., for higher levels.

(c) As $\Delta E \infty 1/a^2$, the energy levels come closer with the increase of the dimension (a) of the box. In the limit $a \to \infty$, $\Delta E \to 0$ and the levels form a continuum corresponding to a free particle.

7.5. ***What is utility of application of the problem of a particle confined in a one-dimensional potential box?***

Ans. **The infinite square well on particle in a box model provides one of the few problems of quantum mechanics that can be solved analytically. These quantum well problems are particularly important in photo electronics and are used in devices such as quantum well laser, the quantum well infrared photo detector, etc.**

7.6. ***What are the phenomena which can be explained by quantum mechanical tunnel effect?***

Ans. **Following phenomena can be satisfactorily explained by quantum mechanical tunnel effect:**

(i) Emission of a α-particle from nucleus.

(ii) Filed emission of electrons from cold metallic surface.

(iii) Switching action of tunnel diode.

(iv) Reverse breakdown of semiconductor diode.

7.7. ***Discuss the significance of zero-point energy.***

Ans. **See Art. 7.7.8.**

7.8. ***What is zero-point energy of a linear harmonic oscillator?***

Ans. **The energy of the *n*th quantum state for the LHO is given by**

$$E_n = (n + \tfrac{1}{2})\, \hbar\omega_0$$

the different letter having their usual meaning.

For $n = 0$, $E_0 = \frac{1}{2}\omega_0$ and is called the *zero point energy.* This is not in accordance with Planck's hypothesis which states $E_n = n\hbar\omega$ and hence $E_0 = 0$ for $n = 0$. This establishes the superiority of new quantum theory over the old quantum theory.

7.9. ***Write down first three wave functions for the linear harmonic oscillator.***

(i) $\psi_0(x) = \left(\frac{\alpha}{\pi}\right)^{1/4} e^{-\alpha x^2/2}, \ (n = 0).$

(ii) $\psi_1(x) = \left(\frac{\alpha}{\pi}\right)^{1/2} \sqrt{2}\sqrt{\alpha}\, x e^{-\alpha x^2/2}, \ (n = 1).$

(iii) $\psi_2(x) = \left(\frac{\alpha}{\pi}\right)^{1/4} \frac{1}{\sqrt{2}}(2a^2x^2 - 1)\, e^{-\alpha x^2/2}, \ (n = 2).$

7.9 Solved problems

» ***Example*** **7.1.** Calculate the value of lowest energy of an electron in a one-dimensional force free region of length 4 Å.

[HPU 1997]

Solution: We know, for a particle in a one-dimensional force free region of length *a* the energy of *n*th state is

$$E_n = \frac{n^2\hbar^2\pi^2}{2ma^2} = \frac{n^2h^2}{8ma^2}$$

For the lowest energy state, $n = 1$.

$$\therefore E_1 = \frac{h^2}{8ma^2} = \frac{(6.63\times10^{-34})^2}{8\times9.1\times10^{-31}\times(4\times10^{-10})^2} = 3.78\times10^{-19}\ \text{J} = \frac{3.78\times10^{-19}}{1.6\times10^{-19}} = \mathbf{2.36\ eV}$$

» ***Example* 7.2.** What is the lowest energy of a neutron confined to a nucleus of size 10^{-14} m? [mass of the neutron = 1.67×10^{-27} kg.]

***Solution*:** Here, we have $a = 10^{-14}$ m, $m = 1.67 \times 10^{-27}$ kg.

∴ lowest energy of the neutron is

$$E_1 = \frac{1^2\pi^2\hbar^2}{2ma^2} = \frac{\pi^2 \times (1.05 \times 10^{-34})^2}{2 \times 1.67 \times 10^{-27} \times (10^{-14})^2} \quad [\because \hbar = 1.05 \times 10^{-34}\ \text{J-s}]$$

$$= 3.28 \times 10^{-13}\ \text{J} = \frac{3.28 \times 10^{-13}}{1.6 \times 10^{-19}} = \mathbf{2.05\ MeV}$$

» ***Example* 7.3.** Show that a particle in a one-dimensional box of length a cannot have a definitely known momentum and that the average value of the momentum is zero.

***Solution*:** The normalised wave function of a particle in a box of one dimension is

$$\psi(x) = \sqrt{\frac{2}{a}} \sin\frac{n\pi x}{a}, \quad n = 1, 2, 3, \ldots$$

The average momentum of the particle in the box is

$$\langle p \rangle = \int_0^a \psi^*(-i\hbar)\frac{\partial \psi}{\partial x}\, dx = \frac{2}{a}(-i\hbar)\int_0^a \sin\frac{n\pi x}{a} \cos\frac{n\pi x}{a} \cdot \frac{n\pi}{a}\, dx = 0$$

$$\langle p_x^2 \rangle = \int_0^a \psi^*(-\hbar^2)\frac{\partial^2 \psi}{\partial x^2}\, dx = \frac{n^2\hbar^2}{4a^2}$$

then the mean deviation $\Delta p = \langle p_x^2 \rangle - \langle p_x \rangle^2 \neq \mathbf{0}$.

∴ it is not possible to know p_x definitely.

» ***Example* 7.4.** The lowest energy of a particle entrapped in a box is 40 eV. What are the next three higher energies the particle can have?

***Solution*:** We know, the energy of a particle in the nth state trapped in a box is given by

$$E_n = \frac{n^2\hbar^2\pi^2}{2ma^2}$$

where m and a are the mass of the particle and width of the potential box respectively.

For the lowest energy, $n = 1$ and

$$E_1 = \frac{\pi^2\hbar^2}{2ma^2} = 40\ \text{eV}$$

The next higher energy state corresponds to $n = 2$.

$$\therefore E_2 = 2^2\frac{\pi^2\hbar^2}{2ma^2} = 4 \times 40 = 160\ \text{eV}$$

For the $n = 3$ state

$$E_3 = 3^2\frac{\pi^2\hbar^2}{2ma^2} = 9 \times 40 = 360\ \text{eV}$$

» ***Example* 7.5.** An electron with kinetic energy of 10 eV at $x = -\infty$ is moving from left to right along the x-axis (Fig 7.19). The potential energy is $V = 0$ for $x < 0$ and $V = 20$ eV for $x > 0$. Treat the electron as a one-dimensional plane wave. (a) Write the Schrödinger equation for $x < 0$ and $x > 0$. (b) Sketch the solutions in the two regions. (c) What is the wavelength for $x < 0$? (d) What are the boundary conditions at $x = 0$? (e) Make a general statement about the possibility of finding the electron at some positive value of x.

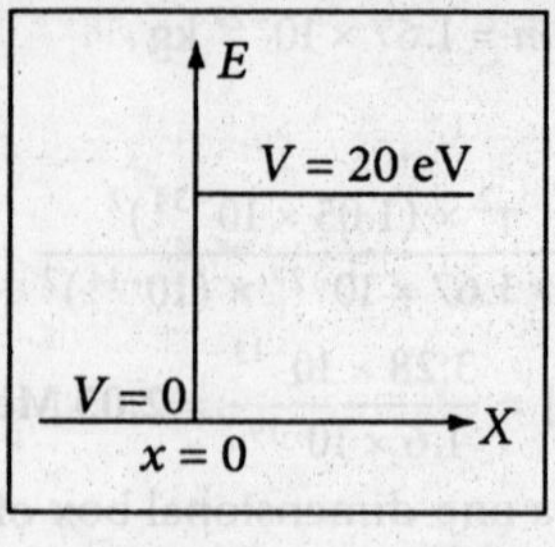

Fig 7.19 Figure for problem 7.5.

Solution:

(a) The Schrödinger equation for the two regions $x > 0$ and $x < 0$ are given by

$$-\frac{\hbar^2}{2m}\frac{d^2\psi(x)}{dx^2} + V_0\psi(x) = E\psi(x), \text{ for } x > 0$$

$$\text{and } -\frac{\hbar^2}{2m}\frac{d^2\psi(x)}{dx^2} = E\psi(x), \text{ for } x < 0$$

According to the problem, $V_0 = 20$ eV and $E = 10$ eV.

(b) The solutions of the problem in region $x < 0$ and $x > 0$ are sketched in Fig 7.20.

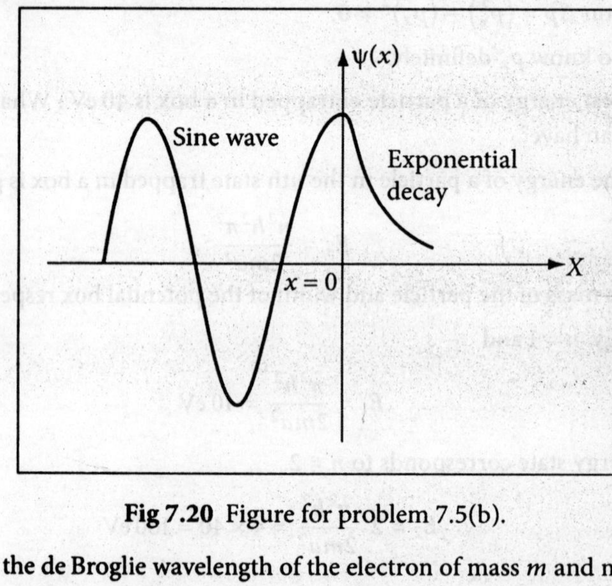

Fig 7.20 Figure for problem 7.5(b).

(c) For the region $x < 0$, the de Broglie wavelength of the electron of mass m and momentum $p = \sqrt{2mE}$, E being the KE of the electron, is given by

$$\lambda = \frac{h}{p} = \frac{2\pi\hbar c}{\sqrt{(2mc^2)E}} = \frac{1.973 \times 2\pi \times 10^{-11}}{3.2 \times 10^{-3}} = 4 \times 10^{-8} \text{ cm},$$

where $m = 0.5$ MeV/C^2; $\hbar c = 1.973 \times 10^{-11}$ MeV-cm and $E = 10$ eV

and $\sqrt{(2mc^2)E} = \left[1.02 \times 10^{-5}\right]^{1/2} \simeq 3.2 \times 10^{-3}$ MeV.

(d) The boundary conditions:

$$\psi(x) \text{ and } \frac{d\psi(x)}{dx} \text{ at } x = 0 \text{ are continuous}$$

(e) The probability of finding the electron at some positive value of x is given by

$$P(x) \propto e^{-2x/\lambda}$$

» ***Example* 7.6.** Estimate the first three energy levels of an electron enclosed in a box of width 10 Å. Compare it with those of glass marble of mass 1 gm, contained in a box of width 20 cm. Can these levels of the marble be measured experimentally? **[PU 1994]**

***Solution*:** We know, the energy of a particle in a one-dimensional box is

$$E_n = \frac{n^2h^2}{8ma^2}$$

Here, $m = 9.1 \times 10^{-31}$ kg, $a = 10 \times 10^{-10}$ m.
The energy corresponding to $n = 1$ is

$$E_1 = \frac{1^2 \times (1.63 \times 10^{-34})^2}{8 \times 9.1 \times 10^{-31} \times 100 \times 10^{-20}} = 0.604 \times 10^{-19}\text{ J} = \frac{0.604 \times 10^{-19}}{1.6 \times 10^{-19}} = 0.377\text{ eV}$$

For $n = 2$

$$E_2 = 2^2 \times \frac{h^2}{8ma^2} = 4 \times 0.377 = 1.508\text{ eV}$$

For $n = 3$

$$E_3 = 3^2 \times \frac{h^2}{8ma^2} = 9 \times 0.377 = 3.393\text{ eV}$$

Now, for glass marble, $n = 1$, $a = 20$ cm $= 0.2$ m, $m = 1$ gm $= 10^{-3}$ kg

$$\therefore E_1 = \frac{1 \times (6.63 \times 10^{-34})^2}{8 \times 10^{-3} \times (0.2)^2} = 1.374 \times 10^{-63}\text{ J} = \frac{1.374}{1.6 \times 10^{-19}} = 0.858 \times 10^{-44}\text{ eV}$$

For $n = 2$

$$E_2 = 2^2 \times \frac{h^2}{8ma^2} = 4 \times 0.858 \times 10^{-44} = 3.432 \times 10^{-44}\text{ eV}$$

For $n = 3$

$$E_3 = 3^2 \times \frac{h^2}{8ma^2} = 9 \times 0.858 \times 10^{-44} = 7.722 \times 10^{-44}\text{ eV}$$

The energy levels of marble cannot be experimentally measured as the energy values are almost zero.

» ***Example* 7.7.** Show that an electron in a potential box is in consistence with Bohr's correspondence principle.

***Solution*:** If a be the length of the square well potential box, the classical probability of the average value of the square of the position coordinate is

$$\langle x^2 \rangle_{cl} = \frac{\int_0^a x^2\,dx}{\int_0^a dx} = \frac{a^2}{3}$$

The quantum mechanical average value of the square of the position coordinate of the same electron is

$$\langle x^2\rangle_q = \frac{\int_0^a \sin^2\frac{n\pi x}{a}x^2\,dx}{\int_0^a \sin^2\frac{n\pi x}{a}\,dx}\quad\left[\because \psi(x) = A\sin\tfrac{n\pi x}{a}\text{ is the wave function representing the electron}\right]$$

$$= \frac{a^2}{3}\left(1-\frac{3}{2n^2\pi^2}\right)$$

When $n\to\infty$, $\frac{3}{2n^2\pi^2}\to 0$.

$$\therefore \langle x^2\rangle_q = a^2/3$$

Hence, $\langle x^2\rangle_{cl} = \langle x^2\rangle_q$, **which is Bohr's correspondence principle.**

» ***Example* 7.8. An electron of mass m is located in a unidimensional square potential well with perfectly impenetrable walls in the range $0 < x < a$. What is (a) the probability of the electron with the lowest energy staying within the region $a/3 < x < 2a/3$? (b) the number of energy levels in the interval E to to $E + dE$?**

***Solution*:**

(a) We know, the energy eigenfunction of the electron confined in the unidimensional square potential well is

$$\psi_n(x) = \sqrt{\frac{2}{a}}\sin\frac{n\pi x}{a}$$

the energy eigenvalues are

$$E_n = \frac{n^2\pi^2\hbar^2}{2ma^2}$$

∴ the probability density in the lowest state with $n = 1$ is

$$|\psi_1(x)|^2 = \frac{2}{a}\sin^2\frac{\pi x}{a}$$

So the probability of finding the electron in the interval is

$$P = \frac{2}{a}\int_{a/3}^{2a/3}\sin^2\frac{\pi x}{a}\,dx = \frac{1}{a}\int_{a/3}^{2a/3}\left(1-\frac{\cos 2\pi x}{a}\right)dx = \frac{1}{3}+\frac{\sqrt{3}}{2\pi} = 0.61$$

(b) The energy

$$E_n = \frac{\pi^2\hbar^2}{2ma^2}n^2;\quad \therefore dE = \frac{\pi^2\hbar^2}{ma^2}n\,dn;\quad \therefore dn = \frac{a}{\pi\hbar}\sqrt{\frac{m}{2E}}\,dE\quad\left[\because n = \frac{a\sqrt{2mE}}{\pi\hbar}\right]$$

is the number of energy levels in the interval E and $E + dE$.

» ***Example* 7.9. Find out the probability of finding a particle in the (a) ground state and (b)nth excited state, the particle being confined in a one-dimensional square potential well ($0 < x < a$). The wave function for the ground state and nth state for the particle are $\psi_1(x) = A\sin^2\left(\frac{n\pi}{a}\right)$ and $\psi_n(x) = Ax(a-x)$ respectively.**

Solution:

(a) According to the condition of normalisation

$$\int_0^a |\psi(x)|^2\, dx = A^2 \int_0^a \sin^4\left(\frac{\pi x}{a}\right) = 1 \text{ or, } A^2 \times \frac{3a}{8} = 1 \text{ or, } A^2 = \frac{8}{3a}$$

The probability of finding the particle in the ground state for

$$\psi_n(x) = \sqrt{\frac{2}{a}} \sin\frac{n\pi x}{a}$$

[$\because$ the normalised wave function for the particle in the one-dimensional square potential well is $\psi_n(x) = \sqrt{\frac{2}{a}} \sin\frac{n\pi x}{a}$]

We have the probability in the nth state $P_n |C_n|^2$ [where C_n is the coefficient in the expansion of $\psi(x)$, i.e., $\psi(x) = \sum\limits_n C_n \psi_n(x)$ and $\psi_n(x)$ is the eigenfunction of the square potential well]

Now,

$$C_n = \int_0^a \psi(x)\,\psi_n(x)\,dx = \int_0^a A\sin^2\left(\frac{\pi x}{a}\right)\sqrt{\frac{2}{a}}\sin\left(\frac{n\pi x}{a}\right) dx$$

$$= A\sqrt{\frac{2}{a}}\int_0^a \sin^2\frac{\pi x}{a}\sin\left(\frac{n\pi x}{a}\right) dx = A\sqrt{\frac{2}{a}}\int_0^a \left(1-\cos\frac{2\pi x}{a}\right)\sin\frac{n\pi x}{a}\,dx$$

$$= \frac{A}{\sqrt{2a}}\left[\frac{a}{n\pi}\left\{\cos\frac{n\pi x}{a}\right\}_0^a - \frac{1}{2}\int_0^a \left\{\sin(n+2)\frac{\pi x}{a} + \sin(n-2)\frac{\pi x}{a}\right\} dx\right]$$

$$\therefore C_1 = \frac{A}{\sqrt{2a}} \times \frac{-8a}{3\pi} \quad \left[\begin{array}{l}\because \text{ for the ground state } n = 1 \text{ and} \\ \psi(x) = \psi(0) = 0\end{array}\right]$$

$$= -\frac{8}{3}\frac{\sqrt{8}}{\sqrt{6\pi}}.$$

$\therefore$ the required probability for the ground state ($n = 1$) is $P_1 = C_1^2 = \dfrac{256}{27\pi^2}$.

(b) To find the probability in the nth state we can show that $A^2 = 30/a^5$ and

$$P_n = |C_n|^2 = \frac{240}{(n\pi)^6}\left[1-(-1)^n\right]^2.$$

If n is odd, then $P_n = \dfrac{240}{(n\pi)^6}$ and if n is even, then $P_n = 0$.

» ***Example 7.10.***

(a) The time-independent Schrödinger equation for a particle moving in a one-dimensional potential $V(x)$ is

$$-\frac{\hbar^2}{2m}\frac{d^2\psi(x)}{dx^2} + V(x)\,\psi(x) = E\psi(x)$$

Suppose, $V(x) = V(-x)$ and $\psi(x)$ is non-degenerate. Prove that $\psi(x)$ has definite parity, i.e.,

$$\psi(x) = +\psi(-x) \text{ (Even parity)}$$

$$\text{or, } \psi(x) = -\psi(-x) \text{ (Odd parity).}$$

(b) Consider the potential well shown in Fig 7.21. Sketch the approximate character of the two lowest energy solutions to the time-independent Schrödinger equation for this potential. Label the two solutions ψ_1 and ψ_2 and find their respective energies E_1 and E_2.

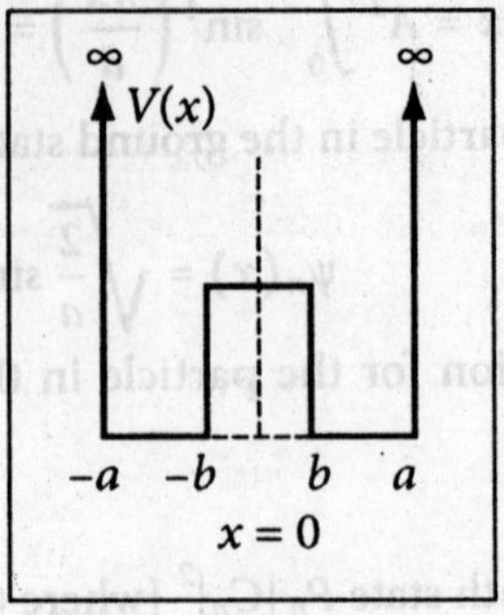

Fig 7.21 Figure for problem 7.10.

(c) A particular solution of time-dependent Schrödinger equation for the potential shown in Fig 7.21 can be constructed by superposing $\psi_1 = e^{-i(E_1/\hbar)t}$ and $\psi_2 = e^{-i(E_2/\hbar)t}$..Construct a wave packet ψ, which at time $t = 0$ is (almost) entirely in the left-hand side well. Describe in detail the motion of this wave packet as a function of time.

Solution:

(a) Substituting $-x$ for x into the time-independent Schrödinger equation

$$-\frac{\hbar^2}{2m}\frac{d^2\psi(x)}{dx^2} + V(x)\psi(x) = E\psi(x)$$

we get

$$-\frac{\hbar^2}{2m}\frac{d^2\psi(-x)}{dx^2} + V(-x)\psi(-x) = E\psi(-x)$$

Using $V(x) = V(-x)$.

$$-\frac{\hbar^2}{2m}\frac{d^2\psi(-x)}{dx^2} + V(x)\psi(-x) = E\psi(-x)$$

Thus, we see that $\psi(-x)$ is also a wave function of the Schrödinger wave equation corresponding to the same energy E. Moreover, $\psi(x)$ is degenerate, so $\psi(x)$ and $\psi(-x)$ must be linearly dependent. Both $\psi(x)$ and $\psi(-x)$ being normalised, we get $\psi(x) = \pm\psi(x)$.

The presence of ± sign means $\psi(x)$ has either positive or negative parity.

(b) We write:

$$\alpha = \frac{n\pi}{2a} \text{ and } \beta = \frac{1}{\hbar}\sqrt{(2mV_0 - \alpha^2\hbar^2)}$$

where $n = 1, 2, \ldots$.

As $\psi(x)$ must vanish at $x = \pm a$

$$\psi(x) = \begin{cases} \sin\alpha(x+a), & \text{in the region } -a < x < -b \\ \sin\alpha(x-a), & \text{in the region } b < x < a \end{cases}$$

Further,more

$$\psi(x) = \begin{cases} e^{-\beta(b+x)}, & -b < x < 0 \\ \pm e^{-\beta(b-x)}, & 0 < x < b \end{cases}$$

Here, the sign ± represents positive (or, negative) parity wave function. The ground state wave function $\psi_1(x)$ has positive parity and the first excited state wave function $\psi_2(x)$ has negative parity.

For the nth quantum state the energy is given by

$$E_n = \frac{n^2\pi^2\hbar^2}{2ma^2}, \; n = 1, 2, \ldots$$

The plots of $\psi_1(x), \psi_2(x)$ and $\psi_1 + \psi_2$ at $t = 0$ are plotted below (Fig 7.22) as a function of x.

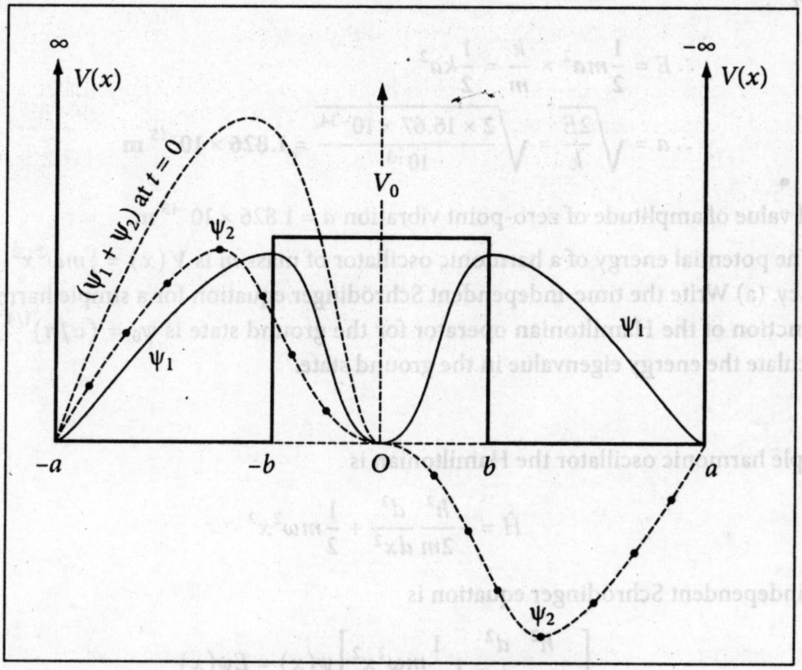

Fig 7.22 Figure for problem 7.10.

(c) We construct a wave packet as $\psi(x, t) = \psi_1(x)\, e^{-iE_1 t/\hbar} + \psi_2(x)\, e^{-iE_2 t/\hbar}$.

The wave function $\psi(x, t)$ is almost entirely in the left-hand well. This is because ψ_1 and ψ_2 are destructive in the right-hand well and constructive in the left-hand well. It is thus entirely in the left-hand well at $t = 0$.

As time goes on, the phase difference between ψ_1 and ψ_2 changes as $\frac{(E_2 - E_1)\,t}{\hbar}$. So the wave packet gradually moves to the right-hand well and, then reflects back.

The time period of oscillation is

$$T = \frac{2\pi}{T} = \frac{2\pi\hbar}{(E_2 - E_1)}$$

» ***Example* 7.11.** A particle of mass 1 mg is attached to a spring of spring constant 10^{-3} N/m. Calculate its zero-point energy and classical value of amplitude of zero-point vibration.

Solution: Here, spring constant $k = 10^{-3}$ N/m and mass of the spring $m = 1\text{ mg} = 10^{-6}$ kg.

$$\therefore \text{ frequency of vibration, } \nu = \frac{1}{2\pi}\sqrt{\frac{k}{m}} = \frac{1}{2\pi}\sqrt{\frac{10^{-3}}{10^{-6}}} = 5.03\text{ Hz.}$$

Zero-point energy

$$E_0 = \tfrac{1}{2}h\nu = \tfrac{1}{2} \times 6.63 \times 10^{-34} \times 5.03 = 16.67 \times 10^{-34}\text{ J} = \frac{16.67 \times 10^{-34}}{1.6 \times 10^{-19}} = 10.4 \times 10^{-15}\text{ eV}$$

Classically, the total energy of the oscillator is $E = \tfrac{1}{2}ma^2\omega^2$.

But, $\omega^2 = k/m$.

$$\therefore E = \frac{1}{2}ma^2 \times \frac{k}{m} = \frac{1}{2}ka^2$$

$$\therefore a = \sqrt{\frac{2E}{k}} = \sqrt{\frac{2 \times 16.67 \times 10^{-34}}{10^{-3}}} = 1.826 \times 10^{-15}\text{ m}$$

$\therefore$ the classical value of amplitude of zero-point vibration $a = 1.826 \times 10^{-15}$ m.

» ***Example* 7.12.** The potential energy of a harmonic oscillator of mass m is $V(x) = \tfrac{1}{2}m\omega^2x^2$, where ω is the angular frequency. (a) Write the time-independent Schrödinger equation for a simple harmonic oscillator. (b) The eigenfunction of the Hamiltonian operator for the ground state is $\psi_0 = (\alpha/\pi)^{1/4}\, e^{-\alpha x^2/2}$, where $\alpha = m\omega/\hbar$. Calculate the energy eigenvalue in the ground state.

Solution:

(a) For the simple harmonic oscillator the Hamiltonian is

$$\hat{H} = -\frac{\hbar^2}{2m}\frac{d^2}{dx^2} + \frac{1}{2}m\omega^2x^2$$

$\therefore$ the time-independent Schrödinger equation is

$$\left[-\frac{\hbar^2}{2m}\frac{d^2}{dx^2} + \frac{1}{2}m\omega^2x^2\right]\psi(x) = E\psi(x)$$

(b) For the ground state the Schrödinger equation is

$$\left[-\frac{\hbar^2}{2m}\frac{d^2}{dx^2} + \frac{1}{2}m\omega^2x^2\right]\psi_0(x) = E_0\psi_0(x) \qquad \text{(i)}$$

Now, $\psi_0 = \left(\dfrac{\alpha}{\pi}\right)^{1/4} e^{-\alpha x^2/2}$.

$$\therefore \frac{d\psi_0}{dx} = \left(\frac{\alpha}{\pi}\right)^{1/4} e^{-\alpha x^2/2}(-\alpha x) = -\alpha x\left(\frac{\alpha}{\pi}\right)^{1/4} e^{-\alpha x^2/2} = -\alpha x\psi_0$$

$$\text{and } \frac{d^2\psi_0}{dx^2} = -\alpha x\frac{d\psi_0}{dx} - \alpha\psi_0 = -\alpha x(-\alpha x\psi_0) - \alpha\psi_0 = \alpha^2x^2\,\psi_0 - \alpha\psi_0$$

Multiplying both sides by $-\hbar^2/2m$, we get

$$-\frac{\hbar^2}{2m}\frac{d^2\psi_0}{dx^2} = -\frac{\hbar^2}{2m}\alpha^2x^2\psi_0 + \frac{\hbar^2}{2m}\alpha\psi_0$$

$$\text{or,}\quad \left[-\frac{\hbar^2}{2m}\frac{d^2}{dx^2} + \frac{1}{2}m\omega^2x^2\right]\psi_0 = \frac{1}{2}\hbar\omega\psi_0 \quad \left[\because \alpha = \frac{m\omega}{\hbar}\right] \qquad \text{(ii)}$$

Comparing (i) and (ii), we get $E_0 = \frac{1}{2}\hbar\omega$.

» ***Example* 7.13.** **The wave function of a particle moving under the influence of a potential $V = \frac{1}{2}kr^2$ is $\psi(\vec{r}, t)$. If the wave function is changed to $\psi(\alpha\vec{x}, t)$ what are the mean kinetic energy and potential energy of the particle?**

***Solution*:** Mean KE is given by

$$\langle T\rangle = \left\langle\psi(\alpha x)\frac{\hbar^2}{2m}\frac{d^2}{dx^2}\psi(\alpha x)\right\rangle = \alpha^2\left\langle\psi(\alpha x)\frac{\hbar^2}{2m}\frac{d^2}{d^2(\alpha x)}\psi(\alpha x)\right\rangle$$

Mean PE is

$$\langle V\rangle = \left\langle\psi(\alpha x), \frac{k}{2}x^2\,\psi(\alpha x)\right\rangle = \frac{1}{\alpha^2}\left\langle\psi(\alpha x), \frac{k}{2}\alpha^2x^2\,\psi(\alpha x)\right\rangle = \frac{1}{\alpha^2}\left\langle\psi(x), \frac{k}{2}x^2\,\psi(x)\right\rangle$$

» ***Example* 7.14.** **If the expectation value of the square of the displacement harmonic oscillator is $\langle x^2\rangle = (n+\frac{1}{2})\,\hbar/m\omega$, what is the expectation value of the potential energy?**

***Solution*:** **For the harmonic oscillator the potential energy is given by $V = \frac{1}{2}m\omega^2x^2$.**

∴ the expectation value of the potential energy in the *n*th state is

$$\langle V\rangle_n = \int_{-\infty}^{+\infty}\psi_n^*(x)\,\hat{V}\,\psi_n(x)\,dx = \int_{-\infty}^{+\infty}\psi_n^*(x)\cdot\tfrac{1}{2}m\omega^2x^2\,\psi_n(x)\,dx$$

$$= \tfrac{1}{2}m\omega^2\int_{-\infty}^{+\infty}\psi_n^*(x)\,x^2\,\psi_n(x)\,dx = \tfrac{1}{2}m\omega^2\langle x^2\rangle = \tfrac{1}{2}m\omega^2\cdot(n+\tfrac{1}{2})\frac{\hbar}{m\omega}$$

$$= \tfrac{1}{2}(n+\tfrac{1}{2})\,\hbar\omega = \frac{E_n}{2}$$

where E_n is the energy in the *n*th state of the oscillator.

» ***Example* 7.15.** **The ground state of a harmonic oscillator is given by $\psi(x) = Ae^{-\alpha^2x^2}$, where $\alpha^2 = \frac{k}{2\hbar\omega}$, k is the quasistatic force constant $\left(U = \frac{1}{2}kx^2\right)$. Calculate the mean values of the kinetic and potential energies of the oscillator.**

***Solution*:** The condition of normalisation is

$$\int_{-\infty}^{+\infty}|\psi|^2\,dx = 1$$

$$\therefore\quad \int_{-\infty}^{+\infty}A^2e^{-2\alpha^2x^2}\,dx = 1 \text{ or, } 2\int_0^{+\infty}A^2e^{-2\alpha^2x^2}\,dx = 1 \text{ or, } A = \sqrt{\frac{2}{\pi}\alpha}$$

∴ the mean kinetic energy $\langle T\rangle$ and potential energy $\langle V\rangle$ are given by

$$\langle T\rangle = \langle V\rangle = \int_{-\infty}^{+\infty}\psi^*\tfrac{1}{2}kx^2\,\psi\,dx = \tfrac{1}{2}k\int_{-\infty}^{+\infty}A^2e^{-2\alpha^2x^2}x^2\,dx = kA\int_0^{\infty}e^{-2\alpha^2x^2}x^2\,dx = \tfrac{1}{4}\hbar\omega$$

» ***Example* 7.16.** If the ground state and the first excited state of an LHO are expressed by the wave functions

$$\psi_0(x,t) = \left(\frac{\alpha}{\sqrt{\pi}}\right)^{1/2} e^{-\alpha^2x^2/2} \text{ and } \psi_1(x,t) = \left(\frac{\alpha}{2\sqrt{\pi}}\right)^{1/2} e^{-\alpha^2x^2/2}2ax$$

where $\alpha = \sqrt{\dfrac{m\omega}{\hbar}}$ and the state function is $\psi(x,t) = 1/\sqrt{2}[\psi_0(x,t) + \psi_1(x,t)]$, find the expectation value of energy of the LHO.

Solution: The expectation value of energy of the LHO is given by $\langle E\rangle = \int \psi^*(x,t)E\psi(x,t)\,dx$.

Now, $\psi^*(x,t)\,\psi(x,t) = \frac{1}{\sqrt{2}}[\psi_0^*(x,t) + \psi_1^*(x,t)] \times \frac{1}{\sqrt{2}}[\psi_0(x,t) + \psi_1(x,t)]$

$= \frac{1}{2}[\psi_0(x,t) + \psi_1(x,t)]^2 = \frac{1}{2}[\psi_0^2(x,t) + \psi_1^2(x,t) + 2\psi_0(x,t)\,\psi_1(x,t)]$.

$$\therefore \langle E\rangle = \tfrac{1}{2}\int_{-\infty}^{+\infty} \psi_0^2(x,t)E\,dx + \tfrac{1}{2}\int_{-\infty}^{+\infty} \psi_1^2(x,t)E\,dx + \tfrac{1}{2}\cdot 2\int_{-\infty}^{+\infty} \psi_0(x,t)\,\psi_1(x,t)E\,dx = \frac{E_0}{2} + \frac{E_1}{2} + 0$$

where $E_0 = \frac{1}{2}\hbar\omega$ and $E_1 = \frac{3}{2}\hbar\omega$ and

$$\int_{-\infty}^{+\infty} \psi_0(x,t)\psi_1(x,t)E\,dx = 0 \text{ [as } \psi_0(x,t) \text{ and } \psi_1(x,t) \text{ are orthogonal]}$$

$$\int_{-\infty}^{+\infty} \psi_0^2(x,t)E\,dx \text{ and } \int_{-\infty}^{+\infty} \psi_1^2(x,t)E\,dx$$

represent the expectation value of energy in the ground state and first excited state respectively.

∴ the expectation value of energy in the given state is $\langle E\rangle = \frac{1}{4}\hbar\omega + \frac{3}{4}\hbar\omega = \hbar\omega$.

» ***Example* 7.17.** If the ground state of the LHO is given by

$$\psi_0(x) = \left(\frac{\alpha}{\sqrt{\pi}}\right)^{1/2} e^{-\alpha^2x^2/2}$$

where $\alpha = \sqrt{\frac{m\omega}{\hbar}}$, then show that the probability of finding the LHO outside the classical limit $(-a < x < a)$ is only 17 %, where a is the amplitude of vibration of the LHO.

Solution: Quantum mechanically, the energy of the oscillator is

$$E = \frac{1}{2}\hbar\omega = \frac{1}{2}\hbar\sqrt{\frac{k}{m}} = \frac{1}{2}ka^2. \qquad \therefore a^2 = \frac{\hbar}{\sqrt{mk}} = \frac{\hbar}{\sqrt{mm\omega^2}} = \frac{\hbar}{m\omega} = \frac{1}{\alpha^2}$$

So quantum mechanically, x varies from $-1/\alpha$ to $1/\alpha$.

The classical and quantum mechanical limits are depicted in Fig 7.24.

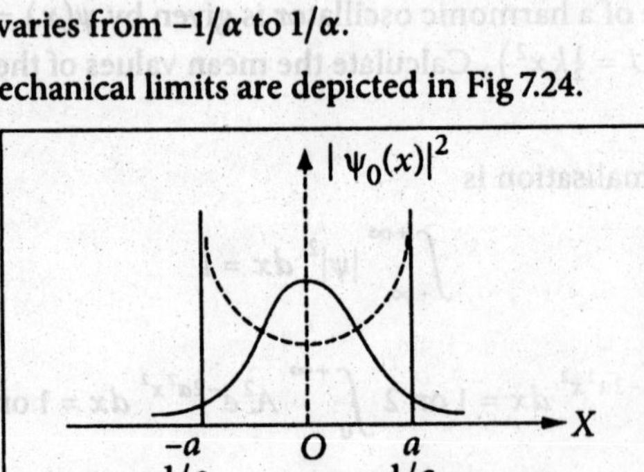

Fig 7.23 Classical and quantum mechanical limits of the LHO

Now, quantum mechanical probability of finding the oscillator in $x = -1/\alpha$ to $x = 1/\alpha$ is

$$P = \int_{-1/\alpha}^{1/\alpha} |\psi_0|^2\, dx = 2\int_0^{1/\alpha} \left(\frac{\alpha}{\sqrt{\pi}}\right) e^{-\alpha^2 x^2}\, dx$$

Putting $z = \alpha x$, $\therefore dx = dz/\alpha$.

$$\therefore P = \frac{2}{\sqrt{\pi}}\int_0^1 e^{-z^2}\, dz = \frac{2}{\sqrt{\pi}}\int_0^1 \left[1 - \frac{z^2}{1!} + \frac{z^4}{2!} - \frac{z^6}{3!} + \frac{z^8}{4!} - \ldots\right] dz$$

$$= \frac{2}{\sqrt{\pi}}\left[z - \frac{z^3}{3(1!)} + \frac{z^5}{5(2!)} - \frac{z^7}{7(3!)} + \frac{z^9}{9(4!)}\right]_0^1 = \frac{2}{\sqrt{\pi}}\left[1 - \frac{1}{3(1!)} + \frac{1}{5(2!)} - \frac{1}{7(3!)} + \frac{1}{9(4!)} - \ldots\right]$$

$= \mathbf{0.83}$

So the probability of finding the LHO within the classical limit is 83 %.

Hence, the probability of finding the LHO outside the classical limit is 17 %.

» ***Example* 7.18.** Show that the ratio of the probability density of the LHO at the classical limits of motion to the probability density at the origin is $1/e$ when the LHO is in the ground state.

***Solution*:** If a be the amplitude of the LHO, then its energy, $E_0 = \frac{m\omega^2 a^2}{2} = \frac{\hbar\omega}{2}$ in the ground state.

$$\text{or, } \frac{m\omega^2 a^2}{2} = \frac{1}{2}\hbar\omega \text{ or, } a = \sqrt{\frac{\hbar}{m\omega}}$$

$$q = \sqrt{\alpha}a = \sqrt{\frac{m\omega}{\hbar}} \times \sqrt{\frac{\hbar}{m\omega}} = 1 \left[\because \alpha = \frac{m\omega}{\hbar}\right]$$

Now, the wave function of the LHO in the ground state is

$$\psi_0 = \left(\frac{1}{\sqrt{\pi}}\right)^{1/4} e^{-q^2/2}$$

Again, at the origin $q = 0$ and at the classical limit $(x = a)$, $q = \sqrt{\alpha}a = 1$.

$\therefore$ the probability density of finding the oscillator in the ground state at the origin is $P(0) = |\psi_0(0)|^2 = 1/\sqrt{\pi}$.

Similarly, the probability density of finding the LHO at the classical limit is

$$P(1) = |\psi_0(1)|^2 = \frac{1}{\sqrt{\pi}}e^{-1} = \frac{1}{e\sqrt{\pi}} \qquad \therefore \frac{P(1)}{P(0)} = \frac{1}{e\sqrt{\pi}} \times \sqrt{\pi} = \frac{1}{e}$$

» ***Example* 7.19.** A particle moving in an infinite potential well of width L in one dimension has the wave function, $\psi(x) = A\sin\frac{2\pi x}{L}$ for $-\frac{L}{2} \leqslant x \leqslant \frac{L}{2}$ and $\phi(x)$ is zero outside. What is the average momentum?

[BU (H) 2002]

***Solution*:** To find out the normalisation constant, we have

$$\int_{-L/2}^{+L/2} \phi^*(x)\phi(x)\, dx = 1 \text{ or, } \int_{-L/2}^{+L/2} a^* \sin\frac{2\pi x}{L} a \sin\frac{2\pi x}{L}\, dx = 1 \text{ or, } a^2\int_{-L/2}^{+L/2} \sin^2\frac{2\pi x}{L}\, dx = 1$$

$$\text{or, } \frac{a^2}{2}\int_{-L/2}^{+L/2}\left(1 - \frac{\cos 2\cdot 2\pi x}{L}\right) dx = 1 \text{ or, } \frac{a^2}{2}\cdot L = 1 \text{ or, } a = \sqrt{\frac{2}{L}}$$

$\therefore$ the normalised wave function is $\phi(x) = \sqrt{\frac{2}{L}} \sin \frac{2\pi x}{L}$.

Now, the average value of momentum is

$$\langle p_x \rangle = \int_{-L/2}^{+L/2} \phi^*(x) \hat{p}_x \phi(x)\, dx = \left(\frac{2}{L}\right)(-i\hbar) \int_{-L/2}^{+L/2} \sin \frac{2\pi x}{L} \frac{d}{dx}\left(\sin \frac{2\pi x}{L}\right) dx$$

$$= (-i\hbar)\frac{2}{L} \cdot \frac{2\pi}{L} \int_{-L/2}^{+L/2} \sin \frac{2\pi x}{L} \cos \frac{2\pi x}{L}\, dx = (-i\hbar)\frac{2\pi}{L^2} \int_{-L/2}^{+L/2} \sin \frac{4\pi x}{L}\, dx$$

$$= +i\frac{2\pi\hbar}{L^2}\left[\frac{L}{4\pi} \cos \frac{4\pi x}{L}\right]_{-L/2}^{+L/2} = 0$$

$\therefore \langle p_x \rangle = 0$

» ***Example* 7.20. A particle moving in one dimension is confined to move within $x = 0$ and $x = a$. Suppose, the particle is in a state represented by the wave function**

$$\psi(x) = A \sin \frac{\pi x}{a}$$

(a) Find the value of A that will normalise ψ. (b) If a measurement of momentum is carried out what are the possible values of p those will be observed? (c) What are the average values of p and p^2? (d) Estimate the uncertainty product $\Delta x\, \Delta p_x$ for the given state. Is the value consistent with the uncertainty principle?

[BU (H) 1998, 2001]

***Solution*:**

(a) From the condition of normalisation, we have

$$\int_0^a \psi^*(x)\, \psi(x)\, dx = 1 \text{ or, } A^2 \int_0^a \sin^2 \frac{\pi x}{a}\, dx = 1$$

$$\text{or, } \frac{A^2}{2} \int_0^a \left[1 - \cos\left(\frac{2\pi x}{a}\right)\right] dx = 1 \text{ or, } \frac{A^2}{2} \cdot a = 1 \text{ or, } A = \sqrt{\frac{2}{a}}$$

$\therefore$ the value of A that will normalise ψ is $\sqrt{2/a}$.

(b) The value of the momentum of the particle is given by

$$p = \frac{h}{\lambda} = \frac{h}{2\pi}\frac{2\pi}{\lambda} = \hbar k$$

where λ is the de Broglie wavelength of the particle of momentum p and $k = 2\pi/\lambda$.

Now, $\psi(x) = A \sin \frac{\pi x}{a} = A \sin kx$, where $k = \pi/a$.

$\therefore$ the momentum of the particle is $\pi\hbar/a$.

(c) The average value of p is given by

$$\langle p \rangle = \int_0^a \psi^*(x) \hat{p} \psi(x)\, dx = (-i\hbar)\frac{2}{a} \int_0^a \sin \frac{\pi x}{a} \frac{d}{dx}\left(\sin \frac{\pi x}{a}\right) dx \quad \left[\because \hat{p} = -i\hbar\frac{\partial}{\partial x}\right]$$

$$= -\frac{2i\hbar}{a}\frac{\pi}{a} \int_0^a \sin \frac{\pi x}{a} \cos \frac{\pi x}{a}\, dx = -\frac{2i\hbar\pi}{a^2}\frac{a}{2\pi}\left[\sin^2 \frac{\pi x}{a}\right]_0^a \quad \left[\because \int \sin mx \cos mx\, dx = \frac{1}{2m} \sin^2 mx\right]$$

$$= -\frac{2i\hbar\pi}{a^2}\frac{a}{2\pi}[0] = 0$$

$\therefore \langle p \rangle = 0$

$$\langle p^2\rangle = \int_0^a \psi^*(x)\hat{p}^2\psi(x)\,dx$$

$$= \frac{2}{a}(-\hbar^2)\int_0^a \sin\frac{\pi x}{a}\frac{\partial^2}{\partial x^2}\left(\sin\frac{\pi x}{a}\right)dx \quad \left[\because \hat{p}^2 = \left(-i\hbar\frac{\partial}{\partial x}\right)^2\right]$$

$$= \frac{2\pi^2\hbar^2}{a^3}\int_0^a \sin^2\frac{\pi x}{a}\,dx = \frac{2\pi^2\hbar^2}{a^3}\left[\int_0^a\left(1-\cos\frac{2\pi x}{a}\right)dx\right] = \frac{\pi^2\hbar^2}{a^3}a = \frac{\pi^2\hbar^2}{a^2}$$

$$\therefore \langle p^2\rangle = \frac{\pi^2\hbar^2}{a^2}$$

(d) The uncertainty relation is $\Delta x\,\Delta p_x \sim \hbar$.

Now, $\Delta x = \sqrt{\langle x^2\rangle - \langle x\rangle^2}$.

$$\langle x^2\rangle = \int_0^a \psi^*(x)\,x^2\,\psi(x)\,dx = \frac{2}{a}\int_0^a x^2\sin^2\frac{\pi x}{a}\,dx = \frac{2}{a}\cdot\frac{a^3}{2}\left(\frac{1}{3}-\frac{1}{2\pi^2}\right) = a^2\left(\frac{1}{3}-\frac{1}{2\pi^2}\right)$$

Also

$$\langle x\rangle = \frac{2}{a}\int_0^a x\sin^2\frac{\pi x}{a}\,dx = \frac{2}{a}\cdot\frac{a^2}{4} = \frac{a}{2}$$

$$\therefore \Delta x = \sqrt{\langle x^2\rangle - \langle x\rangle^2} = \sqrt{a^2\left(\frac{1}{3}-\frac{1}{2\pi^2}\right)-\frac{a^2}{4}} = a\sqrt{\frac{1}{12}-\frac{1}{2\pi^2}}$$

$$\text{and } \Delta p = \sqrt{\langle p^2\rangle - \langle p\rangle^2} = \sqrt{\frac{\pi^2\hbar^2}{a^2}-0} = \frac{\pi\hbar}{a}$$

$$\therefore \Delta x\,\Delta p_x = \sqrt{a^2\left(\frac{1}{12}-\frac{1}{2\pi^2}\right)}\times\frac{\pi\hbar}{a} = \left(\sqrt{\frac{1}{12}-\frac{1}{2\pi^2}}\right)\pi\hbar \sim \hbar$$

So the value of $\Delta x\,\Delta p_x$ is consistent with uncertainty principle.

» ***Example* 7.21.** A one-dimensional harmonic oscillator is a particle of mass m oscillating with the natural frequency ω_0 due to the action of an elastic force acting along x-axis. The potential energy is

$$V(x) = \frac{m\omega_0^2}{2}x^2$$

(a) Write down the classical Hamiltonian and time-independent Schrödinger equation for the system.

(b) Substitute a trial wave function $\psi(x) = Ae^{-\beta x^2/2}$ in the Schrödinger equation and determine the energy of the system for which it is a solution. **[BU (H) 2002, NBU (H) 2003]**

***Solution*:**

(a) For a linear harmonic oscillator (LHO) of mass m and angular frequency ω_0 the potential energy

$$V(x) = \tfrac{1}{2}m\omega_0^2x^2 \tag{i}$$

The Hamiltonian is

$$\hat{H} = \frac{\hat{p}_x^2}{2m} + V(x) = -\frac{\hbar^2}{2m}\frac{d^2}{dx^2} + \frac{1}{2}m\omega_0^2x^2 \tag{ii}$$

The time-independent Schrödinger equation becomes

$$-\frac{\hbar^2}{2m}\frac{d^2\psi}{dx^2} + V(x)\psi = E\psi \text{ or, } \frac{d^2\psi}{dx^2} + \left(\frac{2mE}{\hbar^2}\right)\psi - \left(\frac{2m}{\hbar^2}\right)\frac{1}{2}m\omega_0^2x^2\psi = 0 \tag{iii}$$

(b) Putting

$$\alpha = \frac{2mE}{\hbar^2},\ \beta = \frac{m\omega_0}{\hbar}$$

in (i), we get

$$\frac{d^2\psi}{dx^2} + (\alpha - \beta^2 x^2)\,\psi = 0 \tag{iv}$$

We now find out the value of α for which the solution of (iii) will be finite and unique in all space.

For large value of x we can at first neglect α and take the approximate solution as

$$\psi(x) = Ae^{-\beta x^2/2} \tag{v}$$

The general solution is of the form:

$$\psi(x) = Au(x)\, e^{-\beta x^2/2} \tag{vi}$$

so that the condition, ψ at $x \to \pm\infty$ is zero be satisfied.

From (iii) and (iv), we get

$$\frac{d^2u}{dx^2} - 2\beta x\frac{du}{dx} + (\alpha - \beta)\, x = 0 \tag{vii}$$

The solution of (v) is of the form:

$$u(x) = \sum_r a_r x^r \tag{viii}$$

Substituting (viii) in (vi) and, then equating the coefficients of the various powers of x to zero, we get the recurrence relation

$$(r+2)\,(r+1)\,a_{r+2} + [\alpha - \beta(2r+1)]\,a_r = 0 \tag{ix}$$

Putting $r = 0$, we get a_0 from (ix) and this a_0 value determines the coefficients of all the powers of x.

Similarly, a_1, obtained by putting $r = 1$, determines the coefficients of all the odd powers of x.

Now, $e^{-\frac{1}{2}\beta x^2} \to 0$ as $x \to \infty$ still, $\psi(x) = Au(x)\, e^{-\beta x^2/2}$ tends to ∞ as the power series diverges more rapidly than the exponential series when $x \to \infty$. So $\psi(x)$ does not converge unless the power series terminates. The recurrence relation shows that the series breaks off at $r = n$, if $a_n = 0$, i.e., if

$$\alpha = \beta(2n+1) \text{ or, } E_n = (n + \tfrac{1}{2})\,\hbar\omega_0 \tag{x}$$

Thus, the energy for the ground state ($n = 0$) is $E_0 = \frac{1}{2}\hbar\omega_0$ and is called the *zero-point energy*. This is not in accordance with Planck's hypothesis which states $E_n = n\hbar\omega$ and hence, $E_0 = 0$ for $n = 0$. This establishes the superiority of the new quantum theory over the old quantum theory.

The energy of the LHO is discrete and the energy levels are equispaced with an energy difference

$$\Delta E = E_{n+1} - E_n = \hbar\omega \tag{xi}$$

» ***Example* 7.22.** An oscillator has 1 kg attached to a light rod of length 1 m. Taking amplitude of oscillation to be 0.1 m, calculate (i) frequency of oscillation, (ii) energy of oscillation, (iii) the quantum number n, (iv) energy gap, (v) zero-point energy. **[cf. CU (H) 2003]**

Solution:

(i) Given that, the oscillator has mass $m = 1$ kg; length $l = 1$ m.

So, the angular frequency of the oscillator is

$$\omega = \sqrt{\frac{g}{l}} = \sqrt{\frac{9.8\,\text{ms}^2}{1\,\text{m}}} = 3.13\,\text{s}^{-1}$$

The linear frequency

$$\nu = \frac{\omega}{2\pi} = \frac{3.13\,\text{s}^{-1}}{2\pi} = \mathbf{0.498\,s^{-1}}$$

(ii) The energy of the oscillator is

$$E = \tfrac{1}{2}m\omega^2 A^2 = \tfrac{1}{2}(1\,\text{kg})(3.13\,\text{s}^{-1})^2(0.1\,\text{m})^2 = (0.048\,\text{J})$$

(iii) Putting $E = E_n = (n + \frac{1}{2})\,\hbar\omega$.

$$0.049\,\text{J} = (n+1)\left(\frac{6.625\times 10^{-3}\,\text{Js}}{2\pi}\right)(3.13\,\text{s}^{-1}) \text{ or, } \boldsymbol{n = 1.48\times 10^{12}}$$

(iv) Energy gap $\Delta E = E_{n+1} - E_n = (n + 1 + \frac{1}{2})\,\hbar\omega - (n + \frac{1}{2})\,\hbar\omega = \hbar\omega$

$$= \frac{6.626\times 10^{-34}\,\text{Js}}{2\pi}\times 3.13\,\text{s}^{-1}$$
$$= 3.3\times 10^{-34}\,\text{J} \simeq \mathbf{0}\text{, continuum.}$$

(v) Zero-point energy

$$E_0 = \frac{\hbar\omega}{2\pi} = \frac{1}{2}(3.3\times 10^{-34}\,\text{J}) = 1.65\times 10^{-34}\,\text{J} = \frac{1.65\times 10^{-34}}{1.6\times 10^{-34}}\,\text{eV} = 10^{-15}\,\text{eV} \simeq \mathbf{0}\text{, classical result}$$

» ***Example* 7.23.** A laboratory oscillator consisting of a mass of 1 gm on a spring exhibits a period of 1 sec. The velocity of the mass when it crosses the zero displacement position is 10 cm/s.

(i) Is the oscillator possibly in an eigenstate of the Hamiltonian?

(ii) What is the approximate value of the quantum number n associated with the energy E of the oscillator?

(iii) Has the zero point energy any significance here? **[cf. CU (H) 2003, 2005]**

Solution: KE energy of the mass m of the oscillator at any instant of motion with velocity v is

$$E = \tfrac{1}{2}mv^2 = (n + \tfrac{1}{2})^2\,\hbar\omega = (n + \tfrac{1}{2})\,\hbar\frac{2\pi}{T} = (n + \tfrac{1}{2})\,\frac{h}{T}$$

$$\therefore n = \frac{mv^2 T}{2h} - \frac{1}{2} = \frac{(10^{-3}\,\text{kg})(10\times 10^{-2}\,\text{m s}^{-1})^2}{2\times 6.626\times 10^{-34}\,\text{Js}} - \frac{1}{2} = \mathbf{7.5\times 10^{27}}$$

(i) n = an integer. So, the oscillator is possibly in an eigenstate $\hat{H}$.

(ii) n = very large, ∴ the oscillator is a classical oscillator (by Bohr's correspondence principle).

(iii) Hence, zero point energy has no significance here.

» ***Example* 7.24.** What is the amplitude of zero-point oscillation of a pendulum of length 1 m and mass of bob 1 kg. Can the amplitude be measured?

Solution: The zero-point energy of the oscillator is

$$E_0 = \frac{\hbar\omega}{2} = \frac{1}{2}m\omega^2 A^2$$

the different letter have their usual meanings.

$$\therefore A = \sqrt{\frac{\hbar}{m\omega}} = \sqrt{\frac{\hbar}{m}}\sqrt{\frac{l}{g}} \quad [\because \omega = \sqrt{l/g}]$$

$$= \sqrt{\frac{6.626 \times 10^{-34}\,\text{Js}}{1\,\text{kg}}}\sqrt{\frac{1\,\text{m}}{9.8\,\text{m s}^{-1}}}$$

$$\text{or, } \mathbf{A = 1.45 \times 10^{-17}\ m}$$

The amplitude A is too small to be measured. The zero-point energy here has no consequence.

» ***Example* 7.25.** Find the expectation value of energy if a harmonic oscillator happens to be in a state which is a linear combination of ground state and first excited state as

$$\psi(x) = \frac{1}{\sqrt{2}}[\psi_0(x) + \psi_1(x)]$$

Solution: The expectation value of the total energy $H = T + V$ is

$$\langle \hat{H} \rangle = \int_{-\infty}^{\infty} \psi^* \hat{H}\psi\, dx = \int_{-\infty}^{\infty} \psi^* \hat{H}\left[\frac{1}{\sqrt{2}}(\psi_0 + \psi_1)\right] dx = \int_{-\infty}^{\infty} \psi^* \frac{1}{\sqrt{2}}(\hat{H}\psi_0 + \hat{H}\psi_1)\, dx$$

$$= \frac{1}{\sqrt{2}}\int_{-\infty}^{+\infty} \psi^*(E_0\psi_0 + E_1\psi_1)\, dx = \frac{E_0}{\sqrt{2}}\int_{-\infty}^{\infty} \psi^* \psi_0 dx + \frac{E_1}{\sqrt{2}}\int_{-\infty}^{\infty} \psi^* \psi_1\, dx$$

$$\frac{E_0}{\sqrt{2}}\int_{-\infty}^{\infty} \frac{1}{\sqrt{2}}(\psi_0 + \psi_1)^* \psi_1\, dx$$

$$= \frac{E_0}{\sqrt{2}}\int_{-\infty}^{\infty} \psi_0^* \psi_0\, dx + \frac{E_0}{\sqrt{2}}\int_{-\infty}^{\infty} \psi_1^* \psi_0\, dx + \frac{E_0}{\sqrt{2}}\int_{-\infty}^{\infty} \psi_0^* \psi_1\, dx + \frac{E_0}{\sqrt{2}}\int \psi_1^* \psi_1\, dx$$

By orthonormalisation condition of ψ_n, we get

$$\langle \hat{H} \rangle = \frac{E_0}{2}(1) + \frac{E_0}{2}(0) + \frac{E_1}{2}(0) + \frac{E_1}{2}(1) = \frac{E_0 + E_1}{2} = \frac{1}{2}\frac{\hbar\omega}{2} + \frac{3}{2}\hbar\omega \text{ or, } \langle \hat{H} \rangle = \hbar\omega$$

EXERCISES

7(A): Objective (or MCQ) Type Questions

Select the correct answer:

7.1. The energy of a particle in the nth quantum state in a one-dimensional closed box is proportional to

(a) n (c) $1/n$

(b) n^2 (d) $1/n^2$.

7.2. If a be the length of a one-dimensional closed box in which a particle is enclosed, then its energy is proportional to

(a) a (c) a^2
(b) $1/a$ (d) $1/a^2$.

7.3. The number of antinodes in the ground state wave function of a particle in an infinite square well is

(a) 1 (c) 0
(b) 2 (d) 4.

7.4. The expectation value of momentum $\langle p \rangle$ of a particle trapped in a one-dimensional box of length a is

(a) h/a (c) 1
(b) 0 (d) $h/2a$.

7.5. The eigenvalues of momentum of a particle trapped in a one-dimensional box are

(a) continuous (c) 0
(b) discrete (d) $\pi\hbar/a$.

7.6. The probability of finding the particle trapped in a one-dimensional box is the maximum for $n = 1$ at

(a) $x = 0$ (c) $x = -a$
(b) $x = a/2$ (d) $x = a$.

7.7. If λ be the wavelength of radiation emitted due to transition between energy states of a particle trapped in a one-dimensional box and a the length of the box, then

(a) $\lambda \propto a$ (c) $\lambda \propto a^2$
(b) $\lambda \propto a^n$ (d) $\lambda \propto a/n$.

7.8. Transition between a pair of states is possible for a particle trapped in a one-dimensional box if

(a) $(n + m)$ is odd (c) $(n \pm m)$ is odd
(b) $(n - m)$ is odd (d) $(n \pm m)$ is even.

[m and n are quantum states]

7.9. For a particle trapped in a square well potential, the wave functions $\psi_n(x)$ are alternately of even and odd parity as

(a) $V(x)$ is antisymmetric w.r.t. 0 (c) $V(x) = 0$
(b) $V(x)$ is symmetric w.r.t. 0 (d) $V(x) < 0$.

7.10. Classical probability of finding the LHO is

(a) $\dfrac{dx}{\pi\sqrt{a^2 - x^2}}$ (c) $\dfrac{dx}{\sqrt{x^2 - a^2}}$

(b) $\dfrac{dx}{\pi\sqrt{x^2 - a^2}}$ (d) $\dfrac{dx}{\sqrt{a^2 - x^2}}$.

7.11. In the ground state the energy of the LHO is

(a) ∞ (c) $\frac{3}{2}\hbar\omega$

(b) $\frac{1}{2}\hbar\omega$ (d) 0.

7.12. The energy of an LHO in the nth quantum state is

(a) $(n+\frac{1}{2})\hbar\omega$ (c) $\frac{1}{2}\hbar\omega$

(b) $n\hbar\omega$ (d) $\frac{5}{2}\hbar\omega$.

7.13. The probability of finding the LHO in the classical limit is

(a) 17% (c) 83%

(b) 20% (d) 100%.

7.14. A quantum mechanical oscillator may have displacement

(a) $|x| > a$ (c) $|x| < a$

(b) $|x| = a$ (d) $|x| = 2a$.

7.15. The energy of an LHO is

(a) discrete and equispaced (c) partly discrete and partly continuous

(b) continuous (d) discrete, but not equispaced.

7.16. Match list I with list II and select the correct answer using the code given below:

List I (system)	*List II (Ground state wave function)*
A. Hydrogen atom	1. e^{ikx}
B. Particle in a box	2. $(\pi a_0^3)^{-1/2} e^{-r/a_0}$
C. Harmonic oscillator	3. $\left(\frac{2}{L}\right)^{-1/2} \sin\frac{\pi x}{L}$
D. Potential step	4. $\sqrt{\frac{a}{\sqrt{\pi}}} e^{-\frac{a^2x^2}{2}}$

	A	B	C	D
(a)	2	3	4	1
(b)	3	2	1	4
(c)	3	2	4	1
(d)	2	3	1	4

[IAS(Pre) 2007]

7.17. The figure shown is the energy level diagram corresponding to a

(a) particle in a box (c) harmonic oscillator

(b) particle tunneling (d) hydrogen atom. **[IAS(Pre) 2002]**

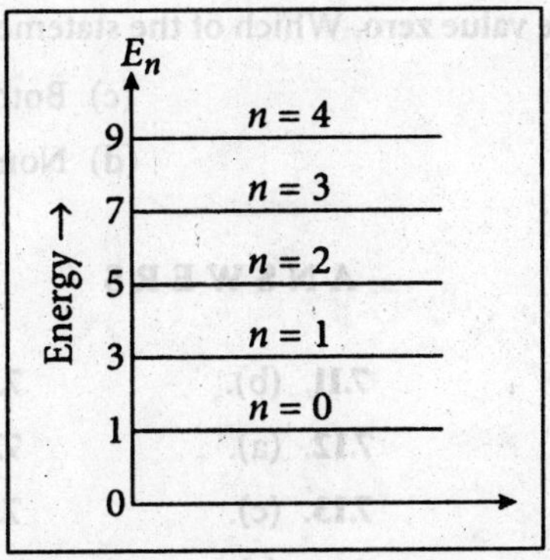

7.18. If E_0 is the zero-point energy of a harmonic oscillator of recency ν and h is the Planck's constant then its energy in the $n = 2$ state will be

(a) $E_0 + h\nu$
(b) $2E_0$
(c) $4E_0$
(d) $E_0 + 2h\nu$. **[IAS(Pre) 2002]**

7.19. For a particle of mass m in a one-dimensional harmonic oscillator potential of the form: $V(x) = \frac{1}{2}m\omega^2x^2$, the first excited energy eigenstate is $\psi(x) \sim xe^{-ax^2}$. The value of a is

(a) $\frac{m\omega}{4\hbar}$
(b) $\frac{m\omega}{3\hbar}$
(c) $\frac{m\omega}{2\hbar}$
(d) $\frac{2m\omega}{3\hbar}$. **[GATE 2007]**

7.20. A one-dimensional harmonic oscillator carrying charge $-q$ is placed in a uniform electric field $\vec{E}$ along $+x$ axis. The corresponding Hamiltonian operator is

(a) $\frac{\hbar^2}{2m}\frac{d^2}{dx^2} + \frac{1}{2}kx^2 + qE_x$
(b) $\frac{\hbar^2}{2m}\frac{d^2}{dx^2} + \frac{1}{2}kx^2 - qE_x$
(c) $-\frac{\hbar^2}{2m}\frac{d^2}{dx^2} + \frac{1}{2}kx^2 + qE_x$
(d) $-\frac{\hbar^2}{2m}\frac{d^2}{dx^2} + \frac{1}{2}kx^2 - qE_x$.

7.21. A particle of mass m in a simple harmonic oscillator potential $V = x^2$. If the ground state wave function is Ae^{-ax^2}. What is the expression for constant a equal to?

(a) $2\pi\left(\frac{m}{2}\right)^{\frac{1}{2}/h}$
(b) $2\pi\left(\frac{1}{m}\right)^{\frac{1}{2}/h}$
(c) $h\left(\frac{m}{2}\right)^{\frac{1}{2}} 2\pi$
(d) $h\left(\frac{m}{2}\right)^{\frac{1}{2}/h}$. **[IAS(Pre) 2007]**

7.22. Consider 5 electrons of mass m each in one-dimensional simple harmonic oscillator potential $V = \frac{1}{2}kx^2$. Ignoring the Colombian interaction between the electrons what is the minimum total energy of the 7 electrons. $[\varepsilon = \hbar\sqrt{k/m}, \hbar = h/2\pi]$

(a) 7ε
(b) $\frac{25}{2}\varepsilon$
(c) $\frac{15}{2}\varepsilon$
(d) $\frac{7}{2}\varepsilon$. **[IAS(Pre) 2006]**

7.23. Consider the following statements associated with quantum mechanical simple harmonic oscillator

(i) Space between energy levels increases with quantum number.

(ii) Lowest energy has the value zero. Which of the statement is/are correct?

(a) 1 only (c) Both

(b) 2 only (d) None. [IAS(Pre) 2006]

ANSWERS

7.1. (b).	7.6. (b).	7.11. (b).	7.16. (a).	7.21. (a).
7.2. (d).	7.7. (c).	7.12. (a).	7.17. (c).	7.22. (b).
7.3. (a).	7.8. (c).	7.13. (c).	7.18. (d).	7.23. (d).
7.4. (b).	7.9. (b).	7.14. (a).	7.19. (c).	
7.5. (b).	7.10. (a).	7.15. (a).	7.20. (d).	

7(B): Short Answer Type Questions

7.1. What is a bound state?

7.2. Write the Schrödinger wave equation for a particle in a square well potential of infinite depth. [WBUT 2006]

7.3. Show that the energy eigenvalues of a particle in a one-dimensional box of infinite depth are discrete and proportional to n^2.

7.4. Show that if the width a of the one-dimensional box tends to infinity, then energy levels leads to a continuum.

7.5. Draw the wave function of a particle in a box of infinite depth.

7.6. Find out an expression for the eigenvalues of momentum of a particle trapped in a one-dimensional box.

7.7. Explain the significance of zero-point energy of a particle in a one-dimensional box.

7.8. Find out an expression for the wavelength of the radiation emitted during the transition of a particle from mth to nth states in a one-dimensional box.

7.9. Show that the energy of a particle trapped in a one-dimensional box of length a is

$$E_n = \frac{n^2\hbar^2\pi^2}{2ma^2}$$

7.10. Show that a particle will remain appreciable beyond the potential well if $(V_0 - E)$ is small.

7.11. Show that for V_0 finite, the energy levels of a particle depend on the product Va^2, a being the width of the potential well.

7.12. Draw the first three wave functions for a particle enclosed in a potential well of depth V_0.

7.13. Distinguish between a classical and a quantum harmonic oscillator.

7.14. Justify that a quantum mechanical LHO is consistent with the uncertainty principle but the classical LHO is not.

7.15. Show that the normalised wave function for an LHO in the nth state is

$$\psi_n(x) = \left[\frac{\sqrt{\alpha}}{2^n n! \sqrt{\pi}}\right]^{1/2} e^{-\alpha x^2/2} H_n(\sqrt{\alpha} x)$$

where $\alpha = \frac{m\omega}{\hbar}$ and H_n is the Hermite polynomial of order n.

7.16. Draw the first five wave functions of the LHO.

7.17. Show that for large quantum numbers, the classical probability and quantum mechanical probability of an LHO become identical.

7.18. What is the ground state energy of a linear harmonic oscillator of angular frequency ω in quantum mechanics? **[BU (H) 2003]**

7.19. A particle moving in an infinite potential well of length L in one dimension has the wave function $\psi(x) = a \sin \frac{2\pi x}{L}$ for $-\frac{L}{2} \leq x \leq \frac{L}{2}$ and $\psi(x) = 0$ outside.

What is the average momentum? **[BU (H) 2002]**

7.20. Write down the Schrödinger equation for the stationary state of a simple harmonic oscillator confined to the x-axis. What is the nature of the solution at $x \to \pm\alpha$? **[CU (H) 2002]**

7.21. The ground state wave function of a one-dimensional harmonic oscillator of mass m frequency ω, $\alpha = \sqrt{\frac{m\omega}{\hbar}}$ is

$$\psi_0 = \sqrt{\frac{\alpha}{\sqrt{\pi}}} e^{-\alpha^2 x^2/2}$$

What is the energy corresponding to this state? Is it an eigenfunction of momentum? **[cf. CU (H) 2000]**

[*Ans.* $E_0 = \frac{\hbar\omega}{2}$; yes]

7(C): Broad Answer Type Questions

7.1. (a) Write down the Schrödinger equation of a one-dimensional harmonic oscillator. What is the energy of this oscillator when it is in the eigenstate associated with the quantum number n? What values may n have?

(b) The lowest energy eigenfunction of the linear harmonic oscillator is

$$\psi_0(x) = \left(\frac{2m\nu}{\hbar}\right)^{1/4} e^{-(\pi m\nu/\hbar)x^2}$$

where m is the mass and ν is the classical frequency of the oscillator.

If $\hat{T}$ and $\hat{V}$ are respectively the expectation values of the kinetic and potential energies of the oscillator in the above state, show that $\hat{T} = \hat{V} = \frac{E_0}{2}$, where E_0 is the zero-point energy. **[BU (H) 1996]**

7.2. A particle is confined in a one-dimensional infinite square well

$$V(x) = \begin{cases} 0, & -a < x < a \\ \infty, & |x| \geq a \end{cases}$$

where a is a positive constant.

(a) Write down the stationary state (time-independent) Schrödinger equation for $-a < x < a$.

(b) State the boundary conditions at $x = \pm a$.

(c) Obtain the energy eigenvalues and the eigenfunction.

(d) Calculate the expectation value of momentum in the ground state. **[BU (H) 2004]**

7.3. Write down Schrödinger's equation for the simple harmonic oscillator. Show how a study of the behaviour of the solution of this equation at large distances can lead to the expression of discrete spectrum of energy levels of the quantum mechanical simple harmonic oscillator. **[BU (H) 1998]**

7.4. The wave function for the ground state of a particle confined to a one-dimensional box with boundaries at $x = 0$ and $x = L$ is given by

$$\psi(x) = \begin{cases} N\sin\left(\frac{\pi x}{L}\right), & 0 \leqslant x \leqslant L \\ 0, & \text{for } x < 0 \text{ and } x > L \end{cases}$$

(a) Calculate N.

(b) Calculate the uncertainties in position (Δx) and momentum (Δp_x) in the state.

(c) Check the validity of the uncertainty relation. **[BU (H) 1998]**

7.5. Consider a linear harmonic oscillator.

(a) Write down the classical Hamiltonian and the time-independent Schrödinger equation for the system.

(b) Substitute a trial solution $\psi(x) = Ae^{-\alpha^2 x^2/2}$ in the Schrödinger equation and determine the value of α and that of the energy of the system for which it is a solution.

(c) Determine A from the normalisation condition. $\left[\text{Use: } \int_{-\infty}^{+\infty} e^{-z^2}\,dz = \sqrt{\pi}\right]$

(d) Is the state represented by this wave function a bound state? State reason for your answer.

[BU (H) 2002]

7.6. Solve the Schrödinger equation for the one-dimensional potential well defined by

$$V(x) = \begin{cases} 0, & x < 0 \text{ and } x > a \\ V_0, & 0 < x < a \end{cases}$$

Show that for a deep well the energy states agree with those obtained by Wilson-Sommerfeld quantisation rule.

7.7. Calculate the discrete energy levels of a particle in one-dimensional square well potential with (a) perfectly rigid walls (b) finite potential step.

7.8. (a) A particle is enclosed in a one-dimensional box. Find (i) the energy eigenvalues (ii) wave functions (iii) expectation value of the momentum operator.

(b) Define parity and discuss the parity of the wave functions for the particle in a box.

7.9. Give an account of quantum mechanical description of a simple harmonic oscillator. In what way is this description different from classical description?

Discuss the properties of the oscillator in the lowest energy states.

7.10. Write down the Schrödinger equation for a harmonic oscillator in momentum representation and obtain its solution.

Ans. The entire mathematical structure of quantum mechanics can be described by the alternative basis of momentum representation of the operators $\hat{x}$ and $\hat{p}$.

For the oscillator, the Hamiltonian operator is given by

$$\hat{H} = \frac{\hat{p}^2}{2m} + \frac{1}{2}m\omega^2\hat{x}^2, \text{ where } \hat{x} = i\hbar\frac{\partial}{\partial p},\ \hat{p} = p$$

Hence,

$$\hat{H} = \frac{p^2}{2m} + \frac{1}{2}m\omega^2\left(i\hbar\frac{\partial}{\partial p}\right)^2 = -\frac{m\omega^2\hbar}{2}\frac{\partial^2}{\partial p^2} + \frac{p^2}{2m} \tag{1}$$

The Schrödinger wave equation in momentum representation is

$$\hat{H}\psi(p) = E\psi(p) \tag{2}$$

where $\psi(p)$ is the wave function

$$\text{or,}\quad -\frac{m\omega^2\hbar^2}{2}\frac{\partial^2\psi}{\partial p^2} + \frac{p^2}{2m}\psi = E\psi$$

Putting

$$q = \frac{p}{\sqrt{m\omega\hbar}},\quad \lambda = \frac{2E}{\hbar^2\omega^2}$$

we get

$$\frac{\partial^2\psi}{\partial q^2} + (\lambda - q^2)\psi = 0 \tag{3}$$

This is the same equation as eq. (7.142).

Hence, the eigenfunction in momentum representation is similar to that of position representation.

$$\therefore \psi_n(q) = N_n e^{-q^2/2}H_n(q)$$

where $H_n(q)$ is the Hermite polynomial of order n and $N_n = \left(\frac{1}{2^n n!\sqrt{\pi}}\right)^{1/2}$.

The energy eigenvalues are

$$E_n = \left(n + \tfrac{1}{2}\right)\hbar\omega \tag{4}$$

The normalised eigenfunctions $\psi(q)$ for different n are the same as in case of coordinate representation.

Physical Significance of $\psi(p)$

The physical significance of the wave function $\psi(p)$ is that $|\psi(p)|^2\,dp$ represents the probability of finding the particle in the momentum range p and $p + dp$. Thus, $|\psi(p)|^2$ is the probability density in the momentum space.

The expectation value of a dynamical variable α in the momentum representation is given by

$$\langle\alpha\rangle = \frac{\int_{-\infty}^{+\infty} \psi^*(p)\hat{\alpha}\psi(p)\,dp}{\int_{-\infty}^{+\infty} \psi^*(p)\,\psi(p)\,dp} \tag{5}$$

The operator $\hat{\alpha}$ obeys the following relation:

$$\hat{\alpha}(x,p) = \alpha(\hat{x},\hat{p}) = \alpha\left(i\hbar\frac{\partial}{\partial p}, p\right) \tag{6}$$

It can also be proved that if a wave function $\psi(x)$ in coordinate space is normalised, then the corresponding wave function $\psi(p)$ in momentum space is also normalised, i.e., if

$$\int_{-\infty}^{+\infty} \psi^*(x)\,\psi(x)\,dx = 1$$

then

$$\int_{-\infty}^{+\infty} \psi^*(p)\,\psi(p)\,dp = 1 \text{ also}$$

7.11. (a) What is the classical Hamiltonian for a linear harmonic oscillator?

(b) Write down Schrödinger equation for this Hamiltonian. What are the eigenvalues and the eigenfunction of the Hamiltonian? What conditions lead to discrete energy levels? Are all the eigenfunctions orthogonal?

(c) Explain the significance of zero-point energy. How does it reconcile with the classical view point?

7.12. Consider an otherwise free particle of mass m to be confined in a one-dimensional box of length L and infinitely hard walls at $x = 0$ and $x = L$.

(a) Find out the wave function of the particle inside the box and show that the energy of the particle inside the box can have only discrete values.

Also show that the energy of the particle cannot be zero.

(b) Normalise the wave function.

(c) Sketch the wave function corresponding to the lowest energy levels.

(d) Show that the energy eigenfunction you have obtained above is not an eigenfunction of the momentum operator, but the eigenfunction can be written as a linear combination of two momentum eigenfunctions.

(e) Find the lowest energy of a neutron when it is confined in a box as described here with $L = 10^{-4}$ m. **[BU (H) 1995]**

7.13. (a) Give the physical interpretation of the wave function $\psi(\vec{x})$.

(b) Obtain the expression for stationary energy levels for a particle of mass m, which is free to move in a region of zero potential between two rigid walls of $x = 0$ and $x = L$. Are the energy levels degenerate?

(c) Find the corresponding wave function pertaining to (b) and normalise. **[WBUT 2005]**

7.14. (a) For a one-dimensional SHO obtained the ground state wave function. **[KU (H) 2011]**

(b) Make a plot of first and second energy eigenfunctions for a LHO.

7.15. (a) Set up the Schrödinger equation for one-dimensional harmonic oscillator and obtain the general wave function. **[KU (H) 2013]**

(b) Write down the wave function of a SHO in *n*th eigenstate. Normalise it. **[cf. KU (H) 2011]**

7(D): Numerical Problems

7.1. Show that

$$\langle x^2 \rangle = \frac{a^2}{3} - \frac{a^2}{2\pi^2 n^2}$$

for the particle in an infinite square well with limits $(0, a)$.

7.2. A particle of mass m is confined by an infinite square well potential of the form:

$$V(x) = 0, \ |x| \leqslant a$$

$$V(x) \to \infty, \ |x| \geqslant a$$

If the particle is in the state

$$\psi(x) = x, \ |x| \leqslant a$$

$$\psi(x) = 0, \ |x| > a$$

what is the relative probability that a measurement of energy will give the result E_2 or E_4.

7.3. Find the probability that a particle in a one-dimensional box of length l can be found between $0.4l$ and $0.6l$.

(a) for the ground state (b) for the first excited state (c) for the second excited state. **[PU 1993]**

7.4. A particle is moving in a one-dimensional potential box of infinite height and width 10 Å. Calculate the probability of finding the particle within an interval of 4 Å at the centre of the box when it is in the state of least energy. **[PU 1996]**

7.5. Show that the lowest ground state energy of a particle in one-dimensional potential box with rigid walls is in agreement with the uncertainty principle.

7.6. Find the expectation value of the energy when the state of the harmonic oscillator is described by the following wave function $\psi(x, t) = \frac{1}{\sqrt{2}}[\psi_0(x, t) + \psi_1(x, t)]$, where $\psi_0(x, t)$ and $\psi_1(x, t)$ are the wave functions for the ground state and the first excited state respectively.

7.7. A particle in an infinitely deep potential well is in a state described by the wave function $\psi(x) = Ax(a - x)$, where a is the well depth and A is constant. Find the probability distribution for the different energies of the particle and also the average value and the dispersion of the energy.

[*Hints*: Probability $P_n = \frac{240}{(\pi n)^2}[1 - (-1)^n]^2$. Average value of energy$= \frac{\pi^2 \hbar^2}{2ma^2} \cdot \frac{10}{\pi^2}$. Dispersion of energy, $\Delta E = \sqrt{5}\frac{\hbar^2}{ma^2}$.]

7.8. Calculate the probability of finding a simple harmonic oscillator within the classical limits if the oscillator is in its normal state.

7.9. A linear harmonic oscillator is defined by the point mass m and a resisting force $-kq$, where k is a positive constant and q is the coordinate of the point. Show that the wave function $\psi(x) =$

$e^{-\frac{x^2}{2}}(4x^2+2)$ is an eigenfunction of the Schrödinger equation of this oscillator where x is defined by the substitution, $q^2 = \frac{x^2\hbar}{\sqrt{km}}$. Find the value of energy belonging to the given eigenfunction.

7.10. Calculate the zero-point energy of a system consisting of a mass of 1 gm connected to a fixed point by a spring which is stretched 1 cm by a force of 10, 000 dynes the particle being constraint to move only along x-axis. [*Ans.* 5.25×10^{20} ergs]

7.11. Using the Schrödinger equation find the energy of a harmonic oscillator of angular frequency ω in the stationary state $\psi(x) = \beta x e^{-\alpha^2 x^2/2}$, where β and α are constants. [BU (H) 1999]

7.12. Calculate the expectation values $\langle x^2\rangle$, $\langle p_x\rangle$ and $\langle p_x^2\rangle$ for the one-dimensional harmonic oscillator in the nth excited state. [*Ans.* $\frac{2n+1}{2\alpha}, 0, (2n+1)\frac{\hbar^2\alpha}{2}$]

7.13. Given that for a particle in a rigid box (spacing from $x=0$ to $x=a$), the eigenfunctions are given by $\psi_n = \sqrt{\frac{2}{a}}\sin\frac{n\pi x}{a}$. What are the energy eigenvalues in these states? [WBUT 2007]

7.14. A particle is confined in a one-dimensional potential well-defined by

$$V(x) = \begin{cases} 0, & \text{for } -a < x < a \\ \infty, & \text{elsewhere} \end{cases}$$

Obtain energy eigenvalues and the normalised eigenfunctions of the system. [CU (H) 2009]

7.15. A particle trapped in a one-dimensional infinite square well within limits $0 \leqslant x \leqslant L$ is described by the wave function

$$\psi(x,0) = A\sin\frac{5\pi x}{L}\cos\frac{2\pi x}{L}$$

Normalise the wave function. [KU (H) 2015]

7.16. For a particle trapped in an infinite square well of width L, sketch the ground state and the first excited state wave functions. End points of the well are at $x = 0$ and $x = L$. [KU (H) 2015]

7.17. A particle of mass m moves in a potential well given by

$$V(x) = \begin{cases} \infty, & \text{for } x < 0 \\ 0, & \text{for } 0 < x < L \\ V_0, & \text{for } x > L \end{cases}$$

Assuming $E < V_0$ find the wave functions in the three regions. Also find the transcendental equation for energy eigenvalue (applying proper boundary conditions). What are the three molecular energy states? Explain their origins? [KU (H) 2013]

7.18. For an one-dimensional rectangular potential well-defined by

$$V(x) = \begin{cases} 0, & \text{for } x < 0 \\ -V_0, & \text{for } 0 < x < a \\ 0, & \text{for } a < x \end{cases}$$

Obtain the corresponding wave functions in such mentioned potential region. [KU (H) 2011]

7.19. A simple harmonic oscillator moves in a potential $V(x) = \frac{1}{2}kx^2 + cx$, where c is a constant. Find energy eigenvalues. [*Ans.* $E_n = (n+\frac{1}{2})\hbar\omega, n = 0, 1, 2, 3, \ldots$]

7.20. Cast Schrödinger equation for LHO in the form

$$\frac{d^2}{dx^2}\psi + (\lambda - x^2)\psi = 0$$

Assuming solution of the form $\psi = H(z)e^{-x^2/2}$ convert it into a differential equation of H.

Now take

$$H = \sum_{n=0}^{\infty} a_n z^n$$

as its solution.

Find condition of termination of the series and hence calculate minimum energy of the oscillation.

[KU (H) 2012]

Chapter 8

Applications of Schrödinger's Equation to Three-dimensional Problems

8.1 Introduction

In chapters 6 and 7, we have applied Schrödinger one-dimensional time-independent equation to solve one-dimensional problems relating to free and bound states respectively. In the present chapter we shall apply Schrödinger three-dimensional time-independent equation to solve the three-dimensional problems like particle in a three-dimensional box, rigid rotor, three-dimensional harmonic oscillator and hydrogen atom with a special emphasis to harmonic oscillator and hydrogen atom problem.

It is to be kept in mind that for most of the problems of physical interest, exact solutions cannot be obtained by analytical method.

However, various approximation methods have been developed in quantum mechanics with a view to obtain satisfactory result.

8.2 The free particle in three dimensions

We consider a particle of mass m moving in three dimensions in a field free space of constant potential energy $V(\vec{r})$. For convenience, we assume that $V(\vec{r}) = 0$ and the motion is non-relativistic.

The Schrödinger time-independent wave equation for such a particle in Cartesian coordinates is given by

$$-\frac{\hbar^2}{2m}\nabla^2 \psi(r) = E\psi(r) \text{ or, } \frac{\partial^2 \psi(r)}{\partial x^2} + \frac{\partial^2 \psi(r)}{\partial y^2} + \frac{\partial^2 \psi(r)}{\partial z^2} + \frac{2m}{\hbar^2}\psi(r) = 0 \tag{8.1}$$

Equation (8.1) is a partial differential equation in three independent variables, viz., x, y and z. The equation may be solved by the methods of separation of variables. So the solution of eq. (8.1) may be expressed as

$$\psi(x, y, z) = X(x)Y(y)Z(z) \tag{8.2}$$

$X(x)$, $Y(y)$ and $Z(z)$ are functions of x, y and z respectively.

Using eq. (8.2) in eq. (8.1), we get

$$YZ\frac{d^2X}{dx^2} + XZ\frac{d^2Y}{dy^2} + XY\frac{d^2Z}{dz^2} + \frac{2m}{\hbar^2}EXYZ = 0$$

Dividing throughout by XYZ, we get

$$\frac{1}{X}\frac{d^2X}{dx^2} + \frac{1}{Y}\frac{d^2Y}{dy^2} + \frac{1}{Z}\frac{d^2Z}{dz^2} + \frac{2m}{\hbar^2}E = 0 \tag{8.3}$$

In eq. (8.3), the first, second and third terms are respectively function of x, y and z alone and their sum is equal to $-\frac{2mE}{\hbar^2}$ for any choice of x, y and z. This is possible only if each term is equal to a constant as follows:

$$\frac{1}{X}\frac{d^2X}{dx^2} = k_x \tag{8.4}$$

$$\frac{1}{Y}\frac{d^2Y}{dy^2} = k_y \tag{8.5}$$

$$\frac{1}{Z}\frac{d^2Z}{dz^2} = k_z \tag{8.6}$$

with the condition that

$$k_x + k_y + k_z = -\frac{2mE}{\hbar^2} \tag{8.7}$$

Now, the kinetic energy E of the particle can be expressed as

$$E = E_x + E_y + E_z \tag{8.8}$$

$$\therefore k_x + k_y + k_z = -\frac{2m}{\hbar^2}(E_x + E_y + E_z) \tag{8.9}$$

or

$$k_x = -\frac{2mE_x}{\hbar^2} \tag{8.10}$$

$$k_y = -\frac{2mE_y}{\hbar^2} \tag{8.11}$$

$$\text{and } k_z = -\frac{2mE_z}{\hbar^2} \tag{8.12}$$

Now, using eqs (8.10), (8.11), (8.12) in eqs (8.4), (8.5), (8.6) respectively, we get

$$\frac{d^2X}{dx^2} + \frac{2mE_x}{\hbar^2}X = 0 \tag{8.13}$$

$$\frac{d^2Y}{dy^2} + \frac{2mE_y}{\hbar^2}Y = 0 \tag{8.14}$$

$$\frac{d^2Z}{dz^2} + \frac{2mE_z}{\hbar^2}Z = 0 \tag{8.15}$$

The general solutions of eqs (8.13), (8.14) and (8.15) are

$$X(x) = N_x \sin \frac{\sqrt{2mE_x}}{\hbar}(x - x_0) \tag{8.16}$$

$$Y(y) = N_y \sin \frac{\sqrt{2mE_y}}{\hbar}(y - y_0) \tag{8.17}$$

$$Z(z) = N_z \sin \frac{\sqrt{2mE_z}}{\hbar}(z - z_0) \tag{8.18}$$

where N_x, N_y, N_z and x_0, y_0, z_0 are constants.

$$\therefore \psi(x, y, z) = X(x)Y(y)Z(z)$$

$$= N_x N_y N_z \sin \frac{\sqrt{2mE_x}}{\hbar}(x - x_0) \times \sin \frac{\sqrt{2mE_y}}{\hbar}(y - y_0) \sin \frac{\sqrt{2mE_z}}{\hbar}(z - z_0)$$

$$\text{or, } \psi(x, y, z) = N \sin \frac{\sqrt{2mE_x}}{\hbar}(x - x_0) \sin \frac{\sqrt{2mE_y}}{\hbar}(y - y_0) \sin \frac{\sqrt{2mE_z}}{\hbar}(z - z_0) \tag{8.19}$$

where $N = N_x N_y N_z$ is the *normalisation constant* of the wave function $\psi(x, y, z)$. As the sine function is continuous, single-valued and finite for all values of its arguments, the only restriction is that E_x, E_y and E_z and hence E must have positive values.

If we put $E_y = E_z = 0$ and $E = E_x$, then

$$\psi(x, y, z, t) = N \sin \frac{\sqrt{2mE}}{\hbar}(x - x_0)\, e^{-\frac{iEt}{\hbar}} \tag{8.20}$$

Equation (8.20) represents a set of standing waves with the wavefronts normal to the x-axis. The wavelength of the standing wave is given by

$$\lambda = \frac{h}{\sqrt{2mE}} \tag{8.21}$$

Also the kinetic energy of a particle moving with velocity v in a field free space is

$$E = \tfrac{1}{2}mv^2 \tag{8.22}$$

From eqs (8.21) and (8.22)

$$\lambda = \frac{h}{\sqrt{2m \times \frac{1}{2}mv^2}} = \frac{h}{mv} \tag{8.23}$$

Here, λ is the de Broglie wavelength of a particle having momentum $p = mv$.

8.3 Particle in a three-dimensional rectangular box (the infinite square well in three dimensions)

We consider a free particle of mass m confined to move non-relativistically in a three-dimensional potential box of sides a, b and c parallel to the x-, y- and z-axes respectively [Fig 8.1].

Since, the particle is free, the potential inside the box is $V(x, y, z) = 0$. So the total energy of the particle inside the box is equal to its kinetic energy which is always a positive quantity.

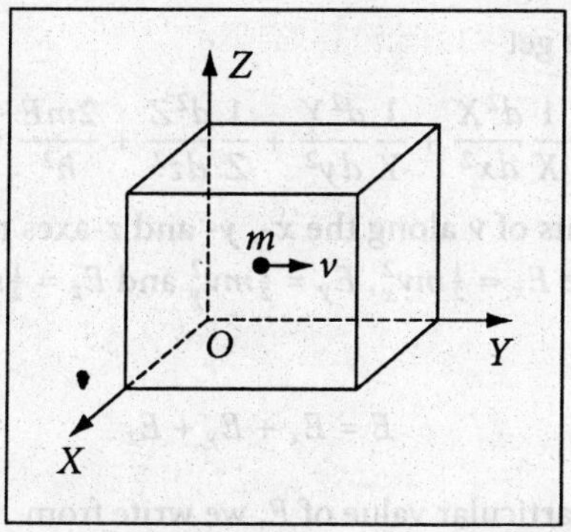

Fig 8.1 Three-dimensional box

If v be the velocity and $p = mv$ the momentum of the particle, then its energy

$$E = \tfrac{1}{2}mv^2 = \frac{p^2}{2m}$$

Since, the potential energy $V(x, y, z)$ is independent of time we here apply Schrödinger time-independent three-dimensional wave equation to obtain the wave function $\psi(x, y, z)$ and the eigenvalue of energy of the particle.

The three-dimensional potential box has the following characteristics:

$$V(x, y, z) = \begin{cases} 0, & \text{for } 0 < x < a,\ 0 < y < b,\ 0 < z < c \\ \infty, & \text{for } \quad x > a, \quad y > b, \quad z > c \end{cases} \tag{8.24}$$

8.3.1 The wave equation for the particle

The wave equation describing the motion of the particle can be written as

$$\hat{H}\psi = -\frac{\hbar^2}{2m}\nabla^2\psi = E\psi \text{ or, } \frac{\partial^2\psi}{\partial x^2} + \frac{\partial^2\psi}{\partial y^2} + \frac{\partial^2\psi}{\partial z^2} + \frac{2mE}{\hbar^2}\psi = 0 \tag{8.25}$$

where $\psi = \psi(x, y, z)$ is the wave function of the particle.

As $V(x, y, z) = \infty$ at the boundaries and outside the box, $\psi(x, y, z) = 0$ at the boundaries and outside the potential box. The boundary conditions can therefore, be written as

$$\left.\begin{aligned} \psi(x, y, z) = 0 \text{ at } x = 0 \text{ and } x &= a \\ y = 0 \text{ and } y &= b \\ z = 0 \text{ and } z &= c \end{aligned}\right\} \tag{8.26}$$

Equation (8.25) can be solved by using the method of separation of variables. We therefore, write:

$$\psi(x, y, z) = X(x)Y(y)Z(z) \tag{8.27}$$

where $X(x)$, $Y(y)$ and $Z(z)$ are functions of x, y and z only.

Substituting eq. (8.27) in eq. (8.25), we get

$$YZ\frac{d^2X}{dx^2} + ZX\frac{d^2Y}{dy^2} + XY\frac{d^2Z}{dz^2} + \frac{2mE}{\hbar^2}XYZ = 0 \tag{8.28}$$

Dividing throughout by XYZ, we get

$$\frac{1}{X}\frac{d^2X}{dx^2}+\frac{1}{Y}\frac{d^2Y}{dy^2}+\frac{1}{Z}\frac{d^2Z}{dz^2}+\frac{2mE}{\hbar^2}=0 \tag{8.29}$$

If v_x, v_y and v_z be the components of v along the x-, y- and z-axes respectively, then the corresponding kinetic energies of the particle are $E_x = \frac{1}{2}mv_x^2$, $E_y = \frac{1}{2}mv_y^2$ and $E_z = \frac{1}{2}mv_z^2$.

So that

$$E = E_x + E_y + E_z \tag{8.30}$$

Since, $\frac{2mE}{\hbar^2}$ is a constant for a particular value of E, we write from

$$\left[\frac{1}{X}\frac{d^2X}{dx^2}+\frac{2mE_x}{\hbar^2}\right]+\left[\frac{1}{Y}\frac{d^2Y}{dy^2}+\frac{2mE_y}{\hbar^2}\right]+\left[\frac{1}{Z}\frac{d^2Z}{dz^2}+\frac{2mE_z}{\hbar^2}\right]=0$$

$$\frac{1}{X}\frac{d^2X}{dx^2}+\frac{2mE_x}{\hbar^2}X=0 \tag{8.31}$$

$$\frac{1}{Y}\frac{d^2Y}{dy^2}+\frac{2mE_y}{\hbar^2}Y=0 \tag{8.32}$$

$$\frac{1}{Z}\frac{d^2Z}{dz^2}+\frac{2mE_z}{\hbar^2}Z=0 \tag{8.33}$$

The solutions of eqs (8.31), (8.32) and (8.33) are

$$X(x) = A_1 \sin k_x x + B_1 \cos k_x x,\ k_x = \frac{\sqrt{2mE_x}}{\hbar} \tag{8.34}$$

$$Y(y) = A_2 \sin k_y y + B_2 \cos k_y y,\ k_y = \frac{\sqrt{2mE_y}}{\hbar} \tag{8.35}$$

$$Z(z) = A_3 \sin k_z z + B_3 \cos k_z z,\ k_z = \frac{\sqrt{2mE_z}}{\hbar} \tag{8.36}$$

where A_1, A_2, A_3 and B_1, B_2, B_3 are constants whose values may be found out by applying the boundary condition.

Now, applying the condition $X(x) = 0$, at $x = 0$ and $x = a$ to eq. (8.34), we get $B_1 = 0$.

$\therefore$ $\sin k_x a = 0$ or, $k_x a = n_x \pi$ where $n_x = 1, 2, 3, \ldots$.

Similarly, applying the condition $Y(y) = 0$ at $y = 0$ and $y = b$ in eq. (8.35), we get $B_2 = 0$.

$\therefore$ $\sin k_y b = 0$ or, $k_y b = n_y \pi$ where $n_y = 1, 2, 3, \ldots$.

Substituting $Z(z) = 0$ for $z = 0$ and $z = c$ in eq. (8.36), we get $B_3 = 0$ and $k_z c = n_z \pi$, where $n_z = 1, 2, 3, \ldots$

So from eqs (8.34), (8.35) and (8.36), we get

$$X(x) = A_1 \sin \frac{n_x \pi}{a} x \tag{8.37}$$

$$Y(y) = A_2 \sin \frac{n_y \pi}{b} y \tag{8.38}$$

$$Z(z) = A_3 \sin \frac{n_z \pi}{c} z \tag{8.39}$$

Using eqs (8.37), (8.38) and (8.39) in eq. (8.27), we get

$$\psi(x, y, z) = A_1 A_2 A_3 \sin \frac{n_x \pi}{a} x \sin \frac{n_y \pi}{b} y \sin \frac{n_z \pi}{c} z$$

$$\text{or, } \psi(x, y, z) = A \sin \frac{n_x \pi}{a} x \sin \frac{n_y \pi}{b} y \sin \frac{n_z \pi}{c} z \tag{8.40}$$

where $A = A_1 A_2 A_3$ and n_x, n_y, n_z are all independent integers.

To find out the value of the normalisation constant A we proceed as follows:

The probability of finding the particle inside the box is equal to one, since the particle must be found out somewhere inside the box, there being no chance of being present anywhere outside the box. So

$$\int_0^a \int_0^b \int_0^c |\psi|^2 \, dx\, dy\, dz = 1$$

$$\text{or, } A^2 \int_0^a \int_0^b \int_0^c \sin^2 \frac{n_x \pi}{a} x \sin^2 \frac{n_y \pi}{b} y \sin^2 \frac{n_z \pi}{c} z \, dx\, dy\, dz = 1$$

$$\text{or, } A^2 \frac{a}{2} \frac{b}{2} \frac{c}{2} = 1 \text{ or, } A = \sqrt{\frac{8}{abc}} = \sqrt{\frac{8}{V}} \tag{8.41}$$

where $V = abc$ the volume of the box

$$\left[\because \int_0^a \sin^2 \frac{n_x \pi}{a} x \, dx = \frac{a}{2}, \int_0^b \sin^2 \frac{n_y \pi}{b} y \, dy = \frac{b}{2}, \int_0^c \sin^2 \frac{n_z \pi}{c} z \, dz = \frac{c}{2} \right]$$

Now, the eigenfunction of the particle is

$$\psi(x, y, z) = \sqrt{\frac{8}{abc}} \sin \frac{n_x \pi}{a} x \sin \frac{n_y \pi}{b} y \sin \frac{n_z \pi}{c} z \tag{8.42}$$

The x-, y-, z-components of linear momentum of the particle are

$$p_x = \hbar k_x = \frac{\hbar n_x \pi}{a} = \frac{n_x \hbar \pi}{a}, \; p_y = \frac{n_y \hbar \pi}{b} \text{ and } p_z = \frac{n_z \hbar \pi}{c}$$

∴ the total linear momentum p is

$$p = \sqrt{p_x^2 + p_y^2 + p_z^2} = \left[\left(\frac{n_x \hbar \pi}{a} \right)^2 + \left(\frac{n_y \hbar \pi}{b} \right)^2 + \left(\frac{n_z \hbar \pi}{c} \right)^2 \right]^{1/2} = \pi \hbar \left[\left(\frac{n_x}{a} \right)^2 + \left(\frac{n_y}{b} \right)^2 + \left(\frac{n_z}{c} \right)^2 \right]^{1/2} \tag{8.43}$$

The eigenvalue of energy of the particle is

$$E = \frac{p^2}{2m} = \frac{\pi^2 \hbar^2}{2m} \left[\left(\frac{n_x}{a} \right)^2 + \left(\frac{n_y}{b} \right)^2 + \left(\frac{n_z}{c} \right)^2 \right] \tag{8.44}$$

The integers n_x, n_y, n_z are called *quantum numbers* which are required to describe the *stationary state* of the particle. All the stationary states are given by positive values of n_x, n_y and n_z, because if we change the sign of n_x, n_y and n_z, E and ψ remain unchanged except a negative sign on the right side of eq. (8.42). This confirms that all the stationary states are given by positive integral values of n_x, n_y, n_z.

Since, for any of n_x, n_y, n_z being zero, $\psi(x, y, z) = 0$, i.e., the particle is absent in the box. But, physically this is not acceptable and hence any of n_x, n_y, n_z cannot be zero.

For the *ground state*, $n_x = n_y = n_z = 1$ and the ground state wave function of the particle is

$$\psi_{1,1,1}(x, y, z) = \sqrt{\frac{8}{abc}} \sin\frac{\pi x}{a} \sin\frac{\pi y}{b} \sin\frac{\pi z}{c} \tag{8.45}$$

and the energy associated with the ground state called the *zero-point energy* is given by

$$E_{1,1,1} = \frac{\pi^2\hbar^2}{2m}\left[\frac{1}{a^2} + \frac{1}{b^2} + \frac{1}{c^2}\right] \tag{8.46}$$

The eigenfunctions and eigenvalues of energy for the first excited state are as follows:

$$\left.\begin{aligned} \psi_{2,1,1} &= \sqrt{\frac{8}{abc}} \sin\frac{2\pi x}{a} \sin\frac{\pi y}{b} \sin\frac{\pi z}{c} \\ \psi_{1,2,1} &= \sqrt{\frac{8}{abc}} \sin\frac{\pi x}{a} \sin\frac{2\pi y}{b} \sin\frac{\pi z}{c} \\ \psi_{1,1,2} &= \sqrt{\frac{8}{abc}} \sin\frac{\pi x}{a} \sin\frac{\pi y}{b} \sin\frac{2\pi z}{c} \end{aligned}\right\} \tag{8.47}$$

and

$$\left.\begin{aligned} E_{2,1,1} &= \frac{\pi^2\hbar^2}{2m}\left[\frac{4}{a^2} + \frac{1}{b^2} + \frac{1}{c^2}\right] \\ E_{1,2,1} &= \frac{\pi^2\hbar^2}{2m}\left[\frac{1}{a^2} + \frac{4}{b^2} + \frac{1}{c^2}\right] \\ E_{1,1,2} &= \frac{\pi^2\hbar^2}{2m}\left[\frac{1}{a^2} + \frac{1}{b^2} + \frac{4}{c^2}\right] \end{aligned}\right\} \tag{8.48}$$

8.3.2 Degeneracy

The degeneracy is the property that two or more quantum states of a particle with different sets of quantum numbers and different eigenfunctions having the same value of momentum and energy.

If the potential box be cubical instead of rectangular, then $a = b = c$ and the eigenfunction corresponding to the state represented by n_x, n_y, n_z is

$$\psi_{n_x n_y n_z}(x, y, z) = \psi(n_x n_y n_z) = \sqrt{\frac{8}{a^3}} \sin\frac{n_x \pi}{a} x \sin\frac{n_y \pi}{a} y \sin\frac{n_z \pi}{a} z \tag{8.49}$$

The momentum eigenvalues are

$$p(n_x, n_y, n_z) = \frac{\pi\hbar}{a}\left[n_x^2 + n_y^2 + n_z^2\right]^{1/2} \tag{8.50}$$

and the corresponding energy eigenvalues are

$$E(n_x, n_y, n_z) = \frac{\pi^2\hbar^2}{2ma^2}\left[n_x^2 + n_y^2 + n_z^2\right]^{1/2} \tag{8.51}$$

The ground state wave function for $n_x = n_y = n_z = 1$ is given by

$$\psi_{1,1,1} = \sqrt{\frac{8}{a^3}} \sin\frac{\pi x}{a} \sin\frac{\pi y}{a} \sin\frac{\pi z}{a} \tag{8.52}$$

The eigenvalues of momentum and energy of the ground state are as follows:

$$p_{1,1,1} = \frac{\pi\hbar}{a}\sqrt{3} \tag{8.53}$$

$$E_{1,1,1} = \frac{\pi^2\hbar^2}{2ma^2}3 \tag{8.54}$$

For the *first excited state* there are three sets of quantum numbers as shown:

n_x	n_y	n_z
2	1	1
1	2	1
1	1	2

The eigenfunctions for the first excited state are

$$\left.\begin{aligned} \psi_{2,1,1} &= \sqrt{\frac{8}{a^3}}\sin\frac{2\pi x}{a}\sin\frac{\pi y}{a}\sin\frac{\pi z}{a} \\ \psi_{1,2,1} &= \sqrt{\frac{8}{a^3}}\sin\frac{\pi x}{a}\sin\frac{2\pi y}{a}\sin\frac{\pi z}{c} \\ \psi_{1,1,2} &= \sqrt{\frac{8}{a^3}}\sin\frac{\pi x}{a}\sin\frac{\pi y}{a}\sin\frac{2\pi z}{a} \end{aligned}\right\} \tag{8.55}$$

The eigenvalues of momentum and energy for the first excited state are given by

$$p_{2,1,1} = \frac{\pi\hbar}{a}\left[2^2+1^2+1^2\right]^{1/2} = \frac{\pi\hbar}{a}\sqrt{6} \tag{8.56}$$

$$p_{1,2,1} = \frac{\pi\hbar}{a}\left[1^2+2^2+1^2\right]^{1/2} = \frac{\pi\hbar}{a}\sqrt{6} \tag{8.57}$$

$$p_{1,1,2} = \frac{\pi\hbar}{a}\left[1^2+1^2+2^2\right]^{1/2} = \frac{\pi\hbar}{a}\sqrt{6} \tag{8.58}$$

$$\therefore p_{2,1,1} = p_{1,2,1} = p_{1,1,2} = \frac{\pi\hbar}{a}\sqrt{6} \tag{8.59}$$

Similarly,

$$E_{2,1,1} = E_{1,2,1} = E_{1,1,2} = \frac{\pi^2\hbar^2}{2ma^2}6 \tag{8.60}$$

For *second excited state*, the eigenfunctions are

$$\left.\begin{aligned} \psi_{2,2,1} &= \sqrt{\frac{8}{a^3}}\sin\frac{2\pi x}{a}\sin\frac{2\pi y}{a}\sin\frac{\pi z}{c} \\ \psi_{1,2,2} &= \sqrt{\frac{8}{a^3}}\sin\frac{\pi x}{a}\sin\frac{2\pi y}{a}\sin\frac{2\pi z}{c} \\ \psi_{2,1,2} &= \sqrt{\frac{8}{a^3}}\sin\frac{2\pi x}{a}\sin\frac{\pi y}{a}\sin\frac{2\pi z}{c} \end{aligned}\right\} \tag{8.61}$$

The corresponding eigenvalues of momentum and energy are

$$p_{2,2,1} = p_{1,2,2} = p_{2,1,2} = \frac{\pi\hbar}{a}\left[2^2 + 2^2 + 1^2\right]^{1/2} = \frac{\pi\hbar}{a}3 \tag{8.62}$$

$$\text{and } E_{2,2,1} = E_{1,2,2} = E_{2,1,2} = \frac{\pi^2\hbar^2}{2ma^3}\left[2^2 + 2^2 + 1^2\right] = \frac{\pi^2\hbar^2}{2ma^2}9 \tag{8.63}$$

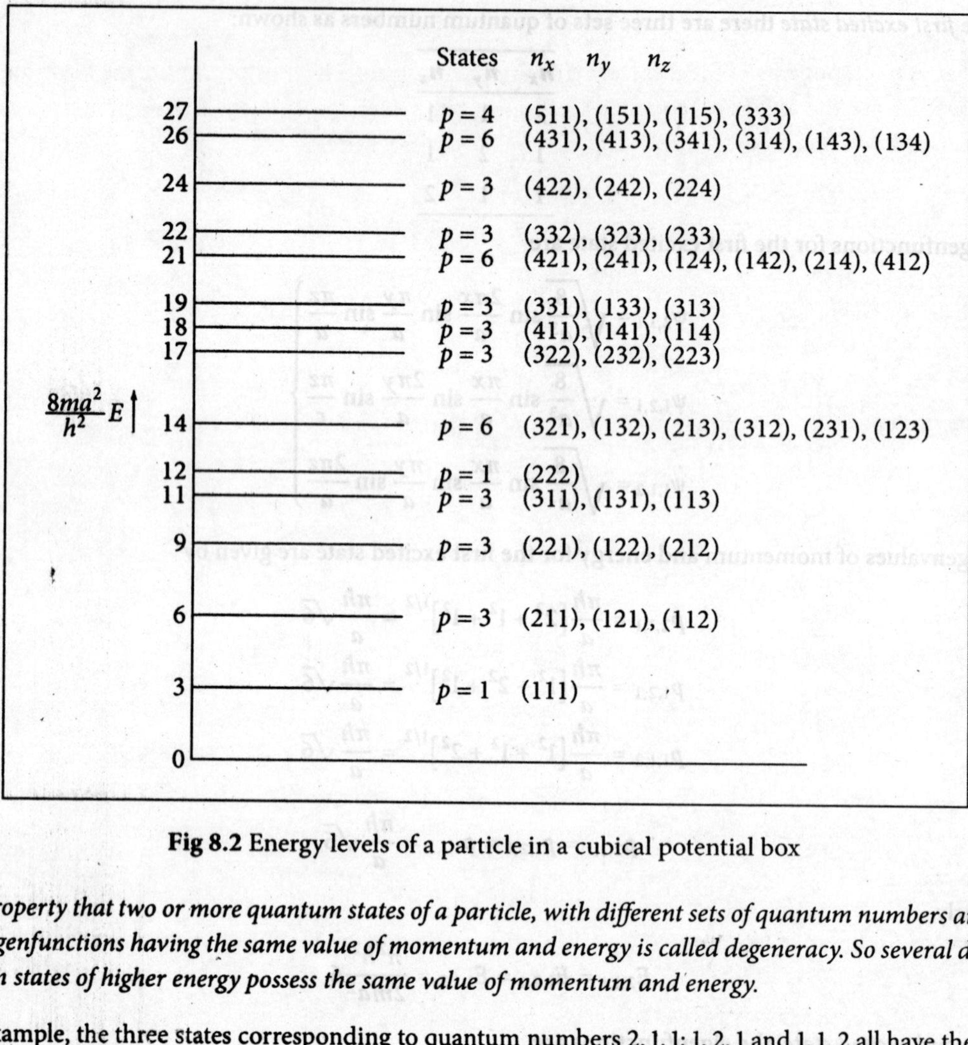

Fig 8.2 Energy levels of a particle in a cubical potential box

The property that two or more quantum states of a particle, with different sets of quantum numbers and different eigenfunctions having the same value of momentum and energy is called degeneracy. So several distinct quantum states of higher energy possess the same value of momentum and energy.

For example, the three states corresponding to quantum numbers 2, 1, 1; 1, 2, 1 and 1, 1, 2 all have the same momentum $\left(\frac{\pi\hbar}{a}\sqrt{6}\right)$ and energy $\left(\frac{\pi^2\hbar^2}{2ma^2}6\right)$. It is one way in which a multidimensional system differs from a one-dimensional system in quantum mechanics.

Fig 8.2 shows the energy levels, degree of degeneracy (p) and the quantum numbers for a particle confined in a cubical potential box of side a.

8.4 Three-dimensional harmonic oscillator (Cartesian coordinates)

We consider a particle of mass m bound to the origin by a force having components

$$F_x = -k_x x = -m\omega_x^2 x \tag{8.64}$$

$$F_y = -k_y y = -m\omega_y^2 y \tag{8.65}$$

$$F_z = -k_z z = -m\omega_z^2 z \tag{8.66}$$

along the x, y, z components of the displacement $\vec{r}$ from the origin. Here, k_x, k_y, k_z are the force constants and $\omega_x, \omega_y, \omega_z$ are the angular frequencies along x-, y-, z-axes respectively. Such a system is called a *three-dimensional harmonic oscillator.*

Again,

$$F_x = -\frac{\partial V}{\partial x},\ F_y = -\frac{\partial V}{\partial y} \text{ and } F_z = -\frac{\partial V}{\partial z}$$

where the potential energy V of the oscillator is

$$V(x,y,z) = \tfrac{1}{2}\left(k_x x^2 + k_y y^2 + k_z z^2\right) \tag{8.67}$$

The time-independent three-dimensional Schrödinger equation for the system is

$$-\frac{\hbar^2}{2m}\nabla^2\psi(x,y,z) + \tfrac{1}{2}\left(k_x x^2 + k_y y^2 + k_z z^2\right)\psi(x,y,z) = E\psi(x,y,z)$$

$$\text{or, } \frac{\partial^2\psi}{\partial x^2} + \frac{\partial^2\psi}{\partial y^2} + \frac{\partial^2\psi}{\partial z^2} + \frac{2m}{\hbar^2}\left[E - \tfrac{1}{2}k_x x^2 - \tfrac{1}{2}k_y y^2 - \tfrac{1}{2}k_z z^2\right]\psi = 0 \tag{8.68}$$

$$\text{or, } \frac{\partial^2\psi}{\partial x^2} + \frac{\partial^2\psi}{\partial y^2} + \frac{\partial^2\psi}{\partial z^2} + \left[\lambda - \left(\alpha_x^2 x^2 + \alpha_y^2 y^2 + \alpha_z^2 z^2\right)\right]\psi = 0 \tag{8.69}$$

where

$$\lambda = \frac{2mE}{\hbar^2},\ \alpha_x = \sqrt{\frac{mk_x}{\hbar^2}},\ \alpha_y = \sqrt{\frac{mk_y}{\hbar^2}} \text{ and } \alpha_z = \sqrt{\frac{mk_y}{\hbar^2}}$$

Equation (8.69) can be solved by the method of separation of variables. Let us try with the solution

$$\psi(x,y,z) = X(x)Y(y)Z(z) \tag{8.70}$$

Using eq. (8.70) in eq. (8.69), we get

$$YZ\frac{d^2X}{dx^2} + XZ\frac{d^2Y}{dy^2} + XY\frac{d^2Z}{dz^2} + \left[\lambda - \left(\alpha_x^2 x^2 + \alpha_y^2 y^2 + \alpha_z^2 z^2\right)\right]XYZ = 0$$

Dividing by XYZ, we get

$$\frac{1}{X}\frac{d^2X}{dx^2} + \frac{1}{Y}\frac{d^2Y}{dy^2} + \frac{1}{Z}\frac{d^2Z}{dz^2} + \left[\lambda - \left(\alpha_x^2 x^2 + \alpha_y^2 y^2 + \alpha_z^2 z^2\right)\right] = 0$$

$$\text{or, } \left[\frac{1}{X}\frac{d^2X}{dx^2} - \alpha_x^2 x^2\right] + \left[\frac{1}{Y}\frac{d^2Y}{dy^2} - \alpha_y^2 y^2\right] + \left[\frac{1}{Z}\frac{d^2Z}{dz^2} - \alpha_z^2 z^2\right] + \lambda = 0 \tag{8.71}$$

The above equation holds good only if each term on the left-hand side is equal to a zero.

So

$$\left[\frac{1}{X}\frac{d^2X}{dx^2} - \alpha_x^2 x^2\right] + \lambda_x = 0$$

$$\left[\frac{1}{Y}\frac{d^2Y}{dy^2} - \alpha_y^2 y^2\right] + \lambda_y = 0$$

$$\left[\frac{1}{Z}\frac{d^2Z}{dz^2} - \alpha_z^2 z^2\right] + \lambda_z = 0$$

where $\lambda = \lambda_x + \lambda_y + \lambda_z$

or

$$\frac{d^2X}{dx^2} + \left(\lambda_x - \alpha_x^2 x^2\right) X = 0 \quad (8.72)$$

$$\frac{d^2Y}{dy^2} + \left(\lambda_y - \alpha_y^2 y^2\right) Y = 0 \quad (8.73)$$

$$\frac{d^2Z}{dz^2} + \left(\lambda_z - \alpha_z^2 z^2\right) Z = 0 \quad (8.74)$$

Each of the above three equations is the same as that of one-dimensional harmonic oscillator. The energy eigenvalues and eigenfunctions of the three-dimensional oscillator are therefore

$$\left.\begin{aligned} E_x &= \left(n_x + \tfrac{1}{2}\right) \hbar\omega_x \\ E_y &= \left(n_y + \tfrac{1}{2}\right) \hbar\omega_y \\ E_z &= \left(n_z + \tfrac{1}{2}\right) \hbar\omega_z \end{aligned}\right\} \quad (8.75)$$

where n_x, n_y and n_z are integers, i.e., $n_x, n_y, n_z = 1, 2, 3, \ldots$ and

$$X(x) = N_{n_x}\, e^{-\frac{\alpha_x^2 x^2}{2}} H_{n_x}\left(\sqrt{\alpha_x}\, x\right) \quad (8.76)$$

$$Y(y) = N_{n_y}\, e^{-\frac{\alpha_y^2 y^2}{2}} H_{n_y}\left(\sqrt{\alpha_y}\, y\right) \quad (8.77)$$

$$Z(z) = N_{n_z}\, e^{-\frac{\alpha_z^2 z^2}{2}} H_{n_z}\left(\sqrt{\alpha_z}\, z\right) \quad (8.78)$$

Therefore, the complete wave function is

$$\psi(x, y, z) = N_{n_x} N_{n_y} N_{n_z}\, e^{-\frac{1}{2}\left(\alpha_x^2 x^2 + \alpha_y^2 y^2 + \alpha_z^2 z^2\right)} H_{n_x}\left(\sqrt{\alpha_x}\, x\right) H_{n_y}\left(\sqrt{\alpha_y}\, y\right) H_{n_y}\left(\sqrt{\alpha_z}\, z\right) \quad (8.79)$$

where $N_{n_x n_y n_z} = N_{n_x} N_{n_y} N_{n_z}$ is the normalising factor and is given by

$$N_{n_x n_y n_z} = \left[\frac{\sqrt{\alpha_x\, \alpha_y\, \alpha_z}}{\pi^{3/2} 2^{(n_x + n_y + n_z)}\, n_x!\, n_y!\, n_z!}\right]^{1/2} \quad (8.80)$$

$$E_{n_x n_y n_z} = \hbar\left[\left(n_x + \tfrac{1}{2}\right)\omega_x + \left(n_y + \tfrac{1}{2}\right)\omega_y + \left(n_z + \tfrac{1}{2}\right)\omega_z\right] \quad (8.81)$$

For the *isotropic oscillator* $\omega_x = \omega_y = \omega_z$ and $\alpha_x = \alpha_y = \alpha_z$.

$$\therefore E_n = \left[n_x + n_y + n_z + \tfrac{3}{2}\right] \hbar\omega = \left(n + \tfrac{3}{2}\right) \hbar\omega \quad (8.82)$$

where $n = n_x + n_y + n_z$ is the total quantum number and $n = 0, 1, 2, 3, \ldots$. It can be shown that except for the lowest energy levels ($n_x = n_y = n_z = 0$, i.e., $n = 0$) all the energy levels are degenerate. Table 8.1 shows a few energy level and their degeneracy.

Table 8.1: A few energy levels and their degeneracy

n_x	n_y	n_z	n	*Wave function*	E_n	*Degeneracy* (p)
0	0	0	0	$\psi_{0,0,0}$	$\frac{3}{2}\hbar\omega$	$p = 0$
1	0	0		$\psi_{1,0,0}$		
0	1	0	1	$\psi_{0,1,0}$	$\frac{5}{2}\hbar\omega$	$p = 3$
0	0	1		$\psi_{0,0,1}$		
1	1	0		$\psi_{1,1,0}$		
1	0	1	2	$\psi_{1,0,1}$		
0	1	1		$\psi_{0,1,1}$	$\frac{7}{2}\hbar\omega$	$p = 6$
2	0	0		$\psi_{2,0,0}$		
0	2	0	2	$\psi_{0,2,0}$		
0	0	2		$\psi_{0,0,2}$		

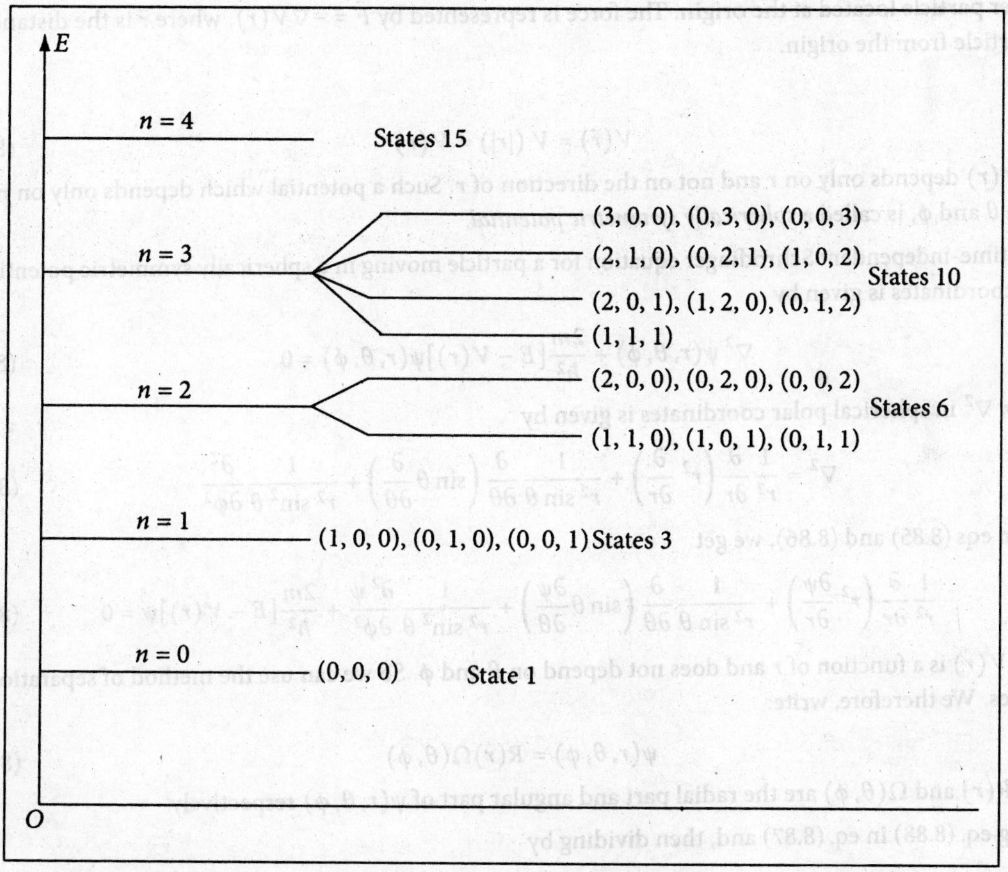

Fig 8.3 Energy levels and degeneracy for the 3-dimensional harmonic oscillator

The *zero-point* (or, *the ground level*) *energy* E_0 for $n_x = n_y = n_z = 0$ of the oscillator is given by [from eq. (8.82)]

$$E_0 = \frac{3}{2}\hbar\omega \tag{8.83}$$

Thus, the zero-point energy of a three-dimensional isotropic harmonic oscillator is three times that for the linear harmonic oscillator. The ground level of the oscillator is thus non-degenerate.

The state with principal quantum number $n = 1$ there are three levels each of energy $\frac{5}{2}\hbar\omega$. So the first excited state is 3-fold degenerate.

Similarly, the state with $n = 2$ is 6-fold degenerate. Thus, all the energy levels of the isotropic three-dimensional oscillator are degenerate and the degree of degeneracy is $p = \frac{1}{2}(n+1)(n+2)$.

Fig 8.3 shows some of the energy levels of the oscillator.

8.5 Particle in spherically symmetric potentials (Spherical harmonics)

We consider a particle of mass m moving in three dimensions under the influence of a force exerted by another particle located at the origin. The force is represented by $\vec{F} = -\nabla V(\vec{r})$, where $\vec{r}$ is the distance of the particle from the origin.

If

$$V(\vec{r}) = V(|\vec{r}|) = V(r) \tag{8.84}$$

then $V(r)$ depends only on r and not on the direction of r. Such a potential which depends only on r and not on θ and ϕ, is called a *spherically symmetric potential.*

The time-independent Schrödinger equation for a particle moving in a spherically symmetric potential in polar coordinates is given by

$$\nabla^2 \psi(r,\theta,\phi) + \frac{2m}{\hbar^2}[E - V(r)]\psi(r,\theta,\phi) = 0 \tag{8.85}$$

Now, ∇^2 in spherical polar coordinates is given by

$$\nabla^2 = \frac{1}{r^2}\frac{\partial}{\partial r}\left(r^2\frac{\partial}{\partial r}\right) + \frac{1}{r^2 \sin\theta}\frac{\partial}{\partial\theta}\left(\sin\theta\frac{\partial}{\partial\theta}\right) + \frac{1}{r^2\sin^2\theta}\frac{\partial^2}{\partial\phi^2} \tag{8.86}$$

From eqs (8.85) and (8.86), we get

$$\frac{1}{r^2}\frac{\partial}{\partial r}\left(r^2\frac{\partial\psi}{\partial r}\right) + \frac{1}{r^2\sin\theta}\frac{\partial}{\partial\theta}\left(\sin\theta\frac{\partial\psi}{\partial\theta}\right) + \frac{1}{r^2\sin^2\theta}\frac{\partial^2\psi}{\partial\phi^2} + \frac{2m}{\hbar^2}[E - V(r)]\psi = 0 \tag{8.87}$$

But, $V(r)$ is a function of r and does not depend on θ and ϕ. So we can use the method of separation of variables. We therefore, write:

$$\psi(r,\theta,\phi) = R(r)\Omega(\theta,\phi) \tag{8.88}$$

where $R(r)$ and $\Omega(\theta,\phi)$ are the radial part and angular part of $\psi(r,\theta,\phi)$ respectively.

Using eq. (8.88) in eq. (8.87) and, then dividing by

$$\frac{R(r)\Omega(\theta,\phi)}{r^2}$$

we get

$$\frac{1}{R(r)}\frac{d}{dr}\left(r^2\frac{dR}{dr}\right)+\frac{1}{\Omega(\theta,\phi)}\frac{1}{\sin\theta}\frac{\partial}{\partial\theta}\left[\sin\theta\frac{\partial\Omega(\theta,\phi)}{\partial\theta}\right]$$
$$+\frac{1}{\Omega(\theta,\phi)}\frac{1}{\sin^2\theta}\frac{\partial^2\Omega(\theta,\phi)}{\partial\phi^2}+\frac{2mr^2}{\hbar^2}[E-V(r)]=0 \tag{8.89}$$

In eq. (8.89), the second and third terms are independent of r, whereas the first and fourth terms dependent on r. So we can write:

$$\frac{1}{\Omega(\theta,\phi)}\left[\frac{1}{\sin\theta}\frac{\partial}{\partial\theta}\left(\sin\theta\frac{\partial\Omega(\theta,\phi)}{\partial\theta}\right)+\frac{1}{\sin^2\theta}\frac{\partial^2\Omega(\theta,\phi)}{\partial\phi^2}\right]=C \tag{8.90}$$

$$\text{and } -\frac{1}{R(r)}\frac{d}{dr}\left(r^2\frac{dR}{dr}\right)-\frac{2mr^2}{\hbar^2}[E-V(r)]=C \tag{8.91}$$

where C is a constant.

Equation (8.90) is independent of $V(r)$, so the solution giving the angular part of the wave function $\psi(r,\theta,\phi)$ will be the same for all spherical potentials. Further, eq. (8.90) is related to angular momentum in the following way:

$$L^2=-\hbar^2\left[\frac{1}{\sin\theta}\frac{\partial}{\partial\theta}\left(\sin\theta\frac{\partial}{\partial\theta}\right)+\frac{1}{\sin^2\theta}\frac{\partial^2}{\partial\phi^2}\right]$$

So

$$L^2\Omega(\theta,\phi)=-\hbar^2\left[\frac{1}{\sin\theta}\frac{\partial}{\partial\theta}\left(\sin\theta\frac{\partial\Omega(\theta,\phi)}{\partial\theta}\right)+\frac{1}{\sin^2\theta}\frac{\partial^2\Omega(\theta,\phi)}{\partial\phi^2}\right]=-\hbar^2C\Omega(\theta,\phi)$$

$$\text{or, } L^2\Omega(\theta,\phi)=\lambda\Omega(\theta,\phi) \tag{8.92}$$

where, we have written λ in place of $-\hbar^2C$.

Clearly, eq. (8.92) is the eigenvalue equation for the operator L^2. $\Omega(\theta,\phi)$ and λ are the *eigenfunction* and *eigenvalue* respectively belonging to the operator L^2.

We can solve eq. (8.92) by using the method of separation of variables by substituting

$$\Omega(\theta,\phi)=\Theta(\theta)\Phi(\phi) \tag{8.93}$$

Then, we get differential equation for $\Theta(\theta)$ and $\Phi(\theta)$ as below:

$$\frac{1}{\Phi(\phi)}\frac{d^2\Phi(\phi)}{d\phi^2}=-\left[\frac{1}{\Theta(\theta)}\frac{d}{d\theta}\left(\sin\theta\frac{d\Theta}{d\theta}\right)+\frac{\lambda}{\hbar^2}\sin^2\theta\right]=\alpha \tag{8.94}$$

where α is a constant. The solutions of eq. (8.92) are expressed in terms of spherical harmonics. The spherical harmonics are standard functions and have found important applications in many physical problems.

Spherical harmonics: From eq. (8.94), we get

$$\frac{1}{\Phi(\phi)}\frac{d^2\Phi(\phi)}{d\phi^2}=\alpha \tag{8.95}$$

and

$$\frac{1}{\Theta(\theta)}\sin\theta\frac{d}{d\theta}\left(\sin\theta\frac{d\Theta}{d\theta}\right)+\frac{\lambda}{\hbar^2}\sin^2\theta=-\alpha \tag{8.96}$$

8.5.1 Solution of Φ-equation

To solve eq. (8.95) we express it in the form:

$$\frac{d^2\Phi(\phi)}{d\phi^2} - \alpha\Phi(\phi) = 0 \tag{8.97}$$

The general solutions of this equation is

$$\Phi(\phi) = A\,e^{\alpha^{1/2}\phi} + B\,e^{-\alpha^{1/2}\phi} \tag{8.98}$$

where A and B are constants and $\alpha^{1/2}$ is a real quantity.

Now, since the wave function is a single-valued function everywhere

$$\Phi(\phi) = \Phi(\phi + 2\pi) \tag{8.99}$$

as the coordinates (r, ϕ, ϕ) and $(r, \phi, \phi + 2\pi)$ represent the same point in coordinate space.

$$\therefore A\,e^{\alpha^{1/2}(\theta+2\pi)} + B\,e^{-\alpha^{1/2}(\theta+2\pi)} = A\,e^{\alpha^{1/2}\phi} + B\,e^{-\alpha^{1/2}\phi} \tag{8.100}$$

Differentiating w.r.t. ϕ

$$\alpha^{1/2}\left[A\,e^{\alpha^{1/2}(\phi+2\pi)} - B\,e^{\alpha^{1/2}(\phi+2\pi)}\right] = \alpha^{1/2}\left[A\,e^{\alpha^{1/2}\phi} - B\,e^{-\alpha^{1/2}\phi}\right]$$

$$\text{or, } \left[A\,e^{\alpha^{1/2}(\phi+2\pi)} - B\,e^{-\alpha^{1/2}(\phi+2\pi)}\right] = \left[A\,e^{\alpha^{1/2}\phi} - B\,e^{-\alpha^{1/2}\phi}\right] \tag{8.101}$$

Now, adding and subtracting eqs (8.100) and (8.101), we get

$$A\,e^{\alpha^{1/2}(\phi+2\pi)} = A\,e^{\alpha^{1/2}\phi} \tag{8.102}$$

$$\text{and } B\,e^{-\alpha^{1/2}(\phi+2\pi)} = B\,e^{-\alpha^{1/2}\phi} \tag{8.103}$$

Equations (8.102) and (8.103) are satisfied if

$$e^{\pm 2\pi\alpha^{1/2}} = 1 \text{ or, } 2\pi\alpha^{1/2} = 2\pi i m, \text{ where } m = 0, \pm 1, \pm 2, \ldots$$

$$\text{or, } \alpha^{1/2} = im$$

Using $\alpha^{1/2} = im$ in eq. (8.98), we get

$$\Phi(\phi) = A\,e^{im\phi} \tag{8.104}$$

where $m = \pm 1, \pm 2, \pm 3, \ldots$.

Equation (8.104) represents the ϕ-dependent part of the spherical harmonics, if $m = m_l$. Then the spherical harmonics is expressed as

$$Y_l^{m_l}(\theta, \phi) = \Theta(\theta)\,e^{im_l\phi} \tag{8.105}$$

8.5.2 Solution of Θ-equation

Now, to find out the solution of $\Theta(\theta)$ equation we put $\alpha = -m^2$ in eq. (8.96) and, then get

$$\sin\theta\frac{d}{d\theta}\left(\sin\theta\frac{d\Theta}{d\theta}\right) + \left(\frac{\lambda}{\hbar^2}\sin^2\theta - m^2\right)\Theta = 0 \tag{8.106}$$

Let $\cos\theta = x$, then

$$\frac{d}{dx} = \frac{d}{d\theta}\frac{d\theta}{dx} = -\frac{1}{\sin\theta}\frac{d}{d\theta}$$

Equation (8.106), then becomes

$$\frac{d}{dx}\left[\left(1-x^2\right)\frac{d\Theta}{dx}\right]+\left(\frac{\lambda}{\hbar^2}-\frac{m^2}{1-x^2}\right)\Theta=0 \tag{8.107}$$

Now, either $m = 0$ or $m \neq 0$.

(a) When $m = 0$, eq. (8.107) takes the form:

$$\frac{d}{dx}\left[\left(1-x^2\right)\frac{d\Theta}{dx}+\frac{\lambda}{\hbar^2}\right]\Theta=0 \tag{8.108}$$

Equation (8.108) is the *Legendre's differential equation*, the solution of which can be written in the form of a series

$$\Theta(x)=\sum_{k=0}^{\infty}a_k x^k \tag{8.109}$$

Substituting eq. (8.109) in eq. (8.108), we get

$$\sum_{k=0}^{\infty}x^k\left[(k+1)(k+2)\,a_{k+2}-\left\{k(k+1)-\frac{\lambda}{\hbar^2}\right\}a_k\right]=0 \tag{8.110}$$

$$\text{or, } a_{k+2}=\frac{k(k+1)-\frac{\lambda}{\hbar^2}}{(k+1)(k+2)}a_k \tag{8.111}$$

Now, $\frac{a_{k+2}}{a_k}\to 1$ if the series represented by eq. (8.109) does not terminate as $k\to\infty$. Then for $x\to\pm1$ the series diverges [as the series, then behaves as $\sum\left(\frac{1}{k}\right)x^k\to\sum\frac{1}{k}=\infty$].

But, physically this cannot be allowed as $\Theta(\cos\theta)$ must remain finite for all $\cos\theta$ values between -1 to $+1$. Let the series terminates for $k = l$, then

$$a_l \neq 0$$

$$a_k = 0, \text{ for } k > l$$

Using $a_l \neq 0$ and $a_{k+2} = 0$ in eq. (8.107), we get

$$\frac{l(l+1)-\frac{\lambda}{\hbar^2}}{(l+1)(l+2)}a_l=0 \text{ or, } l(l+1)-\frac{\lambda}{\hbar^2}=0\ [\because a_l\neq 0]$$

$$\text{or, } \lambda = l(l+1)\,\hbar^2 \tag{8.112}$$

$l = 0, 1, 2, \ldots.$

So $\lambda = -\hbar C = l(l+1)\,\hbar^2$, $\therefore C = -l(l+1)$.

Now, for $l = 0$, we have from eq. (8.112), $\lambda = 0$, $a_0 \neq 0$, $a_i = 0$ (for $i > 0$). Then $\Theta(x) = a_0$.

For $l = 1$, $\lambda = 2\hbar^2$, $a_1 \neq 0$, $a_i = 0$ (for $i > 0$).

Since, $a_2 = 0$, we get $a_0 = 0$. Then

$$\Theta(x) = a_l x = a_l \cos\theta$$

For $l = 2$, we have $\lambda = 6\hbar^2$, $a_2 \neq 0$, $a_i = 0$ (for $i > 2$).

Since, $a_3 = 0$, we get $a_1 = 0$. Then from eq. (8.111) $a_2 = -3a_0$.

Thus,

$$\Theta(x) = a_0(1 - 3x^2) = a_0'(3\cos^2\theta - 1)$$

Here, $\Theta(x)$ is normalised, so $\Theta(x = 1) = 1$.

Then

$$\Theta(x) = \frac{3\cos^2\theta - 1}{2}$$

For any value of l we can easily show that

$$\Theta(x) = P_l(x) = \frac{1}{2^l l!}\frac{d^l}{dx^l}(x^2 - 1)^l \tag{8.113}$$

Here, $P_l(x)$ is *Legendre polynomial in x of degree l*.

(b) When $m \neq 0$.

If $\Theta(x)$ is a solution of eq. (8.108), then

$$(1 - x^2)^{|m|}\frac{d^{|m|}\Theta(x)}{dx^{|m|}}$$

is a solution of eq. (8.107). Therefore, the solution of eq. (8.107) can be expressed in terms of Legendre polynomials in the form:

$$\Theta(x) = P_l^m(x) = (1 - x^2)^{\frac{|m|}{2}}\frac{d^{|m|}P_l(x)}{dx^{|m|}}$$

$$= \frac{(1 - x^2)^{\frac{|m|}{2}}}{2^l l!}\frac{d^{l+|m|}(x^2 - 1)^l}{dx^{l+|m|}} \tag{8.114}$$

If $|m| > l$, then $P_l^m(x) = 0$ because

$$\frac{d^{l+|m|}}{dx^{l+|m|}}(x^2 - 1)^l = \frac{d^{l+|m|}}{dx^{l+|m|}}\left[x^{2l} + a_1x^{2l-1} + a_2x^{2l-2} + \ldots\right]$$

It is evident that if $|m| > l$, then differentiation of $P_l^m(x)$ will be equal to zero.

Hence, $|m|$ is to be restricted to values > or = 0, i.e.,

$$-l \leqslant m \leqslant l \tag{8.115}$$

Thus, the complete solution is

$$\Omega(\theta, \phi) = \Theta(x)\Phi(\phi) = CP_l^m(x)\,e^{im\phi} = CP_l^m\,e^{im\phi}\ (\text{for } m > 0)$$

$$= CP_l^{|m|}\,e^{-i|m|\phi}\ (\text{for } m < 0) \tag{8.116}$$

Here, C is the normalisation constant. Thus, the spherical harmonics are the normalised function $\Omega(\theta, \phi)$ and are given by

$$Y_l^m(\theta, \phi) = \left[\frac{2l + 1}{4\pi}\frac{(l - m)!}{(l + m)!}\right]^{1/2}(-1)^m\,e^{im\phi}P_l^m(\cos\theta)\ \text{when } 0 \leqslant m \leqslant l \tag{8.117}$$

and

$$Y_l^m(\theta, \phi) = (-1)^m Y_l^{-m},\ \text{for } -l \leqslant m < 0 \tag{8.118}$$

Some of spherical harmonics for $l = 0, 1$ and 2 are listed below:

For $l = 0$:

$$Y_0^0(\theta,\phi) = \left(\frac{1}{4\pi}\right)^{1/2} \tag{8.119}$$

For $l = 1$:

$$\left.\begin{aligned} Y_1^1(\theta,\phi) &= -\left(\frac{3}{8\pi}\right)^{1/2} \sin\theta\, e^{i\phi} \\ Y_1^0(\theta,\phi) &= \left(\frac{3}{4\pi}\right)^{1/2} \cos\theta \\ Y_1^{-1}(\theta,\phi) &= \left(\frac{3}{8\pi}\right)^{1/2} \sin\theta\, e^{-i\phi} \end{aligned}\right\} \tag{8.120}$$

For $l = 2$:

$$\left.\begin{aligned} Y_2^2(\theta,\phi) &= \frac{1}{4}\left(\frac{15}{2\pi}\right)^{1/2} \sin^2\theta\, e^{2i\phi} \\ Y_2^1(\theta,\phi) &= -\left(\frac{15}{8\pi}\right)^{1/2} \sin\theta\cos\theta\, e^{i\phi} \\ Y_2^0(\theta,\phi) &= \left(\frac{5}{4\pi}\right)^{1/2} \frac{3\cos^2\theta - 1}{2} \\ Y_2^{-1}(\theta,\phi) &= \left(\frac{15}{8\pi}\right)^{1/2} \sin\theta\cos\theta\, e^{-i\phi} \\ Y_2^{-2}(\theta,\phi) &= \frac{1}{4}\left(\frac{15}{2\pi}\right)^{1/2} \sin^2\theta\, e^{-2i\phi} \end{aligned}\right\} \tag{8.121}$$

Thus, in general for a given l there are $(2l + 1)$ solutions

$$Y_l^l(\theta,\phi),\ Y_l^{l-1}(\theta,\phi),\ Y_l^{l-2}(\theta,\phi),\ \ldots,\ Y_l^{-l}(\theta,\phi)$$

$$\text{or, } \Omega(\theta,\phi) = Y_l^{m_l}(\theta,\phi) \tag{8.122}$$

where for a particular l there are $(2l + 1)$ values of m_l as

$$m_l = l, l-1, l-2, \ldots, (l-1), -l$$

The functions $Y_l^{m_l}(\theta,\phi)$ satisfy the condition of normalisation

$$\int_{\theta=0}^{\pi}\int_{\phi=0}^{2\pi} Y_{m_l}^*(\theta,\phi)\, Y_l^{m_l}(\theta,\phi)\sin\theta\, d\theta\, d\phi = 1 \tag{8.123}$$

Hence, the spherical harmonics $Y_l^{m_l}(\theta,\phi)$ are *normalised functions.*

8.5.3 Solution of $R(r)$-equation

The radial part of the Schrödinger equation is

$$-\frac{1}{R(r)}\frac{d}{dr}\left(r^2\frac{dR}{dr}\right)-\frac{2m}{\hbar^2}r^2[E-V(r)]=C=-l(l+1)\ [\because C=-l(l+1)]$$

$$\text{or, } \frac{1}{r^2}\frac{d}{dr}\left(r^2\frac{dR}{dr}\right)+\left\{\frac{2m}{\hbar^2}[E-V(r)]-\frac{l(l+1)}{r^2}\right\}R=0\ [\text{multiplying both sides by } R/r^2] \tag{8.124}$$

Now,

$$\psi(r,\theta,\phi)=R(r)\Theta(\theta)\Phi(\phi)=R(r)\Omega(\theta,\phi)=R(r)Y_l^{m_l}(\theta,\phi) \tag{8.125}$$

This is the complete three-dimensional wave function of the Schrödinger wave equation for a particle moving in a spherically symmetric potential $V(r)$. For a *bound system*, the wave function must satisfy the condition that

$$R(r)\to 0 \text{ as } r\to\infty \tag{8.126}$$

If $V(r)$ is known, then only an explicit solution of eq. (8.124) can be found out.

However, pending the explicit solution we can predict the nature of the solution of the same equation.

Since, eq. (8.124) depends on l and l can have a discrete set of values, the allowed energy values must be discrete, say, $E_1^l, E_2^l, \ldots, E_n^l$. The complete wave function of the particle corresponding to energy E_n can be written as

$$\psi_{n,l,m_l}=R_n^l(r)Y_l^{m_l}(\theta,\phi) \tag{8.127}$$

As m_l has $(2l+1)$ values, ranging from $-l$ to $+l$ in integral steps, there will be $(2l+1)$ wave functions ψ_{n,l,m_l}, each differing in m_l-values.

Thus, the energy state of the particle is $(2l+1)$ *fold degenerate*. The quantities n, l and m_l are known as the *principal quantum number, orbital quantum number* and *azimuthal quantum number* respectively. These quantum numbers are needed to specify the different states.

8.6 The hydrogen atom

8.6.1 Introduction

The simplest of the atoms is the hydrogen atom, consisting of a single electron moving around the nucleus containing only one nucleon (the proton) under a central force arising out of the Colombian electrostatic attractive force. It forms the basis for the theoretical formulation of more complex atomic systems containing more than one electron. *The old quantum theory* was first proposed by *N. Bohr* (in 1912) on the basis of hydrogen atom. Schrödinger gave the solution of the wave equation for hydrogen atom in 1926 formulating the *wave mechanics*. There was extensive development of quantum theory of the hydrogen atom by the subsequent workers, viz., *W. Heisenberg, Max Born* and *Jordon* and *Pauli*.

The results of hydrogen atom was successfully extended to other *hydrogenic atoms*. The hydrogenic atoms are of two types:

(i) Ionised atoms containing only one electron moving around the nucleus which may consist of more than one proton and neutron. Singly ionised Helium (He^+), doubly ionised Lithium (Li^{++}), triply ionised Beryllium (Be^{+++}) are some of the hydrogenic atoms.

(ii) Atoms or ions with closed shells and only one electron in the outermost shell form the second type of hydrogenic atoms. Atoms of alkali metals and alkaline earth ions are the examples.

8.6.2 The wave equation for the hydrogen atom

The hydrogen or hydrogenic atom is a two-particle system consisting of an electron of charge $-e$ moving round the nucleus of charge +Ze. The electron in this system moves under the central force arising out of the Coulombic electrostatic attraction between the nucleus and the electron along the radius of the circular path of motion [Fig 8.4]. The potential energy of the system in absence of any external field is

$$V(r) = -\frac{Ze^2}{r}$$

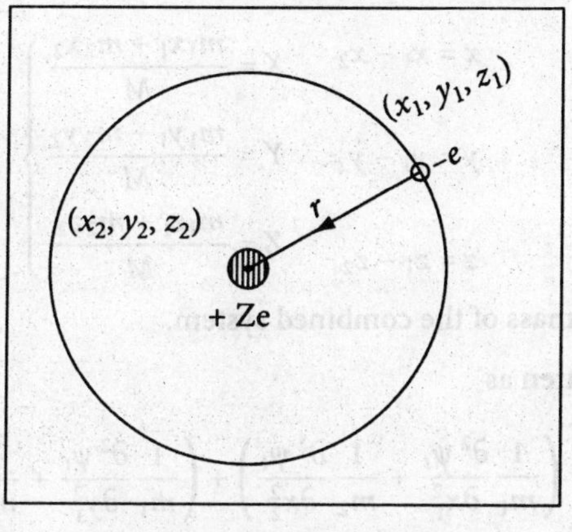

Fig 8.4 The hydrogen atom

The potential energy is independent of θ and ϕ and thus *spherically symmetric*. The three-dimensional wave equation can be separated into total differential equations in terms of each of the three space coordinates.

The total Hamiltonian of the system is

$$H_t = -\frac{\hbar^2}{2}\left(\frac{1}{m_1}\nabla_1^2 + \frac{1}{m_2}\nabla_2^2\right) + V(x_1, y_1, z_1; x_2, y_2, z_2) \tag{8.128}$$

where m_1 and m_2 are the masses of the electron and the nucleus respectively.

The wave equation of the system is

$$H_t\psi_t = E_t\psi_t \tag{8.129}$$

where ψ_t is the wave function of the total system.

Since,

$$\nabla_1^2 = \frac{\partial^2}{\partial x_1^2} + \frac{\partial^2}{\partial y_1^2} + \frac{\partial^2}{\partial z_1^2} \tag{8.130}$$

$$\text{and } \nabla_2^2 = \frac{\partial^2}{\partial x_2^2} + \frac{\partial^2}{\partial y_2^2} + \frac{\partial^2}{\partial z_2^2} \tag{8.131}$$

the explicit form of eq. (8.129) is

$$-\frac{\hbar^2}{2m_1}\left(\frac{\partial^2\psi_t}{\partial x_1^2}+\frac{\partial^2\psi_t}{\partial y_1^2}+\frac{\partial^2\psi_t}{\partial z_1^2}\right)-\frac{\hbar^2}{2m_2}\left(\frac{\partial^2\psi_t}{\partial x_2^2}+\frac{\partial^2\psi_t}{\partial y_2^2}+\frac{\partial^2\psi_t}{\partial z_2^2}\right)$$
$$+\left[V(x_1,y_1,z_1;x_2,y_2,z_2)-E_t\right]\psi_t=0 \qquad (8.132)$$

8.6.3 Reduction to equivalent one-body problem

The potential energy of the system depends upon the relative coordinates of the two particles, i.e.,

$$V=V\left[(x_1-x_2),(y_1-y_2),(z_1-z_2)\right]$$

The effect of external interaction is well-exhibited by the centre of mass having coordinates (X, Y, Z).

Hence, we replace the six coordinates of the electron and the proton by three relative coordinates x, y, z and three coordinates X, Y, Z defined as follows:

$$\left.\begin{aligned} x&=x_1-x_2 & X&=\frac{m_1x_1+m_2x_2}{M}\\ y&=y_1-y_2 & Y&=\frac{m_1y_1+m_2y_2}{M}\\ z&=z_1-z_2, & Z&=\frac{m_1z_1+m_2z_2}{M}\end{aligned}\right\} \qquad (8.133)$$

where $M = m_1 + m_2$ the total mass of the combined system.

Equation (8.132) can be written as

$$-\frac{\hbar^2}{2}\left[\left(\frac{1}{m_1}\frac{\partial^2\psi_t}{\partial x_1^2}+\frac{1}{m_2}\frac{\partial^2\psi_t}{\partial x_2^2}\right)+\left(\frac{1}{m_1}\frac{\partial^2\psi_t}{\partial y_1^2}+\frac{1}{m_2}\frac{\partial^2\psi_t}{\partial y_2^2}\right)\right.$$
$$\left.+\left(\frac{1}{m_1}\frac{\partial^2\psi_t}{\partial z_1^2}+\frac{1}{m_2}\frac{\partial^2\psi_t}{\partial z_2^2}\right)\right]+V(x,y,z)\,\psi_t=E_t\psi_t \qquad (8.134)$$

Now,

$$\frac{\partial\psi_1}{\partial x_1}=\frac{\partial\psi_t}{\partial X}\frac{\partial X}{\partial x_1}+\frac{\partial\psi_t}{\partial x}\frac{\partial x}{\partial x_1}=\frac{m_1}{M}\frac{\partial\psi_t}{\partial X}+\frac{\partial\psi_t}{\partial x}\cdot 1=\left(\frac{m_1}{M}\frac{\partial}{\partial X}+\frac{\partial}{\partial x}\right)\psi_t.$$

$$\therefore\ \frac{\partial}{\partial x_1}=\frac{m_1}{M}\frac{\partial}{\partial X}+\frac{\partial}{\partial x}$$

$$\text{and}\ \frac{\partial^2}{\partial x_1^2}=\left(\frac{m_1}{M}\frac{\partial}{\partial X}+\frac{\partial}{\partial x}\right)^2=\left(\frac{m_1^2}{M^2}\frac{\partial^2}{\partial X^2}+\frac{2m_1}{M}\frac{\partial^2}{\partial X\partial x}+\frac{\partial^2}{\partial x^2}\right)$$

$$\therefore\ \frac{\partial^2\psi_t}{\partial x_1^2}=\left(\frac{m_1^2}{M^2}\frac{\partial^2}{\partial X^2}+\frac{2m_1}{M}\frac{\partial^2}{\partial X\partial x}+\frac{\partial^2}{\partial x}\right)\psi_t \qquad (8.135)$$

Similarly,

$$\frac{\partial^2\psi_t}{\partial x_2^2}=\left(\frac{m_2^2}{M^2}\frac{\partial^2}{\partial X^2}-\frac{2m_2}{M}\frac{\partial^2}{\partial X\partial x}+\frac{\partial^2}{\partial x^2}\right)\psi_t \qquad (8.136)$$

Now, from eqs (8.135) and (8.136), we get by addition

$$\frac{1}{m_1}\frac{\partial^2 \psi_t}{\partial x_1^2} + \frac{1}{m_2}\frac{\partial^2 \psi_t}{\partial x_2^2} = \frac{1}{M}\frac{\partial^2 \psi_t}{\partial X^2} + \frac{1}{\mu}\frac{\partial^2 \psi_t}{\partial x^2} \tag{8.137}$$

where $\mu = \dfrac{m_1 m_2}{m_1 + m_2}$ is the *reduced mass* of the electron-nucleus system.

Similarly,

$$\frac{1}{m_1}\frac{\partial^2 \psi_t}{\partial y_1^2} + \frac{1}{m_2}\frac{\partial^2 \psi_t}{\partial y_2^2} = \frac{1}{M}\frac{\partial^2 \psi_t}{\partial Y^2} + \frac{1}{\mu}\frac{\partial^2 \psi_t}{\partial y^2}$$

$$\text{and } \frac{1}{m_1}\frac{\partial^2 \psi_t}{\partial z_1^2} + \frac{1}{m_2}\frac{\partial^2 \psi_t}{\partial z_2^2} = \frac{1}{M}\frac{\partial^2 \psi_t}{\partial Z^2} + \frac{1}{\mu}\frac{\partial^2 \psi_t}{\partial z^2}$$

Now, eq. (8.134) can be written as

$$-\frac{\hbar^2}{2}\left[\frac{1}{M}\left(\frac{\partial^2 \psi_t}{\partial X^2} + \frac{\partial^2 \psi_t}{\partial Y^2} + \frac{\partial^2 \psi_t}{\partial Z^2}\right) + \frac{1}{\mu}\left(\frac{\partial^2 \psi_t}{\partial x^2} + \frac{\partial^2 \psi_t}{\partial y^2} + \frac{\partial^2 \psi_t}{\partial z^2}\right)\right] + V(x, y, z)\,\psi_t = E_t \psi_t \tag{8.138}$$

Putting $\psi_t = \psi(x, y, z)\Phi(X, Y, Z)$ in eq. (8.138), we get

$$-\frac{\hbar^2}{2}\left[\frac{1}{M}\left(\frac{\partial^2 \Phi}{\partial X^2} + \frac{\partial^2 \Phi}{\partial Y^2} + \frac{\partial^2 \Phi}{\partial Z^2}\right)\psi(x, y, z)\right.$$

$$\left. + \frac{1}{\mu}\left(\frac{\partial^2 \psi}{\partial x^2} + \frac{\partial^2 \psi}{\partial y^2} + \frac{\partial^2 \psi}{\partial z^2}\right)\Phi(X, Y, Z)\right] + V\Phi(X, Y, Z)\,\psi(x, y, z) = E_t \Phi(X, Y, Z)\,\psi(x, y, z) \tag{8.139}$$

$$\text{or, } -\frac{\hbar^2}{2M}\frac{1}{\Phi}\nabla^2_{XYZ}\Phi - \frac{\hbar^2}{2\mu}\frac{1}{\psi}\nabla^2_{xyz}\psi + V(x, y, z) = E_t$$

$$\text{or, } -\frac{\hbar^2}{2M}\frac{1}{\Phi}\nabla^2_{XYZ}\Phi - E_t = -\left[-\frac{\hbar^2}{2\mu}\frac{1}{\psi}\nabla^2_{xyz}\psi + V(x, y, z)\right] = -E \text{ (say)} \tag{8.140}$$

where E is some constant quantity independent of $X, Y, Z; x, y, z$

$$\text{or, } -\frac{\hbar^2}{2\mu}\left(\frac{\partial^2 \psi}{\partial x^2} + \frac{\partial^2 \psi}{\partial y^2} + \frac{\partial^2 \psi}{\partial z^2}\right) + V(x, y, z)\,\psi = E\psi \tag{8.141}$$

$$\text{and } -\frac{\hbar^2}{2M}\left(\frac{\partial^2 \Phi}{\partial X^2} + \frac{\partial^2 \Phi}{\partial Y^2} + \frac{\partial^2 \Phi}{\partial Z^2}\right) - (E_t - E)\Phi = 0 \tag{8.142}$$

Equation (8.142) represents the equation of motion of the centre of mass of the system which behaves like a free particle of mass M and translational energy $(E_t - E)$. The eq. (8.141) involves V and describes the relative motion of the system in which we are actually interested.

Since, the Hamiltonian of the system has spherical symmetry, the variables can be easily separated, if we use spherical polar coordinates of the electron relative to the nucleus.

If (r, θ, ϕ) be the spherical polar coordinates of the electron with the nucleus as origin [Fig 8.5], then we have

$$x = r \sin\theta \cos\phi \tag{8.143}$$
$$y = r \sin\theta \sin\phi \tag{8.144}$$
$$z = r \cos\theta \tag{8.145}$$

and

$$\nabla^2 = \frac{1}{r^2}\frac{\partial}{\partial r}\left(r^2 \frac{\partial}{\partial r}\right) + \frac{1}{r^2 \sin\theta}\frac{\partial}{\partial \theta}\left(\sin\theta \frac{\partial}{\partial \theta}\right) + \frac{1}{r^2 \sin^2\theta}\frac{\partial^2}{\partial \phi^2}$$

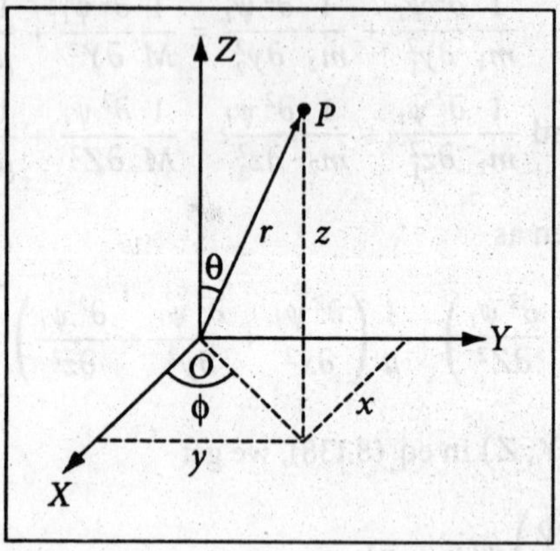

Fig 8.5 Relation between x, y, z and r, θ, ϕ

Then eq. (8.141) becomes

$$\frac{1}{r^2}\frac{\partial}{\partial r}\left(r^2 \frac{\partial \psi}{\partial r}\right) + \frac{1}{r^2 \sin\theta}\frac{\partial}{\partial \theta}\left(\sin\theta \frac{\partial \psi}{\partial \theta}\right) + \frac{1}{r^2 \sin^2\theta}\frac{\partial^2 \psi}{\partial \phi^2} + \frac{2\mu}{\hbar^2}[E - V(r)]\psi = 0 \tag{8.146}$$

Equation (8.146) is the *Schrödinger's wave equation* of the electron in spherical polar coordinates.

8.6.4 Separation of variables

Equation (8.146) can be separated in variables r, θ and ϕ by substituting

$$\psi(r, \theta, \phi) = R(r)\Theta(\theta)\Phi(\phi) \tag{8.147}$$

and, then multiplying the result by $r^2 \sin^2\theta$, we get

$$\therefore \frac{\sin^2\theta}{R}\frac{d}{dr}\left(r^2 \frac{dR}{dr}\right) + \frac{\sin\theta}{\Theta}\frac{d}{d\theta}\left(\sin\theta \frac{d\Theta}{d\theta}\right) + \frac{2\mu r^2 \sin^2\theta}{\hbar^2}[E - V(r)] + \frac{1}{\Phi}\frac{d^2\Phi}{d\phi^2} = 0 \tag{8.148}$$

As the last term of left side of eq. (8.148) depends only on ϕ and the first and the second terms are independent of ϕ we can write:

$$\frac{1}{\Phi}\frac{d^2\Phi}{d\phi^2} = -m^2 \text{ or, } \frac{d^2\Phi}{d\phi^2} + m\Phi = 0 \tag{8.149}$$

where m is a constant.

Hence, eq. (8.141) can be written as

$$\frac{1}{R}\frac{d}{dr}\left(r^2 \frac{dR}{dr}\right) + \frac{1}{\sin\theta}\frac{1}{\Theta}\frac{d}{d\theta}\left(\sin\theta \frac{d\Theta}{d\theta}\right) + \frac{2\mu r^2}{\hbar^2}[E - V(r)] - \frac{m^2}{\sin^2\theta} = 0 \tag{8.150}$$

The first and third terms of this equation depend only on r and the second and fourth terms depend only on θ. It is, therefore, possible to separate the r-part and the θ-part of the eq. (8.150) by putting each part equal to a constant λ (say).

$$\therefore \frac{1}{R}\frac{d}{dr}\left(r^2\frac{dR}{dr}\right)+\frac{2\mu r^2}{\hbar^2}(E-V)=\frac{m^2}{\sin^2\theta}-\frac{1}{\Theta\sin\theta}\frac{d}{d\theta}\left(\sin\theta\frac{d\Theta}{d\theta}\right)=\lambda$$

So

$$\frac{1}{\sin\theta}\frac{d}{d\theta}\left(\sin\theta\frac{d\Theta}{d\theta}\right)+\left(\lambda-\frac{m^2}{\sin^2\theta}\right)\Theta=0 \tag{8.151}$$

and

$$\frac{1}{r^2}\frac{d}{dr}\left(r^2\frac{dR}{dr}\right)+\left[\frac{2\mu}{\hbar^2}\{E-V(r)\}-\frac{\lambda}{r^2}\right]R=0 \tag{8.152}$$

The complete solution of eq. (8.146) depends on the solutions of eqs (8.149), (8.151) and (8.152) known as *azimuthal equation, polar equation* and *radial equation* respectively.

8.6.5 Solution of the ϕ-equation

The azimuthal equation

$$\frac{d^2\Phi}{d\phi^2}+m^2\Phi=0$$

is a second order linear differential equation whose solution may be written as

$$\Phi(\phi)=A\,e^{\pm im\phi} \tag{8.153}$$

where A is a normalisation constant. For $\Phi(\phi)$ to be single-valued, it must have the same value for $\phi=0$ and $\phi=2\pi$.

$$\therefore A\,e^{\pm im\phi}=A\,e^{\pm im(\phi+2\pi)}$$

$$\text{or, } e^{\pm 2\pi im}=1=\cos 2\pi m\pm i\sin 2\pi m$$

Again, $\cos 2\pi m\pm i\sin 2\pi m=1$, if $m=0$ or $\pm n\pi$, where $n=1,2,3,\ldots$.

$$\therefore \Phi(\phi)=A\,e^{im\phi},\ \text{for } m=0,\pm1,\pm2,\ldots \tag{8.154}$$

To find out the normalisation constant A, we have the condition of normalisation

$$\int_0^{2\pi}\Phi^*(\phi)\Phi(\phi)\,d\phi=1 \text{ or, } A^2\int_0^{2\pi}d\phi=1 \text{ or, } A^2 2\pi=1 \text{ or, } A=\frac{1}{\sqrt{2\pi}} \tag{8.155}$$

Therefore,

$$\Phi(\phi)=\frac{1}{\sqrt{2\pi}}e^{im\phi},\ (m=0,\pm1,\pm2,\ldots) \tag{8.156}$$

Here, m is called the *magnetic quantum number*. For $m=0$

$$\Phi(\phi)=\frac{1}{\sqrt{2\pi}} \tag{8.157}$$

For $m=+1$, the real solution is

$$\Phi(\phi)=\frac{1}{\sqrt{2\pi}}\cos\phi$$

and for $m = -1$, the real solution is

$$\Phi(\phi) = \frac{1}{\sqrt{2\pi}} \sin\phi$$

8.6.6 Solution of θ-equation

To solve eq. (8.151) we put $x = \cos\theta$, then x varies from +1 to −1 and

$$\frac{dx}{d\theta} = -\sin\theta$$

Now,

$$\frac{d\Theta}{d\theta} = \frac{dx}{d\theta}\frac{d\Theta}{dx} = -\sin\theta\frac{d\Theta}{dx}$$

$$\therefore \frac{d}{d\theta} = -\sin\theta\frac{d}{dx} \tag{8.158}$$

and

$$\sin\theta\frac{d}{d\theta} = -(1-x^2)\frac{d\Theta}{dx} \quad [\because \sin^2\theta = (1-x^2)] \tag{8.159}$$

From eqs (8.158), (8.159) and (8.151), we get

$$\frac{d}{dx}\left[(1-x^2)\frac{d\Theta}{dx}\right] + \left(\lambda - \frac{m^2}{1-x^2}\right)\Theta = 0$$

$$\text{or, } \frac{d}{dx}\left[(1-x^2)\frac{d\Theta}{dx}\right] + \left[l(l+1) - \frac{m^2}{1-x^2}\right]\Theta = 0 \; [\text{if } \lambda = l(l+1)] \tag{8.160}$$

The *associated Legendre equation of degree l* is

$$\frac{d}{dx}\left[(1-x^2)\frac{d}{dx}P_l^m(x)\right] + \left[l(l+1) - \frac{m^2}{1-x^2}\right]P_l^m(x) = 0 \tag{8.161}$$

Comparing eqs (8.160) and (8.161), we get the solution of eq. (8.159) as

$$\Theta = BP_l^m(x) = BP_l^m(\cos\theta), \; |m| < l \tag{8.162}$$

Here, B is a normalisation constant whose value may be found out from the normalisation condition

$$\int_0^\pi \Theta^*\Theta\sin\theta\, d\theta = 1 \text{ or, } \int_{-1}^{+1} B^2 P_l^{*m}(x)P_l^m(x)\, dx = 1 \text{ or, } B^2\int_{-1}^{+1} |P_l^m(x)|^2\, dx = 1$$

$$\text{or, } B^2\frac{(l+m)!}{(l-m)!}\frac{2}{2l+1} = 1 \text{ or, } B = \sqrt{\frac{2l+1}{2}\cdot\frac{(l-m)!}{(l+m)!}} \tag{8.163}$$

$$\therefore \int_{-1}^{+1} |P_l^m(x)|^2\, dx = \frac{(l+m)!}{(l-m)!}\cdot\frac{2}{2l+1} \tag{8.164}$$

Therefore, the solution of eq. (8.151) is

$$\Theta(\theta) = \left[\sqrt{\frac{2l+1}{2}\frac{(l-m)!}{(l+m)!}}\right]P_l^m(\cos\theta) \tag{8.165}$$

where $P_l^m(\cos\theta)$ is called the *associated Legendre function* and

$$P_l^m(x) = (1-x^2)^{m/2}\frac{d^m}{dx^m}P_l(x) \text{ with } P_l(x) = 2^{1/l}l!\frac{d^l}{dx^l}(x^2-1)^l$$

The values of Legendre functions for different values l and $|m|$ are listed below:

$$P_0^0(\cos\theta) = 1$$
$$P_1^0(\cos\theta) = \cos\theta$$
$$P_2^0(\cos\theta) = \tfrac{1}{2}(3\cos^2\theta - 1)$$
$$P_3^0(\cos\theta) = \tfrac{1}{2}(5\cos^3\theta - 3\cos\theta)$$
$$P_1^1(\cos\theta) = \sin\theta$$
$$P_2^1(\cos\theta) = 3\sin\theta\cos\theta$$
$$P_3^1(\cos\theta) = \tfrac{3}{2}\sin\theta(5\cos^2\theta - 1), \text{ etc.}$$

Table 8.2 shows normalised Θ functions for $l = 0, 1, 2$ and $m = 0, \pm1, \pm2$.

Table 8.2: Normalised Θ functions

n	l	m	$\Theta(\theta)$
1	0	0	$\frac{1}{\sqrt{2}}$
2	0	0	$\frac{1}{\sqrt{2}}$
2	1	0	$\frac{\sqrt{6}}{2}\cos\theta$
2	1	±1	$\frac{\sqrt{3}}{2}\sin\theta$
3	0	0	$\frac{1}{\sqrt{2}}$
3	1	0	$\frac{\sqrt{6}}{2}\cos\theta$
3	1	±1	$\frac{\sqrt{3}}{2}\sin\theta$
3	2	0	$\frac{\sqrt{10}}{4}(3\cos^2\theta - 1)$
3	2	±1	$\frac{\sqrt{15}}{2}\sin\theta\cos\theta$
3	2	±2	$\frac{\sqrt{15}}{4}\sin^2\theta$

The complete solution of the angular part of eq. (8.146) can now be expressed as

$$\begin{aligned}\Omega(\theta,\phi) &= \Theta(\theta)\Phi(\phi)\\ &= (AB)P_l^m(\cos\theta)\,e^{im\phi},\ (m>0)\ \text{[from eqs (8.154) and (8.162)]}\\ &= (AB)P_l^{|m|}\,e^{-i|m|\phi},\ (m<0)\end{aligned} \tag{8.166}$$

The normalised functions $\Omega(\theta,\phi)$ are the spherical harmonics and are given by

$$Y_l^m(\theta,\phi) = \left[\frac{2l+1}{4\pi}\frac{(l-m)!}{(l+m)!}\right]^{1/2}(-1)^m e^{im\phi}P_l^m(\cos\theta),\ 0 \leqslant m \leqslant l \tag{8.167}$$

$$Y_l^m(\theta,\phi) = (-1)^m Y_l^{-m}(\theta,\phi),\ -1 \leqslant m \leqslant 0 \tag{8.168}$$

using the values of A and B from eqs (8.155) and (8.163). Spherical harmonics for $l = 0, 1, 2$ are given by eqs (8.119), (8.120) and (8.121) respectively.

8.6.7 Solution of radial equation

The radial part of the Schrödinger wave equation is

$$\frac{1}{r^2}\frac{d}{dr}\left(r^2\frac{dR}{dr}\right)+\left[\frac{2\mu}{\hbar^2}\{E-V(r)\}-\frac{\lambda}{r^2}\right]R=0$$

The above equation with

$$V(r)=-\frac{Ze^2}{r}$$

is called the *radial equation* for the hydrogenic (or, hydrogen-like) atoms and for $z = 1$ it is called the *radial equation for the hydrogen atom.*

In view of classical mechanics, the states with $E > 0$ correspond to hyperbolic orbits representing *unbound states.* But, the states with $E < 0$ correspond to elliptic orbits and represent *bound states.* In the present case, the electron is bound to the hydrogen nucleus and is $E < 0$, i.e., a *negative energy state.*

Taking

$$V(r)=-\frac{Ze^2}{r}$$

in the radial equation, we get

$$\frac{1}{r^2}\frac{d}{dr}\left(r^2\frac{dR}{dr}\right)+\frac{2\mu}{\hbar^2}\left[E+\frac{Ze^2}{r}-\frac{l\,(l+1)\,\hbar^2}{2\mu r^2}\right]R=0 \tag{8.169}$$

The term $-\frac{Ze^2}{r}$ represents the Coulomb potential and the term

$$\frac{\hbar^2}{2m}\frac{l\,(l+1)}{r^2}$$

is known as the *centrifugal potential.* That is

$$V_{\text{cent}}=\frac{\hbar^2}{2m}\frac{l\,(l+1)}{r^2}\text{ and }V_{\text{coul}}=-\frac{Ze^2}{r}$$

Figs 8.6(a) and (b) show respectively the variation of V_{cent} with r and V_{coul} with r for different values of l.

We now introduce a dimensionless variable ρ such that $\rho = \alpha r$ or, $r = \frac{\rho}{\alpha}$, then the radial equation in terms of ρ takes the form:

$$\alpha^2\frac{1}{\rho^2}\alpha\frac{d}{d\rho}\left[\frac{\rho^2}{\alpha^2}\alpha\frac{dR}{d\rho}\right]+\frac{2\mu}{\hbar^2}\left[-|E|+\frac{Ze^2\alpha}{\rho}-\frac{\alpha^2 l\,(l+1)\,\hbar^2}{2\mu\rho}\right]R=0$$

Here, we have written $-|E|$ in place of E because of the bound state $(E < 0)$ of the electron.

$$\text{or, }\alpha^2\frac{1}{\rho^2}\frac{d}{d\rho}\left[\rho^2\frac{dR}{d\rho}\right]+\frac{2\mu}{\hbar^2}\left[-|E|+\frac{Ze^2\alpha}{\rho}-\frac{\alpha^2 l\,(l+1)\,\hbar^2}{2\rho^2\mu}\right]R=0$$

$$\text{or, }\frac{1}{\rho^2}\frac{d}{d\rho}\left[\rho^2\frac{dR}{d\rho}\right]+\left[-\frac{2\mu|E|}{\hbar^2\alpha^2}+\frac{2\mu Ze^2}{\hbar^2\alpha\rho}-\frac{l\,(l+1)}{\rho^2}\right]R=0 \tag{8.170}$$

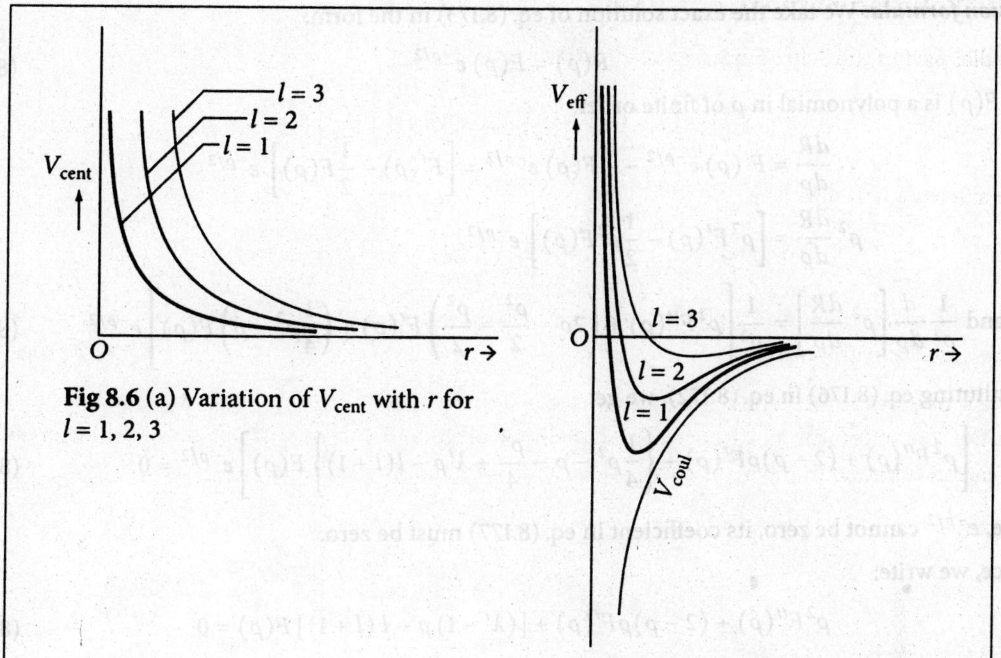

Fig 8.6 (a) Variation of V_{cent} with r for $l = 1, 2, 3$

Fig 8.6 (b) Variation of V_{coul} and V_{eff} with r for $l = 1, 2, 3$

Now, let

$$\frac{2\mu|E|}{\alpha^2\hbar^2} = \frac{1}{4}. \therefore \alpha = \left[\frac{8\pi|E|}{\hbar^2}\right]^{1/2}$$

and

$$\lambda' = \frac{2\mu Ze^2}{\alpha\hbar^2} = \frac{2Ze^2}{\hbar^2}\frac{1}{2}\left(\frac{\mu^2}{2\mu|E|}\right)^{1/2}\hbar = \frac{Ze^2}{\hbar}\left(\frac{\mu}{2|E|}\right)^{1/2} \tag{8.171}$$

Then eq. (8.170) becomes

$$\frac{1}{\rho^2}\frac{d}{d\rho}\left(\rho^2\frac{dR}{d\rho}\right) + \left[\frac{\lambda'}{\rho} - \frac{1}{4} - \frac{l(l+1)}{\rho^2}\right]R = 0 \tag{8.172}$$

Asymptotic behaviour: If $\rho \to \infty$, then $\lambda'/\rho \to 0$ and $[l(l+1)]/\rho^2 \to 0$. Under this condition eq. (8.172) is reduced to

$$\frac{1}{\rho^2}\frac{d}{d\rho}\left(\rho^2\frac{dR}{d\rho}\right) - \frac{R}{4} = 0 \tag{8.173}$$

The solutions of the above equation can be put in the form:

$$R(\rho) = e^{+\rho/2} \text{ and } R(\rho) = e^{-\rho/2} \tag{8.174}$$

As ρ may vary from 0 to ∞, $R(\rho) = e^{+\rho/2} \to \infty$ as $\rho \to \infty$ and is unacceptable. On the other hand, $R(\rho) = e^{-\rho/2} \to 0$ as $\rho \to \infty$, i.e., this solution is *well-behaved* and thus acceptable.

Recursion formula: We take the exact solution of eq. (8.173) in the form:

$$R(\rho) = F(\rho)\, e^{-\rho/2} \tag{8.175}$$

where $F(\rho)$ is a polynomial in ρ of finite order.

$$\therefore \frac{dR}{d\rho} = F'(\rho)\, e^{-\rho/2} - \frac{1}{2}F(\rho)\, e^{-\rho/2} = \left[F'(\rho) - \frac{1}{2}F(\rho)\right] e^{-\rho/2}$$

$$\rho^2 \frac{dR}{d\rho} = \left[\rho^2 F'(\rho) - \frac{1}{2}\rho^2 F(\rho)\right] e^{-\rho/2}$$

and $$\frac{1}{\rho^2}\frac{d}{d\rho}\left[\rho^2 \frac{dR}{d\rho}\right] = \frac{1}{\rho^2}\left[\rho^2 F''(\rho) + \left(2\rho - \frac{\rho^2}{2} - \frac{\rho^2}{2}\right)F'(\rho) + \left(\frac{1}{4}\rho^2 - \rho\right)F(\rho)\right] e^{-\rho/2} \tag{8.176}$$

Substituting eq. (8.176) in eq. (8.172), we get

$$\left[\rho^2 F''(\rho) + (2-\rho)\rho F'(\rho) + \left\{\frac{1}{4}\rho^2 - \rho - \frac{\rho^2}{4} + \lambda'\rho - l(l+1)\right\} F(\rho)\right] e^{-\rho/2} = 0. \tag{8.177}$$

Since, $e^{-\rho/2}$ cannot be zero, its coefficient in eq. (8.177) must be zero.

Hence, we write:

$$\rho^2 F''(\rho) + (2-\rho)\rho F'(\rho) + [(\lambda' - 1)\,\rho - l\,(l+1)]\, F(\rho) = 0 \tag{8.178}$$

Let us now find a solution for $F(\rho)$ in the form: $F(\rho) = \rho^s L(\rho)$.

$\therefore F'(\rho) = s\rho^{s-1}L + \rho^s L'$ and $F''(\rho) = s(s-1)\rho^{s-2}L + 2s\rho^{s-1}L' + \rho^s L''$.

Substituting these values of $F(\rho)$, $F'(\rho)$ and $F''(\rho)$ in eq. (8.178), we get

$$\rho^2 L'' + \left(2s\rho + 2\rho + \rho^2\right) L' + [s(s-1) + 2s - s\rho + (\lambda' - 1)\rho - l(l+1)]L = 0 \tag{8.179}$$

Now, we put

$$L(\rho) = \sum_{k=0}^{\infty} a_k \rho^k$$

then

$$F(\rho) = \rho^s L(\rho) = \sum_{k=0}^{\infty} a_k \rho^{s+k} \tag{8.180}$$

From eqs (8.180) and (8.178), we get

$$\rho^2 \sum_{k=0}^{\infty} a_k (s+k)(s+k-1)\rho^{s+k-2} + (2-\rho)\rho \sum_{k=0}^{\infty} a_k (s+k)\rho^{s+k-1} + [(\lambda'-1)\,\rho - l\,(l+1)] \sum_{k=0}^{\infty} a_k \rho^{s+k} = 0$$

or, $$\sum_{k=0}^{\infty} a_k(s+k)(s+k-1)\rho^{s+k} + 2\sum_{k=0}^{\infty} a_k(s+k)\rho^{s+k} - \sum_{k=0}^{\infty} a_k(s+k)\rho^{s+k-1} + (\lambda'-1)\sum_{k=0}^{\infty} a_k\rho^{s+k-1} - l\,(l+1)\sum_{k=0}^{\infty} a_k\rho^{s+k} = 0$$

or, $$\sum_{k=0}^{\infty}[(s+k)(s+k-1) + 2(s+k) - l\,(l+1)]a_k\rho^{s+k} - [(s+k) - (\lambda'-1)]\sum_{k=0}^{\infty} a_k\rho^{s+k+1} = 0 \tag{8.181}$$

For eq. (8.181) to hold good the coefficients of various powers of ρ must vanish separately.

Now, collecting the coefficients of ρ^s in eq. (8.181) and equating to zero, we get

$$[s(s-1) + 2s - l\,(l+1)]a_0 = 0$$

Since, $a_0 \neq 0, \therefore s(s+1) - l(l+1) = 0$, which yields either

$$s = l \text{ or, } s = -(l+1) \tag{8.182}$$

The boundary condition demands that $R(\rho)$ should be finite at $\rho \to 0$ and for this $s = l$.

Now, comparing the coefficients of ρ^{s+k+1}, we get from eq. (8.181)

$$[(s+k+1)(s+k) + 2(s+k+1) - l(l+1)]a_{k+1} = (s+k+1-\lambda')\,a_k$$

$$\text{or, } \frac{a_{k+1}}{a_k} = \frac{s+k+1-\lambda'}{(s+k+1)(s+k+2) - l(l+1)} \tag{8.183}$$

This is called the *recursion formula* where k is an integer or zero.

Now, the series

$$F(\rho) = \sum_{k=0}^{\infty} a_k \rho^{s+k}$$

behaves as e^{ρ} when $\rho \to 0$. Then $R = e^{-\rho/2}F(\rho) = e^{-\rho/2}.\,e^{\rho} = e^{\rho/2}$ diverges as $\rho \to \infty$ and thus is not acceptable. To satisfy the physical conditions, $F(\rho)$ must terminate after a finite number of terms, i.e., after a definite value of k such that $a_k \neq 0$, but $a_{k+1}, a_{k+2}, a_{k+3}$, etc., are each equal to zero. For this reason

$$s + k + 1 - \lambda' = 0$$

$$\text{or, } \lambda' = l + k + 1 \; (\because s = l)$$

$$\text{or, } n = l + k + 1 \text{ (replacing } \lambda' \text{ by } n)$$

Here, k is called the *radial quantum number* and can have values $k = 1, 2, 3, \ldots$ and n is called the *total quantum number* and $n = 1, 2, 3, \ldots$.

Equation (8.179) can now be written as

$$\rho L'' + [(2l+1) + 1 + \rho]L' + (n - l - 1)L = 0 \tag{8.184}$$

The *associated Laguerre equation* is

$$\rho L_q^{''p} + (p + 1 + \rho)L_q^{'p} + (q - p)L_q^p = 0 \tag{8.185}$$

Comparing eqs (8.184) and (8.185), we get $p = 2l + 1$ and $q = n + l$.

The solution of eq. (8.185) is an associated Laguerre polynomial $L_q^p(\rho)$. The solution of the radial equation now can be expressed as

$$R_n^l(r) = C e^{-\rho/2} \rho^l L_{n+l}^{2l+1}(\rho) \tag{8.186}$$

The normalising constant C can be found out from the condition

$$\int_0^{\infty} R^*(r)R(r)\, r^2\, dr = 1 \tag{8.187}$$

In eq. (8.187), the factor r^2 is introduced to convert the length dr into volume element $r^2\, dr$.

Using eq. (8.186) in eq. (8.187) we can evaluate the value of C as

$$C = \left[(\alpha)^3 \frac{(n-l-1)!}{2n\{(n+l)!\}^3}\right]^{1/2} \tag{8.188}$$

We have

$$\alpha = \left[\frac{8\mu|E|}{\hbar^2}\right]^{1/2}$$

Since,

$$|E| = -\frac{\mu Z^2 e^4}{2\hbar^2 n^2} \text{ (to be proved below)}$$

$$\therefore \alpha = \left[\frac{8\pi}{\hbar^2}\frac{\mu Z^2 e^4}{2\hbar^2 n^2}\right]^{1/2} = \frac{2\mu Z e^2}{n\hbar^2} = \frac{2Z}{n\left(\frac{\hbar^2}{\mu e^2}\right)} = \frac{2Z}{na_0}$$

where $a_0 = \hbar^2/\mu e^2 = 0.529 \times 10^{-8}$ cm = the first Bohr radius.

Therefore, the radial wave equation is

$$R_n^l(r) = \left[\left(\frac{2Z}{na_0}\right)^3 \frac{(n-l-1)!}{2n\{(n+1)!\}^3}\right]^{1/2} e^{\left(\frac{-Zr}{na_0}\right)}\left(\frac{2Zr}{na_0}\right)^l L_{n+l}^{2n+l}\left(\frac{2Zr}{na_0}\right) \qquad (8.189)$$

Note: Proof of eq. (8.188):

$$\because \rho = \alpha r, \therefore r = \frac{\rho}{\alpha} \text{ and } dr = \frac{d\rho}{\alpha} \text{ and } r^2 = \frac{\rho^2}{\alpha^2}$$

$$\therefore \int_0^\infty R^*(r)R(r)r^2\,dr = 1 \text{ or, } C^2\int_0^\infty e^{-\rho}\rho^{2l}\left[L_{n+l}^{2n+l}(\rho)\right]^2\frac{\rho^2}{\alpha^2}\frac{d\rho}{\alpha} = 1$$

$$\text{or, } C^2\frac{1}{\alpha^3}\left[\frac{2n\{(n+1)!\}^3}{(n-l-1)!}\right] = 1 \text{ or, } C = \left[(\alpha)^3\frac{(n-l-1)!}{2n\{(n+1)!\}^3}\right]^{1/2}$$

Table 8.3 shows normalised radial wave function for $n = 1, 2, 3$ and $l = 0, 1, 2$.

Table 8.3: Normalised radial wave functions

n	l	$R_n^l(r)$
1	0	$2\left(\frac{1}{a_0}\right)^{3/2} e^{-r/a_0}$
2	0	$\frac{1}{2\sqrt{2}}\left(\frac{1}{a_0}\right)^{3/2}\left(2-\frac{r}{a_0}\right)e^{-\frac{r}{2a_0}}$
2	1	$\frac{1}{2\sqrt{6}}\left(\frac{1}{a_0}\right)^{3/2}\frac{r}{a_0}e^{-\frac{r}{2a_0}}$
3	0	$\frac{2}{81\sqrt{3}}\left(\frac{1}{a_0}\right)^{3/2}\left(27-18\frac{r}{a_0}+2\frac{r^2}{a_0^2}\right)e^{-\frac{r}{3a_0}}$
3	1	$\frac{4}{81\sqrt{6}}\left(\frac{1}{a_0}\right)^{3/2}\left(6-\frac{r}{a_0}\right)\frac{r}{a_0}e^{-\frac{r}{3a_0}}$
3	2	$\frac{4}{81\sqrt{30}}\left(\frac{1}{a_0}\right)^{3/2}\frac{r^2}{a_0^2}e^{-\frac{r}{3a_0}}$

The variation of the radial wave function for different n and l is depicted in Fig 8.7.

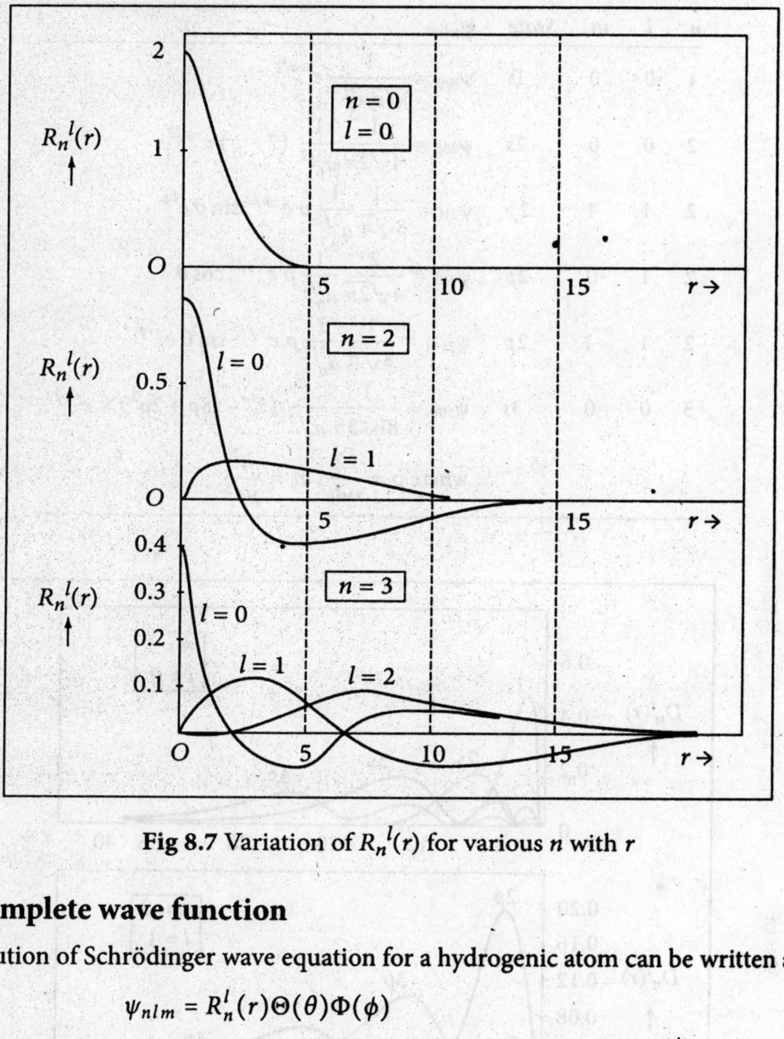

Fig 8.7 Variation of $R_n^l(r)$ for various n with r

8.6.8 The complete wave function

The complete solution of Schrödinger wave equation for a hydrogenic atom can be written as

$$\psi_{nlm} = R_n^l(r)\Theta(\theta)\Phi(\phi)$$

$$\text{or, } \psi_{nlm}(r,\theta,\phi) = \left[\frac{(2l+1)(l-m)!}{4\pi(l+m)!}\left(\frac{2Z}{na_0}\right)^3 \frac{(n-l-1)!}{2n\{(n+l)!\}^3}\right]^{1/2} e^{-\frac{Zr}{na_0}} \times \left(\frac{2Zr}{na_0}\right)^l L_{n+1}^{2n+l}\left\{\frac{2Zr}{na_0}\right\} e^{im\phi} P_l^m(\cos\theta) \quad (8.190)$$

Table 8.4 shows the real form of the complete eigenfunction ψ_{nlm} (excluding spin of the electron) for various values of n, l and m.

8.6.9 Radial probability density

We have the time-independent Schrödinger total wave function

$$\psi_{nlm}(r,\theta,\phi) = R_n^l(r)Y_l^m(\theta,\phi) \quad (8.191)$$

where $Y_l^m(\theta,\phi) = \Omega(\theta,\phi) = \Theta(\theta)\Phi(\theta)$ the spherical harmonics which are normalised wave functions.

Table 8.4: Real form of the complete eigenfunction ψ_{nlm} for various

n	l	m	*State*	ψ_{nlm}
1	0	0	1s	$\psi_{100} = \dfrac{1}{\sqrt{\pi}a_0^{3/2}} e^{-\rho/2}$
2	0	0	2s	$\psi_{200} = \dfrac{1}{4\sqrt{2\pi}} \dfrac{1}{a_0^{3/2}} (2-\rho) e^{-\rho/2}$
2	1	1	2p	$\psi_{211} = \dfrac{-1}{8\sqrt{\pi}} \dfrac{1}{a_0^{3/2}} \rho\, e^{-\rho/2} \sin\theta\, e^{i\phi}$
2	1	0	2p	$\psi_{210} = \dfrac{2}{4\sqrt{2\pi}} \dfrac{1}{a_0^{3/2}} \rho\, e^{-\rho/2} \cos\theta$
2	1	–1	2p	$\psi_{21-1} = \dfrac{1}{8\sqrt{\pi}} \dfrac{1}{a_0^{3/2}} \rho\, e^{-\rho/2} \sin\theta\, e^{-i\phi}$
3	0	0	3s	$\psi_{300} = \dfrac{1}{81\sqrt{3\pi}} \dfrac{1}{a_0^{3/2}} (27 - 18\rho + 2\rho^2) \times e^{-\rho/2}$,
				where $\rho = \dfrac{2Z}{na_0}$, $a_0 = \dfrac{\hbar^2}{\mu e^2}$

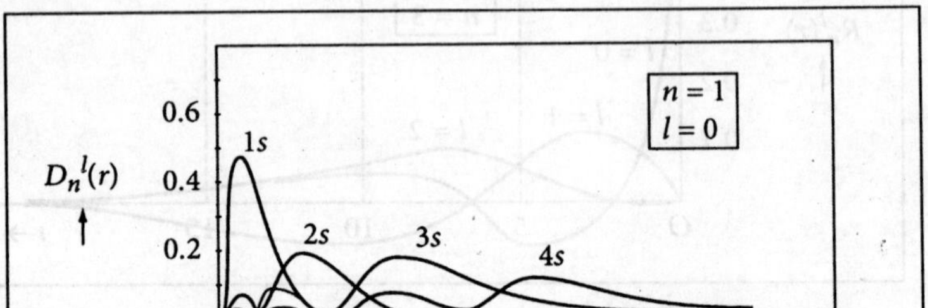

Fig 8.8 Variation of $D_n^l(r)$ with r in units of a_0

The probability of finding the electron in volume element $d\tau$ is given by $|\psi_{nlm}|^2\, d\tau = |\psi_{nlm}|\, r^2\, dr\, d\omega$ where $d\omega$ is the solid angle in the direction of θ, ϕ.

$$d\tau = r^2\, dr\, d\omega = r^2 \sin\theta\, dr\, d\theta\, d\phi.$$

$\therefore$ the probability of finding the electron in a spherical shell of radius r and thickness dr is

$$D_n^l(r) = \int_\omega |\psi_{nlm}|^2 r^2\, dr\, d\omega = \left|R_n^l(r)\right|^2 r^2\, dr \int_\omega |Y_l^m(\theta,\phi)|^2\, d\omega = r^2 \left|R_n^l(r)\right|^2 dr \tag{8.192}$$

Here, $D_n^l(r)$ is called the *radial probability density*.

$D_n^l(r)$ has the maximum value for states $1s(n=1, l=0)$, $2p(n=2, l=1)$, $3d(n=3, l=2)$. Fig 8.8 shows the variation of $D_n'(r)$ with r for different n and l.

8.6.10 Energy values for the hydrogen atom

From eq. (8.171), we have

$$\lambda' = \frac{Ze^2}{\hbar}\left[\frac{\mu}{2|E|}\right]^{1/2}$$

Squaring both sides, we get

$$\lambda'^2 = \frac{Z^2e^4\mu}{2\hbar^2|E|} \quad \text{or,}\ n^2 = \frac{Z^2\mu e^4}{2\hbar^2|E|} \quad [\text{replacing } \lambda' \text{ by } n]$$

So the energy of nth quantum level for the hydrogenic atom is

$$E_n = -|E_n| = -\frac{\mu Z^2 e^4}{2\hbar^2 n^2} \quad \text{or,}\ E_n = -\frac{\mu Z^2 e^4}{2\hbar^2 n^2} \tag{8.193}$$

For the hydrogen atom $Z = 1$ and hence

$$E_n = -\frac{\mu e^4}{2\hbar^2 n^2} \tag{8.194}$$

where $n = 1, 2, 3\ldots$ and is called the *total quantum number.*

Equation (8.193) shows that the energy values for different quantum states are discrete which is in agreement with old quantum theory and experiment.

8.6.11 Degeneracy

We have seen that relation of radial quantum number (k) and the azimuthal quantum number (l) with principal quantum number (n) is $n = k + l + 1$.

The above relation is satisfied for the combination of different k and l values to give a fixed n value. This means, there are several possible wave functions for a given value of n and hence given E_n. This is true for $n > 1$. If this happens, then we, say, that the nth state is *degenerate*, i.e., having different eigenstates corresponding to the same value of energy.

Again, for a particular value of n, $l = (n-1), (n-2), -2, -1, 0$, i.e., $(n-1)$ values and $m = l, (l-1), (l-2), \ldots, 0, \ldots, -(l-1), -l$, i.e., $(2l+1)$ values. Thus, the total number of energy levels or degeneracy is

$$\sum_{l=0}^{(n-1)} (2l+1) = \frac{2(n)(n-1)}{2} + n = n^2 \tag{8.195}$$

Thus, the state $n = 1$ is single, $n = 2$ is 4-fold and $n = 3$ is 9-fold degenerate as shown in Fig 8.9.

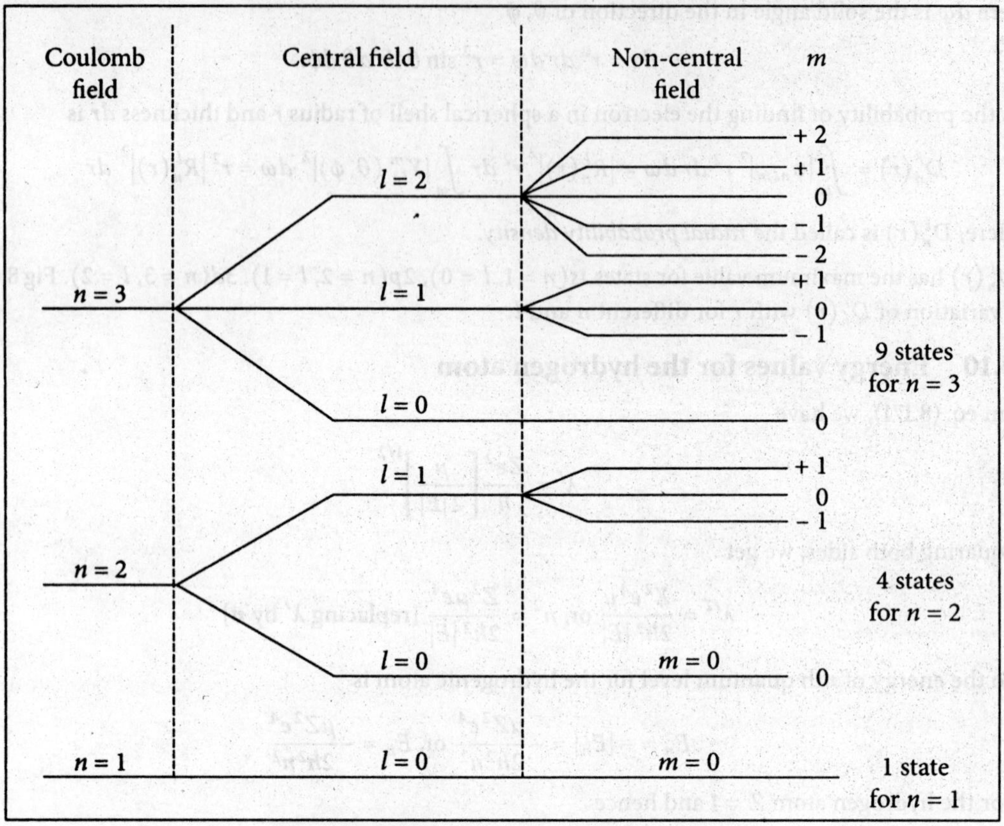

Fig 8.9 Degeneracy of hydrogenic atom

8.6.12 Polar graphs of probability distribution function

The probability distribution of electronic charge around the nucleus is given by

$$|Y_l^m(\theta,\phi)|^2 = |\Theta(\theta)\Phi(\phi)|^2 = \frac{1}{2\pi}|\Theta_{lm}(\theta)|^2$$

$$\because\ |\Phi(\phi)|^2 = \Phi^*(\phi)\Phi(\phi) = \frac{1}{2\pi}$$

So the probability distribution function depends on the value of $|\Theta_{lm}(\theta)|^2$ only.

For the s-electron, $l = 0$ and $m = 0$. $\therefore \Theta_{00} = 1/\sqrt{2}$.

$$\therefore P(\theta) = \text{the polar probability density} = \Theta_{00}^*\Theta_{00} = \tfrac{1}{2} = \text{constant}$$

$$\therefore Y_l^m(\theta,\phi) = \left(\frac{1}{4\pi}\right)^{1/2} = \text{constant} \tag{8.196}$$

The charge distribution for the s-state is thus *spherically symmetric* and independent of θ (Fig 8.10). Hence, probability of locating an electron in s-state has the same value at a given r in all directions.

For the p-state, $l = 1, 0$ and $m = 0, \pm 1$. For $l = 1$ and $m = 0$

$$\Theta_{10} = \sqrt{\tfrac{3}{2}}\cos\theta. \therefore\ P(\theta) = \tfrac{3}{2}\cos^2\theta$$

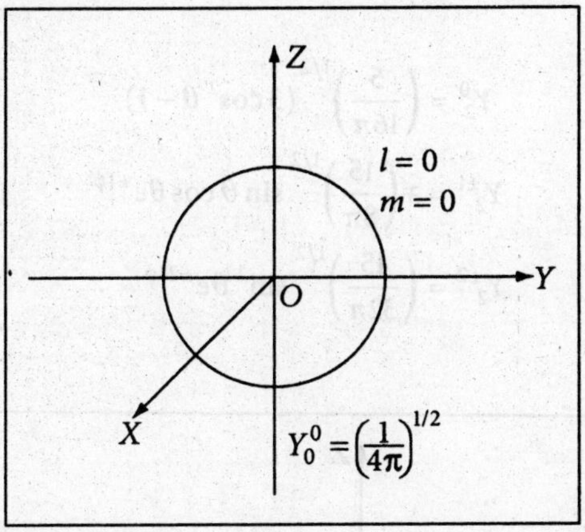

Fig 8.10 Electronic charge distribution for the *s*-state around the nucleus.

So $P(\theta)$ has the maximum value for $\theta = 0$ and $\theta = \pi$ and minimum value for $\theta = \pm\pi/2$.

$\therefore Y_1^0 = (3/4)^{1/2}\cos\theta$.

The variation of $P(\theta)$ for $l = 1$, $m = 0$, *p*-state is shown in Fig 8.11. The figure shows that the probability distribution has two lobes which are symmetrical about the *z*-axis.

When $l = 1$, $m = \pm 1$, *p*-state

$$\Theta_{1\pm 1} = \sqrt{\tfrac{3}{4}}\sin\theta \text{ and } P(\theta) = \Theta_{1\pm 1}^{*}\Theta_{1\pm 1} = \tfrac{3}{4}\sin^2\theta$$

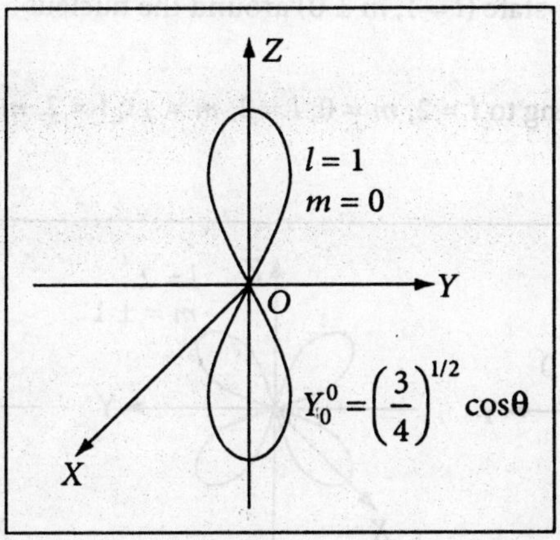

Fig 8.11 Electronic charge distribution for the *p*-state ($l = 1$, $m = 0$) around the nucleus

So when $\theta = \pm\pi/2$, $P(\theta)$ is the maximum and when $\theta = 0$ or π, $P(\theta)$ is minimum.

$\therefore Y_1^{\pm 1}(\theta, \phi) = \mp(3/8\pi)^{1/2}\sin\theta e^{\pm i\phi}$.

The variation of $P(\theta)$ for $l = 1$, $m = \pm 1$ with θ is shown in Fig 8.12.

For the d-state, $l = 2$, $m = 0$; $l = 2$, $m = \pm 1$ and $l = 2$, $m = \pm 2$. We can calculate $P(\theta)$ for d-state for which

$$Y_2^0 = \left(\frac{5}{16\pi}\right)^{1/2} (3\cos^2\theta - 1)$$

$$Y_2^{\pm 1} = \mp\left(\frac{15}{8\pi}\right)^{1/2} \sin\theta\cos\theta e^{\pm 1\phi}$$

$$Y_2^{\pm 2} = \left(\frac{15}{32\pi}\right)^{1/2} \sin^2\theta e^{\pm 2i\phi}$$

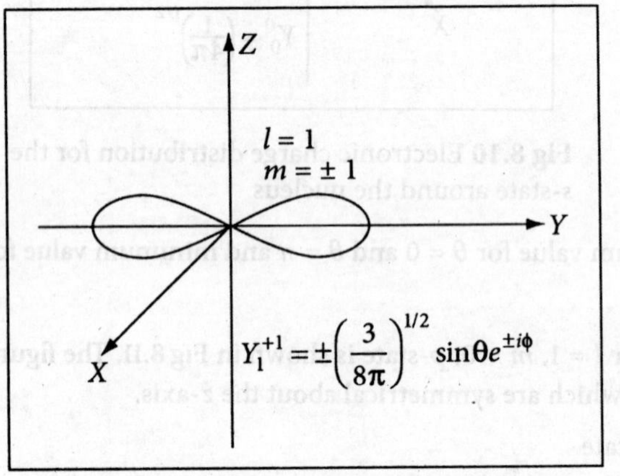

Fig 8.12 Electronic charge distribution for the p-state ($l = 1$, $m \pm 0$) around the nucleus

The distributions corresponding to $l = 2$, $m = 0$; $l = 2$, $m = \pm 1$; $l = 2$, $m = \pm 2$ are shown in Figs 8.13, 8:14 and 8.15 respectively.

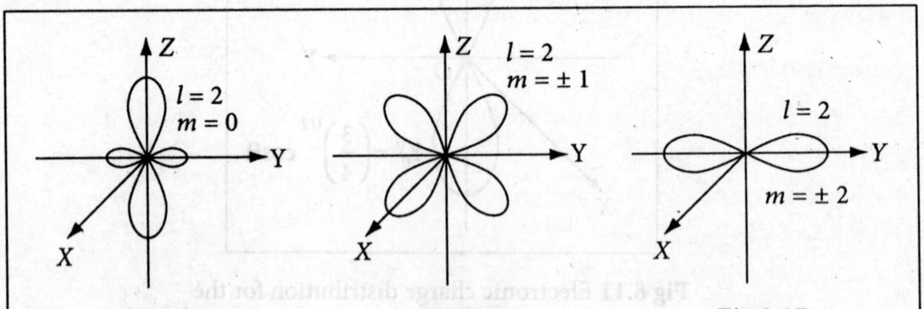

Fig 8.13 **Fig 8.14** **Fig 8.15**

Electronic charge distribution for the d-state around the nucleus.

Likewise, for the *f-state* $l = 3$, $m = 0, \pm1, \pm2, \pm3$ and

$$Y_3^0 = \left(\frac{7}{16\pi}\right)^{1/2} (5\cos^2\theta - 3\cos\theta) \quad Y_3^{\pm1} = \mp\left(\frac{21}{64\pi}\right)^{1/2} \sin\theta(5\cos^2\theta - 1)\, e^{\pm i\phi}$$

$$Y_3^{\pm2} = \left(\frac{105}{32\pi}\right)^{1/2} \sin^2\theta\cos\theta\, e^{\pm2\phi} \quad Y_3^{\pm3} = \mp\left(\frac{35}{64\pi}\right)^{1/2} \sin^2\theta\, e^{\pm3i\phi}$$

The distribution curves for $l = 3$, $m = 0$; $l = 2$, $m = \pm1$, $l = 2$, $m = \pm2$ and $l = 2$, $m = \pm3$ are shown in Figs 8.16, 8.17, 8.18 and 8.19 respectively.

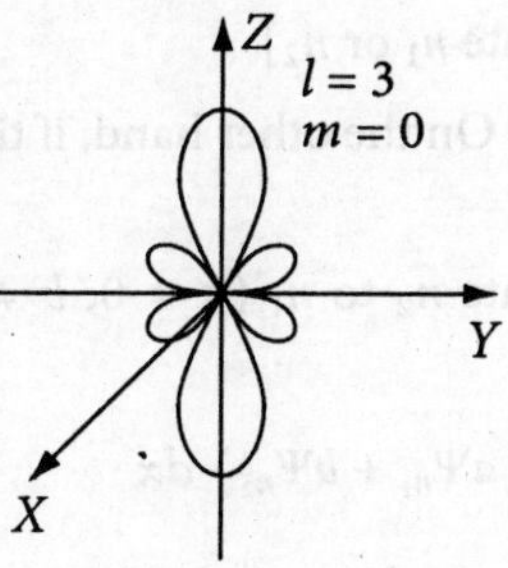

Fig 8.16

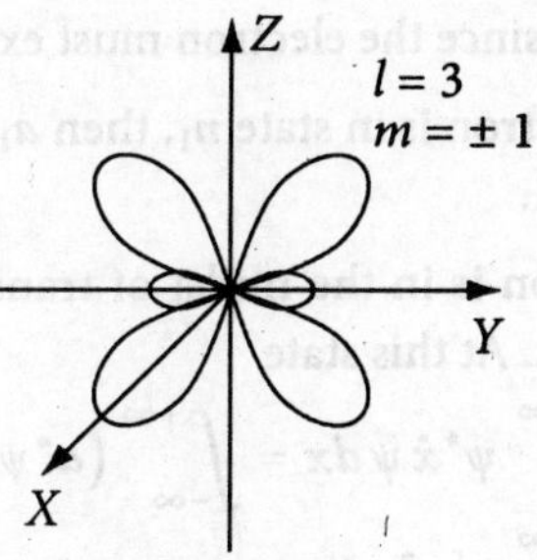

Fig 8.17

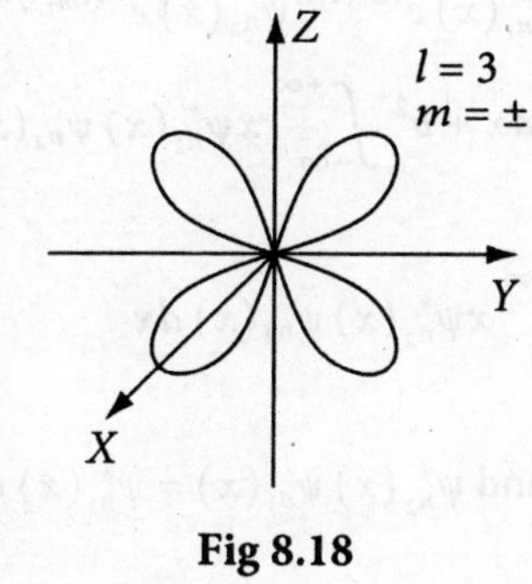

Fig 8.18

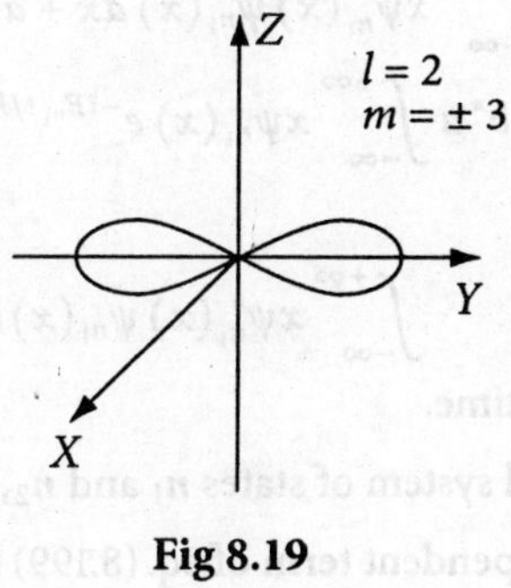

Fig 8.19

Electronic charge distribution for *f*-state around the nucleus.

8.6.13 Selection rules

We know, whenever the average position of an electron relative to the nucleus oscillates, electromagnetic radiations are emitted, the frequency of which is the same as the oscillation of the electron.

Let the wave function of an electron moving along the X-direction in the nth state at time t be represented by $\psi_n(x, t) = \psi_n(x)\, e^{-\frac{iE_n t}{\hbar}}$.

Then the average position of the electron is

$$\langle x \rangle = \int_{-\infty}^{+\infty} \psi_n^*(x, t)\, \hat{x}\psi_n(x, t)\, dx = \int_{-\infty}^{+\infty} \psi_n^*(x)\, e^{iE_n t/\hbar}\, x\psi_n(x)\, e^{-iE_n t/\hbar}\, dx$$

$$= \int_{-\infty}^{+\infty} \psi_n^*(x)\, x\psi_n(x)\, dx \tag{8.197}$$

$\because$ $\langle x \rangle$ is constant in time, the average position of the electron does not oscillate and hence no radiation is emitted.

Hence, according to quantum mechanics *an electron in a specific quantum state does not radiate energy.*

Now, we consider that an electron, by absorbing energy, changes state from n_1 to n_2 $(n_2 > n_1)$. When the electron reverts back from the excited state n_2 to n_1, radiation is emitted. The wave function of the electron capable of existing in states n_1 and n_2 is

$$\psi = a\psi_{n_1} + b\psi_{n_2} \tag{8.198}$$

where $a^* a$ = probability that the electron be in state n_1 and $b^* b$ = probability that the electron be in state n_2.

$\therefore a^* a + b^* b = 1$ [since the electron must exist either in state n_1 or n_2]

If at $t = 0$ the electron is in state n_1, then $a_1 = 1$ and $b = 0$. On the other hand, if the electron be in state n_2, then $a = 0$, $b = 1$.

When the electron is in the midst of transition from state n_2 to n_1 $(a \neq 0,\ b \neq 0)$, electromagnetic radiation comes out. At this state

$$\langle x \rangle = \int_{-\infty}^{+\infty} \psi^* \hat{x}\, \psi\, dx = \int_{-\infty}^{+\infty} (a^* \psi_{n_1}^* + b^* \psi_{n_2}^*)\, x\, (a\Psi_{n_1} + b\Psi_{n_2})\, dx$$

$$= \int_{-\infty}^{+\infty} x \left[a^2 \psi_{n_1}^* \psi_{n_1} + a^* b \psi_{n_1}^* \psi_{n_2} + b^* a \psi_{n_1} \psi_{n_2}^* + b^2 \psi_{n_2}^* \psi_{n_2}\right] dx$$

$$\text{or, } \langle x \rangle = a^2 \int_{-\infty}^{+\infty} x\psi_{n_1}^*(x)\, \psi_{n_1}(x)\, dx + a^* b \int_{-\infty}^{+\infty} x\psi_{n_1}^*(x)\, e^{iE_{n_1} t/\hbar} \psi_{n_2}(x)\, e^{-iE_{n_2} t/\hbar}\, dx$$

$$+ b^* a \int_{-\infty}^{+\infty} x\psi_{n_1}(x)\, e^{-iE_{n_1} t/\hbar} \psi_{n_2}^*\, e^{iE_{n_2} t/\hbar}\, dx + b^2 \int_{-\infty}^{+\infty} x\psi_{n_2}^*(x)\, \psi_{n_2}(x)\, dx \tag{8.199}$$

Here

$$\int_{-\infty}^{+\infty} x\psi_{n_1}^*(x)\, \psi_{n_1}(x)\, dx \text{ and } \int_{-\infty}^{+\infty} x\psi_{n_2}^*(x)\, \psi_{n_2}(x)\, dx$$

are independent of time.

For a finite bound system of states n_1 and n_2, $b^* a = a^* b$ and $\psi_{n_2}^*(x)\, \psi_{n_1}(x) = \psi_{n_1}^*(x)\, \psi_{n_2}(x)$.

Now, the time-dependent term of eq. (8.199) is

$$a^* b \int_{-\infty}^{+\infty} x\psi_{n_1}^* \psi_{n_2} \left[e^{-i\frac{(E_{n_2} - E_{n_1})t}{\hbar}} + e^{i\frac{(E_{n_2} - E_{n_1})t}{\hbar}}\right] dx$$

But,

$$\frac{e^{\frac{-i(E_{n_2} - E_{n_1})t}{\hbar}} + e^{\frac{i(E_{n_2} - E_{n_1})t}{\hbar}}}{2} = \cos\left(\frac{E_{n_2} - E_{n_1}}{\hbar}\right)t$$

$$\therefore a^* b \int_{-\infty}^{+\infty} x\psi_{n_1}^* \psi_{n_1} \left[e^{\frac{-i(E_{n_2} - E_{n_1})t}{\hbar}} + e^{\frac{i(E_{n_2} - E_{n_1})t}{\hbar}}\right] dx$$

$$= 2a^* b \cos\left(\frac{E_{n_2} - E_{n_1}}{\hbar}\right)t \int_{-\infty}^{+\infty} x\psi_{n_1}^*(x)\psi_{n_2}(x)\, dx \tag{8.200}$$

The time-dependent factor of eq. (8.199) is

$$\cos\left(\frac{E_{n_2} - E_{n_1}}{\hbar}\right)t = \cos\left[2\pi\frac{(E_{n_2} - E_{n_1})}{h}t\right] = \cos 2\pi\nu t \tag{8.201}$$

Thus, $\langle x \rangle$ oscillates with a frequency

$$\nu = \frac{E_{n_2} - E_{n_1}}{h} \tag{8.202}$$

The amplitude of oscillation of $\langle x \rangle$ is

$$2a^* b \int_{-\infty}^{+\infty} x\psi_{n_1}^*(x)\,\psi_{n_2}\,dx$$

Hence,

$$\langle x \rangle = a^2 \int_{-\infty}^{+\infty} x\psi_{n_1}^*(x)\,\psi_{n_1}(x)\,dx + b^2 \int_{-\infty}^{+\infty} x\psi_{n_1}^*(x)\,\psi_{n_2}(x)\,dx + 2a^* b \cos 2\pi\nu t \int_{-\infty}^{+\infty} x\psi_{n_1}^*(x)\,\psi_{n_2}(x)\,dx \tag{8.203}$$

Thus, when the electron in a hydrogen atom makes a transition from state n_2 to n_1, $\langle x \rangle$ oscillates with a frequency ν, which is the same as predicted by Bohr.

8.6.14 Forbidden transition and allowed transition

As the motion of the electron can take along any direction $\vec{r}$, the amplitude of oscillation can be written as

$$M_{n_1 n_2} = \int_{\text{all space}} \vec{r}\psi_{n_1}^*(\vec{r})\,\psi_{n_2}(\vec{r})\,d\tau \tag{8.204}$$

where $d\tau$ is the volume element and $M_{n_1 n_2}$ is the matrix element of transition vector $\vec{r}$.

For a transition between states n_1 and n_2

$$M_{n_1 n_2} = \int_{\text{all space}} \vec{r}\psi_{n_1}^*(\vec{r})\psi_{n_2}(\vec{r})\,d\tau = 0$$

and the intensity of the emitted radiation $I \propto M_{n_1 n_2}^2$.

On the other hand, if $M_{n_1 n_2} = 0$, then $I = 0$, i.e., there is no transition. So the *forbidden transition* is characterised by the condition

$$M_{n_1 n_2} = \int_{\text{all space}} \vec{r}\psi_{n_1}^*(\vec{r})\psi_{n_2}(\vec{r})\,d\tau = 0 \tag{8.205}$$

The condition for allowed transition is

$$M_{n_1} M_{n_2} = \int_{\text{all space}} \vec{r}\psi_{n_1}^*(\vec{r})\psi_{n_2}(\vec{r})\,d\tau \neq 0 \tag{8.206}$$

The initial and the final states of an electron in hydrogen atom are specified by three quantum numbers n, l, m_l (replacing m by m_l).

For radiative transition of the electron, the following conditions must hold good:

(a) *The magnetic quantum number m_l should change by unity or remain unchanged, i.e., $\Delta m_l = \pm 1$ or 0.*

(b) *The orbital quantum number l should change by unity, i.e., $\Delta l = \pm 1$.*

Now, for the derivation of selection rule for m_l we proceed as follows:

For the hydrogen atom, we have the complete wave function, $\psi_{nlm_l} = R_n^l(r)\Theta(\theta)\Phi(\phi)$.

$$\therefore M_{n_1 n_2} = \int_{\text{all space}} \vec{r} R_{n_1}^{l_1^*}(r) R_{n_2}^{l_2}(r) \Theta_{l_1}^*(\theta)\Theta_{l_2}(\theta)\Phi_{m_{l_1}}^*(\phi)\Phi_{m_{l_2}}(\phi) \times r^2 \sin\theta\,dr\,d\theta\,d\phi$$

$$= \int_{r=0}^{\infty} R_{n_1}^{l_1^*} R_{n_2}^{l_2} r^2\,dr \int_{\theta=0}^{\pi} \Theta_{l_1}^* \Theta_{l_2} \sin\theta\,d\theta \int_{\phi=0}^{2\pi} \vec{r}\Phi_{m_{l_1}}\Phi_{m_{l_2}}\,d\phi \tag{8.207}$$

For the derivation of selection rule for m_l we consider

$$\int_{\phi=0}^{2\pi} \vec{r}\Phi^*_{m_{l_1}} \Phi_{m_{l_2}}\, d\phi = \hat{i}\int_0^{2\pi} x\Phi^*_{m_{l_1}} \Phi_{m_{l_2}}\, d\phi + \hat{j}\int_0^{2\pi} y\Phi^*_{m_{l_1}} \Phi_{m_{l_2}}\, d\phi + \hat{k}\int_0^{2\pi} z\Phi^*_{m_{l_1}} \Phi_{m_{l_2}}\, d\phi$$

$$= \hat{i}I_x + \hat{j}I_y + \hat{k}I_z \qquad (8.208)$$

where

$$I_x = \int_0^{2\pi} x\Phi^*_{m_{l_1}} \Phi_{m_{l_2}}\, d\phi,\ I_y = \int_0^{2\pi} y\Phi^*_{m_{l_1}} \Phi_{m_{l_2}}\, d\phi \text{ and } I_z = \int_0^{2\pi} z\Phi^*_{m_{l_1}} \Phi_{m_{l_2}}\, d\phi$$

But, $x = r\sin\theta\cos\phi$, $y = r\sin\theta\sin\phi$ and $z = r\cos\theta$.

$$\therefore I_x = r\sin\theta \int_0^{2\pi} \Phi^*_{m_{l_1}} \Phi_{m_{l_2}} \cos\phi\, d\phi = \frac{r\sin\theta}{2\pi}\int_0^{2\pi} e^{-im_{l_1}\phi} \times e^{+im_{l_2}} \times e^{i\phi}\cos\phi\, d\phi$$

where $\Phi^*_{m_{l_1}} = \frac{1}{\sqrt{2\pi}} e^{-im_{l_1}\phi}$ and $\Phi_{m_{l_2}} = \frac{1}{\sqrt{2\pi}} e^{im_{l_2}\phi}$

Again,

$$\because \cos\phi = \frac{e^{i\phi} + e^{-i\phi}}{2}$$

$$\therefore I_x = \frac{r\sin\theta}{4\pi}\left[\int_0^{2\pi} e^{-im_{l_1}\phi}\cdot e^{im_{l_2}\phi}\cdot e^{i\phi}\, d\phi + \int_0^{2\pi} e^{-im_{l_1}\phi}\cdot e^{im_{l_2}\phi}\cdot e^{-i\phi}\, d\phi\right]$$

$$= \frac{r\sin\theta}{4\pi}\left[\int_0^{2\pi} e^{i(m_{l_2}-m_{l_1}+1)\phi}\, d\phi + \int_0^{2\pi} e^{i(m_{l_2}-m_{l_1}-1)\phi}\, d\phi\right] \qquad (8.209)$$

Since,

$$m_{l_1} = 0, \pm 1, \pm 2, \pm 3, \ldots, \pm l_1$$

$$\text{and } m_{l_2} = 0, \pm 1, \pm 2, \pm 3, \ldots \pm l_2$$

$(m_{l_2} - m_{l_1} + 1)$ and $(m_{l_2} - m_{l_1} - 1)$ are each equal to 0 or an integral value, $k = \pm 1, \pm 2, \pm 3$, etc.

Again,

$$\int_0^{2\pi} e^{ik\phi} = \int_0^{2\pi} \cos k\phi\, d\phi + i\int_0^{2\pi} \sin k\phi\, d\phi$$

$$= 0, \text{ for, } k = \pm 1, \pm 2, \pm 3, \ldots$$

$\therefore m_{l_2} - m_{l_1} + 1 = 0$, or $m_{l_2} - m_{l_1} = -1$, or $\Delta m_l = -1$.

Similarly, from $m_{l_2} - m_{l_1} - 1 = 0$, $\Delta m_l = +1$.

So if $\Delta m_l = \pm 1$, then only the integral represented by eq. (8.209) has a definite value and hence non-zero.

Similarly, we can show that I_y leads to the same result, viz., for allowed transition $\Delta m_l = \pm 1$.

But, considering

$$I_z = \int_0^{2\pi} z\Phi^*_{m_{l_1}} \Phi_{m_{l_2}}\, d\phi = r\cos\theta\int_0^{2\pi} \Phi^*_{m_{l_1}} \Phi_{m_{l_2}}\, d\phi = \frac{r\cos\theta}{4\pi}\int_0^{2\pi} e^{i(m_{l_2}-m_{l_1})\phi}\, d\phi \qquad (8.210)$$

The above integral vanishes for all values of Δm_l except $\Delta m_l = 0$. So the condition for allowed transition from this integral comes out to be $\Delta m_l = 0$.

Hence, *the conditions for allowed transition in a hydrogen atom for magnetic quantum number are* $\Delta m_l = 0$ *or* ± 1.

It is to be kept in mind that not only for hydrogen atom, but for all hydrogenic atoms *the selection rule for allowed transition is*

$$\left.\begin{array}{c}\Delta m_l = 0, \pm 1 \\ \Delta l = \pm 1\end{array}\right\} \tag{8.211}$$

8.7 The rigid rotator

A rigid rotator is a system of two spherical masses m_1 and m_2 situated at fixed distance r_1 and r_2 from the centre of mass of the system and capable of rotating about an axis passing through the centre of mass and normal to the plane containing the two particles [Fig 8.20].

If the particles m_1 and m_2 are constrained to move in a plane, then the axis of rotation is fixed and the system is called the *rigid rotator with fixed axis.*

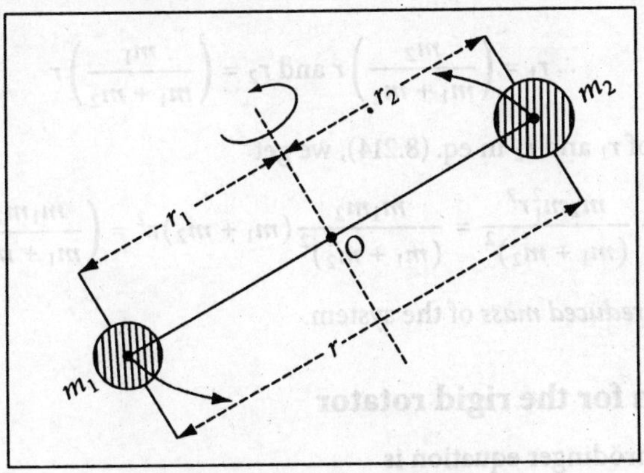

Fig 8.20 The rigid rotator

On the other hand, if the plane of the two particles can move, then the axis of rotation is free to take any orientation in space and the system a *rigid rotator with free axis.* The rigid rotator with free axis has an important role in the study of diatomic molecules like HCl, CO, etc.

8.7.1 Rigid rotator with free axis

Energy: If (x, y, z) be the coordinates of a moving particle at any time t, then its kinetic energy is given by

$$T = \tfrac{1}{2}m\left(\dot{x}^2 + \dot{y}^2 + \dot{z}^2\right)$$

where m is the mass of the particle and $\dot{x}$, $\dot{y}$ and $\dot{z}$ are the components of velocity of the particle along x-, y- and z-axes respectively.

If (r, θ, ϕ) be the polar coordinates of the particle, then $x = r\sin\theta\cos\phi$, $y = r\sin\theta\sin\phi$, $z = r\cos\theta$ and $T = \tfrac{1}{2}m\left(\dot{r}^2 + r^2\dot{\theta}^2 + r^2\dot{\phi}^2\sin^2\theta\right)$.

Since, r is constant, $\dot{r} = 0$.

$$\therefore T = \tfrac{1}{2}m\left(r^2\dot{\theta}^2 + r^2\dot{\phi}^2\sin^2\theta\right) = \tfrac{1}{2}mr^2\left(\dot{\theta}^2 + \dot{\phi}^2\sin^2\theta\right) \tag{8.212}$$

The kinetic energy of the rigid rotator under consideration is given by

$$T = \tfrac{1}{2}m_1r_1^2\left(\dot{\theta}^2 + \dot{\phi}^2\sin^2\theta\right) + \tfrac{1}{2}m_2r_2^2\left(\dot{\theta}^2 + \dot{\phi}^2\sin^2\theta\right) = \tfrac{1}{2}I\left(\dot{\theta}^2 + \dot{\phi}^2\sin^2\theta\right) \tag{8.213}$$

where

$$I = m_1 r_1^2 + m_2 r_2^2 \tag{8.214}$$

Here, I represents the moment of inertia of the rigid rotator about the axis of rotation.

Comparing eqs (8.212) and (8.213) we note that the rigid rotator is equivalent to a system of one particle of mass I situated at a distance $r = 1$ from 0.

In other words, the rigid rotator is just like a single particle of mass I moving on the surface of a sphere of radius unity.

Now, by the definition of centre of mass

$$m_1 r_1 = m_2 r_2 \text{ or, } \frac{r_1}{1/m_1} = \frac{r_2}{1/m_2} = \frac{r_1 + r_2}{1/m_1 + 1/m_2} = \frac{r}{1/m_1 + 1/m_2} \text{ or, } m_1 r_1 = m_2 r_2 = \left(\frac{m_1 m_2}{m_1 + m_2}\right) r$$

$$\therefore r_1 = \left(\frac{m_2}{m_1 + m_2}\right) r \text{ and } r_2 = \left(\frac{m_1}{m_1 + m_2}\right) r$$

Substituting the values of r_1 and r_2 in eq. (8.214), we get

$$I = \frac{m_1 m_2^2 r^2}{(m_1 + m_2)^2} + \frac{m_2 m_1^2 r^2}{(m_1 + m_2)^2} = \frac{m_1 m_2}{(m_1 + m_2)^2}(m_1 + m_2) r^2 = \left(\frac{m_1 m_2}{m_1 + m_2}\right) r^2 = \mu r^2 \tag{8.215}$$

where $\mu = \dfrac{m_1 m_2}{m_1 + m_2}$ is the *reduced mass* of the system.

8.7.2 Wave equation for the rigid rotator

The three-dimensional Schrödinger equation is

$$-\frac{\hbar^2}{2m} \nabla^2 \psi + V\psi = E\psi$$

In polar coordinates

$$\frac{1}{r^2}\frac{\partial}{\partial r}\left(r^2 \frac{\partial \psi}{\partial r}\right) + \frac{1}{r^2 \sin\theta}\frac{\partial}{\partial \theta}\left(\sin\theta \frac{\partial \psi}{\partial \theta}\right) + \frac{1}{r^2 \sin^2\theta}\frac{\partial^2 \psi}{\partial \phi^2} + \frac{2m}{\hbar^2}(E - V)\psi = 0$$

The wave function $\psi(r, \theta, \phi)$ is a function of r, θ and ϕ. Since, for the rigid rotator there is no potential energy, ψ is independent on r.

Therefore, the above equation reduces to

$$\frac{1}{\sin\theta}\frac{\partial}{\partial \theta}\left(\sin\theta \frac{\partial \psi}{\partial \theta}\right) + \frac{1}{\sin^2\theta}\frac{\partial^2 \psi}{\partial \phi^2} = -\frac{2I}{\hbar^2} E\psi \tag{8.216}$$

putting $r = 1$ and $m = I$ for the rigid rotator.

$$\text{or, } -\hbar^2 \left[\frac{1}{\sin\theta}\frac{\partial}{\partial \theta}\left(\sin\theta \frac{\partial}{\partial \theta}\right) + \frac{1}{\sin^2\theta}\frac{\partial^2}{\partial \phi^2}\right] Y(\theta, \phi) = 2IEY(\theta, \phi)$$

$$\text{or, } \hat{L}^2 Y(\theta, \phi) = 2IEY(\theta, \phi) \tag{8.217}$$

replacing ψ by $Y(\theta, \phi)$. In eq. (8.217) $\hat{L}^2$ is the operator for the square of the angular momentum.

8.7.3 Eigenfunctions and eigenvalues

The eigenfunctions $Y_l^m(\theta, \phi)$ for the rotator are the solutions of eq. (8.216) and are given by

$$Y_l^m(\theta, \phi) = \left[\frac{2l+1}{4\pi}\frac{(l-m)!}{(l+m)!}\right]^{1/2}(-1)^m e^{im\phi} P_l^m(\cos\theta), \ (0 \leqslant m \leqslant l)$$

Now, comparing eq. (8.217) with the equation

$$\hat{L}^2 Y(\theta, \phi) = l(l+1)\hbar^2 Y(\theta, \phi) \tag{8.218}$$

We get

$$2IE = l(l+1)\hbar^2 \text{ or, } E_l = \frac{l(l+1)\hbar^2}{2I} \tag{8.219}$$

where $l = 0, 1, 2, \ldots$.

Thus, the energy eigenvalues of the rigid rotator depends on l alone. For each l $(l > 0)$ there are $(2l+1)$ values of m, viz., $m = 0, \pm1, \pm2, \ldots, \pm l$. This means there are $(2l+1)$ eigenfunctions for each energy state and hence the degeneracy of the state is $(2l+1)$-fold. Eq. (8.219) represents the allowed energy values of the rigid rotator with free axis.

8.7.4 Rigid rotator in a fixed plane

A rigid rotator has no potential energy and hence the wave function ψ does not depend on the radial coordinate r. If the motion of the rotator is confined to the XY-plane only, then the polar angle $\theta = \frac{\pi}{2}$. Under this condition the wave equation of the rigid rotator can be written as

$$\frac{\partial^2\Phi}{\partial\phi^2} + \frac{2IE}{\hbar^2}\Phi = 0, \ [I = \mu r^2] \tag{8.220}$$

where Φ is a function of the azimuthal angle ϕ and can be expressed as $\Phi(\phi)$. Now, eq. (8.218) becomes

$$\frac{1}{\Phi}\frac{d^2\Phi}{d\phi^2} + \frac{2IE}{\hbar^2} = 0 \text{ or, } \frac{1}{\Phi}\frac{d^2\Phi}{d\phi^2} = -\frac{2IE}{\hbar^2} = -m^2 \text{ (a constant)}$$

$$\therefore \frac{1}{\Phi}\frac{d^2\Phi}{d\phi^2} + m^2 = 0 \text{ or, } \frac{d^2\Phi}{d\phi^2} + m^2\Phi = 0 \tag{8.221}$$

where

$$\frac{2IE}{m^2} = \hbar^2 \tag{8.222}$$

Equation (8.221) is the time-independent Schrödinger wave equation for the rigid rotator with a fixed axis.

8.7.5 Eigenfunction for the rotator

The solution of eq. (8.221) is

$$\Phi_m = Ae^{im\phi} \tag{8.223}$$

where A is a constant and $m = 0, \pm1, \pm2, \ldots$.

Now, the normalisation condition is

$$\int_0^{2\pi} \Phi_m^* \Phi_m \, d\phi = 1 \text{ or, } A^2 \int_0^{2\pi} e^{im\phi} \cdot e^{-im\phi} \, d\phi = 1 \text{ or, } A^2 \int_0^{2\pi} d\phi = 1 \text{ or, } A = \frac{1}{\sqrt{2\pi}}$$

this is the normalising constant of the wave function of the rigid rotator with a fixed axis.

Hence, the normalised wave function of the rotator is

$$\boxed{\Phi_m = \frac{1}{\sqrt{2\pi}} e^{im\phi}} \quad m = 0, \pm 1, \pm 2, \ldots \tag{8.224}$$

8.7.6 Energy eigenvalues

The energy eigenvalues of the rigid rotator with a fixed axis is given by

$$\frac{2IE_m}{\hbar^2} = m^2 \text{ [cf. (8.222)]}$$

$$\text{or, } \boxed{E_m = \frac{m^2\hbar^2}{2I}} \tag{8.225}$$

Equation (8.225) represents the rotational energy eigenvalues of the rigid rotator with a fixed axis.

8.8 Three-dimensional harmonic oscillator

(a) *Solution in Cartesian rectangular coordinates*: We consider a particle being acted upon a force proportional to displacement and directed towards a fixed point O. If $\vec{r} = \hat{i}x + \hat{j}y + \hat{k}z$ be the displacement of the particle from O at any instant of time t, then its potential energy

$$V(r) = \tfrac{1}{2}kr^2 = \tfrac{1}{2}k\left(x^2 + y^2 + z^2\right) \tag{8.226}$$

where k is the force constant. The potential energy $V(r)$ is spherically symmetric.

The time-independent Schrödinger equation for the oscillator is

$$-\frac{\hbar^2}{2m}\nabla^2 \psi + \tfrac{1}{2}kr^2 \psi = E\psi \tag{8.227}$$

$$\text{or, } \nabla^2 \psi + \frac{2m}{\hbar^2}\left[E - \tfrac{1}{2}\left(k_x x^2 + k_y y^2 + k_z z^2\right)^2\right]\psi = 0$$

$$\text{or, } \nabla^2 \psi + \left[\lambda - \left(a_x^2 x^2 + a_y^2 y^2 + a_z^2 z^2\right)\right]\psi = 0 \tag{8.228}$$

where $\lambda = \frac{2mE}{\hbar^2}$, $\alpha_x = \sqrt{\frac{mk_x}{\hbar^2}}$, $\alpha_y = \sqrt{\frac{mk_y}{\hbar^2}}$ and $\alpha_z = \sqrt{\frac{mk_z}{\hbar^2}}$.

Applying the method of separation of variables, the solution of eq. (8.228) can be written as

$$\psi(x, y, z) = X(x)Y(y)Z(z) \tag{8.229}$$

Using eq. (8.229) in eq. (8.228), we get after dividing throughout by XYZ

$$\frac{1}{X}\frac{d^2X}{dx^2} + \frac{1}{Y}\frac{d^2Y}{dy^2} + \frac{1}{Z}\frac{d^2Z}{dz^2} + \left[\lambda - \left(\alpha_x x^2 + \alpha_y y^2 + \alpha_z z^2\right)\right] = 0 \tag{8.230}$$

$$\left.\begin{aligned} \text{or, } &\left(\frac{d^2X}{dx^2} - \alpha_x^2 x^2\right) + \lambda_x = 0 \\ &\left(\frac{d^2Y}{dy^2} - \alpha_y^2 y^2\right) + \lambda_y = 0 \\ &\left(\frac{d^2Z}{dz^2} - \alpha_z^2 z^2\right) + \lambda_z = 0 \end{aligned}\right\} \tag{8.231}$$

where $\lambda = \lambda_x + \lambda_y + \lambda_z$

$$\text{or, } \frac{d^2X}{dx^2} + \left(\lambda_x - \alpha_x^2 x^2\right) X = 0 \tag{8.232}$$

$$\frac{d^2Y}{dy^2} + \left(\lambda_y - \alpha_y^2 y^2\right) Y = 0 \tag{8.233}$$

$$\frac{d^2Z}{dz^2} + \left(\lambda_z - \alpha_z^2 z^2\right) Z = 0 \tag{8.234}$$

Each of eqs (8.232), (8.233) and (8.234) represents the differential equation of one-dimensional harmonic oscillator having normalised wave functions:

$$X(x) = N_{n_x} e^{-\frac{\alpha_x x^2}{2}} H_{n_x}\left(\alpha_x^{1/2} x\right) \tag{8.235}$$

$$Y(y) = N_{n_y} e^{-\frac{\alpha_y y^2}{2}} H_{n_y}\left(\alpha_y^{1/2} y\right) \tag{8.236}$$

$$Z(z) = N_{n_z} e^{-\frac{\alpha_z z^2}{2}} H_{n_z}\left(\alpha_z^{1/2} z\right) \tag{8.237}$$

where $n_x, n_y, n_z = 0, 1, 2, \ldots.$

Hence, the *normalised wave function* for the three-dimensional harmonic oscillator is

$$\psi\left(n_x, n_y, n_z\right)(x, y, z) = N_{n_x} N_{n_y} N_{n_z} e^{-1/2\left(\alpha_x x^2 + \alpha_y y^2 + \alpha_z z^2\right)} \times H_{n_x}\left(\alpha_x^{1/2} x\right) H_{n_y}\left(\alpha_y^{1/2} y\right) H_{n_z}\left(\alpha_z^{1/2} z\right) \tag{8.238}$$

The corresponding normalisation factor is

$$N_{n_x n_y n_z} = \frac{\alpha_x^{1/2}\,\alpha_y^{1/2}\,\alpha_z^{1/2}}{\pi^{3/2}\, 2^{(n_x+n_y+n_z)}\, n_x!\, n_y!\, n_z!} \tag{8.239}$$

The energy eigenvalues are

$$E_x = \left(n_x + \tfrac{1}{2}\right)\hbar\omega_x,\ E_y = \left(n_y + \tfrac{1}{2}\right)\hbar\omega_y,\ E_z = \left(n_z + \tfrac{1}{2}\right)\hbar\omega_z$$

$$\text{or, } E_{n_x n_y n_z} = \left(n_x + n_y + n_z + \tfrac{3}{2}\right)\hbar\omega \text{ or, } E_n = \left(n + \tfrac{3}{2}\right)\hbar\omega \tag{8.240}$$

with $n = n_x + n_y + n_z$.

Thus, E_n depends on n. For a given value of n the wave functions $\psi_{n_x n_y n_z}$ are a set of *linearly independent functions* each corresponding to a sets of integers n_x, n_y, n_z.

Hence, except the lowest one, all the energy levels are *degenerate,* the degree of degeneracy being $\frac{1}{2}(n+1)(n+2)$. [cf. Quantum Mechanics by J.L. Powell and B. Crasemann, p. 205]

A few energy levels of the three-dimensional harmonic oscillator are shown in Fig 8.21.

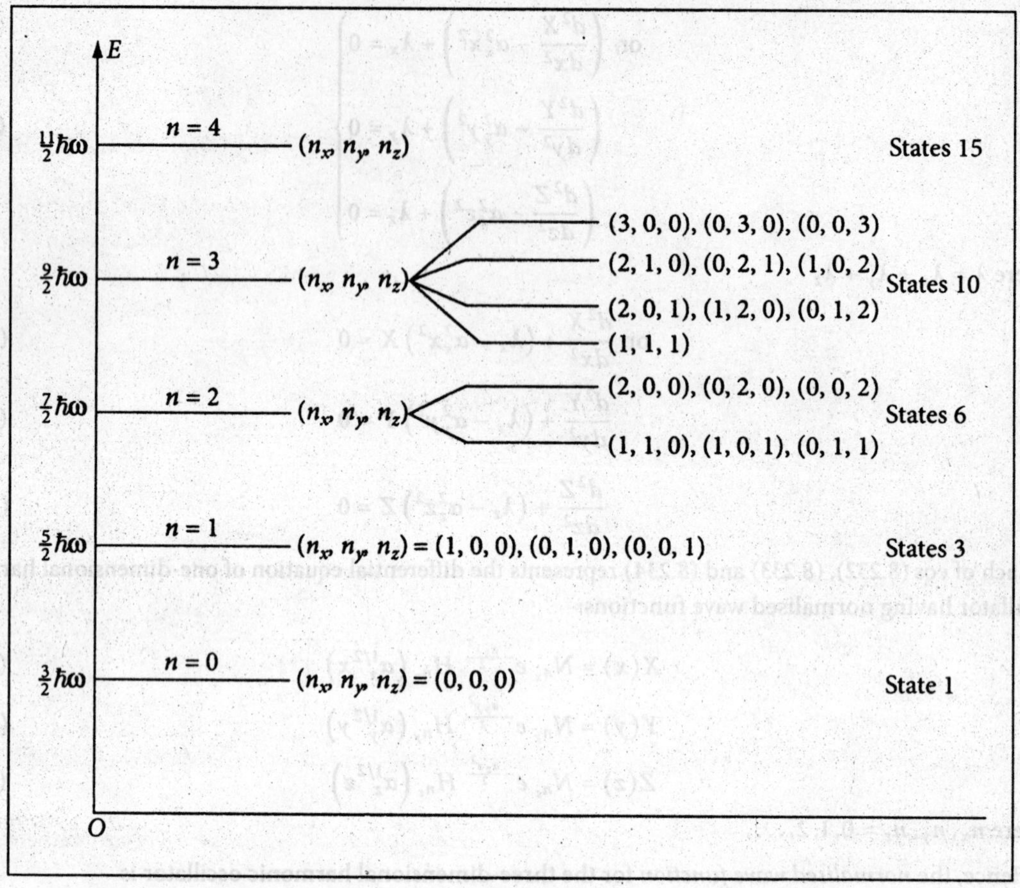

Fig 8.21 Energy levels with degree of degeneracy of a 3-dimensional harmonic oscillator

(b) ***Solution in spherical polar coordinates*:** The time-independent harmonic oscillator in spherical polar coordinates is

$$\frac{1}{r^2}\frac{\partial}{\partial r}\left(r^2\frac{\partial \psi}{\partial r}\right)+\frac{1}{r^2\sin\theta}\frac{\partial}{\partial\theta}\left(\sin\theta\frac{\partial\psi}{\partial t}\right)+\frac{1}{r^2\sin^2\theta}\frac{\partial^2\psi}{\partial\psi^2}+\frac{2m}{\hbar^2}\left(E-\tfrac{1}{2}kr^2\right)\psi=0 \tag{8.241}$$

The angular part of the solution of the above equation will be the spherical harmonics $Y_l^m(\theta,\phi)$. So the wave function is

$$\psi(r,\theta,\phi)=R(r)Y_l^m(\theta,\phi) \tag{8.242}$$

The radial part $R(r)$ of the wave function will satisfy the equation

$$\frac{d^2R}{dr^2}+\frac{2}{r}\frac{dR}{dr}+\frac{2m}{\hbar^2}\left[E-\tfrac{1}{2}kr^2\right]R-\frac{l(l+1)}{r^2}R=0 \tag{8.243}$$

The solution of eq. (8.243) will be of the form:

$$R(r)=e^{-r^2/2}f(r) \tag{8.244}$$

Using eq. (8.244) in eq. (8.243), we get

$$f''(r) + \left(\frac{2}{r} - 2r\right) f'(r) + \left[\lambda - 3 - \frac{l(l+1)}{r^2}\right] f(r) = 0 \quad (8.245)$$

Equation (8.245) can be solved by the method of power series.

Since, the series solution does not converge, it is not acceptable. To make the series solution acceptable we reduce it as a polynomial solution by breaking off the series after a finite number of terms.

From the recurrence relation

$$a_{v+2} = \frac{2(s+v) - \lambda - 3}{(s+v+2)(s+v+1) + 2(s+v+2) - l(l+1)} a_v \quad (8.246)$$

where v is given by the assumed series solution of eq. (2.245)

$$f(r) = \sum a_v r^{s+v} \quad (8.247)$$

The energy level diagram of the three-dimensional oscillator is shown in Fig 8.22.

The condition of termination of the series is

$$2(s+v) - \lambda + 3 = 0 \text{ or, } \lambda = 2s + 2v + 3 \text{ or, } \frac{2mE}{\hbar^2} = 2s + 2v + 3$$

$$\text{or, } E = \hbar\left(n + \frac{3}{2}\right), \; n = 0, 1, 2, \ldots \quad (8.248)$$

where $n = l + v$; $l = 0, 1, 2, \ldots$; $v = 0, 2, 4, \ldots$

Equation (8.248) shows that the *three-dimensional isotropic oscillator has an infinite set of equally spaced levels.* The degree of degeneracy will be the same as in the previous case.

Table 8.5 shows the degeneracy of the three-dimensional harmonic oscillator eigenfunctions characterised by the quantum numbers n and l.

Table 8.5: Degeneracy of the dimensional oscillator

n \ l	0	1	2	3	4	5
5		3		7		11
4	1		5		9	
3		3		7		
2	1		5			
1		3				
0	1					

The wave functions of the three-dimensional oscillator are

$$\psi_{nlm} = N_n e^{-r^2/2} f_{n,l}(r) Y_l^m(\theta, \phi) \quad (8.249)$$

Here, N_n is the normalisation factor.

The energy level diagram of the three-dimensional oscillator is shown in Fig 8.22.

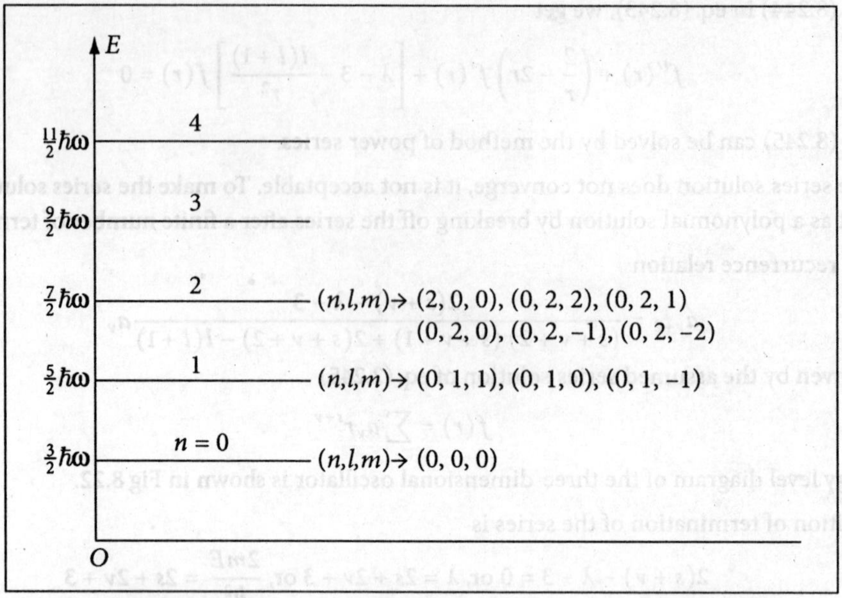

Fig 8.22 Energy level diagram of 3-dimensional harmonic oscillator

8.9 Answers of some important conceptual questions

8.1. *What is a spherically symmetric potential?*

Ans. If the potential $V(r)$ depends only on the radial distance r and not on direction defined by θ and ϕ, then it is called a *spherically symmetric potential.* The value of $V(r)$ is the same at every point on the surface of a sphere of radius r.

8.2. *Mention some applications of spherically symmetric potential in atomic physics.*

Ans. Spherically symmetric potential is applied in the cases of (a) Hydrogen atom (b) singly ionised Helium (He^+) (c) doubly ionised Lithium (Li^{++}) (d) triply ionised Beryllium (Be^{+++}), etc.

8.3. *What is the significance of the magnetic quantum number m?*

Ans. The significance of the magnetic quantum number m (or m_l) is that the *mth represents the z-component of the angular quantum momentum vector*, i.e., $L_z = n\hbar$. The value of $|m| \leqslant l$ and $m = 0, \pm1, \pm2, \ldots$. The magnetic quantum number plays the major role in spatial quantisation of angular momentum and tells us the number and magnitude of different quantised values of L_z for a given value of L.

8.4. *What role magnetic quantum number plays in space quantisation of angular momentum? Show that the z-component of $\vec{L}$ is quantised.* **[Luck U 1996; HPU 1995; PU 1993]**

Ans. The role of orbital magnetic quantum number lies in the fact that the z-component of angular momentum $\vec{L}$ is given by

$$L_z = m_l\hbar \quad \text{(i)}$$

where m_l is the orbital magnetic quantum number. The value of $|m_l| \leqslant l$ the orbital quantum number $m_l = 0, \pm 1, \pm 2$, etc. For a given value of l the total orbital angular momentum is given by

$$L = \sqrt{l(l+1)}\hbar$$

but its component along z-axis is

$$L_z = m_l\hbar = 0, \pm\hbar, \pm 2\hbar \quad \text{(ii)}$$

Eqns (i) and (ii) show that though L, the total angular momentum, is not an angular moment integral multiple of $\hbar$, the z-component of l, is an integral multiple of $\hbar$. The z-component is, therefore, quantised and is said to have *spatial quantisation*. For a given value of L there are $(2l+1)$ different values of L_z. *Thus, magnetic quantum number plays the major role in spatial quantisation of angular momentum and tells us the number and magnitude of different quantised values of L_z for a given value of l.*

8.5. *Why $\vec{L}$ does not have a definite direction?*

Ans. $\vec{L}$ does not have definite direction in space, instead it is continuously changing but it always makes the same angle with the z-axis, i.e., with the direction of the applied magnetic field. The angular momentum vector, therefore, describes a cone, with constant vertex angle θ around the direction of magnetic field. In other words, $\vec{L}$ is continuously precessing around the z-axis. If the angular momentum vector $\vec{L}$ were fixed in space, then its components L_x, L_y and L_z will have definite values. This implies that the electron will be confined to a definite plane. For example, if $\vec{L}$ is always in the z-direction, the electron will have to be in the XY-plane at all times, but according to the uncertainty principle, the electron cannot be confined to a single plane. There is, therefore, a built-in uncertainty in the electron's z-coordinate. Hence, the direction of $\vec{L}$ is continuously changing, such that the average values of L_x and L_y are zero whereas L_z has specific values $m_l\hbar$.

The above explanations state why only one component, i.e., z-component L is quantised and why the angular momentum vector L cannot have definite direction.

8.6. *Give the significance of quantum numbers n, l and m in brief.*

Ans. The three quantum numbers n, l and m were found in the application of Schrödinger's equation to hydrogen atom problem.

The quantum number n is called the *principal quantum number n* can have positive integral values $1, 2, 3, \ldots$. The energy of the electron in the nth state is proportional to $E_n \propto \frac{1}{n^2}$. The radius of the electron is also given by $a_n \propto n^2$. Since, n^2 is also a positive integer, E_n and a_n are *discrete*.

The quantum number l is called the orbital *angular momentum quantum number*. It arises during quantisation of angular momentum L of the electron, i.e., $L = l\hbar$ or, $L = \sqrt{L(l+1)}\hbar$. The values of l are $0, 1, 2, \ldots, (n-1)$. Historically, $l = 0, 1, 2, 3, \ldots$, states of the electron, are represented by the letters s, p, d, f respectively after the classification, i.e., *sharp, principal, diffuse, fundamental*, etc.

Bohr, while formulating his theory on hydrogen atom, introduced as postulate without giving any justification. In quantum mechanics, the quantisation of angular momentum appears as the condition of the solution of $\Theta(\theta)$, which should be finite.

The quantum number m is called the *magnetic orbital quantum number*. It appears in quantisation of the z-component of angular momentum, i.e., $L_z = m\hbar$. m can have values $-l, -(l-1), -l, 0, (l-1), l$, i.e., $(2l+1)$ values.

An atom behaves as a small magnet of moment

$$\vec{\mu} = \left(-\frac{e}{2m'}\right)\vec{L}$$

m' is the mass of the electron and $\vec{L}$, the angular momentum vector. The z-component of $\vec{\mu}$ is

$$\mu_z = -\frac{e}{2m'}L = -\frac{e}{2m'}m\hbar$$

where $L = m\hbar$. When an external magnetic field $\vec{B}$ is applied along the z-axis, the different states relating to different values of m are observed. This justifies the name *magnetic orbital quantum number.*

8.7. *Can the value of $n = 0$ be accepted?*

Ans. No. Because, if $n = 0$, then energy $E_n \propto \frac{1}{n^2} = \infty$, which is absurd. Again, if $n = 0$, then $r_n = 0$, where r_n is the radius of the nth electron orbit. This implies that the electron will be located at the nucleus which is also absurd.

8.8. *What is the role of magnetic quantum number m with reference to space quantisation?*

Ans. For a given value of l the total angular momentum is given by

$$L = \sqrt{l(l+1)}\hbar \tag{i}$$

The z-component of L along z-axis is given by

$$L_z = m\hbar = 0, \pm\hbar, \pm 2\hbar, \pm 3\hbar \tag{ii}$$

where $m = 0, \pm 1, \pm 2, \pm 3, \ldots$.

From eqns (i) and (ii), we note that the total angular momentum L is not an integral multiple of $\hbar$ but the z-component of angular momentum L_z is an integral multiple.

L_z is, therefore, quantised and has spatial quantisation. For a definite L there are $(2l+1)$ values of L_z. Hence, magnetic quantum number m has special role in *space quantisation* and gives the number and magnitude of different quantised values of L_z for a given value of L.

8.10 Solved problems

» ***Example* 8.1.** An electron is confined in a cubical box of side 1 Å. Find the momentum and energy of the electron in the ground state and the first excited state.

Solution: We have for the ground state of the electron, $n_x = n_y = n_z = 1$.

∴ momentum in the ground state is

$$p_{1,1,1} = \frac{\pi\hbar}{l}\sqrt{3} = \frac{\pi h}{2\pi l}\sqrt{3} = \frac{h}{2l}\sqrt{3} = \frac{6.63\times10^{-34}}{2\times10^{-10}}\sqrt{3} = 5.74\times10^{-24}\text{ kg-m/s}$$

Energy of the electron in the ground state

$$E_{1,1,1} = \frac{\pi^2\hbar^2}{2ml^2}3 = \frac{h^2}{8ml^2}3 = \frac{(6.63\times10^{-34})^2\times3}{8\times9.1\times10^{-31}\times10^{-20}} = 60.4\times10^{-19}\text{ J} = 37.7\text{ eV}$$

For the first excited state one of the quantum numbers is 2.

$$\therefore \text{momentum } p_{2,1,1} = p_{1,2,1} = p_{1,1,2} = \frac{\pi\hbar}{l}\sqrt{6} = \frac{h}{2l}\sqrt{6} = 3.315 \times 10^{-24} \times \sqrt{6} = 8.13 \times 10^{-24} \text{ kg-m/s}$$

$$\text{and energy } E_{2,1,1} = E_{1,2,1} = E_{1,1,2} = \frac{\pi^2\hbar^2}{2ml^2}6 = \frac{h^2}{8ml^2}6 = 60.4 \times 6 \times 10^{-19} = 362.4 \times 10^{-19} \text{ J} = \mathbf{226.25\ eV}$$

» ***Example* 8.2.** **Find the lowest energy of an electron in a rectangular box of dimensions 1×10^{-13} cm, 1.5×10^{-13} cm and 2×10^{-13} cm.**

Solution: The energy of the electron in the rectangular potential box for (n_x, n_y, n_z) state is given by

$$E_{n_x,n_y,n_z} = \frac{h^2}{8m}\left(\frac{n_x^2}{a^2} + \frac{n_y^2}{b^2} + \frac{n_z^2}{c^2}\right)$$

where a, b and c are the sides of the box.

For the lowest state, $n_x = n_y = n_z = 1$.

$\therefore$ energy of the lowest state is

$$E_{1,1,1} = \frac{6.6 \times 6.6 \times 10^{-54}}{8 \times 9.1 \times 10^{-28}}\left[\frac{1}{10^{-26}} + \frac{1}{2.25 \times 10^{-26}} + \frac{1}{4 \times 10^{-26}}\right] = 1.014 \text{ ergs} = 1.014 \times 10^{-7} \text{ J}$$

» ***Example* 8.3.** **An electron is confined in a cubical box of side 1 Å. (i) Find its lowest energy. (ii) What is the temperature at which the average energy of the molecule of a perfect gas would be equal to the lowest energy of the electron ($k = 1.38 \times 10^{-23}$ J/K)?**

Solution:

(i) Here, $a = 10^{-10}$ m, the length of a side of the cube.

The energy of the electron in a cubical box of side a in the (n_x, n_y, n_z) state is

$$E_{n_x,n_y,n_z} = \frac{\pi^2\hbar^2}{2ma^2}\left(n_x^2 + n_y^2 + n_z^2\right)$$

For the lowest state, $n_x = n_y = n_z = 1$. So the energy in the lowest state is

$$E_{1,1,1} = \frac{3\pi^2\hbar^2}{2ma^2} = \frac{3\pi^2 \times \left(1.054 \times 10^{-34}\right)^2}{2 \times 9.11 \times 10^{-31} \times \left(10^{-10}\right)^2} = 18.03 \times 10^{-18} \text{ J}$$

(ii) If T K be the required temperature of the molecules, then

$$\frac{3}{2}kT = \frac{3\pi^2\hbar^2}{2ma^2} \quad [k \text{ is the Boltzmann constant}]$$

$$= 18.03 \times 10^{-18} \text{ J}$$

$$\therefore T = \frac{2 \times 18.03 \times 10^{-18}}{3 \times 1.38 \times 10^{-23}} = 8.71 \times 10^5 \text{ K}$$

» ***Example* 8.4.** **Considering the nucleus as a cubical box with side 10^{-14} m, calculate the energy of the neutron confined to the nucleus. $[h = 6.26 \times 10^{-34}$ J-s, $m = 1.6 \times 10^{-27}$ kg$]$** **[PU 1995]**

Solution: Here, we have $a = b = c = 10^{-14}$ m.

The neutron will have the lowest energy in the lowest quantum state given by $n_x = n_y = n_z = 1$.

∴ energy of the neutron in the lowest quantum state is

$$E_{1,1,1} = \frac{\pi^2\hbar^2}{2m}\left[\frac{n_x^2}{a^2} + \frac{n_y^2}{b^2} + \frac{n_z^2}{c^2}\right] = \frac{\pi^2\hbar^2}{2m}\left(\frac{3}{a^2}\right) = \frac{3h^2}{8ma^2} = \frac{3\times\left(6.62\times10^{-34}\right)^2}{8\times1.6\times10^{-27}\times10^{-28}}$$

$$= 10.29\times10^{-13}\ \text{J} = 6.43\ \text{MeV}$$

For the first excited state, energy is

$$E_{2,1,1} = E_{1,2,1} = E_{1,1,2} = \frac{\pi\hbar^2}{2m}\left(\frac{1+4+1}{a^2}\right) = \frac{\pi^2\hbar^2}{2ma^2}\times 6 = \frac{h^2}{8ma^2}6 = \mathbf{12.86\ MeV}$$

∴ the degeneracy of the first excited state is 3.

For the second excited state energy is

$$E_{2,2,1} = E_{2,1,2} = E_{1,2,2} = \frac{\pi^2\hbar^2}{2m}\left[\frac{4+4+1}{a^2}\right] = \frac{\pi^2\hbar^2}{2ma^2}\times 9 = \frac{h^2}{8ma^2}\times 9 = \mathbf{19.29\ MeV}$$

∴ the degeneracy of the second excited state is also 3.

For the third excited state

$$E_{2,2,2} = \frac{\pi\hbar^2}{2m}\left[\frac{4+4+4}{a^2}\right] = \frac{\pi^2\hbar^2}{2ma^2}\times 12 = \frac{h^2}{8ma^2}\times 12 = \mathbf{25.72\ MeV}$$

This state is not degenerate.

» ***Example* 8.5.** (i) Find out the radial probability function for the ground state of hydrogen atom. (ii) Show that the most probable distance of the electron from the nucleus in the ground state of hydrogen atom is equal to the Bohr's radius. **[BU (H) 2002]**

***Solution*:**

(i) For the ground state of hydrogen atom, $n = 1$, $l = 0$ and $m = 0$.

Hence, the ground state wave function is

$$\psi_{1,0,0} = \left(\frac{1}{\pi a_0^3}\right)^{1/2} e^{-\rho/2}$$

where $\rho = \alpha r = \dfrac{2Zr}{na_0} = \dfrac{2r}{a_0}$ [∵ $Z = 1, n = 1$].

$$\therefore \psi_{1,0,0} = \left(\frac{1}{\pi a^3}\right)^{1/2} e^{-r/a_0} \quad \text{(i)}$$

The radial probability density $P(r)$ is independent of θ and ϕ and the probability between r and $r + dr$ is given by

$$P(r)\,dr = \int_\tau \psi_{1,0,0}^*\,\psi_{1,0,0}\,d\tau$$

where $d\tau = r^2 \sin\theta\, dr\, d\theta\, d\phi$ is the volume element in spherical polar coordinates.

$$\therefore P(r)\,dr = \int \frac{1}{\pi a_0^3} e^{-2r/a_0} r^2\, dr \sin\theta\, d\theta\, d\phi = \frac{1}{\pi a_0^3} e^{-2r/a_0} r^2\, dr \int_{\theta=0}^{\pi} \sin\theta\, d\theta \int_{\phi=0}^{2\pi} d\phi$$

$$= \frac{4}{a_0^3} e^{-2r/a_0} r^2\, dr \quad \text{(ii)}$$

This is the probability that the electron be found between the distance r and $r + dr$ around the nucleus.

(ii) Now, for $P(r)$ to be the maximum, $dP/dr = 0$. From eq. (8.225)

$$\frac{dP}{dr} = \frac{4}{a_0^3}\left[r^2\left(-\frac{2}{a_0}\right)e^{-2r/a_0} + 2r\,e^{-2r/a_0}\right]$$

∴ for maximum value of $P(r)$

$$\frac{4}{a_0^3}\left[r^2\left(-\frac{2}{a_0}\right)e^{-2r/a_0} + 2r\,e^{-2r/a_0}\right] = 0 \text{ or, } 2r - \frac{2r^2}{a_0} = 0 \text{ or, } r = a_0$$

Thus, *the most probable distance of the electron from the nucleus of hydrogen atom in the ground state is equal to the Bohr radius* (a_0).

» ***Example* 8.6.** The normalised ground state wave function of a one-electron atom is given by

$$\psi_{(r,\theta,\phi)} = \pi^{-1/2}\left(\frac{Z}{a_0}\right)^{3/2} e^{-Zr/a_0}$$

where the notations have usual meaning. Evaluate the probability of the electron in this state of finding the electron at a distance greater than $2a_0/Z$. **[BU (H) 1997, 2000]**

Solution: Here, the normalised ground state wave function for the electron of a hydrogen-like atom (i.e., one-electron atom) of atomic number Z is

$$\psi(r,\theta,\phi) = \pi^{-1/2}\left(\frac{Z}{a_0}\right)^{3/2} e^{-Zr/a_0}$$

the notations having their usual meaning.

The probability of finding the electron at a distance greater than $2a_0/Z$ is given by

$$P = \int_{2a_0/Z}^{\infty} \psi^*(r,\theta,\phi)\,\psi(r,\theta,\phi)4\pi r^2\,dr = 4\pi\int_{2a_0/Z}^{\infty} |\psi|^2\, r^2\,dr$$

$$= 4\pi\int_0^{\infty} |\psi|^2\, r^2\,dr - 4\pi\int_0^{2a_0/Z} |\psi|^2 r^2\,dr = 1 - 4\left(\frac{Z}{a_0}\right)^3 \int_0^{2a_0/Z} r^2\,e^{-2Zr/a_0}\,dr$$

$$= 1 - 4\left(\frac{Z}{a_0}\right)^3\left(\frac{a_0}{2Z}\right)^3 \int_0^4 \rho^2\,e^{-\rho}\,d\rho, \left[\text{where } \rho = \frac{2Z}{a_0}r\right]$$

$$= 1 - \tfrac{1}{2}\left(2 - 26e^{-4}\right) = 13e^{-4} = \mathbf{0.15}$$

∴ the probability of finding the electron in the normalised ground state of a one-electron atom at a distance greater than $2a_0/Z$ from the nucleus is 15 %.

» ***Example* 8.7.** If the ground state wave function for the hydrogen atom is

$$\psi = \frac{1}{\sqrt{\pi}}\frac{1}{a_0^{3/2}}\,e^{-r/a_0}$$

show that average distance of the electron from the nucleus is $1.5a_0$. **[GNDU 1995]**

Solution: We have the average distance of the electron from the hydrogen nucleus in the ground state is

$$\langle r\rangle = \int_0^{\infty} \psi^*\,\hat{r}\,\psi\,d\tau = \int_0^{\infty} \frac{1}{\sqrt{\pi}}\frac{1}{a_0^{3/2}}\,e^{-r/a_0}\,r\,\frac{1}{\sqrt{\pi}}\frac{1}{a_0^{3/2}}\,e^{-r/a_0}\,d\tau$$

$$= \frac{1}{\pi a_0^3}\int_0^{\infty} r\,e^{-2r/a_0}\,r^2 \sin\theta\,dr\,d\theta\,d\phi$$

[∵ $d\tau$ = the volume element in spherical polar coordinates = $r^2 \sin\theta\,dr\,d\theta\,d\phi$]

Now, $\int_0^\infty r^3 e^{-2r/a_0}\,dr$ can be evaluated by using the standard integral

$$\int_0^\infty e^{-tx}x^n\,dx = \frac{n!}{t^{n+1}}.\ \therefore \int_0^\infty r^3 e^{-2r/a_0}\,dr = \frac{3!}{(2/a_0)^4},\ (n = 3, t = 2/a_0)$$

$$\therefore \langle r\rangle = \frac{1}{\pi a_0^3}\int_0^\infty r^3 e^{-2r/a_0}\,dr \int_0^\pi \sin\theta\,d\theta \int_0^{2\pi} d\phi = \frac{1}{\pi a_0^3}\times\frac{3!}{(2/a_0)^4}\times 2\times 2\pi$$

$$\left[\because \int_0^\pi \sin\theta\,d\theta = 2 \text{ and } \int_0^{2\pi} d\phi = 2\pi\right]$$

$$\therefore \langle r\rangle = \tfrac{3}{2}a_0 = 1.5a_0$$

» ***Example* 8.8.** Find the expectation value of potential energy of the electron in hydrogen atom in the 1*s* state. The wave function for the electron in 1*s* state is given by $\psi_{1,0,0} = 1/\sqrt{\pi a_0^3}\,e^{-r/a_0}$, where $a_0 = \hbar^2/me^2$ is the Bohr radius.

***Solution*:** The potential energy of the electron in hydrogen atom is given by $V = -\frac{e^2}{r}$.

∴ the expectation value of V is

$$\langle V\rangle = +\int \psi_{1,0,0}^*\left(-\frac{e^2}{r}\right)\psi_{1,0,0}\,d\tau = -e^2\int |\psi_{1,0,0}|^2\frac{1}{r}\,d\tau$$

$$= -e^2\left(\frac{1}{\sqrt{\pi a_0^3}}\right)^2\int_0^\pi \sin\theta\,d\theta\int_0^{2\pi}d\phi\int_0^\infty \frac{e^{-2r/a_0}}{r}r^2\,dr = -\frac{e^2}{\pi a_0^3}\times 4\pi\int_0^\infty e^{-2r/a_0}r\,dr$$

By standard integral

$$\int_0^\infty e^{-\alpha x^2}x^n\,dx = \frac{n!}{\alpha^{n+1}}$$

$$\therefore \langle V\rangle = -\frac{4e^2}{a_0^3}\times\frac{1!}{(2/a_0)^2} = -\frac{e^2}{a_0} = -\frac{e^2}{(\hbar^2/me^2)} = -\frac{me^4}{\hbar^2}$$

» ***Example* 8.9.** The ground state wave function of an electron in a hydrogen atom is given by

$$\psi(\vec{r}) = \left(\frac{1}{\pi a_0^3}\right)^{1/2} e^{-r/a_0}$$

where a_0 is the radius of the innermost orbit. Verify that this wave function is normalised. [BU (H) 1997]

***Solution*:** The condition of normalisation of a wave function ψ is

$$\int_{\text{All space}} \psi^*\psi\,d\tau = 1$$

Here

$$\psi = \psi(\vec{r}) = \left(\frac{1}{\pi a_0^3}\right)^{1/2} e^{-r/a_0}$$

$$\therefore \int_{\text{All space}} \psi^*\psi\,d\tau = \int_{\text{All space}}\left(\frac{1}{\pi a_0^3}\right)^{1/2} e^{-r/a_0}\times\left(\frac{1}{\pi a_0^3}\right)^{1/2} e^{-r/a_0}\,4\pi r^2\,dr$$

Here, $d\tau = 4\pi r^2\, dr$ is the volume element.

$$\text{or,} \quad \int_{\text{All space}} \psi^* \psi\, d\tau = \frac{4\pi}{\pi a_0^3} \int_{r=0}^{\infty} e^{-2r/a_0} r^2\, dr$$

Now, put $2r/a_0 = Z$.

$$\therefore r = \frac{a_0 Z}{2} \text{ and } dr = \frac{a_0}{2} dZ$$

$$\therefore \frac{4}{a_0^3} \int_0^{\infty} e^{-2r/a_0} r^2\, dr = \frac{4}{a_0^3} \int_0^{\infty} e^{-Z} \left(\frac{a_0 Z}{2}\right)^2 \left(\frac{a_0}{2}\right) dZ = \frac{4}{a_0^3} \times \frac{a_0^3}{8} \int_0^{\infty} e^{-Z} Z^{3-1}\, dZ$$

$$= \tfrac{1}{2} \int_0^{\infty} e^{-z} z^2 dz = \tfrac{1}{2}\Gamma(3) = \tfrac{1}{2} 2\Gamma(2) = \Gamma(1+1) = 1\Gamma(1) = 1$$

$\therefore \psi(\vec{r})$ *is normalised wave function.*

» ***Example* 8.10.** What is the expectation value of the kinetic energy (E_k) of the electron in the 1s state of the hydrogen atom?

***Solution*:** We have the wave function of the hydrogen atom in the 1s state as

$$\psi_{1,0,0} = R_{1,0,0} Y_0^0 = \frac{1}{\sqrt{4\pi}} \frac{2}{a_0^{3/2}} e^{-r/a_0}$$

The expectation value of the kinetic energy is

$$\langle E_k \rangle = \int_\tau \psi_{1,0,0}^* \hat{E}_k \psi_{1,0,0}\, d\tau = \int_0^{\infty} \int_0^{\pi} \int_0^{2\pi} \psi_{1,0,0}^* \hat{E}_k \psi_{1,0,0} r^2 \sin\theta\, dr\, d\theta\, d\phi$$

But,

$$\hat{E}_k = -\frac{\hbar^2}{2m}\nabla^2 = -\frac{\hbar^2}{2m}\left[\frac{1}{r^2}\frac{\partial}{\partial r}\left(r^2 \frac{\partial}{\partial r}\right) + \frac{1}{r^2 \sin\theta}\frac{\partial}{\partial \theta}\left(\sin\theta \frac{\partial}{\partial \theta}\right) + \frac{1}{r^2 \sin^2\theta}\frac{\partial^2}{\partial \phi^2}\right]$$

Since, $\psi_{1,0,0}$ depends on r only, the terms in $\frac{\partial}{\partial \theta}$ and $\frac{\partial}{\partial \phi}$ are zero.

$$\therefore \hat{E}_k = -\frac{\hbar^2}{2m}\left(\frac{1}{r^2}\frac{\partial^2}{\partial r^2} + 2r\frac{\partial}{\partial r}\right)$$

$$\therefore \langle E_k \rangle = -\frac{\hbar^2}{2m} \cdot 4\pi \left(\frac{1}{\sqrt{\pi a_0^3}}\right)^2 \int_0^{\infty} e^{-r/a_0} \times \left\{\left(\frac{\partial^2}{\partial r^2} + \frac{2}{r}\frac{\partial}{\partial r}\right) e^{-r/a_0}\right\} r^2\, dr$$

$$= -\frac{\hbar^2}{2m}\frac{4}{a_0^3} \int_0^{\infty} \left(\frac{1}{a_0^2} - \frac{2}{r a_0}\right) e^{-2r/a_0} r^2\, dr$$

$$= -\frac{2\hbar^2}{m a_0^3}\left[\frac{1}{a_0^2}\int_0^{\infty} r^2 e^{-2r/a_0}\, dr - \frac{2}{a_0}\int_0^{\infty} r e^{-2r/a_0}\, dr\right]$$

$$= -\frac{2\hbar^2}{m a_0^3}\left[\frac{1}{a_0^2}\frac{2}{\left(\frac{2}{a_0}\right)^3} - \frac{2}{a_0}\frac{1}{\left(\frac{2}{a_0}\right)^2}\right] = \frac{2\hbar^2}{m a_0^3} \cdot \frac{a_0}{4}$$

$$\therefore \langle E_k \rangle = \frac{\hbar^2}{2m a_0^2} = \frac{\hbar^2}{2m}\left(\frac{me^2}{\hbar^2}\right) = \frac{me^4}{2\hbar^2}$$

» ***Example* 8.11.** Show that the orbital angular momentum of the electron in hydrogen atom is $L = \hbar\sqrt{l(l+1)}$, where l is the orbital quantum number. **[GNDU 1991]**

Solution: The radial equation is given by

$$\frac{d^2R}{dr^2} + \frac{2}{r}\frac{dR}{dr} + \frac{2m}{\hbar^2}\left[E + \frac{Ze^2}{r^2}\right]R = 0$$

The above equation depends only on r and is independent on θ and ϕ.

$\therefore$ total energy $E = E_k$ (orbital) + E_k (radial) $-Ze^2/r$.

$$\therefore E + \frac{Ze^2}{r} = E_k(\text{orbital}) + E_k(\text{radial})$$

$$\therefore \frac{d^2R}{dr^2} + \frac{2}{r}\frac{dR}{dr} + \frac{2m}{\hbar^2}[E_k(\text{orbital}) + E_k(\text{radial})]R - \frac{l(l+1)}{r^2}R = 0$$

The equation depends only on radial motion, if

$$E_k(\text{orbital}) - \frac{l(l+1)\hbar^2}{2mr^2} = 0$$

or, $\frac{1}{2}mv^2 = \frac{l(l+1)\hbar^2}{2mr^2}$ [where v = the orbital velocity of the electron]

or, $(mvr)^2 = \hbar^2 l(l+1)$ or, $|\vec{L}|^2 = \hbar^2 l(l+1)$ or, $|\vec{L}| = \hbar\sqrt{l(l+1)}$

where $|\vec{L}| = mvr$ the orbital angular momentum of the electron.

» ***Example* 8.12.** Verify that the average value of $1/r$ for 1*s* electron in hydrogen atom is $1/a_0$.

Solution: We know that the wave function of the 1*s* electron of hydrogen atom is

$$\psi = \left(\frac{1}{\sqrt{\pi a_0^{3/2}}}\right)e^{-r/a_0}$$

where a_0 is the radius of the Bohr orbit.

$\therefore$ the average value of $1/r$ is

$$\left\langle\frac{1}{r}\right\rangle = \int_\tau \psi^*\left(\frac{1}{r}\right)\psi\, d\tau = \frac{1}{\pi a_0^3}\int_0^\infty \frac{1}{r}e^{-2r/a_0} \times r^2 \sin\theta\, dr\, d\theta\, d\phi$$

[$\because$ the volume element $d\tau = r^2\sin\theta\, dr\, d\theta\, d\phi$]

$$= \frac{1}{\pi a_0^3}\int_0^\infty e^{-2r/a_0} r\, dr \times \int_0^\pi \sin\theta\, d\theta \times \int_0^{2\pi} d\phi$$

Now,

$$\int_0^\infty e^{-2r/a_0} r\, dr = \int_0^\infty e^{-z}\left(\frac{a_0 Z}{2}\right)\frac{a_0}{2}dZ \left[\text{putting } \frac{2r}{a_0} = Z; r = \frac{a_0 Z}{2} \text{ and } dr = \frac{a_0}{2}dZ\right]$$

$$= \left(\frac{a_0}{2}\right)^2\int_0^\infty e^{-Z} z\, dz = \left(\frac{a_0}{2}\right)^2\Gamma(2) = \frac{a_0^2}{4}1\Gamma(1) = \frac{a_0^2}{4}$$

$$\int_0^\pi \sin\theta\, d\theta = \left(1 - \cos\theta\right)_0^\pi = 2 \text{ and } \int_0^{2\pi} d\phi = 2\pi$$

$$\therefore \left\langle\frac{1}{r}\right\rangle = \frac{1}{\pi a_0^3} \times \frac{a_0^2}{4} \times 2 \times 2\pi \text{ or, } \left\langle\frac{1}{r}\right\rangle = \frac{1}{a_0}$$

$\therefore$ the average value of $(1/r)$ of 1s electron is $\langle 1/r \rangle = 1/a_0$.

» ***Example* 8.13.** (i) Show that the rotational lines of HCl-molecular spectra are equispaced. (ii) The inter-molecular distance of HCl molecule is 1.29×10^{-8} cm, $m_{\text{Cl}} = 35m_{\text{H}}$ and $m_{\text{H}} = 1.68 \times 10^{-24}$ gm; calculate the separation of the lines in the far infrared region. [$h = 6.625 \times 10^{-27}$ erg-sec, $c = 3 \times 10^{10}$ cm/s]

***Solution*:**

(i) The rotational energy of HCl molecule is

$$E_l = \frac{\hbar^2}{2I} l(l+1)$$

Now,

$$\Delta E = E_{l+1} - E_l = \frac{\hbar^2}{2m}(l+1)(l+2) - \frac{\hbar^2}{2m} l(l+1) = \frac{\hbar^2}{2m}(l+1)(l+2-l) = \frac{\hbar^2}{2m}(l+1)$$

Again, $\Delta E = h\nu$.

$$\therefore \bar{\nu} = \frac{1}{\lambda} = \frac{\nu}{c} \text{ or, } \bar{\nu} = \frac{\Delta E}{hc} = \frac{h(l+1)}{4\pi^2 Ic} \text{ or, } \bar{\nu} = B(l+1), \text{ where } B = \frac{h}{4\pi^2 Ic}$$

$\because l = 0, 1, 2, \ldots, \bar{\nu} = B, 2B, 3B, 4B, \ldots$

In the molecular spectra of HCl, the rotational lines are equispaced.

(ii) For HCl molecule, the reduced mass is

$$\mu = \frac{m_{\text{H}} m_{\text{Cl}}}{m_{\text{H}} + m_{\text{Cl}}} = \frac{35m_{\text{H}} \cdot m_{\text{H}}}{36m_{\text{H}}} = \frac{35}{36} m_{\text{Cl}} = \frac{35}{36} \times 1.68 \times 10^{-24}$$

and moment of inertia of HCl molecule about the axis of rotation is

$$I = \mu r_e^2 = \frac{35}{36} \times 1.68 \times 10^{-24} \times (1.29 \times 10^{-8})^2$$

The separation between the lines is given by

$$B = \frac{h}{4\pi^2 Ic} = \frac{6.625 \times 10^{-27} \times 36}{4 \times (3.14)^2 \times 35 \times 1.68 \times 10^{-24} \times (1.29)^2 \times 10^{-16} \times 3 \times 10^{10}} = 20.68 \text{ cm}^{-1}$$

» ***Example* 8.14.** Find the probability of finding the electron at a distance r from nucleus of hydrogen-like atom. When is this maximum? **[cf. KU (H) 2015]**

***Solution*:** The probability of finding the electron within volume $d\tau$ located at $\vec{r}$ is

$$\psi^* \psi \, d\tau = \left(\sqrt{\frac{z^3}{\pi a_0^3}} e^{-zr/a_0} \right)^* \left(\sqrt{\frac{z^3}{\pi a_0^3}} e^{-zr/a_0} \right)^* \times r^2 dr \sin\theta \, d\theta \, d\phi$$

$$= \frac{z^3}{\pi a_0^3} e^{-2zr/a_0} \cdot r^2 dr \sin\theta \, d\theta \, d\phi$$

where the $\psi = \sqrt{\dfrac{z^3}{\pi a_0^3}} e^{-zr/a_0}$ is the ground state wave function of hydrogen atom.

The probability of finding the electron within distance r and $r + dr$ is obtained by integrating over all angles θ, ϕ. It is called *radial probability.*

$$\int_{\theta,\phi} \psi^* \psi \, d\tau = \frac{z^3}{\pi a_0^3} e^{-2zr/a_0} \int_0^{\pi} \sin\theta \, d\theta \int_0^{2\pi} d\phi = \frac{z^3}{\pi a_0^3} e^{-2zr/a_0} r^2 \, dr \cdot 2 \cdot 2\pi$$

$$= \frac{4z^3}{a_0^3} r^2 e^{-2zr/a_0} \, dr = D \, dr$$

∴ the radial probability density

$$D = \frac{4z^3 r^2}{a_0^3} e^{-2zr/a_0}$$

Now for D is maximum

$$\frac{d}{dr} D = 0 \text{ or, } \frac{d}{dr} \frac{4z^3}{a_0^3} r^2 e^{-2zr/a_0} = 0 \text{ or, } \frac{4z^3}{a_0^3} \cdot 2r \, e^{-2zr/a_0} \left(1 - \frac{zr}{a_0}\right) = 0$$

∴ either $r = 0$ or, $e^{-2zr/a_0} = 0$, i.e., $r = \infty$

$$\text{or, } 1 - \frac{zr}{a_0} = a_0 \text{ or, } r = \frac{a_0}{z}$$

$r = 0$ is the position of the nucleus where the electron cannot stay and hence $r = 0$ is not possible. Also $\psi = \infty$ is not acceptable as the wave function should vanish there. Thus, $r = a_0/z$ is possible.

Further, it can be shown that at $r = a_0/z$

$$\frac{d^2 D}{a_0^3} e^{-2} = \text{-ve}$$

∴ $r = a_0/z$ is acceptable. Thus means $r = a_0/2$ is the position of maximum value D, i.e., $r = a_0/2$ is the most probable distance of electron from the nucleus when the hydrogen-like atom is in the ground state.

The value of the radial probability density in ground state of hydrogen-like atom is

$$D\Big|_{r=r_0/z} = \frac{4z^3}{a_0^3} r^2 e^{-2zr/a_0}\Big|_{r=r_0/z} = \frac{4z}{a_0} e^{-2}$$

For hydrogen atom $z = 1$

$$\therefore D\Big|_{r=a_0} = \frac{4}{a_0 e^{-2}} = \frac{4}{0.53 \text{ Å}} e^{-2} = 1.02 \left(\text{in Å}^{-1}\right)$$

» ***Example* 8.15.** (i) Write down the hydrogen atom ground state wave function. (ii) Show that it is normalised.

***Solution*:**

(i) The ground state wave function of hydrogen atom is

$$\psi_{100} = \frac{1}{\sqrt{\pi a_0^3}} e^{-r/a_0}$$

where the Bohr radius $a_0 = 4\pi\varepsilon_0 \dfrac{h^2}{me^2}$.

(ii) We consider the integral

$$\int_{\text{all space}} \psi_{100}^* \psi_{100}\, d\tau = \int \left[\frac{1}{\sqrt{\pi a_0^3}} e^{-r/a_0}\right]^* \left[\frac{1}{\sqrt{\pi a_0^3}} e^{-r/a_0}\right] \times r^2 \sin\theta\, d\theta\, dr\, d\theta\, d\phi$$

$$= \frac{1}{\pi a_0^3} \int_0^\infty e^{-2r/a_0} r^2\, dr \int_0^\pi \sin\theta\, d\theta \int_0^{2\pi} d\phi$$

$$= \frac{1}{\pi a_0^3}\left[\frac{2!}{(2/a_0)^{2+1}}\right](2)(2\pi) = 1$$

So, ψ_{100} is normalised.

» ***Example* 8.16.** The $2p$ state wave function for hydrogen atom is known be $r\, e^{-r/2a_0} \cos\theta$ ($a_0 = 1$ = the 1st Bohr orbit). Obtain the expectation value of r in this state.

***Solution*:** The hydrogen atom wave function in $2p$ state [$n = 2, l = 1$ and $m_l = 0$] is given by

$$\psi_{200} = Nr\, e^{-r/200} \cos\theta$$

where N is the normalisation constant. To find N we consider the condition of normalisation

$$\int_{\text{all space}} \psi_{210}^* \psi_{210}\, d\tau = 1$$

$$\text{or, } \int [Nre - r/2a_0 \cos\theta]^* \left[Nr\, e^{-r/2a_0} \cos\theta\right] \times (r^2 \sin\theta\, d\theta\, dr\, d\phi) = 1$$

$$\text{or, } |N|^2 \int_0^\infty r^4 e^{-r/a_0}\, dr \int_0^\pi \cos^2\theta \sin\theta\, d\theta \int_0^{2\pi} d\phi = 1$$

$$\text{or, } |N|^2 \left(\frac{4!}{\left(\frac{1}{a_0}\right)^{4+1}}\right)^0 \left(\frac{2}{3}\right)(2\pi) = 1 \left(\because \int_0^\infty r^n e^{-\beta r}\, dr = \frac{n!}{\beta^{n+1}}\right)$$

$$\text{or, } N(32\pi a_0^5) = 1 \text{ or, } N = \frac{1}{\sqrt{32\pi a_0^5}}$$

$$\therefore \psi_{210} = \frac{1}{\sqrt{32\pi a_0^5}} r\, e^{-4/2a_0} \cos\theta$$

The expectation value of r in this state is

$$\langle r\rangle = \int_{\text{all space}} \psi_{210}^* \hat{r}\, \psi_{210}\, d\tau = \int (\psi_{210})^2\, r\, dr = \int \frac{1}{32\pi a_0^5} r^2 e^{-r^2/a_0} \cos^2\theta\, r\, r^2 \sin\theta\, d\theta\, d\phi\, dr$$

$$= \frac{1}{32\pi a_0^5} \int_0^\infty r^5 e^{-r/a_0}\, dr \int_0^\pi \cos^2\theta \sin\theta\, d\theta \int_0^{2\pi} d\phi = \frac{1}{32\pi a_0^5}\left(\frac{5!}{\left(\frac{1}{a_0}\right)^{5+1}}\right)\left(\frac{2}{3}\right)(2\pi)$$

$$= 5a_0 = 5 \times 0.63\ \text{Å}$$

$$\therefore \langle r\rangle = 2.65\ \text{Å}.$$

» ***Example* 8.17.** The ground state wave function of hydrogen atom is

$$\psi(r) = \frac{1}{\sqrt{\pi a_0^3}} e^{-r/a_0} \left[a_0 = 4\pi\varepsilon_0 \frac{\hbar^2}{me^2}\right]$$

What is the energy corresponding to this state? Is it an eigenstate of momentum? **[CU (H) 1994]**

Solution: We have

$$\hat{H}\psi(r) = -\frac{\hbar^2}{2m}\left[\frac{1}{r^2}\frac{d}{dr}\left(r^2\frac{d}{dr}\right)\right]\psi(r) + V(r)\,\psi(r) = -\frac{\hbar^2}{2m}\left[\frac{d^2}{dr^2} + \frac{2}{r}\frac{d}{dr}\right]\psi + V\psi$$

$$= -\frac{\hbar^2}{2m}\frac{d}{dr}\left(-\frac{\psi}{a_0}\right) - \frac{\hbar^2}{mr}\left(-\frac{\psi}{a_0}\right) + V\psi = \frac{\hbar^2}{2ma_0}\frac{d\psi}{dr} + \frac{\hbar^2}{mra_0}\psi + V\psi$$

$$= \frac{\hbar^2}{2ma_0}\cdot\left(-\frac{\psi}{a_0}\right) + \frac{\hbar^2}{mr}\left(\frac{me^2}{4\pi\varepsilon_0\hbar^2}\right)\psi + V\psi = -\frac{\hbar^2}{2ma_0}\cdot\frac{me^2}{4\pi\varepsilon_0\hbar^2}\psi(r) - V\psi(r) + V\psi(r)$$

$$= -\left[\frac{e^2}{4\pi\varepsilon_0 2a_0}\right]\psi(r) = \boldsymbol{E_1\psi(r)}$$

$\therefore \psi(r)$ is eigenfunction of $\hat{H}$.

Now, we consider

$$\vec{p}\psi(r) = -i\hbar\nabla\psi(r) = -i\hbar\hat{r}\frac{d}{dr}\frac{1}{\sqrt{\pi a_0^3}}e^{-r/a_0} = -i\hbar\hat{r}\frac{1}{\sqrt{\pi a_0^3}}e^{-r/a_0}\left(-\frac{1}{a_0}\right) = \frac{i\hbar}{a_0\sqrt{\pi a_0^3}}\hat{r}\psi(r)$$

or, $\hat{\vec{p}}\psi(r) \neq \psi(r)$ is not eigenfunction of $\vec{p}\cdot\psi(r)$ is eigenfunction of $\hat{H}$.

» ***Example* 8.18.** **A hydrogen atom is in the state represented by the normalised wave function**

$$\psi(\vec{r}) = \frac{1}{\sqrt{3}}\psi_{200}(\vec{r}) + \sqrt{\frac{2}{3}}\psi_{210}(\vec{r})$$

(i) Does the state correspond to definite value of energy, orbital angular momentum and its z-component?

[CU (H) 2000]

(ii) Calculate expectation values of these quantities.

***Solution*:**

(i) We have

$$\hat{H}\psi(\vec{r}) = \frac{1}{\sqrt{3}}\hat{H}\psi_{200} + \sqrt{\frac{2}{3}}\hat{H}\psi_{200} = E_2\left[\frac{1}{\sqrt{3}}\psi_{200} + \frac{\sqrt{2}}{\sqrt{3}}\psi_{210}\right] = \left[-\frac{1}{4\pi\varepsilon_0}\frac{1}{2a_0}\left(\frac{e}{2}\right)^2\right]\psi(\vec{r})$$

This implies $\psi(\vec{r})$ is an eigenstate of $\hat{H}$ with eigenvalue E_2.

Here, ψ_{210} have $l = 0$, l is not conserved.

So

$$\hat{l}\psi(\vec{r}) = \frac{1}{\sqrt{3}}\hat{l}\psi_{200} + \sqrt{\frac{2}{3}}\psi_{210} = \frac{1}{\sqrt{3}}\sqrt{0(0+1)}\hbar\psi_{200} + \sqrt{\frac{2}{3}}\sqrt{l(1+1)}\hbar\psi_{210} = \frac{2}{\sqrt{3}}\psi_{210}$$

This shows that $\psi(\vec{r})$ is not reproduced and hence it is not an eigenstate of $\hat{l}$.

ψ_{200}, ψ_{210} both have, $m = 0$ $\therefore$ m_l is conserved.

$$\therefore \hat{l}_z\psi(r) = \frac{1}{\sqrt{2}}\hat{l}_z\psi_{200} + \sqrt{\frac{2}{3}}\hat{l}_z\psi_{210} = \frac{1}{\sqrt{3}}(0\hbar)\psi_{200} + \sqrt{\frac{2}{3}}(0\hbar)\psi_{200} = (0)\,\psi(\vec{r})$$

i.e., $\psi(\vec{r})$ is eigenstate of $\hat{l}_z$ with eigenvalue 0.

(ii) Again $\langle \hat{H} \rangle = \psi^+ \hat{H} \psi = \tau^+ E_2 \psi = E_2 \psi^+ \psi = E_2$.

$$\langle \hat{l} \rangle = \tau^+ \hat{l} \psi = \psi^+ \frac{2}{\sqrt{3}} \hbar \psi_{200} = \frac{2\hbar}{\sqrt{3}} \left[\frac{1}{\sqrt{3}} \psi_{200}^+ + \sqrt{\frac{2}{3}} \psi_{210}^+ \right] \psi_{210} = \sqrt{2}\hbar \frac{2}{3} = \frac{2^{3/2}}{3} \hbar = 0.94\hbar$$

$$\therefore \langle \hat{l} \rangle = \psi^+ \hat{l}_z \psi = \psi^+(0) = 0.$$

EXERCISES

8(A): Objective (or MCQ) Type Questions

Select the correct answer:

8.1. The waves representing a free particle in three dimensions are

(a) progressive waves
(b) standing waves
(c) transverse waves
(d) polarised waves.

8.2. If the quantum numbers n_x, n_y and n_2 are each equal to zero for a particle trapped in a rectangular potential well, then the particle

(a) may be absent from the well
(b) may be present in the well
(c) may or may not be present in the well
(d) none of the three.

8.3. For a particle confined in a three-dimensional cubical potential well, the momentum eigenvalue for the state $(2, 2, 1)$ is

(a) $\frac{\pi\hbar}{3a}$
(b) $\frac{\pi\hbar}{a}$
(c) $\frac{3\pi\hbar}{a}$
(d) $\frac{\pi a}{3\hbar}$

8.4. The energy eigenvalue for a particle trapped in a cubical potential well for $(2, 2, 1)$ state is

(a) $\frac{2\pi^2\hbar^2}{9ma^2}$
(b) $\frac{9\pi^2\hbar^2}{2ma^2}$
(c) $\frac{\pi^2\hbar^2}{9ma^2}$
(d) $\frac{2ma^2}{9\pi^2\hbar^2}$.

8.5. The energy of a three-dimensional LHO is given by

(a) $(n+1)\,\hbar\omega$
(b) $n\hbar\omega$
(c) $(n+\frac{1}{2})\,\hbar\omega$
(d) $(n+3)\,\hbar\omega$.

8.6. The zero-point energy of LHO is given by

(a) $\hbar\omega$
(b) 0
(c) $\frac{1}{2}\hbar\omega$
(d) $3\hbar\omega$.

8.7. For an LHO, the degeneracy for state $n = 1$ is

(a) 3-fold
(b) 4-fold
(c) 2-fold
(d) 5-fold.

8.8. Total number of energy states for $n = 3$ for an LHO is

(a) 3
(b) 5
(c) 10
(d) 15.

8.9. The spherical harmonics are wave functions of θ and ϕ, which are

(a) unnormalised
(b) normalised
(c) some of them are normalised
(d) cannot be normalised.

8.10. The value of Y_1^0 is given by

(a) $\left(\frac{3}{4\pi}\right)^{1/2} \sin\theta$
(b) $\left(\frac{3}{4\pi}\right)^{1/2} e^{im\phi}$
(c) $\left(\frac{3}{4\pi}\right)^{1/2} \cos\theta$
(d) $\left(\frac{3}{4\pi}\right)^{1/2} \sin\theta\cos\theta$.

8.11. B^{+++} is

(a) a hydrogenic atom
(b) an ionised atom
(c) a non-hydrogenic atom
(d) a neutral atom.

8.12. The electron in a hydrogen atom moves in a potential which is regarded as

(a) asymmetric
(b) spherically symmetric
(c) Coulombian
(d) none of them.

8.13. The normalised $\Theta(\theta)$ function in case of hydrogen atom for $n = 2$, $l = 1$ and $m_l = 0$ is given by

(a) $\frac{\sqrt{6}}{2}\cos\theta$
(b) $\frac{\sqrt{6}}{2}\sin\theta$
(c) $\sqrt{\frac{6}{2}}\cos\theta$
(d) $\frac{\sqrt{6}}{2}\sin\theta\cos\theta$.

8.14. The wave function of the electron in hydrogen atom for $n = 1$, $l = 0$, $m = 0$ is given by

(a) $\frac{1}{\sqrt{\pi a_0^{3/2}}} e^{-\rho/2}$
(b) $\frac{1}{\sqrt{\pi}\, a_0^{3/2}} e^{-\rho/2}$
(c) $\sqrt{\frac{\pi}{a_0^{3/2}}}\, e^{-\rho}$
(d) $\sqrt{\pi a_0^{3/2}}\, e^{-\rho/2}$.

8.15. Total number of energy levels (or, degeneracy) for the nth state of hydrogen atom is

(a) $(n+1)$ (c) n^2

(b) n^3 (d) (n^2+1).

8.16. The selection rule for radiative transition of an electron in hydrogen atom is

(a) $\Delta l = 0, \Delta m_l = \pm 1$ (c) $\Delta l = \pm 1, \Delta m_l = 0$

(b) $\Delta l = \pm 1, \Delta m_l = \pm 1$ or 0 (d) $\Delta l = \pm 1, \Delta m_l = \pm 1$.

8.17. In case of the rigid rotator with free axis, the degeneracy for l is

(a) $2l$ (c) $2nl$

(b) $(2l+1)$ (d) $4l^2$.

8.18. The energy of a rigid rotator with fixed axis is given by

(a) $E_m = \dfrac{m^2\hbar^2}{2I}$ (c) $E_m = \dfrac{2I}{m^2\hbar^2}$

(b) $E_m = \dfrac{1}{2}Im^2\hbar^2$ (d) $E_m = \dfrac{m^2 I}{2\hbar^2}$.

8.19. An atomic state of hydrogen is represented by the following wave function

$$\psi(r, \theta\phi) = \frac{1}{\sqrt{\pi}} \frac{1}{(2a_0)^{3/2}} \left(1 - \frac{r}{2a_0}\right) e^{-r/2a_0}$$

The quantum number of the state are

(a) $l = 0, m_l = 0, n = 2$ (c) $l = 1, m_l = 0, n = 2$

(b) $l = 1, m_l = 1, n = 2$ (d) $l = 1, m_l = -1, n = 2$. **[GATE 2007]**

8.20. The expectation value of z-coordinate in the ground stature of hydrogen atom is [a_0 = 1st Bohr radius]

(a) a_0 (c) $\frac{3}{2}a_0$

(b) $\frac{1}{2}a_0$ (d) 0. **[cf. GATE 2005]**

[a_0 = radius of the first Bohr orbit]

8.21. If a three-dimensional harmonic oscillator has an energy $3.5\frac{h\omega}{2\pi}$ in a particular state then the degree of degeneracy is

(a) 7 (c) 4

(b) 6 (d) 3. **[IAS(Pre) 2010]**

8.22. $D(r)\,dr$ denotes the probability of finding the electron between r and $r+dr$. If a_0 and z be respectively the Bohr radius and charge number, then

(a) $\int_0^\infty D(r)\,dr = 1$ (c) $D(r) = \dfrac{z}{\pi a_0^3}$ at $r = \infty$

(b) $D(r) = 0$ at $r = 0$ (d) $D(r)$ has maximum value at $r = \frac{a_0}{z}$.

[IAS(Pre) 2010]

Select the correct answer:

(a) 1 and 2 (b) 3 and 4 (c) 1, 2 and 3 (d) 1, 2 and 4.

8.23. Match list I and list II and select the correct answer using the codes given below the list

List I (Function)	*List II (Quantum system)*
(A) Hermite polynomials	1. Particle in a box
(B) Associated Laugurre	2. Hydrogen atom
(C) Sine function	3. Harmonic oscillator
(D) Sine function	4. Potential step.

	A	B	C	D
(a)	1	4	3	2
(b)	3	4	1	2
(c)	1	2	3	4
(d)	3	2	1	4

[IAS(Pre) 2006]

8.24. The wave function of the ground state of hydrogen is $\psi(r) = Ae^{-r/a_0}$, where a_0 Bohr radius and A = a constant. What is the average value of r?

(a) a_0 (c) $2a_0$

(b) $\frac{3a_0}{2}$ (d) $a_0/2$. **[IAS(Pre) 2007]**

8.25. The expectation wave of z-coordinate in the ground state is

(a) a_0 (c) $\frac{3}{2}a_0$

(b) $a_0/2$ (d) 0. **[GATE 2006]**

8.26. At $r = a_0$ the ground state radial wave function drops by an amount [a_0 = 1st Bohr radius]

(a) $\frac{1}{a_0}$ (c) $e^{-\infty}$

(b) $\frac{1}{a_0^2}$ (d) e^{-1}.

ANSWERS

8.1. (b).	**8.7.** (a).	**8.13.** (a).	**8.19.** (a).	**8.25.** (d).
8.2. (a).	**8.8.** (c).	**8.14.** (b).	**8.20.** (d).	**8.26.** (d).
8.3. (c).	**8.9.** (b).	**8.15.** (c).	**8.21.** (b).	
8.4. (b).	**8.10.** (c).	**8.16.** (b).	**8.22.** (d).	
8.5. (c).	**8.11.** (a).	**8.17.** (b).	**8.23.** (d).	
8.6. (c).	**8.12.** (b).	**8.18.** (a).	**8.24.** (b).	

8(B): Short Answer Type Questions

8.1. Show that the solution of Schrödinger wave equation for a free particle in three dimensions represents a set of standing waves with the wavefronts normal to the x-axis.

8.2. What is a three-dimensional rectangular potential box?

8.3. Show that for a particle in a three-dimensional potential box, all the stationary states are given by positive integral values of n_x, n_y, n_z.

8.4. What do you mean by the *degeneracy* of a state?

8.5. Show that for a cubical potential box, the ground state wave function of a particle is given by

$$\psi_{1,1,1} = \sqrt{\frac{8}{a^3}} \sin\frac{\pi x}{a} \sin\frac{\pi y}{a} \sin\frac{\pi z}{a}$$

8.6. Show that the eigenvalues of momentum and energy of the ground state of a particle in a cubical potential box are $\frac{\pi\hbar}{a}\sqrt{3}$ and $\frac{\pi^2\hbar^2}{2ma^2}3$ respectively, the different symbols have their usual meaning.

8.7. The momentum and energy of a particle in a cubical box for the states 2, 1, 1; 1, 2, 1 and 1, 1, 2 all have the same value $\left(\frac{\sqrt{6}\pi\hbar}{a}\right)$ and $\left(\frac{6\pi^2\hbar^2}{2ma^2}\right)$ respectively. In view of this, can you differentiate a multidimensional and a unidimensional system?

8.8. Show that for a three-dimensional isotropic harmonic oscillator, the energy for the nth state is $E_n = (n + \frac{3}{2})\,\hbar\omega$.

8.9. What do you mean by a spherically symmetric potential?

8.10. What are spherical harmonics?

8.11. What condition must satisfy the wave function for a particle in a spherically symmetrical potential?

8.12. What are principal quantum number, orbital quantum number and azimuthal quantum number? What is their use?

8.13. Write down the Hamiltonian for the hydrogen atom.

8.14. Write down the Schrödinger wave equation for the hydrogen atom.

8.15. Separate the Schrödinger equation for the motion of an electron in hydrogen atom into one radial and two angular parts. **[BU (H) 2000]**

8.16. Are all the wave functions of the LHO orthogonal? Explain.

8.17. Prove that for the hydrogen atom the wave functions ψ_{100} and ψ_{200} are orthogonal.

8.18. Prove that for a three-dimensional LHO, the degree of degeneracy is $\frac{1}{2}(n+1)(n+2)$.

8.19. Show that for all hydrogenic atoms, the selection rule for allowed transition is $\Delta m_l = 0, \pm 1$ and $\Delta l = \pm 1$.

8.20. Explain the zero-point energy. Can you reconcile it with the classical viewpoint?

8.21. Show that the energy values of a rigid rotator are discrete.

8.22. Discuss how the harmonic oscillator eigenfunction approaches those of the free particle in the limit $\omega \to 0$.

8.23. Can the principal quantum number for an electron in hydrogen atom be zero?

[*Hints*: n cannot be zero for in that both E_n and r_n would be zero and the electron would be located at the centre of the nucleus with zero energy which is absurd]

8.24. What is the total number of states in the $n = 2$ state of a hydrogen atom? **[BU (H) 1996]**

8.25. What is the degeneracy of the $n = 2$ level for a three-dimensional isotropic oscillator. [GATE 2005]

8.26. Write down the complete Schrödinger equation for a single electron atom in spherical coordinated. Separate the equation into radial part and angular radial equation. [IAS 2007]

8.27. A hydrogen atom is in the state represented by the normalised wave function

$$\psi(\vec{r}) = \frac{1}{\sqrt{3}}\psi_{200}(r) + \sqrt{\frac{2}{3}}\psi_{210}(\vec{r})$$

Does this state correspond to a definite value of energy, orbital angular momentum and its z-component? [CU (H) 2000]

8(C): Broad Answer Type Questions

8.1. (a) Consider a particle confined to move inside a cubical box of sides L.

(i) Show that stationary state Schrödinger equation is separable in Cartesian system of coordinates.

(ii) State boundary conditions.

(iii) Obtain the energy eigenfunctions and eigenvalues. Is there any degeneracy?

(b) The normalised ground state wave function of hydrogen atom is given to be

$$\psi(r,\theta,\phi) = \frac{1}{\sqrt{\pi a_0^{3/2}}}e^{-r/a_0}$$

Find the distance from the nucleus at which the electron is most likely to be found. [BU (H) 2002]

[*Ans.* a_0]

8.2. The normalised ground state wave function of an electron in a hydrogen atom is given by

$$\psi(\vec{r}) = \left(\frac{1}{\pi a_0^3}\right)^{1/2} e^{-r/a_0}$$

where a_0 is the Bohr radius. Verify that $\psi(\vec{r})$ is normalised. Obtain the probability of finding the ground state electron at a distance greater than a_0 from the nucleus. [BU (H) 1994]

8.3. (a) Write down Schrödinger equation for the hydrogen atom.

(b) Given that in spherical polar coordinates

$$\nabla^2 \equiv \frac{1}{r^2}\frac{\partial}{\partial r}\left(r^2\frac{\partial}{\partial r}\right) + \frac{1}{r^2}\frac{1}{\sin\theta}\frac{\partial}{\partial\theta}\left(\sin\theta\frac{\partial}{\partial\theta}\right) + \frac{1}{r^2\sin^2\theta}\frac{\partial^2}{\partial\phi^2}$$

Show that the Schrödinger equation in (a) can be separated into two angular parts and one radial part.

(c) The normalised ground state wave function of an electron in a hydrogen atom is given by

$$\psi(\vec{r}) = \left(\frac{1}{\pi a_0^3}\right)^{1/2} e^{-r/a_0}$$

where a_0 is the radius of the innermost Bohr orbit. Obtain the probability of finding the ground state electron at a distance greater than a_0 from the nucleus. [BU (H) 1997]

8.4. (a) Write down the time-independent Schrödinger equation for the motion of the electron in hydrogen atom, assuming that the proton is at rest.

Given

$$\nabla^2 \equiv \frac{1}{r^2}\frac{\partial}{\partial r}\left(r^2\frac{\partial}{\partial r}\right) + \frac{1}{r^2\sin\theta}\frac{\partial}{\partial\theta}\left(\sin\theta\frac{\partial}{\partial\theta}\right) + \frac{1}{r^2\sin^2\theta}\frac{\partial^2}{\partial\phi^2}$$

separate the Schrödinger equation into one radial and two angular parts.

(b) The normalised ground state wave function of a one-electron atom is given by

$$\psi(r,\theta,\phi) = \pi^{-1/2}\left(\frac{Z}{a_0}\right)^{3/2} e^{-2r/a_0}$$

where the notations have usual meaning. Evaluate the probability that the electron in this state will be found at a distance greater than $2a_0/Z$. **[BU (H) 2000]**

8.5. Find the normalised eigenfunctions and the allowed energy eigenvalues of a particle of mass m, which is constrained by impenetrable walls to move in a cubical box of side L. Find the degeneracy of the first excited state.

8.6. Consider the hydrogen atom whose wave function at $t = 0$ is the following superposition of energy eigenfunctions $\psi_{nlm}(\vec{r})$:

$$\psi(\vec{r}, t=0) = \frac{1}{\sqrt{14}}\left[2\psi_{100}(\vec{r}) - 3\psi_{200}(\vec{r}) + \psi_{322}(\vec{r})\right]$$

(a) What is the probability of finding the system in the ground state (100), in the state (200) and in the state (322); also in some other energy eigenstate?

(b) What is the expectation value of the energy in the state given above?

[*Ans.* (i) $\frac{4}{14}$, (ii) $\frac{9}{14}$, (iii) $\frac{1}{14}$; 0 (ii) Use the fact that $H\psi_{nlm} = E_n\psi_{nlm}$]

8.7. Solve the Schrödinger's equation for a rigid rotator with free axis and obtain the eigenvalues of energy and the eigenfunctions. How are the results modified if the axis is fixed?

8.8. Discuss the wave mechanics of the electron in a hydrogen atom in a spherically symmetric potential and derive its energy states and eigenfunctions.

8.9. Solve the Schrödinger equations for the case of hydrogen atom. Neglect electron spin and assume the nucleus to be stationary. How do the electron energy levels obtained from the theory agree with experiment?

8.10. Write down the Hamiltonian of a particle of mass m confined to move in a three-dimensional potential box with rigid boundaries, having sides a, b, c parallel to the three rectangular coordinates x, y, z. Solve the Schrödinger equation by the separation of variables and apply the boundary conditions to show that the system can have a discrete set of energy values determined by three integers n_x, n_y and n_z.

What is the energy of the lowest level? Show that if $a = b = c$, some of the states are degenerate. Find the energy of the first three non-degenerate states in this case.

8.11. Solve Schrödinger equation for a three-dimensional harmonic oscillator and obtain the eigenvalues and eigenfunctions. Are the energy levels degenerate? Explain.

8.12. Why is it important to study harmonic oscillator problem? Write down the Schrödinger equation for a three-dimensional harmonic oscillator. Draw pictorially the energy levels of a particle in a harmonic oscillator potential. **[BU (H) 2005]**

8.13. Apply Schrödinger equation in Cartesian coordinates to find out the wave functions and energy eigenvalues of the three-dimensional harmonic oscillator. Draw the energy levels with degeneracy in the above case.

8.14. Solve Schrödinger equation for a three-dimensional harmonic oscillator in spherical polar coordinates. Find out expressions for energy eigenvalues and draw the energy level diagram for such an oscillator.

8.15. (a) Write down the time-independent Schrödinger equation for a simple harmonic oscillator. The eigenfunction of the Hamiltonian operator for the ground state is

$$\psi_0 = \left(\frac{\alpha}{\sqrt{\pi}}\right)^{1/2} e^{-\alpha^2 x^2/2}$$

where $\alpha = \left(\frac{m\omega_0}{\hbar}\right)^{1/2}$.

Find the energy eigenvalue corresponding to the ground state.

(b) Using the fundamental commutation relation $[\hat{x}, \hat{p}] = i\hbar$ find $[\hat{p}, \hat{H}]$, where $\hat{H}$ is the Hamiltonian of a linear harmonic oscillator. Hence, determine whether the ground state wave function ψ_0 of the Hamiltonian is an eigenstate or not.

(c) Find the constant B, which makes $e^{-\alpha x^2}$ an eigenfunction of the operator

$$\left(\frac{d^2}{dx^2} - Bx^2\right).$$

[CU (H) 2012]

8.16. (a) Consider a linear harmonic oscillator for which the total energy is given by

$$E = \frac{p_x^2}{2m} + \frac{1}{2} m\omega^2$$

where the symbols have their usual meanings. The particle is assumed to be confined in a region of width d. Use uncertainty principle to obtain the minimum energy (ground state energy of the oscillator).

(b) Given $\hat{a} = \hat{p}_x + i\hat{X}^2, \hat{a}^+ = \hat{p}_x - i\hat{X}^2$.

Find $[a, a^+]$ using the commutation relation of position and momentum operators. **[CU (H) 2011]**

8.17. (a) Consider a one-dimensional oscillator moving in a potential $V(x) = \frac{1}{2} m\omega^2 x^2$. Given that the ground state wave function is

$$\psi(x) = \left(\frac{\alpha}{\pi}\right)^{1/4} \exp\left(-\tfrac{1}{2}\alpha x^2\right)$$

where $\alpha = \frac{m\omega}{\hbar}$. Find the ground state energy eigenvalue E_0.

(b) If the expectation value of the square of the displacement in case of linear harmonic oscillator is

$$\langle x^2 \rangle = \left(n + \tfrac{1}{2}\right) \hbar / m\omega$$

What is the expectation value of the potential energy?

8.18. (a) Find the eigenfunction of the operator, $x + \frac{d}{dx}$.

(b) Two operators $\hat{a}_+$ and $\hat{a}_-$ are defined as

$$\hat{a}_{\pm} = \frac{1}{\sqrt{2\hbar m\omega}}(\pm i\hat{p} + m\omega\hat{x})$$

where that the Hamiltonian operator $\hat{H}$ for a linear harmonic oscillator is given by $\hat{H} = \hbar\omega(\hat{a}_+\hat{a}_- + \frac{1}{2})$. Find the ground state of energy of the oscillator.

(c) Prove that a Hermitian operator has real eigenvalues and orthogonal eigenvectors.[CU (H) 2014]

8.19. (a) Derive radial and angular parts of Schrödinger equation for an electron bounded by a spherically symmetric potential $V(r)$.

Hence, show that the radial part of the wave equation can be transformed into a form which is formally the same as the Schrödinger equation for a one-dimensional motion in the region $0 < r < \infty$ with an effective potential

$$V_0 = V(r) + \frac{\lambda h^2}{2mr^2}$$

Give an interpretation of the additional term in the expression for the potential. (Here, λ is a constant and all other symbols have their usual meanings).

(b) Write down the Hamiltonian for the hydrogen atom.

Suppose, the electron in the hydrogen atom is in state

$$\psi(r) = \left(\frac{1}{\pi a_0^2}\right)^{1/2} e^{-r/a_0}$$

where $a_0 = \hbar^2/me^2$ is the Bohr radius. Is the state is an eigenstate of the Hamiltonian?[BU (H) 2012]

8.20. (a) Reduce the Schrödinger equation for the hydrogen atom written in spherical coordinates into three equations involving r, θ and ϕ separately.

[*Given*: $\nabla^2 \equiv \frac{1}{r^2}\frac{\partial}{\partial r}\left(r^2\frac{\partial}{\partial r}\right) + \frac{1}{r^2 \sin\theta}\frac{\partial}{\partial\theta}\left(\sin\theta\frac{\partial}{\partial\theta}\right) + \frac{1}{r^2 \sin^2\theta}\frac{\partial^2}{\partial\phi^2}$]

(b) Neglecting spin what are the quantum numbers of the electron that specify the state of hydrogen atom completely and what are their possible values? [BU (H) 2010]

8.21. (a) What is the function in the spherical polar coordinate system which the eigenfunction of the operators $\hat{L}^2$ and $\hat{L}_z^2$? What are the eigenvalues of these operators?

(b) Why is it that only one component of $\vec{L}$ is quantised? [BU (H) 2009]

8.22. (a) Separate the time-independent Schrödinger equation of the hydrogen atom in radial and angular parts.

[*Given*: $\nabla^2 \equiv \frac{1}{r^2}\frac{\partial}{\partial r}\left(r^2\frac{\partial}{\partial r}\right) + \frac{1}{r^2 \sin^2\theta}\frac{\partial}{\partial\theta}\left(\sin\theta\frac{\partial}{\partial\theta}\right) + \frac{1}{r^2 \sin\theta}\frac{\partial^2}{\partial\phi^2}$]

(b) Why is the centrifugal potential so called? [BU (H) 2008]

8.23. (a) Write down the time-independent Schrödinger equation for a spherically symmetric potential.

(b) The ground state wave function of the electron in a hydrogen atom is

$$\psi_{100}(r) = \frac{1}{\sqrt{\pi a_0^{3/2}}} e^{-r/a_0}, \text{ where } a_0 \text{ is the Bohr radius.}$$

(i) Write down radial probability distribution and sketch its graph, mentioning the value of r at which it is the maximum.

(ii) Calculate the expectation values of kinetic energy of the electron in its ground state.

(c) Using the above ground state wave function, show that the average distance of the electron from the nucleus is $1.5a_0$. [CU (H) 2012]

8.24. (a) Write down the Schrödinger equation for hydrogen atom assuming the nucleus to be heavy. Obtain the radial part of the wave function equation.

(b) In the ground state of the hydrogen atom, show that the probability P for the electron to lie within the sphere of radius R is

$$P = 1 - \exp\left(-\frac{2R}{a_0}\right)\left(1 + \frac{2R}{a_0} + \frac{2R^2}{a_0^2}\right)$$

where $\psi(100) = \left(\pi a_0^3\right)^{1/2} \exp\left(-\frac{r}{a_0}\right)$.

(c) Write down the operators for L^2 and L_z in polar coordinates. Hence, verify that $\psi = A \sin\theta e^{i\phi}$, where A is a constant, is an eigenfunction of L^2 and L_z. Find the eigenvalues. [CU (H) 2015]

8.25. (a) Prove that Hermitian operator has real eigenvalues and mutually orthogonal eigenvectors.

(b) Write down Schrödinger equation for the electron of hydrogen atom assuming the nucleus to be stationary. By separation of variables obtain the radial equation.

(c) The normalised wave function for the ground state of hydrogen atom is given by

$$\psi(r) = \frac{1}{\sqrt{\pi}a_0^{3/2}} \exp\left(-\frac{r}{a_0}\right)$$

where a_0 is the Bohr radius.

Find the distance from the nucleus at which the electron is most likely to be found. [BU (H) 2010]

8.26. (a) Verify that $\hat{Q} = \frac{1}{2}(\hat{x}\hat{p}_x + \hat{p}_x\hat{x})$ is a Hermitian operator.

(b) The wave function of the 1s electron in hydrogen atom is $\psi(\pi a_0^3)^{-1/2} e^{-r/a_0}$. Show that the average value of $1/r$ for this electron would be $1/a_0$. [CU (H) 2008]

8.27. (a) Write down time-independent Schrödinger's equation for hydrogen atom in spherical polar coordinates.

(b) Obtain the radial equation for the electron motion and hence show that the angular momentum of the electron is given by $\sqrt{l(l+1)}\hbar$. [NBU (H) 2009]

8.28. (a) Write down Schrödinger wave equation for a hydrogen atom and hence solve the radial part.

(b) The normalised ground state wave function of electron in an hydrogen atom is given by

$$\psi(\vec{r}) = \frac{1}{(\pi a_0^3)^{1/2}} \exp\left(-\frac{r}{a_0}\right)$$

(i) Calculate the expectation value of r, i.e., $\langle r \rangle$ for hydrogen atom is 1s state.

(ii) Show that the radial probability density for hydrogen atom in 1s state is maximum at $r = a_0$. [KU (H) 2010]

8.29. (a) Solve hydrogen atom problem to find the ground state eigenfunction and eigenvalue.

(b) Find the probability of observing an electron at a distance r from the nucleus and explain your answer.

(c) Find the average distance of the electron from the nucleus in the ground state of hydrogen atom taking the first Bohr orbit radius $a_0 = 0.53$ Å and hence deduce the necessary working formula.

[KU (H) 2015]

8.30. (a) For one-dimensional SHO, obtain the ground state wave function.

(b) The normalised ground state wave function of an electron in a hydrogen atom is given by

$$\psi(r) = \left(\frac{1}{\hbar a_0^3}\right)^{1/2} e^{-r/a_0}$$

where a_0 is the radius of the first Bohr orbit. Find the expectation value of position and momentum of the electron. [KU (H) 2010]

8.31. Determine the phase trajectory of a one-dimensional LHO of constant energy E moving along x-axis. What is the phase space area accessible? How many accessible micro states are there if the oscillator is (i) a classical oscillator, (ii) a quantum mechanical oscillator. [CU (H) 2007]

8.32. (a) Make a plot of first and second energy eigenfunction of LHO.

(b) Make a plot of 1s and 2p energy eigenfunctions of hydrogen atom.

8.33. (a) Set-up the Schrödinger wave equation for a one-dimensional quantum mechanical harmonic oscillator and obtain the energy eigenvalues.

(b) Draw the energy level diagram of the harmonic oscillator. [Kakatiya U. May/June 2016]

8.34. (a) Solve Schrödinger's equation for one-dimensional motion of a particle in constant potential and obtain an expression for density of states in terms of energy.

(b) What is zero point energy? What is its significance? [Kakatiya U. May/June 2016]

8.35. (a) Derive Schrödinger's time-department and time-independent wave equations.

(b) An electron is bound in one-dimensional potential well of width 1 Å but of infinite wall height. Find its energy value in the ground and first excited states ($m = 9.11 \times 10^{-31}$ and $e = 1.6 \times 10^{-19}$ C).

[Kakatiya U. May/June 2016]

8(D): Numerical Problems

8.1. Assuming the nucleus as a box with a size of 10^{-14} m, compute the energy of a neutron confined to the nucleus. [*Given*: $m = 1.6 \times 10^{-27}$ kg] [PU 1995]

8.2. A large number of identical Fermi particles are confined in a box of volume V, occupying the lowest accessible levels. By what factor does the maximum momentum of the particles change, if we double the volume of the box, leaving the number of particles unchanged?

[*Hints*: $P_{\max} \propto (V)^{-1/3}(1/2)^{1/3}$.]

8.3. A particle of mass m is confined by an appropriate potential to a spherical region of radius R. Give a rough estimate of the particles kinetic energy in its ground state.

[*Hints*: By uncertainty principle, $\Delta x\, \Delta p_x \sim \hbar$, or $Rp \sim \hbar$, or $p \sim \frac{\hbar}{R}$.

$\therefore T = \frac{p^2}{2m} = \frac{\hbar^2}{2R^2 m}$]

8.4. Compute the value of $\langle r \rangle$ and the most probable value of r in the ground state of hydrogen atom. **[GNDU 1995]**

[*Ans.* $\psi_{1,0,0} = \frac{1}{\pi^{1/2} a_0^{3/2}} e^{-r/a_0}$, a_0 = Bohr radius. $\langle r \rangle = 1.5a_0$, $r = a_0$]

8.5. The radial wave function for the ground state ($n = 1, l = 0$) for the hydrogen atom is

$$R_{100}(r) = \left(\frac{1}{a_0}\right)^{3/2} 2e^{-r/a_0}$$

where $a_0 = \frac{\hbar^2}{me^2}$ is the Bohr radius. If the probability of finding the electron between r and $r + dr$ is $P(r)$, draw a rough sketch to show how $P(r)$ varies with r for the ground state.

8.6. Calculate the probability of finding an electron between the volume enclosed between two spheres of radii r and $r + dr$ for hydrogen atom state defined by

$$\psi_{1,0,0} = \frac{1}{\sqrt{\pi a_0^3}} e^{\left(-\frac{r}{a_0} - \frac{iEt}{\hbar}\right)}$$

where a_0 = Bohr radius. Where is the probability of finding the electron maximum? Also calculate the mean radius of the orbit for this state.

8.7. Find out $\langle L^2 \rangle$ and $\langle L_z \rangle$ for the states ψ_{322} and ψ_{200} of hydrogen atom. [*Ans.* $6\hbar^2, 2\hbar$]

8.8. For an LHO, the mass of the particle is 1 gm and is connected to a fixed point by a spring which is stretched 1 cm by a force of 10 k-dynes. What is the zero-point energy? [*Ans.* 5.25×10^{-26} erg]

8.9. Show that for an electron in the $1s$ state of hydrogen atom

$$\langle E_k \rangle + \langle V \rangle = -\frac{me^4}{2\hbar^2}$$

the symbols have their usual meaning.

8.10. Find the value of r at which the radial probability densities for the states ψ_{210} and ψ_{320} of hydrogen atom become maximum. [*Ans.* $r = 4a_0$ and $r = 9a_0$]

8.11. If the radial function for the electron of the hydrogen atom is represented as

$$R(r) = Ce^{-\rho/2} \rho^l L_{n+l}^{2l+1}(\rho)$$

find the normalisation constant C. [*Ans.* $C = \left(\frac{\rho}{r^3}\right) \frac{(n-l-1)!}{2n\{(n+1)!\}^3}$]

8.12. One of the energy eigenfunction of a particle in a linear harmonic oscillator to potential is $\psi(x) = Axe^{-x^2/x_0^2}$, where A, x_0 are constants. Find plotting the wave function whether the particle is in the ground state, the first excited state or the second excited state. **[BU (H) 2014]**

8.13. Demonstrate the parities of the wave functions of the ground state and the first excited state of a linear harmonic oscillator by drawing those wave functions. **[BU (H) 2009]**

8.14. Consider a linear harmonic oscillator for which the total energy is given by $E = p_x^2/2m + \frac{1}{2}m\omega^2 w^2$, where the symbols have their usual meanings. The particle is assumed to be confined in a region of d. Use the uncertainty principle to obtain the minimum energy (ground state energy) of the oscillator. **[CU (H) 2009, 11]**

8.15. Cast the Schrödinger equation for a linear harmonic oscillator in the form:

$$\frac{d^2u}{d\xi^2} + (\lambda - \xi^2)u = 0$$

Assuming solution of the form: $u = H(\xi)\, e^{-\xi^2/2}$ convert it into a different equation of the series and hence calculate the minimum energy of the oscillator. **[KU (H) 2012]**

8.16. For a simple harmonic oscillator, the wave function is given by

$$\psi_n = A_n e^{-x^2/2}$$

where $H_n(x) = (-1)^n e^{x^2} \frac{d^n}{dx^n}(e^{-x^2})$.

Hence, obtain the orthonormal eigenfunctions of the harmonic oscillator. **[KU (H) 2011]**

8.17. The $n = 2, l = 1, m = 0$ normalised wave function of hydrogen atom is

$$\psi_{210}(r,\theta,\phi) = \frac{1}{\sqrt{32\pi a_0^3}} r\, e^{-r/2a_0 \cos\theta}$$

Calculate the expectation value $\langle r \rangle$, when the atom is in the state described by this wave function and compare it with the value of r, where the radial probability density of the electron is the maximum. **[BU (H) 2015]**

8.18. The eigenvalue of the operator L^{-2} is $l(l+1)\hbar^2$, whereas the maximum eigenvalue of the operator L_z is $l\hbar$ instead of $\sqrt{l(l+1)}\hbar$. Whey this is so? **[BU (H) 2013]**

8.19. The normalised wave function for the ground state of hydrogen atom is

$$\phi_{100} = \frac{1}{\sqrt{\pi a_0^3}} e^{-r/a_0}$$

where a_0 is the Bohr radius.

Calculate (a) the most probable distance (b) average distance of the electron from the nucleus. **[CU (H) 2014]**

• • •

Chapter 9

Dirac Notation

9.1 Dirac's *Bra* and *Ket* notations

In quantum mechanics a dynamical system may have many different states, the superposition of any two of which give rise to other states consisting the set of the states. This conception leads to the possibility of representing every dynamical state by certain type of vector, called a *ket* vector and is represented by the symbol $|\ \rangle$. To distinguish the ket vectors from one another, a label is inserted, viz., the ket of ψ is written as $|\psi\rangle$ and represents a state vector.[1]. The x-component of $|\psi\rangle$ is represented by $\langle x|\psi\rangle$. There may be an infinity of components which together form the state function

$$\psi(x) \equiv \langle x|\psi\rangle \tag{9.1}$$

If $|\psi_m\rangle$ and $|\psi_n\rangle$ be two ket vectors and C_m and C_n any two arbitrary complex numbers, then the linear combination

$$C_m|\psi_m\rangle + C_n|\psi_n\rangle \tag{9.2}$$

is also a vector in the ket vector space. Thus, a ket vector results due to any linear combination of several ket vectors[2].

Now, according to linear algebra to every vector space, there is a dual vector space, so that a scalar product of two vectors, one from each space can be defined. The vectors of the dual space are called the *bra* vectors and are denoted by $\langle\ |$. So, we write:

$$\psi^*(x) \equiv \langle\psi|x\rangle \tag{9.3}$$

The scalar product of $|\psi_m\rangle$ and $\langle\psi_n|$ is given by $\langle\psi_n|\psi_m\rangle$ and is a complex number, i.e.,

$$(\vec{\psi}_m \cdot \vec{\psi}_n) = \langle\psi_n|\psi_m\rangle$$

---

[1]For more details the reader is referred to go through the book: *The Principles of Quantum Mechanics* by P.A.M. Dirac

[2]*Quantum Mechanics* by A. Messiah, Vol 1, North-Holland, 1970

Here, $\langle\psi_n|$ is called the *bra* vector corresponding to ket vector $|\psi_n\rangle$.

Now, if (i) for any $|\psi_m\rangle$, $\langle\psi_n|\psi_m\rangle = 0$, then $\langle\psi_n| = 0$ (ii) for any $|\psi_m\rangle$, $\langle\psi_n|\psi_m\rangle = \langle\psi_p|\psi_m\rangle$, then $\langle\psi_n| = \langle\psi_p|$.

The bra vectors are defined only for the extent that their scalar products with the original ket vectors are a set of numbers.

Here, we assume the following two points:

1. A point to point correspondence exists between kets and bras. Thus, any state of a dynamical system at a certain time may be specified by the direction of a bra vector just as by a ket vector. The bra corresponding to the ket vector $|\psi\rangle$ is denoted by $\langle\psi|$. Further, we have $|\psi_m\rangle + |\psi_n\rangle \rightarrow \langle\psi_m| + \langle\psi_n|$ and $C|\psi\rangle \rightarrow C^*\langle\psi|$.

   That is the bra vectors corresponding to the ket vectors $|\psi_m\rangle + |\psi_n\rangle$ and $C|\psi\rangle$ are respectively $\langle\psi_m| + \langle\psi_n|$ and $C^*\langle\psi|$. Here, C is a complex number and C^* is its complex conjugate.
2. $\langle\psi_m|\psi_n\rangle = \langle\psi_n|\psi_m\rangle^*$.

   If $|\psi_n\rangle = |\psi_m\rangle$, then we see from the above that $\langle\psi_m|\psi_n\rangle$ must be a real.

   We shall further assume that $\langle\psi|\psi\rangle \geqslant 0$. The equality sign holds only when $|\psi\rangle = 0$. The bra vector $\langle\psi|$ and the ket vector $|\psi\rangle$ may be thought as conjugate imaginary of each other.

9.2 Properties of *Bra* and *Ket* notations

1. If an operator $\hat{\alpha}$ operates on a ket vector $|\psi\rangle$ from left, another ket vector is produced $|\psi'\rangle$, i.e.,

$$\hat{\alpha}|\psi\rangle = |\psi'\rangle \tag{9.4}$$

   If the operator $\hat{\alpha}$ operates on a bra vector $\langle\psi|$ from the right another bra vector is produced, i.e.,

$$\langle\psi|\hat{\alpha} = \langle\psi'| \tag{9.5}$$

2. If kets $|A\rangle$ and $|B\rangle$ be multiplied by complex numbers a_1 and a_2 respectively, then

$$a_1|A\rangle + a_2|B\rangle = |C\rangle \tag{9.6}$$

   If ket $|A\rangle$ assumes all values in a certain range, then the addition may be expressed as

$$\int |A\rangle dq = |D\rangle \tag{9.7}$$

   The summation of two bras is defined by the condition that its scalar product with any ket $|A\rangle$ is equal to the sum of the scalar products of $\langle B|$ and $\langle C|$ with $|A\rangle$, i.e.,

$$[\langle B| + \langle C|]|A\rangle = \langle B|A\rangle + \langle C|A\rangle \tag{9.8}$$

3. If $\hat{\alpha}$ be an operator, then its expectation value $\langle\hat{\alpha}\rangle$ in the state ψ can be expressed as

$$\langle\hat{\alpha}\rangle = \int \psi^* \hat{\alpha}\psi\, d\tau = \int \psi^* \psi' d\tau = \langle\psi|\psi'\rangle = \langle\psi|A|\psi\rangle \tag{9.9}$$

4. If an operator $\hat{\alpha}$ operates on the ket $|\psi\rangle$ from the left, then $\hat{\alpha}|\psi\rangle = \lambda|\psi\rangle$, the ket $|\psi\rangle$ is the *eigenket* of $\hat{\alpha}$ and λ is its *eigenvalue*.
5. If $|\psi_1\rangle, |\psi_2\rangle, |\psi_3\rangle, \ldots, |\psi_n\rangle, \ldots, |\psi_m\rangle$ be a set of eigenkets, then the set of eigenkets will be orthonormal, if

$$\langle\psi_n|\psi_m\rangle = \delta_{n,m};\ n, m = 1, 2, 3, \ldots \tag{9.10}$$

where δ_{nm} is the Kronecker δ. So two dynamical states are orthogonal if the vectors corresponding to them are orthogonal.

6. If $|\psi_1\rangle, |\psi_2\rangle, |\psi_3\rangle, \ldots, |\psi_i\rangle, \ldots$ be a complete set of eigenkets, then any arbitrary ket $|\psi\rangle$ can be written as

$$|\psi\rangle = \sum_{i=1}^{\infty} C_i|\psi_i\rangle, \text{ where } C_i = \langle\psi_i|\psi\rangle$$

7. *When a bra is put on the right of a ket, it forms a linear operator*. This can be proved as below: Multiplying the product $|A\rangle\langle B|$ by an arbitrary ket $|C\rangle$ on the right, we have $|A\rangle\langle B|C\rangle$. The ket $\langle B|C\rangle$ is a number.

   Since, the new ket $|A\rangle\langle B|C\rangle$ is the result of operation of $|A\rangle\langle B|$ on the arbitrary ket $|C\rangle$, it depends on $|C\rangle$. Thus, the product $|A\rangle\langle B|$ appears as a *linear operator* and can operate on kets. $|A\rangle\langle B|$ can also operate on bras as $\langle D|A\rangle\langle B|$.

   If, however, a bra is put on the left of a ket, it forms scalar of two operators.

9.3 Condition of normalisation

The normalisation condition is expressed as

$$\int \langle\psi|x\rangle\langle x|\psi\rangle dx = \int |\langle\psi|x\rangle|^2\, dx = 1 \text{ [The integration is overall physical range]} \quad (9.11)$$

The condition of normalisation of the state vector $|\psi\rangle$ can also be expressed as

$$\langle\psi|\psi\rangle = 1 \quad (9.12)$$

The ket vector $|\psi\rangle$ can be multiplied by a phase factor $e^{i\gamma}$ (γ is real) and still the condition of normalisation remains intact.

9.4 Orthogonality condition

The overlap integral between the state vectors $|\phi\rangle$ and $|\psi\rangle$ is given by

$$\int \langle\phi|x\rangle\langle x|\psi\rangle dx = \langle\phi|\psi\rangle \quad (9.13)$$

Now, $\langle\phi|$ and $|\psi\rangle$ will be called *orthogonal to each other*, if

$$\langle\phi|\psi\rangle = 0 \quad (9.14)$$

Proof: Let us have two eigenvalues λ_1 and λ_2 ($\lambda_2 \neq \lambda_1$) with eigenkets $|\psi\rangle$ and $|\phi\rangle$, then $\hat{\alpha}|\psi\rangle = \lambda_1|\psi\rangle$ and $\hat{\alpha}|\phi\rangle = \lambda_2|\phi\rangle$, where $\hat{\alpha}$ is an operator.

The conjugate of the second equation is $\langle\phi|\hat{\alpha} = \lambda_2\langle\phi|$.
Multiplying by $|\psi\rangle$, we get $\langle\phi|\hat{\alpha}|\psi\rangle = \lambda_1\langle\phi|\psi\rangle$ [$\because \alpha|\psi\rangle = \lambda_1|\psi\rangle$].
Also $\langle\phi|\hat{\alpha}|\psi\rangle = \lambda_2\langle\phi|\psi\rangle$ [$\because \langle\phi|\alpha\rangle = \lambda_2\langle\phi|$].

$$\therefore (\lambda_1 - \lambda_2)\langle\phi|\psi\rangle = 0 \quad (9.15)$$

If $\lambda_1 \neq \lambda_2$, then $\langle\phi|\psi\rangle = 0$.

Thus, two eigenvectors of a real dynamical variable belonging to different eigenvalues are orthogonal.

The eigenkets $|\psi_n\rangle$ of a real linear operator satisfy the following conditions

(a) $|\psi_n\rangle$ form a complete set, i.e., any ket $|\psi\rangle$ can be expressed as a linear combination of $|\psi_n\rangle$, i.e.,

$$|\psi\rangle = \sum_n a_n|\psi_n\rangle \tag{9.16}$$

here a_n are constants.

(b) kets $|\psi_n\rangle$ are orthogonal and normalised, i.e.,

$$\langle\psi_n|\psi_m\rangle = \delta_{mn} \tag{9.17}$$

where

$$\delta_{mn} = \begin{cases} 0, & \text{for } m \neq n \\ 1, & \text{for } m = n. \end{cases}$$

(c) kets $|\psi_n\rangle$ are linearly independent, i.e.,

$$\sum_n a_n|\psi_n\rangle = 0 \tag{9.18}$$

then we can show that

$$a_n = 0 \tag{9.19}$$

and

$$\sum_n |\psi_n\rangle\langle\psi_n| = 1 \tag{9.20}$$

where 1 is the *unit operator*.

Proof: We have

$$\sum_n a_n|\psi_n\rangle = 0$$

Multiplying by $\langle\psi_m|$ on both sides, we get

$$0 = \sum_n a_n\langle\psi_m|\psi_n\rangle = \sum_n a_n\delta_{mn} = \sum_m a_m\delta_{mn} = a_m \; [\because \delta_{mm} = 1]$$

$$\text{i.e., } a_m = 0 \text{ or, } a_n = 0 \tag{9.21}$$

Again, multiplying both sides of eq. (9.16) by $\langle\psi_m|$, we get

$$\langle\psi_m|\psi\rangle = \sum_n a_n\langle\psi_m|\psi_n\rangle = \sum_n a_n\delta_{mn} = a_m, \text{ i.e., } a_n = \langle\psi_n|\psi\rangle \tag{9.22}$$

$$\therefore |\psi\rangle = \sum_n a_n|\psi_n\rangle = \sum_n [\langle\psi_n|\psi\rangle]|\psi_n\rangle \; [\because a_n = \langle\psi_n|\psi\rangle]$$

$$\text{or, } |\psi\rangle = \sum^n [|\psi_n\rangle\langle\psi_n|]|\psi\rangle \tag{9.23}$$

$$\therefore \sum_n |\psi_n\rangle\langle\psi_n| = 1 \tag{9.24}$$

9.5 Eigenvalues and eigenvectors

If λ be an eigenvalue of the linear operator $\hat{\alpha}$ and $|\psi\rangle$ be an eigenket ($|\psi\rangle \neq 0$) of the operator $\hat{\alpha}$ belonging to this eigenvalue, then

$$\hat{\alpha}|\psi\rangle = \lambda|\psi\rangle \tag{9.25}$$

1. If $|\phi_1\rangle$ and $|\phi_2\rangle$ be two eigenkets belonging to the same eigenvalue λ, then any linear combination of them will also be an eigenket belonging to the same eigenvalue.

   Proof: If $|\psi\rangle = C_1|\phi_1\rangle + C_2|\phi_2\rangle$, C_1 and C_2 are complex numbers and $\hat{\alpha}|\phi_1\rangle = \lambda|\phi_1\rangle$ and $\hat{\alpha}|\phi_2\rangle = \lambda|\phi_2\rangle$, then $\hat{\alpha}|\psi\rangle = \hat{\alpha}[C_1|\phi_1\rangle + C_1|\phi_2\rangle] = C_1\hat{\alpha}|\phi_1\rangle + C_2\hat{\alpha}|\phi_2\rangle = C_1\lambda|\phi_1\rangle + C_2\lambda|\phi_2\rangle = \lambda[C_1|\phi_1\rangle + C_2|\phi_2\rangle]$ or, $\hat{\alpha}|\psi\rangle = \lambda|\psi\rangle$.

   But, the formulation is symmetric w.r.t. bras and kets.

   $\therefore \langle\theta|\hat{\alpha} = \lambda'\langle\theta|$, where $\langle\theta| \neq 0$ and λ' is a number.

   Hence, for any linear operator $\hat{\alpha}$ there are two types of eigenvalues relating to bras and kets respectively.

2. Now, we can prove that *the eigenvalues of a real linear operator are real numbers.*

   Proof: Let $\hat{\alpha}|\psi\rangle = \lambda|\psi\rangle$.

   Multiplying both sides by $\langle\psi|$

   $$\langle\psi|\hat{\alpha}|\psi\rangle = \lambda\langle\psi|\psi\rangle$$

   $\langle\psi|\psi\rangle$ is real and $\neq 0$.

   Since, $\langle\psi|\hat{\alpha}|\psi\rangle = (\langle\psi|\hat{\alpha}^*|\psi\rangle)^* = (\langle\psi|\hat{\alpha}|\psi\rangle)^*$.

   $\therefore \langle\psi|\hat{\alpha}|\psi\rangle$ is real.

   Hence, λ must be a real number.

   Again, the conjugate equation of $\hat{\alpha}|\psi\rangle = \lambda|\psi\rangle$ is

   $$\langle\psi|\hat{\alpha}^* = \lambda\langle\psi| \text{ or, } \langle\psi|\hat{\alpha} = \lambda\langle\psi| \quad [\because \hat{\alpha}^* = \hat{\alpha}, \alpha \text{ being real}] \tag{9.26}$$

   Thus, eigenbras are the conjugates of the eigenkets. The eigenvalues are the same for both $|\psi\rangle$ and $\langle\psi|$.

9.6 Observable operator

The fact that the average value of repeated observations, $\hat{\alpha}$, on a normalised state $|\psi\rangle$ is real and imposes some restrictions on the operator $\hat{\alpha}$. *A necessary condition is*

$$\langle\psi|\hat{\alpha}|\psi\rangle = \langle\psi|\hat{\alpha}\psi\rangle^* \tag{9.27}$$

for any state represented by ket $|\psi\rangle$.

Again, *a sufficient condition is represented as follows:*

The operator $\hat{\alpha}$ can represent a physical observation, only if

$$\langle\phi|\hat{\alpha}|\psi\rangle = \langle\psi|\hat{\alpha}|\phi\rangle^* \tag{9.28}$$

The above restriction for any two arbitrary states $|\phi\rangle$ and $|\psi\rangle$ can be expressed as

$$\int \langle\phi|x\rangle\hat{\alpha}\langle x|\psi\rangle dx = \left[\int \langle\psi|x\rangle\,\hat{\alpha}\langle x|\phi\rangle dx\right]^* \tag{9.29}$$

The operator satisfying condition expressed by eq. (9.29) is called an *observable operator*.

9.7 The Hermitian adjoint

$\hat{\alpha}^\dagger$ (α-dagger) represents the Hermitian adjoint of $\hat{\alpha}$. The Hermitian adjoint $\hat{\alpha}^\dagger$ of $\hat{\alpha}$ is defined by

$$\langle\phi|\hat{\alpha}^\dagger|\psi\rangle = \langle\psi|\hat{\alpha}|\phi\rangle^* \tag{9.30}$$

Here, $|\psi\rangle$ and $|\phi\rangle$ are any two arbitrary kets.

Generally, the operator $\hat{\alpha}$ has a matrix representation in that case $\hat{\alpha}^\dagger$ is its transposed conjugate as

$$\left(\hat{\alpha}^\dagger\right)_{mn} = (\hat{\alpha})^*_{nm} \tag{9.31}$$

where m, n refer to the elements of the corresponding matrix.

(a) *We can show that $\langle\hat{\alpha}\hat{\alpha}^\dagger\rangle$ is never negative.*

We have the average value of $\hat{\alpha}$ given by

$$\langle\alpha\rangle = \int \phi^*(x)\left[\hat{\alpha}\phi(x)\right]dx = \int \phi^*(x)\left[\sum_n C_n\hat{\alpha}\,\psi_n(x)\right]dx$$

Now, substituting $\hat{\alpha}\hat{\alpha}^\dagger$ in place of $\hat{\alpha}$ in the above equation, we get

$$\langle\hat{\alpha}\hat{\alpha}^\dagger\rangle = \int \phi^*(x)\left[\hat{\alpha}\hat{\alpha}^\dagger\phi(x)\right]dx = \int \phi^*(x)\hat{\alpha}\left[\hat{\alpha}^\dagger\phi(x)\right]dx = \int \left|\hat{\alpha}^\dagger\phi(x)\right|^2 dx \geqslant 0$$

since the square of a quantity is always positive.

$$\therefore \langle\hat{\alpha}\hat{\alpha}^\dagger\rangle \text{ is } \geqslant 0 \tag{9.32}$$

(b) We can show that the *adjoint operator* $(i\hat{\alpha})^\dagger = -i\hat{\alpha}^\dagger$.

We have according to the definition of $\hat{\alpha}^\dagger$

$$\int \phi^*(x)\left[\hat{\alpha}\psi(x)\right]dx = \int \left[\hat{\alpha}^\dagger\phi(x)\right]^*\psi(x)\,dx$$

where $\phi^*(x)$ is the complex conjugate of $\phi(x)$ and the integral extends to cover the whole domain of the coordinate space.

Multiplying the above equation by i, we get

$$i\int \phi^*(x)\left[\hat{\alpha}\psi(x)\right]dx = i\int \left[\hat{\alpha}^\dagger\phi(x)\right]^*\psi(x)\,dx$$

$$\text{or,} \quad \int \phi^*(x)\left[i\hat{\alpha}\psi(x)\right] = \int \left[-i\hat{\alpha}^\dagger\phi(x)\right]^*\psi(x)\,dx \tag{9.33}$$

$$\text{and} \quad \int \phi^*(x)\left[i\hat{\alpha}\psi(x)\right]dx = \int \left[(i\hat{\alpha})^\dagger\phi(x)\right]^*\psi(x)\,dx \tag{9.34}$$

In eq. (9.34), $(i\hat{\alpha})$ has been considered as an operator. Comparing eqs (9.33) and (9.34), we get

$$(i\hat{\alpha})^\dagger = -i\hat{\alpha}^\dagger \tag{9.35}$$

In this way, if $\hat{\alpha}$ and $\hat{\beta}$ be two operators, then

$$\left(\hat{\alpha} + i\hat{\beta}\right)^\dagger = \hat{\alpha}^\dagger - i\hat{\beta}^\dagger \tag{9.36}$$

9.8 The linear harmonic oscillator in Dirac notation

The theory of radiation, quantisation of lattice vibrations and many other topics are based on the principle of linear harmonic oscillator. That is why the problem of the linear harmonic oscillator plays an important role in quantum mechanics.

If m be the mass, ω be the angular frequency of oscillation of the linear harmonic oscillator, then its Hamiltonian operator is given by

$$\hat{H} = \frac{1}{2m}\left(\hat{p}^2 + m^2\omega^2\hat{x}^2\right) \tag{9.37}$$

We want to find out the wave vector ψ_E and the corresponding eigenvalues E by solving the eigenvalue equation

$$H|\psi_E\rangle = E|\psi_E\rangle \tag{9.38}$$

With the dynamical variables $\hat{x}$ and $\hat{p}$, which obey commutation relation, viz.,

$$[\hat{x}, \hat{p}] = \hat{x}\hat{p} - \hat{p}\hat{x} = i\hbar \tag{9.39}$$

Here, we introduce the *dimensionless* complex dynamical variables $\hat{a}$ and $\hat{a}^\dagger$, where

$$\hat{a} = \frac{1}{\sqrt{2m\hbar\omega}}(m\omega\hat{x} + i\hat{p}) \tag{9.40}$$

$$\text{and } \hat{a}^\dagger = \frac{1}{\sqrt{2m\hbar\omega}}(m\omega\hat{x} - i\hat{p}) \tag{9.41}$$

$\hat{a}^\dagger$ is the adjoint of $\hat{a}$, since $\hat{p}$ and $\hat{x}$ are Hermitian.

Now,

$$\hbar\omega\hat{a}\hat{a}^\dagger = \frac{1}{2m}\left[m^2\omega^2\hat{x}^2 + \hat{p}^2 - im\omega\left(\hat{p}\hat{x} - \hat{x}\hat{p}\right)\right] = \frac{1}{2m}\left[\hat{p}^2 + m^2\omega^2\hat{x}^2\right] + \tfrac{1}{2}\hbar\omega = \hat{H} + \tfrac{1}{2}\hbar\omega \tag{9.42}$$

In a similar way we can show that

$$\hbar\omega\hat{a}^\dagger\hat{a} = \hat{H} - \tfrac{1}{2}\hbar\omega \tag{9.43}$$

$$\therefore \hat{H} = \tfrac{1}{2}\hbar\omega\left(\hat{a}^\dagger\hat{a} + \hat{a}\hat{a}^\dagger\right) \tag{9.44}$$

and

$$\hat{a}\hat{a}^\dagger - \hat{a}^\dagger\hat{a} = \left[\hat{a}, \hat{a}^\dagger\right] = 1 \tag{9.45}$$

Now, multiplying by $\hat{a}$ on both sides of eq. (9.42), we get

$$\hbar\omega\hat{a}\hat{a}^\dagger\hat{a} = \hat{H}\hat{a} + \tfrac{1}{2}\hbar\omega\hat{a} \tag{9.46}$$

Also, we get from eq. (9.43)

$$\hbar\omega\hat{a}\hat{a}^\dagger\hat{a} = \hat{a}\hat{H} - \tfrac{1}{2}\hbar\omega\hat{a} \tag{9.47}$$

From eqs (9.46) and (9.47), we get

$$\hat{a}\hat{H} - \tfrac{1}{2}\hbar\omega\hat{a} = \hat{H}\hat{a} + \tfrac{1}{2}\hbar\omega\hat{a} \text{ or, } \hat{a}\hat{H} - \hat{H}\hat{a} = \hbar\omega\hat{a} \text{ or, } \left[\hat{a}, \hat{H}\right] = \hbar\omega\hat{a} \tag{9.48}$$

Similarly,

$$\left[\hat{a}^\dagger, \hat{H}\right] = -\hbar\omega\hat{a}^\dagger \tag{9.49}$$

Now, according to the definition of scalar product of two vectors, we have

$$\hbar\omega\,\langle\hat{a}\psi_E|\hat{a}\psi_E\rangle = \hbar\omega\,\langle\psi_E|\hat{a}^\dagger\hat{a}|\psi_E\rangle = \langle\psi_E|\hbar\omega\hat{a}^\dagger\hat{a}|\psi_E\rangle = \langle\psi_E|\left(\hat{H}-\tfrac{1}{2}\hbar\omega\right)|\psi_E\rangle$$
$$= \left(E-\tfrac{1}{2}\hbar\omega\right)\langle\psi_E|\psi_E\rangle \quad \left[\because \hbar\omega\hat{a}^\dagger a = \hat{H}-\tfrac{1}{2}\hbar\omega\right] \tag{9.50}$$

where $|\psi_E\rangle$ is an eigenket of $\hat{H}$ belonging to the eigenvalue E.

The quantity $\langle\hat{a}\psi_E|\hat{a}\psi_E\rangle$ represents the square of the length of the vector $\hat{a}\psi_E$.

$$\therefore\ \langle\hat{a}\psi_E|\hat{a}\psi_E\rangle \geqslant 0 \tag{9.51}$$

which holds for $\hat{a}\psi_E = 0$.

Also, we have

$$\langle\psi_E|\psi_E\rangle \geqslant 0 \tag{9.52}$$

and hence

$$E \geqslant \tfrac{1}{2}\hbar\omega \tag{9.53}$$

Considering the operator $\hat{H}\hat{a}$ operating on $|\psi_E\rangle$, we have

$$\hat{H}\hat{a}|\psi_E\rangle = \left(\hat{a}\hat{H}-\hbar\omega\hat{a}\right)|\psi_E\rangle \quad \left[\because \hat{a}\hat{H}-\hat{H}\hat{a} = \hbar\omega\hat{a}\right]$$
$$= \left(\hat{a}E-\hbar\omega\hat{a}\right)|\psi_E\rangle$$
$$\text{or, } \hat{H}\hat{a}|\psi_E\rangle = (E-\hbar\omega)\hat{a}|\psi_E\rangle \tag{9.54}$$

Equation (9.54) shows that $\hat{a}|\psi_E\rangle$ is an eigenvector of $\hat{H}$ belonging to the eigenvalue $(E-\hbar\omega)$, if $\hat{a}|\psi_E\rangle \neq 0$.

Similarly, $\hat{a}\hat{a}|\psi_E\rangle = \hat{a}^2|\psi_E\rangle$ would be an eigenvector belonging to the eigenvalue $(E-2\hbar\omega)$ and so on.

Now, we note from eq. (9.50) that $\hat{a}|\psi_E\rangle$ is non-zero and an eigenket belonging to the eigenvalue $(E-\hbar\omega)$, if $E \neq \frac{1}{2}\hbar\omega$.

Hence, if $(E-\hbar\omega) \neq \hbar\omega$, $(E-2\hbar\omega)$ is another eigenvalue of the operator $\hat{H}$. Proceeding in this way, we obtain a series of eigenvalues E, $(E-\hbar\omega)$, $(E-2\hbar\omega)$, $(E-3\hbar\omega)$, etc. The series of eigenvalues so obtained cannot extend up to infinity because in that case it would contradict $E \geqslant \frac{1}{2}\hbar\omega$. The eigenvalues can terminate only with the value $\frac{1}{2}\hbar\omega$.

From equation

$$\left[\hat{a}^\dagger,\hat{H}\right] = \hat{a}^\dagger\hat{H}-\hat{H}\hat{a}^\dagger = -\hbar\omega\hat{a}^\dagger \text{ or, } \hat{H}\hat{a}^\dagger = \hat{a}^\dagger\hat{H}+\hbar\omega\hat{a}^\dagger$$
$$\text{or, } \hat{H}\hat{a}^\dagger|\psi_E\rangle = \left(\hat{a}^\dagger\hat{H}+\hbar\omega\hat{a}^\dagger\right)|\psi_E\rangle = (E+\hbar\omega)\hat{a}^\dagger|\psi_E\rangle \tag{9.55}$$

Equation (9.55) shows that if $\hat{a}^\dagger|\psi_E\rangle \neq 0$, then $(E+\hbar\omega)$ is another eigenvalue of the Hamiltonian operator $\hat{H}$ with $\hat{a}^\dagger|\psi_E\rangle$ as the eigenket belonging to it. Now, we can show that $\hat{a}^\dagger|\psi_E\rangle$ cannot be zero. If it were so, then

$$0 = \hbar\omega\hat{a}\hat{a}^\dagger|\psi_E\rangle = \hbar\omega\left(\hat{a}^\dagger\hat{a}+1\right)|\psi_E\rangle \text{ [by eq. (9.48)]}$$
$$= \left(\hat{H}+\tfrac{1}{2}\hbar\omega\right)|\psi_E\rangle \quad \left[\because \hbar\omega\hat{a}^\dagger\hat{a} = \hat{H}-\tfrac{1}{2}\hbar\omega\right]$$
$$= \left(E+\tfrac{1}{2}\hbar\omega\right)|\psi_E\rangle \tag{9.56}$$

So $E+\frac{1}{2}\hbar\omega = 0$ and $E = -\frac{1}{2}\hbar\omega$. But, this value of E contradicts eq. (9.53). So we can say, that if E is an eigenvalue, then $(E+\hbar\omega)$, $(E+2\hbar\omega)$, $(E+3\hbar\omega)$, etc. are other eigenvalues of $\hat{H}$. Therefore, the eigenvalues of the Hamiltonian operator $\hat{H}$ of the linear harmonic oscillator are

$$\tfrac{1}{2}\hbar\omega,\ \tfrac{3}{2}\hbar\omega,\ \tfrac{5}{2}\hbar\omega,\ \tfrac{7}{2}\hbar\omega, \ldots \text{ to } \infty \tag{9.57}$$

or,

$$E = (n + \tfrac{1}{2})\,\hbar\omega,\ n = 0, 1, 2, 3, \ldots \tag{9.58}$$

Writing ψ_n in place of ψ_E, we have

$$\hat{H}|\psi_n\rangle = (n + \tfrac{1}{2})\,\hbar\omega|\psi_n\rangle \tag{9.59}$$

where $|\psi_n\rangle$ represents the eigenket corresponding to the eigenvalue $(n + \frac{1}{2})\,\hbar\omega$.

Now, assuming $|\psi_n\rangle$ are normalised, so that

$$\langle\psi_m|\psi_n\rangle = \delta'_{mn} \tag{9.60}$$

where δ_{mn} is the *Kronecker delta*, i.e.,

$$\delta_{mn} = \begin{cases} 0, & \text{for } m \neq n \\ 1, & \text{for } m = n' \end{cases}$$

then

$$\hat{a}|\psi_n\rangle = \alpha_n|\psi_{n-1}\rangle \tag{9.61}$$

$$\hat{a}^\dagger|\psi_n\rangle = \beta_n|\psi_{n+1}\rangle \tag{9.62}$$

The value of the constants α_n and β_n can be determined from the condition of normalisation. Let us now consider the scalar product

$$\langle\psi_n|\hat{a}^\dagger\hat{a}|\psi_n\rangle = \langle\alpha\psi_n|\hat{a}\psi_n\rangle = |\alpha_n|^2\langle\psi_{n-1}|\psi_{n-1}\rangle = |\alpha_n|^2 \tag{9.63}$$

Again, $\hbar\omega\,\langle\psi_n\,|\hat{a}^\dagger\hat{a}|\,\psi_n\rangle = \langle\psi_n\,|\{\hat{H} - \frac{1}{2}\hbar\omega\}|\,\psi_n\rangle = \langle\psi_n\,|\{(n + \frac{1}{2})\,\hbar\omega - \frac{1}{2}\hbar\omega\}|\,\psi_n\rangle = n\hbar\omega\,\langle\psi_n|\psi_n\rangle$.

$$\therefore \langle\psi_n\,|\hat{a}^\dagger\hat{a}|\,\psi_n\rangle = n \tag{9.64}$$

From eqs (9.63) and (9.64), we get $\alpha_n = \sqrt{n}$.

Proceeding exactly in the same way we can find out the value of β_n as $\beta_n = \sqrt{n+1}$.

$$\therefore \hat{a}|\psi_n\rangle = (\sqrt{n})\,|\psi_{n-1}\rangle \tag{9.65}$$

$$\text{and } \hat{a}^\dagger|\psi_n\rangle = (\sqrt{n+1})\,|\psi_{n+1}\rangle \tag{9.66}$$

Now, the relations

$$\hat{a} + \hat{a}^\dagger = \left(\sqrt{\frac{m\omega}{2\hbar}}\right)\hat{x} \text{ and } \hat{a} - \hat{a}^\dagger = \left(\sqrt{\frac{2}{m\hbar\omega}}\right)ip$$

give

$$\sqrt{n}|\psi_{n-1}\rangle - \sqrt{\left(\frac{m\omega}{2\hbar}\right)}\hat{x}|\psi_n\rangle + (\sqrt{n+1})\,|\psi_{n+1}\rangle = 0 \text{ [adding eqs (9.65) and (9.66)]} \tag{9.67}$$

$$\sqrt{n}|\psi_{n-1}\rangle - \sqrt{\left(\frac{2}{m\hbar\omega}\right)}i\hat{p}|\psi_n\rangle - (\sqrt{n+1})\,|\psi_{n+1}\rangle = 0 \text{ [subtracting eq. (9.66) from eq. (9.65)]}$$

$$\text{or, } \sqrt{n}|\psi_{n-1}\rangle - \sqrt{\left(\frac{2\hbar}{m\omega}\right)}\frac{d|\psi_n\rangle}{dx} - (\sqrt{n+1})\,|\psi_{n+1}\rangle = 0 \quad \left(\because \hat{p} = \frac{\hbar}{i}\frac{d}{dx}\right) \tag{9.68}$$

Taking the sum of eqs (9.67) and (9.68), we get

$$2\sqrt{n}|\psi_{n-1}\rangle = \sqrt{\left(\frac{m\omega}{2\hbar}\right)}\hat{x}|\psi_n\rangle + \sqrt{\left(\frac{2\hbar}{m\omega}\right)}\frac{d|\psi_n\rangle}{dx}$$

$$\text{or, } |\psi_{n-1}\rangle = \frac{1}{2\sqrt{n}}\left[\sqrt{\left(\frac{2m\omega}{\hbar}\right)}\hat{x}|\psi_n\rangle + \sqrt{\left(\frac{2\hbar}{m\omega}\right)}\frac{d|\psi_n\rangle}{dx}\right] \quad (9.69)$$

Again, we write eqs (9.67) and (9.68) as

$$(\sqrt{n-1})|\psi_{n-2}\rangle - \sqrt{\left(\frac{m\omega}{2\hbar}\right)}\hat{x}|\psi_{n-1}\rangle + \sqrt{n}|\psi_n\rangle = 0 \quad (9.70)$$

$$(\sqrt{n-1})|\psi_{n-2}\rangle - \sqrt{\left(\frac{2\hbar}{m\omega}\right)}\frac{d|\psi_{n-1}\rangle}{dx} - \sqrt{n}|\psi_n\rangle = 0 \quad (9.71)$$

Substracting eq. (9.69) from eq. (9.70) and, then using the value of $|\psi_{n-1}\rangle$ from eq. (9.71), we get

$$|\psi_n\rangle = \frac{1}{2\sqrt{n}}\left[\left(\sqrt{\frac{m\omega}{2\hbar}}\right)\frac{1}{2\sqrt{n}}\left\{\left(\sqrt{\frac{m\omega}{2\hbar}}\right)\hat{x}^2|\psi_n\rangle + \sqrt{\frac{2\hbar}{m\omega}}\hat{x}\frac{d|\psi_n\rangle}{dx}\right\}\right.$$

$$\left. -\left(\sqrt{\frac{2\hbar}{m\omega}}\right)\frac{1}{2\sqrt{n}}\left\{\sqrt{\frac{m\omega}{2\hbar}}\left(|\psi_n\rangle + \hat{x}\frac{d|\psi_n\rangle}{dx}\right) + \sqrt{\frac{2\hbar}{m\omega}}\frac{d^2|\psi_n\rangle}{dx^2}\right\}\right]$$

$$\text{or, } \frac{d^2|\psi_n\rangle}{dx^2} + \frac{2m}{\hbar^2}\left\{\left(n+\tfrac{1}{2}\right)\hbar\omega - \tfrac{1}{2}m\omega^2x^2\right\}|\psi_n\rangle = 0 \quad (9.72)$$

This is the *Schrödinger equation for one-dimensional harmonic oscillator*. The operator $\hat{a}^\dagger$ operating on an excited state $|\psi_n\rangle$, containing n quanta of energy $\hbar\omega$, creates a quantum of energy. That is why $\hat{a}^\dagger$ is called the *creation operator*. The operator $\hat{a}$ annihilates a quantum of energy and hence, called the *annihilation operator*. [cf. eqs (9.65) and (9.66)]

9.9 Answers of some important conceptual questions

9.1. *What are ket vectors and bra vectors?*

Ans. In quantum mechanics a dynamical system may have many different states, the superposition of any two of which give rise to other states consisting of the set of the states. This conception leads to the possibility of representing every dynamical state by certain type of vector, called a *ket vector* and is represented by the symbol | ⟩. To distinguish the *ket vectors* from one another a level is inserted, viz., the ket of ψ is written as $|\psi\rangle$ and represents a state vector. The x-component of $|\psi\rangle$ is represented by $\langle x|\psi\rangle$.

A ket vector results due to any *linear* combination of of several ket vectors.

According to linear algebra to every vector space, there is a dual vector space, so that a scalar product of two vectors, one from each space can be defined. The vectors of the dual space are called the *bra vectors* and are denoted by ⟨ |. So, we write $\psi^*(x) = \langle\psi|x\rangle$.

The scalar product of $|\psi_m\rangle$ and $\langle\psi_n|$ is given by $\langle\psi_n|\rangle$ and is a complex number

$$(\vec{\psi}_m \cdot \vec{\psi}_n) = \langle\vec{\psi}_n|\vec{\psi}_m\rangle$$

Here, $\langle\psi_n|$ is called the *bra vector* corresponding to *ket vector* $|\psi_n\rangle$.

9.2. *What led Dirac to introduce bra and ket notations?*

Ans. The scalar product of two state vectors $\vec{\psi}_m$ and $\vec{\psi}_n$ can be defined as

$$(\vec{\psi}_m \cdot \vec{\psi}_n) = \int \psi_m^* \psi_n \, d\tau \text{ and } (\vec{\psi}_n \cdot \vec{\psi}_m) = \int \psi_n^* \psi_m d\tau = (\vec{\psi}_m \cdot \vec{\psi}_n)^*$$

This shows that $(\vec{\psi}_n \cdot \vec{\psi}_m) \neq (\vec{\psi}_m \cdot \vec{\psi}_n)$.

Further, $(\vec{\psi}_m \cdot C\vec{\psi}_n) = C(\vec{\psi}_m \cdot \vec{\psi}_n)$ and $(C\vec{\psi}_m \cdot \vec{\psi}_n) = C^*(\vec{\psi}_m \cdot \vec{\psi}_n)$.

This shows that the scalar product is linear w.r.t. post-factor while it depends upon pre-factor in another anti-linear fashion. To avoid this confusion, Dirac introduced two very important notations, called the *bra* and *ket*. He considered all possible state vectors ψ in the *Hilbert space* and denoted the typical vector by symbol $|\,\rangle$. Into the *ket* $|\,\rangle$ we can introduce symbols denoting eigenvalues, quantum numbers, etc. which specify the space in questions. All the ket vectors of a system together form a linear vector space which may be called as *ket space*.

The scalar product of two state vectors $\vec{\psi}_m$ and $\vec{\psi}_n$ denoted by kets $|\psi_m\rangle$ and $|\psi_n\rangle$ is written as

$$(\vec{\psi}_m \cdot \vec{\psi}_n) = \langle \psi_m | \psi_n \rangle$$

where the symbol $\langle \psi_n|$ is called the bra vector corresponding to ket vector $|\psi_m\rangle$.

In this way, *bra* and *ket* notations distinguish very clearly between pre-factor and post-factor wave vectors in a scalar product. Merely writing ψ_m or ψ_n one cannot know whether it is to be pre-factor or post-factor but writing $\langle \psi_m|$ and $|\psi_n\rangle$ clearly tells that ψ_m will be a pre-factor and ψ_n will be a post-factor.

9.3. *Prove that when bra is put on the right of a ket, it forms a linear operator.*

Ans. We consider the product $|Q\rangle\langle R|$. Multiplying this product by an arbitrary ket $|A\rangle$, we get $|Q\rangle\langle R|A\rangle$.

This another ket $\langle R|A\rangle$ constitutes a number. The new ket must depend on the arbitrary ket $|A\rangle$ because it is the result of operation of the product $|Q\rangle\langle R|$ on the arbitrary ket $|A\rangle$. Thus, the product $|Q\rangle\langle R|$ appears as a linear operator which can operate on kets. This product can also operate on bras. Its operation upon bras is written as $\langle B|Q\rangle\langle R|$.

9.4. *Express the condition of normalisation.*

Ans. The normalisation condition is expressed as

$$\int \langle \psi|x\rangle \langle x|\psi\rangle dx = \int |\langle \psi|x\rangle|^2 \, dx$$

The integration is over all physical range. The condition of normalisation of the state vector $|\psi\rangle$ can also be expressed as $\langle\psi|\psi\rangle = 1$.

The ket vector $|\psi\rangle$ can be multiplied by a phase factor $e^{i\gamma}$ (γ is real) and still the condition of normalisation remains in tact.

9.5. *Show that Hermitian operators possess real eigenvalues and the eigenfunctions of a Hermitian operator corresponding to different eigenvalues are orthogonal to each other.*

Ans. We have $\langle \psi_n|A|\psi_n\rangle = a_n\langle\psi_n|\psi_n\rangle$ [since $A|\psi_n\rangle = a_n|\psi_n|$].

Similarly, $\langle A\psi_n|\psi_n\rangle = a_n^*\langle\psi_n|\psi_n\rangle$.

When A is Hermitian, $\langle\psi_n|A|\psi_n\rangle = \langle A\psi_n|\psi_n\rangle$ and hence, $a_n = a_n^*$, i.e., eigenvalues are real.

To prove the second part, if ψ_i and ψ_j are eigenfunctions of A having distinct eigenvalues a_i and a_j, we have $\langle A\psi_i|\psi_j\rangle - \langle\psi_i|A|\psi_j\rangle = a_j^*\langle\psi_i|\psi_j\rangle - a_j\langle\psi_i|\psi_j\rangle = (a_i - a_j)\langle\psi_i|\psi_j\rangle = 0$ as A is Hermitian and a_i is real from the first part. Then if $a_i \neq a_j$ (distinct eigenvalues), it follows that $\langle\psi_i|\psi_j\rangle = 0$.

Hence, the two eigenfunctions corresponding to distinct eigenvalues are orthogonal to each other.

9.10 Solved problems

» ***Example* 9.1.** If $\hat{H}_1$ and $\hat{H}_2$ be two Hamiltonians differing by a constant independent of r and t, potential V_0, then compare the eigenvalues and eigenvectors of $\hat{H}_1$ and $\hat{H}_2$.

Solution: If $|\psi\rangle$ be an eigenket of $\hat{H}_2$, then $\hat{H}_2|\psi\rangle = E_2|\psi\rangle$.

Now, let $\hat{H}_1 - \hat{H}_2 = V_0$ or $\hat{H}_1 = \hat{H}_2 + V_0$, then $\hat{H}_1|\psi\rangle = \hat{H}_2|\psi\rangle + V_0|\psi\rangle = (\hat{H}_2 + V_0)|\psi\rangle = (E_2 + V_0)|\psi\rangle$.

This shows that the eigenvalue of $\hat{H}_1 = E_2 + V_0$.

Hence, $\hat{H}_1$ and $\hat{H}_2$ have the same eigenvector and their eigenvalues differ by V_0.

» ***Example* 9.2.** Prove that $(|A\rangle\langle B|)^* = |B\rangle\langle A|$.

Solution: We multiply $|A\rangle\langle B|$ by an arbitrary bra $\langle Q|$ to get $\langle Q|A\rangle\langle B|$.

The conjugate imaginary ket of the above is given by $(\langle Q|A\rangle)^*|B\rangle = \langle A|Q\rangle|B\rangle = |B\rangle\langle A|Q\rangle$.

$\therefore (|A\rangle\langle B|)^* = |B\rangle\langle A|$.

» ***Example* 9.3.** If a Hermitian operator possesses a complete set of eigenfunction ψ_n, then show that the identity operator I can be expressed as

$$I = \sum_n |\psi_n\rangle\langle\psi_n|$$

Solution: We know that the relation of completeness for the eigenfunction ψ_n of a Hermitian operator is

$$\sum_n \psi_n^*(\vec{r}')\,\psi_n(\vec{r}) = \delta(\vec{r} - \vec{r}')$$

where $\delta(\vec{r} - \vec{r}')$ is the Dirac delta function.

If ϕ and ψ be two-wave functions, then their scalar product is expressed as

$$\langle\phi|\psi\rangle = \int \phi^*(\vec{r},t)\,\psi(\vec{r},t)\,d\vec{r} = \iint \phi^*(\vec{r}',t)\,\delta(\vec{r} - \vec{r}')\,\psi(\vec{r} - \vec{r}')\,d\vec{r},d\vec{r}'$$

$$\therefore \langle\phi|\psi\rangle = \sum_n \int \phi^*(\vec{r},t)\,\psi_n(\vec{r})\,d\vec{r} \int \psi^*(\vec{r}')\,\psi(\vec{r}',t)\,d\vec{r}' = \sum_n \langle\phi|\psi_n\rangle\langle\psi_n|\psi\rangle = \sum_n \langle\phi|\psi\rangle$$

where $\sum_n |\psi_n\rangle\langle\psi_n| = I$.

» ***Example* 9.4.** If $\hat{\alpha}$ is a real linear operator and $\hat{\alpha}^m|P\rangle = 0$ for a particular ket $|P\rangle$, m being a positive integer, then prove that $\hat{\alpha}|P\rangle = 0$.

Solution: We first consider $m = 2$, then from $\hat{\alpha}^m|P\rangle = 0$, we have $\langle P|\hat{\alpha}^2|P\rangle = 0$ or, $\langle Q|Q\rangle = 0$, where $|Q\rangle = \hat{\alpha}|P\rangle$ [$\langle Q| = \langle P|\hat{\alpha}^\dagger$ as $\hat{\alpha}$ is real].

$\therefore 0 = |Q\rangle = \hat{\alpha}|P\rangle$.

Thus, the theorem is proved for the case $m = 2$. For $m > 2$, we put $\hat{\alpha}^{m-2}|P\rangle = |R\rangle$, then $\hat{\alpha}^2|R\rangle = 0$.

Then, we have $0 = \hat{\alpha}|R\rangle = \hat{\alpha}^{m-1}|P\rangle$.

Now, we put $\hat{\alpha}^{m-1}|P\rangle = |S\rangle$, then we get $\hat{\alpha}^{m-2}|P\rangle = 0$.

Proceeding in this way we can prove that $\hat{\alpha}^{m-3}|P\rangle = 0, \hat{\alpha}^{m-2}|P\rangle = 0, \ldots, \hat{\alpha}^2|P\rangle = 0, \hat{\alpha}|P\rangle = 0. \therefore \hat{\alpha}|P\rangle = 0.$

» **Example 9.5.** Prove that (i) $(CP)^\dagger = C^* P^\dagger$ (ii) $(PQ)^\dagger = Q^\dagger P^\dagger$.

Solution:

(i) According to the definition of adjoint

$$\langle\Phi|(CP)^\dagger|\Psi\rangle = \langle CP\Phi|\Psi\rangle = C^*\langle P\Phi|\Psi\rangle = C^*\langle\Phi|P^\dagger|\Psi\rangle = \langle\Phi|C^* P^\dagger|\Psi\rangle. \therefore (CP)^\dagger = C^* P^\dagger.$$

(ii) $\langle\Phi|(PQ)^\dagger|\Psi\rangle = \langle PQ\Phi|\Psi\rangle = \langle Q\Phi|P^\dagger|\Psi\rangle = \langle\Phi|Q^\dagger P^\dagger|\Psi\rangle. \therefore (PQ)^\dagger = Q^\dagger P^\dagger.$

» **Example 9.6.** If $|\psi_1\rangle$ and $|\psi_2\rangle$ are two linearly independent eigenfunctions of an operator $\hat{\alpha}$ belonging to the eigenvalue λ (same for both), then prove that a linear combination of these functions will also be an eigenfunction of $\hat{\alpha}$ and it will have an eigenvalue λ.

Solution: By the problem, $\hat{\alpha}|\psi_1\rangle = \lambda|\psi_1\rangle$ and $\hat{\alpha}|\psi_2\rangle = \lambda|\psi_2\rangle$.

The linear combination, $\hat{\alpha}|\psi\rangle = \hat{\alpha}\,(C_1|\psi_1\rangle + C_2|\psi_1\rangle) = C_1\hat{\alpha}|\psi_1\rangle + C_2\hat{\alpha}|\psi_2\rangle = C_1\lambda|\psi_1\rangle + C_2\lambda|\psi_2\rangle$
$= \lambda\,(C_1|\psi_1\rangle + C_2|\psi_1\rangle) = \lambda|\psi\rangle.$

$\therefore \psi$ is also an eigenfunction of $\hat{\alpha}$ having eigenvalue λ.

» **Example 9.7.** Prove that the eigenfunctions of a Hermitian operator corresponding to different eigenvalues are orthogonal to each other.

Solution: We know, $\langle\psi_n|\hat{\alpha}|\psi_n\rangle = \lambda_n\langle\psi_n|\psi_n\rangle$ $[\because \hat{\alpha}|\psi_n\rangle = \lambda_n|\psi_n\rangle]$.

Also $\langle\hat{\alpha}\,\psi_n|\psi_n\rangle = \lambda_n^*\,\langle\psi_n|\psi_n\rangle$.

If $\hat{\alpha}$ is Hermitian, then $\langle\psi_n|\hat{\alpha}|\psi_n\rangle = \langle\hat{\alpha}\,\psi_n|\psi_n\rangle$.

$\therefore \lambda_n = \lambda_n^*$, i.e., eigenvalues are equal.

Again, $\langle\hat{\alpha}\,\psi_m|\psi_n\rangle - \langle\psi_m|\hat{\alpha}|\psi_n\rangle = \lambda_m^*\langle\psi_m|\psi_n\rangle - \lambda_n\langle\psi_m|\psi_n\rangle = (\lambda_m - \lambda_n)\langle\psi_m|\psi_n\rangle$ $[\because \lambda_m^* = \lambda_m]$
$= 0$

If $\lambda_m \neq \lambda_n$, then $\langle\psi_m|\psi_n\rangle = 0$.

Hence, two eigenfunctions having distinct eigenvalues are orthogonal to each other.

» **Example 9.8.** If $\hat{P}$ be a momentum operator, then prove that (i) $\hat{P} = \int d^3pP|P\rangle\langle P|$ (ii) $\langle r|\hat{P}|P\rangle = \frac{\hbar}{i}\nabla\langle r|P\rangle$.

Solution:

(i) If p be an eigenvalue of $\hat{P}$, then $\hat{P}|P\rangle = p|P\rangle$.

If $\hat{P}$ operates on an arbitrary ket $|\psi\rangle$, then

$$\hat{P}|\psi\rangle = \int d^3pp|P\rangle\langle P|\psi\rangle$$

But, $|\psi\rangle$ is arbitrary

$$\therefore \hat{P} = \int d^3pp|P\rangle\langle P|$$

(ii) We have

$$\langle r|\hat{p}|P\rangle = p\langle r|P\rangle = p\frac{1}{(2\pi\hbar)^{3/2}}e^{(i\vec{P}\cdot\vec{r}/\hbar)}$$

$$\because \langle r|P\rangle = \frac{1}{(2\pi\hbar)^{3/2}}e^{i(\vec{P}\cdot\vec{r})/\hbar} \text{ or, } \langle r|\hat{p}|P\rangle = \frac{\hbar}{i}\nabla(\langle r|P\rangle)$$

» ***Example* 9.9.** If ϕ and ψ be any two state functions, then show that $|\langle\phi|\psi\rangle|^2 \leqslant \langle\psi|\psi\rangle\langle\phi|\phi\rangle$.

Solution: We consider a vector $\psi + \lambda\phi$, where λ is in general a complex number.

Since, the scalar product of $\psi + \lambda\phi$ with itself should always be greater than or equal to zero, we write:

$$\langle\psi + \lambda\phi|\psi + \lambda\phi\rangle \geqslant 0 \qquad \text{(i)}$$

Let

$$f(\lambda) = \langle\psi + \lambda\phi|\psi + \lambda\phi\rangle \geqslant 0$$

$$\text{or, } f(\lambda) = \langle\psi|\psi\rangle + \lambda\langle\psi|\phi\rangle + \lambda^* \langle\phi|\psi\rangle + \lambda\lambda^*\langle\phi|\phi\rangle \qquad \text{(ii)}$$

Differentiating w.r.t. λ^*

$$\frac{d}{d\lambda^*}[\langle\psi|\psi\rangle + \lambda\langle\psi|\phi\rangle + \lambda^* \langle\phi|\psi\rangle + \lambda^*\lambda\langle\phi|\phi\rangle] = \langle\phi|\psi\rangle + \lambda\langle\phi|\phi\rangle$$

But, $f(\lambda)$ is minimum or maximum, if

$$\frac{df(\lambda)}{d\lambda} = 0 \text{ or, } \frac{df(\lambda)}{d\lambda^*} = 0$$

$$\therefore \langle\phi|\psi\rangle + \lambda\langle\phi|\phi\rangle = 0 \text{ or, } \lambda = -\frac{\langle\phi|\psi\rangle}{\langle\phi|\phi\rangle}. \qquad \text{(iii)}$$

From (ii) and (iii)

$$f(\lambda)_{\min} = \langle\psi|\psi\rangle + \left(-\frac{\langle\phi|\psi\rangle}{\langle\phi|\phi\rangle}\right)\langle\psi|\phi\rangle + \lambda^* \langle\phi|\psi\rangle + \lambda^*\left(-\frac{\langle\phi|\psi\rangle}{\langle\phi|\phi\rangle}\right)\langle\phi|\phi\rangle \geqslant 0$$

$$\text{or, } \langle\psi|\psi\rangle - \frac{\langle\phi|\psi\rangle\langle\psi|\phi\rangle}{\langle\phi|\phi\rangle} + \lambda^* \langle\phi|\psi\rangle - \lambda^* \langle\phi|\psi\rangle \geqslant 0 \text{ or, } \langle\psi|\psi\rangle \geqslant \frac{\langle\phi|\psi\rangle\langle\phi|\psi\rangle^*}{\langle\phi|\phi\rangle}$$

$$\text{or, } \langle\psi|\psi\rangle\langle\phi|\phi\rangle \geqslant |\langle\phi|\psi\rangle|^2 \text{ or, } |\langle\phi|\psi\rangle|^2 \leqslant \langle\psi|\psi\rangle\langle\phi|\phi\rangle \qquad \text{(iv)}$$

This is known as *Schwartz inequality*.

» ***Example* 9.10.** Show that $\hat{\pi}$ is Hermitian.

Solution: Consider the scalar product $\pi\psi$ and a function ϕ, i.e., $+\infty$.

$$\langle\hat{\pi}|\phi\rangle = \int_{-\infty}^{+\infty} \psi^*(-x)\phi(x)\,dx \int_{-\infty}^{+\infty} \psi^*(x')\phi(-x')\,dx'$$

where $x' = -x$. Thus, on remaining the variable, the value of integral is unaffected and, we get

$$\langle\hat{\pi}\psi|\phi\rangle = \int_{-\infty}^{+\infty} \psi^*(x)\phi(-x)\,dx = \langle\psi|\hat{\pi}\phi\rangle$$

This shows that $\hat{\pi}$ is Hermitian operator.

» ***Example* 9.11.** If a Hermitian operator possesses a complete set of eigenfunction $\{\psi_n\}$, then show that the identity operator I can also be expressed as

$$I = \sum_n |\psi_n\rangle\langle\psi_n|$$

Solution: The completeness relation for the eigenfunctions $\{\psi_n\}$ of a Hermitian operator is

$$\sum_n \psi_n^*(\vec{r}')\,\psi_n(\vec{r}) = \delta(\vec{r}-\vec{r}') \tag{i}$$

where $\delta(\vec{r}-\vec{r}')$ is the delta function.

Now, consider the scalar product of two-wave functions ϕ and ψ :

$$\langle\phi|\psi\rangle = \int \phi^*(\vec{r},t)\,\psi(\vec{r},t)\,d\vec{r} = \iint \phi^*(\vec{r},t)\delta(\vec{r}-\vec{r}')\,\psi(\vec{r}',t)\,d\vec{r},d\vec{r}'$$

by the substitution property of the delta function.

Using relation (i)

$$\langle\phi|\psi\rangle = \sum_n \int \phi^*(\vec{r},t)\,\psi_n(\vec{r})\,d\vec{r}\int \psi_n^*(\vec{r}')\,\psi(\vec{r}',t')\,d\vec{r}' = \sum_n \langle\psi|\psi_n\rangle\langle\psi_n|\psi\rangle \tag{ii}$$

in Dirac *bra* and *ket* notation. Comparing LHS and RHS of (ii), we conclude

$$\sum_n |\psi_n\rangle\langle\psi_n| = I = \text{Identity operator.}$$

» ***Example* 9.12.** Let the state of a system be represented by $|\psi\rangle = a_1|\psi_1\rangle + a_2|\psi_2\rangle$. If the probability that the system is in state $|\psi_1\rangle$ is double in state $|\psi_2\rangle$, find a_1 and a_2. **[CU (H) 2013]**

Solution: The probability that the system is in states $|\psi_1\rangle$ and $|\psi_2\rangle$ is $|a_1|^2$ and $|a_2|^2$ respectively, i.e., $|a_1|^2 + |a_2|^2 = 1$.

Here, $|a_1|^2 = 2|a_2|^2$.

So $2|a_2|^2 + |a_2|^2 = 1$ or, $3|a_2|^2 = 1$ or, $a_2 = \frac{1}{\sqrt{3}}$. $\therefore\ \boldsymbol{a_1 = \sqrt{\frac{2}{3}}}$.

» ***Example* 9.13.** Write condition of Hermiticity of an operator $\hat{A}$. Check Hermiticity operators $\hat{x}$, $\frac{d}{dx}$, $i(d/dx)$, $\hat{p}$, e^x, $e^{i\hat{x}}$.

Solution: We have $\langle\psi|\hat{A}\,\psi\rangle = \langle\hat{A}\,\psi|\psi\rangle$ or, $\hat{A} = \hat{A}^\dagger$ for Hermiticity.

Now

(i) $\langle\psi|\hat{x}\,\psi\rangle = \int_{-\infty}^{\infty}\psi^*(x\psi)\,dx = \int_{-\infty}^{\infty}x\psi^*\psi\,dx = \int_{-\infty}^{\infty}(x\psi)^*\psi\,dx$ (as $x = x^*$)

$= \langle x\psi^*|\psi\rangle$ i.e., $\hat{x}$ is Hermitian (i.e., $\hat{x}^\dagger = \hat{x}$)

(ii) $$\left\langle\psi\left|\frac{d\psi}{dx}\right.\right\rangle = \int_{-\infty}^{\infty}\psi^*\frac{d\psi}{dx}\,dx = \psi^*\psi\Big|_{-\infty}^{\infty} - \int_{-\infty}^{\infty}\frac{d\psi^*}{dx}\psi\,dx$$

$$= 0 - \int_{-\infty}^{\infty}\left(\frac{d\psi}{dx}\right)^*\psi\,dx \quad (\text{as } \psi \to 0 \text{ at } x \to \pm\infty)$$

$$= -\left\langle\frac{d\psi}{dx}\right|\psi\left.\right\rangle \text{ or, i.e., } \frac{d}{dx} \text{ is anti-Hermitian } \left[\text{i.e., } \left(\frac{d}{dx}\right)^\dagger = -\frac{d}{dx}\right]$$

[$\because$ $\langle\psi|\hat{A}\,\psi\rangle = -\langle\hat{A}\,\psi^*|\psi\rangle$ or $\hat{A} = -\hat{A}^\dagger$ is the condition of anti-Hermiticity]

(iii) $\left(i\frac{d}{dx}\right)^\dagger = -i\left(\frac{d}{dx}\right)^\dagger = -i\left(-\frac{d}{dx}\right) = i\frac{d}{dx}$, i.e., $i\frac{d}{dx}$ is Hermitian.

(iv) $\hat{p} = -i\hbar\frac{\partial}{\partial x}$; $\hat{p}^\dagger = \left(-i\hbar\frac{\partial}{\partial x}\right)^\dagger = i\hbar\left(\frac{\partial}{\partial x}\right)^\dagger = i\hbar\frac{d}{\partial x} = -i\hbar\frac{\partial}{\partial x} = \hat{p}$, i.e., $\hat{p}$ is Hermitian.

(v) $e^\dagger = e^{\hat{x}^\dagger} = e^{\hat{x}}$, i.e., e^x is Hermitian.

(vi) $\left(e^{i\frac{d}{dx}}\right)^{\dagger} = i\left(\frac{d}{dx}\right)^{\dagger} = e^{-i\left(-\frac{d}{dx}\right)} = e^{i\frac{d}{dx}}$, i.e., $e^{i\frac{d}{dx}}$ is Hermitian.

» *Example* 9.14.

(a) Express a generalised spinor $|\chi\rangle = \begin{pmatrix} a \\ v \end{pmatrix}$ in terms of spin up and spin down.

(b) What is the probability of getting values $\pm\frac{\hbar}{2}$ in a measurement of $\hat{s}_2$?

(c) What is the probability of getting $\pm\frac{\hbar}{2}$ in a measurement of $\hat{s}_x$?

(d) What is the probability of getting values $+\frac{\hbar}{2}$ in a measurement of $\hat{s}_y$?

***Solution*:**

(a) Let us denote spin up state as $|\chi_+\rangle = |s=\frac{1}{2}, m_s = \frac{1}{2}\rangle = |\frac{1}{2},\frac{1}{2}\rangle = |\text{up}\rangle$, i.e., $|\alpha\rangle \equiv |4\rangle = \begin{pmatrix} 1 \\ 0 \end{pmatrix}$ and spin down state as $|\chi_-| = |s=\frac{1}{2}, m_s = -\frac{1}{2}\rangle = |\frac{1}{2},-\frac{1}{2}\rangle = |\text{down}\rangle = |\beta\rangle| \downarrow\rangle = \begin{pmatrix} 0 \\ 1 \end{pmatrix}$.

We assume $|\chi_+\rangle$ and $|\chi_+\rangle$ as basis (orthogonalised + complete)

$$|\chi\cdot\rangle = \hat{I}|\chi\rangle = |\chi_+\rangle\langle\chi_+|\chi\rangle + |\chi_-\rangle\langle\chi-|\chi\rangle = \langle\chi_+|\chi\rangle||\chi_+\rangle\rangle + \langle\chi_-|\chi\rangle|\chi_-\rangle = (1,0)\begin{pmatrix} a \\ b \end{pmatrix}|\chi_+\rangle + (0,1)\begin{pmatrix} a \\ b \end{pmatrix}|\chi_-\rangle$$

(b) Probability to get $\frac{\hbar}{2}$ in a measurement of $\hat{s}_2$ in a state χ is $|a|$ and to get $-\frac{\hbar}{2}$ is $|b|^2$. And $|a|^2 + |b|^2 = 1$ (χ is normalised).

(c) $\hat{s}_x = \frac{5}{3}\hat{\sigma}_x$ has eigenvalue $+\frac{\hbar}{2}$ with eigenspinor $|\chi_1\rangle = \frac{1}{\sqrt{2}}\begin{pmatrix} 1 \\ 1 \end{pmatrix}$ and $-\frac{\hbar}{2}$ with eigenspinor $|\chi_1\rangle = \frac{1}{\sqrt{2}}\begin{pmatrix} 1 \\ 1 \end{pmatrix}^2$ and $-\frac{\hbar}{2}$ with eigenspinor $|\chi_2\rangle = \frac{1}{\sqrt{2}}\begin{pmatrix} 1 \\ -1 \end{pmatrix}$.

Now, we express $|\chi\rangle$ in the basis of $|\chi_1\rangle, |\chi_2\rangle$.

$$|\chi\rangle = \hat{I}|\chi\rangle = (|\chi_1\rangle\langle\chi_1| + |\chi_2\rangle\langle\chi_2|)|\chi\rangle = \langle\chi_1|\chi|\chi_1\rangle + \langle\chi_2|\chi\rangle|\chi_2\rangle$$

$$= \frac{1}{\sqrt{2}}(1\quad 1)\begin{pmatrix} a \\ b \end{pmatrix}|\chi_1\rangle + \frac{1}{\sqrt{2}}(1\quad -1)\begin{pmatrix} a \\ b \end{pmatrix}|\chi_2\rangle = \frac{a+b}{\sqrt{2}}|\chi_1\rangle + \frac{a-b}{\sqrt{2}}|\chi_2\rangle$$

So, probability to get $\frac{\hbar}{2}$ in a measurement of $\hat{s}_x$ in state χ is $\left|\frac{a+b}{\sqrt{2}}\right|^2$ and probability to get $-\frac{\hbar}{2}$ is $\left|\frac{a+b}{\sqrt{2}}\right|^2$.

They add up to $a^2 + b^2 = 1$.

$\therefore$ χ is normalised.

(d) $\hat{S}_y = \frac{\hbar}{2}\hat{\sigma}_y$ has eigenvalues $\frac{\hbar}{2}$ with eigenspinor $|\chi_L\rangle = \frac{1}{\sqrt{2}}\begin{pmatrix} 1 \\ i \end{pmatrix}$ and $-\frac{\hbar}{2}$ with eigenspinor $||\chi_R\rangle = \frac{1}{\sqrt{2}}\begin{pmatrix} i \\ 1 \end{pmatrix}$.

Now, let us express $|\chi\rangle$ in the basis of $|\chi_L\rangle, |\chi_R\rangle$.

$$|\chi\rangle = \hat{I}|\chi\rangle = (|\chi_L\rangle\langle\chi_L| + |\chi_R\rangle\langle\chi_R|)|\chi\rangle = |\chi_L\rangle\langle\chi|\chi_L\rangle + \langle\chi_R|\chi\rangle|\chi_R\rangle$$

$$= \frac{1}{\sqrt{2}}(1 \quad -i)\begin{pmatrix} a \\ b \end{pmatrix}|\chi_L\rangle + \frac{1}{\sqrt{2}}(-i \quad 1)\begin{pmatrix} a \\ b \end{pmatrix}|\chi_R\rangle = \frac{a-ib}{\sqrt{2}}|\chi_L\rangle + \frac{-ia+b}{\sqrt{2}}|\chi_K\rangle$$

So, probability to get $\frac{\hbar}{2}$ in a measurement of $\hat{s}_y$ in state χ is

$$\left|\frac{a-ib}{\sqrt{2}}\right|^2 = \frac{a^2+b^2}{2}$$

and to get $-\frac{\hbar}{2}$ is

$$\left|\frac{-ia+b}{\sqrt{2}}\right| = \frac{a^2+b^2}{2} \quad \text{or, } a^2+b^2 = 1 \text{ (since } \chi \text{ is normalised)}$$

» ***Example* 9.15.** What will be the normalised singlet and triplet spin states formed by 2 electrons?

[CU (H) 2014]

Solution: State of the 1st electron, i.e., $s_1 = \frac{1}{2}$, $m_{s_1} = \pm\frac{1}{2}$ or, $|sm_s\rangle = |\frac{1}{2} \pm \frac{1}{2}\rangle$ [$m_s = \frac{1}{2}$ means up state $|\uparrow\rangle$, $m_s = -\frac{1}{2}$] means down state $|\downarrow\rangle$]

Combined system spin state, $\vec{S} = \vec{s}_1 + \vec{s}_2 = \frac{1}{2} + \frac{1}{2} \to 0, 1, S$ values.

For $S = 0, M_s = 0$, i.e., $|SM_s\rangle = |00\rangle$.

Multiplicity $= 2S + 1 = 2(0) + 1 = 1$ state.

So, it is a singlet state.

For $S = 1, M_s = 1, 0, -1$, i.e., $|SM_s\rangle = |11\rangle, |10\rangle, |1, -1\rangle$.

Multiplicity $= 2S + 1 = 2(1) + 1 = 3$ states.

So it is a triplet state.

Connection of singlet state ($|SM_s\rangle = |00\rangle$) and triplet state ($|SM_s\rangle$) $= |11\rangle|10\rangle, |1 \quad -1\rangle$ with the original electron state ($|sm_s\rangle = |\frac{1}{2} \pm \frac{1}{2}\rangle, |\frac{1}{2} \pm \frac{1}{2}\rangle$).

Construction of single state ($S = 0$).

$M_s = 0$ is possible if $m_{s_1} = \frac{1}{2}(\uparrow), m_{s_2} = -\frac{1}{2}(\downarrow)$ or, if $m_{s_1} = -\frac{1}{2}(\downarrow), m_{s_2} = \frac{1}{2}(\uparrow)$ [$\because M_S = m_{s_1} + m_{s_2}$]

So, $|SM_s\rangle = |00\rangle$ will be a linear combination of $|\uparrow\downarrow\rangle$ and $|\downarrow\uparrow\rangle$ in the representation $|s_{s_1} m_{s_2}\rangle$.

Again, singlet $S = 0$ state is antisymmetric [as determined from a factor $(-1)^{1+S} = (-1)^{1+0}$, which is negative]. So, the normalised singlet state will be $|00\rangle = \frac{1}{\sqrt{2}}(|\downarrow\rangle - |\downarrow\uparrow\rangle = |\text{singlet}| = |\text{one down on up}\rangle)$

$\frac{1}{\sqrt{2}}$ is the normalisation factor and the intermediate negative sign makes $|00\rangle$ antisymmetric.

Construction of triplet state ($S = 1$)

$M_s = 1$ is possible if $m_{s_1} = \frac{1}{2}(\uparrow), m_{s_2} = \frac{1}{2}(\uparrow)$.

So, $|SM_S\rangle = |11\rangle$ state is $|11\rangle = |\uparrow\uparrow\rangle = |\text{both up}\rangle$.

$M_S = 0$ is a linear combination of $|\uparrow\downarrow\rangle$ in the representation $|m_{s_1} m_{s_2}\rangle$.

Again triplet state $S = 1$ is symmetric [as $(-1)^{1+S} = (-1)^{1+1}$, which is positive].

So, the normalised state $|SM_S\rangle = |10\rangle$ is $|10\rangle = \frac{1}{\sqrt{2}}(|\uparrow\downarrow\rangle + |\downarrow\uparrow\rangle) = |\text{one down one up}\rangle$.

$\frac{1}{\sqrt{2}}$ is the normalisation factor and the intermediate +ve sign makes $|10\rangle$ symmetries $M_S = -1$ is possible if $m_{s_1} = -\frac{1}{2}(\downarrow), m_{s_2} = -\frac{1}{2}(\downarrow)$.

So, $|SM_S\rangle = |1 \quad -1\rangle$ state is $|1 \quad -1\rangle = |\downarrow\downarrow\rangle = |\text{both down}\rangle$.

Clearly, $|11\rangle = |\uparrow\uparrow\rangle = |10\rangle = \frac{1}{\sqrt{2}}(|\uparrow\downarrow\rangle + |\downarrow\uparrow\rangle)$, triplet, $|1 \quad 1\rangle = |\downarrow\downarrow\rangle$.

EXERCISES

9(A): Objective (or MCQ) Type Questions

Select the correct answer:

9.1. The operation on $|\psi\rangle$ from the left with an operator $\hat{\alpha}$ produces

(a) $|\psi\rangle$ (c) $\langle\psi'|$

(b) $|\psi'\rangle$ (d) $\langle\psi|$.

9.2. The scalar product of two state vectors ψ_m and ψ_n, $\langle\psi_m|\psi_n\rangle$ is linear w.r.t.

(a) post-factor (c) $\vec{r}$

(b) pre-factor (d) none of these.

9.3. If $|\psi_m\rangle$ and $|\psi_n\rangle$ be two eigenvectors corresponding to the operator $\hat{\alpha}$ belonging to different eigen λ_m and λ_n, then

(a) $\langle\psi_m|\psi_n\rangle = 1$ (c) $\langle\psi_m|\psi_n\rangle = 0$

(b) $\langle\psi_m|\psi_n\rangle = -1$ (d) $\langle\psi_m|\psi_n\rangle \geqslant 0$.

9.4. If $\hat{\alpha}|\psi_1\rangle = \lambda|\psi_1\rangle$, $\hat{\alpha}|\psi_2\rangle = \lambda|\psi_2\rangle$ and $|\psi\rangle = C_1|\psi_1\rangle + C_2|\psi_2\rangle$, then

(a) $\hat{\alpha}|\psi\rangle \neq \lambda|\psi\rangle$ (c) $\hat{\alpha}|\psi\rangle = 0$

(b) $\hat{\alpha}|\psi\rangle = \lambda|\psi\rangle$ (d) $\hat{\alpha}|\psi\rangle = \lambda'|\psi\rangle$.

9.5. The condition of orthogonality of $|\psi_m\rangle$ and $|\psi_n\rangle$ is

(a) $\langle\psi_m|\psi_n\rangle = 1$ (c) $\langle\psi_m|\psi_n\rangle = \delta_{mn}$

(b) $\langle\psi_m|\psi_m\rangle = 0$ (d) $\langle\psi_m|\psi_n\rangle = -1$.

9.6. If the expectation value of a dynamical variable be represented by $\langle\alpha\rangle = \langle\psi|\alpha|\psi\rangle$, then $|\psi\rangle$ is

(a) normalised (c) may be both

(b) unnormalised (d) none of these.

9.7. The condition of normalisation of the state vector $|\psi\rangle$ is

(a) $\langle\psi|\psi\rangle = 0$ (c) $\langle\psi|\psi\rangle = -1$

(b) $\langle\psi|\psi\rangle = 1$ (d) $\langle\psi|\psi\rangle^* = 0$.

ANSWERS

9.1. (b). **9.2.** (a). **9.3.** (c). **9.4.** (b). **9.5.** (c). **9.6.** (a). **9.7.** (b).

9(B): Short Answer Type Questions

9.1. Why Dirac introduced *bra* and *ket* notations?

Ans. If $\vec{\psi}_m$ and $\vec{\psi}_n$ be two arbitrary state vectors, then their scalar product is given by

$$(\vec{\psi}_m \cdot \vec{\psi}_n) = \int \vec{\psi}_m^* \psi_n \, d\tau \text{ and } (\vec{\psi}_n \cdot \vec{\psi}_n) = \int \vec{\psi}_n^* \vec{\psi}_m d\tau = (\vec{\psi}_m \cdot \vec{\psi}_n)^*$$

This shows that $(\vec{\psi}_n \cdot \vec{\psi}_m) \neq (\vec{\psi}_m \cdot \vec{\psi}_n)$.

Also $(\vec{\psi}_m \cdot C\vec{\psi}_n) = C(\vec{\psi}_m \cdot \vec{\psi}_n)$ and $(C\vec{\psi}_m \cdot \vec{\psi}_n) = C^*(\vec{\psi}_m \cdot \vec{\psi}_n)$.

It, therefore, transpires that the scalar product is *linear* w.r.t. post-factor $(\vec{\psi}_n)$ and *anti-linear* w.r.t. pre-factor $(\vec{\psi}_m)$. This leads to a confusion.

In order to avoid this confusion Dirac introduced *bra* and *ket* notations. He considered all possible state vectors ψ in *Hilbert space* (a linear vector space) and represented by the ket vector $|\psi\rangle$. The linear vector space formed by all ket vectors is called a *ket-space*.

The scalar product of $\vec{\psi}_m$ and $\vec{\psi}_n$ is represented as $(\vec{\psi}_m \cdot \vec{\psi}_n) = \langle \psi_m | \psi_n \rangle$.

Here, $\langle \psi_m|$ is the bra vector corresponding to $|\psi_n\rangle$. By this consideration, the pre-factor and post-factor are clearly distinguished in a scalar product. In $\langle \psi_m | \psi_n \rangle$, $\langle \psi_m |$ is the pre-factor and $|\psi_n\rangle$ is the post-factor, but one cannot make distinction so as to pre-factor or post-factor merely by writing ψ_m and ψ_n.]

9.2. Write down the condition of orthonormality of state vector $|\psi_m\rangle$ and $|\psi_n\rangle$.

9.3. If $\psi(x)$ be a state function, then write $\psi(x)$ and its complex conjugate in Dirac notation.

9.4. Use Dirac notation to show that the eigenvalues of an observable operator are real.

9.5. Using Dirac notation show that the momentum operator $\hat{p} = -i\hbar\nabla$ is Hermitian.

9.6. Show that the product of the Hermitian operators is Hermitian, if and only if they commute.

9.7. If $a^\dagger$ and a be respectively the creation and annihilation operators and

$$x = \sqrt{\frac{\hbar}{2m\omega}}\,(a + a^\dagger) \text{ and } p = \sqrt{\frac{m\hbar\omega}{2}}\, i\,(a^\dagger - a)$$

then show that $\Delta p \Delta x = \hbar/2$.

9(C): Broad Answer Type Questions

9.1. Explain the properties of *bra* and *ket* notations.

9.2. Using Dirac notation find out the eigenvalue and eigenfunction of the linear harmonic oscillator. Prove that the eigenfunctions are normalised.

9.3. The normalised eigenfunction of the ground state of a system of energy E_0 is

$$\langle x|\psi_0\rangle = \left(\frac{\alpha}{\sqrt{\pi}}\right)^{1/2} e^{-\frac{1}{2}\alpha^2 x^2}$$

If the same system is in the state represented by

$$\langle x|\psi\rangle = \left(\frac{\sigma}{\sqrt{\pi}}\right)^{1/2} e^{-\frac{1}{2}\sigma^2 x^2}$$

then find out the probability that a measurement of energy gives the result E_0.

9(D): Numerical Problems

9.1. If $\hat{p}$ be an operator corresponding to the momentum, then prove that $\langle\phi|\hat{p}|\psi\rangle = (\langle\psi|\hat{p}|\phi\rangle)^*$.

9.2. If $\langle x\rangle = \langle\psi_n|x|n\rangle = 0$ and $\langle p\rangle = \langle\psi|p|\psi_n\rangle = 0$, then show that $\Delta x\,\Delta p_x = (n + \frac{1}{2})\,\hbar$.

9.3. Prove that $\left[\hat{a}, e^{\alpha\hat{a}^\dagger}\right] = \alpha e^{\alpha\hat{a}^\dagger}$, where α is an arbitrary complex number. Using the above equation show that $e^{\alpha\hat{a}^\dagger}|0\rangle$ is an eigenket of the operator $\hat{a}$. What is the corresponding eigenvalue?

[*Ans.* (α)]

9.4. A system is described by the time-dependent Hamiltonian H, having non-degenerate eigenvalues E_n,

$$H|\psi_n\rangle = E_n|\psi_n\rangle.$$

The system at $t = 0$ is in the state $|\psi_n\rangle$ and let a measurement of the observable α be performed on this system. Assuming that the eigenvalues α_m of α is non-degenerate, show that the probability of obtaining the eigenvalue α_m would be $|\langle\alpha_m|\psi_n\rangle|^2$.
Assume that the eigenvalue α_m is indeed obtained, show that if the measurement is repeated at $t = \tau$, then the probability of obtaining α_m is

$$P_m = \left|\sum_n e^{-\frac{iE_n t}{\hbar}} |\langle\alpha_m|\psi_n|^2\right|^2$$

Also prove that $\sum_m P_m = 1$.

9.5. In the harmonic oscillator problem express x and p in terms of $\hat{a}$ and $\hat{a}^\dagger$.

9.6. Prove the identity $\langle|A|y\rangle = \langle A^\dagger x|y\rangle$ for any matrix A and its Hermitian conjugate. Also prove that the scalar product (x, y) does not depend on the representation used.

Chapter 10

Angular Momentum and Spin

10.1 The operator method for solving the angular momentum problems

In Chapter 5, we have discussed the operator representation of the components L_x, L_y, L_z of the orbital angular momentum L of a particle rotating around an axis. We have also established the commutation relations $[\hat{L}_x, \hat{L}_y]$, $[\hat{L}_y, \hat{L}_z]$, $[\hat{L}_z, \hat{L}_x]$ and $[\hat{L}^2, \hat{L}_x]$, $[\hat{L}^2, \hat{L}_y]$, $[\hat{L}^2, \hat{L}_z]$.

In the present chapter we shall find the eigenvalues of the square of the angular momentum ($\hat{L}^2$) and of the components $\hat{L}_x$, $\hat{L}_y$ and $\hat{L}_z$. Subsequently, we shall pass over to the angular momentum in general which includes spin angular momenta along with orbital angular momentum.

10.2 Eigenvalues of L_z

Let c be the eigenvalue of $\hat{L}_z$ corresponding to the wave function ψ, then

$$L_z\psi = c\psi \tag{10.1}$$

$$\text{or, } \frac{\hbar}{i}\frac{\partial\psi}{\partial\phi} = c\psi \quad \left[\because L_z = \frac{\hbar}{i}\frac{\partial}{\partial\phi}\right] \tag{10.2}$$

where $L_z = \dfrac{\hbar}{i}\dfrac{\partial}{\partial\phi}$.

The solution of eq. (10.2) is

$$\psi = f(r,\theta)\, e^{ic\phi/\hbar} \tag{10.3}$$

where $f(r, \theta)$ is an arbitrary function of r, θ.

Since, ψ is a single-valued function, if ϕ be changed to $\phi + 2\pi$, ψ should remain unchanged.

$$\therefore f(r,\theta)\, e^{ic\phi/\hbar} = f(r,\theta)\, e^{\frac{ic(\phi+2\pi)}{\hbar}} \text{ or, } e^{ic2\pi/\hbar} = 1 \text{ or, } \frac{2\pi c}{\hbar} = 2m\pi \text{ or, } c = m\hbar$$

where m is an integer.

Hence, the eigenvalues of L_z are

$$L_z = m\hbar \tag{10.4}$$

the eigenfunctions of L_z are given by

$$\psi = f(r,\theta)\, e^{\frac{im\phi\hbar}{\hbar}} \text{ or, } \psi = f(r,\theta)\, e^{im\phi} \tag{10.5}$$

10.3 Generation of eigenfunctions

► ***Theorem* 1.** If ψ_m is an eigenfunction of L_z with eigenvalue $m\hbar$, then $(L_x \pm iL_y)\,\psi_m$ is also an eigenfunction of L_z having eigenvalue $(m \pm \nu)\hbar$, where ν is equal to a non-negative integer.

Proof: We have

$$L_z(L_x \pm iL_y) = L_zL_x \pm iL_zL_y \tag{10.6}$$

But, $[L_z, L_x] = L_zL_x - L_xL_z = i\hbar L_y$ and $[L_y, L_z] = L_yL_z - L_zL_y = i\hbar L_x$.

Now, from (10.6)

$$L_z(L_x \pm iL_y) = L_zL_x \pm iL_zL_y = (i\hbar L_y + L_xL_z) \pm i(L_yL_z - i\hbar L_x)$$
$$= i\hbar L_y + L_xL_z \pm iL_yL_z \pm \hbar L_x = \pm\hbar(L_x \pm iL_y) + (L_x \pm iL_y)L_z$$

$$\therefore L_z(L_x \pm iL_y)\,\psi_m = (L_x \pm iL_y)\,(L_z \pm \hbar)\,\psi_m \quad (\because L_z\psi_m = m\hbar\psi_m)$$

$$\therefore L_z[(L_x \pm iL_y)\,\psi_m] = (m \pm 1)\hbar[(L_x \pm iL_y)\,\psi_m] \tag{10.7}$$

Thus, if ψ_m be an eigenfunction of L_z with eigenvalues $m\hbar$, then (ii) shows that $(L_x \pm iL_y)\,\psi_m$ is also an eigenfunction of L_z with eigenvalues $(m \pm 1)\hbar$. Repeating this process r times we can easily prove the theorem.

If the operator $(L_x + iL_y)$ be repeatedly applied, then we get eigenfunctions of fixed L^2 belonging to indefinitely large eigenvalues of L_z, unless $(L_x + iL_y)\,\psi_m$ vanishes for some value of m. In the same way repeated application of $(L_x - iL_y)$ will give indefinitely large negative values of m unless $(L_x - iL_y)\,\psi_m$ vanishes for some m values.

Since,

$$L^2 = L_x^2 + L_y^2 + L_z^2 = L_x^2 + L_y^2 + m^2\hbar^2 \tag{10.8}$$

$|m|$ cannot be indefinitely large for a fixed L^2. As $\langle L_x\rangle^2$ and $\langle L_y\rangle^2$ are always positive for a state in which L^2 and L_z have definite values, we must have $\hbar^2m^2 \leqslant L^2$ and as such $|\hbar m|$ must not be greater than $\sqrt{L^2}$.

10.4 Eigenvalues of L^2

Let m_1 be the maximum positive value of $L_z/\hbar$ for a definite L^2, then

$$L_+\,\psi_{m_1} = (L_x + iL_y)\,\psi_{m_1} = 0 \tag{10.9}$$

Similarly, if m_2 be the maximum negative value of $L_z/\hbar$ for a definite L^2, then

$$L_- \psi_{m_2} = (L_x - iL_y)\,\psi_{m_2} = 0 \tag{10.10}$$

Now,

$$\begin{aligned} L^2\,\psi_{m_1} &= (L_x^2 + L_y^2 + L_z^2)\,\psi_{m_1} = [(L_x - iL_y)(L_x + iL_y) + L_z^2 + \hbar L_z]\psi_{m_1} \\ &= [(L_x - iL_y)(L_x + iL_y) + \hbar^2(m_1^2 + m_1)]\psi_{m_1} \\ \text{or, } L^2\,\psi_{m_1} &= \hbar^2 m_1(m_1+1)\,\psi_{m_1} \quad \text{[by eq. (10.9)]} \qquad (10.11) \\ \text{and } L^2\,\psi_{m_2} &= (L_x^2 + L_y^2 + L_z^2)\,\psi_{m_2} = [(L_x + iL_y)(L_x - iL_y) + L_z^2 - \hbar L_z]\psi_{m_2} \\ &= [(L_x + iL_y)(L_x - iL_y) + \hbar^2(m_2^2 - m_2)]\psi_{m_2} \\ \text{or, } L^2\,\psi_{m_2} &= \hbar^2 m_2(m_2 - 1)\,\psi_{m_2} \qquad (10.12) \end{aligned}$$

If eqs (10.11) and (10.12) are true simultaneously, then

$$m_2(m_2-1) = m_1(m_1-1) \text{ or, } m_2^2 - m_1^2 = m_2 - m_1$$
$$\text{or, } (m_2 + m_1)(m_2 - m_1 - 1) = 0$$

Now, either $m_2 = -m_1$ or, $m_2 = m_1 + 1$. But, as m_1 is the largest value of $L_z/\hbar$, $(m_2 - m_1 - 1)$ is unacceptable. Hence, $m_2 = -m_1$.

Writing l for m, where l is an integer, we have

$$L^2 = \hbar^2 l(l+1) \tag{10.13}$$

This gives the eigenvalues of L^2.

The total angular momentum L has magnitude $\sqrt{l(l+1)}$. The Z-component of L is $l\hbar$.

Since, L_z does not commute with L_x and L_y, the values of L_x and L_y are not known when L_z is determined.

Hence, the orbital angular momentum vector lies on the surface of the cone as shown in Fig 2010.1.

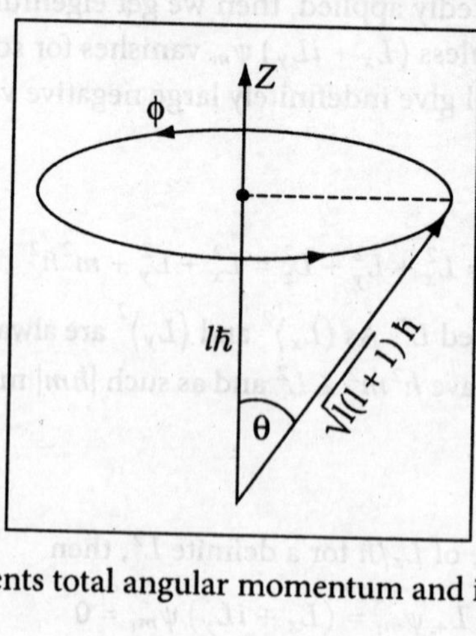

Fig 10.1 Represents total angular momentum and its Z-component

10.5 Eigenfunctions of L^2 and L_z

Suppose, we have obtained a single eigenfunction of L^2 and L_z. Then other eigenfunctions relating to the same value of L^2 can be obtained by repeating the application of $L_+ = (L_x + iL_y)$ and $L_- = (L_x - iL_y)$.

Now, the first eigenfunction for $m = l$ may be obtained from

$$(L_x + iL_y)\,\psi_l^l = 0 \tag{10.14}$$

But, in polar coordinates

$$L_x + iL_y = \hbar e^{i\phi}\left(\frac{\partial}{\partial\theta} + i\cot\theta\frac{\partial}{\partial\phi}\right) \tag{10.15}$$

$\therefore$ from eqs (10.14) and (10.15)

$$\hbar e^{i\phi}\left(\frac{\partial}{\partial\theta} + i\cot\theta\frac{\partial}{\partial\phi}\right)\psi_l^l = 0 \tag{10.16}$$

Again, for $m = l$, $\psi_l^l = f_l^l(\theta)\,e^{il\phi}$.

$$\therefore \frac{\partial\psi_l^l}{\partial\phi} = f_l^l(\theta)\,(il)\,e^{il\phi} = il\psi_l^l \text{ or, } i\frac{\partial\psi_l^l}{\partial\phi} = -l\psi_l^l$$

$$\therefore \frac{\partial}{\partial\phi} = il\psi_l$$

Then from eq. (10.16), we get

$$\hbar e^{i\phi}\left[\frac{\partial\psi_l^l}{\partial\theta} - l\cot\theta\psi_l^l\right] = 0 \text{ or, } \frac{\partial\psi_l^l}{\partial\theta} = l\cot\theta\psi_l^l \text{ or, } \frac{\partial\psi_l^l}{\psi_l^l} = l\cot\theta\,d\theta \tag{10.17}$$

Integrating, we get $\ln\psi_l^l = l\ln(\sin\theta) + k(\phi)$, where $k(\phi)$ is a constant of integration and may be written as $k(\phi) = \ln g(\phi)$.

$\therefore \ln\psi_l^l = l\ln(\sin\theta) + \ln g(\phi) = \ln[g(\phi)(\sin\theta)^l]$.

Taking exponential on both sides, we get

$$\psi_l^l = g(\phi)(\sin\theta)^l \tag{10.18}$$

Here, the function $g(\phi)$ must be such that ψ_l^l be an eigenfunction of L_z with eigenvalue $l\hbar$.

So

$$\psi_l^l = e^{il\phi}(\sin\theta)^l \tag{10.19}$$

The eigenfunction ψ_l^m for smaller value of m may be obtained by operating with $L_- = (L_x - iL_y)$. But,

$$-(L_x - iL_y) = \hbar e^{-i\phi}\left(\frac{\partial}{\partial\theta} - i\cot\theta\frac{\partial}{\partial\phi}\right) \tag{10.20}$$

$$\therefore -(L_x - iL_y)\,\psi_l^l = \hbar e^{-i\phi}\left[\frac{\partial}{\partial\theta} + l\cot\theta\right]\psi_l^l \quad \left[\because -i\frac{\partial\psi_l^l}{\partial\phi} = l\psi_l^l\right] \tag{10.21}$$

Again,

$$\frac{\partial}{\partial\theta}\left[(\sin\theta)^l \psi_l^l\right] = (\sin\theta)^l \frac{\partial\psi_l^l}{\partial\theta} + \psi_l^l l(\sin\theta)^{l-1}\cot\theta = (\sin\theta)^l\left[\frac{\partial}{\partial\theta} + l\cot\theta\right]\psi_l^l$$

$$\therefore \left[\frac{\partial}{\partial\theta} + l\cot\theta\right]\psi_l^l = \frac{1}{(\sin\theta)^l}\frac{\partial}{\partial\theta}\left[(\sin\theta)^l\psi_l^l\right] \tag{10.22}$$

Substituting eq. (10.22) in eq. (10.21), we get

$$\psi_l^{l-1} = -\hbar e^{-i\phi}\frac{1}{(\sin\theta)^l}\frac{\partial}{\partial\theta}\left[(\sin\theta)^l e^{il\phi}(\sin\theta)^l\right]\left[\because -(L_x - iL_y)\psi_l^l = \psi_l^{l-1} \text{ and } \psi_l^l = e^{il\phi}(\sin\theta)^l\right]$$

$$\text{or, } \psi_l^{l-1} = C\frac{1}{(\sin\theta)^l}e^{i(l-1)\phi}\frac{\partial}{\partial\theta}(\sin\theta)^{2l} \tag{10.23}$$

where $C = -\hbar$ is a constant.

Again,

$$(L_x - iL_y)\psi_l^{l-1} = -\hbar e^{-i\phi}\left[\frac{\partial}{\partial\theta} + (l-1)\cot\theta\right]\psi_l^{l-1} = -\frac{\hbar e^{-i\phi}}{(\sin\theta)^{l-1}}\frac{\partial}{\partial\theta}\left[(\sin\theta)^{l-1}\psi_l^{l-1}\right]$$

$$\therefore \psi_l^{l-2} = C\frac{1}{(\sin\theta)^{l-1}}e^{i(l-1)\phi}\left[\frac{1}{\sin\theta}\frac{\partial}{\partial\theta}(\sin\theta)^{2l}\right]$$

Applying $(L_x - iL_y)$ repeatedly, we get

$$\psi_l^{l-s} = C\frac{e^{i(l-s)\phi}}{(\sin\theta)^{l-s}}\left[\frac{1}{\sin\theta}\frac{\partial}{\partial\theta}\right]^s(\sin\theta)^{2l} \tag{10.24}$$

Now, putting $\cos\theta = x$ and $\frac{\partial}{\partial\theta} = -\sin\theta\frac{\partial}{\partial x}$, eq. (10.24) reduces to

$$\psi_l^{-s} = C\frac{e^{i(l-s)\phi}}{(1-x^2)^{\frac{l-s}{2}}}\frac{\partial^s}{\partial x^s}(1-x^2)^l \tag{10.25}$$

Equations (10.24) and (10.25) represent normalised wave functions of L_z and L^2 with normalisation constant C.

10.6 Spin angular momentum

Before the introduction of the concept of electron spin, the following experimental facts could not be accounted for by the quantum theory developed so far:

1. Fine structure of spectral lines
2. The results of Stern-Gerlach experiment
3. The anomalous Zeeman effect

4. The permanent magnetic moment of the electron

$$\vec{\mu}_s = -\frac{e}{2m_0 c} 2\vec{S}$$

where $\vec{S}$ is the spin angular momentum associated with the spin motion of the electron.

In order to resolve the above difficulties, *Goudsmit and Uhlenbeck* (1925) proposed that each electron spins, while revolving around the nucleus and has a *quantised spin angular momentum.* The electron being a charged body also possesses *spin magnetic moment.*

According to them, the electron has a spin angular momentum, $S = \frac{1}{2}\hbar$, so that the spin quantum number is $\frac{1}{2}$. According to the rules of space quantisation, only two orientations of the spin vector are possible w.r.t. special direction in space. Then the possible values of the components of the spin angular momentum will be $S_z = \frac{1}{2}\hbar$ and $S_z = -\frac{1}{2}\hbar$.

Hence, we conclude that each orbital electron in an atom provides a contribution to the angular momentum in an assigned direction equal to $\pm\frac{1}{2}\hbar$.

10.7 Pauli's theory of spin

Spin of the electron is a purely quantum mechanical concept having no classical analogue. As $S = \hbar/2$, in classical limit when $\hbar \to 0$, $S \to 0$. But, for orbital angular momentum $p_l = l\hbar$, p_l remains finite in classical limit of $\hbar \to 0$ and $l \to \infty$.

If S_x, S_y and S_z be the x-, y- and z-components of spin angular momentum, then $S^2 = S_x^2 + S_y^2 + S_z^2$.

The following commutation relations hold:

$$\left.\begin{aligned} \hat{S}_x\hat{S}_y - \hat{S}_y\hat{S}_x &= i\hbar\hat{S}_z \\ \hat{S}_z\hat{S}_x - \hat{S}_x\hat{S}_z &= i\hbar\hat{S}_y \\ \hat{S}_y\hat{S}_z - \hat{S}_z\hat{S}_y &= i\hbar\hat{S}_x \end{aligned}\right\} \tag{10.26}$$

Since, $\hat{S}_x$, $\hat{S}_y$ and $\hat{S}_z$ have values $\pm\frac{1}{2}\hbar$

$$\therefore \hat{S}_x^2 + \hat{S}_y^2 + \hat{S}_z^2 = \frac{\hbar^2}{4}\hat{I} \tag{10.27}$$

where $\hat{I}$ is the identity operator.

10.7.1 Pauli's spin matrices

The spin operators are expressed as matrices with two rows and two columns

$$\left.\begin{aligned} \hat{S}_x &= \begin{bmatrix} a_x & b_x \\ b_x^* & c_x \end{bmatrix} \\ \hat{S}_y &= \begin{bmatrix} a_y & b_y \\ b_y^* & c_y \end{bmatrix} \\ \hat{S}_z &= \begin{bmatrix} \hbar/2 & 0 \\ 0 & -\hbar/2 \end{bmatrix} \end{aligned}\right\} \tag{10.28}$$

Now, operators expressed as matrices are *Hermitian* and hence the corresponding coefficients on opposite sides of the diagonal are *complex conjugate of each other.*

$$\therefore \hat{S}_y\hat{S}_z - \hat{S}_z\hat{S}_y = \begin{bmatrix} a_y & b_y \\ b_y^* & c_y \end{bmatrix}\begin{bmatrix} \hbar/2 & 0 \\ 0 & -\hbar/2 \end{bmatrix} - \begin{bmatrix} \hbar/2 & 0 \\ 0 & -\hbar/2 \end{bmatrix}\begin{bmatrix} a_y & b_y \\ b_y^* & c_y \end{bmatrix}$$

$$= \hbar/2\begin{bmatrix} a_y & -b_y \\ b_y^* & -c_y \end{bmatrix} - \hbar/2\begin{bmatrix} a_y & b_y \\ -b_y^* & -c_y \end{bmatrix} = \begin{bmatrix} 0 & -\hbar b_y \\ \hbar b_y^* & 0 \end{bmatrix} = i\hbar\begin{bmatrix} a_x & b_x \\ b_x^* & c_x \end{bmatrix} \tag{10.29}$$

Hence,

$$a_x = c_x = 0 \text{ and } i\hbar b_x = -\hbar b_y \text{ or, } b_y = -ib_x \tag{10.30}$$

Again, from eq. (10.26), we have

$$\hat{S}_z\hat{S}_x - \hat{S}_x\hat{S}_z = \begin{bmatrix} \hbar/2 & 0 \\ 0 & -\hbar/2 \end{bmatrix}\begin{bmatrix} a_x & b_x \\ b_x^* & c_x \end{bmatrix} - \begin{bmatrix} a_x & b_x \\ b_x^* & c_x \end{bmatrix}\begin{bmatrix} \hbar/2 & 0 \\ 0 & -\hbar/2 \end{bmatrix}$$

$$= \hbar/2\begin{bmatrix} a_x & b_x \\ -b_x^* & -c_x \end{bmatrix} - \hbar/2\begin{bmatrix} a_x & -b_x \\ b_x^* & -c_x \end{bmatrix} = \begin{bmatrix} 0 & \hbar b_x \\ -\hbar b_x^* & 0 \end{bmatrix}$$

$$\therefore \hat{S}_z\hat{S}_x - \hat{S}_x\hat{S}_z = \begin{bmatrix} 0 & \hbar b_x \\ -\hbar b_x^* & 0 \end{bmatrix} = i\hbar\begin{bmatrix} a_y & b_y \\ b_y^* & c_y \end{bmatrix} \tag{10.31}$$

From eq. (10.31), we get $a_y = c_y = 0$ and $b_x = ib_y$.

We have from eq. (10.27)

$$\hat{S}_z^2 = \hat{S}_x^2 = \hbar^2/4 \tag{10.32}$$

$$\hat{S}_z^2 = \begin{bmatrix} a_x^2 & b_x^2 \\ b_x^{*2} & c_x^2 \end{bmatrix} = \begin{bmatrix} \hbar^2/4 & 0 \\ 0 & \hbar^2/4 \end{bmatrix}$$

$$\text{or, } \hat{S}_z^2 = \begin{bmatrix} 0 & |b_x|^2 \\ |b_x|^2 & 0 \end{bmatrix} = \begin{bmatrix} \hbar^2/4 & 0 \\ 0 & \hbar^2/4 \end{bmatrix} \quad [\text{as } a_x = c_x = 0]$$

$$\text{or, } S_z^2 = \begin{bmatrix} |b_x|^2 & 0 \\ 0 & |b_x|^2 \end{bmatrix} = \begin{bmatrix} \hbar^2/4 & 0 \\ 0 & \hbar^2/4 \end{bmatrix} \tag{10.33}$$

From eq. (10.32) it is evident that

$$|b_x|^2 = \hbar^2/4, \therefore b_x = \hbar/2 \tag{10.34}$$

Using eqs (10.30), (10.32) and (10.34) in eq. (10.28), we get

$$\hat{S}_x = \begin{bmatrix} 0 & \hbar/2 \\ \hbar/2 & 0 \end{bmatrix}, \hat{S}_y = \begin{bmatrix} 0 & -i\hbar/2 \\ +i\hbar/2 & 0 \end{bmatrix}, \hat{S}_z = \begin{bmatrix} \hbar/2 & 0 \\ 0 & -\hbar/2 \end{bmatrix}$$

$$\text{or, } \hat{S}_x = \hbar/2\begin{bmatrix} 0 & 1 \\ 1 & 0 \end{bmatrix}, \hat{S}_y = \hbar/2\begin{bmatrix} 0 & -i \\ +i & 0 \end{bmatrix}, \hat{S}_z = \hbar/2\begin{bmatrix} 1 & 0 \\ 0 & -1 \end{bmatrix} \tag{10.35}$$

Putting

$$\hat{\sigma}_x = \begin{bmatrix} 0 & 1 \\ 1 & 0 \end{bmatrix}, \ \hat{\sigma}_y = \begin{bmatrix} 0 & -i \\ +i & 0 \end{bmatrix}, \ \hat{\sigma}_z = \begin{bmatrix} 1 & 0 \\ 0 & -1 \end{bmatrix} \tag{10.36}$$

we get

$$\hat{S}_x = \frac{\hbar}{2}\hat{\sigma}_x, \ \hat{S}_y = \frac{\hbar}{2}\hat{\sigma}_y \text{ and } \hat{S}_z = \frac{\hbar}{2}\hat{\sigma}_z \tag{10.37}$$

$\hat{\sigma}_x$, $\hat{\sigma}_y$ and $\hat{\sigma}_z$ are *the Pauli's spin matrices.*

Now, $\hat{\sigma}^2 = \hat{\sigma}_x^2 + \hat{\sigma}_y^2 + \hat{\sigma}_z^2 = 3$ and $\hat{S}^2 = \hat{S}_x^2 + \hat{S}_y^2 + \hat{S}_z^2 = \frac{3}{4}\hbar^2$.

Since, the eigenvalues of each $\hat{S}$ are just $\frac{1}{2}\hbar$ and $-\frac{1}{2}\hbar$ the eigenvalues of each σ must be +1 and −1.

So $\sigma_x^2 = \sigma_y^2 = \sigma_z^2 = 1$, where 1 is the unit operator.

10.7.2 Further, Pauli operators

Further, the Pauli's operators associated with x, p_x and L_z are

$$x = x\begin{bmatrix} 1 & 0 \\ 0 & 1 \end{bmatrix} = \begin{bmatrix} x & 0 \\ 0 & x \end{bmatrix}, \ p_x = \begin{bmatrix} p_x & 0 \\ 0 & p_x \end{bmatrix} \text{ and } L_z = \begin{bmatrix} L_z & 0 \\ 0 & L_z \end{bmatrix}$$

Again, we know that the Schrödinger operators associated with $\hat{x}$, $\hat{p}_x$ and $\hat{L}_z$ are x, $\frac{\hbar}{i}\frac{\partial}{\partial x}$ and $\frac{\hbar}{i}\frac{\partial}{\partial \phi}$ respectively.

$$\therefore x = \begin{bmatrix} x & 0 \\ 0 & x \end{bmatrix}, \ p_x = \begin{bmatrix} \frac{\hbar}{i}\frac{\partial}{\partial x} & 0 \\ 0 & \frac{\hbar}{i}\frac{\partial}{\partial x} \end{bmatrix}, \ L_z = \begin{bmatrix} \frac{\hbar}{i}\frac{\partial}{\partial \phi} & 0 \\ 0 & \frac{\hbar}{i}\frac{\partial}{\partial \phi} \end{bmatrix} \tag{10.38}$$

10.8 Commutation relations satisfied by $\hat{\sigma}_x$, $\hat{\sigma}_y$ and $\hat{\sigma}_z$

We have

$$[\hat{\sigma}_x, \hat{\sigma}_y] = [\hat{\sigma}_x\hat{\sigma}_y - \hat{\sigma}_y\hat{\sigma}_x] = \begin{bmatrix} 0 & 1 \\ 1 & 0 \end{bmatrix}\begin{bmatrix} 0 & -i \\ i & 0 \end{bmatrix} - \begin{bmatrix} 0 & -i \\ i & 0 \end{bmatrix}\begin{bmatrix} 0 & 1 \\ 1 & 0 \end{bmatrix}$$

$$= \begin{bmatrix} i & 0 \\ 0 & -i \end{bmatrix} - \begin{bmatrix} -i & 0 \\ 0 & i \end{bmatrix} = \begin{bmatrix} 2i & 0 \\ 0 & -2i \end{bmatrix} = 2i\begin{bmatrix} 1 & 0 \\ 0 & -1 \end{bmatrix}$$

$$\therefore [\hat{\sigma}_x, \hat{\sigma}_y] = 2i\hat{\sigma}_z$$

Similarly,

$$[\hat{\sigma}_y, \hat{\sigma}_z] = \hat{\sigma}_y\hat{\sigma}_z - \hat{\sigma}_z\hat{\sigma}_y = \begin{bmatrix} 0 & -i \\ i & 0 \end{bmatrix}\begin{bmatrix} 1 & 0 \\ 0 & 1 \end{bmatrix} - \begin{bmatrix} 1 & 0 \\ 0 & 1 \end{bmatrix}\begin{bmatrix} 0 & -i \\ i & 0 \end{bmatrix} = \begin{bmatrix} 0 & i \\ i & 0 \end{bmatrix} - \begin{bmatrix} 0 & -i \\ -i & 0 \end{bmatrix}$$

$$= \begin{bmatrix} 0 & i \\ i & 0 \end{bmatrix} + \begin{bmatrix} 0 & i \\ i & 0 \end{bmatrix} = 2\begin{bmatrix} 0 & i \\ i & 0 \end{bmatrix} = 2i\begin{bmatrix} 0 & 1 \\ 1 & 0 \end{bmatrix}$$

$$\therefore [\hat{\sigma}_y, \hat{\sigma}_z] = 2i\sigma_z$$

Also

$$[\hat{\sigma}_z, \hat{\sigma}_x] = \hat{\sigma}_z\hat{\sigma}_x - \hat{\sigma}_x\hat{\sigma}_z = \begin{bmatrix} 1 & 0 \\ 0 & -1 \end{bmatrix}\begin{bmatrix} 0 & 1 \\ 1 & 0 \end{bmatrix} - \begin{bmatrix} 0 & 1 \\ 1 & 0 \end{bmatrix}\begin{bmatrix} 1 & 0 \\ 0 & -1 \end{bmatrix} = \begin{bmatrix} 0 & 1 \\ -1 & 0 \end{bmatrix} - \begin{bmatrix} 0 & -1 \\ 1 & 0 \end{bmatrix}$$

$$= \begin{bmatrix} 0 & 1 \\ -1 & 0 \end{bmatrix} + \begin{bmatrix} 0 & 1 \\ -1 & 0 \end{bmatrix} = 2\begin{bmatrix} 0 & 1 \\ -1 & 0 \end{bmatrix} = -2\begin{bmatrix} 0 & 1 \\ 1 & 0 \end{bmatrix} = 2i^2\begin{bmatrix} 0 & -1 \\ 1 & 0 \end{bmatrix} = 2i\begin{bmatrix} 0 & -i \\ i & 0 \end{bmatrix} = 2i\hat{\sigma}_y$$

$$\therefore [\hat{\sigma}_z, \hat{\sigma}_x] = 2i\hat{\sigma}_y$$

Commutation relation satisfied by $\hat{\sigma}^2$ and $\hat{\sigma}_x$, $\hat{\sigma}_y$ and $\hat{\sigma}_z$

We have $\hat{\sigma}^2 = \hat{\sigma}_x^2 + \hat{\sigma}_y^2 + \hat{\sigma}_z^2$.

We consider $[\hat{\sigma}^2, \hat{\sigma}_x] = [\hat{\sigma}_x^2 + \hat{\sigma}_y^2 + \sigma_z^2, \hat{\sigma}_x] = [\hat{\sigma}_x^2, \hat{\sigma}_x] + [\hat{\sigma}_y^2, \hat{\sigma}_x] + [\hat{\sigma}_z^2, \hat{\sigma}_x]$.

Again,

$$[\hat{\sigma}_x^2, \hat{\sigma}_x] = \hat{\sigma}_x^3 - \hat{\sigma}_x = 0$$

$$\therefore [\hat{\sigma}^2, \hat{\sigma}_x] = [\hat{\sigma}_y^2, \hat{\sigma}_x] + [\hat{\sigma}_z^2, \hat{\sigma}_x] = [\hat{\sigma}_y\hat{\sigma}_y, \hat{\sigma}_x] + [\hat{\sigma}_z\hat{\sigma}_z, \hat{\sigma}_x]$$

$$= \hat{\sigma}_y[\hat{\sigma}_y, \hat{\sigma}_x] + [\sigma_y, \hat{\sigma}_x]\hat{\sigma}_y + \hat{\sigma}_z[\hat{\sigma}_z, \hat{\sigma}_x] + [\hat{\sigma}_z, \hat{\sigma}_y]\sigma_z \; (\because [ab, c] = a[b, c] + [a, c]b)$$

$$= \hat{\sigma}_y(-2i\hat{\sigma}_z) + (-2i\hat{\sigma}_z)\hat{\sigma}_y + \hat{\sigma}_z(2i\hat{\sigma}_y) + (2i\hat{\sigma}_y)\hat{\sigma}_z$$

$$= -2i\hat{\sigma}_y\hat{\sigma}_z - 2i\hat{\sigma}_z\hat{\sigma}_y + 2i\hat{\sigma}_z\hat{\sigma}_y + 2i\hat{\sigma}_y\hat{\sigma}_z = 0$$

$$\therefore [\hat{\sigma}^2, \hat{\sigma}_x] = 0$$

Likewise, $[\hat{\sigma}^2, \hat{\sigma}_y] = 0$ and $[\hat{\sigma}^2, \hat{\sigma}_z] = 0$.

Hence, $\hat{\sigma}^2$ commutes with each component $\hat{\sigma}_x$, $\hat{\sigma}_y$ and $\hat{\sigma}_z$.

10.9 Two-component wave functions

The Pauli operators $\hat{\sigma}_x$, $\hat{\sigma}_y$, $\hat{\sigma}_z$ are 2×2 matrices. Therefore, Pauli operands must be two-component column matrix. Further, Pauli's theory retains the form of Schrödinger theory, so Pauli's operand must retain the form of Schrödinger operand.

Hence, the wave function must be the function of (x, y, z) or (r, θ, ϕ). Therefore, the Pauli's operand ψ can be expressed in the form:

$$\psi = \begin{pmatrix} \psi_1(x, y, z) \\ \psi_2(x, y, z) \end{pmatrix} \text{ or, } \psi = \begin{pmatrix} \psi_1 \\ \psi_2 \end{pmatrix} \tag{10.39}$$

If at least one of the components of Pauli's wave function ψ is non-zero, then ψ is called *well-behaved*. In view of Schrödinger theory, each of the non-zero component of ψ is well-behaved.

The complex conjugate of ψ is $\psi^* = (\psi_1^* \; \psi_2^*)$.

So

$$\psi^* \psi = (\psi_1^* \; \psi_2^*)\begin{pmatrix}\psi_1 \\ \psi_2\end{pmatrix} = \psi_1^* \psi_1 + \psi_2^* \psi_2 \tag{10.40}$$

ψ is said to be normalised, if

$$\int_\tau \psi^* \psi \, d\tau = 1 \text{ or, } \int_\tau (\psi_1^* \psi_1 + \psi_2^* \psi_2)\, d\tau = 1 \tag{10.41}$$

The integration is over all space.

10.10 Properties of Pauli operators

The Pauli operators have the following important properties:

(a) The Pauli operators are added together by the standard rule of matrix addition:

$$L_z + S_z = L_z + \frac{1}{2}\hbar\sigma_z = \frac{\hbar}{i}\frac{\partial}{\partial\phi}\begin{bmatrix}1 & 0\\ 0 & 1\end{bmatrix} + \frac{\hbar}{2}\begin{bmatrix}1 & 0\\ 0 & -1\end{bmatrix} = \begin{bmatrix}\left(\frac{\hbar}{i}\frac{\partial}{\partial\phi} + \frac{\hbar}{2}\right) & 0\\ 0 & \left(\frac{\hbar}{i}\frac{\partial}{\partial\phi} - \frac{\hbar}{2}\right)\end{bmatrix}$$

$$= \hbar\begin{bmatrix}\left(-i\frac{\partial}{\partial\phi} + \frac{1}{2}\right) & 0\\ 0 & \left(i\frac{\partial}{\partial\phi} - \frac{1}{2}\right)\end{bmatrix} \tag{10.42}$$

(b) Pauli's operators follow the standard rule of matrix multiplication:

$$p_x S_x = \begin{bmatrix}p_x & 0\\ 0 & p_x\end{bmatrix}\frac{1}{2}\hbar\begin{bmatrix}1 & 0\\ 0 & -1\end{bmatrix}\left(\because S_x = \frac{1}{2}\hbar\sigma_x = \frac{\hbar}{2}\begin{bmatrix}1 & 0\\ 0 & -1\end{bmatrix}\right)$$

$$= \frac{\hbar}{2}\begin{bmatrix}0 & p_x\\ p_x & 0\end{bmatrix} = \frac{\hbar}{2}\begin{bmatrix}0 & \frac{\hbar}{i}\frac{\partial}{\partial x}\\ \frac{\hbar}{i}\frac{\partial}{\partial x} & 0\end{bmatrix} = \frac{1}{2}\frac{\hbar}{i}\begin{bmatrix}0 & \frac{\partial}{\partial x}\\ \frac{\partial}{\partial x} & 0\end{bmatrix} \tag{10.43}$$

(c) In the multiplication of Pauli matrices with non-commuting elements, the order of the matrix factors and that of the element factors must be preserved, e.g.,

$$x p_x = \begin{bmatrix}x & 0\\ 0 & x\end{bmatrix}\begin{bmatrix}p_x & 0\\ 0 & p_x\end{bmatrix} = \begin{bmatrix}x & 0\\ 0 & x\end{bmatrix}\begin{bmatrix}\frac{\hbar}{i}\frac{\partial}{\partial x} & 0\\ 0 & \frac{\hbar}{i}\frac{\partial}{\partial x}\end{bmatrix} = \frac{\hbar}{i}\begin{bmatrix}x\frac{\partial}{\partial x} & 0\\ 0 & x\frac{\partial}{\partial x}\end{bmatrix} \tag{10.44}$$

$$\text{and } p_x x = \begin{bmatrix}p_x & 0\\ 0 & p_x\end{bmatrix}\begin{bmatrix}x & 0\\ 0 & x\end{bmatrix} = \begin{bmatrix}\frac{\hbar}{i}\frac{\partial}{\partial x} & 0\\ 0 & \frac{\hbar}{i}\frac{\partial}{\partial x}\end{bmatrix}\begin{bmatrix}x & 0\\ 0 & x\end{bmatrix} = \frac{\hbar}{i}\begin{bmatrix}\frac{\partial}{\partial x}x & 0\\ 0 & \frac{\partial}{\partial x}x\end{bmatrix} \tag{10.45}$$

(d) The Pauli operators operate on the two-component wave functions according to standard rule, e.g.,

$$\sigma_x \begin{bmatrix} \psi_1 \\ \psi_2 \end{bmatrix} = \begin{bmatrix} 0 & 1 \\ 1 & 0 \end{bmatrix} \begin{bmatrix} \psi_1 \\ \psi_2 \end{bmatrix} = \begin{bmatrix} \psi_2 \\ \psi_1 \end{bmatrix} \tag{10.46}$$

$$\sigma_y \begin{bmatrix} \psi_1 \\ \psi_2 \end{bmatrix} = \begin{bmatrix} 0 & -i \\ i & 0 \end{bmatrix} \begin{bmatrix} \psi_1 \\ \psi_2 \end{bmatrix} = \begin{bmatrix} -i\psi_2 \\ i\psi_1 \end{bmatrix} \tag{10.47}$$

$$\sigma_z \begin{bmatrix} \psi_1 \\ \psi_2 \end{bmatrix} = \begin{bmatrix} 1 & 0 \\ 0 & -1 \end{bmatrix} \begin{bmatrix} \psi_1 \\ \psi_2 \end{bmatrix} = \begin{bmatrix} \psi_1 \\ -\psi_2 \end{bmatrix} \tag{10.48}$$

Likewise

$$S_x \begin{bmatrix} \psi_1 \\ \psi_2 \end{bmatrix} = \frac{1}{2}\hbar\sigma_x \begin{bmatrix} \psi_1 \\ \psi_2 \end{bmatrix} = \frac{1}{2}\hbar \begin{bmatrix} 0 & 1 \\ 1 & 0 \end{bmatrix} \begin{bmatrix} \psi_1 \\ \psi_2 \end{bmatrix} = \frac{\hbar}{2} \begin{bmatrix} \psi_2 \\ \psi_1 \end{bmatrix}$$

$$S_y \begin{bmatrix} \psi_1 \\ \psi_2 \end{bmatrix} = \frac{1}{2}\hbar\sigma_y = \frac{1}{2}\hbar \begin{bmatrix} 0 & -i \\ i & 0 \end{bmatrix} \begin{bmatrix} \psi_1 \\ \psi_2 \end{bmatrix} = \frac{\hbar}{2} \begin{bmatrix} -i\psi_2 \\ i\psi_1 \end{bmatrix}$$

$$\text{and } S_z \begin{bmatrix} \psi_1 \\ \psi_2 \end{bmatrix} = \frac{1}{2}\hbar\sigma_z = \frac{1}{2}\hbar \begin{bmatrix} 1 & 0 \\ 0 & -1 \end{bmatrix} \begin{bmatrix} \psi_1 \\ \psi_2 \end{bmatrix} = \frac{\hbar}{2} \begin{bmatrix} \psi_1 \\ -\psi_2 \end{bmatrix}$$

Also

$$p_x\psi = \begin{bmatrix} p_x & 0 \\ 0 & p_x \end{bmatrix} \begin{bmatrix} \psi_1 \\ \psi_2 \end{bmatrix} = \begin{bmatrix} \frac{\hbar}{i}\frac{\partial}{\partial x} & 0 \\ 0 & \frac{\hbar}{i}\frac{\partial}{\partial x} \end{bmatrix} \begin{bmatrix} \psi_1 \\ \psi_2 \end{bmatrix} = \frac{\hbar}{i} \begin{bmatrix} \frac{\partial \psi_1}{\partial x} \\ \frac{\partial \psi_2}{\partial x} \end{bmatrix}$$

A Pauli operator can be written in various forms, e.g.

$$p_x \begin{bmatrix} p_x & 0 \\ 0 & p_x \end{bmatrix} \begin{bmatrix} \frac{\hbar}{i}\frac{\partial}{\partial x} & 0 \\ 0 & \frac{\hbar}{i}\frac{\partial}{\partial x} \end{bmatrix} \text{ or, } \frac{\hbar}{i} \begin{bmatrix} \frac{\partial}{\partial x} & 0 \\ 0 & \frac{\partial}{\partial x} \end{bmatrix} \tag{10.49}$$

10.11 Pauli eigenvalues and eigenfunctions

According to Pauli theory, the eigenvalue equation of the operator σ_x is given by

$$\sigma_x\psi = \lambda\psi \tag{10.50}$$

where ψ is an eigenfunction of σ_x and λ is the eigenvalue of σ_x.

Equation (10.47) can be written as

$$\begin{pmatrix} 1 & 0 \\ 0 & 1 \end{pmatrix} \begin{bmatrix} \psi_1 \\ \psi_2 \end{bmatrix} = \lambda \begin{bmatrix} \psi_1 \\ \psi_2 \end{bmatrix} \left[\because \psi = \begin{pmatrix} \psi_1 \\ \psi_2 \end{pmatrix}, \sigma_x = \begin{pmatrix} 0 & 1 \\ 1 & 0 \end{pmatrix} \right]$$

$$\text{or, } \begin{bmatrix} \psi_1 \\ \psi_2 \end{bmatrix} = \begin{bmatrix} \lambda\psi_1 \\ \lambda\psi_2 \end{bmatrix} \tag{10.51}$$

We thus get $\lambda\psi_1 = \psi_1$ and $\lambda\psi_2 = \psi_2$.

From the above equations, we get $\lambda^2 = 1$ or, $\lambda = \pm 1$.

So σ_x has two eigenvalues +1 and −1. This is true for σ_y and σ_z.

In the present case, the eigenfunctions are $(2s+1)$ row and one column matrices. For electrons $s = \frac{1}{2}$, so the electron eigenfunction matrices have 2 rows and 1 column.

The normalised wave functions are given by

$$\psi_x\left(\frac{1}{2}\right) = \frac{1}{\sqrt{2}}\begin{bmatrix}1\\1\end{bmatrix} \text{ and } \psi_x\left(-\frac{1}{2}\right) = \begin{bmatrix}1\\-1\end{bmatrix} \tag{10.52}$$

$$\psi_y\left(\frac{1}{2}\right) = \frac{1}{\sqrt{2}}\begin{bmatrix}1\\i\end{bmatrix} \text{ and } \psi_y\left(-\frac{1}{2}\right) = \frac{1}{\sqrt{2}}\begin{bmatrix}1\\-i\end{bmatrix} \tag{10.53}$$

$$\psi_z\left(\frac{1}{2}\right) = \frac{1}{\sqrt{2}}\begin{bmatrix}1\\0\end{bmatrix} \text{ and } \psi_z\left(-\frac{1}{2}\right) = \frac{1}{\sqrt{2}}\begin{bmatrix}1\\0\end{bmatrix} \tag{10.54}$$

We can express two-component ψ in the form:

$$\psi = \psi_1\alpha + \psi_2\beta = \psi_1\begin{bmatrix}1\\0\end{bmatrix} + \psi_2\begin{bmatrix}0\\1\end{bmatrix} \tag{10.55}$$

Here, $\psi_1 = \psi_1(\vec{r})$ and $\psi_2 = \psi_2(\vec{r})$ are two functions of x, y, z only. The spin eigenvectors are

$$\alpha = \begin{bmatrix}1\\0\end{bmatrix}, \beta = \begin{bmatrix}0\\1\end{bmatrix} \tag{10.56}$$

α and β represent two spin states having values $+\frac{\hbar}{2}$ and $-\frac{\hbar}{2}$ of the z-component of the spin vector and are called *spinors*.

Now,

$$\hat{S}_z\alpha = \frac{\hbar}{2}\begin{bmatrix}1 & 0\\0 & -1\end{bmatrix}\begin{bmatrix}1\\0\end{bmatrix} = \frac{\hbar}{2}\begin{bmatrix}1\\0\end{bmatrix} = \frac{\hbar}{2}\alpha$$

$$\hat{S}_z\beta = \frac{\hbar}{2}\begin{bmatrix}1 & 0\\0 & -1\end{bmatrix}\begin{bmatrix}0\\1\end{bmatrix} = \frac{\hbar}{2}\begin{bmatrix}0\\-1\end{bmatrix} = -\frac{\hbar}{2}\begin{bmatrix}0\\1\end{bmatrix} = -\frac{\hbar}{2}\beta \tag{10.57}$$

Hence, α and β represent the eigenvectors of $\hat{S}_z$ for the two eigenstates having eigenvalues $\hbar/2$ (spin up) and $-\hbar/2$ (spin down) respectively. For state α the spin quantum number $s = 1/2$ and spin magnetic quantum number $m_s = +1/2$. For state β, $s = 1/2$ and $m_s = -1/2$.

Now, by operating on spinors α and β by $\hat{\sigma}_x$, $\hat{\sigma}_y$ and $\hat{\sigma}_z$, we get

$$\hat{\sigma}_x\alpha = \begin{bmatrix}0 & 1\\1 & 0\end{bmatrix}\begin{bmatrix}1\\0\end{bmatrix} = \begin{bmatrix}0\\1\end{bmatrix} = \beta \quad \hat{\sigma}_y\alpha = \begin{bmatrix}0 & -i\\i & 0\end{bmatrix}\begin{bmatrix}1\\0\end{bmatrix} = \begin{bmatrix}0\\i\end{bmatrix} = i\begin{bmatrix}0\\1\end{bmatrix} = i\beta \quad \hat{\sigma}_z\alpha = \begin{bmatrix}1 & 0\\0 & -1\end{bmatrix}\begin{bmatrix}1\\0\end{bmatrix} = \begin{bmatrix}1\\0\end{bmatrix} = \alpha$$

$$\hat{\sigma}_x\beta = \begin{bmatrix}0 & 1\\1 & 0\end{bmatrix}\begin{bmatrix}0\\1\end{bmatrix} = \begin{bmatrix}1\\0\end{bmatrix} = \alpha \quad \hat{\sigma}_y\beta = \begin{bmatrix}0 & -i\\i & 0\end{bmatrix}\begin{bmatrix}0\\1\end{bmatrix} = \begin{bmatrix}-i\\0\end{bmatrix} = -i\begin{bmatrix}1\\0\end{bmatrix} = -i\beta \quad \hat{\sigma}_z\beta = \begin{bmatrix}1 & 0\\0 & -1\end{bmatrix}\begin{bmatrix}0\\1\end{bmatrix} = \begin{bmatrix}0\\-1\end{bmatrix} = -\begin{bmatrix}0\\1\end{bmatrix} = -\beta$$

We can show that $\hat{\sigma}_x^2$, $\hat{\sigma}_y^2$ and $\hat{\sigma}_z^2$ are unit matrices. So $\hat{\sigma}_x^2 = \hat{\sigma}_y^2 = \hat{\sigma}_z^2 = \begin{bmatrix}1 & 0\\0 & 1\end{bmatrix}$.

Then $S_x^2 = S_y^2 = S_z^2 = \frac{\hbar^2}{4}\begin{bmatrix} 1 & 0 \\ 0 & 1 \end{bmatrix}$ and $S^2 = S_x^2 + S_y^2 + S_z^2 = \frac{3\hbar^2}{4}\begin{bmatrix} 1 & 0 \\ 0 & 1 \end{bmatrix}$

Again, operating α and β by S^2, we get

$$S^2\alpha = \frac{3\hbar^2}{4}\begin{pmatrix} 1 & 0 \\ 0 & 1 \end{pmatrix}\begin{pmatrix} 1 \\ 0 \end{pmatrix} = \frac{3\hbar^2}{4}\begin{pmatrix} 1 \\ 0 \end{pmatrix} = \frac{3\hbar^2}{4}\alpha \tag{10.58}$$

$$\text{and } S^2\beta = \frac{3\hbar^2}{4}\begin{pmatrix} 1 & 0 \\ 0 & 1 \end{pmatrix}\begin{pmatrix} 0 \\ 1 \end{pmatrix} = \frac{3\hbar^2}{4}\begin{pmatrix} 0 \\ 1 \end{pmatrix} = \frac{3\hbar^2}{4}\beta \tag{10.59}$$

Hence, both α and β are eigenvectors of S^2 having the same eigenvalue $3\hbar^2/4$.

Now,

$$\frac{3\hbar^2}{4} = \hbar^2 \times \tfrac{1}{2} \times \tfrac{3}{2} = \hbar^2 \times \tfrac{1}{2}\left(\tfrac{1}{2}+1\right) = s(s+1)\hbar^2 \quad (\because s = \tfrac{1}{2}) \tag{10.60}$$

It is to be noted that α *is the simultaneous eigenfunction of* S^2 *and* S_z *belonging to the eigenvalues* $3\hbar^2/4$ *and* $\hbar/2$ *respectively.*

Similarly, β is the simultaneous eigenfunction of S^2 and S_z belonging to respective eigenvalues $3\hbar^2/4$ and $-\hbar^2/2$.

It is to note that *relativistic quantum mechanics developed by P.A.M. Dirac,* however established the ad hoc assumption regarding the existence of electron spin by *Uhlenbeck and Goudsmit. For relativistic particles like electron, Dirac theory is applicable, but for non-relativistic particles like neutrons and protons within the nucleus (spin 1/2) Pauli's theory is not applicable.*

10.12 The total angular momentum operator

The total angular momentum of a particle is given by

$$\vec{J} = \vec{L} + \vec{S} \tag{10.61}$$

where $\vec{L}$ is the orbital angular momentum and $\vec{S}$ is the spin angular momentum. From classical point of view, this generalised angular momentum is defined as

$$\vec{J} = \vec{r} \times \vec{p} \tag{10.62}$$

and the x-, y- and z-components of $\vec{J}$ are given by J_x, J_y and J_z, where

$$\left.\begin{aligned} J_x &= \frac{\hbar}{i}\left(y\frac{\partial}{\partial z} - z\frac{\partial}{\partial y}\right) \\ J_y &= \frac{\hbar}{i}\left(z\frac{\partial}{\partial x} - x\frac{\partial}{\partial z}\right) \\ \text{and } J_z &= \frac{\hbar}{i}\left(x\frac{\partial}{\partial y} - y\frac{\partial}{\partial x}\right) \end{aligned}\right\} \tag{10.63}$$

10.12.1 Commutation relations followed by J_x, J_y and J_z

We have $[J_x, J_y] = [J_x J_y - J_y J_x]$.

Now,

$$J_xJ_y = \frac{\hbar}{i}\left(y\frac{\partial}{\partial z} - z\frac{\partial}{\partial y}\right)\frac{\hbar}{i}\left(z\frac{\partial}{\partial x} - x\frac{\partial}{\partial z}\right)$$

$$= \frac{\hbar^2}{i^2}\left\{y\frac{\partial}{\partial z}\left(z\frac{\partial}{\partial x}\right) - y\frac{\partial}{\partial z}\left(x\frac{\partial}{\partial z}\right) - z\frac{\partial}{\partial y}\left(z\frac{\partial}{\partial x}\right) + z\frac{\partial}{\partial y}\left(x\frac{\partial}{\partial z}\right)\right\}$$

$$= -\hbar^2\left\{y\frac{\partial}{\partial x} + yz\frac{\partial^2}{\partial z\partial x} - yx\frac{\partial^2}{\partial z^2} - z^2\frac{\partial^2}{\partial y\partial z} + zx\frac{\partial^2}{\partial y\partial z}\right\} \tag{10.64}$$

and $J_yJ_x = \frac{\hbar}{i}\left(z\frac{\partial}{\partial x} - x\frac{\partial}{\partial z}\right)\frac{\hbar}{i}\left(y\frac{\partial}{\partial z} - z\frac{\partial}{\partial y}\right)$

$$= -\hbar^2\left\{z\frac{\partial}{\partial x}\left(y\frac{\partial}{\partial z}\right) - z\frac{\partial}{\partial x}\left(z\frac{\partial}{\partial y}\right) - x\frac{\partial}{\partial z}\left(y\frac{\partial}{\partial z}\right) + x\frac{\partial}{\partial z}\left(z\frac{\partial}{\partial y}\right)\right\}$$

$$= -\hbar^2\left\{zy\frac{\partial^2}{\partial x\partial z} - z^2\frac{\partial^2}{\partial x\partial y} - xy\frac{\partial^2}{\partial z^2} + xz\frac{\partial^2}{\partial z\partial y}\right\} \tag{10.65}$$

Subtracting eq. (10.65) from eq. (10.64), we get

$$[J_x, J_y] = -\hbar^2\left(y\frac{\partial}{\partial x} - x\frac{\partial}{\partial y}\right) = i\hbar\left\{\frac{\hbar}{i}\left(x\frac{\partial}{\partial y} - y\frac{\partial}{\partial x}\right)\right\} = i\hbar J_z \tag{10.66}$$

Similarly, it can be proved that

$$[J_y, J_z] = i\hbar J_x \text{ and } [J_z, J_x] = i\hbar J_y$$

$$[J_y, J_x] = -i\hbar J_z,\ [J_z, J_y] = -i\hbar J_x \text{ and } [J_x, J_z] = -i\hbar J_y$$

$$\text{and } [J_x, J_x] = [J_y, J_y] = [J_z, J_z] = 0 \tag{10.67}$$

10.12.2 Commutation relation of total angular momentum (J) with components (J_x, J_y, J_z)

We have

$$J^2 = J_x^2 + J_y^2 + J_z^2 \tag{10.68}$$

$$\therefore [J^2, J_x] = [J_x^2 + J_y^2 + J_z^2; J_x] = (J_x^2 + J_y^2 + J_z^2)J_x - J_x(J_x^2 + J_y^2 + J_z^2)$$

$$= [J_x^2J_x + J_y^2J_x + J_z^2J_x] - [J_xJ_x^2 + J_xJ_y^2 + J_xJ_z^2] = [J_y^2J_x - J_xJ_y^2] + [J_z^2J_x - J_xJ_z^2]$$

$$= [J_y^2, J_x] + [J_z^2, J_x]$$

But,

$$[ab, c] = a[b, c] + [a, c]b$$

$$\therefore [J^2, J_x] = [J_yJ_y, J_x] + [J_zJ_z, J_x] = J_y[J_y, J_x] + [J_y, J_x]J_y + J_z[J_z, J_x] + [J_z, J_x]J_z$$

$$= J_y(-i\hbar J_z) + (-i\hbar J_z)J_y + J_z(i\hbar J_y) + (i\hbar J_y)J_z$$

$$\therefore [J^2, J_x] = 0 \tag{10.69}$$

Similarly,

$$[J^2, J_y] = 0 \tag{10.70}$$

$$\text{and } [J^2, J_z] = 0 \tag{10.71}$$

10.12.3 Commutation relation of J_z with J_+ and J_-

Let $J_+ = J_x + iJ_y$ and $J_- = J_x - iJ_y$, where J_+ and J_- so defined are called *ladder operators.*

Now,

$$\begin{aligned}[J_z, J_+] &= J_zJ_+ - J_+J_z = J_z\left(J_x + iJ_y\right) - \left(J_x + iJ_y\right)J_z = J_zJ_x - J_xJ_z - i\left(J_yJ_z - J_zJ_y\right)\\ &= [J_z, J_x] - i\left[J_y, J_z\right] = i\hbar J_y - i(i\hbar J_y)\ (\because [J_z, J_x] = i\hbar J_y \text{ and } \left[J_y, J_z\right] = i\hbar J_x)\\ &= i\hbar J_y + \hbar J_x = \hbar\left(J_x + iJ_y\right)\end{aligned}$$

$$\therefore [J_z, J_+] = \hbar J_+ \tag{10.72}$$

Similarly,

$$\begin{aligned}[J_z, J_-] &= J_zJ_- - J_-J_z = J_z\left(J_x - iJ_y\right) - \left(J_x - iJ_y\right)J_z\\ &= J_zJ_x - J_xJ_z - i\left(J_zJ_y - J_yJ_z\right) = [J_z, J_x] - i\left[J_z, J_y\right]\\ &= i\hbar J_y - i\left(-i\hbar J_x\right) = -\hbar\left(J_x - iJ_y\right)\end{aligned}$$

$$\therefore [J_z, J_-] = -\hbar J_- \tag{10.73}$$

10.12.4 Commutation relation of J_+ and J_-

We have

$$\begin{aligned}[J_+, J_-] &= J_+J_- - J_-J_+ = \left(J_x + iJ_y\right)\left(J_x - iJ_y\right) - \left(J_x - iJ_y\right)\left(J_x + iJ_y\right)\\ &= J_x\left(J_x - iJ_y\right) + iJ_y\left(J_x - iJ_y\right) - J_x\left(J_x + iJ_y\right) + iJ_y\left(J_x + iJ_y\right)\\ &= J_x^2 - iJ_xJ_y + iJ_yJ_x - i^2J_y - J_x^2 - iJ_xJ_y + iJ_yJ_x - i^2J_y^2\\ &= 2i\left(J_yJ_x - J_xJ_y\right) = 2i\left[J_y, J_x\right] = 2i\left(-i\hbar J_z\right)\end{aligned}$$

$$\therefore [J_+, J_-] = 2\hbar J_z \tag{10.74}$$

10.12.5 Commutation relation of J^2 with J_+ and J_-

We have

$$\begin{aligned}J_+J_- &= \left(J_x + iJ_y\right)\left(J_x - iJ_y\right) = J_x^2 + J_y^2 - i\left(J_xJ_y - J_yJ_x\right)\\ \text{and } J_-J_+ &= \left(J_x - iJ_y\right)\left(J_x + iJ_y\right) = J_x^2 + J_y^2 + i\left(J_xJ_y - J_yJ_x\right)\end{aligned}$$

So

$$\tfrac{1}{2}\left(J_+J_- + J_-J_+\right) = J_x^2 + J_y^2$$

$$\therefore J^2 = J_x^2 + J_y^2 + J_z^2 = \tfrac{1}{2}\left(J_+J_- + J_-J_+\right) + J_z^2 \tag{10.75}$$

Hence,

$$\begin{aligned}\left[J^2, J_+\right] &= \left[\tfrac{1}{2}\left(J_+J_- + J_-J_+\right) + J_z^2;\ J_+\right] = \tfrac{1}{2}[J_+J_-, J_+] + \tfrac{1}{2}[J_-J_+, J_+] + \left[J_z^2, J_+\right]\\ &= \tfrac{1}{2}J_+[J_-, J_+] + \tfrac{1}{2}[J_+, J_+]J_- + \tfrac{1}{2}J_-[J_+, J_+] + \tfrac{1}{2}[J_-, J_+]J_+ + J_z[J_z, J_+] + [J_z, J_+]J_z\\ &\qquad (\text{using the identity } [ab,\ c] = a[b,\ c] + [a,\ c]b)\\ &= \tfrac{1}{2}J_+\left(-2\hbar J_z\right) + 0 + 0 + \tfrac{1}{2}\left(-2\hbar J_z\right)J_+ + J_z\left(\hbar J_+\right) + \hbar J_+J_z = 0\end{aligned} \tag{10.76}$$

Similarly,

$$\left[J^2, J_-\right] = 0 \tag{10.77}$$

10.13 Eigenvalues of J^2 and J_z

If b be the eigenvalue and $|\psi\rangle$ be an eigenfunction corresponding to the operator J_z, then the eigenvalue equation is

$$J_z|\psi\rangle = b|\psi\rangle \tag{10.78}$$

Now, if $J_+|\psi\rangle = |\phi\rangle$, then $J_z|\phi\rangle = J_zJ_+|\psi\rangle$.

But, $[J_z, J_+] = J_zJ_+ - J_+J_z$, or, $J_zJ_+ = [J_z, J_+] + J_+J_z$.

$$\therefore J_z|\phi\rangle = \{[J_z, J_+] + J_+J_z\}|\psi\rangle = \{[J_z, J_+]|\psi\rangle + J_+J_z\}|\psi\rangle$$

$$= \hbar J_+|\psi\rangle + J_+J_z|\psi\rangle \; [\because J_z, J_+ = \hbar J_+]$$

$$\text{or, } J_z|\phi\rangle = (\hbar J_+ + J_+J_z)|\psi\rangle$$

Again,

$$J_z|\phi\rangle = \hbar J_+|\psi\rangle + J_+J_z|\psi\rangle = \hbar J_+|\psi\rangle + bJ_+|\psi\rangle \text{or, } J_z|\phi\rangle = (b+\hbar)|\phi\rangle \tag{10.79}$$

If $|L\rangle$ represents a new state function such that $J_+|\phi\rangle = |L\rangle$

$$\therefore J_z|L\rangle = J_zJ_+|\phi\rangle = \{[J_z, J_+] + J_+J_z\}|\phi\rangle = [J_z, J_+]|\phi\rangle + J_+J_z|\phi\rangle = \hbar J_+|\phi\rangle + J_+J_z|\phi\rangle$$

$$= \hbar J_+|\phi\rangle + (b+\hbar)J_+|\phi\rangle = (b+2\hbar)J_+|\phi\rangle$$

$$\text{or, } J_z|L\rangle = (b+2\hbar)|L\rangle \; [\because J_+|\phi\rangle = |L\rangle] \tag{10.80}$$

This shows that operating $|\psi\rangle$ several times by J_+ we simply get the addition of $\hbar$ to the term b several times.

Hence, we can write:

$$J_zJ_+^n|\psi\rangle = (b+n\hbar)J_+^n|\psi\rangle \tag{10.81}$$

Similarly,

$$J_zJ_-^n|\psi\rangle = (b-n\hbar)J_-^n|\psi\rangle \tag{10.82}$$

Equations (10.81) and (10.82) show that depending on n, J_z has a *discrete set of eigenvalues*, i.e., $(b-l\hbar)$, $\ldots$, $(b-2\hbar)$, $(b-\hbar)$, b, $(b+\hbar)$, $(b+2\hbar)$, $\ldots$, $(b+n\hbar)$ for $n = -l, \ldots, 2, -1, 0, 1, 2, \ldots, n$.

Hence, we note that the eigenvalues of J_z tend from $-\infty$ to $+\infty$, as $n \to -\infty$ to $+\infty$.

Now, if a be an eigenvalue of J^2, then

$$J^2|\psi\rangle = a|\psi\rangle \tag{10.83}$$

where a has a fixed value.

Again, since $[J^2, J_z] = 0$, i.e., J^2 and J_z commute each other, $|\psi\rangle$ will be an eigenfunction for both J^2 and J_z.

Since, J^2 has a fixed eigenvalue a and $J^2 \geqslant J_z^2$, the eigenvalues of J_z must be terminated after a certain terms so that $J^2 \geqslant J_z^2$ holds good.

Hence, we write:

$$\left.\begin{aligned} J_zJ_+^{l+1}|\psi\rangle &= 0 \\ \text{and } J_zJ_-^{k+1}|\psi\rangle &= 0 \end{aligned}\right\} \tag{10.84}$$

i.e., the maximum and minimum values of J_z are l and k respectively. So the eigenvalues are limited to the region $(b - k\hbar), \ldots, (b - \hbar), b, (b + \hbar), \ldots, (b + l\hbar)$ for $n = -k, -1, 0, 1, \ldots, l$.

Again, since $J_+ = J_x + iJ_y$ and $J_- = J_x - iJ_y$.

$$\therefore J_-J_+ = (J_x - iJ_y)(J_x + iJ_y) = J_x^2 + J_y^2 + i[J_x, J_y]$$

$$\text{or, } J_x^2 + J_y^2 = J_-J_+ - i[J_x, J_y] = J_-J_+ + \hbar J_z$$

$\therefore J^2 = J_x^2 + J_y^2 + J_z^2.$

$$J^2 = J_z^2 + J_-J_+ + \hbar J_z \tag{10.85}$$

Also

$$J^2 = J_z^2 + J_+J_- - \hbar J_z \tag{10.86}$$

Now, we are in a position to find out the eigenvalues of J^2 and J_z.

As J^2 and J_z commute each other, i.e., $[J_z, J^2] = 0$, J^2 and J_z must have the same eigenvalue. So if $J_+^l|\psi\rangle$ is an eigenfunction of J_z, it will also be an eigenfunction of J^2.

$$\therefore J^2J_+^l|\psi\rangle = (J_z^2 + J_-J_+ + \hbar J_z)J_+^l|\psi\rangle = J_z^2J_+^l|\psi\rangle + J_-J_+J_+^l|\psi\rangle + \hbar J_zJ_+^l|\psi\rangle$$

$$\text{or, } J^2J_+^l|\psi\rangle = J_zJ_zJ_+^l|\psi\rangle + J_-J_+^{l+1}|\psi\rangle + \hbar J_zJ_+^l|\psi\rangle = J_z(b + l\hbar)J_+^l|\psi\rangle + 0 + \hbar(b + l\hbar)J_+^l|\psi\rangle$$

[by eqs (10.81) and (10.84)]

$$= \{(b + l\hbar)(b + l\hbar) + \hbar(b + l\hbar)\}J_+^l|\psi\rangle$$

$$\text{or, } J^2J_+^l|\psi\rangle = (b + l\hbar)(b + l\hbar + \hbar)J_+^l|\psi\rangle \tag{10.87}$$

Thus, the eigenvalue of J^2 is $(b + l\hbar)(b + l\hbar + \hbar)$.

Similarly,

$$J^2J_-^k|\psi\rangle = (J_x^2 + J_+J_- - \hbar J_z)J_-^k|\psi\rangle = \{(b - k\hbar)(b - k\hbar) + 0 - \hbar(b - k\hbar)\}J_-^k|\psi\rangle$$

$$\text{or, } J^2J_-^k|\psi\rangle = (b - k\hbar)(b - k\hbar - \hbar)J_-^k|\psi\rangle \tag{10.88}$$

From this equation the eigenvalue of J^2 is

$$(b - k\hbar)(b - k\hbar - \hbar)$$

Since, the eigenvalue of J^2 has been assumed to be a, we write:

$$a = (b + l\hbar)(b + l\hbar + h) = (b - k\hbar)(b - k\hbar - \hbar) \tag{10.89}$$

Simplifying, we get

$$b = \frac{\hbar}{2}(k - l) \tag{10.90}$$

Substituting this value of b in eq. (10.86), we have

$$a = \hbar^2\left(\frac{k + l}{2}\right)\left(\frac{k + l}{2} + 1\right)$$

Putting $\frac{k + l}{2} = j$, we get

$$a = j(j + 1)\hbar^2 \tag{10.91}$$

From the relation, $j = \dfrac{k+l}{2}$, we have

$$\left.\begin{aligned} k_{\max} &= 2j;\ l_{\max} = 2j \\ k_{\min} &= 0;\ l_{\min} = 0 \end{aligned}\right\}$$

So when $k = k_{\max}$, $l = l_{\max}$ and $k = k_{\min}$, $l = l_{\min}$,

$\therefore \dfrac{[\hbar(k-l)]}{2}$ have values which lie in

$$-j\hbar, (-j+1)\hbar, \ldots, -\hbar, 0, \hbar, \ldots, (j-1)\hbar, j\hbar \tag{10.92}$$

So J_z has eigenvalues from $-j$ to $+j$, i.e., total values $(2j+1)$.

Putting $b = m_z$, where m_z is the magnetic quantum number and has $(2j+1)$ values,

$$\therefore J^2 = j(j+1)\hbar \text{ and } J_z = m_z\hbar \tag{10.93}$$

Again, the eigenvalues of J_z corresponding to the eigenfunctions J_+ and J_- are respectively $(b+\hbar)$ and $(b-\hbar)$.

$$\therefore (b+\hbar) = m_z\hbar + \hbar = (m_z+1)\hbar$$

$$\text{and } (b-\hbar) = m_z\hbar - \hbar = (m_z-1)\hbar \ [\text{where } J_z = m_z\hbar]$$

Hence, if ψ is an eigenfunction of J_z belonging to eigenvalues m_z, then it follows that J_+ is also an eigenfunction of J_z belonging to eigenvalue $(m \pm 1)\hbar$.

10.14 Eigenvalues of J_+ and J_-

Let a state function be represented by $\psi(r, j, m)$, where j and m are respectively angular momentum and magnetic momentum quantum numbers.

Now, let

$$\psi(r, j, m \pm 1) = J_\pm\, \psi(r, j, m) N_\pm \tag{10.94}$$

where $N_\pm$ is the normalisation constant.

Now,

$$\begin{aligned} \psi^*(r, j, m \pm 1)\, \psi(r, j, m \pm 1) &= N_\pm^* N_\pm J_\pm J_\pm\, \psi^*(r, j, m)\, \psi(r, j, m) \\ &= |N_\pm|^2 \left[J^2 - J_z(J_z \pm \hbar)\right] \psi^*(r, j, m)\, \psi(r, j, m) \end{aligned} \tag{10.95}$$

To apply the condition of normalisation

$$\int \psi^*(r, j, m)\, \psi(r, j, m)\, d\tau = 1$$

Now, we integrate eq. (10.95) for all values of r.

$$\therefore 1 = |N_\pm|^2 \left[j(j+1)\hbar^2 - m_z\hbar(m_z\hbar \pm \hbar)\right] \text{ or, } N_\pm = \frac{e^{i\delta}}{\hbar[\{j(j+1) - m_z(m_z+1)\}]^{1/2}} \tag{10.96}$$

where $e^{i\delta}$ is the phase factor.

Putting $e^{i\delta} = 1$ in eq. (10.96), we get

$$N_\pm = \frac{1}{\hbar[\{j(j+1) - m_z(m_z+1)\}]^{1/2}} \tag{10.97}$$

$\therefore$ eq. (10.91) becomes

$$J_\pm\, \psi(r, j, m) = \frac{1}{N_\pm}\, \psi(r, j, m \pm 1)$$

So

$$J_+\, \psi(r, j, m) = \frac{1}{N_\pm}\, \psi(r, j, m + 1) \tag{10.98}$$

$$\text{and } J_-\, \psi(r, j, m) = \frac{1}{N_\pm}\, \psi(r, j, m - 1) \tag{10.99}$$

The matrix elements of J_+ are

$$\begin{aligned}\langle j'm'|J_+|jm\rangle &= \langle j'm'|jm+1\rangle\, \hbar[(j-m)\,(j+m+1)]^{1/2} \\ &= \hbar[(j-m)\,(j+m+1)]^{1/2}\, \delta_{j,\,j'},\ \delta_{m',\,m+1}\end{aligned} \tag{10.100}$$

Similarly, the matrix elements of J_- are

$$\langle j'm'|J_-|jm\rangle = \hbar[(j+m)\,(j-m+1)]^{1/2}\, \delta_{j,\,j'},\ \delta_{m',\,m-1} \tag{10.101}$$

10.15 Eigenvalues of J_x and J_y

We have $J_+ = J_x + iJ_y$ and $J_- = J_x - iJ_y$.

$$\therefore J_x = \tfrac{1}{2}(J_+ + J_-) \text{ and } J_y = \tfrac{1}{2i}(J_+ - J_-) = -\tfrac{1}{2}i\,(J_+ - J_y) \tag{10.102}$$

$\therefore\ \langle j'm'|J_x|jm\rangle = \langle j'm'|\tfrac{1}{2}(J_+ + J_-)|\,jm\rangle = \tfrac{1}{2}\langle j'm'|(J_+ + J_-)|jm\rangle = \tfrac{1}{2}\langle j'm'|J_+|jm\rangle + \tfrac{1}{2}\langle j'm'|J_-|jm\rangle$

or, $\langle j'm'|J_x|jm\rangle = \tfrac{1}{2}\hbar[\{(j-m)\,(j+m+1)\}]^{1/2}\,\delta_{j,j'}\,\delta_{m',\,m+1} + \tfrac{1}{2}\hbar[(j+m)\,(j-m+1)]^{1/2}\,\delta_{j,j'}\,\delta_{m',\,m-1}$

[by eqs (10.100) and (10.101)] (10.103)

The matrix elements of J_y are

$$\begin{aligned}\langle j'm'|J_y|jm\rangle &= \langle j'm'|-\tfrac{1}{2}i(J_+ - J_-)|\,jm\rangle = -\tfrac{i}{2}\langle j'm'|J_+|jm\rangle + \tfrac{i}{2}\langle j'm'|J_-|jm\rangle \\ &= -\tfrac{i\hbar}{2}[(j-m)\,(j+m+1)]^{1/2}\,\delta_{j',\,j}\,\delta_{m',\,m+1} + \tfrac{i\hbar}{2}[(j+m)\,(j-m+1)]^{1/2}\,\delta_{j',\,j}\,\delta_{m',\,m-1}\end{aligned} \tag{10.104}$$

The simplified forms of eqs (10.103) and (10.104) are as follows:

$$J_x\psi(jm) = \tfrac{1}{2}\hbar[(j-m)\,(j+m+1)]^{1/2}\,\psi(j, m+1) + \tfrac{1}{2}\hbar[(j+m)\,(j-m+1)]^{1/2}\,\psi(j, m-1) \tag{10.105}$$

$$\text{and } J_y\psi(jm) = -\tfrac{i\hbar}{2}[(j-m)\,(j+m+1)]^{1/2}\,\psi(j, m+1) + \tfrac{i\hbar}{2}[(j+m)\,(j-m+1)]^{1/2}\,\psi(j, m-1) \tag{10.106}$$

If $j = j'$, then eq. (10.103) becomes

$$\langle jm'|J_x|jm\rangle = \tfrac{1}{2}\hbar[(j-m)\,(j+m+1)]^{1/2}\,\delta_{j,\,j}\,\delta_{m',\,m+1} + \tfrac{1}{2}\hbar[(j+m)\,(j-m-1)]^{1/2}\,\delta_{j,\,j}\,\delta_{m',\,m-1} \tag{10.107}$$

Similarly, for $j = j'$, eq. (10.104) reduces to

$$\langle jm'|J_y|jm\rangle = -\tfrac{i\hbar}{2}[(j-m)\,(j+m+1)]^{1/2}\,\delta_{j,\,j}\,\delta_{m',\,m+1} + \tfrac{i\hbar}{2}[(j+m)\,(j-m+1)\,\delta_{j,\,j}\,\delta_{m',\,m-1} \tag{10.108}$$

Since, $\delta_{j,j} = 1$

$$\therefore \langle jm'|J_x|jm\rangle = \tfrac{\hbar}{2}[(j-m)(j+m+1)]^{1/2}\delta_{m',m+1} + \tfrac{\hbar}{2}[(j+m)(j-m+1)]^{1/2}\delta_{m',m-1} \quad (10.109)$$

Now,

$$\delta_{m',m+1} = 1, \text{ when } m' = m+1$$
$$\text{and } \delta_{m',m-1} = 1, \text{ when } m' = m-1.$$

So the matrix elements of J_x will be finite.

On the other hand

$$\delta_{m',m+1} = 0, \text{ if } m' \neq m+1$$
$$\text{and } \delta_{m',m-1} = 0, \text{ if } m' \neq m-1.$$

The matrix elements of J_x will vanish.

Again, eq. (10.108) for $\delta_{j,j} = 1$ becomes

$$\langle jm'|J_y|jm\rangle = -\tfrac{i}{2}\hbar[(j-m)(j+m+1)]^{1/2}\,\delta_{m',m+1} + \tfrac{i\hbar}{2}[(j+m)(j-m+1)]^{1/2}\,\delta_{m',m-1} \quad (10.110)$$

But, for $m' \neq m \pm 1$, the matrix elements will vanish.

If $m' = m \pm 1$, then the matrix elements will have finite values.

So for $m' = m+1$

$$\langle j, m+1|J_x|jm\rangle = \tfrac{\hbar}{2}[(j-m)(j+m+1)]^{1/2} \quad (10.111)$$

and for $m' = m-1$

$$\langle j, m-1|J_x|jm\rangle = \tfrac{\hbar}{2}[(j+m)(j-m+1)]^{1/2} \quad (10.112)$$

Likewise, the matrix elements of J_y are as below:

For $m' = m+1$

$$\langle j, m+1|J_y|jm\rangle = -\tfrac{i\hbar}{2}[(j-m)(j+m+1)]^{1/2} \quad (10.113)$$

and for $m' = m-1$

$$\langle j, m-1|J_y|jm\rangle = \tfrac{i\hbar}{2}[(j+m)(j-m+1)]^{1/2} \quad (10.114)$$

10.16 Angular momentum matrices

The matrices of J^2 and J_z are $J_z = m\hbar$ and $J^2 = j(j+1)\hbar^2$.

As m varies from $-j$ to $+j$, it has $(2j+1)$ values. So the dimension of the diagonal matrices will be $(2j+1)$. This means the matrices have $(2j+1)$ rows and $(2j+1)$ columns. The angular momentum matrices are explicitly shown below:

$$J_z = \hbar \begin{vmatrix} j & 0 & \cdots\cdots\cdots & 0 \\ 0 & j-1 & \cdots\cdots\cdots & 0 \\ \cdots & \cdots & \cdots\cdots\cdots & \cdots \\ \cdots & \cdots & \cdots\cdots\cdots & \cdots \\ 0 & 0 & \cdots\cdots\cdots & -j \end{vmatrix} \tag{10.115}$$

$$\text{and } J^2 = \hbar^2 \begin{vmatrix} j(j+1) & \cdots & \cdots\cdots & 0 \\ 0 & j(j+1) & \cdots\cdots & 0 \\ \cdots & \cdots & \cdots\cdots & \cdots \\ \cdots & \cdots & \cdots\cdots & \cdots \\ 0 & \cdots & \cdots\cdots & j(j+1) \end{vmatrix} \tag{10.116}$$

Each of J_z and J^2 matrices has the form:

j	m \ j':	0	$\frac{1}{2}$	$\frac{1}{2}$	
	m':	0	$\frac{1}{2}$, $-\frac{1}{2}$	$\frac{1}{2}$, $-\frac{1}{2}$	
0	0	(Element present)	(0)	(0)	...
$\frac{1}{2}$	$\frac{1}{2}$, $-\frac{1}{2}$	(0)	(Element present)	(0)	...
1	1, 0, -1	(0)	(0)	(Element present)	...

Hence, J_z and J^2 matrices contain only principal diagonals.

The matrix elements of J_+ are given by

$$\langle j, m+1|J_+|jm\rangle = \hbar[(j-m)\,(j+m+1)]^{1/2} \tag{10.117}$$

for $\delta_{j,j} = 1$ and $\delta_{m+1,m+1} = 1$.

All other elements of J_+ will be zero.

For $m' = m - 1$, $j = j'$ the matrix elements of J_- are

$$\langle j, m-1|J_-|jm\rangle = \hbar[(j+m)\,(j-m+1)]^{1/2} \tag{10.118}$$

for $\delta_{j,j} = 1$ and $\delta_{m-1,m-1} = 1$.

All the other elements of J_- will be zero.

So J_+ will be of the type:

$m \backslash m'$	0	1	2	3	$\cdots$
m					$\cdots$
0	0	(Element present)	0	0	$\cdots$
1	0	0	(Element present)	0	$\cdots$
$J_+ \equiv 2$	0	0	0	(Element present)	$\cdots$
3	$\vdots$				

This shows that if

$$m = 0,\ m' = m + 1 = 1, \text{ element will be present}$$
$$m = 1,\ m' = m + 1 = 2, \text{ element will be present}$$
$$\text{and } m = 2,\ m' = m + 1 = 3, \text{ element will be present}$$
$$\cdots\cdots\cdots\cdots\cdots\cdots\cdots\cdots$$

All other elements of the matrix will be zero. These will be only upper diagonal for the J_+ matrix.

Likewise, the form of J_- is as below:

$m \backslash m'$	0	1	2	0	$\cdots$
m					
0	0	0	0	0	$\cdots$
1	(Element present)	0	0	0	$\cdots$
$J_+ \equiv 2$	0	(Element present)	0	0	$\cdots$
3	0	(Element present)	0	0	$\cdots$
$\vdots$	$\cdots$				

i.e., if $m = 1,\ m' = 1 - 1 = 0,$ element present

$m = 2,\ m' = 2 - 1 = 1,$ element present

and $m = 3,\ m' = 3 - 1 = 2,$ element present

......, ..., ...

All other elements of the matrix will be zero.

$\therefore J_-$ matrix contains only the lower diagonal.

Examples

1. For $j = \frac{1}{2}$, $m = \frac{1}{2}, -\frac{1}{2}$

$$J^2 = \hbar^2 j(j+1) = \tfrac{1}{2}\hbar^2\left(\tfrac{1}{2}+1\right) = \tfrac{3}{4}\hbar^2$$

J^2 matrix is of dimension $(2j+1) = (2 \times \frac{1}{2} + 1) = 2$.

Hence, J^2 matrix will be of two rows and two columns.

$$J^2 = \hbar^2 \begin{pmatrix} \frac{3}{4} & 0 \\ 0 & \frac{3}{4} \end{pmatrix} \text{ and } J_z = m_z\hbar = \begin{cases} \frac{1}{2}\hbar, & \text{for } m_z = \frac{1}{2} \\ \frac{1}{2}\hbar, & \text{for } m_z = -\frac{1}{2} \end{cases}$$

$$\therefore J_z = \hbar \begin{pmatrix} \frac{1}{2} & 0 \\ 0 & -\frac{1}{2} \end{pmatrix}$$

and $J_+ = \hbar[(j-m)(j+m+1)]^{1/2}$ [All elements are zero except those of upper diagonal]

$= \hbar[(\frac{1}{2}-\frac{1}{2})(\frac{1}{2}+\frac{1}{2}+1)]^{1/2}$, for $j = \frac{1}{2}$, $m = \frac{1}{2}$

$= 0$

Again, $J_+ = \hbar[(\frac{1}{2}+\frac{1}{2})(\frac{1}{2}-\frac{1}{2}+1)]^{1/2}$, for $j = \frac{1}{2}$, $m = -\frac{1}{2} = \hbar$

$$\therefore J_+ = \hbar \begin{bmatrix} 0 & 1 \\ 0 & 0 \end{bmatrix}$$

Also $J_- = \hbar[(j+m)(j-m+1)]^{1/2}$ [All elements except those of lower diagonal are zero]

$= \hbar[(\frac{1}{2}+\frac{1}{2})(\frac{1}{2}-\frac{1}{2}+1)]^{1/2}$

$= \hbar$ [for $j = \frac{1}{2}$, $m = \frac{1}{2}$]

$J_- = 0$ [for $j = \frac{1}{2}$, $m = -\frac{1}{2}$]

$$\therefore J_- = \hbar \begin{bmatrix} 0 & 0 \\ 1 & 0 \end{bmatrix}$$

2. For $j = 1$, $m = 1, 0, -1$, we can show that

$$J^2 = \hbar \begin{bmatrix} 2 & 0 & 0 \\ 0 & 2 & 0 \\ 0 & 0 & 2 \end{bmatrix}, \quad J_z = \hbar \begin{bmatrix} 1 & 0 & 0 \\ 0 & 0 & 0 \\ 0 & 0 & -1 \end{bmatrix},$$

$$J_+ = \hbar \begin{bmatrix} 0 & \sqrt{2} & 0 \\ 0 & 0 & \sqrt{2} \\ 0 & 0 & 0 \end{bmatrix}, \quad J_- = \hbar \begin{bmatrix} 0 & 0 & 0 \\ \sqrt{2} & 0 & 0 \\ 0 & \sqrt{2} & 0 \end{bmatrix}$$

3. For $j = \frac{3}{2}$

$$J^2 = \hbar^2 \begin{bmatrix} \frac{15}{4} & 0 & 0 & 0 \\ 0 & \frac{15}{4} & 0 & 0 \\ 0 & 0 & \frac{15}{4} & 0 \\ 0 & 0 & 0 & \frac{15}{4} \end{bmatrix}, \quad J_z = \hbar \begin{bmatrix} \frac{3}{2} & 0 & 0 & 0 \\ 0 & \frac{1}{2} & 0 & 0 \\ 0 & 0 & \frac{1}{2} & 0 \\ 0 & 0 & 0 & -\frac{3}{2} \end{bmatrix}$$

$$J_+ = \hbar \begin{bmatrix} 0 & \sqrt{3} & 0 & 0 \\ 0 & 0 & 2 & 0 \\ 0 & 0 & 0 & \sqrt{3} \\ 0 & 0 & 0 & 0 \end{bmatrix}, \quad J_- = \begin{bmatrix} 0 & 0 & 0 & 0 \\ \sqrt{3} & 0 & 0 & 0 \\ 0 & 2 & 0 & 0 \\ 0 & 0 & \sqrt{3} & 0 \end{bmatrix}$$

10.17 Eigenfunctions of J^2 and J_z

If a single eigenfunction of the operators J^2 and J_z be obtained, then all other eigenfunctions corresponding to the value J^2 can be found out by the repeated application of the operators $(J_x - iJ_y)$.

If ψ_m be an eigenfunction of J^2, then we have

$$J^2 \psi_m = (J_x^2 + J_y^2 + J_z^2)\psi_m = [(J_x - iJ_y)(J_x + iJ_y) + J_z^2 + \hbar J_z]\psi_m$$

$$= [(J_x - iJ_y)(J_x + iJ_y + \hbar^2(m_z^2 + m_z)]\psi_m \tag{10.119}$$

$$\text{or, } J^2 \psi_m = \hbar^2 m_z(m_z + 1)\psi_m \; [\because (J_x + iJ_y)\psi_m = 0] \tag{10.120}$$

Also

$$J^2 \psi'_m = \hbar^2 m'_z(m'_z - 1)\psi'_m \; [\text{when } (J_x - iJ_y)\psi'_m = 0] \tag{10.121}$$

As eqs (10.120) and (10.121) hold good simultaneously, we have $m'_z = m_z$; taking $m_z = j$, we get

$$J^2 = \hbar^2 j(j+1) \text{ so that } (J_x - iJ_y)\psi'_j = 0 \tag{10.122}$$

Now, we know that

$$\left.\begin{aligned} J_x &= -\frac{\hbar}{i}\left(\sin\phi\frac{\partial}{\partial\theta} + \cot\theta\cos\phi\frac{\partial}{\partial\phi}\right) \\ J_y &= \left(\cos\phi\frac{\partial}{\partial\theta} - \cot\theta\sin\phi\frac{\partial}{\partial\phi}\right) \\ \text{and } J_z &= \frac{\hbar}{i}\frac{\partial}{\partial\phi} \end{aligned}\right\} \tag{10.123}$$

$$\therefore J_x + iJ_y = \hbar e^{i\phi}\left(\frac{\partial}{\partial\phi} + i\cot\theta\frac{\partial}{\partial\phi}\right) \tag{10.124}$$

Again, $m_z = j$ and $\psi_j^i = f_j^i(\theta)\, e^{ij\phi}$.

$$\therefore i\frac{\partial\phi_j^i}{\partial\theta} = -j\psi_j^i$$

Then eq. (10.124) takes the form:

$$\frac{\partial \psi_i^j}{\partial \theta} = j \cot \theta \psi_j^i \tag{10.125}$$

Integrating eq. (10.125), we get

$$\ln \psi_j^i = j \ln(\sin \theta) + k(\phi)$$

$$\text{or, } \psi_j^i = g(\phi)(\sin \theta)^j \tag{10.126}$$

Here, the function $g(\phi)$ is such that ψ_j^i is an eigenfunction of J_z with eigenvalue $j\hbar$, i.e., $g(\phi) = e^{ij\phi}$.
Hence,

$$\psi_j^i = e^{ij\phi}(\sin \theta)^j \tag{10.127}$$

To obtain values of ψ_j^m relating to similar values of m, we operate with $(J_x - iJ_y)$.
From eq. (10.123)

$$-(J_x - iJ_y) = \hbar e^{-i\phi}\left[\frac{\partial}{\partial \theta} - i \cot \theta \frac{\partial}{\partial \phi}\right]$$

Using the relation $-i\frac{\partial \psi_i^j}{\partial \phi} = j\psi_i^j$ we get

$$-(J_x - iJ_y)\,\psi_j^i = \hbar e^{-i\phi}\left(\frac{\partial}{\partial \theta} + j \cot \theta\right)\psi_j^i \tag{10.128}$$

Again, making use of the relation

$$\left(\frac{\partial}{\partial \theta} + j \cot \theta\right)\psi_j^i = \frac{1}{(\sin \theta)^j}\left[\frac{\partial}{\partial \theta}(\sin \theta)^j \psi_j^i\right]$$

we obtain

$$\psi^{j-1} = \frac{1}{(\sin \theta)^j} e^{i(j-1)\phi} \frac{\partial}{\partial \theta}(\sin \theta)^{2j} \tag{10.129}$$

Now, since

$$(J_x - iJ_y)\,\psi_j^{j-1} = e^{-i\phi}\left[\frac{\partial}{\partial \theta} + (j-1)\cot \theta\right]\psi_j^{i-1}$$

we get

$$\psi_j^{j-2} = \frac{1}{(\sin \theta)^{j-2}} e^{j(j-1)\phi} \frac{\partial}{\partial \theta}\left[\frac{\frac{\partial}{\partial \theta}(\sin \theta)^{2j}}{\sin \theta}\right]$$

Applying $(J_x - iJ_y)$ repeatedly, we obtain

$$\psi_j^{j-s} = \frac{e^{j(j-s)\phi}}{(\sin \theta)^{j-s}}\left(\frac{1}{\sin \theta}\frac{\partial}{\partial \theta}\right)^s (\sin \theta)^{2j} \tag{10.130}$$

Substituting $\cos \theta = \xi$, $\frac{\partial}{\partial \theta} = -\sin \theta \frac{\partial}{\partial \xi}$, then

$$\psi_j^{j-s} = \frac{e^{j(j-s)\phi}}{(1-\xi)^{\frac{(j-s)}{2}}} \frac{\partial^j}{\partial \xi^j}(1-\xi^2)^j \tag{10.131}$$

Here, ψ_j^{j-s} are unnormalised wave functions of J_z and J^2, which can be normalised by the application of normalisation condition.

10.18 Addition of angular momenta: Clebsch-Gordan coefficients

There are systems in which the total angular momentum is composed of two or more independent parts, viz., the orbital momenta of two electrons in an atom or the spin and angular momenta of the same electron, etc. The problem is to find out how the total angular momentum is related to its components.

We suppose that the total angular momentum $\vec{J}$ of a particular system is the sum of two parts $\vec{J}_1$ and $\vec{J}_2$. Then

$$\vec{J} = \vec{J}_1 + \vec{J}_2 \tag{10.132}$$

where the components of $\vec{J}_1$ commute with the components of $\vec{J}_2$ and each satisfies the fundamental commutation relations for angular momenta.

In the system of units in which $\hbar = 1$, we define angular momentum as

$$\vec{J} = (\vec{r} \times \vec{p}) \tag{10.133}$$

The commutation relations of $\vec{J}_1$ and $\vec{J}_2$ are

$$\left.\begin{aligned} \vec{J}_1 \times \vec{J}_1 &= i\vec{J}_1 \\ \vec{J}_2 \times \vec{J}_2 &= i\hbar\vec{J}_2 \end{aligned}\right\} \tag{10.134}$$

Now,

$$\begin{aligned} \left(\vec{J} \times \vec{J}\right)_x &= J_yJ_z - J_zJ_y = \left(J_{1y} + J_{2y}\right)\left(J_{1z} + J_{2z}\right) - \left(J_{1z} + J_{2z}\right)\left(J_{1y} + J_{2y}\right) \\ &= \left(J_{1y}J_{1z} - J_{1z}J_{1y}\right) + \left(J_{2y}J_{2z} - J_{2z}J_{2y}\right) = i\left(J_{1x} + J_{2x}\right) = iJ_x \end{aligned}$$

In general,

$$\vec{J} \times \vec{J} = i\vec{J} \tag{10.135}$$

This shows that $\vec{J}$ has the required properties for an angular momentum operator. Thus, $\vec{J}_1$ and $\vec{J}_2$ may represent the orbital angular momentum $\vec{L}$ and intrinsic spin angular momentum $\vec{S}$ of the same electron of an atom or $\vec{J}_1$ and $\vec{J}_2$ may represent the orbital angular momenta $\vec{L}_1$ and $\vec{L}_2$ of two spinless particles. Let us denote the eigenvalues of J_1^2 by $j_1(j_1+1)$, those of J_2^2 by $j_2(j_2+1)$and those of J^2 by $j(j+1)$.

Similarly, the eigenvalues of J_{1z}, J_{2z} and J_z are respectively m_1, m_2 and m.

Let the simultaneous eigenvectors of J_1^2 and J_{1z} be $\psi_1(j_1m_1)$ and $\vec{J}_2$ has no effect on them.

Similarly, the simultaneous eigenvectors of J_2^2 and J_{2z} are $\psi_2(j_2m_2)$ and $\vec{J}_1$ has no effect on them.

$$\therefore J_1^2\,\psi_1(j_1\,m_1) = j_1(j_1+1)\,\psi_1(j_1\,m_1) \tag{10.136}$$

$$J_{1z}\,\psi_1(j_1\,m_1) = m_1\psi_1(j_1\,m_1) \tag{10.137}$$

$$J_2^2\,\psi_2(j_2\,m_2) = j_2(j_2+1)\,\psi_2(j_2\,m_2) \tag{10.138}$$

$$\text{and } J_{2z}\,\psi_2(j_2\,m_2) = m_2\,\psi_2(j_2\,m_2) \tag{10.139}$$

The products

$$\psi(j_1\,j_2\,m_1\,m_2) = \psi_1(j_1\,m_1)\,\psi_2(j_2\,m_2) \tag{10.140}$$

form a complete set of simultaneous eigenvectors of the mutually commuting set of operators J_1^2, J_{1z}, J_2^2 and J_{2z}.

The eigenvectors $\psi(j_1\, j_2\, m_1\, m_2)$ defines the eigenstate in which the eigenvalues of the four observables J_1^2, J_{1z}, J_2^2 and J_{2z} are respectively $j_1\,(j_1+1)$, m_1, $j_2\,(j_2+1)$ and m_2.

Again, for given values of j_1 and j_2 there will be $(2j_1+1)$ values of m_1 and $(2j_2+1)$ values of m_2, i.e., there will be $(2j_1+1)\,(2j_2+1)$ possible eigenvectors $\psi\,(j_1\, j_2\, m_1\, m_2)$. Here, $\psi\,(j_1\, j_2\, m_1\, m_2)$ are the simultaneous eigenvectors of J_1^2 and J_2^2, but not the eigenvectors of J^2 and J_z. As $\psi\,(j_1\, j_2\, m_1\, m_2)$ form a complete set, a linear combination of them can give eigenvectors $\phi(j_1\, j_2\, j\, m)$ of J^2 and J_z, j and m being the quantum numbers related to the values of J^2 and J_z respectively.

Now, we are in a position to find the unitary transformation [rotation of axes in *Hilbert space*, cf., *Quantum Mechanics* by A. Messiah, North Holland Publishing Co., Amsterdam (1970)] that changes the vectors from $(j_1\, j_2\, m_1\, m_2)$ to $(j_1\, j_2\, m)$ representation.

Let us assume that $j_1 \geqslant j_2$ and the eigenvector $\psi(j_1\, j_2\, m_1\, m_2)$ be the eigenvector of J_z belonging to the eigenvalue (m_1+m_2). Then

$$\begin{aligned} J_z\psi\,(j_1 j_2,\, m_1 m_2) &= (J_{1z}+J_{2z})\,\psi_1\,(j_1 m_1)\,\psi_2\,(j_2 m_2) \\ &= (m_1+m_2)\,\psi_1\,(j_1 m_1)\,\psi_2\,(j_2 m_2) \\ &= (m_1+m_2)\,\psi_1\,(j_1\, j_2\, m_1\, m_2) \end{aligned}$$

As $|m_1| \leqslant j_1, |m_2| \leqslant j_2$

$$\therefore |m_1+m_2| \leqslant j_1+j_2 \tag{10.141}$$

This means, j_1 and j_2 are the largest values of m_1 and m_2. So the largest value of $m = m_1+m_2 = j_1+j_2 = j$.

We now consider the case $j_1 = 1$, $j_2 = \frac{1}{2}$. There are three possible values of m_1 and two possible values of m_2. So, there will be $3 \times 2 = 6$ possible vectors as follows:

$$\begin{aligned} &\psi\,(1, \tfrac{1}{2}, 1, \tfrac{1}{2}) && \text{for } m_1 = 1, \quad m_2 = \tfrac{1}{2} \\ &\psi\,(1, \tfrac{1}{2}, 1, -\tfrac{1}{2}) && \text{for } m = 1, \quad m_2 = -\tfrac{1}{2} \\ &\psi\,(1, \tfrac{1}{2}, 0, \tfrac{1}{2}) && \text{for } m_1 = 0, \quad m_2 = \tfrac{1}{2} \\ &\psi\,(1, \tfrac{1}{2}, 0, -\tfrac{1}{2}) && \text{for } m_1 = 0, \quad m_2 = -\tfrac{1}{2} \\ &\psi\,(1, \tfrac{1}{2}, -1, \tfrac{1}{2}) && \text{for } m_1 = -1, \, m_2 = \tfrac{1}{2} \\ &\psi\,(1, \tfrac{1}{2}, -1, -\tfrac{1}{2}) && \text{for } m_1 = -1, \, m_2 = -\tfrac{1}{2} \end{aligned}$$

Since, $J_z = J_{1z} + J_{2z}$ the corresponding m values will be

$$\tfrac{3}{2}, \tfrac{1}{2}, \tfrac{1}{2}, -\tfrac{1}{2}, -\tfrac{1}{2}, -\tfrac{3}{2} \quad (\because m = m_1 + m_2)$$

So, there will be four states with $m = \frac{3}{2}, \frac{1}{2}, -\frac{1}{2}, -\frac{3}{2}$ for $j = \frac{3}{2}$ and two states, $m = \frac{1}{2}$ or $-\frac{1}{2}$ for $j = \frac{1}{2}$. We thus conclude that the possible values of j in the present case are $\frac{3}{2}$ and $\frac{1}{2}$.

Now, it is quite possible, an alternative description in terms of the six vectors may be as follows:

$$\phi\left(1, \tfrac{1}{2}, \tfrac{3}{2}, \tfrac{3}{2}\right) \text{ for } j = \tfrac{3}{2},\ m = \tfrac{3}{2}$$
$$\phi\left(1, \tfrac{1}{2}, \tfrac{3}{2}, \tfrac{1}{2}\right) \text{ for } j = \tfrac{3}{2},\ m = \tfrac{1}{2}$$
$$\phi\left(1, \tfrac{1}{2}, \tfrac{3}{2}, -\tfrac{1}{2}\right) \text{ for } j = \tfrac{3}{2},\ m = -\tfrac{1}{2}$$
$$\phi\left(1, \tfrac{1}{2}, \tfrac{3}{2}, -\tfrac{3}{2}\right) \text{ for } j = \tfrac{3}{2},\ m = -\tfrac{3}{2}$$
$$\phi\left(1, \tfrac{1}{2}, \tfrac{1}{2}, +\tfrac{1}{2}\right) \text{ for } j = \tfrac{1}{2},\ m = \tfrac{1}{2}$$
$$\phi\left(1, \tfrac{1}{2}, \tfrac{1}{2}, -\tfrac{1}{2}\right) \text{ for } j = \tfrac{1}{2},\ m = -\tfrac{1}{2}$$

We now consider the general case and let $j_1 \geqslant j_2$. The numbers j_1 and j_2 are the largest values of m_1 and m_2 respectively.

Since, $j \geq |m|$ the greatest value of j will be $j_{\max} = (m_1 + m_2)_{\max} = j_1 + j_2 = j$.

The corresponding vector is

$$\phi(j_1 j_2 j_1 + j_2 j_1 + j_2) = \psi(j_1 j_2 j_1 j_2) \tag{10.142}$$

This is because there is only one function out of $\psi(j_1\, j_2\, m_1\, m_2)$ for which $m_1 + m_2 = j_1 + j_2$.

The next largest value of m is $j_1 + j_2 - 1$, which occurs twice when $m_1 = j_1$ and $m_2 = j_2 - 1$ and when $m_1 = j_1 - 1$ and $m_2 = j_2$, if $j_1 \neq 0$, $j_2 \neq 0$. So, there will be two distinct vectors for $m = j_1 + j_2 - 1$ as below:

$$\phi(j_1 j_2, j_1 + j_2, j_1 + j_2 - 1)$$
$$\phi(j_1 j_2, j_1 + j_2 - 1, j_1 + j_2 - 1)$$

because m can take the values $j_1 + j_2 - 1$, when $j = j_1 + j_2$ and when $j = j_1 + j_2 - 1$.

For $m = j_1 + j_2 - 2$, there will be three vectors

$$\psi(j_1, j_2, j_1 - 2, j_2)$$
$$\psi(j_1, j_2, j_1 - 1, j_2 - 1)$$
$$\text{and } \psi(j_1, j_2, j_1, j_2 - 2)$$

A linear combination of the above gives three vectors in ϕ's, viz.,

$$\phi(j_1 j_2, j_1 + j_2, j_1 + j_2 - 2)$$
$$\phi(j_1 j_2, j_1 + j_2 - 1, j_1 + j_2 - 2)$$
$$\phi(j_1 j_2, j_1 + j_2 - 2, j_1 + j_2 - 2)$$

In this way, for $m = j_1 + j_2 - k$, we have

$$\psi(j_1 j_2, j_1 - k, j_2)$$
$$\psi(j_1 j_2, j_1 - k + 1, j_2 - 1)$$
$$\cdots\ \cdots\ \cdots\ \cdots\ \cdots\ \cdots\ \cdots$$
$$\cdots\ \cdots\ \cdots\ \cdots\ \cdots\ \cdots\ \cdots$$
$$\psi(j_1 j_2, j_1, j_2 - k)$$

and we obtain the vectors

$$\phi(j_1 j_2, j_1 + j_2, j_1 + j_2 - k)$$
$$\phi(j_1 j_2, j_1 + j_2 - 1, j_1 + j_2 - k)$$
$$\cdots\cdots\cdots\cdots\cdots\cdots$$
$$\cdots\cdots\cdots\cdots\cdots\cdots$$
$$\phi(j_1 j_2, j_1 + j_2 - k, j_1 + j_2 - k)$$

As $j_1 \geqslant j_2$ the smallest value of m_2 is $-j_2$. The largest value of k will be corresponding to

$$j_2 - k = -j_2 \text{ or, } k = 2j_2$$

corresponding to which the smallest value of j is $j_{\min} = j_1 + j_2 - 2j_2 = j_1 - j_2$.

So the possible j values are such that $j_1 + j_2 \geqslant j \geqslant j_1 - j_2$.

Thus, there are $(2j_2 + 1)$ possible values of j and for each value of j there are $(2j + 1)$ values of m. Then the total number of eigenstates are given by

$$\sum_{j=j_1-j_2}^{j=j_1+j_2} (2j+1) = (2j_1 + 1)(2j_2 + 1)$$

This represents the number of independent ψ's.

Therefore, the vector $\phi(j_1 j_2 jm)$ may be expressed as a linear combination of vector $\psi(j_1\, j_2\, m_1\, m_2)$

$$\phi(j_1 j_2 jm) = \sum_{m_1}\sum_{m_2} C^{j_1 j_2}_{jm\, m_1\, m_2}\psi(j_1\, j_2\, m_1\, m_2) = \sum_{m_1+m_2=m} C^{j_1 j_2}_{jm\, m_1\, m_2}\psi(j_1\, j_2\, m_1\, m_2)$$

Now, $C^{j_1 j_2}_{jm\, m_1\, m_2} = 0$ if $m \neq m_1 + m_2$.

The set of eigenvectors $\psi(j_1\, j_2\, m_1\, m_2)$ and $\phi(j_1 j_2 jm)$ are each orthogonal set. In addition to this, if the vectors are normalised, then the coefficients $C^{j_1 j_2}_{jm\, m_1\, m_2}$ define unitary transformation between two angular momentum representations. $C^{j_1 j_2}_{jm\, m_1\, m_2}$ are called *Clebsch-Gordan vector coupling or Wigner coefficient.*

In Dirac's notation

$$C^{j_1 j_2}_{jm\, m_1\, m_2} = \langle j_1\, j_2\, m_1\, m_2 \,|\, j_1 j_2 jm\rangle \tag{10.143}$$

Table 10.1: Clebsch-Gordan coefficient for $j_1 = 1$ and $j_2 = \frac{1}{2}$, i.e., $\langle j_1 \tfrac{1}{2} m_1 m_2 | j_1 \tfrac{1}{2} jm\rangle$ [Ref. E.U. Condon and G.H. Shortley's *The Theory of Atomic Spectra.*]

j	$m_2 = \frac{1}{2}$	$m_2 = -\frac{1}{2}$
$j_1 + \frac{1}{2}$	$\sqrt{\dfrac{j_1 + m + \frac{1}{2}}{2j_1 + 1}}$	$\sqrt{\dfrac{j_1 - m + \frac{1}{2}}{2j_1 + 1}}$
$j_1 - \frac{1}{2}$	$-\sqrt{\dfrac{j_1 - m + \frac{1}{2}}{2j_1 + 1}}$	$\sqrt{\dfrac{j_1 + m + \frac{1}{2}}{2j_1 + 1}}$

Table 2010.2: Clebsch-Gordan coefficients for $j = 1$, i.e., $\langle j_1 m_1 m_2 | j_1 | jm \rangle$

$j = 1$	$m_2 = 1$	$m_2 = 0$	$m_2 = -1$
$j_1 + 1$	$\left[\frac{(j_1+m)(j_1+m+1)}{(2j_1+1)(2j_1+2)}\right]^{1/2}$	$\left[\frac{(j_1-m+1)(j_1+m+1)}{(2j_1+1)(j_1+1)}\right]^{1/2}$	$\left[\frac{(j_1-m)(j_1-m+1)}{(2j_1+1)(2j_1+2)}\right]^{1/2}$
$j_1 + 0$	$-\left[\frac{(j_1+m)(j_1-m+1)}{2j_1(j_1+1)}\right]^{1/2}$	$\left[\frac{m^2}{j_1(j_1+1)}\right]^{1/2}$	$\left[\frac{(j_1-m)(j_1+m+1)}{2j_1(j_1+1)}\right]^{1/2}$
$j_1 - 1$	$\left[\frac{(j_1-m)(j_1-m+1)}{2j_1(2j_1+1)}\right]^{1/2}$	$-\left[\frac{(j_1-m)(j_1+m)}{j_1(2j_1+1)}\right]^{1/2}$	$\left[\frac{(j_1+m+1)(j+m)}{2j_1(2j_1+1)}\right]^{1/2}$

10.19 Addition of orbital and spin angular momentum

If $\vec{L}$ and $\vec{S}$ be the orbital and spin angular momentum of a system, then the total angular momentum $\vec{J}$ is given by $\vec{J} = \vec{L} + \vec{S}$.

The total number of combined states is

$$J = \sum_{|l-s|}^{l+s} (2j+1) = (2l+1)(2s+1) \tag{10.144}$$

and the number of eigenvector is

$$\sum_{|L-s|}^{|l+s|} (2j+1) = (2l+1)(2s+1)$$

Now, we have

$$\phi(j_1+j_2, j_1+j_2) = \psi_1(j_1, j_1)\,\psi_2(j_2, j_2) \tag{10.145}$$

and

$$\begin{aligned}&\phi(j_1+j_2, j_1+j_1-1)\\ &= [j_1+j_2]^{-1/2}\left[j_1^{1/2}\psi_1(j_2, j_1-1)\,\psi_2(j_2, j_2) + j_2^{1/2}\psi_1(j_1, j_1)\,\psi_2(j_2, j_2-1)\right]\end{aligned} \tag{10.146}$$

Now, for $j_1 = l$ and $j_2 = s = \frac{1}{2}$, which is the case for fermions, eqs (10.145) and (10.146) become

$$\phi\left(l+\tfrac{1}{2}, l+\tfrac{1}{2}\right) = \psi_1(l, l)\,\psi_2\left(\tfrac{1}{2}, \tfrac{1}{2}\right) \tag{10.147}$$

$$\phi\left(l+\tfrac{1}{2}, l-\tfrac{1}{2}\right) = \left[l+\tfrac{1}{2}\right]^{-1/2}\left[(2l)^{1/2}\,\psi_1(l, l-1)\psi_2\left(\tfrac{1}{2}, \tfrac{1}{2}\right) + \psi_1(l, l)\,\psi_2\left(\tfrac{1}{2}, \tfrac{1}{2}\right)\right] \tag{10.148}$$

or in general

$$\begin{aligned}\phi\left(l-\tfrac{1}{2}, m\right) = (2l+1)^{-1/2}&\left[\left(l+m+\tfrac{1}{2}\right)^{1/2}\psi_2\left(l, m-\tfrac{1}{2}\right)\psi_2\left(\tfrac{1}{2}, \tfrac{1}{2}\right)\right.\\ &\left.+\left(l-m+\tfrac{1}{2}\right)^{1/2}\psi_2\left(l, m+\tfrac{1}{2}\right)\psi_2\left(\tfrac{1}{2}, -\tfrac{1}{2}\right)\right]\end{aligned} \tag{10.149}$$

Putting $m = l - \frac{1}{2}$ in eq. (10.149), we get back eq. (10.148).

Now, the orthonormal states to eq. (10.148) is

$$\phi\left(l-\tfrac{1}{2}, l-\tfrac{1}{2}\right) = (2l+1)^{-1/2}\left[(2l)^{1/2}\,\psi_1(l, l)\,\psi_2\left(\tfrac{1}{2}, -\tfrac{1}{2}\right) - \psi_1(l, l-1)\,\psi_2\left(\tfrac{1}{2}, \tfrac{1}{2}\right)\right] \tag{10.150}$$

The general form of the above equation is

$$\phi\left(l-\tfrac{1}{2},m\right)=(2l+1)^{-1/2}\Big[\left(l+m+\tfrac{1}{2}\right)^{1/2}\psi_1\left(l,m+\tfrac{1}{2}\right)\psi_2\left(\tfrac{1}{2},-\tfrac{1}{2}\right)$$
$$-\left(l-m+\tfrac{1}{2}\right)^{1/2}\psi_1\left(l,m-\tfrac{1}{2}\right)\psi_2\left(\tfrac{1}{2},\tfrac{1}{2}\right)\Big] \quad (10.151)$$

All the states of the system are represented by the above equations.

10.20 The *p*-states of an electron

The *p*-state of an electron is represented by $l = 1$ and $s = \pm 1/2$. The orbital functions for $l = 1$ states of an electron are the three spherical harmonics Y_1^1, Y_1^0, Y_1^{-1} and the spin states are $S_{1/2}$ and $S_{-1/2}$. The two possible values of the total angular momentum are $l+s=1+\frac{1}{2}=\frac{3}{2}$ and $l-s=1-\frac{1}{2}=\frac{1}{2}$.

Now, the states corresponding to $j = 3/2$ and $m = 3/2$ can be written as

$$\phi\left(\tfrac{3}{2},\tfrac{3}{2}\right)=Y_1^1 S_{+1/2} \quad (10.152)$$

Using the properties of the lowering operator $J_- = L_- + S_-$, we can calculate the remaining states for $j = 3/2$.

$\therefore J_-\psi(j,m)=[j(j+1)-m(m-1)]^{1/2}\phi(j,m-1)$ so that

$$J_-\phi\left(\tfrac{3}{2},m\right)=\left[\tfrac{15}{4}-m(m-1)\right]^{1/2}\phi\left(\tfrac{3}{2},m-1\right) \quad (10.153)$$

$$L_-Y_1^{m_l}=[2-m_l(m_l-1)]Y_1^{m_l-1} \quad (10.154)$$

$$S_-S_{m_s}=\left[\tfrac{3}{4}-m_s(m_s-1)\right]S_{m_s-1} \quad (10.155)$$

Now, from eq. (10.152)

$$J_-\phi\left(\tfrac{3}{2},\tfrac{3}{2}\right)=(L_-+S_-)Y_1^1S_{+1/2},\ \text{i.e., } \sqrt{3}\,\phi\left(\tfrac{3}{2},\tfrac{1}{2}\right)=\sqrt{2}\,Y_1^0S_{+1/2}+Y_1^1S_{-1/2}$$

$$\text{or, } \phi\left(\tfrac{3}{2},\tfrac{1}{2}\right)=\sqrt{\tfrac{2}{3}}\,Y_1^0S_{+1/2}+\tfrac{1}{\sqrt{3}}\,Y_1^1S_{-1/2} \quad (10.156)$$

Again, the application of lowering operator J_- yields

$$\phi\left(\tfrac{3}{2},-\tfrac{1}{2}\right)=\tfrac{1}{\sqrt{3}}\,Y_1^{-1}S_{+1/2}+\sqrt{\tfrac{2}{3}}\,Y_1^0S_{-1/2} \quad (10.157)$$

$$\text{and } \phi\left(\tfrac{3}{2},-\tfrac{3}{2}\right)=Y_1^{-1}S_{-1/2} \quad (10.158)$$

The state represented by $\phi\left(\frac{1}{2},\frac{1}{2}\right)$ is a linear combination of $Y_1^0S_{+1/2}$ and $Y_1^1S_{-1/2}$ and $\phi\left(\frac{1}{2},\frac{1}{2}\right)$ is orthogonal to the state $\phi\left(\frac{3}{2},\frac{1}{2}\right)$.

$$\therefore \phi\left(\tfrac{1}{2},\tfrac{1}{2}\right)=\tfrac{1}{\sqrt{3}}\,Y_1^0S_{+1/2}+\sqrt{\tfrac{2}{3}}\,Y_1^1S_{-1/2} \quad (10.159)$$

Again, applying the lowering operator to eq. (10.156), we get

$$\phi\left(\tfrac{1}{2},-\tfrac{3}{2}\right)=\sqrt{\tfrac{2}{3}}\,Y_1^{-1}S_{+1/2}-\tfrac{1}{\sqrt{3}}\,Y_1^0S_{-1/2} \quad (10.160)$$

Thus, all the six states $\phi(j,m)$ are $\phi\left(\frac{3}{2},\frac{3}{2}\right)$, $\phi\left(\frac{3}{2},+\frac{1}{2}\right)$, $\phi\left(\frac{1}{2},-\frac{1}{2}\right)$, $\phi\left(\frac{3}{2},-\frac{3}{2}\right)$, $\phi\left(\frac{1}{2},\frac{1}{2}\right)$ and $\phi\left(\frac{1}{2},-\frac{3}{2}\right)$.

These states are mutually orthogonal and normalised.

Hence, we conclude that if spherical symmetry is maintained, the *p*-state for $j=\frac{3}{2}$ is *four-fold* and the state for $j=\frac{1}{2}$ is *two-fold* degenerate w.r.t. energy.

10.21 Selection rules for j and m

When electromagnetic radiation is incident on an atom, it undergoes a transition from one energy level to another. The condition which governs the process of transition is called the *selection rule.*

We now consider two operators, a scalar and a vector operator. A rotation operator is a vector operator represented by $U = e^{imJ}$. An operator which remains invariant under the rotation of coordinates is called a *scalar operator* (S) and obeys the condition

$$USU^{\dagger} = S \text{ or, } US - SU = 0$$

S also commutes with each component of J, i.e., $[S, J] = 0$, $[S, J_z] = 0$ and $[S, J^2] = 0$.

We can represent matrix element in jm notation in the following way:

$$\begin{aligned}\langle j'm' |[S, J_z]| jm\rangle &= \langle J'm' |SJ_z - J_zS| jm\rangle = \langle j'm' |SJ_z| jm\rangle - \langle j'm' |J_zS| jm\rangle \\ &= (m - m')\langle j'm' |S| jm\rangle \end{aligned} \quad (10.161)$$

Likewise

$$\begin{aligned}\left\langle j'm' \left|SJ^2 - J^2S\right| jm\right\rangle &= \left\langle j'm' \left|SJ^2\right| jm\right\rangle \left\langle j'm' \left|J^2S\right| jm\right\rangle \\ &= \langle j'm' |S| jm\rangle\, j(j+1) - j'(j'+1)\langle j'm' |S| jm\rangle \\ &= [j(j+1) - j'(j'+1)]\langle j'm' |S| jm\rangle \end{aligned} \quad (10.162)$$

If $j' \neq j$ and $m' \neq m$, then $\langle j'm' |S| jm\rangle = 0$.

$\therefore$ the selection rules from eqs (10.161) and (10.162) are

$$\Delta j = j' - j = 0 \text{ and } \Delta m = m' - m = 0 \quad (10.163)$$

We now consider a vector operator T satisfying the following relations with J:

$$\begin{array}{lll} [J_x, T_x] = 0 & [J_y, T_x] = -iT_z & [J_z, T_z] = iT_y \\ [J_x, T_y] = iT_z & [J_y, T_y] = 0 & [J_z, T_y] = -iT_x \\ [J_x, T_z] = -iT_y & [J_y, T_z] = iT_x & [J_z, T_z] = 0 \end{array}$$

The corresponding lowering and raising operators are $T_- = T_x - jT_y$, $T_+ = T_x + iT_y$ so that we can write:

$$\begin{aligned}\langle j'm' |J_zT_+ - T_+J_z| jm\rangle &= \langle j'm' |J_zT_+| jm\rangle - \langle j'm' |T_+J_z| jm\rangle \\ &= m'\langle j'm' |T_+| jm\rangle - \langle j'm' T_+|jm\rangle\, m \\ &= (m' - m)\langle j'm' |T_+| jm\rangle \end{aligned}$$

Now, $\langle j'm' |T_+| jm\rangle = 0$, if $m' \neq m + 1$. T_+ is a ladder matrix whose off-diagonal elements are zero and the upper diagonal elements are non-zero.

So

$$m' - m = 1 \text{ i.e., } \Delta m = 1 \quad (10.164)$$

Likewise

$$\begin{aligned}\langle j'm'|J_zT_- - T_-J_z|\,jm\rangle &= \langle j'm'|J_zT_-|\,jm\rangle - \langle j'm'|T_-T_z|\,jm\rangle \\ &= m'\langle j'm'|T_-|\,jm\rangle - \langle j'm'|T_-|\,jm\rangle\, m \\ &= (m'-m)\langle j'm'|T_-|\,jm\rangle\end{aligned}$$

Since, T_- matrix has only lower diagonal operator and other elements are zero, hence $\langle jm'|T_-|\,jm\rangle = 0$ unless

$$m' = m-1 \text{ or, } m-m' = -1 \text{ or, } \Delta m = -1 \qquad (10.165)$$

Similarly,

$$\begin{aligned}\langle j'm'|J_zT_z - T_zJ_z|\,jm\rangle &= \langle j'm'|J_zT_z|\,jm\rangle - \langle j'm'|T_zJ_z|\,jm\rangle \\ &= m'\langle j'm'|T_z|\,jm\rangle - \langle j'm'|T_z|\,jm\rangle\, m \\ &= (m'-m)\langle j'm'|T_z|\,jm\rangle\end{aligned}$$

Now, since T_z is a diagonal matrix, $m' = m$. So the matrix $\langle j'm'|T_z|\,jm\rangle \neq 0$, if $m' = m$

$$\text{i.e., } m'-m = 0 \text{ or, } \Delta m = 0 \qquad (10.166)$$

Thus, the selection rule for m so that a transition can occur is

$$\Delta m = 0, \pm 1 \qquad (10.167)$$

To obtain a selection rule for j we consider the following general commutation relation as below:

$$\left[J^2, [J^2, \vec{T}]\right] = 2(J^2\vec{T} + \vec{T}J^2) - 4\vec{J}[\vec{J}\cdot\vec{T}] \qquad (10.168)$$

$$\therefore \vec{J}\cdot\vec{T} = \vec{T}\cdot\vec{J} = T_xJ_x + T_yJ_y + T_zJ_z$$

Again,

$$\left[J^2, [J^2, \vec{T}]\right] = \left[J^2, (J^2\vec{T} - \vec{T}J^2)\right] = J^4\vec{T} + 2J^2\vec{T}J^2 + \vec{T}J^4 \qquad (10.169)$$

From eqs (10.168) and (10.169), we get

$$J^4\vec{T} - 2J^2\vec{T}J^2 + \vec{T}J^4 = 2(J^2\vec{T} + \vec{T}J^2) - 4\vec{J}(\vec{J}\cdot\vec{T})$$

$$\text{or, } J^4\vec{T} - 2J^2\vec{T}J^2 + \vec{T}J^4 - 2(J^2\vec{T} + \vec{T}J^2) + 4\vec{J}(\vec{J}\cdot\vec{T}) = 0 \qquad (10.170)$$

$\left[J^2, (\vec{J}\cdot\vec{T})\right] = 0$, so $(\vec{J}\cdot\vec{T})$ will be a diagonal matrix. Hence, the only non-zero matrix element of $(\vec{J}\cdot\vec{T})$ is for $j = j'$.

For $j' \neq j$, the general matrix elements of the operator of eq. (10.170) can be written as under:

$$\left\langle jm\left|\left\{J^4\vec{T} - 2J^2\vec{T}J^2 + \vec{T}J^4 - 2(J^2\vec{T} + \vec{T}J^2) + 4\vec{J}(\vec{J}\cdot\vec{T})\right\}\right| j'm'\right\rangle = 0$$

$$\text{or, } \left\{[j(j+1)]^2 - 2j(j+1)\,j'(j'+1) + [j'(j'+1)]^2 - 2[j(j+1) + j'(j'+1)^2]\right\}\left\langle jm\left|\vec{T}\right| j'm'\right\rangle = 0 \quad (10.171)$$

On simplification the above equation reduces to

$$[\{(j+j'+1)^2 - 1\}\{(j-j')^2 - 1\}]\left\langle jm\left|\vec{T}\right| j'm'\right\rangle = 0$$

The matrix elements of T will be zero except

$$(j-j')^2 = 1 \text{ or, } (j-j')^2 = -1 \text{ or, } 4j = \pm 1 \qquad (10.172)$$

Hence, the non-zero matrix elements of T are those for which

$$j = 0, \pm 1 \tag{10.173}$$

The selection rule can be interpreted as

The photon having many of the properties of a particle of spin one, during transition of the system absorbs it from one energy level to another.

The angular momentum of the system is affected in the following three ways:

1. It increases the total angular momentum by one unit.
2. It leaves the angular momentum unchanged.
3. It decreases the angular momentum of the system by one unit.

The above three effects are controlling the selection rule for transition.

Any other transitions except those indicated above are *forbidden* (i.e., not allowed).

10.22 Answers of some important conceptual questions

10.1. *Justify that the quantum-mechanical theory of angular momentum is founded on the theory of rotations in three dimensions.*

Ans. We know that the operators corresponding to angular momentum are infinitesimal rotation operators. The quantity

$$\frac{i}{\hbar} L_z \psi(\vec{r})\, d\phi = \frac{\partial \psi}{\partial \phi}\, d\phi$$

is the change in the function $\psi(\vec{r})$, which is produced by rotation of the vector $\vec{r}$ through the angle $d\phi$ about the z-axis. This fact explains the importance of angular momentum operators in describing the rotational symmetry properties of a physical system. These operators are completely characterised by the fact that they are infinitesimal rotation operators. Thus, it can be said that the quantum mechanical theory of angular momentum is founded on the theory of rotations in three-dimensional space.

10.2. *How can the intrinsic degeneracy of the quantum states of an isolated system be removed?*

Ans. The intrinsic degeneracy of the quantum states of an isolated system can only be removed by subjecting the system to an external force which is not spherically symmetric. For a system of charged particles, the forces arising from an externally applied magnetic field are of this nature and provide a means for the complete removal of the degeneracy. The study of magnetic effects is, therefore, important for understanding the physical nature and significance of angular momentum states.

10.3. *What is Goudsmit and Uhlenbeck's proposal regarding electron spin? Indicate its success in the development of quantum theory.*

Ans. Goudsmit and Uhlenbeck in 1925, proposed that each electron spins while revolving around the nucleus and has a *quantised spin angular momentum*. The electron being a charged body also possesses spin magnetic moment. According to them, the electron has a spin angular momentum $S = \frac{1}{2}\hbar$, so that the spin quantum number $s = \frac{1}{2}$. According to the rules of space quantisation, only two orientations of the spin vector $\vec{S}$ are possible w.r.t. special direction in space. Then the possible values of the components of the spin angular momentum will be $S_z = \frac{1}{2}\hbar$ and $S_z = -\frac{1}{2}\hbar$.

Hence, each orbital electron in an atom provides a contribution to the angular momentum in an assigned direction equal to $\pm\frac{1}{2}\hbar$.

The introduction of the concept of electron spin has tremendous success in accounting the following experimental results:

1. Fine structure of spectral lines.
2. The results of Stern-Gerlach experiment.
3. The anomalous Zeeman effect.
4. The permanent magnetic moment of the electron

$$\vec{\mu}_s = -\frac{e}{2m_0c}2\vec{S}$$

where $\vec{S}$ is the spin angular momentum associated with the spin motion of the electron.

10.23 Solved problems

» ***Example* 10.1.** Given $\hat{H} = \frac{p^2}{2\mu} + V(r)$, calculate $\sum_m (E_m - E_n)\,|x_{mn}|^2$.

Solution: We have the commutation relation of x and H

$$[x,H] = \left[x, \left\{\frac{p^2}{2\mu} + V(r)\right\}\right] = \left[x, \frac{p^2}{2\mu}\right] + [x, V(r)] = \left[x, \frac{p^2}{2\mu}\right]$$

Since, x and $V(r)$ commute

$$[x, V(x)] = 0$$

$$\text{or, } [x,H] = \frac{1}{2\mu}[x,p^2] = \frac{1}{2\mu}\{p[x,p] + [x,p]p\} = \frac{2i\hbar}{2\mu}p_x = \frac{i\hbar}{\mu}p_x$$

$$\because [x,p_y] = 0 \text{ and } [x,p_z] = 0, [x,p_x] = i\hbar$$

Now, taking double commutator

$$[x,[x,H]] = \frac{i\hbar}{\mu}[x,p_x] \text{ or, } x[x,H] - [x,H]x = \frac{i\hbar}{\mu}i\hbar = -\frac{\hbar^2}{\mu}$$

$$\therefore \langle n|x[x,H] - [x,H]x|m\rangle = \left\langle n\left|-\frac{\hbar^2}{\mu}\right|m\right\rangle$$

Putting $n = m$, we obtain

$$\langle n|x[x,H] - [x,H]x|n\rangle = -\frac{\hbar^2}{\mu}$$

Here, $x[x,H]$ is product of two matrices A, B as

$$A\cdot B = \sum_j A_{ij}B_{jk}.$$

$$\therefore \text{LHS} = \sum_m \langle n|x|m\rangle\langle m|[x,H]|n\rangle - \sum_m \langle n|[x,H]|m\rangle\langle m|x|n\rangle$$
$$= \sum_m \langle n|x|m\rangle\langle m|xH - Hx|n\rangle - \sum_m \langle n|xH - Hx|m\rangle\langle m|x|n\rangle$$
$$= \sum_m \langle n|H|m\rangle\langle xE_n - E_m x\rangle|n\rangle - \sum_m \langle n|xE_m - E_n x|m\rangle\langle m|x|n\rangle$$
$$= \sum_m 2(E_n - E_m)\langle n|x|m\rangle\langle m|x|n\rangle$$
$$= \sum_m 2(E_n - E_m)\, x_{nm}x_{mn}$$
$$\text{or, } \sum_m (E_n - E_m)|x_{nm}|^2 = -\frac{\hbar^2}{2\mu} \quad [\because x_{mn} = x^*_{nm}, \text{ if } x \text{ is Hermitian}]$$

» ***Example* 10.2.** Suppose, that the total angular momentum of a system $\vec{J} = \vec{J}_1 + \vec{J}_2$. Assuming that the system is in a state in which J_1^2, J_2^2, J_{1z} and J_{2z} have the values $j_1(j_1+1)$, $j_2(j_2+1)$, m_1 and m_2, show that $\bar{J}^2 = j_1(j_1+1) + j_2(j_2+1) + 2m_1m_2$.

Solution: We have

$$\bar{J}^2 = \overline{(\vec{J}_1 + \vec{J}_2)^2} \text{ or, } \bar{J}^2 = J_1^2 + J_2^2 + 2(J_{1x}J_{2x} + J_{1y}J_{2y} + J_{1z}J_{2z}) = J_1^2 + J_2^2 + 2\left[\overline{J_{1x}J_{2x}} + \overline{J_{1y}J_{2y}} + \overline{J_{1z}J_{2z}}\right].$$

In the given state, J_1^2 and J_2^2 have definite values equal to $j_1(j_1+1)$ and $j_2(j_2+1)$ respectively.

And $\overline{J_{1x}J_{2x}} = \overline{J_{1y}J_{2z}}$, etc. $\bar{J}_{1x} = \bar{J}_{2x} = \bar{J}_{1y} = \bar{J}_{2y} = 0$ and $\bar{J}_{1z} = m_1$, $\bar{J}_{2z} = m_2$.

$\therefore \bar{J}^2 = j_1(j_1+1)\, j_2(j_2+1) + 2m_1m_2$.

» ***Example* 10.3.** A particle is moving in the field of potential $V(r)$. Show that for any eigenstate of energy and angular momentum $\hbar^2[l(l+1)]\langle r^{-3}\rangle = m\left\langle \frac{\partial V}{\partial r}\right\rangle$.

Solution: We have the Hamiltonian of the particle, $H = \frac{p^2}{2m} + V(r)$, where $p^2 = -\hbar^2\nabla^2$. If $p_r = -i\hbar\left(\frac{\partial}{\partial r} + \frac{1}{r}\right)$ and L^2 = square of angular momentum operator $= L_x^2 + L_y^2 + L_z^2$, then $p^2 = p_r^2 + \frac{L^2}{r^2}$.

Hence,

$$[p_r, H] = \left[p_r, \frac{p^2}{2m}\right] + [p_r, V(r)] = \frac{1}{2m}\left[p_r, p_r^2 + \frac{L^2}{r^2}\right] + [p_r, V(r)]$$

Now,

$$\frac{1}{2m}\left[p_r, \frac{L^2}{r^2}\right] = \frac{L^2}{2m}\left[p_r, \frac{1}{r^2}\right] = -i\hbar\frac{L^2}{2m}\left[\frac{\partial}{\partial r} + \frac{1}{r}, \frac{1}{r^2}\right] = -i\hbar\frac{L^2}{2m}\left(-\frac{2}{r^3}\right) = i\hbar\frac{L^2}{m}\frac{1}{r^3}$$

Again, $[p_r, V(r)] = -i\hbar\frac{\partial V}{\partial r}$.

Then $[p_r, H] = i\hbar\frac{L^2}{mr^3} - i\hbar\frac{\partial V}{\partial r}$.

Hence, expectation value of both sides are

$$\langle [p_r, H]\rangle = \frac{i\hbar}{m}\left\langle\frac{L^2}{r^3}\right\rangle - i\hbar\left\langle\frac{\partial V}{\partial r}\right\rangle$$

But, $\langle[p_r, H]\rangle = 0$.

$$\therefore \frac{i\hbar}{m}\left\langle\frac{L^2}{r^3}\right\rangle = i\hbar\left\langle\frac{\partial V}{\partial r}\right\rangle \text{ or, } l(l+1)\hbar^2\left\langle\frac{1}{r^3}\right\rangle = \left\langle\frac{\partial V}{\partial r}\right\rangle$$

» ***Example* 10.4.** If the Hamiltonian for a two-particle system $H(\vec{r}_1, \vec{r}_2)$ is to be invariant under translation, then show that H cannot depend on centre of mass coordinate.

Solution: Let $\vec{r}_1 \to \vec{r}_1 + \varepsilon, \vec{r}_2 \to \vec{r}_2 + \varepsilon$, then $H \to H + \delta H$ $[\because H = H(\vec{r}_1, \vec{r}_2)]$

$$\therefore \delta H = \frac{\partial H}{\partial r_1}\varepsilon + \frac{\partial H}{\partial r_2}\varepsilon = \varepsilon\left(\frac{\partial H}{\partial r_1} + \frac{\partial H}{\partial r_2}\right)$$

For the Hamiltonian H to be invariant, then $\frac{\partial H}{\partial r_1} + \frac{\partial H}{\partial r_2} = 0$.

Now, let $\vec{R} = \dfrac{m_1\vec{r}_1 + m_2\vec{r}_2}{m_1 + m_2}$ be the position vector of the CM of m_1 and m_2 at $\vec{r}_1$ and $\vec{r}_2$ respectively and $\vec{r} = \vec{r}_1 - \vec{r}_2$ the relative coordinate of m_1 and m_2.

Then

$$dH = \frac{\partial H}{\partial \vec{R}}d\vec{R} + \frac{\partial H}{\partial \vec{r}}d\vec{r}$$

$$\frac{\partial H}{\partial \vec{r}_1} = \frac{\partial H}{\partial \vec{R}}\frac{\partial \vec{R}}{\partial \vec{r}_1} + \frac{\partial H}{\partial \vec{r}}\frac{\partial \vec{r}}{\partial \vec{r}_1} = \frac{\partial H}{\partial R}\frac{m_1}{m_1 + m_2} + \frac{\partial H}{\partial \vec{r}}$$

$$\frac{\partial H}{\partial \vec{r}_2} = \frac{\partial H}{\partial \vec{R}}\frac{\partial \vec{R}}{\partial \vec{r}_2} + \frac{\partial H}{\partial \vec{r}}\frac{\partial \vec{r}}{\partial \vec{r}_2} = \frac{\partial H}{\partial \vec{R}}\frac{m_2}{m_1 + m_2} - \frac{\partial H}{\partial \vec{r}}$$

$$\therefore \frac{\partial H}{\partial \vec{r}_1} + \frac{\partial H}{\partial \vec{r}_2} = \frac{\partial H}{\partial \vec{R}}\left(\frac{m_1}{m_1 + m_2} + \frac{m_2}{m_1 + m_2}\right) = \frac{\partial H}{\partial \vec{R}}$$

But,

$$\frac{\partial H}{\partial \vec{r}_1} + \frac{\partial H}{\partial \vec{r}_2} = 0, \therefore \frac{\partial H}{\partial \vec{R}} = 0.$$

Hence, H cannot depend on CM variable R.

» ***Example* 10.5.** If we neglect the spin-orbit interaction for the electron moving in a central force field, show that it is possible to obtain eigenstates in which the observables H, L^2, L_z and σ_z are all well-defined. Obtain the eigenfunctions.

Solution: We can show that σ_z commutes with each of the operators H, L^2, L_z.

Now,

$$H\sigma_z\psi = H\begin{pmatrix}1 & 0\\ 0 & -1\end{pmatrix}\begin{pmatrix}\psi_+\\ \psi_-\end{pmatrix} = H\begin{pmatrix}\psi_+\\ -\psi_-\end{pmatrix} = \begin{pmatrix}H\psi_+\\ -H\psi_-\end{pmatrix}$$

$$\text{and } \sigma_z H\psi = \begin{pmatrix}1 & 0\\ 0 & -1\end{pmatrix}\begin{pmatrix}H\psi_+\\ H\psi_-\end{pmatrix} = \begin{pmatrix}H\psi_+\\ -H\psi_-\end{pmatrix}$$

ψ being an arbitrary vector.

The eigenstates of H must be represented by vectors satisfying the eigenvalue equation

$$H\begin{pmatrix}\psi_+\\ \psi_-\end{pmatrix} = E_n\begin{pmatrix}\psi_+\\ \psi_-\end{pmatrix}$$

Thus, $H\psi_+ = E_n\psi_+, H\psi_- = E_n\psi_-$.

So ψ_+ and ψ represent energy eigenfunction corresponding to the energy value E_n when spin was neglected. Also ψ_+ and ψ_- are eigenfunctions of L^2 and L_z as for the case of no spin.

For the eigenstate corresponding to $\sigma_z = +1$

$$\begin{pmatrix} 1 & 0 \\ 0 & -1 \end{pmatrix}\begin{pmatrix} \psi_+ \\ \psi_- \end{pmatrix} = \begin{pmatrix} \psi_+ \\ \psi_- \end{pmatrix} \text{ or, } \psi_- = 0$$

Thus, the eigenfunctions for which the operators H, L^2, L_z and σ_z have the eigenvalues $E_n, l(l+1)\hbar, m\hbar$ and +1 respectively are

$$\psi_{nlm}\Big\uparrow = \begin{pmatrix} R_{nl}(r)\,Y_{lm}(\theta,\phi) \\ 0 \end{pmatrix}$$

where ↑ represents that the spin is up ($\sigma_z = +1$).

For $\sigma_z = -1$

$$\psi_{nlm}\Big\downarrow = \begin{pmatrix} 0 \\ R_{nl}(r)\,Y_{lm}(\theta,\phi) \end{pmatrix}$$

» ***Example* 10.6.** A system consisting of two weakly interacting systems (1) and (2) with angular momentum $j_1 = 1$ and $j_2 = 1/2$ is in a state with a total angular momentum $j = 1/2$ and a component of angular momentum along a given z-axis $m = 1/2$; show that $\bar{m}_1 = 2/3$ and $\bar{m}_2 = 1/6$.

Solution: We have $\phi\,(1, \frac{1}{2}, \frac{1}{2}, \frac{1}{2}) = \sqrt{(\frac{2}{3})}\,\psi_{1,-\frac{1}{2}} - \sqrt{(\frac{1}{3})}\,\psi_{0,\frac{1}{2}}$.

For a state with $j = \frac{1}{2}, m = \frac{1}{2}$ the probabilities of the values

$$m_1 = 1, \quad m_2 = -\tfrac{1}{2} \text{ equal } \tfrac{2}{3}$$

$$m_1 = 0, \quad m_2 = \tfrac{1}{2} \text{ equal } \tfrac{1}{3}$$

$\therefore \bar{m}_1 = 1\cdot\frac{2}{3} + 0\cdot\frac{1}{3} = \frac{2}{3}$ and $\bar{m}_2 = -\frac{1}{2}\cdot\frac{2}{3} + \frac{1}{2}\cdot\frac{1}{3} = -\frac{1}{6}$.

$\therefore m = \bar{m}_1 + \bar{m}_2 = \frac{1}{2}$.

» ***Example* 10.7.** Calculate the Clebsch-Gordan coefficients for $j_1 = 1, j_2 = \frac{1}{2}$. Use these results to write the wave functions for the p-states of an electron ($l = 1, s = \frac{1}{2}$) for $j = \frac{3}{2}$; $m_j = -\frac{3}{2}, -\frac{1}{2}, \frac{1}{2}, \frac{3}{2}$ and for $j = \frac{1}{2}$; $m_j = -\frac{1}{2}, \frac{1}{2}$ states in terms of spherical harmonics and spin states.

Solution: *Clebsch-Gondan coefficients for* $j_1 = 1, j_2 = \frac{1}{2}$.

Let

$$\phi(j_1 j_2 jm) = \phi_{jm} \tag{i}$$

$$\text{and } \psi(j_1\, j_2\, m_1\, m_2) = \psi_{m_1 m_2}$$

The six independent vectors are

$$\psi_{1,\frac{1}{2}};\ \psi_{1,-\frac{1}{2}};\ \psi_{0,\frac{1}{2}};\ \psi_{0,-\frac{1}{2}};\ \psi_{-1,-\frac{1}{2}};\ \psi_{-1,-\frac{1}{2}}$$

$$\text{or, } \phi_{\frac{3}{2},\frac{3}{2}};\ \phi_{\frac{3}{2},\frac{1}{2}};\ \phi_{\frac{3}{2},-\frac{1}{2}};\ \phi_{\frac{3}{2},-\frac{3}{2}};\ \phi_{\frac{1}{2},\frac{1}{2}};\ \phi_{\frac{1}{2},-\frac{1}{2}}$$

Obviously

$$\phi_{\frac{3}{2},\frac{3}{2}} = \psi_{1,\frac{1}{2}} \tag{ii}$$

$$\text{and } \phi_{\frac{3}{2},-\frac{3}{2}} = \psi_{-1,-\frac{1}{2}} \tag{iii}$$

For the remaining ϕ vectors, let

$$\phi_{\frac{3}{2},\frac{1}{2}} = \alpha_1\psi_{1,-\frac{1}{2}} + \alpha_2\psi_{0,\frac{1}{2}} \tag{iv}$$

$$\phi_{\frac{3}{2},-\frac{1}{2}} = \alpha_3\psi_{0,-\frac{1}{2}} + \alpha_4\psi_{-1,\frac{1}{2}} \tag{v}$$

$$\phi_{\frac{1}{2},\frac{1}{2}} = \alpha_5\psi_{1,-\frac{1}{2}} + \alpha_6\psi_{0,\frac{1}{2}} \tag{vi}$$

$$\phi_{\frac{1}{2},-\frac{1}{2}} = \alpha_7\psi_{0,-\frac{1}{2}} + \alpha_8\psi_{-1,\frac{1}{2}} \tag{vii}$$

The coefficients α_i are elements of the unitary transformation matrix. It is to be kept in mind that all those coefficients are zero for which $m \neq m_1 + m_2$.

Now,

$$J_-\phi_{jm} = [(j+m)(j-m+1)]^{1/2}\phi_{j,m-1} \tag{viii}$$

where $J_- = J_x - iJ_y = [J_{1x} + J_{2x}] - i[J_{1y} + J_{2y}] = (J_1)_- + (J_2)_-$.

Also

$$(J_1)_-\psi_{m_1m_2} = \begin{cases} [(1+m_1)(2-m_1)]^{1/2}\psi_{m_1-1,m_2}, & \text{if } m_1 - 1 \geq -1 \\ 0, & \text{if } m_1 - 1 < -1 \end{cases} \tag{ix}$$

$$\text{and } (J_2)_-\psi_{m_1m_2} = \begin{cases} [(\frac{1}{2}+m_2)(\frac{3}{2}-m_2)]^{1/2}\psi_{m_1,m_2-1}, & \text{if } m_2 - 1 \geq -\frac{1}{2} \\ 0, & \text{if } m_2 - 1 < -\frac{1}{2} \end{cases} \tag{x}$$

So

$$J_-\phi_{\frac{3}{2},\frac{3}{2}} = [(\tfrac{3}{2}+\tfrac{3}{2})(\tfrac{3}{2}-\tfrac{3}{2})+1]^{1/2}\phi_{\frac{3}{2},\frac{1}{2}}$$

$$\phi_{\frac{3}{2},\frac{1}{2}} = \frac{1}{\sqrt{3}}J_-\psi_{1,\frac{1}{2}} = \frac{1}{\sqrt{3}}\left[\sqrt{(1+1)(2-1)}\,\psi_{0,\frac{1}{2}} + \sqrt{(\tfrac{1}{2}+\tfrac{1}{2})(\tfrac{3}{2}-\tfrac{1}{2})}\,\psi_{1,-\frac{1}{2}}\right] = \frac{1}{\sqrt{3}}\psi_{1,-\frac{1}{2}} + \sqrt{\tfrac{2}{3}}\,\psi_{0,\frac{1}{2}} \tag{xi}$$

which gives the value of α_1 and α_2.

Applying the lowering operator again, we get

$$J_-\phi_{\frac{3}{2},\frac{1}{2}} = \sqrt{2\times 2}\,\phi_{\frac{3}{2},\frac{1}{2}}$$

$$\text{or, } \phi_{\frac{3}{2},-\frac{1}{2}} = \tfrac{1}{2}[(J_1)_- + (J_2)_-]\left[\frac{1}{\sqrt{3}}\psi_{1,-\frac{1}{2}} + \sqrt{\tfrac{2}{3}}\,\psi_{0,\frac{1}{2}}\right]$$

$$= \tfrac{1}{2}\left[\frac{1}{\sqrt{3}}\left\{\sqrt{(1+1)(2-1)}\,\psi_{0,-\frac{1}{2}+0}\right\} + \sqrt{\tfrac{2}{3}}\left\{\sqrt{(1+0)(2-0)}\,\psi_{-1,\frac{1}{2}} + \sqrt{(\tfrac{1}{2}+\tfrac{1}{2})(\tfrac{3}{2}-\tfrac{1}{2})}\,\psi_{0,-\frac{1}{2}}\right\}\right]$$

$$\text{or, } \phi_{\frac{3}{2},-\frac{1}{2}} = \sqrt{\tfrac{2}{3}}\,\psi_{0,-\frac{1}{2}} + \sqrt{\tfrac{1}{3}}\,\psi_{-1,\frac{1}{2}} \tag{xii}$$

$$\text{and } \phi_{\frac{3}{2},-\frac{3}{2}} = \psi_{-1,-\frac{1}{2}} \tag{xiii}$$

Now, the state $\phi_{\frac{1}{2},\frac{1}{2}}$ is a linear combination of $\psi_{1,-\frac{1}{2}}$ and $\psi_{0,\frac{1}{2}}$, which is orthogonal to $\phi_{\frac{3}{2},\frac{1}{2}}$.

$$\therefore \alpha_1^*\alpha_5 + \alpha_2^*\alpha_6 = 0 \text{ or, } \frac{\alpha_6}{\alpha_5} = -\frac{\alpha_1^*}{\alpha_2^*} = -\frac{1/\sqrt{3}}{\sqrt{2/3}} = -1/\sqrt{2}.$$

$$\therefore \phi_{\frac{1}{2},\frac{1}{2}} = \alpha\left[\psi_{1,-\frac{1}{2}} - \frac{1}{\sqrt{2}}\psi_{0,\frac{1}{2}}\right].$$

The condition for normalisation gives

$$\phi_{\frac{1}{2},\frac{1}{2}} = \frac{1}{\sqrt{3}}\left[\sqrt{2}\,\psi_{1,\frac{1}{2}} - \psi_{0,\frac{1}{2}}\right] \qquad \text{(xiv)}$$

The application of lowering operator J_- again leads to

$$\phi_{\frac{1}{2},-\frac{1}{2}} = \frac{1}{\sqrt{3}}\left[\psi_{0,-\frac{1}{2}} - \sqrt{2}\,\psi_{-1,\frac{1}{2}}\right] \qquad \text{(xv)}$$

Thus,

$$\begin{bmatrix} \phi_{\frac{3}{2},\frac{3}{2}} \\ \phi_{\frac{3}{2},\frac{1}{2}} \\ \phi_{\frac{3}{2},-\frac{1}{2}} \\ \phi_{\frac{3}{2},-\frac{3}{2}} \\ \phi_{\frac{1}{2},\frac{1}{2}} \\ \phi_{\frac{1}{2},-\frac{1}{2}} \end{bmatrix} = \begin{bmatrix} 1 & 0 & 0 & 0 & 0 & 0 \\ 0 & \sqrt{\frac{1}{3}} & \sqrt{\frac{2}{3}} & 0 & 0 & 0 \\ 0 & 0 & 0 & \sqrt{\frac{2}{3}} & \sqrt{\frac{1}{3}} & 0 \\ 0 & 0 & 0 & 0 & 0 & 1 \\ 0 & \sqrt{\frac{2}{3}} & -\sqrt{\frac{1}{3}} & 0 & 0 & 0 \\ 0 & 0 & 0 & \sqrt{\frac{1}{3}} & -\sqrt{\frac{2}{3}} & 0 \end{bmatrix} \begin{bmatrix} \psi_{1,\frac{1}{2}} \\ \psi_{1,-\frac{1}{2}} \\ \psi_{0,\frac{1}{2}} \\ \psi_{0,-\frac{1}{2}} \\ \psi_{-1,\frac{1}{2}} \\ \psi_{-1,-\frac{1}{2}} \end{bmatrix}$$

Wave functions for the *p*-states of an electron

The p-state of the electron has angular momentum $l = 1$ and is represented by three spherical harmonics Y_1^1, Y_1^0 and Y_1^{-1}. The spin states are $|\uparrow\rangle$ and $|\downarrow\rangle$ corresponding to $S_z = +\frac{\hbar}{2}$ and $-\frac{\hbar}{2}$ respectively.

$$\therefore \psi_{1,\frac{1}{2}} = Y_1^1\,|\uparrow\rangle; \qquad \psi_{1,-\frac{1}{2}} = Y_1^1\,|\downarrow\rangle$$
$$\psi_{0,\frac{1}{2}} = Y_1^0\,|\uparrow\rangle; \qquad \psi_{0,-\frac{1}{2}} = Y_1^0\,|\downarrow\rangle$$
$$\psi_{-1,\frac{1}{2}} = Y_1^{-1}\,|\uparrow\rangle; \qquad \psi_{-1,-\frac{1}{2}} = Y_1^{-1}\,|\downarrow\rangle$$

» ***Example* 10.8.** Calculate the Clebsch-Gondan coefficients for $j_1 = j_2 = \frac{1}{2}$.

Solution: This is the case of two particles of spin $\frac{1}{2}$. These will be four independent states and possible values of j are 1 and 0.

$$\psi_{m_1 m_2} = \psi_{\frac{1}{2},\frac{1}{2}};\ \psi_{\frac{1}{2},-\frac{1}{2}};\ \psi_{-\frac{1}{2},\frac{1}{2}};\ \psi_{-\frac{1}{2},-\frac{1}{2}}$$
$$\text{and } \phi_{jm} = \phi_{1,1};\ \phi_{1,0};\ \phi_{0,0};\ \phi_{-1,-1}$$

Obviously, $\phi_{1,1} = \psi_{\frac{1}{2},\frac{1}{2}}$ because $\psi_{\frac{1}{2},\frac{1}{2}}$ is the state for which $m_1 + m_2 = 1$.

Similarly, $\phi_{1,-1} = \psi_{-\frac{1}{2},-\frac{1}{2}}$.

The remaining states are

$$\phi_{1,0} = a_1\psi_{\frac{1}{2},-\frac{1}{2}} + a_2\psi_{-\frac{1}{2},\frac{1}{2}}$$
$$\phi_{0,0} = a_3\psi_{\frac{1}{2},-\frac{1}{2}} + a_4\psi_{-\frac{1}{2},\frac{1}{2}}$$

Now, $J_-\phi_{1,1} = \sqrt{2}\phi_{1,0}$.

$\therefore \phi_{1,0} = \frac{1}{\sqrt{2}}\left[(J_1)_- + (J_2)_-\right]\psi_{\frac{1}{2},\frac{1}{2}} = \frac{1}{\sqrt{2}}\left[\psi_{\frac{1}{2},-\frac{1}{2}} + \psi_{-\frac{1}{2},\frac{1}{2}}\right]$ and $\phi_{0,0}$ must be orthogonal to $\phi_{1,0}$; so we write, $\phi_{0,0} = \frac{1}{\sqrt{2}}\left[\psi_{\frac{1}{2},-\frac{1}{2}} - \psi_{-\frac{1}{2},\frac{1}{2}}\right]$.

The matrix transformation is given below:

$$\begin{bmatrix} \phi_{1,1} \\ \phi_{1,0} \\ \phi_{0,0} \\ \phi_{1,-1} \end{bmatrix} = \begin{bmatrix} 1 & 0 & 0 & 0 \\ 0 & \frac{1}{\sqrt{2}} & \frac{1}{\sqrt{2}} & 0 \\ 0 & \frac{1}{\sqrt{2}} & -\frac{1}{\sqrt{2}} & 0 \\ 0 & 0 & 0 & 1 \end{bmatrix} \begin{bmatrix} \psi_{\frac{1}{2},\frac{1}{2}} \\ \psi_{\frac{1}{2},-\frac{1}{2}} \\ \psi_{-\frac{1}{2},\frac{1}{2}} \\ \psi_{-\frac{1}{2},-\frac{1}{2}} \end{bmatrix}$$

For $j = 0$, there is only one vector $\phi_{0,0}$ and this is called the *singlet state*. The vectors $\phi_{1,1}$; $\phi_{1,0}$; $\phi_{1,-1}$ belong to triplet state with $j = 1$.

» ***Example* 10.9.** Show that $[\sigma_x, \sigma_y] = 2i\sigma_z$.

Solution: We have $\sigma_x = \begin{bmatrix} 0 & 1 \\ 1 & 0 \end{bmatrix}$, $\sigma_y = \begin{bmatrix} 0 & -i \\ i & 0 \end{bmatrix}$, $\sigma_z = \begin{bmatrix} 1 & 0 \\ 0 & -1 \end{bmatrix}$.

$$\therefore [\sigma_x, \sigma_y] = \begin{bmatrix} 0 & 1 \\ 1 & 0 \end{bmatrix}\begin{bmatrix} 0 & -i \\ i & 0 \end{bmatrix} - \begin{bmatrix} 0 & -i \\ i & 0 \end{bmatrix}\begin{bmatrix} 0 & 1 \\ 1 & 0 \end{bmatrix}$$

$$= \begin{bmatrix} i & 0 \\ 0 & -i \end{bmatrix} - \begin{bmatrix} -i & 0 \\ 0 & i \end{bmatrix} \quad (\because [\sigma_x, \sigma_y] = \sigma_x\sigma_y - \sigma_y\sigma_x)$$

$$= \begin{bmatrix} i & 0 \\ 0 & -i \end{bmatrix} + \begin{bmatrix} i & 0 \\ 0 & -i \end{bmatrix} = 2\begin{bmatrix} i & 0 \\ 0 & -i \end{bmatrix} = 2i\begin{bmatrix} 1 & 0 \\ 0 & -1 \end{bmatrix} = 2i\sigma_z$$

Similarly, $[\sigma_y, \sigma_z] = 2i\sigma_x$ and $[\sigma_z, \sigma_x] = 2i\sigma_y$.

» ***Example* 10.10.** A hydrogen atom is in the p-state ($l = 1, s = \frac{1}{2}$). Obtain the wave functions in terms of the spherical harmonics and the spin states.

Solution: Let us take $j_1 = l = 1$ and $j_2 = s = \frac{1}{2}$. Hence, the state of the atom can be written as

$$|jm\rangle = \psi_{jm} \tag{i}$$

Here, j is the total angular momentum of the atom and m is its z-component.

Now, the states $|jm\rangle$ can be expanded in terms of the states $|j_1m_1\rangle$ and $|-j_2m_2\rangle$ using CG coefficients as

$$|jm\rangle = \sum_{m_1,m_2} |j_1j_2m_1m_2\rangle\langle j_1j_2m_1m_2|jm\rangle \tag{ii}$$

where we have used the abbreviation

$$|j_1j_2m_1m_2\rangle = |j_1m_1\rangle \otimes |j_2m_2\rangle$$

For $j_1 = 1$, $m_1 = 1, 0, -1$ and for $j_2 = \frac{1}{2}$, $m_2 = \frac{1}{2}$; and $-\frac{1}{2}$.

$\therefore$ possible $|j_1m_1\rangle$ states are $|1,1\rangle$, $|1,0\rangle$ and $|1,-1\rangle$ and $|j_2m_2\rangle$ states are $|\frac{1}{2},\frac{1}{2}\rangle$ and $|\frac{1}{2},-\frac{1}{2}\rangle$.

j can take values from $|j_1 - j_2|$ to $j_1 + j_2$, i.e., $j = \frac{3}{2}$ and $\frac{1}{2}$.

Corresponding to $j = \frac{3}{2}$, $m = \frac{3}{2}, \frac{1}{2}, -\frac{1}{2}, -\frac{3}{2}$ and corresponding to $j = \frac{1}{2}$, $m = \frac{1}{2}$ and $-\frac{1}{2}$.

$\therefore$ possible $|jm\rangle$ states are $|\frac{3}{2},\frac{3}{2}\rangle$, $|\frac{3}{2},\frac{1}{2}\rangle$, $|\frac{1}{2},\frac{1}{2}\rangle$, $|\frac{3}{2},-\frac{1}{2}\rangle$, $|\frac{1}{2},-\frac{1}{2}\rangle$ and $|\frac{3}{2},-\frac{3}{2}\rangle$.

Using (i) and (ii), coefficients for $j_1 = 1$ and $j_2 = \frac{1}{2}$, we get

$$|\tfrac{3}{2},\tfrac{3}{2}\rangle = |1,1\rangle \otimes |\tfrac{1}{2},\tfrac{1}{2}\rangle$$

$$|\tfrac{3}{2},\tfrac{1}{2}\rangle = \frac{1}{\sqrt{3}}|1,1\rangle \otimes |\tfrac{1}{2},-\tfrac{1}{2}\rangle + \sqrt{\tfrac{2}{3}}|1,0\rangle \otimes |\tfrac{1}{2},\tfrac{1}{2}\rangle$$

$$|\tfrac{1}{2},\tfrac{1}{2}\rangle = \sqrt{\tfrac{3}{2}}|1,1\rangle \otimes |\tfrac{1}{2},-\tfrac{1}{2}\rangle - \frac{1}{\sqrt{3}}|1,0\rangle \otimes |\tfrac{1}{2},\tfrac{1}{2}\rangle$$

$$|\tfrac{3}{2},-\tfrac{1}{2}\rangle = \sqrt{\tfrac{2}{3}}|1,0\rangle \otimes |\tfrac{1}{2},-\tfrac{1}{2}\rangle + \frac{1}{\sqrt{3}}|1,-1\rangle \otimes |\tfrac{1}{2},\tfrac{1}{2}\rangle$$

$$|\tfrac{1}{2},-\tfrac{1}{2}\rangle = \frac{1}{\sqrt{3}}|1,0\rangle \otimes |\tfrac{1}{2},-\tfrac{1}{2}\rangle - \sqrt{\tfrac{2}{3}}|1,-1\rangle \otimes |\tfrac{1}{2},\tfrac{1}{2}\rangle$$

$$|\tfrac{3}{2},-\tfrac{3}{2}\rangle = |1,-1\rangle \otimes |\tfrac{1}{2},-\tfrac{1}{2}\rangle$$

Now, the wave functions for the hydrogen atoms can be written as $\psi_{nlm} = R_{nl}(r)Y_l^m(\theta,\phi) \times$ spin state, where R_{nl} is the radial wave function and Y_l^m's are the spherical harmonics. The spin up state $|\frac{1}{2},\frac{1}{2}\rangle$ can be represented by $\begin{pmatrix}0\\1\end{pmatrix}$. Thus, we can write:

$$|\tfrac{3}{2},\tfrac{3}{2}\rangle = R_{nl}(r)Y(\theta,\phi)\begin{pmatrix}1\\0\end{pmatrix} = \begin{vmatrix}R_{nl}(r)Y_1^1(\theta,\phi)\\0\end{vmatrix}$$

$$|\tfrac{3}{2},\tfrac{1}{2}\rangle = \frac{1}{\sqrt{3}}R_{nl}(r)Y_1^1\begin{pmatrix}\theta\\\phi\end{pmatrix}\begin{pmatrix}0\\1\end{pmatrix} + \sqrt{\tfrac{2}{3}}R_{nl}Y_1^0(\theta,\phi)\begin{pmatrix}1\\0\end{pmatrix} = \begin{bmatrix}\frac{1}{\sqrt{3}}R_{nl}(r)Y_1^1(\theta,\phi)\\\sqrt{\tfrac{2}{3}}R_{nl}(r)Y_1^0(\theta,\phi)\end{bmatrix}$$

In a similar way we can write all the other states in terms of the spherical harmonics as a two-component spinors.

» ***Example* 10.11.** Consider an electron with orbital angular momentum l and spin $\frac{1}{2}$. Under spin orbit interaction the two possible values of j are $(l+\frac{1}{2})$ and $(l-\frac{1}{2})$. Show that the expectation values of S_z in the states $j = (l+\frac{1}{2})$ and $(l-\frac{1}{2})$ are $\left(\frac{m\hbar}{2l+1}\right)$ and $\left(\frac{-m\hbar}{2l+1}\right)$ respectively.

Solution: Taking $j_1 = 1$ and $j_2 = \frac{1}{2}$, we have $j = (j_1 + j_2)$ to $|j_1 - j_2| = (l+\frac{1}{2})$ and $(l-\frac{1}{2})$.

The z-component of J has $(2j+1)$ values, $(-j$ to $+j)$ denoted by m, then we have two sets of states for the electron as $|jm\rangle \equiv |l+\frac{1}{2}, m\rangle$ and $|l-\frac{1}{2}, m\rangle$.

The C.G. coefficient matrix for these states is given in the table below:

	j	$l+\frac{1}{2}$	$l-\frac{1}{2}$
		m	
m_1		m	m
	m_2		
$m-\frac{1}{2}$	$\frac{1}{2}$	$\sqrt{\dfrac{l+m+\frac{1}{2}}{2l+1}}$	$-\sqrt{\dfrac{l-m+\frac{1}{2}}{2l+1}}$
$m+\frac{1}{2}$	$-\frac{1}{2}$	$\sqrt{\dfrac{l-m+\frac{1}{2}}{2l+1}}$	$\sqrt{\dfrac{l+m+\frac{1}{2}}{2l+1}}$

$$\therefore |l+\tfrac{1}{2}, m\rangle = \sqrt{\left(\frac{l+m+\frac{1}{2}}{2l+1}\right)}|m-\tfrac{1}{2},\tfrac{1}{2}\rangle$$

$$\therefore \langle S_z\rangle_{j=l+\frac{1}{2}} = \left(\frac{l+m+\frac{1}{2}}{2l+1}\right)\tfrac{1}{2}\hbar + \left(\frac{l-m+\frac{1}{2}}{2l+1}\right)(-\tfrac{1}{2}\hbar)$$

$$\therefore \langle S_z\rangle_j = \frac{m\hbar}{2l+1}$$

Similarly,

$$\langle S_z\rangle_{j=l-\frac{1}{2}} = -\frac{m\hbar}{2l+1}$$

» ***Example* 10.12. Use the C.G. coefficients for $j_1 = j_2 = \frac{1}{2}$ to write the spin states for two particles of spin $\left(\frac{1}{2}\right)$ and determine the nature of symmetry of these functions.**

Solution: For $j_1 = j_2 = \left(\frac{1}{2}\right)$, we have $j = 1$ and 0. Corresponding to $j_1 = 1$, $m = 1, 0, -1$ and for $j = 0$, $m = 0$. Thus, we have four $|jm\rangle$ states, viz., $|1,1\rangle, |1,0\rangle, |1,-1\rangle$ and $|0,0\rangle$.

Using the value of C.G. coefficients, these can be written in terms of the states $|j_1, m_1\rangle \otimes |j_2, m_2\rangle$ as

$$|1,1\rangle = \alpha\alpha \tag{i}$$

$$|1,0\rangle = \frac{1}{\sqrt{2}}(\alpha\beta + \beta\alpha) \tag{ii}$$

$$|1,-1\rangle = \beta\beta \tag{iii}$$

and

$$|0,0\rangle = \frac{1}{\sqrt{2}}(\alpha\beta - \beta\alpha) \tag{iv}$$

where we have used the notations:

$$|j_1m_1\rangle \equiv |\tfrac{1}{2},\tfrac{1}{2}\rangle = \alpha \text{ (spin up) and } |\tfrac{1}{2},-\tfrac{1}{2}\rangle = \beta \text{ (spin down)}$$

$$|j_2m_2\rangle \equiv |\tfrac{1}{2},\tfrac{1}{2}\rangle = \alpha \text{ (spin up) and } |\tfrac{1}{2},-\tfrac{1}{2}\rangle = \beta \text{ (spin down)}$$

On interchanging α and β in the above four states, we note that (i), (ii) and (iii) remain unchanged while the state (iv) changes sign. Thus, the first three states are symmetric and the last one is antisymmetric. The state with $j = 1$ is a triplet one while the state with $j = 0$ is a singlet state.

» ***Example* 10.13. Show that the operator $(\vec{\sigma}_1 \cdot \vec{\sigma}_2)^n$, where $\vec{\sigma}_1$ and $\vec{\sigma}_2$ are Pauli matrices, depends linearly on the product $(\vec{\sigma}_1 \cdot \vec{\sigma}_2)$. Find the explicit form of this dependence.**

Solution: In Example 10.12, we note that if $j_1 = s_1 = \frac{1}{2}$ and $j_2 = s_2 = \frac{1}{2}$ are the spin values for two electrons, then (i), (ii), (iii) and (iv) are the states of the combined system of the electrons. If $\vec{S}$ is the spin operator for the combined system, then $\vec{S} = \vec{S}_1 + \vec{S}_2$ or, $S^2 = \vec{S}_1^2 + \vec{S}_2^2 + 2\vec{S}_1 \cdot \vec{S}_2$.

$$\therefore (\vec{S}_1 \cdot \vec{S}_2) = |0,0\rangle = -\frac{3\hbar}{2}(\vec{S}_1^2 + \vec{S}_2^2)|0,0\rangle = -\frac{\hbar^2}{2}\left(\frac{3}{4}+\frac{3}{4}\right)|0,0\rangle = -\frac{3\hbar^2}{4}|0,0\rangle \tag{i}$$

Hence, from (i), we have

$$(\vec{\sigma}_1 \cdot \vec{\sigma}_2)|0,0\rangle = -3|0,0\rangle \tag{ii}$$

Similarly, we can show that

$$\left.\begin{aligned}(\vec{\sigma}_1\cdot\vec{\sigma}_2)|1,1\rangle &= 1|1,1\rangle\\ (\vec{\sigma}\cdot\vec{\sigma}_2)|1,0\rangle &= 1|1,0\rangle\end{aligned}\right\} \quad \text{(iii)}$$

and $(\vec{\sigma}_1\cdot\vec{\sigma}_2)|1,-1\rangle = 1|1,-1\rangle$.

Thus, we see that

$$\left.\begin{aligned}(\vec{\sigma}_1\cdot\vec{\sigma}_z)|s=0,s_z\rangle &= -3|s=0,s_z\rangle\\ \text{and } (\vec{\sigma}_1\cdot\vec{\sigma}_z)|s=1,s_z\rangle &= 1|s=1,s_z\rangle\end{aligned}\right\} \quad \text{(iv)}$$

From (iv) we can write:

$$(\vec{\sigma}_1\cdot\vec{\sigma}_2)^n|s=0,s_z\rangle = (-3)^n|s=0,s_z\rangle$$

$$(\vec{\sigma}_1\cdot\vec{\sigma}_2)^n|s=1,s_z\rangle = |s=1,s_z\rangle$$

As the three triplet states and the singlet state together form a complete orthonormal basis in the space of the spin states of two electrons, it is sufficient to verify the linear dependence of $(\vec{\sigma}_1\cdot\vec{\sigma}_2)$ on $(\vec{\sigma}_1\cdot\vec{\sigma}_2)$ for these states.

Then let us suppose that

$$(\vec{\sigma}_1\cdot\vec{\sigma}_2)^n = C_1(\vec{\sigma}_1\cdot\vec{\sigma}_2) + C_2 \quad \text{(v)}$$

where the coefficients C_1 and C_2 are to be determined.

Applying both sides of the operator relation (v) to the triplet states and the singlet state, we get $C_1 + C_2 = 1$ and $-3C_1 + C_2 = (-3)^n$ so that $C_1 = \frac{1}{4}[1-(-3)^n]$ and $C_2 = \frac{1}{4}[3+(-3)^n]$.

For $n = 2$, $(\vec{\sigma}_1\cdot\vec{\sigma}_2)^2 = -2(\vec{\sigma}_1\cdot\vec{\sigma}_2) + 3$.

EXERCISES

10(A): Objective (or MCQ) Type Questions

Select the correct answer:

10.1. Eigenvalues of L_z are

(a) $(m+1)$ (c) $(m-1)\hbar$

(b) $m\hbar$ (d) $(m-1)$.

10.2. Eigenvalues of L^2 are

(a) $l^2\hbar^2$ (c) $l(l+1)\hbar^2$

(b) $(l+1)\hbar^2$ (d) $l(l+1)$.

10.3. Possible values of the z-component of spin angular momentum are given by

(a) $\pm h$ (c) $\pm\hbar$

(b) $\pm\frac{\hbar}{2}$ (d) $\pm 2\hbar$.

10.4. Which is correct

(a) $[\hat{\sigma}^2, \hat{\sigma}_x] = 2i\sigma_z$ (c) $[\hat{\sigma}^2, \hat{\sigma}_x] \neq 0$

(b) $[\hat{\sigma}^2, \hat{\sigma}_x] = 0$ (d) $[\hat{\sigma}^2, \hat{\sigma}_x] = \sigma_z$.

10.5. The value of $[J_z, J_-]$ is

(a) $\hbar J_-$ (c) $-\hbar J_-$

(b) J_- (d) $-\hbar J_+$.

10.6. $\vec{J} \times \vec{J}$ is given by

(a) $\vec{J}$ (c) $-i\vec{J}$

(b) $i\vec{J}$ (d) $-j$.

10.7. The p-state for $j = \frac{3}{2}$ has degeneracy

(a) two-fold (c) three-fold

(b) four-fold (d) non-degenerate.

10.8. The selection rule for m so that a transition can occur is

(a) $\Delta m = 0$ (c) $\Delta m = 0, \pm 1$

(b) $\Delta m = \pm 1$ (d) $\Delta m = -1$.

10.9. The selection rule for j so that a transition can occur is

(a) $\Delta j = \pm 1$ (c) $\Delta j = 0$

(b) $\Delta j = -1$ (d) $\Delta j = 0, \pm 1$.

10.10. If the state $|jm\rangle$ is a simultaneous eigenstate of J^2 and J_z, then $\langle J_x^2 \rangle$ is

(a) $\frac{1}{2}[j(j+1) - m^2]$ (c) $[j(j+1) - m^2]$

(b) $\frac{\hbar^2}{2}[j(j+1) - m^2]$ (d) $\frac{\hbar^2}{2}[j(j+1) - m]$.

ANSWERS

10.1. (b).	**10.3.** (b).	**10.5.** (c).	**10.7.** (b).	**10.9.** (d).
10.2. (c).	**10.4.** (b).	**10.6.** (b).	**10.8.** (c).	**10.10.** (b).

10(B): Short Answer Type Questions

10.1. Show that $[J^2, J_z] = 0$.

10.2. Show that $\psi(j_1m_1)\,\psi(j_2m_2)$ is an eigenvector of J^2 only for particular values of m_1 and m_2.

10.3. If L_x, L_y and L_z are orbital angular momentum operators, show that $[L_x, L_y] = i\hbar L_z$.

10.4. If J_x, J_y, J_z are angular momentum operators, then show that $[J_+, J_-] = 2\hbar J_z$, where $J_+ = J_x - iJ_y$ an $J_- = J_x - iJ_y$.

10.5. What are Clebsch-Gordan coefficients.

10.6. Mention the properties and selection rules for Clebsch-Gordan coefficients.

0.7. Show that an operator which commutes with J_x and J_y commutes with J_z also.

0.8. Show that if the angular momentum of a particle is $\frac{3}{2}$, then the angular momentum vector is of length $\sqrt{\left(\frac{3}{2}\cdot\frac{5}{2}\right)}\hbar$.

0.9. Prove that σ^2 commutes with each of the matrices σ_x, σ_y and σ_z.

0.10. Show that σ_x, σ_y and σ_z anti-commute when taken in pair.

Prove that $[\hat{x}, \hat{p}_x] = i\hbar 1$, where $1 = \begin{pmatrix} 1 & 0 \\ 0 & 1 \end{pmatrix}$ is the unit matrix.

0.11. Determine the eigenvalues of L^2 and L_z.

10(C): Broad Answer Type Questions

0.1. Derive the commutation relation for the components of angular momentum L_x, L_y, L_z and show that all the three components commute with $L^2 = L_x^2 + L_y^2 + L_z^2$. Derive eigenvalues of L^2 and L_z.

0.2. Derive the eigenvalues of operator J^2 and J_z, where J^2 and J_z represent the square and the z-component of angular momentum operator.

0.3. If ψ_m is an eigenfunction of L_z with eigenvalue $m\hbar$, then show that $(L_x \pm iL_y)\psi_m$ is also an eigenfunction of L_z having eigenvalue $(m \pm \nu)\hbar$, where ν is equal to a non-negative integer.

0.4. Derive eigenfunctions of L^2 and L_z. Discuss the reason which shows that L is not the total angular momentum in case of the electron.

0.5. Derive Pauli eigenvalues and eigenfunctions. Discuss the commutation relations satisfied by $\hat{\sigma}^2$ and $\hat{\sigma}_x$, $\hat{\sigma}_y$ and $\hat{\sigma}_z$.

0.6. Find out eigenvalues of J_+ and J_-. Obtain eigenfunctions of J^2 and J_z.

0.7. Define angular momentum operator in terms of the commutation relations between its components. Use this definition to find out the various allowed values that this operator can take.

0.8. Calculate eigenvalue spectrum of raising and lowering operators $L_\pm = L_x \pm iL_y$ in a representation in which both L^2 and L_z are diagonal.

0.9. If $J_+ = J_x + iJ_y$ and $J_- = J_x - iJ_y$, then show that $[J^2, J_\pm] = 0$, $[J_z, J_\pm] = \pm\hbar J_\pm$ and $[J_+, J_-] = 2\hbar J_z$. Interpret $J_+ (J_-)$ as raising (lowering) operator and derive the result, $\langle J_\pm | jm\rangle = [j(j+1) - m(m \pm 1)]^{1/2}\hbar | j, m \pm 1\rangle$.

0.10. What are Clebsch-Gordan coefficients? Determine all the coefficients for $j_1 = j_2 = \frac{1}{2}$.

0.11. Find the matrix representation for spin angular momentum. Show that the spin angular moments of two electrons may be combined. Obtain the eigenvalues and eigenfunction of total angular momentum operators.

0.12. (a) Prove that no two of the three components of angular momentum L commute with each other, but all of them commute with L^2.

(b) Deduce the eigenvalues of total angular momentum operator and its components along z-axis.

10.13. Derive the selection rules governing transition of energy from one level to another.

10.14. For a hydrogen-like atom, neglecting the spin-orbit interaction, construct eigenstates which are simultaneously eigenstates of H, L^2, L_z and S_z.

10.15. How does quantised angular momentum vector differ from a classical angular momentum vector? What is the length of the angular momentum vector in a state with $L = 2$? What are the possible z-components of $\vec{L}$? **[BU (H) 1994]**

10(D): Numerical Problems

10.1. Show that $x_{nm} \neq 0$ unless $E_n - E_m = \pm\hbar\omega$ for harmonic oscillator whose Hamiltonian is $H = \frac{p^2}{2\mu} + \frac{1}{2}kx^2$ and $[x, p] = i\hbar$.

10.2. In the electron spin-orbit interaction the two possible values of j are $l + \frac{1}{2}$ and $l - \frac{1}{2}$. Show that the expectation values of s_z in the states $j = l + \frac{1}{2}$ and $j = l - \frac{1}{2}$ are $+\frac{m_j\hbar}{(2l+1)}$ and $-\frac{m_j\hbar}{(2l+1)}$ respectively.

10.3. A spin $\left(\frac{1}{2}\right)$ particle moves in a central field. Find the wave function of this particle which is simultaneously an eigenfunction of the three commutating operators $J_z = L_z + S_z$, L^2 and J^2.

10.4. A particle of spin $\frac{1}{2}\hbar\sigma$ and magnetic moment $\mu\hbar\sigma$ is placed in a uniform magnetic field $(Be^{-\alpha t}, 0, 0)$, where B and $\alpha > 0$ constant. If at time $t = 0$, the electron is certainly in the state $+\frac{\hbar}{2}$, show that after a very long time the probability of a transition to a state with z-component of spin $-\frac{1}{2}\hbar$ is $\sin^2\left(\frac{\mu B}{\alpha}\right)$.

10.5. An electron has orbital angular momentum $\vec{L}$ in addition to the intrinsic spin momentum $\vec{S} = \frac{1}{2}\vec{\sigma}$. Show that the projection operating to the state $j = l + \frac{1}{2}$ is

$$\frac{(l+1) + \vec{L}\cdot\vec{\sigma}}{(2l+1)}$$

10.6. Let the two spin $(-\frac{1}{2})$ states of an electron be $|\alpha\rangle, |\beta\rangle$. What will be the normalised singlet and triplet spin states formed by two electrons?

10.7. (a) Consider a single electron atom with orbital angular momentum $L = \sqrt{2}\hbar$. What are the possible values of a measurement of the z-component of L?

(b) Find out the different states in L-S coupling scheme for a two-electron atom with $l_1 = 2$ and $l_2 = 1$.

10.8. (a) Using vector diagram determine the possible values of the total angular momentum of an electron system for which $L = 3, S = 5/2$.

(b) Calculate the total angle and orbital angular momentum vector for an electron in $4D_{3/2}$ state. **[CU (H) 2010]**

(c) The quantum numbers of two electrons in a two-valence electron atom are $n_1 = 6$, $l_1 = 3$, $s_1 = \frac{1}{2}$, $n_2 = 5$, $l_2 = 1$, $s_2 = \frac{1}{2}$.

(i) Assuming L-S coupling, find the possible values of L and hence J.

(ii) Assuming J-J coupling, find possible values of J.

10.9. (a) Find the eigenvalues and normalised eigenvectors of the matrix $\begin{pmatrix} 5 & 2 \\ 2 & 2 \end{pmatrix}$.

(b) Consider the matrices $\sigma_x = \begin{pmatrix} 0 & 1 \\ 1 & 0 \end{pmatrix}$ and $\sigma_y = \begin{pmatrix} 0 & -i \\ i & 0 \end{pmatrix}$. Which of the matrices is/are Hermitian? Which of the matrices is/are unitary? Find $\sigma_x\sigma_y - \sigma_y\sigma_x$. **[CU (H) 2013]**

10.10. **Establish $[\hat{L}_x, \hat{L}_y] = i\hbar\, \hat{L}_z; \hat{L}^2, \hat{L}_z = 0$.** **[BU (H) 2000; KU (H) 2012, 2014, 2015]**

10.11. **Show that $[\hat{L}_x^2 + \hat{L}_y^2, \hat{L}_z] = [\hat{L}_x^2, \hat{L}_z] + \hat{L}_y^2, \hat{L}_z = 0$.** **[CU (H) 2006]**

10.12. **Write down angular momentum commutation relations. What is the implication of these results?** **[cf. CU (H) 2007, 2008]**

10.13. (a) **Find $\hat{L}_+\hat{L}_-$, $\hat{L}_-\hat{L}_+$**

(b) **Prove that $\hat{L}^2 = \frac{1}{2}[\hat{L}_+\hat{L}_- \hat{L}_-\hat{L}_+] + \hat{L}_z^2$.** **[cf. CU (H) 2015]**

10.14. **Find $[\hat{L}_x^2 + \hat{L}_y^2, \hat{L}_z^2]$, $[\hat{L}^2, \hat{L}_x]$, $[\hat{L}^2, \hat{L}_-]$, $[\hat{L}_z, \hat{L}_+]$, $[\hat{L}_z, \hat{L}_-]$, $[\hat{L}_x, \hat{L}_-]$.** **[cf. CU (H) 2012, 2014]**

10.15. **Establish $[\hat{L}_x, \hat{L}_y] = i\hbar\, \hat{L}_z$; $[\hat{L}^2, \hat{L}_z] = 0$. What conclusions about the eigenfunctions of the operators involved can be drawn from these relations?** **[BU (H) 2000; KU (H) 2012, 2014, 2015]**

Chapter 11

Identical Particles and Spin

11.1 Introduction

In a quantum mechanics, *identical particles* are those for which there will be no change in the system by interchanging the positions and spin coordinates. So *the word identical means to describe the particles that can be substituted for each other in the most general possible circumstances with no change in physical situation.* The wave packets of these particles do not overlap.

A pair of identical particles can be identified by the continuity of their trajectories which enables an observer to follow each particle. This is because the particles move in separate distinguishable, continuous orbits and their wave packets do not overlap. For example, volume available for a gas molecule at NTP is $\sim 10^{-25}\ \text{m}^3$ and the volume of a single molecule is $\sim 10^{-30}\ \text{m}^3$. So the volume of a molecule being much smaller than the volume available to it, we can in principle identify every molecule of the gas. Hence, the gas molecules are *distinguishable.*

Quantum mechanically, a particle is represented by a wave packet of finite size and spread. That is why the exact specification of the position and momentum of the particle simultaneously is restricted by Heisenberg's uncertainty principle $\Delta q \Delta p \simeq \hbar$.

Hence, there is no way of following the track of individual particles separately even if they interact with each other to an appreciable extent. The wave functions of the quantum mechanical particles overlap considerably making the particles indistinguishable.

For example, the volume available for each electron taking part in conduction within the metals is $\sim 10^{-28}\ \text{m}^3$, but the volume of a single electron is $\sim 10^{-27}\ \text{m}^3$. This shows that the electron wave functions overlap considerably and hence cannot be identified separately, i.e., they are *indistinguishable.*

In conclusion, particles are of two general categories, viz.,

1. Classical particles that are identical, but *distinguishable.*
2. Quantum particles that are identical and *indistinguishable.*

In *elastic collision* of particles, the component of spin along a particular axis is assumed unchanged and these particles can be distinguished provided they have different spin components.

11.2 Indistinguishability of identical particles

If two particles are identical, then one can be substituted for the other without any change left in the system.

For example, if two electrons in an atom have the same value of energy and angular momentum (i.e., same n and l), then they are said to be identical.

According to the principle of indistinguishability, the Hamiltonian of a system remains unchanged for a pair of identical particles. Taking the case of Helium atom, the Hamiltonian is given by

$$H = \sum_{i=1}^{2}\left(\frac{p_i^2}{2m} + V_i\right) = \frac{p_1^2}{2m} + \frac{p_2^2}{2m} - \frac{2e^2}{4\pi\varepsilon_0 r_1} - \frac{2e^2}{4\pi\varepsilon_0 r_2} + \frac{e^2}{4\pi\varepsilon_0 r_{12}} \tag{11.1}$$

where r_1, r_2 are the position coordinates of the two electrons w.r.t. the nucleus and r_{12} is their separation. If two particles are indistinguishable, then

$$H(1,2) = H(2,1) \tag{11.2}$$

In quantum mechanics, the electron is represented by a wave function ψ, which is again the solution of the wave equation. That is

$$H\psi = i\hbar\frac{\partial\psi}{\partial t} \tag{11.3}$$

$$\text{or, } H(1,2)\,\psi(1,2) = i\hbar\frac{\partial}{\partial t}\psi(1,2) \tag{11.4}$$

Interchanging the pair of particles, we have

$$H(2,1)\,\psi(2,1) = i\hbar\frac{\partial}{\partial t}\psi(2,1) \tag{11.5}$$

$$\therefore H(1,2)\,\psi(2,1) = i\hbar\frac{\partial}{\partial t}\psi(2,1)\ [\because H(1,2) = H(2,1)] \tag{11.6}$$

From eqs (11.5) and (11.6) it is clear that $\psi(1,2)$ and $\psi(2,1)$ are the wave functions of the same Hamiltonian. Further, the linear combination of the two functions is also a solution of the wave equation.

Now,

$$\psi(1,2) = \tfrac{1}{2}\,[\psi(1,2) + \psi(2,1)] + \tfrac{1}{2}\,[\psi(1,2) - \psi(2,1)] \tag{11.7}$$

suggests that each state is the superposition of symmetric state

$$\psi_S = \tfrac{1}{2}\,[\psi(1,2) + \psi(2,1)] \tag{11.8}$$

and antisymmetric state

$$\psi_A = \tfrac{1}{2}\,[\psi(1,2) - \psi(2,1)] \tag{11.9}$$

The separation of the two states ψ_S and ψ_A is invariant with time in view of eq. (11.3). Eq. (11.3) indicates that $\frac{\partial\psi}{\partial t}$ (i.e., the increment in ψ w.r.t. t) is of the same nature as ψ and as such ψ_S and ψ_A are orthogonal.

11.3 The exchange symmetry of wave functions

We consider a system of N identical particles with Hamiltonian $H(1,2,\ldots,N)$. Let the non-interacting N particle systems be described by the wave function

$$\psi(1,2,3,\ldots,i,\ldots,j,\ldots,N) \tag{11.10}$$

where 1 stands for all coordinates including position and spin, similarly for $2,3,\ldots,i,\ldots,j,\ldots,N$.

Now, an operator $\hat{P}_{ij}$, which interchanges all the coordinates of the ith and jth particles is defined as

$$\hat{P}_{ij}\psi(1,2,\ldots,i,\ldots,j,\ldots,N) = \psi(1,2,\ldots,j,\ldots,i,\ldots,N) \tag{11.11}$$

The effect of the operator $\hat{P}_{ij}$ is to interchange the particles i and j. Here $\hat{P}_{ij}$ is a Hermitian operator. The eigenvalue equation for the operator $\hat{P}_{ij}$ is expressed as

$$\hat{P}_{ij}\psi(1,2,\ldots,i,\ldots,j,\ldots,N) = P\psi(1,2,\ldots,i,\ldots,j,\ldots,N) \tag{11.12}$$

where P is the eigenvalue of $\hat{P}_{ij}$. Operating $\hat{P}_{ij}$ again, we get

$$\hat{P}_{ij}\left[\hat{P}_{ij}\psi(1,2,\ldots,i,\ldots,j,\ldots,N)\right] = P\hat{P}_{ij}\psi(1,2,\ldots,i,\ldots,j,\ldots,N)$$

$$\text{or, } \hat{P}_{ij}\left[\hat{P}_{ij}\psi(1,2,\ldots,i,\ldots,j,\ldots,N)\right] = P^2\psi(1,2,\ldots,i,\ldots,j,\ldots,N) \tag{11.13}$$

Operating $\hat{P}_{ij}$ once more eq. (11.11) becomes

$$\hat{P}_{ij}\left[\hat{P}_{ij}\psi(1,2,\ldots,i,\ldots,j,\ldots,N)\right] = \hat{P}_{ij}\psi(1,2,\ldots,i,\ldots,j,\ldots,N)$$

$$\text{or, } \hat{P}_{ij}\left[\hat{P}_{ij}\psi(1,2,\ldots,i,\ldots,j,\ldots,N)\right] = \psi(1,2,\ldots,i,\ldots,j,\ldots,N) \tag{11.14}$$

Now, comparing eqs (11.13) and (11.14), we get

$$P^2 = 1 \text{ or, } P = \pm 1 \tag{11.15}$$

So the eigenvalues of P are $+1$ and -1.

Hence,

$$\hat{P}_{ij}\psi(1,2,\ldots,i,\ldots,j,\ldots,N) = \pm\psi(1,2,\ldots,j,\ldots,i,\ldots,N) \tag{11.16}$$

11.4 Symmetric and antisymmetric wave functions

The Schrödinger equation for N identical particles is

$$H(1,2,\ldots,i,\ldots,j,\ldots,N)\,\psi(1,2,\ldots,i,\ldots,j,\ldots,N) = i\hbar\frac{\partial}{\partial t}\psi(1,2,\ldots,i,\ldots,j,\ldots,N) \tag{11.17}$$

where each of the numbers $1,2,\ldots,i,\ldots,j,\ldots,N$ denotes all the coordinates (position and spin) of one of the particles. Due to the identity of the particles, the Hamiltonian is symmetrical in its arguments. The solution of eq. (11.17) is of two types, viz.,

(1) Symmetric wave function (ψ_S)

A wave function is said to be symmetric if the interchange of any pair of particles among its arguments leaves the wave function unchanged.

The wave function ψ is said to be symmetric if it satisfies the condition:

$$\psi_S(1,2,\ldots,i,\ldots,j,\ldots,N) = +\psi_S(1,2,\ldots,j,\ldots,i,\ldots,N) \tag{11.18}$$

The particles which are described by symmetric wave function obey *Bose-Einstein* (B-E) *statistics* and hence are termed as *bosons.*

Example: Photons (spin 1), neutral helium atom (spin 0), α-particles (spin 0), deuteron (spin 1), π-meson (spin 0), etc.

(2) Antisymmetric wave function (ψ_A)

A wave function is said to be antisymmetric if the interchange of any pair of particles among its arguments changes the sign of the wave function.

A wave function ψ is said to be antisymmetric, if it satisfies the condition:

$$\psi(1,2,\ldots,i,\ldots,j,\ldots,N) = -\psi(1,2,\ldots,j,\ldots,i,\ldots,N) \tag{11.19}$$

The particles described by antisymmetric wave function obey *Fermi-Dirac* (FD) *statistics* and hence are called *Fermions.*

Example: Electrons, protons, neutrons, muons all have spin $\left(\frac{1}{2}\right)$, but Ω-particle is also a fermion having spin $\left(\frac{3}{2}\right)$.

The fermions described by antisymmetric wave functions are concerned with *Pauli principle.*

The symmetry properties represented by eqs (11.18) and (11.19) give rise to macroscopic phenomena like *super-fluidity of liquid helium at low temperature, blackbody radiation, magnetism, stimulated emission of radiation, shell structure in atoms and in nuclei,* etc.

11.5 The effect of exchange operation

The Hamiltonian of a system of two particles is expressed as

$$H = \frac{1}{2m}\left(p_1^2 + p_2^2\right) + V(1,2)$$

where m is the mass of each particle, $p_i^2 = p_{ix}^2 + p_{iy}^2 + p_{iz}^2$ and $V(1,2)$ is the potential energy of the system. The related operator for the system is

$$\hat{H}(1,2) = -\frac{\hbar^2}{2m}\left(\nabla_1^2 + \nabla_2^2\right) + V(1,2) \tag{11.20}$$

$$\text{and } \hat{H}(2,1) = -\frac{\hbar^2}{2m}\left(\nabla_2^2 + \nabla_1^2\right) + V(2,1)$$

$$= \frac{\hbar^2}{2m}\left(\nabla_2^2 + \nabla_1^2\right) + V(1,2) \text{ [since } V(1,2) = V(2,1)] \tag{11.21}$$

Hence, the Hamiltonian is symmetric w.r.t. the interchange of any pair of particles.

So $\hat{H}(1,2) = \hat{H}(2,1)$.

Now, the Schrödinger equation is

$$\hat{H}(1,2)\,\psi(1,2) = E\psi(2,1)$$

where E is the energy eigenvalue of $\hat{H}(1,2)$.

So

$$\hat{H}(2,1)\,\psi(2,1) = E\psi(2,1)$$

$$\hat{H}(1,2)\,\psi(1,2) = E\psi(2,1)$$

$\therefore \psi(2,1)$ is also a solution of the wave equation.

Hence, $\psi(1,2)$ and $\psi(2,1)$ are eigenfunctions of $\hat{H}$ having the same eigenvalue E, i.e., the system is *degenerate*. A linear combination of $\psi(1,2)$ and $\psi(2,1)$ is also an eigenfunction of belonging to the energy eigenvalue E. That is $\psi'(1,2) = A\psi(1,2) + B\psi(2,1)$ and $\hat{H}\psi'(1,2) = E\psi'(1,2)$.

11.6 Construction of symmetric and antisymmetric wave functions

The wave function obtained by the permutation of the arguments of ψ is also a solution of the Schrödinger wave equation. This is due to the fact that the same permutation applied throughout eq. (11.17) only relabel the particles without impairing its validity. Thus, the N arguments of ψ can be permuted in $N!$ ways and hence $N!$ solutions can be obtained from any one solution of eq. (11.17). Any linear combination of these $N!$ solutions is also a solution of eq. (11.17).

Now, in case of a two-particle system the Schrödinger *time-independent equation* is written as

$$\hat{H}\psi(1,2) = E\psi(1,2) \tag{11.22}$$

The two degenerate solutions of eq. (11.22) are

$$\psi(1,2) \text{ and } \psi(2,1)$$

Then the symmetric wave function is

$$\psi_S = \psi(1,2) + \psi(2,1) \tag{11.23}$$

and antisymmetric wave function is

$$\psi_A = \psi(1,2) - \psi(2,1) \tag{11.24}$$

It is to be kept in mind that $\psi(1,2)$ and $\psi(2,1)$ are unnormalised.

For a system consisting of three particles, the Schrödinger's equation is written as

$$H(1,2,3)\,\psi(1,2,3) = E\psi(1,2,3) \tag{11.25}$$

This equation has $3! = 6$ solutions relating to the same eigenvalue. The six possible functions are

$$\psi(1,2,3),\ \psi(2,3,1),\ \psi(3,2,1),\ \psi(1,3,2),\ \psi(2,1,3),\ \psi(3,1,2)$$

The functions arising out of even number of interchanges of the pair of particles in $\psi(1,2,3)$ are

$$\psi(1,2,3),\ \psi(2,3,1),\ \psi(3,1,2)$$

The functions arising out of odd number of interchanges are

$$\psi(2,1,3),\ \psi(1,3,2),\ \psi(3,2,1)$$

Now, the symmetric and antisymmetric wave functions can be written as

$$\psi_S = \psi(1,2,3) + \psi(2,3,1) + \psi(3,2,1) + \psi(1,3,2) + \psi(2,1,3) + \psi(3,1,2) \tag{11.26}$$

$$\text{and } \psi_A = [\psi(1,2,3) + \psi(2,3,1) + \psi(3,1,2)] - [\psi(1,3,2) + \psi(2,1,3) + \psi(3,2,1)] \tag{11.27}$$

Here, ψ_S and ψ_A are *unnormalised*.

For ψ_S an exchange of coordinates of the particles leaves both ψ and ψ^* unchanged.

Hence, the average value of p_s remains unchanged, since

$$\langle p_S \rangle = \int \psi_S^* \hat{p} \psi_S d\tau$$

In case of ψ_A the exchange of coordinates of the two particles changes the sign of both ψ_A and ψ_A^*. So

$$\langle p_A \rangle = \int \psi_A^* \hat{p} \psi_A d\tau$$

remains unchanged.

Hence, any interchange of two particles leaves the average of the observable property of the system unchanged. That is why the similar particles cannot be distinguished in quantum mechanics.

11.7 Particle exchange operator

In Art. 11.3, we have seen that the effect of operation of the particle exchange operator $\hat{P}_{ij}$ on a wave function ψ_{ij} representing a system consisting a number (N) of particles is to interchange of the subscripts of the spin (j) and position (i) variables of the wave function, i.e.,

$$\hat{P}_{ij}\psi(1,2,\ldots,i,\ldots,j,\ldots,N) = \psi(1,2,\ldots,j,\ldots,i,\ldots,N)$$

Here, we shall discuss the commutation relations obeyed by $\vec{P}_{ij}$.

11.7.1 Commutation Relation with Hamiltonian

If the system is consisted of two particles 1 and 2 and is described by $\psi(1,2)$, then

$$\hat{P}_{12}\psi(1,2) = \psi(2,1)$$
$$\text{and } \hat{P}_{12}\hat{H}(1,2)\,\psi(1,2) = \hat{H}(2,1)\,\psi(2,1)$$
$$\text{or, } \hat{P}_{12}\hat{H}(1,2)\,\psi(1,2) = \hat{H}(1,2)\,\psi(2,1)$$

[$\because$ $\hat{H}$ is symmetric for identical particles, i.e., $\hat{H}(1,2) = \hat{H}(2,1)$]

$$\therefore \hat{P}_{12}\hat{H}(1,2)\,\psi(1,2) = \hat{H}(1,2)\hat{P}_{12}\psi(1,2)$$
$$\text{or, } \left\{\hat{P}_{12}\hat{H}(1,2) - \hat{H}(1,2)\hat{P}_{12}\right\}\psi(1,2) = 0$$
$$\text{or, } \left\{\hat{P}_{12}\hat{H}(1,2) - \hat{H}(1,2)\hat{P}_{12}\right\} = 0, \; [\because \psi(1,2) \neq 0]$$
$$\therefore \left[\hat{P}_{12}, \hat{H}(1,2)\right] = 0 \tag{11.28}$$

Thus, $\hat{P}_{12}$ and $\hat{H}(1,2)$ commutes with each other and both have simultaneous eigenfunctions.

11.7.2 Commutation of Exchange Operators

Let the system have three identical particles 1, 2 and 3 and is described by the wave function $\psi(1,2,3)$. Then

$$\hat{P}_{12}\psi(1,2,3) = \psi(2,1,3)$$
$$\text{and } \hat{P}_{13}\hat{P}_{12}\psi(1,2,3) = \hat{P}_{13}\psi(2,1,3) = \psi(2,3,1) \tag{11.29}$$

Also

$$\hat{P}_{12}\hat{P}_{13}\psi(1,2,3) = \hat{P}_{12}\,(3,2,1) = \psi(3,1,2) \tag{11.30}$$

Hence,

$$\hat{P}_{12}\hat{P}_{13} \neq \hat{P}_{13}\hat{P}_{12} \tag{11.31}$$

$$\text{or, } \left[\hat{P}_{12}, \hat{P}_{13}\right] \neq 0$$

This shows that the particle exchange operators do not commute each other.

Again,

$$\hat{P}_{23}\hat{P}_{13}\psi(1,2,3) = \hat{P}_{23}\psi(3,2,1) = \psi(2,3,1) \tag{11.32}$$

From eqs (11.29) and (11.32)

$$\hat{P}_{13}\hat{P}_{12} = \hat{P}_{23}\hat{P}_{13} \tag{11.33}$$

In general,

$$\hat{P}_{ij}\hat{P}_{ik} = \hat{P}_{kj}\hat{P}_{ij} \tag{11.34}$$

Thus, it is impossible to find a complete set $N!$ degenerate eigenfunction of $\hat{H}$ which are all simultaneous eigenfunction of P (except the case $N = 2$).

Again, we can show that $\hat{P}_{ij}\,\psi_S = \psi_S$ and $\hat{P}_{ij}\,\psi_A = -\psi_A$ as below:

Since,

$$\psi_S = \psi(1,2) + \psi(2,1), \ [\text{for } i = 1 \text{ and } j = 2]$$

$$\therefore \hat{P}_{ij}\,\psi_S = \hat{P}_{ij}\psi(1,2) + \hat{P}_{ij}\psi(2,1) = \psi(2,1) + \psi(1,2) = \psi(1,2) + \psi(2,1) = \psi_S \tag{11.35}$$

Also

$$\hat{P}_{ij}\,\psi_A = \psi(2,1) - \psi(1,2) \ [\because \psi_A = \psi(1,2) - \psi(2,1)]$$

$$= -[\psi(1,2) - \psi(2,1)] = -\psi_A \tag{11.36}$$

11.8 Distinguishability of identical particles

Two identical particles in a system can be distinguished from each other if the sum of the probabilities of individual wave function in two states is equal to probability derived by the symmetries wave function.

For example, in an atom two electrons moving in different orbits are well distinguished but two electrons in the same orbits are not. Quantum mechanically two particles are distinguished if their wave functions do not overlap, i.e., the overlap terms $\psi(1,2)\,\psi^*(2,1) = 0$ and $\psi^*(1,2)\,\psi(1,2) = 0$.

According to the statement of distinguishability

$$|\psi(1,2)|^2 + |\psi(2,1)|^2 = |[\psi(1,2) \pm \psi(2,1)]|^2 = |\psi(1,2)|^2 + |\psi(2,1)|^2 \pm 2\text{Re}\,[\psi(1,2)\,\psi^*(2,1)] \tag{11.37}$$

where Re $[\psi(1,2)\,\psi^*(2,1)]$ denotes the real part of $[\psi(1,2)\,\psi^*(2,1)]$.

Now, from eq. (11.37), for distinguishability, $|\psi(1,2)|^2 + |\psi(2,1)|^2 = |[\psi(1,2) \pm \psi(2,1)]|^2$ if the overlap of wave functions $\psi(1,2)$ and $\psi(2,1) = 0$ or the interference term 2Re $[\psi(1,2)\,\psi^*(1,2)] = 0$.

Thus, if the space and spin coordinates of two particles are not the same, the interference term

$$2\text{Re}\,[\psi(1,2)\,\psi^*(2,1)] = 0$$

and the coordinates of the particles do not overlap.

11.9 The Pauli's exclusion principle

If in a system of many particles, the interaction between the particles be neglected, then the approximate (or, unperturbed) Hamiltonian becomes equal to the Hamiltonians of individual particles. Thus, for a system of n particles

$$H_0(1, 2, \ldots, n) = H_0'(1) + H_0'(2) + \cdots + H_0'(n) \tag{11.38}$$

The approximate eigenfunction is the product of individual particle eigenfunction of H_0', i.e.,

$$\psi(1, 2, \ldots, n) = \phi_a(1)\phi_b(2)\ldots\phi_k(n) \tag{11.39}$$

The total energy is given by

$$E = E_a + E_b + \cdots + E_k$$

where $E_a, E_b, \ldots, E_k$ being the energy eigenvalues corresponding to the eigenfunctions $\phi_a(1)$, $\phi_b(2), \ldots,$ $\phi_k(n)$ belonging to the operator $H_0'(1), H_0'(2), \ldots, H_0'(n)$ respectively.

For the electrons the wave function is antisymmetric $\psi_A(1, 2, \ldots, n)$, which can be constructed from the $\psi(1, 2, \ldots, n)$ of eq. (11.39). This antisymmetric wave function is expressed as the *Slater's determinant* of ϕ's as below:

$$\psi_A(1, 2, \ldots, n) = \begin{vmatrix} \phi_a(1) & \phi_a(2) & \ldots & \phi_a(n) \\ \phi_b(1) & \phi_b(2) & \ldots & \phi_b(n) \\ \ldots\ldots & \ldots\ldots & \ldots & \ldots\ldots \\ \phi_k(1) & \phi_k(2) & \ldots & \phi_k(n) \end{vmatrix} \tag{11.40}$$

From eq. (11.40) it is clear that $\psi_A(1, 2, \ldots, n)$ is *unnormalised* and is a solution of the approximate wave equation

$$H_0\psi_A = E\psi_A \text{ or, } (H_0 - E)\,\psi_A = 0 \tag{11.41}$$

Now, from eq. (11.40) it is evident that if $\phi_a(1) = \phi_b(1)$ or, $\phi_a(1) = \phi_a(2)$, then the determinant of eq. (11.40) reduces to zero. This means that an antisymmetric wave function $\psi_A(1, 2, \ldots, n)$ cannot be constructed from a solution which remains unchanged by the interchange of any pair of particles. Thus, H_0 has no solutions if there are more than one electron in any one of the states represented by $a, b, \ldots, k$. If the system is consisted of two identical noninteracting fermions, then this means that no two identical noninteracting fermions can be in states described by the same set of quantum numbers. This rule was first stated by Pauli (in 1925) and is known as *Pauli's exclusion principle* which states:

No two electrons in an atom can have the same set of quantum numbers n, l, m_l and m_s.

The principle forms the basis of understanding the structure of the atom.

11.10 Physical significance of Pauli's exclusion principle

The Pauli's exclusion principle along with *the principle of lowest possible energy* in the ground state forms the basis of *Aufbau principle* which helps to build up *Mendeleev's periodic table* and to explain the course of the ground state of the elements. The periodicity in the properties of the elements is understood from the repetition of the ground state configuration.

The group behaviour of the elementary particles having spin $\hbar/2$ is governed by the exclusion principle. As a consequence of this principle, new forces of non-dynamical character arise during the interaction of identical particles.

Illustration I. If x_1, p_1 and x_2, p_2 be the position and momentum coordinates of two electrons moving in the same line and ψ_A be the antisymmetric wave function of the two-electron system, then it can be shown that the probability of finding the first electron at x_1 and the second at x_2 is

$$\psi_A^* \psi_A = 1 - \cos\left[\frac{(p_1 - p_2)(x_1 - x_2)}{\hbar}\right]$$

If $x_1 \to x_2$, then $\psi_A^* \psi_A \to 0$, i.e., in addition to Coulomb repulsion a new type of repulsion is found. For a given smaller separation and definite difference in momenta, the repulsive force is large.

According to this principle, when atoms come to very short distances so that their electronic structure interpenetrates, then a strong repulsion is developed. This accounts the extremely small compressibility of liquids and solids.

Pauli's principle also explains the difference in behaviour of ordinary gas and an electron gas.

11.11 Shells and subshells in atoms

In an atom the nucleus is situated at the centre and the extranuclear electrons are moving in different stationary orbits around the nucleus. For such a system the total Hamiltonian of the electrons is given by

$$H = H_0 + H_1 + H_2 + H_3, \tag{11.42}$$

where

$$H_0 = \sum_i \left[\frac{p_i^2}{2m} - \frac{Ze^2}{4\pi\varepsilon_0 r_i} + V(r_i)\right] \tag{11.43}$$

$$H_1 = \frac{1}{2}\sum_{i\neq j} \frac{e^2}{4\pi\varepsilon_0 r_{ij}} - \sum_i V(r_i) \tag{11.44}$$

$$H_2 = \sum_i \frac{1}{2m^2c^2 r_i}\left(\frac{dV}{dr_i}\right)\vec{l}_i \cdot \vec{s}_i \tag{11.45}$$

The term H_3 arises due to *spin-spin interaction, relativistic correction,* etc. and is less important in the present case.

H_0 contains the potential due to the nucleus $\frac{Ze^2}{4\pi\varepsilon_0 r_i}$ at the position of the ith electron and some average potential $V(r_i)$ due to the electrons which screen the nuclear charge. $V(r_i)$ contributes significantly for $r_i \to \infty$. It can be shown that the screening contribution is given by

$$V(r) = \frac{(Z-1)e^2}{4\pi\varepsilon_0 r}\left(1 - e^{-\frac{r}{b}}\right) \tag{11.46}$$

$V(r)$ is accurate for the limiting values of r, i.e., for $r \to 0$ and $r \to \infty$. As $b > r$, we have

$$1 - e^{-\frac{r}{b}} = \left(\frac{1}{b} - \frac{1}{2b^2}r_i + \ldots\right) \tag{11.47}$$

$$\text{and } V(r) \simeq \frac{(Z-1)^2 e^2}{4\pi\varepsilon_0}\left[\frac{1}{b} - \frac{r_i}{2b^2} + \ldots\right] \tag{11.48}$$

H_1 given by eq. (11.44) expresses the difference between the actual repulsive potential $\frac{e^2}{4\pi\varepsilon_0 r_{ij}}$ and $V(r_i)$.

H_2 arises due to the spin-orbit interaction of the atomic electrons. H_0 determines the general structure of the atom. In the determination of the fine structure of the energy levels, the terms H_1 and H_2 play an important role.

Determination of the Structure of the Atom

Considering only the kinetic energy of the electrons and the electrostatic interaction of the electrons with the nucleus, the energy of the electrons is given by

$$E_n^{(0)} = -\frac{m}{2\hbar^2}\left(\frac{Ze^2}{4\pi\varepsilon_0}\right)^2 \frac{1}{n^2} \tag{11.49}$$

If the atom contains N electrons, then the total energy is given by

$$E^{(0)} = \sum_{i=1}^{N} E_n^{(0)}(i) \tag{11.50}$$

The distribution of electrons in different states is governed by the Pauli's exclusion principle. The ground state energy is obtained by placing electrons one by one in the lowest unoccupied states. For each vlaue of n (the principal quantum number), there are actually

$$\sum_{0}^{n-1} 2(2l+1) = 2n^2$$

states having same energy. Thus, for the states $n = 1, 2, 3$, etc., the number of electrons is 2, 8, 18, etc.

Thus, electrons with the same value of n form shell. The electrons belong to the same n are on the average at the same distance from the atomic nucleus. The shells are designated as K, L, M, N, etc. for $n = 1, 2, 3, 4$, etc.

Now, the average potential $V(r_i)$ will remove the degeneracy of the different l states with $l \leqslant n-1$. Then the states with same n and different l will have different energies. As $V(r_i)$ is positive and has significant contribution for increasing r_i the states with larger l values will be raised more than those with smaller l.

Now, it can be shown that

$$E_{n,l} \simeq E_n^{(0)} + \frac{(Z-1)\,e^2}{4\pi\varepsilon_0}\left[\frac{1}{b} - \frac{a_1}{4b^2 Z}\{3n^2 - l(l+1)\}\right] \tag{11.51}$$

where, we have used the relation

$$\langle r \rangle = \frac{a_1 n^2}{Z}\left[1 + \frac{1}{2}\left\{1 - \frac{l(l+1)}{n^2}\right\}\right]$$

a_1 being the radius of the first Bohr orbit with $Z = 1$. Then the screening effect due to other electrons removes the l-degeneracy and the energies $E_{n,l}$ increase as l increases. This means that for the same n, there are different *subshells* with different l-values. The energy of an electron in a particular shell depends to some extent on l.

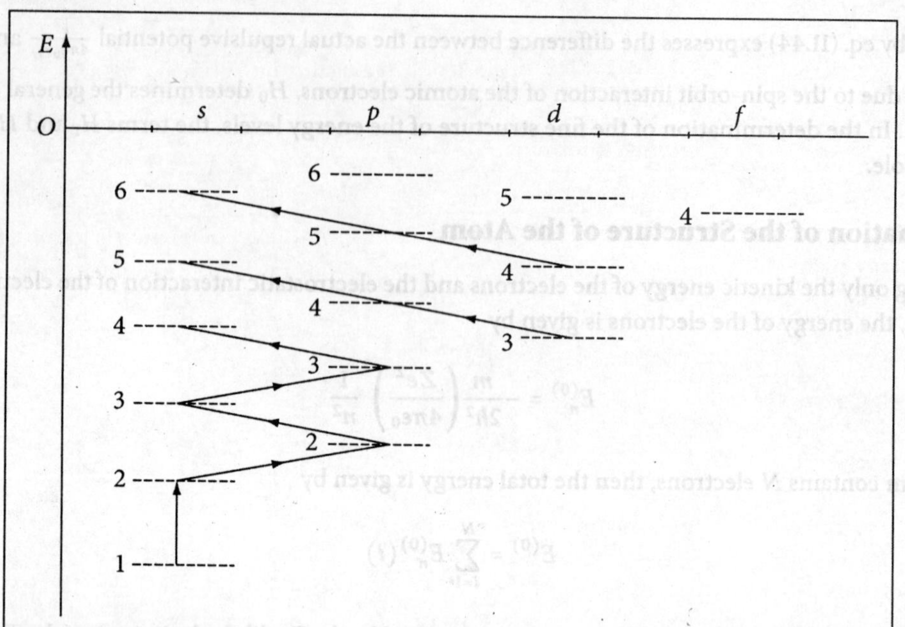

Fig 11.1 Energy levels of atomic subshells and the order of filling the subshells (shown by the arrows)

In general, the electron energy in each shell increases with l. However, the electrons in subshell have almost identical energies as the dependence of electron energy on m_l and m_s is extremely small. Energy of an electron in a subshell with sufficiently larger l may be higher than that of another electron with larger n, but lower l-values. Thus, for a given n, i.e., shell, there are different l and hence different subshells. The subshells are denoted by s, p, d, f, g, etc. for l = 1, 2, 3, 4, 5, etc.

The relative positions of the energy levels of atomic subshells and the order of filling the subshells is shown in Fig 11.1.

11.12 Electron configuration

If Z be the nuclear charge of an atom, then the charge neutrality demands that Z-electrons must be arranged outside the nucleus. But, according to the principle of mechanics, a system of particle is stable if its total energy is the minimum. In that view, the Z-electrons in the atom should occupy the lowest quantum state. But, Pauli's exclusion principle does not allow such a distribution of electrons in an atom. The *Building-up Principle* (*Aufbau principle*), however provides a way for the distribution of electrons in various shells and subshells in an atom. The above principle is based on two basic points, viz.,

1. the principle of the minimum energy of the system
2. the Pauli's exclusion principle.

Now, the electrons with total quantum number n belong to the same shell which are roughly on the same distance from the nucleus. These electrons interact with the same electric field and hence have similar energies. For a given n, electrons with same value of l belong to the same subshell. Subshells and the number of electrons in a particular subshell that can be occupied is shown in Table 11.1.

Table 11.1: No. of electrons in a particular subshell

Value of l	0	1	2	3	4	...
Subshell notations	*s*	*p*	*d*	*f*	*g*	...
No. of electrons	2	6	10	14	18	...

The energy of an electron for a particular n depends on l, but the dependence is less as compared to n. The energy of an electron (for a particular n) though depends on l, the dependence on m_l and m_s is extremely small.

Permitted number of electrons in a shell

We have for a given n, l can have the values

$$l = 0, 1, 2, 3, \ldots, (n-1)$$

and for a given l

$$m_l = 0, \pm 1, \pm 2, \pm 3, \cdots \pm l$$

So m_l can have $(2l+1)$ different values.

Again, the spin magnetic quantum number has values $m_s = \pm 1$.

Hence, in accordance with Pauli's principle, the maximum number of electrons in a shell is

$$N(n) = \sum_{l=0}^{n-1} 2(2l+1) = 2n^2$$

Thus, for s-, p- and d-subshell the number of electrons will be 2, 6 and 10 respectively.

We shall now find out the possible number of electrons and their distribution in a shell.

***K*-shell**: For K-shell, $n = 1$, $l = 0$, $m_l = 0$ and $m_s = \pm\frac{1}{2}$.

Hence, K-shell can contain a maximum of two electrons with $m_s = +\frac{1}{2}$ and $m_s = -\frac{1}{2}$. These electrons are called *s-electrons* and designated as $1s^2$.

***L*-shell**: For L-shell, $n = 2$, $l = 0, 1$. For $l = 0$, there will be $2s$-electrons and for $l = 1$, there will be six p-electrons as shown in Table 11.2.

Table 11.2: No. of electrons in the L-shell

n	*l*	m_l	m_s	*No. of electrons*	*Type*
2	0	0	$\pm\frac{1}{2}$	2	*s*-electrons
		1	$\pm\frac{1}{2}$		
2	1	0	$\pm\frac{1}{2}$	6	*p*-electrons
		−1	$\pm\frac{1}{2}$		

Thus, the total number of electrons in the L-shell is 8. The 6 electrons in p-state are designated as $2p^6$.

***M*-shell**: For M-shell, $n = 3$, $l = 0, 1, 2$. So these are two s-electrons for $l = 0$, six p-electrons for $l = 1$.

For $l = 2$, $m_l = -2, -1, 0, +1, +2$. For each value of m_l there are two electrons with $m_s = \pm\frac{1}{2}$. So, there are ten d-electrons for $l = 2$.

The distribution of electrons in different subshells under M-shell and the total number of electrons therein is shown in Table 11.3 below:

Table 11.3: No. of electrons in the M-shell

n	l	m_l	m_s	No. of electrons	Type
		2	$\pm\frac{1}{2}$		
		1	$\pm\frac{1}{2}$	$5 \times 2 = 10$	d-electrons
3	2	0	$\pm\frac{1}{2}$		
		−1	$\pm\frac{1}{2}$		
		−2	$\pm\frac{1}{2}$		

So, there can be 2 + 6 + 10 = 18 electrons in the M-shell. The 10 d-electrons are designated as $3d^{10}$.

Example: The electronic configuration of Na(11) is expressed as $1s^2 2s^2 2p^6 3s^1$. The electron configurations of first 30 elements of the periodic table are shown in Table 11.4.

Fig 11.1 shows the order in which shells are filled according to the increasing values of their energies. The order is $1s, 2s, 2p, 3s, 3p, 4s, 3d, 4p, 5s, 4d, 5p, 6s, 4f, 5d, 6p, 7s$.

Table 11.4: Order of filling the shells according to the increasing energy

		K ($n = 1$)	L ($n = 2$)		M ($n = 3$)			N ($n = 4$)	
Elements	Z	$l = 0$ (1s)	$l = 0$ (2s)	$l = 1$ (2p)	$l = 0$ (3s)	$l = 1$ (3p)	$l = 2$ (3d)	$l = 0$ (4s)	$l = 1$ (4p)
H	1	1							
He	2	2							
Li	3	2	1						
Be	4	2	2						
B	5	2	2	1					
C	6	2	2	2					
N	7	2	2	3					
O	8	2	2	4					
F	9	2	2	5					
Ne	10	2	2	6					
Na	11		There are		1				
Mg	12		10 electrons		2				
Al	13		in 1s to 2p		2	1			
Si	14		shells		2	2			
P	15				2	3			
S	16				2	4			
Cl	17				2	5			
Cr	18				2	6			

K	**19**	**There are 18 electrons**	—	**1**
Ca	**20**	**in 1*s* to 3*p* shells**	—	**2**
Sc	**21**		**1**	**2**
Ti	**22**		**2**	**2**
V	**23**		**3**	**2**
Cr	**24**		**5**	**1**
Mn	**25**		**5**	**2**
Fe	**26**		**6**	**2**
Co	**27**		**7**	**2**
Ni	**28**		**8**	**2**
Cu	**29**		**10**	**1**
Zn	**30**		**10**	**2**

11.13 The periodic table

According to *Mendeleev*, if the elements are arranged in order of atomic number (Z), then the similar chemical and physical properties repeat at regular intervals. This *periodic law* and the tabular arrangement of the elements exhibiting this property which repeats periodically constitute a *periodic table* (Fig 11.2).

	IA	IIA		IIIA	IVA	VA	VIA	VIA	VIIIA			IB	IIB		IIIB	IVB	VB	VIB	VIIB	VIIIB
1*s*	1 H	2 He																		
2*s*	3 Li	4 Be												2*p*	5 B	6 C	7 N	8 O	9 F	10 Ne
3*s*	11 Na	12 Mg												3*p*	13 Al	14 Si	15 P	16 S	17 Cl	18 Ar
4*s*	19 K	20 Ca	3d	21 Sc	22 Ti	23 V	24 Cr	25 Mn	26 Fe	27 Co	28 Ni	29 Cu	30 Zn	4*p*	31 Ga	32 Ge	33 As	34 Se	35 Br	36 Kr
5*s*	37 Rb	38 Sr	4d	39 Y	40 Zr	41 Nb	42 Mo	43 Tc	44 Ru	45 Rh	46 Pd	47 Ag	48 Cd	5*p*	49 In	50 Sn	51 Sb	52 Te	53 I	54 Xe
6*s*	55 Cs	56 Ba	5d	71 Lu	72 Hf	73 Ta	74 W	75 Re	76 Os	77 Ir	78 Pt	79 Au	80 Hg	6*p*	81 Tl	82 Pb	83 Bi	84 Po	85 At	86 Rn
7*s*	87 Fr	88 Ra	6d	89 Ac	90 Th	91 Pa	92 U	93 Np	94 Pu											
	1	2		1	2	3	4	5	6	7	8	9	10		1	2	3	4	5	6

	1	2	3	4	5	6	7	8	9	10	11	12	13	14
41	57 La	58 Ce	59 Pr	60 Nd	61 Pm	62 Sm	63 Eu	64 Gd	65 Tb	66 Dy	67 Ho	68 Er	69 Tm	70 Vb

Fig 11.2 Periodic Table of elements

The periodic table consists of *eight groups* and *seven periods*. The vertical columns containing the elements having similar properties are called *groups* of the periodic table. *Group* I consists of hydrogen and alkali metals. These elements have valency +1 and are chemically highly active. Halogens are included in *Group* VII which are all having valency −1 and are highly chemically active. These elements form diatomic molecules in the gaseous state. *Group* VIII consists of inert gases which are chemically inactive. The periodic table of elements is shown in Fig 11.2.

The horizontal rows in the periodic table are called *period* which are seven in number. We note the following transition along each period of the periodic table as follows:

Active metal ⟶ less active metal ⟶ weakly active non-metal ⟶ highly active non-metal ⟶ inert gas.

The change in property of elements within each period is much more prominent than that within each group. Due to this reason, the alkali metals are chemically more active with increasing atomic number. But, the halogens behave just in opposite order.

A series of *transition elements* appear in each period just after the third between *Group* II and *Group* III. These are all metals and have similar chemical properties, but do not show any similarity in behaviour with other elements in the groups. These transition elements are virtually indistinguishable. Fifteen of these occur in 6th period and are called *lanthanides* (or, rare earths). A similar group of fifteen elements in 7th period beyond uranium are called *actinides*.

Explanation of the arrangement in the periodic table

According to *Pauli's principle*, we have seen that (i) each shell contains a maximum of $2n^2$ electrons; (ii) each subshell can contain a maximum of $2(2l+1)$ electrons.

If the maximum possible number of electrons be present in a shell or subshell, then it is called *a closed shell* or *closed subshell*.

For example, a closed s-subshell has 2 electrons, a closed p-shell has 6 electrons, etc.

The total orbital and spin angular momenta in a closed shell vanish and the effective charge distribution of the electrons is perfectly symmetric.

Since, the positive nuclear charge is large compared to the inner shielding electrons in a closed shell, the electrons are very tightly bound. As an atom with closed shell is devoid of dipole moment, it has no attractive force on other electrons. The electrons of such an atom cannot be easily detached as well.

Consequently, the atoms with closed shell are chemically inert as, for example, the inert gases.

A single-valence electron in the outermost shell is relatively at a large distance from the nucleus of the atom and is shielded by the inner electrons from all, but an effective nuclear charge $+e$. Such atoms are of valency +1, like hydrogen and alkali metals. If, on the other hand, an atom lacks a single electron to have closed shell, then for the sake of stability it will have a tendency to acquire an electron through electrostatic attraction. These atoms are of valency −1, like the halogens. Similar considerations can be extended to other members of various groups in the periodic table.

The origin of transition elements lies in the fact that the s-electrons in a complex atom are more tightly bound than the d- or f-electrons.

For example, in potassium the outermost electron is in a $4s$ subshell instead of $3d$ subshell. In chromium and copper the binding energy between $3d$ and $4s$ subshells does not differ much because an additional $3d$-electron is present and a vacancy exists in the $4s$ subshell.

It has been indicated earlier that the order of filling the subshells is $1s, 2s, 2p, 3s, 3p, 4s, 3d, 5p, 5s, 4d, 4p, 6s, 4f, 5d, 6p, 7s, 6d$.

Now, all the *lanthanides* have configuration $5s^2 5p^6 6s^2$, but differ in complete $4f$ subshells whose filling does not affect the chemical properties. The *actinides* also have identical configuration $6s^2 6p^6 7s^2$, but the $5f$- and $5d$-electrons differ in number. This explains the similarity in chemical behaviour of *lanthanides* and *actinides*.

11.14 Ionisation potential

The least energy required to ionise the atom is called the *ionisation potential*. It is also equal to the binding energy of the least strongly bound electron. This electron generally belongs to last filled or unfilled shell.

We now consider an atom having Z electrons revolving around its nucleus and i of these Z electrons belong to the last subshell. The potential at the position of an electron in the last shell will be due to the nucleus with charge $+Ze$ screened by $(Z - i)$ electrons. The attractive potential is, therefore, equal to the charge $[Z - (Z - i)]e = +ie$. This shows that the ionisation potential increases with i. The ionisation potential is minimum for atoms with single electron in the outermost shell, e.g., Li, Na, K, Ga, Rb, and a maximum for atoms with the last shell being complete, e.g., He, Ne, Ar, Kr and Zn.

If some of the subshells have approximately the same energy or interpenetrate, then a rigorous analysis will be needed.

11.15 Chemical properties of elements

The origin of chemical properties in atoms is the forces between them.

(a) The ionisation energy of atoms is small in case of atoms with slightly filled outer shells. These atoms can form positive ions by losing electrons. These elements are chemically active.

(b) If there is small number of vacancies in the outer shell, then atoms can easily accommodate electrons in the unfilled shells although net charge on the atom is zero.

For example, Cl can accommodate an additional electron with a binding energy of 3.80 eV. By accepting an additional electron they form negative ions which are chemically active.

(c) The chemical activity is related to the number of electrons in the outer shell and there are groups of elements whose atoms have similar outer shells. These elements have similar chemical properties.

Examples:

1. Noble gases, like He, Ne, Ar, Kr, Xe and Rn have closed outer shells. Their ionisation energy is large and they cannot easily form positive ions. They also cannot bind an additional electron and hence hardly experience any attraction. These are, therefore, *chemically inactive* and are known as *inert gases*.
2. Alkali metals Li, Na, K, Rb, Cs, Fr have only one s-electron in the outer shell whose ionisation energy is small. Due to small binding energy, the outermost electron can be lost forming positive monovalent ions. These atoms, therefore, are *chemically active*.

3. The halogens, like F, Cl, Br, I, Ac all have five electrons in the outer p-shell. Naturally, they have a tendency to acquire an additional electron which would complete the outer shell. By capturing the additional electron the atom becomes a negative ion and chemically active.
4. The three series of elements, viz., $21 \leqslant Z \leqslant 30$ for $n = 3$, $39 \leqslant Z \leqslant 48$ for $n = 4$ and $71 \leqslant Z \leqslant 80$ for $n = 5$ are of special importance. They corresponds to progressive filling of an nd subshell, but have a complete $(n + 1)$ s shell. Because of their similar outer shell, all have similar chemical properties.
   Some of the elements, like Fe, Co and Ni have large magnetic moments due to their partially filled inner $3d$ shell.
5. The elements, $57 \leqslant Z \leqslant 70$ form a group of 14 elements which correspond to a progressive filling of the $4f$ subshell, $6s$ subshell being completed. These elements are called *rare earths* and have similar chemical properties. They have *high paramagnetic susceptibilities* also.
6. The third transition group from Hf $(Z = 72)$ to last three elements Os $(Z = 76)$, Ir $(Z = 77)$ and Pt $(Z = 78)$ all have the $5d$ subshell with gradual filling. These are placed in the eighth column.
7. In elements having $Z = 79$ (Au) up to $Z = 86$ (Rn), the $6s$ and $6p$ subshells of shell P are gradually filled up and at the end of the sixth period $Z = 86$ (Rn) is present which is inert chemically.
8. In Fr $(Z = 87)$ and Ra $(Z = 88)$ the extra electron goes to the $7s$ subshell of Q-shell. Then the *actinide series* begins with Ac $(Z = 89)$ and ends in *transuranic element* No $(Z = 102)$, the inner $5f$ subshell being gradually filled up.

In conclusion, the electronic structure consideration of the atoms of the elements presents an insight into their physical and chemical properties which led the Russian chemist *Dimitri Mendeleev* to build up the periodic table of elements.

11.16 The relative importance of residual Coulomb interaction and spin-orbit interaction

The characterisation of different energy levels for a given shell configuration (in particular, the ground state) can be made by considering the mutual interaction of the electrons represented by the Hamiltonian H_1 and the spin-orbit interaction represented by the Hamiltonian H_2 of an atom. This will be useful for predicting the multiplicity of the spectral lines and describing the behaviour of the atom in presence of an external magnetic field, etc.

The total angular momentum as well as spin angular momentum remain no longer conserved if the mutual interaction and spin-orbit interaction terms are taken into account. But, the total angular momentum $\vec{J} = \vec{L} + \vec{S}$ is conserved, i.e., $[L, H] \neq 0$, $[S, H] \neq 0$, but $[J, H] = 0$. In this case, the energy eigenstates can be represented by their total angular momentum.

Since, the Z-component of total orbital angular momentum and the total spin angular momentum are both zero for the assembly, it follows that

$$\sum_{m_l}^{l} m_l = 0 \text{ and } \sum_{m_s=-\frac{1}{2}}^{1/2} m_s = 0$$

$$\therefore \sum_{\text{Shell}} m_j = \sum_{\text{Shell}} (m_l + m_s) = 0$$

Hence, the total angular momentum is zero. The unfilled shells, therefore, contribute to the total angular momentum.

The contribution of H_1 and H_2 leads to the fact that all the atoms can be divided into two categories:

(i) The *residual mutual interaction* between the electron (i.e., H_1) is more important than *spin-orbit interaction* (i.e., H_2) for most of the atoms. This case is treated as *L-S coupling.*

(ii) For many heavy atoms with large nuclear charges, the *spin-orbit interaction* (i.e., H_2) is more important than the *residual mutual interaction* between the electrons (i.e., H_1). This is treated as *j-j coupling.*

11.17 Russell-Saunders (or *L-S* coupling)

The Hamiltonian due to residual mutual interaction between the electrons (H_1) is independent of S and commutes with S and also L. So the energy states of the atom can be represented by L and S. To deduce the final multiplicity of the energy state, the effect of spin-orbit interaction (H_2) is, then incorporated after finding the different states and their qualitative ordering.

Now, the state having largest spin, the spins are essentially parallel to each other. The associated wave function is, then antisymmetric. The minimum energy state is the state having the largest total spin S, as the electrons under such circumstance have a tendency to repel each other.

In view of the above, as the total spin decreases, the energy of the state increases. Similarly, the state having the largest L will correspond to the electrons moving in the same direction and they do not come together. Thus, we get *Hund's rule* which states:

The ground state has the largest possible L value compatible with the largest possible S value.

For the states with different J, but same L and S, the degeneracy is removed by the action of H_2 given by

$$H_2 = C_{LS}\,(\vec{L}\cdot\vec{S}) \tag{11.52}$$

where C_{LS} is a constant. H_2 gives a perturbative contribution to energy expressed as

$$\Delta E_J = \tfrac{1}{2}\, C_{LS}\,[J(J+1) - L(L+1) - S(S+1)] \tag{11.53}$$

Here, C_{LS} is positive for multiplets formed from a subshell which is half or less than half-filled. C_{LS} is, however, negative for the multiplets formed out of subshells which is more than half-filled. In this connection we can state the *multiplet rule* as below:

In case of a subshell, that is half or less filled, the energy within the multiplet increases as J increases, but for a subshell which is more than half-filled, the energy within the multiplet decreases as J increases.

The meaning of the multiplet rule is that the ground state of an atom has the smallest J value obeying Hund's rule provided the subshell is half or less filled, but the largest J value, if the subshell is more than half-filled. The separation of energy between the levels with $J+1$ and J with same value of L and S is given by

$$\begin{aligned} E_{J+1} - E_J &= \tfrac{1}{2}\, C_{LS}\,[(J+2)(J+1) - L(L+1) - S(S+1) - J(J+1) + L(L+1) + S(S+1)] \\ &= \tfrac{1}{2}\, C_{LS}\,[(J+1)(J+2-J)] = C_{LS}\,(J+1) \end{aligned} \tag{11.54}$$

Thus, the spacing between consecutive levels of a fine structure multiplet is proportional to the larger of the two J values of the levels. This is known as Landé interval rule.

Example 1. We consider two valence electrons in ns state and $n'l$ state $(l \neq 0)$. For the electron the degeneracy is $2(2l+1)$ and that due to s-electron is 2.

Hence, the total degeneracy of the state is $2 \cdot 2(2l+1) = 4(2l+1)$.

Here, the total wave function is expressed by the spatial part and the radial part given by

$$\psi_{n,n',L}(\vec{r}_1, \vec{r}_2) = R_n(r_1) Y_0^0(\theta_1, \phi_1) R_{n'}(r_2) Y_l^m(\theta_2, \phi_2) \tag{11.55}$$

$$\text{and } \nu = \nu^{m_s}(1)\nu^{m_s'}(2) \tag{11.56}$$

respectively.

For state ψ the total orbital angular momentum is for $L = l$. Thus, w.r.t. two electrons the state can be either symmetric or antisymmetric. The two states for $L = l$ are thus represented by the wave functions

$$\psi^{+}_{n,n',l} = \frac{1}{\sqrt{2}} [\psi_{n,n',l}(\vec{r}_1, \vec{r}_2) + \psi_{n,n',l}(\vec{r}_2, \vec{r}_1)] \tag{11.57}$$

$$\text{and } \psi^{-}_{n,n',l} = \frac{1}{\sqrt{2}} [\psi_{n,n',l}(\vec{r}_1, \vec{r}_2) + \psi_{n,n',l}(\vec{r}_2, \vec{r}_1)] \tag{11.58}$$

Again, since both the electrons have $s = 1/2$ the values of S are $S = 1, 0$. We can show that the three symmetric states

$$\nu_1^{+} = \nu^{m_s}(1)\nu^{m_s'}(2) + \nu^{m_s}(2)\nu^{m_s'}(1) \tag{11.59}$$

relating to $S = 1$ states. The antisymmetric state relating to $S = 0$ is

$$\nu_0^{-} = \nu^{1/2}(1)\nu^{-1/2}(2) - \nu^{1/2}(2)\nu^{-1/2}(1) \tag{11.60}$$

The total symmetric and antisymmetric wave functions ψ_{LS} are given by

$$\psi_{l,0} = \psi^{+}_{n,n',l}\nu_0^{-} \tag{11.61}$$

$$\text{and } \psi_{l,1} = \psi^{-}_{n,n',l}\nu_1^{+} \tag{11.62}$$

where $\psi_{l,0}$ and $\psi_{l,1}$ are $(2l+1)$ and $3(2l+1)$ in number respectively.

Now, $\psi_{l,1}$ states with largest S value, $S = 1$ have lower energy.

$\psi_{l,1}$ can have three J values, viz., $J = l+1, l, l-1$. Since, the degeneracy of these states has been removed due to spin-orbit interaction, the states with larger J value have greater energy. The total number of energy levels is $[(2l+1) + 3(2l+1)] = 4(2l+1)$ as shown in Fig 11.3.

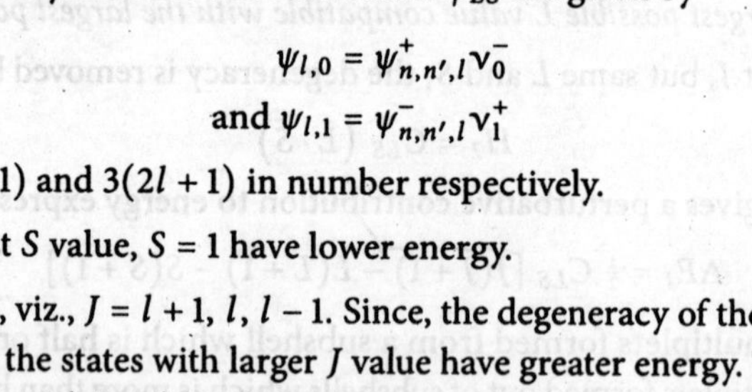

Fig 11.3 Schematic representation of fine structure splitting of a level with *ns* and *n′s* ($l \neq 0$) electrons

Since, the energy states are represented by $^{(2S+1)}L_J$ the singlet state, for $l = 1$, is represented by $^1P_1$. The triplet states are represented by $^3P_2$, $^3P_1$, $^3P_0$.

Now, there are only $S = 0$ and $S = 1$ levels for $l = 0$ and $n \neq n'$ with $J = 0$ and 1 respectively. The states are ${}^1S_0$ and ${}^3S_1$ respectively. For $l = 0$ and $n = n'$, there is only one level ${}^1S_0$ with $S = J = 0$, as the state $S = 1$ is not allowed by Pauli's exclusion principle.

Example 2. We now consider an atom with one valence electron in the np state and the other in the $n'l$ state. The total degeneracy in this case, is $12(2l + 1)$. Assuming $l \neq 0$ and $l \neq 1$ the wave functions are same as represented by eqs (11.61) and (11.62) with the allowed values of angular momentum quantum numbers $l + 1, l, l - 1$. The J degeneracy is removed by spin-orbit interaction. Fig 11.4 shows the final energy levels in the case under consideration. The states are represented by the notation ${}^{(2S+1)}L_J$.

If $l = 1$, but $n \neq n'$, there is only one level relating to $L = l - 1$ for each S with $J = 0$ for $S = 0$, and $J = 1$ for $S = 1$. The other levels for $L = l, l + 1$ are singlet or triplet as shown in Fig 11.4.

The case with $l = 1$ and $n = n'$, has the allowed values of $L = 2, 0$ for $\psi^+_{n,n,L}$ and $L = 1$ for $\psi^-_{n,n,L}$. So $S = 0$ state is associated with $L = 2, 0$ states and $S = 1$ state is associated with $L = 1$ state. But, $S = 1$ state splits into $J = 2, 1, 0$ state due to spin-orbit interaction.

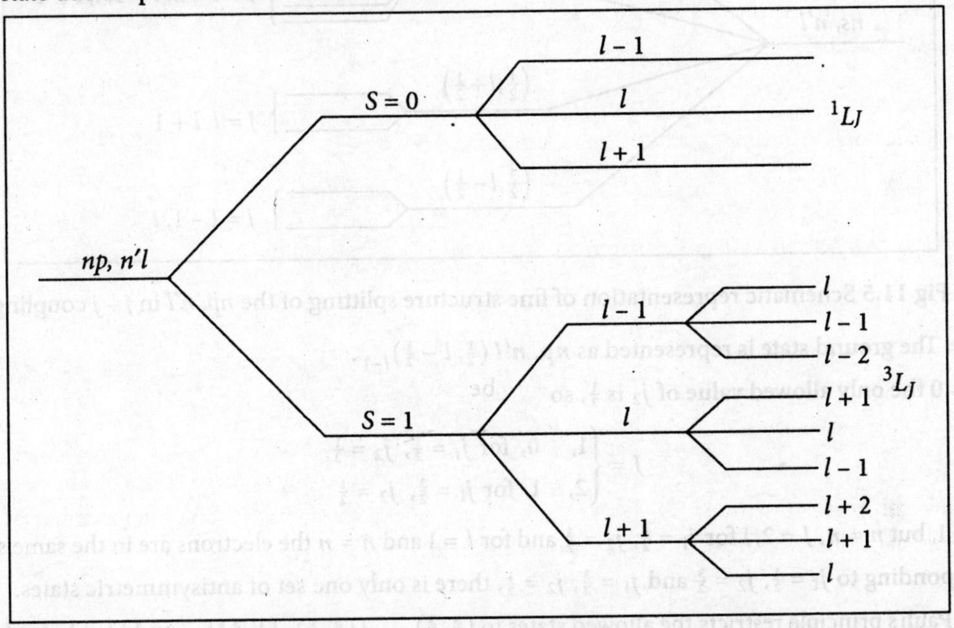

Fig 11.4 Schematic representation of the fine structure splitting of a level with np, $n'l$ ($l > 1$) electrons

11.18 *j-j* Coupling

There are some heavy atoms for which the spin-orbit interaction (H_2) is large compared to the residual electrostatic interaction (H_1). The electrostatic correlation energy splits the states of different J, where $J = \sum_i j_i$. For these heavy atoms the large $V(r)$ makes the spin-orbit energy for the individual electrons, the dominant perturbation. The *j-j* coupling scheme is used for such atoms. The number of levels depends on number of valence electrons, individual orbital angular momenta and Pauli's exclusion principle.

The separation energy between two levels into which the original single particle level is split up corresponding to $j_i = l_i \pm 1/2$ (except for $l_i = 0$) by the action of H_2 is given by

$$\Delta E_i = \tfrac{1}{2} C\left[j_i(j_i+1) - l_i(l_i+1) - \tfrac{3}{4}\right] \tag{11.63}$$

$$j_i = l_i \pm \tfrac{1}{2},\ l_i \neq 0$$

Here, C represents the expectation value of the coefficient of $(\vec{l}_i \cdot \vec{s}_i)$ in H_2.

For a given set of j_i the many-electron state contains states with different J, which are degenerate and this degeneracy can be removed by the residual interaction (electrostatics) between the electrons.

Fig 11.5 shows the fine structure splitting of the levels with one electron in the np state and another electron in the $n'l$ state ($l \geqslant 2$).

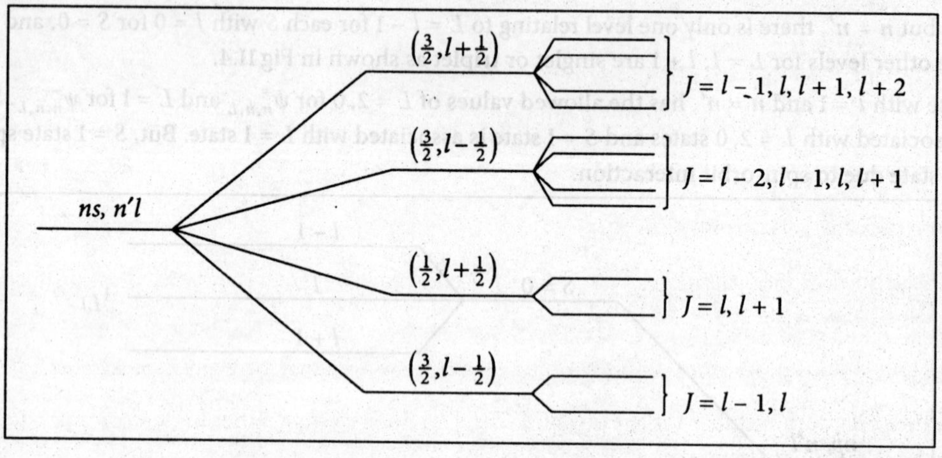

Fig 11.5 Schematic representation of fine structure splitting of the *np*, *n'l* in *j* – *j* coupling

Example: The ground state is represented as $np, n'l\left(\tfrac{1}{2}, l - \tfrac{1}{2}\right)_{l-1}$.

For $l = 0$ the only allowed value of j_2 is $\tfrac{1}{2}$, so

$$J = \begin{cases} 1, & 0, \text{ for } j_i = \tfrac{1}{2},\ j_2 = \tfrac{1}{2} \\ 2, & 1, \text{ for } j_1 = \tfrac{3}{2},\ j_2 = \tfrac{1}{2} \end{cases}$$

For $l = 1$, but $n \neq n$, $J = 2, 1$ for $j_1 = \tfrac{3}{2}, j_2 = \tfrac{1}{2}$ and for $l = 1$ and $n = n$ the electrons are in the same subshell. Corresponding to $j_1 = \tfrac{1}{2}, j_2 = \tfrac{3}{2}$ and $j_1 = \tfrac{3}{2}, j_2 = \tfrac{1}{2}$, there is only one set of antisymmetric states. Again, Pauli's principle restricts the allowed states to $\left(\tfrac{3}{2}, \tfrac{3}{2}\right)_{2,0}$, $\left\{\left(\tfrac{3}{2}, \tfrac{1}{2}\right), \left(\tfrac{1}{2}, \tfrac{3}{2}\right)\right\}_{2,1}$ and $\left(\tfrac{1}{2}, \tfrac{1}{2}\right)_0$ for $(j_1, j_2)_J$. The state $\left(\tfrac{1}{2}, \tfrac{1}{2}\right)_0$ represents the ground state. It is evident that the allowed J values as well as the number of final levels for both *L-S* coupling and *j-j* coupling schemes are the same.

It is to be noted that the *L-S* and *j-j* coupling schemes are applicable for lighter and heavier elements respectively.

For example, a lighter element carbon is described by *L-S* coupling scheme whereas for heavy element lead, *j-j* coupling scheme is applied. An intermediate coupling is suitable for silicon, germanium and tin.

11.19 Selection rules

The emerging spectral lines arise due to transition of the electric dipole moment. These transitions follow the selection rules.

Selection rules for *L-S* Coupling

1. Only one electron jumps in the transition between the levels in most common type of spectra.
2. The orbital angular momentum quantum number of the jumping electrons must change by unity, i.e., $\Delta l = \pm 1$.
3. For the atoms as a whole, the other quantum numbers change according to the following rule:

$\Delta S = 0$

$\Delta L = 0, \pm 1$

$\Delta J = 0, \pm 1$ but $J = 0$ to $J = 0$ not allowed

$\Delta M_J = 0, \pm 1$ but $M_J = 0$ to $M_J = 0$ not all allowed.

In this case, $\Delta J = 0$.

Selection rules for *j-j* coupling

1. The transitions in most common type of spectra take place between the levels when only one electron jumps at a time.
2. The orbital quantum number of the jumping electron must change by unity, i.e., $\Delta l = \pm 1$.
3. The inner quantum number for the jumping electron must change either by zero or unity, for the other electrons it must not change, i.e.,

$$\Delta j = \begin{cases} 0, \pm 1, & \text{for jumping electron} \\ 0, & \text{for all other electrons.} \end{cases}$$

4. For the atom as a whole, the other quantum numbers have the following change:

$$\Delta J = 0, \pm 1, \qquad J = 0 \text{ to } J = 0 \text{ not allowed}$$

$$\Delta M_J = 0, \pm 1, \quad M_J = 0 \text{ to } M_J = 0 \text{ not allowed, if } \Delta J = 0.$$

Illustration I. The mercury atom provides an illustration of the energy levels and the allowed transitions. For the ground state the atom has both the valence electrons in 6*s* shell. One of these electrons goes to $n'l$ state in the excited state. The energy levels for each configuration are the same as shown in Fig 11.3 except for $l = 0$ and $n' \neq n$. For this situation there are only two states $S = J = 0$ and $S = J = 1$. For $l = 0$ and $n' = n$, only $S = J = 0$ state exists.

In Fig 11.6, the energy levels according to *L-S* coupling scheme and the observed transitions are shown.

The transition $(6s)(6p)\,^3P_1 \to (6s)(6s)\,^1S_0$ does not follow the selection rule, $\Delta S = 0$ for *L-S* coupling scheme. The reason behind this observation is that all the atoms with $S = 1$ go down to $6\,^3P_1$ lowest energy triplet state and hence highly populated. Then the level becomes a mixture of $S = 0$ and $S = 1$ terms. There are some observed lines which violate the selection rule and are known as *intercombination or intersystem lines*.

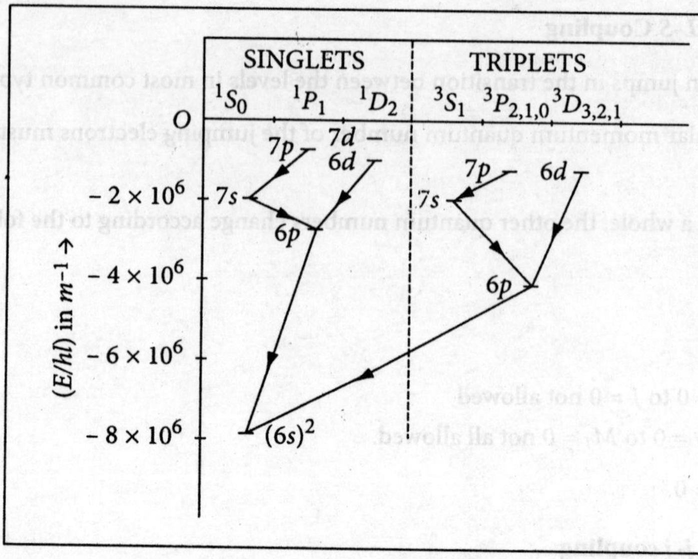

Fig 11.6 Some energy levels and allowed transitions for the mercury atom

11.20 X-rays

X-rays were discovered by W.C. Röntgen in 1895, while carrying out investigations on the discharge of electricity through rarefied gases. He found that when the pressure in the discharge tube is reduced to 0.001 mm Hg and an electric discharge is passed between the anode and the cathode, the glass wall behind the cathode begins to glow with a greenish-yellow colour. Performing a series of experiments Röntgen concluded that *when a beam of fast-moving electrons strikes a solid target an invisible high-penetrating radiation is produced.* Because of the unknown nature, Röntgen called these radiations as *X-rays* (also known as *Röntgen rays* after its discoverer).

Actually X-rays are *electromagnetic waves of very short wavelength*. The discovery of X-rays opened a new era in the history of Physics.

11.21 Production of X-rays

When a beam of fast-moving electrons strikes a target of suitable material, X-rays are produced. So the essential requirements for the production of X-rays are (a) a source of high energy electron (b) a target of high atomic weight.

X-ray tubes are of two types: (i) the gas-filled or Röntgen X-ray tube and (ii) the hot filament tube or Coolidge tube.

The modern X-ray tube was designed by the American physicist W.D. Coolidge in 1913. It consists essentially of a highly evacuated hard glass bulb (B) containing a cathode F and an anode (or anticathode) T. The cathode is a spiral of tungsten and emits electrons when heated to ~ 2000 °C by electric current [Fig 11.7]. The metallic reflector is a molybdenum cylinder kept at a negative potential to F. It focuses the thermoelectrons onto the anode A, which is made of copper with a target of tungsten or molybdenum. The requirements of the target are as follows:

(a) high atomic weight so that hard X-rays are reduced

(b) high melting point so that it does not melt easily

(c) high thermal conductivity so that the generated heat is conducted away.

The kinetic energy of the electron beam after traversing through a P. D. of ~ 50 to 200 kV becomes enormous and produces X-rays after the impact with the target (T). By regulating the filament current, the intensity of the emitted X-rays can be controlled. The target is cooled by the cooling arrangement to save it from the hand of damage.

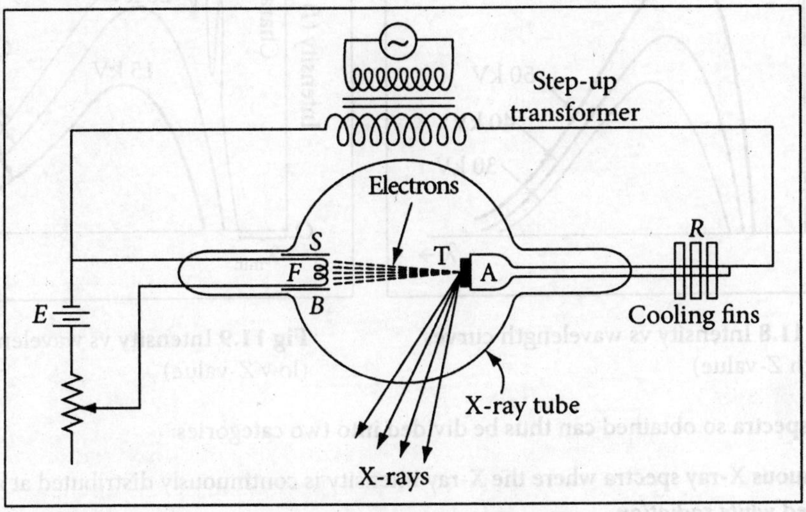

Fig 11.7 Coolidge X-ray tube

The Coolidge tube has the advantages over the gas-filled tube:

(i) the intensity of X-rays can be controlled by regulating the filament current;

(ii) the Coolidge tube is more stable than the gas-filled tube.

11.22 Variation of X-ray intensity with wavelength

The wavelength (λ) of the emitted X-rays from X-ray tube can be measured by a Bragg spectrometer and the corresponding intensity (I) by an ionisation chamber. If the intensity (I) be plotted against wavelength λ, we get curves as shown in Figs 11.8 and 11.9.

From these curves the following features are apparent:

(i) The intensity (I) increases with the increase of potential difference (V) between the anode and the cathode.

(ii) For the same potential difference, the intensity increases with the increase of the atomic number Z of the target. For a particular Z the peak increases with the increase of V.

(iii) There is a minimum value of $\lambda = \lambda_m$ below which the intensity $I = 0$. This λ_m depends on V, the applied voltage.

(iv) Sharp intensity peaks appear above a minimum limiting value of $V = V_m$ (say). V_m increases with the increase of Z.

(v) The wavelengths (λ) of peaks are *characteristic* of the target material and independent on V. These are *characteristic X-rays.*

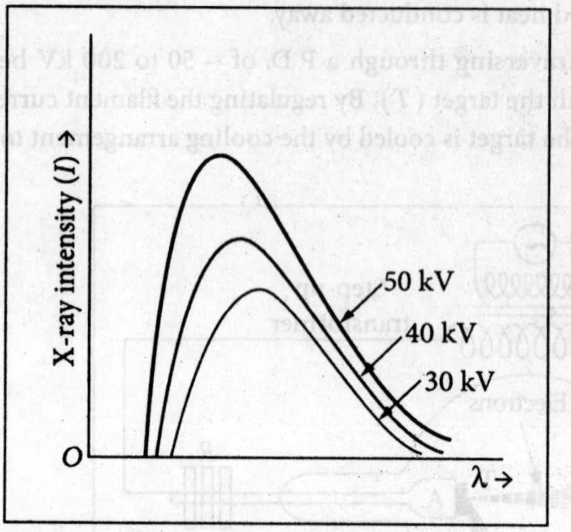

Fig 11.8 Intensity vs wavelength curves (high Z-value)

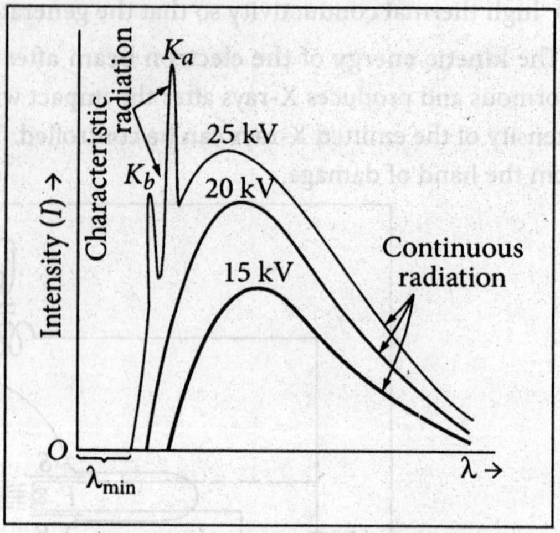

Fig 11.9 Intensity vs wavelength curve (low Z-value)

The X-ray spectra so obtained can thus be divided into two categories:

1. The continuous X-ray spectra where the X-ray intensity is continuously distributed at all wavelengths. It is also called *white radiation.*
2. The characteristic X-ray spectra where the sharp intensity peaks are found to appear at certain wavelengths depending on Z.

11.23 The origin of continuous X-ray spectra

The origin of the continuous X-ray spectra and the existence of the minimum wavelength can be understood from the quantum theory of radiation.

When a fast-moving electron penetrates deep into the interior of the target atom, it experiences an attractive force due to the nucleus. As a result the path of the electron is deviated from the original path. The electron, then undergoes deceleration and according to *classical electromagnetic theory* suffers a loss of energy. This loss of energy is given off in the form of electromagnetic radiation, i.e., X-rays of continuously varying wavelength. The X-rays so emitted are of continuous frequencies up to a minimum ν_m and minimum wavelength λ_{min}. These are continuous X-rays [Fig 11.10].

If ν_1 and ν_2 be the velocities of the electron of mass m before entering and leaving the nuclear electrostatic field respectively, then the loss in kinetic energy is $\frac{1}{2}m\nu_1^2 - \frac{1}{2}m\nu_2^2$. If this lost energy is converted to the energy of a photon of frequency ν, then we can write:

$$h\nu = \frac{1}{2}m\nu_1^2 - \frac{1}{2}m\nu_2^2 \tag{11.64}$$

where h is the Planck's constant.

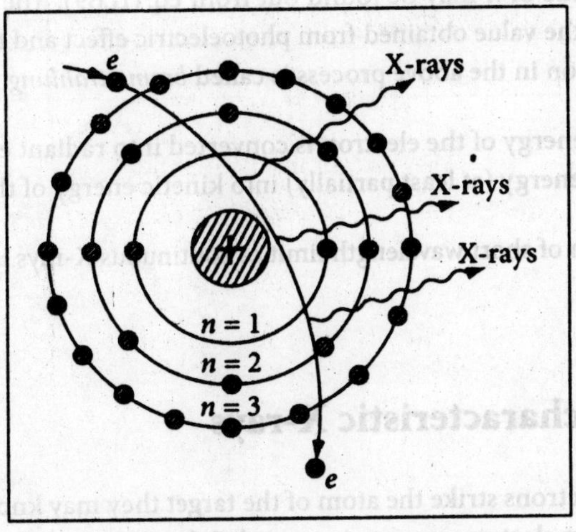

Fig 11.10 Emission of continuous X-rays

If the electron moves very close to the nucleus and loses whole of its energy in a single encounter, then the liberated photon has the maximum energy given by

$$\tfrac{1}{2} mv^2 = h\nu_{max} = \frac{ch}{\lambda_{min}} \quad [\because c = \nu\lambda] \tag{11.65}$$

If *e* be the charge and *V* the accelerating potential of the electron, then

$$\tfrac{1}{2} mv^2 = eV \tag{11.66}$$

From eqs (11.65) and (11.66)

$$\frac{ch}{\lambda_{min}} = eV \tag{11.67}$$

$$\text{or, } h\nu_{max} = eV \tag{11.68}$$

From eq. (11.67)

$$\lambda_{min} = \frac{ch}{eV} \tag{11.69}$$

Equation (11.68) is called the *Duane and Hunt law.*

Substituting the standard values of

$$c = 3 \times 10^8 \text{ m/s}$$
$$h = 6.62 \times 10^{-34} \text{ J-s}$$
$$\text{and } e = 1.602 \times 10^{-19} \text{ C}$$

we get from eq. (11.69)

$$\lambda_{min} = \frac{3 \times 10^8 \times 6.62 \times 10^{-34}}{1.602 \times 10^{-9} \times V} \text{ or, } \lambda_{min} = \frac{12413}{V \text{ (volt)}} \text{ Å} \tag{11.70}$$

Equation (11.70) shows that $\lambda_{min} \propto 1/V$ and is independent on the target material. If V = 10,000 volts, then λ_{min} = 1.24 Å.

Measuring λ_{min} the value of h may be found out from eq. (11.69). The value of h so determined is in excellent agreement with the value obtained from photoelectric effect and thermal radiation. The emission of electromagnetic radiation in the above process is called *bremsstrahlung*.

In this process kinetic energy of the electron is converted into radiant energy (X-rays), while in photoelectric effect the radiant energy (at least partially) into kinetic energy of the electron.

Hence, the phenomenon of short wavelength limit of continuous X-rays is called *the inverse photoelectric effect*.

11.24 Origin of characteristic X-rays

When the fast-moving electrons strike the atom of the target they may knock out an electron from one of the inner shells and thus an electron vacancy is created. Subsequently, an electron from an outer shell may jump into the vacancy by emitting a photon of energy equal to the energy difference of the two levels.

If an electron in the L-shell of energy (E_L) undergoes a transition to the vacancy created in the K-shell of energy (E_K), then the energy of the emitted X-ray photon is given by $h\nu = E_K - E_L$.

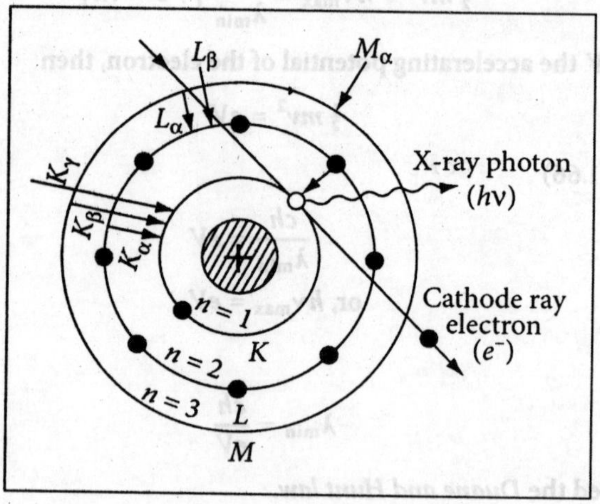

Fig 11.11 Origin of characteristic X-rays

The X-ray lines due to the transitions between the different X-ray levels are usually denoted by indices α, β, γ, etc. [Figs 11.11 and 11.12]. For example

K_α line : Transition from K to L level

K_β line : Transition from K to M level

L_α line : Transition from L to M level

L_β line : Transition from L to N level, etc.

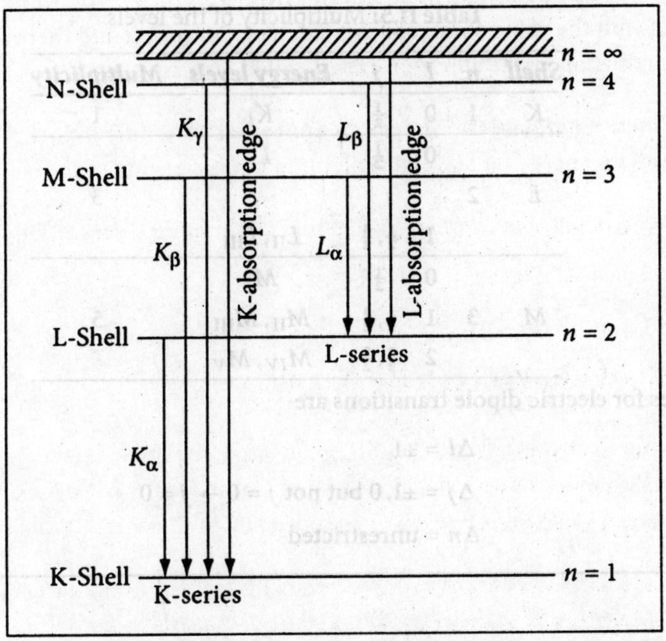

Fig 11.12 Energy level diagram for X-ray series spectra

The corresponding frequencies are as follows:

$$\text{For } K_\alpha : \nu(K_\alpha) = \frac{E_K - E_L}{h}$$

$$\text{For } K_\beta : \nu(K_\beta) = \frac{E_K - E_M}{h}$$

$$\text{For } L_\alpha : \nu(L_\alpha) = \frac{E_L - E_M}{h}$$

$$\text{For } L_\beta : \nu(L_\beta) = \frac{E_L - E_M}{h} \text{ and so on}$$

The characteristic X-ray spectral lines are independent of V, but depend on the material of the anode (i.e., on Z). The characteristic X-radiation is also termed as *fluorescent radiation*.

Note: $\because E_K > E_L > E_M, \ldots, \nu_K > \nu_L > \nu_M, \ldots$ or, $\lambda_K < \lambda_L < \lambda_M$.

11.24.1 Energy level diagram

A close look into the structure of the characteristic X-ray lines reveals that they are not single line, but each line is composed of a closely lying components. This *fine structure* of the lines can be understood on the basis of quantum numbers n, l and j associated with each energy level.

For a given l, the total quantum number j can have only two values, $j = l + \frac{1}{2}$ and $j = l - \frac{1}{2}$. Table 11.5 shows the multiplicity of the levels for different values of n, l and j.

Table 11.5: Multiplicity of the levels

Shell	*n*	*l*	*j*	*Energy levels*	*Multiplicity*
K	1	0	$\frac{1}{2}$	K_I	1
L	2	0	$\frac{1}{2}$	L_{I}	3
		1	$\frac{1}{2}, \frac{3}{2}$	$L_{\text{II}}, L_{\text{III}}$	
M	3	0	$\frac{1}{2}$	M_I	5
		1	$\frac{1}{2}, \frac{3}{2}$	$M_{\text{II}}, M_{\text{III}}$	
		2	$\frac{3}{2}, \frac{5}{2}$	M_{IV}, M_V	

The selection rules for electric dipole transitions are

$$\Delta l = \pm 1$$

$$\Delta j = \pm 1, 0 \text{ but not } j = 0 \rightarrow j = 0$$

$$\Delta n = \text{unrestricted}$$

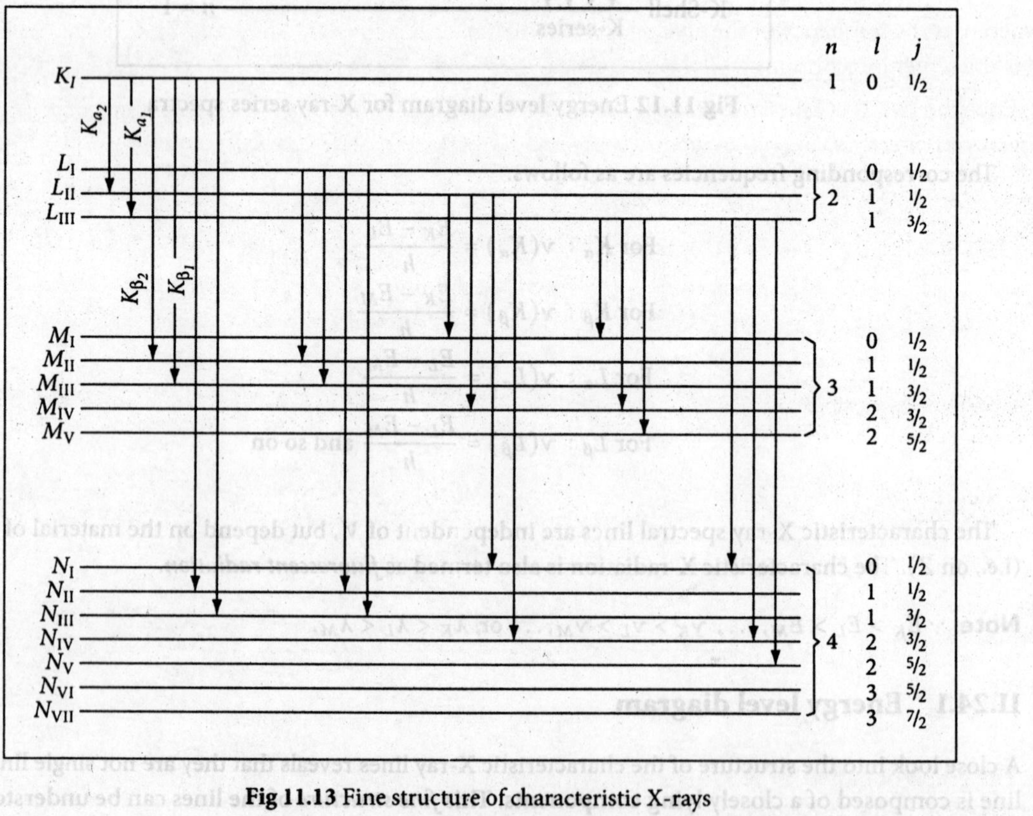

Fig 11.13 Fine structure of characteristic X-rays

The fine structure of X-ray characteristic lines is shown in Fig 11.13. From Fig 11.13 it is evident that *K*-line is a doublet K_{α_1} and K_{α_2}, and K_{β_1} and K_{β_2}. K_{α_1} is for $K \rightarrow L_{\text{III}}$, K_{α_2} for $K \rightarrow L_{\text{II}}$. The first few allowed transitions are shown in Fig 11.13. The binding energy of an electron in *K*-shell is much more higher than that of an L-shell. The energy and hence the frequency of K-lines are much more higher than those of L-lines. So

$$\nu_K > \nu_L > \nu_M > \nu_N > \ldots \text{ or, } \lambda_K < \lambda_L < \lambda_M < \lambda_N < \ldots$$

11.24.2 Moseley's law

An English physicist *H.G.J. Moseley* (1913–14) carried out systematic study of characteristic X-ray spectra on nearly 40 elements from aluminium to gold as targets in X-ray tube. He observed that

(i) the characteristic X-ray spectra of different elements are similar in the sense that each consists of K, L and M series;

(ii) in every series, the frequency of spectral lines emitted from an element of higher atomic number (Z) is greater than that produced by an element of lower atomic number.

Mathematically,

$$\nu \propto (Z - a)^2$$

$$\text{or, } \nu = b(Z - a)^2 \tag{11.71}$$

where ν is the frequency of the characteristic radiation, b is a constant, different for different elements and a is the screening constant which is also different for different elements.

Equation (11.71) is called the *Moseley's law* which states that *The frequency of a characteristic X-ray spectral line varies directly as the square of the atomic number of the emitting element.*

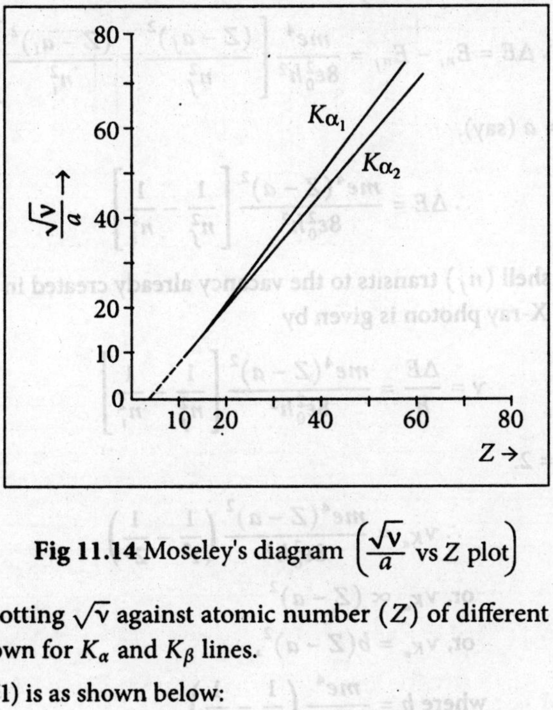

Fig 11.14 Moseley's diagram $\left(\frac{\sqrt{\nu}}{a} \text{ vs } Z \text{ plot}\right)$

The graph obtained by plotting $\sqrt{\nu}$ against atomic number (Z) of different elements is known as the *Moseley's diagram* and is shown for K_α and K_β lines.

The exact form of eq. (11.71) is as shown below:

$$\frac{1}{\lambda} = R(Z - \sigma)^2 \left(\frac{1}{n_i^2} - \frac{1}{n_f^2}\right) \tag{11.72}$$

where n_i and n_f are the principal quantum numbers of the energy levels between which transition takes place and σ is a correction factor. The separation between K_{α_1} and K_{α_2} lines, being related to spin-relativity separation, increases rapidly as Z.

11.24.3 Derivation of Moseley's law from Bohr's theory

According to Bohr's theory, the energy of an electron moving in n_i orbit of an atom of atomic number Z is

$$E_{n_i} = -\frac{mZ^2e^4}{8\varepsilon_0^2 n_i^2 h^2} \tag{11.73}$$

Similarly, the energy of an electron in n_f orbit of the same atom is

$$E_{n_f} = -\frac{mZ^2e^4}{8\varepsilon_0^2 n_f^2 h^2} \tag{11.74}$$

Due to the screening effect, the positive charge of the nucleus is reduced from Ze to $(Z-a)\,e$, where a is the screening constant. If a_i and a_f be the screening constants for the electrons in orbits n_i and n_f respectively, then

$$E_{n_i} = -\frac{m(Z-a_i)^2e^4}{8\varepsilon_0^2 n_i^2 h^2} \tag{11.75}$$

$$\text{and } E_{n_f} = -\frac{m(Z-a_f)^2e^4}{8\varepsilon_0^2 n_f^2 h^2} \tag{11.76}$$

$$\therefore \Delta E = E_{n_i} - E_{n_f} = \frac{me^4}{8\varepsilon_0^2 h^2}\left[\frac{(Z-a_f)^2}{n_f^2} - \frac{(Z-a_i)^2}{n_i^2}\right]$$

If Z is very large, $a_i = a_f = a$ (say).

$$\therefore \Delta E = \frac{me^4(Z-a)^2}{8\varepsilon_0^2 h^2}\left[\frac{1}{n_f^2} - \frac{1}{n_i^2}\right] \tag{11.77}$$

If an electron from higher shell (n_f) transits to the vacancy already created in the lower shell (n_i), then the frequency of the emitted X-ray photon is given by

$$\nu = \frac{\Delta E}{h} = \frac{me^4(Z-a)^2}{8\varepsilon_0^2 h^3}\left[\frac{1}{n_f^2} - \frac{1}{n_i^2}\right] \tag{11.78}$$

For K_α line, $n_f = 1$ and $n_i = 2$.

$$\therefore \nu_{K_\alpha} = \frac{me^4(Z-a)^2}{8\varepsilon_0^2 h^3}\left(\frac{1}{1^2} - \frac{1}{2^2}\right)$$

$$\text{or, } \nu_{K_\alpha} \propto (Z-a)^2$$

$$\text{or, } \nu_{K_\alpha} = b(Z-a)^2, \tag{11.79}$$

$$\text{where } b = \frac{me^4}{8\varepsilon_0^2 h^3}\left(\frac{1}{1^2} - \frac{1}{2^2}\right) \tag{11.80}$$

Equation (11.79) is *the Moseley's law.*

11.24.4 Importance of Moseley's law

Moseley's law has the following importances:

1. According to Moseley's law, the arrangements of the elements in the periodic table should be according to the atomic number (Z) and not according to the atomic weight (A) of the element.

   Previously, ${}_{18}A^{20}$ as placed latter than ${}_{19}K^{39}$, and ${}_{27}Co^{58.9}$ latter than ${}_{28}Ni^{58.7}$. But, according to Moseley's law the correct arrangement must be as ${}_{18}A^{40}$ before ${}_{19}K^{39}$ and ${}_{27}Co^{58.9}$ before ${}_{28}N^{58.7}$.

2. The law helped in the discovery of new elements in the missing gaps of the periodic table such as hafnium (57), promethium (61), rhenium (75) and technetium (43), etc.

3. By observing the X-ray spectra of rare earths, their atomic numbers were also determined by Moseley's law.

4. The law proves that a fundamental relationship existing among all the elements is that all elements are constructed with the same building units, complexity arises with increasing Z.

5. Characteristic X-ray lines arise due to the transition in the innermost part of the atom, the inner arrangement of all atoms is the same whatever be their outer electron distribution.

11.24.5 Absorption of X-rays

When a parallel beam of X-rays be incident on a slab of some material, major part of it is transmitted through the slab with reduced intensity and a small part is absorbed by the material. The absorbed part of the radiation removed from the incident beam gives rise to secondary radiation consisting chiefly of the (i) photoelectrons (ii) scattered X-rays (iii) characteristic X-rays.

***Absorption coefficient*:** Let I be the intensity of a monochromatic X-ray beam after penetrating a thickness x of the material and $-\frac{dI}{dx}$ be the rate of decrease of intensity with thickness. The negative sign before $\frac{dI}{dx}$ represents that I decreases with the increase of x [Fig 11.15].

The diminution in I is proportional to I, i.e., $-\frac{dI}{dx} \propto I$.

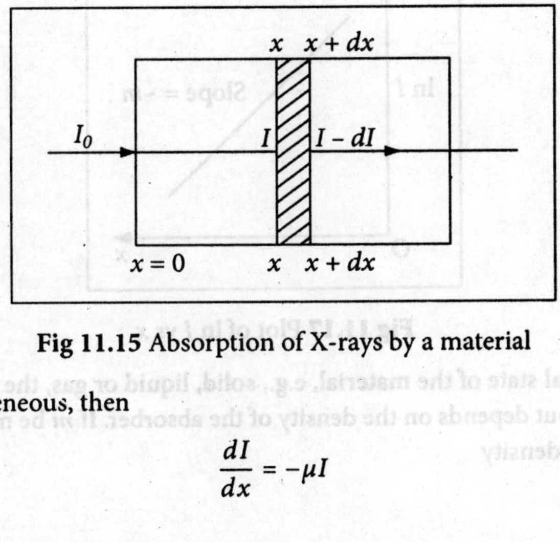

Fig 11.15 Absorption of X-rays by a material

If the material be homogeneous, then

$$\frac{dI}{dx} = -\mu I \tag{11.81}$$

Here, μ is a constant of proportionality and is known as *the linear absorption coefficient.*

$$\text{or, } \frac{dI}{I} = -\mu dx$$

Integrating, $\ln I = -\mu x + A$, where A is a constant of integration.

$\because$ at $x = 0$, $I = I_0$, we get from the above equation, $A = \ln I_0$.

$$\therefore \ln\left(\frac{I}{I_0}\right) = -\mu x \text{ or, } I = I_0\, e^{-\mu x} \tag{11.82}$$

Equation (11.82) shows that the intensity of the X-ray beam decreases exponentially with x [Fig 11.16].

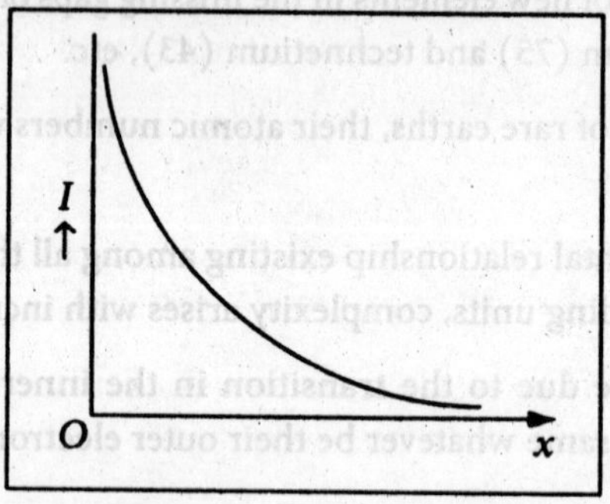

Fig 11.16 Exponential decrease of I with x

The linear absorption coefficient μ can be determined by finding the thickness x at which I_0 reduces to $I_0/2$.

$$\therefore \frac{I}{I_0} = \frac{1}{2} = e^{-\mu x} \text{ or, } \mu = \frac{\ln 2}{x} = \frac{0.693}{x} \tag{11.83}$$

To measure μ, $\ln I$ is plotted against x, which gives a straight line with negative slope. The slope of the curve [Fig 11.17] gives the value of μ. The μ-value of a material determines the quality of the incident X-rays. If μ is large, then x is small and the X-rays are, then called *soft*; on the other hand, if μ is small, then x is large and the X-rays are *hard*.

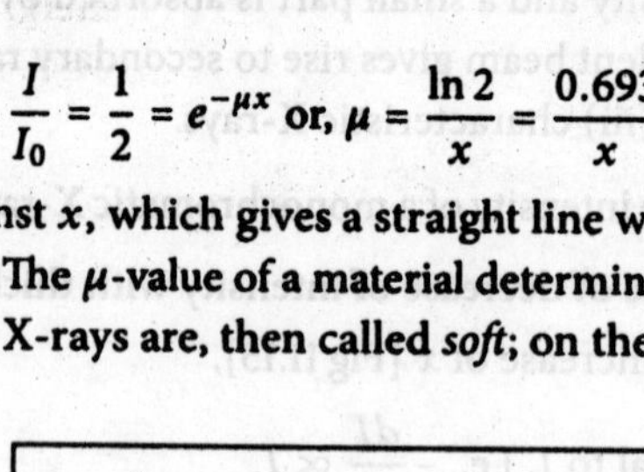

Fig 11.17 Plot of $\ln I$ vs x

Depending on the physical state of the material, e.g., solid, liquid or gas, the value of μ is found to vary. So μ is not a true constant, but depends on the density of the absorber. If m be mass of a slab of thickness x and surface area s, then the density

$$\rho = \frac{m}{sx} \text{ or, } x = \frac{m}{\rho s}. \therefore I = I_0 e^{\frac{\mu m}{\rho s}} \tag{11.84}$$

The quantity $\mu_m = \mu/\rho$ is called the *mass absorption coefficient* of the absorber which increases rapidly with atomic weight (A).

Since, A for Pb is greater than A for Al, $(\mu_m)_{Pb} > (\mu_m)_{Al}$, which justifies that Pb-*screens shield X-rays better than Aluminium screens*. Eq. (11.84) has practical applications in X-ray and γ-ray radiography. The same equation is also used to detect the presence of cavity in cast metal.

Frequency dependence of μ (X-ray absorption spectra): The variation of $\mu(\nu)$ with the frequency ν of the incident radiation is shown in Fig 11.18(a). The curve shows that as ν is lowered, $\mu(\nu)$ increases gradually and at a certain value of $\nu = \nu_K$ it suddenly decreases to a low value. If ν is further lowered $\mu(\nu)$ increases again and, then drops to a low value at $\nu = \nu_{L_1}$. This process continues and becomes complicated as $\nu \to 0$. Fig 11.18(a) shows the K-absorption edge, L_I-absorption edge, etc., at $\nu = \nu_K, \nu_{L_1}$, etc. respectively. Fig 11.18(a) shows three absorption edges, viz., L_I, L_{II} and L_{III} for L-shell for which $\mu(L_{III}) < \mu(L_{II}) < \mu(L_I)$.

Similarly, there are five absorption edges corresponding to M-shell and so on. The K-absorption edge has a low value of $\mu(\nu)$. The frequencies of these absorption edges are given by

$$\nu_a = |E|/h \tag{11.85}$$

where $|E|$ is the magnitude of energy of K-shell, L_I-shell, L_{II}-shell, etc.

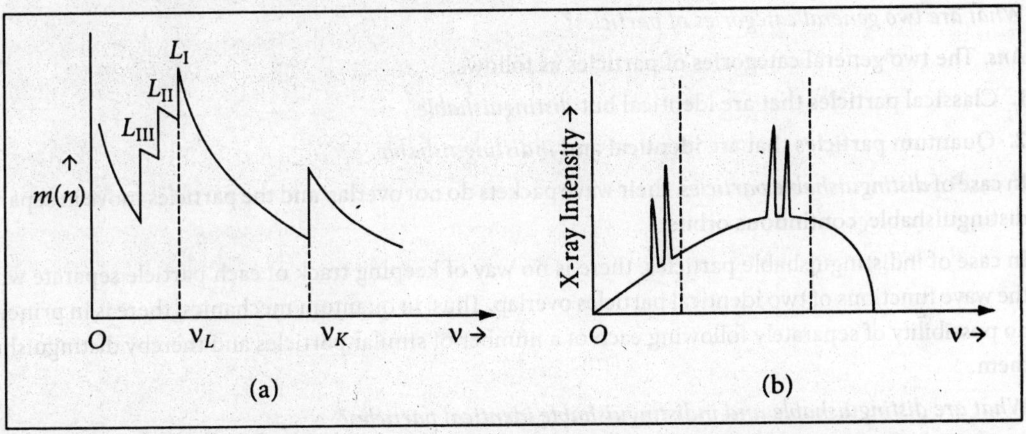

Fig 11.18 A comparison of absorption and emission characteristics of X-rays

Since, the frequency of characteristic emission line is $\nu_e = \dfrac{E_i - E_f}{h}$, it is evident that $\nu_a > \nu_e$. This is because the emission lines arises due to transition between two shells (or subshells), but absorption edge arise due to transition between the lower shell (or, subshell) and the continuum states.

11.24.6 Auger effect

If N_p and N_e be the number of K-photons and K-electrons respectively, then the *fluorescent yield* W is defined by

$$W = N_p/N_e \tag{11.86}$$

Experimentally, W is found to have values ranging from 0.1 (for light elements) to 0.95 (for uranium). This shows that $N_e > N_p$.

***The explanation of the above fact was put forward by Auger in* 1925:** When a K-electron is knocked out from an atom either by high energy electron impact or absorption of an X-ray photon, a vacancy is created in the K-shell. This vacancy is filled up by an L-electron through transition. The energy of the photon thus emitted is $h\nu_{KL} = E_K - E_L$.

But, sometimes it may so happen that the excess energy in such transition may be directly absorbed by another L-electron which gets emitted and thus two vacancies are created in the L-shell. If this level be repressed as LL-level with energy E_{LL}, then the KE of the later emitted electron is $\frac{1}{2}mv^2 = E_K - E_{LL}$.

This phenomenon in which there is no emission of electromagnetic radiation, but emission of two electrons occurs is called *Auger effect*. It is also known as *radiation-less transition*. Thus, $N_e > N_p$ and $W < 1$.

Previously, it was thought that the K_α-radiation emitted from K-shell due to transition of L-electron is absorbed by a second electron in the L-shell of the same atom. That is why it was called *inner photoelectric effect*. But, later on, it was realised that in this process there is no emission of photon at all and hence the name *radiation-less transition*. *Radiation-less transition* has also been detected in the transition of electrons in the outer shells of atoms and in nuclear transitions leading to the emission of *internal conversion electrons*.

11.25 Answers of some important conceptual questions

11.1. *What are two general categories of particles?*

Ans. The two general categories of particles as follows:

1. Classical particles that are identical but *distinguishable*.
2. Quantum particles that are identical and *indistinguishable*.

In case of *distinguishable particles*, their wave packets do not overlap and the particles move in separate, distinguishable, continuous orbits.

In case of indistinguishable particles, there is no way of keeping track of each particle separate when the wave functions of two identical particles overlap. Thus, in quantum mechanics, there is in principle, no possibility of separately following each of a number of similar particles and thereby distinguishing them.

11.2. *What are distinguishable and indistinguishable identical particles?*

Ans. ***Distinguishable particles*:** In classical mechanics identical particles do not lose their individuality despite the identity of their physical properties. If a pair particles are completely equivalent (i.e., if each has the same form of Hamiltonian under all conditions) even, then it is possible to identify them by the continuity of their trajectories because this property enables an observer to follow each particle. This is due to the fact that their wave packets do not overlap and the particles move in separate, *distinguishable*, continuous orbits.

***Example*:** The volume available for a gas molecule at NTP is ~ 10^{-25} m^3 and the volume of a single molecule is ~ 10^{-3} m^3. So the volume of a molecule being much smaller than the volume to it we can in principle identify every molecule of the gas. Hence, the gas molecules are *distinguishable*.

***Indistinguishable particles*:** Quantum mechanically, a particle is represented by a wave packet of finite size and spread. That is why the exact specification of the position and momentum of the particle simultaneously restricted by Heisenberg's uncertainty principle, $\Delta q \Delta p \simeq \hbar$. Hence, there is no way of following the track of individual particles separately even if they interact with each other to an appreciable extent. The wave functions of the quantum mechanical particles overlap considerably making the particles indistinguishable.

Example: The volume available for each electron taking part in conduction within the metal is ~ 10^{-28} m^3 but the volume of a single electron is ~ 10^{-27} m^3. This shows that the electron wave functions overlap considerably and hence cannot be identified separately, i.e., they are *indistinguishable.*

11.3. *Explain with example the meaning of symmetric and antisymmetric wave functions.*

Ans. See Art. 11.14.

11.4. *State Pauli's exclusion principle. Give its physical significance.*

Ans. Statement: Pauli's exclusion principle states that *no two electrons in an atom can have the same set of quantum numbers n, l, m_l and m_s.*

The principle forms the basis of understanding the structure of the atom.

Physical significance of Pauli's Exclusion Principle.

The *Pauli's exclusion principle* along the *principle of lowest possible energy* in the ground state forms the basis of *Aufbau principle* which helps to build up Mendeleev's periodic table and to explain the course of the ground state of the elements. The periodicity in the properties of the elements is understood from the ground state configuration.

The group behaviour of the elementary particles having spin $\hbar/2$ is governed by the exclusion principle. As a consequence of this principle new forces of non-dynamical character raised during the interaction of identical particles.

11.5. *State the selection rules of L-S coupling.*

Ans. The emerging spectral lines arise due to transition of the electric dipole moment. These transitions follow the selection rules. The selection rules are

1. Only one electron jumps in the transition between the levels in most common type of spectra.
2. The orbital angular momentum quantum number of the jumping electrons must change by unity, i.e., $\Delta l = \pm 1$.
3. For the atoms as a whole, the other quantum numbers change according to the following rule:

   $\Delta S = 0$.

   $\Delta L = 0, \pm 1$.

   $\Delta J = 0, \pm 1$, but $J = 0$ to $J = 0$ not allowed.

   $\Delta M_J = 0, \pm 1$, but $M_J = 0$, to $M_J = 0$ not all allowed; in this case, $\Delta J = 0$.

11.6. *State the selection rules for j-j coupling.*

Ans.

1. The transition in most common type of spectrum take place between the levels when only one electron jumps at a time.
2. The orbital quantum number of the jumping electron must change by unity, i.e., $\Delta l = \pm 1$.

3. The inner quantum number for the jumping electron must change either by zero or unity, for the other electrons it must not change, i.e.,

$$\Delta j = \begin{cases} 0, \pm 1, & \text{for jumping electron} \\ 0, & \text{for all other electrons.} \end{cases}$$

4. For the atom as a whole, the other quantum numbers have the following change:

$\Delta J = 0, \pm 1, J = 0$ to $J = 0$ not allowed

$\Delta M_J = 0, \pm 1, M_J = 0$ to $M_J = 0$ not allowed, if $\Delta J = 0$.

11.7. ***What is Duane and Hunt law?***

Ans. X-rays emitted by all the target materials have a continuous distribution of intensity above a certain minimum wavelength $\lambda_{\min}$ given by

$$\lambda_{\min} = \frac{Ch}{eV}$$

Since, $C = 3 \times 10^{10}$ m/s, $h = 6.62 \times 10^{-34}$ J-s and $e = 1.602 \times 10^{-19}$ C

$$\lambda_{\min} = \frac{3 \times 10^{10} \times 6.62 \times 10^{-34}}{1.602 \times 10^{-19} \times V} = \frac{12413}{V(\text{volt})} \text{ Å i.e., } \lambda_{\min} \propto \frac{1}{V}$$

where V is the applied voltage. This relation is known as *Duane-Hunt law*. The minimum wavelength is known as *continuous spectrum boundary*. Duane-Hunt law is independent of the nature of the target but depends on the anode voltage (V).

11.8. ***Why the characteristic X-ray is so called?***

Ans. The characteristic X-ray spectra consist of sharp lines, the wavelength of which depends only upon the element. The characteristic wavelength is governed by Moseley's law: $\lambda(z - b)$ = constant, where z is the atomic number of the target element.

As the wavelength is characteristic of the target element it is known as characteristic X-ray. Characteristic X-ray spectrum is independent of applied voltage to the anode of the X-ray tube.

11.9. ***Both line and continuous spectrum emitted by the same target are of different origin. Explain.***

[Bang U. 1992]

Ans. ***Continuous X-ray spectrum***: The continuous spectrum is emitted due to the loss of energy of the incident electron during collision with the nucleus. The frequency of X-rays product is given by

$$h\nu = \tfrac{1}{2}mv_i - \tfrac{1}{2}mv_f^2$$

where v_i and v_f are the initial and final velocities of the electron. The loss of energy may have any value and hence the spectrum is *continuous*.

There is, however, a maximum frequency or minimum wavelength when $v_f = 0$, i.e., the incident electron loses energy to produce X-rays. The lowest wavelength limit is given by

$$\lambda_{\min} = \frac{c}{\nu_{\max}} = \frac{ch}{eV}$$

where V is the applied voltage.

Line spectrum: Line spectrum is the characteristic X-ray spectrum emitted when the incident electron knocks out an electron of the target atom from the K, L, M, etc. level thereby causing a vacancy in the particular orbit, say, K-level. At this state an electron from a higher level, say, L-level jumps into the K-level thereby giving out an X-ray spectral line of frequency ν given by

$$\nu = \frac{E_k - E_L}{h}$$

where E_k is the energy required to extract the electron from the K-level and E_L the energy required to extract the electron from L-level. The spectral line corresponding to this frequency is called the K_α-line.

Similarly, if a vacancy is caused in the L-level and an electron jumps into it from the M-level, we get the L_α-line and such line spectrum is characteristic of the element.

11.10. *What is Bremsstrahlung?*

Ans. In an X-ray tube, electrons emitted from the cathode and accelerated towards the anode or target are *slowed down* or undergo *braking* due to the strong electrostatic interaction of the nucleus. In the process, electron loses some of its energy. This energy appears as an X-ray photon. As, we get photons of almost of all energies up to a certain maximum, it gives rise to continuous X-ray spectrum having a maximum frequency or minimum wavelength.

As these continuous X-rays are produced due to *slowing down* or *braking* of the fast moving electron, these are known as *Bremsstrahlung* which is a German word for *slowing down* or *braking*.

11.11. *What do you mean by inverse photoelectric effect in relation to X-ray production?*

Ans. In photoelectric effect, the energy of the incident photon is absorbed by an electron in the metal and it is knocked out carrying the energy of the incident photon. If v be the velocity of the ejected electron, E its kinetic energy and ν the frequency of the incident photon, then $h\nu = \frac{1}{2}mv^2 = E$.

The production of continuous X-ray spectra is the result of inverse photoelectric effect in which the kinetic energy of the incident fast moving electron is carried by the emitted X-ray photon. If E_i and E_f be the initial and final kinetic energies respectively, then $h\nu = E_i - E_f = E$.

Hence, the production of continuous X-rays is appropriate to regard as *inverse photoelectric effect*.

11.12. *Discuss the importance of Moseley's law.* **[PU 1996; Cal.U. 1992; Luck.U. 1995]**

Ans. See Art. 12.24.2.

11.26 Solved problems

» ***Example* 11.1.** Consider an atom with electronic configuration $1s^2 2s^2 2p^1$ and find possible values of l, s and j.

Also find the magnitude of angular momenta L, S, J. **[HPU 1995]**

Solution: The electronic configuration of the atom is $1s^2 2s^2 2p^1$. Here, $1s$ and $2s$ subshells are closed. So their angular momentum is zero.

Hence, the angular momentum of the electron is due to the $2p^1$ electron.

For this p-electron, $n = 2$, $l = 1$, $s = \frac{1}{2}$. The allowed values of j are $j = l + s = 1 + \frac{1}{2} = \frac{3}{2}$ and $j = l - s = 1 - \frac{1}{2} = \frac{1}{2}$.

So

$$|L| = \sqrt{l(l+1)}\,\hbar = \sqrt{1(1+1)}\,\hbar = \sqrt{2}\,\hbar$$

$$|S| = \sqrt{s(s+1)}\,\hbar = \sqrt{\tfrac{1}{2}\left(\tfrac{1}{2}+1\right)}\,\hbar = \tfrac{\sqrt{3}}{2}\,\hbar$$

For $j = \frac{3}{2}$

$$|J| = \sqrt{j(j+1)}\,\hbar = \sqrt{\tfrac{3}{2}\left(\tfrac{3}{2}+1\right)}\,\hbar = \tfrac{\sqrt{15}}{2}\,\hbar$$

For $j = \frac{1}{2}$

$$|J| = \sqrt{\tfrac{1}{2}\left(\tfrac{1}{2}+1\right)}\,\hbar = \tfrac{\sqrt{3}}{2}\,\hbar$$

» ***Example* 11.2.** Represent the states of an atom in spectral notation. **[KU 1996]**

Solution: We know, the electronic configuration of an electron is represented by small letters as

l	= 0	1	2	3	4	5
Letter	= s	p	d	e	f	g

For an atom, the entire electronic state of the system is represented by capital letters as

L	= 0	1	2	3	4	5
Letter	= S	P	D	F	G	H

The state of an atom is represented by three quantum numbers L, S, J, where

$$L = l_1 + l_2 + l_3 + \ldots$$

$$S = s_1 + s_2 + s_3 + \ldots$$

For all the individual electrons of the atom, $J = L + S$.

J has possible values, $J = (L+S), (L+S-1), \ldots, (L-S)$.

The state of the atom according to L-S coupling scheme is $(2s+1)_{L_j}$.

This term is called the *spectroscopic term.* $(2S+1)$ is the multiplicity of the states when $L > S$, because J ranges from $(L+S)$ to $(L-S)$ through O. Thus, when $S = 0$, $(2S+1) = 1$ (singlet state) and $J = L$. When $S = \frac{1}{2}$, $(2S+1) = 2$ (doublet state) and $J = L \pm \frac{1}{2}$.

When $S = 1$, $(2S+1) = 3$ (triplet state) and $J = L+1, L, L-1$.

When $S > L$, the multiplicity is given by $(2L+1)$.

The state of an atom with $L = 1$, $S = \frac{1}{2}$ and $J = \frac{3}{2}$ is ${}^2P_{3/2}$.

For $L = 2$, $S = 1$ and $J = 3, 2, 1$ (taking all possible values of J). The corresponding terms are ${}^3D_3$, ${}^3D_2$, ${}^3D_1$.

» ***Example* 11.3.** (a) Explain the meaning of the term ${}^{2S}L_J$ (b) find the S, L and J values corresponding to ${}^1P_1$, ${}^1S_0$, ${}^2P_2$, ${}^3P_2$, ${}^2D_{3/2}$, ${}^2S_{1/2}$, ${}^5F_5$, ${}^6H_{5/2}$. **[GNDU 1991, HPU 1997]**

Solution:

(a) The spectroscopic term ${}^{2S+1}L_J$ contains S, L and J, which represent the state of an atom.

Now, S is the total spin quantum number and is given by $S = s_1 + s_2 + s_3 + \ldots$ for all the electrons in the atom.

L represents the total orbital quantum number given by $L = l_1 + l_2 + l_3 + \ldots$ for all the electrons in the atom.

J represents the total angular momentum quantum number of the atom and is given by $J = L + S$.

J has the possible values, $J = |L + S|, |L + S - 1|, \ldots, (L - S)$.

$(2S + 1)$ represents multiplicity of the states.

(b) (i) For $^1P_1$, $L = 1$, $2S + 1 = 1$. $\therefore S = 0$, so $J = 1$.

[Since, $J = 1$ is an admissible value for $L = 1$ and $S = 0$, $^1P_1$ represents an allowed state with $J = 1$.]

(ii) For $^1S_0$, $L = 0$, $(2S + 1) = 1$. $\therefore S = 0$, so $J = 0$.

[As $J = 0$ is an admissible value for $L = 0$, $S = 0$, $^1S_0$ represents an allowed state with $J = 0$.]

(iii) For $^2P_2$, $L = 1$, $2S + 1 = 1$. $\therefore S = \frac{1}{2}$.

Hence, $J = \frac{3}{2}, \frac{1}{2}$.

[As $J = 2$ is not an admissible value of J with $L = 1$ and $S = \frac{1}{2}$, $^2P_2$ term does not represent an allowed state.]

(iv) For $^3P_2$, $L = 1$, $2S + 1 = 3$. $\therefore S = 1$, so $J = 2, 0$.

[$J = 2$ is admissible when $L = 1$, $S = 1$. $\therefore {}^3P_2$ represents an allowed state.]

(v) For $^2D_{3/2}$, $L = 2$, $2S + 1 = 2$. $\therefore S = \frac{1}{2}$. $\therefore J = \frac{3}{2}$, so $^2D_{3/2}$ represents an allowed state with $L = 2$, $S = \frac{1}{2}$.

(vi) For $^2P_{3/2}$, $L = 1$, $2S + 1 = 2$. $\therefore S = \frac{1}{2}$. $\therefore J = \frac{3}{2}, \frac{1}{2}$.

[Now, $J = \frac{3}{2}$ is admissible. $\therefore {}^2P_{3/2}$ is an allowed state with $L = 1$, $S = \frac{1}{2}$.]

(vii) For $^2S_{1/2}$, $L = 0$, $2S + 1 = 2$. $\therefore S = \frac{1}{2}$.

Hence, $J = \frac{1}{2}$.

[As $J = \frac{1}{2}$ is an admissible value for $L = 0$ and $S = \frac{1}{2}$, the $^2S_{1/2}$ is admissible.]

(viii) For $^5F_5$, $L = 3$, $2S + 1 = 5$. $\therefore S = 2$, $J = 5, 4, 3, 2, 1$.

[$\therefore J = 5$ is admissible for $L = 3$, $S = 2$. $\therefore {}^5F_5$ is an admissible state.]

(ix) For $^6H_{5/2}$, $L = 5$, $2S + 1 = 6$. $\therefore S = \frac{5}{2}$, so $J = \frac{15}{2}, \frac{13}{2}, \frac{11}{2}, \frac{9}{2}, \frac{7}{2}, \frac{5}{2}$. $\because J = \frac{5}{2}$ for $L = 5$ and $S = \frac{5}{2}$ is admissible, so $^6H_{5/2}$ is an admissible state.

» **Example 11.4.** In a two-valence electron, the quantum numbers of the electrons are $n_1 = 6$, $l_1 = 3$, $s_1 = \frac{1}{2}$ and $n_2 = 5$, $l_2 = 1$, $s_2 = \frac{1}{2}$ respectively. Find the values of L and J in L-S coupling and J in j-j coupling.

Solution: *L-S coupling scheme*:

Here $l_1 = 3$, $l_2 = 1$, $\therefore L = (l_1 + l_2), (l_1 + l_2 - 1), \ldots, |l_1 - l_2| = 4, 3, 2$

$s_1 = \frac{1}{2}$, $s_2 = \frac{1}{2}$, $\therefore S = (s_1 + s_2), (s_1 + s_2 - 1), \ldots, |s_1 - s_2| = 1, 0$.

For $S = 1$, $L = 4$, the allowed values of $J = 5, 4, 3$

For $S = 1$, $L = 3$, the allowed values of $J = 4, 3, 2$

For $S = 1$, $L = 2$, the allowed values of $J = 3, 2, 1$

For $S = 0$, $L = 4$, the allowed values of $J = 4$

For $S = 0$, $L = 3$, the allowed values of $J = 3$

For $S = 0$, $L = 2$, the allowed values of $J = 2$.

j-j coupling scheme: $l_1 = 3,\ s_1 = \frac{1}{2}.\ \therefore\ j_1 = 3 + \frac{1}{2} = \frac{7}{2}$ and $j_1 = 3 - \frac{1}{2} = \frac{5}{2}.\ \therefore\ j_1 = \frac{7}{2}, \frac{5}{2}$

$l_2 = 1,\ s_2 = \frac{1}{2}.\ \therefore\ j_2 = 1 + \frac{1}{2} = \frac{3}{2}$ and $j_2 = 1 - \frac{1}{2} = \frac{1}{2}.\ \therefore\ j_2 = \frac{3}{2}, \frac{1}{2}.$

For $j_1 = \frac{7}{2}$ and $j_2 = \frac{3}{2}$, $J = (j_1 + j_2), (j_1 + j_2 - 1), \ldots, |j_1 - j_2| = 5, 4, 3, 2$

For $j_1 = \frac{7}{2}$ and $j_2 = \frac{1}{2}$, $J = 4, 3$

For $j_1 = \frac{5}{2}$ and $j_2 = \frac{3}{2}$, $J = 4, 3, 2, 1$

For $j_1 = \frac{5}{2}$ and $j_2 = \frac{1}{2}$, $J = 3, 2.$

» ***Example* 11.5.** Show that the number of terms for *pd* configuration is the same for *L-S* and *j-j* coupling. **[PU 1993]**

Solution: Let n_p and n_d be the outer configuration of an atom with two valence electrons.

L-S coupling scheme: For *p*-electron, $l_1 = 1,\ s_1 = \frac{1}{2}$.

For *d*-electron, $l_2 = 2,\ s_2 = \frac{1}{2}$.

$\therefore L = 3, 2, 1$ and $S = 1, 0$.

For $S = 0$, i.e., singlet state, $L = 1,\ J = 1$, the term is $^1P_1$

$L = 2,\ J = 2$, the term is $^1D_2$

$L = 3,\ J = 3$, the term is $^1F_3$

For $S = 1$, i.e., triplet state, $L = 1,\ J = 2, 1, 0$, the spectroscopic terms are $^3P_2, {}^3P_1, {}^3P_0$

$L = 2,\ J = 3, 2, 1$, the terms are $^3D_3, {}^3D_2, {}^3D_1$

$L = 3,\ J = 4, 3, 2$, the terms are $^3F_4, {}^3F_3, {}^3F_2$

There are, thus, $3 + 9 = 12$ terms.

j-j coupling scheme:

For *p*-electron, $l_1 = 1,\ s_1 = \frac{1}{2},\ \therefore j_1 = (1 + \frac{1}{2}) = \frac{3}{2}$ and $j_1 = (1 - \frac{1}{2}) = \frac{1}{2}$

For *d*-electron, $l_2 = 2,\ s_2 = \frac{1}{2},\ \therefore j_2 = 2 + \frac{1}{2} = \frac{5}{2}$ and $j_2 = 2 - \frac{1}{2} = \frac{3}{2}$

$\therefore$ for $j_1 = \frac{1}{2},\ j_2 = \frac{3}{2},\ J = 2, 1$

$j_1 = \frac{1}{2},\ j_2 = \frac{5}{2},\ J = 3, 2$

$j_1 = \frac{3}{2},\ j_2 = \frac{3}{2},\ J = 3, 2, 1, 0$

$j_1 = \frac{3}{2},\ j_2 = \frac{5}{2},\ J = 4, 3, 2, 1$

The four states are $(\frac{1}{2}, \frac{3}{2})_{2,1}\ (\frac{1}{2}, \frac{5}{2})_{3,2}\ (\frac{3}{2}, \frac{3}{2})_{3,2,1,0}\ (\frac{3}{2}, \frac{5}{2})_{4,3,2,1}$.

» ***Example* 11.6.** Find the possible values of resultant angular momentum for two electrons with $j_1 = \frac{3}{2}$ and $j_2 = \frac{5}{2}$. **[PU 1994]**

Solution: We have $j_1 = \frac{3}{2}$ and $j_2 = \frac{5}{2}$.

$\therefore j_2 + j_1 = \frac{5}{2} + \frac{3}{2} = 4$ and $j_2 - j_1 = \frac{5}{2} - \frac{3}{2} = 1$.

$\therefore$ the possible values of $J = 4, 3, 2, 1$.

The resultant angular momentum, $|\vec{J}| = \sqrt{J(J+1)}\,\hbar$.

When $J = 4, |\vec{J}| = \sqrt{4(4+1)}\,\hbar = \sqrt{20}\,\hbar$

$J = 3, |\vec{J}| = \sqrt{3(3+1)}\,\hbar = \sqrt{12}\,\hbar$

$J = 2, |\vec{J}| = \sqrt{2(2+1)}\,\hbar = \sqrt{6}\,\hbar$

$J = 1, |\vec{J}| = \sqrt{1(1+1)}\,\hbar = \sqrt{2}\,\hbar.$

» ***Example* 11.7. What are non-equivalent and equivalent electrons? Is it possible to have two equivalent electrons in the same atom? Calculate the spectral terms for non-equivalent (s, s), (s, p) and (p, p) electrons and for two equivalent (s^2) and (p^2) electrons.** **[PU 1993]**

Solution: *Non-equivalent electrons*: **Non-equivalent electrons are those which have different principal quantum numbers.**

Actually these electrons belong to (n, l) subgroups.

Terms for non-equivalent electrons:

1. ***Two non-equivalent s-electrons*: Let the atom has the configuration $(ns)(n's)$. Then $l_1 = 0, l_2 = 0$.**

$\therefore L = (l_1 + l_2), (l_1 + l_2 - 1), \ldots, (l_2 - l_1) = 0.$

Again, $s_1 = \frac{1}{2}, s_2 = \frac{1}{2}$.

$\therefore \quad S = (s_1 + s_2), (s_1 + s_2 - 1), \ldots, (s_2 - s_1) = 1, 0$

$J = (L + S), (L + S - 1), \ldots, (L - S)$

For $L = 0, S = 1, J = 1$

$L = 0, S = 0, J = 0.$

$\therefore$ the ground state configurations for $J = 1$ and $J = 0$ are $^3S_1$ and $^3S_0$.

2. ***Two non-equivalent (s, p) electrons*: Let $(ns)(n'p)$ be the configuration of the atom.**

Here, for *s*-electron, $l_1 = 0, s_1 = \frac{1}{2}$ and

for *p*-electron, $l_2 = 1, s_2 = \frac{1}{2}$.

$\therefore L = 1$ and $S = 1, 0$ and hence

$$J = \begin{cases} 2, 1, 0, & \text{for } L = 1, S = 1 \\ 1, & \text{for } L = 1, S = 0 \end{cases}$$

Hence, for $J = 2, 1, 0$ and $J = 1$, the possible ground state configurations for $(ns)(n'p)$ are $^3P_0$, $^3P_1$, $^3P_2$ and $^1P_1$.

3. ***Two non-equivalent p-electrons*: Let the configuration of the atom be represented by $(np)(n'p)$.**

Then $l_1 = 1, s_1 = \frac{1}{2}$

$l_2 = 1, s_2 = \frac{1}{2}$.

$\therefore L = (l_1 + l_2), (l_1 + l_2 - 1), \ldots, (l_2 - l_1) = 2, 1, 0$ and $S = 1, 0$.

Hence, $J = (L + S), (L + S - 1), \ldots, (L - S) = 3, 2, 1$ for $L = 2, S = 1$.

$\therefore$ the possible terms $^3D_3, {}^3D_2, {}^3D_1$ for $L = 2, S = 0, J = 2$.

Possible term is $^1D_2$ for $L = 1, S = 1, J = 2, 1, 0$.

Possible terms are $^3P_2, {}^3P_1, {}^3P_0$ for $L = 1, S = 0, J = 1$.

Possible term is $^1P_1$ for $L = 0$, $S = 1$, $J = 1$.

For $L = 0$, $S = 0$, $J = 0$ the possible term is $^3S_1$.

The possible term is $^1S_0$.

Hence, for $(np)(n'p)$ configurations there are 10 possible terms: $^1S_0$, $^1P_1$, $^1D_2$, $^3S_1$, $^3P_0$, $^3P_1$, $^3P_2$, $^3D_1$, $^3D_2$, $^3D_3$.

Equivalent electrons: Two electrons with same principal and orbital quantum numbers are called *equivalent electrons*. They belong to the same (n, l) subgroup.

For equivalent electrons (n, l) values are the same. All the electrons have $s = \frac{1}{2}$. To satisfy Pauli's exclusion principle, at least one of the remaining quantum numbers m_l or m_s must be different. So certain terms are not possible for two equivalent electrons though they are possible for two non-equivalent electrons.

Terms for equivalent electrons:

1. *Two equivalent s-electrons*: Let the atom has the configuration ns^2. $\therefore l_1 = 0, l_2 = 0. \therefore L = 0$.

   Since, both the electrons have same n, l, s, they must have opposite m_s, i.e., $m_{s_1} = \frac{1}{2}$ and $m_{s_2} = -\frac{1}{2}$. The spins must be antiparallel. $\therefore S = s_1 - s_2 = 0$.

   Hence, $J = 0$. The possible term is $^1S_0$.

2. *Two equivalent p-electrons*: Let the configuration of the atom be np^2.

   Here, $l_1 = 1$, $l_2 = 1$, $s_1 = \frac{1}{2}$, $s_2 = \frac{1}{2}$. $\therefore L = 2, 1, 0$ and $S = 1, 0$.

   For p-electron, $l = 1$. $\therefore m_l = 1, 0, -1$ and $m_s = +\frac{1}{2}$ and $-\frac{1}{2}$.

   $\therefore$ possible combinations of m_l and m_s for a single p-electron are as follows:

States	***a***	***b***	***c***	***d***	***e***	***f***
m_l	1	0	−1	1	0	−1
m_s	$\frac{1}{2}$	$\frac{1}{2}$	$\frac{1}{2}$	$-\frac{1}{2}$	$-\frac{1}{2}$	$-\frac{1}{2}$

   Thus, the p-electron can exist in an atom in six possible states, viz., a, b, c, d, e and f. By taking all possible combinations of the above six states taking two at a time with no two combinations alike we can obtain the possible states for the two equivalent electrons as follows:

$$\begin{array}{ccccc} ab & ac & ad & ae & af \\ & bc & bd & be & bf \\ & & cd & ce & cf \\ & & & de & df \\ & & & & ef \end{array}$$

   The possible values of M_L and M_S for the above 15 combinations are given below:

	ab	*ac*	*ad*	*ae*	*af*	*bc*	*bd*	*be*	*bf*	*cd*	*ce*	*cf*	*de*	*df*	*ef*
M_L	1	0	2	1	0	−1	1	0	−1	0	−1	−2	1	0	−1
M_S	1	1	0	0	0	1	0	0	0	0	0	0	−1	−1	−1

From above it is evident that the highest value of $M_L = 2$, this is from $L = 2$ giving the D-term. The other values of M_L for $L = 2$ are 2, 1, 0, −1, −2. All these arise only with $M_S = 0$ for which $S = 0$, which is ^1D-term.

For $L = 2$, $S = 0$, the following combinations are formed. They form: ^1D-term:

	ad	*ae*	*af*	*bf*	*cf*
M_L	2	1	0	−1	−2
M_S	0	0	0	0	0

Again, for $L = 2$, $S = 0$, $J = 2$. The above five combinations give $^1D_2$.

The next highest value of $M_L = 1$. This comes from $L = 1$ giving P-term. For $L = 1$, the other values of M_L are 1, 0, −1. The highest value of $M_S = 1$ for which $S = 1$ gives ^3P-term.

For $S = 1$, $M_S = 1, 0, -1$.

For $L = 1$, $S = 1$, the following combinations are produced and form a ^3P-term:

	ab	*ac*	*bc*	*bd*	*cd*	*ce*	*de*	*df*	*ef*
M_L	1	0	−1	1	0	−1	1	0	−1
M_S	1	1	1	0	0	0	−1	−1	−1

Also for $L = 1$, $S = 1$, $J = 2, 1, 0$. ∴ the above nine combinations give rise to three terms: $^3P_2$, $^3P_1$ and $^3P_0$.

There is only one combination with $M_L = 0$, $M_S = 0$. This combination arises from $L = 0$, $S = 0$ and gives ^1S-term.

Again, for $L = 0$, $S = 0$, $J = 0$. This gives the term $^1S_0$.

Hence, two equivalent p-electrons give the five terms, viz., $^1S_0$, $^1D_2$, $^3P_2$, $^3P_1$, $^3P_0$.

Thus, though ten terms are possible for two non-equivalent p-electrons, only five terms are formed for two equivalent p-electrons due to Pauli's exclusion principle.

» ***Example* 11.8.** What is the angle between the total angular momentum vectors for the state represented by $^4D_{3/2}$?

Solution: Since, the state is represented by $^4D_{3/2}$, we have

$$2S + 1 = 4, \therefore S = \tfrac{3}{2}$$

$$L = 2, \because J = \tfrac{3}{2}$$

Again, $\vec{J} = \vec{L} + \vec{S}$ or $\vec{J} - \vec{L} = \vec{S}$.

So

$$\vec{S}\cdot\vec{S} = (\vec{J}-\vec{L})\cdot(\vec{J}-\vec{L}) = J^2 + L^2 - 2\vec{J}\cdot\vec{L}$$

$$\text{or, } \vec{J}\cdot\vec{L} = \frac{J^2 + L^2 - S^2}{2}$$

$$\vec{J}\cdot\vec{L} = |\vec{J}||\vec{L}|\cos\theta = \sqrt{J(J+1)}\sqrt{L(L+1)}\,\hbar^2\cos\theta$$

$$\text{and } J^2 + L^2 - S^2 = J(J+1)\hbar^2 + L(L+1)\hbar^2 - S(S+1)\hbar^2$$

$$\therefore \cos\theta = \frac{J(J+1)\hbar^2 + L(L+1)\hbar^2 - S(S+1)\hbar^2}{2\sqrt{J(J+1)}\sqrt{L(L+1)}}$$

Putting $L = 2$, $S = \frac{3}{2}$ and $J = \frac{3}{2}$, we get

$$\cos\theta = \frac{\frac{3}{2}\left(\frac{3}{2}+1\right)\hbar^2 + 2(2+1)\hbar^2 - \frac{3}{2}\left(\frac{3}{2}+1\right)\hbar^2}{2\sqrt{\frac{3}{2}\left(\frac{3}{2}+1\right)}\sqrt{2(2+1)}\,\hbar^2} = \frac{\frac{3}{2}\times\frac{5}{2} + 2\times 3 - \frac{3}{2}\times\frac{5}{2}}{2\sqrt{\frac{3}{2}\times\frac{5}{2}\times 2\times 3}} = \frac{2\times 3}{\sqrt{3\times 5\times 2\times 3}} = \sqrt{\tfrac{2}{5}}$$

$$\text{or, } \theta = \cos^{-1}\left(\sqrt{\tfrac{2}{5}}\right) = 50°52'$$

» ***Example* 11.9.** An X-ray tube operates at the voltage 100 kV. Find the maximum speed of electrons striking the anticathode and shortest wavelength of X-rays produced. **[PU 1994; GNDU 1996]**

Solution: An electron traversing through a potential difference V volt acquires an energy eV. If v be the velocity of the electron, then its KE is $= \frac{1}{2}mv^2$, m being the mass of the electron.

$$\therefore \frac{1}{2}mv^2 = eV \text{ or, } v = \sqrt{\frac{2eV}{m}} = \sqrt{V}\sqrt{\frac{2e}{m}} = \sqrt{V}\sqrt{\frac{2\times 1.6\times 10^{-19}}{9\times 10^{-31}}} = \sqrt{V}\times 0.596\times 10^6 \text{ m/s}$$

$$\because V = 100\text{ kV} = 100{,}000\text{ V}$$

$$\therefore v = \sqrt{100000}\times 0.596\times 10^6 = 188.45\times 10^6 \text{ m/s}$$

Shortest wavelength of X-rays produced is

$$\lambda_{\min} = \frac{ch}{eV} = \frac{12420}{V}\times 10^{-10}\text{ m} = \frac{12420}{V}\text{ Å} = \frac{12420}{100000} = 0.124\text{ Å}$$

» ***Example* 11.10.** An X-ray tube operating at 200 kV emits a continuous spectrum with shortest wavelength 6.2×10^{-12} m. Calculate the Planck's constant $[e = 1.6\times 10^{-19}\text{C}]$. **[LU 1995]**

Solution: The minimum wavelength of emitted X-rays is $\lambda_{\min} = \frac{ch}{eV}$.

$$\therefore h = \frac{eV}{c}\lambda_{\min} \text{ or, } h = \frac{1.6\times 10^{-19}\times 200000}{3\times 10^8}\times 6.2\times 10^{-12} = 6.61\times 10^{-34}\text{ J-s}$$

» ***Example* 11.11.** An X-ray tube is operated at an anode potential of 10 kV and anode current of 15 mA. Calculate (i) the number of electrons hitting the anode per second (ii) the minimum wavelength of the X-rays produced. **[LU 1993]**

Solution:

(i) Anode current = 15 mA = 15×10^{-3} A.

$\therefore$ charge conveyed to the anode/sec = 15×10^{-3} C. But, charge on the electron = 1.6×10^{-19} C.

$\therefore$ number of electrons hitting the anode per second

$$= \frac{15\times 10^{-3}}{1.6\times 10^{-19}} = 9.4\times 10^{16} \text{ electrons}$$

(ii) Again,

$$\lambda_{min} = \frac{ch}{eV} = \frac{3\times 10^8 \times 6.62\times 10^{-34}}{1.6\times 10^{-19}\times 10\times 10^3} = 1.24\times 10^{10}\ \text{m} = 1.24\ \text{Å}$$

» ***Example* 11.12.** Calculate the frequency of K_α-line when the atomic number of the anticathode is $Z = 79$ $[R_H = 1.097\times 10^7/\text{m}]$. **[HPU 1991]**

***Solution*:** The frequency of K_α-line is given by

$$\nu_{K_\alpha} = R_H c(Z-1)^2\left[\frac{1}{1^2}-\frac{1}{2^2}\right] = \frac{3}{4}cR(h-1)^2$$

$$= \frac{3\times(3\times 10^{10})\times(1.097\times 10^7)(79-1)^2}{4} = 1.502\times 10^{19}/\text{s}$$

» ***Example* 11.13.** Calculate the ionisation potential of K-shell electron of copper. [*Given*: $Z = 29$ and $R_H = 1.097\times 10^7/\text{m}$.] **[HPU 1992]**

***Solution*:** According to Moseley's law, the frequency of K_α X-ray spectral line of copper is

$$\nu = \frac{3}{4}lR_H(Z-1)^2 = \frac{3}{4}\times(3\times 10^8)\times(1.097\times 10^7)\times(29-1)^2 = 1.936\times 10^{18}\ \text{Hz}$$

∴ energy of the K_α-line, $E = h\nu$

$$\text{or, } E = 4.136\times 10^{-15}\times 1.936\times 10^{18} = 8.011\times 10^3\ \text{eV}$$

$$= 8.011\ \text{keV}\ \left[\because h = 6.626\times 10^{-34}\ \text{J-s} = 4.136\times 10^{-15}\ \text{eV}\right]$$

Now, $E = E_K - E_L$ [where E_K = ionisation energy of the K-shell]

$$\text{or, } E_K = E + E_L = 8.011 + 0.931\ [\text{taking } E_L = 0.931\ \text{keV}]$$

$$= 8.942\ \text{keV}$$

» ***Example* 11.14.** The wavelength of L_α X-ray line of platinum ($Z = 78$) is 1.321 Å. An unknown substance emits L_α X-rays of wavelength 4.174 Å. Calculate the atomic number of the unknown substance [$b = 7.4$ for L_α-line]. **[BU 1995]**

***Solution*:** According to Moseley's law, the frequency of an X-ray spectral line is $\nu = a(Z-b)^2$.

Let ν_1 and ν_2 be the frequencies of the L_α-line of platinum and the unknown substance respectively;, then $\nu_1 = a(Z_1-b)^2$ and $\nu_2 = a(Z_2-b)^2$.

$$\therefore \frac{\nu_1}{\nu_2} = \frac{(Z_1-b)^2}{(Z_2-b)^2} \text{ or, } \sqrt{\frac{\nu_1}{\nu_2}} = \frac{Z_1-b}{Z_2-b}$$

$$\text{or, } (Z_2-b) = (Z-b)\sqrt{\frac{\nu_2}{\nu_1}} = (Z_1-b)\sqrt{\frac{\lambda_1}{\lambda_2}}\ [\because c = \nu\lambda]$$

Here, $b = 7.4$, $Z_1 = 78$; $\lambda_1 = 1.321$ Å and $\lambda_2 = 4.174$ Å.

$$\therefore (Z_2-b) = (78-7.4)\sqrt{\frac{1.321}{4.174}} = 39.72.$$

$$\therefore Z_2 = 39.72 + 7.4 = 47.12.$$

» ***Example* 11.15.** The K-absorption edge in tungsten is $\lambda_\alpha = 0.178$ Å and the average wavelengths of some lines in K-series are $\lambda_\alpha = 0.210$ Å, $\lambda_\beta = 0.184$ Å, $\lambda_\gamma = 0.179$ Å. If a tungsten target is bombarded with electrons of 120 keV, find the value of the maximum kinetic energies of electrons emitted from $n = 1, 2$ and 3 levels.

Solution: The energies of the K-series lines are

$$\frac{hc}{\lambda_\alpha} = W_2 - W_1, \quad \frac{hc}{\lambda_\beta} = W_3 - W_1, \quad \frac{hc}{\lambda_\gamma} = W_4 - W_1, \quad \frac{hc}{\lambda_\infty} = W_\infty - W_1$$

But, $W_\infty = 0$.

$$\therefore W_1 = -\frac{hc}{\lambda_\infty} = -\frac{6.62 \times 10^{-34} \times 3 \times 10^8}{0.178 \times 10^{-10} \times 1.6 \times 10^{-19}} = -69.7\,\text{keV}$$

$$W_2 = \frac{hc}{\lambda_\infty} + W_1 = \frac{6.62 \times 10^{-34} \times 3 \times 10^8}{0.210 \times 10^{-10} \times 1.6 \times 10^{-19}} - 69.7$$

$$= 59.1 - 69.7 = -10.6\,\text{keV}$$

Similarly, $W_3 = \dfrac{hc}{\lambda_\beta} + W_1$ and $W_4 = \dfrac{hc}{\lambda_\gamma} + W_1$ can be found out.

Now, the maximum kinetic energy of the electrons from each level is

$$E_{max} = 120 - |W_n|$$

$\therefore E_{max}$ for $n = 1$ is $(E_{max})_1 = 120 - 69.7 = 50.3\,\text{keV}$.

Similarly, $(E_{max})_1$, $(E_{max})_2$ and $(E_{max})_3$ may be found out.

» ***Example* 11.16.** The intensity of a beam of X-rays is reduced to 90 % of its initial value after passing through a slab of beryllium 3 mm thick and density $1.84 \times 10^3\,\text{kg/m}^3$. Calculate the mass absorption coefficient.

Solution: We have $\dfrac{I}{I_0} = \dfrac{90}{100} = 0.9$, $x = 3\,\text{mm} = 3 \times 10^{-3}$ m and $\rho = 1.84 \times 10^3\,\text{kg/m}^3$.

$\because \dfrac{I}{I_0} = e^{-\mu x}$, $\therefore 0.9 = e^{-3\times10^{-3}\mu}$ or $\mu = 35$.

$\therefore$ mass absorption coefficient, $\mu_m = \dfrac{\mu}{\rho} = \dfrac{35}{(1.84 \times 10^3)} = 19 \times 10^{-3}\,\text{m}^2/\text{kg}$.

» ***Example* 11.17.** The ionisation energies of K and L shells in Zn ($Z = 30$) are respectively 9.659 keV and 1.021 keV. Determine the kinetic energy of an Auger electron emitted from L-shell by K_α X-rays.

Solution: The kinetic energy of K_α X-rays is

$$E_{K_\alpha} = E_L - E_K = -1.021\,\text{keV} - (-9.659\,\text{keV}) = 8.638\,\text{keV}$$

The kinetic energy of the Auger electron = (K_α-photon energy) – (energy of the electron in L-shell)

$$K = E_{K_\alpha} - E_{B_L} = 8.638 - 1.021 = 7.617\,\text{keV}.$$

» ***Example* 11.18.** Calculate, using Moseley's law with the screening constant assumed to be 1, the wavelength of the K_α-line of Ni ($Z = 28$) if that of the K_α-line of Mo ($Z = 42$) is given to be 0.71 Å. **[BU (H) 2004]**

Solution: According to Moseley's law, K_α-line has frequency $\nu_{K_\alpha} = a(Z - b)^2$, where a = a general constant and b = the screening constant.

For Ni ($Z = 28$),

$$\nu_{K_\alpha} = a(28 - 1)^2 \;(\because b = 1) \text{ or, } \nu_{K_\alpha} = 27^2 a.$$

Similarly, for Mo ($Z = 42$), $\nu'_{K_\alpha} = 41^2 a$.

$$\therefore \frac{\nu'_{K_\alpha}}{\nu_{K_\alpha}} = \left(\frac{41}{27}\right)^2 \text{ or, } \frac{(c/\lambda'_{K_\alpha})}{(c/\lambda_{K_\alpha})} = \left(\frac{41}{27}\right)^2 \text{ or, } \lambda_{K_\alpha} = \lambda'_{K_\alpha}\left(\frac{41}{27}\right)^2$$

$$\text{or, } \lambda_{K_\alpha} = 0.71 \times \left(\frac{41}{27}\right)^2 = 0.71 \times 2.3059 = 1.637\text{ Å } [\because \lambda'_{K_\alpha} = 0.71\text{ Å}]$$

» ***Example* 11.19.** **Using vector atom model determine the possible terms corresponding to principal quantum number 3 and compute the angle between l and $\vec{s}$ vectors for the term $^2D_{5/2}$. What are the possible values of j and m_j?**

Solution: For $n = 3, l = 0, 1, 2$. For s-electron, $s = \frac{1}{2}$; $2s + 1 = 2 \times \frac{1}{2} + 1 = 2$.

So, j-values are given by

l	0	1		2	
j	$\frac{1}{2}$	$1 + \frac{1}{2} = \frac{3}{2}$	$1 - \frac{1}{2} = \frac{1}{2}$	$2 + \frac{1}{2} = \frac{5}{2}$	$2 - \frac{1}{2} = \frac{3}{2}$
Term	$3\,^2S_{1/2}$	$3\,^2P_{3/2}$	$3\,^2P_{1/2}$	$3\,^2D_{5/2}$	$3\,^2D_{3/2}$

For the term $^2D_{5/2}, l = 2, s = \frac{1}{2}, j = \frac{5}{2}$.

$$\therefore \cos(\vec{l}\cdot\vec{s}) = \frac{j(j+1) - l(l+1) - s(s+1)}{2\sqrt{l(l+1)}\sqrt{s(s+1)}} = \frac{\frac{5}{2}(\frac{5}{2}+1) - 2(2+1) - \frac{1}{2}(\frac{1}{2}+1)}{2\sqrt{2(2+1)}\sqrt{\frac{1}{2}(\frac{1}{2}+1)}} = 4718 \text{ or, } \vec{l}\cdot\vec{s} = 61.9°$$

The j value is $j = \frac{5}{2}$. So $m_j = -\frac{5}{2}, -\frac{3}{2}, -\frac{1}{2}, \frac{1}{2}, \frac{3}{2}, \frac{5}{2}$.

Orientation of $\vec{j}$ in space w.r.t. field are

$$\cos\theta = \frac{m_j}{\sqrt{j(j+1)}} = \frac{-5/2}{\sqrt{\frac{5}{2}(\frac{5}{2}+1)}}, \frac{-3/2}{\sqrt{\frac{5}{2}(\frac{5}{2}+1)}}, \frac{-1/2}{\sqrt{\frac{5}{2}(\frac{5}{2}+1)}}$$

$$= \frac{1/2}{\sqrt{\frac{5}{2}(\frac{5}{2}+1)}}, \frac{3/2}{\sqrt{\frac{5}{2}(\frac{5}{2}+1)}}, \frac{5/2}{\sqrt{\frac{5}{2}(\frac{5}{2}+1)}}, \frac{5/2}{\sqrt{\frac{5}{2}(\frac{5}{2}+1)}}$$

$\therefore \boldsymbol{\theta = 147°.69, 120°.47, 99.73, 80.27, 59°.53, 32°.31.}$

» ***Example* 11.20.** **What are the angles between angular momentum and magnetic field vectors for $l = 2$?**

[KU (H) 2009]

Solution: If magnetic field $\vec{B} = B\vec{z}, \vec{l}$, orbital angular momentum and θ = angle between $\vec{l}$ and $\vec{B}$, then

$$\cos\theta = \frac{m_l}{\sqrt{l(l+1)}}$$

Now $l = 2, m_l = -2, -1, 0, +1, +2$.

$$\therefore \cos\theta = \frac{-2}{\sqrt{2(2+1)}}; \frac{-1}{\sqrt{2(2+1)}}; 0; \frac{1}{\sqrt{2(2+1)}}; \frac{2}{\sqrt{2}}(2+1)$$

$$= -0.82, -0.41, 0, +0.41, 0.82$$

$$\text{or, } \theta = 145°, 114°, 90°, 66°, 35°$$

» ***Example* 11.21.** **Determine the state which can be formed from a 2 electron configuration in the L-S coupling scheme. [Given: $l_1 = 3, l_2 = 2$]** **[cf. CU (H) 2000, 2001; KU (H) 1995, 1996, 2012, 2013, 2014, 2015]**

Solution: Given $l_1 = 3, l_2 = 2; \therefore L = 1, 2, 3, 4, 5.$

$s_1 = \frac{1}{2}, s_2 = \frac{1}{2}. \therefore \frac{1}{2} - \frac{1}{2} = 0, \frac{1}{2} + \frac{1}{2} = 1.$

So, $s_1 - s_2 = 0, 1.$

L	S	J	$2S+1$	Term
1	0	1	1	$^1P_1$
	1	2, 1, 0	3	$^3P_{2,1,0}$
2	0	2	1	$^1D_2$
	1	3, 2, 1	3	$^3D_{3,2,1}$
3	0	3	1	$^3F_3$
	1	4, 3, 2	3	$^1F_{4,3,2}$
4	0	4	1	$^3G_4$
	1	5, 4, 3	3	$^1G_{5,4,3}$
5	0	5	1	$^1H_5$
	1	6, 5, 4	3	$^3H_{6,5,4}$

» ***Example* 11.22. Explain whether the level $^4D_{1/2}$ can exhibit Zeeman splitting.** **[BU (H) 2007]**

Solution: For the term $^4D_{1/2}$, $2S+1 = 4; S = 3/2; D$ state $\Rightarrow L = 2, K = \frac{1}{2}$.

Now Landé g-factor

$$g = 1 + \frac{J(J+1) + S(S+1) - K(L+1)}{2J(J+1)} = 1 + \frac{\frac{1}{2}(\frac{1}{2}+1) + \frac{3}{2}(\frac{3}{2}+1) - 2(2+1)}{2\frac{1}{2}(\frac{1}{2}+1)}$$

$\therefore g = 0$. Hence, the level $^4D_{1/2}$ cannot exhibit Zeeman splitting.

» ***Example* 11.23. For sodium the energy difference between two p energy levels is 0.00214 eV. Calculate the separation of D_1 and D_2 lines taking wavelength of line before spin-orbit splitting to be 589.3 nm.**

Solution: We know that, $E_{3P} - E_{3S} = h\nu$.

For D-line

$$\nu = \frac{E_{3P} - E_{3S}}{h}, \ \lambda = \frac{C}{\nu} = 589.3\,\text{nm}$$

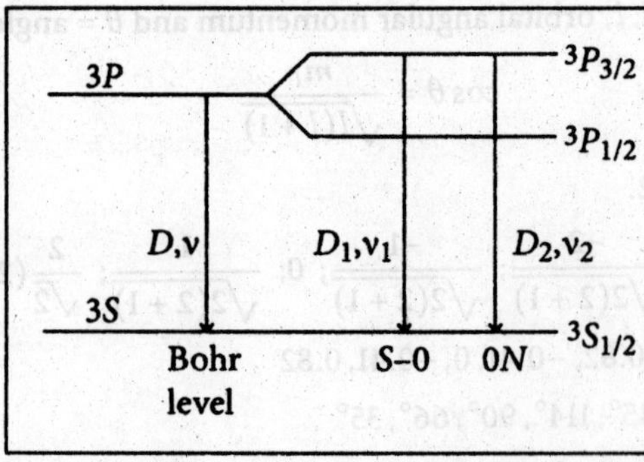

Fig 11.19 Sodium energy levels

$^{E_3}P_{3/2} - {}^{E_3}S_{1/2} = h\nu_2$; D_2 line.

$^{E_3}P_{1/2} - {}^{E_3}S_{1/2} = h\nu_1$; D_1 line.

Subtracting, $^{E_3}P_{3/2} - {}^{E_3}P_{1/2} = h(\nu_1 - \nu_1) = h\Delta\nu$.

$\therefore 0.0214\text{ eV} = h\Delta\nu$.

$$\therefore \nu = \frac{c}{\lambda}. \therefore d\nu = -\frac{C}{\lambda^2}d\lambda. \therefore \Delta\nu = -\frac{C}{\lambda^2}\Delta\lambda = \frac{C}{\lambda^2}\Delta\lambda \text{ (considering modulus)}$$

$$\therefore 0.00214\text{ eV} = h\frac{C}{\lambda^2}\Delta\lambda \text{ or, } \Delta\lambda = \frac{\lambda^2}{hC} \times 0.00214\text{ eV} = \frac{(589.3\text{ nm})^2 \cdot (0.00214\text{ eV})}{1242\text{ eV}}$$

$$\text{or, } \Delta\lambda = 5.98\text{ Å} \simeq 6\text{ Å}$$

So, the wavelength separation of D_1, D_2 lines is 6 Å. [D_1 = 5896 Å; D_2 = 5890 Å]

» ***Example* 11.24.** Discuss the fine structure of $3d \to 2p$ line Balmer series of hydrogen spectrum.

Solution: We know that for $3d$ state $n = 3, l = 2, s = \frac{1}{2}$.

$\therefore j = 2 + \frac{1}{2}, 2 - \frac{1}{2} = \frac{5}{2}, \frac{3}{2}; 2s_+ = 2$.

Term: $3\,^2d_{5/2,3/2}$.

For $2p$-state, $n = 2, l = 1, s = \frac{1}{2}, j = 1 + \frac{1}{2}, 1 - \frac{1}{2} - \frac{3}{2}, \frac{1}{2}; 2s + 1 = 2$.

Term: $2\,^2P_{3/2,1/2}$.

So, due to spin-orbit interaction the $2p$ and $3d$ levels split up into doublets. Selection rule $\Delta l = \pm 1, \Delta j = 0, \pm 1$ allows three transitions. The $3\,^2d_{5/2} \to 2\,^2p_{1/2}$ transition does not occur since if it occurs then $\Delta j = \frac{5}{2} - \frac{1}{2} = 2$ which violets selection rule $\Delta j = 0, \pm 1$.

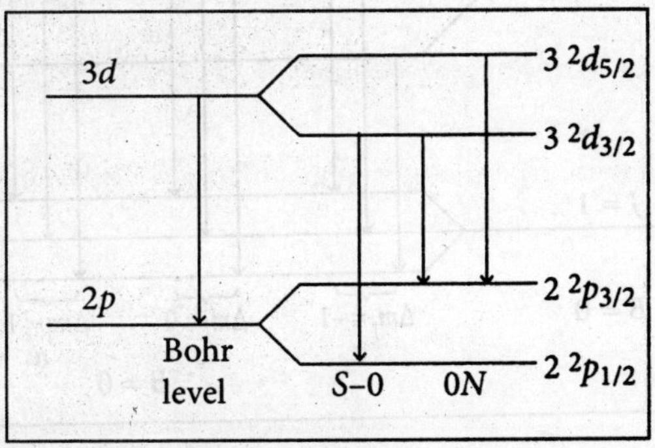

Fig 11.20

Fine structure of $3d \to 2p$ line of Balmer series of hydrogen spectrum.

» ***Example* 11.25.** Write the spectral symbol (term) with $S = \frac{1}{2}, J = \frac{5}{2}$, and $g = \frac{6}{7}$. **[IAS 2002]**

Solution: We know, for $S = \frac{1}{2}, J = \frac{5}{2}$,

$$g = 1 + \frac{J(J+1) + S(S+1) - L(L+1)}{2J(J+1)} = \frac{6}{7} = 1 + \frac{\frac{5}{2}(\frac{5}{2}+1) + \frac{1}{2}(\frac{1}{2}+1) - L(L+1)}{2 \cdot \frac{5}{2} \cdot \frac{7}{2}}$$

$$= 1 + \frac{\frac{5}{2} \cdot \frac{7}{2} + \frac{1}{2} \cdot \frac{3}{2}}{2 \cdot \frac{5}{2} \cdot \frac{7}{2}} - \frac{L(L+1)}{2 \cdot \frac{5}{2} \cdot \frac{7}{2}}$$

$L(L+1) = 12$.

$\therefore L = 3$, i.e., F-state; $2S + 1 = 2$.

$\therefore$ term $= {}^2F_{5/2}$.

» ***Example* 11.26.** In a many electron atom, the orbital, spin and total angular momenta are denoted by L, S and J respectively. If $L = 2$, $S = 1$ and $J = 2$, find the angle between L and S using the vector atom model.

[CU (H) 1995, 2004]

Solution: In this case, the relative orientation of $\vec{L}$, $\vec{S}$ and $\vec{J}$ is given by

$$\cos(\vec{L},\vec{S}) = \frac{J(J+1) - L(L+1) - S(S+1)}{2\sqrt{L(K+1)}\sqrt{S(S+1)}} = \frac{2(2+1) - 2(2+1) - 1(1+1)}{2\sqrt{2(2+1)}\sqrt{1(1+1)}} = -\frac{1}{\sqrt{12}}$$

$\therefore \vec{L}\cdot\vec{S} = 106.8°$.

» ***Example* 11.27.** For the electronic transition ${}^1D_2 \to {}^1P_1$. (a) Draw the energy level diagram and show Zeeman splitting of the energy levels in the presence of a magnetic field. (b) Show allowed transitions. (c) How many distinct lines are obtained in the Zeeman spectrum.

[GATE 1997]

Solution: For ${}^1D_2$, $2S + 1 = 1 \therefore S = 0, L = 2, J = 2, m_J = \pm 2, \pm 1, 0, g = 1$.

${}^1P_1$, $2S + 1 = 1$ or, $S = 0, L = 1, J = 1, m_J = \pm 1, 0, g = 1$.

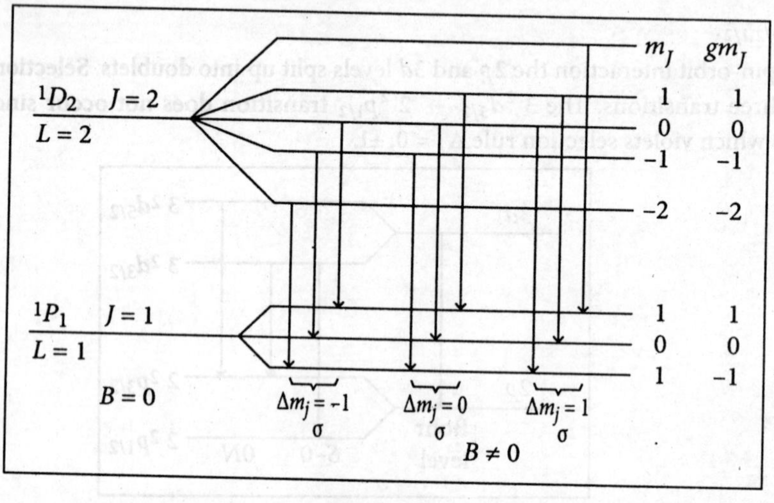

Fig 11.21

Normal Zeeman splitting and the transitions between ${}^1D_2$ and ${}^1P_1$ levels.

Selection rule, $\Delta m_J = 0, \pm 1$ gives 3 lines.

» ***Example* 11.28.** What are the values of L, S and J and the multiplicities of the levels having spectrum terms: ${}^1S_0, {}^3D_2, {}^4P_{5/2}$.

Solution: For ${}^1S_0$: S-state, $L = 0$, multiplicity $2S + 1 = 1$.

$\therefore S = 0, J = 0$.

For ${}^3D_2$: D-state, $L = 2$, multiplicity $2S + 1 = 3$.

$\therefore S = 1, J = 2$.

For $^4P_{5/2}$: *P*-state, $L = 1$, multiplicity $2S + 1 = 4$.

$\therefore S = \frac{3}{2}, J = \frac{5}{2}$.

» ***Example* 11.29.** Using vector diagram determine the possible values of the total angular momentum of f-electron according to vector atom model. How will the values be modified according to wave mechanics? Determine the angle between s and l in the latter case. **[WBCS 2005; KU (H) 2009]**

Solution: For f-electron, $l = 3, s = \frac{1}{2}, 2s + 1 = 2$ and j values are $3 - \frac{1}{2} = \frac{5}{2}, 3 + \frac{1}{2} = \frac{7}{2}$.

$$\text{Eigenvalue of } \hat{i} = \begin{cases} j\hbar = \frac{5}{2}; \frac{7}{2}\hbar \text{ or } j = 2.5\hbar, 3.5\hbar \text{ (by old quantum theory)} \\ \sqrt{j(j+1)}\hbar = \frac{5}{2}\left(\frac{5}{2}+1\right)\hbar \\ \sqrt{\frac{7}{2}\left(\frac{7}{2}+1\right)}\hbar = 2.96\hbar, 3.97\hbar \end{cases}$$

Again,

$$\cos(\vec{l}\cdot\vec{s}) = \frac{j(j+1) - l(l+1) - s(s+1)}{2\sqrt{l(l+1)}\sqrt{s(s+1)}} = \begin{cases} -0.67 \text{ or, } \vec{l}\cdot\vec{s} = 132°, \text{ for } j = \frac{5}{2}, l = 3, s = \frac{1}{2} \\ 0.5 \text{ or, } (\vec{l}\cdot\vec{s}) = 60°, \text{ for } j = \frac{7}{2}, l = 3, s = \frac{1}{2} \end{cases}$$

» ***Example* 11.30.** Write the spectral symbol (term) with $S = \frac{1}{2}, J = \frac{5}{2}$ and $g = \frac{6}{7}$.

Solution: We know that $S = \frac{1}{2}, J = \frac{5}{2}$

$$g = 1 + \frac{J(J+1) + S(S+1) - L(L+1)}{2J(J+1)} = \frac{6}{7} = 1 + \frac{\frac{5}{2}\left(\frac{5}{2}+1\right) + \frac{1}{2}\left(\frac{1}{2}+1\right) - L(L+1)}{2\cdot\frac{5}{2}\cdot\frac{7}{2}} - \frac{L(L+1)}{2\cdot\frac{5}{2}\cdot\frac{7}{2}}$$

or, $L(L+1) = 12$ or, $L = 3$

For *F*-state, $(2S + 1) = 2$.

$\therefore$ the term value $= {}^2F_{5/2}$.

» ***Example* 11.31.** For the electronic transition ${}^1D_2 \to {}^1P_1$ (a) Draw the energy level diagram and show Zeeman splitting of energy levels in the presence of magnetic field (b) Show allowed transitions. (c) How many distinct lines are obtained in the Zeeman spectrum? **[GATE 1997]**

Solution: Ref. Fig. 8.19 of Art. 8.14.2.

Section rule $\Delta m_j = 0, \pm 1$, total number of lines 3.

» ***Example* 11.32.** X-rays emitted from a copper target and a molybdenum target are found to contain a line of wavelength 2.285 Å attributed to K_α-line of an impurity element. The K_α-line of copper ($Z = 29$) and Molybdenum ($Z = 42$) have wavelengths 1.542 Å and 0.712 Å respectively. Using Moseley's law find the atomic number of the impurity element present in the target sample. **[GATE 1992]**

Solution: Moseley's law is

$$\sqrt{\nu} = C_0(z - b) = \sqrt{\frac{c}{\lambda}} \text{ or, } \frac{1}{\sqrt{\lambda}} = C_0'(Z - b).$$

$$\therefore \frac{1}{\sqrt{1.542}} = C_0'(29 - B), \text{ for Cu}$$

$$\frac{f}{\sqrt{0.712}} = C_0'(42 - B), \text{ for Mo}$$

$$\frac{1}{\sqrt{2.285}} = C_0'(Z - B), \text{ for impurity element}$$

$$\therefore \frac{\sqrt{1.542}}{\sqrt{0.712}} = \frac{C_0'(42-B)}{C_0'(29-B)} \text{ or, } 1.5716(29-B) = 42 - B \text{ or, } B = 1.434$$

$$\therefore \frac{1}{\sqrt{1.542 \times 10^{-10}_m}} = C_0'(29-1.434) \text{ or, } C_0' = 2.92 \times 10^{-3}\ \text{m}^{-1/2}$$

$$\text{or, } \frac{1}{\sqrt{2.285 \times 10^{-10}\ \text{m}}} = 2.92 \times 10^{3}\ \text{m}^{-1/2}\ (Z - 1.434)$$

$$\text{or, } Z = 24$$

∴ *atomic number of the impurity element is* $Z = 24$.

EXERCISES

11(A): Objective (or MCQ) Type Questions

Select the correct answer:

11.1. Classical particles are

(a) identical
(b) distinguishable
(c) identical and distinguishable
(d) identical, but indistinguishable.

11.2. Quantum particles are

(a) identical
(b) indistinguishable
(c) distinguishable
(d) identical and indistinguishable.

11.3. For two indistinguishable particles

(a) $H(1,2) \neq H(2,1)$
(b) $H(1,2) = H(2,1)$
(c) $T(1,2) = T(2,1)$
(d) $V(1,2) = V(2,1)$.

11.4. Particles described by symmetric wave functions are

(a) electrons
(b) photons
(c) μ-meson
(d) H^+.

11.5. The maximum number of electrons in a shell of principal quantum number n is

(a) n^2
(b) $(n+1)^2$
(c) $2n^2$
(d) $2n$.

11.6. The chemical property of the elements is related to the number of electrons in the

(a) innermost shell
(b) number of closed shells
(c) outer shell
(d) number of protons in the nucleus.

11.7. *j-j* coupling schemes is applicable to

(a) light elements
(b) heavy elements
(c) all elements
(d) medium elements.

11.8. The quality of X-rays is determined by

(a) size of the anode
(b) filament voltage
(c) filament current
(d) PD between the cathode and anode.

11.9. X-ray intensity is determined by

(a) filament voltage
(b) filament current
(c) pressure in the X-ray tube
(d) PD between the anode and cathode.

11.10. Continuous X-rays are produced due to

(a) loss of KE of the incident electron
(b) transition of electron of the target from higher to lower orbit
(c) attraction of the nucleus to the incident electron
(d) collision of the incident electron with the nucleus.

11.11. The minimum frequency limit of X-ray spectrum depends on

(a) PD between the anode and cathode
(b) nature of the target material
(c) KE of the incident electron
(d) pressure inside the X-ray tube.

11.12. For characteristic X-ray lines, the frequency depends on

(a) Z of the target material
(b) Z^2 of the target material
(c) A^2 of the target material
(d) $\sqrt{Z}$ of the target material.

11.13. X-rays are

(a) elastic waves
(b) electromagnetic waves of short wavelength
(c) electromagnetic waves of low frequency
(d) elastic waves of short wavelength.

11.14. The operating voltage of a Coolidge tube is 50 kV. The energy of the most energetic X-ray photon produced is

(a) 5000 eV
(b) 50 eV
(c) 50 keV
(d) 50000 keV.

11.15. Which of the following wavelengths falls in X-ray region?

(a) 10^{-4} Å
(b) 1 Å
(c) 1000 Å
(d) 10000 Å.

11.16. When a beam of accelerated electrons hits the target, a continuous X-ray spectrum is emitted from the target. Which of the following wavelengths is absent in the X-ray spectrum, if the X-ray tube operates at 40,000 V?

(a) 1.5 Å
(b) 0.5 Å
(c) 0.25 Å
(d) 1.0 Å.

11.17. If λ_L and λ_K are the wavelengths of L and K X-rays respectively, then

(a) $\lambda_L > \lambda_K$
(b) $\lambda_L < \lambda_K$
(c) $\lambda_L = \lambda_K$
(d) $\lambda_K = 4\lambda_L$. **[WBUT 1995]**

11.18. The degeneracy of the spectral term ^3F is

(a) 7
(b) 9
(c) 15
(d) 21. **[GATE 2004]**

11.19. The Landé g-factor for the level $^3D_3$ is

(a) $\frac{2}{3}$
(b) $\frac{3}{2}$
(c) $\frac{3}{4}$
(d) $\frac{4}{3}$. **[GATE 2004]**

11.20. The sodium doublet lines are due to transitions from $^2P_{3/2}$ and $^2P_{1/2}$ levels to $^3S_{1/2}$ level. On application of a weak magnetic field the total number of allowed transitions are:

(a) 4
(b) 6
(c) 8
(d) 10. **[GATE 2005]**

11.21. A spectral line results from the transition $n = 2$ to $n = 1$ in the atoms/species given below. Which one of these will produce the shortest wavelength emission?

(a) hydrogen atom
(b) singly ionised helium
(c) doubly ionised lithium
(d) deuterium atom.

11.22. If an electron has orbital angular momentum quantum number $l = 7$, then it will have an angular momentum equal to

(a) $7\hbar$
(b) $42\hbar$
(c) $\sqrt{7}\hbar$
(d) $\sqrt{56}\hbar$.

11.23. If the magnetic quantum number m_l of an electron in a particular excited state of hydrogen atom is 5, then the possible choice of quantum numbers n, l, m_l and m_s would respectively

(a) $8, 6, 5, \frac{1}{2}$
(b) $6, 6, 5, -\frac{1}{2}$
(c) $7, 7, 5, \frac{1}{2}$
(d) $6, 6, 6 - \frac{1}{2}$. **[IAS(Pre) 1998]**

11.24. According to Pauli's principle $2n^2$ electrons can occupy the major shell with principal quantum number n. Out of 50 electrons in the major shell $n = 5$, the number of electrons having magnetic quantum huber $m_l = 1$ would be

(a) 5
(b) 10
(c) 8
(d) 2. **[IAS(Pre) 1998]**

11.25. Stern-Gerlach experiment gives a direct and convincing confirmation of

(a) space quantisation
(b) wave nature of electron
(c) spin of electron
(d) for quantised atomic magnetic moment.

Which of the above statements are true?

(a) 3 and 4
(b) 1, 2 and 4
(c) 1, 3 and 4
(d) 1 and 2. **[IAS(Pre) 2001]**

11.26. Match list I (property) with list II (Associated experiment) and selected the correct answer:

List I	List II
A. Particle nature of light	1. Davisson-Germer experiment
B. Spin of the electron	2. Zeeman effect
C. Wave nature of particle	3. Compton scattering
D. Magnetic moment of atom	4. Stern-Gerlach experiment.

	A	B	C	D
(a)	4	2	1	3
(b)	3	1	2	4
(c)	4	1	2	3
(d)	3	2	1	4

[IAS(Pre) 2003]

11.27. The selection rule for transition between various energy levels of an atom is given by (l is the orbital angular momentum and j is the total angular momentum)

(a) $\Delta L = \pm\Delta j = 0$
(b) $\Delta l = \pm 1, \Delta j = 0$ or, ± 1
(c) $\Delta l = 0, \Delta j = \pm 1$
(d) $\Delta l = 0$ or, ± 1; $\Delta j = 0$ or, ± 1. **[cf. IAS(Pre) 2004]**

11.28. Match list I and list II and select the correct answer.

List I	List II
A. Goudsmit and Uhlenbeck	1. Antisymmetric state
B. Moseley	2. Space quantisation
C. Pauli's exclusion principle	3. Bohr's theory
D. Stern-Gerlach	4. Electron spin.

	A	B	C	D
(a)	1	3	4	2
(b)	4	3	1	2
(c)	1	2	4	3
(d)	4	2	1	3

[IAS(Pre) 2005]

11.29. For He^+, which has one electron and a nuclear charge $2|e|$, which is the binding energy of the first excited state with principal quantum number $n = 2$.

(a) −13.6 eV
(b) −54.42 eV
(c) −10.2 eV
(d) −20.3 eV. **[IAS(Pre) 2006]**

11.30. The possible values of the total angular momentum J resulting from the addition of two angular momenta $J_1 = 1$ and $J_2 = 2$ are

(a) 1, 2 (c) 0, 1, 2

(b) 1, 3 (d) 1, 2, 3. [GATE 1995]

11.31. Degeneracy of the first excited state of an isolated hydrogen atom is

(a) 2 (c) 6

(b) 4 (d) 8. [GATE 1997]

11.32. Which one of the following groups of particle is a boson group?

(a) photon, muon, pion, proton (c) pion, neutron, graviton, photon

(b) photon, pion, graviton, kaon (d) electron, muon, neutron, proton.

[IAS(Pre) 2004]

11.33. A hydrogen atom is in the $n = 2, j = \frac{3}{2}, m_j = \frac{3}{2}$ state. What is the change in energy of the state when it is placed in a magnetic field B in the z-direction?

(a) $\frac{|e|\hbar B}{m}$ (c) $\frac{e\hbar B}{\pi m}$

(b) $\frac{3\hbar B}{2\pi m}$ (d) $\frac{e\hbar B}{2m}$. [IAS(Pre) 2006]

11.34. What can be the value of total angular momentum of p electron?

(a) $\frac{\sqrt{3}h}{4\pi}$ (c) $\frac{h}{4\pi}$

(b) $\frac{3h}{4\pi}$ (d) $\frac{\sqrt{2}h}{2m}$. [IAS(Pre) 2007]

11.35. An atomic beam in a Stern-Gerlach experiment splits into 4 components when it is subjected to a spatially varying magnetic field. What is the total angular momentum quantum number for the atom?

(a) 3 (c) 4

(b) $\frac{3}{2}$ (d) $\frac{1}{2}$. [IAS(Pre) 2007]

11.36. How is the de Broglie wavelength (λ) of an electron in the nth Bohr orbit related to the radius of the orbit?

(a) $n\lambda = \pi r$ (c) $n\lambda = 2\pi r$

(b) $n\lambda = \frac{3\pi r}{2}$ (d) $n\lambda = 4\pi r$. [IAS(Pre) 2007]

11.37. What type of magnetic field is needed in a Stern-Gerlach experiment to act upon the beam of silver atoms?

(a) highly uniform (c) uniform but varying with time

(b) non-uniform but slowly varying (d) highly non-uniform. [IAS(Pre) 2008]

11.38. If an atom is in the $^3D_3$ state, the angle between its orbital and spin angular momentum vectors $\vec{L}$ and $\vec{S}$ is

(a) $\cos^{-1}\frac{1}{\sqrt{3}}$ (c) $\cos^{-1}\frac{1}{2}$

(b) $\cos^{-1}\frac{2}{\sqrt{3}}$ (d) $\cos^{-1}\frac{\sqrt{3}}{2}$. [GATE 2004]

11.39. The typical wavelength emitted by diatomic molecules in purely vibrational and purely rotational transitions are respectively in the region of

(a) infrared and visible
(b) visible and infrared
(c) infrared and microwave
(d) microwave and infrared. **[GATE 2005]**

11.40. The L, S and J quantum numbers corresponding to the ground state electronic configuration of Boron ($Z = 5$) are

(a) $L = 1, S = \frac{1}{2} m J = \frac{3}{2}$
(b) $L = 1, S = \frac{1}{2} m J = \frac{1}{2}$
(c) $L = 1, S = \frac{3}{2} m J = \frac{1}{2}$
(d) $L = 0, S = \frac{3}{2} m J = \frac{3}{2}$. **[GATE 2004]**

11.41. The degeneracies of the J-states arising from ^3P term with spin orbit interaction are

(a) 1, 3, 5
(b) 1, 2, 3
(c) 3, 5, 7
(d) 2, 6, 10. **[GATE 2004]**

11.42. Assuming L-S coupling scheme is valid, the number of permitted transitions from $^2P_{3/2}$ to $^2S_{1/2}$ due to weak magnetic field is

(a) 2
(b) 4
(c) 6
(d) 10. **[GATE 2004]**

11.43. Sodium atom has 11 electrons. If the sequence in which the energy levels are filled is $1s$, $2s$, $2p$, $3s$, $4p$, $4s$, $3d$, . . . , the ground state of sodium to

(a) $^3P_{1/2}$
(b) $^2P_{1/2}$
(c) $^1P_{1/2}$
(d) $^2S_{1/2}$. **[GATE 1998]**

11.44. Stern Gerlach experiment gives a direct and convincing confirmation of

1. space quantisation
2. wave nature of electron
3. spin of electron
4. quantised atomic magnetic moment.

Which of the above statements are true?

(a) 3 and 4 (b) 1, 2 and 4 (c) 1, 3 and 4 (d) 1 and 2. **[IAS Pre. 2001]**

ANSWERS

11.1. (c).	**11.9.** (b).	**11.17.** (a).	**11.25.** (a,b,d).	**11.33.** (b).	**11.41.** (a).
11.2. (d).	**11.10.** (a).	**11.18.** (d).	**11.26.** (d).	**11.34.** (a).	**11.42.** (c).
11.3. (b).	**11.11.** (c).	**11.19.** (d).	**11.27.** (b).	**11.35.** (b).	**11.43.** (d).
11.4. (b).	**11.12.** (b).	**11.20.** (d).	**11.28.** (b).	**11.36.** (c).	**11.44.** (c).
11.5. (c).	**11.13.** (b).	**11.21.** (c).	**11.29.** (b).	**11.37.** (d).	
11.6. (c).	**11.14.** (c).	**11.22.** (d).	**11.30.** (b).	**11.38.** (a).	
11.7. (b).	**11.15.** (b).	**11.23.** (a).	**11.31.** (b).	**11.39.** (c).	
11.8. (d).	**11.16.** (c).	**11.24.** (c).	**11.32.** (b).	**11.40.** (a).	

11(B): Short Answer Type Questions

11.1. What are symmetric and antisymmetric wave functions?

11.2. Show that $\left[\hat{P}_{12}, \hat{H}(1,2)\right] = 0$, but $\left[\hat{P}_{12}, \hat{P}_{13}\right] \neq 0$.

11.3. Show that antisymmetric wave function $\psi_A(1, 2, \ldots, n)$ cannot be constructed from a solution which remains unchanged by the interchange of any pair of particles.

11.4. Write down the total Hamiltonian of the extranuclear electrons of an atom.

11.5. Explain the concept of *shell* and *subshell* as applied to an atom.

11.6. What is the building up or *aufbau* principle?

11.7. Write down the electronic configuration: Al (13), Ca (20), Fe (26), Cu (29) and Zn (30).

11.8. State and explain Hund's rule. What is Landé interval rule? Find the ground state of nitrogen, $Z = 7$. **[cf. BU (H) 1995]**

11.9. Why the chemical properties of lanthanides and actinides are similar?

11.10. What is meant by the ionisation potential energy of elements? Show that the ionisation potential energy is minimum for atoms with single electron in the outermost shell.

11.11. Why argon is an inert gas but chlorine is a chemically active gas?

11.12. State the selection rules for (a) *L-S* coupling (b) *j-j* coupling.

11.13. State the difference between continuous and characteristic X-rays.

11.14. State Moseley's law regarding characteristic X-rays and explain it. State two remarkably important applications of this law. **[BU (H) 1993, 2001, 2006]**

11.15. What is the origin of continuous X-ray spectra? **[BU (H) 1995, 1997, 2002, 2004]**

11.16. What is Auger effect? Why is it called a *non-radiative transition process*?

11.17. What is *L-S* coupling? **[BU (H) 2003]**

11.18. Discuss the variation of characteristic X-ray frequency if the atomic number (Z) of the target varies systematically from Al to Cu. **[WBUT 2006]**

11.19. The ground state of sodium atom is described by $3^2S_{1/2}$. Explain the meaning of this notation. **[BU (H) 2009]**

11.20. How do the continuous and characteristic X-ray spectra originate? **[BU (H) 2009]**

11.21. (a) Spectral notation of an atomic state is 2^3D_2. Find the quantum numbers.

(b) Determine the state formed by 2 electron $(l_1 = 3, l_2 = 2)$ configuration in the *L-S* coupling scheme. **[KU (H) 2015, 2014]**

11.22. Taking screening effect into consideration prove Moseley's law from Bohr's theory. **[KU (H) 2013]**

11.23. What is Moseley's law? Explain it from the Bohr model of the atom. **[CU (H) 2008]**

11.24. State and explain Pauli's exclusion principle. Prove that if n be the principal quantum number of a shell, it can at best accommodate $2n^2$ electrons. **[VU (H) 2014]**

11.25. Why does the continuous X-ray spectrum emitted by the target of an X-ray tube have a limit at a minimum wavelength? **[NBU (H) 2008]**

11(C): Broad Answer Type Questions

11.1. Show that the sum of the degeneracies for the $(ns)\,(n'l)$ system is $4l(2l+1)$ and for the $(np)\,(n'l)$ system is $12(2l+1)$ (assume that the electrons are equivalent). What are the sums of the degeneracies for equivalent electrons?

11.2. Show the energy levels of carbon and indicate the first few allowed transitions.

11.3. Using Hund's rule show that the ground states of the electronic configurations nd^3 and np^5 are $^4F_{3/2}$ and $^2P_{1/2}$ respectively.

11.4. Define Landé splitting factor and derive expressions for its values for *L-S* and *j-j* coupling in a two-electron system.

11.5. Represent schematically the fine structure splitting of the np, $n'l$ in *j-j* coupling scheme.

11.6. (a) How does the continuous X-ray spectrum arise? Why does it have a short wavelength limit? What is Daune and Hunt law?

(b) State and explain Moseley's law. Derive Moseley's law.

11.7. Discuss fine structure characteristic of X-ray lines with reference to energy level diagram.

11.8. Find two-particle symmetric and antisymmetric wave functions and demonstrate Pauli's exclusion principle. **[BU (H) 2006]**

11.9. (a) Give the order of magnitude of the wavelength of X-rays.

(b) Explain the nature of the continuous X-ray spectra considering the X-rays as showers of photons. **[BU (H) 2000]**

11.10. On the basis of Bohr's theory deduce an expression for the frequency of the K_α X-ray line of an element of atomic number Z.

Calculate the wavelength of K_α-line for a uranium atom. **[BU (H) 1998]**

11.11. What is Moseley's law? Discuss how this law can be understood in terms of the one-electron energy levels including screening effect. **[BU (H) 1995]**

11.12. State Pauli's principle. How does it help in understanding the periodic table of elements? Discuss briefly. **[BU (H) 2004]**

11.13. Show that the maximum number of electrons in a shell is $2n^2$. What is the periodic table of chemical elements? Point out the most important features of the table. **[BU (H) Old Syllabus 2006]**

11.14. State Pauli's exclusion principle. What should be the form of a many-electron wave function in order to comply with this principle? Write down and explain such wave function for a two-electron system as an example. **[BU (H) 2000]**

11.15. (a) State Pauli's exclusion principle.

(b) The ground state of sodium atom is $^3S_{1/2}$. What is the value of l? **[BSc (H) 2002]**

11.16. (a) Write the approximate range of wavelength of X-rays.

(b) Give brief qualitative explanation of the origin of continuous and characteristic X-ray spectra.

(c) Show that the frequency of the k_α X-ray line of an element of atomic number Z is given by

$$\nu = \frac{3CR(Z-1)}{4}$$

where R is the Rydberg constant. [BU (H) 2015]

11.17. (a) Write down (i) the electronic configuration and (ii) the spectroscopic notation for the ground state of Na ($Z = 11$).

(b) Indicate the transitions between the levels, giving rise to the sharp, the principal and the diffuse series of alkali metals.

(c) State the parity selection rules for the electric dipole transitions in atoms. [BU (H) 2012]

11.18. (a) Show that if we neglect the spin of the electron, the number of allowed states of hydrogen atom, with principal quantum number n is n^2.

(b) What are the shell-model predictions of the ground state angular momenta of ^{16}O and ^{17}Al?

(c) The ground state of chlorine atom (in which L-S coupling holds) is $2P_{3/2}$. Find the magnitude of its magnetic moment. [*Given*: the Bohr magneton = 9.27×10^{-24} joule/weber/m^2.] [BU (H) 2008]

11.19. (a) Write down Pauli's exclusion principle. Write down the electronic configuration of sodium ($Z = 11$) atom in the ground state.

(b) Given the energy of a hydrogen-like atom taking into account the fine structure correction in the form:

$$E_{nl} = E_n \left[1 + \frac{Z^2\alpha^2}{n^2}\left(\frac{n}{l+\frac{1}{2}} - \frac{3}{4}\right)\right]$$

where n and l represent the principal and orbital angular momentum quantum numbers respectively and $\alpha = \frac{1}{137}$, show that for fixed n the difference between the maximum and minimum energy as l varies from 0 to $n-1$ is equal to $E_n Z^2\alpha^2 4(n-1)/n(n-1)$. [CU (H) 2013]

11.20. (a) Calculate the angle between the total and orbital angular momentum vectors for an electron in $4D_{3/2}$-state.

(b) The quantum numbers of two electrons in a two-valence electron atom are $n_1 = 6$, $l_1 = 3$, $s_1 = \frac{1}{2}$; $n_2 = 5$, $l_2 = 1$, $s_2 = \frac{1}{2}$.

(i) Assuming L-S coupling, find the possible values of L and hence J.

(ii) Assuming J-J coupling, find the possible values of L and hence J. Write down the ground state electronic configuration of copper ($Z = 29$). [CU (H) 2012]

11.21. (a) State the origin of continuous and characteristic X-ray spectra.

(b) In an X-ray tube containing tungsten as target metal, a voltage of 60 kV is applied. Calculate the minimum X-ray wavelength that will be emitted. If the material of the target is replaced by molybdenum, will there be any change in minimum wavelength? Justify your answer. (1 eV = 1.6×10^{-19} J, $h = 6.63 \times 10^{-34}$ J-s). [CU (H) 2009]

11.22. (a) X-rays are produced in an X-ray tube operated at 66 kV. What is the range of wavelength of the continuous X-rays?

(b) Explain the origin of continuous X-ray spectra and characteristic X-ray spectra.

(c) The minimum wavelength found in an X-ray tube is $\lambda_{\min}$ and the wavelength of k_α-line is λ_k. If the accelerated voltage is increased, what will be the effect on $\lambda_{\min}$ and λ_k? [VU (H) 2013]

11.23. (a) Explain the origin of continuous and characteristic X-ray spectra. [NBU (H) 2009]

(b) If the K, L energy levels of a certain element are 78 keV and 12 keV respectively, calculate the wavelength of k_α-line from the element. [VU (H) 2011]

11.24. (a) What are continuous and characteristic X-rays?

(b) Derive Moseley's law on the basis of Bohr's theory.

(c) Estimate the value of k_α-line of silver ($Z = 47$). [*Given*: $R = 10973700\ \text{m}^{-1}$).] [VU (H) 2009]

11.25. Why does the continuous X-ray spectrum emitted by the target of an X-ray tube have a limit at a minimum wavelength? [NBU (H) 2008]

11.26. (a) Derive a simplified expression for spin-orbit interaction energy.

(b) Using Pauli's exclusion principle deduce the maximum number of electrons that can occupy a given shell for principal quantum number n. [BU (H) 2012]

11.27. (a) What is the binding energy of electron in the ground state of Mn^+ ion? [*Given*: the binding energy is 13.6 eV].

(b) A carbon atom in its ground state has four electrons in the s-wave orbit and two electrons in the p-wave orbitals. How many states are allowed for these two electrons by Pauli's exclusion principle? [BU (H) 2010]

11(D): Numerical Problems

11.1. Calculate the isonisation potential of K-shell electron of copper. [*Given*: Z for copper is 29 and $R_H = 1.097 \times 10^7/\text{m}$.]

11.2. The mass absorption coefficient of silver is $3.8\ \text{m}^2/\text{kg}$ for X-rays of $\lambda = 0.4$ Å. Find the atomic mass absorption coefficient of silver. [*Ans.* $6.087 \times 10^{-25}\ \text{m}^2/\text{atom}$]

11.3. The K_α and L_α absorption edges of copper occur at wavelength 1.380 Å and 13.288 Å respectively. Calculate the atomic number of copper. [*Ans.* 26.27]

11.4. If a potential difference of 30,000 V is applied between the anode and cathode of an X-ray tube, what is the minimum wavelength of the emitted X-rays? [*Ans.* 0.4136 Å]

11.5. An X-ray tube is operated at 18 kV; calculate the minimum velocity of electron bombarding at anticathode. [*Ans.* 8×10^7 m/s]

11.6. Calculate using Moseley's law with the screening constant assumed to be 1, the wavelength of K_α-line of Ni ($Z = 28$) if that of K_α-line of Mo ($Z = 42$) is given to be 0.71 Å. [BU (H) 2004]

11.7. What are the angles between $\vec{L}$ and the z-axis for $l = 2$? [BU (H) 2000]

11.8. Find the wavelength of short wave limit for continuous X-ray spectrum if it shifts by 0.5 Å when the voltage applied to the X-ray tube is doubled. [VU (H) 2011]

11.9. Which of the following wave functions satisfy the principle of indistinguishability of electrons?

(a) $\psi_{1s}(1)\,\psi_{2s}(2)\alpha(1)\alpha(2)$ (b) $\frac{1}{\sqrt{2}}\,\psi_{1s}(1)\,\psi_{2s}(2)[\alpha(1)\beta(2) + \beta(1) + \alpha(2)]$

(c) $\psi_{2s}(1)\,\psi_{1s}(2)\beta(1)\alpha(2)$.

11.10. Which of the following functions are symmetric, antisymmetric or neither in coordinates of particles 1 and 2?

(a) $\alpha(1)\alpha(2)$ (b) $\beta(1)\alpha(2)$

(c) $r_{12}\exp[-(r_1 + r_2)]$ (d) $\cos(x_1 + y_2)\sin(x_2 - y_1)$.

11.11. If σ_x, σ_y and σ_z are Pauli spin matrices and $\vec{A}$ and $\vec{B}$ any constant vectors, show that $(\vec{\sigma}\cdot\vec{A})(\vec{\sigma}\cdot\vec{B}) = \vec{A}\cdot\vec{B} + i\vec{\sigma}(\vec{A}\times\vec{B})$.

11.12. An X-ray tube is operated at an anode potential of 20 kV and anode current of 12 mA. Calculate

(a) the number of electrons hitting the anode per second

(b) the minimum wavelength of the X-ray produced. **[Manipur Univ (Elective) 2016]**

Chapter 12

Theory of Scattering

12.1 Introduction

In a scattering experiment, a beam of particles is directed at a target containing the scattering material, and the energy and angular distribution of the deflected beam or of the recoil particles are measured. The experimental data can be interpreted in terms of a model of the microscopic details of the singular collision and conclusions conceding the interaction between the scattered particle and the scatterer can be drawn. The results of this theory also find application in the study of matter in bulk.

The problem of studying the scattering process is important for the following two reasons:

(i) many interesting effects, such as the stopping of electrons in gaseous discharges, the collision of gas molecules and the stopping radioactive and cosmic ray particles, mesons are all determined, at least in part, by the probability of scattering.

(ii) even more important is the fact that from a detailed study of the result of scattering much can be learned about the nature of the particles that are being scattered as well as of those which are doing scattering.

In some collisions the energy of the incident particle does not change. Such collisions are called *elastic collisions* and the corresponding scattering is called *elastic scattering.* If the energy of the incident particle suffers a change, after its interaction, then it is an *inelastic scattering.*

Importance of scattering experiments:

The following examples reveal the importance of scattering experiments:

(i) scattering of nucleons of various energies from a nucleus gives the idea of the strength and range of nuclear forces.

(ii) scattering of high energy electrons gives charge distribution in a nucleus and even in a nucleon.

(iii) scattering of mesons and other hadrons give an idea about the strength of strong interaction process.

In the consideration of the scattering process it will be assumed that the interaction between the projectile particles and the scatterer can be represented by a potential energy function $V(\vec{r})$, where $\vec{r}$ is the vector representing the relative coordinate of the particles and the mass m of the scatterer. In general, $V(\vec{r})$ decreases in magnitude as the distance $r = |\vec{r}|$ from the scattering centre becomes large. Therefore, $V(r) \to 0$ as $r \to \infty$ and total energy of the scattered particle is mostly kinetic. The particle wave functions will be those of a free particle at $r \to \infty$.

12.2 Scattering cross-section

It is defined as the effective area presented by the target to the incident beam of particles, such that when particles hit within this area scattering is bound to occur. If they hit outside the area the particles go undeflected after traversing the target.

It is also defined as the probability that a particle will be scattered as it traverses a given thickness of the target material.

Let a beam of particles of flux N (i.e., the number of incident particles crossing per unit time per unit surface area placed perpendicular to the direction of propagation) be incident on a scattering centre, particles are scattered in all directions. If dN be the number of particles scattered per unit time into the solid angle $d\omega$ located in the direction θ and ϕ w.r.t. the bombarding direction as polar axis, then dN will be proportional to the incident flux and the solid angle.

Thus,

$$dN = \sigma(\theta, \phi) N d\omega \tag{12.1}$$

where $\sigma(\theta, \phi)$ is a proportionality constant and is a function of θ and ϕ. $\sigma(\theta, \phi)$ as is evident from eq. (12.1) has the dimension of area. All the particles of the incident beam which pass through $\sigma(\theta, \phi)$ will be scattered through solid angle $d\omega$ about θ and ϕ. For this reason $\sigma(\theta, \phi)$ is called the *differential cross-section.*

The *total cross-section* (σ_t) is obtained by integrating eq. (12.1) over entire solid angle.

$$N_{\text{scatt}} = \int dN = \int \sigma(\theta, \phi) N d\omega = N \int \sigma(\theta, \phi)\, d\omega$$

$$= N\sigma_t, \left[\text{where } \sigma_t = \int \sigma(\theta, \phi)\, d\omega\right]$$

$$\therefore \frac{N_{\text{scatt}}}{N} = \sigma_t \tag{12.2}$$

When some particles of the incident beam are absorbed in some reaction, then we can define *total absorption cross-section* in a similar fashion.

In nuclear physics, the scattering centres have linear dimensions of the order of 10^{-12} to 10^{-13} cm, the cross-sections are usually measured in barns (1 barn = 10^{-24} cm^2) or millibar.

12.3 Quantum mechanical description of scattering

In a scattering experiment, a beam of particles impinges on a target which scatters them mostly isotropically in a sphere around the scattering centre [Fig 12.1]. In quantum mechanical description of scattering, the total wave function describing the *trajectory* of the particle can be regarded as consisting of two parts—one representing the incoming part and the other, the particle after scattering. The complete wave function is given by

$$\psi = e^{ikz} + f(\theta, \phi)\frac{e^{ikr}}{r} \tag{12.3}$$

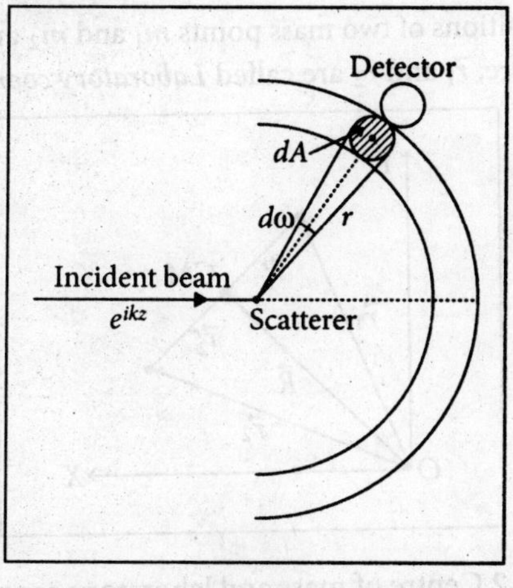

Fig 12.1 Scattering of a beam of particles from a scatterer

The first term represents a particle moving in the positive z-direction or along the polar axis $\theta = 0$, since $z = \cos\theta$, it is an infinite plane wave. The second term represents a particle that is moving radially outward whose amplitude depends upon θ and ϕ and is inversely proportional to r.

The magnitude of the incident flux density along Z-axis is

$$v|\psi_{\text{inc}}|^2 = v|e^{ikz}|^2 = v \tag{12.4}$$

The magnitude of the scattered flux is

$$v|\psi_{\text{scatt}}| = v\left|f(\theta)\frac{e^{ikr}}{r}\right|^2 = \frac{v}{r^2}|f(\theta)|^2 \tag{12.5}$$

Now, if the scattered particles are intercepted by a detector of area dA at a distance r, then the solid angle $(d\omega)$ subtended by the detector at the scattering centre

$$d\omega = \frac{dA}{r^2} \text{ or, } dA = d\omega r^2 \tag{12.6}$$

$\therefore$ the number of particles crossing area dA per seconc is

$$\frac{v}{r^2}|f(\theta)|^2 dA = v|f(\theta)|^2\frac{dA}{r^2} = v|f(\theta)|^2 d\omega \tag{12.7}$$

Now, the differential scattering cross-section $\sigma(\theta)$ is defined as

$$\sigma(\theta)\, d\omega = \frac{\text{no. of particles scattered into } d\omega \text{ per unit time}}{\text{no. of particles scattered per unit area per unit time}} = \frac{v|f(\theta)|^2\, d\omega}{v}$$

$$\text{or, } \sigma(\theta) = |f(\theta)|^2 \tag{12.8}$$

Equation (12.8) represents the *differential scattering cross-section.*

12.4 Centre of mass and laboratory coordinates: Laboratory frame of reference

It is a reference frame in which the bombarding particle or the target particle is at rest.

In this frame of reference the positions of two mass points m_1 and m_2 are located with the help of position vector $\vec{r}_1$ and $\vec{r}_2$ w.r.t. origin O. Here, $\vec{r}_1$ and $\vec{r}_2$ are called *Laboratory coordinates* [Fig 12.2].

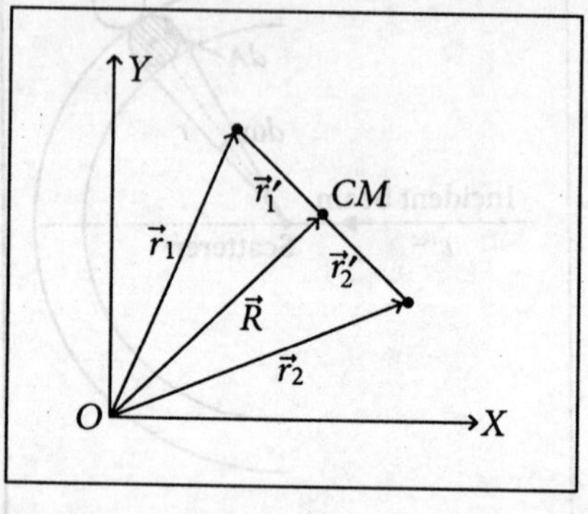

Fig 12.2 Centre of mass and laboratory coordinates

Centre of mass system

The centre of mass of colliding particles in this coordinate system is initially and always at rest.

In this frame the position of the mass points is referred by $\vec{r}_1'$ and $\vec{r}_2'$ w.r.t. the centre of mass (CM). The two situations are shown in Fig 12.2. The centre of mass coordinate in laboratory frame is

$$\vec{R} = \frac{m_2\vec{r}_1' + m_1\vec{r}_2'}{m_1 + m_2} \tag{12.9}$$

The calculation of a collision experiment is easier in centre of mass system than in Laboratory system. This is due to the fact that the six degree of freedom in centre of laboratory system are reduced to three in case of centre of mass system. Practically, the observations are made in centre of mass system and calculations are made in centre of mass system. In the centre of mass system, the collision process may be thought as one in which a particle having reduced mass

$$\mu = \frac{m_1 m_2}{(m_1 + m_2)}$$

and an initial velocity v collides with a fixed scattering centre.

12.4.1 Relation between angles in laboratory and centre of mass system

The relation between angles in the L-system and in the C-system can be found by translating the L-system in the direction of the incident particle with sufficient speed to bring the centre of mass to rest.

Let a particle of mass m_1 and initial velocity v strikes a particle of mass m_2 at rest in L-system as shown in Fig 12.3(a). If the centre of mass moves to right with speed v', then

$$(m_1 + m_2)\, v' = m_1 v \text{ or, } v' = \frac{m_1 v}{m_1 + m_2} \tag{12.10}$$

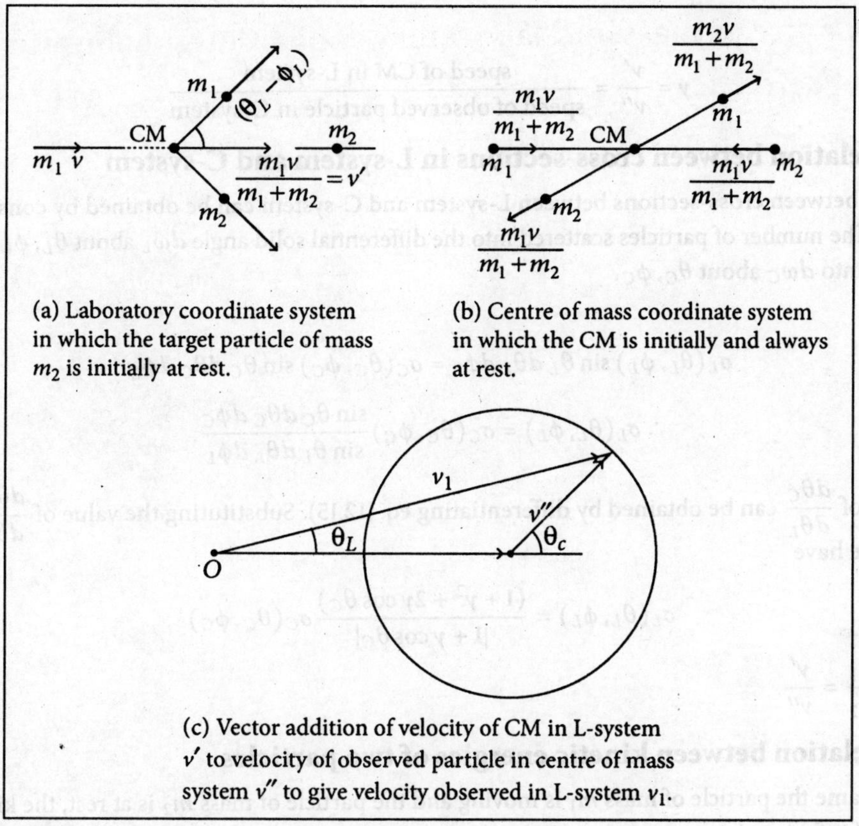

(a) Laboratory coordinate system in which the target particle of mass m_2 is initially at rest.

(b) Centre of mass coordinate system in which the CM is initially and always at rest.

(c) Vector addition of velocity of CM in L-system v' to velocity of observed particle in centre of mass system v'' to give velocity observed in L-system v_1.

Fig 12.3

Thus, in centre of mass system the particles of mass m_1 and m_2 approach the CM with speed $v'' = (v - v')$ and v'.

$$\therefore v'' = v - v' = v - \frac{m_1 v}{m_1 + m_2} \tag{12.11}$$

$$v' = \frac{m_2 v}{m_1 + m_2}$$

When the collision is elastic, the speed remains the same after collision [Fig 12.3(b)].

From figure it is clear that θ_c and ϕ_c are related to θ_L and ϕ_L by the relation

$$v'' \cos\theta_c + v' = v_1 \cos\theta_L \tag{12.12}$$

$$v'' \sin\theta_c \quad = v_L \sin\theta_L \tag{12.13}$$

If the collision is symmetric, then

$$\phi_C = \phi_L \tag{12.14}$$

From eqs (12.12) and (12.13), we get

$$\theta_L = \frac{v'' \sin\theta_c}{v'' \sin\theta_c + v'} = \frac{\sin\theta_c}{\cos\theta_c + (v'/v'')} = \frac{\sin\theta_c}{\cos\theta_c + \gamma} \tag{12.15}$$

where

$$\gamma = \frac{v'}{v''} = \frac{\text{speed of CM in L-system}}{\text{speed of observed particle in C-system}}$$

12.4.2 Relation between cross-sections in L-system and C-system

The relation between cross-sections between L-system and C-system can be obtained by considering their definitions. The number of particles scattered into the differential solid angle $d\omega_L$ about θ_L, ϕ_L are the same as scattered into $d\omega_C$ about θ_C, ϕ_C.

Hence,

$$\sigma_L(\theta_L, \phi_L) \sin\theta_L \, d\theta_L \, d\phi_L = \sigma_C(\theta_C, \phi_C) \sin\theta_C \, d\theta_C \, d\phi_C$$

$$\therefore \sigma_L(\theta_L, \phi_L) = \sigma_C(\theta_C, \phi_C) \frac{\sin\theta_C \, d\theta_C \, d\phi_C}{\sin\theta_L \, d\theta_L \, d\phi_L} \tag{12.16}$$

The value of $\frac{d\theta_C}{d\theta_L}$ can be obtained by differentiating eq. (12.15). Substituting the value of $\frac{d\theta_C}{d\theta_L}$ and using eq. (12.14), we have

$$\sigma_L(\theta_L, \phi_L) = \frac{(1+\gamma^2+2\gamma\cos\theta_C)}{|1+\gamma\cos\theta_C|} \sigma_C(\theta_C, \phi_C) \tag{12.17}$$

where $\gamma = \frac{m_1}{m_2} = \frac{v'}{v''}$.

12.4.3 Relation between kinetic energies of two particles

Since, in L-frame the particle of mass m_1 is moving and the particle of mass m_2 is at rest, the kinetic energy in L-frame of reference is

$$T_L = \frac{1}{2} m_1 v^2 = \frac{p_i^2}{2m_1} \tag{12.18}$$

where $m_1 v = p_i$, the momentum of the incident particle kinetic energy in the C-system is

$$T_C = \frac{p^2}{2m_1} + \frac{p^2}{2m_2} = \frac{p^2}{2\mu} \tag{12.19}$$

where

$$\mu = \frac{m_1 m_2}{m_1 + m_2} \tag{12.20}$$

Here, p is the momentum of the incident particle in C-system. This means that if p is the momentum of incident particle, then $-p$ is the momentum of target particle.

The relation between the magnitude momenta in the two frames of reference is

$$p = p_i - m_1 v = p_i - \frac{m_1 p_i}{m_1 + m_2} = \frac{m_2 p_i}{m_1 + m_2} \quad \left[\because v = \frac{p_i}{m_1 + m_2}\right]$$

$$\text{or, } p^2 = \left(\frac{m_2}{m_1 + m_2}\right) p_i^2 \tag{12.21}$$

Substituting the value of p^2 in eq. (12.19)

$$T_c = \left(\frac{m_2}{m_1 + m_2}\right)^2 \frac{p_i^2}{2\mu} = \frac{p_i^2}{2m_1} \cdot \frac{m_2}{m_1 + m_2} \quad \text{or, } T_c = \frac{m_2}{m_1 + m_2} T_L \tag{12.22}$$

12.5 Stationary (steady) state quantum theory of scattering

In wave mechanics, an incident beam of particles is represented by a plane wave in incident channel. We consider the scattering of a particle of mass m by a central potential $V(\vec{r})$ such that $V(r) \to 0$ as $r \to \infty$. The Schrödinger equation for the potential $V(\vec{r})$ is

$$\left[-\frac{\hbar^2}{2m}\nabla + V(r)\right]\psi = E\psi$$

Since, the scattering eigenfunctions represent unbound state, we do not require them to vanish at infinity.

The region of interest in scattering experiment is far from the target. In asymptotic region, the incident flux due to plane wave travelling along z-axis is represented as e^{ikz} and the scattered wave with outgoing spherical wave as $f(\theta,\phi)\frac{e^{ikr}}{r}$. Thus, the wave function is given by

$$\psi_{r\to\infty} = e^{ikz} + f(\theta,\phi)\frac{e^{ikr}}{r} \tag{12.23}$$

with energy

$$E = \frac{\hbar^2 k^2}{2m} \text{ or, } k = \left(\frac{2mE}{\hbar^2}\right)^{1/2} \tag{12.24}$$

Though there are several solutions of Schrödinger equation stated above, our interest is to find out a solution that fits well in asymptotic region. For this purpose, the incident flux is obtained from the incident wave as follows:

$$J_{\text{inc}} = -\frac{i\hbar}{2m}\left[\left(e^{ikz}\right)^* \frac{\partial}{\partial z}e^{ikz} - e^{ikz}\frac{\partial}{\partial z}e^{-ikz}\right] = \frac{\hbar k}{m} \tag{12.25}$$

Similarly, scattered flux is obtained from the relation

$$\begin{aligned} J_{\text{sc}} &= -\frac{i\hbar}{2m}|f(\theta,\phi)|^2\left[\left(\frac{e^{ikr}}{r}\right)^* \frac{\partial}{\partial r}\left(\frac{e^{ikr}}{r}\right) - \left(\frac{e^{ikr}}{r}\right)\frac{\partial}{\partial r}\left(\frac{e^{ikr}}{r}\right)^*\right] \\ &= \frac{\hbar k}{m}\cdot\frac{|f(\theta,\phi)|^2}{r^2} \end{aligned} \tag{12.26}$$

Now, the number of particles scattered per unit solid angle is

$$\frac{dN}{d\Omega} = J_{\text{sc}}r^2 = \frac{\hbar^2 k}{m}|f(\theta,\phi)|^2 \tag{12.27}$$

Now, the scattering cross-section is

$$\sigma(\theta,\phi) = \frac{1}{J_{\text{inc}}}\cdot\frac{dN}{d\Omega} \text{ or, } \sigma(\theta,\phi) = |f(\theta,\phi)|^2 \tag{12.28}$$

This relation is exact. The angular momentum factor is called the *scattering amplitude.*

Note:

1. In classical approach, Newton's laws of motion are solved for trajectory of the particle of given energy, impact parameter and scattering potential.

   In quantum approach Schrödinger equation is solved for asymptotic solution for a given energy and scattering problem.

2. A relation is, then established between scattering angle θ and impact parameter in classical theory. But, in quantum theory scattering amplitude is obtained.
3. In classical theory scattering cross-section is obtained by using

$$\sigma(\theta) = \frac{-b d\theta}{f m\theta\, d\theta} \; [b = \text{impact parameter}]$$

But, in quantum theory, $\sigma(\theta, \phi) = |f(\theta, \phi)|^2$.

12.6 Quantum theory for Rutherford scattering

The scattering of one charged particle by another was given by *Rutherford* classically to investigate the charge distribution in atoms.

The classical Rutherford formula for the scattering cross-section of particles with charge $2e$, and Ze in the centre of mass system is

$$\sigma(\theta) = \left(\frac{2Ze^2}{16\pi\varepsilon_0 E}\right)^{1/2} \operatorname{cosec}^4 \theta \tag{12.29}$$

where E and θ are respectively the energy of the particle (α-particle of charge $2e$) and the angle of scattering.

The Schrödinger equation is

$$\left[-\frac{\hbar^2}{2m}\nabla^2 + V(r)\right]\psi = E\psi \tag{12.30}$$

where

$$V(r) = \frac{2Ze^2}{4\pi\varepsilon_0 r} = \frac{c}{r} \text{ with } C = \frac{2Ze^2}{4\pi\varepsilon_0} \tag{12.31}$$

Now, eq. (12.30) can be written as

$$\left[-\frac{\hbar^2}{2m}\nabla^2 + \frac{c}{r}\right]\psi = E\psi \tag{12.32}$$

$$\text{or, } \left[-\frac{\hbar^2}{2m}\nabla^2 + \frac{c}{r}\right]\psi = \frac{\hbar^2 k^2}{2m}\psi \tag{12.33}$$

$$\text{with } E = \frac{\hbar^2 k^2}{2m} \tag{12.34}$$

$$\therefore \left[-\nabla^2 + \frac{2mC}{\hbar^2 r}\right]\psi = k^2\psi \tag{12.35}$$

Now, the problem of *pure Coulomb scattering* is best solved using parabolic coordinates

$$\left.\begin{aligned} \xi &= r(1+\cos\theta) = r + z \\ \eta &= r(1-\cos\theta) = r - z \\ \phi &= \tan^{-1}\frac{y}{x} \end{aligned}\right\} \tag{12.36}$$

The matrix coefficients, Laplacian and volume element in parabolic coordinates are

$$h_\xi = -\left[\frac{\eta+\xi}{4\xi}\right]^{1/2},\ h_\eta = \left[\frac{\eta+\xi}{4\eta}\right]^{1/2},\ h_\phi = (\xi\eta)^{1/2} \tag{12.37}$$

$$\nabla^2 = \frac{4}{\xi+\eta}\left[\frac{\partial}{\partial\xi}\left(\xi\frac{\partial}{\partial\xi}\right) + \frac{\partial}{\partial\eta}\left(\eta\frac{\partial}{\partial\eta}\right) + \frac{1}{\xi\eta} + \frac{1}{\xi\eta}\cdot\frac{\partial^2}{\partial\phi^2}\right] \tag{12.38}$$

$$d\tau = \frac{1}{4}(\xi+\eta) + d\xi\, d\eta\, d\phi \tag{12.39}$$

Substituting in eq. (12.35), we get

$$\frac{d^2\Phi}{d\phi^2} = -m_l^2\phi \tag{12.40}$$

$$\frac{d}{d\xi}\left(\xi\frac{d}{d\xi}F\right) + \left(\frac{m_l^2}{4\xi} + \frac{k^2}{4\xi} - \beta\right)F = 0 \tag{12.41}$$

$$\text{and } \frac{d}{d\eta}\left(\eta\frac{d}{d\eta}G\right) + \left(-\frac{m_l^2}{4\eta} + \frac{k^2\eta}{4} + \beta - \frac{Cm}{\hbar^2}\right)G = 0 \tag{12.42}$$

where m_l and β are separation constants. The values of m_l and β are so chosen that ψ has asymptotic form:

$$\psi_{r-\infty} = e^{ikz} + f(\theta)\frac{e^{ikr}}{r} \tag{12.43}$$

The solution of eq. (12.40) can be written as

$$\Phi = e^{im_l\phi} \tag{12.44}$$

Since, the scattering amplitude does not contain Φ function, $m_l = 0$ is taken so that $\Phi = 1$.

∴ the solution takes the form:

$$\psi = F(\xi)G(\eta) \tag{12.45}$$

where F and G happen to be the solution of the following equations

$$\frac{d}{d\xi}\left(\xi\frac{dF}{d\xi}\right) + \left(\frac{k^2}{4}\xi - \beta\right)F = 0 \tag{12.46}$$

$$\text{and } \frac{d}{d\eta}\left(\eta\frac{d}{d\eta}G\right) + \left(\frac{k^2\eta}{4} + \beta - \frac{Cm}{\hbar^2}\right)G = 0 \tag{12.47}$$

The required asymptotic form can be written as

$$\psi(\xi,\eta) = e^{ikz} + f\frac{e^{ikr}}{r}$$

$$r - z = \eta \to \infty (\text{as } r > z)$$

$$\psi(\xi,\eta) = e^{ikz}\left(1 + f\frac{e^{ik(r-z)}}{r}\right)$$

$$\text{or, } \psi(\xi,\eta) = e^{ik\left(\frac{\xi+\eta}{2}\right)}\left(1 + \frac{fe^{ik\eta}}{r}\right) \tag{12.48}$$

This suggests that the exact solution is of the form:

$$\psi(\xi,\eta) = e^{ik\left(\frac{\xi-\eta}{2}\right)}\left(1+\frac{fe^{ik\eta}}{r}\right)W(\eta) = e^{(ik\xi)/2}\cdot e^{-(ik\eta)/2}W(\eta) \tag{12.49}$$

$$\text{with } W(\eta) = \left(1+f\frac{e^{ik\eta}}{r}\right) \tag{12.50}$$

We get from eq. (12.49)

$$F(\xi) = e^{ik\xi/2} \tag{12.51}$$

$$\therefore \frac{d}{d\xi}\left(\xi\frac{dF}{d\xi}\right) = \frac{d}{d\xi}\left(\frac{ik}{2}\xi e^{ik\xi/2}\right) = \frac{ik}{2}e^{(ik\xi)/2} - \frac{k^2}{4}\xi e^{(ik\xi)/2} \tag{12.52}$$

Substituting in eq. (12.46), we get

$$\beta = \frac{ik}{2} \tag{12.53}$$

Thus, from eq. (12.47)

$$\eta = \frac{d^2W}{d\eta^2} + (1-ik\eta)\frac{dW}{d\eta} + \gamma W = 0 \tag{12.54}$$

$$\text{with } \gamma = \frac{Cm}{k\hbar^2} \text{ and } G = e^{-(ik\eta)/2}W(\eta) \tag{12.55}$$

Equation (12.54) is recognised as *confluent hypergeometric* equation. The solution that is well-behaved at $\eta = 0$ is called *confluent series*. The asymptotic form of solution of Schrödinger equation, then becomes

$$\psi_{\eta\to\infty} = \exp i[kz + \gamma \ln kr(1-\cos\theta)] + f(\theta)\exp\left[\frac{ikr - \gamma\ln 2kr}{r}\right]$$

$$\text{with } f(\theta) = \frac{-[\Gamma(1+i\gamma)\exp(-i\gamma\ln\sin^2\theta/2)]^*}{\Gamma(1-ir)\,k2\sin^2\theta/2}$$

The scattering cross-section

$$\sigma(\theta) = |f(\theta)|^2 = \frac{\gamma^2}{4k^2\sin^4\theta/2}\left|\frac{\Gamma(1+i\gamma)}{\Gamma(1-i\gamma)}\right|^2$$

$\because |\Gamma(i+i\gamma)| = |\Gamma(1-i\gamma)|$.

Finally

$$\sigma(\theta) = \frac{\gamma^2}{4k^2\sin^4\frac{\theta}{2}} = \left(\frac{2Ze^2}{16\pi\varepsilon_0 E}\right)^2 \operatorname{cosec}^4\frac{\theta}{2}$$

Hence, finding solution of Schrödinger equation in the present case is not an easy affair. That is why to solve such problems *Born approximation* or method of partial wave analysis is employed.

12.7 Expansion of plane wave in spherical harmonics (Bauer's formula)

We consider a plane wave propagating along z-axis, i.e., the axis of quantisation for the angular momentum. This wave is given by

$$e^{ikz} = e^{ikr\cos\theta} = e^{i\zeta\cos\theta}$$

where $\zeta = kr$.

The function $\exp(i\zeta\cos\theta)$ is independent of ϕ, hence in expansion of $\nabla^2\psi$ in Schrödinger equation no ϕ term will occur.

Now, the Schrödinger equation for a free particle is

$$\nabla^2\psi + k^2\psi = 0$$

i.e.,

$$\frac{1}{r^2}\cdot\frac{\partial}{\partial r}\left(r^2\frac{\partial\psi}{\partial r}\right) + \frac{1}{r^2\sin\theta}\cdot\frac{\partial}{\partial\theta}\left(\sin\theta\frac{\partial\psi}{\partial\theta}\right) + k^2\psi = 0 \tag{12.56}$$

Putting $\psi = R(r)\Theta(\theta)$ and separating the radial and angular parts, we get

$$\frac{1}{R}\cdot\frac{d}{dr}\left(r^2\frac{dR}{dr}\right) + k^2r^2 = -\frac{1}{\Theta\sin\theta}\cdot\frac{d}{d\theta}\left(\sin\theta\frac{d\Theta}{d\theta}\right) = \lambda = l(l+1) \tag{12.57}$$

$\therefore$ the angular part is

$$\frac{1}{\sin\theta}\cdot\frac{d}{d\theta}\left(\sin\theta\frac{d\Theta}{d\theta}\right) + l(l+1)\Theta = 0 \tag{12.58}$$

The solutions of eq. (12.58) are Legendre polynomials

$$\Theta(\theta) = P_l(\cos\theta) \tag{12.59}$$

The radial part is given by

$$\frac{1}{R}\cdot\frac{d}{dr}\left(r^2\frac{dR}{dr}\right) + k^2r^2 - l(l+1) = 0 \tag{12.60}$$

Putting $\zeta = kr$, the above equation takes the form:

$$\frac{1}{R}\cdot\frac{d}{d\zeta}\left(\zeta^2\frac{dR}{d\alpha}\right) + \zeta^2 - l(l+1)$$

$$\text{or, } \frac{1}{\zeta^2}\cdot\frac{d}{d\zeta}\left(\zeta^2\frac{dR}{d\zeta^2}\right) + \left[1 - \frac{l(l+1)}{\zeta^2}\right]R = 0 \tag{12.61}$$

Let $R = v/\sqrt{\zeta}$, then from eq. (12.6), we get

$$\zeta^2\frac{d^2v}{d\zeta^2} + \zeta\frac{dv}{d\zeta^2} + \left[\zeta^2 + \left(l + \frac{1}{2}\right)^2\right]v = 0 \tag{12.62}$$

The solution of eq. (12.62) is

$$v = J_{l+\frac{1}{2}}(\zeta)$$

$$\therefore R = \frac{J_l + \frac{1}{2}(\zeta)}{\sqrt{\zeta}} \tag{12.63}$$

Now, the solution of Schrödinger equation will contain two terms $P_l(\cos\theta)$ and $J_l + \frac{1}{2}(\zeta)/\sqrt{\zeta}$, i.e., the solution is

$$\psi = \sum_i A_l P_l(\cos\theta) \frac{1}{\sqrt{\zeta}} J_{l+\frac{1}{2}(\zeta)} \tag{12.64}$$

where A_l are constants.

As the function $\psi = \exp(i\zeta\cos\theta)$ represents a wave which is also denoted by eq. (12.64), we write:

$$\exp(i\zeta\cos\theta) = \sum_l A_l P_l(\cos\theta) \frac{1}{\sqrt{\zeta}} J_{l+\frac{1}{2}(\zeta)} \tag{12.65}$$

Now, substituting the values of $J_{l+\frac{1}{2}(\zeta)}$, we get

$$\exp(i\zeta\cos\theta) = \sum_l A_l P_l(\cos\theta) \frac{1}{\sqrt{\zeta}} \sqrt{\frac{2}{\pi\zeta}} f_l(\zeta) = \sum a_l P_l(\cos\theta) \frac{f_l(\zeta)}{\zeta} \tag{12.66}$$

$$\left[\because J_{l+\frac{1}{2}(\zeta)} = \sqrt{\left(\frac{2}{\pi\zeta}\right)} f_l(\zeta)\right] \tag{12.67}$$

where $a_l = A_l \sqrt{\frac{2}{\pi}}$

To find out the values of the constants a_l, we use the orthogonality property. Multiplying eq. (12.67) by $P_l'(\cos\theta)$ and integrating between −1 to +1, we get

$$\int_{-1}^{+1} P_l'(\cos\theta) \exp(i\zeta\cos\theta)\, d(\cos\theta) = \sum a_l \frac{f_l(\zeta)}{\zeta} \int_{-1}^{+1} P_l'(\cos\theta) P_l(\cos\theta) d(\cos\theta) \tag{12.68}$$

Now, the Legendre polynomial have the orthogonality relation of the form:

$$\int_{-1}^{+1} P_l(\cos\theta) P_l'(\cos\theta)\, d(\cos\theta) = \frac{2}{2l+1} \delta_{ll'} \tag{12.69}$$

where δ'_{ll} is *Kronecker delta function.*

From eqs (12.68) and (12.69), we get

$$\int_{-1}^{+1} P_l(\cos\theta) \exp(i\zeta\cos\theta)\, d(\cos\theta) = \frac{2}{2l+1} \left[\frac{f_l(\zeta)}{\zeta}\right] a_l \tag{12.70}$$

substituting $l = l'$.

Now, integrating by parts the LHS of eq. (12.70)

$$\int_{-1}^{+1} P_l(\cos\theta)\exp(i\zeta\cos\theta)\,d(\cos\theta) = \frac{1}{i\zeta}\left[P_l(\cos\theta)\exp\left((i\zeta\cos\theta)\right)\right]_{-1}^{+1}$$

$$= \frac{1}{i\zeta}\left[\exp(i\zeta)(-1)^l\exp(-i\zeta)\right]$$

$$\left[\because P_l(+1) \text{ and } P_l(-1) = (-1)^l\right]$$

$$= \frac{1}{i\zeta}\exp(il\pi/2)\left[e^{i(\zeta-l\pi/2)} - (-1)^l e^{-i(\zeta+l\pi/2)}\right]$$

$$= \frac{1}{i\zeta}\exp(il\pi/2)\left[e^{i(\zeta-l\pi/2)} - e^{il\pi}\cdot e^{-i(\zeta+l\pi/2)}\right]$$

$$\left[\because (-1)^l = e^{ilx}\right]$$

$$= \frac{1}{i\zeta}e^{il\pi/2}\left[e^{i(\zeta-l\pi/2)}e^{-i(\zeta-l\pi/2)}\right]$$

$$= \frac{2}{\zeta}e^{il\pi/2}\sin(\zeta - l\pi/2)$$

$$\therefore \int_{-1}^{+1} P_l(\cos\theta)\,e^{i\zeta\cos\theta}\,d(\cos\theta) = \frac{2}{\zeta}(i)^l\sin\left(\zeta - \frac{i\pi}{2}\right) \tag{12.71}$$

where $(i)^l = e^{il\pi/2}$.

$$\left[\because e^{il\pi/2} = \cos\frac{l\pi}{2} + i\sin\frac{l\pi}{2}\right.$$

$$\left.\therefore \begin{rcases} \text{when } l = 0 = \cos\theta + i\sin\theta = 1 \\ l = 1 = \cos\frac{\pi}{2} + i\sin\frac{\pi}{2} = i \\ l = 2 = \cos\pi + i\sin\pi = -1 \end{rcases} = (i)^l\right]$$

From eqs (12.70) and (12.71)

$$\frac{2i^l}{\zeta}\sin\left(\zeta - \frac{l\pi}{2}\right) = \frac{2}{2l+1}\left[\frac{f_l(\zeta)}{\zeta}\right]a_l \tag{12.72}$$

$$\text{or, } \frac{2i^l}{\zeta}\sin\left(\zeta - \frac{l\pi}{2}\right) = \frac{2}{2l+1}\frac{(\sin\zeta - l\pi/2)}{\zeta}a_l$$

whence

$$a_l = i^l(2l+1) \tag{12.73}$$

Substituting the value of a_l in eq. (12.67), we get

$$\exp(i\zeta\cos\theta) = \sum_{l=0} i^l(2l+1)\left[\frac{f_l(\zeta)}{\zeta}\right]P_l(\cos\theta) \tag{12.74}$$

This is *Bauer's formula*.

Equation (12.74) shows that there is no azimuthal dependence. It is because of spherical symmetry about z-axis. Further, it implies that the scattering amplitude must be independent of ϕ as there is nothing in the potential due to spherical symmetry so that it can discriminate between different values of ϕ.

12.8 Scattering by spherically symmetric potentials: Partial wave analysis

Partial wave analysis is applicable to spherically symmetric potentials where *Born approximation fails.* For spherically symmetric potential, angular momentum of scattered particle is constant of motion. When there exists a short range potential, only the few eigenfunctions of small angular momenta are affected by this potential. In partial wave analysis problems, the scattering of particles with angular momenta satisfying the condition $l \ll ka$, is considered where a is the range of potential.

The wave function ψ for a particle of energy E, which moves in a central, spherically symmetric field of force is a solution of the Schrödinger equation

$$\nabla^2 \psi + \frac{2m}{\hbar^2}[E - V(r)]\psi = 0 \tag{12.75}$$

where V is a function of r only. We shall see the asymptotic form of the solution of the above equation. Since, in case of spherically symmetric potential, the wave function is independent of angle ϕ, the general solution of eq. (12.75) is

$$\psi = R_l(r)P_l(\cos\theta) \tag{12.76}$$

where $R_l(r) = \dfrac{\chi_l(r)}{r}$ and $P_l(\cos\theta)$ is the Legendre polynomial of order l. χ_l satisfies the equation

$$\frac{d^2\chi_l}{dr^2} + \left[k^2 - U(r) - \frac{l(l+1)}{r^2}\right]\chi_l = 0 \tag{12.77}$$

$$k = \left[\frac{2mE}{\hbar^2}\right]^{1/2} \text{ and } U(r) = \frac{2mV(r)}{\hbar^2}$$

To find the general nature of the asymptotic behaviour, we consider r to be so large that the V and l terms in eq. (12.77) can be neglected. Then the solution of eq. (12.77) will be of the form: $e^{\pm ikr}$.

$$\therefore \psi = \sum_l P_l(\cos\theta)\frac{\chi_l}{r} = \sum_l P_l(\cos\theta)\frac{e^{\pm ikr}}{r} \tag{12.78}$$

Hence, the general solution is

$$\psi = \sum_l P_l(\cos\theta)\left[C_l\frac{e^{-ikr}}{r} + D_l\frac{e^{-ikr}}{r}\right] \tag{12.79}$$

where C_l and D_l are constants.

Now, the form of ψ at a large distance from the scattering centre must be such as to describe an incident plane wave with energy E and an outgoing spherical wave representing the scattered particle.

Considering the direction of incidence as the positive z-axis, we have

$$\psi = e^{ikz} + f(\theta)\frac{e^{ikr}}{r} \tag{12.80}$$

From eqs (12.79) and (12.80), we get

$$e^{ikz} + f(\theta)\frac{e^{ikr}}{r} = \sum_l P_l(\cos\theta)\left[C_l\frac{e^{+ikr}}{r} + D_l\frac{e^{-ikr}}{r}\right]$$

$$\text{or, } e^{ikz} + f(\theta)\frac{e^{ikr}}{r} = \sum_l P_l(\cos\theta)[b_l]\sin\frac{[\zeta - \frac{l\pi}{2} + \delta_l]}{\zeta} \quad (12.81)$$

Here, we have written $l\pi/2$ in place of $\pi/2$ for convenience. Also from eq. (12.79), we get on expanding $\cos\zeta$, i.e., $-\sin(\zeta - \pi/2)$, which has been written as $\sin(\zeta - l\pi/2)$, the negative sign and constants have been generalised in new constant b_l. The quantity δ_l is the phase-shift of the lth partial wave. It measures the amount by which the regular function $\sin[\zeta - \frac{l\pi}{2} + \delta_l]$ is displace in ζ, relative to the free particle function $f_l(\zeta)$. If $V(r) = 0$, the function $\sin\left(\zeta - \frac{l\pi}{2} + \delta_l\right)$ and $f_l(\zeta)$ are identical.

Thus, the effect of the scattering potential is to shift the phase of each outgoing partial wave.

Calculation of b_l

We substitute the value of e^{ikr} in eq. (12.81) from equation

$$e^{ikr} = \sum_{l=0}^{\infty} i^l(2l+1)\left[\frac{f_l(\zeta)}{\zeta}\right]P_l(\cos\theta) \quad (12.82)$$

we have

$$\sum i^l(2l+1)P_l(\cos\theta)\frac{\sin(\zeta - l\pi/2)}{\zeta} + kf(\theta)\frac{e^{i\zeta}}{\zeta}$$

$$= \sum b_l P_l(\cos\theta)\sin\left(\zeta - \frac{l\pi}{2} + \delta_l\right)$$

$$\text{or, } \sum_l i^l(2l+1)P_l(\cos\theta)\sin\left(\zeta - \frac{l\pi}{2}\right) + kf(\theta)\,e^{i\zeta}$$

$$= \sum_l b_l P_l(\cos\theta)\sin\left(\zeta - \frac{l\pi}{2} + \delta_l\right)$$

$$\text{or, } \sum_l i^l(2l+1)P_l(\cos\theta)\frac{e^{i(\zeta - \frac{l\pi}{2})} - e^{-i(\zeta - \frac{l\pi}{2})}}{2i} + kf(\theta)\,e^{i\zeta}$$

$$= \sum_l b_l P_l(\cos\theta)e^{i(\zeta - \frac{l\pi}{2} + \delta_l)} - \frac{e^{-i(\zeta - \frac{1}{2}l\pi + \delta_l)}}{2i} + kf(\theta)\,e^{i\zeta}\left[\because \sin x = \frac{e^{ix} - e^{-ix}}{2i}\right]$$

$$\text{or, } e^{i\zeta}\left[\left\{\sum_l i^l(2l+1)P_l(\cos\theta)\,e^{-\frac{il\pi}{2}}\right\} - \left\{\sum_l b_l P_l(\cos\theta)\,e^{-\frac{il\pi}{2} + i\delta_l}\right\} + 2ikf(\theta)\right]$$

$$= e^{-i\zeta}\left[\left\{\sum_l i^l(2l+1)\,e^{il\pi/2}P_l(\cos\theta)\right\} - \left\{\sum b_l P_l(\cos\theta)\,e^{\frac{il\pi}{2} - i\delta_l}\right\} + 2ikf(\theta)\right]$$

This is an identity in ζ. Since, $e^{+i\zeta}$ and $e^{-i\zeta}$ are linearly independent, each of the quantities in square brackets must vanish. Therefore, from the coefficient of $e^{-i\zeta}$, we have

$$\sum_l \{(i^l)(2l+1) - b_l e^{i\delta_l}\}P_l(\cos\theta)\,e^{\frac{il\pi}{2}} = 0 \text{ or, } b_l = i^l(2l+1)\,e^{i\delta_l} \quad (12.83)$$

The constant b_l are thus determined by the condition that the scattered wave part of the solution shall contain outgoing wave only, i.e., a term in $\frac{\exp(-i\zeta)}{\zeta}$ is not allowed in scattered wave part of eq. (12.80).

Using eq. (12.83) in eq. (12.82) and, then equating the coefficients of $e^{i\zeta}$ equal to zero

$$2ikf(\theta) = \sum_l i^l(2l+1)\,e^{2i\delta_l}e^{il\pi/2}P_l(\cos\theta) - \sum_l i^l(2l+1)P_l(\cos\theta)\,e^{-il\pi/2}$$

$$= \sum_l (2l+1)P_l(\cos\theta)\left[e^{2i\delta_l}-1\right]\left[\because e^{il\pi/2}=i^{-l}\right] \tag{12.84}$$

$$= \sum_l (2l+1)P_l(\cos\theta)\,e^{i\delta l}\left[e^{i\delta_l}-e^{-i\delta_l}\right]$$

$$\therefore f(\theta) = \frac{1}{k}\sum_{l=0}^{\infty}(2l+1)P_l(\cos\theta)\,e^{i\delta_l}\left[\frac{e^{i\delta_l}-e^{-i\delta_l}}{2i}\right] = \frac{1}{k}\sum_{l=0}^{\infty}(2l+1)P_l(\cos\theta)\,e^{i\delta_l}\sin\delta_l \tag{12.85}$$

So the differential cross-section is $\sigma(\theta) = |f(\theta)|^2$

$$= \frac{1}{k^2}\sum_{l=0}^{\infty}\left|(2l+1)P_l(\cos\theta)\,e^{i\delta_l}\sin\delta_l\right|^2$$

$$\text{or, } \sigma(\theta) = \sum_{l=0}^{\infty}\frac{(2l+1)^2P_l^2\cos(\theta)\sin^2\delta_l}{k^2} \tag{12.86}$$

The total scattering cross-section is the integral of eq. (12.86) over a unit sphere. Because of the orthogonality of the Legendre polynomials it contains no products of the factors involving different values of l.

$\therefore$ the total scattering cross-section is

$$\sigma_l = \int_\omega \sigma(\theta)\,d\omega = 2\pi\int_{-1}^{+1}|f(\theta)|^2 d(\cos\theta) \tag{12.87}$$

$$= 2\pi\int_{-1}^{+1}\sigma(\theta)\,d(\cos\theta) = \frac{2\pi}{k^2}\sum_l\int_{-1}^{+1}(2l+1)^2P_l^2(\cos\theta)\sin^2\delta_l d(\cos\theta)$$

$$= \frac{2\pi}{k^2}\sum_l(2l+1)^2\sin^2\delta_l\int_{-1}^{+1}P_l(\cos\theta)\,d(\cos\theta) = \frac{2\pi}{k^2}\sum_l(2l+1)^2\sin^2\theta_l\times\frac{2}{2l+1}$$

$$\left[\because \text{from the orthogonality relation } \int_{-1}^{+1}P_l^2(\cos\theta)\,d(\cos\theta) = \frac{2}{2l+1}\right]$$

$$\therefore \sigma_t = \frac{4\pi}{k^2}\sum_l(2l+1)\sin^2\delta_l \tag{12.88}$$

From eq. (12.88) it is evident that the different partial waves contribute independently to the total cross-section. Eq. (12.88) gives the total cross-section of scattering. The scattering is completely determined by the phase-shift. The scattering cross-section vanishes when each of the δ_l is 0° or 180°. The scattering cross-section is the maximum for the values of $\delta_l = \pm\frac{\pi}{2}, \pm\frac{3\pi}{2}, \ldots$.

The relation of phase-shifts and potential energy function is given by

$$\sin\delta_l = -\int_0^a \frac{V}{E}F_l(\zeta)\,f_l(\zeta)\,d\zeta \tag{12.89}$$

where $f_l(\zeta)$ and $F_l(\zeta)$ are the functions in the differential equations:

$$\frac{d^2 f_l}{d\zeta^2} + \left[1 - \frac{l(l+1)}{\zeta^2}\right] f_l = 0 \tag{12.90}$$

$$\text{and } \frac{d^2 F_l}{d\zeta^2} + \left[1 + \frac{V}{E} - \frac{l(l+1)}{\zeta^2}\right] F_l = 0 \tag{12.91}$$

Derivation of δ_l

Multiplying eq. (12.91) by F_l and eq. (12.90) by f_l and, then subtracting

$$\frac{d}{d\zeta}[f_l' F_l - f_l F_l'] + \frac{V}{E} F_l f_l = 0 \tag{12.92}$$

when $\zeta \to \infty$.

$$[f_l' F_l - f_l F_l'] = \cos\left(\zeta - \frac{l\pi}{2}\right) \sin\left(\zeta - \frac{l\pi}{2} + \delta_l\right) - \sin\left(\zeta - \frac{1}{2} l\pi\right) \cos\left(\zeta - \frac{l\pi}{2} + \delta_l\right) = \sin \delta_l$$

$\therefore$ from eq. (12.92), we get

$$\frac{d}{d\zeta} \sin \delta_l + \frac{V}{E} F_l \delta_l = 0$$

We know that

$$\frac{1}{r^2} \frac{d}{dr}\left\{r^2 \frac{dF_l}{dr}\right\} + \left\{k^2 - U(r) - \frac{l(l+1)}{r^2}\right\} F_l(r) = 0$$

$$\text{or, } \frac{1}{\zeta^2} \frac{1}{d\zeta}\left\{\zeta^2 \frac{dF_l}{d\zeta}\right\} + \left\{1 - \frac{V(E)}{E} - \frac{l(l+1)}{\zeta^2}\right\} F_l(\zeta) = 0$$

$$\text{or, } \frac{d^2 F_l}{d\zeta^2} + \left[\left(1 - \frac{V}{E} - \frac{l(l+1)}{\zeta^2}\right)\right] F_l(\zeta) = 0$$

When $V = 0$

$$\frac{d^2 F_l}{d\zeta^2} + \left\{1 - \frac{l(l+1)}{\zeta^2}\right\} f_l(\zeta) = 0 \text{ or, } \sin \delta_l = -\int \frac{V}{E} F_l f_l d\zeta$$

If the phase-shifts are small, i.e., $F_l(\zeta) \to f_l(\zeta)$

$$\delta_l = -\int \frac{V}{E} |f_l(\zeta)|^2 d\zeta \tag{12.93}$$

For low energies approaching zero, only phase-shift for $l = 0$, partial waves are important and scattering becomes isotropic.

Optical theorem: This theorem relates the total scattering cross-section to the scattering amplitude in the forward direction, i.e., the scattering amplitude for $\theta = 0$.

Putting $\theta = 0$ in eq. (12.85), we get

$$f(0) = \frac{1}{k} \sum_{l=0}^{\infty} (2l+1) P_l(\cos 0)\, e^{i\delta_l} \sin \delta_l$$

But, $P_l \cos \theta = P_l(1) = 1$ for all l.

$$\therefore f(0) = \frac{1}{k} \sum_{l=0}^{\infty} (2l+1)\,(\cos \delta_l + i \sin \delta_l) \sin \delta_l$$

The imaginary part of $f(0)$, i.e., $I_m f(0)$ is given by

$$I_m f(0) = \frac{1}{k}\sum_{l=0}^{\infty}(2l+1)\sin^2\delta_l \tag{12.94}$$

According to eq. (12.88), we have

$$\sigma_t = \frac{4\pi}{k^2}\sum_{l=0}^{\infty}(2l+1)\sin^2\delta_l = \frac{4\pi}{k}\frac{1}{k}\sum_{l=0}^{\infty}(2l+1)\sin^2\delta_l \tag{12.95}$$

$$\text{or, } \sigma_t = \frac{4\pi}{k} I_m f(0) \tag{12.96}$$

The relation (12.96) is called the *optical theorem*, in view of which the imaginary part of the forward scattering amplitude measures the intensity which the incident beam suffers due to scattering.

12.9 Scattering by a perfectly rigid sphere

The potential associated with a perfectly rigid sphere is

$$V(r) = \begin{cases} +\infty, & \text{for } r < a \\ 0, & \text{for } r < a \end{cases}$$

where a is the radius of the rigid sphere.

The Schrödinger equation is

$$\nabla^2\psi + \frac{2m}{\hbar^2}[E - V(r)]\psi = 0 \tag{12.97}$$

For $r > a$, $V(r) = 0$.

∴ eq. (12.97) reduces to

$$\nabla^2\psi + \frac{2m}{\hbar^2}E\psi = 0 \text{ or, } \nabla^2\psi + k^2\psi = 0, \left(\text{where } k^2 = \frac{2mE}{\hbar^2}\right) \tag{12.98}$$

The radial part of eq. (12.98) is

$$\frac{1}{r^2}\cdot\frac{d}{dr}\left(r^2\frac{dR}{dr}\right) + \left[k^2 - \frac{l(l+1)}{r^2}\right]R = 0 \tag{12.99}$$

Putting $\zeta = kr$ or, $r = \zeta/k$, we get

$$\frac{d}{d\zeta}\left(\zeta^2\frac{dR}{d\zeta}\right) + [\zeta^2 - l(l+1)]R = 0 \tag{12.100}$$

Now, let

$$R(\zeta) = \frac{V(\zeta)}{\sqrt{\zeta}} \tag{12.101}$$

Substituting this value of $R(\zeta)$ in eq. (12.100)

$$\zeta^2\frac{d^2v}{d\zeta^2} + \zeta\frac{dv}{d\zeta} + \left[\zeta^2 - \left(l + \frac{1}{2}\right)^2\right]v = 0 \tag{12.102}$$

This is *Bessel's equation* of order $(l+\frac{1}{2})$. Its solutions are the linear combinations of $J_{l+\frac{1}{2}}(\zeta)$ and $Y_{l+\frac{1}{2}}(\zeta)$. The solution of $Y_{l+\frac{1}{2}}(\zeta)$ is rejected, since it diverges at $\zeta = 0$, so the solution remains $v = J_{l+\frac{1}{2}}(\zeta)$.

$$\therefore R(\zeta) = \frac{v}{\sqrt{\zeta}} = \frac{J_{l+\frac{1}{2}}(\zeta)}{\sqrt{\zeta}}$$

This can be written as

$$R(\zeta) = \sqrt{\frac{\pi}{2}} \cdot \frac{J_{l+\frac{1}{2}}(\zeta)}{\sqrt{\zeta}} \tag{12.103}$$

In the above equation, $\sqrt{\pi/2}$ is a normalisation constant.

$$R(\zeta) = j_l(\zeta) \tag{12.104}$$

where $j_l(\zeta)$ is the spherical Bessel function

$$j_l(\zeta) = \sqrt{\left(\frac{\pi}{2\zeta}\right)} J_{l+\frac{1}{2}}(\zeta)$$

Again, if J_n is a solution, then J_{-n} is also a solution.

The *spherical Neumann function* is defined by

$$n_l(\zeta) = (-1)^{l+1}\left(\frac{\pi}{2\zeta}\right)^{1/2} J_{-l-\frac{1}{2}}(\zeta) \tag{12.105}$$

Hence, the required solution of eq. (12.100)

$$R_l = Aj_l(\zeta) + Bn_l(\zeta) \tag{12.106}$$

Now, we like to find out the asymptotic behaviour of R_l.

$$j_l(\zeta) = \sqrt{\left(\frac{\pi}{2\zeta}\right)} J_{l+\frac{1}{2}}(\zeta)$$

When $\zeta \to \infty$

$$j_{l+\frac{1}{2}}(\zeta) = \sqrt{\left(\frac{2}{\pi\zeta}\right)} \sin\left(\zeta - \tfrac{1}{2}l\pi\right)$$

$$\therefore j_l(\zeta) = \frac{\sin\left(\zeta - \frac{l\pi}{2}\right)}{\zeta} \tag{12.107}$$

Also

$$n_l(\zeta) = (-1)^{l+1}\sqrt{\left(\frac{\pi}{2\zeta}\right)} J_{-l-\frac{1}{2}}(\zeta)$$

where $J_{-l-\frac{1}{2}}(\zeta) = (-1)^l \sqrt{\frac{2}{\pi\zeta}} \cos\left(\zeta - \frac{1}{2}l\pi\right)$.

$\therefore$ when $\zeta \to \infty$

$$n_l(\zeta) = (-1)^{l+1}(-1)^l \frac{\cos\left(\zeta - \frac{l\pi}{2}\right)}{\zeta} \tag{12.108}$$

$$\text{or, } n_l(\zeta) = -\frac{\cos\left(\zeta - \frac{1}{2}l\pi\right)}{\zeta} \tag{12.109}$$

$\therefore$ from eq. (12.106)

$$R_l = \frac{A\sin\left(\zeta - \frac{l\pi}{2}\right) B\cos\left(\zeta - \frac{l\pi}{2}\right)}{\zeta}$$

$$\text{or, } R_l = \frac{C}{\zeta}\left[\sin\left(\zeta - \frac{l\pi}{2}\right)\cos\delta_l + \cos\left(\zeta - \frac{l\pi}{2}\right)\sin\delta_l\right] \tag{12.110}$$

where $A = C\cos\delta_l$ and $B = -C\sin\delta_l$.

Equation (12.110) can be expressed as

$$R_l = \frac{C}{\zeta}[j_l(\zeta)\cos\delta_l - n_l(\zeta)\sin\delta_l] \tag{12.111}$$

$$\text{and } R_l(r) = \frac{C}{kr}[j_l(kr)\cos\delta_l - n_l(kr)\sin\delta_l] \tag{12.112}$$

Equation (12.112) represents the solution of wave equation, when $r > a$. When $r = a$, the wave function vanishes.

$\therefore R_l(r) = 0$ at $r = a$.

From eq. (12.112)

$$0 = \frac{C}{ka}[j_l(ka)\cos\delta_l - n_l(ka)\sin\delta_l] \text{ or, } j_l(ka)\cos\delta_l - n_l\sin\delta_l = 0$$

$$\text{or, } \tan\delta_l = \frac{j_l(ka)}{n_l(ka)} \tag{12.113}$$

Equation (12.112) gives the tangent of the phase-shift δ_l.

$$\tan\delta_l = \frac{j_l(ka)}{n_l(ka)} = \frac{\left(\frac{1}{2}\frac{\pi}{ka}\right)J_{l+\frac{1}{2}}(ka)}{(-1)^{l+1}\sqrt{\frac{\pi}{2ka}}J_{-l-\frac{1}{2}}(ka)} = \frac{J_{l+\frac{1}{2}}(ka)}{(-1)^{l+1}J_{-l-\frac{1}{2}}(ka)}$$

When $l = 0$

$$\tan\delta_0 = -\frac{J_{1/2}(ka)}{J_{-\frac{1}{2}}(ka)} = -\frac{\sqrt{\left(\frac{2}{\pi ka}\right)}\sin ka}{\sqrt{\left(\frac{2}{\pi ka}\right)}\cos ka} = -\tan(ka) \text{ i.e., } \delta_0 = -ka$$

The scattering cross-section is

$$\sigma = \frac{4\pi}{k^2}\sum_l (2l+1)\sin^2\delta_l \tag{12.114}$$

When $l = 0$

$$\sigma = \frac{4\pi}{k^2}\sin^2 ka = \frac{4\pi}{k^2}k^2a^2 \; (ka \ll 0)$$

$$\therefore \sigma \simeq 4\pi a^2 \tag{12.115}$$

Equation (12.115) gives the expression for σ for *low energy scattering* $(ka \ll 1)$.

From eq. (12.115), $\sigma \propto a^2$, implying that the scattering is spherically symmetric.

Equation (12.115) shows that this is four times the result obtained on the basis of classical theory. It is to be remembered that classical theory applies when we can follow the trajectory of the particle.

In the *high energy limit* ($ka \gg 1$), we might expect to get the classical result, since it is, then possible to make wave packets that are small compared to the size of the scattering region and these can follow the classical trajectories without spreading appreciably.

Now, from eq. (12.113)

$$\tan\delta_l = \frac{j_l(ka)}{n_l(ka)}$$

Using eqs (12.107) and (12.113), we have

$$\tan\delta_l = -\tan\left(ka - \frac{l\pi}{2}\right) = \tan\left(-ka + \frac{l\pi}{2}\right)$$

Substituting eq. (12.113)

$$\sigma = \frac{4\pi}{k^2}\sum_{l=0}^{l=ka}(2l+1)\sin^2\left(-ka + \frac{l\pi}{2}\right) \simeq \frac{4\pi}{k^2}\int_0^{ka}(2l+1)\sin^2\left(-ka + \frac{\pi x}{2}\right)dx$$

$$\text{or, } \sigma \simeq 2\pi a^2 \tag{12.116}$$

Again, the scattering cross-section is seen to be twice of that obtained on the basis of classical theory. Further, it suggests that from the region $\theta \simeq 0$, there is contribution of πa. *This contribution is due to Poisson's diffraction phenomenon, according to which when a plane wave is incident normally on an opaque disc, there is a bright spot on the centre of the geometric shadow.*

12.10 Scattering by an attractive square potential well

We consider a square well of radius a and depth V_0 as shown in Fig 12.4. The potential energy function of an attractive square potential well is of the form:

$$V(r) = \begin{cases} -V_0, & (r < a) \\ 0, & (r > a) \end{cases} \tag{12.117}$$

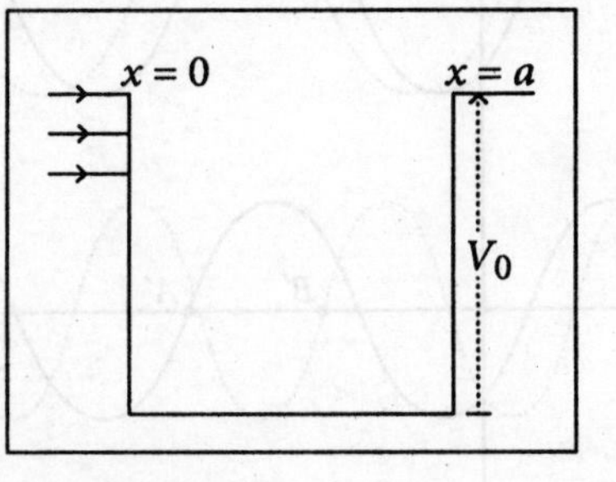

Fig 12.4 Attractive potential well

Phase-shift: The radial part of the wave equation satisfies for $l = 0$.

$$\frac{1}{R^2}\cdot\frac{d}{dr}\left(r^2\frac{dR}{dr}\right) + \frac{2\mu}{\hbar^2}[E - V(r)]R = 0$$

$$\text{or, } \frac{d^2G}{dr^2} + \frac{2\mu}{\hbar^2}[E - V(r)]G = 0 \text{ where } R(r) = \frac{G(r)}{r}.$$

Applying the above condition, we have

$$\frac{d^2G}{dr^2} + k_1^2 G(r) = 0, \; (r < a)$$

$$\frac{d^2G}{dr^2} + k^2 G(r) = 0, \; (r > a)$$

where $k^2 = \frac{2\mu E}{\hbar^2}$, $k_0^2 = \frac{2\mu V_0}{\hbar^2}$ and $k_1^2 = k^2 + k_0^2 = \frac{2m}{\hbar^2}[E + V_0]$.

Since, $\frac{G(r)}{r}$ must be bounded at $r = 0$, the solutions which vanish at the origin are given by

$$A \sin k_1 r, \; (r < a)$$

$$\text{and } \sin(kr + \delta_0), \; (r > a)$$

The constant A and δ_0 must be chosen in such a way that $G(r)$ and $\frac{dG}{dr}$ are continuous at $r = a$

$$\text{i.e., } A \sin k_1 a = \sin(ka + \delta_0)$$

$$\text{and } Ak_1 \cos k_1 a = k \cos(ka + \delta_0)$$

$$\text{or, } \tan(ka + \delta_0) = \frac{k}{k_1} \tan k_1 a$$

Thus,

$$\delta_0 = \tan^{-1}\left[\frac{k}{k_1} \tan k_1 a\right] - ka \tag{12.118}$$

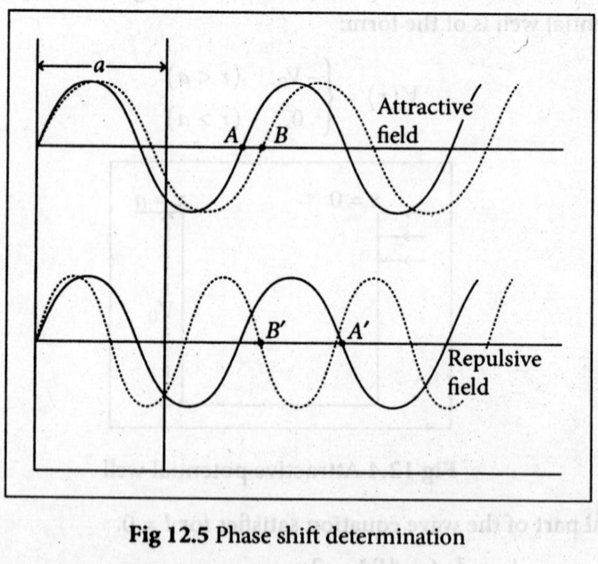

Fig 12.5 Phase shift determination

Fig 12.5 shows the wave function by full line and the function $G = \sin kr$ is denoted by dotted line. The phase-shift δ_0 is represented by the length AB or $A'B'$ multiplied by k.

We can show that δ_0 is positive for an attractive field and negative for a repulsive field.

We know that when $\zeta \to \infty$

$$J_{1/2}(\zeta) = \sqrt{\frac{2}{\pi\zeta}}\sin\zeta \text{ and } J_{-\frac{1}{2}}(\zeta) = \sqrt{\frac{2}{\pi\zeta}}\cos\zeta.$$

The recurrence formula is

$$J_{n-1} = \left(\frac{2n}{\zeta}\right)J_n - J_{n+1}$$

For $n = -\frac{1}{2}$

$$J_{-\frac{1}{2}-1} - \left(\frac{2}{2\zeta}\right)J_{-\frac{1}{2}}(\zeta) - J_{1/2}(\zeta) \text{ or, } J_{-\frac{3}{2}}(\zeta) = -J_{1/2}(\zeta) - \left(\frac{1}{\zeta}\right)J_{-\frac{1}{2}}(\zeta)$$

When $\zeta \to \infty$

$$J_{3/2}(\zeta) = -J_{1/2}(\zeta) = -\sqrt{\frac{2}{\pi\zeta}}\sin\zeta$$

$$J_{-\frac{3}{2}}(\zeta) = -\sqrt{\left(\frac{2}{\pi\zeta}\right)}\cos\left(\zeta - \frac{\pi}{2}\right)$$

For $n = -\frac{3}{2}$

$$J_{-\frac{3}{2}-1}(\zeta) = -\left\{\frac{2.3}{2\zeta}\right\}J_{-\frac{3}{2}}(\zeta) - J_{-\frac{1}{2}}(\zeta)$$

$$J_{-5/2}(\zeta) = -\left(\frac{3}{\zeta}\right)J_{-\frac{3}{2}}(\zeta) - J_{-\frac{1}{2}}(\zeta)$$

When $\zeta \to \infty$

$$J_{-5/2}(\zeta) = -J_{-\frac{1}{2}}(\zeta) = -\sqrt{\frac{2}{\pi/\zeta}}\cos\zeta = (-1)^l\sqrt{\frac{2}{\pi\zeta}}\cos\left(\zeta - \frac{\pi}{2}\right)$$

Proceeding in the same way, we have

$$J_{-l-\frac{1}{2}}(\zeta) = (-1)^l\sqrt{\frac{2}{\pi\zeta}}\cos\left(\zeta - \frac{l\pi}{2}\right),\ l = 1, 2, 3, \ldots$$

Again,

$$\sin(\tan^{-1}x) = \frac{x}{\sqrt{1+x^2}} \text{ and } \cos(\tan^{-1}x) = \frac{x}{\sqrt{1+x^2}}$$

$$\therefore \sin\delta_0 = \sin\left[\tan^{-1}\left(\frac{k}{k_1}\tan k_1 a\right) - ka\right]$$

$$= \sin\left[\tan^{-1}\left(\frac{k}{k_1}\tan k_1 a\right)\right]\cos ka - \cos\tan^{-1}\left(\frac{k}{k_1}\tan ka\right)\sin ka$$

$$= \frac{(k/k_1)\tan k_1 a}{\sqrt{\left\{1+\left(\frac{k}{k_1}\right)^2\tan^2 k_1 a\right\}}}\cos ka - \frac{\sin ka}{\sqrt{\left\{1+\left(\frac{k}{k_1}\right)^2\tan^2 k_1 a\right\}}}$$

$$\text{or, } \sin\delta_0 = \frac{\sin ka}{\sqrt{(k^2 + k_1^2\cot^2 k_1 a)}}[k\cot ka - k_1\cot k_1 a] \qquad (12.119)$$

For the existence of a level of energy $-E_1$, the radial part of the equation must be

$$\left.\begin{aligned} \frac{d^2G}{dr^2} + \frac{2\mu}{\hbar}[-E_1 + V_0]G = 0, r < a \\ \text{and } \frac{d^2G}{dr^2} - \frac{2\mu}{\hbar^2}E_1 G = 0, r > a \end{aligned}\right\} \quad (12.120)$$

Applying boundary conditions, we get

$$A \sin k_2 a = e^{-\gamma a}$$

$$A k_2 \cos k_2 a = -\gamma e^{-\gamma a}$$

where $\gamma = \frac{2\mu E_1}{\hbar^2}$ and $k_2^2 = \frac{2\mu}{\hbar^2}(V_0 - E_1) = k_0^2 - \gamma^2$.

$$\therefore \cot k_2 a = -\frac{\gamma}{k_2}$$

$$\text{or, explicitly } \cot\left[\sqrt{\left(\frac{2\mu}{\hbar^2}(v_0 - E)\,a\right)}\right] = -\sqrt{\frac{E_1}{(V_0 - E_1)}}$$

Assuming $V_0 \gg E_1$, we get

$$\cot\left[\sqrt{\left(\frac{2\mu V_0}{\hbar^2}a\right)}\right] = -\sqrt{\left(\frac{E_1}{V_0}\right)} \quad (12.121)$$

[Actually this case prevails in neutron-proton scattering where the binding energy of the deuteron is (2.223) MeV (E_1) and $V_0 \simeq 50$ MeV]

Phase-shift: We have from eq. (12.118)

$$\delta_0 = \tan^{-1}\left(\frac{k}{k_1}\tan k_1 a\right) - ka \text{ or, } \tan\delta_0 = \frac{\left[\frac{k}{k_1 \cot k_1 a} - \frac{1}{\cot ka}\right]}{1 + \frac{k}{k_1 \cot k_1 a}\frac{1}{\cot ka}}$$

$$\text{or, } k\cot\delta_0 = \frac{k\left(1 + \frac{k}{L\cot ka}\right)}{\left(\frac{k}{L} - \frac{1}{\cot ka}\right)}$$

[where $L = k_1 \cot k_1 a$]

Thus,

$$k\cot\delta_0 = \frac{(L\cos ka + k\sin ka)}{(\cos ka - \frac{L}{k}\sin ka)} \quad (12.122)$$

As the numerator and denominator are even functions of k, we obtain

$$k\cot\delta_0 = \frac{L}{1 - La} = \frac{k_1 \cot k_1 a}{1 - k_1 a \cot k_1 a}$$

$$\text{or, } k_1^2 = k_0^2 + k^2 = \frac{2\mu}{\hbar^2}(V_0 + E) \quad (12.123)$$

For small values of E

$$k_1^2 = \frac{2\mu V_0}{\hbar^2}$$

Thus,

$$k\cot\delta_0 = \frac{\sqrt{\left(\frac{2\mu V_0}{\hbar^2}\right)}\cot\sqrt{\left(\frac{2\mu V_0}{\hbar^2}\right)}a}{1-\sqrt{\left(\frac{2\mu V_0}{\hbar^2}\right)}a\cot\sqrt{\left(\frac{2\mu V_0}{\hbar^2}\right)}a} = \frac{-\sqrt{\left(\frac{2\mu V_0}{\hbar^2}\right)}\sqrt{\frac{E_1}{V_0}}}{1+\sqrt{\left(\frac{2\mu V_0}{\hbar^2}\right)}a\sqrt{\frac{E_1}{V_0}}}$$

$$= -\sqrt{\left(\frac{2\mu V_0}{\hbar^2}\right)} = -\gamma \tag{12.124}$$

where it is assumed that $\sqrt{\left(\frac{2\mu E_0}{\hbar^2}\right)}a \ll 1$.

Then the expression for total cross-section is

$$\sigma = \frac{4\pi}{k^2}\sin^2\delta_0 = \frac{4\pi}{k^2(1+\cot\delta_0)} = \frac{4\pi}{k^2+\gamma^2} \text{ or, } \sigma = \frac{4\pi\hbar^2}{2\mu}\cdot\frac{1}{E+E_1} \tag{12.125}$$

Ramsaur effect: We have eq. (12.118)

$$\tan(ka+\delta_0) = \frac{k}{k_1}\tan k_1a \text{ or, } \frac{\tan ka+\tan\delta_0}{1-\tan ka\tan\delta_0} = \frac{k}{k_1}\tan k_1a$$

$$\text{or, } \tan ka+\tan\delta_0 = \frac{k}{k_1}\tan k_1a - \frac{k}{k_1}\tan k_1a\times\tan ka\tan\delta_0$$

$$\text{or, } \tan\delta_0\left[1+\frac{k}{k_1}\tan k_a\tan ka\right] = \frac{k}{k_1}\tan k_1a - \tan ka$$

Now, $\tan\delta_0$ would vanish, if

$$\frac{\tan ka}{k} = \frac{\tan k_1a}{k_1} \tag{12.126}$$

The condition is true for $ka \ll 1$. It was observed by Ramsaur that for low energy (~ 0.7 eV) electrons scattering by noble atoms is extremely small. This is known as *Ramsaur effect*. It is an important property that cross-section vanishes for a non-zero potential.

Application to neutron proton scattering

Experiments shows that the angular distribution of scattering of neutrons with energies up to 7 MeV or more is still isotropic and as such must involve neutrons with $l = 0$.

If $E = 7$ MeV, then

$$k = \sqrt{\left(\frac{2\mu E}{\hbar^2}\right)} \text{ or, } k\left[2\times\frac{940}{2}\text{ MeV}\times 7\text{ MeV}\right]^{1/2}\frac{1}{\hbar c}$$

where $\mu = \frac{m_n m_p}{m_n+m_p} \simeq \frac{m}{2}$, [$m_n$ and m_p being the mass of the neutron and proton respectively].

$$\text{or, } k = \frac{\sqrt{6880}\text{ MeV}}{1.042\times10^{-27}\text{ (erg-sec)}\times 0.6285\times10^6\frac{\text{MeV}}{\text{ergs}}\times 3\times10^{10}\frac{\text{cm}}{\text{sec}}} \simeq 3\times10^{12}\text{ cm}^{-1}$$

As the value of k must correspond to $ka \simeq 1$, the range a cannot be greater than 3×10^{-13} cm.

The depth of the well may be found out from the relation

$$\cot\left[\sqrt{\frac{2\mu}{\hbar^2}(V_0 - E)}\, a\right] = -\sqrt{\frac{E}{V_0 - E_1}}$$

We assume, $V_0 \ll E_1$, then

$$\sqrt{\left(\frac{2\mu V_0}{\hbar^2}\right)}\, a \simeq \frac{\pi}{2} \text{ or, } V_0^2 \simeq 1.03 \times 10^{-21} \text{ MeV cm}^2$$

For $a \simeq 3 \times 10^{-13}$ cm, $D \simeq 100$ MeV, which is much greater than the binding energy of the deuteron ($E_1 = 2.23$ MeV).

Now, for small energies

$$\sigma \simeq \frac{4\pi\hbar^2}{2\mu} \cdot \frac{1}{E_1 + E} \simeq \frac{4\pi\hbar^2}{2\mu E_1} = \frac{4\pi\hbar^2 c^2}{(4\pi c^2)E_1}$$

$$\text{or, } \sigma = \frac{4 \times 3.14 \times [1.04 \times 10^{-27}\,(\text{erg-sec}) \times 0.6285 \times 10^6 (\text{MeV/erg})^2 (3 \times 10^{10})^2]}{940\,(\text{MeV}) \times 2.23\,(\text{MeV})}$$

$$\simeq 2.4 \times 10^{-24} \text{ cm}^2 \simeq 2.4 \text{ barns}$$

The experimentally observed value of $\sigma = 2.1 \times 10^{-24}$ cm^2.

Alternative treatment of scattering by an attractive potential square well

Here, somewhat more complicated problem of the scattering from the spherically symmetric square well potential will be considered.

The potential energy function is of the form:

$$V(r) = \begin{cases} -V_0, & (r < a) \\ 0, & (r > a) \end{cases}$$

The radial wave function $R_0(kr) = R$ for the s state ($l = 0$) obeys the differential equation

$$\left.\begin{aligned} &\frac{d^2R}{dr^2} + \frac{2\mu}{\hbar}(E + V_0)R = 0, (r < a) \\ \text{and } &\frac{d^2R}{dr^2} + \frac{2\mu}{\hbar^2}(E)R = 0, (r > a) \end{aligned}\right\}$$

The boundary conditions are $R = 0$ at $r = 0$.

R and $\frac{dR}{dr}$ should be continuous at $r = a$. Solution of the first equation of eq. (12.126) for $r < a$ is given by $R = A \sin \alpha r$, where

$$\alpha = \sqrt{\left(\frac{2m(E + V_0)}{\hbar^2}\right)}$$

Outside the well region, i.e., $r > a$, the situation can be expressed in the form of a linear combination of functions f_0 and g_0, which are $f_0 = \sin kr$ and $g_0 = \cos kr$ or, $R = \sin(kr + \delta_0)$, where $k = \sqrt{\frac{2\mu E}{\hbar^2}}$.

The continuity at $r = a$ is expressed independently of the normalisation of R, by the requirement that the logarithmic derivative $\frac{1}{R}\left(\frac{dR}{dr}\right)$ shall be continuous, i.e.,

$$\alpha \cos \alpha a = k \cot(ka + \delta_0)$$

Let

$$\alpha \cos \alpha a = k \cot(ka + \delta_0) = \beta$$

Putting $k\cot(ka + \delta_0) = \beta$, we have

$$\cot(ka + \delta_0) = \beta/k \text{ or, } \frac{1 - \tan ka \tan \delta_0}{\tan ka + \tan \delta_0} = \beta/k$$

$$\text{or, } \frac{\beta}{k}\tan ka + \frac{\beta}{k}\tan \delta_0 = 1 - \tan ka \tan \delta_0$$

$$\text{or, } \tan \delta_0 = \frac{1 - \frac{\beta}{k}\tan ka}{\frac{\beta}{k} + \tan ka} = \frac{\frac{k}{\beta} - \tan ka}{1 + \frac{k}{\beta}\tan ka} = \frac{k\left(\frac{1}{\beta} - \frac{\tan ka}{k}\right)}{1 + \frac{k}{\beta}\tan ka}$$

The total cross-section is

$$\sigma = \frac{4\pi}{k^2}\sum(2l+1)\sin^2\delta_l \tag{12.127}$$

For $l = 0$

$$\sigma = \frac{4\pi}{k^2}\sin^2\delta_0 = \frac{4\pi}{k^2(1 + \cot^2\delta_0)} \text{ or, } \sigma = \frac{4\pi}{k^2 + \dfrac{(\beta + k\tan ka)^2}{\left(1 - \dfrac{\beta \tan ka}{k}\right)}} \tag{12.128}$$

By calculating β from equation $\alpha \cdot \cot \alpha a = k\cot(ka + \delta_0) = \beta$, we can obtain the cross-section of scattering.

Approximation for small k : For small k we can expand the expression for $\tan \delta_0$. Retaining only terms up to order of k^2, we get

$$\tan \delta_0 \simeq k\left[\frac{\left(\frac{1}{\beta} - a\right) - \frac{k^2 a^3}{3}}{1 + \frac{k^2 a}{\beta}}\right]$$

We see that as $k \to 0$, the phase-shift also approaches to zero. The sign of the phase at small k depends on the sign of $\left(\frac{1}{\beta} - a\right)$.

If k is so small that $k^2 a^3 \ll 1$ and $\frac{k^2 a}{\beta} \ll 1$, then the above expression reduces to

$$\tan \delta_0 \simeq k\left(\frac{1}{\beta} - a\right)$$

In this approximation, the cross-section of scattering is

$$\sigma = \frac{4\pi}{k^2 + \dfrac{\beta^2}{(1-\beta a)^2}} = \left(\frac{2\hbar^2}{\mu}\right) \frac{\pi}{E + \dfrac{\hbar^2\beta^2}{2\mu} \cdot \dfrac{1}{(1-\beta)^2}} \tag{12.129}$$

To have a better idea of low energy cross-section, one need to obtain β.

12.11 Scattering by a Coulomb field

Coulomb field has an infinite range. The result already derived under the assumption of a potential with a finite range is not immediately applicable without a separate investigation.

If Z and Z' be the atomic number of the scattering nucleus and that of the scattered particle, then the potential at a distance r from the scattering centre is

$$V(r) = \frac{ZZ'e^2}{r}$$

Here, we propose to present a particular method of solving this problem because of its intrinsic interest. The method is based on the observation that the Schrödinger equation with a Coulomb potential is separable in parabolic coordinates. The transformation to the parabolic coordinates is as follows:

$$\left.\begin{aligned} \xi &= r + z, \quad r = \frac{\xi + n}{2} \\ \eta &= r - z, \quad z = \frac{-\eta + \xi}{2} \end{aligned}\right\} \tag{12.130}$$

In these coordinates, the Schrödinger equation is

$$-\frac{\hbar^2}{2\mu}\left[\frac{4}{\xi+\eta}\frac{d}{d\xi}\left(\xi\frac{d}{d\xi}\right) + \frac{4}{\xi+\eta}\frac{\partial}{\partial\eta}\left(\eta\frac{\partial}{\partial\eta}\right) + \frac{1}{\partial\eta}\frac{\partial^2}{\partial\phi^2}\right]\psi + \frac{2ZZ'e^2}{(\xi+\eta)}\psi = E\psi \tag{12.131}$$

Here, we are interested in a solution with axial symmetry about the direction of incidence (say, z-axis). $\therefore$ the solutions which are independent of ϕ have the form:

$$\psi(\xi, \eta) = f_1(\xi)\, f_2(\eta) \tag{12.132}$$

Equation (12.130) is separated into variable ξ and η as

$$\frac{d}{d\xi}\left(\xi\frac{df_1}{d\xi}\right) + \frac{k^2}{4}\xi f_1 - C_1 f_1 = 0 \tag{12.133}$$

and
$$\frac{d}{d\eta}\left(\xi\frac{df_2}{d\eta}\right) + \frac{k^2}{4}\eta f_2 - C_2 f_2 = 0 \tag{12.134}$$

where $k^2 = \dfrac{2\mu E}{\hbar^2}$ and C_1 and C_2 are the two separation constants.

They are related as

$$C_1 + C_2 = \frac{ZZ'\mu e^2}{\hbar^2} \tag{12.135}$$

Scattering is described by a wave function which asymptotically has an incident plane wave part.

$$e^{ikz} = e^{ik/2(\xi-\eta)} \tag{12.136}$$

and a radially outgoing scattering part

$$e^{ikr} = e^{ik/2(\xi+\eta)} \tag{12.137}$$

The behaviour of these two parts suggests that the solution $f_1(\xi)$ is of the form:

$$f_1(\xi) = e^{ik/2\xi} \tag{12.138}$$

Now, substituting eq. (12.138) in eq. (12.133)

$$C_1 = \frac{ik}{2}$$

$$\therefore C_2 = \frac{ZZ'\mu e^2}{\hbar^2} - C_1 = \frac{ZZ'\mu e^2}{\hbar^2} - \frac{ik}{2} \tag{12.139}$$

Equation (12.134) now becomes

$$\frac{d}{d\eta}\left(\eta\frac{df_2}{d\eta}\right) - \left(\frac{ZZ'\mu e^2}{\hbar^2} - \frac{ik}{2}\right)f_2 + \frac{k^2}{4}\eta f_2 = 0 \tag{12.140}$$

Let

$$f_2(\eta) = e^{-ik\eta/2}g(\eta) \tag{12.141}$$

Using

$$n = -\frac{ZZ'\mu e^2}{\hbar^2 k} = -\frac{ZZ'e^2}{\hbar\omega} \tag{12.142}$$

where v is the relative velocity and μ is the reduced mass.

Now, we obtain the equation

$$\eta\frac{d^2g}{d\eta^2} + (1 - ik\eta)\frac{dg}{d\eta} + nkg = 0 \tag{12.143}$$

By comparing eq. (12.143) with the equation below, its solution can be obtained. The said equation is

$$z\frac{d^2F}{dz^2} + (a - c)\frac{dF}{dz} - aF = 0 \tag{12.144}$$

which has a solution $[F(a, c, z)]$ in terms of the confluent hypergeometric function. The asymptotic expansion of this function has the leading terms

$$F(a, c, z) \sim \frac{\Gamma(c)}{\Gamma(c-a)}|z|^{-a}e^{ia\pi/2} = \frac{\Gamma(c)}{\Gamma(a)}|z|^{a-c}\exp[i(a-c)\pi/2]e^x, \text{ for } |z| \text{ large} \tag{12.145}$$

Applying the approximation to $g(\eta)$, we obtain for large values of η

$$g(\eta) \sim \frac{e^{-n\pi/2}}{[\Gamma(1+in)]}\left\{\exp[-i(n\log k\eta - \sigma_n)] + \frac{n}{k\eta}\exp[i(k\eta + n\log\eta - \sigma_n)]\right\} \tag{12.146}$$

where $\Gamma(1 + in) = |\Gamma(1 + in)|e^{i\alpha_n}$.

The particular solution is given by

$$\psi(\xi, \eta, \phi)\, e^{(1/2)\,ik(\xi-\eta)}g(\eta) = e^{ik(\zeta-\eta)/2}F(in, 1, ikn) \tag{12.147}$$

$$\therefore \psi \sim \frac{e^{-n\pi/2}}{|\Gamma(1+in)|}\left\{ \exp[i(kz-n)\log k(r-z)+\sigma_n] + \frac{n}{k(r-z)} \right.$$

$$\left. \exp[i\{kr + n\log k(r-z) - \sigma_n]\right\} \tag{12.148}$$

Equation (12.147) represents the asymptotic form for large values of $r - z$, i.e., at large distances from the scattering centre except in the forward direction.

It is to be noted that the Coulomb wave function does not approach asymptotically the form:

$$\psi = e^{ikr} + f(\theta)\frac{e^{ikr}}{r} \tag{12.149}$$

The first term of eq. (12.147) contains a factor e^{ikz}, which is modified by another factor $e^{-in\log k(r-z)} + \sigma_n$. This factor indicates that the incident plane wave is slightly distorted, no matter how far away from the origin one chooses to go. The second factor is also modified indicating that the outgoing wave does not approach a definite phase.

The differentiating scattering cross-section is

$$\sigma = |f_k(\theta)|^2 = \frac{x^2}{(2k\sin^2\theta/2)} = \left(\frac{ZZ'e^2}{2\mu}\right)^2 \operatorname{cosec}^4\frac{\theta}{2} \tag{12.150}$$

Here

$$f_k(\theta) = \frac{n}{k(1-\cos\theta)}\exp[in\log(1-\cos\theta)] \tag{12.151}$$

but angle independent addition to the phase has been omitted in eq. (12.150).

Equation (12.150) is the formula obtained by Rutherford from classical dynamics. This was verified experimentally for the collision of α-particles with heavier nuclei.

The following properties are note worthy:

1. At a given angle, the cross-section falls of like $1/E^2$ as the energy (E) increases.
2. It depends upon the absolute value of potential and not upon its sign.
3. The angular distribution is independent on the energy.
4. The total cross-section is finite; $\int \sigma d\omega$ diverges at small angles.

12.12 General formulation of scattering theory

The analysis of scattering by partial-wave method is limited to *scattering at relatively low energy*. Further,more, the expansion of the wave function in spherical harmonics is generally useful only for spherically symmetric systems. We shall now present a more general analysis which is not subjected to these analysis.

Green's function

We shall describe a method for the solution of the inhomogeneous partial differential equation in the form:

$$-\nabla^2 v(\vec{r}) - k^2 v(\vec{r}) = -U(\vec{r})\, e^{ikz} = F(\vec{r}),\ \text{say} \tag{12.152}$$

where $F(\vec{r})$ is a known function.

Now, we shall expand $v(\vec{r})$ in terms of $\psi_{k'}(\vec{r})$. k' has been written in place of k to show a fixed value of k.

$$v(\vec{r}) = \int A_{k'}\,\psi_{k'}(\vec{r})\,dk' \tag{12.153}$$

Substituting eq. (12.153) in eq. (12.152)

$$(-\nabla^2 - k^2)\,v(\vec{r}) = \int A_{k'}[-\nabla^2\,\psi_{k'}(\vec{r}) - k^2\,\psi_{k'}(\vec{r})]dk' = F(\vec{r}) \tag{12.154}$$

$$\text{or, } \int A_{k'}[k'^2 - k^2]\psi_{k'}(\vec{r})\,dk' = F(\vec{r}) \tag{12.155}$$

$$\text{or, } \int A_{k'}[k'^2 - k^2]\delta(k' - k'')\,dk' = \int \psi^*_{k''}(\vec{r})F(\vec{r})d\vec{r}$$

where

$$\int \psi^*_{k''}(\vec{r})\,\psi_{k'}(\vec{r})\,d\vec{r} = \delta(k - k'') \text{ or, } A_{k''}[k'' - k^2] = \int \psi^*_{k''}(\vec{r})F(\vec{r})\,d\vec{r}$$

$$\text{or, } A_{k''} = \frac{\int \psi^*_{k''}(\vec{r})F(\vec{r})\,d\vec{r}}{(k''^2 - k^2)} \tag{12.156}$$

Since, we have a definite integral, we can replace double prime by a single prime.

Now, substituting the value of A'_k in eq. (12.153)

$$v(\vec{r}) = \int A_{k'}\,\psi_{k'}(\vec{r})\,dk' \tag{12.157}$$

$$= \int \frac{\int \psi^*_{k'}(\vec{r}')F(\vec{r}')\,d\vec{r}'\psi_{k'}(\vec{r})}{(k'^2 - k^2)}\,dk' \tag{12.158}$$

To avoid confusion in two $\vec{r}$'s in eqs (12.157) and (12.158), $\vec{r}$ has been replaced by $\vec{r}'$.

$v(\vec{r})$ can be expressed in another form:

$$v(\vec{r}) = \int G_{k'}(\vec{r},\vec{r}')F(\vec{r}')\,d\vec{r}' \tag{12.159}$$

where

$$G_{k'}(\vec{r},\vec{r}') = \int \frac{\psi^*_{k'}(\vec{r}')\,\psi_{k'}(\vec{r})\,dk'}{(k'^2 - k^2)} \tag{12.160}$$

$G_{k'}(\vec{r},\vec{r}')$ given by eq. (12.160) is known as *Green's function.*

Green's function for a free particle

For a free particle, the wave function is

$$\psi_{k'}(\vec{r}) = \frac{1}{(2\pi)^{3/2}}e^{i\vec{k}'\vec{r}} \tag{12.161}$$

where $\vec{k}'$ is another vector having magnitude k'.

∴ the green's function is

$$G = \frac{1}{(2\pi)^3}\int \frac{e^{-i\vec{k}'.\vec{r}'}\cdot e^{i\vec{k}'.\vec{r}'}}{(k'^2 - k^2)}\,dk' \tag{12.162}$$

$$\text{or, } G = \frac{1}{(2\pi)^3}\int \frac{e^{i\vec{k}'\cdot(\vec{r}-\vec{r}')}}{(k'^2 - k^2)}\,dk' \tag{12.163}$$

The integration is over the entire $\vec{k}'$ space and to carry out the integration we choose $(\vec{r}-\vec{r}')$ to lie along the polar axis.

$$\therefore \vec{k}'\cdot(\vec{r}-\vec{r}') = k'\rho\cos\theta = k'\rho\omega$$

where $\rho = |(\vec{r}-\vec{r}')|$, $\omega = \cos\theta$.

$$\therefore G = \frac{1}{(2\pi)^3}\int_0^\infty\int_{-1}^{+1}\int_0^{2\pi}\frac{e^{ik'\rho\omega}}{(k'^2-k^2)}k'^2\,dk'd\omega d\phi \tag{12.164}$$

$$= \frac{1}{(2\pi)^3}\int_0^\infty\int_{-1}^{+1}\frac{e^{ik'\rho\omega}}{(k'^2-k^2)}k'^2\,d\omega dk' \tag{12.165}$$

Now, considering the integral

$$\int_{-1}^{+1}e^{ik'\rho\omega}\,d\omega = \frac{1}{ik'\rho}\left[e^{ik'\rho\omega}\right]_{-1}^{+1} = \frac{1}{ik'\rho}\left[e^{ik'\rho}-e^{-ik'\rho}\right]$$

$$= \frac{2}{k'\rho}\left[\frac{e^{ik'\rho-e^{-k'\rho}}}{2i}\right] = \frac{2}{k'\rho}\sin k'\rho$$

Substituting this value in eq. (12.165)

$$G = \frac{2}{(2\pi)^3}\int_0^3\frac{\sin k'\rho}{k'\rho(k'^2-k^2)}k'^2\,dk' \tag{12.166}$$

Now, putting $k'\rho = x$, we have

$$G = \frac{2}{4\pi^2}\int_0^\infty\frac{\sin x}{\dfrac{x(x^2-\sigma^2)}{\rho^2}}\cdot\frac{x^2\,dx}{\rho^3},\text{ where } \sigma = k\rho \tag{12.167}$$

$$\text{or, } G = \frac{1}{2\pi^2\rho}\int_0^\infty\frac{x\sin x}{(x^2-\sigma^2)}dx \tag{12.168}$$

Equation (12.166) can be solved by contour integration method. For this purpose, we write:

$$G = \frac{1}{4\pi^2\rho}\int_{-\infty}^{+\infty}\frac{x\sin x}{(x^2-\sigma^2)}dx = \frac{1}{4\pi^2\rho}[\text{I-II}] \tag{12.169}$$

where $\text{I} = \dfrac{1}{2i}\displaystyle\int\frac{xe^{ix}\,dx}{(x+\sigma)(x-\sigma)}$ and $\text{II} = \dfrac{1}{2i}\displaystyle\int\frac{xe^{-ix}\,dx}{(x+\sigma)(x-\sigma)}$.

The integral used in eq. (12.169) has the poles at $x = \pm\sigma$. We can have four solutions, when

(i) both the poles are below the semicircle

(ii) both the poles are within the semicircle

(iii) one pole $x = -\sigma$ is within the semicircle

(iv) one pole $x = +\sigma$ is within the semicircle.

Only those solutions are to be chosen which satisfy the boundary condition, i.e., the function $v(r)$ should represent an outward wave, when $r \to \infty$

$$v(r) = \frac{e^{ikr}}{r}f(\theta,\phi)$$

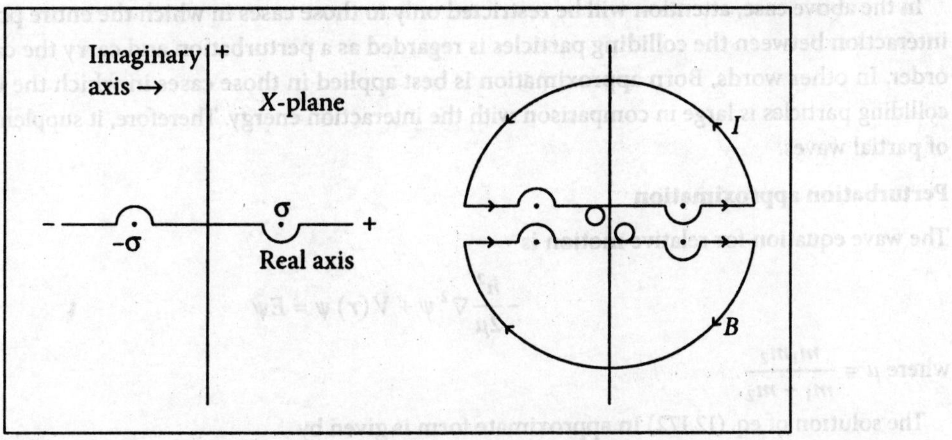

Fig 12.6 Contours to evaluate two integrals

The value of I integral can be evaluated by closing the contour with an infinite semicircle in positive imaginary half-plane, since the exponential becomes vanishingly small there and contributes nothing to the integral. Then the value of I is

$$2\pi i(\text{Residue at } x = +\infty) = \frac{2\pi i}{2i}\frac{\sigma e^{i\sigma}}{2\sigma}$$

$$\therefore I = \frac{\pi}{2}e^{i\sigma}$$

The second integral II can be evaluated by closing the contour with an infinite semicircle in the negative imaginary half-plane. The value of second integral is

$$-2\pi i(\text{Residue at } x = -\sigma) = -\frac{2\pi i}{2i}\left(\frac{-\sigma e^{i\sigma}}{-2\sigma}\right)$$

$$\therefore \text{II} = -\frac{\pi}{2}e^{i\sigma}$$

∴ eq. (12.169) becomes

$$G = \frac{1}{4\pi^2\rho}\left[\frac{\pi}{2}e^{i\sigma} + \frac{\pi}{2}e^{i\sigma}\right] = \frac{1}{4\pi}e^{i\sigma} \tag{12.170}$$

$$\text{or, } G = \frac{1}{4\pi|\vec{r} - \vec{r}'|}e^{ik|\vec{r}-\vec{r}'|} \tag{12.171}$$

This is Green's function whose each term can be treated as propagator which carry the waves from their last point of scattering to the next.

12.13 Born approximation

If the scattering takes place from the scattering centres which are localised but are weak such that the scattering does not take place at large distances from the scatterer and the scattered wave is weak in amplitude, then Born approximation can be used to evaluate the scattering amplitude $f(\theta, \phi)$ and hence the differential cross-section $\sigma(\theta, \phi) = |f(\theta, \phi)|^2$.

In the above case, attention will be restricted only to those cases in which the entire potential energy of interaction between the colliding particles is regarded as a perturbation and carry the calculation to first order. In other words, Born approximation is best applied in those cases in which the kinetic energy of colliding particles is large in comparison with the interaction energy. Therefore, it supplements the method of partial waves.

Perturbation approximation

The wave equation for relative motion is

$$-\frac{\hbar^2}{2\mu}\nabla^2\psi + V(r)\psi = E\psi \tag{12.172}$$

where $\mu = \dfrac{m_1 m_2}{m_1 + m_2}$.

The solution of eq. (12.172) in approximate form is given by

$$\psi(r,\theta,\phi) \xrightarrow{r\to\infty} e^{ikz} + f(\theta,\phi)\frac{e^{ikr}}{r} \tag{12.173}$$

Here, e^{ikz} represents particles moving in the z-direction and the second term represents the particles which are moving radially outwards.

Applying a perturbation approach such that

$$\psi(\vec{r}) = e^{ikz} + v(\vec{r}) \tag{12.174}$$

Here, the scattered wave $v(\vec{r})$ may be regarded as small addition to the unperturbed plane wave e^{ikz}. The term $v(\vec{r})$ will be found only to first order, otherwise Born approximation becomes tedious for higher orders.

Equation (12.172) can be written as

$$\nabla^2\psi + k^2\psi = U(\vec{r})\psi \tag{12.175}$$

where $k^2 = \dfrac{2\mu E}{\hbar^2}$ and $U(\vec{r}) = \dfrac{2\mu V(r)}{\hbar^2}$.

Substituting eq. (12.174) in eq. (12.175), we have

$$\nabla^2\left[e^{ikz} + v(\vec{r})\right] + k^2\left[e^{ikz} + v(\vec{r})\right] = U(\vec{r})[e^{ikz} + v(\vec{r})]$$

$$\text{or, } \nabla^2\left[e^{ikz} + v(\vec{r})\right] + k^2\left[e^{ikz} + v(\vec{r})\right] = U(\vec{r})e^{ikz} + U(\vec{r})v(\vec{r}) \tag{12.176}$$

Now, in view of our assumption that $v(\vec{r})$ is small in comparison with e^{ikz} or roughly that $U(\vec{r})$ is small in comparison with k^2 leads us to neglect the term $U(\vec{r})v(\vec{r})$ in eq. (12.176). Hence, eq. (12.176) reduces to the form:

$$\nabla^2 e^{ikz} + \nabla^2 v(\vec{r}) + k^2 e^{ikz} + k^2 v(\vec{r}) = U(\vec{r})e^{ikz} \tag{12.177}$$

Now,

$$\nabla^2 e^{ikz} = -k^2 e^{ikz}$$

$$\therefore \nabla^2 v(\vec{r}) + k^2 v(\vec{r}) = U(\vec{r})e^{ikz} \tag{12.178}$$

Thus, we have to solve the above inhomogeneous wave equation where the right-hand side is known.

A sufficient condition for the validity of our solution is

$$|v(\vec{r})| \ll |e^{ikz}| = 1, \text{ for all } \vec{r} \tag{12.179}$$

Expression for scattering cross-section

The solution of eq. (12.178) can be written as

$$v(\vec{r}) = \int G_{k'}(\vec{r}, \vec{r}')F(\vec{r}')\, d\vec{r}' \tag{12.180}$$

where $G_{k'}(\vec{r}, \vec{r}')$ is the Green's function.

Now, for a free particle, the value of Green's function is

$$G_{k'}(\vec{r}, \vec{r}') = \frac{1}{4\pi(\vec{r} - \vec{r}')} e^{ik|\vec{r}-\vec{r}'|}$$

$$\therefore v(\vec{r}) = \int \frac{e^{ik|(\vec{r}-\vec{r}')|}}{4\pi|\vec{r} - \vec{r}'|} F(\vec{r}')\, d\vec{r}' \tag{12.181}$$

$$\text{or, } v(\vec{r}) = -\int e^{ik|(\vec{r}-\vec{r}')|} \tag{12.182}$$

We have to find out asymptotic form of eq. (12.182). We assume that $U(\vec{r}')$ falls off rapidly enough for large r', so that there is an asymptotic region in which r is large in comparison with those values of r' for which there is a significant contribution to the integrand.

For $r \to \infty$ the following asymptotic relations are valid [Fig 12.7].

$$(\vec{r} - \vec{r}')^2 = \vec{r}^2 + \vec{r}'^2 - 2\vec{r}\cdot\vec{r}' \cos\theta = \vec{r}^2 + \vec{r}^2 - 2\vec{r}\cdot\vec{r}'\omega, \text{ where } \omega = \cos\theta$$

$$= (\vec{r})^2 \left[1 + \frac{\vec{r}'^2}{\vec{r}^2} - \frac{2\vec{r}\cdot\vec{r}'\omega}{r^2}\right]$$

$$\text{or, } |(\vec{r} - \vec{r}')|_{r\to\infty} = r\left[1 - \frac{2r'\omega}{r}\right]^{1/2} = r\left[1 - \frac{r'\omega}{r}\right] = r - r'\omega$$

Similarly,

$$\left|\frac{1}{(\vec{r} - \vec{r}')}\right|_{r\to\infty} = \frac{1}{r\left[1 - \dfrac{2r'\omega}{r}\right]^{1/2}} = \frac{1}{r}\left[1 + \frac{r'\omega}{r}\right] = \frac{1}{r} + \frac{r'\omega}{r^2} = \frac{1}{r}$$

where $\omega = \cos\theta$, θ being the angle between the vectors $\vec{r}$ and $\vec{r}'$.

$\therefore$ eq. (12.182) becomes

$$v(\vec{r}) = -\frac{1}{4\pi}\int \left(\frac{1}{r} + \frac{\omega r'}{r^2}\right) e^{ikr} \cdot e^{-ik\omega r'} U(r')\, e^{ikz'}\, d\vec{r}$$

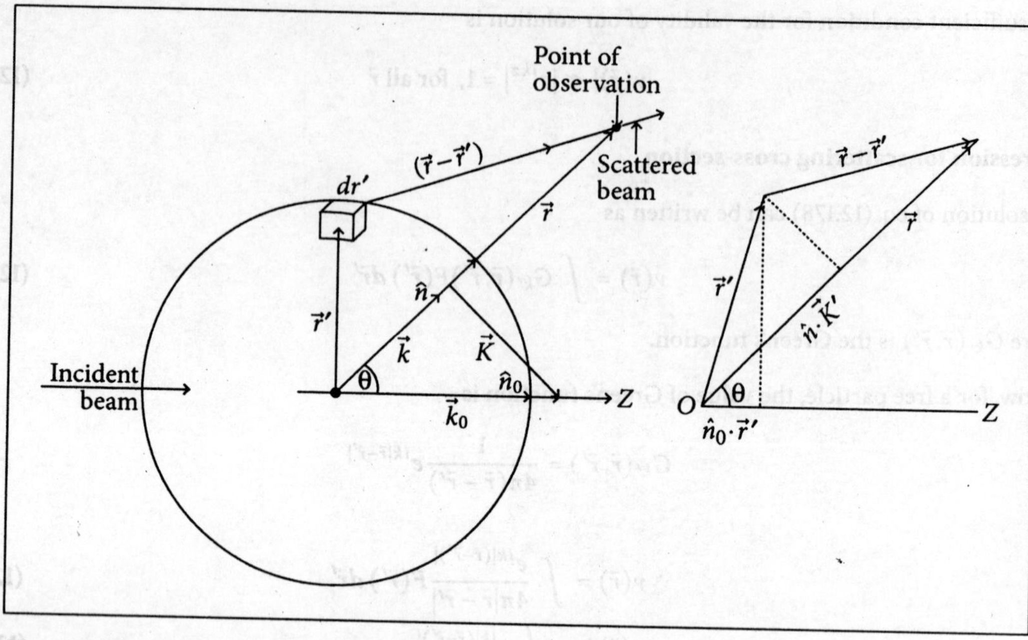

Fig 12.7 The geometric relation between $\vec{r}$ and $\vec{r}'$ used in computing the scattering of particles in Born approximation. $\vec{r}$ is the vector from the centre of the coordinate system, taken at the centre of the scatterer to the point at which the scatterred wave is being computed; $\vec{r}'$ is the position vector of a point in the scattering potential.

When $r \to \infty$ we can neglect $\omega r'/r^2$.

$$\therefore V(\vec{r})_{r\to\infty} = -\frac{1}{4\pi r} e^{ikr} \int U(\vec{r}')\, e^{ik(z'-\omega r')}\, d\vec{r}'$$

Now,

$$\psi(\vec{r}) = e^{ikz} + v(\vec{r}) = e^{ikz} - \frac{1}{4\pi r} e^{ikr} \int u(r')\, e^{ik(z'-\omega r')}\, dr' \tag{12.183}$$

Comparing eq. (12.183) with the equation

$$\psi(\vec{r}) = e^{ikz} + \frac{e^{ikr}}{r} f(\theta, \phi) \tag{12.184}$$

we get

$$f(\theta, \phi) = -\frac{1}{4\pi} \int U(\vec{r}')\, e^{ik(z-r'\omega)} \tag{12.185}$$

If the wave vector $\vec{k}_0$ denotes the incoming beam having direction of the incident beam (polar axis) and that of the scattered beam by $\vec{k}$, then

$$kz' = k\hat{n}_0 \cdot \vec{r}' = \vec{k}_0 \cdot \vec{r}' \ [\hat{n}_0 \text{ is the unit vector } \vec{k}_0]$$

$$\text{or, } k\omega r' = k\hat{n} \cdot \vec{r}' = \vec{k} \cdot \vec{r}' \ [\hat{n} \text{ is the unit vector along } \vec{k}]$$

$$\therefore f(\theta, \phi) = \frac{1}{4\pi} \int U(\vec{r}')\, e^{i(k_0-k)\hat{r}'}\, d\vec{r}'$$

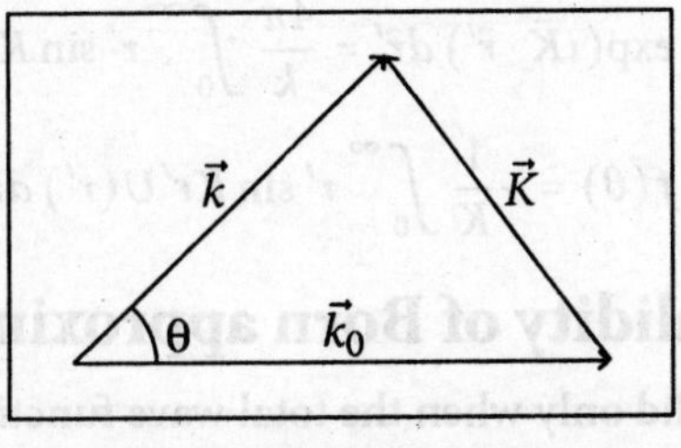

Fig 12.8 Relation between the propagation $\vec{k}_0$ for the incident particle, $\vec{k}$ for scattered particle and the angle of scattering θ

We define vector $\vec{K} \equiv \vec{k}_0 - \vec{k}$, which represents the change in the wave vector of the incident particle caused by scattering, then

$$f(\theta,\phi) = -\frac{1}{4\pi}\int U(\vec{r}')\, e^{i\vec{K}\cdot\vec{r}'}\, d\vec{r} \tag{12.186}$$

Equation (12.186) represents the scattering amplitude. The corresponding differential scattering cross-section is

$$\sigma(\theta,\phi) = \frac{1}{16\pi^2}\left|\int U(\vec{r}')\, e^{i(\vec{K}\cdot\vec{r})}\, d\vec{r}'\right|^2 \tag{12.187}$$

$$\because \sigma(\theta,\phi) = |f(\theta,\phi)|^2$$

Substituting

$$U(\vec{r}') = \frac{2\mu V(r')}{\hbar^2}$$

$$\sigma(\theta,\phi) = \left(\frac{\mu}{2\pi\hbar^2}\right)^2 \int \left|V(r')\, e^{i\vec{K}\cdot\vec{r}}\, d\vec{r}'\right|^2 \tag{12.188}$$

Equation (12.188) shows that *the Born scattering amplitude is proportional to the spherical Fourier transform of scattering potential w.r.t. momentum transfer.*

Special case: Considering $U(\vec{r}')$ spherically symmetric, i.e., $U(\vec{r}') = U(r')$, then the scattering amplitude

$$f(\theta,\phi) = -\frac{1}{4\pi}\int U(r')\exp(i\vec{K}\cdot\vec{r}')\, d\vec{r}' \tag{12.189}$$

Choosing the direction of $\vec{K}$ along the polar axis we can integrate eq. (12.189) as

$$\int U(\vec{r}')\exp(i\vec{K}\cdot\vec{r}')\, d\vec{r}' = \int_0^\infty \int_{-1}^{+1} \int_0^{2\pi} U(\vec{r}')\exp(i\vec{K}\cdot\vec{r}'\omega)\vec{r}'^2 d\vec{r}'\, d\omega\, d\phi,$$

where $\omega = \cos\theta$

$$= 2\pi \int_0^\infty r'^2 U(r')\, d\vec{r}' \left[\frac{\exp(i\vec{K}r')\omega}{ikr'}\right]_{-1}^{+1}$$

$$= 2\pi \int_0^\infty r'^2 U(r')\, d\vec{r}' \left[\frac{\exp(iKr') - \exp(-iKr')}{2i}\right]\left[\frac{2}{kr'}\right]$$

$$= 2\pi \int_0^\infty r'^2 U(r')\sin Kr' \frac{2}{kr'}\, d\vec{r}'$$

$$\therefore \int U(\vec{r}') \exp(i\vec{K}\cdot\vec{r}')\, d\vec{r}' = \frac{4\pi}{k}\cdot\int_0^\infty r' \sin Kr' U(r')\, d\vec{r}' \tag{12.190}$$

$$\therefore f(\theta) = -\frac{1}{K}\int_0^\infty r' \sin Kr' U(r')\, dr' \tag{12.191}$$

12.14 Condition for validity of Born approximation

The Born approximation will be valid only when the total wave function is not greatly different from the incident wave function. Therefore, it will be valid whenever the scattered wave or interaction energy is small compared to $\exp[i\vec{k}\cdot\vec{r}]$ in the region where $V(\vec{r})$ is large. Hence, the sufficient condition for the validity of Born approximation is

$$|v(\vec{r})| \ll |e^{ikz}| = 1 \text{ or, } |v(\vec{r})|^2 \ll 1$$

Now, eq. (12.181)

$$v(\vec{r}) = -\int \frac{e^{ik|\vec{r}-\vec{r}'|}}{4\pi|\vec{r}-\vec{r}'|} U(\vec{r}')\, e^{ikz'}\, d\vec{r}'$$

$$= -\frac{\mu}{2\pi\hbar^2}\int \frac{e^{ik|\vec{r}-\vec{r}'|}}{|\vec{r}-\vec{r}'|} V(\vec{r}')\, e^{ikz'}\, d\vec{r}' \quad \left[\because U(\vec{r}') = \frac{2\mu V(\vec{r}')}{\hbar^2}\right]$$

$$= \lim_{r\to 0} -\frac{\mu}{2\pi\hbar^2}\int \frac{e^{ikr'}}{r'} V(\vec{r}')\, e^{ikz'}\, d\vec{r}'$$

Dropping the primes

$$|v(0)|^2 = \left|\frac{\mu}{2\pi\hbar^2}\int \frac{e^{ikr}}{r} e^{ikr\omega} V(r)\, dr\right|^2 \ll 1 \tag{12.192}$$

Equation (12.192) represents the validity of *Born approximation*.
If $V(r)$ is spherically symmetric, choosing $\vec{k}_0$ as polar axis, we can write:

$$|v(0)|^2 = \left(\frac{\mu}{2\pi\hbar^2}\right)\left|\int_0^\infty\int_{-1}^{+1}\int_0^{2\pi} \frac{e^{ikr}\cdot e^{ikr\omega}}{r} V(r)\times r^2\, dr\, d\omega\, d\phi\right|^2$$

$$= \left(\frac{\mu}{\hbar^2}\right)\left|\int_0^\infty\int_{-1}^{+1} e^{ikr}\cdot e^{ikr\omega} V(r)\, r\, dr\, d\omega\right|^2$$

Now,

$$\int_{-1}^{+1} e^{ikr\omega}\, d\omega = \left[\frac{e^{ikr}-e^{-ikr}}{ikr}\right]$$

$$\therefore |v(0)|^2 = \left(\frac{\mu}{\hbar^2 k}\right)^2\left|\int_0^\infty e^{ikr}\left(e^{ikr}-e^{-ikr}\right) V(r)\, dr\right|^2$$

$$\text{or, } |v(0)|^2 = \left(\frac{\mu}{\hbar^2 k}\right)^2\left|\int_0^\infty \left(e^{2ikr}-1\right) V(r)\, dr\right|^2 \ll 1 \tag{12.193}$$

Here, for elastic scattering, we have $k = k_0$.

12.15 Application of Born approximation

Here, we shall discuss some applications of Born approximation, viz.,

1. Scattering by a square well potential
2. Scattering by a screened Coulomb field
3. Scattering from an exponential potential
4. Scattering by a Gaussian potential
5. Scattering by a potential $V(r) = -V_0 r^2 \exp\left(-\frac{r^2}{a^2}\right)$.

12.15.1 Scattering by a square well potential

We have for a square well potential

$$V(r) = \begin{cases} -V_0, & r < a \\ 0, & r > a \end{cases} \tag{12.194}$$

According to Born approximation

$$f(\theta) = -\frac{1}{k}\int_0^\infty r \sin kr U(r)\, dr \tag{12.195}$$

But,

$$U(r) = \frac{2\mu}{\hbar^2} V(\vec{r}) = \frac{2\mu}{\hbar^2} V(r)$$

$$f(\theta) = -\frac{1}{k}\int_0^\infty r \sin kr \frac{2\mu}{\hbar^2} V(r)\, dr$$

$$\text{or, } f(\theta) = \frac{2\mu}{\hbar^2}\frac{V_0}{k}\int_0^a r \sin kr\, dr \tag{12.196}$$

$$= \frac{2\mu V_0}{\hbar^2 k}\left[-\frac{a\cos ka}{k} + \frac{\sin ka}{k^2}\right]$$

evaluating the integral in eq. (12.194) by parts.

Putting $ka = x$

$$f(\theta) = \frac{2\mu V_0 a^3}{\hbar^2}\left[\frac{\sin x - x\cos x}{x^3}\right]$$

∴ the difference scattering cross-section is

$$\sigma(\theta) = |f(\theta)|^2 = \left(\frac{2\mu V_0 a^3}{\hbar^2}\right)^3 \left[\frac{\sin x - x\cos x}{x^3}\right] = \left(\frac{2\mu V_0 a^3}{\hbar^2}\right)^2 g(x)$$

where $g(x) = \left[\dfrac{\sin x - x\cos x}{x^3}\right]^2$.

Low energy limit ($ka \ll 1$) : We have

$$\left[\frac{\sin x - x\cos x}{x^3}\right] = \left[\frac{\left(x - \frac{x^3}{6} + \frac{x^5}{120} + \ldots\right) - x\left(1 - \frac{x^2}{2} + \frac{x^4}{24} + \ldots\right)}{x^3}\right]$$

$$= \frac{x^3\left(-\frac{1}{6} + \frac{1}{2}\right) + x^5\left(\frac{1}{120} - \frac{1}{24}\right)}{x^3} = \frac{1}{3} - \frac{1}{30}x^2 + \cdots \approx \frac{1}{3} \tag{12.197}$$

$$\therefore g(x) = \left[\frac{\sin x - x\cos x}{x^3}\right]^2 = \left(\frac{1}{3}\right)^2 = \frac{1}{9}$$

Here, $x = 2ka \sin\frac{\theta}{2} \simeq 0$.

$$\therefore \sigma(\theta) = \left(\frac{2\mu V_0 a^3}{\hbar^2}\right)^2 \frac{1}{9} = \frac{4\mu^2 V_0^2 a^6}{9\hbar^4}$$

Now, the total scattering cross-section is

$$\sigma_t = \int_0^\pi \int_0^{2\pi} \sigma(\theta) \sin\theta \, d\theta \, d\phi = 2\pi \int_0^\pi \sigma(\theta) \sin\theta \, d\theta$$

$$= 2\pi \int_0^\pi \frac{4\mu^2 V_0^2 a^6}{9\hbar^2} \sin\theta \, d\theta = \frac{16\pi\mu V_0^2 a^6}{9\hbar^2} \qquad (12.198)$$

High energy limit $ka \gg 1$

Equation (12.193) shows that for large value of x, $g(x)$ falls off rapidly which shows a forward peak scattering. The plot of $g(x)$ against x gives a graph as in Fig 12.9.

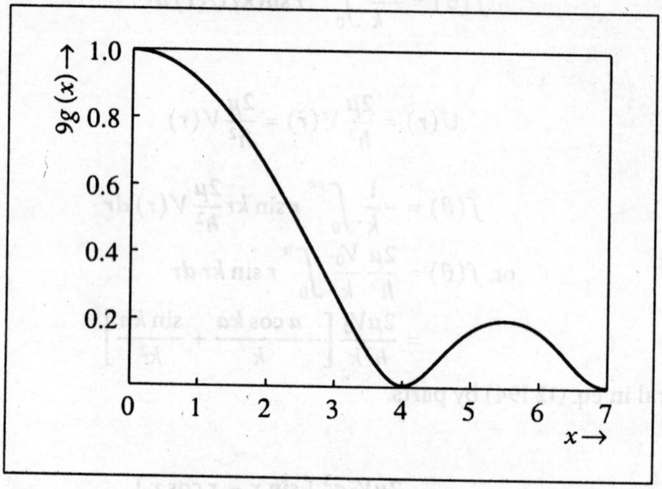

Fig 12.9 Plot of $9g(x)$ against x

At high energies ($ka \gg 1$) the scattering shows a strong maximum in the forward direction so that most of the scattered particles are in a cone whose angular opening is of the order $\frac{1}{ka}$.

The total cross-section is given by

$$\sigma_t = \int \sigma(\theta) \, d\Omega \text{ or, } \sigma_t = 2\pi \int_0^\pi \sigma(\theta) \sin\theta \, d\theta \qquad (12.199)$$

Since,

$$k = 2k \sin\frac{\theta}{2}, \; x = ka = 2ka \sin\frac{\theta}{2}$$

$$\therefore dx = 2ka \frac{1}{2} \cos\frac{\theta}{2} \, d\theta$$

$$\text{or, } x\,dx = 2ka \sin\frac{\theta}{2} ka \cos\frac{\theta}{2} \cdot d\theta = 2k^2 a^2 \sin\frac{\theta}{2} \cos\frac{\theta}{2} \cdot d\theta = k^2 a^2 \sin\theta \, d\theta$$

$$\text{or, } \sin\theta \, d\theta = \frac{x\,dx}{k^2 a^2}$$

$$\therefore \sigma_t = 2\pi \int_0^{2ka} \sigma(\theta) \frac{x\,dx}{k^2a^2}$$

When $\theta = 0, x = 0; \theta = \pi, x = 2ka$.

$\therefore$ from eq. (12.199), substituting the value of $\alpha(\theta)$

$$\sigma_t = 2\pi \int_0^{2ka} \left(\frac{2\mu V_0 a^3}{\hbar^2}\right) g(x) \frac{x\,dx}{k^2a^2} = \frac{2\pi \cdot 4\mu^2 V^2 a^6}{\hbar^4} \cdot \frac{1}{(ka)^2} \int_0^{2ka} g(x)\,x\,dx$$

$$= \frac{8\pi\mu^2 V_0^2 a^6}{\hbar^4 (ka)^2} \int_0^{2ka} \frac{(\sin x - x\cos x)^2}{x^6} x\,dx$$

$$= \frac{32\pi\mu^2 V_0^2 a^6}{\hbar^4} \left[\frac{1}{(2ka)^2} \int_0^{2ka} \frac{\sin x - x\cos x}{x^5}\,dx\right]$$

$$= \frac{32\pi\mu^2 V_0^2 a^6}{\hbar^4} [y(2ka)]$$

$$= \frac{32\pi\mu^2 V^2 a^6}{\hbar^4} y(y) \tag{12.200}$$

where $y = 2ka$.

Now,

$$y(y) = \frac{1}{y^2} \int_0^{y} \frac{(\sin x - x\cos x)^2}{x^5}\,dx$$

$$= \frac{1}{4y^2}\left[1 - \frac{1}{y^2} + \frac{\sin^2 y}{y^3} - \frac{\sin^2 y}{y^4} + \dots\right] \tag{12.201}$$

When $y \to \infty$

$$y(y) = \frac{1}{4y^2}$$

$\therefore$ total scattering cross-section

$$\sigma_t = \frac{32\pi\mu^2 V_0^2 a^6}{\hbar^4} \cdot \frac{1}{4y^2} = \frac{32\pi\mu^2 V_0^2 a^6}{\hbar^4 4 \cdot (4k^2a^2)} = \frac{2\pi\mu^2 V_0^2 a^4}{\hbar^4 k^2}$$

$$= \frac{\pi\mu V_0^2 a^4}{\hbar^2} \cdot \frac{2\mu}{\hbar^2 k^2} = \frac{\pi\mu V_0^2 a^4}{\hbar^2 E}, \left[\text{where } E = \frac{\hbar^2 k^2}{2\mu}\right] \tag{12.202}$$

Validity of Born's approximation in square well potential

We have seen in eq. (12.193) that

$$|V(0)|^2 = \left(\frac{\mu}{\hbar^2 k}\right)^2 \left|\int_0^{\infty} (e^{2ikr} - 1)\,V(r)\,dr\right|^2$$

Again, for a square well potential

$$V(r) = \begin{cases} -V_0 = |V_0|, & r < a \\ 0, & r > a \end{cases}$$

$$\therefore |v(0)| = \left(\frac{\mu V_0}{\hbar^2 k}\right)\left|\int_0^a (e^{2ikr}-1)\,dr\right| = \left(\frac{\mu V_0}{\hbar^2 k}\right)\left|\left[\frac{e^{2ikr}}{2ikr} - r\right]_0^a\right|$$

$$= \left(\frac{\mu V_0}{\hbar^2 k}\right)\left|\left[\frac{e^{2ikr}}{2ikr} - a - \frac{1}{2ik}\right]\right|$$

$$= \frac{\mu V_0}{\hbar^2 k}\frac{1}{2k}\left|\left[e^{2ika} - 2ika - 1\right]\right| = \frac{\mu V_0}{2\hbar^2 k^2}\left|\left[e^{2ika} - 1 - 2ika\right]\right|$$

$$= \frac{\mu V_0}{2\hbar^2 k^2}\left|\left[(\cos 2ka - 1) + i(\sin 2ka - 2ka)\right]\right| \tag{12.203}$$

Low energy limit $ka \ll 1$

$$|v(0)| = \left(\frac{\mu V_0}{2\hbar^2 k^2}\right)\left|\left[\left(1 - \frac{(2ka)^2}{2!} + \frac{(2ka)^4}{4!} - 1\cdots - 1\right) + i\left(2ka - \frac{(2ka)^3}{3!} + \cdots - 2ka\right)\right]\right|$$

$$= \left(\frac{\mu V_0}{2\hbar^2 k^2}\right)\left|\left[-\frac{(2ka)^2}{2!} - i\frac{(2ka)^3}{3!} + \frac{(2ka)^4}{4!}\right]\right|$$

$$= \left(\frac{\mu V_0}{2\hbar^2 k^2}\right)\cdot\frac{4k^2a^2}{2} = \frac{\mu V_0 a^2}{\hbar^2} \tag{12.204}$$

Hence, in low energy limit, the validity of Born approximation is

$$\frac{\mu V_0 a^2}{\hbar^2} \ll 1 \tag{12.205}$$

Equation (12.202) shows that if the potential energy V_0 is strong enough to bind a particle, i.e., $\frac{\mu V_0 a^2}{\hbar^2} \geqslant 1$, then Born approximation cannot be used in low energy limit.

High energy limit

In high energy limit $ka \gg 1$.

$$\left|e^{2ika} - 1 - 2ika\right| = 2ka$$

$$\therefore |v(0)| = \frac{\mu V_0}{2\hbar^2 k^2}\cdot 2ka = \frac{\mu V_0 a}{\hbar^2 k}$$

For the validity of Born's approximation

$$\frac{\mu V_0 a}{\hbar^2 k} \ll 1 \text{ or, } \frac{V_0 a}{\hbar\omega} \ll 1 \tag{12.206}$$

The above condition is satisfied for high energy limit and as such Born approximation can be safely used in high energy limit. Born approximation improves with increasing energy of the bombarding particle. Again, for small energy the method of partial waves is applicable. Thus, the two methods supplement each other.

12.15.2 Scattering by a screened Coulomb field

In case of elastic scattering of a particle of charge $-Z'e$ by a nucleus of charge Ze, the simple from of screened Coulomb potential is given by

$$V(r) = -\frac{ZZ'e^2}{r}e^{-\alpha r} \tag{12.207}$$

where the length $1/\alpha = a$ is the range of the potential. Hence, a is the atomic radius.

We have

$$U(r) = \frac{2\mu V(r)}{\hbar^2} = -\frac{2\mu}{\hbar^2}\frac{ZZ'e^2}{r}e^{-\alpha r} \tag{12.208}$$

$$\text{and } f(\theta) = -\frac{1}{k}\int_0^\infty r \sin kr U(r)\, dr \tag{12.209}$$

Substituting the value of $U(r)$ from eq. (12.208) in eq. (12.209), we get

$$f(\theta) = \frac{2\mu}{\hbar^2}\frac{ZZ'e^2}{k}\int_0^\infty r\sin kr\frac{e^{-\alpha r}}{r}\,dr = \frac{2\mu ZZ'e^2}{\hbar^2 k}\int_0^\infty \sin kr\, e^{-\alpha r}$$

$$= \frac{2\mu ZZ'e^2}{\hbar^2 k}\int_0^\infty\left(\frac{e^{ikr}+e^{-ikr}}{2i}\right)e^{-\alpha r}\,dr = \frac{\mu ZZ'e^2}{\hbar^2 ik}\int_0^\infty\left[e^{-(\alpha-ik)r}-e^{-(\alpha+ik)r}\right]dr$$

$$= \frac{\mu ZZ'e^2}{\hbar^2 ik}\left[\frac{1}{(\alpha-ik)}-\frac{1}{(\alpha+ik)}\right]\left[\because \int_0^\infty e^{-\alpha r}\,dr = -\frac{1}{\alpha}\left[e^{-\alpha r}\right]_0^\infty\right] = \frac{1}{\alpha}$$

$$\therefore f(\theta) = \frac{\mu ZZ'e^2}{\hbar^2 ik}\left[\frac{2ik}{\alpha^2+k^2}\right] \text{ or, } f(\theta) = \frac{2\mu ZZ'e^2}{\hbar^2\left[\alpha^2+4k^2\sin^2\frac{\theta}{2}\right]} \tag{12.210}$$

The differential cross-section of scattering is

$$\sigma(\theta) = |f(\theta)|^2 = \left[\frac{2\mu ZZ'e^2}{\hbar^2\left\{\alpha^2+4k^2\sin^2\frac{\theta}{2}\right\}}\right]^2 \tag{12.211}$$

$[\because |\vec{k}|^2 = k_0^2 + k^2 - 2kk_0 = k^2 + k^2 - 2k^2\cos\theta, \therefore |k_0| = |k| = 2k^2(1-\cos\theta) = 4k^2\sin^2\frac{\theta}{2}]$

When $\alpha \to 0$, i.e., the potential $V(r)$ becomes the Coulomb potential

$$f(\theta) = \frac{2ZZ'e^2\mu}{\hbar^2(4k^2\sin^2\frac{\theta}{2})} \tag{12.212}$$

Again,

$$\lambda = \frac{h}{p} = \frac{h}{\mu v'}$$

$$\therefore k = \frac{2\pi}{\lambda} = \frac{2\pi\mu v}{h} \text{ or, } kh = 2\pi\mu v \text{ i.e., } \hbar^2 k^2 = \mu^2 v^2$$

$$\therefore f(\theta) = \frac{2ZZ'e^2\mu}{4\sin^2\frac{\theta}{2}\mu^2 v^2} = \frac{ZZ'e^2}{2\mu v^2\sin^2\frac{\theta}{2}} \tag{12.213}$$

Hence, the differential cross-section of scattering is

$$\sigma(\theta) = \left(\frac{ZZ'e^2}{2\mu v'}\right)^2\frac{1}{\sin^4\frac{\theta}{2}} \tag{12.214}$$

Thus, the result is in exact agreement with both classical Rutherford cross-section and the exact quantum mechanical evaluation of the Coulomb scattering cross-section.

The validity criterion for screened Coulomb field

We have

$$v(0) = \left(\frac{\mu}{\hbar^2 k}\right)\left|\int_0^\infty \frac{ZZ'e^2}{r}e^{-\alpha r}\left(e^{2ikr}-1\right)dr\right| = \left(\frac{\mu ZZ'e^2}{\hbar^2 k}\right)\left|\int_0^\infty e^{-\alpha r}\left(\frac{e^{2ikr}}{r}\right)dr\right|$$

Now, let

$$I = -\int_0^\infty e^{-\alpha r}\frac{(e^{2ikr}-1)}{r}\,dr$$

$$\therefore \frac{dI}{d\alpha} = \int_0^\infty e^{-\alpha r}\left(e^{2ikr}-1\right)dr = \int_0^\infty \left[e^{-r}(\alpha-2ik) - e^{-\alpha r}\right]dr$$

$$= \left[\frac{e^{-r}(\alpha-2ik)}{-(\alpha-2ik)} - \frac{e^{-\alpha r}}{-\alpha}\right] = \frac{1}{\alpha-2ik} - \frac{1}{\alpha}$$

$$\therefore I = \ln(\alpha-2ik) - \ln\alpha + C$$

where C is a constant of integration.

To evaluate the value of C, we use the condition that in the limit $\alpha \to \infty$, $I \to 0$.

$$\therefore 0 = \ln\left(1-\frac{2ik}{\alpha}\right) + C = \ln 1 + C = C$$

and hence

$$I = \ln(\alpha-2ik) - \ln\alpha = \ln\left(1-\frac{2ik}{\alpha}\right)$$

Thus, the condition for validity of screened Coulomb field is

$$\frac{\mu ZZ'e^2}{\hbar^2 k}\left|\log\left(1-\frac{2ik}{\alpha}\right)\right| \ll 1$$

But,

$$\left(1-\frac{2ik}{\alpha}\right) = \sqrt{\left(1+\frac{4k^2}{\alpha^2}\right)}\,e^{i\tan^{-1}(2k/a)}$$

$$\therefore \frac{\mu ZZ'e^2}{\hbar^2 k}\left|\ln\left\{\sqrt{(1+4k^2/a^2)}\,e^{i\tan^{-1}(2k/a)}\right\}\right| \ll 1$$

$$\text{or,}\quad \frac{\mu ZZ'e^2}{\hbar^2 k}\left|\ln\sqrt{\frac{4k^2}{\alpha^2}} + i\tan^{-1}\frac{2k}{\alpha}\right| \ll 1$$

$$\text{or,}\quad \frac{\mu ZZ'e^2}{\hbar^2 k}\left|\left\{\ln\sqrt{1+\frac{4k^2}{\alpha^2}}\right\}^2 + \left\{\tan^{-1}\frac{2k}{\alpha}\right\}^2\right|^{1/2} \ll 1 \qquad (12.215)$$

Low energy limit

For low energy scattering $k < a \ll 1$. Hence, $\ln\left(\sqrt{1+4k^2/\alpha^2}\right) \simeq \ln 1 = 0$ and $\tan^{-1}(2k/\alpha)$ is limited to $\pi/2$.

Thus, the condition of validity is

$$\frac{\mu ZZ'e^2}{\hbar^2 v} \ll 1 \text{ or, } \frac{ZZe^2}{\hbar v} \ll 1\ (\because v = \hbar k/\mu) \qquad (12.216)$$

For a particular case of electron's energy 10 keV, $v = 6 \times 10^9$ cm/sec and $z = 10$

$$\frac{ZZ'e^2}{\hbar v} \simeq 4$$

Thus, Born approximation is barely satisfied in this case. When $Z = 1$, Born approximation will be fairly satisfied.

Hence, for the applicability of the Born approximation, the energy of incident particle should be very high and the scatterer should have low atomic number.

High energy limit

In this case $ka \gg 1$.

Hence, the validity of condition is

$$\frac{ZZ'e^2}{\hbar v}\ln(2ka) \ll 1 \tag{12.217}$$

12.15.3 Scattering from an exponential potential

An exponential potential is given by

$$V(r) = -V_0 e^{-r/a} \tag{12.218}$$

Here, V_0 and a are constants representing the strength and range of the potential respectively.

We have

$$f(\theta) = -\frac{1}{k}\int_0^\infty r \sin kr U(r)\, dr \tag{12.219}$$

where

$$U(r) = \frac{2\mu V(r)}{\hbar^2} = -\frac{2\mu}{\hbar^2}V_0 e^{-r/a}$$

$$\therefore f(\theta) = \frac{2\mu V_0}{\hbar^2 K}\int_0^\infty r\left(\frac{e^{iKr} - e^{-iKr}}{2i}\right)e^{-r/a}\, dr \tag{12.220}$$

$$= \frac{2\mu V_0}{\hbar^2 K2i}\int_0^\infty r\left[e^{-r(\frac{1}{a}-iK)} - e^{-r(\frac{1}{a}+iK)}\right] dr$$

$$= \frac{2\mu V_0}{\hbar^2 K2i}\left[\int_0^\infty r\, e^{-r(\frac{1}{a}-iK)}\, dr - \int_0^\infty r\, e^{-r(\frac{1}{a}+iK)}\right] dr \quad \left[\because \int_0^\infty x^n e^{-ax}\, dx = \frac{n!}{a^{n+1}}\right]$$

$$\therefore f(\theta) = \frac{2\mu V_0}{\hbar^2 K2i} \times \left[\frac{1}{\left(\frac{1}{a} - iK\right)^2} - \frac{1}{\left(\frac{1}{a} + iK\right)^2}\right] = \frac{2\mu V_0}{\hbar^2 K2i}\left[\frac{\left(\frac{1}{a} + iK\right)^2 - \left(\frac{1}{a} - iK\right)^2}{\left(\frac{1}{a} + K^2\right)^2}\right]$$

$$= \frac{2\mu V_0}{\hbar^2 K2i}\frac{(4iK/a)}{\left(\frac{1}{a^2} + K^2\right)} = \frac{4\mu V_0 a^3}{\hbar^2(1 + K^2a^2)^2}$$

$$\text{or, } f(\theta) = \frac{4\mu V_0 a^3}{\hbar^2\left(1 + 4k^2a^2\sin^2\frac{\theta}{2}\right)^2}\quad\left[\because K = 2k\sin\frac{\theta}{2}\right] \tag{12.221}$$

Now,

$$\sigma(\theta) = |f(\theta)|^2 \text{ or, } \sigma(\theta) = \frac{16\mu^2 V_0^2 a^6}{\hbar^4}\left(1 + 4k^2a^2\sin^2\frac{\theta}{2}\right)^{-4}$$

Total scattering cross-section is

$$\sigma_t = \int \sigma(\theta)\sin\theta\, d\theta\, d\phi = \frac{16\mu^2 V_0^2 a^6}{\hbar^4}\int_0^{\pi}\left(1+4k^2a^2\sin^2\frac{\theta}{2}\right)^{-4} \times 2\sin\frac{\theta}{2}\cos\frac{\theta}{2}\int_0^{2\pi} d\phi$$

$$= \frac{16\pi\mu^2 V_0^2 a^4}{3\hbar^4 k^2}\left[-\left(1-4k^2a^2\sin^2\frac{\theta}{2}\right)^{-3}\right]_0^{\pi} = \frac{16\pi\mu^2 V_0^2 a^4}{3\hbar^4 k^2}\left[-\frac{1}{(1+4k^2a^2)^3}+1\right]$$

$$= \frac{64\pi\mu^2 V_0^2 a^6}{3\hbar^4}\left[\frac{(4k^2a^2)+3(4k^2a^2)+3}{(1+4k^2a^2)^3}\right] \quad (12.222)$$

Condition for the validity of Born approximation

We have

$$|v(0)|^2 = \left(\frac{\mu}{\hbar^2 k}\right)^2 \left|\int_0^{\infty} V(r)\left(e^{2ikr}-1\right) dr\right|^2$$

$$\text{or, } |v(0)| = \frac{\mu}{\hbar^2 k}\left|\int_0^{\infty} V(r)\left(e^{2ikr}-1\right) dr\right| \quad (12.223)$$

Here, $V(r) = -V_0 e^{-r/a}$.

$$\therefore |v(0)| = \frac{\mu V_0}{\hbar^2 k}\left|\int_0^{\infty} e^{-r/a}\left(e^{2ikr}-1\right) dr\right| = \frac{\mu V_0}{\hbar^2 k}\left|\int_0^{\infty}\left[e^{(2ik-1/a)r}-e^{-r/a}\right] dr\right|$$

$$= \frac{\mu V_0}{\hbar^2 k}\left|\left[\frac{e^{(2ik-1/a)r}}{(2ik-1/a)}-\frac{e^{-r/a}}{(-1/a)}\right]_0^{\infty}\right| = \frac{\mu V_0}{\hbar^2 k}\left|-\left[\frac{1}{(2ik-1/a)}+\frac{1}{1/a}\right]\right|$$

$$= \frac{\mu V_0}{\hbar^2 k}\left|-\left[\frac{a}{(2ika-1)}+\frac{a}{1}\right]\right| = \frac{\mu V_0 a}{\hbar^2 k}\left|-\left[\frac{2ika}{(2ika-1)}\right]\right| = \frac{2\mu V_0 a^2}{\hbar^2}\left|\frac{i}{(2ika-1)}\right|$$

$$\text{or, } |v(0)| = \frac{2\mu V_0 a^2}{\hbar^2\sqrt{1+4k^2a^2}} \quad (12.224)$$

For low energy limit $ka \ll 1$.

$$\therefore |v(0)| \approx \frac{2\mu V_0 a^2}{\hbar^2} \ll 1 \quad (12.225)$$

For high energy limit, $ka \gg 1$, i.e., $\sqrt{1+4k^2a^2} \simeq ka$.

$$\therefore |v(0)| \simeq \frac{2\mu V_0 a^2}{\hbar^2 2ka} = \frac{\mu V_0 a}{\hbar^2 k} = \frac{V_0 a}{\hbar v} \ll 1 \quad (12.226)$$

12.15.4 Scattering by a Gaussian potential

Mathematically, the Gaussian potential is expressed as

$$V(r) = -V_0 \exp -\frac{r^2}{a^2} \quad (12.227)$$

where V_0 and a represent the strength and range of the potential respectively, both being constant.

According to Born approximation, the scattering amplitude is

$$f(\theta) = -\frac{1}{k}\int_0^{\infty} r\sin Kr \cdot U(r)\, dr$$

where

$$V(r) = \frac{2\mu}{\hbar^2}V(r) \text{ and } K = 2k\sin\frac{\theta}{2}$$

$$\therefore f(\theta) = -\frac{2\mu}{\hbar^2 k}\int_0^\infty r\sin kr V(r)\,dr \tag{12.228}$$

If the potential function is an even function of r as in case of Gaussian potential, the scattering amplitude can be written as

$$f(\theta) = -\frac{\mu}{\hbar^2 Ki}\int_{-\infty}^{\infty} r e^{iKr} V(r)\,dr \tag{12.229}$$

$\therefore$ for Gaussian potential

$$f(\theta) = \frac{\mu V_0}{\hbar^2 Ki}\int_{-\infty}^{+\infty} r e^{iKr} e^{-r^2/a^2}\,dr = \frac{\mu V_0}{\hbar^2 Ki}\int_{-\infty}^{+\infty} r\exp\left\{-\left(\frac{r}{a}-\frac{iKa}{2}\right)^2\right\}\exp\left\{\frac{-K^2a^2}{4}\right\}dr \tag{12.230}$$

Putting $\left(\frac{r}{a}-\frac{iKa}{2}\right) = t$, we get

$$\frac{dr}{a} = dt \text{ or, } dr = a\,dt \text{ and } r = a\left(t+\frac{iKa}{2}\right)$$

Hence, eq. (12.230) becomes

$$f(\theta) = \frac{\mu V_0}{\hbar^2 Ki}\exp\left\{-\frac{K^2a^2}{4}\right\}\int_{-\infty}^{\infty} a\left(t+\frac{iKa}{2}\right)e^{-r^2}\,a\,dt$$

$$= \frac{\mu V_0 a^2}{\hbar^2 Ki}\exp\left[-\frac{K^2a^2}{4}\right]\left[\int_{-\infty}^{+\infty} t\cdot e^{-t^2}\,dt + \frac{iKa}{2}\int_{-\infty}^{+\infty}\right]e^{-t^2 dt}$$

$$= \frac{\mu V_0 a^2}{\hbar^2 Ki}\exp\left[-\frac{K^2a^2}{4}\right]\left[0+\frac{iKa}{2}\sqrt{\pi}\right]\left[\because \int_{-\infty}^{+\infty} e^{-r^2 dt} = \sqrt{\pi}\right]$$

$$= \frac{\mu V_0 a^3\sqrt{\pi}}{2\hbar^2}\exp\left\{\left(K^2a^2\sin^2\frac{\theta}{2}\right)\right\}$$

The differential scattering cross-section is given by

$$\sigma(\theta) = |f(\theta)|^2 \text{ i.e., } \sigma(\theta) = \left(\frac{\mu V_0}{\hbar^2}\right)^2\cdot\frac{\pi a^6}{4}\exp\left\{\left(-2k^2a^2\sin^2\frac{\theta}{2}\right)\right\}$$

Condition of validity of Born approximation

The condition of validity of Born approximation is given by

$$|V(0)| = \frac{\mu}{\hbar^2 k}\left|\int_0^\infty V(r)\left(e^{2ikr}-1\right)dr\right| \ll 1 \tag{12.231}$$

For low energy, $ka \ll 1$, where a is the range of the potential. Hence, the first two terms in the series expansion of $\exp(2ikr)$ will give very good approximation. Hence, for low energy scattering the condition expressed by eq. (12.231) can be written as

$$\frac{2\mu}{\hbar^2}\left|\int_0^\infty V(r)\cdot r\,dr\right| \ll 1 \tag{12.232}$$

If the kinetic energy of the particle is very large $(ka \gg 1)$, the contribution of the oscillating term can be neglected, then eq. (12.231) becomes

$$\frac{\mu}{\hbar^2 k}\left|\int_0^\infty V(r)\,dr\right| \ll 1 \tag{12.233}$$

Thus, for Gaussian potential, the validity condition in low energy region is

$$\frac{2\mu V_0}{\hbar^2}\left|\int_0^\infty e^{-r^2/a^2}\, r\, dr\right| \ll r \tag{12.234}$$

Now,

$$\int_0^\infty e^{-r^2/a^2}\, r\, dr = \frac{a^2}{2}\int_0^\infty e^{-t}\, dt = \frac{a^2}{2}\Big[-e^{-t}\Big]_0^\infty = \frac{a^2}{2} \quad [\text{putting } r^2/a^2 = t]$$

∴ the validity condition for low energy is

$$\frac{2\mu V_0 a^2}{2\hbar^2} \ll 1 \tag{12.235}$$

Equation (12.235) shows that to satisfy the above condition V_0 should be negligibly small. Hence, *the Born approximation for low energy is a good approximation for very weak potential only.*

For *high energy*, the condition of validity of Born approximation is

$$\frac{\mu V_0}{\hbar^2 k}\left|\int_0^\infty e^{-r^2/a^2}\, dr\right| \ll 1 \text{ or, } \frac{\mu V_0 a\sqrt{\pi}}{2\hbar^2 k} \ll 1 \tag{12.236}$$

Thus, *for Gaussian potential, Born approximation is a very good approximation for high incident energy and weak potential.*

12.15.5 Scattering by potential $V(r) = -V_0 r^2 \exp(-r^2/a^2)$

In this case, the potential is an even function of r.

Hence,

$$f(\theta) = -\frac{\mu}{\hbar^2 ki}\int_{-\infty}^{+\infty} r\, e^{ikr}\, V(r)\, dr = \frac{\mu V_0}{\hbar^2 ki}\int_{-\infty}^{+\infty} r^3 e^{ikr} e^{-r^2/a^2}\, dr$$

$$= \frac{\mu V_0}{\hbar^2 ki} e^{-k^2a^2/4}\int_{-\infty}^{+\infty} r^2 \exp\left\{-\left(\frac{r}{a}-\frac{ika}{2}\right)^2\right\} dr \tag{12.237}$$

Now, putting

$$\left(\frac{r}{a}-\frac{ika}{2}\right) = t \text{ or, } dr = a\, dt$$

$$\therefore \int_{-\infty}^{+\infty} r^3 \exp\left\{-\left(\frac{r}{a}-\frac{ika}{2}\right)^2\right\} dr = a^4\int_{-\infty}^{+\infty}\left(t+\frac{ika}{2}\right)^3 e^{-t^2}\, dt$$

$$= a^4\int_{-\infty}^{\infty} t^3 e^{-t^2}\, dt - \frac{ik^3a^7}{8}\int_{-\infty}^{\infty} e^{-t^2}\, dt + \frac{3ika^5}{2}\int_{-\infty}^{+\infty} t^2 e^{-t^2}\, dt - \frac{3k^2a^6}{4}\int_{-\infty}^{+\infty} t e^{-t^2}\, dt \tag{12.238}$$

Now, since the first and the last integrands are odd functions of t, they vanish. The second integral is standard integral having value $\sqrt{\pi}$. The third integral can be evaluated by using the standard integral

$$\int_{-\infty}^{+\infty} x^{2n} e^{-x^2}\, dx = \frac{1.3.5\ldots(2n-1)}{2^n}\sqrt{\pi} = \frac{\sqrt{\pi}}{2}$$

$$\therefore f(\theta) = \frac{\mu V_0 a^5\sqrt{\pi}}{8\hbar^2}(6-k^2a^2)\, e^{-k^2a^2/4}$$

$$\text{or, } f(\theta) = \frac{\mu V_0 a^5\sqrt{\pi}}{4\hbar^2}\left(3-2k^2a^2\sin^2\frac{\theta}{2}\right) e^{-k^2a^2\sin^2\frac{\theta}{2}} \tag{12.239}$$

Now, the differential scattering cross-section is given by

$$\sigma(\theta)=|f(\theta)|^2=\frac{\mu^2V_0^2a^{10}\pi}{16\hbar^2}-\left(3\ 2k^2a^2\sin^2\frac{\theta}{2}\right)^2e^{-2k^2a^2\sin^2\frac{\theta}{2}} \tag{12.240}$$

Condition for validity of Born approximation

For *low energy limit,* the condition of validity of Born approximation is given by

$$\frac{2\mu V_0}{\hbar^2}\left|\int_0^\infty r^3e^{-r^2/a^2}dr\right|\ll 1$$

Putting $r^2/a^2=t$, we get

$$\frac{2\mu V_0a^4}{\hbar^2 2}\left|\int_0^\infty te^{-t}\,dt\right|\ll 1\ \text{or,}\ \frac{2\mu V_0a^4}{2\hbar^2}\ll 1 \tag{12.241}$$

For high energy limit, the validity condition is

$$\frac{\mu V_0}{\hbar^2k}\left|\int_0^\infty r^2e^{-r^2/a^2}dr\right|\ll 1\ \text{or,}\ \frac{\mu V_0}{\hbar^2k}\left|\int_0^\infty \frac{r^2}{a^2}e^{-r^2/a^2}d\left(\frac{r}{a}\right)\right|\ll 1$$

$$\text{or,}\ \frac{\mu V_0a^3\sqrt{\pi}}{4\hbar^2k}\ll 1 \tag{12.242}$$

Equations (12.241) and (12.242) show that in the high energy limit Born approximation is a good approximation for week potential and high incident energy.

12.16 Answers of some important conceptual questions

12.1. *Discuss the phenomenon of scattering. What is its importance?*

Ans. In a scattering experiment, a beam of particles is directed at a target containing the scattering material and the energy, and angular distribution of the deflected beam of the recoil particles are measured. The experimental data can be interpreted in terms of a model of the microscopic details of a single collision and conclusions concerning the interaction between the scattered particles and the scatterer can be drawn.

The scattering experiments with atoms and interactions between the scattered particles and the target materials have given very useful information about the forces and nuclei. The study of scattering process is important for the following reasons:

Firstly, many interesting effects, such as the stopping of electrons in gaseous discharges, the collision of gas molecules and the stopping radio active and cosmic ray particles mesons are all determined at least in part by the probability of scattering.

Secondly, from a detailed study of the result of scattering much can be learned about the nature of the particles that are being scattered as well as of those which are doing the scattering.

In particular, following examples reveal the importance of scattering experiments:

(i) scattering of nucleus of various energies from a nucleus gives an idea of the strength and range of nuclear forces

(ii) scattering of high energy electrons gives charge distribution in a nucleus and even in a nucleon.

(iii) scattering of mesons and other hadrons give an idea about the strong interaction process.

12.2. ***Discuss the different points of view in which scattering process may be described.***

Ans. **In monochromatic wave treatment, an incident influx as particles represented by monochromatic wave, is considered. Then the continuous stream of particles emerge from interaction region. The probability that a single particle is transmitted or reflected from such an interaction region is given by the ratio of transmitted or reflected flux to the incident flux in this continuous model.**

In wave packet approach, a single particle approaching the interaction region from left is represented by the superposition of waves grouped about the velocity corresponding to that of incident particle. As the incident particle approaches the interaction region the wave packet also approaches the region and from this region reflected and transmitted wave packets start. The probability that a single particle being transmitted is interpreted by the ratio of transmittedand to incident packet. In classical limit for very heavy particles and potentials that vary very slowly with distance only one wave packet is found. Therefore, in this approach it is necessary to evaluate the ratio of amplitudes.

12.3. ***What is scattering cross-section. What are differential scattering cross-section and total scattering cross-section?***

Ans. Scattering cross-section is defined as the area presented by the target to the incident beam of particles, such that, particles when hit within this area while scattering is bound to occur. If they hit outside the area the particles go undeflected after traversing the target.

Differential scattering cross-section: **Let a beam of particles of flux N, i.e., the number of particles crossing per unit time a unit surface placed perpendicular to the direction of propagation be incident on a scattering centre, particles are scattered in all directions. If dN be the number of particles scattered per unit time into the solid angle $d\omega$ located in the direction θ and ϕ w.r.t. the bombarding direction as polar axis, then dN will be proportional to the incident flux and the solid angle taken.**

Thus,

$$dN = \sigma(\theta,\phi) N\, d\omega \tag{i}$$

where $\sigma(\theta,\phi)$ is the proportionality coefficient and a function of θ and ϕ. Equation (i) shows that $\sigma(\theta,\phi)$ has the dimension of an area. This can be regarded as the cross-section of the incident beam through which all the particles pass that are scattered into the solid angle $d\omega$ about θ and ϕ. For this reason $\sigma(\theta,\phi)$ is known as the *differential cross-section.*

Total scattering cross-section: **The total number of particles scattered in unit time is obtained by integrating (i) over all angles. It is equal to**

$$N_{\text{scatt}} = \int dN = \int \sigma(\theta,\phi) N\, d\omega = N \int \sigma(\theta,\phi)\, d\omega = N\sigma_t$$

Thus,

$$\frac{N_{\text{scatt}}}{N} = \sigma_t \tag{ii}$$

σ_t is known as the *total scattering cross-section.*

In nuclear physics, the cross-sections are measured in barns (1 barn = 10^{-24} cm^2) or millions.

12.4. ***State optical theorem. Show that the imaginary part of the forward scattering amplitude measures the intensity which the incident beam suffers due to scattering.***

Ans. **See Art. 12.8.**

12.5. *Write the expression of Green's function for a free particle. What is its significance?*

Ans. If $\vec{k}$ be the propagation vector in the direction of incidence and $\vec{k}'$ the vector along the direction of observation, then Green's function for a free particle is given by

$$G(\vec{r},\vec{r}') = \frac{\exp(ik|\vec{r}-\vec{r}'|)}{|\vec{r}-\vec{r}'|}$$

It represents the solution of scattering problem for a source of unit intensity at the point $\vec{r}'$. The *Green's function in each term can be fought of as 'propagators' which carry the waves from their last point of scattering to the next.*

12.6. *Discuss the applicability of Born approximation.*

Ans. If the scattering takes place from scattering centres which are localised but are weak such that the scattering does not take place at large distance from the scatterer and the scattering wave is weak in amplitude, then Born approximation can be used to evaluate the scattering amplitude $f(\theta,\phi)$ and hence the differential cross-section.

Born approximation is applicable whenever V is fairly small. The idea is simply that of successive approximation. *The Born approximation simply amounts neglecting the re-scattering of the scattered wave provided the scattered wave is small compared with incident wave.* It, therefore, supplements the method of partial waves.

12.7. *Mention the condition of validity of Born approximation.*

Ans. The condition of validity of Born approximation for scattering by a Gaussian potential is

$$\frac{\mu V_0 a\sqrt{\pi}}{2\hbar^2 k} \ll 1$$

The different letters have their usual significances.

In this case, Born approximation is a very good approximation for high incident energy and weak potential.

12.17 Solved problems

» ***Example* 12.1.** Assuming the mass of neutron and proton to be equal, prove that there cannot be any back scattering in the scattering of neutrons and protons.

Solution: If θ_L and θ_C be the angles of scattering in the L-system and C-system respectively, then

$$\tan\theta_L = \frac{\sin\theta_C}{\frac{m_1}{m_2} + \cos\theta_C}$$

Since, $m_1 = m_2 = m$, where m_1 and m_2 the masses of neutron and proton respectively, then

$$\tan\theta_L = \frac{\sin\theta_C}{1+\cos\theta_C} = \frac{2\sin\frac{\theta_C}{2}\cos\frac{\theta_C}{2}}{2\cos^2\frac{\theta_C}{2}} = \tan\frac{\theta_C}{2}$$

$$\therefore \theta_L = \frac{\theta_C}{2}$$

when $\theta_L \to 0$ to π, $\theta_C \leqslant \frac{\pi}{2}$. This implies that there cannot be any back scattering.

» ***Example* 12.2.** In the neutron proton scattering, if range of attacking force is ~ 3×10^{-13} cm, show that the scattering will be isotropic in the centre of mass system for neutron energy $\leqslant$ 10 MeV.

Solution: We have

$$T_{\text{Lab}} = 2T_{\text{cm}} = 2\left(\frac{\hbar^2 k^2}{2\mu}\right)$$

when $ka \leqslant 1$, this is purely s-wave scattering, i.e., isotropic.

$$\therefore T_{\text{Lab}} = \frac{\hbar^2}{2\mu}\cdot\frac{1}{a^2} = \frac{(1.05\times 10^{-34})}{2\times(3\times 10^{-15})^2\times(0.83\times 10^{-27})} = 1.6\times 10^{-12} = \mathbf{10\ MeV}$$

» ***Example* 12.3.** In elastic scattering experiment, there is no absorption of incident particles by the scatterer. This means that incoming and outgoing intensities are equal and phase-shift is real. Prove that reaction cross-section which is defined as the number of particles removed from the beam per unit flux is given by

$$\sigma_r = \sum_l \sigma_{r,l} = \sum_l \frac{\pi}{k}(2l+1)\,(1-|\eta_l|^2)$$

where $\eta_l = \exp(2i\delta_l)$ = collision amplitude.

Also show that there can be scattering without absorption, but no absorption without scattering.

Solution: Through a sphere of radius R, the net flux of particles is given by

$$J_R = \frac{\hbar}{2\mu i}\int\left[\psi^*\frac{\partial\psi}{\partial r} - \psi\frac{\partial\psi^*}{\partial r}\right]R^2\,d\tau \qquad \text{(i)}$$

Now,

$$\frac{\partial\psi}{\partial r} = \sum\frac{2l+1}{2ikR}P_l\left[e^{2i\delta_l}ike^{ikr} + (-1)^l(ik)\,e^{-ikR}\right]$$

$$= \sum\frac{2l+1}{2R}P_l\left[\eta_l e^{ikr} + (-1)^l e^{-ikR}\right] \qquad \text{(ii)}$$

$$\text{and } \psi^* = \sum -\frac{2l+1}{2ikR}P_l\left[\eta_l^* e^{-ikr} - (-1)^l e^{ikR}\right] \qquad \text{(iii)}$$

$$\therefore \left[\psi^*\frac{\partial\psi}{\partial r}\right] = \sum_l\sum_{l'}\frac{(2l+1)P_l(2l'+1)P_{l'}}{-4ikR^2}\left[\eta_l\eta_{l'}^* - 1\right] + \text{terms involving } e^{\pm 2ikR}.$$

The terms which oscillate rapidly and average out to zero have been neglected. Thus, net flux inward

$$J_R = \frac{\hbar}{4\mu k}\sum_l\sum_{l'}\int(2l+1)\,(2l'+1)P_lP_{l'}\left[1-\eta_l\eta_{l'}^*\right]d\tau \qquad \text{(v)}$$

Again, from orthogonality condition

$$\int P_l(\cos\theta)P_{l'}(\cos\theta)\,d(\cos\theta) = \frac{2}{(2l+1)}\delta_{ll'} \qquad \text{(vi)}$$

From (v), we get

$$J_R = \frac{4\pi\hbar}{4\mu k}\sum(2l+1)\,(1-|\eta_l|^2) \qquad \text{(vii)}$$

The current in lth partial wave is

$$J_R = \frac{\mu\hbar}{\mu k}(2l+1)\,(1-|\eta_l|^2) \qquad \text{(viii)}$$

Thus, the reaction cross-section is

$$\sigma_r^l = \frac{\pi}{k^2}(2l+1)(1-|\delta_l|^2) \tag{ix}$$

In the above equation, $|\eta_l| < 1$, for otherwise the outgoing current will exceed the incoming current. The scattering cross-section is

$$\sigma_{sc}^l = \frac{\pi}{k^2}(2l+1)|(1-\eta_l)|^2 \tag{x}$$

Obviously, σ_{sc}^l is larger for $\eta_l = -1$ and σ_r^l vanishes. On the other hand, $\eta_l = 0$ gives the maximum reaction cross-section for which

$$\sigma_{sc}^l = \sigma_{rl} = \frac{2l+1}{k^2}\pi \tag{xi}$$

If there is no scattering ($\eta_l = 1$), there is no reaction either.

» ***Example* 12.4.** Considering that a change of potential acts like a change in refractive index in optics, obtain the condition of validity of Born approximation. Apply the condition in case of scattering of high energy particles by the quare potential well of radius a and depth V_0, when $V_0 \ll E$.

***Solution*:** We suppose that the total wave function does not differ greatly from the initial wave function if the phase of the incident wave is not much altered as it passes through the region in which it is influenced by the perturbing potential.

The magnitude of the wave vector at large distance is

$$k = \sqrt{\left(\frac{2mE}{\hbar^2}\right)}$$

The magnitude of the wave vector near the centre of the force

$$= \sqrt{\frac{2m(E-V)}{\hbar^2}}$$

Now, the first order Born approximation will be valid, if

$$|\Delta\phi| = \left|\int_0^\infty \sqrt{\left\{\left(\frac{2m}{\hbar^2}\right)\right\}}\left[\sqrt{(E-V)} - \sqrt{(E)}\right] dr\right| \ll 1$$

where $V \ll E$ the criterion may be simplified by expressing the above equation as a function of the ratio V/E and expanding the square root.

Hence,

$$\sqrt{\left\{\left(\frac{2mE}{\hbar^2}\right)\right\}}\left|\int_0^\infty \left[\left(1-\frac{V}{E}\right)^{1/2} - 1\right] dr\right| \ll 1$$

$$\text{or, } \sqrt{\left\{\left(\frac{2mE}{\hbar^2}\right)\right\}}\left|\int_0^\infty \frac{V}{2E}\, dr\right| \ll 1 \quad \left[\because \left(1-\frac{V}{E}\right)^{1/2} = 1 - \frac{V}{2E}\right]$$

$$\text{or, } \sqrt{\left\{\left(\frac{m}{2\hbar^2 E}\right)\right\}}\left|\int_0^\infty V dr\right| \ll 1 \text{ or, } \sqrt{\left\{\left(\frac{m}{2\hbar^2 E}\right)\right\}}\bar{V}\bar{r} \ll 1 \tag{i}$$

where $\bar{V}$ is the average potential and $\bar{r}$ is the mean range.

In square potential well, $\bar{V} = V_0$ and $\vec{r} = a$.

$$\therefore \sqrt{\left\{\left(\frac{m}{2\hbar^2 E}\right)\right\}} V_0 a \ll 1$$

EXERCISES

12(A): Objective (or MCQ) Type Questions

Select the correct answer:

12.1. If the energy of the incident particle does not change then the scattering is called

(a) inalstic (b) elastic (c) semielastic (d) none of these.

12.2. Scattering of nucleons of various energies by a nucleus gives the idea of the

(a) strength (b) range (c) strength and range (d) none of the nuclear forces.

12.3. The dimension of scattering cross-section is that of

(a) volume (b) area (c) momentum (d) mass.

12.4. The relation between the angles of scattering in L- and C-system is

(a) $\tan\theta_L = \dfrac{\tan\theta_c}{\cos\theta_c + \gamma}$

(b) $\tan\theta_L = \dfrac{\cos\theta_c}{\tan\theta_c + \gamma}$

(c) $\tan\theta_L = \dfrac{\cos\theta_c}{\sin\theta_c}$

(d) $\tan\theta_L = \dfrac{\sin\theta_c}{\cos\theta_c + \gamma}$

where the different letters have their usual meaning.

12.5. Relation between the kinetic energies of the incident and target particles in L- and C-system is

(a) $T_c = \dfrac{m_2}{m_1} T_L$

(b) $T_c = \dfrac{m_1}{m_2} T_c$

(c) $T_c = \dfrac{m_1}{m_1 + m_2} T_c$

(d) $T_c = \dfrac{m_2}{m_1 + m_2} T_c$.

12.6. Green's function for a free particle is

(a) $G(\vec{r}, \vec{r}') = \dfrac{\exp(k|\vec{r} - \vec{r}'|)}{|\vec{r} - \vec{r}'|}$

(b) $G(\vec{r}, \vec{r}') = \dfrac{\exp(ik|\vec{r} - \vec{r}'|)}{|\vec{r} - \vec{r}'|}$

(c) $G(\vec{r}, \vec{r}') = \dfrac{\exp(ik|\vec{r}|)}{|\vec{r} - \vec{r}'|}$

(d) $G(\vec{r}, \vec{r}') = \dfrac{\exp(ik|\vec{r} - \vec{r}'|)}{|\vec{r}'|}$.

12.7. Born approximation is applicable whenever potential energy $V(\vec{r})$ is

(a) small (b) large (c) medium (d) none of these.

12.8. The method of partial wave analysis is applicable when the potential has symmetry.

(a) non-spherical

(b) spherical

(c) both (a) and (b)

(d) none of these.

12.9. In scattering by square well potential, the validity of Born approximation in high energy limit is given by

(a) $\frac{V_0 a}{\hbar v} = 1$

(b) $\frac{V_0 a}{\hbar v} \ll 1$

(c) $\frac{V_0 a}{\hbar v} \gg 1$

(d) $\frac{V_0 \hbar}{a v} \ll 1$.

12.10. In case of scattering by Gaussian potential, the condition for validity of Born approximation for high incident energy and weak potential is

(a) $\frac{\mu V_0 a \sqrt{\pi}}{2\hbar^2 k} = 1$

(b) $\frac{V_0 a \sqrt{\pi}}{2\hbar^2 k \mu} \ll 1$

(c) $\frac{\mu V_0 a \sqrt{\pi}}{2\hbar^2 k} \ll 1$

(d) $\frac{\mu V_0 a \sqrt{\pi}}{2\hbar^2 k} \gg 1$.

ANSWERS

12.1. (b).	**12.3.** (b).	**12.5.** (d).	**12.7.** (a).	**12.9.** (b).
12.2. (c).	**12.4.** (d).	**12.6.** (b).	**12.8.** (b).	**12.10.** (c).

12(B): Short Answer Type Questions

12.1. What do you mean by scattering? Define the scattering cross-section.

12.2. Compare the ratio between the differential cross-section for α-scattering and the Rutherford cross-section.

12.3. Define total and differential cross-section. How are these related to the angle of scattering? Discuss on the basis of classical theory the scattering process when the angle of deflection is small.

12.4. What are the suitable conditions for the study of scattering problem by the method of partial waves? The total scattering cross-section is given by

$$\sigma = \frac{4\pi}{k^2} \sum_{l=0}^{\infty} (2l+1) \sin^2 \delta_l$$

What will be the value of this expression for (i) low and (ii) high energy particles scattered by a rigid sphere of radius a.

12.5. Discuss the method of partial waves for the determination of the rigid scattering amplitude for scattering from the spherically symmetric potential

$$V(r) = \begin{cases} V_0, & r < a \\ 0, & r > a \end{cases}$$

12.6. What is Green's function? Explain its significance with reference to scattering problem.

12.7. Show that the Born approximation will not give correct scattering amplitude for the forward direction.

12(C): Broad Answer Type Questions

12.1. **Show that when a particle of mass m_1 collides elastically with a particle of mass m_2 that is initially at rest, all the recoil (mass m_2) particles are scattered in the forward hemisphere in the laboratory coordinate system. If the angular distribution is spherically symmetrical in the centre of mass of system what is it for m_2 in the laboratory system.**

12.2. **Outline the main features of partial wave analysis. Find out the scattering length for an S-state.**

12.3. **Give the theory of Born approximation in scattering calculations. Use it to find the cross-section for scattering of the electrons by nuclei.**

12.4. **Show that the amplitude of scattering of a particle by a central potential $V(r)$ in the Born approximation is proportional to the three-dimensional Fourier transform: $V(r)$ w.r.t. momentum transfer. Deduce the Rutherford formula.**

12.5. **Deduce an expression for the scattering cross-section of the particles by a spherically symmetric potential. Explain the significance of the *phase-shift* term that appears in the formula.**

12.6. **Obtain the Rutherford cross-section for the scattering of charged particles by the Coulomb field of nuclei using the first Born approximation.**

12.7. **Analyse the p-scattering of a particle by a spherically symmetric potential well.**

12.8. **Give the theory of Born approximation in the scattering calculations. Use it to find the cross-section for scattering for the electrons by nuclei.**

12.9. **Find a general expression for phase-shift produced by scattering potential**

$$V(r) = \frac{A}{r^2}, \text{ where } A > 0$$

Is the total cross-section finite? If not, does the divergence come from small or large scattering angles and why?

12.10. **Give the theory of Born approximation and discuss its validity for the scattering by a square well potential.**

12.11. **(a) Distinguish between incoming and outgoing waves.**

(b) Deduce an expression for the total scattering cross-section of particles by spherically symmetric potential

$$V(r) = \begin{cases} -V_0, & r = a \\ 0, & r > a \end{cases}$$

Find the S-wave phase-shift when energy tends to zero.

12.12. **Obtain the scattering amplitude $f(k^2, \theta)$ in the following form: $f(k^2, \theta) = \Sigma(2l+1)\, a_l(k^2) P_l(\cos\theta)$ and determine the form of $a_l(k^2)$. What is optical theorem? Is there any usefulness of this theorem in physics?**

12.13. **Apply the method of partial waves to derive expression for phase-shifts for various partial waves for the scattering of a particle by hard sphere. Obtain an expression for the scattering cross-section in the zero energy limit. Compare the classical and quantum results.**

12.14. Obtain an expression for the scattering amplitude from a spherically symmetric potential in terms of phase-shifts. Use the result to prove the optical theorem.

12.15. (a) Show that the amplitude of scattering of a particle by central potential $V(r)$ in Born approximation is proportional to the three-dimensional Fourier transform of $V(r)$ w.r.t. momentum transfer.

(b) Find the condition of validity of Born approximation for scattering by a square well potential. Deduce the Rutherford scattering formula.

12.16. Determine in the Born approximation, the effective cross-section of scattering for a spherical potential well:

$$V(r) = \begin{cases} -V_0, & \text{for } r < a \\ 0, & \text{for } r > a \end{cases}$$

12.17. Find Green's function for an outgoing wave in the scattering problem and use it to calculate the scattering amplitude in the first Born approximation.

12.18. Using Born approximation, calculate the differential scattering cross-section for the Gaussian potential

$$V(r) = -V_0 \exp\left[-\frac{1}{2}\left(\frac{r}{a}\right)^2\right]$$

and show that scattering by a square well potential of depth V_0 and range a_0 has the following characteristics:

(a) scattering is peaked in the forward direction

(b) at large energies the total scattering cross-section is inversely proportional to the energy.

12.19. Construct Green's function for a free particle. Use it to write the total wave function in the first Born approximation. Develop this equation by iteration and interpret the series.

12.20. Discuss the method of Born approximation for evaluating scattering cross-section for a short range scattering potential. Calculate the total scattering cross-section for the scattering potential $V(r) = -V_0 e^{-\alpha r}/\alpha r$.

12.21. Give the theory of Born approximation for the scattering process and calculate the differential cross-section in the first Born approximation for the scattering of an election by the potential

$$V(r) = g\frac{e^{-\alpha r}}{r}$$

12.22. Derive an expression for scattering amplitude for a particle scattered by a central potential. Show that the imaginary part of scattering amplitude in the forward direction is related to the total scattering cross-section.

12.23. Use the method of partial waves for the scattering amplitude in terms of phase whist. How is the imaginary part of scattering amplitude in the forward direction is related to the total scattering cross-section.

12.24. Discuss Born approximation in scattering calculations. Use it to find cross-section for scattering of electrons by nuclei.

12.25. Derive the formula of Fascen and Holtsmark for the scattering amplitude on the basis of partial wave analysis. Discuss the optical theorem.

12(D): Numerical Problems

12.1. Show that if $r \neq 0$

$$(\nabla^2 + k^2)\frac{e^{ikr}}{r} = 0$$

12.2. The potential energy of scattering of an electron by an atom can be represented approximately as a *shielded* Coulomb field given by

$$V(r) = \frac{ze^2}{r}e^{-r/a}$$

where a is the *shielding radius*. Show that in Born approximation the scattering amplitude is

$$f(\theta) = -\frac{2mze^2}{\hbar k^2}\left[1 - \frac{1}{(1-ka)^2}\right]$$

12.3. Show that for low energies, the total s-wave cross-section in the potential $V = V_0$ for $r \leqslant R$ is

$$\sigma_0 = 4\pi R^2\left[\frac{\tanh(k_0 R)}{k_0 R} - 1\right]$$

where $k_0 = \left[\frac{2mV_0}{\hbar^2}\right]^{1/2}$. What happens to the expression as $V_0 \to 0$?

12.4. For the rigid sphere of radius a, show that the scattering cross-section is given by $\sigma = 4\pi a^2$.

12.5. Show that Born scattering amplitude is proportional to the spatial Fourier transform of the scattering potential w.r.t. the momentum transfer.

12.6. Using Born approximation calculate the cross-section for scattering of electrons by nuclei.

12.7. Find the expression for phase shift produced by scattering potential $V(r) = \frac{A}{r^2}$, where $A > 0$.

Chapter 13

Approximate Methods in Quantum Mechanics (Time-independent Perturbations)

13.1 Introduction

In quantum mechanics, the number of problems which can be solved by analytical methods is very limited. Most of the problems of physical interest require different types of approximate methods for solutions as the exact solution of the Schrödinger equation presents mathematical difficulties. Various approximate methods of solving wave equation have been proposed leading to more or less accurate evaluation of energy values and wave functions. Out of these methods a simple and beautiful one is *wave mechanical perturbation theory,* developed by *Schrödinger* in 1926.

Perturbation theories are of two kinds: (i) time-independent (or stationary) perturbation theory (ii) time-dependent perturbation theory.

In this chapter we shall discuss time-independent (stationary) perturbation theory and its application in various physical problems.

13.2 Stationary perturbation theory

We consider a physical system subjected to perturbation. The effect of this perturbation is to shift the energy levels; of course the arrangement of energy levels remains the same. Thus, the perturbation introduces some additional terms in the Hamiltonian of unperturbed system.

Now, the perturbed Hamiltonian can be written as

$$H = H^{(0)} + H^{(1)} \tag{13.1}$$

where $H^{(0)}$ is the unperturbed Hamiltonian and $H^{(1)}$ corresponds to perturbed effect.

In the perturbed state, the Schrödinger equation can be written as

$$H\psi = E\psi \tag{13.2}$$

If $\psi_n^{(0)}$ and $E_n^{(0)}$ be the orthonormal eigenfunctions and eigenvalues of unperturbed Hamiltonian $H^{(0)}$, then

$$H^{(0)}\,\psi_n^{(0)} = E_n^{(0)}\cdot\psi_n^{(0)} \tag{13.3}$$

13.3 Time-independent perturbation theory for non-degenerate levels

If in a system there is only one eigenfunction corresponding to each energy level or eigenvalue, then it is *non-degenerate*. If on the other hand, there are two or more eigenfunctions corresponding to one energy level, then the system is called *degenerate system*.

Let the true wave function in perturbed stated be ψ, then

$$H\psi = E\psi \tag{13.4}$$

where H is the operator given by

$$H = -\left(\frac{\hbar^2}{2m}\nabla^2 + V\right) \tag{13.5}$$

In stationary system, the Hamiltonian H does not depend on time and it is possible to expand H in terms of some parameter λ, yielding the equation

$$H = H^{(0)} + \lambda H^{(1)} + \lambda^2 H^{(2)} + \ldots \tag{13.6}$$

where $H^{(0)}$ is large compared to $H^{(1)}$. Here, $H^{(0)}$ is the unperturbed and other terms are due to perturbation.

Let the eigenvalues and eigenfunctions of the unperturbed problem be $E_1^{(0)}$, $E_2^{(0)}$, $E_3^{(0)}$, ... and $\psi_1^{(0)}$, $\psi_2^{(0)}$, $\psi_3^{(0)}$, respectively.

Then

$$H^{(0)}\,\psi_n^{(0)} = E_n^{(0)}\,\psi_n^{(0)},\quad n = 1, 2, 3, \ldots \tag{13.7}$$

Equation (13.7) is the wave equation for the unperturbed system. $E_n^{(0)}$ denotes the energy of the nth level of the system.

The eigenfunction ψ_n for the perturbed system satisfies the equation

$$H\psi_n = E_n\psi_n \tag{13.8}$$

where E_n are the energy eigenvalues of the modified Hamiltonian.

We assume that it is possible to expand the eigenfunctions and eigenvalues of the total Hamiltonian H of eq. (13.8) in a power series in λ

$$E_n = E_n^{(0)} + \lambda E_n^{(1)} + \lambda^2 E_n^{(2)} + \ldots \tag{13.9}$$

$$\psi_n = \psi_n^{(0)} + \lambda\psi_n^{(1)} + \lambda^2\psi_n^{(2)} + \ldots \tag{13.10}$$

where $E_n^{(1)}, E_n^{(2)}, \ldots$ and $\psi^{(1)}, \psi_n^{(2)}, \ldots$ are quantities to be determined.

Substituting eqs (13.6), (13.9) and (13.10) in eq. (13.8), we get

$$\left(H^{(0)} + \lambda H^{(1)}\right)\left(\psi_n^{(0)} + \lambda\psi_n^{(1)} + \lambda^2 \psi_n^{(2)} + \ldots\right)$$
$$= \left(E_n^{(0)} + \lambda E_n^{(1)} + \lambda^2 E_n^{(2)} + \ldots\right)\left(\psi_n^{(0)} + \lambda\psi_n^{(1)} + \lambda^2 \psi_n^{(2)} + \ldots\right) \tag{13.11}$$

$$\text{or, } H^{(0)} \psi_n^{(0)} + \lambda\left(H^{(0)} \psi^{(1)} \psi_n^{(0)}\right) + \lambda^2\left(H^{(0)} \psi_n^{(2)} + H^{(1)} \psi_n^{(1)} + \ldots\right) + \ldots$$
$$= E_n^{(0)} \psi_n^{(0)} + \lambda\left(E_n^{(0)} \psi_n^{(1)} + E_n^{(1)} \psi_n^{(0)}\right) + \lambda^2\left(E_n^{(0)} \psi_n^{(2)} + E_n^{(2)} \psi_n^{(0)} + E_n^{(1)} \psi_n^{(1)}\right) + \ldots \tag{13.12}$$

In the above substitution, we have limited H to two terms, viz., $H^{(0)}$, the unperturbed and major part and $H^{(1)}$, the perturbed and small part.

In order that eq. (13.12) be satisfied for all values of λ, the coefficients of the same powers of λ on both sides of eq. (13.12) must equal each other.

So

$$H^{(0)} \psi_n^{(0)} = E_n^{(0)} \psi_n^{(0)} \tag{13.13}$$

$$H^{(0)} \psi_n^{(1)} \psi_n^{(0)} = E_n^{(0)} \psi_n^{(1)} + E_n^{(1)} \psi_n^{(0)} \tag{13.14}$$

$$H^{(0)} \psi_n^{(2)} + H_n^{(1)} \psi_n^{(1)} = E_n^{(0)} \psi_n^{(2)} + E_n^{(1)} \psi_n^{(1)} + E_n^{(2)} \psi_n^{(0)} \tag{13.15}$$

Equations (13.13), (13.14) and (13.15) represent respectively the unperturbed, first order perturbation and second order perturbation equations.

Evaluation of first order energy $E_n^{(1)}$

The first order perturbation equation is

$$H^{(0)} \psi_n^{(1)} + H^{(1)} \psi_n^{(0)} = E_n^{(0)} \psi_n^{(1)} + E_n^{(1)} \psi_n^{(0)}$$

For the solution of this equation, the expansion theorem will be used. We consider that the unknown wave function $\psi_n^{(1)}$ can be expanded in terms of known functions

$$\psi_1^{(0)}, \psi_2^{(0)}, \psi_3^{(0)}, \ldots, \psi_m^{(0)} \text{ as } \psi_n^{(1)} = \sum_{n=0}^{\infty} C_m \psi_m^{(0)} \tag{13.16}$$

To determine the perturbation of the first order we now solve eq. (13.14). Substituting eq. (13.16) in eq. (13.14), we get

$$\sum_m C_m H^{(0)} \psi_m^{(0)} + H_1^{(1)} \psi_n^{(0)} = \sum_m C_m E_n^{(0)} \psi_m^{(0)} + E_n^{(1)} \psi_n^{(0)} \tag{13.17}$$

But,

$$H^{(0)} \psi_m^{(0)} = E_m^{(0)} \psi_m^{(0)}$$

$$\therefore \sum_m C_m E_m^{(0)} \psi_m^{(0)} - \sum_m C_m E_n^{(0)} \psi_m^{(0)} + H^{(1)} \psi_n^{(0)} = E_n^{(1)} \psi_n^{(0)}$$

$$\text{or, } \sum_m C_m\left[E_m^{(0)} - E_n^{(0)}\right] \psi_m^{(0)} + H^{(1)} \psi_n^{(0)} = E_n^{(1)} \psi_n^{(0)} \tag{13.18}$$

Now, multiplying eq. (13.18) and integrating over the space variables

$$\int \sum_m \psi_n^{(0)*} C_m\left[E_m^{(0)} - E_n^{(0)}\right] \psi_m^{(0)} d\tau + \int \psi_n^{(0)*} H^{(1)} \psi_n^{(1)} d\tau = \int E_n^{(1)} \psi_n^{(0)*} \psi_n^{(0)} d\tau \tag{13.19}$$

Again,

$$\int \psi_i^{(0)*} \psi_j^{(0)} d\tau = \begin{cases} 0, & \text{if } i \neq 0 \\ 1, & \text{if } i = j \end{cases} \tag{13.20}$$

Using eq. (13.20) in eq. (13.19), we have

$$0 + \int \psi_n^{(0)*} H^{(1)} \psi_n^{(0)} d\tau = E_n^{(1)} \text{ or, } E_n^{(1)} = \int \psi_n^{(0)*} H^{(1)} \psi_n^{(0)} d\tau \tag{13.21}$$

Equation (13.21) gives the first order perturbation of the energy eigenvalue.

Equation (13.21) can be written as

$$E_n^{(1)} = \langle n|H'|n\rangle \tag{13.22}$$

The above eq. (13.22) shows that the first order energy correction for a non-degenerate system is just the expectation value of first order Hamiltonians $H^{(1)}$ over the unperturbed state, here the nth state of the system.

Evaluation of the first order wave function $\psi_n^{(1)}$

To calculate the wave functions we proceed as follows:

Multiplying eq. (13.18) by $\psi_m^{(0)*}$ and integrating over all space, we have

$$\int \sum_m C_m \left[E_m^{(0)} - E_n^{(0)}\right] \psi_m^{(0)*} \psi_n^{(0)} d\tau + \int \psi_m^{(0)*} H^{(1)} \psi_n^{(0)} d\tau = \int E_n^{(1)} \psi_m^{(0)*} \psi_n^{(0)} d\tau \tag{13.23}$$

Applying eq. (13.20), we get

$$C_m \left[E_m^{(0)} - E_n^{(0)}\right] + \int \psi_m^{(0)*} H^{(1)} \psi_n^{(0)} d\tau = 0 \text{ or, } C_m = -\frac{\int \psi_m^{(0)*} H^{(1)} \psi_n^{(0)} d\tau}{E_m^{(0)} - E_n^{(0)}}, \quad m \neq n \tag{13.24}$$

∴ we can denote the first order energy and wave functions by the expressions

$$E_n = E_n^{(0)} + \lambda \int \psi_n^{(0)*} H^{(1)} \psi_n^{(0)} d\tau \tag{13.25}$$

$$\text{and } \psi_n = \psi_n^{(0)} - \lambda \sum_{m=0}^{\infty} \frac{\int \psi_m^{(0)} H^{(1)} \psi_n^{(0)} d\tau}{E_m^{(0)} - E_n^{(0)}} \psi_m^{(0)} \tag{13.26}$$

$$\text{or, } \psi_n = \psi_n^{(0)} - \lambda \sum_{m=0}^{\infty}{}' \frac{\langle m|H^{(1)}|n\rangle}{E_m^{(0)} - E_n^{(0)}} \psi_m^{(0)} \tag{13.27}$$

Here, Σ' indicates summation over the values of m from 0 to ∞, excluding $m \neq n$.

Evaluation of second order energy $E_n^{(2)}$

We suppose that the unknown function $\psi_n^{(2)}$ can be expanded in terms of the known wave function $\psi_m^{(0)}$ as

$$\psi_n^{(2)} = \sum_m g_m \psi_m^{(0)} \tag{13.28}$$

Substituting eqs (13.28) and (13.14) in eq. (13.15), we get

$$\sum_m g_m H^{(0)} \psi_m^{(0)} + \sum_m C_m H^{(1)} \psi_m^{(0)} = \sum_m g_m E_n^{(0)} \psi_m^{(0)} + \sum_n C_m E_n^{(1)} \psi_m^{(0)} + E_m^{(2)} \psi_n^{(0)} \qquad (13.29)$$

$$\text{or, } \sum_m g_m H^{(0)} \psi_m^{(0)} - \sum_m g_m E_m^{(0)} \psi_m^{(0)} = \sum_m C_m E_n^{(1)} \psi_m^{(0)} - \sum_m C_m H^{(1)} \psi_m^{(0)} + E_n^{(2)} \psi_n^{(0)}$$

$$\text{or, } \sum_m g_m E_m^{(0)} \psi_m^{(0)} - \sum_m g_m E_n^{(0)} \psi_m^{(0)} = \sum_m C_m E_n^{(1)} \psi_m^{(0)} - \sum_m C_m H^{(1)} \psi_m^{(0)} + E_n^{(2)} \psi_n^{(0)}$$

$$\text{or, } \sum_m g_m \left[E_m^{(0)} - E_m^{(0)}\right] \psi_m^{(0)} = \sum_m C_m \left[E_n^{(1)} - H^{(1)}\right] \psi_m^{(0)} + E_n^{(2)} \psi_n^{(0)} \qquad (13.30)$$

Now, multiplying eq. (13.30) by $\psi_n^{(0)*}$ and, then integrating over the space variables

$$\int \sum_m g_m \left[E_m^{(0)} - E_n^{(0)}\right] \psi_n^{(0)*} \psi_m^{(0)} d\tau$$

$$= \int \sum_m \psi_n^{(0)*} C_m \left[E_n^{(1)} - H^{(1)}\right] \psi_m^{(0)} d\tau + \int E_n^{(2)} \psi_n^{(0)*} \psi_n^{(0)} d\tau \qquad (13.31)$$

Applying eqs (13.20) to (13.31), we get

$$0 = -\sum_m C_m \int \psi_n^{(0)*} H^1 \psi_m^{(0)} d\tau + E_n^{(2)} \text{ or, } E_n^{(2)} = \sum_m C_m \int \psi_n(0)^* H^{(1)} \psi_m^{(0)} d\tau \qquad (13.32)$$

Putting the value of C_m from eq. (13.24) in eq. (13.32), we get

$$E_n^{(2)} = -\sum_m{}' \frac{\int \psi_m^{(0)*} H^{(1)} \psi_n^{(0)} d\tau \int \psi_n^{(0)*} H^{(1)} \psi_m^{(0)} d\tau}{E_m^{(0)} - E_n^{(0)}}$$

$$\text{or, } E_n^{(2)} = \sum_m{}' \frac{\psi_m^{(0)*} H^{(1)} \psi_n^{(0)} d\tau \int \psi_n^{(0)*} H^{(1)} \psi_m^{(0)} d\tau}{E_n^{(0)} - E_m^{(0)}} \qquad (13.33)$$

If $H^{(1)}$ is Hamiltonian operator, then

$$E_n^{(2)} = \sum_m{}' \frac{\int \psi_m^{(0)*} H^{(1)} \psi_n^{(0)} d\tau \int \psi_m^{(0)*} H^{(1)} \psi_n^{(0)} d\tau}{E_n^{(0)} - E_m^{(0)}} \qquad (13.34)$$

$$\text{or, } E_n^{(2)} = \sum_m{}' \frac{\left\langle m|H^{(1)}|n\right\rangle^2}{E_n^{(0)} - E_m^{(0)}} \qquad (13.35)$$

Evaluation of second order wave function $\psi_n^{(2)}$

We multiply eq. (13.30) by $\psi_m^{(0)*}$ $(m \neq n)$ and, then integrating, we get

$$\int \sum_m g_m \int \left[E_m^{(0)} - E_n^{(0)}\right] \psi_m^{(0)*} \psi_m^{(0)} d\tau$$

$$= \sum_m C_m \int \psi_m^{(0)*} \left[E_n^{(1)} - H^{(1)}\right] \psi_m^{(0)} d\tau + \int E_n^{(2)} \psi_m^{(0)*} \psi_n^{(0)} d\tau \tag{13.36}$$

$$\text{or, } g_m \left[E_m^{(0)} - E_n^{(0)}\right] = \sum_m C_m \left(E_n^{(1)} \int \psi_m^{(0)*} H^{(1)} \psi_m^{(0)} d\tau\right)$$

$$\text{or, } g_m = \sum_m{}' \frac{C_m}{E_m^{(0)} - E_n^{(0)}} \left(E_n^{(1)} - \int \psi_m^{(0)*} H^{(1)} \psi_m^{(0)} d\tau\right) \tag{13.37}$$

$$\text{or, } g_m = \sum_m{}' \frac{C_m}{E_m^{(0)} - E_n^{(0)}} \left(E_n^{(1)} - \langle m|H^{(1)}|m\rangle\right) \tag{13.38}$$

Substituting the value of g_m from eq. (13.38) in eq. (13.27) we can calculate the value of second order perturbed wave functions.

13.4 The perturbed harmonic oscillator

(a) ***Interaction energy is proportional to x*** **:** The case of perturbed harmonic oscillator is *a* and example of application of the time-independent perturbation theory to non-degenerate state. The Hamiltonian of one-dimensional harmonic oscillator is given by

$$H = \text{KE} + \text{PE} = \frac{p^2}{2m} + \frac{1}{2}kx^2 + bx \tag{13.39}$$

$$\text{or, } \hat{H} = -\frac{\hbar^2}{2m}\nabla^2 + \frac{1}{2}kx^2 + bx \tag{13.40}$$

$$\text{or, } \hat{H} = \hat{H}_0 + \hat{H}' \tag{13.41}$$

where $\hat{H}_0$ = unperturbed Hamiltonian.

$$\hat{H}^{(0)} = -\frac{\hbar^2}{2m}\nabla^2 + \frac{1}{2}kx^2 \tag{13.42}$$

$\hat{H}'$ = perturbing Hamiltonian = bx.

The unperturbed Schrödinger equation is

$$H_0\psi_n^{(0)} = E_n\,\psi_n^{(0)} \text{ or, } -\frac{\hbar^2}{2m}\frac{\partial^2}{\partial x^2}\psi_n^{(0)} + \frac{1}{2}kx^2\,\psi_n^{(0)} = E_n\,\psi_n^{(0)} \tag{13.43}$$

This is simple harmonic oscillator equation having the energy eigenvalue

$$E_n = \left(n + \tfrac{1}{2}\right)\hbar\omega,\ n = 0, 1, 2, \ldots \tag{13.44}$$

$$\omega = \text{angular frequency of the oscillator} = \sqrt{k/m} \tag{13.45}$$

The first order energy correction is given by

$$E_n' = \langle n|\hat{H}'|n\rangle \tag{13.46}$$

$$\text{or, } E_n' = \int \psi_n^{(0)*}\hat{H}'\psi_n^{(0)}\,d\tau \tag{13.47}$$

with

$$\psi_n^{(0)} = N_m H n e^{-\alpha^2 x^2/2} \tag{13.48}$$

$$\alpha^4 = \frac{mk}{\hbar^2} \tag{13.49}$$

H_n = Hermite polynomial of degree n.

N_n = normalisation constant = $\left[\dfrac{\alpha}{\pi^{1/2} 2^n n!}\right]^{1/2}$.

Now, from eq. (13.47)

$$E_n' = N_n^2 \int e^{-(\alpha^2 x^2)/2} H_n^2 b x e^{(-\alpha^2 x^2)/2} dx \tag{13.50}$$

$$\text{or, } E_n' = N_n^2 \int e^{-\alpha^2 x^2} H_n^2 b x \, dx \tag{13.51}$$

The integrand being an odd function of x, it vanishes which implies that *the first order energy correction is zero.*

Second order energy correction

The second order energy correction is given by

$$E_n'' = -\sum_m{}' \frac{|\langle n|\hat{H}'|m\rangle|^2}{E_n^0 - E_m^{(0)}} \tag{13.52}$$

Summation over prime indicates that $n = m$ is excluded.

The term

$$\langle n|\hat{H}'|m\rangle = \int_{-\infty}^{\infty} \psi_n \hat{H}' \psi_m d\tau$$

$$\langle n|\hat{H}'|m\rangle = b \int_{-\infty}^{\infty} \psi_n x \psi_m dx$$

$$\langle n|\hat{H}'|m\rangle = \frac{b}{\alpha}\left(\frac{n+1}{2}\right)^2, \text{ for } m = n+1$$

$$\langle n|\hat{H}'|m\rangle = \left(\frac{b}{\alpha}\left(\frac{n}{2}\right)\right)^{1/2}, \text{ for } m = n-1$$

$$\langle n|\hat{H}'|n\rangle = 0, \text{ for all other values}$$

Now, for $m = n+1$

$$\langle n|\hat{H}'|m\rangle = \frac{b}{\alpha}\left(\frac{m}{2}\right)^{1/2}$$

Hence,

$$\langle \psi_0|\hat{H}'|\psi_1\rangle = \frac{b}{\alpha}\left(\frac{1}{2}\right)^{1/2}; \langle \psi_0|\hat{H}'|\psi_1\rangle = \frac{b^2}{2\alpha^2}$$

$$E'' = \frac{E_0 + \frac{b^2}{2\alpha^2}}{E_1 - E_0} = \frac{\frac{b^2}{2\alpha^2}}{\frac{3}{2}\hbar\omega - \frac{1}{2}\hbar\omega} = \frac{b^2}{2\alpha^2\hbar\omega} \text{ and } E = E_0 + \frac{b^2}{2\alpha^2\hbar\omega}$$

$$E = \frac{1}{2}\hbar\omega + \frac{b^2}{2\alpha^2\hbar\omega} \tag{13.53}$$

The correction in energy value of nth level

$$E_n'' = \sum_m{}' \frac{\frac{b^2}{\alpha^2}\left(\frac{n+1}{2}\right)}{E_m^0 - E_n^0}, \text{ for } m = n+1$$

$$E_n'' = \sum_m{}' \frac{\frac{b^2}{\alpha^2}\frac{n}{2}}{E_m^0 - E_n^0}, \text{ for } m = n-1$$

$$E_n'' = 0, \text{ for all other values}$$

For $m = n+1, E_m^0 - E_n^0 = \hbar\omega$.

For $m = n-1, E_m^0 - E_n^0 = -\hbar\omega$.

$$\therefore E_n'' = \sum_m{}' \frac{b^2}{\alpha^2}\left(\frac{n+1}{2}\right)\frac{1}{\hbar\omega}, \text{ for } m = n+1$$

$$E_n'' = -\sum_m{}' \frac{b^2}{\alpha^2}\left(\frac{n}{2}\right)\frac{1}{\hbar\omega}, \text{ for } m = n-1$$

$$E_n'' = 0, \text{ for all other values}$$

(b) ***Interaction is proportional to x^2*** : In this case the Hamiltonian is given by

$$H = \frac{p^2}{2m} + \frac{1}{2}kx^2 + \frac{1}{2}bx^2 \tag{13.54}$$

with

$$H^0 = \frac{p^2}{2m} + \frac{1}{2}kx^2 \tag{13.55}$$

$$H' = \frac{bx^2}{2} \tag{13.56}$$

The angular frequency is given by

$$\omega = \left(\frac{k+b}{m}\right)^{1/2} = \sqrt{\frac{k}{m}\left(1+\frac{b}{k}\right)} \tag{13.57}$$

$$\omega = \omega_0\left(1 + \frac{b}{2k} - \frac{b^2}{8k^2}\right) \tag{13.58}$$

with $\omega_0 = \left(\frac{k}{m}\right)$.

The energy eigenvalue is given by

$$E_n = \left(n+\frac{1}{2}\right)\hbar\omega\left(1+\frac{b}{2k} - \frac{b^2}{8k^2}\right)$$

Then the first order correction is given by

$$E_n' = \langle n|\hat{H}'|n\rangle \tag{13.59}$$

$$= \frac{b}{2}\langle n|x^2|n\rangle \tag{13.60}$$

with

$$\psi_n = N_n H_n e^{-\alpha^2 x^2/2} \tag{13.61}$$

and normalisation constant

$$N_n = \left(\frac{\alpha}{\pi^{1/2}2^n n!}\right)^{1/2} \tag{13.62}$$

$$E'_n = N_n^2 \frac{b}{2}\int H_n^2 e^{-\alpha^2 x^2} x^2\, dx \tag{13.63}$$

$$E'_n = \frac{N_n^2}{\alpha^3}\frac{b}{2}\int H_n^2 e^{-\eta^2}\eta^2 d\eta;\ \eta = \alpha x \tag{13.64}$$

Now, the recurrence formula

$$H_{n+1} = 2\xi H_n - 2nH_{n-1}$$

we have

$$\eta H_n = \tfrac{1}{2}H_{n+1} + nH_{n-1} \tag{13.65}$$

Using in eq. (13.64), we get

$$E'_n = \frac{N_n^2 b}{2\alpha^3}\int_{-\infty}^{+\infty} e^{-\eta^2}\left[\tfrac{1}{2}H_{n+1} + nH(n-1)\right]^2 d\eta \tag{13.66}$$

Using

$$\int_{-\infty}^{+\infty} e^{-\eta^2} H_n H_m\, d\eta = 0,\ \text{for } m \neq n$$

$$\int_{-\infty}^{+\infty} e^{-\eta^2} H_n H_m\, d\eta = 2^n n!\,\pi^{1/2},\ \text{for } m = n$$

we have

$$E'_n = \frac{N_m^2 b}{2\alpha^3}\left[\tfrac{1}{4}2^{n+1}(n+1)!\,\pi^{1/2} + n^2 2^{n-1}(n-1)!\,\pi^{1/2}\right]$$

$$= \frac{N_m^2 b}{2\alpha^3}\left[\pi^{1/2}2^{n-1}(n-1)!\{(n+1)n + n^2\}\right]$$

$$E'_n = \frac{b}{4\alpha^2}(2n+1)$$

Thus, the energy eigenvalue in the nth state is given by

$$E_n = E_n^0 + E'_n = \left(n + \frac{1}{2}\right)\hbar\omega_0 + \frac{b}{4\alpha^2}(2n+1)$$

But,

$$\omega_0 = \left(\frac{k}{m}\right)^{1/2};\ \alpha^2 = \left(\frac{km}{\hbar^2}\right)^{1/2}$$

$$\therefore E_n = \left(n + \frac{1}{2}\right)\hbar\left(\frac{k}{m}\right)^{1/2} + \frac{b}{n}\hbar\frac{1}{(mk)^{1/2}}\left(n + \frac{1}{2}\right)$$

$$\text{or, } E_n = \left(n + \frac{1}{2}\right)\hbar\left(\frac{k}{m}\right)^{1/2}\left(1 + \frac{b}{2k}\right)$$

(c) ***Interaction energy proportional to*** x^4 : We consider the approximate energy levels of the system having we equation

$$\frac{d^2\psi}{dx^2} + \frac{2m}{\hbar^2}\left[E - \frac{1}{2}kx^2 - ax^3 - bx^4\right]\psi = 0 \tag{13.67}$$

If a and b are zero, then the above equation reduces to a wave equation for harmonic oscillator. If a and b are small we can treat these terms as perturbation

$$H^{(1)} = ax^3 + bx^4 \tag{13.68}$$

The first order perturbed energy is given by

$$E_n^{(1)} = \int \psi_n^{(0)*} H^{(1)} \psi_n^{(0)} d\tau = \int_{-\infty}^{+\infty} \psi_n^{(0)} ax^3 \psi_n^{(0)} dx + \int_{-\infty}^{+\infty} \psi_n^{(0)} x^4 b \psi_n^{(0)} dx$$

$$\therefore E_n^{(1)} = a\int_{-\infty}^{+\infty} \psi_n^{(0)*} x^3 \psi_n^{(0)} dx + b\int_{-\infty}^{+\infty} \psi_n^{(0)*} x^4 \psi_n^{(0)} dx \tag{13.69}$$

Since, x^3 is an odd function and $\psi_n^{(0)*}\ \psi_n^{(0)}$ is an even function, the value of the first integral is zero. This means the first order perturbation due to ax^3 is zero.

So

$$E_n^{(1)} = b\int_{-\infty}^{+\infty} \psi_n^{(0)*} x^4 \psi_n^{(0)} dx \tag{13.70}$$

Now, let

$$I = \int_{-\infty}^{+\infty} \psi_n^{(0)*} x^4 \psi_n^{(0)} dx \tag{13.71}$$

But, in case of linear harmonic oscillator

$$\psi_n(x) = N_n H_n(\xi) \exp\left(-\frac{\xi^2}{2}\right), \text{ where } \xi = \alpha x \tag{13.72}$$

Substituting eq. (13.72) in eq. (13.71), we get

$$I = \int_{-\infty}^{+\infty} N_n^2 H_n^2(\xi) e^{-\xi^2}\left(\frac{\xi^4}{\alpha^4}\right)\left(\frac{d\xi}{\alpha}\right) = \frac{N_n^2}{\alpha^5}\int_{-\infty}^{+\infty} H_n^2(\xi) e^{-\xi^2} \xi^4 d\xi \tag{13.73}$$

Now, from recurrence relation

$$\xi H_n(\xi) = \tfrac{1}{2} H_{n+1}(\xi) + nH_{n-1}(\xi) \tag{13.74}$$

$$\text{or, } \xi^2 H_n(\xi) = \tfrac{1}{2}\xi H_{n+1}(\xi) + n\xi H_{n-1}(\xi) \tag{13.75}$$

Substituting $n = (n+1)$ and $n = n-1$ in eq. (13.74)

$$\xi H_{n+1}(\xi) = \tfrac{1}{2} H_{n+2}(\xi) + (n+1) H_n(\xi)$$

$$\xi H_{n+1}(\xi) = \tfrac{1}{2} H_n(\xi) + (n+1) H_{n-2}(\xi)$$

Substituting these values in eq. (13.75), we get

$$\xi^2 H_n(\xi) = \tfrac{1}{2}\left[\tfrac{1}{2} H_{n+2}(\xi) + (n+1) H_n(\xi)\right] + n\left[\tfrac{1}{2} H_n(\xi) + (n-1) H_{n-2}(\xi)\right]$$

$$\text{or, } \xi^2 H_n(\xi) = \tfrac{1}{4} H_{n+2}(\xi) + (n+\tfrac{1}{2}) H_n(\xi) + n(n-1) H_{n-2}(\xi) \tag{13.76}$$

Now, substituting eq. (13.76) in eq. (13.73)

$$I = \frac{N_n^2}{\alpha^5}\int_{-\infty}^{+\infty} e^{\xi^2}\left[\tfrac{1}{4}H_{n+2}(\xi) + (n+\tfrac{1}{2})\,H_n(\xi) + n(n-1)\,H_{n-2}(\xi)^2\right]d\xi \tag{13.77}$$

But,

$$\int_{-\infty}^{+\infty} e^{-\xi^2} H_n(\xi)\,H_m(\xi)\,d\xi = \begin{cases} 0, & \text{if } m \neq n \\ 2^n n!\,\pi^{1/2}, & \text{if } m = n \end{cases} \tag{13.78}$$

Using eq. (13.78) in eq. (13.77), we get

$$\begin{aligned} I &= \frac{N_n^2}{\alpha^5}\sqrt{\pi}\left\{\tfrac{1}{16}(n+2)\,!2^{n+2} + (n+\tfrac{1}{2})^2\,2^n n! + n^2(n-1)^2 2^{n-2}(n-2)\,!\right\} \\ &= \frac{3}{4\alpha^4}(2n^2+2n+1) \end{aligned} \tag{13.79}$$

From eqs (13.79) and (13.70)

$$E_n^{(1)} = \frac{3}{4\alpha^4}(2n^2+2n+1) \tag{13.80}$$

∴ the total energy to first order perturbation is

$$E_n = E_n^{(0)} + E_n^{(1)} = \left(n+\frac{1}{2}\right)\hbar\omega_0 + \frac{3b}{4}\frac{\hbar^2}{mk}(2n^2+2n+1) \tag{13.81}$$

where $\omega_0 = \sqrt{\dfrac{k}{m}}$ and $\alpha^4 = \dfrac{mk}{\hbar^2}$.

13.5 The normal helium atom

The normal state of helium atom contains a nucleus of charge Ze and two electrons each having charge $-e$. The potential energy of such a system is given by

$$V = \left(-\frac{ze^2}{r_1} - \frac{ze^2}{r_2} + \frac{e^2}{r_1 r_2}\right)\frac{1}{4\pi\varepsilon_0} \tag{13.82}$$

where r_1 and r_2 are the distances of electrons 1 and 2, respectively from the nucleus and r_{12} is the separation of two electrons as shown in Fig 13.1.

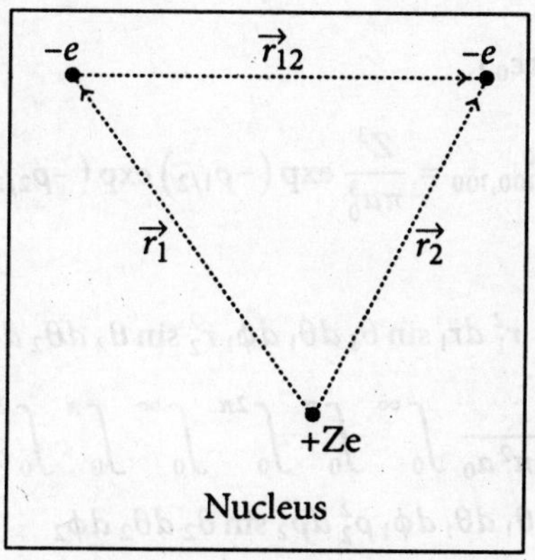

Fig 13.1 The Helium atom

Considering the nucleus at rest, the wave equation for the two electrons is

$$\left(\frac{\partial^2 \psi}{\partial x_1^2}+\frac{\partial^2 \psi}{\partial y_1^2}+\frac{\partial^2 \psi}{\partial z_1^2}+\frac{\partial^2 \psi}{\partial x_2^2}+\frac{\partial^2 \psi}{\partial y_2^2}+\frac{\partial^2 \psi}{\partial z_2^2}\right)+\frac{2m}{\hbar^2}\left[E+\frac{1}{4\pi\varepsilon_0}\left(\frac{Ze^2}{r_1}+\frac{Ze^2}{r_2}-\frac{e^2}{r_{12}}\right)\right]\psi=0 \tag{13.83}$$

Here, x_1, y_1, z_1 and x_2, y_2, z_2 are the Cartesian coordinates of the two electrons respectively. If we neglect the term $\frac{e^2}{4\pi\varepsilon_0 r_{12}}$, then the exact solution of the wave equation is possible. So this term may be regarded as the perturbation term

$$H^{(1)}=\frac{e^2}{r_{12}}\cdot\frac{1}{4\pi\varepsilon_0} \tag{13.84}$$

The unperturbed wave function can be separated into two parts by substituting

$$\psi^{(0)}(x_1y_1z_1;x_2y_2z_2)=u_1(x_1y_1z_1)\,u_2(x_2y_2z_2) \tag{13.85}$$

In polar coordinates r_1, θ_1, ϕ_1 and r_2, θ_2, ϕ_2, the wave function in normal state

$$\psi^{(0)}_{100,100}=\psi^{(0)}_{100}(r_1,\theta_1,\phi_1)\psi^{(0)}_{100}(r_2,\theta_2,\phi_2)=u^{(0)}_{1s}(r_1,\theta_1,\phi_1),u^{(0)}_{1s}(r_2,\theta_2,\phi_2)$$

and the corresponding eigenvalue is

$$E^{(0)}_{100,100}=E^{(0)}_1+E^{(0)}_2=-2Z^2E_{\text{eff}} \tag{13.86}$$

The first order perturbed energy $E^{(1)}$ is the average value of the perturbation function

$$H^{(1)}=\frac{e^2}{4\pi\varepsilon_0 r_{12}}$$

over the unperturbed state of the system.

Hence,

$$E^{(1)}=\int \psi_n^{(0)*}\,H^{(1)}\,\psi_n^{(0)}\,d\tau=\frac{1}{4\pi\varepsilon_0}\int\frac{e^2}{r_{12}}\,\psi^{(0)}_{100,100}\,d\tau \tag{13.87}$$

We know that

$$u^{(0)}_{1s}=\psi^{(0)}_{100}=\sqrt{\left(\frac{Z^3}{\pi a_0^3}\right)}e^{-\rho/2} \tag{13.88}$$

where $\rho=2Zr/a_0$ and $a_0=\frac{\hbar^2}{me^2}(4\pi\varepsilon_0)$.

$$\therefore\ \psi^0_{100,100}=\frac{Z^3}{\pi a_0^3}\exp\left(-\rho_{1/2}\right)\exp\left(-\rho_{2/2}\right) \tag{13.89}$$

The volume element is

$$d\tau=r_1^2\,dr_1\sin\theta_1\,d\theta_1\,d\phi_1\,r_2^2\sin\theta_2\,d\theta_2\,d\phi_2 \tag{13.90}$$

$$\therefore\ E^{(1)}=\frac{Ze^2}{(4\pi\varepsilon_0)\,2^5\pi^2 a_0}\int_0^\infty\int_0^\pi\int_0^{2\pi}\int_0^\infty\int_0^\pi\int_0^{2\pi}\frac{\exp(-\rho_1-\rho_2)}{\rho_{12}}$$
$$\times\,\rho_1^2\,d\rho_1\sin\theta_1\,d\theta_1\,d\phi_1\,\rho_2^2\,d\rho_2\sin\theta_2\,d\theta_2\,d\phi_2 \tag{13.91}$$

in which $\rho_{12}=\frac{2Z_{12}}{a_0}$.

Considering the electrostatic interaction energy of two spherically symmetrical distribution of electricity with density functions $\exp(-\rho_1)$ and $\exp(-\rho_2)$ respectively we can solve the above integral to get

$$E^{(1)} = \left(\frac{5}{4}\right) ZE_{\text{eff}} \tag{13.92}$$

Hence, the total energy of the system is

$$E = -\left(2Z^2 - \frac{5}{4}Z\right) E_{\text{eff}} \tag{13.93}$$

Similarly, we can calculate the energy of He, Li^+, Be^{++}, etc. with $Z = 1, 2, 3$, respectively.

13.6 The Stark effect of the plane rotator

The wave equation of a plane rotator under the influence of electric field E' is

$$\frac{d^2\psi}{d\phi^2} + \frac{2I}{\hbar^2}(E + \mu E' \cos\phi)\psi = 0 \tag{13.94}$$

where I is the moment of inertia about the axis of rotation, μ the electric moment. It rotates in a plane about an axis passing through its centre of mass.

If $E' = 0$, the wave equation is

$$\frac{d^2\psi}{d\phi^2} + \frac{2I}{\hbar^2} E\psi = 0 \tag{13.95}$$

The normalised solution of eq. (13.85) is

$$\psi_m^{(0)} = \left(\frac{1}{\sqrt{2\pi}}\right) e^{im\phi}, \quad m = 0, \pm 1, \pm 2, \pm 3 \tag{13.96}$$

and energy value

$$E^{(0)} = \frac{m^2\hbar}{8\pi^2 I} \tag{13.97}$$

Here, $-\mu E' \cos\phi$ may be regarded as the perturbed term and hence the perturbation energy will be

$$E_m^{(1)} = -\mu E' \int \psi_m^{(0)*} \cos\phi \psi_m^{(0)} d\phi = \frac{\mu}{2\pi} E' \int_0^{2\pi} e^{i(m'-m)\phi} \cos\phi \, d\phi$$

$$\text{or, } E_m^{(1)} = -\frac{\mu E'}{4\pi} \int_0^{2\pi} e^{i(m'-m+2)\phi} d\phi - \frac{\mu E'}{2\pi} \int_0^{2\pi} e^{i(m'-m-1)} d\phi \tag{13.98}$$

$$\left.\begin{aligned} &= 0, && \text{for } m' \neq m \pm 1 \\ &= -\frac{\mu E'}{2}, && \text{for } m' = m \pm 1 \end{aligned}\right\} \tag{13.99}$$

Hence, the first order energy correction is zero, i.e.,

$$E_m^{(1)} = 0 \tag{13.100}$$

The second order energy is given by

$$E_m^{(2)} = \frac{\left[\int \psi_m^{(0)*}\ \mu\cos\phi\psi_{(m-1)}^{(0)}\ d\phi\right]^2}{E_m^{(0)} - E_{(m-1)}^{(0)}} + \frac{E'^2\left[\int \psi_m^{(0)*}\ \cos\phi\psi_{(m+1)}^{(0)}\ d\phi\right]^2}{E_m^{(0)} - E_{(m+1)}^{(0)}}$$

$$= \frac{4\pi^2 I\mu^2 E'^2}{h^2(4m^2-1)}$$

Hence, the total energy to the second order is

$$E = E^{(0)} + E_m^{(1)} + E_m^{(2)} \text{ or, } E = \frac{m^2h^2}{8\pi^2 I} + \frac{4\mu^2 I\mu^2 E'^2}{h^2(4m^2-1)} \tag{13.101}$$

13.7 First order Zeeman effect in H-Atom (without electron spin)

The phenomenon of splitting the energy levels in presence of a magnetic field is known as the Zeeman effect. This is due to change in energy value of a level due to interaction between applied field and magnetic moment of electron. The unperturbed Hamiltonian is

$$\hat{H}^{(0)} = -\frac{\hbar^2}{2m}\nabla^2 - \frac{e^2}{4\pi\varepsilon_0 r} \tag{13.102}$$

The first order perturbation of Hamiltonian

$$\hat{H}' = -\frac{ie\hbar}{2\mu}B\frac{\partial^2}{\partial\phi^2} \tag{13.103}$$

The unperturbed wave function of nth state is

$$\psi_{nlm}^{(0)} = A_{nlm}R_{nl}(r)\hat{H}_{lm}\Theta e^{im\phi} \tag{13.104}$$

∴ the first order change in energy value

$$E^{(1)} = \int \psi_{nlm}^{(0)*} - \frac{ie\hbar B}{2\mu}\frac{\partial}{\partial\phi}\psi_{nlm}\, d\tau \tag{13.105}$$

$$= -\frac{ie\hbar B}{2\mu}(im)\int \psi_{nlm}^{(0)*}\psi_{nlm}^{(0)}\, d\tau \tag{13.106}$$

$$\text{or, } E^{(1)} = \frac{\hbar B}{2\mu}m_l \tag{13.107}$$

replacing m by m_l the orbital magnetic quantum number.

$$\therefore E_n = E_n^{(0)} + \frac{B\hbar}{2\mu}m_l \tag{13.108}$$

The frequency of radiation is given by

$$\nu = \frac{E_{n_2} - E_{n_1}}{h} - 1$$

$$\nu = \nu_0 + \frac{eB}{4\pi\mu}\Delta m_l \tag{13.109}$$

But, $\Delta m_l = 0, \pm 1$.

$$\left.\begin{aligned} \therefore \nu &= \nu_0 \\ \nu &= \nu_0 + \frac{eB}{4\pi\mu} \\ \nu &= \nu_0 - \frac{eB}{4\pi\mu} \end{aligned}\right\} \tag{13.110}$$

So in absence of electron spin, a given transition splits into three transitions as shown in Fig 13.2 for transition $l_1 = 2$ to $l_2 = 1$.

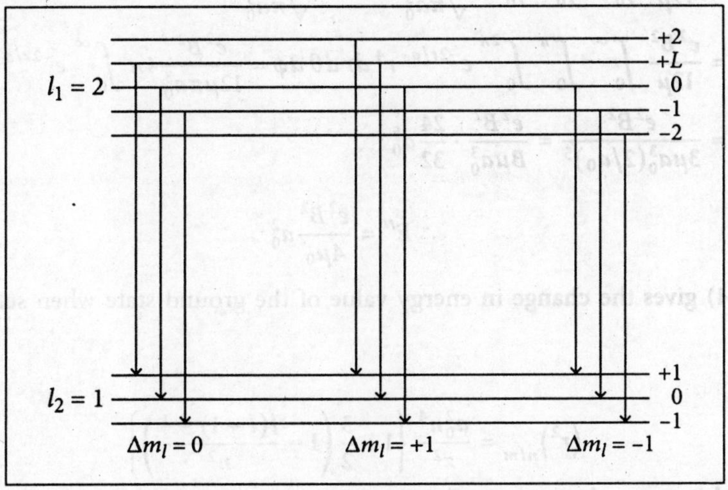

Fig 13.2 Possible transitions between two levels with $l_1 = 2$ and $l_2 = 1$

The transition corresponding to same value of Δm_l coincides as they represent the same value of energy.

Second order Zeeman effect in hydrogen

The second order perturbing Hamiltonian is

$$\hat{H}'' = \frac{q^2 A^2}{2m} = \frac{e^2 A^2}{2\mu} \tag{13.111}$$

But,

$$A^2 = Ax^2 + Ay^2 + Az^2 \tag{13.112}$$

Again, from the relation

$$\vec{A} = \tfrac{1}{2}\vec{B} \times \vec{r} \; [\vec{A} = \text{vector potential}]$$

$$A_x = -\tfrac{1}{2}By$$

$$A_y = -\tfrac{1}{2}B_x$$

$$A_z = 0$$

$$\therefore \hat{H} = \frac{e^2 A^2}{2\mu} = \frac{e^2 B^2}{8\mu}(x^2 + y^2)$$

Considering the potential due to hydrogen as spherically symmetric

$$\hat{H} = \frac{e^2 B^2}{8\mu} \cdot \frac{2}{3} r^2$$

Now, the change in energy value of the first state due to this perturbation

$$E'' = \int \psi_{100}^* \frac{2e^2 B^2}{24\mu} r^2 \psi_{100} \, d\tau \tag{13.113}$$

$$= \frac{e^2 B^2}{12\mu} \int_0^\infty \int_0^\pi \int_0^{2\pi} \frac{1}{\sqrt{\pi a_0^3}} e^{-r/a_0} r^2 \frac{1}{\sqrt{\pi a_0^3}} e^{-r/a_0} \, dr \sin\theta \, d\theta \, d\phi$$

$$= \frac{e^2 B^2}{12\mu} \int_0^\infty \int_0^\pi \int_0^{2\pi} e^{-2r/a_0} r^4 \, dr \, d\theta \, d\phi = \frac{e^2 B^2}{12\mu\pi a_0^3} 4\pi \int_0^\infty e^{-2r/a_0} r^4 \, dr$$

$$= \frac{e^2 B^2}{3\mu a_0^3 (2/a_0)^5} = \frac{e^2 B^2}{B\mu a_0^3} \cdot \frac{24}{32} a_0^5$$

$$\therefore E'' = \frac{e^2 B^2}{4\mu_0} a_0^2 \tag{13.114}$$

Equation (13.114) gives the change in energy value of the ground state when second order effect is considered.

In general,

$$\langle r^2 \rangle_{nlm} = \frac{a_0^2 n^4}{z^2} \left\{ 1 + \frac{3}{2} \left(1 - \frac{l(l+1) - \frac{1}{3}}{n^2} \right) \right\} \tag{13.115}$$

and the second order energy change is

$$E'' = \frac{e^2 B^2}{8\mu} \cdot \frac{2}{3} \frac{a_0^2 n^4}{Z^2} \left\{ 1 + \frac{3}{2} \left(1 - \frac{l(l+1) - \frac{1}{3}}{n^2} \right) \right\} \tag{13.116}$$

13.8 The first order perturbation theory for a degenerate level

If the energy level of unperturbed system is *degenerate,* first order non-degenerate perturbation method is not applicable. Because in the treatment of non-degenerate perturbation theory we assumed that the perturbed wave function differs slightly from one function $\psi_m^{(0)}$, which is the solution of unperturbed wave equation for a given value whereas in case of degenerate level, there are such several functions which corresponds to the same energy.

An energy level is called α-fold degenerate when there exist α linearly independent wave functions such as $\psi_{m1}, \psi_{m2}, \ldots, \psi_{m\alpha}$ satisfying the wave equation.

Let, in the perturbed state, the Schrödinger wave equation be given by

$$H\psi = E\psi \tag{13.117}$$

where H, E and ψ are respectively the perturbed Hamiltonian, perturbed energy and perturbed wave function.

The perturbed Hamiltonian H can be expressed in terms of unperturbed Hamiltonian $H^{(0)}$ as below:

$$H = H^{(0)} + \lambda H^{(1)} + \lambda^2 H^{(2)} \tag{13.118}$$

The wave equation for the unperturbed system is

$$H^{(0)}\,\psi^{(0)} = E^{(0)}\,\psi^{(0)} \tag{13.119}$$

The solutions of eq. (13.119) are

$$\psi_{01}^{(0)}, \psi_{02}^{(0)}, \psi_{03}^{(0)}, \psi_{04}^{(0)}, \ldots, \psi_{0\alpha}^{(0)} \text{ corresponding to } E_0^{(0)}$$
$$\psi_{11}^{(0)}, \psi_{12}^{(0)}, \psi_{13}^{(0)}, \psi_{14}^{(0)}, \ldots, \psi_{1\alpha}^{(0)} \text{ corresponding to } E_1^{(0)}$$
$$\ldots\ldots\ldots\ldots\ldots\ldots\ldots\ldots\ldots\ldots\ldots\ldots$$
$$\psi_{m1}^{(0)}, \psi_{m2}^{(0)}, \psi_{m3}^{(0)}, \psi_{m4}^{(0)}, \ldots, \psi_{m\alpha}^{(0)} \text{ corresponding to } E_m^{(0)}$$

For a degenerate system, the problem is to determine the set of unperturbed wave functions to which the perturbed functions reduce when the perturbation vanishes, i.e., the evaluation of coefficients in the linear transformation converting the initially chosen wave functions into the correct zeroth order wave functions. These combinations are as below:

$$\chi_{ml}^{(0)} = \sum_{l'=1}^{\alpha} k_{ll'}\,\psi_{ml'}^{(0)},\ l = 1, 2, \ldots, \alpha \tag{13.120}$$

$$= k_{l1}\,\psi_{m1}^{(0)} + k_{l2}\,\psi_{m2}^{(0)} + \cdots + k_{l\alpha}\,\psi_{m2}^{(0)}$$

$$\text{or, } \chi_{m1}^{(0)} = k_{11}\,\psi_{m1}^{(0)} + k_{12}\,\psi_{m2}^{(0)} + \cdots + k_{1\alpha}\,\psi_{m\alpha}^{(0)}$$

$$\chi_{m2}^{(0)} = k_{21}\,\psi_{m1}^{(0)} + k_{22}\,\psi_{m2}^{(0)} + \cdots + k_{2\alpha}\,\psi_{m\alpha}^{(0)} \tag{13.121}$$

provided the function ψ_{ml} is given by

$$\psi_{ml} = \chi_{ml}^{(0)} + \lambda\psi_{ml}^{(1)} + \lambda^2\psi_{ml}^{(2)} \tag{13.122}$$

$$\text{and } E_{ml} = E_m^{(0)} + \lambda E_{ml}^{(1)} + \lambda^2 E_{ml}^{(2)} \tag{13.123}$$

Now, the perturbed wave function is given by

$$H\psi_{ml} - E_{ml}\,\psi_{ml} = 0 \tag{13.124}$$

Substituting the values of H, ψ_{ml} and E_{ml} from eqs (13.118) and (13.121) in eq. (13.124), we get

$$\left(H^{(0)} + \lambda H^{(1)} + \lambda^2 H^{(2)} + \ldots\right)\left(\psi_{ml}^{(0)} + \lambda\psi_{ml}^{(1)} + \lambda^2\psi_{ml}^{(2)} + \ldots\right)$$
$$-\left(E_m^{(0)} + \lambda E_{ml}^{(1)} + \lambda^2 E_{ml}^{(2)} + \ldots\right)\left(\chi_{ml}^{(0)} + \lambda_{ml}^{(1)} + \lambda^2\psi_{ml}^{(2)} + \ldots\right) = 0 \tag{13.125}$$

Solving eq. (13.125), we get

$$\left(H^{(0)}\chi_{ml}^{(0)} - E_m^{(0)}\chi_{ml}^{(0)}\right) + \chi\left(H^{(0)}\psi_{ml}^{(1)} + H^{(1)}\chi_{ml}^{(0)}\right) - E_m^{(0)}\psi_{ml}^{(1)} - E_{ml}^{(1)}\chi_{ml}^{(0)} + \cdots = 0 \tag{13.126}$$

The first order perturbation can be obtained by equating the coefficients of χ equal to zero, i.e.,

$$H^{(0)}\psi_{ml}^{(1)} + H^{(1)}\chi_{ml}^{(0)} - E_n^{(0)}\psi_{ml}^{(1)} - E_{ml}^{(1)}\chi_{ml}^{(0)} = 0 \tag{13.127}$$

We now expand $\psi_{ml}^{(1)}$ as

$$\psi_{ml}^{(1)} = \sum_{m'l'} a_{ml\,m'l'}\,\psi_{m'l'}^{(0)} \tag{13.128}$$

Substituting eqs (13.128) and (13.120) in eq. (13.127), we get

$$\sum a_{ml\,m'l'}\, H^{(0)}\, \psi^{(0)}_{m'l'} + \sum_{l'=1}^{\alpha} k_{ll'}\, H^{(1)}\, \psi^{(0)}_{ml'} - \sum_{m'l'} a_{ml\,m'l'}\, E^{(0)}_{m}\, \psi^{(0)}_{m'l'} - \sum_{l'=1}^{\alpha} k_{ll'}\, E^{(1)}_{m}\, \psi^{(0)}_{m'l'} = 0$$

$$\text{or,}\ \sum_{m'l'} a_{ml\,m'l'}\, E^{(0)}_{m'}\, \psi^{(0)}_{m'l'} + \sum_{l'=1}^{\alpha} k_{ll'}\, H^{(1)}\, \psi^{(0)}_{m'l'} - \sum a_{ml\,m'l'}\, E^{(0)}_{m}\, \psi^{(0)}_{m'l'} - \sum_{l'=1}^{\alpha} k_{ll'}\, E^{(1)}_{ml}\, \psi^{(0)}_{ml'} = 0$$

$$\text{or,}\ \sum_{m'l'} a_{ml\,m'l'} \left(E^{(0)}_{m'} - E^{(0)}_{m}\right) \psi^{(0)}_{m'l'} = \sum_{l'=1}^{\alpha} k_{ll'} \left(E^{(0)}_{ml} - H^{(1)}\right) \psi^{(0)}_{ml'} \tag{13.129}$$

Multiplying both sides of eq. (13.129) by ψ^{*}_{mj} and, then integrating over the configuration space, we get

$$\sum_{m'l'} a_{ml\,m'l'} \left(E^{(0)}_{m'} - E^{(0)}_{m}\right) \int \psi^{(0)*}_{mj}\, H^{(1)}\, \psi^{(0)}_{ml'}\, d\tau$$
$$= \sum_{l'=1}^{\alpha} k_{ll'} \left[E^{(1)}_{ml} \int \psi^{(0)*}_{ml'}\, d\tau - \int \psi^{(0)*}_{mj}\, H^{(1)}\, \psi^{(0)}_{ml'}\, d\tau\right] \tag{13.130}$$

Since, $\psi^{(0)}_{mj}$ and $\psi^{(0)}_{m'l'}$ are orthogonal, the left-hand side of eq. (13.130) is equal to zero, if $m \neq m'$.

Also $E^{(0)}_{m'} - E^{(0)}_{m}$ is zero, if $m = m'$.

Putting

$$H^{(1)}_{jl'} = \int \psi^{(0)*}_{mj}\, H^{(1)}\, \psi^{(0)}_{ml'}\, d\tau \tag{13.131}$$

$$\text{and}\ \Delta_{jl'} = \int \psi^{(0)}_{mj}\, \psi^{(0)}_{ml'}\, d\tau \tag{13.132}$$

Using eqs (13.131) and (13.132) in eq. (13.130), we get

$$\sum_{l'=1}^{\alpha} k_{ll'} \left[H^{(1)}_{jl'} - \Delta_{jl'}\, E^{(1)}_{ml}\right] = 0 \tag{13.133}$$

where $j = 1, 2, 3, \ldots \alpha$.

Equation (13.133) represents a system of α homogeneous linear simultaneous equations in the α unknown quantities $k_{l1}, k_{l2}, \ldots, k_{l\alpha}$.

These equations are

$$\left.\begin{aligned}
&\left(H^{(1)}_{11} - \Delta_{11}\, E^{(1)}_{ml}\right) k_{l1} + \left(H^{(1)}_{12} - \Delta_{12}\, E^{(2)}_{ml}\right) k_{l2} + \cdots + \left(H^{(1)}_{1\alpha} - \Delta_{1\alpha}\, E^{(1)}_{ml}\right) k_{l\alpha} = 0\\
&\left(H^{(1)}_{21} - \Delta_{21}\, E^{(1)}_{ml}\right) k_{l1} + \left(H^{(1)}_{22} - \Delta_{22}\, E^{(1)}_{ml}\right) k_{l2} + \cdots + \left(H^{(1)}_{2\alpha} - \Delta_{2\alpha}\, E^{(1)}_{ml}\right) k_{l\alpha} = 0\\
&\cdots\cdots\cdots\cdots\cdots\cdots\cdots\cdots\cdots\cdots\cdots\cdots\\
&\left(H^{(1)}_{\alpha 1} - \Delta_{\alpha 1}\, E^{(1)}_{ml}\right) k_{l1} + \left(H^{(1)}_{\alpha 2} - \Delta_{\alpha 2}\, E^{(2)}_{ml}\right) k_{l2} + \cdots + \left(H^{(1)}_{\alpha\alpha} - \Delta_{\alpha\alpha}\, E^{(1)}_{ml}\right) k_{l\alpha} = 0
\end{aligned}\right\} \tag{13.134}$$

The condition which must be satisfied so that such a set of homogeneous linear equation is to have non-zero solutions is that the determinant of the coefficients of the unknown quantities vanished.

$$\begin{vmatrix} \left(H_{11}^{(1)} - \Delta_{11} E_{ml}^{(1)}\right) & \left(H_{12}^{(1)} - \Delta_{12} E_{ml}^{(2)}\right) & \dots & \left(H_{1\alpha}^{(1)} - \Delta_{1\alpha} E_{ml}^{(1)}\right) \\ \left(H_{21}^{(1)} - \Delta_{21} E_{ml}^{(1)}\right) & \left(H_{22}^{(1)} - \Delta_{22} E_{ml}^{(1)}\right) & \dots & \left(H_{2\alpha}^{(1)} - \Delta_{2\alpha} E_{ml}^{(1)}\right) \\ \dots & \dots & \dots & \dots \\ \left(H_{\alpha 1}^{(1)} - \Delta_{\alpha 1} E_{ml}^{(1)}\right) & \left(H_{\alpha 2}^{(1)} - \Delta_{\alpha 2} E_{ml}^{(2)}\right) & \dots & \left(H_{\alpha\alpha}^{(1)} - \Delta_{\alpha\alpha} E_{ml}^{(1)}\right) \end{vmatrix} = 0 \qquad (13.135)$$

Applying the condition

$$\nabla_{jl'} = \begin{cases} 1, & \text{if } j = l' \\ 0, & \text{if } j = l' \end{cases}$$

we get

$$\begin{vmatrix} \left(H_{11}^{(1)} - E_{ml}^{(1)}\right) & H_{22}^{(1)} & \dots & H_{1\alpha}^{(1)} \\ H_{21}^{(1)} & \left(H_{22}^{(1)} - E_{ml}^{(1)}\right) & \dots & H_{2\alpha}^{(1)} \\ \dots & \dots & \dots & \dots \\ H_{\alpha 1}^{(1)} & H_{\alpha 2}^{(1)} & \dots & \left(H_{\alpha\alpha}^{(1)} - E_{ml}^{(1)}\right) \end{vmatrix} = 0 \qquad (13.136)$$

Equations (12.135) and (13.136) are often called *scalar equations.*

It is to be noted that in case the secular equation has the form:

$$\begin{vmatrix} \left(H_{11}^{(1)} - E_{ml}^{(1)}\right) & 0 & \dots & 0 \\ 0 & \left(H_{22}^{(1)} - E_{ml}^{(1)}\right) & \dots & 0 \\ \dots & \dots & \dots & \dots \\ 0 & 0 & \dots & \left(H_{\alpha\alpha}^{(1)} - E_{ml}^{(1)}\right) \end{vmatrix} = 0 \qquad (13.137)$$

then the initially assumed functions $\psi_{m1}^{(1)}, \psi_{m2}^{(1)}, \dots, \psi_{m\alpha}^{(1)}$ are the correct zero order functions for the perturbation $H^{(1)}$. The secular equation in which all the elements are zero except along the principal diagonal is said to be in the diagonal form as the algebraic equation equivalent to it is

$$\left(H_{11}^{(1)} - E_{ml}^{(1)}\right)\left(H_{22}^{(1)} - E_{ml}^{(1)}\right)\dots\left(H_{\alpha\alpha}^{(1)} - E_{ml}^{(1)}\right) = 0 \qquad (13.138)$$

with the roots $E_{ml}^{(1)} = H_{11}^{(1)}, H_{22}^{(1)}, \dots, H_{\alpha\alpha}^{(1)}$.

13.9 The first order Stark effect in hydrogen

The splitting of energy levels of an atom caused by a uniform external electric field of strength F directed along the z-axis is called *Stark effect*. It was discovered by a *German Physicist Johannes Start in* 1913 while observing Balmer lines of hydrogen with a field of 10^5 volt/cm. Here, we shall discuss the energy shift of Stark effect by perturbation theory.

The unperturbed Hamiltonian for the hydrogen atom is given by

$$H^{(0)} = -\frac{\hbar^2}{2\mu}\nabla^2 - \frac{e^2}{r}\cdot\frac{1}{4\pi\varepsilon_0} \tag{13.139}$$

where μ is the reduced mass. The perturbation $H^{(1)}$, which is the extra energy of nucleus and electron due to external field F is

$$H^{(1)} = -eFz = -eFr\cos\theta \tag{13.140}$$

where e is the electronic charge and $z = r\cos\theta$ in polar coordinates.

Here, the potential energy and wave functions are spherically symmetric. We know that the wave functions for any spherically symmetric potential energy when expressed in spherical harmonics, have even parity when azimuthal quantum number l is even, and odd when l is odd.

We consider the ground state ($n = l$, $l = 0$, $m = 0$) of hydrogen atom. Since, in hydrogen atom the wave function is spherically symmetric and has the same form for all orientations of the polar axis, there is no degeneracy. For such a non-degenerate, the wave function is

$$\psi_{100} = R_{10}(r)Y_{00}(\theta,\phi) = R_{10}(r)\frac{1}{\sqrt{4\pi}} \tag{13.141}$$

From eq. (13.140) we note that the perturbation $H^{(1)}$ has odd parity. Hence, for the ground state of hydrogen

$$H^{(1)}_{100,100} = \int_{-\infty}^{+\infty} \psi^*_{100}\, H^{(1)}\, \psi_{100}\, d\tau = 0 \text{ or, } H^{(1)}_{100,100} = 0$$

because when ψ_m and ψ_n have the same parity, even or odd and $H^{(1)}$ has the definite parity, then

$$\int \psi^*_m H^{(1)}\, \psi_n\, d\tau = 0$$

Also it can be understood as $\lambda H^{(1)} = -eFr\cos\theta$, i.e., $\lambda = eFH^{(1)} = -r\cos\theta$.

$$H^{(1)}_{nlmn'l'm'} = \iiint \psi^*_{nlm}(-r\cos\theta)\,\psi_{n'l'm'}r^2\sin\theta\, dr\, d\theta\, d\phi$$

The first order perturbation energy in ground state is

$$\lambda H^{(1)}_{100,100} = -eF\iiint r\cos\theta\frac{1}{\pi a_0^3}e^{-2r/a}r^2\sin\theta\, dr\, d\theta\, d\phi = 0 \tag{13.142}$$

Hence, for the ground sate hydrogen atom, there is no first order Stark effect. Hence, we may, say, that only the non-zero matrix element of H' are those that have odd parity, since the perturbation term H' is odd w.r.t. inversion (odd parity).

We now consider the first excited state $n = 2$ for hydrogen atom. Since, $n = 2$, $l = 0, 1$ and $m = 1, 0, -1$. Thus, the quantum numbers l and m have the following combinations: (0, 0), (1, 0), (1, 1), (1, −1), i.e., when $n = 2$ the state of hydrogen is four-fold degenerate. The corresponding wave functions are as follows:

$$\psi_{200} = R_{20}Y_{00} = \sqrt{4\pi}R_{20}(r) \quad \text{even parity} \tag{13.143}$$

$$\psi_{210} = R_{21}Y_{10} = \sqrt{\frac{3}{4\pi}}R_{21}(r)\cos\theta \quad \text{odd parity} \tag{13.144}$$

$$\psi_{211} = R_{21}Y_{11} = \sqrt{\frac{3}{8\pi}}R_{21}(r)\sin\theta e^{i\theta} \quad \text{odd parity} \tag{13.145}$$

$$\psi_{2,1,-1} = R_{21}Y_{1,-1} = \sqrt{\frac{3}{8\pi}}R_{21}(r)\sin\theta e^{-i\theta} \quad \text{odd parity} \tag{13.146}$$

As there is four-fold degeneracy, the secular determinant contains 4 × 4 = 16 elements ψ_{200}, ψ_{211}, ψ_{210}, $\psi_{2,1,-1}$—all have the same energy.

Hence,

$$\begin{vmatrix} H^{(1)}_{200,200} - E^{(1)} & H^{(1)}_{200,211} & H^{(1)}_{200,210} & H^{(1)}_{200,21,-1} \\ H^{(1)}_{211,200} & H^{(1)}_{211,211} - E^{(1)} & H^{(1)}_{211,210} & H^{(1)}_{211,21,-1} \\ H^{(1)}_{210,200} & H^{(1)}_{210,211} & H^{(1)}_{210,210} - E^{(1)} & H^{(1)}_{210,21,-1} \\ H^{(1)}_{21,-1,200} & H^{(1)}_{21,-1,211} & H^{(1)}_{21,-1,210} & H^{(1)}_{21,-1,21,-1} - E^{(1)} \end{vmatrix}$$

We now calculate the elements of the secular determinant

$$H^{(1)}_{200,200} = \int_{-\infty}^{\infty} \psi^*_{200} H^{(1)} \psi_{200}\, d\tau = 0 \tag{13.147}$$

since $\psi^*_{200}\psi_{200}$ has the even parity.

Similarly,

$$H^{(1)}_{211,211}, H^{(1)}_{210,210}, H^{(1)}_{21,-1,21,-1} \text{ are zero.}$$

On the same ground

$$H^{(1)}_{210,210}, H^{(1)}_{211,21,-1}, H^{(1)}_{210,211}, H^{(1)}_{210,21,-1}, H^{(1)}_{21,-1,211} \text{ and } H^{(1)}_{21,-1,210} \text{ are all zero.}$$

Now, we consider the integral

$$H^{(1)}_{200,211} = -\iiint \frac{1}{\sqrt{4\pi}}R_{20}(r)\, r\cos\theta\sqrt{\frac{3}{8\pi}}\sin\theta e^{i\phi}R_{21}(r)\,\psi r^2\, dr\sin\theta\, d\theta\, d\phi \tag{13.148}$$

But,

$$\int_0^{2\pi} e^{i\phi}\, d\phi = 0$$

$\therefore H^{(1)}_{200,211} = 0.$

Similarly,

$$H^{(1)}_{211,210}, H^{(1)}_{200,21,-1} \text{ and } H^{(1)}_{21,-1,200} \text{ are zero.}$$

Thus, the remaining elements are

$$H^{(1)}_{200,200} \text{ and } H^{(1)}_{210,200}.$$

Now,

$$H^{(1)}_{200,210} = \int \psi^*_{200} H^{(1)} \psi_{210}\, d\tau \tag{13.149}$$

where

$$\psi_{200} = \frac{1}{\sqrt{4\pi}} R_{20}(r) = \frac{1}{\sqrt{4\pi}}\left(\frac{1}{2a_0}\right)^{3/2}\left(2-\frac{r}{a_0}\right)\exp\left(-\frac{r}{2a_0}\right)$$

$$\text{and } \psi_{210} = \sqrt{\left(\frac{3}{4\pi}\right)}\cos\theta R_{21}(r) = \sqrt{\left(\frac{3}{4\pi}\right)}\cos\theta\left(\frac{1}{2a_0}\right)^{3/2}\frac{r}{a_0\sqrt{3}}\exp\left(-\frac{r}{2a_0}\right)$$

where a_0 is the radius of the first Bohr orbit.

Now, substituting these values in eq. (13.149), we get

$$H^{(1)}_{200,210} = -\frac{1}{\sqrt{4\pi}}\sqrt{\frac{3}{4\pi}}\left\{\left[\int_0^\infty \left(\frac{1}{2a_0}\right)^{3/2}\left(2-\frac{r}{a_0}\right)\right]\exp\frac{r}{2a_0}\right.$$

$$\left.\times\left(\frac{1}{2a_0}\right)^{3/2}\left[\frac{r}{a_0\sqrt{3}}\exp\left(-\frac{r}{2a_0}\right)r^3 dr\right]\times\left[\int_0^\pi \cos^2\theta\sin\theta\, d\theta\right]\left[\int_0^{2\pi} d\phi\right]\right\} \tag{13.150}$$

where $d\tau = r^2 \sin\theta\, dr\, d\theta\, d\phi$ and $H = -r\cos\theta$.

Again,

$$\int_0^\pi \cos^2\theta\sin\theta\, d\phi = -\frac{1}{3}\left[\cos^3\theta\right]_0^\pi = \frac{2}{3}$$

$$\int_0^{2\pi} d\phi = 2\pi$$

and

$$\int_0^\infty \left(\frac{1}{2a_0}\right)^{3/2}\left(2-\frac{r}{a_0}\right)\exp\left(-\frac{r}{2a}\right)\left(\frac{1}{2a_0}\right)^{3/2}\frac{1}{a_0\sqrt{3}}\times\exp\left(-\frac{r}{a_0}\right)r^4\, dr$$

$$= \left(\frac{1}{2a_0}\right)^3 \frac{1}{a_0\sqrt{3}}\int_0^\infty \left(2-\frac{r}{a_0}\right)\exp\left(-\frac{r}{a_0}\right)r^4\, dr$$

(Let $r/a_0 = \eta$)

$$= \left(\frac{1}{2a_0}\right)^3 \frac{1}{a_0\sqrt{3}}\int_0^\pi (2-\eta)\, e^{-\eta} a_0^4 \delta^4 a_0\, d\eta$$

$$= \frac{1}{8}\frac{a_0^5}{a_0^4\sqrt{3}}\left[2\int_0^\infty \eta^4 e^{-\eta}\, d\eta - \int_0^\infty \eta^5 e^{-\eta}\, d\eta\right]$$

$$= \frac{a_0}{8\sqrt{3}}[2\Gamma(5)-\Gamma(6)]\left[\because \int_0^\infty x^{n-1}e^{-x}\, dx = \Gamma(n)\right]$$

$$= \frac{a_0}{8\sqrt{3}}[48-120] = -\frac{a_0}{8\sqrt{3}}\times 72 = -3\sqrt{3}a_0$$

Substituting these values in eq. (13.143), we get

$$H^{(1)}_{200,210} = -\frac{1}{\sqrt{4\pi}}\sqrt{\frac{3}{4\pi}}\times(-3\sqrt{3}a_0)\times\frac{2}{3}\times 2\pi = 3a_0 \tag{13.151}$$

Similarly,

$$H^{(1)}_{210,100} = 3a_0 \tag{13.152}$$

Substituting all these values in the secular determinant, we have

$$\begin{vmatrix} -E^{(1)} & 0 & 3a_0 & 0 \\ 0 & -E^{(1)} & 0 & 0 \\ 3a_0 & 0 & -E^{(1)} & 0 \\ 0 & 0 & 0 & -E^{(1)} \end{vmatrix} = 0 \tag{13.153}$$

The four roots of the above equation are

$$E^{(1)} = 0, 0, +3a_0 - 3a_0 \tag{13.154}$$

The corresponding perturbed energy values are

$$\lambda E^{(1)} = 0, 0, 3a_0 eF, -3a_0 eF \tag{13.155}$$

Hence, for $n = 2$ state of hydrogen, the energy level is raised by $3a_0 eF$ for one function and is lowered by the same amount for another function. For the remaining two-wave functions the energy levels are the same.

The splitting of energy level is as shown in Fig 13.3.

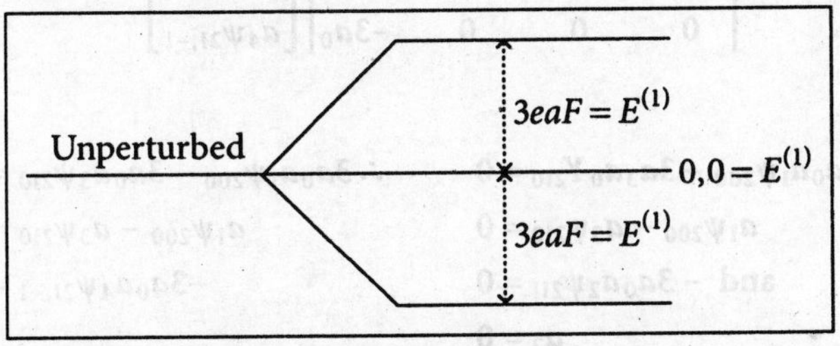

Fig 13.3 Splitting of energy level in electric field

Further, when an electric dipole is placed in electrical field, the change in energy value is $-\vec{p}\cdot\vec{E}$. It suggests that hydrogen atom in $n = 2$ state (i.e., in the first excited state) behaves like an electric dipole of dipole moment $3ea_0$. The dipole in electric field gets itself oriented in three different ways: (i) parallel to electric field (ii) anti-parallel to electric field (iii) two states perpendicular to the field.

Eigenfunctions

The following secular equation is used to evaluate the eigenfunctions

$$\begin{vmatrix} 0 & 0 & 3a_0 & 0 \\ 0 & 0 & 0 & 0 \\ 3a_0 & 0 & 0 & 0 \\ 0 & 0 & 0 & 0 \end{vmatrix} \begin{bmatrix} a_1\psi_{200} \\ a_2\psi_{211} \\ a_3\psi_{210} \\ a_4\psi_{21,-1} \end{bmatrix} = 0 \tag{13.156}$$

$$3a_0 a_3 \psi_{210} = 0 \tag{13.157}$$

$$\text{and } 3a_0 a_1 \psi_{200} = 0 \tag{13.158}$$

From these two, we get $a_1 = 0, a_3 = 0$.

As there is no restriction on a_2 and a_4. The linear combination of wave functions is

$$\psi = a_2\psi_{211} + a_4\psi_{21,-1} = 0 \tag{13.159}$$

Using

$$\int \psi^* \psi \, d\tau = 1$$

we get $a_2^2 + a_4^2 = 0$, $a_2 = 1/\sqrt{2}$, $a_4 = 1\sqrt{2}$ or, $a_2 = 1/\sqrt{2}$, $a_4 = -1/\sqrt{2}$.

Hence, the wave functions corresponding to $E^{(1)} = 0$ are

$$\left.\begin{aligned} \psi &= \frac{1}{\sqrt{2}}[\psi_{211} + \psi_{21,-1}] \\ \psi &= \frac{1}{\sqrt{2}}[\psi_{211} - \psi_{21,-1}] \end{aligned}\right\} \tag{13.160}$$

For $E^{(1)} = -3eFa_0$, we have

$$\begin{vmatrix} -3a_0 & 0 & 3a_0 & 0 \\ 0 & -3a_0 & 0 & 0 \\ 3a_0 & 0 & -3a_0 & 0 \\ 0 & 0 & 0 & -3a_0 \end{vmatrix} \begin{bmatrix} a_1\psi_{200} \\ a_2\psi_{211} \\ a_3\psi_{210} \\ a_4\psi_{21,-1} \end{bmatrix} = 0 \tag{13.161}$$

Solving

$$-3a_0a_1\psi_{200} + 3a_3a_0Y_{210} = 0 \qquad \therefore 3a_0a_1\psi_{200} - 3a_0a_3\psi_{210} = 0$$

$$a_1\psi_{200} - a_3\psi_{210} = 0 \qquad a_1\psi_{200} - a_3\psi_{210} = 0$$

$$\text{and } -3a_0a_2\psi_{211} = 0 \qquad -3a_0a_4\psi_{21,-1} = 0$$

$$a_2 = 0$$

$\therefore a_4 = 0.$

From the condition of normalisation, $a_1^2 + a_3^2 = 1$ or, $a_1 = a_3 = 1/\sqrt{2}$.

$$\psi = \frac{1}{\sqrt{2}}[\psi_{200} - \psi_{210}]$$

corresponding to $E^{(1)} = 3ea_0F$.

Similarly,

$$\psi = \frac{1}{\sqrt{2}}[\psi_{200} + \psi_{210}]$$

corresponding $E^{(1)} = 3ea_0F$.

Only in the degenerate state, energy change proportional to F^2 is possible, every state will have non-vanishing matrix element. In the first case, the permanent dipole moments are affected but in the later case, the dipole moments are induced in the atom in any state. Since, the ground states of all it is expected that an atom or a nucleus in its ground state will not possess a permanent electric dipole moment and none has ever been found experimentally.

13.10 Heitler-London theory of hydrogen molecules

The exchange degeneracy provides us a conception of interaction, which causes atom to be held up together in a stable molecule providing to develop a picture of chemical bond. *Heitler and London* made the highest contribution to the classification of the chemical conception of valency.

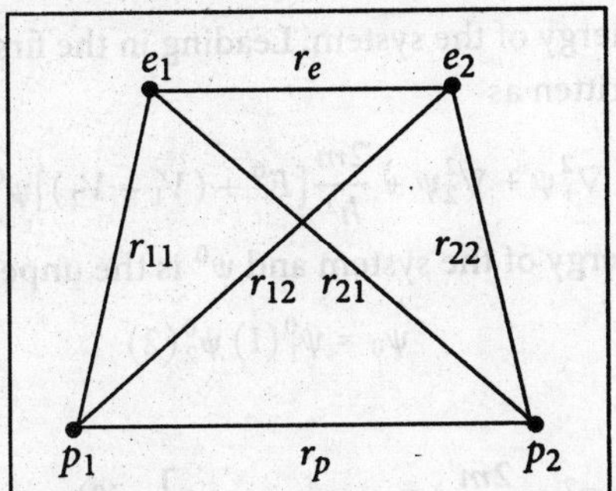

Fig 13.4 The hydrogen molecule

The hydrogen molecule may be pictured as shown in Fig 13.4 where p_1 and p_2 are the positions of two electrons. The separation between them are as marked in the figure. The attractive force between two atoms of molecules is due to sharing of electrons between two atoms. The shared electrons spend more time in between the atoms than anywhere else and as such produces an attractive force.

The covalent bond formation is purely a quantum mechanical process and can be explained using perturbation theory together with the Pauli's exclusion principle.

The electrostatic energy of the hydrogen molecule is given by

$$V = \frac{e^2}{4\pi\varepsilon_0}\left[-\frac{1}{r_{12}}-\frac{1}{r_{22}}-\frac{1}{r_{12}}-\frac{1}{r_{21}}+\frac{1}{r_e}+\frac{1}{r_p}\right] \tag{13.162}$$

$$= -\frac{e^2}{4\pi\varepsilon_0}\left(\frac{1}{r_{11}}+\frac{1}{r_{22}}\right)+\frac{e^2}{4\pi\varepsilon_0}\left(-\frac{1}{r_{12}}-\frac{1}{r_{21}}+\frac{1}{r_e}+\frac{1}{r_p}\right) \tag{13.163}$$

$$= (V_1 + V_2) + V_3 \tag{13.164}$$

where

$$V_1 + V_2 = -\frac{e^2}{4\pi\varepsilon_0}\left(\frac{1}{r_{11}}+\frac{1}{r_{22}}\right) \tag{13.165}$$

representing unperturbed term.

$$V_3 = \frac{e^2}{4\pi\varepsilon_0}\left(-\frac{1}{r_{12}}-\frac{1}{r_{21}}+\frac{1}{r_e}+\frac{1}{r_p}\right) \tag{13.166}$$

representing the perturbed terms.

Now, the Schrödinger equation for hydrogen molecule is given by

$$\nabla_1^2\psi + \nabla_2^2\psi + \frac{2m}{\hbar^2}(E - V)\,\psi = 0 \tag{13.167}$$

$$\text{or, } \nabla_1^2\psi + \nabla_2^2\psi + \frac{2m}{\hbar^2}[E - V_3 - (V_1 + V_2)]\psi = 0 \tag{13.168}$$

with

$$\nabla_1^2 = \frac{\partial^2}{\partial x_1^2} + \frac{\partial^2}{\partial y_1^2} + \frac{\partial^2}{\partial z_1^2}, \quad \nabla_2^2 = \frac{\partial^2}{\partial x_2^2} + \frac{\partial^2}{\partial y_2^2} + \frac{\partial^2}{\partial z_2^2}$$

where (x_1, y_1, z_1) and (x_2, y_2, z_2) are the position coordinates of the two electrons.

In eq. (13.160) E is the total energy of the system. Leading in the first instance, the perturbation term, the Schrödinger equation can be written as

$$\nabla_1^2\psi + \nabla_2^2\psi + \frac{2m}{\hbar^2}[E^0 + (V_1 + V_2)]\psi^0 = 0 \tag{13.169}$$

where E^0 is the unperturbed energy of the system and ψ^0 is the unperturbed state wave function where

$$\psi_0 = \psi_1^0(1)\,\psi_2^0(2) \tag{13.170}$$

Substituting in eq. (13.169)

$$\left[\nabla_1^2 + \nabla_2^2 + \frac{2m}{\hbar^2}\{E - (V_1 + V_2)\}\right]\psi_1^{(0)}(1)\,\psi_2^{(0)}(2) = 0 \tag{13.171}$$

$$\psi_2^{(0)}\nabla_1^2\psi_1^{(0)} + \psi_1^{(0)}\nabla_2^2\psi_2^{(0)} + \frac{2m}{\hbar^2}[E - (V_1 + V_2)]\,\psi_1^{(0)}\psi_2^{(0)} = 0 \tag{13.172}$$

The Schrödinger equation for two atoms independently can be written as

$$\nabla_1^2\psi_1^{(0)} + (E_1 - V_1)\,\psi_1^{(0)} = 0 \tag{13.173}$$

$$\nabla_2^2\psi_2^{(0)} + (E_2 - V_1)\,\psi_2^{(0)} = 0 \tag{13.174}$$

Multiplying eqs (13.173) and (13.174) by $\psi_2^{(0)}$ and $\psi_1^{(0)}$ respectively and, then adding, we get

$$\psi_2^{(0)}\nabla_1^2\psi_1^{(0)} + \psi_1^{(0)}\nabla_2^2\psi_2^{(0)} + [(E_1 + E_2) - (V_1 + V_2)]\,\psi_1^{(0)}\psi_2^{(0)} = 0 \tag{13.175}$$

Comparing eqs (13.172) and (13.175), we get

$$E_1 + E_2 = E^{(0)} = 2E_{\mathrm{H}} \tag{13.176}$$

As the two electrons are indistinguishable, the unperturbed wave function may be written as

$$\psi_s^{(0)} = \psi_1^{(0)}(2)\,\psi_2^{(0)}(1) \tag{13.177}$$

The linear combinations of wave functions given in eqs (13.170) and (13.177) are also solutions of Schrödinger equation and are as follows:

$$\psi_s^{(0)} = N_s\left[\psi_1^{(0)}(1)\,\psi_2^{(0)}(2) + \psi_1^{(0)}(2)\,\psi_2^{(0)}(1)\right] \tag{13.178}$$

$$\psi_A^{(0)} = N_A\left[\psi_1^{(0)}(1)\,\psi_2^{(0)}(2) - \psi_1^{(0)}(2)\,\psi_2^{(0)}(1)\right] \tag{13.179}$$

where $\psi_s^{(0)}$ and $\psi_A^{(0)}$ are symmetric and antisymmetric wave functions with N_s and N_A are the corresponding normalisation constant.

N_s and N_A come out to be

$$N_s = \frac{1}{\sqrt{2 + 2\Delta}} \tag{13.180}$$

$$N_A = \frac{1}{\sqrt{2 - 2\Delta}} \tag{13.181}$$

with

$$\Delta = \int \psi_1^{(0)}(1)\,\psi_2^{(0)}(2)\,\psi_1^{(0)}(2)\,\psi_2^{(0)}(2)\,d\tau_1\,d\tau_2 \tag{13.182}$$

Now, taking the contribution of perturbation into account, we get

$$E' = \frac{\iint \left[\psi_1^{(0)}(1)\,\psi_2^{(0)}(2) \pm \psi_1^{(0)}(2)\,\psi_2^{(0)}(1)\right] \hat{H} \left[\psi_1^{(0)}(1)\,\psi_2^{(0)}(2) \pm \psi_1^{(0)}(2)\,\psi_2^{(0)}(1)\right] d\tau}{\iint \left[\psi_1^{(0)}(1)\,\psi_2^{(0)}(2) \pm \psi_1^{(0)}(2)\,\psi_2^{(0)}(1)\right]\left[\psi_1^{(0)}(1)\,\psi_2^{(0)}(2) \pm \psi_1^{(0)}(2)\,\psi_2^{(0)}(1)\right] d\tau}$$

This gives

$$E' = \frac{C \pm \Delta}{1 \pm \Delta} \tag{13.183}$$

where

$$C = \iint \psi_1(1)\,\psi_2(2)\hat{H}'\psi_1(1)\,\psi_2(2)\,d\tau$$

$$A = \iint \psi_1(2)\,\psi_2(1)\hat{H}'\psi_1(2)\,\psi_2(1)\,d\tau = \text{exchange integral}$$

With this, energy in two cases can be written as

$$E_s = 2E_{\mathrm{H}} + \frac{C + A}{1 + \Delta} \tag{13.184}$$

$$\text{and } E_A = 2E_{\mathrm{H}} + \frac{C - A}{1 - \Delta} \tag{13.185}$$

The Coulomb integral

$$C = \frac{e^2}{4\pi\varepsilon_0} \iint \psi_1^{(0)}(1)\,\psi_2^{(0)} \left(\frac{1}{r_p} + \frac{1}{r_p} - \frac{1}{r_{12}} - \frac{1}{r_{21}}\right) \psi_1^{(0)}(1)\,\psi_2^{(0)}(2)\,d\tau \tag{13.186}$$

Physically, the terms in above integral represent the average potential energy of two protons, two electrons, first proton and second electron and second proton and first electron, respectively.

The exchange integral is

$$A = \int \psi_1^{(0)}(1)\,\psi_2^{(0)}(2)\,H\psi_1^{(0)}(2)\,\psi_2^{(0)}(1)\,d\tau \tag{13.187}$$

This implies that the electron (1) is sometimes in state (1) and sometimes in state (2). This can be possible when electrons form a cloud-like structure around the nucleus.

Physical interpretation

The hydrogen molecule contains two electrons instead of single election H_2^+. Both the electrons can share the same orbit i.e, described by the same wave function when they have ***antiparallel spins.***

With two electrons to contribute to the bond, H_2 appears to be twice as stable as H_2^+ with a bond energy of 2 × 2.65 = 5.63 eV . But, due to electron-electron repulsion the bond energy comes out to 4.5 eV instead of 5.3 eV. The bond length is 0.074 nm.

Further, when the total wave function is antisymmetric (= spatially symmetric and spin antisymmetric) the wave function is more stable [Fig 13.5]. Physically, when spins are parallel it leads to more dominating repulsion in accordance with exclusion principle making the state less stable.

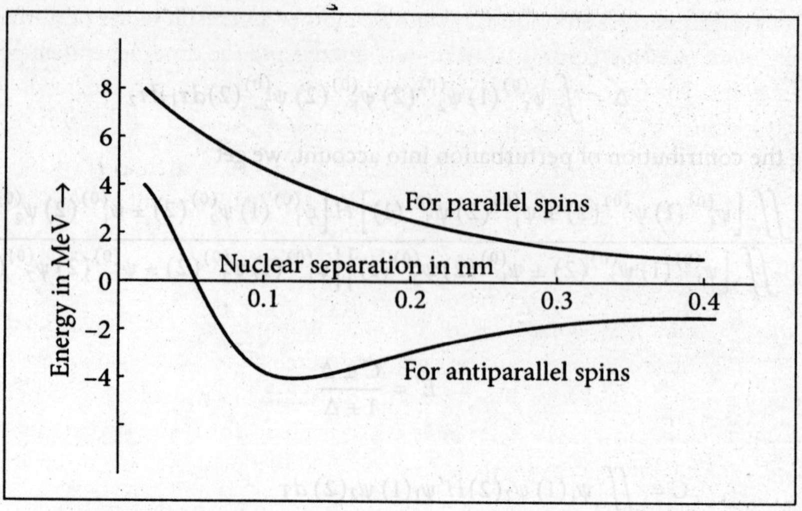

Fig 13.5 Potential energy vs nuclear separation

13.11 Answers of some important conceptual questions

13.1. *What do you mean by perturbation theory?*

Ans. In quantum mechanics (as in classical mechanics) exactly solvable problems are rare and one must frequently resolve approximation. Approximations are therefore expected to play an important part. That is why various methods of approximate solution of the wave equation have been devised, leading to the more or less accurate approximate evaluation of energy values and wave functions. Of these methods the first and in many respects the most interesting is simple wave mechanical perturbation theory developed by *Schrödinger*. Perturbation theories are of two kinds:

(i) time-independent or stationary perturbation theory

(ii) time-dependent perturbation theory.

The theory of dependent perturbation was developed by *Dirac* and is often called the theory of the variation of constants.

13.2. *Explain time-independent and time-dependent perturbation theories.*

Ans. Time-independent or stationary perturbation theory is concerned finding the changes in the energy levels and eigenfunctions of a system when a small disturbance is applied. The Hamiltonian may be regarded as broken up into two parts, one of which is large and characterises a system for which wave equation can be solved exactly while the other part is small and can be treated as perturbation. The effect of the perturbation is to shift the energy levels, of course the arrangement of energy levels remains the same. Thus, the perturbation introduces some additional terms in the Hamiltonian of the unperturbed system. The perturbed Hamiltonian can be written as $H = H^{(0)} + H^{(1)}$, where $H^{(0)}$ and $H^{(1)}$ are respectively the unperturbed and perturbed Hamiltonian.

The perturbation theory of a system for which Hamiltonian depends explicitly on time is called the *time-dependent perturbation theory or method of variation of constant*. They theory of time-dependent perturbation was developed by Dirac and permits us to discuss the probability of transition of a system from one unperturbed stationary state to another as a result of the perturbation. In spectral analysis the main question is the quantisation of energy and the time-dependent perturbation finds wide applications. The time-dependent perturbation theory is frequently applied in nuclear reactions and disintegrations.

The Hamiltonian of the system is time-dependent and can be written as $H' = H_0 + XH'(t)$, where H^0 is the original time-independent Hamiltonian and $\lambda M'(t)$ is a small perturbation depending one time.

13.3. *What is exchange degeneracy?*

Ans. We can get source of degeneracy if there is more than one particle of given kind. As helium atom has two similar electrons, if these two electrons are interchanged, we obtain a wave function which in general not the same as one with which we stored. But, the energy of the system cannot change in interchanging of electrons while all electrons are equivalent. The two-wave functions must, therefore, compound to degenerate energy levels.

13.12 Solved problems

» ***Example* 13.1.** A particle is confined in an infinitely deep potential well inside a cube of dimension L. It is given a perturbation $H' = kx^2$. Apply the first order perturbation theory to first excited state.

Solution: We have the energy eigenvalue and eigenfunctions of the nth state are as under:

$$E_n^{(0)} = \frac{\pi^2\hbar^2}{2mL^2}\left(n_x^2 + n_y^2 + n_z^2\right),\ n_x, n_y, n_z = 1.2.3.\ldots$$

$$\psi = \left(\frac{8}{L^3}\right)^{1/2} \sin\frac{n_x\pi}{L}x \sin\frac{n_y\pi}{L}y \sin\frac{n_z\pi}{L}z$$

In the first excited state [for first excited state, either of three values is 2 and others are 1]

$$E_1^{(0)} = \frac{6\pi^2\hbar^2}{8mL^2}$$

$$\psi_1 = \left(\frac{8}{L^3}\right)^{1/2} \sin\frac{2\pi x}{L} \sin\frac{2\pi y}{L} \sin\frac{2\pi z}{L}\ [n_x = 2, n_y = 1, n_z = 1]$$

$$\psi_2 = \left(\frac{8}{L^3}\right)^{1/2} \sin\frac{\pi x}{L} \sin\frac{2\pi y}{L} \sin\frac{2\pi z}{L}\ [n_x = 1, n_y = 2, n_z = 1]$$

$$\psi_3 = \left(\frac{8}{L^3}\right)^{1/2} \sin\frac{\pi x}{L} \sin\frac{\pi y}{L} \sin\frac{2\pi z}{L}\ [n_x = 1, n_y = 1, n_z = 2]$$

Now, according to perturbation theory, matrix element

$$H'_{ij} = \int \psi_i^* H' \psi_j d\tau = k \iiint \psi_i^* x^2 \psi_j dx\,dy\,dz$$

$$H'_{ij} = 0,\ \text{for } i \neq j$$

$$H'_{11} = kL^2\left(\frac{1}{3} - \frac{1}{8\pi^2}\right) = h_1$$

$$H'_{22} = kL^2\left(\frac{1}{3} - \frac{1}{2\pi^2}\right) = h_2$$

Hence, energy values can be calculated by

$$\begin{vmatrix} h_1 - E^{(1)} & 0 & 0 \\ 0 & H_2 - E^{(1)} & 0 \\ 0 & 0 & h_2 - E^{(1)} \end{vmatrix} = 0$$

The roots of this determinant are h_1, h_2 and h_2. It thus suggests that the degeneracy is partially lifted. The energy values in the perturbed state are

$$E_1^{(0)} + h_1,\ E_1^{(0)} + h_2,\ E_1^{(0)} + h_2$$

» ***Example* 13.2.** A linear harmonic oscillator of charge e is perturbed by an electric field F in the positive x-direction. The first order change in the energy level is zero. Why? Calculate the energy up to second order.

Solution: Here, the potential energy of the oscillator is $\frac{1}{2}kx^2$ + a small quantity eFx, where e is the charge of the oscillator. The Schrödinger equation is

$$\frac{\partial^2 \psi}{\partial x^2} + \frac{2m}{\hbar^2}\left(E - \tfrac{1}{2}kx^2 + eFx\right)\psi = 0 \tag{i}$$

The term eFx can be regarded as perturbation.

Thus,

$$\lambda H^{(1)} = -eFx \tag{ii}$$

Setting

$$\lambda = -eFx \tag{iii}$$

then $H^{(1)} = x$.

The unperturbed wave function in a linear harmonic are given as

$$\psi_n(\xi) = N_n\, H_n(\xi)\exp\left(-\frac{\xi^2}{2}\right) \tag{iv}$$

where

$$N_n = \left\{\frac{\alpha}{\pi^{1/2}2^n n!}\right\}^{1/2},\ \alpha = \sqrt{\left(\frac{m\omega}{\hbar}\right)} \text{ and } \omega = \sqrt{\frac{k}{m}}$$

When perturbation is applied, the matrix elements $H_{mn}^{(1)}$ are given as

$$H_{mn}^{(1)} = \int_{-\infty}^{+\infty}\left\{\frac{\alpha}{\pi^{1/2}2^m m!}\right\}^{1/2} H_m(\xi)\exp\left(-\frac{\xi^2}{2}\right)\left\{\frac{\alpha}{\pi^{1/2}2^n n!}\right\}^{1/2} \times H_n(\xi)\exp\left(-\frac{\xi^2}{2}\right)dx \tag{v}$$

$$= \int_{-\infty}^{+\infty}\frac{1}{\alpha}\left\{\frac{1}{\pi 2^{m+n} n!\, m!}\right\}^{1/2} H_m(\xi)\,\xi H_n(\xi)\exp(-\xi^2)\,d\xi$$

Putting

$$\xi H_n(\xi) = \tfrac{1}{2}H_{n+1}(\xi) + nH_{n-1}(\xi)$$

we get

$$H_{mn}^{(1)} = \frac{1}{\alpha}\left\{\frac{1}{\pi 2^{m+n} m!\, n!}\right\}^{1/2}\int_{-\infty}^{+\infty} H_m(\xi)\times\left[\tfrac{1}{2}H_{n+1}(\xi) + nH_{n-1}(\xi)\right]\exp(-\xi^2)\,d\xi \tag{vi}$$

The right-hand side of (vi) vanishes except when $m = n + 1$ or $m = n - 1$.

Also

$$\int_{-\infty}^{+\infty} \left\{ H_{n+1}(\xi) \exp\left(-\frac{\xi^2}{2}\right) \right\}^2 d\xi = (\sqrt{\pi}) 2^{n+1} (n+1)! \tag{vii}$$

Thus,

$$H^{(1)}_{n+1,n} = \frac{1}{\alpha} \left\{ \frac{1}{\pi 2^{2n+1} (n+1)!\, n!} \right\}^{1/2} \times \frac{1}{2} (\sqrt{\pi}) 2^{n+1} (n+1)! = \frac{1}{\alpha} \sqrt{\left(\frac{n+1}{2}\right)} \tag{viii}$$

For $m = n - 1$, we have

$$H^{(1)}_{n-1,n} = \frac{1}{\alpha} \left\{ \frac{1}{\pi 2^{n-1} (n-1)!\, n!} \right\}^{1/2} n (\sqrt{\pi}) 2^{n-1} (n-1)! = \frac{1}{\alpha} \sqrt{\frac{n}{2}} \tag{ix}$$

So

$$H^{(1)}_{m,n} = \begin{cases} \frac{1}{\alpha}\sqrt{\left(\frac{n+1}{2}\right)}, & \text{for } m = n+1 \\ 0, & \text{for } m \neq n+1 \\ \frac{1}{\alpha}\sqrt{\left(\frac{n}{2}\right)}, & \text{for } m = n-1 \end{cases} \tag{x}$$

Using (x), we get $\boldsymbol{E_n^{(1)} = 0}$.

Here, the first order energy correction is zero as the perturbing Hamiltonian $H^{(1)}$ is an odd function of x and as such the total integrand giving the value of $E_n^{(1)}$ is also an odd function of x. Hence, its value comes out to be zero. Physically we can also, say, that as the first order correction depends only on x, its expectation value, i.e., the first order energy correction is zero.

Now,

$$E_n^{(2)} = \frac{-\frac{1}{\alpha^2}\left(\frac{n+1}{2}\right) + \frac{1}{\alpha^2}\left(\frac{n}{2}\right)}{\hbar\omega} \tag{xi}$$

$$\text{or, } E_n^{(2)} = \sum_m{}' \frac{\int \psi_m^{(0)*} H^{(1)} \psi_n^{(0)}\, d\tau \int \psi_n^{(0)*} H^{(1)} \psi_m^{(0)}\, d\tau}{E_n^{(0)} - E_m^{(0)}} \tag{xii}$$

$$= -\frac{1}{2\alpha^2\hbar^2\omega} = -\frac{1}{2m\omega^2} \tag{xiii}$$

As

$$E_n^{(0)} = \left(n + \frac{1}{2}\right)\hbar\omega \tag{xiv}$$

$$E_n = E_n^{(0)} + \lambda^r E_n^{(2)} \tag{xv}$$

We have

$$\boldsymbol{E_n = \left(n + \frac{1}{2}\right)\hbar\omega - \frac{e^2F^2}{2m\omega^2}} \tag{xvi}$$

» ***Example* 13.3.** When the perturbing term in linear harmonic oscillator problem is $\left(\frac{1}{2}\right) ax^2$, use perturbation theory to obtain correction in energy eigenvalue up to first order. Show that this problem can also be solved exactly.

Solution: The Hamiltonian in one-dimensional harmonic oscillator is given by

$$H = \frac{p^2}{2m} + \frac{1}{2}kx^2 + \frac{1}{2}ax^2 \quad \text{(i)}$$

$$\text{or, } H = H^0 + H^{(1)} \quad \text{(ii)}$$

where $H^{(1)} = \frac{1}{2}ax^2$.

First order correction

$$E_n^{(1)} = \langle n|H^{(1)}|n\rangle \quad \text{(iii)}$$

$$\text{or, } E_n^{(1)} = \int \psi_n^{(0)} \frac{1}{2}ax^2 \, \psi_n^{(0)} \, d\tau \quad \text{(iv)}$$

The unperturbed wave function in nth state is

$$\psi_n = N_n \, H_n(\alpha x) \, e^{-\alpha^2 x^2/2} \quad \text{(v)}$$

$$\text{and } N_n = \left(\frac{\alpha}{\pi^{1/2} 2^n n!}\right)^{1/2} \quad \text{(vi)}$$

$$\therefore E_n^{(1)} = \frac{N_n^2 \alpha}{2} \int H_n(\alpha x) \, e^{-\alpha^2 x^2/2} H_{n(\alpha x)} e^{-\alpha^2 x^2/2} \, dx \quad \text{(vii)}$$

Let us take

$$I = \frac{1}{\alpha^3} \int \eta^2 H_n^2(\eta) e^{-\eta^2} \, d\eta \quad \text{(viii)}$$

Putting $\eta = \alpha x$, we get

$$I = \frac{1}{\alpha^3} \int \left[\frac{1}{4} H_n(\eta) \, H_{n+2}(\eta) + \left(n + \frac{1}{2}\right) H_n^2(\eta) + n(n-1) \, H_n(\eta) \, H_{n-2}(\eta)\right] e^{-\eta^2} \, d\eta$$

$$\text{or, } I = \frac{1}{\alpha^3}\left(n + \frac{1}{2}\right) 2^n n! \, \pi^{1/2} \quad \text{(ix)}$$

Now,

$$E_n^{(1)} = \frac{\alpha}{\pi^{1/2} 2^n n!} \frac{a}{2} \frac{1}{\alpha^3} \left(n + \frac{1}{2}\right) 2^n n! \, \pi^{1/2}$$

$$\left[\because E_n' = \langle n|\hat{H}'|n\rangle = \int \psi_n^{(0)*} \hat{H}' \psi_n^{(0)} \, d\tau \text{ and with } \psi_n^{(0)} = N_n \, H_n e^{-\alpha^2 x^2/2}\right]$$

$$\therefore E_n^{(1)} = \frac{\alpha}{\pi^{1/2} 2^n n!} \frac{a}{2} \cdot \frac{1}{\alpha^3} \left(n + \frac{1}{2}\right) 2^n n! \, \pi^{1/2}$$

$$E_n^{(1)} = \frac{a(2n+1)}{4\alpha^2}$$

$$\text{or, } E_n^{(1)} = \frac{a}{4}(2n+1)\left(\frac{\hbar}{mk}\right)^{1/2} \quad \text{(x)}$$

Energy corrected up to first order

$$E_n = \left(n + \frac{1}{2}\right)\hbar\omega + \frac{a}{4}(2n+1)\left(\frac{\hbar^2}{mk}\right)^{1/2}$$

$$\text{or, } E_n = \left(n + \frac{1}{2}\right)\hbar\omega\left[1 + \frac{a}{2k}\right] \quad \text{(xi)}$$

Exact solution possible

We have

$$H^{(0)} = \frac{p^2}{2m} + \frac{1}{2}kx^2 \tag{xii}$$

$$H = \frac{p^2}{2m} + \frac{1}{2}kx^2 + \frac{1}{2}ax^2 \tag{xiii}$$

$$H = \frac{p^2}{2m} + \frac{1}{2}(k+a)x^2 \tag{xiv}$$

Thus,

$$\omega = \sqrt{\frac{k+a}{m}} = \left(\frac{k}{m}\right)^{1/2}\left(1+\frac{a}{k}\right)^{1/2} \tag{xv}$$

$$\omega = \omega_0\left(1+\frac{a}{2k} - \frac{a^2}{8k^2} + \ldots\right) \tag{xvi}$$

where $\omega_0 = \sqrt{k/m}$.

The corrected value of energy is

$$E_n = \left(n+\frac{1}{2}\right)\hbar\omega \tag{xvii}$$

$$= \left(n+\frac{1}{2}\right)\hbar\left(1+\frac{a}{2k} - \frac{1}{8}\frac{a^2}{k^2}\right) \tag{xviii}$$

The energy corrected up to first order is

$$E_n = \left(n+\frac{1}{2}\right)\hbar\left(1+\frac{a}{2k}\right) \tag{xix}$$

and up to second order

$$E_n = \left(n+\frac{1}{2}\right)\hbar\left(1+\frac{a}{2k} - \frac{a^2}{8k^2}\right) \tag{xx}$$

Equation (xix) is exactly the same as derived on the basis of perturbation theory.

Example 13.4. Calculate the first order energy correction for one-dimensional harmonic whose Hamiltonian is

$$H = \frac{p^2}{2m} + \frac{1}{2}kx^2 + ax^4$$

and the ground state wave function is

$$\psi_0 = \left(\frac{k}{\pi\hbar\omega}\right)^{1/4}\exp\left(-\frac{kx^2}{2\hbar\omega}\right)$$

Further, show that the value of the ground state energy can be written as

$$E = \frac{\hbar\omega}{2}\left(1+\frac{3\alpha}{2}\frac{\hbar\omega}{k^2}\right)$$

***Solution*: For the one-dimensional harmonic oscillator, the perturbing Hamiltonian is given by**

$$H^{(1)} = ax^4$$

$$\therefore E_0^{(1)} = \int_{-\infty}^{+\infty} \psi_0^{(0)*} ax^4 \psi_0^{(0)} dx$$

Substituting for ψ_0

$$E_0^{(1)} = \left(\frac{k}{\pi\hbar\omega}\right)^{1/2} \int_{-\infty}^{+\infty} \exp\left(-\frac{kx^2}{2\hbar\omega}\right) ax^4 \exp\left(-\frac{kx^2}{2\hbar\omega}\right) dx$$

$$= \left(\frac{k}{\pi\hbar\omega}\right)^{1/2} a \int_{-\infty}^{+\infty} \exp\left(-\frac{kx^2}{\hbar\omega}\right) x^4 dx$$

$$= 2\left(\frac{k}{\pi\hbar\omega}\right)^{1/2} a \int_{0}^{\infty} \exp\left(-\frac{kx^2}{\hbar\omega}\right) x^4 dx$$

Using the standard integral, the value of

$$E_n^{(1)} = 2\left(\frac{k}{\pi\hbar\omega}\right)^{1/2} a \frac{1}{2} \frac{\Gamma\left(\frac{4+1}{2}\right)}{\left(\frac{k}{\hbar\omega}\right)^{5/2}}$$

$$\left[\because \int_0^{\infty} x^m e^{-\alpha x^n} dx = \frac{1}{n} \frac{\Gamma\left(\frac{m+1}{n}\right)}{\alpha^{(m+1)/n}},\ m \geqslant 0 \text{ or, } n \geqslant 0\right]$$

$$= \frac{2}{\pi^{1/2}} \left(\frac{\hbar\omega}{k}\right)^2 a \frac{1}{2} \cdot \frac{3}{2} \frac{1}{2} \pi^{1/2}$$

$$\text{or, } E_n^{(1)} = \left(\frac{\hbar\omega}{k}\right)^2 \frac{3a}{4}$$

» ***Example* 13.5.** For two harmonic oscillators placed along the same line, the combined Hamiltonian is given by

$$H = -\frac{\hbar^2}{2m}\left(\frac{\partial^2}{\partial x_1^2} + \frac{\partial^2}{\partial x_2^2}\right) + \frac{1}{2}k\left(x_1^2 + x_2^2\right)$$

The oscillators interact and as such the change in Hamiltonian is due to interaction is

$$H^{(1)} = ax_1x_2.$$

Calculate the first order shift in the energy level of the first excited state. Show that this problem can be solved exactly.

Solution: The composite eigenfunction of the two oscillators is

$$\psi_{n_1 n_2}^{(0)} = \psi_{n_1}^{(0)} \psi_{n_2}^{(0)} \qquad \text{(i)}$$

$$= NH_{n1}(\eta_1) H_{n2}(\eta_2) \exp\left(-\frac{\eta_1^2 + \eta_2^2}{2}\right) \qquad \text{(ii)}$$

where $n_1, n_2 = 0, 1, 2, 3, \ldots$

$$\eta_1 = \alpha x_1,\ \eta_2 = \alpha x_2$$

$$\alpha = \left(\frac{mk}{\hbar}\right)^{1/4} \text{ and } N = \frac{\alpha}{\pi^{1/2}}\left[\frac{1}{2^{n_1} n_1!} 2^{n_2} n_2!\right]^{1/2}$$

The eigenvalues are

$$E^{(0)}_{n_1 n_2} = E^{(0)}_{n_1} + E^{(0)}_{n_2} = \left(n_1 + \frac{1}{2}\right)\hbar\omega + \left(n_2 + \frac{1}{2}\right)\hbar\omega = (n_1 + n_2 + 1)\,\hbar\omega \qquad \text{(iii)}$$

The first excited state is two-fold degenerate as $n_1 = 0,\ n_2 = 1,\ n_1 = 1,\ n_2 = 0$ and eigenvalue is $2\hbar\omega$ from (iii).

Now, we assume

$$u_1 = \psi^{(0)}_{01} \qquad \text{(iv)}$$

$$u_2 = \psi^{(0)}_{2} \qquad \text{(v)}$$

Now,

$$H^{(1)}_{mn} = \int u^*_m H^{(1)} u_n \, dx_1\, dx_2 \text{ with } m, n = 1, 2 \qquad \text{(vi)}$$

$$H^{(1)}_{11} = H^{(1)}_{22} = 0 \qquad \text{(vii)}$$

and

$$H^{(1)}_{12} = \int_{-\infty}^{+\infty}\int_{-\infty}^{+\infty} u^*_1\, a\frac{\eta_1\eta_2}{\alpha^2} u_2 \frac{d\eta_1\, d\eta_2}{\alpha^2} \qquad \text{(viii)}$$

$$H^{(1)}_{12} = \frac{a}{\alpha^4}\left(\frac{\alpha}{\sqrt{2\pi}}\right)^2 \int_{-\infty}^{\infty} 2\eta_1^2 e^{-\eta_1^2}\, d\eta_1 \int_{-\infty}^{\infty} 2\eta_2^2\, e^{-\eta_2^2}\, d\eta_2 \qquad \text{(xi)}$$

$$= \frac{a}{\alpha^4}\left(\frac{\alpha}{\sqrt{2\pi}}\right)^2 4\int_{-\infty}^{+\infty} \eta_1^2 e^{-\eta_1^2}\, d\eta_1 \int_{-\infty}^{+\infty} \eta_2^2 e^{-\eta_2^2} \qquad \text{(x)}$$

$$= \frac{a}{\alpha^4}\left(\frac{\alpha}{\sqrt{2\pi}}\right)^2 8\int_{0}^{\infty} \eta_1^2 e^{-\eta_1^2}\, d\eta_1 \int_{0}^{\infty} \eta_2^2\, e_2^{-\eta_2^2}\, d\eta_2 \qquad \text{(xi)}$$

Now, from

$$\int_0^\infty x^m e^{-ax^2} dx = \frac{\frac{1}{n}\Gamma\left(\frac{m+1}{n}\right)}{\alpha^{(m+n)/n}} = \frac{a}{\alpha^4}\left(\frac{a}{\sqrt{2\pi}}\right)^2 4\frac{1}{2}\pi^{1/2}\frac{1}{2}\pi^{1/2}$$

$$\left.\begin{aligned} H^{(1)}_{12} &= \frac{a}{2\alpha^2} \\ \text{Similarly, } H^{(1)}_{21} &= \frac{a}{2\alpha^2} \end{aligned}\right\} \qquad \text{(xii)}$$

$\therefore$ the secular equation becomes

$$\begin{vmatrix} H_{11} & H_{12} \\ H_{21} & H_{22} \end{vmatrix} = 0 \text{ or, } \begin{vmatrix} E^{(1)} & \dfrac{a}{2\alpha^2} \\ \dfrac{a}{2\alpha^2} & E^{(1)} \end{vmatrix} = 0 \qquad \text{(xiii)}$$

$$\therefore E^{(1)} = \pm\frac{a}{2\alpha^2},\ E^{(1)}_1 = \pm\frac{a}{2k}\hbar\omega \qquad \text{(xiv)}$$

Hence,

$$E_1 = E^{(0)}_0 + E^{(1)} \text{ or, } E_1 = 2\hbar\omega \pm \frac{a}{2k}\hbar\omega = \left(2 \pm \frac{a}{2k}\right)\hbar\omega \qquad \text{(xv)}$$

Exact solution

The Hamiltonian is

$$H = -\frac{\hbar^2}{2m}\left(\frac{\partial}{\partial x_1^2} + \frac{\partial^2}{\partial x_2^2}\right) + \frac{1}{2}k(x_1^2 + x_2^2) + a x_1 x_2$$

Putting

$$y_1 = \frac{x_1 + x_2}{\sqrt{2}},\ y_2 = \frac{x_1 - x_2}{\sqrt{2}} \tag{xvi}$$

$$\text{and } x_1 = \frac{y_1 + y_2}{\sqrt{2}},\ x_2 = \frac{y_1 - y_2}{\sqrt{2}}$$

we get

$$H = -\frac{\hbar^2}{2m}\left(\frac{\partial^2}{\partial x_1^2} + \frac{\partial^2}{\partial x_2^2}\right) + \frac{1}{2}(k+a)y_1^2 + \frac{1}{2}(k-a)y_2^2 \tag{xvii}$$

In eq. (xii), k has been replaced by $(k + a)$ and $(k - 1)$ respectively and hence the two frequencies are

$$\omega_1 = \sqrt{\left(\frac{k+a}{m}\right)} = \omega_0\left(1 + \frac{a}{k}\right)^{1/2}$$

$$\omega_2 = \sqrt{\left(\frac{k+a}{m}\right)} = \omega_0\left(1 - \frac{a}{k}\right)^{1/2}$$

and the energy values will be

$$E = \left(n_1 + \frac{1}{2}\right)\hbar\omega_0\left(1 + \frac{a}{k}\right)^{1/2} + \left(n_2 + \frac{1}{2}\right)\hbar\omega_0\left(1 - \frac{a}{k}\right)^{1/2}$$

This gives the same result as obtained on the basis of perturbation theory. Thus, exact solution is possible.

» ***Example* 13.6.** Assume that the nuclear charge is distributed uniformly in a sphere of radius R equal to nuclear radius. Use first order perturbation theory to calculate the change in energy value.

Solution: The charge density inside the nucleus is given by

$$\rho = \frac{Ze}{\frac{4\pi}{3}R^3} \tag{i}$$

The electric field-inside and outside the nucleus are

$$E_i = \frac{Zer}{4\pi\varepsilon_0 R^3} \tag{ii}$$

$$E_o = \frac{Ze}{4\pi\varepsilon_0 r^2} \tag{iii}$$

The forces on the electron are

$$F_i = -\frac{Ze^2 r}{4\pi\varepsilon_0 R^3},\ 0 \leqslant r \leqslant R \tag{iv}$$

$$F_o = -\frac{Ze^2}{4\pi\varepsilon_0 r^2},\ r > 0 \tag{v}$$

Potential energy

$$V = -\int F dr = -\frac{Ze^2}{4\pi_0 R}\left(\frac{3}{2} - \frac{1}{2}\frac{r^2}{R^2}\right), \; 0 \leqslant r \leqslant R$$

$$= -\frac{Ze^2}{4\pi\varepsilon_0 r}, \; r \geqslant R$$

Assume that the potential energy is zero at infinity and it is continuous at the boundary of the nucleus. The perturbation is given by

$$H' = -\frac{Ze^2}{4\pi\varepsilon_0 R}\left(\frac{3}{2} - \frac{r^2}{R^2}\right) + \frac{Ze^2}{4\pi\varepsilon_0 r}, \text{ for } 0 \leqslant r \leqslant R \tag{vi}$$

$$= 0, \text{ for } r \geqslant R \tag{vii}$$

Hence,

$$E_n^{(1)} = \int_0^k r^2\, dr \int_0^\pi \sin\theta\, d\theta \int_0^{2\pi} d\phi\, \psi_n^{(0)} H' \psi_n^{(0)} d\tau$$

As the radius of the nucleus is much smaller than the Bohr radius we can assume that wave function remains constant over entire region. For convenience, we can replace it by its value at $r = 0$.

Consequently, we have

$$E_n^{(1)} = R_{nl}^{2(0)} \int_0^R \frac{1}{4\pi\varepsilon_0}\left\{-\frac{Ze^2}{R}\left(\frac{3}{2} - \frac{1}{2}\frac{r^2}{R^2}\right) + \frac{Ze^2}{r}\right\} r^2\, dr \iint Y_{lm}(\theta,\phi)\sin\theta\, d\theta\, d\phi \tag{viii}$$

$$= \frac{2}{5}\frac{Z^4}{a_0^3}\frac{e^2}{n^3}\frac{R^2\delta_{l,0}}{(4\pi\varepsilon_0)} \tag{ix}$$

$$R_{nl}^2(0) = \left(\frac{Z}{a_0}\right)^3 \frac{4}{n^3}\delta_{l,0} \tag{x}$$

$$E_n = E_n^{(0)} + E_n^{(1)} = -\frac{Z^2 me^4}{2n^2\hbar^2}\frac{1}{(4\pi\varepsilon_0)^2}\left[1 - \frac{4Z^2}{5}\frac{R_0^2 A^{2/3}}{a_0^2 n}\delta_{l,0}\right] \tag{xi}$$

with $R = R_0 A^{1/3}$, $E_n^{(0)} = -13.6$ eV, $E_n^{(1)} = 4\psi 10^{-9}$ eV.

Thus, there is a small correction.

» ***Example* 13.7.** In the Stark effect from the ground state of the hydrogen atom ($n = 1$), show that $E_1^{(1)}$ and $E_1^{(2)}$, which represent the first and second order energy corrections are given by $E_1^{(1)} = 0; E_1^{(2)} = -\frac{9}{4}a_0^2 F$, where $a_0 = \dfrac{\hbar^2}{me^2}$ and ε is the strength of the electric field.

***Solution*:** If the z-axis of coordinates is taken in the direction of the electric field ε, the perturbation to the Hamiltonian is given by $H' = -e\varepsilon Z = -e\varepsilon r\cos\theta$. For the ground state, the unperturbed eigenfunction is given by

$$\psi_0^{(0)} = u_0 = \frac{1}{\sqrt{\pi}}\exp(-r) \; [\text{taking } 6\hbar = 1, m = 1, e = 1]$$

and the unperturbed eigenvalue is

$$E_0^{(0)} = \tfrac{1}{2} \text{ atomic unit}$$

$$\therefore \int_0^\pi \cos\theta \sin\theta \, d\theta = 0$$

the first order energy correction $E_0^{(1)}$ vanishes.

Now, the first order perturbed wave function $\psi_0^{(1)}$ satisfies

$$\left[H_0 - E_0^{(0)}\right]\psi_0^{(1)} + H'\psi_0 = W_0^{(1)}\psi_0^{(0)} = 0$$

or in atomic units

$$\left[-\frac{1}{2}\nabla^2 - \frac{1}{r} + \frac{1}{2}\right]\psi_0^{(1)} = Fr\cos\theta\frac{1}{\sqrt{\pi}}e^{-r}, \text{ where } F = \frac{\hbar^4}{m^2e^5}\varepsilon$$

If we assume, $\psi_0^{(1)} = R(r)\cos\theta$, we would obtain

$$\left[-\frac{1}{2}\left\{\frac{1}{r^2}\frac{d}{dr}\left(r^2\frac{dR}{dr}\right) - \frac{2R}{r^2}\right\} + \left\{-\frac{1}{r} + \frac{1}{2}\right\}R\right] = F\frac{1}{\sqrt{\pi}}r\,e^{-r}$$

We define $G(r)$ by the equation $R(r) = rG(r)\,e^{-r}$ and obtain

$$rG'' + (4-2r)G' - 2G = -\frac{2Fr}{\sqrt{\pi}}$$

If we assume $G(r) = G_0(r) + \alpha + \beta r$, we get $rG'' + (4-2r)G_0' - 2G_0 = 0$, where

$$\alpha = 2\beta = \frac{F}{\sqrt{\pi}} \tag{i}$$

We need to worry about the solution to the homogeneous part of the equation, because it will not vanish when $\varepsilon \to 0$; thus we obtain

$$\psi_0^{(1)} = Fr\left(\frac{r}{2} + 1\right)\frac{1}{\sqrt{\pi}}e^{-r}$$

Now,

$$E_0^{(2)} = \int \psi_0^{(0)} H' \psi_0^{(1)} d\tau = -W_0^{(1)} \int \psi_0^{(0)}\psi_0^{(1)} d\tau$$

The integration is easy to carry out and the result is

$$E_0^{(2)} = -\frac{9}{4}F^2 \text{ atomic units} = -\frac{9}{4}\left[\frac{\hbar^4}{m^2e^5}\varepsilon\right]\frac{me^4}{\hbar^2} = -\frac{9}{4}a_0^3\varepsilon^2$$

where a_0 is the Bohr radius $\left(= \dfrac{\hbar^2}{me^2}\right)$.

Thus, we may write:

$$E = W_0^{(0)} - \frac{9}{4}a_0^3\varepsilon^2$$

which corresponds to the value $\alpha = \frac{9}{2}a_0^3 = 0.677 \times 10^{-24}$ cm^3 for the polarisability of the normal hydrogen atom.

It can be shown that $E_0^{(3)} = 0$ (as indeed do all odd order energies in this case).

Further,

$$\psi_0^{(2)}(r) = \left[\frac{3!}{8} - \frac{7r^2}{16} - \frac{r^3}{24} - \frac{z^2}{16}(15 + 10r + 2r)^2\right]F^2 \text{ and } E_0^{(4)} = -\frac{3555}{64}F^4$$

» ***Example* 13.8.** Calculate the expectation value of kinetic energy for hydrogen atom in ground state.

Solution: If p be the momentum, then KE of the hydrogen atom is KE $= p^2/2m$.

$$\therefore \langle \text{KE} \rangle = \langle T \rangle = \left\langle \frac{p^2}{2m} \right\rangle$$

The wave function in the ground state us ψ_0, i.e., $\psi_0 \propto N_0 e^{-a/n_0}$, where $N_0 = \frac{1}{\sqrt{\pi a_0^3}}$, a_0 is the Bohr radius.

$$\langle T \rangle = \left\langle \psi_0^* \left| \frac{p^2}{2m} \right| \psi_0 \right\rangle = \int \psi_0^* \frac{p^2}{2m} \psi_0 d^3r = \frac{1}{2m} \int \psi_0^* (-i\hbar\nabla)^2 \psi_0 d^3r$$

Since, momentum operator is $-i\hbar\nabla$

$$= -\frac{\hbar^2 |N_0|^2}{2m} \int e^{-n/a_0} \nabla^2 e^{-r/a_0} d^3r = -\frac{\hbar^2}{2\pi a_0^3 m} \int e^{-r/a_0} \nabla^2 e^{-r/a_0} d^3r$$

Now, we put $r/a_0 = \rho$, $\therefore dr = a_0 d\rho$ and

$$\nabla^2 = \frac{\partial^2}{\partial x^2} + \frac{\partial^2}{\partial y^2} + \frac{\partial^2}{\partial z^2}$$

$$\therefore \langle T \rangle = \frac{\hbar^2 \times a_0}{2m\pi a_0^3} \int e^{-\rho} \nabla^2 e^{-\rho} d^3\rho = \frac{\hbar^2}{2\pi m a_0^2} \int e^{-\rho} \nabla^2 e^{-\rho} d^3\rho$$

Since, in polar coordinates

$$\nabla^2 = \frac{1}{r^2}\left(r^2 \frac{\partial}{\partial r}\right) + \frac{1}{r^2 \sin\theta} \frac{\partial}{\partial\theta}\left(\sin\theta \frac{\partial}{\partial\theta}\right) + \frac{1}{r^2 \sin^2\theta} \frac{\partial^2}{\partial\phi^2}$$

$$\text{and } \nabla^2 e^{-r} = \frac{1}{r} \frac{\partial}{\partial r}\left(r^2 \frac{\partial}{\partial r} e^{-r}\right) = \left(e^{-r} - \frac{2}{r} e^{-r}\right)$$

Hence,

$$\langle T \rangle = -\frac{\hbar^2}{2m\pi a_0^2} \int e^{-\rho}\left(e^{-\rho} - \frac{2}{\rho} e^{-\rho}\right) d^3\rho = -\frac{\hbar^2}{2m\pi a_0^4} \int \left(e^{-2\rho} - \frac{2}{\rho} e^{-2\rho}\right) d^3\rho$$

where $d^3\rho$ is volume element and we know that the volume element is given by

$$d^3\rho = \rho^2 \sin\theta \, d\theta \, d\theta \, d\phi \, d\rho$$

$$\therefore \langle T \rangle = -\frac{\hbar^2}{2m\pi a_0^2} \int_0^{\alpha} \int_0^{\pi} \int_0^{2\pi} \left(e^{-2\rho} - \frac{2}{\rho} e^{-2\rho}\right) \rho^2 d\rho \sin\theta \, d\theta \, d\phi$$

But,

$$\int_0^{\pi} \int_0^{2\pi} \sin\theta \, d\theta \, d\phi = 4\pi$$

$$\therefore \langle T \rangle = -\frac{\hbar^2}{2m\pi a_0^2} \left[\int_0^{\infty} e^{-2\rho} d\rho - 2\int_0^{\infty} \frac{e^{-2\rho}}{\rho} d\rho\right] \rho^2$$

$$= \hbar^2 \frac{2\hbar^2}{m a_0^2} \left[\int_0^{\pi} \rho^2 e^{-2\rho} d\rho - 2\int_0^{\infty} \rho e^{-2\rho} d\rho\right]$$

On integrating by parts, we get

$$\langle T\rangle = \frac{\hbar^2}{a_0^2 m}\int_0^\infty e^{-2\rho}\,d\rho = -\frac{\hbar^2}{2a_0^2 m}\left[e^{-2\rho}\right]_0^\infty = -\frac{\hbar^2}{2a_0^2 m}[0-1] = \frac{\hbar^2}{2a_0^2 m}$$

But, $a_0 = \dfrac{\hbar^2}{2me^2}$ (Bohr radius).

$$\therefore \langle T\rangle = \frac{\hbar^2}{2a_0^2 m} = \frac{\hbar^2}{2a_0 m}\times\frac{me^2}{\Gamma^2} \text{ or, } \langle T\rangle = \frac{e^2}{2a_0}$$

Also the KE of hydrogen atom in the ground state is $e^2/2a_0$.

We can express $\langle T\rangle$ in other terms as under:

$$\langle T\rangle = \frac{\hbar^2}{2m}\times\frac{m^2e^4}{\hbar^2} = \frac{me^4}{2\hbar^4}$$

EXERCISES

13(A): Objective (or MCQ) Type Questions

Select the correct answer:

13.1. For a perturbed harmonic oscillator when interaction is proportional to x the first order energy correction is

(a) positive
(b) negative
(c) zero
(d) none of these.

13.2. For the normal helium atom the first order perturbed energy $E^{(1)}$ is

(a) ZE_H
(b) $\frac{3}{4}ZE_H$
(c) $\frac{5}{4}ZE_H$
(d) zero.

13.3. For first order Zeeman effect in H-atom, the change in energy value E^{11} of first state due to perturbation is given by

(a) $\dfrac{e^2B}{4\mu_0}a_0^2$
(b) $\dfrac{e^2B^2}{4\mu_0}a_0^2$
(c) $\dfrac{e^2B^2}{4\mu_0}a_0$
(d) $\dfrac{e^2\mu_0}{4B^2}a_0^2$.

the different letter have their usual significances.

13.4. In first order Stark effect in hydrogen, the hydrogen atom in $n = 2$ state behaves like an electric dipole of moment

(a) $\dfrac{3e}{a_0}$
(b) $3ea_0$
(c) $\dfrac{ea_0}{3}$
(d) $\dfrac{3a_0}{e}$.

13.5. The bond energy in H_2 molecule considering electron-electron repulsion is

(a) 4.5 eV (c) 4.3 eV

(b) 5.3 eV (d) 3.5 eV.

13.6. The ground state energy of the He-atom calculated by perturbation method is

(a) $-2.75\ e^2/a_0$ (c) $-2.75\frac{a_0}{e^2}$

(b) $2.75\ e^2/a_0$ (d) $-2.75\ e^2 a_0$.

where $a_0 = \hbar^2/m_0 e^2$.

13.7. The expectation value of kinetic energy $\langle T \rangle$ for hydrogen atom in the ground state is given by

(a) $\dfrac{2e^2}{a_0}$ (c) $2e^2 a_0$

(b) $\dfrac{2a_0}{e^2}$ (d) $\dfrac{e^2}{2a_0}$.

ANSWERS

13.1. (c). **13.2.** (c). **13.3.** (b). **13.4.** (b). **13.5.** (a). **13.6.** (a). **13.7.** (d).

13(B): Short Answer Type Questions

13.1. Explain the meaning of perturbation in quantum mechanics. What are its different kinds?

13.2. Discuss the field of applicability of time-independent perturbation and time-dependent perturbation theory.

13.3. A plane rigid rotor of charge e and mass μ is acted on by a weak electric field X along the X-axis. Show that the first order correction to its energy correction to its energy (for the nth state) is

$$E_n^{(1)} = \frac{X^2 e a^4 \mu}{\hbar^2} \cdot \frac{1}{4n^2 - 1}$$

where a is the distance of the particle from its centre.

13.4. Calculate energy of the ground state of the atom corrected to first order. Suggest further improvement in the calculations.

13.5. Outline the way in which absorption and emission of radiation is explained in quantum mechanics.

13.6. A particle moves in a potential field given by $V(x) = \frac{1}{2}kx^2 + ax^4$. Treating the term ax^4 as a perturbation, calculate the ground state energy for the particle.

13.7. Outline Schrödinger perturbation theory for non-degenerate levels.

13.8. Discuss the quantum theory of excited helium atom and explain the role of exchange integral.

13.9. Explain Heitler-London model of hydrogen molecule. Comment on the exchange phenomenon and saturation properties of the chemical bond.

13.10. Discuss time-independent perturbation theory for a degenerate state.

13.11. Assume that the nucleus has finite size, use the perturbation of volume effect to calculate the energy change in hydrogen ground state.

13(C): Broad Answer Type Questions

13.1. What do you mean by perturbation theory? Discuss the perturbation theory for non-degenerate levels in the first and second orders.

13.2. (a) Discuss time-independent perturbation theory for non-degenerate stationary state.

(b) A harmonic oscillator of natural frequency ω is placed in a small external potential $\frac{1}{2}kx^2$. By how much is the energy of the ground state changed.

13.3. Discuss time-independent perturbation theory for the non-degenerate stationary state. Obtain the corrected eigenfunctions and eigenvalues.

13.4. Use Schrödinger perturbation theory for non-degenerate levels and apply it to explain first order Stark effect in hydrogen.

13.5. Use perturbation theory to find out the energy levels of the ground state of helium atom. How are results modified if one uses variation technique.

13.6. A one-dimensional potential barrier of arbitrary shape can be divided into a large number of narrow rectangular potential barrier (width δx) by drawing lines parallel to the ordinate in the V vs. x graph. Show that in the limit of $\Delta x \to 0$, the probability of penetration through the barrier of a particle of mass μ and energy E can be obtained by multiplying the probabilities of penetration through the successive rectangular barriers of width Δx each and is given by

$$T = \exp\left[-\frac{2}{\hbar}\int_{x_1}^{x_2}\sqrt{2\mu(V-E)\,dx}\right]$$

where x_1 and x_2 are the limits of the barriers each region.

13.7. In the free electron theory of the metal, the potential energy of a free electron in the metal acted upon by an electric X perpendicular to the metal surface can be written as $V(x) = V_0 - X\,ex$, if we neglect the image force. V_0 is the height of the potential well, assumed rectangular in which the electrons are disturbed in the metal. Show that the probability of penetration of the potential barrier at the metal surface by an electron of energy E (cold emission) is given by

$$T = \exp\left[-\frac{4}{3}\sqrt{2\mu}\frac{(V_0-E)^{3/2}}{Xe\hbar}\right]$$

13.8. A one-dimensional potential barrier has the shape

$$V = 0, \text{ for } x < 0$$

$$V(x) = \frac{V_0}{1+x/l}, \text{ for } x \geqslant 0$$

Show by substitution

$$\sin^2\phi = \frac{E}{V(x)}$$

that the probability of penetration through the barrier for a particle of mass μ and energy E is

$$T = \exp\left[-G(\pi - 2\phi_0 - \sin 2\phi_0)\right]$$

where $\sin\phi_0 = \sqrt{E/V_0}$ and $G = \frac{Vol}{\hbar}\sqrt{\frac{2\mu}{E}}$.

13.9. A linear harmonic oscillator is perturbed by an additional potential = λx^3. Determine the eigenvalues and eigenfunctions. Repeat the calculation for λx^4.

13.10. Discuss the perturbation of isotropic three-dimensional harmonic oscillator by an additional potential λr. Repeat for λr^3 and λr^4.

13.11. A particle moves in a potential field given by $V(x) = \frac{1}{2}kx^2 + ax^4$. Treating the term ax^4 as perturbation, calculate ground state energy for the particle.

13.12. Develop the time-independent perturbation theory up to the second order.

13.13. Give the theory of the time-independent first order perturbation and apply it to the case of Zeeman effect.

13.14. Develop the stationary perturbation theory for the non-degenerate case up to second order.

13.15. Develop the stationary perturbation theory for a degenerate case. Calculate the perturbed energies for a weak field Zeeman effect.

13.16. Calculate the first order Stark effect for the first excited state ($n = 2$) of the hydrogen atom. Show that our result would imply that a hydrogen atom in its $n = 2$ state behaves as though it has a permanent electric dipole moment of magnitude $3ea_0$. Sketch the arrangement of the energy levels and indicate the wave functions associated with each.

13.17. Calculate energy levels for an anharmonic oscillator using perturbation theory.

13.18. Outline Heitler-London theory of the hydrogen molecule and discuss the results.

13.19. Discuss the time-independent perturbation theory for degenerate levels. Consider the hydrogen atom in a weak electric field having perturbed energy $-eEr\cos\theta$. Indicate how the level $n = 2$ will split due to the electric field, without going into detailed calculation of matrix element.

13.20. Explain Heitler-London model of hydrogen molecule. Comment on the exchange phenomenon and saturation properties of the chemical bond.

13.21. Use perturbation theory to calculate the change in energy value of hydrogen level using second order effect. What is exchanged integral? What is implication?

13(D): Numerical Problems

13.1. A perturbation $H^{(1)}$ acts on a system in stationary state n for which the energy is E_n. Obtain the second order perturbation energy in the usual form:

$$\sum_{r \neq n} \frac{H_{nr}^{(1)} H_{rn}^{(1)}}{E_n - E_r}$$

13.2. Apply time-independent perturbation theory to obtain ground state energy of the helium atom correct to first order.

13.3. Explain Stark splitting of $n = 2$ level of hydrogen atom in presence of an electric field using first order time-independent perturbation theory.

13.4. The Hamiltonian of a one-dimensional harmonic oscillator has the form:

$$H = \frac{p_x^2}{2\mu} + \frac{1}{2}kx^2 + bx^3$$

where μ is the mass, k and b are constants. Taking the anharmonicity term bx^3 to be a small perturbation, show that the first order correction to the nth eigenstate is

$$E_n^{(2)} = -\frac{b^2}{8\alpha^2\mu\omega^2}(30n^2 + 30n + 11)$$

where $\alpha = \frac{\mu\omega}{\hbar}$ (ω is the angular frequency). Hence, find three states of the oscillator.

13.5. A charged linear harmonic oscillator is acted upon by a uniform electric field F. Show by the application of the stationary perturbation theory, that the first order correction to the energy of the oscillator vanishes while the second order correction to the energy of the nth eigenstate is

$$E_r^{(2)} = -\frac{F^2e^2}{2\mu\omega^2}$$

where e and μ are the charge and mass of oscillator and ω is the circular frequency.

13.6. If ϕ is a single valued, continuous function and

$$\int_\tau \phi^*\phi\, d\tau = 1$$

show if the operator H has only discrete spectrum

$$\int_\tau \phi^* H\phi\, d\tau \geqslant E_0$$

where E_0 is the lowest energy value of H. Assume that

$$\phi = \frac{Z^2}{\pi a_0^3}e^{-2(r_1+r_2)/a_0}$$

obtain an expression for the energy of the ground state of a helium atom (r_1 and r_2 being the distances of the electrons from the nucleus).

13.7. Explain briefly what is meant by degeneracy? Illustrate this in terms of the isotropic oscillator containing a particle of mass M moving in a potential field $V = \frac{1}{2}\alpha(x^2 + y^2 + z^2)$ and find the degeneracy of the nth energy level.

13.8. Show that the ground state energy E_0 of a Hamiltonian H is always less or at best equal to F when

$$F = \frac{(\psi, H\gamma)}{(\phi, \gamma)}$$

where ϕ is reasonable good choice to ϕ_0, the ground state wave function. Use a trial function $\psi = e^{-x}$ to find the ground state energy of the hydrogen atom.

13.9. If trial function ϕ differs from an eigenfunction u_e by a small amount, so that $\phi = u_e + \varepsilon\psi_1$, where u_e and ψ_1 are normalised and $\varepsilon < 1$. Show that $\langle H\rangle$ differs from E only by a term of order e^2 and find this term.

13.10. Consider $H = H_0 + H'$, where

$$H_0 = \begin{pmatrix} 1 & 0 \\ 0 & -1 \end{pmatrix} \text{ and } \begin{pmatrix} \varepsilon & \varepsilon \\ \varepsilon & \varepsilon \end{pmatrix}$$

assuming $\varepsilon \ll 1$, find the first perturbation theory to obtain the eigenvalues and eigenfunctions of H.

Chapter 14

Time-Dependent Perturbations

14.1 Introduction

The perturbation theory of a system for which Hamiltonian depends on time is sometimes called the *method of variation of constants*. The theory of *time-dependent perturbation* was developed by Dirac.

It permits us to discuss such questions as the probability of transition of the system from one unperturbed stationary state to another as a result of the perturbation. In atomic physics, particularly in spectral analysis, the main concern is the quantisation of energy and the time-dependent perturbation finds wide application. In nuclear physics particularly in nuclear reactions and disintegrations, the time-dependent method is frequently used.

14.2 Time-dependent perturbation

We consider an unperturbed system with wave equation including time as

$$H^{(0)}\,\psi^{(0)} = -\frac{h}{2\pi i}\frac{\partial \tau^{(0)}}{\partial t} \tag{14.1}$$

The general solution of eq. (14.1) is

$$\psi^{(0)} = \sum_{n=0}^{\infty} a_n(0)\, e^{-(2\pi i/h)E_n^{(0)}}\, \psi_n^{(0)} \tag{14.2}$$

where $a_n(0)$ are constants and the functions $\psi_n^{(0)}$ are eigenfunctions of the following time-independent Schrödinger equation

$$H^{(0)}\,\psi_n^{(0)} = E_n^{(0)}\,\psi_n^{(0)} \tag{14.3}$$

where $\psi_n^{(0)}$ and $E_n^{(0)}$ are assumed to be known.

Let the Hamiltonian of a system be expressed as

$$H = H^{(0)} + \lambda H^{(0)} \tag{14.4}$$

where $H^{(0)}$ is the unperturbed Hamiltonian and $H^{(1)}$ is the perturbation depending on time and is very small, i.e., $H^{(1)}$ is slowly varying function of time. The perturbed equation is

$$\left(H^{(0)} + \lambda H^{(1)}\right) \psi = -\frac{h}{2\pi i}\frac{\partial \psi}{\partial t} \tag{14.5}$$

A general solution of the above equation can be written as

$$\psi(x_1, \ldots, t) = \sum a_n(t)\, \psi_n^{(0)}(x_1, \ldots, t) \tag{14.6}$$

where $a_n(t)$ is a function of t alone.

Substituting eq. (14.6) in eq. (14.5), we get

$$\sum a_n(t)\, H^{(0)}\, \psi_n^{(0)} + \lambda \sum a_n(t)\, H^{(1)}\, \psi_n^{(0)} = -\frac{h}{2\pi i}\sum \psi_n^{(0)} \frac{\partial a_n(t)}{\partial t} - \frac{h}{2\pi i}\sum a_n(t) \frac{\partial \psi_n^{(0)}}{\partial t} \tag{14.7}$$

But,

$$H^{(0)}\, \psi_n^{(0)} = -\frac{h}{2\pi i}\frac{\partial \psi_n^{(0)}}{\partial t}$$

Hence, the first and the last terms of eq. (14.7) are cancelled.

$$\therefore \lambda \sum_n a_n(t)\, H^{(1)}\, \psi_n^{(0)} = -\frac{h}{2\pi i}\sum \psi_n^{(0)} \frac{\partial a_n(t)}{\partial t} \tag{14.8}$$

Now, multiplying eq. (14.8) by $\psi_m^{(0)*}$ and integrating over the configuration space, we have

$$\lambda \sum_n \int a_n(t)\, \psi_m^{(0)*}\, H^{(1)}\, \psi_n^{(0)}\, d\tau = -\frac{h}{2\pi i}\sum_n \int \frac{\partial a_n(t)}{\partial t}\, \psi_m^{(0)*}\, \psi_n^{(0)}\, d\tau$$

$$\therefore \frac{\partial a_n(t)}{\partial t} = -\frac{2\pi i}{h}\lambda \sum_n a_n(t) \int \psi_m^{(0)*}\, H^{(1)}\, \psi_n^{(0)}\, d\tau$$

$$\text{or, } a_m(t) = -\frac{2\pi i}{h}\lambda \sum_n a_n(t) \int \psi_m^{(0)+} H^{(1)} \psi_n^{(0)}\, d\tau, \quad m = 0, 1, 2 \tag{14.9}$$

The above equation can also be written as

$$\dot{a}_m(t) = -\frac{2\pi i}{h}\lambda \sum_n a_n(t) \int \psi_m^{(0)*} \exp\left(\frac{2\pi i}{h}E_m t\right) H^{(1)}\, \psi_n^{(0)} \times \exp\left(-\frac{2\pi i}{h}E_n t\right) d\tau$$

$$\text{or, } \dot{a}_m(t) = -\frac{2\pi i}{h}\lambda \sum_n a_n(t)\, H_{mn}^{(1)} \exp\left[\frac{2\pi i}{h}(E_m - E_n)\, t\right]$$

$$\text{or, } \dot{a}_n(t) = -\frac{2\pi i}{h} \times \sum_n a_n(t)\, H_{mn}^{(1)} \exp(i\omega_{mn} t) \tag{14.10}$$

where

$$H_{mn}^{(1)} = \int \psi_m^{(0)*}\, H^{(1)}\, \psi_n^{(0)}\, d\tau \text{ and } \omega_{mn} = \frac{E_m - E_n}{\hbar}$$

Equation (14.10) can be solved by using perturbation theory. We shall assume that the perturbation is sufficiently small, so that the amplitudes $a_n(t)$ are not very much different from their value at $t = 0$. The perturbation approximation consists in expanding a_n as power series in λ and equating coefficients of various powers of λ for different order calculations.

Hence,

$$a_n(t) = a_n^{(0)}(t) + \lambda a_n^{(1)}(t) + \lambda^2 a_n^{(2)}(t) + \ldots \tag{14.11}$$

Now, substituting eq. (14.11) in eq. (14.10), we get

$$\frac{\partial}{\partial t}\left(a_m^{(0)} + \lambda a_m^{(1)} + \lambda^2 a_m^{(2)} + \ldots\right) = -\frac{2\pi i}{h}\lambda \sum_n \left(a_n^{(0)} + \lambda a_n^{(1)} + \lambda^2 a_n^{(2)} + \ldots\right) H_{mn}^{(1)} \exp(i\omega_{mn} t) \tag{14.12}$$

Zero order calculation

Equating the coefficient of λ^0 in eq. (14.12), we get

$$\frac{\partial a_m^{(0)} t}{\partial t} = 0 \text{ i.e., } a_m^{(0)} t = \text{constant} \tag{14.13}$$

This constant gives the initial state of the system and can be chosen properly if the system is completely known before the application of perturbation.

First order calculation

Equating the coefficient λ on both sides of eq. (14.12), we get

$$\frac{\partial}{\partial t} a_m^{(1)}(t) = -\frac{2\pi i}{h}\sum_n a_n^{(0)}(t)\, H_{mn}^{(1)} \exp(i\omega_{mn} t) \tag{14.14}$$

$$\text{or, } a_m^{(1)}(t) = -\frac{2\pi i}{h}\int_{-\infty}^{t}\left[\sum_n a_n^{(0)}(t)\, H_{mn}^{(1)} \exp(i\omega_{mn} t)\right] dt$$

$$= -\frac{2\pi i}{h}\int_{-\infty}^{t}\left[\sum_n \delta_{nl}\, H_{mn}^{(1)} \exp(i\omega_{ml} t)\right] dt$$

$$\text{or, } a_m^{(1)}(t) = -\frac{2\pi i}{h}\int_{-\infty}^{t} H_{ml}^{(1)} \exp(i\omega_{ml} t)\, dt \tag{14.15}$$

Here, the constant of integration is taken to be zero so that $a_m^{(t)}(t)$ be zero at $t = -\infty$ (before perturbation is applied). Eq. (14.15) can be simplified if $H^{(1)}$ is independent of time except for being turned at $t = 0$ and off at $t = t$ respectively.

So

$$a_m^{(1)}(t) = -\frac{2\pi i}{h} H_{ml}^{(1)} \int_0^t \exp(i\omega_{ml} t)\, dt = -\frac{2\pi i}{h} H_{ml}^{(1)} \left[\frac{\exp(i\omega_{ml})}{i\omega_{ml}}\right]_0^t \tag{14.16}$$

This is the *first order perturbation theory.*

The probability of finding the particle in mth state

The probability of finding the particle in the mth at any time t is

$$= a_m^{(1)}(t)\, a_m^{(1)}(t)^* = \left|a_m^{(1)}(t)\right|^2 = \frac{1}{\omega_{ml}^2}\left(e^{i\omega_{ml} t} - 1\right)\left(e^{-i\omega_{ml} t} - 1\right) \times \frac{1}{\hbar^2}\left|H_{ml}^{(1)}\right|^2 \tag{14.17}$$

$$= \frac{1}{\omega_{ml}^2}\left[1 + 1 - 2\frac{e^{i}\omega_{ml} t + e^{-i}\omega_{ml} t}{2}\right] \times \frac{1}{\hbar^2}\left|H_{ml}^{(1)}\right|^2$$

$$= \frac{1}{\omega_{ml}^2} 2(1 - \cos\omega_{ml} t)\frac{\left|H_{ml}^{(1)}\right|^2}{\hbar^2} = \frac{4}{\omega_{ml}^2}\sin^2 \omega_{ml}^{t}/2 \frac{\left|H_{ml}^{(1)}\right|^2}{\hbar^2}$$

$$\therefore \left|a_{ml}^{(1)}(t)\right|^2 = \frac{4\left|H_{ml}^{(1)}\right|^2 \sin^2 \omega_{ml}^t/2}{\hbar^2\omega_{ml}^2} \tag{14.18}$$

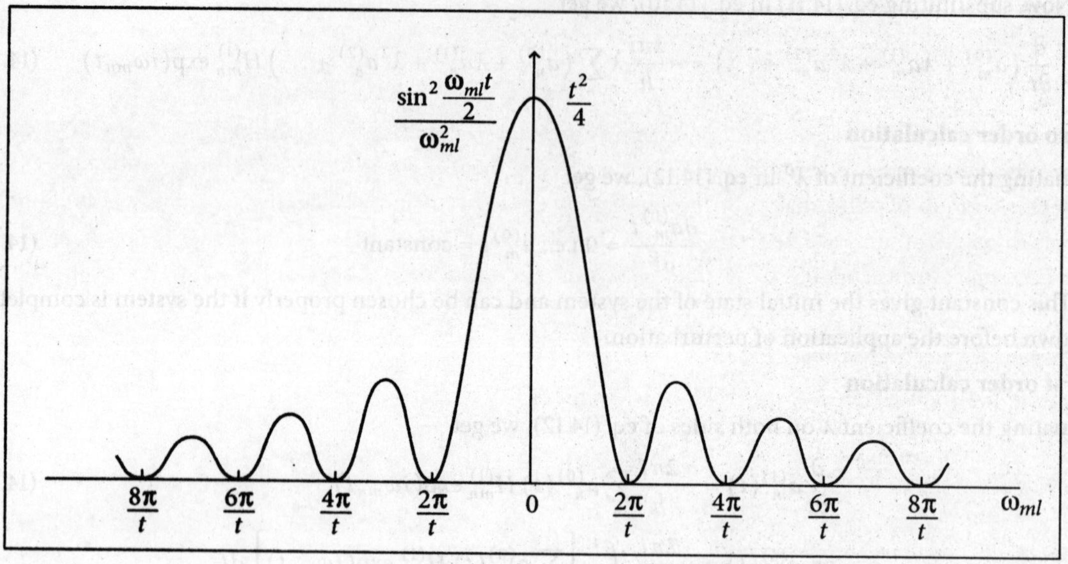

Fig 14.1 Plot of $\sin^2\left|\frac{\omega_{ml}t/2}{\omega_{ml}^2 t}\right|$ against ω_{ml}

Fig 14.1 shows the plot of $\sin(\omega_{ml}/2)/\omega_{ml}^2$ as a function of ω_{ml}.

Physical interpretation of the curve

The above figure shows that

$$\sin^2 \frac{\omega_{ml}t/2}{\omega_{ml}^2}$$

becomes maximum when $\omega_{ml} = 0$. Taking $\omega_{ml} = x$

$$\frac{\sin^2(xt/2)}{x^2} = \frac{1}{x^2}\left[\left(\frac{xt}{2}\right) - \frac{(xt)^3}{3.2^2} + \ldots\right]^2 = \frac{1}{x^2}\left[\frac{xt}{2}\right]^2 \text{ neglecting higher powers of } x$$

$$= \frac{t^2}{4} \tag{14.19}$$

Thus, the highest peak value is $t^2/4$, which can be seen in the graph.

The value of $\frac{\sin^2(xt/2)}{x^2}$ will be zero when

$$\frac{\sin^2(xt/2)}{x^2} = 0 \text{ or, } \frac{xt}{2} = \pm n\pi \text{ or, } x = \pm\frac{2n\pi}{t}$$

Thus,

$$\omega_{ml} = \pm\frac{2\pi n}{t} \tag{14.20}$$

For $n = 1$, $\omega_{ml} = \pm\frac{2\pi}{t}$

$n = 2$, $\omega_{ml} = \pm\frac{4\pi}{t}$

$n = 3$, $\omega_{ml} = \pm\frac{6\pi}{t}$.

It is evident from the Fig 14.1 that the height of the main peak increases proportional to t^2 eq. (14.19) while its breadth decreases inversely as t eq. (14.20). Therefore, the area under the curve is proportional to t. Thus, if there is a group of states that have energies nearly equal to the initial state, the probability of finding the system in one or another of these states is proportional to t. This statement implies that the probability per unit time that a transition has taken place when the perturbation has been on for a time t is proportional to t.

Second order perturbation

Equating the coefficients of λ^2 on both sides of eq. (14.12), we get

$$\frac{\partial}{\partial t}a_m^{(2)}t = -\frac{2\pi i}{h}\sum_n a_n^{(1)}(t)\,H_{mn}^{(1)}\exp(i\omega_{mn}t) = -\frac{2\pi i}{h}\sum_n\left[-\frac{1}{\hbar}H_{nl}^{(1)}\left\{\frac{e^{i\omega_{nl}t}-1}{\omega_{nl}}\right\}H_{mn}^{(1)}e^{i\omega_{mn}t}\right]$$

$$= \frac{i}{\hbar^2}\sum_n\frac{H_{nl}^{(1)}H_{mn}^{(1)}}{\omega_{nl}}\left[e^{i(\omega_{nl}+\omega_{mn})t}-e^{i\omega_{mn}t}\right] = \frac{i}{\hbar^2}\sum_n\frac{H_{nl}^{(1)}H_{mn}^{(1)}}{\omega_{nl}}\left[e^{i\omega_{ml}t}-e^{i\omega_{mn}t}\right]$$

$$\therefore a_m^{(2)}(t) = \frac{i}{\hbar^2}\sum_n\frac{H_{nl}^{(1)}H_{mn}^{(1)}}{\omega_{nl}}\int_0^t\left[e^{i\omega_{ml}t}-e^{i\omega_{mn}t}\right]dt$$

Integrating the above equation so that $a_m^{(2)}(t) = 0$ at $t = 0$, we get

$$a_m^{(2)}t = \frac{i}{\hbar^2}\sum_n\frac{H_{nl}^{(1)}H_{mn}^{(1)}}{\omega_{nl}}\left[\frac{e^{i\omega_{ml}t}-1}{i\omega_{ml}}-\frac{e^{i\omega_{ml}t}-1}{i\omega_{mn}}\right]$$

$$\text{or, } a_m^{(2)}t = \frac{1}{\hbar^2}\sum_n\frac{H_{nl}^{(1)}H_{mn}^{(1)}}{\omega_{nl}}\left[\frac{e^{i\omega_{ml}t}-1}{\omega_{ml}}-\frac{e^{i\omega_{mn}t}-1}{\omega_{mn}}\right] \tag{14.21}$$

The above equation gives the constant in second order perturbation.

14.3 Transition to the continuum: Fermi golden rule

We shall now calculate an explicit expression for transition probability per unit time.

Let the system be enclosed in a cubical box of side L that has boundary conditions at its walls. The stationary states of the system will be discrete, but separated in energy by an interval which is inversely proportional to the volume of the box. If we pass to the limit of a box of infinite size, the levels within a given energy interval increase in number and merge into a continuum. Of course, it being only a formal limiting process, permits direct application of the result in calculating the probability for transition that involves states in the continuum.

The probability of transition of mth state is given by

$$W = |a_m(t)|^2 = \frac{1}{\hbar^2}\left|H_{ml}^{(1)}\right|^2\frac{4\sin^2(\omega_{ml}^t/2)}{\omega_{ml}^2} \tag{14.22}$$

Due to the factor $1/\omega_{ml}^2$, the probability is the largest for those states in which the unperturbed energies $E_m^{(0)}$ are closer to $E_l^{(0)}$. If $E_m^{(0)}$ is far away from $E_l^{(0)}$, the probability is small. Since the levels are very close to one another, they form a cluster around $E_m^{(0)} = E_l^{(0)}$. Since, all levels in the cluster represent nearly the same physical probabilities we may add up the probabilities of all levels in the cluster and consider them as a whole.

∴ total probability of transition = $\Sigma|a_m(t)|^2$. We may define the density of final state $\rho(m)$ such that $\rho(m)\,dE$ is the number of such states in the energy range E_m and $E_m + dE$. The total probability for transition into these states is obtained by multiplying eq. (14.22) by $\rho(m)\,dE$ and, then integrating w.r.t. E. Thus,

$$\sum |a_m(t)|^2 = \int |a_m(t)|^2 \rho(m)\,dE \tag{14.23}$$

Rather than considering transition to a particular state we may consider transitions of state of continuous spectrum. One may, then under limiting conditions, define a probability of transition per unit time (to any group of states).

If the central peak of

$$\frac{4\sin^2(\omega_{ml}t/2)}{\omega_{ml}}$$

is the domain of integration (i.e., energy conserving transitions); the main contribution comes from this peak and only a small error is involved in extending the limit of this integral to $-\infty$ to $+\infty$. The transition probability per unit time is

$$T = \frac{1}{\hbar^2}\left|H_{ml}^{(1)}\right|^2 \rho(m)\int_{-\infty}^{\infty}\frac{4\sin^2(\pi_{ml}t/2)}{\omega_{ml}^2}\,dE$$

But, $dE = \hbar d\omega$ as $E_m - E_l = \hbar\omega_m$.

$$\therefore T = \frac{1}{\hbar^2}\left|H_{ml}^{(1)}\right|^2 \rho(m)\int_{-\infty}^{+\infty}\frac{4\sin^2(\omega_{ml}t/2)}{\omega_{ml}^2}\,d\omega$$

$$= \frac{1}{\hbar}\left|H_{ml}^{(1)}\right|^2 \rho(m)\int_{-\infty}^{+\infty}\frac{4\sin^2(\omega_{ml}t/2)}{\omega_{ml}^2}\,d\omega \tag{14.24}$$

Now, consider the integral

$$4\int_{-\infty}^{+\infty}\frac{\sin^2(\omega_{ml}t/2)}{\omega_{ml}^2}\,d\omega$$

Put $\left(\frac{\omega_{ml}t}{2}\right) = x$

$$4\int_{-\infty}^{+\infty}\frac{\sin^2(\omega_{ml}t/2)}{a_{ml}^2}\,d\omega = 4\int_{-\infty}^{+\infty}\frac{\sin^2 x}{\frac{4x^2}{t}}\cdot\frac{2}{t}\,dx = 2t\int_{-\infty}^{+\infty}\frac{\sin^2 x}{x^2}\,dx = 2t\pi$$

where $\int_{-\infty}^{+\infty}\frac{\sin^2 x}{x^2}\,dx = \pi$.

∴ from eq. (14.24), we have

$$T = \frac{1}{\hbar}\left|H_{ml}^{(1)}\right|^2 \rho(m)2\pi t$$

Hence, the transition probability per unit time is

$$T = \frac{2\pi}{\hbar^2}\left|H_{ml}^{(1)}\right|^2 \rho(m) \tag{14.25}$$

This formula has wide applications in quantum physics. The theory of scattering can be explained with the help of this formula.

Equation (14.25) is known as the *Fermi golden rule* and states that the transition probability per unit time is

(i) non-zero only between quantum states of the same energy.

(ii) proportional to the square of $\left|H_{ml}^{(1)}\right|$ of the perturbation connecting the states.

(iii) proportional to the density of states.

14.4 Harmonic perturbation

Transitions between stationary states of a quantum system can be induced by applying a harmonic perturbation. For example, the oscillating electric and magnetic fields associated with light can induce transition between stationary states in atoms.

We consider a situation in which perturbation is an explicit function of time (e.g., simple harmonic function) except for being turned on at $t = 0$ and turned off at a later time at $t = t_0$, we write

$$H^{(1)} = 2A^{(0)} \sin \omega t^*.$$

Now,

$$H_{ml}^{(1)} = \int \psi_m^{(0)*} 2A^{(0)} \sin \omega t \psi_l^{(0)} \, d\tau = 2H_{ml}^{(0)} \sin \omega t \tag{14.26}$$

In case of harmonic perturbation

$$\begin{aligned} a_m^{(1)}(t \geqslant t_0) &= -\frac{2\pi i}{h} 2H_{ml}^{0(1)} \int_0^{t_0} \sin \omega t \exp(i\omega_{ml} t)\, dt \\ &= -\frac{2\pi}{h} 2iH_{ml}^{0(1)} \int_0^{t_0} \left\{\frac{e^{i\omega t} - e^{-i\omega t}}{2i}\right\} e^{i\omega_{ml} t}\, dt \\ &= -\frac{2\pi}{h} H_{ml}^{0(1)} \int_0^{t_0} \left[e^{i(\omega_{ml}+\omega)t} - e^{i(\omega_{ml}+\omega)t}\right] dt \\ &= -\frac{2\pi}{h} H_{ml}^{0(1)} \left[\frac{e^{i(\omega_{ml}+\omega)t}}{i(\omega_{ml}+\omega)} - \frac{e^{i(\omega_{ml}-\omega)t}}{i(\omega_{ml}-\omega)}\right]_0^{t_0} \\ &= \frac{H_{ml}^{0(1)}}{i\hbar}\left[\frac{e^{i(\omega_{ml}-\omega)t_0} - 1}{(\omega_{ml}-\omega)} - \frac{e^{i(\omega_{ml}+\omega)t_0} - 1}{\omega_{ml}+\omega}\right] \end{aligned} \tag{14.27}$$

This shows that the amplitude depends upon perturbation duration to and does not depend upon time t. The probability of transition, i.e., the probability of finding the system in mth state is appreciable only when the denomination of first or second term in eq. (14.27) is almost equal to zero.

If we consider the mth level above ith level, then

$$\omega_{ml} = \frac{\left(E_m^{(0)} - E_l^{(0)}\right)}{\hbar} = \text{positive quantity}$$

In this case, the first term of eq. (14.27) will be zero due to larger denominator and the first term is important.

If considered the lth level above mth level, then

$$\omega_{ml} = \frac{\left(E_m^{(0)} - E_l^{(0)}\right)}{\hbar} = \text{negative quantity}$$

Here, the first term of eq. (14.27) will be zero due to large denominator and the second term is important.

Thus, the second term of eq. (14.27) is important when the effect of perturbation is to transfer an energy $\hbar\omega$ to the system on which it acts. The process is known as *absorption*, i.e., energy is transferred from the perturbing source to the system. The first term will be important if the effect of perturbation is to receive energy from the system. This is known as *emission*.

When $E_m > E_l$, the first term needs to be considered and the probability is given by

$$|a_m(t \geqslant t_0)|^2 = \frac{4\left|H_{ml}^{0(1)}\right|^2}{\hbar^2}\left[\frac{\sin^2(\omega_{ml} - \omega)\, t_0/2}{(\omega_{ml} - \omega)^2}\right] \tag{14.28}$$

When $E_l > E_m$, the second term needs to be considered and the probability is given by

$$|a_m(t \geqslant t_0)|^2 = \frac{4\left|H_{ml}^{(0)\,(1)}\right|^2}{\hbar^2}\left[\frac{\sin^2(\omega_{ml} + \omega)\, t_0/2}{(\omega_{ml} + \omega)^2}\right] \tag{14.29}$$

14.5 Scattering of free particle by potential energy $V(r)$

The theory of perturbation can be applied to the theory of scattering provided the scattering potential is regarded as a perturbation of the states of the free particle. Here, we shall calculate the transition probability when the initial and final statures are free particle momentum eigenfunctions (i.e., plane waves).

Let for initial and final states, the free particle eigenfunctions be

$$\psi_I^{(0)} = \frac{1}{L^{3/2}} e^{(i\vec{k}_0\cdot\vec{r})} \tag{14.30}$$

$$\text{and } \psi_F^{(0)} = \frac{1}{L^{3/2}} e^{(i\vec{k}'\cdot\vec{r})} \tag{14.31}$$

where $\vec{k}_0$ and $\vec{k}'$ are the initial and final propagation vectors respectively.

Then the perturbation matrix element will be

$$H_{FI}^{(1)} = \int \psi_F^{(0)*} H^{(1)} \psi_I^{(0)}\, d\tau = \int \frac{1}{L^{3/2}} e^{-i\vec{k}'\cdot\vec{r}} V(r) \frac{1}{L^{3/2}} e^{i\vec{k}_0\cdot\vec{r}}\, d\tau$$

$$\therefore \left|H_{FI}^{(1)}\right|^2 = \frac{1}{L^6}\left|\int e^{ik\cdot\vec{r}} V(r)\, d\tau\right|^2 \tag{14.32}$$

where $\vec{k} = (\vec{k}_0 - \vec{k}')$.

Calculation of energy density $\rho(E)$

In case of cubical box of length L, the normalised wave functions and eigenvalues are as follows:

$$\psi_{n_x n_y n_z} = \frac{1}{(L/2)^3} \sin\frac{n_x \pi x}{L} \sin\frac{n_y \pi y}{L} \sin\frac{n_z \pi z}{l} \tag{14.33}$$

and

$$E = \frac{\hbar^2\pi^2}{2mL^2}n^2, \text{ where } n^2 = n_x^2 + n_y^2 + n_z^2 \tag{14.34}$$

All the energy states will be in the octant of the sphere of radius n. The number of energy states would be given by volume of the octant divided by the volume of one energy state (unit cell). Thus, the number of energy states with energy less than E is given by

$$\frac{1}{8}\frac{4\pi}{3}\left(\frac{2mEL^2}{\pi^2\hbar^2}\right)^{3/2}$$

The energy density is given by

$$\rho(E) = \frac{d}{dE}\left[\frac{1}{8}\frac{4\pi}{3}\left(\frac{2mEL^2}{\pi^2\hbar^2}\right)^{3/2}\right] = \frac{\pi}{6}\frac{(2m)^{3/2}L^3}{\pi^3\hbar^3}\frac{d}{dE}(E)^{3/2}$$

$$= \frac{\pi}{6}\frac{(2m)^{3/2}}{\pi^3\hbar^3}\frac{3}{2}(E)^{1/2} = V\frac{(2m)^{3/2}}{4\pi^2\hbar^3}E^{1/2}, \text{ where } V = L^3 \tag{14.35}$$

In terms of momentum p and velocity v, we have

$$\rho(E) = \frac{d}{dE}\left[\frac{\pi}{6}\left(\frac{p^2L^2}{\pi^2\hbar^2}\right)^{3/2}\right] \quad \left[\because 2mE = 2m\cdot\tfrac{1}{2}mv^2 = p^2\right]$$

$$= \frac{\pi L^3}{6\pi^3\hbar^3}\frac{d}{dE}(p)^3 = \frac{\pi L^3}{6\pi^3\hbar^3}3p^3\frac{dp}{dE}$$

$$= \frac{L^3p^2}{2\pi^3\hbar^3 v}\quad\left[\because E = \frac{p^2}{2m}, \therefore \frac{dp}{dE} = \frac{1}{v}\right] \tag{14.36}$$

Now, the energy level density corresponding to scattering direction on (θ,ϕ) per unit solid angle is given by

$$\rho(E,\theta,\phi)\,d\omega = \frac{\rho(E)\,d\omega}{4\pi} = \frac{1}{4\pi}\frac{L^3p^2}{2\pi^2\hbar^3 v}d\omega = \frac{L^3p^2}{8\pi^2\hbar^3 v}d\omega \tag{14.37}$$

Now, $T_{F\to I}\,d\omega$ is the probability per unit time that a system makes a transition to final state in solid angle $d\omega$.

$$T_{F\to I}\,d\omega = \frac{2\pi}{\hbar}\frac{1}{L^6}\left|\int \exp(i\vec{k}\cdot\vec{r})V(r)\,d\tau\right|^2 \rho(E,\theta,\phi)\,d\omega$$

$$= \frac{2\pi}{\hbar}\frac{1}{L^6}\left|\int \exp(i\vec{k}\cdot\vec{r})V(r)\,d\tau\right|\frac{L^3p^2}{8\pi^3\hbar^3 v}d\omega$$

$$= \frac{p^2}{L^3\cdot 4\pi\hbar^4 v}\left|\int \exp(i\vec{k}\cdot\vec{r})\,V(r)\,d\tau\right|^2 d\omega$$

$$\text{or, } \sigma(\theta,\phi) = \left(\frac{m}{2\pi\hbar^2}\right)\left|\int \exp(i\vec{k}\cdot\vec{r})\,V(r)\,d\tau\right|^2 d\omega \tag{14.38}$$

This is equal to the number of particles scattered into the solid angle $d\omega$ per unit time when there is one incident particle in the volume L^3. If v be the velocity of the incident particle then

$$\text{Incident flux} = \text{number of particles} \times \text{velocity per unit volume} = \frac{1}{L^3}v = \frac{p}{mL^3} \tag{14.39}$$

Differential cross-section

The differential cross-section is defined as

$$\sigma(\theta,\phi)\,d\omega = \frac{\text{Scattered flux}}{\text{Incident flux}} = \frac{I_{F\to I}\,d\omega}{p/mL^3}$$

$$\therefore \sigma(\theta,\phi) = \frac{p^2}{L^3 4\pi^2\hbar^4 v}\left|\int \exp(i\vec{k}\cdot\vec{r})V(r)\,d\tau\right|^2 \frac{mL^3}{p}$$

$$= \frac{pm}{4\pi^2\hbar^4 v}\left|\int \exp(i\vec{k}\cdot\vec{r})V(r)\,d\tau\right|^2$$

$$= \frac{m^2}{4\pi^2\hbar^4}\left|\int \exp(i\vec{k}\cdot\vec{r})V(r)\,d\tau\right|$$

$$\text{or, } \alpha(\theta\phi) = \left(\frac{m}{4\pi\hbar^2}\right)^2\left|\int \exp(i\vec{k}\cdot\vec{r})V(r)\,d\tau\right|^2 \tag{14.40}$$

This is the *Born approximation formula.*

14.6 Application of time-dependent perturbation theory to alpha scattering

If α-particles of charge $Z_1 e$ with moderate energy bombard a nucleus of charge $Z_2 e$, the α-particles are deflected by mutual Coulomb repulsion and the Rutherford scattering takes place. Here, the nucleus experiences a rapidly varying electric field which can cause transition from the ground state to some excited state of the nucleus.

In CGS unit, the potential energy of α-scattering is given by

$$V(r) = \frac{Z_1 Z_2 e^2}{r} \tag{14.41}$$

Now, the Schrödinger time-dependent equation is

$$-\frac{\hbar^2}{2m}\nabla^2\psi + \frac{Z_1 Z_2 e^2}{r}\psi = -\frac{\hbar}{i}\frac{\partial\psi}{\partial t} \tag{14.42}$$

Solution of the above equation can be obtained by considering the potential as a perturbation, i.e.,

$$H^{(1)} = \frac{Z_1 Z_2 e^2}{r}$$

We assume that the scattering foil is inserted in position at $t = 0$. Before $t = 0$ the α-particle is described by a plane wave along the X-axis

$$\psi_l^{(0)} = \frac{1}{L^{3/2}}\exp\left[\left(\frac{i}{\hbar}\right)\vec{p}_0\cdot\vec{r}\right] = \frac{1}{\sqrt{V}}\exp\left[\left(\frac{i}{\hbar}\right)\vec{p}_0\cdot\vec{r}\right] \tag{14.43}$$

where p is the magnitude of the momentum vector $\vec{p}_0$ of α-particle. At $t = 0$ the perturbation is switched on. Now, transitions may occur to clusters of the levels—the unperturbed energy level $E_m^{(0)}$ of which are close to $E_l^{(0)}$. We now examine these levels. They are all plane waves (unperturbed eigenfunctions), the momentum vectors of which have nearly the same magnitude but different directions, i.e.,

$$\psi_m^{(0)} = \frac{1}{\sqrt{V}}\exp\left[\left(\frac{i}{\hbar}\right),\ \vec{p}'\cdot\vec{r}\right],\ |\vec{p}'| = p \tag{14.44}$$

These levels may be grouped into sub-clusters and take a typical level $\psi_l'^{(0)}$ for which $|\vec{p}'| = p$. The momentum vector for this level may be denoted by $\vec{p}$. Thus,

$$\psi_{l'}^{(0)} = \frac{1}{\sqrt{V}} \exp\left[\left(\frac{i}{\hbar}\right) \vec{p} \cdot \vec{r}\right] \tag{14.45}$$

The matrix element for transition is given by

$$H_{ll'}^{(0)} = \frac{Z_1 Z_1 e^2}{V} \iiint \frac{\exp\left[\left(\frac{i}{\hbar}\right)(\vec{p}_0 - \vec{p}) \cdot \vec{r}\right]}{r} d\tau \tag{14.46}$$

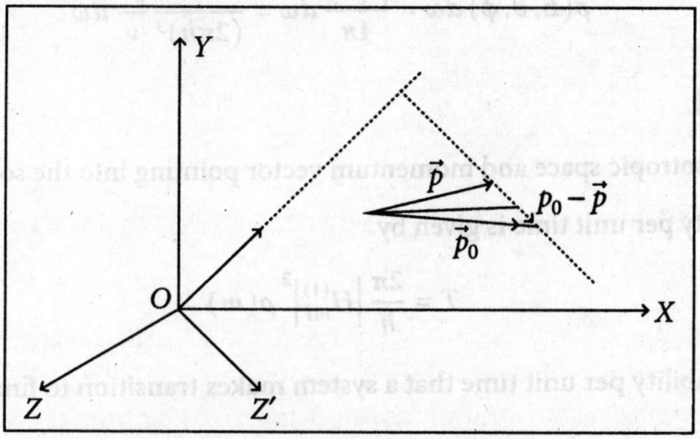

Fig 14.2 Direction of $\vec{p}_0 - \vec{p}$

This matrix element may be determined graphically. The direction of the vector $(\vec{p}_0 - \vec{p})$ is shown in Fig 14.2 and it is $2p \sin\frac{\theta}{2}$, where θ is the angle of scattering. The radius vector $\vec{r}$ is $2p \sin\theta'$ so that

$$[(\vec{p}_0 - \vec{p})]^2 = (\vec{p}_0)^2 + (\vec{p})^2 - 2\vec{p}_0 \vec{p} \cos\theta = 2p^2[1 - \cos\theta] = 4p^2 \sin^2\frac{\theta}{2}$$

$$\therefore (\vec{p}_0 - \vec{p}) \cdot \vec{r} = 2pr \sin\frac{\theta}{2} \cos\theta'$$

θ' is the angle between $(\vec{p}_0 - \vec{p})$ and $\vec{r}$.

Spherical coordinates (r, θ', ϕ) are employed in order to find out the volume integral so that the polar axis is parallel to $(\vec{p}_0 - \vec{p})$.

$$H_{ll'}^{(1)} = \frac{Z_1 Z_2 e^2}{V} \int_0^\infty \int_0^\pi \frac{\exp\left[\left(\frac{i}{\hbar}\right) 2pr \sin\frac{\theta}{2}\right] \cos\theta'}{r} 2\pi r^2\, dr \sin\theta' d\theta'$$

$$= \frac{2\pi Z_1 Z_2 e^2}{V} \int_0^\infty \int_0^\pi \left[\frac{\exp\left[\left(\frac{i}{\hbar}\right) 2pr \sin\frac{\theta}{2} \cos\theta'\right]}{\left(\frac{i}{\hbar}\right) 2pr \sin\frac{\theta}{2}}\right]_{\theta'=0}^{\theta'=\pi} r\, dr$$

$$= \frac{2\pi Z_1 Z_2 e^2 \hbar^2}{V p \sin\frac{\theta}{2}} \int_0^\infty \sin\left(\frac{2pr}{\hbar} \sin\frac{\theta}{2}\right) dr$$

$$= \frac{2\pi Z_1 Z_2 e^2 \hbar^2}{2V p^2 \sin^2\left(\frac{\theta}{2}\right)} \int_0^\infty \sin x\, dx, \left[\text{where } x = \frac{2pr}{\hbar} \sin\frac{\theta}{2}\right] \tag{14.47}$$

Using the limiting value 1 for

$$\int_0^\infty \sin x \, dx$$

we have

$$H_{ll'}^{(1)} = \frac{2\pi Z_1 Z_2 e^2 \hbar^2}{2V p^2 \sin^2 \frac{\theta}{2}} \tag{14.48}$$

Now, the level density of the sub-cluster of level corresponding to scattering in direction (θ, ϕ) is

$$\rho(E, \theta, \phi)\, d\omega = \frac{\rho(E)}{4\pi}\, d\omega = \frac{V}{(2\pi\hbar)^3} \frac{p^2}{v}\, d\omega \tag{14.49}$$

$\left[\text{since } \rho(E) = \dfrac{L^3 p^2}{2\pi^3 \hbar^3 v}\right]$

This is valid for an isotropic space and momentum vector pointing into the solid angle $d\omega$.

Transition probability per unit time is given by

$$T = \frac{2\pi}{\hbar} \left|H_{ml}^{(1)}\right|^2 \rho(m)$$

If $T_{l\to l'}$ be the probability per unit time that a system makes transition to final state in solid angle $d\omega$, then

$$T_{l\to l'}\, d\omega = \frac{2\pi}{\hbar}\left[\frac{2\pi Z_1 Z_2 e^2 \hbar^2}{2V p^2 \sin^2 \frac{\theta}{2}}\right]^2 \frac{V p^2}{(2\pi\hbar)^3 v}\, d\omega = \frac{Z_1 Z_2^2 e^4}{4m^2 v^4} \frac{1}{\sin^4(\theta/2)} \frac{v}{V}\, d\omega \tag{14.50}$$

Now, the differential cross-section is

$$\sigma(\theta, \phi)\, d\omega = \frac{\text{Scattered flux}}{\text{Incident flux}} = \frac{T_{l\to l'}\, d\omega}{p/mL^3} \tag{14.51}$$

$$\therefore \sigma(\theta, \phi) = \frac{Z_1^2 Z_2^2 e^4}{4m^2 v^4} \cdot \frac{1}{\sin^4(\theta/2)} \frac{v}{V} \frac{mL^3}{p} \quad \text{or, } \sigma(\theta, \phi) = \frac{Z_1^2 Z_2^2 e^4}{4m^2 v^4} \cdot \frac{1}{\sin^4 \frac{\theta}{2}} \tag{14.52}$$

This is the Rutherford's scattering formula.

This is an important problem where quantum mechanics and classical mechanics give the same result. This is due to the fact that $\hbar$ does not appear in the expression and when $\hbar \to 0$, the classical picture corresponds to quantum picture.

14.7 Einstein transition probabilities relating to emission and absorption of radiation

We consider an atom exposed to monochromatic electromagnetic wave for which magnetic field H is equal in magnitude to the electric field E as shown in Fig 14.3. Both E and H act on the electron.

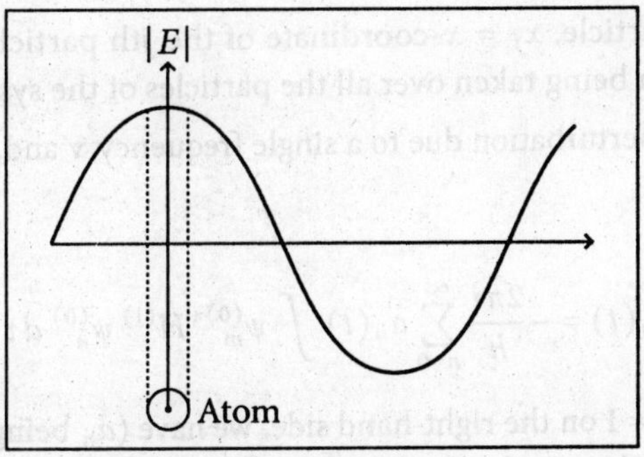

Fig 14.3 Atom exposed to electromagentic wave

The effect of H may be neglected for electrons moving slowly compared to the velocity of light. The time variation of monochromatic electromagnetic wave may be given as

$$E_x(\nu) = 2A\cos\omega t \tag{14.53}$$

where $2A$ represents the amplitude of the wave.

The dimension of the atom is of the order of 10^{-8} cm, while the wavelength of light in the visible region is of the order of 10^{-5} cm which is much greater than the dimension of the atom and as such the field may be regarded as constant. Thus,

$$E_x(\nu) = 2A\cos\omega t = 2A\cos 2\pi\nu t = A[\exp(i2\pi\nu t) + \exp(-i2\pi\nu t)] \tag{14.54}$$

The energy density of radiation of frequency ν in space, with unit dielectric constant and magnetic permeability is given by

$$\rho(\nu) = \overline{\frac{E^2 + H^2}{8\pi}} = \frac{1}{4\pi}\overline{E^2(\nu)} \tag{14.55}$$

where $\overline{E^2(\nu)}$ represents the average value of the square of the electric field strength corresponding to this radiation. The distribution being isotropic we can write:

$$\tfrac{1}{3}\overline{E^2(\nu)} = \overline{E_x^2(\nu)} = \overline{E_y^2(\nu)} = \overline{E_z^2(\nu)} \tag{14.56}$$

$E_x(\nu)$ represents the component of electric field in X-direction, etc.

Now,

$$\rho(\nu) = \frac{1}{4\pi}\overline{E^2(\nu)} = \frac{3}{4\pi}\overline{E_x^2(\nu)} = \frac{6}{4\pi}A^2 \tag{14.57}$$

since the average value of $\cos^2 2\pi\nu$ is $\frac{1}{2}$.

Now, we consider two stationary states m and n of an unperturbed system having energies E_m and E_n $(E_m > E_n)$ and wave functions $\psi_m^{(0)}$ and $\psi_n^{(0)}$ respectively. Let at $t = 0$ the system be in the lowest state n and come under the perturbing influence of radiation, the electric field strength of each frequency is given by eq. (14.53).

We shall now calculate the probability of the transition to the state m. The perturbation energy for a system of electrically charged particles in electric field E_x, parallel to x-axis is given by

$$H^{(1)} = E_x F_j e_j x_j \tag{14.58}$$

where e_j = charge of the jth particle, x_j = x-coordinate of the jth particle, $\sum_j e_j x_j$ = x-component of electric dipole moment (the sum being taken over all the particles of the system).

Temporarily, we consider the perturbation due to a single frequency ν and, then finally summing over all the frequencies.

We know that

$$\dot{a}_m(t) = -\frac{2\pi i}{h}\sum_{n=0}^{\infty} a_n(t)\int \psi_m^{(0)*} H^{(1)}\,\psi_n^{(0)}\,d\tau$$

Introducing $a_m^{(0)} = 0$ and $a_n^{(0)} = 1$ on the right-hand side, we have (a_n being the coefficient of a particular state and all other coefficients in the sum being zero).

$$\dot{a}_m(t) = -\frac{2\pi i}{h}\int \psi_m^{(0)*} H^{(1)}\,\psi_n^{(0)}\,d\tau$$

$$= -\frac{2\pi i}{h}\int \psi_m^{(0)*} e^{2\pi i E_m t/\hbar} \times A(\nu)\left(e^{2\pi i\nu t} + e^{-2\pi i\nu t}\right)\sum_j e_j x_j \times \psi_n^{(0)} e^{-2\pi i E_n t/h}\,d\tau \tag{14.59}$$

Let

$$(\mu_x)_{mn} = \int \psi_m^{(0)*}\sum_j e_j x_j \psi_n^{(0)}\,d\tau \tag{14.60}$$

Putting in eq. (14.60), we get

$$\frac{da_m}{dt} = -\frac{2\pi i}{h}(\mu)_{mn} A(\nu) e^{2\pi i(E_m - E_n + h\nu)t/h} + e^{2\pi i(E_m - E_n - h\nu)t/h} \tag{14.61}$$

Integrating between 0 and t

$$a_m(t) = -\frac{2\pi i}{h}(\mu_x)_{mn} A(\nu)\left[\left\{\frac{\exp\{2\pi i(E_m - E_n + h\nu)t/h\}}{2\pi i(E_m - E_n + h\nu)/h}\right\}_0^t + \left\{\frac{\exp\{2\pi i(E_m - E_n - h\nu)t/h\}}{2\pi i(E_m - E_n - h\nu)/h}\right\}_0^t\right]^0$$

$$= -\frac{2\pi i}{h}(\mu_x)_{mn} A(\nu)\left[\left\{\frac{\exp\{2\pi i(E_m - E_n + h\nu)t/h - 1}{2\pi i(E_m - E_n - h\nu)/h}\right\} + \left\{\frac{\exp\{2\pi i(E_m - E_n - h\nu)t/h}{2\pi i(E_m - E_n - h\nu)/h}\right\}\right]$$

$$= (\mu_x)_{mn} A(\nu)\left[\frac{1 - \exp\{2\pi i(E_m - E_n + h\nu)t/h}{E_m - E_n + h\nu} + \frac{1 - \exp\{2\pi i(E_m - E_n - h\nu)t/h\}}{E_m - E_n - h\nu}\right] \tag{14.62}$$

Now, for a single frequency the term $(\mu_x)_{mn} A$ is always small. The denominator $E_m - E_n - h\nu$ causes the influence of the perturbation in changing the system from the state n to a state m; hence neglecting the first term

$$a_m(t) = (\mu_x)_{mn} A(\nu)\left[\frac{1 - \exp\{2\pi i(E_m - E_n - h\nu)t/h\}}{E_m - E_n - h\nu}\right] \tag{14.63}$$

(In case of induced emission, the first term plays the same role).

Now,

$$a_m^*(t)\,a_m(t) = (\mu_x^2)_{mn} A^2(\nu)\frac{1}{(E_m - E_n - h\nu)^2}$$

$$\times [1 - \exp\{2\pi i(E_m - E_n - h\nu)t/\hbar\}][1 - \exp\{-2\pi i(E_m - E_n - h\nu)t/h\}]$$

$$\text{or, } a_m(t)^* a_m(t) = (\mu_x^2)_{mn} A^2(\nu)\frac{4\sin^2\{(\pi/\hbar)(E_m - E_n - h\nu)t}{(E_m - E_n - h\nu)^2} \tag{14.64}$$

[It can be proved that $[1-\exp\{2\pi i(E_m - E_n - h\nu)\,t/h\}]\times[1-\exp\{-2\pi i(E_m - E_n - h\nu)\,t/h\}]$
$= 4\sin^2\{\pi/h(E_m - E_n - h\nu)\,t\}]$

Equation (14.64) is only due to a single frequency and hence in order to include a range of frequencies we integrate it.

Hence,

$$a_m(t)^* a_m(t) = 4\left(\mu_x^2\right)_{mn} A^2(\nu_{mn}) \int \frac{\sin^2\{(\pi/h)(E_m - E_n - h\nu)\,t\}}{(E_m - E_n - h\nu)^2} \quad (14.65)$$

The integration can be taken from $-\infty$ to $+\infty$.

$$\therefore a_m(t)^* a_m(t) = 4\left(\mu_x^2\right)_{mn} A^2(\nu_{mn}) \int_{-\infty}^{+\infty} \frac{\sin^2\{(\pi/h)(E_m - E_n - h\nu)\,t\}}{(E_m - E_n - h\nu)^2}\, d\nu \quad (14.66)$$

Now, consider

$$I = \int_{-\infty}^{\infty} \frac{\sin^2\{(\pi/h)(E_m - E_n - h\nu)\,t\}}{(E_m - E_n - h\nu)^2}\, d\nu$$

Let

$$\{(\pi/h)(E_m - E_n - h\nu)\,t\} = x$$

Hence,

$$I = \left(\frac{\pi}{h}t\right)^2 \int_{-\infty}^{+\infty} \frac{\sin^2 x}{x^2}\left(-\frac{dx}{\pi t}\right) = \frac{\pi t}{h^2}\int_{-\infty}^{+\infty} \frac{\sin^2 x}{x^2}\, dx$$

$$= \frac{\pi^2}{h^2}t \left[\because \int_{-\infty}^{+\infty} \frac{\sin^2 x}{x^2}\, dx = \pi\right] \quad (14.67)$$

Substituting the value of this integral in eq. (14.65), we get

$$a_m(t)^* a_m(t) = \frac{4\pi^2}{h^2}\left(\mu_x^2\right)_{mn} A^2(\nu_{mn})\, t \quad (14.68)$$

The probability of transition per unit time from state n to m is

$$\frac{4\pi^2}{h^2}\left(\mu_x^2\right)_{mn} A^2(\nu_{mn})$$

Substituting the value of A from eq. (14.67), we get

$$a_m(t)^* a_m(t) = \frac{4\pi^2}{h^2}\left(\mu_x^2\right)_{mn} \frac{4\pi}{6}\rho(\nu_{mn}) \quad (14.69)$$

From (14.24), we obtain the Einstein coefficient of absorption $B_{n\to m}$ as

$$B_{n\to m} = \frac{8\pi^3}{3h^2}\left[\left(\mu_x^2\right)_{mn} + \left(\mu_y^2\right)_{mn} + \left(\mu_z^2\right)_{mn}\right] \quad (14.70)$$

Using the classical relation between $A_{m\to n}$ and $B_{n\to m}$ the Einstein coefficient of spontaneous emission is

$$A_{m\to n} = \frac{64\pi^4\nu_{mn}^3}{3c^3 h}\left[\left(\mu_x^2\right)_{mn} + \left(\mu_y^2\right)_{mn} + \left(\mu_z^2\right)_{mn}\right] \quad (14.71)$$

14.8 Selection rules for simple harmonic oscillator

Let a particle execute a harmonic motion along X-axis and possess a charge $+e$, the instantaneous dipole moment is ex. At the equilibrium, the charge is $-e$ for an oscillator type behaviour.

The transition between two states can be represented by the integral

$$e\int \psi_m^{(0)*} x \psi_n^{(0)}\, d\tau = ex_{mn} \tag{14.72}$$

This should be non-vanishing for a transition to take place between states m and n.

Now,

$$x_{mn} = \int_{-\infty}^{+\infty} \frac{N_n N_m}{\alpha} H_n\, H_m \exp(-\xi^2)\, \xi\, d\xi \tag{14.73}$$

(in case of linear harmonic oscillator)

$$= \frac{N_n N_m}{\alpha} \int_{-\infty}^{+\infty} H_n\, H_m \exp(-\xi^2)\, \xi\, d\xi$$

But,

$$\exp\left(-s^2 + 2s\xi\right) = \sum_{n=0}^{\infty} \frac{H_n(\xi)\, s^n}{n!}$$

$$\text{and } \exp\left(-t^2 + 2t\xi\right) = \sum_{n=0}^{\infty} \frac{H_n(\xi)\, t^m}{m!}$$

$$\therefore \int_{-\infty}^{+\infty} \exp(-s^2 + 2s\xi) \exp(-t^2 + 2t\xi) = \sum_{n=0}^{\infty}\sum_{m=0}^{\infty} \frac{s^m\, t^n}{n!\, m!} \int_{-\infty}^{+\infty} H_n(\xi)\, H_m(\xi)\, d\xi$$

Multiplying by $\xi \exp(-\xi^2)$, we get

$$\int_{-\infty}^{+\infty} \exp(-s^2 + 2s\xi) \exp(-t^2 + 2t\xi) \exp(-\xi^2)\, d\xi$$

$$= \sum_{r=0}^{\infty}\sum_{m=0}^{\infty} \frac{s^m\, t^m}{n!\, m!} \int_{-\infty}^{+\infty} H_n(\xi)\, H_m(\xi) \exp(-\xi^2)\, \xi\, d\xi \tag{14.74}$$

We now consider the integral

$$I = \int_{-\infty}^{+\infty} \exp(-s^2 + 2s\xi) \exp(-t^2 + 2t\xi)^{-\xi^2} e\, \xi\, d\xi = \int_{-\infty}^{+\infty} \exp\left[-(\xi - s - t^2)\right] \xi\, d\xi$$

$$= e^{2st} \int_{-\infty}^{+\infty} \exp\left[-(\xi - s - t)^2\right] \xi\, d\xi = e^{2st} \int_{-\infty}^{+\infty} \exp\left[-(\xi - s - t)^2\right] (\xi - s - t)\, d(\xi - s - t)$$

$$+ e^{2st}(s + t) \int_{-\infty}^{+\infty} \exp[-(\xi - s - t)^2] d(\xi - s - t) \tag{14.75}$$

The first integral vanishes and the second integral gives a value $\sqrt{\pi}$. Thus,

$$I = \sqrt{(\pi)}\, (s + t)\, e^{2st}$$

$$I = \sqrt{\pi}(s+t)\left\{1 + \frac{2st}{1!}\frac{(2st)^2}{2!} + \cdots + \frac{(2st)^n}{n!} + \cdots\right\}$$

$$= \sqrt{\pi}\left[s + \frac{2s^2t}{1!} + \frac{2s^3t^2}{2!} + \cdots + \frac{2^n s^{n+1} t^n}{n!} + \cdots + t + \frac{2st^2}{1!} + \frac{2^2 s^2 t^3}{2!} + \cdots + \frac{2^n s^n t^{n+1}}{n!} + \cdots\right]$$

$$\therefore I = \sqrt{\pi}\left[s + \frac{2s^2t}{1!} + \cdots + \frac{2^n s^{n+1} t^n}{n!} + \cdots + t + \frac{2st^2}{1!} + \cdots + \frac{2^n s^n t^{n+1}}{n!}\right]$$

$$= \sum_{n=0}^{\infty}\sum_{m=0}^{\infty} \frac{s^n t^m}{n!\,m!}\int_{-\infty}^{+\infty} H_m(\xi)H_n(\xi)\xi \exp(-\xi^2)\,d\xi \tag{14.76}$$

Now, comparing the coefficient of $s^n t^n$, we see that x_{mn} is zero except for $m = n+1$. This shows that m will be equal to $n-1$ and $n+1$ for non-vanishing integral.

Thus, $m = n \pm 1$ or, $(m-n) = \pm 1$, i.e., $\Delta n = \pm 1$.

Hence, the only transitions which this system can undergo with the emission or absorption of radiations is from a given stationary state to the two adjacent states.

14.9 Adiabatic and sudden perturbations

1. General adiabatic perturbation

This case has been treated for a slowly varying potential when the potential is so small that perturbation can be applied. Here, we shall extend this treatment to a more general problem in which the perturbing potential undergoes large changes.

The Schrödinger equation

$$i\hbar\frac{\partial\psi}{\partial t} = H(t)\,\psi \tag{14.77}$$

will be solved for $H(t)$ varying slowly with time. When $H(t)$ varies slowly enough we may expect that a good approximation should be given by solving Schrödinger's equation at each instant of time under the assumption that H is a constant and equal to its instantaneous value $H(t')$, where t' is the value of t at which we wish to evaluate H. The stationary state wave functions obtained by setting $t = t' =$ constant, would satisfy the equation:

$$H(t')\,\psi_n(x,t') = E_n(t')\,\psi_n(x,t) \tag{14.78}$$

In case H is a slowly varying function of t', a good approximate solution is

$$\psi_n = \psi_n(x,t)\exp\left[-\frac{i}{\hbar}\int_0^t E_n(t')\,dt'\right] \tag{14.79}$$

If the wave function is known at $t = 0$, then at later time we put

$$\psi = \sum a_n(t)\,\psi_n(x,t)\exp\left[-\frac{i}{\hbar}\int_0^t E_n(t')\,dt'\right] \tag{14.80}$$

Substituting eq. (14.80) in eq. (14.77), we get

$$i\hbar\sum_n\left(\dot{a}_n\psi_n + a_n\frac{\partial\psi_n}{\partial t}\right)\exp\left(-\frac{i}{\hbar}\int_0^t E_n(t')\,dt'\right) + \sum_n a_n\psi_n\exp\left(-\frac{i}{\hbar}\int_0^t E_n(t')\,dt'\right)$$

$$= H(t)\sum_n a_n\psi_n\exp\left(-\frac{i}{\hbar}\int_0^t E_n(t')\,dt'\right)$$

$$\text{or, } i\hbar\sum_n\left(\dot{a}_n\psi_n + a_n\frac{\partial\psi_n}{\partial t}\right)\exp\left(-\frac{i}{\hbar}\int_0^t E_n(t')\,dt'\right) + \sum_n a_n\psi_n E_n\exp\left(-\frac{i}{\hbar}\int_0^t E_n(t')\,dt'\right)$$

$$= \sum a_n E_n\psi_n\exp\left(-\frac{i}{\hbar}\int_0^t E_n(t')\,dt'\right)$$

$$\text{or, } i\hbar\sum_n\left(\dot{a}_n\psi_n + a_n\frac{\partial\psi_n}{\partial t}\right)\exp\left(-\frac{i}{\hbar}\int_0^t E_n(t')\,dt'\right) = 0 \tag{14.81}$$

Now, multiplying eq. (14.81) by

$$\psi_n^*\exp\left(\frac{i}{\hbar}\int_0^t E_m(t')\,dt'\right)$$

and integrating over all space, we get

$$\int\sum_n\dot{a}_n\psi_n\psi_m^*\exp\left(-\frac{i}{\hbar}\int_0^t E_n(t')\,dt'\right)\exp\left(\frac{i}{\hbar}\int_0^t E_m(t')\,dt'\right)d\tau$$

$$+\sum_n a_n\int\psi_m^*\frac{\partial\psi_n}{\partial t}\exp\left(-\frac{i}{\hbar}\int_0^t (E_n - E_m)\,dt'\right)d\tau = 0$$

Orthogonality and normalisation conditions give

$$\dot{a}_m = -\sum_n a_n\psi_m^*\frac{\partial\psi_n}{\partial t}\exp\left(-\frac{i}{\hbar}\int_0^t (E_n - E_m)\,dt'\right)d\tau$$

$$\text{or, } \dot{a}_m = -\sum_n a_n\int\psi_m^*\frac{\partial\psi_n}{\partial t}\exp\left(-\frac{i}{\hbar}\int_0^t (E_n - E_m)\,dt'\right)d\tau \tag{14.82}$$

To obtain the value of the integral

$$\int\psi_m^*\frac{\partial\psi_n}{\partial t}\,d\tau$$

we differentiate the equation

$$H(t)\,\psi_n(t) = E_n(t)\,\psi_n \tag{14.83}$$

with respect to t.

Thus,

$$\frac{\partial H}{\partial\psi}\psi_n + H\frac{\partial\psi_n}{\partial t} = \frac{\partial E_n}{\partial t}\psi_n + E_n\frac{\partial\psi_n}{\partial t}$$

Multiplying the above equation by ψ_m^* and integrating over the coordinates

$$\int\psi_m^*\frac{\partial H}{\partial t}\psi_n\,d\tau + \int\psi_m^* H\frac{\partial\psi_n}{\partial t}\,d\tau = \frac{\partial E_n}{\partial t}\int\psi_m^*\psi_n\,d\tau + E_n\int\psi_m^*\frac{\partial\psi_n}{\partial t}\,d\tau \tag{14.84}$$

Now,

$$\int\psi_m^*\psi_n\,d\tau = 0$$

and remembering that H is Hermitian

$$\int \psi_m^* H \frac{\partial \psi_n}{\partial t} d\tau = \int (G\psi_m)^* \frac{\partial \psi_n}{\partial t} dt = E_m \int \psi_m^* \frac{\partial \psi_n}{\partial t} d\tau$$

Substituting in eq. (14.84), we get

$$\int \psi_m^* \frac{\partial H}{\partial t} \psi_n \, d\tau + E_m \int \psi_m^* \frac{\partial \psi_n}{\partial t} d\tau = E_n \int \psi_m^* \frac{\partial \psi_n}{\partial t} d\tau$$

$$\text{or, } (E_m - E_n) \int \psi_m^* \frac{\partial \psi_n}{\partial t} d\tau = -\int \psi_m^* \frac{\partial H}{\partial t} \psi_n \, d\tau$$

$$\text{or, } \int \psi_m^* \frac{\partial \psi_n}{\partial t} d\tau = -\frac{\int \psi_m^* \left(\frac{\partial H}{\partial t} \psi_n \, d\tau\right)}{(E_m - E_n)}, \quad m \neq n \tag{14.85}$$

Substituting this in eq. (14.82)

$$\dot{a}_m = -\sum_{n \pm m} \frac{a_n \left[-\int \psi_m^* \left(\frac{\partial H}{\partial t}\right) \psi_n \, d\tau\right] \exp\left(-\frac{i}{\hbar}\int_0^t (E_n - E_m)\, dt'\right)}{(E_m - E_n)}$$

$$\text{or, } \dot{a}_m + \sum_{m \neq n} \frac{\int \psi_m^* \left(\frac{\partial H}{\partial t}\right) \psi_n \, d\tau \exp\left(-\frac{i}{\hbar}\int_0^t (E_n - E_m)\, dt'\right)}{(E_n - E_m)} \tag{14.86}$$

Now, to proceed as in the method of variation of constants, let the system starts with $a_s = 1$ and $a_n = 0$ for $n \pm s$, then we can solve for a_m by successive approximation

$$\dot{a}_{ms} + \frac{\left(\frac{\partial H}{\partial t}\right)_{ms}}{(E_s - E_n)} \exp\left(-\frac{i}{\hbar}\int_0^t (E_s - E_m)\, dt'\right) = 0 \tag{14.87}$$

where $\left(\frac{\partial H}{\partial t}\right)_{ms} = \int \psi_m^* \left(\frac{\partial H}{\partial t}\right) \psi_s d\tau.$

To estimate the matrix element we can neglect the slow change of $\left(\frac{\partial H}{\partial t}\right)_{ms}$. Then

$$a_{ms} = \frac{\hbar}{i(E_s - E_m)^2} \left(\frac{\partial H}{\partial t}\right)_{ms} \left[\exp\left(-\frac{i}{\hbar}(E_s - E_m)\, t\right) - \exp\left(-\frac{i}{\hbar}(E_s - E_m)\, t_0\right)\right] \tag{14.88}$$

The exponential factor in eq. (14.88) is at most of the order of unity. Hence, the total probability of transmission to the mth level is less than

$$|a_{ms}|^2 = \frac{4\hbar^2}{(E_m - E_s)^4} \left|\left(\frac{\partial H}{\partial t}\right)_{ms}\right|^2 \tag{14.89}$$

If $\left(\frac{\partial H}{\partial t}\right)_{ms}$ is small enough, a negligible error is made by neglecting $|a_{ms}|^2$ and saying that the system remains in the state $\psi_s(x, t)$ even through ψ_s itself is changing with time. This is known as *adiabatic approximation.*

An application of adiabatic approximation

We consider the collisions of the gas molecules. If the gas atoms approach each other, then the question is whether the force resulting from their interaction can cause transition among the electronic states. It is known that the molecular velocities are ~ 10^4 cm/s whether velocities of electrons in atoms are much higher ~ 10^8 cm/s. As a result in an electronic period of rotation in the atom, the molecule does not move very far, so the interaction energy does not change appreciably. Hence, the collision may be regarded as adiabatic process in which the electron remains in its original quantum state.

Note: *Adiabatic approximation is valid only for smallness of* $\left(\frac{\partial H}{\partial t}\right)$ *as well as* $(E_m - E_n)$ *not too small.*

2. Sudden approximation

When the change in the Hamiltonian occupies a very short but finite interval of time, we use sudden approximation. Let us suppose that at the time $t = 0$, the Hamiltonian suddenly changes from H_0 to H_1 and remains constant. Then the wave functions are given by $\psi_n \exp[(-i/\hbar)E_n^{(0)}(t)]$, where $H_0\psi_n \simeq E_0^{(0)}\psi_n$.

After $t = 0$, the eigenfunction and Hamiltonian operator are denoted as $\psi_m e^{(-i/\hbar)E_m t}$ and $H_1\psi_m = E_m\psi_m$.

To find how ψ changes after $t = 0$, we adopt the procedure of expanding it in series solution of Schrödinger's equation, i.e.,

$$\psi_n(x) = \sum_m C_{mn}\,\psi_m(x) \tag{14.90}$$

The values of C_{mn} are obtained by multiplying eq. (14.90) by $\psi_m^*(x)$ and integrating. We have

$$C_{mn} = \int \psi_m^*(x)\,\psi_n(x)\,d\tau$$

After $t = 0$, the wave function becomes

$$\psi_n(y) = \sum_m C_{mn}\,\psi_m(y)\,e^{-(i/\hbar)E_m t} = \sum_m \int \psi_m^*(x)\,\psi_n(x)\,d\tau\psi_m(y)\,e^{-(i/\hbar)E_m t} \tag{14.91}$$

The above derivation holds for instantaneous changes of the Hamiltonian and must, therefore, hold for sufficiently rapid, but not instantaneous changes. The essential simplification resulting from the instantaneous change was that the wave function did not change, while the Hamiltonian was changing.

14.10 Answers of some important conceptual questions

14.1. *What is time-dependent perturbation?*

Ans. The perturbation theory of a system for which the Hamiltonian depends on time is the time-dependent perturbation. The energy eigenstates of such a system are stationary. It is sometimes called the method of variation of constant. The theory of time-dependent perturbation was developed by Dirac. The theory permits us to discuss such questions as the probability of transition of the system from one perturbed stationary state to another as a result of perturbation.

In atomic physics, particularly in spectral analysis, the main concern is the quantisation of energy and the time-dependent perturbation finds wide application. In nuclear physics particularly in nuclear reactions and disintegrations, the time-dependent method is frequently used.

14.2. *What is Fermi Golden Rule?*

Ans. Fermi Golden Rule deals with the transitions of states of continuous spectrum. Under certain limiting conditions we can define a probability of transition per unit time (to any group of states).

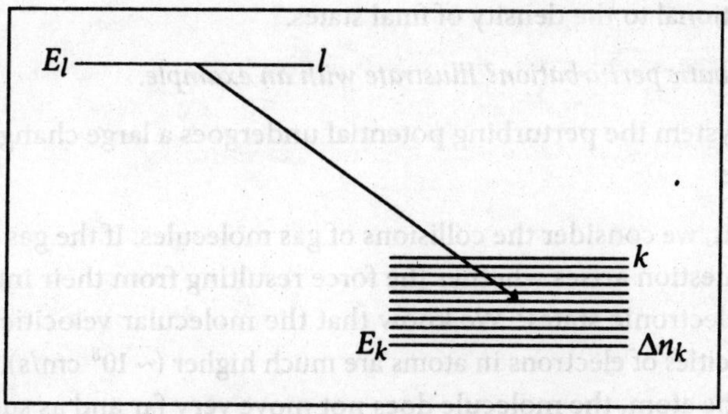

Fig 14.4 Transition to final state

Consider a range of energy ΔE for the final states in which the number of levels is Δn_k (Fig 14.4). The probability of transition to these Δn_k states is obtained by summing over all these states:

$$T_{kl}\Delta n_k = T_{kl}\frac{\nabla n_k}{\Delta E}\Delta E \tag{14.92}$$

If $\rho(E)$ denotes the density of sates we can write:

$$\rho(E) = \lim_{\Delta E \to 0}\frac{\Delta n_k}{\Delta E} = \frac{dn_k}{dE} \tag{14.93}$$

So the total probability of transition to any of the final states becomes

$$T_{kl} = \int_{-\infty}^{+\infty} T'_{kl}\rho(E)\,dE = \frac{\left|H_{kl}^{(1)}\right|^2 t^2}{\hbar^2}\int \frac{\sin^2 x}{x^2}\rho(E)\,dE \tag{14.94}$$

where T'_{kl} is the first order probability of transition from state l to the state k and $x = \omega_{kl}t/2$ so that

$$dx = \frac{1}{2}d\omega_{kl} = \frac{t}{2\hbar}dE \tag{14.95}$$

Because of the sharpness of the peak in Fig 14.1, we can take $\rho(E)$ outside the integral and write:

$$\begin{aligned} T_{kl} &= \frac{\left|H_{kl}^{(1)}\right|^2 t^2}{\hbar^2}\rho(E)\int_{-\infty}^{+\infty}\frac{\sin^2 x}{x^2}dx\frac{2\hbar}{t} = \frac{2t}{\hbar}\left|H_{kl}^{(1)}\right|^2\rho(E)\,\pi \\ &= \frac{2\pi}{\hbar}\left|H_{kl}^{(1)}\right|^2\rho(E)\,t \end{aligned} \tag{14.96}$$

Hence, the rate of transition per second is

$$T = \frac{dT_{kl}}{dt} = \frac{2\pi}{\hbar}\left|H_{kl}^{(1)}\right|^2\rho(E) \tag{14.97}$$

This formula has wide applications in quantum physics. The theory of scattering can be explained with the help of this formula.

Equation (14.97) is the *Fermi's golden rule* which states that the probability of transition per unit time is

(i) non-zero only between continuum states of the same energy

(ii) proportional to the square of $\left|H_{ml}^{(1)}\right|$ of the perturbation connecting the states

(iii) proportional to the density of final states.

14.3. *What is adiabatic perturbation? Illustrate with an example.*

Ans. If in a system the perturbing potential undergoes a large changes, then the perturbation is an adiabatic one.

To illustrate it, we consider the collisions of gas molecules. If the gas atoms approach each other, an important question arises whether the force resulting from their interaction can cause transitions among the electronic states? We know that the molecular velocities are usually low (~ 10^4 cm/s) whereas velocities of electrons in atoms are much higher (~ 10^8 cm/s). During an electronic period of rotation in the atom, the molecule does not move very far and as such the interaction energy does not change much. Hence, the collision may be regarded as an adiabatic process in which the electron remains in the original quantum state.

14.11 Solved problems

» ***Example* 14.1.** A system with discrete eigenstates $|\psi_n|$ and eigenenergies $|E_n|$ is exposed to the perturbation $H'(r,t) = H_0'(r)\, e^{-r^2/\tau^2}\sqrt{\pi}\tau$. The perturbation is turned on at $t = -\infty$, when the unperturbed system is in its ground state ψ_0. What is the probability that at $t = +\infty$, the system suffers a transition to the state $\psi_k, k > 0$?

Solution: For the first order perturbation, the transition probability is given by

$$P_{mi}^{(1)}(t) = h^{-2}\left|\int_{t_0}^{t} H'_{mi}(t')\, e^{i\omega_{mi}t'}\, dt'\right|^2, \quad m \neq i'$$

where $|i>$ and $|m>$ represent the initial and ground state respectively. Since, $H'(r,t)$ is separate

$$P_{ni}(t) = \left|\frac{H'_{ni}}{h}\right|^2 \left|\int_{-\infty}^{\infty} \frac{e^{i\omega t}\, e^{-t^2/\tau^2}}{\sqrt{\pi}}\, d(t/\tau)\right|^2$$

$$= \frac{|H'_{ni}|}{\pi\hbar^2}\left|\int_{-\infty}^{\infty} e^{i\omega' x - x^2}\, dx\right|^2, \quad \left[x = \frac{t}{\tau}, \omega' = \rho\omega\right]$$

$$= \frac{|H'_{ni}|}{\pi\hbar^2}\left|\int_{-\infty}^{\infty} \exp\left[-\left(x - \frac{i\omega'}{2}\right)^2 - \frac{\omega'^2}{4}\right] dx\right|^2$$

$$= \frac{|H'_{ni}|^2}{\pi\hbar^2}\left|e^{-\frac{\omega'^2}{4}}\int_{-\infty}^{+\infty} e^{-z^2} dz\right|^2, \quad \left(z = x - \frac{i\omega'}{2}\right)$$

$$\text{or, } P_{k0} = \frac{|H_{k0}|^2}{\hbar^2}\, e^{-(E_0 - E_k)\tau^2}/2\hbar^2$$

$$\left[\text{since} \int_{-\infty}^{\infty} e^{-x^2}\, dx = 2\int_0^{\infty} e^{-x^2}\, dx = 2 \times \frac{\sqrt{\pi}}{2} = \sqrt{\pi}\right]$$

» ***Example* 14.2.** A particle of charge e and mass m executes a simple harmonic motion along X-axis with a force constant k. An electric field $E(t) = E_0(x)\, e^{-t/\tau}$, where $E_0(x)$ and τ are time-independent, is switched on at $t = 0$, so that the system is perturbed by an interaction $H'(t) = -ex\, E(t)$. If the oscillator is in the ground state for $t \leqslant 0$, find the probability that the oscillator will be found in an excited state as $t \to \infty$.

Solution: The normalised wave function of the oscillator is

$$\psi_n(x) = e^{i\delta_n}\left(\frac{2\nu m}{h}\right)^{1/4}\frac{1}{\sqrt{2^n n!}}H_n\left(x\sqrt{\frac{2\pi\nu m}{h}}\right)e^{-\frac{2\pi\nu m}{h}\frac{x^2}{2}}$$

the letters have their usual significance.

Here, the oscillator wave function in terms of Hermite polynomials, can be written as

$$\psi_i = N_i e^{-\alpha^2x^2/2}H_i(\alpha x)$$
$$\psi_f = N_f e^{-\alpha^2x^2/2}H_f(\alpha x)$$

Since, $H'_{fi}(t) = -eE(t)\langle\psi_f|x|\psi_i\rangle$, we get on substitution $\xi = \alpha x$

$$x_{fi} = \frac{N_i N_f}{\alpha^2}\int_{-\infty}^{\infty} e^{-\xi^2}\xi H_i(\xi)\,d\xi$$

which has the solution

$$x_{fi} = \begin{cases} \frac{1}{\alpha}\left(\frac{f+1}{2}\right)^{1/2}, & i = f+1 \\ \frac{1}{\alpha}\left(\frac{f}{2}\right)^{1/2}, & i = f-1 \\ 0, & \text{otherwise} \end{cases}$$

In the present problem, $i = 0$ for the ground state. Hence, the only non-vanishing matrix is H'_{10}.

∴ the probability that the oscillator will be found in an excited state as $t \to \infty$ is

$$P_{10}(t=+\infty) = \frac{e^2E_0^2}{2m\hbar\omega}\left|\int_0^{\infty} e^{i\omega t}e^{-t/\tau}\,dt\right|^2 = \frac{e^2E_0^2}{2m\hbar\omega}\frac{\tau^2}{(\omega\tau)^2+1}$$

where $\omega = \sqrt{k/m}$.

***Example* 14.3.** A charge e of mass m is confined to a rectangular parallelepiped of sides $2a, 2b$ and $2c$. It is acted on by an electric field along X-axis $E(x,t) = E_0e^{-\alpha t}$, $\alpha > 0$, where α is a constant. Calculate the transition probability from the ground state at $t = 0$ to the first excited state at $t = \infty$.

Solution: The energy eigenvalue and eigenfunctions are

$$E_{jkl} = \frac{\pi^2\hbar^2}{8ma^2}(j^2+k^2+l^2),\ j = k, l = 1, 2, 3, \ldots$$

$$\psi_{jkl} = |jkl\rangle = a^{-3/2}\sin\frac{j\pi x}{2a}\sin\frac{k\pi y}{2a}\sin\frac{l\pi z}{2a}$$
$$= b^{-3/2}\sin\frac{j\pi x}{2b}\sin\frac{k\pi y}{2b}\sin\frac{l\pi z}{2b}$$
$$= c^{-3/2}\sin\frac{j\pi x}{2c}\sin\frac{k\pi y}{2c}\sin\frac{l\pi z}{2c}$$

The ground state is (111) and the first excited states $|211\rangle$, $|121\rangle$ and $|112\rangle$. Since, the electric field is along X-axis the dipole moment is ex and the perturbed Hamiltonian is

$$H' = -\vec{\mu}\cdot\vec{E} = -ex\,E_0e^{-\alpha t}$$

This gives rise to transition probability in first order.

$$P = |C_m^{(1)}|^2 = \frac{1}{\hbar^2}\left|\int_0^\infty H'_{mn} e^{i\omega_{mn}t}\,dt\right|^2, \quad \omega_{mn} = (E_m - E_n)\,l\hbar$$

The transition moment between $|111\rangle$ and $|211\rangle$ is

$$H'_{mn} = \langle 111|H'|211\rangle = \langle 111| - eE_0 x e^{-\alpha t}|211\rangle$$

$$= -\frac{eE_0 e^{-\alpha t}}{a^3}\int_0^{2a} x \frac{\sin \pi x}{2a} \sin\frac{\pi x}{a}\,dx$$

$$= -\frac{eE_0 e^{-\alpha t}}{a^3}\left(-\frac{32a^2}{9\pi^2}\right) a\cdot a = \frac{32aeE_0 e^{-\alpha t}}{9\pi^2}$$

Similarly,

$$\langle 111|H'|121\rangle = -\frac{eE_0 e^{-\alpha t}}{b^3}\int_0^{2b} x\sin^2\frac{\pi x}{2b}\,dx \times \int_0^{2b}\sin\frac{\pi y}{2b}\sin\frac{\pi y}{b}\,dy \times \int_0^{2b}\sin^2\frac{\pi z}{2b}\,dz$$

$$= 0$$

and $\langle 111|H'|112\rangle = 0$.

The angular frequency is given by

$$\omega_{21} = \frac{E_2 - E_1}{\hbar} = \frac{E_{211} - E_{111}}{\hbar} = \frac{\pi^2\hbar}{8ma^2}(2^2 + 1^2 + 1^2 - 1^2 - 1^2 - 1^2) = \frac{3\pi^2\hbar}{8ma^2}$$

Similarly, the transition probability from the ground state $|111\rangle$ at $t = 0$ to the first excited state $|211$ $t = \infty$ is

$$P = \left(\frac{32aeE_0}{9\pi^2\hbar}\right)^2\left|\int_0^\infty e^{-\alpha t + \omega_{21}t}\,dt\right|^2 = \left(\frac{32aeE_0}{9\pi^2\hbar}\right)^2\frac{1}{\alpha^2 + \omega_{21}^2}$$

» ***Example* 14.4.** A system in an unperturbed state n is suddenly disturbed by a constant perturbation H'(which exists during $0 \to t$. Find the transition probability from state m to state n and also its amplitude angular frequency.

Solution: When the perturbation is independent of time we can directly write the transition probabili

$$P_{mn} = |C_m(t)|^2 = \frac{4}{\hbar^2}\frac{|H'_{mn}|^2}{\omega_{mn}^2}\sin^2\frac{\omega_{mn}t}{2}$$

Thus, the transition probability varies simple harmonically with amplitude

$$\frac{4|H'_{mn}|^2}{\hbar^2\omega_{mn}^2} = \frac{4|H'_{mn}|^2}{(E_m - E_n)^2}$$

and the angular frequency

$$\frac{\omega_{mn}}{2} = \frac{E_m - E_n}{2\hbar}$$

EXERCISES

14(A): Objective (or MCQ) Type Questions

1. Fermi Golden rule deals with

(a) transition probability per unit time
(b) total energy of the system
(c) arrangement of energy levels
(d) frequency of spectral lines.

2. The theory of time dependent perturbation was developed by

(a) Max Born
(b) P.A.M. Dirac
(c) Sommerfeld
(d) Zeemann.

3. The Born approximation formula is

(a) $\sigma(\theta,\phi) = \left(\frac{m}{2\pi}\hbar^2\right)\left|\int e^{i\vec{k}\cdot\vec{r}}V(\vec{r})\,d\vec{r}\right|$

(b) $\sigma(\theta,\phi) = \left(\frac{m}{2\pi}\hbar^2\right)\left|\int e^{i\vec{k}\cdot\vec{r}}V(\vec{r})\,d\vec{r}\right|^2$

(c) $\sigma(\theta,\phi) = \left(\frac{\pi\hbar^2}{2m}\right)\left|\int e^{i\vec{k}\cdot\vec{r}}V(\vec{r})\,d\vec{r}\right|^2$

(d) $\sigma(\theta,\phi) = \left(\frac{m}{2\pi\hbar^2}\right)\left|\int e^{i\vec{k}\cdot\vec{r}}V(\vec{r})\,d\vec{r}\right|^2$.

4. The Rutherford scattering formula

$$\sigma(\theta,\phi) = \frac{z_1^2 z_2^2 e^4}{4m^2 v^4}\frac{1}{\sin^4\theta/2}$$

is the same in classical as well as quantum mechanics because

(a) the formula contains $z_1^2 z_2^2$
(b) the formula does not contain $\hbar$
(c) the formula contains m
(d) none of these.

5. Stark effect is the splitting of spectral lines in presence of

(a) magnetic field
(b) external electric field
(c) high energy radiation
(d) either by electric or magnetic field.

6. A LHS of charge $\bar{e}$ is perturbed by an electric field F in the positive x-direction the first order change in energy level is

(a) $(n + \frac{1}{2})\hbar\omega$
(b) 0
(c) $(n + \frac{1}{2})\hbar\omega$
(d) $\hbar\omega$.

7. The transition probability from ground state $|111\rangle$ at $t = 0$ to the first excited state $|211\rangle$ at $t = \infty$ is

(a) $\left(\frac{32aeE_0}{9\pi^2\hbar}\right)^2 \frac{1}{a^2 + \omega_{21}^2}$

(b) $\left(\frac{32aeE_0}{\pi^2\hbar^2}\right)^2 \frac{1}{a^2 + \omega_{12}^2}$

(c) $\left(\frac{aeE_0}{9\pi^2\hbar}\right)^2 \frac{1}{a^2 + \omega_{12}^2}$

(d) $\left(\frac{16aeE_0}{9\pi^2\hbar}\right)^2 \frac{1}{a^2 + \omega_{12}^2}$.

ANSWERS

14.1. (a). **14.2.** (b). **14.3.** (d). **14.4.** (b). **14.5.** (b). **14.6.** (b). **14.7.** (a).

14(B): Short Answer Type Questions

14.1. A perturbation H' is sinusoidal in time $H' = V(\vec{r})\cos\omega t$ and the unperturbed Hamiltonian H_0 is time-independent. Show that the first order transition probability of the system from its initial state $|a\rangle$ to final sate $|b\rangle$ can be written as

$$P_{ab}(t) = \frac{1}{2}\frac{|V_{ab}|^2}{\hbar^2}F(t, \Delta\omega)$$

in the approximation when the driving frequency ω_0 and $\Delta\omega$ is defined by $\Delta\omega = \omega_0 - \omega$. Show the nature of the function F by depicting graphically its variation with ω and t. **[NBU MSc 2012]**

14.2. A spin $\frac{1}{2}$ particle is in an eigenstate of S_y with eigenvalue $\hbar/2$ at time $t = 0$. At what time it is placed in a constant magnetic field B in the z-direction. The spin is allowed to precess for a time T. At that instant the magnetic field is very quickly switched to the X-direction. After another time interval J a measurement of the y-component of the spin is made. What is the probability that the value $\hbar/2$ will be found?

14.3. Consider a system of spin $\frac{1}{2}$. What are the eigenstates and eigenvalues of the operator $S_s + S_y$? Suppose, a measurement of this quantity is made and the system is found to be in the eigenstate with the larger eigenvalues. What is the probability that a subsequent measurement of S_y yields $\hbar/2$?

14.4. Prove that the transition probability per unit time is $\frac{2\pi}{\hbar}\rho(k)\left|H'_{km}\right|^2$, where $\rho(k)$ denotes the density of the final states and H'_{km} is the matrix element of the perturbation H'_{kn}.

14.5. Calculate the transition probability per unit time and per unit of radiation intensity for a harmonic perturbation. Evaluate Einstein coefficient for spontaneous emission. **[Rohilkhand 1980, 77]**

14(C): Broad Answer Type Questions

14.1. Write about the sudden approximations. Show that the sudden approximation can be applied to the extranuclear electron initially present in the hydrogen atom and is superior to the other approximation methods that might be used.

14.2. Discuss briefly the time-dependent perturbation theory and derive an expression for the transition probability to a group of states per unit time. **[Agra U 1971]**

14.3. Give the time-dependent perturbations for a harmonic perturbation. Discuss the electric dipole approximation. **[Rohilkhand 1980, 77]**

14.4. (a) A quantum mechanical system is in a state $\psi(r)$ at $t = 0$. It is subjected to a large impulse $F(t) = P\delta(t)$. Show that immediately after impulse, the wave function of the system $\psi_i(\vec{r}) = \psi_0 e^{i\vec{p}\cdot\vec{v}/\hbar}$.

(b) Also obtain an expression for the probability that following the impulse, the system is found in the state n.

14.5. A small time-dependent perturbation $H_1(t)$ is applied to a system in the state ψ. Obtain the expansion coefficient $C_j(t)$ for the new state function when perturbation is kept on for a small time t.

Calculate the transition probability per unit time when $H_1(t) = A\cos\omega t$. **[Agra U 1981]**

14.6. Give time-dependent perturbation theory for case of perturbation which is constant in time except that it is switched on at time $t = 0$ and switched off at time t. **[Agra U 1981]**

14.7. Use time-dependent perturbation theory to obtain an expression for the transition probability in first order approximation.

Use this to obtain Rutherford scattering cross-section. **[Agra U 1973]**

14.8. Give time-dependent perturbation theory for a constant perturbation acting for a short interval of time. Relate the transition probability per unit time with the differential cross-section for scattering. **[Hohilkhand U 1988]**

14.9. Discuss the time-dependent perturbation theory. Apply the theory to obtain Fermi's golden rule.

14.10. Apply time-dependent perturbation theory to obtain Einstein's transition probabilities. With the help of the above theory arrive at the selection rules for simple harmonic oscillator.

14.11. State and prove Fermi-Golden Rule for the rate of transitions induced by a constant perturbation and apply it to the scattering problem, considering the scattering potential as the perturbation which induces transition from an initial momentum eigenstate $|\vec{k}\rangle$ to the final state $|\vec{k}'\rangle$.

14.12. Discuss briefly the time-dependent perturbation theory and derive an expression for the transition probability to a group of states per unit time.

14.13. Calculate the transition probability per unit time and per unit of radiation. Evaluate Einstein's coefficients for spontaneous emission.

14.14. (a) Show how the first order effect of a time-dependent perturbation varying sinusoidally in time leads to the emission and absorption of energy.

(b) Find the transition rates for emission and absorption.

14(D): Numerical Problems

14.1. Show that in case of transitions of states of continuous spectrum Fermi Golden Rule is given by

$$T = \frac{2\pi}{\hbar}\left|H_{kl}^{(1)}\right|^2 \rho(E)$$

14.2. A system is in unperturbed state n is suddenly disturbed by a constant perturbation $H'(r)$, which exists during $0 \to t$. Find the transition probability from stature m to state n and also its amplitude, and its angular frequency.

14.3. Obtain an expression for the transition probability in first order approximating using first order approximation.

Chapter 15

WKB Approximation and Variational Method

15.1 The WKB approximation

The WKB approximation is applicable in cases where the potential is a slowly varying function and the change in potential over the distance of the de Broglie wavelength is small compared to the KE of the particle.

These methods are useful in almost ideal situations. But, in practical situations, such as in chemical phenomenon the ideal conditions are not fulfilled. Under such circumstances, we apply *variational technique* (also also *variation method, the Rayleigh-Ritz method,* etc.) This method is widely used in atomic and nuclear physics, quantum chemistry and also is the genesis of more elaborate methods of *Hartree* and *Hartree-Fock* approximations.

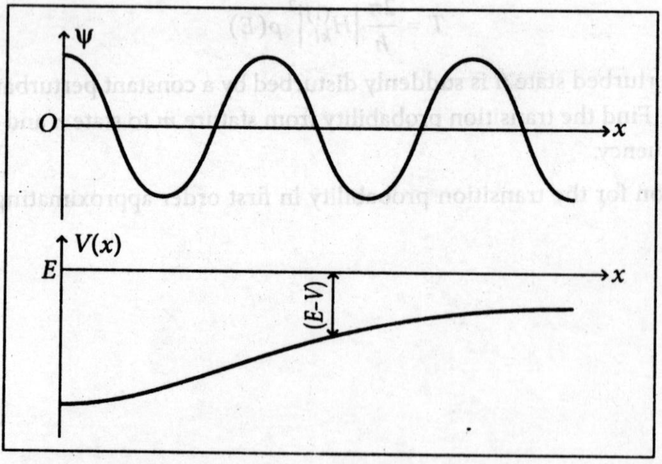

Fig 15.1 Slowly changing one-dimensional potential and its associated wave function

A slowly changing potential means the variation of potential energy $V(r)$ slightly over several de Broglie wavelengths of the as shown in Fig 15.1.

Phase integral method

Phase integral method is the approximate method of solving the wave equation. This method is based on the expansion, the first term of which leads to the classical result, the second term to the old quantum theory results and the higher terms, to the characteristic of the new mechanics.

The de Broglie wavelength of a particle of mass m having energy E moving in a potential V is given by

$$\lambda = \frac{h}{mv} = \frac{h}{[2m(E-V)]^{1/2}}$$

Since, $\frac{1}{2}mv^2 = E - V$ and $m^2v^2 = 2m(E-V)$.

The propagation constant k is given by

$$k = \frac{2\pi}{\lambda} = \frac{[2m(e-V)]^{1/2}}{\hbar} \tag{15.1}$$

Mathematically, slowly varying potential can be expressed by the condition

$$\left|\frac{1}{k^2}\frac{dk}{dx}\right| \ll 1 \tag{15.2}$$

Substituting the value of k in eq. (15.1), we get

$$\left|\frac{m\hbar\dfrac{\partial v}{\partial x}}{[2m(E-V)]^{3/2}}\right| \ll 1 \text{ or, } \frac{\lambda\left|\dfrac{\partial V}{\partial x}\right|}{4\pi(E-V)} \ll 1 \tag{15.3}$$

Equation (15.2) gives the validity of WKB *approximation.*

So the change in the potential V in a distance λ must be small compared to the KE $(E-V)$ of the particle.

15.2 Method of classical approximation

The method of classical approximation has been developed, among others by *G. Wentzel, H.A. Kramers* and *L. Brillouin* and is commonly known as WKB approximation.

For one-dimensional problem, Schrödinger equation is given by

$$-\frac{\hbar^2}{2m}\frac{\partial^2\psi}{\partial x^2} + (V-E)\psi = 0 \tag{15.4}$$

We have seen above that the method is applicable when the potential $V(x)$ is very slowly varying. Hence, the wave function $\psi(x)$ is not much different from the case in which V is constant. In latter case, the wave function is given by

$$\psi(x) = \exp\left(\frac{ip_2x}{\hbar}\right), \text{ where } p_2 = \sqrt{2m(E-V)}$$

This suggests that when $E > V$, it will be convenient to write the wave function

$$\psi = \exp[iS/\hbar] \tag{15.5}$$

where S is a function of x. In general, S may be complex.

If V is very nearly constant, we may expect that S is roughly equal to $p_2 x$. S can be expanded in a series as

$$S = S_0(x) + \hbar S_1(x) + \frac{\hbar^2}{2} S_2(x) + \dots \tag{15.6}$$

Now, from eq. (15.5)

$$\frac{\partial \psi}{\partial x} = \frac{i}{\hbar} \exp(iS/\hbar) \frac{\partial S}{\partial x}$$

$$\frac{\partial^2 \psi}{\partial x^2} = \frac{i}{\hbar}\frac{i}{\hbar} \exp(iS/\hbar) \left(\frac{\partial S}{\partial x}\right)^2 + \frac{i}{\hbar} \exp(iS/\hbar) \frac{\partial^2 S}{\partial x^2}$$

$$\therefore -\frac{\hbar^2}{2m}\frac{\partial^2 \psi}{\partial x^2} + (V-E)\psi = \left[\frac{1}{2m}\left\{\left(\frac{\partial S}{\partial x}\right)^2 - i\hbar \frac{\partial^2 S}{\partial x^2}\right\} + (V-E)\right] \exp(iS/\hbar) \tag{15.7}$$

To solve for S, we substitute the value of $\frac{\partial^2 \psi}{\partial x^2}$ from eq. (15.5) in eq. (15.4)

$$\therefore 0 = -\frac{\hbar^2}{2m}\frac{\partial^2 \psi}{\partial x^2} + (V-E)\psi = \left[\frac{1}{2m}\left\{\left(\frac{\partial S}{\partial x}\right)^2 - i\hbar \frac{\partial^2 S}{\partial x^2}\right\} + (V-E)\right] \exp(iS/\hbar)$$

$$\text{or, } \frac{1}{2m}\left(\frac{\partial S}{\partial x}\right)^2 + (E-V) - \frac{i\hbar}{2m}\frac{\partial^2 S}{\partial x^2} = 0 \tag{15.8}$$

Substituting eq. (15.6) in eq. (15.7) (up to second order in $\hbar$), we have

$$\frac{1}{2m}\left[\left(\frac{\partial S_0}{\partial x} + \hbar \frac{\partial S_1}{\partial x} + \frac{\hbar^2}{2}\frac{\partial S_2}{\partial x}\right)^2\right] + (V-E) - \frac{i\hbar}{2m}\left[\frac{\partial^2 S_0}{\partial x^2} + \hbar \frac{\partial^2 S_1}{\partial x^2} + \frac{\hbar^2}{2}\frac{\partial^2 S_2}{\partial x^2}\right] = 0$$

$$\text{or, } \frac{1}{2m}\left[\left(\frac{\partial S_0}{\partial x}\right)^2 + \hbar^2 \left(\frac{\partial S_1}{\partial x}\right)^2 + 2\hbar \left(\frac{\partial S_0}{\partial x}\right)\left(\frac{\partial S_1}{\partial x}\right) + \hbar^2 \left(\frac{\partial S_2}{\partial x}\right)\left(\frac{\partial S_0}{\partial x}\right)\right] + (V-E) - \frac{i}{2m}\left[\hbar\left(\frac{\partial^2 S_0}{\partial x^2}\right) + \hbar^2 \left(\frac{\partial^2 S_1}{\partial x^2}\right)\right] = 0$$

$$\text{or, } \frac{1}{2m}\left[\left(\frac{\partial S_0}{\partial x}\right)^2\right] + (V-E) + \frac{\hbar}{m}\left(\frac{\partial S_0}{\partial x}\frac{\partial S_1}{\partial x} - \frac{i}{2}\frac{\partial^2 S_0}{\partial x^2}\right) + \frac{\hbar^2}{2m}\left[\frac{\partial S_0}{\partial x} + \frac{\partial S_2}{\partial x} + \frac{\partial S_1}{\partial x} - i\frac{\partial^2 S_1}{\partial x^2}\right] = 0 \tag{15.9}$$

As eq. (15.9) must be satisfied independent of the value of $\hbar$, the coefficient of each power of $\hbar$ be separately equal to zero. This requirement leads to the following series of equations:

$$\frac{1}{2m}\left(\frac{\partial S_0}{\partial x}\right)^2 + (V-E) = 0 \tag{15.10}$$

$$\frac{\partial S_0}{\partial x}\frac{\partial S_1}{\partial x} - \frac{i}{\hbar^2}\frac{\partial^2 S_0}{\partial x^2} = 0 \tag{15.11}$$

$$\frac{\partial S_0}{\partial x} + \frac{\partial S_2}{\partial x} + \frac{\partial S_1}{\partial x} - i\frac{\partial^2 S_1}{\partial x^2} = 0 \tag{15.12}$$

Now, the solution of eq. (15.10) is

$$\frac{\partial S_0}{\partial x} = \pm\sqrt{[2m(E-V)]}$$

$$\text{or, } S_0 = \pm \int_{x_0}^{x} \sqrt{[2m(E-V)]}\, dx \tag{15.13}$$

The constant of integration is adjusted while normalising the wave function.

Solution of eq. (15.11) is

$$\frac{\partial S_1}{\partial x} = \frac{i}{2}\frac{1}{(\partial S_0/\partial x)}\frac{\partial^2 S_0}{\partial x^2} = \frac{i}{2}\frac{\partial}{\partial x}\log\frac{\partial S_0}{\partial x}$$

$$\therefore S_1 = \frac{i}{2}\log\frac{\partial S_0}{\partial x}$$

and

$$iS_1 = -\frac{1}{2}\log\frac{\partial S_0}{\partial x} = \log\left(\frac{\partial S_0}{\partial x}\right)^{-1/2} \quad \text{or, } e^{iS_1} = \frac{1}{\sqrt{\partial S_0/\partial x}} = \frac{1}{[2m(E-V)]^{1/4}}$$

Similarly, from eq. (15.12)

$$S_2 = \frac{1}{2}\frac{m(\partial V/\partial x)}{[2m(E-V)]^{3/2}} - \frac{1}{2}\int \frac{m^2(\partial V/\partial x)^2\,dx}{[2m(E-V)]^{5/2}} \tag{15.14}$$

Neglecting S_2 the solution of eq. (15.4) is given by

$$\psi = \exp[i/\hbar(S_0 + \hbar S_1)] = \exp(iS_0/\hbar)\exp(iS_1)$$

As S_0 has positive as well as neglecting values, hence

$$\psi = A\exp(iS_0/\hbar)\exp(iS_1) + B\exp(-iS_0/\hbar)\exp(iS_1)$$

$$= \frac{A}{(2m)^{1/4}(E-V)^{1/4}}\exp\left[i\int_{x_0}^{x}\sqrt{2m(E-V)\,dx/\hbar}\right]$$

$$+ \frac{B}{(2m)^{1/4}(E-V)^{1/4}}\exp\left[-i\int_{S_0}^{S}\sqrt{2m(E-V)\,dx/\hbar}\right]$$

$$= \frac{A_2}{\sqrt{p_2}}\exp\left[i\int_{x_0}^{x}p_2\frac{dx}{\hbar}\right] + \frac{B}{\sqrt{p_2}}\exp\left[-i\int_{x_0}^{x}p_2\frac{dx}{\hbar}\right] \tag{15.15}$$

where A and B are arbitrary constants and $p_2 = \sqrt{2m(E-V)}$. The positive exponential represents a wave moving in the positive direction, while the negative exponential a wave moving in the negative direction. For the special case where V is a constant, these reduce exponentially to the plane waves $\exp(ip_2x/\hbar)$ and $\exp(-ip_2x/\hbar)$.

For $V > E$, the WKB approximation possesses real exponential solution

$$\psi = \frac{A}{\sqrt{p_1}}\exp\left[\int_{x_0}^{x}p_1\frac{dx}{\hbar}\right] + \frac{B}{\sqrt{p_1}}\exp\left[-\int_{x_0}^{x}p_1\frac{dx}{\hbar}\right] \tag{15.16}$$

where $p_1 = \sqrt{2m(V-E)}$.

15.3 Validity of WKB approximation

The validity condition of WBK approximation

$$\left|\frac{dk}{dx}\right| \leqslant k^2 \tag{15.17}$$

But, $\lambda = \dfrac{2\pi}{k}$.

$$\therefore \frac{d\lambda}{dx} = -\frac{2\pi}{k^2}\frac{dk}{dx} \tag{15.18}$$

From eqs (15.17) and (15.18)

$$\frac{d\lambda}{dx} \leqslant 2\pi \tag{15.19}$$

A small change of wavelength over some interval of Δx of x may be expressed as

$$\Delta\lambda = \frac{d\lambda}{dx}\Delta x$$

If $\Delta x = \lambda$, then

$$\frac{d\lambda}{dx} = \frac{\Delta\lambda}{\lambda} \tag{15.20}$$

$$\text{or,} \quad \frac{\Delta\lambda}{\lambda} \leqslant 2\pi \tag{15.21}$$

This implies that the fractional change in wavelength $\frac{\Delta\lambda}{\lambda}$ must be small compared to 2π. Hence, if the condition eq. (15.19) is satisfied, condition eq. (15.17) is automatically satisfied.

WKB approximation gives satisfactory result where potential varies slowly. But, the turning is not valid.

15.4 Turning points and connection formulae

The points at which the velocity of the particle vanishes and changes sign are called *turning point*. They are defined by the equation

$$E = V(x) \tag{15.22}$$

As the wavelength at turning points becomes infinite, WKB approximation is not valid at these points. To obtain complete solution, Schrödinger equation is solved in a region of suitable extension surrounding the turning points and, then this solution is suitably joined to the solution on two sides. The solutions are the wave functions in regions where WKB approximation is valid.

If the connection formulae (the method of connecting the solution) are known, then it is immaterial to know the exact form of solutions in the regions of turning points.

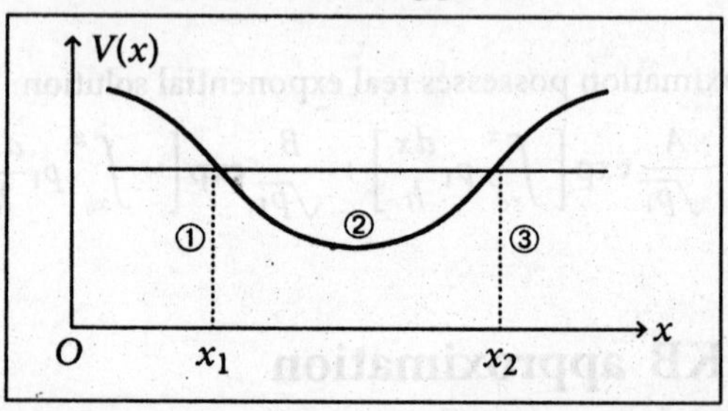

Fig 15.2 Potential well for WKB approximation

Now, we shall consider the problem of finding the wave functions of a particle in a given potential well $V(x)$ as shown in Fig 15.2. In the first region $x < x_1$, the wave function decreases exponentially for $x \to -\infty$. For $V(x) > E$, the wave function is given by

$$\psi_1 = \frac{A}{\sqrt{p_1}} \exp\left[\int_{x_0}^{x} p_1 \frac{dx}{\hbar}\right], \text{ where } p_1 = \sqrt{2m(V - E)} \tag{15.23}$$

In region (2), $x_1 < x < x_2$, $V(x) < E$

$$\psi_2 = \frac{B}{\sqrt{p_2}} \exp\left[i \int_{x_0}^{x} p_2 \frac{dx}{\hbar}\right] + \frac{c}{\sqrt{p_2}} \exp\left[-i \int_{x_0}^{x} p_2 \frac{dx}{\hbar}\right] \tag{15.24}$$

where $p_2 = \dfrac{x_0}{\sqrt{2m(E-V)}}$.

In region (3), $x > x_3$, $V(x) > E$, the wave function decreases exponentially for $x \to \infty$

$$\psi_3 = \frac{D}{\sqrt{p_1}} \exp\left[-\int_{x_0}^{x} p_1 \frac{dx}{\hbar}\right] \tag{15.25}$$

We now mention the restriction on the applicability of these solutions. The solutions fail in the immediate neighbourhood of the points $x = x_1$ and $x = x_2$ (turning points) of these are valid sufficiently remote from x_1 and x_2. So we are controlled with two problems:

(i) What is the expression of ψ in the immediate neighbourhood of x_1 and x_2?

(ii) How the solutions in the region to the left of x_1, to the right of x_1 and in the immediate neighbourhood of x_1, be matched to form a continuous solution (the same applies to point x_2)?

The first problem may be approximated by considering the potential function by a straight line (linear function) near x_1 (or x_2) and by shifting the origin to the coordinate x_1.

Thus, at x_1

$$V(x) = E - k_1(x - x_1) \tag{15.26}$$

and at x_2

$$V(x) = E + k_2(x - x_2) \tag{15.27}$$

The Schrödinger wave function

$$\frac{d^2\psi}{dx^2} + \frac{2m}{\hbar^2}[E - V(x)]\psi = 0$$

in region becomes

$$\frac{d^2\psi}{dx^2} + \frac{2mk_1}{\hbar^2}(x - x_1)\,\psi = 0 \tag{15.28}$$

and near x_2

$$\frac{d^2\psi}{dx^2} - \frac{2mk_2}{\hbar^2}(x - x_2)\,\psi = 0 \tag{15.29}$$

Now, changing the variable in eq. (15.28) as

$$z = -\left(\frac{2mk_1}{\hbar^2}\right)^{1/3}(x - x_1)$$

then

$$\frac{d^2\psi}{dx^2} = \frac{d}{dz}\left[\frac{d\psi}{dz}\left\{-\frac{2mk_1}{\hbar^2}\right\}^{1/3}\right]\frac{dz}{dx} = \left(\frac{2mk_1}{\hbar^2}\right)^{2/3}\frac{d^2\psi}{dz^2}$$

$$\therefore \left(\frac{2mk_1}{\hbar^2}\right)^{2/3}\frac{d^2\psi}{dz^2}+\left(\frac{2mk_1}{\hbar^2}\right)(x-x_1)\,\psi=0 \text{ or, } \frac{d^2\psi}{dx^2}+\left(\frac{2mk_1}{\hbar^2}\right)^{1/2}(x-x_1)\,\psi=0$$

$$\text{or, } \frac{d^2\psi}{dz^2}-z\psi=0 \tag{15.30}$$

In a similar way on substituting

$$z=\left(\frac{2mk_2}{\hbar^2}\right)^{1/3}(x-x_2)$$

in eq. (15.29), we get eq. (15.30).

The solutions of the differential equation (15.30) are the *Airy functions.* We require a function which vanishes asymptotically for large positive $z(z>0)$ corresponds to $x<x_1$ and $x>x_2$. Such a function is

$$A\,iz-\frac{1}{\pi}\int_0^\infty \cos\left(\frac{s^3}{3}+sz\right)dx \tag{15.31}$$

which for large $|z|$, has the asymptotic forms:

$$A\,iz-\frac{1}{2\sqrt{\pi}z^{1/4}}\exp\left(-\frac{z}{3}z^{3/2}\right),\ z>0 \tag{15.32}$$

$$A\,iz-\frac{1}{2\sqrt{\pi}(-z)^{1/4}}\sin\left[\frac{2}{3}(-z)^{3/2}+\frac{\pi}{4}\right],\ z<0 \tag{15.33}$$

The Airy function $Ai(z)$, which passes smoothly through the turning point, provides the required connection among the approximate forms of the wave function given in eqs (15.23), (15.25) and (15.25).

In the neighbourhood of x_1

$$p_2^2=2mk_1(x-x_1)=-(2mk_1\hbar)^{2/3}z$$

and

$$\int_{x_1}^{x} p_1\frac{dx}{\hbar}=\left(\frac{2mk_1}{\hbar}\right)^{1/3}\int_{x_1}^{x} z^{1/2}\,dx=-\int_0^z \sqrt{z}\,dz=-\frac{2}{3}z^{2/3} \tag{15.34}$$

and

$$\int_{x_1}^{x_2} p_1\frac{dx}{\hbar}=\left(\frac{2mk_1}{\hbar}\right)^{1/3}\int_{x_1}^{x} z^{1/2}\,dx=-\int_0^z \sqrt{z}\,dz=-\frac{2}{3}z^{2/3} \tag{15.35}$$

Similarly,

$$\int_{x_1}^{x} p_2\frac{dx}{\hbar}=\left(\frac{2mk_1}{\hbar^2}\right)^{1/3}\int_{x_1}^{x}\sqrt{-z}\,dx=-\int_0^z\sqrt{-z}\,dz=\frac{2}{3}(-z)^{3/2} \tag{15.36}$$

$$\therefore Ai(z)\sim\frac{1}{2\sqrt{\pi}z^{1/4}}\exp\left[\int_{x_1}^{x} p_1\frac{dx}{\hbar}\right] \text{ at } z>0 \text{ or } x<x_1 \tag{15.37}$$

and

$$Ai(z)\sim\frac{1}{2\sqrt{\pi}(-z)^{1/4}}\sin\left[\int_{x_1}^{x} p_1\frac{dx}{\hbar}+\frac{\pi}{4}\right] \text{ at } z<0 \text{ or } x>x_1 \tag{15.38}$$

Now, the function is approximated to the left of x_1 by

$$\psi_1=\frac{1}{\sqrt{p_1}}\exp\left(\int_{x_1}^{x} p_1\frac{dx}{\hbar}\right)(x<x_1) \tag{15.39}$$

and on the right the approximation is

$$\psi_2 = \frac{2}{\sqrt{p_2}} \sin\left(\int_{x_1}^{x} p_2 \frac{dx}{\hbar} + \frac{\pi}{4}\right), \quad x < x_2 \tag{15.40}$$

A similar analysis in the neighbourhood of point x_2 shows that the function is approximated to the right of x_2 by

$$\psi_3 = \frac{1}{\sqrt{p_1}} \exp\left(-\int_{x_2}^{x} p_1 \frac{dx}{\hbar}\right), \quad x > x_2$$

is approximated in region (2) by

$$\psi = \frac{2}{\sqrt{p_2}} \sin\left(\int_{x}^{x_2} p_2 \frac{dx}{\hbar} + \frac{\pi}{4}\right), \quad x < x_2 \tag{15.41}$$

Again, the solution of eq. (15.29) is Airy function

$$Bi(z) = \frac{1}{\pi} \int_0^{\infty} \left[e^{-sz-\left(\frac{1}{3}\right)s^3} + \sin\left(\frac{x^3}{3} + zs\right) \right] ds \tag{15.42}$$

with the asymptotic forms

$$Bi(z) - \frac{1}{\sqrt{\pi} z^{1/4}} \exp\left(\frac{2}{3} z^{3/2}\right), \quad z > 0 \tag{15.43}$$

$$Bi(z) - \frac{1}{\sqrt{\pi}(-z)^{1/4}} \cos\left[\frac{2}{3}(-z)^{3/2} + \frac{\pi}{4}\right], \quad z > 0 \tag{15.44}$$

In a similar way the connection formula linking an increasing exponential solution in region (1) to an oscillatory solution in region (2) is

$$\psi_{\text{WKB}} = \begin{cases} \dfrac{1}{(\sqrt{p_1})} \exp\left(\displaystyle\int_{x_1}^{x} p_1 \frac{dx}{\hbar}\right), & x < x_1 \\ \dfrac{1}{(\sqrt{p_2})} \cos\left(\displaystyle\int_{x_1}^{x} p_2 \frac{dx}{\hbar} + \frac{\pi}{4}\right); & x < x_1 \end{cases} \tag{15.45}$$

Thus, we have *Barrier to the left of* $x = x_0$, *where* $V > E$.

For decreasing exponential

$$\frac{1}{\sqrt{p_1}} \exp\left[\int_{x}^{x_0} \frac{p_1\, dx}{\hbar}\right] = \frac{1}{\sqrt{p_2}} \sin\left[\int_{x_0}^{x} \frac{p_2\, dx}{\hbar} + \frac{\pi}{4}\right] \tag{15.46}$$

$$\frac{1}{\sqrt{p_2}} \exp\left[-\int_{x}^{x_0} \frac{p_1\, dx}{\hbar}\right] = \frac{1}{\sqrt{p_2}} \cos\left[\int_{x_0}^{x} \frac{p_2\, dx}{\hbar} - \frac{\pi}{4}\right] \tag{15.47}$$

For increasing exponential

$$\frac{1}{\sqrt{p_1}} \exp\left[\int_{x}^{x_0} \frac{p_1\, dx}{\hbar}\right] = \frac{1}{\sqrt{p_2}} \cos\left[\int_{x_0}^{x} \frac{p_2\, dx}{\hbar} + \frac{\pi}{4}\right] \tag{15.48}$$

$$\frac{1}{\sqrt{p_2}} \exp\left[-\int_{x}^{x_1} \frac{p_1\, dx}{\hbar}\right] = -\frac{1}{\sqrt{p_2}} \sin\left[\int_{x_0}^{x} \frac{p_2\, dx}{\hbar} - \frac{\pi}{4}\right] \tag{15.49}$$

Barrier to the right: For decreasing exponential

$$\frac{1}{\sqrt{p_1}}\exp\left[-\int_{x_0}^{x}\frac{p_1\,dx}{\hbar}\right] = \frac{2}{\sqrt{p_2}}\sin\left[\int_{x_0}^{x}\frac{p_2\,dx}{\hbar}+\frac{\pi}{4}\right] \tag{15.50}$$

$$\frac{1}{\sqrt{p_1}}\exp\left[-\int_{x_0}^{x}\frac{p_1\,dx}{\hbar}\right] = \frac{2}{\sqrt{p_2}}\cos\left[\int_{x_0}^{x}\frac{p_2\,dx}{\hbar}+\frac{\pi}{4}\right] \tag{15.51}$$

For increasing exponential

$$\frac{1}{\sqrt{p_1}}\exp\left[-\int_{x_0}^{x}\frac{p_1\,dx}{\hbar}\right] = \frac{2}{\sqrt{p_2}}\cos\left[\int_{x_0}^{x}\frac{p_2\,dx}{\hbar}+\frac{\pi}{4}\right] \tag{15.52}$$

$$\frac{1}{\sqrt{p_1}}\exp\left[-\int_{x_0}^{x}\frac{p_1\,dx}{\hbar}\right] = -\frac{2}{\sqrt{p_2}}\sin\left[\int_{x_0}^{x}\frac{p_2\,dx}{\hbar}-\frac{\pi}{4}\right] \tag{15.53}$$

15.5 Potential barrier penetration (by WKB method)

We consider a potential barrier as shown in Fig 15.3. Let the particles be incident from left with sufficient energy to pass to the other side classically.

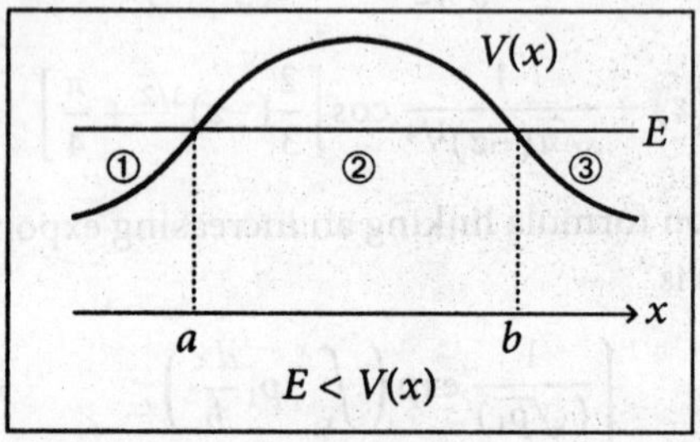

Fig 15.3 Barrier penetration of a particle through potential barrier

If the energy E of the particle is such that $E = V$ (at $x = a$), classically the particle would slow down to zero velocity at this point and, then turn back. Quantum mechanically, however, we know that the wave penetrates some distance further into the barrier. The points where $E = V$ are called *turning points.* In this barrier the turning points are at $x = a$ and $x = b$. Due to the wave nature of the particles, the beam of particles which is incident from left side is partly reflected and partly transmitted at $x = a$. The beam transmitted in region (2) is again partly reflected and transmitted at $x = b$. In region (3), there is only transmitted wave because it continues to be propagated without getting reflected anywhere.

The condition to apply WKB approximation is that the barrier is thick and high enough so that

$$\int_a^b \frac{[2m(E-V)]^{1/2}}{\hbar}\,dx$$

is considerably greater than unity.

The Schrödinger equation is

$$\frac{d^2\psi}{dx^2}+\frac{2m}{\hbar^2}(E-V)\,\psi = 0$$

The solution of this equation in region (3) is

$$\psi_3 = \frac{A}{\sqrt{p_2}} \exp\left(i \int_b^x p_2 \frac{dx}{\hbar} + \frac{\pi}{4}\right), \ x > b, \text{ where } p_2 = \sqrt{2m(E-V)} \tag{15.54}$$

Here, the phase factor $e^{\varepsilon\pi/4}$ has been included to facilitate the application of eq. (15.45). Eq. (15.54) can also be written as

$$\psi_3 = \frac{A}{\sqrt{p_2}}\left[\cos\left(\int_b^x p_2\frac{dx}{\hbar} + \frac{\pi}{4}\right) + i\sin\left(\int_b^x p_2 \frac{dx}{\hbar} + \frac{\pi}{4}\right)\right] \tag{15.55}$$

Now, the connecting wave function of exponential type in region (2) can be obtained with the help of eqs (15.35) and (15.40).

$$\psi_2 = \frac{A}{\sqrt{p_2}}\left[\exp\left(\int_x^b p_2 \frac{dx}{\hbar}\right) + \frac{i}{2}\exp\left(-\int_x^b p_1 \frac{dx}{\hbar}\right)\right] \tag{15.56}$$

In order to obtain the wave function in region (1) we rewrite the integrals of the above equation as below

$$\int_x^b p_1 \frac{dx}{\hbar} = \int_x^b p_1 \frac{dx}{\hbar} - \int_a^b p_1 \frac{dx}{\hbar}$$

$$\therefore \psi_2 = \frac{A}{\sqrt{p_2}}\left[\exp\left(\int_a^b p_1\frac{dx}{\hbar} - \int_a^x p_1 \frac{dx}{\hbar}\right) + \frac{i}{2}\exp\left(-\int_a^b p_1\frac{dx}{\hbar} + \int_a^x p_1 \frac{dx}{\hbar}\right)\right]$$

Again, applying the connection formulae, we obtain the wave function in region (1) as

$$\begin{aligned}
\psi_1 &= \frac{A}{\sqrt{p_2}}\left[\exp\left(\int_a^b p_1\frac{dx}{\hbar}\right)\cdot 2\sin\left(\int_x^a p_2\frac{dx}{\hbar} + \frac{\pi}{4}\right)\right. \\
&\quad \left. + \frac{i}{2}\exp\left(-\int_a^b p_1\frac{dx}{\hbar}\right)\cos\left(p_2\frac{dx}{\hbar}\right) + \frac{\pi}{4}\right] \\
&= \frac{A}{\sqrt{p_2}}\left[\exp\left(\int_a^b p_1 \frac{dx}{\hbar}\right) + \frac{1}{i}\left\{\exp\left(i\int_x^a p_2\frac{dx}{\hbar} + \frac{\pi}{4}\right)\right.\right. \\
&\quad \left. - \exp\left(-\int_x^a p_2\frac{dx}{\hbar} + \frac{\pi}{4}\right)\right\} + \frac{i}{2}\exp\left(-\int_a^b p_1\frac{dx}{\hbar}\right)\Big\} \\
&\quad \left. + \frac{1}{2}\left\{\exp\left(i\int_x^a p_2\frac{dx}{\hbar} + \frac{\pi}{4}\right) + \exp\left(-i\int_x^b p_2\frac{dx}{\hbar} + \frac{\pi}{4}\right)\right\}\right] \\
&= \frac{A}{i\sqrt{p_2}}\left[\exp\left\{i\int_x^a\left(p_2\frac{dx}{\hbar} + \frac{\pi}{4}\right)\right\}\left\{\exp\left(\int_a^b p_1\frac{dx}{\hbar}\right)\right.\right. \\
&\quad - \frac{1}{4}\exp\left(-\int_a^b p_1\frac{dx}{\hbar}\right) - \exp\left\{-i\left(\int p_2\frac{dx}{\hbar} + \frac{\pi}{4}\right)\right\} \\
&\quad \left.\left\{\exp\left(\int_a^b p_1\frac{dx}{\hbar}\right) + \frac{1}{4}\exp\left(-\int_a^b p_1\frac{dx}{\hbar}\right)\right\}\right]
\end{aligned} \tag{15.57}$$

Now, the transmission coefficient T is the ratio of the transmitted flux to the incident flux, i.e.,

$$T = \frac{\text{transmitted intensity}}{\text{incident intensity}} = \frac{\psi_3\psi_3^*}{\psi_1\psi_1^*} = \left[\exp\left(\int_a^b p_1\frac{dx}{\hbar}\right) - \frac{1}{4}\exp\left(-\int_a^b p_1\frac{dx}{\hbar}\right)\right]^{-2} \tag{15.58}$$

If WKB approximation is to be applicable, then

$$\int_a^b p_1 \frac{dx}{\hbar} \gg 1$$

and the negative exponential may be a neglected in comparison to positive exponential.

Hence,

$$T \simeq \exp\left(\int_a^b p_1 \frac{dx}{\hbar}\right)^{-2} = \exp\left[-2\int_a^b \sqrt{2m(V-E)}\frac{dx}{\hbar}\right] \tag{15.59}$$

15.6 Theory of α-disintegration

An α-particle is bound to the nucleus by a strong attractive force of short range. When the particle leaves the nucleus, it is repelled by the Coulomb force as shown in Fig 15.4. The probability of crossing the nucleus boundary is called the transmission probability given by the equation

$$T = \exp\left[-2\int_a^b \frac{\sqrt{2m(E-V)}}{\hbar}\,dx\right]$$

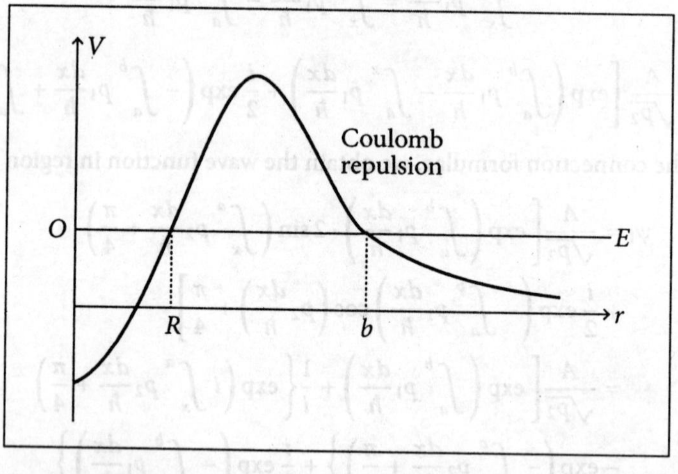

Fig 15.4 Showing nuclear attraction and Coulomb repulstion

We suppose that the potential of the α-particle (charge ze) which has been emitted from a nucleus of atomic number $(Z+z)$ be $V(r)$. For large distances from the nucleus there is a Coulomb potential (energy) given by

$$V(r) = \frac{Zze^2}{r} \tag{15.60}$$

where $-e$ is the electronic charge and r is the distance between the nucleus and α-particle.

The transmission probability of the α-particle is given by

$$T = \exp\left[-2\int_R^b \sqrt{\frac{2\mu}{\hbar^2}[V(r)-E]}\,dr\right]$$

Here, the turning points are $r = R$ and $r = b$ and in place of m, the reduced mass μ has been used. Substituting the value of $V(r)$, we get

$$T = \exp\left[-2\int_R^b \sqrt{\frac{2\mu}{\hbar^2}\left(\frac{Zze^2}{r} - \frac{Zze^2}{b}\right)}\,dr\right]\left[\because V_b = \frac{Zze^2}{b} = E\right] \tag{15.61}$$

$$= \exp(-2G) \tag{15.62}$$

where

$$G = \int_R^b \sqrt{\frac{2\mu}{\hbar^2}\left(\frac{Zze^2}{r} - \frac{Zze^2}{b}\right)}\,dr = \sqrt{\left(\frac{2\mu}{\hbar^2}Zze^2\right)}\int_R^b \sqrt{\left(\frac{1}{r} - \frac{1}{b}\right)}\,dr \tag{15.63}$$

We shall now evaluate the value of the integral

$$I = \int_R^b \sqrt{\left(\frac{1}{r} - \frac{1}{b}\right)}\,dr$$

Put $r = b\cos^2\theta$. $\therefore\ dr = -2b\cos\theta\sin\theta\,d\theta$.

$$\therefore I = \int \sqrt{\left(\frac{1}{r} - \frac{1}{b}\right)} = -\int\left(\sqrt{\frac{1}{b\cos^2\theta} - \frac{1}{b}}\right)2b\cos\theta\sin\theta\,d\theta$$

$$= -2\sqrt{b}\int\sqrt{\left(\frac{1-\cos^2\theta}{\cos^2\theta}\right)}\cos\theta\sin\theta\,d\theta = -2\sqrt{b}\int\sin^2\theta\,d\theta$$

$$= -2\sqrt{b}\int\frac{(1-\cos 2\theta)}{2}\,d\theta = -2\sqrt{b}\left[\frac{1}{2}\left\{\theta - \frac{1}{2}\sin 2\theta\right\}\right]$$

$$\therefore \int\sqrt{\left(\frac{1}{r} - \frac{1}{b}\right)}\,dr = -\sqrt{b}[\theta - \sin\theta\cos\theta]$$

Now, put $\cos\theta = \sqrt{r/b}$, then

$$\int_R^b \sqrt{\left(\frac{1}{r} - \frac{1}{b}\right)}\,dr = -\sqrt{b}\left[\cos^{-1}\left(\frac{r}{b}\right) - \sqrt{\left(1 - \frac{r}{b}\right)}\sqrt{\left(\frac{r}{b}\right)}\right]_R^b$$

$$= -\sqrt{b}\left[\cos^{-1}1 - \cos^{-1}\sqrt{\frac{R}{b}} + \sqrt{\left(1 - \frac{R}{b}\right)}\sqrt{\frac{R}{b}}\right]$$

Now, $\cos^{-1} = 0, 2\pi, 4\pi, \ldots$ For simplicity we shall take $\cos^{-1}1 = 0$, since in other cases we shall have to add an additional factor 2π in the value of $\cos^{-1}\sqrt{R/b}$.

$$\therefore \int_R^b \sqrt{\left(\frac{1}{r} - \frac{1}{b}\right)}\,dr = \sqrt{b}\left[\cos^{-1}\left(\frac{R}{b}\right) - \sqrt{\left(1 - \frac{R}{b}\right)}\sqrt{\left(\frac{R}{b}\right)}\right]$$

Substituting this value in eq. (15.63)

$$G = \sqrt{\left(\frac{2\mu}{\hbar^2}Zze^2\right)}\sqrt{b}\left[\cos^{-1}\sqrt{\left(\frac{R}{b}\right)} - \sqrt{\left(1 - \frac{R}{b}\right)}\sqrt{\left(\frac{R}{b}\right)}\right]$$

$$= \sqrt{\left(\frac{2\mu}{\hbar^2}Zze^2 b\right)}\left[\cos^{-1}\sqrt{\left(\frac{R}{b}\right)} - \sqrt{\left(\frac{R}{b} - \frac{R^2}{b^2}\right)}\right]$$

Let

$$\sqrt{\left(\frac{R}{b}\right)} = X,\ X \ll 1$$

then

$$\cos^{-1} X - X\sqrt{(1-x^2)} = \frac{\pi}{2} - x - X(1-X^2)^{1/2} = \frac{\pi}{2} - 2\pi$$

For small values of

$$X = \frac{\pi}{2}\left(1 - \frac{4X}{\pi}\right)$$

approximately.

$$\therefore G = \sqrt{\left(\frac{2\mu}{\hbar^2} Z ze^2 b\right)} \cdot \frac{\pi}{2}\left(1 - \frac{4X}{\pi}\right) \tag{15.64}$$

The kinetic energy E is given by

$$E = \frac{1}{2}\mu v^2 = \frac{Z ze^2}{b}, \therefore b = \frac{2Z ze^2}{\mu v^2} \tag{15.65}$$

Substituting this value of b in eq. (15.64), we get

$$G = \sqrt{\left(\frac{2\mu Z ze^2}{\hbar^2} \cdot \frac{2Z ze^2}{\mu v^2}\right)} \frac{\pi}{2}\left[1 - \frac{4}{\pi}\sqrt{\left(\frac{R\mu v^2}{2zZe^2}\right)}\right] = \frac{\pi Z ze^2}{\hbar v}\left[1 - \frac{4}{\pi}\sqrt{\left(\frac{R\mu v^2}{2zZe^2}\right)}\right]$$

For an α-particle, $z = 2$.

$$\therefore G = \frac{4\pi^2 Z ze^2}{hv}\left[1 - \frac{4}{\pi}\sqrt{\left(\frac{R\mu v^2}{Ze^2}\right)}\right] \quad \text{or, } G = \frac{4\pi^2 Ze^2}{hv} - \frac{8\pi e}{h}\sqrt{(\mu ZR)}$$

Now,

$$T = \exp\left[-2\left\{\frac{4\pi^2 Ze^2}{hv}\right\} - \frac{8\pi e}{h}\sqrt{(\mu ZR)}\right] \tag{15.66}$$

This is known as *Gamow factor* which successfully explains the dependence of half-life of radioactive nuclei on energy of α-particle. The disintegration constant λ is given by

$$\lambda = nT = \frac{\hbar}{4mR^2}\exp\left[-2\left\{\frac{4\pi^2 Ze^2}{hv} - \frac{8\pi e}{h}\sqrt{(\mu ZR)}\right\}\right]$$

$$\text{or, } L = \log\frac{h}{2mR^2} - \frac{8\pi^2 Ze^2}{hv} + \frac{16\pi e}{h}\sqrt{(\mu ZR)} \tag{15.67}$$

This is known as *Geiger-Nuttal law.*

15.7 The variation (Rayleigh-Ritz) method

The exact solution of Schrödinger equation is possible if the system contains a single electron as in case of hydrogen-like atom. But, if the system contains several electrons as in case of chemical phenomena the exact solution is not possible. In these cases *variation method* or *Rayleigh-Ritz method* are used. The variation method is most useful in the treatment of chemical bonding. This method is widely used in atomic and nuclear physics, quantum chemistry and also in the genesis of more elaborate methods of *Hartree* and *Hartree-Fock approximations.*

The Schrödinger time-independent wave equation is

$$\nabla^2 \psi + \frac{2m}{\hbar^2}(E - V)\psi = 0 \tag{15.68}$$

$$\text{or,} \quad \left(-\frac{\hbar^2}{2m}\nabla^2 + V\right)\psi = E\psi \tag{15.69}$$

The above equation can be written in the form:

$$\hat{H}\psi = E\psi \tag{15.70}$$

where $\hat{H} = \left(-\frac{\hbar^2}{2m}\nabla^2 + V\right)$ is known as *Hamiltonian operator.*

Multiplying both sides of eq. (15.70) by ψ^* and integrating over the configuration space, we have

$$\int \psi^* \hat{H}\psi \, d\tau = E \int \psi^* \psi \, d\tau \text{ or, } E = \frac{\int \psi^* \hat{H}\psi}{\int \psi^* \psi \, d\tau} \tag{15.71}$$

Knowing the value of ψ and $\hat{H}$ we can find the value of energy E. The approximate wave function ψ can be obtained by variation principle.

Now, we can guess a wave function and calculate the energy. Fortunately, if the guess is correct, the energy of the system will be correct. On the other hand, for a poor wave function, the energy will be poor. According to the variation principle it can be shown that if ψ_0 is the correct wave function for a system whose correct energy is E_0, then any other acceptable wave function ψ_i will give an energy greater than E_0. So any energy E_i obtained from a wave function ψ_i will be greater than E. Though it is not correct wave function it will, at least, tell us which of the several trial wave functions is close to the true wave function in the sense of giving the best energy.

The secular equations

In most instances, the variation function can be expressed in terms of functions ϕ_i is

$$\psi = a_1\phi_1 + a_2\phi_2 + \cdots + a_n\phi_n \tag{15.72}$$

where $a_1, a_2, \ldots, a_n$ are arbitrary parameters that can be varied to minimum in the energy. For convenience, we express it as

$$\psi = a_1\phi_1 + a_2\phi_2 \tag{15.73}$$

Substituting this value of ψ in eq. (15.71), we get

$$E = \frac{\int (a_1\phi_1^* + a_2\phi_2^*)\hat{H}(a_1\phi_1 + a_2\phi_2)\,d\tau}{\int (a_1\phi_1^* + a_2\phi_2^*)(a_1\phi_1 + a_2\phi_2)\,d\tau}$$

$$\text{or, } E\int (a_1\phi_1^* + a_2\phi_2^*)(a_1\phi_1 + a_2\phi_2)\,d\tau = \int (a_1\phi_1^* + a_2\phi_2^*)\hat{H}(a_1\phi_1 + a_2\phi_2)\,d\tau$$

$$\text{or, } E\left[a_1^2\int \phi_1^*\phi_1\,d\tau + 2a_1a_2\int \phi_1^*\phi_2\,d\tau + a_2^2\int \phi_2^*\phi_2\,d\tau\right]$$

$$= a_1^2\left[\phi_1^*\hat{H}\phi_1\,d\tau + 2a_1a_2\int \phi_1^*\hat{H}\phi_2\,d\tau + a_2^2\int \phi_2^*\hat{H}\phi_2\,d\tau\right] \tag{15.74}$$

As we desire the minimum value of E, it is necessary to minimise E w.r.t. both a_1 and a_2.

Now, differentiating w.r.t. a_1, we have

$$E\left[2a_1\int \phi_2^*\phi_2\,d\tau + 2a_2\int \phi_2^*\phi_2\,d\tau\right] + \frac{\partial E}{\partial a_1}\left[a_1^2\int \phi_1^*\phi_1\,d\tau + 2a_1a_2\int \phi_1^*\phi_2\,d\tau + a_2^2\int \phi_2^*\phi_2\,d\tau\right]$$

$$= 2a_1\int \phi_1^*\hat{H}\phi_1\,d\tau + 2a_2\int \phi_1^*\hat{H}\phi_2\,d\tau \tag{15.75}$$

Similarly, differentiating w.r.t. a_2, we get

$$E\left[2a_1\int \phi_1^*\phi_2\,d\tau + 2a_2\int \phi_2^*\phi_2\,d\tau\right] + \frac{\partial E}{\partial a_2}\left[a_1^2\int \phi_1^*\phi_1\,d\tau + 2a_1a_2\int \phi_1^*\phi_2\,d\tau + a_2^2\int \phi_2^*\phi_2\,d\tau\right]$$

$$= 2a_1\int \phi_1^*\hat{H}\phi_2\,d\tau + 2a_2\int \phi_2^*\hat{H}\phi_2\,d\tau \tag{15.76}$$

In order to minimise E w.r.t. a_1 and a_2, it is necessary that

$$\left(\frac{\partial E}{\partial a_1}\right)_{a_2} = \left(\frac{\partial E}{\partial a_2}\right)_{a_1} = 0 \tag{15.77}$$

For simplicity we introduce the symbolism

$$\left.\begin{aligned} H_{ij} &= \int \phi_i^*\hat{H}\phi_j d\tau \\ \text{and } \Delta_{ij} &= \int \phi_i^*\phi_j d\tau \end{aligned}\right\} \tag{15.78}$$

Applying condition $\partial E/\partial a_1 = 0$ in eq. (15.75) and expressing in terms of eq. (15.78) after rearranging, we get

$$(H_{11} - E\Delta_{11})\,a_1 + (H_{12} - E\Delta_{12})\,a_2 = 0 \tag{15.79}$$

Similarly, applying the condition $\partial E/\partial a_2 = 0$ in eq. (15.79), we get

$$(H_{21} - E\Delta_{21})\,a_1 + (H_{22} - E\Delta_{22})\,a_2 = 0 \tag{15.80}$$

Equations (15.79) and (15.80) are known together as the *secular equations*.

Equations (15.79) and (15.80) can be solved for a_1 and a_2 if the determinant

$$\begin{vmatrix} H_{11} - E\Delta_{11} & H_{12} - E\Delta_{12} \\ H_{21} - E\Delta_{21} & H_{22} - E\Delta_{22} \end{vmatrix} = 0 \tag{15.81}$$

In more general way, where ψ is expressed in terms of n independent functions, secular equations become

$$\begin{vmatrix} H_{11}-E\Delta_{11} & H_{12}-E\Delta_{12} & \ldots & H_{1n}-E\Delta_{1n} \\ H_{21}-E\Delta_{21} & H_{22}-E\Delta_{22} & \ldots & H_{2n}-E\Delta_{2n} \\ \ldots\ldots & \ldots\ldots & \ldots & \ldots\ldots \\ H_{n1}-E\Delta_{n1} & H_{n2}-E\Delta_{n2} & \ldots & H_{nn}-E\Delta_{nn} \end{vmatrix} \tag{15.82}$$

15.8 Applications of variational method

1. *Ground state energy of the hydrogen atom*: For the ground sate of hydrogen atom energy will be minimum.

For the ground state of hydrogen atom, the potential term will be $-\dfrac{e^2}{4\pi\varepsilon_0 r}$ and hence the form of the Hamiltonian operator $\hat{H}$ will be

$$\hat{H} = -\frac{h^2}{8\pi^2 m}\nabla^2 - \frac{e^2}{4\pi\varepsilon_0 r} \tag{15.83}$$

Since, the energy occurs in the radial equation we need only to consider the radial part of ∇^2.

Hence,

$$\nabla^2 = \frac{1}{r^2}\frac{\partial}{\partial r}\left(r^2\frac{\partial}{\partial r}\right)$$

The value of $\nabla^2\psi$ will be given by

$$\begin{aligned}\nabla^2\psi &= \left[\frac{1}{r^2}\frac{\partial}{\partial r}\left(r^2\frac{\partial}{\partial r}\right)\right]e^{-ar} \\ &= \frac{1}{r^2}\left[2r\frac{\partial}{\partial r} + r^2\frac{\partial^2}{\partial^2 r}\right]e^{-ar} \quad [\because \psi = e^{-ar} \text{ is the wave function in general form in this case}] \\ &= \frac{1}{r^2}\left[2r\frac{\partial}{\partial r}(e^{-ar}) + r^2\frac{\partial^2}{\partial r^2}(e^{-ar})\right] = \frac{1}{r^2}\left[2r(-a)\,e^{-ar} + r^2a^2e^{-ar}\right] \\ &= e^{-ar}\left\{a^2 - \frac{2a}{r}\right\}\end{aligned} \tag{15.84}$$

In view of variational method, the energy E is given by

$$E = \frac{\int \psi^*\hat{H}\psi\,d\tau}{\int \psi^*\psi\,d\tau} = \frac{\int_0^\infty e^{-ar}\left[-\frac{h^2}{8\pi^2 m}\left\{a^2 - \frac{2a}{r}\right\}e^{-ar} - \frac{e^2}{4\pi\varepsilon_0 r}e^{-ar}\right]4\pi r^2\,dr}{\int_0^\infty e^{-2ar}4\pi r^2\,dr}$$

where $4\pi r^2\,dr$ is the volume of a spherical shell at a distance r and $r + dr$ from the nucleus. The limit of integration varies from 0 to ∞.

$$\int_0^\infty -\frac{h^2a^2}{8\pi^2 m}r^2e^{-2ar}\,dr + \int_0^\infty \frac{2h^2a}{8\pi^2 m}r\,e^{-2ar}\,dr$$

$$\therefore E = \frac{-\int_0^\infty r^2 e^{-2ar}\,dr}{\int_0^\infty r^2 e^{-2ar}\,dr}$$

$$= \frac{-\frac{h^2a^2}{8\pi^2 m}\left\{\frac{2}{8a^3}\right\} + \frac{h^2a^2}{4\pi^2 m}\left\{\frac{1}{4a^2}\right\} e^2\left\{\frac{1}{4a^2}\right\}\frac{1}{4\pi\varepsilon_0}}{\left\{\frac{2}{8a^3}\right\}} \tag{15.85}$$

$$= \frac{h^2a^2}{8\pi^2 m} - \frac{e^2 a}{4\pi\varepsilon_0}\left[\because \int_0^\infty x^n x^{-ax}\,dx = \frac{n!}{a^{n+1}}\right] \tag{15.86}$$

Now, for minimum energy

$$\frac{\partial E}{\partial a} = 0 \text{ or, } \frac{\partial E}{\partial a} = \frac{2h^2}{8\pi^2 m}a - \frac{e^2}{4\pi\varepsilon_0} = 0 \text{ or, } a = \frac{4\pi^2 me^2}{h^2(4\pi\varepsilon_0)} \tag{15.87}$$

This is the value of a for which E is minimum.

We substitute this value of a in eq. (15.86).

Thus,

$$E_{\min} = \frac{h^2}{8\pi^2 m}\left\{\frac{4\pi^2 me^2}{h^2(4\pi\varepsilon_0)}\right\}^2 - e^2\left\{\frac{4\pi^2 me^2}{h^2}\right\}\frac{1}{(4\pi\varepsilon_0)^2} \text{ or, } E_{\min} = -\frac{2\pi^2 me^4}{h^2(4\pi\varepsilon_0)^2} \tag{15.88}$$

This is the same value of $E_{\min}$ as has been obtained by other methods.

2. *The normal states of helium atom*: We here apply variational method to obtain the energy of Helium atom in the normal state. The helium atom consists of a nucleus of charge +Ze (where $Z = 2$) and two electrons each of charge $-e$ [Fig 15.5].

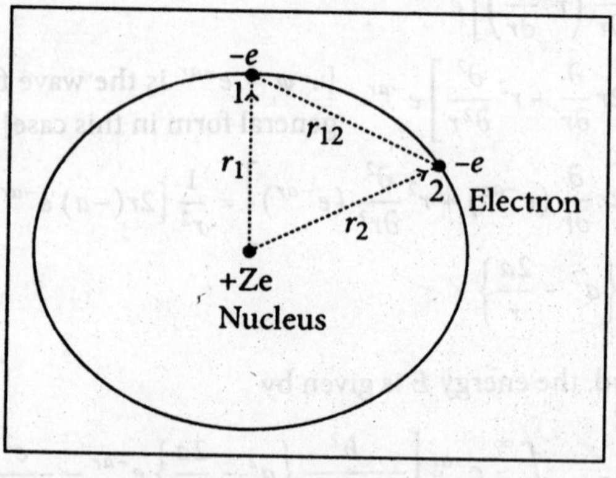

Fig 15.5 Helium atom

The potential energy of the system is given by

$$V = \left(-\frac{Ze^2}{r_1} - \frac{Ze^2}{r_2} + \frac{e^2}{r_{12}}\right)\frac{1}{4\pi\varepsilon_0} \tag{15.89}$$

where r_1, r_2 and r_{12} are respectively the distances of electrons (1), (2) and the separation between them.

The Hamiltonian operator is given by

$$\hat{H} = -\frac{\hbar^2}{2m}\left(\nabla_1^2 + \nabla_2^2\right) + \left(\frac{Z'e^2}{r_1} + \frac{Z'e^2}{r_2} + \frac{e^2}{r_{12}}\right)\frac{1}{4\pi\varepsilon_0} \tag{15.90}$$

The value of Z' is determined with the help of variation method.

The wave function ψ is given by

$$\psi = \psi_1\psi_2 = \left\{\frac{Z^3}{\pi a_0^3}\right\}\exp\left[\frac{Z'r_1}{a_0}\right]\exp\left[-\frac{Z'r_2}{a_0}\right] \tag{15.91}$$

where Z' is the effective atomic number. This is possible when the interaction between two electrons is ignored. When it is done, Helium atom can be treated as combination of two hydrogen atoms. The factors ψ_1 and ψ_2 of ψ are hydrogen-like wave functions for the nuclear charge $Z'e$ so that

$$-\frac{\hbar^2}{2m}\nabla_1^2\psi_1 = \frac{Z'e^2}{4\pi\varepsilon_0 r_1}\psi_1 - Z'^2 W_H\psi_1 \tag{15.92}$$

$$\text{and } -\frac{\hbar^2}{2m}\nabla_2^2\psi_2 = \frac{Z'e^2}{4\pi\varepsilon_0 r_2}\psi_1 - Z'^2 W_H\psi_2 \tag{15.93}$$

where

$$W_H = \frac{e^2}{(4\pi\varepsilon_0)2a_0}, \quad a_0 \text{ being the first Bohr radius}$$

$$= \frac{\hbar^2}{m_0 e^2}(4\pi\varepsilon_0)$$

Now, according to variational method, energy of the system is

$$E = \frac{\int \psi^*\hat{H}\psi\, d\tau}{\int \psi^*\psi\, d\tau}$$

$$\text{or, } E = \int \psi_1^*\psi_2^*\left[-\frac{\hbar^2}{2m}\left(\nabla_1^2 + \nabla_2^2\right) + \frac{1}{4\pi\varepsilon_0}\left(-\frac{Ze^2}{r_1} - \frac{Ze^2}{r_2} + \frac{e^2}{r_{12}}\right)\right]\psi_1\psi_2\, d\tau \tag{15.94}$$

Now, using eqs (15.92) and (15.93), we get

$$E = -2Z^2 W_H + \frac{(Z'-Z)e^2}{4\pi\varepsilon_0}\int \psi^*\left(\frac{1}{r_1} - \frac{1}{r_2}\right)\psi\, d\tau + \frac{1}{4\pi\varepsilon_0}\int \psi^*\frac{e^2}{r_{12}}\psi\, d\tau \tag{15.95}$$

Now, consider the integral

$$\frac{e^2}{4\pi\varepsilon_0}\int \psi^*\left(\frac{1}{r_1} + \frac{1}{r_2}\right)\psi\, d\tau$$

$$\frac{e^2}{4\pi\varepsilon_0}\int \psi^*\left(\frac{1}{r_1} + \frac{1}{r_2}\right)\psi\, d\tau = \frac{2e^2}{4\pi\varepsilon_0}\left(\int \frac{\psi^2}{r_1}\, d\tau_1\right)$$

$$= \frac{2e^2}{4\pi\varepsilon_0}\frac{Z^3}{\pi a_0^3}\int_0^\infty\int_0^\pi\int_0^{2\pi}\frac{1}{r_1}\exp\left(-2Z'r_1/a_0\right)\times r_1^2\, dr_1 \sin\theta\, d\theta\, d\phi$$

$$= \frac{8Z'^3e^2}{a_0^3}\int_0^\infty r_1\exp\left(\frac{2Z'r_1}{a_0}\right)dr_1 = \frac{2Z'e^2}{a_0 4\pi\varepsilon_0} = 4Z'W_H \tag{15.96}$$

Considering the integral

$$\int \frac{\phi^*}{4\pi\varepsilon_0} \frac{e^2}{r_{12}} \psi \, d\tau$$

the interaction integral,

$$\frac{1}{4\pi\varepsilon_0} \int \psi^* \frac{e^2}{r_{12}} \psi \, d\tau = \iint \left(\frac{Z^3}{\pi a_0^3}\right)^2 \frac{e^2}{4\pi\varepsilon_0 r_{12}} \exp\left[\frac{-2Z'}{a_0(r_1+r_2)}\right] d^3r_1 \, d^3r_2$$

$$= \left(\frac{Z'^3}{\pi a_0^3}\right)^2 \frac{e^2}{4\pi\varepsilon_0} \iint \frac{1}{r_{12}} \exp\left[\frac{-2Z'}{a_0(r_1+r_2)}\right] d^3r_1 \, d^3r_2$$

$$= \left(\frac{Z'^3}{\pi a_0^3}\right)^2 \frac{e^2}{4\pi\varepsilon_0} \left(\frac{2Z'}{a_0}\right)\left(\frac{a_0}{2Z'}\right)^6 \iint \frac{\exp\left[2Z'/a_0(r_1+r_2)\right]}{r_{12}(2Z'/a_0)} \times \left(\frac{2Z'}{a_0}\right)^6 d^3r_1 \, d^3r_2$$

$$= \left(\frac{Z'^3}{\pi a_0^3}\right) \frac{e^2}{4\pi\varepsilon_0} \left(\frac{2Z'}{a_0}\right)\left(\frac{a_0}{2Z'}\right)^6 \iint \frac{\exp[-(R_1+R_2)]}{R_{12}} d^3r_1 \, d^3r_2$$

where $\frac{2Z'}{a_0} r_1 = R_1$ **and** $\frac{2Z'}{a_0} r_2 = R_2$.

Hence,

$$\int \psi^* \frac{e^2}{4\pi\varepsilon_0 r_{12}} \psi \, d\tau = \frac{Z'e^2}{32\pi^2 a_0} \cdot \frac{20\pi^2}{4\pi\varepsilon_0} \left[\because \iint \exp\left[-\frac{(R_1+R_2)}{R_{12}}\right] \times d^3R_1 \, d^3R_2 = 20\pi^2\right]$$

$$= \frac{5}{8}\frac{Z'e^2}{a_0} \cdot \frac{1}{4\pi\varepsilon_0} = \frac{5}{4} Z' W_4 \tag{15.97}$$

Substituting the value of the integrals from eqs (15.96) and (15.97) in eq. (15.95), we get

$$E = -2Z'^2 W_H + (Z'-Z)4Z'W_H + \frac{5}{4}Z'W_H = \left[-2Z'^2 + 4Z'(Z'-Z) + \frac{5}{4}Z'\right] W_H$$

Now, E will be minimum when $\frac{\partial E}{\partial Z'} = 0$

$$\therefore \frac{\partial E}{\partial Z'} = 0 = \left(-4Z' + 8Z' - 4Z + \frac{5}{4}\right) W_H \text{ or, } 0 = -4Z' = -4Z + 8Z' - 4Z + \frac{5}{4}$$

$$\text{or, } -4Z' = -4Z + \frac{5}{4} \text{ or, } Z' = Z - \frac{5}{16} \text{ or, } Z' = 2 - \frac{5}{16} = \frac{27}{16} = 1.69 \tag{15.98}$$

which gives

$$E = -2\left(Z - \frac{5}{16}\right)^2 W_H = -\frac{2.85e^2}{a_0} \tag{15.99}$$

The hydrogen like wave functions give the best energy value when $Z' = 27/16 = 1.69$ rather than 2. This shows that each electron screens the nucleus fro the other electron and the electron nuclear charge is reduced by $(2 - 27/16) = 5/16$ of an electronic charge. Thus, the effective charge in the nucleus is less than 2.

3. *Zero-point energy of one-dimensional harmonic oscillator*: The Hamiltonian of one-dimensional harmonic oscillator is

$$H = \frac{p_x^2}{2m} + \frac{1}{2}kx^2 \tag{15.100}$$

$$\text{or, } \hat{H} = -\frac{\hbar^2}{2m}\frac{\partial^2}{\partial x^2} + \frac{1}{2}m\omega^2 x^2 \tag{15.101}$$

$$\text{or, } \sqrt{\frac{k}{m}} = \omega_0 \tag{15.102}$$

We choose a trial wave function which can be normalised (even function) and vanishes at infinity. Let

$$\psi(x) = Ae^{-ax^2} \tag{15.103}$$

where A is normalisation constant and a is a variational parameter. The condition of normalisation of $\psi(x)$ is

$$\int_{-\infty}^{\infty} \psi(x)^* \psi(x)\, dx = 1 \text{ or, } A^2 \int e^{-2ax^2} dx = 1 \tag{15.104}$$

Now, since

$$\int_0^{\infty} x^m e^{-ax^n} dx = \frac{1}{n}\Gamma\left(\frac{m+1}{n}\right) \Big/ a^{(m+1)/n}$$

$$\therefore A^2 \cdot 2\frac{1}{2}\Gamma\left(\frac{1}{2}\right)/(2a)^{1/2} = 1 \text{ or, } A = \left(\frac{2a}{\pi}\right)^{1/4} \tag{15.105}$$

The expectation value of energy is

$$\langle E \rangle = \frac{\int \psi^*(x) \hat{H} \psi(x)\, dx}{\int \psi^*(x)\, \psi(x)\, dx} \tag{15.106}$$

Now,

$$H\psi = \left(-\frac{\hbar^2}{2m}\frac{\partial^2}{\partial x^2} + \frac{1}{2}m\omega_0^2 x^2\right) Ae^{-ax^2} = -\frac{A\hbar^2}{2m}\frac{\partial^2}{\partial x^2}\left(e^{-ax^2}\right) + \frac{A}{2}m\omega_0^2 x^2 e^{-ax^2}$$

$$= -\frac{A\hbar^2}{2m}\left[\left(4a^2 - \frac{m^2\omega_0^2}{\hbar^2}\right)x^2 - 2a\right] e^{-ax^2} \tag{15.107}$$

Now, expectation value of energy is

$$\langle E \rangle = -\frac{A^2\hbar^2}{2m}\int e^{-ax^2}\left[\left(4a^2 - \frac{m^2\omega_0^2}{\hbar^2}\right)x^2 - 2a\right] e^{-ax^2} dx \tag{15.108}$$

$$= -\frac{A^2\hbar^2}{2m}\left(4a^2 - \frac{m^2\omega_0^2}{\hbar^2}\right)\int x^2 e^{-2ax^2} dx + \frac{A^2\hbar^2}{2m}2a\int e^{-2ax^2} dx \tag{15.109}$$

$$= -\frac{A^2\hbar^2}{2m}\left(4a^2 - \frac{m^2\omega_0^2}{\hbar^2}\right) 2\frac{1}{2}\Gamma\left(\frac{2+1}{2}\right)/(2a)^{3/2} + \frac{A^2\hbar^2}{2m}2a \cdot 2\frac{1}{2}\Gamma\left(\frac{1}{2}\right)/(2a)^{1/2}$$

$$= -\frac{A^2\hbar^2}{2m}\left(4a^2 - \frac{m^2\omega_0^2}{\hbar^2}\right)\frac{\pi^{1/2}}{2(2a)^{3/2}} + \frac{A^2\hbar^2}{2m}2a\frac{\pi^{1/2}}{(2a)^{1/2}}$$

$$\text{or, } \langle E \rangle = \frac{A^2\hbar^2}{2m}\left(\frac{\pi}{2a}\right)^{1/2}\left[-a + \frac{m^2\omega_0^2}{4\hbar^2 a} + 2a\right] \tag{15.110}$$

Substituting the value of $A = \left(\frac{2a}{\pi}\right)^{1/4}$

$$\langle E \rangle = \frac{\hbar^2}{2m}\left[a + \frac{m^2\omega_0^2}{4\hbar^2 a}\right] \tag{15.111}$$

Differentiating $\langle E \rangle$ w.r.t. a

$$\frac{d\langle E \rangle}{da} = \frac{\hbar^2}{2m}\left[1 - \frac{m^2\omega_0^2}{4\hbar^2 a^2}\right] \tag{15.112}$$

According to variational principle

$$\langle E \rangle = \frac{\hbar^2}{2m}\left[\frac{m\omega_0}{2\hbar} + \frac{m_0^2\omega_0^2}{4\hbar^2}\right]\frac{2\hbar}{m\omega_0} \text{ or, } \langle E \rangle = \frac{\hbar\omega_0}{4} + \frac{\hbar\omega_0}{4} = \frac{\hbar\omega_0}{2} \tag{15.113}$$

Equation (15.113) gives the *zero-point energy of the harmonic oscillator.*

The eigenfunction is

$$\psi(r) = -\frac{V_0 e^{-r/r_0}}{r/r_0} \tag{15.114}$$

4. *The Deuteron problem*: We know that the force between a neutron and a proton is described by a potential of the form:

$$V(r) = -\frac{V_0 e^{-r/r_0}}{r/r_0} \tag{15.115}$$

where V_0 and r_0 are the strength of potential and the range of the potential respectively.

The Hamiltonian operator of the system is given by

$$\hat{H} = -\frac{\hbar^2}{2\mu}\nabla^2 + V(r)$$

where $\mu = \frac{M_p M_n}{M_p + M_n}$ is the reduced mass of the system.

In s-state, $\hat{H}$ is given by

$$\hat{H} = -\frac{\hbar^2}{2m}\left[\frac{1}{r^2}\frac{d}{dr}\left(r^2\frac{d}{dr}\right)\right] + V(r) \text{ or, } \hat{H} = -\frac{\hbar^2}{2m}\left[\frac{d^2}{dr^2} + \frac{2}{r}\frac{d}{dr}\right] + V(r) \tag{15.116}$$

Using the following trial solution we shall calculate the binding energy of the system

$$\psi = Ce^{-\alpha r/r_0} \tag{15.117}$$

where C is the normalisation constant and can be calculate from

$$C^2\int \psi^*\psi\, d\tau = \int_0^\infty C^2 e^{-\alpha r/r_0} e^{-\alpha r/r_0} 4\pi r^2\, dr = 1$$

$$\text{or, } 4\pi C^2 \int_0^\infty e^{-2\alpha r/r_0} r^2\, dr = 1$$

$$\text{or, } 4\pi C^2 \frac{2!}{(2\alpha/r_0)^3} = 1 \left[\because \int_0^\infty r^n e^{-\alpha r}\, dr = \frac{n!}{\alpha^{n+1}}\right]$$

$$\text{or, } C = \left(\frac{\alpha^3}{\pi r_0^2}\right)^{1/2}$$

The trial function is

$$\psi = \left(\frac{\alpha^3}{\pi r_0^3}\right)^{1/2} e^{-\alpha r/r_0} \tag{15.118}$$

According to variational method

$$E = \frac{\int \psi^* \hat{H}\psi\, d\tau}{\int \psi^*\psi\, d\, d\tau} = \int \psi^*\hat{H}\psi\, d\tau \tag{15.119}$$

$$\text{or, } E = \int_0^\infty \left(\frac{\alpha^3}{\pi r_0^3}\right)^{1/2} e^{-\alpha r/r_0}\left[-\frac{\hbar^2}{2\mu}\left\{\frac{d^2}{dr^2} + \frac{2}{r}\frac{d}{r} + V(r)\right\}\right] \times \left(\frac{\alpha^3}{\pi r_0^3}\right)^{1/2} e^{-\alpha r/r_0} 4\pi r^2\, dr$$

$$= \left(\frac{\alpha^3}{\pi r_0^3}\right) 4\pi \int_0^\infty e^{-\alpha r/r_0}\left[-\frac{\hbar^2}{2\mu}\left\{\frac{d^2}{dr^2} + \frac{2}{r}\frac{d}{dr}\right\}\right] + \left\{-V_0 \frac{e^{-r/r_0}}{r/r_0}\right\} e^{--\alpha r/r_0} r^2\, dr$$

$$= \left(\frac{\alpha^3}{\pi r_0^3}\right) 4\pi\left(-\frac{\hbar^2}{2\mu}\right)\int_0^\infty e^{-\alpha r/r_0}\left\{\frac{d^2}{dr^2} + \frac{2}{r}\frac{d}{dr}\right\} e^{-\alpha r/r_0} r^2\, dr$$

$$+ \frac{\alpha^3}{\pi r_0^3} 4\pi(-V_0 r_0)\int_0^\infty e^{-\alpha r/r_0}\frac{e^{-r/r_0}}{r} e^{-\alpha r/r_0} r^2\, dr \tag{15.120}$$

We now solve the integral

$$\int_0^\infty e^{-\alpha r/r_0}\left\{\frac{d^2}{dr^2} + \frac{2}{r}\frac{d}{dr}\right\} e^{-\alpha r/r_0} r^2\, dr = \int_0^\infty e^{-\alpha r/r_0}\frac{\alpha^2}{r_0^2} e^{-\alpha r/r_0} r^2\, dr + \int_0^\infty r^2 e^{-\alpha r/r_0}\left(-\frac{2\alpha}{r r_0}\right) e^{-\alpha r/r_0} r^2\, dr$$

$$= \frac{\alpha^2}{r_0^2}\int_0^\infty r^2 e^{-2\alpha r/r_0}\, dr - \frac{2\alpha}{r_0}\int_0^\infty r\, e^{-2\alpha r/r_0}\, dr = \frac{\alpha^2}{r_0^2}\left[\frac{2!}{(2\alpha/\alpha_0)^3}\right] - \frac{2\alpha}{r_0}\left[\frac{1!}{(2\alpha/r_0)^2}\right]$$

$$= \left(\frac{r_0}{4\alpha}\right) - \left(\frac{r_0}{2\alpha}\right) = -\frac{r_0}{4\alpha}$$

Again, the integral

$$\int_0^\infty e^{-\alpha r/r_0}\frac{e^{-r/r_0}}{r} e^{-\alpha r/r_0} r^2\, dr = \int_0^\infty r\, e^{-(2\alpha+1)r/r_0}\, dr = \frac{1}{\{(2\alpha+1)/r_0\}^3} = \frac{r_0^2}{(2\alpha+1)^2}$$

Substituting these values of the integrals, we get

$$E = \left(\frac{\alpha^3}{\pi r_0^3}\right) 4\pi\left(-\frac{\hbar^2}{2\mu}\right)\left(-\frac{r_0}{4\alpha}\right) + \left(\frac{\alpha^3}{\pi r_0^3}\right) 4\pi(-V_0 r_0)\frac{r_0^2}{(2\alpha+1)^2} \text{ or, } E = \frac{\hbar^2\alpha^2}{2\mu r_0^2} - \frac{4V\alpha^3}{(2\alpha+1)} \tag{15.121}$$

Now, for E to be the minimum, $\frac{\partial E}{\partial \alpha} = 0$

$$\text{or, } \frac{\hbar^2 \alpha}{\mu r_0^2} - \frac{4V\alpha^2(2\alpha+3)}{(2\alpha+1)^3} = 0 \text{ or, } \frac{4V\alpha^3}{(2\alpha+1)^2} = \frac{\hbar^2\alpha^2(2\alpha+1)}{\mu r_0^2(2\alpha+3)}$$

Substituting in eq. (15.121), we get

$$E_{\min} = \frac{\hbar^2\alpha^3}{4\alpha r_0^2} - \frac{\hbar^2\alpha^2(2\alpha+1)}{\mu r_0^2(2\alpha+3)} = \frac{\hbar^2\alpha^2[(2\alpha+3)-2(2\alpha+1)]}{2\mu r_0^2(2\alpha+3)} \text{ or, } E_{\min} = \frac{\hbar^2\alpha^2\,(1-2\alpha)}{2\mu r_0^2\,(2\alpha+3)} \quad (15.122)$$

5. *van der Waal's interaction*: In order to calculate the van der Waal' interaction between two hydrogen atoms in their ground state, we consider two nuclei A and B of two hydrogen atoms along (Z-axis) which are at a distance R apart. 1 and 2 are two electrons which are at a distance r_1 and r_2 from A and B respectively, r_{12} represents the distance between two electrons. r_{1B} is the distance of electron 1 from B and r_{2A} is the distance of electron 2 from A.

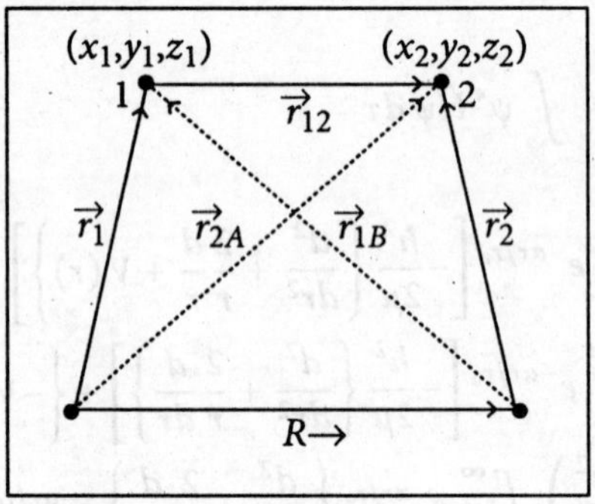

Fig 15.6 Two-hydrogen atoms

The potential energy of the system is

$$V = \left(-\frac{e^2}{r_1} - \frac{e^2}{r_2} + \frac{e^2}{R} + \frac{e^2}{r_{12}} - \frac{e^2}{r_{2A}} - \frac{e^2}{r_{1B}}\right)\frac{1}{4\pi\varepsilon_0}$$

and Hamiltonian is

$$H = -\frac{\hbar^2}{2m}(\nabla_1^2 + \nabla_2^2) + \left(-\frac{e^2}{r_1} - \frac{e^2}{r_2} + \frac{e^2}{R} + \frac{e^2}{r_{12}} - \frac{e^2}{r_{2A}} - \frac{e^2}{r_{1B}}\right)\frac{1}{4\pi\varepsilon_0} \quad (15.123)$$

$$\text{or, } H = H_0 + H'$$

where

$$H_0 = -\frac{\hbar^2}{2m}(\nabla_1^2 + \nabla_2^2) - \left(\frac{e^2}{r_1} + \frac{e^2}{r_2}\right)\frac{1}{4\pi\varepsilon_0}$$

$$\text{and } H' = \left(\frac{e^2}{R} + \frac{e^2}{r_{12}} - \frac{e^2}{r_{2A}} - \frac{e^2}{r_{1B}}\right)\frac{1}{4\pi\varepsilon_0} \quad (15.124)$$

Here, H_0 is the unperturbed Hamiltonian and H' is the perturbed term.

Since, r is much smaller than R

$$\frac{1}{r_{1B}} = \frac{1}{|r_1 - R|} = \frac{1}{(r_1^2 + R^2 - 2r_1 R\cos\theta_1)^{1/2}} = \frac{1}{R\left(1 + \frac{r_1^2}{R} - \frac{2Z_1}{R}\right)^{1/2}}$$

$$= \frac{1}{R}\left(1 - \frac{2Z_1}{R} + \frac{r_1^2}{R^2}\right)^{-1/2} \quad (\because Z_1 = r_1\cos\theta_1) \tag{15.125}$$

Similarly,

$$\frac{1}{r_{2A}} = \frac{1}{|r_2 - R|} = \frac{1}{R}\left(1 + \frac{2Z_2}{R} + \frac{r_2^2}{R^2}\right)^{-1/2} \quad [\because Z_2 = r_2\cos(180° - \theta_2)] \tag{15.126}$$

and
$$\frac{1}{r_{12}} = \frac{1}{R}\left(1 + \frac{2(Z_2 - Z_1)}{R} + \frac{(x_2 - x_1^2) + (y_2 - y_1)^2 + (Z_2 - Z_1)^2}{R^2}\right)^{1/2}$$

$$-\left\{1 + \frac{2Z_2}{R} + \frac{r_2^2}{R^2}\right\}^{-1/2} - \left\{1 - \frac{2Z_1}{R} + \frac{r_1^2}{R^2}\right\}^{-1/2} \tag{15.127}$$

Substituting eqs (15.125), (15.126) and (15.124) in eq. (15.121), we get

$$H' = \frac{e^2}{4\pi\varepsilon_0 R} + \frac{e^2}{4\pi\varepsilon_0 R}\left[1 + \frac{2(Z_2 - Z_1)}{R} + \frac{(x_2 - x_1)^2 + (y_2 - y_1)^2 + (Z_2 - Z_1)^2}{R^2}\right]^{-1/2}$$

$$-\frac{e^2}{4\pi\varepsilon_0 R}\left[1 + \frac{2Z_2}{R} + \frac{r_2^2}{R^2}\right]^{1/2} - \frac{e^2}{R}\left[1 - \frac{2Z_1}{R} + \frac{r_1^2}{R^2}\right]^{-1/2}$$

$$= \frac{e^2}{4\pi\varepsilon_0 R}\left[1 + \left\{\frac{2(Z_2 - Z_1)}{R} + \frac{(x_2 - x_1)^2 + (y_2 - y_1)^2 + (Z_2 - Z_1)^2}{?}\right\}^{1/2}\right.$$

$$\left. - \left\{1 + \frac{2Z_2}{R} + \frac{r_2^2}{R^2}\right\}^{-1/2} - \left\{1 - \frac{2Z_1}{R} + \frac{r_1^2}{R^2}\right\}^{-1/2}\right] \tag{15.128}$$

An approximate solution of eq. (15.128) is

$$H' = \frac{e^2}{4\pi\varepsilon_0 R^3}(x_1x_2 + y_1y_2 - 2Z_1Z_2) \tag{15.129}$$

The unperturbed Hamiltonian H_0 has a solution

$$\psi_0(r_1, r_2) = U_{100}(r_1)U_{100}(r_2) \tag{15.130}$$

It has the form like two non-interaction hydrogen atoms in ground state. With the wave function in eqs (15.130) and (15.118). We obtain the ground state energy E_0.

Now, the interaction energy

$$E_1 = \int \psi_0^*(r_1, r_2)H_0'(r_1 r_2)\, d\tau$$

$$= \int \psi_0^*(r_1, r_2)\left\{\frac{e^2}{4\pi\varepsilon_0 R}(x_1x_2 + y_1y_2 - 2Z_1Z_2)\right\}\psi_0(r_1, r_2)\, d\tau$$

$$= 0 \tag{15.131}$$

[$\because \psi_0(r_1, r_2)$ is an even function of r_1 and ψ_2 and H' is odd function of r_1 and r_2].

We now consider a trial function as

$$\psi(r_1, r_2) = \psi_0(r_1, r_2)(1 + AH') \tag{15.132}$$

where A is a variational parameter.

According to variational principle

$$E_0 + E \leqslant \frac{\int \psi^* H\psi\, d\tau}{\int \psi^* \psi\, d\tau} \tag{15.133}$$

From the right-hand side eq. (15.133)

$$\frac{\int \psi^* H\psi\, d\tau}{\int \psi^* \psi\, d\tau} = \frac{\int \psi^*(1 + AH')(H_0 + H')\psi_0(1 + H')\, d\tau}{\int \psi_0^*(1 + AH')\psi_0(1 + AH')\, d\tau}$$

$$= \frac{E_0 + 2A\int \psi_0^* H'^2 \psi_0 d\tau + A^2 \int \psi_0^* H' H_0 H' \psi_0 d\tau}{1 + A^2 \int \psi_0^* H'^2 \psi_0 d\tau}$$

But,

$$\int \psi_0^* H' H_0 H' \psi_0 d\tau = 0 = \frac{\left[E_0 + 2A\int \psi_0^* H'^2 \psi_0 d\tau\right]\left[1 - A^2 \int \psi_0^* H^2 \psi_0 d\tau\right]}{\left[1 + A^2 \int \psi_0^* H'^2 \psi_0 d\tau\right]\left[1 - A^2 \int \psi_0^* H'^2 \psi_0 d\tau\right]}$$

$$= E_0 + \int \psi_0^* H'^2 \psi_0 d\tau \left[2A - E_0 A^2\right] = E_0 + \int \psi_0^* H'^2 \psi_0 d\tau \cdot \frac{1}{E_0} \left[\because A = \frac{1}{E_0}\right]$$

$$\therefore E_0 + E_2 \leqslant E_0 + \frac{\int \psi_0^* H'^2 \psi_0 d\tau}{E_0} \tag{15.134}$$

Again,

$$H' = \frac{e^2}{4\pi\varepsilon_0 R^3}(x_1 x_2 + y_1 y_2 - 2Z_1 Z_2)$$

$$\therefore H'^2 = \frac{e^4}{(4\pi\varepsilon_0)^2 R^6}\left(x_1^2 x_2^2 + y_1^2 y_2^2 + 4Z_1^2 Z_2^2 + 2x_1 x_2 y_1 y_2 + \ldots\right)$$

We now consider the value of the integral in eq. (15.134)

$$\int \psi_0^* H'^2 \psi_0 d\tau = \int \psi_0^* \frac{e^4}{R^6}\left(x_1^2 x_2^2 + y_1^2 y_2^2 + 4Z_1^2 Z_2^2 + 2x_1 x_2 y_1 y_2 + \ldots\right) \times \frac{1}{(4\pi\varepsilon_0)^2} \psi_0 d\tau$$

$$= \frac{e^4}{(4\pi\varepsilon_0)^2 R^6}\left[\int \psi_0^* x_1^2 x_2^2 \psi_0 d\tau + \int \psi_0^* y_1^2 y_2^2 \psi_0 d\tau + \int \psi_0^* 4Z_1^2 Z_2^2 \psi_0 d\tau + \oint \psi_0^* 2x_1 x_2 y_1 y_2 \psi_0\, d\tau + \ldots\right]$$

$$= \frac{e^4}{(4\pi\varepsilon_0)^2 R^6}\left[a_0^4 + a_0^4 + 4a_0^4 + 0 + \ldots\right] = \frac{6e^4 a_0^4}{R^6} \cdot \frac{1}{(4\pi\varepsilon_0)^2} \tag{15.135}$$

Substituting this value in eq. (15.134), we get

$$E_0 + E_2 \leqslant E_0 + \frac{6e^4 a_0^4}{R^6} \cdot \frac{1}{E_0} \cdot \frac{1}{(4\pi\varepsilon_0)^2}$$

$$\text{or, } E_2 \leqslant \frac{6a_0^4 e^4}{R^6} + \frac{1}{E_0} \cdot \frac{1}{(4\pi\varepsilon_0)^2} \leqslant \frac{6a_0^4 e^4}{R^6}\left(-\frac{4\pi\varepsilon_0 a_0}{e^2}\right)\frac{1}{(4\pi\varepsilon_0)^2} E_0 = -\frac{e^2}{a_0(4\pi\varepsilon_0)}$$

$$\text{or, } E_2 \leqslant \frac{6a_0^5 e^2}{R^6} \cdot \frac{1}{(4\pi\varepsilon_0)} \tag{15.136}$$

According to perturbation theory, the lower limit of E_2 is obtained as $-8a_0^5 e^2/4\pi\varepsilon_0 R^6$.

$$E_2 \geqslant \frac{1}{4\pi\varepsilon_0}\left(-8\frac{a_0^5 e^2}{R^6}\right) \quad \text{or,} \quad -\frac{8a_0^5 e^2}{R^6} \leqslant E_2 \leqslant \frac{6a_0^5 e^2}{R^6} \tag{15.137}$$

The term

$$\frac{e^4}{(4\pi\varepsilon_0)R^2}(-2Z_1 Z_2)$$

in eq. (15.129) represents the interaction energy of two electric dipoles that correspond to the instantaneous configuration of two atoms and the energy varies as $1/R^6$.

15.9 Answers of some important conceptual questions

15.1. *What is* WKB *approximation? Arrive at its validity equation.*

Ans. WKB is an approximate method of solving the wave equation. The method is based on the expansion, the first term of which leads to the classical results, the second term to the old quantum theory result and the higher terms to the characteristic of the new mechanics.

When the potential function is a slowly varying function of position, the problem can be solved by approximation method known as Wentzel-Kramers-Brillouin or WKB approximation method.

Validity equation

For a particle of mass m and kinetic energy E moving in potential, the de Broglie wavelength is

$$\lambda = \frac{h}{mv} = \frac{h}{[2m(E-V)]^{1/2}}$$

$$\therefore \tfrac{1}{2}mv^2 = E - V. \therefore m^2 v^2 = 2m(E-V)$$

The propagation constant k is given by

$$k = \frac{2\pi}{\lambda} = \frac{[2m(E-V)]^{1/2}}{\hbar}$$

The condition for slowly varying potential is

$$\left|\frac{1}{k^2}\frac{dR}{dx}\right| \ll 1$$

$$\therefore \frac{m\hbar\dfrac{\partial V}{\partial x}}{[2m(E-V)]^{1/2}} \ll 1 \text{ or, } \frac{\lambda\dfrac{dV}{dx}}{4\pi(E-V)} \ll 1$$

This is the validity equation of WKB approximation method.

15.2. *What are turning points? Mention connection formulae.*

Ans. Turning points of classical motion are the points where the velocity of the particle vanishes and changes sign. These are defined by the equation $E = V(x)$.

At turning points wavelength becomes infinity. As a result WKB approximation is not valid in the vicinity of turning points. To obtain complete solution Schrödinger equation is solved in a region of suitable extension surrounding the turning points and, then the solution is suitably joined to the solution on two sides. These solutions are the wave functions in regions where WKB approximation is valid.

The connection formula linking an increasing exponential solution in region (1) to an oscillatory solution in region (2) [Fig 15.2] is

$$\psi_{\mathrm{WKB}} = \begin{cases} \dfrac{1}{(p_1)^{1/2}} \exp\left(\displaystyle\int_{x_1}^{x} p_1 \frac{dx}{\hbar}\right), & (x < x_1) \\ \dfrac{1}{(p_2)^{1/2}} \cos\left(\displaystyle\int_{x_1}^{x} p_2 \frac{dx}{\hbar} + \frac{\pi}{4}\right), & (x > x_1) \end{cases}$$

where the symbols have their usual significances.

15.3. *What is variation method? Mention its field of applicability.*

Ans. Approximation methods are applied in cases where the Hamiltonian of the system differs only by small amount from the Hamiltonian of a similar system which is exactly solvable. The WKB method is applied only when the potential is a slowly varying function and the change in the potential over the distance of the de Broglie wavelength is small compared to KE of the particle. These methods are useful in almost ideal situations.

But, in the real world such as in chemical phenomenon these almost ideal conditions are not fulfilled. Under such circumstances, variation method (Rayleigh-Ritz method) is applied. This methods is widely used in atomic and nuclear physics, quantum chemistry also is the genesis of more elaborate methods of Hartree and Hartree-Fock approximation.

The expectation value of energy is given by

$$\langle E \rangle = \int \psi_0^* H \psi_0 \, d\tau = E_0$$

Here, ψ is any normalised function of the coordinates of the system. This integral is known as *variation integral* and gives an upper limit to the energy E_0 of the lowest state of the system. The function ψ is the variation function and its choice may be quite arbitrary but more wisely, it is chosen such that E approaches, more closely to E_0.

15.4. *What is the fact on which variational method is based?*

Ans. The simple fact on which the variational method is based is that the average energy of a system (i.e., $\langle H \rangle$) is in a state represented by any arbitrary wave function ψ which has got to be greater than or equal to the true ground state energy E_0 of the system.

$$\langle H \rangle = \frac{\int \psi^* H \psi \, d\tau}{\int \psi^* \psi \, d\tau} \geqslant E_0 \text{ (Eckart's theorem)}$$

As the ground state energy acts as a lower bound on an expectation value, calculated with any arbitrary wave function we can choose a trial function by an intelligent guess with a number of parameters and, then minimise $\langle H \rangle$ as a function of these parameters. The beauty of this method is that we need not solve any differential equation. The successful use of the variation method depends on the ability to make a judicious choice for the trial function as has been done in solved problem.

15.5. *What is the mathematical statement of the variational theorem due to Eckart?*

Ans. The expectation value of H in the state is given as

$$\langle H \rangle = \frac{\langle \psi | H | \psi \rangle}{\langle \psi | \psi \rangle} = \frac{\sum_n |a_n|^2 E_n}{\sum_n |a_m|^2} \geqslant \frac{E_0 \sum_n |a_n|^2}{\sum_n |a_n|^2} = E_0 \qquad \text{(i)}$$

(Here, $|\psi\rangle$ may not be normalised)

where the inequality follows from the fact as E_0 is the ground state energy, we must have $E_n > E_0$ for $n \neq 0$ and replacing all the E_n's by E_0 gives us a smaller number in general. (i) is the mathematical statement of the variational theorem due to Eckart. It tells us the fact that the energy expectation value of a trial function is an upper limit to the exact ground state energy.

It is to be noted that

(i) intelligently a trial function is to be chosen that contains a whole bunch of parameters λ_i

$$\psi = \psi(\lambda_1, \lambda_2, \ldots) \qquad \text{(ii)}$$

(ii) the parameter λ_i is to be varied until the expectation value of H is minimised and

$$\frac{\partial \langle H \rangle}{\partial \lambda_i} = 0 \qquad \text{(iii)}$$

To prove Eckart's theorem we suppose that $\{u_n\}$ are the complete set of orthonormal eigenfunctions of H for the statement at hand, viz., $Hu_n = E_n U_n$ and $\int u_n^* u_n \, d\tau = \delta_{mn}$.

Since, these form a complete set, any arbitrary trial function ψ can, of course, be expanded in terms of u_n.

$$|\psi\rangle = \sum_n a_n \{u_n\} \qquad \text{(iv)}$$

15.6. *What is field emission? Apply* WKB *approximation method to find an expression of transmission probability of electrons from the metal surface.*

Ans. The emission of electrons in presence of electric field from the metal surface is known as *field emission* or *cold emission*. The electric field imparts sufficient energy to the electrons in the metal to lift it from the potential well. The energy imparted is known as the work function. In the language of quantum mechanics work function effectively changes the shape of potential well reducing thereby the width of the potential as shown in Fig 15.7 below.

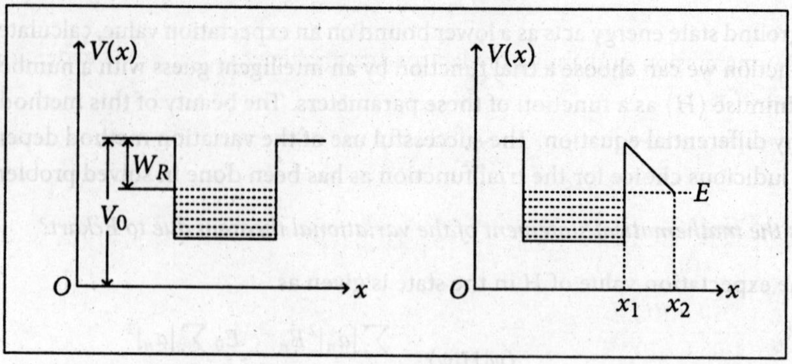

Fig 15.7 Field emission and width of potential

The transmission probability of electrons can be calculated using

$$T = \exp\left\{-\frac{2}{\hbar}\int_{x_1}^{x_2}\left[2m(V-E)^{1/2}\right]dx\right\} = e^{-2G}$$

where

$$G = \exp\left\{\frac{1}{\hbar}\int_{x_1}^{x_2}\left[2m(V-E)^{1/2}\right]dx\right\}$$

If X is the electric field applied, then the work done to liberate electron

$$W = eX\,x_2$$

This gives

$$x_2 = \frac{W}{eX} = x$$

Further, $V(x) = V_0 - eX\,x$.

Hence, $V(x) - E = V_0 - eX\,x - E = W - eX\,x$, as

$$W = V_0 - E$$

Now,

$$G = \frac{1}{\hbar}\int_0^x \{2m(V-E)\}^{1/2}\,dx \text{ or, } G = \frac{1}{\hbar}\int_0^x [2m(W - eX\,x)]^{1/2}\,dx$$

Put $W - eX\,x = t$, then $-dx = -\dfrac{dt}{eX}$.

$$\therefore G = -\int_W^0 \frac{(2m)^{1/2}\,t^{1/2}}{eX}\,dt = -2\frac{(2m)^{1/2}\,t^{3/2}}{3eX} = \frac{2}{3}\frac{(2m)^{1/2}\,W^{3/2}}{eX}$$

Thus, the transmission probability becomes

$$T = \exp\left[-\frac{4}{3}\frac{(2m)^{1/2}\,W^{3/2}}{\hbar eX}\right]$$

15.10 Solved problems

» ***Example* 15.1.** **A stream of electrons, each of energy $E = 3$ eV is incident on a barrier of height $V_0 = 4$ eV. The width of the barrier is 20 Å. Calculate the percentage transmission of the beam through the barrier.**

Solution: The transmission probability of electron beam of energy $E = 3$ eV incident on a potential barrier of height $V_0 = 4$ eV is

$$T = \frac{16E}{V_0}\left(1 - \frac{e}{V_0}\right)e^{-2G}$$

where $2G = 2a_0\left[\frac{2m(V_0 - E)}{\hbar^2}\right]^{1/2}$ and $a_0 = 20$ Å $= 20 \times 10^{-10\,\text{m}}$, $E = 3 \times 1.6 \times 10^{-19}$ J, $V_0 = 4 \times 1.6 \times 10^{-19}$ J.

$$\therefore 2G = \frac{40 \times 10^{-10}}{1.054 \times 10^{-34}}\left[2 \times 9.1 \times 10^{-31}\,(1.6 \times 10^{-19})\right]^{1/2} = \frac{40}{1.054}(2 \times 9.1 \times 1.6)^{1/2} = 20.49$$

Then

$$T = \frac{16 \times 3}{4}\left(1 - \frac{3}{4}\right)e^{-20.49} = \frac{16 \times 3}{4} \times \frac{1}{4}e^{-20.49} = \frac{3}{(2.71)^{20.49}} \text{ or, } \mathbf{T = 3.797 \times 10^{-17}\ \%}$$

» ***Example* 15.2.** An α-particle of energy 10 MeV approaches a potential barrier of height equal to 30 MeV. What is the width of potential if transmission coefficient is 2×10^{-3}? [mass of α-particle is $4 \times 1.67 \times 10^{-31}$ kg and $\hbar = 1.034 \times 10^{-34}$ J-s]

Solution: The transmission coefficient of an α-particle of energy E approaching a potential barrier V_0 is

$$T = \frac{16E}{V_0}\left(1 - \frac{E}{V_0}\right)e^{-2G}$$

$$\text{or, } e^{2G} = \frac{16E}{V_0 T}\left(1 - \frac{E}{V_0}\right) = \frac{16 \times 10}{30T}\left(1 - \frac{10}{30}\right) = \frac{16}{3} \times \frac{2}{3T} = \frac{32}{9T} = \frac{32}{9 \times 2 \times 10^{-3}}$$

$$\beta = \frac{1}{\hbar}\left[2m(V_0 - E)^{1/2}\right]\frac{1}{1.054 \times 10^{-34}} = \left[2 \times 4 \times 1.6 \times 10^{-31} \times 20 \times 1.6 \times 10^{-13}\right]^{1/2} = 19.62 \times 10^{14}$$

$$\therefore a_0 = \frac{1}{2\beta} \times 2.303 \log_{10}\frac{32}{9 \times 2 \times 10^{-3}}$$

$$\text{or, } \mathbf{a_0 = 1.91 \times 10^{-15}\ m}$$

» ***Example* 15.3.** Find the energy eigenvalues of a linear harmonic oscillator by WKB method.

Solution: If E be the total energy and V is potential energy V at a position x of the oscillator, then

$$\text{KE} = \tfrac{1}{2}mv^2 = E - V$$

moment $p = mv = \sqrt{2m(E - V)}$

$$V = \tfrac{1}{2}kx^2 = \tfrac{1}{2}m\omega^2x^2\,[\because \omega^2 = k/m]$$

At the extremities of the oscillator, $V = E = \frac{1}{2}m\omega^2A^2$, where A = amplitude of the oscillator $= \left(\frac{2E}{m\omega^2}\right)^{1/2}$.

Thus, the turning points in connection formulae are $x = -A$ and $x = A$.

The energy levels are obtained by Bohr's quantisation condition

$$\int_a^b p\,dx = (n + \tfrac{1}{2})\,\pi\hbar$$

Here, $a = -A$, $b = A$.

$$\therefore \int_{-A}^{A}\sqrt{2m(E - V)}\,dx = (n + \tfrac{1}{2})\,\pi\hbar \text{ or, } \int_{-A}^{A}(2mE)^{1/2}\left(1 - \frac{E}{V}\right)^{1/2}dx = (n + \tfrac{1}{2})\,\pi\hbar \qquad \text{(i)}$$

$$x = \left(\frac{2E}{m\omega^2}\right)^{1/2} t, \therefore dx = \left(\frac{2E}{m\omega^2}\right) dt$$

$$\frac{V}{E} = \frac{\frac{1}{2}kx^2}{\frac{1}{2}kA^2} = t^2$$

$\therefore$ (i) becomes

$$\frac{2E}{\omega}\int_{-1}^{+1}(1-t^2)^{1/2}\,dt = \left(n+\frac{1}{2}\right)\pi\hbar$$

Put $t = \sin\theta$, $\therefore dt = \cos\theta\, d\theta$.

So

$$\frac{2E}{\omega}\int_{-\pi/2}^{\pi/2}(1-\sin^2\theta)\cos\theta\, d\theta = (n+\tfrac{1}{2})\,\pi\hbar \text{ or, } \frac{2E}{\omega}\int_{-\pi/2}^{\pi/2}\cos^2\theta\, d\theta = (n+\tfrac{1}{2})\,\pi\hbar$$

$$\text{or, } \frac{2E}{\omega}\int_{-\pi/2}^{\pi/2}(1+2\cos\theta)\, d\theta = (n+\tfrac{1}{2})\,\pi\hbar \text{ or, } \frac{E}{\omega}\left[\theta + \frac{\sin 2\theta}{2}\right]_{-\pi/2}^{\pi/2} = (n+\tfrac{1}{2})\,\pi\hbar$$

$$\text{or, } \frac{\pi E}{\omega} = (n+\tfrac{1}{2})\,\pi\hbar \text{ or, } \boldsymbol{E = (n+\tfrac{1}{2})\,\hbar\omega} \qquad \text{(ii)}$$

This gives the required equation for energy of harmonic oscillator.

» ***Example* 15.4.** Apply WKB method to the bound states of a particle in potential well to obtain

$$2\int_0^a \sqrt{2m(E-V)}\,dx = (n+\tfrac{1}{2})\,\hbar,\ n = 0,1,2,\ldots$$

Solution: In Fig 15.3, the wave function in the region (1), where $x < a$ is given by WKB method

$$\psi_1 = \frac{1}{p_1^{1/2}}\exp\left[\left(\int_x^a \frac{p_1\,dx}{\hbar}\right)\right] \text{ from eq. (15.39)}$$

$$\psi_2 = \frac{2}{p_2^{1/2}}\sin\left(\int_a^x \frac{p_2\,dx}{\hbar} + \frac{\pi}{4}\right) \text{ from eq. (15.40)}$$

$$= \frac{2}{p_2^{1/2}}\cos\left(\int_a^x \frac{p_2\,dx}{\hbar} - \frac{\pi}{4}\right)$$

This equation may be rewritten as

$$\psi_2 = \frac{1}{p_2^{1/2}}\cos\left[\int_a^b \frac{p_2\,dx}{\hbar} + \int_b^x \frac{p_2\,dx}{\hbar} - \frac{\pi}{4}\right]$$

Adding and subtracting $\pi/4$, we get

$$\psi_2 = \frac{2}{p_2^{1/2}}\cos\left[\int_a^b \frac{p_2\,dx}{\hbar} - \int_x^b \frac{p_2\,dx}{\hbar} - \frac{\pi}{4}\right]$$

$$\text{or, } \psi_2 = \frac{2}{p_2^{1/2}}\cos\int_a^b \frac{p_2\,dx}{\hbar}\cos\left(\int_a^b \frac{p_2\,dx}{\hbar} + \frac{\pi}{4}\right) + \frac{2}{p_2^{1/2}}\sin\left(\int_a^b \frac{p_2\,dx}{\hbar} + \frac{\pi}{4}\right)$$

From the connection formula

$$\cos\left(\int_x^b \frac{p_2\,dx}{\hbar} + \frac{\pi}{4}\right)$$

will go to an exponential increasing solution in region (3) and for the bound state

$$\cos \int_a^b \frac{p_2\,dx}{\hbar} = 0 \text{ or, } \cos \int_a^b \frac{p_2\,dx}{\hbar} = \cos\left(n+\tfrac{1}{2}\right)\pi \text{ or, } \int_a^b \frac{p_2\,dx}{\hbar} = \left(n+\tfrac{1}{2}\right)\pi$$

$$\therefore \int_0^b [2m(E-V)]^{1/2}\,dx = \left(n+\frac{1}{2}\right)\hbar,\quad n = 0, 1, 2, \ldots$$

» ***Example* 15.5.** Use variation method to estimate the ground state energy of a particle in the potential

$$V = \infty,\ x < 0$$
$$V = kx, x > 0$$

Choose $xe^{-\alpha x}$ as the trial solution.

Solution: As in region $x < 0$, $V = \infty$, wave function vanishes there, i.e., $\psi = 0$, when $x < 0$. This is satisfied by xe^{-ax}, which at the boundary $x = 0$ is zero. Now, the expectation value of energy is given by

$$\langle E\rangle = \int_0^\infty \psi^* H\psi\,dx \Big/ \int_0^\infty \psi^*\psi\,dx = \frac{\int_0^\infty xe^{-\alpha x}\left[-\frac{\hbar^2}{2m}\frac{d^2}{dx^2} + kx\right]xe^{-\alpha x}\,dx}{\int_0^\infty x^2e^{-2\alpha x}\,dx}$$

$$= \frac{\int_0^\infty xe^{-\alpha x}\left[-\frac{\hbar^2}{2m}\left\{\frac{d}{dx}\left(e^{-\alpha x} + x(-a)\,e^{-\alpha x}\right)\right\}\right]dx + \int_0^\infty kx^3e^{-2ax}\,dx}{\int_0^\infty x^2e^{-2ax}\,dx}$$

$$= \frac{\int_0^\infty kx^3e^{2\alpha x}\,dx + \int_0^\infty \frac{\hbar^2}{2m}\alpha xe^{-2\alpha x}\,dx + \int_0^\infty xe^{-\alpha x}\left[\frac{\hbar}{2m}\alpha e^{-\alpha x} - \frac{\hbar^2}{2m}\alpha^2xe^{-\alpha x}\right]dx}{\int_0^\infty x^2e^{-2\alpha x}\,dx}$$

$$= \left(\frac{\hbar^2}{4m\alpha} - \frac{\hbar^2}{8m\alpha} + \frac{3k}{8\alpha^4}\right)\Big/\left(\frac{1}{4\alpha^3}\right) = \frac{3k}{2\alpha} + \frac{\hbar^2\alpha^2}{2m}$$

This is a minimum for $\alpha \simeq (2\,\text{km}/2\hbar)^{1/3}$.

Putting this value of α in the above equation, we get

$$\langle E\rangle \geqslant \frac{9}{4}\left(\frac{2\hbar^2k^2}{3m}\right)^{1/3}$$

» ***Example* 15.6.** A particle of mass m is bounded by a potential $V(r) = -V_0e^{\alpha r/a}$ with $\frac{\hbar^2}{mV_0\alpha^2} = 0.50$. Assume trial function as $e^{-\alpha r/a}$, use variation method to find out lowest energy eigenvalue.

Solution: According to normalisation condition

$$\int_0^\infty \psi^*\psi\,d\tau = 1 \text{ or, } A^2\int_0^\infty e^{-2\alpha r/a}4\pi r^2\,dr \text{ or, } 4\pi^2A^2\int_0^\infty r^2e^{-2\alpha r/a}\,dr = 1$$

$$\left[\text{using} \int_0^\infty e^{-kx}x^n\,dx = \frac{n!}{(k)^{n+1}}\right]$$

We have

$$4\pi A^2 \frac{2!}{(2\alpha/a)^3} = 1 \text{ or, } A^2 = \frac{\alpha^2}{\pi a^3}. \ \therefore\ A = \left(\frac{\alpha^3}{\pi a^3}\right)^{1/2}$$

Thus, the normalised function becomes

$$\psi = \left(\frac{\alpha^3}{\pi a^3}\right)^{1/2} e^{-\alpha x/a}$$

Now, the Hamiltonian of the system is

$$H = -\frac{\hbar^2}{2m}\nabla^2 + V(r) = \frac{\hbar^2}{2m}\left[\frac{\partial^2}{\partial r^2} + \frac{2}{r}\frac{\partial}{\partial r}\right] + V(r) = -\frac{\hbar^2}{2m}\left[\frac{\partial^2}{\partial r^2} + \frac{2}{r}\frac{\partial}{\partial r}\right] - V_0 e^{-r/a}$$

$\therefore$ the expectation value of energy is

$$\langle E\rangle = \int \psi^* H\psi\, d\tau$$

$$= \int \left(\frac{\alpha^3}{\pi a^3}\right)^{1/2} e^{-\alpha r/a}\left[-\frac{\hbar^2}{2m}\left(\frac{\partial^2}{\partial r^2} + \frac{2}{r}\frac{\partial}{\partial r}\right) - V_0 e^{-r/a}\right] \times \left(\frac{\alpha^3}{\pi a^3}\right)^{1/2} e^{-\alpha r/a} 4\pi r^2\, dr$$

$$= -\frac{4\pi\alpha^3}{\pi a^3}\left[-\frac{\hbar^2}{2m}\int e^{-\alpha r/a}\left(\frac{\partial^2}{\partial r^2} + \frac{2}{r}\frac{\partial}{\partial r}\right)e^{-\alpha r/a} r^2\, dr + V_0 \int e^{-2\alpha r/a} e^{-r/a} r^2\, dr\right] \quad \text{(i)}$$

Now,

$$\int_0^\infty e^{-\alpha r/a}\left[\left(\frac{\partial^2}{\partial r^2} + \frac{2}{r}\frac{\partial}{\partial r}\right)e^{-\alpha r/a}\right] r^2\, dr = \int_0^\infty e^{-\alpha r/a}\left(\frac{\alpha^2}{a^2} - \frac{2\alpha}{ra}\right)e^{-\alpha r/a} r^2\, dr$$

$$= \int_0^\infty \frac{\alpha^2}{a^2} e^{-2\alpha r/a} r^2\, dr - \frac{2\alpha}{a}\int e^{-2\alpha r/a} r\, dr = \frac{\alpha^2}{a^2}\frac{2!}{(2\alpha/a)^3} - \frac{2\alpha}{a}\frac{1}{(2\alpha/a)^2} = \frac{a}{4\alpha} - \frac{a}{2\alpha} = -\frac{a}{4\alpha} \quad \text{(ii)}$$

and

$$\int e^{-2\alpha r/a} e^{-r/a} r^2\, dr = \int_0^\infty e^{-(2\alpha+1)r/a} r^2\, dr = \frac{2!}{\left(\frac{2\alpha+1}{a}\right)^3} = \frac{2a^3}{(2\alpha+1)^3} \quad \text{(iii)}$$

Substituting (ii) and (iii) in (i)

$$\langle E\rangle = -\left(\frac{4\pi\alpha^3}{a^3}\right)\left[\frac{\hbar^2}{2m}\left(-\frac{a}{4\alpha}\right) + \frac{2V_0 a^3}{(2\alpha+1)^3}\right] = \frac{\alpha^2\hbar^2}{2ma^2} - \frac{8V_0\alpha^3}{(2\alpha+1)^3} \quad \text{(iv)}$$

For $\langle E\rangle$ to be minimum, then

$$\frac{\partial}{\partial\alpha}\langle E\rangle = 0 \text{ or, } \frac{\partial}{\partial\alpha}\left[\frac{\alpha^2\hbar^2}{2ma^2} - \frac{8V_0\alpha^3}{(2\alpha+1)^3}\right]$$

$$\text{or, } \frac{\hbar^2}{2m}(2\alpha) - 8V_0\{3\alpha^2(2\alpha+1)^{-3}3\alpha^3(2\alpha+1)^{-4}2\} = 0$$

$$\text{or, } \frac{\alpha\hbar^2}{ma^2} - \frac{24V_0\alpha^2}{(2\alpha+1)^4} = 0 \text{ or, } \frac{(2\alpha+1)^4}{\alpha} = \frac{24V_0 ma^2}{\hbar^2}$$

But, given $\dfrac{\hbar^2}{mV_0 a^2} = 48$.

Solving $\alpha \simeq 0.7$.

Hence,

$$E_{min} = \frac{\hbar^2}{8ma^2} \frac{8V_0(0.7)^3}{(2.4)^3} - 8V_0\frac{(0.343)}{13.824} = \frac{\hbar^2}{8ma^2} - 0.2V_0 = \frac{0.5}{8} - 0.02V_0$$

or, E_{min} = $\mathbf{0.138\,V_0}$

» ***Example* 15.7.** Find out the energy levels of a ball bouncing off a perfectly reflecting plane in a uniform gravitational field (energy levels for free fall).

Solution: Let $z = 0$ be the reflecting plane and z the coordinate of the bounding ball at any time. Then the potential energy of the ball is

$$V(z) = \alpha,\ z \leqslant 0$$
$$= mgh,\ z \geqslant 0$$

On the turning points, $z_1 = z = 0$.

At a height where KE = 0, $E - V = 0$ or, $E - mgz = 0$ or, $z_2 = z = E/mg$.

Now,

$$\psi_{WKB} = \frac{1}{\sqrt{p(z)}} \sin\left(\int_{z_1}^{z} \frac{p(z)}{\hbar} dz + \frac{\pi}{4}\right),\ z > z_1$$
$$= \frac{1}{\sqrt{p(z)}} \sin\left(\int_{z}^{z_2} \frac{p(z)\,dz}{\hbar}\right),\ z < z_2$$

These two expressions become equivalent in the region $z_1 \leqslant z \leqslant z_2$ if

$$\frac{1}{\hbar}\int_{z_1}^{z_2} p(z)\,dz = \frac{\pi}{4} = (n+1)\,\pi,\ n = 0, 1, 2, \ldots$$

In this case, the lower turning point is distinct from the upper, since at lower point, a phase-factor $\pi/4$ does not appear in the wave function as V is finite there (no matching of wave function).

∴ the condition now is

$$\frac{1}{\hbar}\int_{z}^{z_2} p(z)\,dz + \frac{\pi}{4} = (n+1)\,\pi \text{ or, } \frac{1}{\hbar}\int_{z}^{z_2} d(z) = \left(n + \frac{3}{4}\right)\pi$$

But,

$$p(z) = [2m(E-V)]^{1/2} = [2m(E - mgz)]^{1/2}$$

Solving, we get energy levels

$$E_n = \left(\frac{9}{8} mg^2\hbar^2\pi^2\right)^{1/3} \left(n + \frac{3}{4}\right)^{2/3},\ n = 0, 1, 2, \ldots$$

» ***Example* 15.8.** Find out which of the two trial wave functions (a) $\psi_1 = A(1+\alpha r)\,e^{-\alpha r}$ (b) $\psi_2 = Be^{-\alpha r^2/2}$ is better for the ground state of hydrogen atom and why?

Solution: Out of the two trial wave functions one will be better that gives the better estimate of the ground state energy of the hydrogen atom (–13.6 eV), i.e., which is close to this value. Generally, a chosen wave function, trial wave function, will be give energy greater than this.

(a) Here, we choose parameters A and α in ψ_1 in such a way that the expectation value of energy

$$\langle E\rangle = \int \psi_1^* H\psi_1\, d\tau \Big/ \int \psi_1^* \psi_1\, d\tau \text{ is minimum}$$

We shall first normalise ψ_1 to estimate the denominator

$$\int |\psi_1|^2 d\tau = 1 = A^2 \int (1+\alpha r)\, e^{-2\alpha r}\, d\tau$$

In spherical polar coordinates, $d\tau = r^2\, dr \sin\theta\, d\theta\, d\phi$.

$$\therefore \int |\psi_1|^2 d\tau = A^2 \int_0^\infty \int_0^\pi \int_0^{2\pi} (1+\alpha r)^2 e^{-2\alpha r} \times r^2\, dr \sin\theta\, d\theta\, d\phi$$

$$= A^2 \int_0^{2\pi} d\phi \int_0^\pi \sin\theta\, d\theta \int_0^\infty (1+\alpha r)^2 e^{-2\alpha r} r^2\, dr$$

$$= 4\pi A^2 \left[\int_0^\infty e^{-2\alpha r} r^2\, dr + 2\int_0^\infty \alpha r\, e^{-2\alpha r} r^2\, dr + \int_0^\infty \alpha^2 r^2 e^{-2\alpha r} r^2\, dr\right]$$

By standard integral

$$\int_0^\infty x^m e^{-\alpha x^n}\, dx = \frac{1}{2}\Gamma\left(\frac{m+1}{m}\right) \Big/ \alpha^{\frac{m+1}{m}},\; m \geqslant 0, n \geqslant 0$$

we get

$$\int |\psi_1|^2 d\tau = 4\pi A^2 \left[\frac{1}{4\alpha^3} + \frac{3}{4}\frac{1}{\alpha^3} + \frac{3}{4}\frac{1}{\alpha^3}\right] = 1$$

$$\therefore A = \left(\frac{\alpha^3}{7\pi}\right)^{1/2}$$

For the hydrogen atom, the Hamiltonian is

$$H = -\frac{\hbar^2}{2m}\nabla^2 - \frac{e^2}{4\pi\varepsilon_0 r}$$

Expectation of energy is

$$\langle E\rangle = \int \psi_1^* H\psi\, d\tau = -\frac{\hbar^2}{2m}\int \psi_1^* \nabla^2 \psi_1\, d\tau - \frac{e^2}{4\pi\varepsilon_0}\int \psi_1^* \frac{1}{r}\psi\, d\tau$$

$$= \frac{\hbar^2}{2m}\int |\text{grad}\,\psi_1|^2 d\tau - \frac{e^2}{4\pi\varepsilon_0}\int |\psi_1|^2 \frac{1}{r}\, d\tau = \frac{3\pi A^2}{\alpha}\cdot\frac{\hbar^2}{2m} - \frac{e^2}{4\pi\varepsilon_0}\frac{9\pi A^2}{2\alpha^2}$$

$$= \frac{3\pi}{\alpha}\cdot\frac{\alpha^3}{7\pi}\cdot\frac{\hbar^2}{2m} - \frac{9}{2}\cdot\pi\cdot\frac{e^2}{4\pi\varepsilon_0}\frac{\alpha^3}{7\pi}\cdot\frac{1}{\alpha^2} = \frac{3}{14}\cdot\frac{\hbar^2}{m}\alpha^2 - \frac{9}{14}\frac{e^2\alpha}{(4\pi\varepsilon_0)}$$

For this to be minimum, $\dfrac{d\langle E\rangle}{d\alpha} = 0$, whence

$$\frac{\hbar^2}{m}2\alpha - \frac{3e^2}{4\pi\varepsilon_0} = 0 \text{ or, } \alpha = \frac{3e^2 m}{2\hbar^2(4\pi\varepsilon_0)}$$

Putting this value, we get

$$\langle E\rangle_{\min} = \frac{3}{14}\left(\frac{\hbar^2}{m}\right)\cdot\frac{9e^4m^2}{4\hbar^4(4\pi\varepsilon_0)^2} - \frac{9}{14}e^2\frac{3e^2m}{2\hbar^2}\cdot\frac{1}{(4\pi\varepsilon_0)^2} = \frac{me^4}{\hbar^2}\cdot\frac{1}{(4\pi\varepsilon_0)^2}\left[\frac{27}{56}-\frac{27}{28}\right]$$

$$\text{or, } \langle E\rangle_{\min} = -0.435\frac{me^4}{\hbar^2}\frac{1}{(4\pi\varepsilon_0)^2}$$

Clearly, for ψ_1 this energy is less than that for ψ_2. Hence, ψ_1 is a better trial function.

EXERCISES

15(A): Objective (or MCQ) Type Questions

Select the correct answer:

15.1. WKB method is used when the potential function of a particle is a

(a) slowly varying function of position
(b) constant
(c) rapidly varying function of position
(d) none of these.

15.2. The condition of validity of WKB method is

(a) $\dfrac{\lambda\dfrac{dV}{dx}}{4\pi(E-V)} \ll 1$
(b) $\dfrac{\lambda\dfrac{dV}{dx}}{4\pi(E-V)} \gg 1$
(c) $\dfrac{\lambda\dfrac{dV}{dx}}{4\pi(E-V)} = 0$
(d) $\dfrac{\lambda\dfrac{dV}{dx}}{4\pi(E-V)} = \infty$.

15.3. Turning points of classical motion are the points, where the velocity of the particle changes sign and is equal to

(a) ∞
(b) 0
(c) constant
(d) none of these.

15.4. Turning points are defined by energy E is

(a) $E = 0$
(b) $E = \alpha$
(c) $E = V(x)$
(d) $E \geqslant 0$.

15.5. At turning points de Broglie wavelength of the particle is

(a) λ = constant
(b) $\lambda = 0$
(c) $\lambda = \infty$
(d) $\lambda = -\infty$.

15.6. To apply WKB approximation, the condition is that the barrier is thick and high enough so that

$$\int_a^b \frac{[2m(E-V)]^{1/2}}{\hbar}\,dx$$

is considerably.

(a) less than unity (c) constant

(b) greater than unity (d) none of these.

15.7. The variation method among others is used if the system contains

(a) single electron (c) extremely large number of electrons

(b) several electrons (d) none of these.

15.8. According to variation method the minimum energy $E_{\min}$ of the hydrogen atom is

(a) $-\dfrac{2\pi^2 me^4}{\hbar^2(4\pi\varepsilon_0)^2}$ (c) $+\dfrac{2\pi^2 me^4}{\hbar^2(4\pi\varepsilon_0)^2}$

(b) $-\dfrac{2\pi^2 me^2}{\hbar^2(4\pi\varepsilon_0)^2}$ (d) $-\dfrac{2\pi^2 me^4}{\hbar^2(4\pi\varepsilon_0)^2}$.

15.9. The variation of method shows that the minimum energy E of the helium atom is

(a) $-2e^2/a_0$ (c) $-2.85e^2/a_0$

(b) $2.85e^2/a_0$ (d) $1.69e^2/a_0$.

15.10. According to variation, method the minimum energy of the deuterium is

(a) $E_{\min} = \dfrac{\hbar^2(2\alpha-1)}{2\mu r_0^2\alpha^2(2\alpha+3)}$ (c) $\dfrac{\hbar^2\alpha^2(2\alpha-1)}{2\mu r_0^2(2\alpha+3)}$

(b) $E_{\min} = \dfrac{\hbar^2\alpha^2(2\alpha+1)}{2\mu r_0^2(2\alpha+3)}$ (d) $\dfrac{\alpha^2 r_0^2(2\alpha-1)}{2\mu\hbar^2(2\alpha+3)}$.

15.11. In view of variation method the interaction energy of two electric dipoles in van der Waal's interaction is

(a) $\dfrac{e^4}{(4\pi\varepsilon_0)R^2}(2z_1z_2)$ (c) $\dfrac{e^4}{(4\pi\varepsilon_0)^2R}(-2z_1z_2)$

(b) $\dfrac{e^4}{(4\pi\varepsilon_0)R^2}(-2z_1z_2)$ (d) $\dfrac{R^2}{(4\pi\varepsilon_0)\,e^4}(-2z_1z_2)$.

15.12. According to variation method the zero-point energy of harmonic oscillator is

(a) $\langle E\rangle = \dfrac{\hbar\omega_0}{2}$ (c) $\langle E\rangle = \dfrac{\hbar}{2\omega_0}$

(b) $\langle E\rangle = \dfrac{(n+1)\hbar\omega_0}{2}$ (d) $\dfrac{(n-1)\hbar\omega_0}{2}$.

ANSWERS

15.1. (a).	**15.3.** (b).	**15.5.** (c).	**15.7.** (b).	**15.9.** (b).	**15.11.** (a).
15.2. (a).	**15.4.** (b).	**15.6.** (b).	**15.8.** (a).	**15.10.** (c).	**15.12.** (a).

15(B): Short Answer Type Questions

15.1. What is WKB approximation method? Mention its field of applicability.

15.2. Arrive at the condition of validity of WKB approximation method.

15.3. Define turning points. What connection formulae?

15.4. What do you mean by field emission? Find out the transmission probability of electrons from metal surface.

15.5. *Each electron screens the nucleus from the other electrons and effective nuclear charge is reduced by 5/16 of an electronic charge.*—Justify.

15.6. Deduce Eckart's theorem in connection with variational method.

15.7. Applying WKB approximation method arrive at semi-classical (Bohr-Sommerfeld) quantum condition

$$\oint p\,dq = nh$$

15(C): Broad Answer Type Questions

15.1. Describe the WKB approximation method for the solution of Schrödinger wave equation in a potential field and discuss its validity.

15.2. (a) Outline WKB method for a one-dimensional case and derive connection formula.

(b) Apply the method to obtain the quantisation connection for boundary state.

15.3. Show that WKB approximation method gives exact energy value for a linear harmonic oscillator.

15.4. Discuss WKB approximation method in quantum mechanics and illustrate its use by applying to the problem of α-decay of radioactive nucleus.

15.5. Describe WKB approximation method and give an application of this method.

15.6. Describe the method of variational constants as applied to time-dependent perturbations. Illustrate it by simple examples.

15.7. Give the theory of WKB approximation and enumerate its application. Discuss one application in detail.

15.8. (a) What is the principle of WKB method.

(b) Apply WKB method to calculate the energy eigenvalues of a one-dimensional harmonic oscillator.

15.9. Discuss WKB method to explain α-decay. Hence, derive Geiger-Nuttal law.

15.10. (a) State the condition under which WKB approximation is applicable.

(b) Apply WKB method to the bound states of a particle in a potential well to obtain

$$2\int_a^b \sqrt{2m(E-V)}\,dx = \left(n+\tfrac{1}{2}\right)\hbar,\; n = 0, 1, 2, \ldots$$

the symbols have their usual meaning, a and b are turning points.

(c) With the help of above equations work out the energy eigenvalues for the harmonic oscillator.

15.11. Give the theory of WKB method.From the asymptotic WKB solution near turning points of a potential well of arbitrary shape, deduce Bohr-Sommerfeld quantisation rule.

15.12. Using the variational method show that the ground state energy of a three-dimensional oscillator is given by

$$\langle H\rangle_{\min} = \tfrac{9}{7}\sqrt{\tfrac{3}{2}}\,\hbar\omega$$

[*Hints*: Choose a trial function, $\psi = a(1 + pr)\, e^{-pr}$, where p is the variational parameter and a is a constant.]

15.13. (a) What is variational method. Discuss its principle.

(b) Using the above method find the normal state of helium atom.

15.14. (a) Discuss variational method.

(b) Using variational method find out interaction energy of two electric dipoles.

15.15. Using variational method find the ground state of hydrogen atom.

15(D): Numerical Problems

15.1. Solve the problem of linear harmonic oscillator by WKB method and compare the results with those obtained by classical method.

15.2. A particle of mass m is bound by the potential $V(r) = -V_0 e^{-r/a}$, where $\dfrac{\hbar^2}{mV_0 a^2} = 0.75$. Use the variation method with the trial function $e^{-\alpha r}$ to get a good limit to the lowest energy eigenvalue.

15.3. Use the variational method to find the energy eigenvalues and eigenfunctions for the found 1s and first excited state 2s states of hydrogen atom assuming the trial wave function $\psi_{1s} = Ae^{-\beta r}$ and $\psi_{2s} = B\left[1 + \gamma\left(\frac{r}{a}\right)\right] e^{-\alpha r/a}$ respectively.

15.4. Obtain energy values of harmonic oscillator by the WKB method.

15.5. Explain variational method of finding the energy of the ground state of a system and discuss how the method can be extended to find the energy of an excited state.

15.6. Show that if a particle has the wave function $\psi = e^{ikz}$, the z-component of its angular momentum is zero.

[*Hints*: $L_z e^{ikz} = -i\hbar\left(x\dfrac{\partial}{\partial y} - y\dfrac{\partial}{\partial x}\right) e^{ikx} = 0.$

Hence, e^{ikz} is an eigenstate of L_z with zero eigenvalues.]

15.7. Derive a trial variation function for a free particle in a one-dimensional box of length l.

[*Hints*: The wave function is zero outside the box and $\psi = 0$ at $x = 0$ and $x = l$. So let us try the trial function $\psi = x(l - x)$ for $0 \leqslant x \leqslant l$ and ψ outside.

$$E \geqslant \frac{\langle\psi|H|\psi\rangle}{\langle\psi|\psi\rangle} = \frac{\displaystyle\int_0^l (lx - x^2)\left(-\frac{\hbar^2}{2m}\right)\frac{d^2}{dx^2}(lx - x^2)\, dx}{\displaystyle\int_0^l x^2(l - x)^2\, dx} = \frac{\hbar^2 l^3/6m}{l^5/30} = \frac{5\hbar^2}{ml^2}\Bigg]$$

15.8. Use variational principle to estimate the ground state energy of a particle moving in a potential well given by

$$V = \begin{cases} \infty, & \text{for } x < 0 \\ cx, & \text{for } x > 0,\ c = \text{constant} \end{cases}$$

[*Hints*: Put the trial wave function as $\psi = xe^{-\alpha x}$, α = variational parameter. Now, by Eckart's theorem

$$\langle H \rangle = \frac{\int \psi^* H\psi\, d\tau}{\int \psi^* Y d\tau} \geqslant E_0$$

the expectation value of energy is

$$\langle E \rangle = \langle H \rangle = \frac{\int_0^\infty \psi^* H\psi\, d\tau}{\int_0^\infty \psi^* \psi\, d\tau} = \frac{\int_0^\infty xe^{-\alpha x}\left[-\frac{\hbar^2}{2m}\frac{\partial^2}{\partial x^2} + cx\right]xe^{-\alpha x}\, dx}{\int_0^\infty x^2 e^{-2\alpha x}\, dx} = \frac{\hbar^2 \alpha^2}{2m} + \frac{3c}{2\alpha}$$

Minimising

$$\frac{\partial \langle H \rangle}{\partial \alpha} = 0 \text{ or, } \alpha = \left(\frac{3cm}{2\hbar^2}\right)^{1/3}$$

Thus, the expectation value of energy is

$$\langle E^0 \rangle = \frac{9}{4}\left(\frac{2\hbar^2 c^2}{3m}\right)^{1/3}$$]

15.9. Use the WKB approximation to show that an attractive three-dimensional potential that falls off like r^{-n} for large r has infinite number of bound station if $n \geqslant 2$.

Chapter 16

Molecular Spectra

16.1 Introduction

The *line spectra* emitted from a gas discharge tube generally consist of a number of discrete lines. The origin of this line spectra lies in the electronic transition in atoms of different elements. The origin of line spectra can be understood with the help of Bohr-Sommerfeld theory and quantum mechanics.

There are other types of spectra emitted by substances in the *molecular state*. The molecules may contain two or more atoms. These molecular spectra appear as continuous bands under low dispersion and are known as *band spectra*. The band has a sharp intense edge on one side and is called *band head*. The intensity of the band gradually decreases as one moves towards the other side of it.

A spectral band, when observed through a high resolving power instrument, is found to consist of the *three-fold* structure as below:

(a) At the long wavelength side, a large number of lines are found to be crowded together. These lines are arranged in a regular fashion. Generally, the lines spread out from a fixed point in the band which is called the *band origin*.

(b) Several bands in the spectrum follow one another in a regular sequence and thus constitute a group of bands.

(c) The closely spaced different bands, because of their regular arrangement form a *band system*

16.2 Category of molecular spectra

The band spectra of molecules may be divided into the following three classes:

(i) *Rotational spectra*: Rotational spectra arise due to the radiation absorbed or emitted when a transition between two rotational energy levels occurs. During this transition the vibrational state of the molecules and the electronic configuration of the atoms remain unchanged.

The order of energy associated with the above transition is ~ 10^{-3} eV and as such the emitted radiation lies in the *far infrared* or *microwave region* of wavelength $\lambda \sim 10^7$ Å. This wavelength is much longer than that of visible light.

(ii) *Rotational-vibrational spectrum*: This spectrum arises due to radiation emitted of absorbed when a transition between two vibrational levels takes place, while the electronic configuration of the atoms remains fixed. Under this condition, rotational levels may or may not undergo a change. A pure vibrational spectrum is obtained if the rotation levels remain unchanged.

The transition energy being ~ 0.1 eV the emitted radiation is in the *near infrared region* of wavelength $\lambda \sim 10^5$ Å.

(iii) *Electronic spectrum*: The electronic spectrum arises due to radiation emitted or absorbed when a transition between two electronic energy levels of an atom in a molecule takes place.

The wavelength of this spectrum usually ranges from 1000 to 7000 Å and hence lies in the *visible* or *in the ultraviolet region*

16.3 Complexity of the molecular spectrum

A molecule may have various allowed energy states. The transition between these energy states gives rise to the emission or absorption of a photon resulting in the molecular spectrum.

We consider two possible energy states of a system, viz., E_1 and E_2 as shown in Fig 16.1. Transition can take place between the levels E_1 and E_2, provided the appropriate amount of energy, $\Delta E = E_2 - E_1$, can be either absorbed or emitted by the system. According to Max Planck, such absorbed or emitted energy can take the form of electromagnetic radiation whose frequency is

$$\nu = \frac{\Delta E}{h}$$

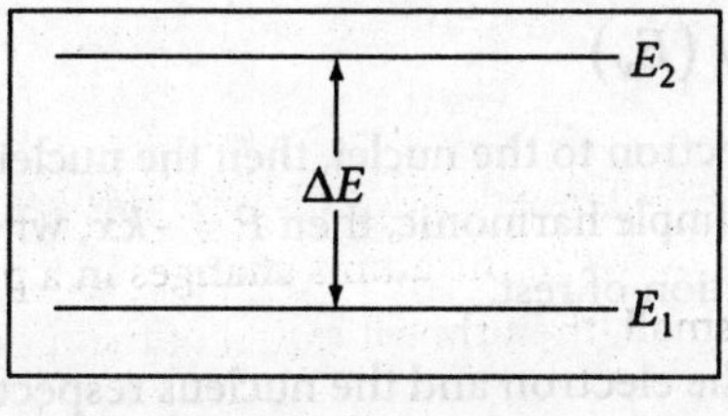

Fig 16.1 Two possible energy states of a molecule

If $E_2 > E_1$, then $\nu = \dfrac{E_2 - E_1}{h}$, energy will be absorbed and if $E_1 > E_2$, then $\nu = \dfrac{E_1 - E_2}{h}$, energy will be emitted.

The molecular spectrum thus obtained is more *complicated* than that of an atom. This is because:

(a) a molecule having many more degrees of freedom, it has many more energy levels as compared to that of an atom.

(b) the electrons in a molecule do not move in a central force field, that is why the method of separation of angular motion from the radial is not possible.

Hence, the orbital angular moment is not conserved and this makes it impossible to level the energy states with orbital quantum number.

(c) In a molecule the nuclei can rotate about its centre of mass or vibrate about their mean position of rest or execute complex type of motion including rotation and vibration. That is why solving Schrödinger equation of motion becomes extremely difficult for such a complicated system

In view of the above impasse, simple models of molecules have been proposed with a view to explain the molecular spectra satisfactorily. Max Born and J.R. Oppenheimer suggested that an approximate solution of Schrödinger equation can be obtained by neglecting the coupling between the different types of motion. Then to a first approximation, the energy of a molecule may be expressed as

$$E = E_e + E_v + E_r \tag{16.1}$$

The electronic energy E_e is of the order of a few electron-volts. The vibrational energy $E_v \ll E_e$ and the rotational energy $E_r < E_v$.

16.3.1 Electronic energy (E_e)

To have an idea of the order of magnitude of electronic energy, let a be the separation between the nuclei of the two atoms of a molecule. According to the uncertainty principle, the momentum of the valence electron is $p_e = \Delta p$.

Since, $a\Delta p = \hbar$, $\therefore \Delta p = \hbar/a$.

The kinetic energy of the electron is, then of the order of

$$E_e = \frac{p_e^2}{2m_e} = \frac{\hbar^2}{2m_e a^2} \tag{16.2}$$

But, $a \sim 1$ Å. Substituting the standard values of $\hbar$ and m_e, we get $E_e \sim 3.80$ eV.

Hence, E_e comes out to be of the same order of magnitude as that of a valence electron in an atom.

16.3.2 Vibrational energy (E_v)

If F be the binding force of the electron to the nuclei, then the nuclei are also bound to the electrons by an equal and opposite force. If F be simple harmonic, then $F = -kx$, where k is the force constant and x is the displacement from the mean position of rest.

If m_e and M be the masses of the electron and the nucleus respectively, then their angular frequency of vibration are

$$\omega_e = \sqrt{\frac{k}{m_e}} \text{ and } \omega_N = \sqrt{\frac{k}{M}} \text{ respectively}$$

Hence,

$$\frac{E_v}{E_e} = \frac{\hbar\omega_N}{\hbar\omega_e} = \sqrt{\frac{m_e}{M}} \tag{16.3}$$

But, $m_e/M \sim 10^{-3}$ to 10^{-5}. $\therefore E_v/E_e \sim 10^{-2}$.

E_e is ~ 1 eV and this produces spectral lines in the *visible* or *ultraviolet regions*. $E_v \sim 10^{-2} \times 1 = 10^{-2}$ eV, which would produce spectral lines in the *infrared region* of the electromagnetic spectrum.

16.3.3 Rotational energy (E_r)

We consider a diatomic molecule with two nuclei each of mass M and separated by a distance a, then the rotational energy of the molecule is

$$E_r = \frac{\hbar}{2I} = \frac{\hbar}{Ma^2} \tag{16.4}$$

where $I = \frac{1}{2} Ma^2$ is the moment of inertia of the diatomic molecule about an axis perpendicular to the line joining the nuclei and passing through the centre of mass of the system.

$$\therefore \frac{E_r}{E_e} = \frac{\hbar}{Ma^2} \Big/ \frac{\hbar^2}{2m_e a^2} = \frac{2}{\hbar}\left(\frac{m_e}{M}\right) \text{ or, } E_r = \frac{2}{\hbar}\left(\frac{m_e}{M}\right) E_e \tag{16.5}$$

Thus, E_r of a nucleus is less than E_e by a factor ~ (m_e/M).

Also the E_r is less than E_v by a factor ~ $(m_e/M)^{1/2}$. So E_r is ~ 10^{-3} eV or less and E_v is 10^{-1} to 10^{-2} eV, less than E_e.

16.4 Pure rotational energy levels of diatomic molecules

If the two atoms of a diatomic molecule are supposed to be joined by a *rigid rod* so that the distance between them remains unchanged, then electronic energy (E_e) and the vibrational energy (E_v) of the molecule remain unchanged as the molecule rotates as a whole. This model of the diatomic molecule is known as a *rigid rotator model.* It will be assumed that the rotation of the molecule takes place only about an axis perpendicular to the line joining the two atoms and the atoms are point masses.

If the two atoms of masses m_1 and m_2 rotate with an angular velocity ω about an axis perpendicular to the line joining the two atoms passing through the centre of mass of the system [Fig 16.2], then the moment of inertia (MI) of the system about the axis is

$$I = m_1 r_1^2 + m_2 r_2^2 \tag{16.6}$$

where r_1 and r_2 are the distances from the centre of mass and R is the distance between the molecules.

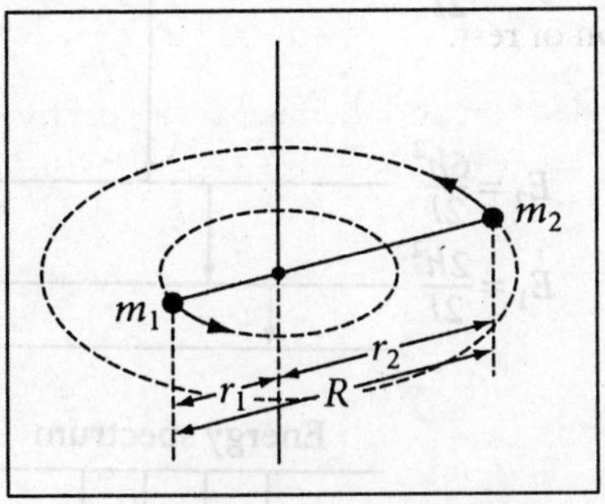

Fig 16.2 Rotation of diatomic molecule

Now, in the centre of mass coordinates

$$m_1 r_1 = m_2 r_2 \tag{16.7}$$

and $R = r_1 + r_2$.

From eq. (16.7), $r_1 = \dfrac{m_2}{m_1} r_2$.

$$\therefore R = r_1 + r_2 = \frac{m_2}{m_1} r_2 + r_2 = r_2 \left(\frac{m_1 + m_2}{m_1} \right) \text{ or, } r_2 = \frac{m_1 R}{m_1 + m_2}$$

Similarly,

$$r_1 = \frac{m_2 R}{m_1 + m_2}$$

$$\therefore I = \frac{m_1 m_2^2}{(m_1 + m_2)^2} R^2 + \frac{m_2 m_1^2}{(m_1 + m_2)^2} R^2 = m_1 m_2 \frac{(m_1 + m_2)}{(m_1 + m_2)^2} R^2$$

$$= \frac{m_1 m_2}{m_1 + m_2} R^2 = \mu R^2 \tag{16.8}$$

where $\mu = \dfrac{m_1 m_2}{m_1 + m_2}$ is the reduced mass of the system.

The kinetic energy of the molecule is

$$E = \frac{1}{2} I \omega^2 = \frac{1}{2I} (I\omega)^2 = \frac{J^2}{2I} \tag{16.9}$$

where J is the angular momentum of the system.

According to quantum mechanics, the orbital angular momentum is quantised and is given by

$$J = \sqrt{j(j+1)}\, \hbar$$

where j is called the *rotational quantum number* which can take values 0, 1, 2, 3, ...

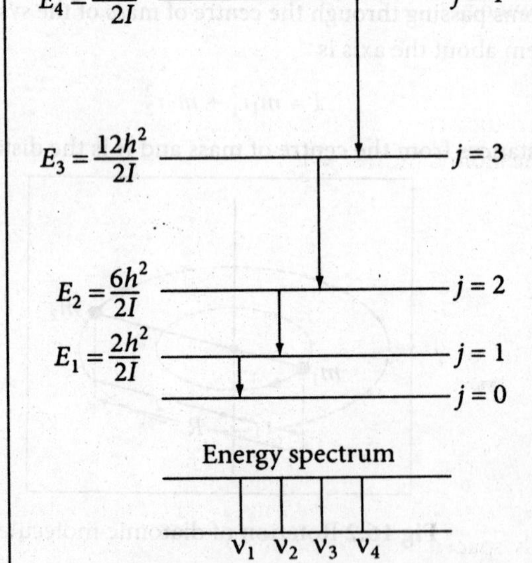

Fig 16.3 Energy levels of rotational spectrum and the spectral lines due to transition

The rotational energy levels of the diatomic molecule are given by

$$E_j = \frac{j(j+1)\hbar^2}{2I} = Bj(j+1),\ j = 0, 1, 2, \ldots \tag{16.10}$$

where $B = \hbar^2/2I$ and is called the *rotational constant.*

Thus, the molecules can have a *discrete* set of rotational energy levels with different j values. These levels constitute the pure rotational levels as shown in Fig 16.3. The difference between two rotational energy levels goes on increasing as j increases, e.g.,

$$E_1 - E_0 = \frac{2\hbar^2}{2I},\ \text{but}\ E_2 - E_1 = \frac{4\hbar^2}{8I}$$

16.4.1 Rotational spectra

Due to the transition between two rotational energy levels, rotational spectra of diatomic molecules arise. The selection rule for such a transition between rotational levels is

$$\Delta j = j_f - j_i = \pm 1 \tag{16.11}$$

Here, j_f and j_i represent respectively the final and initial rotational energy levels. A transition can occur only between two consecutive energy levels, viz., E_j and E_{j-1}.

If ν_j represents the frequency of the emitted photon due to this transition, then the energy of the photon is

$$h\nu_j = E_j - E_{j-1} = \frac{\hbar^2}{2I}j[j+1-(j-1)] = 2Bj \tag{16.12}$$

$$\therefore \nu_j = \frac{2B}{h}j\ (j = 1, 2, 3, \ldots) \tag{16.13}$$

The spectrum has the frequency

$$\nu_1 = \frac{2B}{h},\ \nu_2 = \frac{4B}{h},\ \nu_3 = \frac{6B}{h},\ \ldots$$

The value of B can be determined from the experimental values of ν. B can also be calculated from the internuclear distance and the atomic masses.

Now, for $j = 0$, $\nu_0 = 0$

$j = 1$, $\nu_1 = 2B/h$

$j = 2$, $\nu_2 = 4B/h$

$j = 3$, $\nu_3 = 6B/h$

$j = 4$, $\nu_4 = 8B/h$.

$$\therefore \nu_2 - \nu_1 = \frac{2B}{h},\ \nu_3 - \nu_2 = \frac{2B}{h},\ \nu_4 - \nu_3 = \frac{2B}{h},\ \text{etc.}$$

This shows that the frequencies of lines in rotational spectra are integral multiples of $\nu_1 = \frac{2B}{h}$ and the separation consists of equally spaced lines with spacing ν_1. Hence, here the *frequencies are equispaced, but the energy levels are not.*

In deriving the relation (16.10), we have considered the rotation about an axis perpendicular to the line of the two nuclei. The rotation of the molecule about the line has been neglected. This is due to the fact that the MI about this line is very small as the line passes through the nuclei where most of the mass of the molecule is concentrated.

Since, the radius of the nucleus is 10^{-4} times the radius of the atom, the MI will be smaller by a factor $\sim 10^{-8}$. So E_j would be very large ($\sim 10^5$ eV). The bond energy being of the same order of magnitude, the molecule will dissociate if such a high energy is applied to it. As a result *the energy level arising out of the rotation of the molecule is out of expectation.*

The *importance* of pure rotational spectra is that from the study of the spectra of a molecule, its moment of inertia (I) can be estimated and, then from the knowledge of the masses of the constituent atoms the bond length (R_0) can be determined. Valuable information regarding the molecular structure can be obtained from the discrete pure lines observed in gaseous substances.

16.4.2 All diatomic molecules do not exhibit rotational spectra. Why?

We know, rotational spectra arise from the transition between the rotational energy levels.

Since, the emission of radiation is due to the *rotation of electric dipole moment,* only the molecules having electric dipole moment can absorb or emit electromagnetic radiation and hence can exhibit rotational spectra.

The molecules like HCl or CO consisting of *unlike* atoms possess permanent dipole moment as the centres of gravity of positive and negative charges do not coincide.

For example, in HCl molecule an electron is transferred to Cl from H and the molecules take the form: H^+Cl^-. The centre of gravity of H^+ and Cl^- does not coincide and an electric dipole is formed. The component of dipole moment along a fixed direction changes periodically as the molecule rotates giving rise to rotational spectra.

In case of symmetric molecules consisting of *like* atoms, viz., N_2, H_2, O_2, CO_2 (O=C=O) and CH_4, no such dipole moment is present and hence cannot emit or absorb radiation. Though the rotational levels exist in such atoms, transitions between them normally do not occur and hence rotation spectra are not observed.

16.4.3 Intensities of the rotational lines

The intensities of the rotational spectral lines depend on the relative probabilities of transitions. The transitions between the rotational levels are controlled by the selection rule $\Delta j = \pm 1$, i.e., for a transition between two rotational levels $j = 0$ to 1, $j = 1$ to 2, $j = 2$ to 3, etc. The transition and hence the intensity depends on the following:

(a) Boltzmann factor giving the relative number of molecules in the different initial rotational energy states and

(b) the statistical weights of these states.

If N_0 be the number of molecules in the lowest energy state $j = 0$, then the number of molecules in any higher state j according to Boltzmann factor is given by

$$N_j = N_0\, e^{-E_j/kT} = N_0\, e^{-\frac{Bch\, j(j+1)}{kT}} \tag{16.14}$$

where

$$E_j = Bch\; j(j+1) \quad : (16.15)$$

where B is the rotational constant and k is the Boltzmann constant = 1.38×10^{-23} J/K.

Using $B = 200/\text{m}$, $T = 300$ K and $j = 1$, we get

$$N_j = N_0 \exp\left[-\frac{(2\times 10^2)\times(6.63\times 10^{-34})(3\times 10^8)\times 2}{1.38\times 10^{-23}\times 300}\right] = N_0\, e^{-0.019} = 0.98\, N_0 \qquad (16.16)$$

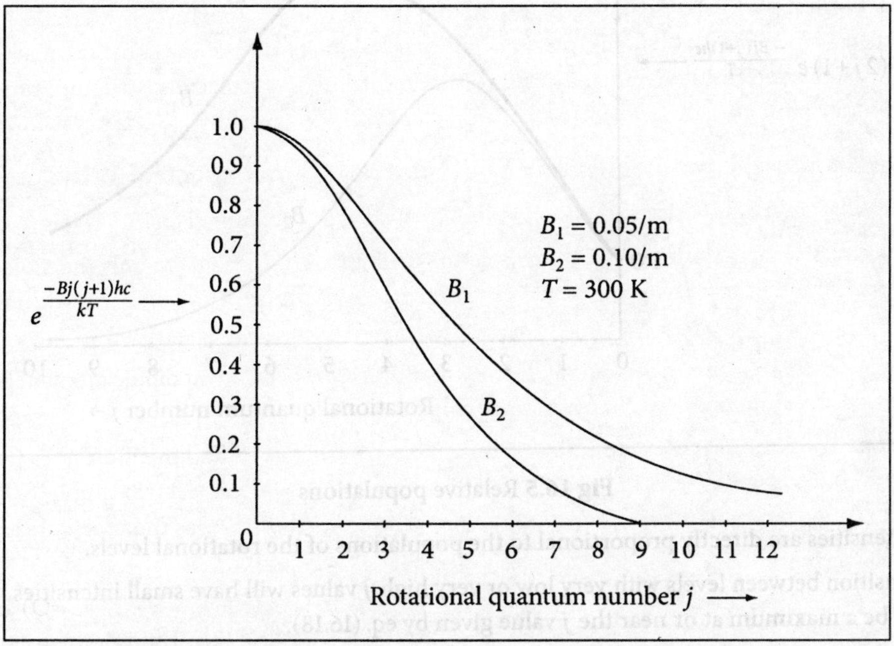

Fig 16.4 Boltzmann population of the rotational energy levels

Here, there are almost as many molecules in the $j = 1$ state at equilibrium as in the $j = 0$ state. But, N_j decreases more rapidly with increasing j and with larger B as shown in Fig 16.4.

The statistical factor of a level of a given j is $(2j+1)$ so that the total population of the jth level is given by the factor

$$\text{population} \propto (2j+1)\, e^{-E_j/kT} \qquad (16.17)$$

When $(2j+1)\, e^{-E_j/kT}$ is plotted against j, we get a curve of the type as shown in Fig 16.5. This curve shows that the population rises to a maximum and, then decreases. By differentiating eq. (16.17) we can show that the population is the maximum at the nearest integral j value to

$$j = \sqrt{\frac{kT}{2hcB}} - \frac{1}{2}$$

So for maximum population

$$j = \sqrt{\frac{kT}{2hcB}} - \frac{1}{2} \qquad (16.18)$$

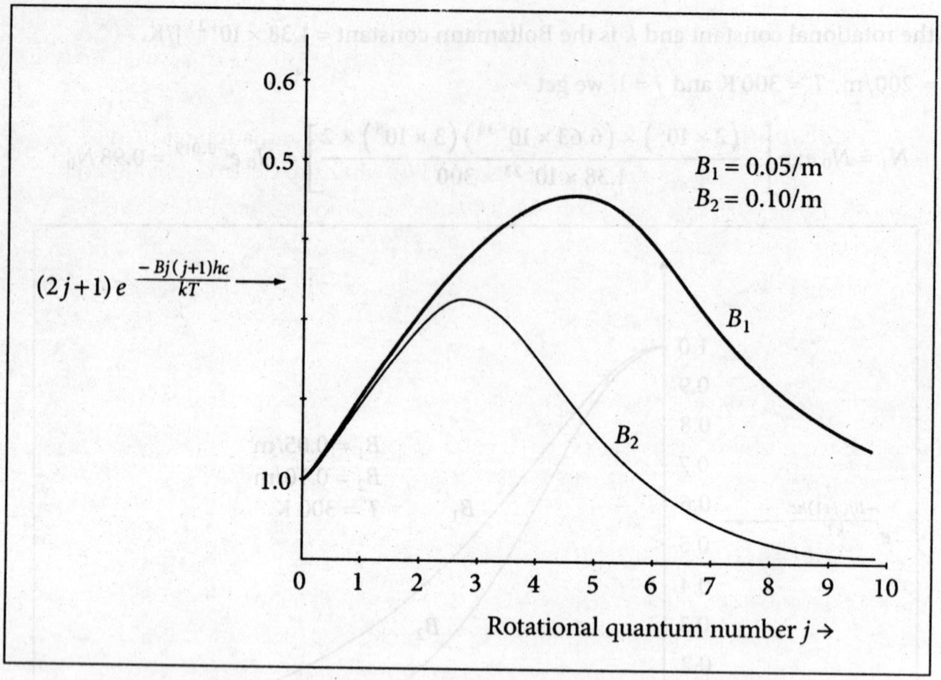

Fig 16.5 Relative populations

The line intensities are directly proportional to the populations of the rotational levels.

Hence, transition between levels with very low or very high j values will have small intensities, while the intensity will be a maximum at or near the j value given by eq. (16.18).

16.5 Vibrational energy levels of diatomic molecules

In the rigid rotator model of a diatomic molecule the vibration of two atoms along the line joining them has been neglected. But, actually the atoms vibrate along the line joining them, as if they are connected by a massless spring. The distance R between the atoms changes in a periodic manner about a mean value R_0. If the amplitude of vibration is small, then the vibration of the molecule may be assumed to be simple harmonic.

The variation of potential energy between the atoms with R is shown in Fig 16.6. In equilibrium position the bond length of a molecule is R_0. The molecule can absorb the energy supplied to it through the vibration of the atoms in it.

Considering the vibration of the atoms around R_0 to be small, the potential energy $V(R)$ about R_0 can be expanded by Taylor series as

$$V(R) = V(R_0) + \frac{1}{2}(R - R_0)^2 \times \left(\frac{d^2V(R)}{dR^2}\right)_{R=R_0} \tag{16.19}$$

since

$$\left(\frac{dV(R)}{dR}\right)_{R=R_0} = 0$$

as R_0 represents the distance corresponding to the minimum potential energy. Putting $(R - R_0) = x$, we have

$$V(R) = V(R_0) + \tfrac{1}{2}V''(R_0)\,x^2 \tag{16.20}$$

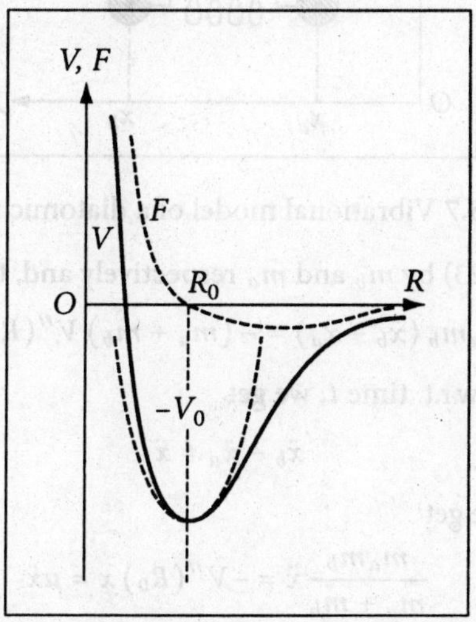

Fig 16.6 Potential energy between two atoms as a function of *R*

In eq. (16.20), $V(R_0)$ is a constant and the second term $\frac{1}{2}\,V''(R_0)\,x^2$ represents the potential energy of a simple harmonic oscillator with a force constant $V''(R_0)$. Calculating the slope of the potential energy curve at every point we can find out the force $F = -\frac{\partial V}{\partial R}$. The nature of variation of $F(R)$ with R is shown in Fig 16.6. It is evident from Fig 16.6 that at

(i) $R = R_0$, $V = -V_0$, the minimum value of potential energy, the force $F = 0$

(ii) $R > R_0$, F is negative, i.e., F is *attractive*

(iii) $R < R_0$, F is positive, i.e., F is *repulsive.*

If the system is distributed slightly, they will vibrate about their equilibrium position (R_0). If the supplied energy is greater than $-V_0$, the atoms will fly apart and the molecule will be *dissociated.*

The potential energy curve can be approximated to a parabolic potential about R_0 corresponding to a linear harmonic oscillator as shown by dotted curve in Fig 16.6.

To find out the vibrational energy levels of a diatomic molecule, we consider two atoms a and b in a molecule having masses m_a and m_b respectively. The vibration in the molecule can be thought to be due to a spring of spring constant $V''(R_0)$ between the atoms a and b.

If the motion is only along x-direction, then let x_a and x_b be the distances of the atoms a and b respectively from the origin O. Then from Fig 16.7

$$x_b - x_a = R_0 + x \tag{16.21}$$

$$m_a\ddot{x}_a = V''(R_0)\,x \tag{16.22}$$

$$m_b\ddot{x}_b = -V''(R_0)\,x \tag{16.23}$$

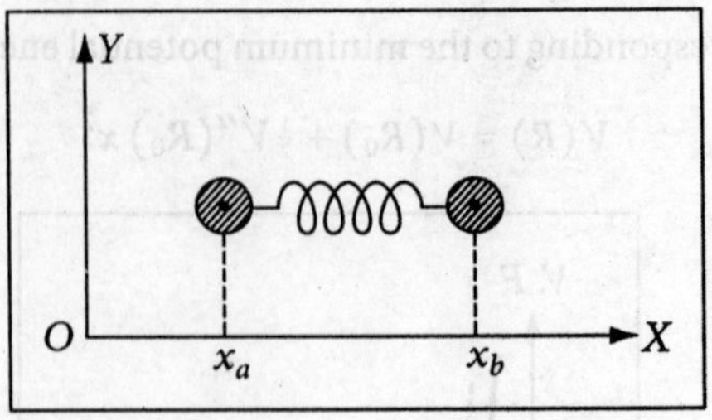

Fig 16.7 Vibrational model of a diatomic molecule

Multiplying eqs (16.22) and (16.23) by m_b and m_a respectively and, then subtracting the result, we get

$$m_a m_b (\ddot{x}_b - \ddot{x}_a) = -(m_a + m_b) V''(R_0)\, x \tag{16.24}$$

Differentiating eq. (16.21) twice w.r.t. time t, we get

$$\ddot{x}_b - \ddot{x}_a = \ddot{x} \tag{16.25}$$

From eqs (16.24) and (16.25), we get

$$\frac{m_a m_b}{m_a + m_b}\ddot{x} = -V''(R_0)\, x = \mu \ddot{x} \tag{16.26}$$

where $\mu = \dfrac{m_a m_b}{m_a + m_b}$ is the *reduced mass* of the system

$$\text{or, } \mu \ddot{x} + V''(R_0)\, x = 0 \tag{16.27}$$

Equation (16.27) represents a simple harmonic motion with frequency

$$\nu_0 = \frac{1}{2\pi}\left[\frac{V''(R_0)}{\mu}\right]^{1/2} \tag{16.28}$$

Quantum mechanical approach

The total vibrational energy of the molecule is

$$E_v = E_k + V = \frac{1}{2}\mu x^2 + \frac{1}{2}V''(R_0)\, x = \frac{p^2}{2\mu} \tag{16.29}$$

where p is the momentum of the system.

The Hamiltonian of the system is

$$\hat{H} = -\frac{p^2}{2\mu} + V = -\frac{\hbar^2}{2\mu}\frac{\partial^2}{\partial x^2} + \frac{1}{2}\mu\omega_0^2 x^2 \tag{16.30}$$

So the Schrödinger equation can be written as

$$\hat{H}\psi = -\frac{\hbar^2}{2\mu}\frac{\partial^2 \psi}{\partial x^2} + \frac{1}{2}\mu\omega_0^2 x^2\, \psi = E_v \psi \tag{16.31}$$

The solution of eq. (16.31) is

$$\psi = \left(\frac{\sqrt{\alpha}}{2^v v!\, \sqrt{\pi}}\right)^{1/2} e^{-\alpha x^2/2} H_v\left(\sqrt{\alpha}\, x\right) \tag{16.32}$$

where $\alpha = \dfrac{m\omega_0}{\hbar}$. $H_v\left(\sqrt{\alpha}\, x\right)$ is the Hermite polynomial of order v.

Putting $\sqrt{\alpha}\, x = q$, we get

$$H_\nu(q) = (-1)^\nu e^{q^2}\left(\frac{d}{dq}\right)^\nu e^{-q^2} \tag{16.33}$$

The energy eigenvalues are given by

$$E_\nu = (\nu + \tfrac{1}{2})\, \hbar\omega_0 = (\nu + \tfrac{1}{2})\, h\nu_0 \tag{16.34}$$

where $\nu = 0, 1, 2, 3, \ldots$. For $\nu = 0$, we get

$$E_0 = \tfrac{1}{2}\, \hbar\omega_0 = \tfrac{1}{2} h\nu_0 \tag{16.35}$$

where ν_0 is called the *proper frequency* of oscillation.

16.5.1 Vibrational spectra

The vibrational energy levels of the diatomic molecule corresponding to vibrational quantum numbers $\nu = 0, 1, 2, 3, \ldots$ are shown in Fig 16.8.

From eq. (16.34)

$$\text{for } \nu = 0,\ E_\nu = E_0 = \tfrac{1}{2}\, h\nu_0$$

$$\nu = 1,\ E_\nu = E_1 = \tfrac{3}{2}\, h\nu_0$$

$$\nu = 2,\ E_\nu = E_2 = \tfrac{5}{2}\, h\nu_0$$

$$\nu = 3,\ E_\nu = E_3 = \tfrac{7}{2}\, h\nu_0$$

$\therefore E_1 - E_0 = E_2 - E_1 = E_3 - E_2 = h\nu_0.$

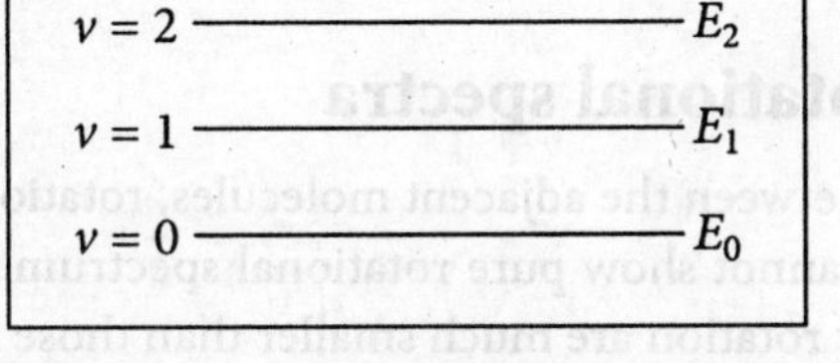

Fig 16.8 Vibrational energy levels of a diatomic molecule

Hence, the vibrational energy levels are equally spaced with energy interval $h\nu_0$.

16.5.2 Selection rule

The selection rule can be stated as follows:

Only those transitions are allowed between vibrational energy levels for which the vibrational quantum number ν changes by one unit, i.e., $\Delta\nu = \pm 1$.

This could be understood as When electromagnetic radiation is incident on the oscillating dipole, it can absorb energy or emit energy of frequency ν_0, which is the frequency of oscillation of it. By absorbing a quantum of energy $h\nu_0$ the energy of the oscillation of dipole increases from $(\nu + \frac{1}{2})\, h\nu_0$ to $(\nu + \frac{1}{2} + 1)\, h\nu_0$. Similarly, it can emit energy $h\nu_0$ at a time when its energy decreases from $(\nu + \frac{1}{2})\, h\nu_0$ to $(\nu + \frac{1}{2} - 1)\, h\nu_0$. This explains the justification of the selection rule, $\Delta\nu = \pm 1$.

16.5.3 Frequency of lines in vibrational spectra

When a transition takes place from an energy level with quantum number $(v+1)$ to an energy level v, then the energy of the emitted photon is

$$h\nu = E_{v+1} - E_v = (v + 1 + \tfrac{1}{2})\,h\nu_0 - (v + \tfrac{1}{2})\,h\nu_0 = h\nu_0$$

$$\therefore \nu = \nu_0 = \frac{1}{2\pi}\sqrt{\frac{V''(R_0)}{\mu}}$$

This is the frequency of the spectral line in vibrational emission spectra. It can be similarly shown that the frequency of the vibrational absorption spectra is also ν_0.

Hence, whatever be the vibrational quantum number, the frequency of the radiation emitted or absorbed is equal to the classical frequency ν_0 of the oscillator.

16.5.4 Behaviour of homonuclear and heteronuclear diatomic molecules

Homonuclear diatomic molecules like O_2, N_2, etc., do not show vibrational spectra. This is because when electromagnetic radiation is incident on a diatomic molecule, its electric vector may cause transition if the electric dipole moment of the molecule varies with the vibrational coordinate x. As a homonuclear molecule does not possess permanent electric dipole moment, there will be no such transition. That is why such molecules do not have any vibrational infrared spectrum. These molecules also do not have any pure rotational infrared spectrum.

The *heteronuclear* molecules like HCl, CO, etc., possess permanent dipole moment which is a function of internuclear distance. If the dipole associated with such a molecule executes linear vibration along the distance between the nuclei, then it can be shown that transitions between the vibrational levels occur. The selection rule for such transition is $\Delta v = \pm 1$.

16.6 Vibrational-rotational spectra

In liquids, due to interaction between the adjacent molecules, rotational motion of the molecules is not possible. That is why liquids cannot show pure rotational spectrum. In a gas or vapour, the excitation energies involved in molecular rotation are much smaller than those involved in vibration. So the freely moving molecules in gas or vapour are almost always rotating, irrespective of their vibrational states. The spectra of such molecule do not show isolated spectral lines due to transition between various vibrational states. Instead, a number of closely spaced rotational lines due to transitions between various rotational states of one vibrational level and the rotational states of other vibrational states, are observed. These spectra when observed with a spectrometer appear as broad *streak* and are called *vibration-rotation band.* It should be noted that like rotational spectra, *rotation-vibration spectra* are observed only for molecules having permanent dipole moment.

16.6.1 Vibration-rotation energy levels

Since, the energies of rotation and vibration of a diatomic molecule are quite different, we may, as a first approximation, consider that the diatomic molecule can execute rotations and vibrations quite independently. This is *the Born-Oppenheimer approximation* according to which the combined rotational-vibrational energy is simply the sum of separate energies, E_j and E_v. The allowed energy levels are specified by the two quantum numbers v and j. So the total energy of the molecule is

$$E_{v,j} = E_v + E_j = (v + \tfrac{1}{2})\, h\nu_0 + j(j+1)\frac{\hbar^2}{2I} \tag{16.36}$$

where $v = 0, 1, 2, \ldots$ and $j = 0, 1, 2, \ldots$.

When $v = 0$, $j = 0$, the total energy of the molecule is $E_{0,0} = \frac{1}{2} h\nu_0$, which is the minimum energy of the diatomic molecule and is called the *ground state* energy.

For $v = 0$, $j = 1$; $v = 0$, $j = 2$, etc., states are called the *excited states*. Thus, the energy of the first excited state is

$$E_{0,1} = \frac{1}{2} h\nu_0 + \frac{\hbar^2}{I}$$

The energy of the second excited state is

$$E_{0,2} = \frac{1}{2} h\nu_0 + \frac{3\hbar^2}{I}$$

Similarly, for the third, fourth, ... excited states of the molecule, the energies are

$$E_{0,3} = \frac{1}{2} h\nu_0 + \frac{6\hbar^2}{I}$$

$$E_{0,4} = \frac{1}{2} h\nu_0 + \frac{10\hbar^2}{I} \quad \text{and so on}$$

$E_{0,0}, E_{0,1}, E_{0,2}, \ldots$ are known as the *vibration-rotation energy levels* of the diatomic molecule.

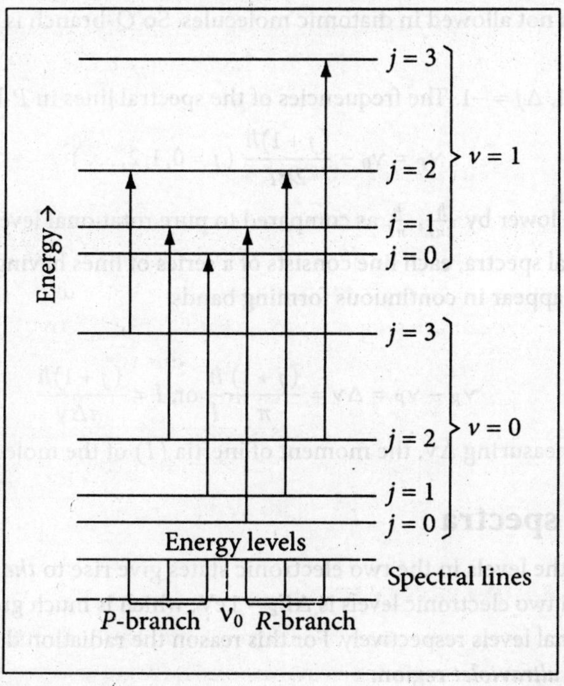

Fig 16.9 Vibrational-rotational levels of a diatomic molecule. The *P*-branch and *R*-branch of the spectrum are shown.

Now, $E_{0,0} > kT$ showing that only a few molecule at room temperature can go to the excited states, $E_{0,1}$, $E_{0,2}$, etc.

Again, since $E_v \gg E_j$, only the vibrational states corresponding to $v = 0$ and $v = 1$ are excited even at high temperatures.

Fig 16.9 shows the spectral lines for $v = 0$, $j = 0, 1, 2, \ldots$ and $v = 1$, $j = 0, 1, 2, \ldots$.

16.6.2 The *R*-branch, *Q*-branch and *P*-branch

Due to the selection rules $\Delta v = \pm 1$ and $\Delta j = \pm 1$, discrete lines are observed in both sides of the band origin at proper frequency ν_0. The selection rules $\Delta v = +1$ and $\Delta v = -1$ correspond respectively for absorption and emission of spectral lines. The transition with $\Delta j = 0$ is not allowed.

The spectrum for higher frequency side, i.e., $\nu > \nu_0$ is known as *R-branch,* while that on the lower frequency side, i.e., $\nu < \nu_0$ is known as the *P-branch.*

1. ***R-branch***: Here, $\Delta v = +1$, $\Delta j = +1$. The frequency of the spectral line in R-branch is given by

$$\nu_R = \frac{1}{h}\left[E_{1,j+1} - E_{0,j}\right] = \nu_0 + \frac{(j+1)}{2\pi}\frac{\hbar}{I}\ (j = 0, 1, 2, \ldots) \tag{16.37}$$

Since, the frequency of a pure vibrational spectrum is ν_0, we can say, the frequency ν_R is greater than the frequency (ν_0) of pure vibrational spectrum by $\frac{\hbar}{2\pi I}, \frac{\hbar}{\pi I}, \ldots$.

2. ***Q-branch***: Here, $\Delta v = 1$, $\Delta j = 0$. The frequencies of the spectral lines in Q-branch are given by

$$E_{1,j} - E_{0,j} = h\nu_Q \text{ or, } h\nu_0 = h\nu_Q \text{ or, } \nu_Q = \nu_0. \tag{16.38}$$

Transition for $\Delta j = 0$ is not allowed in diatomic molecules. So Q-branch is missing from the vibration-rotation spectra.

3. ***P-branch***: Here, $\Delta v = 1$, $\Delta j = -1$. The frequencies of the spectral lines in P-branch are given by

$$\nu_P = \nu_0 - \frac{(j+1)\hbar}{2\pi I}\ (j = 0, 1, 2, \ldots) \tag{16.39}$$

The frequencies ν_P are lower by $\frac{\hbar}{2\pi I}, \frac{\hbar}{\pi I}$ as compared to pure rotational level.

In rotational-vibrational spectra, each line consists of a series of lines having different values of j. These are closely spaced and appear in continuous forming bands.

We have

$$\nu_R - \nu_P = \Delta\nu = \frac{(j+1)}{\pi}\frac{\hbar}{I} \text{ or, } I = \frac{(j+1)\hbar}{\pi\Delta\nu} \tag{16.40}$$

Thus, knowing j and measuring $\Delta\nu$, the moment of inertia (I) of the molecule can be found out.

16.7 Electronic spectra

The transitions between the levels in the two electronic states give rise to *the electronic band spectra.* The energy difference between two electronic levels is $\Delta E_e \sim 1\,\text{eV}$, which is much greater than ΔE_j and ΔE_v due to rotational and vibrational levels respectively. For this reason the radiation due to transition is emitted or absorbed in the *visible* or *ultraviolet* region.

Since, the electronic energy of a molecule is large compared to its vibrational energy $(E_e > E_v)$, there may be a large number of vibrational levels associated with each electronic level.

Again, each vibrational level has a large number of rotational energy levels, since $E_v > E_j$. In a transition between the two electronic levels of a molecule the initial and final energy states in general are different. The vibrational levels of a molecule, due to anharmonicity of the oscillator, are not equally spaced.

The total energy E of a molecule may be written as

$$E = E_e + E_v + E_j \tag{16.41}$$

The energy changes due to electronic, vibrational and rotational transition are

$$\Delta E_e = E_{e_i} - E_{e_f}$$

$$\Delta E_v = E_{v_i} - E_{v_f}$$

$$\text{and } \Delta E_j = E_{j_i} - E_{j_f}$$

So the total energy change in a transition is given by

$$\Delta E = \Delta E_e + \Delta E_v + \Delta E_j \tag{16.42}$$

The order of magnitudes of ΔE_e, ΔE_v and ΔE_j are respectively 1 eV, 0.1 eV and 0.001 eV.

During a transition between electronic levels, transition between vibrational and rotational levels occurs.

Since, ΔE_v and ΔE_j are very small, the spectral lines are close to each other in frequency. This is responsible for the band structure of electronic spectra of molecules. The transitions between two electronic levels having different vibrational quantum numbers are shown in Fig 16.10. The selection rule for lines in a given electronic band is $\Delta J = 0$ or ± 1.

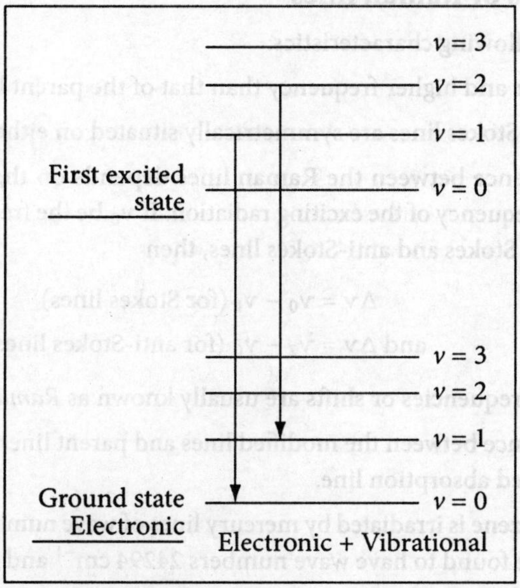

Fig 16.10 Transitions between two electronic levels having different vibrational quantum numbers

16.8 Raman effect

Sir C.V. Raman (in 1928), while studying the phenomenon of scattering of light, found that when a beam of *monochromatic light* is passed through a transparent medium (solid, liquid or gas) and the scattered light at right angles to the incident beam is examined spectroscopically, it is found to contain lines of other frequencies in addition to that of incident light (parent line). This type of *incoherent scattering* was first discovered in liquids and is known as the *Roman effect.* For this important discovery Raman was awarded the Nobel Prize in Physics in the year 1930.

The new lines (or, the satellite lines) present on the low frequency side are called *Stokes lines* and those on the high frequency side, the *anti-Stokes lines.* All these satellite lines taken together are called *Raman lines.* The Raman spectra with Stokes and anti-Stokes lines in Benzene is shown in Fig 16.11.

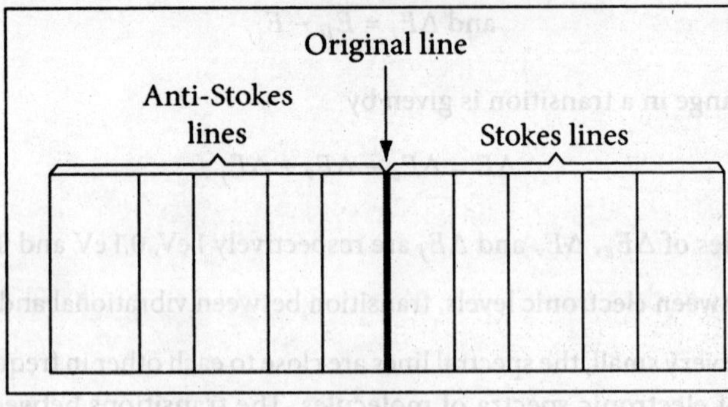

Fig 16.11 Raman spectra in Benzene

16.8.1 Characteristics of Raman lines

Raman spectrum has the following characteristics:

(i) Spectral lines of lower and higher frequency than that of the parent line are observed.

(ii) Both Stokes and anti-Stokes lines are symmetrically situated on either side of the parent line.

(iii) The frequency difference between the Raman lines depends on the nature of the scatterer, but is independent of the frequency of the exciting radiation. If ν_0 be the frequency of the exciting radiation and ν_1, ν_2 that of the Stokes and anti-Stokes lines, then

$$\Delta\nu = \nu_0 - \nu_1 \text{ (for Stokes lines)}$$

$$\text{and } \Delta\nu = \nu_2 - \nu_0 \text{ (for anti-Stokes lines)}$$

These differences in frequencies or shifts are usually known as *Raman frequencies.*

(iv) The frequency difference between the modified lines and parent lines represents the frequency of the corresponding infrared absorption line.

As for example, if benzene is irradiated by mercury lines of wave numbers 27354 cm^{-1} and 27290 cm^{-1} the scattered lines are found to have wave numbers 24294 cm^{-1} and 24231 cm^{-1} respectively.

(v) The intensities of Raman lines are ~ 1% of that of parent lines in solids. For liquids and gases, it is still smaller. The Stokes lines are always more intense than the anti-Stokes lines.

(vi) The degree of polarisation of different Raman lines is different.

(vii) There is no simple relationship between intensity (I) and wavelength (λ) in Raman scattering whereas in case of Rayleigh scattering the relation is $I \propto \frac{1}{\lambda^4}$. Thus, Raman scattering is basically different from *fluorescent scattering*

16.8.2 Quantum theory of Raman effect

According to the quantum theory, a photon of frequency ν has energy $h\nu$, where h is the Planck's constant. When a photon of energy $h\nu$ is incident on a molecule of solid, liquid or gas, the following three processes may occur:

(i) The photon may collide with the molecule elastically and scatter without any change in energy and hence frequency. This explains the presence of *unmodified line* in the Raman spectrum.

(ii) The photon may collide in such a way that a small fraction of its energy is transferred to the molecule. As a result the photon is scattered with reduced energy or frequency and, we get the *Stokes lines.*

(iii) If the photon collides a molecule in the excited state, then it may gain some energy from the molecule. Then the energy and hence the frequency of the scattered photon is increased giving rise to the *anti-Stokes lines.*

Let a photon of energy $h\nu$ collides with a molecule of kinetic energy $\frac{1}{2}mv^2$ and intrinsic energy E. If ν_1 be the frequency of the scattered photon, v_1 be the velocity and E_1 the intrinsic energy of the molecule after collision, then from the law of conservation of energy

$$h\nu + \left(\tfrac{1}{2}mv^2 + E\right) = h\nu_1 + \left(\tfrac{1}{2}mv_1^2 + E_1\right)$$

If there is no change in kinetic energy, i.e., $\frac{1}{2}mv^2 = \frac{1}{2}mv_1^2$, then

$$h\nu + E = h\nu_1 + E_1 \text{ or, } h\nu = h\nu_1 + (E_1 - E_2) = h\nu_1 + \Delta E$$

$$\therefore \nu = \nu_1 + \frac{\Delta E}{h} \tag{16.43}$$

Stokes lines: If the collision of the photon with the molecule is such that the molecule gains some energy from the photon, then the molecule goes from lower energy E to higher energy E_1 and $(E_1 - E) = \Delta E$ is positive.

Then from eq. (16.43)

$$\nu_1 = \nu - \frac{\Delta E}{h} \text{ or, } \nu_1 = \nu - \Delta\nu \tag{16.44}$$

where $\Delta\nu = \dfrac{\Delta E}{h}$.

Thus, the frequency of scattered photon (ν_1) is less than that of the incident photon. As a result the Stokes lines in Raman spectra occur.

Anti-Stokes lines: When the photon collides with the molecules in the excited state, then the photon gains some energy from the molecule. The molecule, then goes from higher energy (E) to the lower energy state (E_1) and $\Delta E = (E_1 - E)$ is negative.

From eq. (16.43)

$$\nu_1 = \nu + \frac{\Delta E}{h} = \nu + \Delta E \tag{16.45}$$

Since, the frequency of the scattered photon is greater than that of the incident photon, anti-Stokes lines appear in the Raman spectrum.

The above explanation justifies the symmetrical distribution of Raman lines about the line of frequency ν_0. It is to be noted that Raman effect also occurs in atomic spectra, but Raman effect in relation to molecules is of more interest.

16.8.3 Intensities in Raman lines

The transition probability of the atoms, molecules, etc., responsible for the origin of the spectral line is the key factor which determines the intensity of the same line. If the transition probability is more, then the intensity of the corresponding spectral line will be more.

According to Maxwell-Boltzmann statistics, the number of molecules in the energy level E is given by

$$n = n_0 e^{-\frac{E}{kT}} \tag{16.46}$$

where k is the Boltzmann constant and T is the absolute temperature.

From eq. (16.46) it is evident that n decreases as E increases. When $E = 0$, $n = n_0 = (n)_{max}$, i.e., the number of molecules is the maximum in the ground state corresponding to $E = 0$ and decreases as E increases. The ratio of the number of molecules in the excited state to that in the ground state is

$$\frac{n}{n_0} = e^{-\frac{E}{kT}} \tag{16.47}$$

At $T = 300$ K the rotational state energy E_r, just above the ground state is

$$E_r \sim 5 \times 10^{-3} \text{ eV}$$

At $T = 300$ K

$$kT = \frac{1.38 \times 10^{-23} \times 300}{1.6 \times 10^{-19}} = 2.6 \times 10^{-2} \text{ eV}$$

$\therefore E_r \ll kT$ and

$$\frac{n_r}{n_0} = e^{-\frac{E_r}{kT}} = e^{\left(\frac{-5\times10^{-3}}{2.6\times10^{-2}}\right)} \simeq e^{-0.2} = 0.82$$

So at room temperature nearly 82 % of the molecules are in the low-lying excited rotational states.

Again, the energy of the low-lying vibrational level, $E_v = 0.36$. $\therefore E_v \gg kT$.

$$\therefore \frac{n_v}{n_0} = e^{-\frac{E_v}{kT}} = e^{-\left(\frac{36\times10^{-2}}{2.6\times10^{-2}}\right)} = e^{-13.85} = 9.7 \times 10^{-7}$$

Thus, the number of molecules in the low-lying vibrational state is only 9.7×10^{-5} %, which is very small.

Since, at $T = 300$ K the number of molecules (n_0) in the ground state is the maximum, the transition probability from the ground state to the higher excited state is greater than the transition probability from higher state to the ground state. This explains the greater intensity of the Stokes lines than the anti-Stokes lines. The intensity of anti-Stokes lines of rotational Raman spectrum may increase with the increase of T.

At any temperature T, the ratio of the intensity of anti-Stokes and Stokes lines is

$$\frac{I_A}{I_S} = e^{-\frac{ch\Delta\bar{\nu}}{kT}} \tag{16.48}$$

The ratio I_A/I_S is the same as the ratio of number of molecules in the excited state to that in the ground state.

16.8.4 Rotational structure of Raman spectra

The Raman lines, viz., Stokes and anti-Stokes lines arise due to the transition of the molecules between vibrational and rotational levels. If the vibrational state of molecule undergoes a change due to Raman transition, then transition between different rotational levels of the involved vibrational states may occur. For this reason, the vibrational Raman bands usually show rotational structure.

To explain the Raman effect, we consider a molecule in the rotational state A with rotational quantum number j. The selection rule governing the transition in ordinary rotational spectrum is $\Delta j = \pm 1$.

According to quantum mechanical theory of Raman effect, under the influence of the incident photon, the molecule first makes a *virtual transition* from the initial state A to an intermediate state C. From C it makes a transition to the final state B [Fig 16.12]. Hence, the molecule first goes from A to C and, then from C to B. So we may, say, that Raman effect is a *two-step process.*

If j_a, j_b and j_c be the rotational quantum numbers of states A, B and C respectively, then for a transition from state A to C the rotational quantum number must change by ±1 in accordance with the selection rule, i.e., $j_c = j_a \pm 1$.

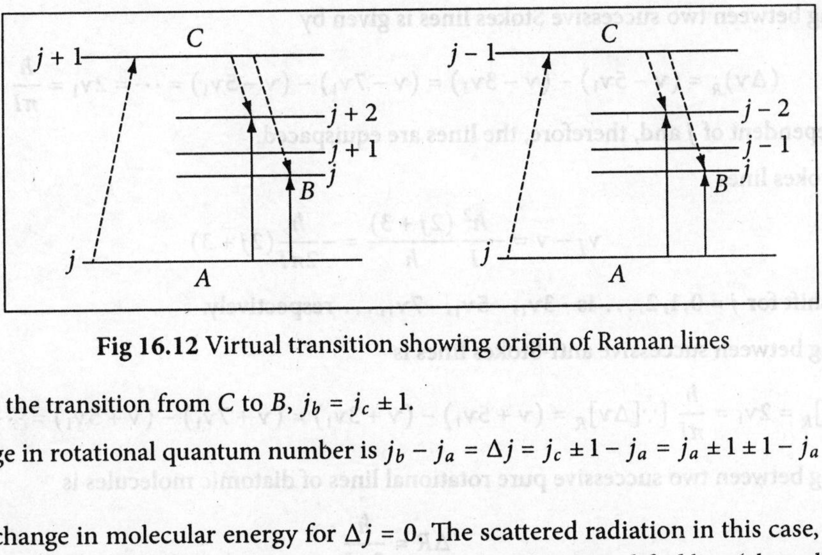

Fig 16.12 Virtual transition showing origin of Raman lines

Similarly, for the transition from C to B, $j_b = j_c \pm 1$.

∴ total change in rotational quantum number is $j_b - j_a = \Delta j = j_c \pm 1 - j_a = j_a \pm 1 \pm 1 - j_a = \pm 1 \pm 1 = \pm 2$ or 0.

There is no change in molecular energy for $\Delta j = 0$. The scattered radiation in this case, has the same frequency as that of the incident one. This explains the origin of the unmodified line (also called the *parent line*).

For $\Delta j = \pm 2$, Raman scattering occurs and the selection rule for Raman scattering is $\Delta j = \pm 2$. It is interesting to note that the transitions corresponding to $\Delta j = \pm 2$ are not allowed in ordinary rotational spectra, but are allowed in Raman spectra.

For $\Delta j = \pm 2$, we get Stokes lines and for $\Delta j = -2$, the anti-Stokes lines. The energy of the rotational quanta being very small, a large number of rotational energy levels will be occupied by many molecules at ordinary temperatures. This leads to several Raman transitions with initial value of $j = 0, 1, 2, \ldots, 10$.

Pure rotational Raman spectra: Stokes lines arise due to the transition from rotational energy level E_j to E_{j+2}. The frequency of the line is

$$\nu_j = \nu - \frac{E_{j+2} - E_j}{h} = \nu - \frac{\hbar^2}{2I}\left[\frac{(j+2)(j+3) - j(j+1)}{h}\right] = \nu - \frac{\hbar^2}{2I}\frac{(4j+6)}{h}$$

$$= \nu - \frac{\hbar^2}{2\pi I}(2j+3),\ (j = 0, 1, 2, \ldots)\left[\because E_j = \frac{\hbar^2}{2I}j(j+1)\right]$$

$$\therefore \Delta\nu = \nu - \nu_j = \frac{\hbar^2}{2\pi I}(2j+3) \text{ is the } \textit{Raman shift}$$

Now,

$$\text{for } j = 0,\ \Delta\nu = \frac{3\hbar}{2\pi I} = 3\nu_1,\ \because \nu_1 = \frac{\hbar}{2\pi I}$$

$$j = 1,\ \Delta\nu = \frac{5\hbar}{2\pi I} = 5\nu_1$$

$$j = 2,\ \Delta\nu = \frac{7\hbar}{2\pi I} = 7\nu_1$$

Hence, the spacing between the parent line and the first Raman line is $3\nu_1$.

The spacing between two successive Stokes lines is given by

$$(\Delta\nu)_R = (\nu - 5\nu_1) - (\nu - 3\nu_1) = (\nu - 7\nu_1) - (\nu - 5\nu_1) = \cdots = 2\nu_1 = \frac{\hbar}{\pi I}$$

which is independent of j and, therefore, the lines are equispaced.

For anti-Stokes lines

$$\nu_j - \nu = -\frac{\hbar^2}{I}\frac{(2j+3)}{h} = -\frac{\hbar}{2\pi I}(2j+3)$$

$\therefore$ Raman shift for $j = 0, 1, 2, \ldots$ is $-3\nu_1, -5\nu_1, -7\nu_1, \ldots$ respectively.

The spacing between successive anti-Stokes lines is

$$[\Delta\nu]_R = 2\nu_1 = \frac{\hbar}{\pi I}\ [\because [\Delta\nu]_R = (\nu + 5\nu_1) - (\nu + 3\nu_1) = (\nu + 7\nu_1) - (\nu + 5\nu_1) = \cdots = 2\nu_1]$$

The spacing between two successive pure rotational lines of diatomic molecules is

$$\Delta R = \frac{\hbar}{2\pi I}$$

Thus, the frequency difference in Raman lines $[\Delta\nu]_R$ is twice the frequency difference between successive lines in pure rotational spectrum.

The origin of pure rotational Raman lines is shown in Fig 16.13.

The diatomic molecules like H_2, O_2, N_2, etc. (which are homonuclear) do not exhibit vibrational and rotational spectra, but these molecules show Raman spectra. Hence, by studying their Raman spectra we can find out their vibrational and rotational energy levels, bond length, bond force constant, etc. and molecular structure.

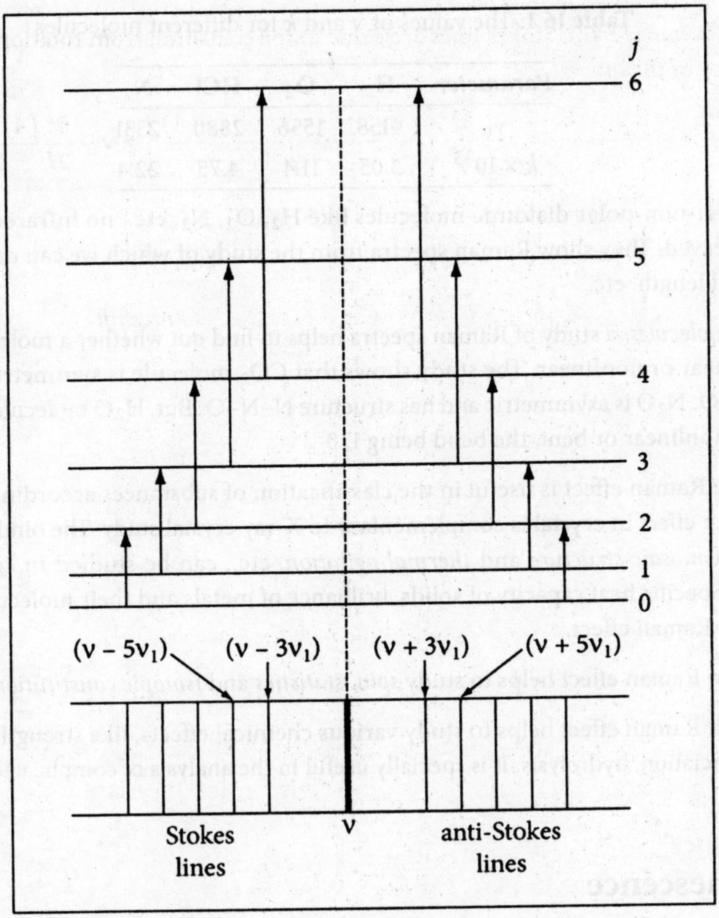

Fig 16.13 Raman lines due to pure rotational transitions

16.8.5 Applications of Raman effect

Raman effect has many important applications as follows:

1. ***Molecular structure in gases, liquids and solids*: Raman spectra include the vibrational spectra of the molecules. In a simple diatomic molecule, the natural frequency of vibration ν_0 is given by**

$$\nu_0 = \frac{1}{2\pi}\sqrt{\frac{k}{\mu}}$$

where k is the force constant and μ is the reduced mass of the molecule. The frequency and force constant of H_2, O_2 and HCl can be determined from the study of Raman spectra. Table 16.1 shows the values of ν and k for different molecules. From the table, k is in the ratio 1 : 2 : 1 : 3 for H_2, O_2, HCl and N_2 indicating the presence of single bond, double bond and triple bond.

Table 16.1: The values of ν and *k* for different molecules

Parameter	H_2	O_2	HCl	N_2
ν_0	4158	1556	2880	2331
$k \times 10^{-5}$	5.05	11.4	4.75	22.4

For homonuclear non-polar diatomic molecules like H_2, O_2, N_2, etc., no infrared rotation-vibration spectra are observed. They show Raman spectra from the study of which we can determine vibrational frequency, bond length, etc.

For *tri-atomic molecules,* a study of Raman spectra helps to find out whether a molecule is symmetric or asymmetric; linear or nonlinear. The study shows that CO_2 molecule is symmetric and linear having structure O–C–O. N_2O is asymmetric and has structure N–N–O. But, H_2O molecule, though symmetric, its structure is nonlinear or bent, the bend being 120°.

2. ***Crystal physics*:** Raman effect is useful in the classification of substances according to their molecular structure. Raman effect in crystal is complementary to X-ray crystal study. The binding force in crystals, the existence of *mosaic structure* and *thermal agitation*, etc., can be studied in crystal using Raman effect as a tool. Specific heat capacity of solids, brilliance of metals and their molecular structure can be explained using Raman effect.
3. ***Nuclear physics*:** Raman effect helps to study *spin, statistics* and *isotopic constitution* of the molecules.
4. ***Chemical effect*:** Raman effect helps to study various chemical effects, like strength of chemical bonds, electrolytic dissociation, hydrolysis. It is specially useful in the analysis of complicated mixture of organic molecules

16.9 Luminescence

Luminescence is the phenomenon in which a molecule absorbs energy from a high frequency radiation and re-emits in the form of visible light. It is, therefore, the reverse process of absorption.

If the absorption and emission of radiation by the molecule take place simultaneously, then it is called *fluorescence.*

If the emission of light is delayed by 10^{-7} second from absorption of radiation by the molecule, then it is called *phosphorescence.* The substance showing phosphorescence is known as *phosphor.*

16.9.1 Fluorescence

There are number of ways in which a molecule in an excited state can lose energy in the excited state and, then come to the ground state. A molecule can return to the ground state in a *single step* by emitting a photon of the same frequency as that of the absorbed photon. In another process, a molecule in the excited state can share its energy to other molecules through collision and, then move to the lower vibrational energy states. Then it may emit a photon of lower energy and return to the ground state as shown in Fig 16.14.

***Applications*:** Fluorescence is used in (i) fluorescent tubes (ii) biochemistry fluorescence excited by UV radiation is employed in the identification of minerals and compounds.

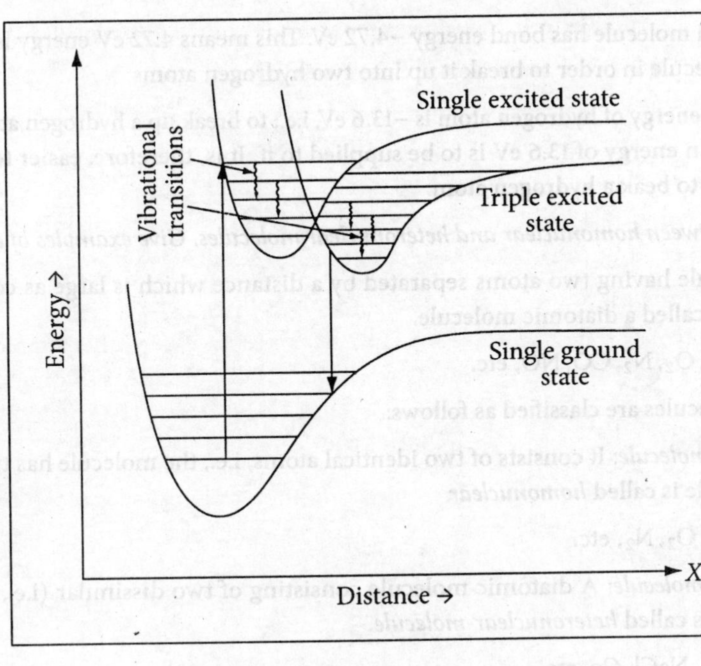

Fig 16.14 Shows molecular transitions

16.9.2 Phosphorescence

The selection rule for spin quantum number S in molecular spectra is the same as that in atomic spectra, viz., $\Delta S = 0$. This means no transition (radiative) takes place between singlet state with $S = 0$ and triplet state with $S = 1$.

A molecule in the singlet ground state transits to the singlet excited state by absorbing a photon. The molecules in the singlet excited state may collide with other molecules and undergo radiation-less transition to lower vibrational level. If this lower vibrational level in the singlet excited state coincides with one of the vibrational levels of the triplet excited state, then there is a certain probability that a shift to the triplet excited state with same vibrational energy may occur. The molecule may finally shift to $v = 0$ level in the triplet state through collisions with other molecules [Fig 16.14].

Though the transition from $S = 1$ to $S = 0$ level is forbidden, there is a little probability of radiative transition between these levels.

Since, the half-life of such transitions is very large, the molecule may stay in the excited state for a considerable long time. The radiation emitted from such a molecular transition usually takes place after minutes or even hours after the absorption of the photon. This delayed radiation emitted due to molecular transition is called *phosphorescence*.

Materials which emit phosphorescence radiation are called *phosphor*. Phosphors have wide applications in fluorescent tubes and electronic appliances.

16.10 Answers of some important conceptual questions

16.1. *Why is it easier to break a hydrogen molecule than to beak up a hydrogen atom?* **[PU 1992, 1991]**

Ans. Hydrogen molecule has bond energy –4.72 eV. This means 4.72 eV energy is to be supplied to a hydrogen molecule in order to break it up into two hydrogen atoms.

The ionisation energy of hydrogen atom is –13.6 eV, i.e., to break up a hydrogen atom into an electron and a proton an energy of 13.6 eV is to be supplied to it. It is, therefore, easier to break a hydrogen molecule than to beak a hydrogen atom.

16.2. *Distinguish between homonuclear and heteronuclear molecules. Give examples of each.*

Ans. A molecule having two atoms separated by a distance which is large as compared to atomic dimensions is called a diatomic molecule.

Examples: H_2, O_2, N_2, CO, NO, etc.

Diatomic molecules are classified as follows:

Homonuclear molecule: It consists of two identical atoms, i.e., the molecule has two identical nuclei. Such a molecule is called *homonuclear.*

Examples: H_2, O_2, N_2, etc.

Heteronuclear molecule: A diatomic molecule consisting of two dissimilar (i.e., two non-identical nuclei) atoms is called *heteronuclear-molecule.*

Examples: NO, NaCl, Co, etc.

16.3. *Show that a diatomic molecule do not rotate in the ground state.* **[GNDU 1997; Pbi. U 1993]**

Ans. The rotational energy of a diatomic molecule is given by

$$E_J = \frac{J(J+1)\hbar^2}{2I}$$

where J is the rotational quantum number and I is the moment of inertia of the molecule about the axis of rotation.

For the ground state $J = 0$. Hence, $E_J = 0$. *Since, the rotational energy E_J of the molecule in the ground state is zero, it cannot rotate in the ground state.*

16.4. *All diatomic molecules do not exhibit rotational spectra. Why?* **[GNDU 1992; BU (H) 1993]**

Ans. The rotational spectra arise from transitions between rotational energy states. Only molecules which have electric dipole moment can absorb or emit electromagnetic radiation (or photons) in such transitions. The emission of radiation is due to the rotation of electric dipole moment.

According to classical electrodynamics, the rotation of molecules leads to radiation of energy only, if a changing dipole moment is associated with its motion. In case of molecules consisting of *unlike* atoms like HCl or CO, such a permanent dipole moment does not exist, since centres of gravity of positive and negative charges do not coincide. The component of dipole moment along a fixed direction changes periodically as the molecules rotate, giving rise to a rotational spectrum. For molecules consisting of like atoms—say, N_2, H_2, etc. no such dipole moment is present and no radiation takes place due to pure rotation. Conversely, energy in the far infrared region is absorbed by molecules having permanent dipole moment.

In general, non-polar diatomic molecules with identical atoms like H_2, O_2, N_2 etc. and symmetric polyatomic molecules like $CO_2(O = C = O)$ and CH_4 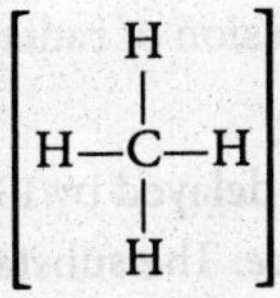 do not exhibit rotational spectra.

16.5. *H_2, N_2 and O_2 do not show vibrational spectra. Why?* **[GNDU 1991]**

Ans. We know that the vibrational spectra are observed only in absorption. The electromagnetic radiation can induce transition among vibrational energy levels. For transfer of radiation energy to vibrating molecule, an electrical coupling should be present. If the vibrating molecule produces an oscillating dipole moment, then this dipole moment interacts with electric field of radiation and thus gives rise to coupling. Molecules like HCl, HBr having dissimilar atoms, have a resultant dipole moment which is a function of intermolecular distance and as such show vibrational spectra. Molecules like H_2, O_2, N_2, which have similar atoms, do not possess net dipole moment. Such molecules cannot interact with radiation and do not show vibrational spectra.

16.6. *Discuss the origin of line spectra and band spectra.*

Ans. The *line spectra* emitted from a gas discharge tube consists of discrete lines. The origin of line spectra lies in the electronic transition in the *atoms* of different elements. The origin of line spectra can be understood with the help of Bohr-Sommerfeld theory and quantum mechanics.

The band spectra are emitted by substances in the *molecular state*. The molecules may contain two or more atoms. These molecular spectra appear as continuous band under low dispersion and are known as *band spectra*. The band has a sharp intense edge on one side and is called *band head*. The intensity of the band gradually decreases as one moves towards the other side.

16.7. *What is Raman effect? How does it differ from Rayleigh scattering?*

Ans. Raman effect: Sir C.V. Raman, while studying the phenomenon of scattering of light, found that when a beam of *monochromatic* light is passed through a transparent medium (solid, liquid or gas) and the scattered light at right angles to the incident beam is examined spectroscopically, it is found to contain lines of other frequencies in addition to that of the incident light (parent line). This type of *incoherent scattering* was first discovered in liquids and is known as the *Raman effect.*

The new lines (or, the satellite lines) present on the low frequency side are called *Stokes lines* and those on the high frequency side, the *anti-Stokes lines.* All these satellite lines taken together are called *Raman lines.*

In Raman effect, there is interchange of energy and the collision is *inelastic*. The scattered photon may have an energy less than or greater than that of the incident photon giving rise to *Stokes lines* on the lower frequency (longer wavelength) side or anti-Stokes lines on the higher (shorter wavelength) side of the unmodified spectral line in the Raman spectrum.

Rayleigh scattering: Rayleigh scattering is *coherent scattering*, i.e., the scattered light has the same frequency as that of the incident light. In other words, there is no exchange of energy between the incident photon and the molecule of the scattering substance or the *collision is elastic.*

16.8. *What are Luminescence, Fluorescence and Phosphorescence?* **[PU 1993; GNDU 1992]**

Ans. Luminescence is the phenomenon in which a molecule absorbs energy from a high frequency radiation and re-emits in the form of visible light. It is, therefore, the reverse process of absorption.

If the absorption and emission of radiation by the molecule take place simultaneously, then it is called *fluorescence*.

If the emission of light is delayed by 10^{-7} second from absorption of radiation by the molecule, then it is called *phosphorescence*. The substance showing phosphorescence is called *phosphor*.

16.9. ***State the characteristics of Raman lines.***

Ans. **The main characteristics of Raman lines are as follows:**

(i) The Stokes lines are always more intense than the anti-Stokes lines.

(ii) The Raman lines are symmetrically placed w.r.t. the parent line.

(iii) The frequency difference between the modified lines and parent line represents the frequency of the corresponding infrared absorption lines.

16.11 Solved problems

» *Example* 16.1. Find out the rotational energy of CO molecule when its rotational quantum number $j = 1$. [*Given*: the masses of oxygen and carbon atoms are 2.66×10^{-26} kg and 1.99×10^{-26} kg respectively and the distance between the two nuclei is 1.13 Å.] **[BU (H) 1997]**

***Solution*: The rotational energy of a diatomic molecule in the jth rotational level is**

$$E_j = j(j+1)\frac{\hbar^2}{2I}$$

where $\hbar = h/2\pi$ and I is the MI of the molecule about an axis perpendicular to the line joining the two nuclei and passing through the CM.

The reduced mass (μ) of the molecule is

$$\mu = \frac{(2.66 \times 10^{-26}) \times (1.99 \times 10^{-26})}{(2.66 + 1.99) \times 10^{-26}} = \frac{(2.66 \times 1.99)}{4.65} \times 10^{-26} = 11.384 \times 10^{-27} \text{ kg}$$

The distance between the two nuclei, $r = 1.13$ Å $= 1.13 \times 10^{-10}$ m.

∴ MI of the molecule, $I = \mu r^2 = (11.384 \times 10^{-27}) \times (1.13 \times 10^{-10})^2 = 11.384 \times 1.13 \times 1.13 \times 10^{-47} = 14.536 \times 10^{-47}$ kg-m^2.

Now,

$$E_j = j(j+1)\frac{\hbar^2}{2I} = \frac{1(1+1)6.62 \times 6.62 \times 10^{-68}}{2 \times (14.536 \times 10^{-47}) \times (4 \times 3.14 \times 3.14) \times 1.6 \times 10^{-19}}$$

$$[\because j = 1 \text{ and } 1 \text{ eV} = 1.6 \times 10^{-19} \text{ J}]$$

$$= 0.0478 \times 10^{-2} \text{ eV} = \mathbf{4.78 \times 10^{-4} \text{ eV}}$$

» *Example* 16.2. The force constant of HF molecule is 966 newtons/metre. Calculate the minimum energy of vibration of this molecule in eV. [*Given*: $m_H = 1.007825$ u, 1 u $= 1.66043 \times 10^{-27}$ kg, atomic number of F = 19, $h = 6.626 \times 10^{-34}$ J-s and 1 eV $= 1.602 \times 10^{-19}$ J.] **[BU (H) 1999]**

***Solution*: The minimum energy of vibration of a diatomic molecule is given by**

$$(E_v)_0 = \frac{h\nu_0}{2}$$

ν_0 is the classical frequency of the molecule.

$$\nu_0 = \frac{1}{2\pi}\sqrt{\frac{k}{\mu}}$$

Here, $k = 966$ N-m and

$$\mu = \text{the reduced mass of the system}$$
$$= \frac{1.007825 \times 19}{1.007825 + 19} \times 1.66043 \times 10^{-27} = \frac{1.007825 \times 19}{20.007825} \times 1.66043 \times 10^{-27}$$
$$= 1.589 \times 10^{-27}\ \text{kg}$$

$$\therefore \nu_0 = \frac{1}{2\pi}\sqrt{\frac{996}{1.589 \times 10^{-27}}} = \frac{1 \times 10^{13}}{2 \times 3.14}\sqrt{\frac{9660}{1.589}} = \frac{10^{23}}{6.28} \times \sqrt{6079.295159} = \frac{10^{13}}{6.28} \times 77.9698$$
$$= 12.41557881 \times 10^{13}\ \text{Hz}$$

$$(E_\nu)_0 = \frac{h\nu_0}{2} = \frac{6.626 \times 10^{-34}}{2} \times 12.41557881 \times 10^{13} = 41.13281258 \times 10^{-21}\ \text{J}$$
$$= \frac{41.13281258}{1.602 \times 10^{-19}}\ \text{eV} = \mathbf{25.6759 \times 10^{-2}\ eV \simeq 0.257\ eV}$$

» ***Example* 16.3.** The lines in the rotational spectra of HCl molecule appear at the wave numbers 83.5 cm^{-1}, 104.1 cm^{-1}, 124.7 cm^{-1}, 145.4 cm^{-1} and 166.0 cm^{-1}. Calculate the rotational energy for the state $j = 4$. **[BU (H) 1993]**

Solution: We have the wave number difference between jth and $(j+1)$th level is given by

$$\left(\frac{1}{\lambda}\right)_{j+1} - \left(\frac{1}{\lambda}\right)_j = \frac{2B}{\hbar c}$$

where B is the rotational constant and c is the velocity of the radiation emitted. From the given wave numbers we see that the difference between two consecutive wave numbers is almost 20.6 cm^{-1}.

$$\therefore \frac{2B}{hc} = 20.6\ \text{cm}^{-1} = 2060\ \text{m}^{-1} \text{ or, } B = \frac{6.63 \times 10^{-34} \times 3 \times 10^{8}}{2} \times 2.06 \times 10^{3} \simeq 2 \times 10^{-22}\ \text{J} \approx 1.25 \times 10^{-3}\ \text{eV}$$

Now,

$$E_j = \text{the rotational energy for the } j\text{th level} = Bj(j+1)$$
$$\text{or, } E_4 = B4(4+1)\ [\text{for } j = 4]$$
$$= 1.25 \times 10^{-3} \times 4 \times 5 = 20 \times 1.25 \times 10^{-3} = \mathbf{2.50 \times 10^{-2}\ eV}$$

» ***Example* 16.4.** HCl has a force constant of 516 N-m^{-1}. Estimate the frequency and wavelength of its vibrational line. [*Given*: m(Cl) = 5.81×10^{-26} kg, m(H) = 1.67×10^{-27} kg.] **[BU (H) 1994]**

Solution: The reduced mass of the system is

$$\mu = \frac{1.67 \times 58.1}{1.67 + 58.1} \times 10^{-27} = \frac{97.027}{59.77} \times 10^{-27} = 1.62 \times 10^{-27}\ \text{kg}$$

Now, frequency

$$\nu = \frac{1}{2\pi}\sqrt{\frac{k}{\mu}} = \frac{1}{2\pi}\sqrt{\frac{516}{1.62 \times 10^{-27}}} = \frac{1 \times 10^{13}}{2 \times 3.14}\sqrt{\frac{5160}{1.62}} = \frac{10^{13}}{6.24} \times \sqrt{3185.185}$$
$$= \frac{10^{13}}{6.24} \times 56.43744 = \mathbf{9.044 \times 10^{13}\ s^{-1}}$$

and wavelength

$$\lambda = \frac{c}{\nu} = \frac{3 \times 10^8}{9.044 \times 10^{13}} = \frac{3}{9.044} \times 10^{-5} = 0.3317 \times 10^{-5} \text{ m} = \mathbf{33170\ Å}$$

The wavelength is in infrared region.

» ***Example* 16.5.** Calculate the moment of inertia of an HCl molecule about an axis passing through the centre of mass and perpendicular to the line joining the two nuclei H and Cl and obtain the rotational energy levels of HCl in eV. [*Given*: R = fixed distance between the two nuclei = 1.3 Å.]

Solution: We have the reduced mass of the system

$$\mu = \frac{m_H \times m_{Cl}}{m_H + m_{Cl}} = \frac{1 \times 35}{1 + 35} m_H = \frac{35}{36} m_H$$

MI of the molecule about an axis passing through the CM and perpendicular to the line joining the two nuclei is

$$I = \mu R^2 = \frac{35}{36} \times m_H \times \left(1.3 \times 10^{-10}\right)^2 = \frac{35}{36} \times 1.0078\,\text{u} \times \left(1.3 \times 10^{-10}\right)^2$$

$$= \frac{35}{36} \times 1.0078 \times 1.66 \times 10^{-27} \times \left(1.3 \times 10^{-10}\right)^2 = 2.75 \times 10^{-47} \text{ kg-m}^2$$

The rotational energy levels are

$$E_j = \frac{\hbar^2}{2I} j(j+1) = \frac{\left(1.05 \times 10^{-34}\right)^2}{2 \times 2.75 \times 10^{-47}} j(j+1)$$

$$= 2 \times 10^{-22} j(j+1) \text{ J}$$

$$= \frac{2 \times 10^{-22} j(j+1)}{1.6 \times 10^{-19}} \text{ eV}$$

$$= \mathbf{1.25 \times 10^{-3}\ j(j+1)\ eV}$$

» ***Example* 16.6.** The interatomic distance in CO molecule is 1.13 Å. Calculate (i) the reduced mass (ii) the moment of inertia (iii) the energy gap (in joule) between the ground state and the first excited rotational level.

[BU (H) 1995, 2002]

Solution:

(i) The reduced mass of the molecule is

$$\mu = \frac{m_C \times m_O}{m_C + m_O}$$

But, $m_C = 1.99 \times 10^{-26}$ kg, $m_O = 2.66 \times 10^{-26}$ kg.

(ii) $$\therefore \mu = \frac{\left(1.99 \times 10^{-26}\right) \times \left(2.66 \times 10^{-26}\right)}{1.99 \times 10^{-26} + 2.66 \times 10^{-26}} = \mathbf{11.384 \times 10^{-27}\ kg.}$$

The moment of inertia, $I = \mu r^2 = \left(11.384 \times 10^{-27}\right) \times \left(1.14 \times 10^{-10}\right)^2 = 14.536 \times 10^{-47} \text{ kg-m}^2$.

(iii) The energy gap between the ground state $(j = 0)$ and the first excited state $(j = 1)$ is

$$\Delta E = E_1 - E_0 = E_1 = 1(1+1) \frac{6.62 \times 6.62 \times 10^{-68}}{2 \times (14.536 \times 10^{-47}) \times (4 \times 3.14 \times 3.14)}$$

$$= \mathbf{7.648 \times 10^{15}\ J} \left[\because E_j = j(j+1)\frac{\hbar^2}{2I} \text{ and } E_0 = 0\right]$$

» ***Example* 16.7.** The wave number of the first line in the rotation spectra of CO is 3.84235 cm^{-1}. Find the bond length. [*Given*: the reduced mass of CO = 11.384 × 10^{-27} kg.]

Solution: The first line corresponds to the transition from $j = 0$ to $j = 1$.

$$\therefore \bar{\nu} = 3.84235 = 2B\ cm^{-1}$$

$$\therefore B = 1.92118\ cm^{-1}$$

Now, the MI of CO molecule about an axis passing through its centre of mass and perpendicular to the line joining the two nuclei is

$$I = \frac{h}{8\pi^2 Bc}$$

$$\therefore I_{CO} = \frac{6.626 \times 10^{-34}}{8 \times 9.87 \times 2.99793 \times 10^{10} \times B} = \frac{27.9907 \times 10^{-47}}{B} = 14.56954 \times 10^{-47}\ \text{kg-m}^2$$

The bond length is given by

$$R^2 = \frac{I}{\mu} = \frac{(14.56954 \times 10^{-47})}{11.384 \times 10^{-27}}\ m^2 = 1.2799 \times 10^{-20}\ m^2$$

$$\therefore R = 0.1131\ m = \mathbf{1.31\ Å}$$

» ***Example* 16.8.** Show that the bond length of HF molecule is about 0.93 Å, given that its rotational constant is 2025 m^{-1}.

Solution: The rotational constant B for a diatomic molecule is given by

$$B = \frac{\hbar}{4\pi Ic}\ \text{or,}\ I = \frac{\hbar}{4\pi Bc}\ \text{or,}\ I = \frac{1.05 \times 10^{-34}}{4\pi \times 2025 \times 3 \times 10^{8}} = 1.38 \times 10^{-47}\ \text{kg-m}^2$$

The reduced mass of the molecule is

$$\mu = \frac{1 \times 19}{1 + 19} u = \frac{19}{20} \times 1.66 \times 10^{-27}\ kg = 1.58 \times 10^{-27}\ kg$$

$$\because I = \mu R^2,\ \therefore R = \sqrt{\frac{I}{\mu}} = \sqrt{\frac{1.38 \times 10^{-47}}{1.58 \times 10^{-27}}} = 0.93 \times 10^{-10}\ m = \mathbf{0.93\ Å}$$

∴ the bond length of the HF molecule is $R = 0.93$ Å.

» ***Example* 16.9.** The spacing between vibrational levels of CO molecule is 0.08 eV. Calculate the value of force constant. [*Given*: mass of carbon atom = 12 and that of oxygen atom = 16 times the mass of proton = 1.67 × 10^{-27} kg.] **[GNDU 1993; PU 1992]**

Solution: The reduced mass of CO molecule is

$$\mu = \frac{12 \times 16}{12 + 16} \times 1.67 \times 10^{-27} = 1.14 \times 10^{-26}\ kg$$

$$\Delta E = E_{v+1} - E_v = h\nu_0 = h\left(\frac{1}{2\pi}\sqrt{\frac{k}{\mu}}\right)\left[\because \nu_0 = \frac{1}{2\pi}\sqrt{\frac{k}{\mu}}\right]$$

$$= 0.08 \text{ eV} = 0.08 \text{ eV} \times 1.6 \times 10^{-19} \text{ J}$$

$$\text{or, } \frac{k}{\mu} = \left(\frac{0.08 \times 1.6 \times 10^{-19} \times 2\pi}{h}\right)^2$$

$$\text{or, } k = \left(\frac{0.08 \times 1.6 \times 10^{-19} \times 2\pi}{6.63 \times 10^{-34}}\right)^2 \times 1.14 \times 10^{-26} = \mathbf{167.8 \text{ N/m}}$$

» ***Example* 16.10.** The force constant of CO bond is 187 N/m. Find the frequency of vibration of CO molecule and the spacing between vibrational level. [*Given*: mass of ^{12}C atom = 1.99×10^{-26} kg and of ^{16}O atom = 2.66×10^{-26} kg.] **[GNDU 1991]**

Solution: The reduced mass of the CO molecule is

$$\mu = \frac{1.99 \times 10^{-26} \times 2.66 \times 10^{-26}}{(1.99 + 2.66) \times 10^{-26}} = 1.14 \times 10^{-26} \text{ kg}$$

The frequency of vibration is

$$\nu = \frac{1}{2\pi}\sqrt{\frac{k}{\mu}} = \frac{1}{2\pi}\sqrt{\frac{187}{1.14 \times 10^{-26}}} = \mathbf{2.04 \times 10^{13} \text{ Hz}}$$

Now, the separation between the vibrational energy levels in CO is given by

$$\Delta E = E_{v+1} - E_v = h\nu_0 = 6.63 \times 10^{-34} \times 2.04 \times 10^{13}$$

$$= 1.35 \times 10^{-20} \text{ J} = \mathbf{8.44 \times 10^{-2} \text{ eV}} \; [\because 1 \text{ eV} = 1.6 \times 10^{-19} \text{ J}]$$

» ***Example* 16.11.** In an experiment in the study of Raman effect using mercury green radiation of $\lambda = 546.1$ nm a Stokes line of wavelength 554.3 nm was observed. Find Raman shift and wavelength corresponding to anti-Stokes lines. **[Bang U 1994]**

Solution: Here, the wavelength of the incident radiation is

$$\lambda = 546.1 \text{ nm} = 546.1 \times 10^{-9} \text{ m} = 5461 \times 10^{-10} \text{ m}$$

Corresponding frequency

$$\nu = \frac{c}{\lambda} = \frac{c}{5461 \times 10^{-10}} = c \times 1.8312 \times 10^{-4} \times 10^{10} = c \times 18312 \times 10^2/\text{s}$$

Wavelength of the Stokes lines, $\lambda_1 = 554.3 \text{ nm} = 5543 \times 10^{-10}$ m.

∴ frequency of this line

$$\nu_1 = \frac{c}{\lambda_1} = \frac{c}{5543 \times 10^{-10}} = c \times 18041 \times 10^2/\text{s}$$

∴ Raman shift, $\Delta\nu = \nu - \nu_1 = c[18312 - 18041] \times 10^2 = c \times 10^2 \times 271 = 3 \times 10^8 \times 10^2 \times 271 = 813 \times 10^{10}/\text{s}$.

As the Stokes and anti-Stokes lines are equally shifted from the parent line of frequency ν, the frequency of the anti-Stokes line, $\nu_1' = \nu + c \times 10^2 \times 271 = c \times 10^2[18312 + 271] = c \times 10^2 \times 18583/\text{s}$.

Wavelength of the anti-Stokes line is

$$\lambda_1' = \frac{c}{c \times 10^2 \times 18583} = 5.381 \times 10^{-5} \times 10^{-2} = 5381 \times 10^{-10} \text{ m} = \mathbf{5381 \text{ Å}}$$

» ***Example* 16.12.** With exciting line 2536 Å, a Raman line for a sample is observed at 2612 Å. Calculate the Raman shift in cm^{-1}.

***Solution*:** The wave number of the exciting radiation is

$$\bar{\nu} = \frac{1}{\lambda} = \frac{1}{2536 \times 10^{-8}\,\text{cm}} = 39432\,\text{cm}^{-1}$$

Wave number for the Raman line

$$\bar{\nu}' = \frac{1}{2618 \times 10^{-8}\,\text{cm}} = 38285\,\text{cm}^{-1}$$

$\therefore$ Raman shift, $\Delta\nu = \bar{\nu} - \bar{\nu}' = (39432 - 38285)\,\text{cm}^{-1} = \mathbf{1157\,cm^{-1}}$.

» ***Example* 16.13.** The first line in the rotational spectrum of CO is 3.84235 cm^{-1}. Find bond length. Also find the region of electromagnetic spectrum, where the spectra lies. **[CU (H) 2015]**

***Solution*:** For CO, the first line in the rotational spectrum is given by

$$\bar{\nu}_{0\to1} = 2B = 3.84235\,\text{cm}^{-1}$$

$$\text{or, } B = \tfrac{1}{2}(3.84235)^{-1} = 1.92115\,\text{cm}^{-1} = 1.921175 \times (10^{-2}\,\text{m})^{-1} = 192.1175\,\text{m}^{-1}$$

Again

$$B = \frac{h}{8\pi^2 IC} \text{ or, } I \equiv I_{CO} = \frac{h}{8\pi^2 Bc}$$

$$\therefore I_{CO} = \frac{6.626 \times 10^{-34}\,\text{J-s}}{8\pi^2(1.92.1175\,\text{m}^{-1})(3 \times 10^8\,\text{m/s})} = 1.456 \times 10^{-46}\,\text{kg-m}^2$$

Now, $I_{CO} = \mu r^2$

$$\mu = \frac{m_c m_O}{m_c + m_O} = \frac{12m_p \times 16m_p}{12m_p + 12m_p}$$

$$\text{or, } \mu = \frac{12 \times 16}{12 + 16} m_p,\ m_p = 1.67 \times 10^{-27}\,\text{kg} = \text{mass of proton}$$

$$\therefore I_{CO} = 1.456 \times 10^{-16}\,\text{kg-m}^2 = \frac{12 \times 16}{12 + 16}(1.67 \times 10^{-27}\,\text{kg})r^2$$

$$\text{or, } r^2 = 1.27 \times 10^{-20}\,\text{m}^2 \text{ or, } r = 1.127 \times 10^{-10}\,\text{m} = 1.13\,\text{Å}$$

$$\therefore E_J = BhCJ(J+1) \simeq BhC = (192.1175\,\text{m}^{-1})(6.626 \times 10^{-34}\,\text{J-s})(3 \times 10\,\text{ms}^{-1})$$

$$\text{or, } E_J = 3.82 \times 10^{-23}\,\text{J} = \frac{3.82 \times 10^{-23}}{1.6 \times 10^{-19}}\,\text{eV}$$

$$\text{or, } E_J = \mathbf{2.39 \times 10^{-4}\,eV}$$

Thus, the rotational energy spectrum lies in the micro-wave range.

» ***Example* 16.14.** The bond length of NO is 1.15 Å. Find the moment of inertia.

***Solution*:** We know that

$$I = \mu r^2 = \frac{m_N m_O}{m_N + m_O} = \frac{14m_p \times 16m_p}{14m_p + 16m_p} \times (1.15\,\text{Å}) = \frac{14 \times 16}{14 + 16}(1.67 \times 10^{-27}\,\text{kg})(1.15 \times 10^{-10})^2$$

$$= \mathbf{1.65 \times 10^{-46}\,kg\text{-}m^2}$$

» ***Example* 16.15.** Given spacing between vibrating levels of a CO molecule is 8.45×10^{-2} eV. Calculate the value of the force constant of the bond in a CO molecule. **[CU (H) 2007]**

Solution: We know that $h\nu \sim \Delta E = 8.45 \times 10^{-2}\ \text{eV} = (8.45 \times 10^{-2})(1.6 \times 10^{-19}\ \text{J}) = 1.352 \times 10^{-20}\ \text{J}$.

$$\therefore \nu = \frac{\Delta E}{h} \text{ or, } \nu = \frac{1.352 \times 10^{20}\ \text{J}}{6.626 \times 10^{-34}\ \text{J-s}} = 2 \times 10^{13}\ \text{s}^{-1}$$

$$\text{or, } \omega = 2\pi\nu = \sqrt{k/\mu}$$

$$\text{or, } k = 4\pi^2\nu^2\mu = 4\pi^2\nu^2 \frac{m_c m_O}{m_c + m_O} = 4\pi^2\nu^2 \frac{12m_p \times 16m_p}{12m_p + 16m_p}$$

$$= 4\pi^2\nu^2(6.86m_p) = 4\pi^2(2 \times 10^{13}\ \text{s}^{-1}) \times (6.86 \times 10^{-27}\ \text{kg}) = \mathbf{181\ N/m}$$

» ***Example* 16.16.** Assume that H_2 molecule has a force constant $k = 573$ N/m. What is the vibrational quantum number corresponding to dissociation energy of 4.5 eV?

Solution: In this case,

$$h\omega\left(v + \frac{1}{2}\right) = 4.5\ \text{eV or, } \hbar\sqrt{\frac{k}{\mu}}\left(v + \frac{1}{2}\right) = 4.5\ \text{eV}$$

$$\mu = \frac{m_H m_H}{m_H + m_H} = \frac{m_H}{2} = \frac{m_p}{2} = \frac{1.67 \times 10^{-19}}{2} = 8.35 \times 10^{-28}\ \text{kg}$$

$$\text{or, } v + \frac{1}{2} = \frac{4.5\ \text{eV}}{\hbar}\sqrt{\frac{\mu}{k}} \frac{4.5 \times 1.6 \times 10^{-19}}{\frac{6.626 \times 10^{-34}}{2\pi}\ \text{J-s}} \times \sqrt{\frac{8.35 \times 10^{-28}\ \text{kg}}{573\ \text{N/m}}} = 8.24$$

$\therefore v = 8.24 - \frac{1}{2} = 7.74 \sim 8.$

$\therefore \boldsymbol{v = 8}$.

Hence, the vibrational quantum number corresponding to dissociation energy of 4.5 eV is 8.

» ***Example* 16.17.** The rotational constant of HF molecule is 2025 m^{-1}. Find the bond length.

Solution: The rotational constant of HF molecule is

$$B = \frac{h}{8\pi^2 Ic} \text{ or, } I = \frac{h}{8\pi^2 Bc} = \frac{6.626 \times 10^{-34}\ \text{J-s}}{8\pi^2(2025\ \text{m}^{-1})(3 \times 10^8\ \text{ms}^{-1})} \text{ or, } I = 1.38 \times 10^{-47}\ \text{kg-m}^2$$

$$\mu = \frac{m_H + m_F}{m_H + m_F} = \frac{m_p \times 19m_p}{m_p + 19m_p} = \frac{19}{20}(1.67 \times 10^{-27}\ \text{kg}) = 1.59 \times 10^{-27}\ \text{kg}$$

Now, $I = \mu r^2$.

$$\therefore r = \sqrt{\frac{I}{\mu}} = \frac{1.38 \times 10^{-47}\ \text{kg-m}^2}{1.59 \times 10^{-27}\ \text{kg}} = 0.93 \times 10^{-10}\ \text{m or, } \boldsymbol{r = 0.93\ \text{Å}}$$

the required bond length.

» ***Example* 16.18.** Given that force constant of HCl molecule is 470 Nm^{-1}. Estimate the energy difference between the lowest and the first vibrational states of HCl. **[KU (H) 2012]**

Solution: The energy difference between the lowest and the first state of HCl is given by

$$\Delta E_{v' \to v'+1} = \hbar\omega = \Gamma\sqrt{\frac{\hbar}{\mu}} = \frac{1}{2\pi}(6.626 \times 10^{-34}\ \text{J-s})\sqrt{\frac{470\ \text{N-m}^{-1}}{1.6236 \times 10^{-27}\ \text{kg}}}$$

$$= 5.67 \times 10^{-20}\ \text{J} = \mathbf{0.35\ eV\ (infrared\ region)}$$

» ***Example* 16.19.** In case of HCl molecule, the first three vibrational lines of vibration spectrum has wave numbers 2886 cm^{-1}, 5668 cm^{-1}, and 8347 cm^{-1} (where transition starts from ground state). Find the anharmonicity coefficient and force constant. Show that the vibration spectrum lies in IR region.

Solution: Here, the first three vibrational lines are

$$\tilde{\nu}_{0\to 1} = \frac{\omega}{2\pi c}(1-2\xi) = 2886\ \text{cm}^{-1} \quad \text{(i)}$$

$$\tilde{\nu}_{0\to 2} = \frac{\omega}{2\pi c}2(1-3\xi) = 5668\ \text{cm}^{-1} \quad \text{(ii)}$$

$$\tilde{\nu}_{0\to 3} = \frac{\omega}{2\pi c}3(1-4\xi) = 8347\ \text{cm}^{-1} \quad \text{(iii)}$$

Now,

$$\frac{\text{Eqn (ii)}}{\text{Eqn (i)}} = \frac{5668\ \text{cm}^{1}}{2886\ \text{cm}^{-1}} = \frac{2(1-3\xi)}{1-2\xi} \text{ or, } 0.98 = \frac{1-3\xi}{1-2\xi} \text{ or, } 0.98(1-2\xi) = 1-3\xi \text{ or, } \xi = 0.0192$$

Similarly,

$$\frac{\text{Eqn (iii)}}{\text{Eqn (ii)}} = \frac{8347\ \text{cm}^{1}}{5668\ \text{cm}^{-1}} = \frac{3(1-4\xi)}{2(1-3\xi)} \text{ or, } \xi = 0.0189$$

Also, from eqn (iii) and (i), we get $\xi = 0.0192$.

$\therefore$ the average value of anharmonicity constant $\xi = 0.0191$.

Again eqn (i) gives

$$\frac{\omega}{2\pi c}[1-2(0.0191)] = 2886 \times (10^{-2}\ \text{m})^{-1}$$

$$\text{or, } \omega = \frac{2886 \times (10^{-2})^{-1}}{1-2\times 0.0191} \times 2\pi \times (3\times 10^{8}\ \text{ms}^{-1}) = 5.656 \times 10^{14}\ \text{s}^{-1}$$

$\therefore \omega = \sqrt{k/\mu} = 5.656 \times 10^{14}\ \text{s}^{-1}$ or $k = \mu(5.656 \times 10^{14}\ \text{s}^{-1})^2$.

$\therefore$ for HCl

$$\mu = \frac{m_H m_{Cl}}{m_H + m_{Cl}} = \frac{m_p(35m_p)}{m_p + 35m_p} = \frac{35}{36}m_p = \frac{35}{36} \times (1.67 \times 10^{-27} k) = 1.6235 \times 10^{-27} k$$

where $m_p = 1.67 \times 10^{-27}$ kg, mass of the proton.

$\therefore$ force constant $k = (1.6236 \times 10^{-27}\ \text{kg})(5.656 \times 10^{14}\ \text{s}^{-1})^2 = 519\ \text{kg-s}^{-1} = \dfrac{519\ \text{kg-ms}^{-1}}{m}$ or, $\mathbf{k = 519\dfrac{N}{m}}$.

Also, energy of the vibration spectrum for HCl

$$\Delta E = \hbar\omega = \frac{(6.62 \times 10^{34}\ \text{J-s})(5.656 \times 10^{14}\ \text{s}^{-1})}{2\pi} = 5.97 \times 10^{-20}\ \text{J} = \frac{5.97 \times 10^{-20}}{1.6 \times 10^{-19}}\ \text{eV}$$

or, $\mathbf{\Delta E = 37 \times 10^{-2}\ eV}$

So, the energy of vibration spectrum lies in the IR region.

» **Example 16.20.** The frequency of absorption line for HCl in the rotational spectra is 6.36×10^{11} Hz for transition $J = 0$ to $J = 1$. Calculate the bond length of HCl molecule. **[IFS 2007]**

Solution: The wave number for transition in rotational spectra of HCl is

$$\bar{\nu}_{J\to J+1} = 2B(J+1) = 2\frac{h}{2\pi^2 Ic} \text{ or, } \frac{\nu}{c}\bigg|_{0\to 1} = 2B(0+1) = 2B \text{ or, } \nu = 2CB$$

$$\text{or, } \nu = 2c\left(\frac{h}{8\pi^2 Ic}\right) = \frac{h}{4\pi^2 \mu R^2} \text{ or, } R^2 = \frac{h}{4\pi^2 \nu \mu}$$

But

$$\mu = \mu_{\text{HCl}} = \frac{m_{\text{H}} + m_{Cl}}{m_{\text{H}} + m_{Cl}} = \frac{m_p \times 35 m_p}{m_p + 35 m_p} = \frac{35}{36} m_p$$

∴ R = the bond length of HCl is

$$= \left(\frac{h}{4\pi^2 \nu} \cdot \frac{36}{35 m_p}\right)^{1/2} = \frac{6.656 \times 10^{-34} \text{ J-s}}{4\pi^2 (6.36 \times 10^{11} \text{ s}^{-1})} \times \frac{36}{35(1.67 \times 10^{-27} \text{ kg})}$$

$$\text{or, } \mathbf{R = 1.28 \times 10^{-10} m = 1.28 \text{ Å}}$$

» **Example 16.21.** The molecules H_2O and D_2O both show rotational spectrum. Which will have smaller separation of lines in rotational spectrum?

Solution: We know that H^2 is heavier than H^1 as a result mass of D_2O is greater than that of H_2O.

$$\therefore I_{H_2O} < I_{D_2O} \text{ or, } B_{H_2O} > B_{D_2O} \left[\because B = \frac{h}{8\pi^2 Ic}\right].$$

∴ the energy of speculation of lines in rotational spectrum of the two molecules is

$$\Delta E_{J'-J'H} = 2Bhc(J+1) \simeq 2B \text{ or, } (\delta E)_{H_2O} > (\Delta E)_{D_2O}$$

Hence, for D_2O, the rotational lines will have smaller separation.

EXERCISES

16(A): Objective (or MCQ) Type Questions

Select the correct answer:

16.1. Molecular spectra consists of

(a) discrete lines
(b) bands
(c) mixture of lines and bands
(d) none of them.

16.2. The order of energy associated with rotational spectra is

(a) ~ 10^{-3} eV
(b) 2 eV
(c) 10^{-6} eV
(d) 10^{-2} eV.

16.3. The radiation emitted in rotational-vibrational spectrum is in

(a) visible region
(b) near infrared region
(c) ultraviolet region
(d) audible region.

16.4. The wavelength of electronic spectrum lies between

(a) 100 to 1000 Å
(b) 1000 to 7000 Å
(c) 10000 to 15000 Å
(d) 10^5 to 10^7 Å.

16.5. If the rotational, vibrational and electronic energy of a molecule be represented by E_r, E_v and E_e respectively, then

(a) $E_v > E_r > E_e$
(b) $E_e > E_r > E_v$
(c) $E_e > E_v > E_r$
(d) $E_r = E_v = E_e$.

16.6. The rotational energy levels of diatomic molecules are

(a) equispaced
(b) closely spaced at higher values
(c) closely spaced at lower j-values
(d) widely spaced at higher j-values.

16.7. The frequency separation of rotational level

(a) is constant
(b) increases with ν
(c) decreases with ν
(d) none of them.

16.8. Molecular spectra is more complicated as compared to atomic spectra because

(a) the molecule has many more energy levels
(b) the electron in a molecule does not move in a central force field like an atom
(c) the molecular mass is usually greater than atomic mass,
(d) the molecule can dissociate into atoms.

16.9. The importance of rotational spectral study is that we can determine the

(a) mass of the atoms forming the molecule
(b) bond length
(c) angular momentum
(d) molecular structure.

16.10. All diatomic molecules do not show rotational spectra because

(a) they possess very heavy mass
(b) they do not have permanent dipole moment
(c) their angular momentum is constant
(d) they always occupy ground state though they have higher rotational level.

16.11. The minimum vibrational energy of a diatomic molecule is given by

(a) $(\nu + \frac{1}{2})\, h\nu_0$
(b) $\nu h\nu_0$
(c) $\frac{1}{2}h\nu_0$
(d) $h\nu_0$.

16.12. The vibrational energy levels of a diatomic molecule are

(a) equispaced
(b) unequispaced
(c) irregularly spaced
(d) none of them.

16.13. If ν_1 and ν_2 be the frequencies of the *R*-branch and *P*-branch of vibrational-rotational spectra, then

(a) $\nu_1 = \nu_2$
(b) $\nu_1 > \nu_2$
(c) $\nu_2 > \nu_1$
(d) $\nu_2 \geqslant \nu_1$.

16.14. The origin of band structure of electronic spectra of molecules is due to the fact that

(a) $\Delta E_e \gamma E_j$
(b) $E_v > E_j$
(c) $\Delta E_v \simeq \Delta E_j$
(d) $\Delta E_e > \Delta E_v \Delta E_j$.

16.15. The bond strength of homonuclear molecules can be determined from the study of

(a) rotational spectra
(b) vibrational-rotational spectra
(c) electronic spectra
(d) Raman spectra.

16.16. The spacing between two successive Stokes lines is

(a) dependent on rotational quantum number *j*
(b) independent of *j*
(c) dependent directly on the MI of the molecule
(d) dependent on the frequency of the lines.

16.17. In phosphorescence

(a) the absorption and emission of light take place simultaneously
(b) emission is delayed by ~ 10^7 s
(c) emission is not always delayed
(d) delay time may be several years.

16.18. The selection rule for transition in rotational spectra is

(a) $\Delta J = 0$
(b) $\Delta J = \pm 1$
(c) $\Delta J = \pm 2$
(d) $\Delta J = \pm 1, \pm 2$.

16.19. The Stokes or anti-Stokes lines in Raman spectra are

(a) equally spaced
(b) unequally spaced
(c) irregularly spaced
(d) none of them.

16.20. In a system of diatomic molecules some atoms of one element are replaced by a heaver isotope such that the reduced mass is changed by 1.05. The shift in spectral line will by a factor.

(a) 0.475
(b) 0.5
(c) 0.95
(d) 1. **[cf. GATE 2007]**

16.21. For a diatomic molecule the vibrational energy level spacing ΔE_v and rotational energy level spacing (ΔE_j) are as follows:

(a) ΔE_v increases with v, ΔE_J increases with J (c) ΔE_v decreases with v, ΔE_J increases with J
(b) ΔE_v increases with v, ΔE_J decreases with J (d) ΔE_v decreases with v, ΔE_J decreases with J.

16.22. The spacing between two successive pure rotational lines of diatomic molecule is

(a) $\Delta R = \dfrac{h}{2\pi I}$ (c) $\dfrac{I}{2\pi\hbar}$

(b) $\Delta R = \dfrac{\hbar}{2\pi E}$ (d) $\dfrac{\pi}{2I\hbar}$.

ANSWERS

16.1. (b).	**16.6.** (c), (d).	**16.11.** (c).	**16.16.** (b).	**16.21.** (c).
16.2. (a).	**16.7.** (a).	**16.12.** (a).	**16.17.** (b).	**16.22.** (b).
16.3. (b).	**16.8.** (b).	**16.13.** (b).	**16.18.** (b).	
16.4. (b).	**16.9.** (b).	**16.14.** (c).	**16.19.** (a).	
16.5. (c).	**16.10.** (b).	**16.15.** (d).	**16.20.** (c).	

16(B): Short Answer Type Questions

16.1. How do the molecular energy states arise? **[BH (H) 2006]**

16.2. What is the Raman effect? **[BH (H) 2004]**

16.3. What is the origin of phosphorescence? **[BH (H) 2006]**

16.4. Explain the origin of Stokes and anti-Stokes lines. **[BU (H) 2004]**

16.5. Explain the origin of vibrational levels of a diatomic molecule with a diagram showing the variation of the potential energy of the molecule with internuclear distance. **[BH (H) 2003]**

16.6. Do all diatomic molecules exhibit rotational spectra? Give reasons for your answer. **[BU (H) 1993, 2012]**

16.7. Why do molecules show band spectra rather than line spectra? **[GNDU 1995, 1996; VU (H) 2013]**

16.8. H_2, N_2 and O_2 do not show vibrational spectra. Why?

16.9. Show that in rotational spectra, the energy levels are not equally spaced whereas the frequencies are equally spaced. **[HPU 1994, 1997]**

16.10. Why molecules show band spectra?

16.11. A diatomic molecule does not rotate in the ground state. Why? **[GNDU 1997]**

16.12. In which region of the electromagnetic spectra do the vibrational spectra of molecules lie?

16.13. Estimate the magnitude of rotational, vibrational and electronic energy of a diatomic molecule.

16.14. State the selection rule of pure rotational spectra and Raman spectra.

16.15. How the bond strength of a diatomic molecule can be determined from the study of molecular spectral study?

16.16. Write down an expression for the vibrational-rotational energy of a diatomic molecule.

16.17. For the emission of rotational spectra and vibration-rotation spectra, the diatomic molecule must have a permanent dipole moment. Why?

16.18. Outline the main features of Raman scattering. [CU (H) 2012]

16.19. Which of the following molecule can give rise to rotation and vibration spectra? Justify you answer with reference to H_2 and CO.

16.20. Why is pure vibrational spectra observed only in liquids? Explain. [CU (H) 2010]

16.21. Pure rotational spectrum is almost always seen as absorption lines and not as emission lines. Explain. [CU (H) 2009]

16.22. (a) Why all molecules do not show rotational spectra?

(b) Why homonuclear molecules (H_2, N_2, O_2, etc.) do not show rotational spectra? [VU (H) 2013]

16(C): Broad Answer Type Questions

16.1. (a) Discuss the quantum states of a molecule and those of an atom with reference to the types of radiation emitted in atomic and molecular transitions.

(b) Derive an expression for the rotational energy levels of a diatomic molecule when the rotation takes place about an axis perpendicular to the line joining the two nuclei and passing through the centre of mass. [BU (H) 1995, 1997]

16.2. Assuming a small one-dimensional vibration about the equilibrium position, obtain an expression for the frequency of vibration of a diatomic molecule. Write down the minimum energy of such vibration and interpret it. [BU (H) 1999]

16.3. Derive an expression for the rotational energy spectrum of a diatomic molecule in terms of its moment of inertia. [BU (H) 2000]

16.4. Give theory and obtain an expression for energies of various vibrational levels. Show that the energy levels are equispaced. How is the frequency of various lines obtained in vibrational spectra of molecules? What are the selection rules for vibrational states? In which region of the electromagnetic spectra do the vibrational spectra of molecules lie? [GNDU 1997; PU 1996]

16.5. (a) Discuss the vibrational-rotational spectra of diatomic molecules. Draw energy level diagram to explain the spectra. [PU 1997; HPU 1994]

(b) Show the vibrational-rotational levels of a diatomic molecule on a potential energy versus interatomic distance curve. Explain the formation of these levels. [GNDU 1991; PU 1992]

16.6. What is Raman effect? Discuss the characteristics of Raman lines. What are Stokes and anti-Stokes lines in Raman spectra? Discuss the practical uses of Raman effect. Compare Raman spectra with infrared spectra.

16.7. What is luminescence? Discuss the difference between fluorescence and phosphorescence.

16.8. Enumerate briefly the theory of Raman effect. Why are Stokes lines brighter than the anti-Stokes lines. [CU (H) 2014, 2009; VU (H) 2012]

16.9. (a) Show that in a pure rotational spectrum of a diatomic gas the spectrum lines are equally spaced in frequency.

(b) Can hydrogen molecules give rise to pure vibration spectra? Justify your answer. [CU (H) 2013]

16.10. (a) Discus briefly the basic difference between Compton effect and Raman effect.

(b) Show that in a pure rotational spectrum of a diatomic gas the spectral lines are equally spaced in frequency. **[CU (H) 2011]**

16.11. (a) Which of the following substances can give rise to pure rotational-vibrational spectra? Justify your answer. H_2, HF, O_2, CO.

(b) What are Stokes and anti-Stokes lines in Raman spectra? **[CU (H) 2010]**

16.12. (a) What is the basic difference between atomic spectra and molecular spectra?

(b) Why is pure vibrational spectra observed only in liquids? **[VU (H) 2011]**

16(D): Numerical Problems

16.1. The separation between the rotational levels $j = 1$ and $j = 0$ of an HI molecule results in a transition of wave number $(\frac{1}{\lambda})$ 1310 m^{-1}. Calculate the bond length of the molecule. **[BU (H) 2000]**

16.2. The atoms of CsI are separated by 3.31 Å. Calculate the reduced mass and moment of inertia of CsI molecule. [*Ans.* 64.96 amu; 1.18×10^{-44} kg-m^2]

16.3. The values of the constants B and ν_0 for a KCl molecule are 1.43×10^{-5} eV and 8.4×10^{12} sec^{-1} respectively. Find the number of rotational levels between the vibrational levels $v = 0$ and $v = 1$.

[*Ans.* No. of levels = 48]

16.4. The most intense vibrational bands of CO and HCl molecules have the wave numbers 2.143×10^5 m^{-1} and 2.886×10^5 m^{-1} respectively. Calculate the force constants of these molecules.

[*Ans.* 1855 N/m; 480.7 N/m]

16.5. Find the ratio of the intensities of the violet and red secondary lines in the Raman scattering of CCl_4 at 300 K, if the frequencies of the normal vibrations of CCl_4 expressed in Hz are 217, 315, 457 and 774.

[*Ans.* 0.35, 0.22, 0.11, 0.024]

16.6. Calculate the energy in joule per molecule to raise NO from its lowest rotational level to the level with $j = 4$. [*Ans.* 97.47]

16.7. The wave numbers of the vibrational transitions occurring in the HF, HCl and HI molecules are 4141.3 cm^{-1}, 2988.9 cm^{-1} and 2309.5 cm^{-1} respectively. Compare the force constants of the three molecules. [*Ans.* 1 : 0.51 : 30]

16.8. The spacing between vibrational level of CO molecule is 0.08 eV. Calculate the value of the force constant of CO bond. [*Given*: the masses of C and O atoms are 1.99×10^{-26} kg and 2.66×10^{-26} kg respectively.]

16.9. Three consecutive lines are observed at 82.19 cm^{-1}, 123.15 cm^{-1} and 164 cm^{-1}. Find the value of the rotational constant. **[VU (H) 2013]**

16.10. The spacing between vibrational levels of CO molecule is 0.08 eV. Calculate the value of the force constant of the CO bond. [*Given*: Mass of C atom = 2.0×10^{-26} kg and mass of O atom = 2.7×10^{-26} kg]

[VU (H) 2011]

16.11. Given that the force constant of HCl nucleus is 470 N/m. Estimate the energy difference between the lowest and the first vibrational states of HCl (assume atomic mass of H and Cl to be 1u and 35u respectively).

[KU (H) 2012]

[*Ans.* $\Delta E_{v' \to v'+1} = 0.35$ eV (IR region)]

16.12. The $J = 0 \rightarrow J = 1$ rotational absorption line occurs at 1.153×10^{11} Hz in $^{12}C^{16}O$ and at 1.102×10^{11} Hz in $^{X}C^{16}O$. What is the value of X? **[BU (H) 2010]**

16.13. Calculate the frequency of vibration of the CO molecule and the spacing between the vibrational energy level in eV. [*Given*: the force constant of the bond in CO = 187 n/m. Mass of C = 12u, mass of O = 15.994916u]. **[BU (H) 2009]**

16.14. Given that the force constant of HCl molecule is 470 Nm^{-1}. Estimate the energy difference between the lowest and the first vibrational states of HCl. **[KU (H) 2012]**

16.15. The observation spectrum of HCl molecules consists of 4 consecutive lines with wave numbers 83.03, 103.73, 124.30, 145.03 cm^{-1}. What J value do they correspond? Calculate the moment of inertia and the inter-nuclear separation. Also, find the region of electromagnetic spectrum where the spectra lies.

[*Ans.* $I_{HCl} = 2.7 \times 10^{-47}$ kg-m^2; $E_{J=1.13\,eV}$ (in micro wave region)]

16.16. The first three vibrational lines of vibration spectra of HCl molecule has wave numbers 2886 cm^{-1}, 5668 cm^{-1}, and 8347 cm^{-1}, where transition starts from ground state. Find the anharmonicity coefficient and the force constant. Show that the vibration spectrum lies in IR region.

[*Ans.* Anharmonicity constant $\xi = 0.0191$; energy = 37×10^{-2} eV $\therefore$ the vibration spectra lies in IR region]

16.17. Considering the diatomic molecule as a rigid rotator, show that the molecule has a discrete set of energy levels and that these levels are equispaced. **[CU (H) 2001]**

16.18. (a) Discuss the vibrational spectra of a diatomic molecule with mass of a nucleus bringing in the concept of reduced mass.

(b) Homonuclear diatomic molecule do not show vibrational spectra. Why? **[Kakatia U. 2015]**

16.19. Give an account of Raman effect and discuss how Raman spectra can be used together with IR spectra for structure determination. **[Kakatia U 2015]**

16.20. (a) Obtain an expression for the rotational energy levels of a diatomic molecule and the frequency of rotational spectra.

(b) A radiation of wavelength 546.1 nm excites a substance to emit Raman line of wavelength 538.2 nm. Calculate the Raman frequency and the wavelength of the corresponding Stokes' line.

[Kakatia U 2016]

Appendix A

Dirac Delta Function

The Dirac delta function is defined by

$$\delta(x-a) = 0 \text{ if } x \neq a \tag{A.1}$$

$$\text{and} \int_{a-\varepsilon}^{a+\varepsilon} \delta(x-a)\,dx = 1,\ \varepsilon > 0 \tag{A.2}$$

Here, the region of integration includes the point $x = a$.

The delta function is also defined as

$$\int_{a-\varepsilon}^{a+\varepsilon} f(x)\delta(x-a)\,dx = f(a) \tag{A.3}$$

The function $f(x)$ in this case, is arbitrary and is continuous at $x = a$. In the special case, $f(x) = 1$. Then (3) takes the form of (2).

Hence, the delta function $\delta(x-a)$ can be taken as the limit of a sequence of functions each having unit area which sharply peaks around the point $x = a$.

The integral representation of delta function is given by

$$\delta(x-a) = \frac{1}{2\pi}\int_{-\infty}^{+\infty} e^{ik(x-a)}\,dx \tag{A.4}$$

This can be proved as follows:

$$\frac{1}{2\pi}\int_{-\infty}^{+\infty} e^{ik(x-a)}\,dk = \lim_{k\to 0}\left[\frac{e^{ik(x-a)}}{2\pi i(x-a)}\right]_{-k}^{+k} = \lim_{k\to 0}\frac{1}{\pi(x-a)}\frac{e^{ik(x-a)} - e^{-ik(x-a)}}{2i}$$

$$= \lim_{k\to 0}\frac{1}{\pi(x-a)}\sin k(x-a) = \lim_{k\to 0}\frac{k}{\pi}\frac{\sin y}{y} \tag{A.5}$$

where $y = k(x-a)$.

Again,

$$\lim_{k\to\infty}\frac{k}{\pi}\int_{-\infty}^{+\infty}\frac{\sin y}{y}\,dx = \lim_{k\to\infty}\frac{1}{\pi}\int_{-\infty}^{+\infty}\frac{\sin y}{y}\,dy\ [\because y = k(x-a),\ \therefore dy = k\,dx] \tag{A.6}$$

$$= 1\left[\because \int_{-\infty}^{+\infty}\frac{\sin y}{y}\,dy = \pi \text{ by complex integration}\right] \tag{A.7}$$

That $\int_{-\infty}^{+\infty}\frac{\sin y}{y}\,dy = \pi$ can be shown as follows:

We plot $\frac{\sin y}{y}$ against y and get a curve as shown in Fig A-1. As

$$\lim_{y\to 0}\frac{\sin y}{y} = 1$$

we note that the main contribution is from the central part of the curve where $\frac{\sin y}{y} = 1$. We may, therefore, approximate the area to that of the triangle shown by dotted line and having vertices at $(-\pi, 0)$, $(+\pi, 0)$ and $(0, 1)$. The area of the triangle is

$$\Delta = \tfrac{1}{2}\times 2\pi \times 1 = \pi$$

Since, the central part of the Fig A-1 is represented by $y = k(x-a)$, the width $(x-a) = y/k \to 0$ as $k\to\infty$.

This establishes the definition of delta function in the integral form expressed by (A-4).

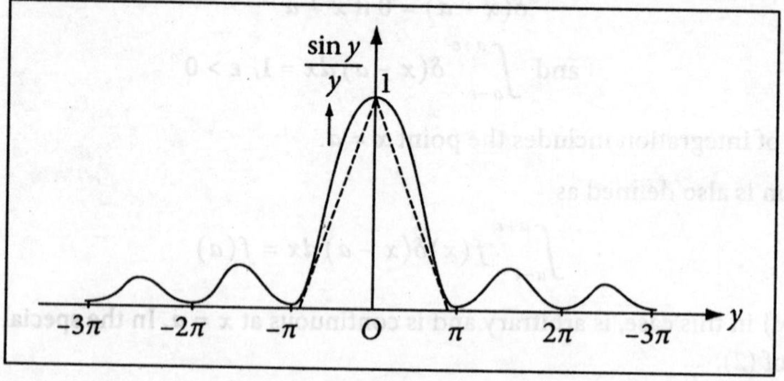

Fig A-1 $\frac{\sin y}{y}$ versus y curve peaks around $y \to 0$

Another representation of delta function is the Gaussian

$$G_\sigma = \frac{1}{\sigma\sqrt{2\pi}}e^{\left[-\frac{(x-x')^2}{2\sigma^2}\right]} \tag{A.8}$$

having unit area

$$\int_{-\infty}^{+\infty} G_\sigma(x, x')\,dx = 1 \text{ (independent of } \sigma)$$

peaking around $x = x'$ having with $\sim\sigma$. At $\sigma\to\infty$, $G_\sigma\to\infty$ at $x = x'$ and $\sigma_a\to 0$ elsewhere [Fig A-2].

Hence,

$$\delta(x-x') = \lim_{\sigma\to 0}\frac{1}{\sigma\sqrt{2\pi}}e^{\left[-\frac{(x-x')^2}{2\sigma^2}\right]} \tag{A.9}$$

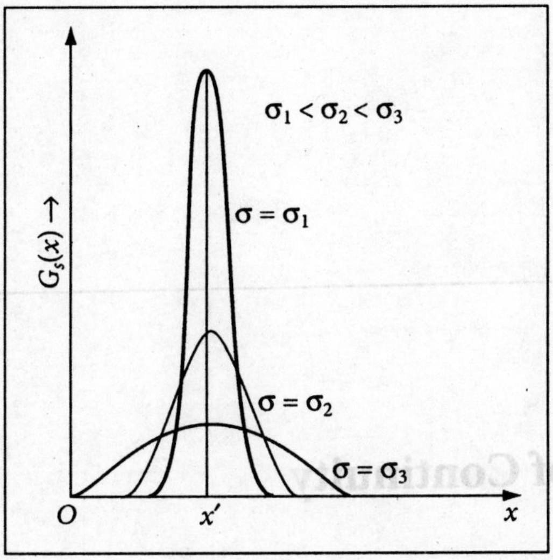

Fig A-2 The delta function as the limit of Gaussian function having half-width $s \rightarrow 0$

Properties of Delta Function:

1. $\delta(x - x') = \delta(x' - x)$
2. $\delta(ax) = \frac{1}{|a|}\,\delta(x)$
3. $x\delta(x) = 0$
4. $\delta(x^2 - a^2) = \frac{1}{2a}\,[\delta(x - a) + \delta(x + a)],\ a > 0$
5. $\int \delta(a - x)\delta(x - b)\,dx = \delta(a - b)$
6. $f(x)\delta(x - a) = f(a)\delta(x - a)$.

Appendix B

The Equation of Continuity

If ρ and $\vec{J}$ represent respectively the number of particles per unit volume and current density of a moving gas, then the equation of continuity is given by

$$\frac{\partial \rho}{\partial t} + \nabla \cdot \vec{J} = 0 \tag{B.1}$$

To understand the physical significance of the above equation, we consider a rectangular parallelopiped of dimensions dx, dy and dz parallel to the x-, y- and z-axes respectively [Fig B-1].

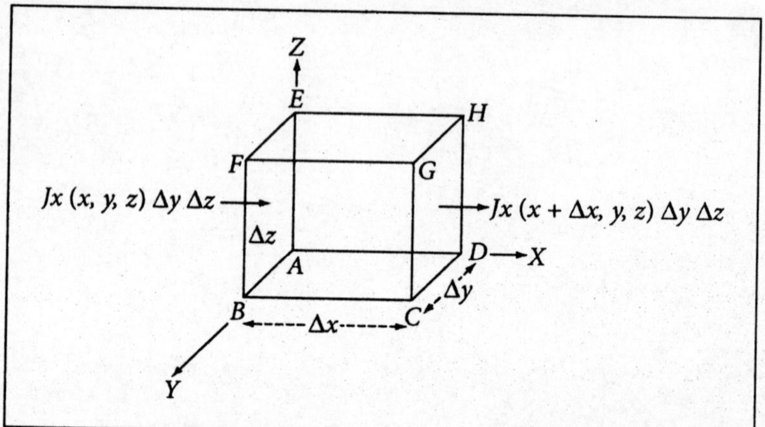

Fig B-1

The number of particles crossing the are a $ABFE$ per unit time is $J_x(x, y, z)\Delta y\, \Delta z$ and that coming out from the right surface $CDHG$ is $J_x(x + \Delta x, y, z)\, \Delta y\, \Delta z$.

$\therefore$ the net influx per unit time through the above two surfaces is

$$J_x(x, y, z)\Delta y\,\Delta z - J_x(x + \Delta x, y, z)\Delta y\,\Delta z = -\frac{\partial J_x}{\partial x}\Delta x\,\Delta y\,\Delta z$$

Similarly, considering the surfaces $ADHE$ and $BCGF$, we get the net influx per unit time as

$$-\frac{\partial J_y}{\partial y}\Delta x\,\Delta y\,\Delta z$$

and that for the surfaces $ABCD$ and $EFGH$ is

$$-\frac{\partial J_z}{\partial z}\Delta x\,\Delta y\,\Delta z$$

Hence, the net influx per unit time is

$$-\left[\frac{\partial J_x}{\partial x} + \frac{\partial J_y}{\partial y} + \frac{\partial J_z}{\partial z}\right]\Delta x\,\Delta y\,\Delta z = -\left(\hat{i}\frac{\partial}{\partial x} + \hat{j}\frac{\partial}{\partial y} + \hat{k}\frac{\partial}{\partial z}\right)\cdot(\hat{i}J_x + \hat{j}J_y + \hat{k}J_z)\,\Delta x\,\Delta y\,\Delta z$$

$$= -(\nabla\cdot\vec{J})\,\Delta x\,\Delta y\,\Delta z$$

Now, the rate of increase of particles in volume $\Delta x\,\Delta y\,\Delta z$ is

$$\frac{\partial \rho}{\partial t}\Delta x\,\Delta y\,\Delta z$$

Thus, in the steady state

$$\frac{\partial \rho}{\partial t}\Delta x\,\Delta y\,\Delta z = -(\nabla\cdot\vec{J})\,\Delta x\,\Delta y\,\Delta z \text{ or, } \frac{\partial \rho}{\partial t} = -(\nabla\cdot\vec{J}) \text{ or, } \frac{\partial \rho}{\partial t} + \nabla\cdot\vec{J} = 0$$

which is the equation of continuity. If we put $\rho = \psi^*\psi$, the position probability density, then

$$\vec{J} = \frac{i\hbar}{2m}[\psi\nabla\psi^* - \psi^*\nabla\psi] \tag{B.2}$$

is the probability current density. Here, $\hbar = h/2\pi$, m is the mass of the particle and ψ the wave function representing the particle.

Appendix C

Some Relations of Hermite Polynomials Used in Quantum Mechanics

1. The Hermite polynomial is defined as

$$H_n(x) = (-1)^n e^{x^2} \frac{d^n}{dx^n}\left(e^{-x^2}\right) \tag{C.1}$$

where $n = 0, 1, 2, \ldots$.

2. The Hermite polynomials satisfy the following differential equation:

$$\frac{d^2 H_n(x)}{dx^2} - 2x\frac{d}{dx}H_n(x) + 2nH_n(x) = 0 \tag{C.2}$$

3. The generating function is

$$e^{(-t^2+2xt)} = \sum_{n=0}^{\infty} H_n(x)\frac{t^n}{n!} \tag{C.3}$$

4. The recurrence relations of Hermite polynomials of different orders are as follows:

(a) $H_{n+1}(x) = 2xH_n(x) - 2nH_{n-1}(x)$; (C.4)

(b) $H_n'(x) = 2nH_{n-1}(x)$; (C.5)

(c) $H_n''(x) = 4n(n-1)\, H_{n-2}(x)$; (C.6)

(d) $H_n(-x) = (-1)^n H_n(x)$ (C.7)

5. Integrals involving Hermite polynomials:

(a)

$$\int_{-\infty}^{+\infty} e^{-x^2} H_n(x)\, H_m(x)\, dx = \begin{cases} 2^n n!\sqrt{\pi}, & \text{if } m = n \\ 0, & \text{for } m \neq n. \end{cases} \tag{C.8}$$

(b) $$\int_{-\infty}^{+\infty} e^{-x^2} H_n(x)\, x H_m(x)\, dx = \begin{cases} 0, & \text{for } m = n \\ 2^{n-1} n!\sqrt{\pi}, & \text{for } m = n-1 \\ 2^n (n+1)!\sqrt{\pi}, & \text{for } m = n+1. \end{cases} \quad \text{(C.9)}$$

(c) $$\int_{-\infty}^{+\infty} e^{-x^2} H_n(x)\, x^2 H_m(x)\, dx = (2n+1)2^{n-1} n!\sqrt{\pi}, \text{ for } m = n \quad \text{(C.10)}$$

6. Approximate value of $H_n(x)$, when $n \to \infty$

$$H_n(x) \sim \frac{2^{n+1}\left(\frac{n}{2e}\right)^{n/2}}{\sqrt{2\cos\alpha}} e^{n\alpha^2} \cos\left[\left(2n+\frac{1}{2}\right)\alpha - \frac{n\pi}{2}\right]$$

where α is the smallest positive angle given by

$$\alpha = \sin^{-1}\left(\frac{x}{\sqrt{2n}}\right)$$

7. A few Hermite polynomials: $H_0(x) = 1$

$$H_1(x) = 2x$$
$$H_2(x) = 4x^2 - 1$$
$$H_3(x) = 8x^3 - 12x$$
$$H_4(x) = 16x^4 - 48x^2 + 12$$
$$H_5(x) = 32x^5 - 160x^3 + 120x$$
$$H_6(x) = 64x^6 - 480x^4 + 720x^3 - 120$$
$$H_7(x) = 128x^7 - 1344x^5 + 3360x^3 - 1680x$$

Appendix D

Legendre Polynomials $[P_l(x)]$ and Associated Legendre Functions $[P_l^m(x)]$

1. Legendre polynomials are defined by

$$P_l(x) = \frac{1}{2^l l!}\frac{d^l}{dx^l}(x^2-1)^l,\ l = 0, 1, 2, \ldots \tag{D.1}$$

2. Associated Legendre polynomials are defined by

$$P_l^m(x) = (1-x^2)^{\frac{m}{2}}\frac{d^m}{dx^m}P_l(x),\quad l = 0, 1, 2, \ldots \quad m = 0, 1, \ldots, l \tag{D.2}$$

This is defined for positive integral values of m, $m \leqslant l$.

For $m = 0$, $P_l^{\circ}(x) = P_l(x)$.

3. The generating function is given by

$$(1-2tx+x^2)^{-1/2} = \sum_{l=0}^{\infty} t^l P_l(x),\ (|t|<1) \tag{D.3}$$

$$(2m-1)\,!(1-x^2)^{\frac{m}{2}}\frac{t^m}{(1-2tx+t^2)^{m+\frac{1}{2}}} = \sum_{l=m}^{\infty} t^l P_l^m(x),\ (|t|<1) \tag{D.4}$$

4. Differential equations satisfied by $P_l(x)$ and $P_l^m(x)$:

$$(1-x^2)\frac{d^2P_l(x)}{dx^2} - 2x\frac{dP_l(x)}{dx} + l(l+1)P_l(x) = 0 \tag{D.5}$$

$$(1-x^2)\frac{d^2P_l^m(x)}{dx^2} - 2x\frac{dP_l^m(x)}{dx} + \left[l(l+1) - \frac{m^2}{1-x^2}\right]P_l^m(x) = 0 \tag{D.6}$$

5. Orthogonality relations of Legendre polynomials:

$$\int_{-1}^{+1} P_l(x)P_m(x)\,dx = \frac{2}{2l+1}\delta_{lm} \tag{D.7}$$

$$\int_{-1}^{+1} P_l^m(x)P_{l'}^m(x)\,dx = \frac{2}{(2l+1)}\frac{(l+m)!}{(l-m)!}\delta_{ll'} \tag{D.8}$$

6. Recurrence relations:

(a) $(l+1)P_{l+1} = (2l+1)\,xP_l - lP_{l-1}$

(b) $(l+1-m)P_{l+1}^m = (2l+1)\,xP_l^m - (l+m)P_{l-1}^m$

(c) $(1-x^2)\frac{d}{dx}P_l^m = -lxP_l^m + (l+m)P_{l-1}^m$.

7. A few Legendre polynomials:

(a) $P_0(x) = 1$

(b) $P_1(x) = x$

(c) $P_2(x) = \frac{1}{2}(3x^2 - 1)$

(d) $P_3(x) = \frac{1}{2}(5x^3 - 3x)$

(e) $P_4(x) = \frac{1}{8}(35x^4 - 30x^2 + 3)$

(f) $P_5(x) = \frac{1}{8}(63x^5 - 70x^3 + 15x)$.

Particular values:

(g) $P_l(1) = 1$

(h) $P_l(-1) = (-1)^l$

(i) $P_l^m(1) = P_l^m(-1) = 0,\ m \neq 0$

(j) $P_l^m(0) = \begin{cases} (-1)^n \dfrac{(2n+2m)!}{2^l n!(n+m)!}, & \text{if } l - m = 2n \\ 0, & \text{if } l - m = 2n+1. \end{cases}$

Appendix E

Laguerre Polynomials and Associated Laguerre Polynomials

1. Laguerre polynomials are defined as

$$L_n(\rho) = e^{\rho}\left(\frac{d}{d\rho}\right)^n \left[e^{(-\rho)}\rho^n\right] = (-1)^n\left[\rho^n - \frac{n^2}{1!}\rho^{n-1} + \frac{n^2(n-1)^2}{2!}\rho^{n-2} + \cdots + (-1)^n n!\right]$$

where n is an integer.

2. Associated Laguerre polynomials of degree $(n-k)$ are defined as

$$L_n^k(\rho) = \left(\frac{d}{d\rho}\right)^k L_n(\rho) = \left(\frac{d}{d\rho}\right)^k\left[e^{\rho}\left(\frac{d}{d\rho}\right)^n \{e^{-\rho}\rho_n\}\right] \tag{E.1}$$

3. Generating functions are given by

$$\frac{e^{-\frac{\rho z}{(1-z)}}}{(1-z)} = \sum_{n=0}^{\infty} \frac{L_n(\rho)}{n!} z^n \quad |z| < 1 \tag{E.2}$$

$$\frac{(-z)^k e^{-\frac{\rho z}{(1-z)}}}{(1-z)^{k+1}} = \sum_{n=k}^{\infty} \frac{L_n^k(\rho)}{n!} z^n \tag{E.3}$$

4. Differential equations satisfied by $L_n(\rho)$ and $L_n^k(\rho)$ are as given below:

$$\rho\frac{d^2L_n(\rho)}{d\rho^2} + (1-\rho)\frac{dL_n(\rho)}{d\rho} + nL_n(\rho) = 0 \tag{E.4}$$

$$\rho\frac{d^2L_n^k(\rho)}{d\rho^2} + (k+1-\rho)\frac{dL_n^k(\rho)}{d\rho} + (n-k)L_n^k(\rho) = 0 \tag{E.5}$$

Relation involving integration of associated Laguerre polynomials:

$$\int_0^{\infty} e^{-\rho}\rho^{2l}\left[L_{n+1}^{2l+1}(\rho)\right]^2\rho^2 d\rho = \frac{2n[(n+l)!]^2}{(n-l-1)!}, \tag{E.6}$$

$$L_{n+1}^{2l+1} = \sum_{k=0}^{n-l-1}(-1)^{k+1}\frac{\{(n+l)!\}^2\rho^k}{(n-l-1-k)!(2l+1+k)!k!} \tag{E.7}$$

Recurrence relations:

(a) $L_{n+1}(x) + (x - 2n - 1)L_n(x) + n^2 L_{n-1}(x) = 0.$

(b) $xL_n''(x) + (1 - x)L_n'(x) + nL_n(x) = 0.$

A few Laguerre polynomials:

(a) $L_0(x) = 1$

(b) $L_1(x) = 1 - x$

(c) $L_2(x) = 2 - 4x + x^2$

(d) $L_3(x) = 6 - 18x + 9x^2 + x^3$

(e) $L_4(x) = 24 - 96x + 72x^2 - 16x^3 + x^4.$

A few associated Laguerre polynomials:

(a) $L_0^0(x) = 1$

(b) $L_1^0(x) = 1 - x$

(c) $L_2^0(x) = 2 - 4x + x^2$

(d) $L_1^1(x) = -1$

(e) $L_2^1(x) = -4 + 2x.$

Also $L_n^m(x) = 0$, if $m > n$.

Appendix F

Spherical Harmonics

1. The Spherical Harmonics in θ, ϕ is defined as

$$Y_l^m(\theta, \phi) = \frac{1}{2^l l!}\sqrt{\frac{(2l+1)(l-m)!}{4\pi(l+m)!}}(-1)^m e^{im\phi} P_l^n(\cos\theta) \tag{F.1}$$

The relation is true for positive integer, $m \leqslant l$, where $l = 0, 1, 2, \ldots$; and $m = -l, -l+1, \ldots, +l$.

For negative integer, $|m| \leqslant l$, we have

$$Y_l^m(\theta, \phi) = (-1)^m Y_l^{m^*}(\theta, \phi) \tag{F.2}$$

where $Y_l^{m^*}$ is the complex conjugate of Y_l^m.

2. Differential equation satisfied by Y_l^m is given by

$$\frac{1}{\sin\theta}\frac{\partial}{\partial\theta}\left\{\sin\theta\frac{\partial Y_l^m(\theta,\phi)}{\partial\theta}\right\} + \frac{1}{\sin^2\theta}\frac{\partial^2 Y_l^m(\theta,\phi)}{\partial\phi^2} = -l(l+1)Y_l^m(\theta,\phi) \tag{F.3}$$

3. The normalisation condition is

$$\int_0^{2\pi} d\phi \int_0^{\pi} Y_l^{m^*}(\theta,\phi) Y_l^{m'}(\theta,\phi)\sin\theta\, d\theta = \delta_{ll'}\delta_{mm'} \tag{F.4}$$

where $\delta_{ll'} = +1$, for $l = l'$ and $\delta_{ll'} = 0$, for $l \neq l'$

$\delta'_{mm'} = +1$, for $m = m'$ and $\delta_{mm'} = 0$, for $m = m'$

$Y_l^m(\theta, \phi)$ form an *orthonormal set*. For the expansion of the wave function they also form a complete set

4. First few Spherical Harmonics are as follows:

$$l = 0, \quad 2l + 1 = 1, \quad Y_0^0 = \frac{1}{\sqrt{4\pi}}$$

$$l = 1, \quad 2l + 1 = 2, \quad Y_1^1 = -\sqrt{\frac{3}{8\pi}} \sin\theta e^{i\phi}$$

$$Y_1^0 = \sqrt{\frac{3}{4\pi}} \cos\theta$$

$$Y_1^{-1} = \sqrt{\frac{3}{8\pi}} \sin\theta e^{-i\phi}$$

$$l = 2, \quad 2l + 5, \quad Y_2^2 = \sqrt{\frac{15}{32\pi}} \sin^2\theta e^{2i\phi}$$

$$Y_2^1 = -\sqrt{\frac{15}{8\pi}} \sin\theta \cos\theta e^{i\phi}$$

$$Y_2^0 = \sqrt{\frac{5}{16\pi}} (3\cos^2\theta - 1)$$

$$Y_2^{-1} = \sqrt{\frac{15}{8\pi}} \sin\theta \cos\theta e^{-i\phi}$$

$$Y_2^{-2} = \sqrt{\frac{15}{32\pi}} \sin^2\theta e^{-2i\phi}$$

5. Recurrence relation of Spherical Harmonics:

$$\text{(a)} \cos\theta Y_l^m = \sqrt{\frac{(l+1+m)(l+1-m)}{(2l+1)(2l+3)}} Y_{l+1}^m + \sqrt{\frac{(l+m)(l-m)}{(2l+1)(2l-1)}} Y_{l-1}^m; \tag{F.5}$$

$$\text{(b)} \sin\theta Y_l^m = -\sqrt{\frac{(l+1-m)(l+2-m)}{(2l+1)(2l+3)}} Y_{l+1}^{m-1} + \sqrt{\frac{(l+m)(l+m-1)}{(2l+1)(2l-1)}} Y_{l-1}^{m-1} \tag{F.6}$$

In particular, for $m = 0$

$$Y_l^0 = \sqrt{\frac{2l+1}{4\pi}} P_l(\cos\theta) \tag{F.7}$$

and for $m = 1$

$$Y_l^1 = (-1)^l \sqrt{\frac{2l+1}{4\pi}} \frac{(2l)!}{2^{2l}(l!)^2} (\sin\theta e^{i\phi})^l \tag{F.8}$$

Appendix G

The Hydrogen Atom Wave Functions

1. The total hydrogenic normalised eigenfunctions are

$$\psi_{nlm}(r,\theta,\phi) = R_{nl}(r)Y_l^m(\theta,\phi) \tag{G.1}$$

where $Y_l^m(\theta,\phi)$ are the spherical harmonics and

$$R_{nl}(r) = -\left[\left(\frac{2Z}{na_0}\right)^3 \frac{(n-l-1)!}{2n[(n+l)!]^3}\right]^{1/2} e^{-\rho/2}\rho^l L_{n+l}^{2l+1}(\rho) \tag{G.2}$$

where $a_0 = \frac{\hbar}{\mu e^2}$ and $\rho = \frac{2Z}{na_0}$ and

$$L_{n+l}^{2l+1}(\rho) = \sum_{k=0}^{n-l-1} (-1)^{k+2l+1} \frac{[(n+l)!]^2 \rho^k}{(n-l-1-k)!(2l+1+k)!k!} \tag{G.3}$$

are the associated *Laguerre polynomials.*

2. First few R_{nl} are given by

$$R_{1,0} = 2\left(\frac{Z}{a_0}\right)^{3/2} e^{-\sigma}$$

$$R_{2,0} = \frac{1}{\sqrt{2}}\left(\frac{Z}{a_0}\right)^{3/2} e^{-\frac{\sigma}{2}}\left(1-\frac{\sigma}{2}\right)$$

$$R_{2,0} = \frac{1}{2\sqrt{6}}\left(\frac{Z}{a_0}\right)^{3/2} e^{-\frac{\sigma}{2}}\sigma$$

$$R_{3,0} = \frac{2}{3\sqrt{3}}\left(\frac{Z}{a_0}\right)^{3/2} e^{-\frac{\sigma}{3}}\left(1-\frac{2\sigma}{3}+\frac{2\sigma^2}{27}\right)$$

$$R_{3,1} = \frac{8}{27\sqrt{6}}\left(\frac{Z}{a_0}\right)^{3/2} e^{-\frac{\sigma}{3}}\left(\sigma - \frac{\sigma^2}{6}\right)$$

$$R_{3,2} = \frac{4}{81\sqrt{30}}\left(\frac{Z}{a_0}\right)^{3/2} e^{-\frac{\sigma}{3}}\sigma^2$$

3. First few ψ_{nlm} are as follows:

$$\psi_{100} = \frac{1}{\sqrt{\pi}}\left(\frac{Z}{a_0}\right)^{3/2} e^{-\sigma}$$

$$\psi_{200} = \frac{1}{4\sqrt{2\pi}}\left(\frac{Z}{a_0}\right)^{3/2} (2-\sigma)\, e^{-\frac{\sigma}{2}}$$

$$\psi_{210} = \frac{1}{4\sqrt{2\pi}}\left(\frac{Z}{a_0}\right)^{3/2} \sigma e^{-\frac{\sigma}{2}} \cos\theta$$

$$\psi_{21\pm1} = \mp\frac{1}{8\sqrt{\pi}}\left(\frac{Z}{a_0}\right)^{3/2} \sigma e^{-\frac{\sigma}{2}} \sin\theta e^{\pm i\phi}$$

$$\psi_{300} = \frac{1}{81\sqrt{3\pi}}\left(\frac{Z}{a_0}\right)^{3/2} \left(27 - 18\sigma + 2\sigma^2\right) e^{-\frac{\sigma}{3}}$$

$$\psi_{310} = \frac{1}{81}\sqrt{\frac{2}{\pi}}\left(\frac{Z}{a_0}\right)^{3/2} \sigma(6-\sigma)e^{-\frac{\sigma}{3}} \cdot \cos\theta$$

$$\psi_{31\pm1} = \pm\frac{1}{81\sqrt{\pi}}\left(\frac{Z}{a_0}\right)^{3/2} \sigma(6-\sigma)e^{-\frac{\sigma}{3}} \sin\theta e^{\pm i\phi}$$

$$\psi_{320} = \frac{1}{81\sqrt{6\pi}}\left(\frac{Z}{a_0}\right)^{3/2} \sigma^2 e^{-\frac{\sigma}{3}}(3\cos\theta - 1)$$

$$\psi_{32\pm1} = \mp\frac{1}{81\sqrt{\pi}}\left(\frac{Z}{a_0}\right)^{3/2} \sigma^2 e^{-\frac{\sigma}{3}} \sin\theta\cos\theta e^{\pm i\phi}$$

$$\psi_{32\pm2} = \frac{1}{162\sqrt{\pi}}\left(\frac{Z}{a_0}\right)^{3/2} \sigma^2 e^{-\frac{\sigma}{3}} \sin^2\theta e^{\pm 2i\phi}$$

where $\sigma = Zr/a_0$.

Appendix H

Fourier Integral and Fourier Transform

Fourier Series

If $f(x)$ be a single-valued periodic function in any arbitrary interval satisfying the following conditions:

(i) $f(x)$ is continuous having finite number of maxima and minima and

(ii) $f(x)$ is differentiable at every point within the interval of expansion on them

$$f(x) = \frac{a_0}{2} + \sum_{n=1}^{\infty}\left[a_n \cos\left(\frac{2n\pi x}{l}\right) + b_n \sin\left(\frac{2n\pi x}{l}\right)\right] \tag{H.1}$$

where

$$a_n = \frac{2}{l}\int_{-\frac{l}{2}}^{+\frac{l}{2}} f(x') \cos\left(\frac{2n\pi x'}{l}\right) dx'$$

$$b_n = \frac{2}{l}\int_{-\frac{l}{2}}^{+\frac{l}{2}} f(x') \sin\left(\frac{2n\pi x'}{l}\right) dx'$$

with $n = 0, 1, 2, \ldots.$

The series represented by (H-1) is called *Fourier series* and the conditions imposed on $f(x)$ are known as *Dirichlet conditions.*

Fourier Integral and Fourier Transform

If $f(x)$ be non-periodic, then also it can be expanded in terms of sine and cosine functions if the limit of expansion be from $-\infty$ to $+\infty$. For this

$$\int_{-\infty}^{+\infty} f(x)\, dx$$

must exist and is single-valued in the interval $-\infty \leqslant x \leqslant +\infty$. Then we can express $f(x)$ as

$$f(x) = \frac{1}{\sqrt{2\pi}}\int_{-\infty}^{+\infty} g(k)\, e^{ikx}\, dx \tag{H.2}$$

where

$$g(k) = \frac{1}{\sqrt{2\pi}} \int_{-\infty}^{+\infty} f(x)\, e^{-ikx}\, dx \tag{H.3}$$

Equations (H-2) and (H-3) are known as *Fourier integral* and *Fourier transform* respectively.

In three dimensions

$$f(\vec{r}) = \left(\frac{1}{2\pi}\right)^{3/2} \iiint g(\vec{k})\, e^{i\vec{k}\cdot\vec{r}}\, dk_x\, dk_y\, dk_z \tag{H.4}$$

where

$$g(\vec{k}) = \left(\frac{1}{2\pi}\right)^{3/2} \iiint f(\vec{r})\, e^{-i\vec{k}\cdot\vec{r}}\, dx\, dy\, dz \tag{H.5}$$

Appendix I

Some Fundamental Physical Constants

No. Name of Quantity	Symbol	Value
1. Gravitational constant	G	6.6720×10^{-11} N-m^2/kg^2
2. Velocity of light in vacuum	c	2.997925×10^{8} m/s
3. Avogadro number	N_A	6.02205×10^{23} molecules/mole
4. Universal gas constant	R	8.314 J/(K mol)
5. Boltzmann constants	k	1.3807×10^{-23} J/K = 8.62×10^{-5} eV/K
6. Stefan-Boltzmann constant	$\sigma = \frac{\pi^2 k^2}{60\hbar^3 c^2}$	5.67×10^{-8} W-m^{-2} K^{-4}
7. Rydberg constant	$R = \frac{e^4 m_e}{8h^3 \varepsilon_0 c}$	1.0973731×10^{7} m^{-1}
8. Permeability (vacuum)	μ_0	$4\pi \times 10^{-7} = 1.26 \times 10^{-6}$ H/m
9. Permittivity (vacuum)	$\varepsilon_0 = \frac{1}{\mu_0 c^2}$	8.85×10^{-12} F/m; $\frac{1}{4\pi\varepsilon_0} = 8.99 \times 10^{9}$ m/F
10. Faraday constant	$F = N_A e$	9.65×10^{4} C/mol
11. Electronic charge	$-e$	-1.60×10^{-19} c 4.80×10^{-10} e.s.u
12. Electronic radius	$r_e = \frac{e^2}{mc^2}$	2.82×10^{-15} m
13. Bohr radius	a_0	0.529177×10^{-10} m
14. Specific charge of electron	$\left\lvert \frac{e}{m_e} \right\rvert$	1.76×10^{11} c/kg
15. Electron rest mass	m_0	9.10953×10^{-31} kg
16. Electron rest energy	$m_0 c^2$	0.511003 MeV
17. Electron-volt	eV	1.60219×10^{-19} J
18. Fine structure constant	$\alpha = \frac{e^2}{2\varepsilon_0 hc}$	$7.29735 \times 10^{-3} = \frac{1}{137.03605}$
19. Proton-electron mass ratio	$\frac{m_p}{m_e}$	1837
20. Electron Compton wavelength	λ_c	2.43×10^{-12} m
21. Bohr magneton	$\lvert \mu_B \rvert = \frac{eh}{2m_e}$	9.28×10^{-24} J/T
22. Proton rest mass	m_p	1.673×10^{-27} kg = 1.007 u = 938.3 MeV/c^2
23. Neutron rest mass	m_n	1.675×10^{-27} kg = 1.009 u = 939.6 MeV/c^2
24. Electron magnetic moment	μ_e	$-\frac{(1.00eh)}{2m_e} = 9.28 \times 10^{-24}$ J/T
25. Neutron magnetic moment	μ_n	$1.91\frac{eh}{2m_n} = 9.66 \times 10^{-27}$ J/T
26. Proton magnetic moment	μ_p	$2.79\frac{eh}{2m_p} = 1.41 \times 10^{-26}$ J/T
27. One atomic mass	1 u	931 MeV = 1.66×10^{-27} kg
28. Wavelength associated with 1 eV/c		12,400 Å

List for References

1. Anderson, E.E., *Modern Physics and Quantum Mechanics,* Macmillan Co. Ltd, 1979.
2. Baym, G., *Lectures on Quantum Mechanics,* W.A. Benjamin Inc., New York, 1969.
3. Bethe, H.A. and Jackiw, R.W., *Intermediate Quantum Mechanics,* W.A. Benjamin, New York, 1968.
4. Bohm, D., *Quantum Theory,* Prentice-Hall, New York, 1951.
5. Davydov, A.S., *Quantum Mechanics,* Pergamon Press, Oxford, 1965.
6. Dirac, P.A.M., *The Principles of Quantum Mechanics,* Clarendon Press, Oxford, 1947.
7. Feynman, R.P., Leighton, R.B. and Sands, M., *The Feynman Lectures on Physics,* Vol. III, Addison Wesley Publishing Co., Reading, Mass, 1965.
8. Flugge, S., *Practical Quantum Mechanics,* Springer Verlag, Berlin, 1971.
9. Fock, V., *Quantum Mechanics,* Mir Publisher, Moscow, 1986.
10. Ghatak, A.K. and Loknathan, S., *Quantum Mechanics,* The Macmillan Company of India Limited, 1975.
11. Ghoshal, S.N., *Introductory Quantum Mechanics,* Calcutta Book House, 1987.
12. Gropper, W.H., *The Quantum Physics and an Introduction to their Physics,* Oxford University Press, New York, 1970.
13. Heisenberg, W., *The Physical Principles of Quantum Theory,* Dover Publications, London, 1949.
14. Heitler, W., *Elementary Wave Mechanics,* Oxford University Press, 1956.
15. Hertzberg, G., *Spectra of Diatomic Molecules,* Van-Norstand Reinhold Co. by Litton Publishing Inc., 1950.
16. Jammer, M., *The Conceptual Development of Quantum Mechanics,* McGraw-Hill, 1966.
17. Kogan, V.I. and Galitsky, V.M., *Problems in Quantum Mechanics,* Prentice Hall, Inc., Englewood Cliffs, N.J. 1963.
18. Lawden, D., *The Mathematical Principles of Quantum Mechanics,* Methuen, 1967.
19. Ludwig, G., *Wave Mechanics,* Pergamon Press, Oxford, 1968.
20. Merzbacher, E., *Quantum Mechanics,* John Wiley and Sons Inc., New York, 1970.
21. Messiah, A., *Quantum Mechanics,* Vol. I, North Holland, London, 1970.
22. Morse, P.M. and Feshbach, H., *Methods of Theoretical Physics,* McGraw-Hill, 1953.
23. Pauling, L. and Wilson, E.B., *Introduction to Quantum Mechanics,* McGraw-Hill Book Co. Inc., New York, 1935.
24. Powell, J.L. and Craseman, B., *Quantum Mechanics,* Addison Wesley Publishing Co. Inc., Reading, Mass., 1961.

25. Rojansky, V., *Introductory Quantum Mechanics,* Prentice Hall Inc., 1938.

26. Saxon, D.S., *Elementary Quantum Mechanics,* Holden Day Inc., San Francisco, 1964.

27. Schiff, L.I., *Quantum Mechanics,* McGraw-Hill, 1968.

28. Slater, J.C., *Quantum Theory of Atomic Structures,* Vols. I and II, McGraw-Hill, 1960.

29. Sokolov, A.A., Loskutov, Y.M. and Ternov, I.M., *Quantum Mechanics,* Holt, Rinehart and Winston, 1966.

30. Wichmann, E.H., *Quantum Physics,* Berkley Physics Course, Vol. 5, McGraw-Hill, 1967.

31. Wollrab, J.E., *Rotational Spectra and Molecular Structure,* Academic Press, 1967

Index